POLYMER SCIENCE DICTIONARY

POLYMER SCIENCE DICTIONARY

MARK S. M. ALGER

London School of Polymer Technology,
Polytechnic of North London, UK

ELSEVIER APPLIED SCIENCE
LONDON and NEW YORK

ELSEVIER SCIENCE PUBLISHERS LTD
Crown House, Linton Road, Barking, Essex IG11 8JU, England

Sole Distributor in the USA and Canada
ELSEVIER SCIENCE PUBLISHING CO., INC.
655 Avenue of the Americas, New York, NY10010, USA

© 1989 ELSEVIER SCIENCE PUBLISHERS LTD

First edition 1989
Reprinted 1990

British Library Cataloguing in Publication Data

Alger, Mark S. M.
 Polymer science dictionary.
 1. Polymers. Encyclopaedias
 I. Title
 547.7′03′21

ISBN 1-85166-220-0

Library of Congress Cataloging in Publication Data

Alger, Mark S. M.
  Polymer science dictionary/Mark S. M. Alger.
   p. cm.
  ISBN 1-85166-220-0
  1. Polymers and polymerization—Dictionaries. I. Title.
 QD380.3.A52 1989
 547.7′03′21—dc19   88-11034
                                                    CIP

Printed in Great Britain by Galliard (Printers) Ltd, Great Yarmouth

PREFACE

Polymer science is often regarded as a new science, although the fundamental molecular principle of the long chain molecule, which underlies much of the subject, was first proposed by Staudinger in 1920. Polymer technology is of course much older, dating back to the early nineteenth century. Today several important industries (those of plastics, rubbers, fibres, coatings and adhesives) are based on polymers, and scientists and engineers working in even wider fields are concerned with polymers. This is not to mention the importance of the concepts of polymer science in the biological area—much of molecular biology may be considered as part of the broader subject of polymer science.

It is therefore perhaps surprising that there is no publication devoted solely to explaining the terminology of polymer science, though several dictionaries covering the related technologies do exist. This dictionary is an attempt to fill this gap, a need which I have felt many times during my 20 years of teaching polymers. Furthermore many scientists and technologists working with polymers have little polymer experience in their formal education.

The scope of the dictionary is restricted to polymer science and deliberately excludes technological areas. Specifically, polymer processing terms are not included. Within the fields of polymer chemistry and physics, an attempt has been made to be as comprehensive as possible, with coverage of polymerisation, polymer structure, properties and individual polymer materials. However, given the multidisciplinary nature of polymer science, it is difficult to decide where to draw the limits of the scope. A particular problem has been to decide how much coverage to give to biopolymers, which are often, with little justification, not regarded as being within the mainstream of polymer science. I have included proteins and polysaccharides, partly since they form the basis of many useful materials, but, reluctantly, I have excluded nucleic acids. In all about 1000 specialist polymer monographs and numerous reviews have been consulted. I hope that this has provided a good coverage of the terms used in the polymer literature, both the older literature and the more recent. It has proved impossible to be as up-to-date as is desirable in such a rapidly developing subject.

In writing the individual entries the approach has been to provide an explanation of what the terms mean, together with any necessary background, as they are used in the polymer literature, rather than to provide definitions of the terms. Thus the entries are longer than is usual in a dictionary strictly concerned with definitions, and this dictionary could perhaps be considered more as an encyclopaedic dictionary. Many terms have been included that are not specifically polymer terms, but which have a special relevance in polymer science. I have attempted to place such terms in the context of polymer science.

Increasingly, SI units are being used in the polymer literature and have been used throughout this dictionary. However, where other units are frequently used, these have been included. Appendix A gives a listing of SI and other units and Appendix B lists some conversion factors. The symbols used have been selected as those most frequently used elsewhere and are collected together following this preface.

For the more important polymers some values are given for the main mechanical properties. Since these properties are very variable, the values given are merely indicative and are only to provide a rough comparison between different polymers.

The compilation of the dictionary has stretched over many (too many!) years and during this time, but especially during the last stages, my family has had to put up with my prolonged absences from participation in normal family life. For this forbearance I am very grateful. I would also like to thank Hazel Fryer and my wife Jill Alger for considerable help in the preparation of the final manuscript and the publishers for their patience in waiting so long for the manuscript. The extensive use of the material in the libraries of the Polytechnics of the South Bank and North London is also gratefully acknowledged.

MARK ALGER

CONTENTS

NOTATION

a	activity	$\bar{e}$	equivalent strain
a_1	solvent activity	e^*	complex strain
a_{11}, etc.	direction cosine	e_0	maximum value of a cyclically varying strain
a_2	solute activity	e_1, e_2, e_3	principal strains
a_T	shift factor	e_{12}, etc.	strain components
A	area	e_{ij}, etc.	finite strain components
A_2	second virial coefficient	e_{xx}, e_{yy}, e_{zz}	principal strains
A_3	third virial coefficient	e_{xy}, e_{yz}, etc.	strain components
b	form factor	$e(t)$	time-dependent strain (creep strain)
b	scattering length	E	Young's modulus
B	second virial coefficient	E	internal energy
B'	bulk storage compliance	E	dielectric strength
B''	bulk loss compliance	E	electric field strength
B_e	equilibrium bulk compliance	E'	tensile storage modulus
B_r	relaxed bulk compliance	E''	tensile loss modulus
B_u	unrelaxed bulk compliance	E_e	equilibrium tensile modulus
$B(t)$	bulk creep compliance	E_r	relaxed tensile modulus
Bu	butyl group	E_r	tensile relaxation modulus
n-Bu	normal butyl group	E_u	unrelaxed tensile modulus
t-Bu	tertiary butyl group	Et	ethyl group
$B(\omega)$	complex bulk compliance	$E(t)$	tensile stress relaxation modulus
c	weight concentration	$E_r(t)$	tensile relaxation modulus
c_2	solute concentration	$E^*(\omega)$	complex tensile modulus
C	elastic modulus	f	initiator efficiency factor
C	third virial coefficient	f	normal stress in tension
C	characteristic ratio	f	friction coefficient
C	chain transfer constant	f	functionality
C'	storage compliance	f	atomic form factor
C''	loss compliance	f	orientation factor
C_1	constant of the Rivlin–Saunders equation	f_A	mole fraction of monomer A in a copolymerisation feed
C_I	chain transfer constant to initiator	f_{av}	average functionality
C_M	chain transfer constant to monomer	F	force
C_p	specific heat	F_A	mole fraction of comonomer A units in a copolymer
C_P	chain transfer constant to polymer		
C_S	chain transfer constant to solvent	F_{hkl}	structure factor
$C^*(\omega)$	complex compliance	F_N	Fourier number
D	diffusion coefficient	$F(\dot{e})$	constant strain rate modulus
D'	tensile storage compliance	g	branching parameter
D''	tensile loss compliance	G	shear modulus
De	equilibrium tensile compliance	G	branching factor
D_e^0	steady state compliance in tension	G	strain energy release rate
D_r	relaxed tensile compliance	G'	shear storage modulus
D_u	unrelaxed tensile creep compliance	G''	shear loss modulus
D_e	Deborah number	G_c	critical strain energy release rate
$D(t)$	tensile creep compliance	G_e	equilibrium shear modulus
$D^*(\omega)$	complex tensile compliance	ΔG_m	free energy of mixing
e	monomer polarity factor		
e	engineering strain		

G_r	relaxed shear modulus
G_r	shear relaxation modulus
G_u	unrelaxed shear modulus
$G(t)$	stress relaxation modulus
$G_r(t)$	shear relaxation modulus
$G^*(\omega)$	complex shear modulus
$\mathbf{h}$	scattering vector
H	enthalpy
ΔH	enthalpy change
ΔH_m	enthalpy of mixing
ΔH_p	enthalpy of polymerisation
$H(\tau)$	relaxation spectrum
i	individual polymer molecular species (or its size)
I	light intensity
$[I]$	initiator concentration
I_1, I_2, I_3	strain invariants
I_a	absorbed light intensity
$\mathscr{I}$	rate of initiation
J	shear compliance
J'	shear storage compliance
J''	shear loss compliance
J_e	equilibrium shear compliance
J_e^0	steady state compliance in shear
J_r	relaxed shear compliance
J_u	unrelaxed shear compliance
$J(t)$	shear creep compliance
$J^*(\omega)$	complex shear compliance
k	degree of coupling
k	consistency index
k'	Huggins constant
k''	Kraemer's constant
k_d	rate constant for initiator decomposition
k_E	Einstein coefficient
k_p	rate constant for propagation
k_t	rate constant for termination
k_{tr}	rate constant for transfer
K	bulk modulus
K	ebullioscopic constant
K	stress intensity factor
K'	bulk storage modulus
K''	bulk loss modulus
K_c	critical stress intensity factor
K_e	equilibrium bulk modulus
K_r	relaxed bulk modulus
K_u	unrelaxed bulk modulus
$K(t)$	bulk stress relaxation modulus
$K^*(\omega)$	complex bulk modulus
l	length
l_0	initial length
l_c	critical fibre length
l, m, n	direction cosines
$L(\tau')$	retardation spectrum
$\mathscr{L}$	Langevin function
M	molecular weight
M	absolute modulus
M	monomer
$[\mathrm{M}]$	monomer concentration
$[\mathrm{M}^*]$	active centre concentration
M'	storage modulus

M''	loss modulus
M_c	complex modulus
M_c	network parameter
$[\mathrm{M}]_c$	equilibrium monomer concentration
M_i	molecular weight of species i
$\bar{M}_n$	number average molecular weight
$\bar{M}_{SD}$	sedimentation–diffusion average molecular weight
$\bar{M}_v$	viscosity average molecular weight
$\bar{M}_w$	weight average molecular weight
$\bar{M}_z$	z average molecular weight
$\bar{M}_{z+1}$	$z+1$ average molecular weight
$[\mathrm{M}]_\lambda$	molar rotation at wavelength λ
$M^*(\omega)$	complex modulus
n	refractive index
n	number of chain links
n^*	number of scattering centres
n_1	number of moles of solvent
n_2	number of moles of solute
N_1	number of molecules of solvent
N_2	number of molecules of solute
N_A	Avogadro's number
N_i	number of molecules of type i
p	fractional conversion
p_c	critical conversion
P	pressure
P	electric polarisation
P	permeability coefficient
P	colligative property
P_1, P_2, P_3	principal stresses
$P(\theta)$	particle scattering function
P_θ	particle scattering factor
Pe	Peclet number
Ph	phenyl group
$P(x)$	probability of x
q	crosslink density
$\mathbf{q}$	scattering vector
Q	volume flow rate
Q	dynamic amplification factor
Q	monomer reactivity factor
r	radial distance
r	chain end-to-end distance
r_A	reactivity ratio of monomer A
$\langle r^2 \rangle$	mean square end-to-end distance
$\langle r^2 \rangle_0$	unperturbed mean square end-to-end distance
R	alkyl group
R	gas constant
Re	Reynolds number
R_G	radius of gyration
R_i	rate of initiation
R_p	rate of propagation
R_θ	reduced scattering factor (Rayleigh ratio)
s	radius of gyration
S	compliance
S	sedimentation constant
S	conductance
S	entropy
$S(q), S(\kappa)$	structure factors

S_{conf}	configurational entropy	ε	relative permittivity
ΔS_m	entropy of mixing	$\Delta \varepsilon$	relaxation strength
t	trans configuration	ε'	relative permittivity
T	temperature (kelvin)	ε''	dielectric loss factor
T_2	spin–lattice relaxation time	ε^*	complex relative permittivity
ΔT	boiling point elevation, freezing point depression	ε_0	static relative permittivity
		ε_∞	high frequency relative permittivity
T_B	brittleness temperature	ε_s	static relative permittivity
T_c	ceiling temperature	$\varepsilon_{xy}, \varepsilon_{yz}$, etc.	tensor strain components
T_g	glass transition temperature	η	coefficient of viscosity
$T_{1,1}$	liquid–liquid transition	η	efficiency of reinforcement
T_m	melting temperature	$[\eta]$	limiting viscosity number
u_i, u_j	velocity components	η'	apparent viscosity, dynamic viscosity
u, v, w	displacements	η^*	complex viscosity
U	inhomogeneity	η_0	Newtonian viscosity at zero shear rate
v	velocity component		
$\bar{v}$	partial molar volume	η_∞	upper Newtonian viscosity
$\bar{v}_2$	solute partial specific volume	η_r	relative viscosity, viscosity ratio
v_e	effective network chain	η_{rel}	relative viscosity, viscosity ratio
V	volume	η_{sp}	specific viscosity
V	voltage	η_{sp}/c	viscosity number
ΔV	change in volume	η_t	viscous traction
V_f	free volume	η'_0	apparent viscosity, dynamic viscosity
V_R	retention volume	θ	temperature
w_A	work of adhesion	θ	theta temperature
w_C	work of cohesion	κ	thermal conductivity
w_i	weight of molecular species i	κ	compressibility
W	strain energy function	λ	deformation ratio
W_s	Weissenberg number	λ	Lamé constant
$W(r)$	differential weight distribution function	λ	relaxation time
x	mole fraction	λ	elongational viscosity
x_1	mole fraction of solvent	λ	branching density
x_2	mole fraction of solute	λ	radiation wavelength
$\bar{x}_n$	number average degree of polymerisation	$\lambda_1, \lambda_2, \lambda_3$	principal extension ratios
$\bar{x}_w$	weight average degree of polymerisation	μ	coefficient of viscosity
$\bar{x}_z$	z average degree of polymerisation	μ	coefficient of friction
z	inhibition constant	μ	dipole moment
z	dissymmetry ratio	v	kinetic chain length
α	polarisability	v	Poisson ratio
α	branching coefficient	ξ	monomer friction coefficient
α	degree of degradation	Π	osmotic pressure
α	volume expansion coefficient	ρ	density
α_c	critical branching coefficient	ρ	branching density
α_R	expansion factor	ρ_1	density of solvent
γ	crosslinkining index	ρ_2	density of solute
γ	shear strain	ρ_r	atomic radial distribution function
$\dot{\gamma}$	shear strain rate	ρ_u	scattering depolarisation for unpolarised light
γ_c	critical surface tension		
$\dot{\gamma}_c$	critical shear strain rate	σ	stress
Γ_2	second virial coefficient	σ	scattering cross-section
Γ_3	third virial coefficient	σ	steric hindrance parameter
δ	crosslinking coefficient	$\bar{\sigma}$	equivalent stress
δ	solubility parameter	$\hat{\sigma}$	tensile strength
δ	loss angle	σ^*	complex stress
δ_{ij}	Kronecker delta	σ_0	maximum value of a cyclically varying stress
Δ	dilatation		
Δ	rate of deformation tensor	σ_c	initial yield stress
Δ	logarithmic decrement	σ_N	stress normal to a plane
ε	extinction coefficient	$\sigma_{xx}, \sigma_{yy}, \sigma_{zz}$	normal stresses
ε	tensor strain	$\sigma(t)$	time-dependent stress

σ_θ	hoop stress	ϕ	quantum yield
τ	coefficient of viscous traction	ϕ	torsional angle
τ	shear stress	ϕ	volume fraction
τ	turbidity	ϕ_1	volume fraction of solvent
τ	relaxation time	ϕ_2	volume fraction of solute
τ'	retardation time	ϕ_f	volume fraction of filler
$\hat{\tau}$	shear strength	ϕ_{max}	maximum packing fraction
τ_c	critical shear stress	χ	Flory–Huggins interaction parameter
τ_{oct}	octahedral shear stress	ψ	torsional angle
τ_s	lifetime of a kinetic chain	Ψ	excess partial molar entropy of mixing
τ_{xy}, τ_{yz}, etc.	shear stress component	ω	angular velocity
τ_y	yield value for flow	∇	nabla operator

A

A Symbol for α-alanine.

AABB NYLON Alternative name for dyadic nylon.

AABB POLYAMIDE Alternative name for dyadic nylon.

AABB POLYMERISATION A step-growth polymerisation in which the two types of functional group (A and B), which react together in the polymer forming reaction, are attached to different monomer molecules. Thus, for example, in polyesterification using a diol and a diacid:

$$n\,HO—R—OH + n\,HOOC—R'—COOH \rightarrow$$
$$\text{⦅O—R—OCO—R'—CO⦆}_n + 2n\,H_2O$$

ABA BLOCK COPOLYMER Alternative name for triblock polymer.

ABACA (Manila hemp) A leaf fibre from the plant *Musa textilis*, consisting of strands of up to 400 cm in length, which can be almost white. Its composition is about 80% cellulose with about 10% lignin. It is useful for ropes and cordage.

ABALYN Tradename for methyl abietate.

AB BLOCK COPOLYMER Alternative name for diblock polymer.

AB CROSSLINKED POLYMER A block/graft copolymer which consists of chains of poly-A crosslinked through chains of poly-B. Prepared by generating active centres at specific sites on preformed poly-A in the presence of monomer B. The polymer formed initially resembles an ABA block copolymer but, as more initiating sites are generated, a network structure is gradually built up.

ABHESION The reduction in adhesion to a solid surface caused by the presence on that surface of a layer of an anti-stick material, such as a fluorocarbon, of low surface tension.

ABIBN Abbreviation for azobisisobutyronitrile.

ABIETIC ACID

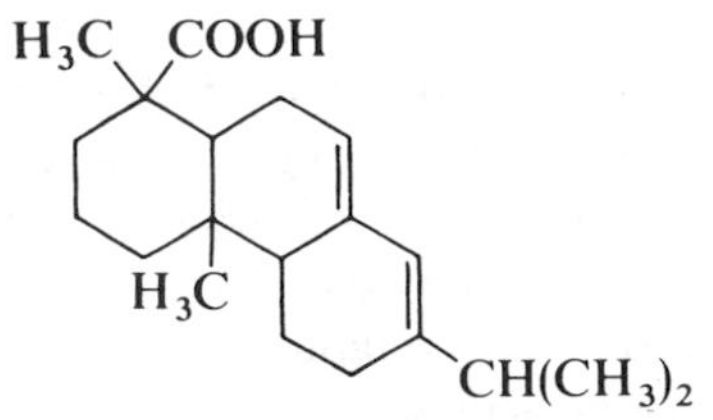

M.p. 170–172°C.

The major, about 90%, component of commercial rosin, whose structure is related to phenanthrene. The pine oleoresin from which rosin is produced by distillation contains the isomeric acid laevopimeric acid which is converted to abietic acid by the heat treatment during distillation. Simple esters, e.g. methyl abietate, are useful plasticisers, whilst more complex esters, e.g. of glycerol and pentaerythritol, form the ester gums. When rosin is used as a modifier in alkyd resins it is active in ester exchange. It forms an adduct with maleic anhydride via a Diels–Alder reaction which is also used in alkyd resins.

ABLATION The sacrificial loss of material when used as a heat shield to protect space vehicles during re-entry to the earth's atmosphere. The loss is due to decomposition and volatilisation caused by the frictional heating in the earth's upper atmosphere. Thermal protection is provided by the material acting both as a heat sink (the loss processes being endothermic) and as a thermal insulant.

ABLATIVE POLYMER A polymer composition, which under the pyrolytic conditions encountered in use as space vehicle heat shields, has useful ablation properties. Effective materials degrade endothermically to yield volatile products and a thermally insulating char. In practice nylon and phenolic/glass composites are widely used.

AB NYLON Alternative name for monadic nylon.

AB POLYAMIDE Alternative name for monadic nylon.

AB POLYMERISATION A step-growth polymerisation in which the two types of functional group (A and B), which react together in the polymer forming reaction, are attached to the same monomer molecule. Thus, for example, in the polymerisation by self-condensation of a hydroxyacid to a polyester:

$$n\,HO—R—COOH \rightarrow \text{⦅O—R—CO⦆}_n + n\,H_2O$$

ABS Abbreviation for acrylonitrile–butadiene–styrene copolymer.

ABSOLUTE MODULUS Symbol $|M|$. The absolute value of the complex modulus for dynamic mechanical behaviour. $|M| = \sigma_0/e_0$ where σ_0 and e_0 are the maximum amplitudes of the cyclically varying stress and strain respectively. $|M|$ is related to the storage (M') and loss (M'') components by $|M| = [(M')^2 + (M'')^2]^{1/2}$.

ABSOLUTE PERMITTIVITY A measure of the ability of a material to store electric charge, as energy per unit volume, and having units of farads per metre, when the material is polarised due to an applied electric field. Its value is dependent on both sample geometry and field strength. Strictly, it is the ratio of the electric displacement (which is the electric flux in a material in a uniform field compared with that in free space) to the field strength. The term is sometimes used to mean the relative permittivity, which is a material property independent of field strength.

Any confusion is limited since absolute permittivities have values of about 10^{-11} to $10^{-9}\,F\,m^{-1}$, whereas relative permittivities have dimensionless values of 1 to about 50.

ABSON Tradename for acrylonitrile–butadiene–styrene copolymer.

ABSORPTION OPTICS An optical system used in an ultracentrifuge for determination of the solute concentration in any part of the cell. Particularly useful for proteins and nucleic acids which absorb ultraviolet light strongly. A slit beam of light passes through the cell and produces an image of the column of liquid on a photographic plate. The degree of blackening at any point in the image is proportional to the solute concentration at the corresponding point in the cell. Alternatively, the image may be scanned with a photoelectric scanner. The optical system is simpler than Schlieren or other optics and with some biopolymers is so sensitive that solution concentrations down to 0·001% may be used.

ABSORPTOMETER An instrument for measuring the structure of carbon black by determining the volume of the air spaces between the aggregated black particles. This is done by measuring the volume of dibutyl phthalate required to fill the voids in 100 g of black. A weighed amount of black is stirred in a chamber and dibutyl phthalate is added. When all the voids are just filled the mixture noticeably stiffens.

AC Abbreviation for azodicarbonamide.

ACCELERATED WEATHERING Attempted speeding up of natural weathering processes in materials by exposure to a radiation source (carbon arc, xenon lamp, fluorescent lamp or concentrated sunlight) simulating sunlight, such that the sample receives a higher radiation flux than during natural weathering. Owing to the complex participation of factors other than the radiation itself, that bring about degradation during weathering, accelerated weathering is only partially successful in predicting actual weathering behaviour.

ACCELERATOR (1) A component of a rubber compound which is usually present when the compound is to be sulphur vulcanised. An accelerator is also sometimes used for non-sulphur vulcanisation of some chlororubbers such as chloroprene rubber and chlorosulphonated polyethylene. Its function is not only to speed up the vulcanisation, but also to increase the efficiency of vulcanisation by encouraging useful mono- and disulphide crosslinks to form, rather than to waste sulphur in forming ineffective polysulphide crosslinks and cyclic structures. Choice of accelerator is determined by both the required vulcanisation behaviour (especially freedom from scorch—a delayed action being desirable) and vulcanisate properties. A very wide range of materials is available; sometimes they are used in combination as a mixture of primary and secondary accelerators. Particu-

larly active accelerators are termed ultraaccelerators. Important types of accelerator are the thiazoles, sulphenamides, dithiocarbamates, guanidines and thiuram disulphides.

Accelerators are usually used in combination with an activator, most commonly a mixture of stearic acid and zinc oxide. Synthetic rubbers vulcanise more slowly than natural rubber and need a higher amount of accelerator. A typical recipe for natural rubber in phr, might be: zinc oxide (6), stearic acid (0·5), cyclohexylbenzothiazylsulphenamide (0·4), sulphur (3·0), giving a cure time of about 30 min. For styrene–butadiene rubber, for a similar cure time, a suitable recipe might be: zinc oxide (3·0), stearic acid (3·0), cyclohexylbenzothiazylsulphenamide (2·0) and sulphur (2·0). For synthetic rubbers with only very few double bonds, such as butyl rubber and EPDM, even more active accelerator systems are needed.

The chemistry of action of accelerators is complex. In sulphur vulcanisation reaction occurs between a zinc salt of an accelerator (formed from the accelerator and the zinc oxide) with sulphur to give a perthiosalt of the type XS_xZnS_xX, where X is a group derived from the accelerator. This salt then reacts with rubber (R—H):

$$XS_xZnS_xX + R-H \rightarrow XS_xR + ZnS + HS_{x-1}X$$

The perthioaccelerator will react with more zinc oxide to form more zinc perthiosalt, but of lower sulphur content than before. This will similarly again react with more R—H. In this way rubber intermediates of the type $X-S_{x-1}-R$, $X-S_{x-2}-R$, etc., of varying polysulphidity are formed. Crosslinks are then formed when these react with a further rubber molecule:

$$X-S_x-R + R-H \rightarrow R-S_{x-1}-R + X-SH$$

Many other reaction steps must be involved in the overall process since, for example, as vulcanisation proceeds the crosslinks formed decrease in polysulphidity. Furthermore, crosslinks initially formed α to the double bond in the rubber appear to shift to other positions. Thus in natural rubber:

$$\sim\!CH_2-C(CH_3)\!=\!CH-\underset{\underset{R}{\overset{|}{S_x}}}{\overset{|}{CH}}\!\sim \rightarrow$$

$$\sim\!CH_2-\underset{\underset{R}{\overset{|}{S_x}}}{\overset{|}{C}}(CH_3)-CH\!=\!CH\!\sim$$

(2) An activator for an organic peroxide free radical initiator, which enables polymerisation to be readily performed at ambient temperature. Accelerators are used especially for the room temperature curing of unsaturated polyester resins, using methylethyl ketone or cyclohexanone peroxide. Accelerators are variable valency metal soaps. The most commonly used metal is cobalt, but tin, iron, manganese, cerium and vanadium are also used. The most commonly used acid group is the complex industrial product naphthenic acid. Thus cobalt naphthenate is a typical example. Metal octoates are also widely

used. Accelerators are capable of catalytically decomposing hydroperoxides by the reactions:

$$ROOH + M^{2+} \rightarrow RO\cdot + OH^- + M^{3+}$$

and

$$ROOH + M^{3+} \rightarrow ROO\cdot + H^+ + M^{2+}$$

ACCESSIBILITY   The percentage of a polymer accessible to direct chemical reaction. It applies largely to crystalline polymers in which chemically reactive groups, in the crystalline or other regions, do not readily react due to the inability of reagent to diffuse to the reaction sites. The term applies especially to cellulose, in which accessibility of the hydroxyl groups is measured by deuterium or tritium exchange. When these groups are reacted (e.g. acetylated) with the cellulose merely swollen with solvent, accessibility will be < 100%. Accessibility is altered by any treatment which affects morphology such as gross or fine structure, crystallite size or extent of crystallinity. Thus in cotton accessibility is about 35%, increased to about 55% on mercerisation and to about 75% in regenerated viscose film.

ACCORDION MODE   (Longitudinal acoustic mode) A vibrational mode of a chain molecule consisting of a symmetrical elongation and contraction of the molecular chain along its length. The frequencies of such vibrations have been studied, for example, by Raman spectroscopy, especially in polyethylene. The frequency may be related to the elastic constants of the material.

ACENAPHTHYLENE

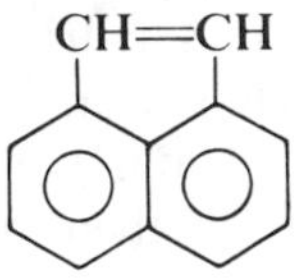

CH=CH

M.p. 93°C.

Obtained by the dehydrogenation of acenaphthene, itself obtained from heavy coal tar fractions. It polymerises by a free radical mechanism to polyacenaphthylene.

ACETA   Tradename for cellulose acetate rayon.

ACETAL   In general the chemical group

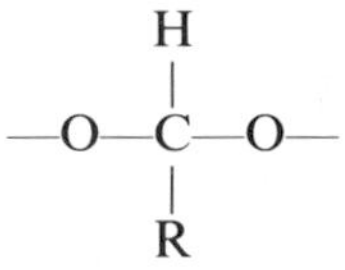

H
|
—O—C—O—
|
R

where R is an alkyl group or hydrogen; polymers containing this group in the repeat unit are polyacetals. However, the term is also used as an alternative name for the commercially important homo- and copolymer polyoxymethylenes.

2-ACETAMIDO-2-DEOXY-D-GALACTOSE   Alternative name for *N*-acetyl-D-galactosamine.

2-ACETAMIDO-2-DEOXY-β-D-GLUCOSE   Alternative name for *N*-acetyl-D-glucosamine.

ACETATE   Alternative name for cellulose acetate rayon fibres of degree of substitution about 2·5.

ACETATE RAYON   Alternative name for cellulose acetate rayon.

ACETONE   (Propan-2-one) CH_3COCH_3. B.p. 56·3°C. A solvent for most cellulose esters and ethers, many natural resins, polyvinyl acetate, low molecular weight silicones and polystyrene. Somewhat swells polyvinyl chloride, polyethylene, polymethylmethacrylate and some rubbers. Widely used in lacquers and in cellulose acetate fibre production. In general, it is a powerful solvent (but not particularly so for polymers) of high volatility, low cost and low toxicity. Miscible with water in all proportions.

N-ACETYL-2-AMINO-2-DEOXY-D-GALACTOSE Alternative name for *N*-acetyl-D-galactosamine.

N-ACETYL-2-AMINO-2-DEOXY-D-GLUCOSE Alternative name for *N*-acetyl-D-glucosamine.

ACETYLCYCLOHEXANESULPHONYL PEROXIDE

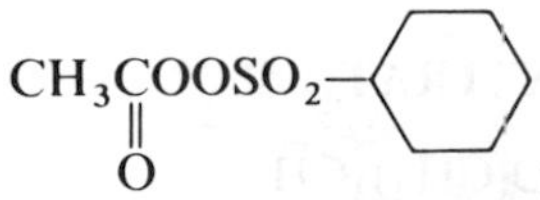

$$CH_3COOSO_2{-}\bigcirc$$
          ‖
          O   M.p. 33°C.

A particularly active diacyl peroxide with half-life times of 0·3 h/50°C and 0·04 h/70°C and half-life temperatures of 30°C/10 h, 46°C/1 h and 75°C/1 min. Therefore it is useful for the initiation of free radical polymerisations at lower than usual temperature (around 40°C), especially of vinyl chloride.

ACETYLENE BLACK   A form of carbon black produced by high temperature thermal decomposition of acetylene. It has a very high structure, with particle sizes of about 50 nm and unusually high electrical and thermal conductivity. Its main use is in dry cell batteries.

N-ACETYL-D-GALACTOSAMINE   (2-Acetamido-2-deoxy-D-galactose) (*N*-Acetyl-2-amino-2-deoxy-D-galactose)

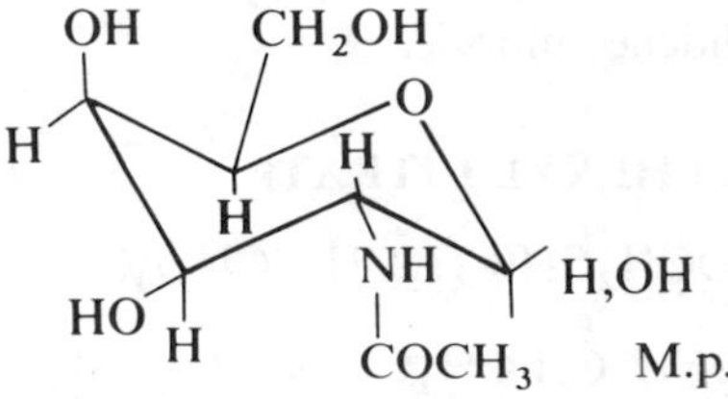

OH   CH₂OH
                    O
H   H
       H
              NH   H,OH
HO   H |
         COCH₃   M.p. 172–173°C. α_D^{27} +86°.

An aminosugar occurring as a monosaccharide unit in several aminopolysaccharides, e.g. chondroitin, chondroitin sulphate and blood-group substances. Acid hydrolysis of these polymers yields the parent monosaccharide D-galactosamine.

N-ACETYL-D-GLUCOSAMINE

(2-Acetamido-2-deoxy-D-glucose)
(N-Acetyl-2-amino-2-deoxy-D-glucose)

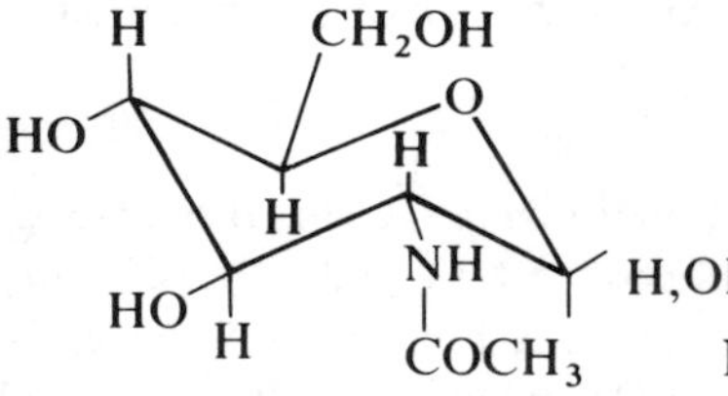

M.p. 210°C. α_D^{25} + 41·3°.

An aminosugar occurring as a monosaccharide unit in several aminopolysaccharides, e.g. hyaluronic acid, blood-group substances, keratosulphate, heparin and peptidoglycans and as a linear homopolysaccharide (chitin). Acid hydrolysis of these polymers yields the parent D-glucosamine. Obtained by controlled acid or enzyme hydrolysis of chitin.

N-ACETYLHEPARAN SULPHATE
A group of mucopolysaccharides similar in structure to heparin but with a lower sulphate content and in which some amino groups are acetylated. Heparitin sulphate and heparin monosulphate are of this type.

N-ACETYLNEURAMINIC ACID
Alternative name for sialic acid.

ACETYLTRIBUTYL CITRATE

$$CH_2COO(CH_2)_3CH_3$$
$$CH_3COO-C-COO(CH_2)_3CH_3 \quad \text{B.p. } 173°C/1\,\text{mm.}$$
$$CH_2COO(CH_2)_3CH_3$$

A plasticiser for polyvinyl chloride, polyvinyl acetate, polyvinyl acetals and polystyrene. Also a stabiliser for polyvinylidene chloride. It is also compatible with many cellulose derivatives, but not with cellulose acetate. It has a low toxicity.

ACETYLTRIETHYL CITRATE

$$CH_2COOC_2H_5$$
$$CH_3COO-C-COOC_2H_5 \quad \text{B.p. } 127°C/1\,\text{mm.}$$
$$CH_2COOC_2H_5$$

A plasticiser for cellulose derivatives, notably ethyl cellulose, vinyl chloride polymers and copolymers and a stabiliser for polyvinylidene chloride.

ACETYLTRI-2-ETHYLHEXYL CITRATE

$$CH_2COOCH_2CH(CH_2CH_3)(CH_2)_3CH_3$$
$$CH_3COO-C-COOCH_2CH(CH_2CH_3)(CH_2)_3CH_3$$
$$CH_2COOCH_2CH(CH_2CH_3)(CH_2)_3CH_3$$

B.p. 225°C/1 mm.

A low volatility plasticiser for polyvinyl chloride and a stabiliser for polyvinylidene chloride.

α_1-ACID GLYCOPROTEIN
An animal serum glycoprotein, of which mammalian sera, especially human plasma α-glycoprotein, have been widely studied. Typically of molecular weight about 45 000 and having a very high carbohydrate content of 4–12 groups linked to asparagine residues of the protein and containing, in one example, 18 units each of D-galactose and D-mannose, 36 of 2-acetamido-2-deoxy-G-glucose, 3 of L-fucose and 16 sialic acid.

ACIDIC MUCOPOLYSACCHARIDE
(Acid mucopolysaccharide) A glycosoaminoglycan, usually found in animal connective tissues. It consists of alternating sequences of hexosamine (D-galactosamine or D-glucosamine) and hexuronic acid (D-glucuronic or L-iduronic) units, i.e. formally a glycosoaminoglycurono-glycan, and may contain N-acetyl or O- or N-sulphate groups. Examples include hyaluronic acid, the chondroitin sulphates, keratan sulphate (in which D-galactose replaces uronic acid) and heparin. In the native state it is often linked covalently to protein, i.e. it is a glycoprotein, from which it may be separated by mild degradative procedures.

ACID MUCOPOLYSACCHARIDE
Alternative name for acidic mucopolysaccharide.

ACID NUMBER
Alternative name for acid value.

ACID VALUE
(AV) (Acid number) A measure of the concentration of carboxyl end groups in a polymer, often used in the characterisation of polyesters. Defined as the weight in milligrams of potassium hydroxide required to neutralise 1 g of the polymer. Thus, if x g of polymer, when titrated with base (KOH or NaOH) of normality N, required y cm³ for neutralisation, then the acid value is $5·61\,yN/x$.

ACLAR
Tradename for polychlorotrifluoroethylene.

ACM
Abbreviation for acrylic elastomer.

AC POLYETHYLENE
Tradename for low density polyethylene.

AC POLYMER
Tradename for ethylene–acrylic acid copolymer.

ACRIBEL
Tradename for polyacrylonitrile fibre.

ACRILAN
Tradename for a polyacrylonitrile fibre containing about 10% vinyl pyridine comonomer units to improve dyeability.

ACROLEIN
(Acrylaldehyde) (Acrylic anhydride) $CH_2{=}CHCHO$. B.p. 52·6°C. Obtained by the dehydration of glycerol, e.g. with potassium hydrogen sulphate, at 190°C or, commercially, by the vapour phase catalysed oxidation of propylene with oxygen at 300–550°C. It is mostly used as a chemical intermediate, especially for

conversion to glycerol, but also to 1,2,6-hexanetriol. It may be polymerised by free radical, cationic or anionic mechanisms to polyacrolein. However, the polymer has a complex structure, being crosslinked and containing 10–20 mol% of aldehyde groups. Polymeric condensation products with phenol are useful as adhesives and those with formaldehyde as lacquers.

AC RUBBER   Alternative name for anticrystallising rubber.

ACRYL-ACE   Tradename for polymethylmethacrylate.

ACRYLALDEHYDE   Alternative name for acrolein.

ACRYLAMIDE CH_2=$CHCONH_2$. M.p. 85°C. Produced by the partial hydrolysis of acrylonitrile by its dissolution in concentrated sulphuric acid, followed by dilution with water. The resultant acrylamide sulphate is then decomposed with alkali. It is useful for the production of polyacrylamide, for copolymers with N,N'-methylenebisacrylamide as chemical grouts and for chromatographic column and electrophoresis materials and with various acrylates for surface coatings. Various N-substituted acrylamides, e.g. N-methylolacrylamide, are also made from acrylamide.

ACRYLATE   An ester of acrylic acid, i.e. a compound of structure CH_2=$CHCOOR$, where R is an alkyl group.

ACRYLATE RUBBER   Alternative name for acrylic elastomer.

ACRYLIC   (1) Generic name for a fibre containing at least 85% by weight of acrylonitrile repeating units in its polymer chains. Frequently small amounts of comonomer are also incorporated to improve certain properties, notably dyeability. If more than 15% of comonomer is present then the fibre is described as modacrylic. Examples are Acrilan, Courtelle, Creslan, Orlon and Zefran.

(2) Technological name for polymethylmethacrylate in plastics technology.

ACRYLIC ACID CH_2=$CHCOOH$. B.p. 142°C. M.p. 13°C. Produced either by dehydration followed by hydrolysis of ethylene cyanohydrin, itself obtained by reaction of ethylene oxide with hydrogen cyanide:

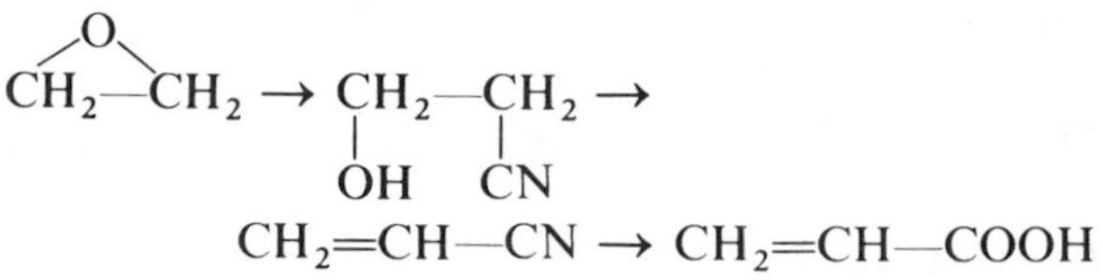

$$CH_2\overset{O}{-}CH_2 \rightarrow \underset{OH}{CH_2}-\underset{CN}{CH_2} \rightarrow$$

$$CH_2=CH-CN \rightarrow CH_2=CH-COOH$$

by reaction of acetylene with carbon monoxide and water, or by aerial oxidation of propylene at 400–500°C. Useful as the monomer for polyacrylic acid and as a comonomer in thermosetting acrylic coatings.

ACRYLIC ALDEHYDE   Alternative name for acrolein.

ACRYLIC ELASTOMER   (Acrylate rubber) (Polyacrylate rubber) (Polyacrylic elastomer) (ACM) Tradenames: Cyanacril, Elaprim AR, Hycar, Krynac, Lactoprene BN, Lactoprene EV, Paracril, Thiacryl. A rubber based on a polyacrylate, usually polyethylacrylate or poly-n-butylacrylate. The former has the better oil resistance but poorer low temperature resistance than the latter. The vulcanisates show better high temperature resistance (to about 200°C) than most rubbers and also good oil resistance. The polymers may be crosslinked by heating with peroxides or with alkali or, better, through reactive sites on a suitable incorporated comonomer. Usually about 5% of 2-chloroethylvinyl ether, vinyl chloroacetate or allylglycidyl ether may be used as the comonomer. Better oil resistance combined with good low temperature properties are obtained by the incorporation of 20–50% of an alkoxyacrylate comonomer such as ethoxy- or methoxyethylacrylate.

ACRYLIC POLYMER   A polymer with repeat units which can be considered as derivatives of acrylic or of a substituted acrylic acid, i.e. containing units of the type

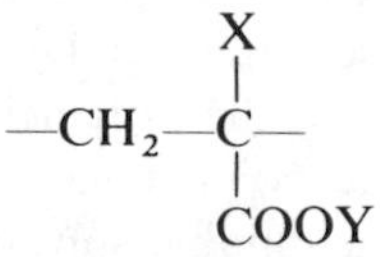

$$-CH_2-\underset{COOY}{\overset{X}{C}}-$$

where, most commonly, X = H (acrylic acid derivative) or X = CH_3 (methacrylic acid derivative). The most important group of such polymers is the esters, both the polyacrylates and the polymethacrylates, where Y is an alkyl group. Other important polymers are the acrylic amides

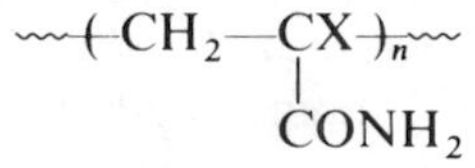

$$\sim(CH_2-\underset{CONH_2}{CX})_n\sim$$

and the nitriles

$$\sim(CH_2-\underset{CN}{CX})_n\sim$$

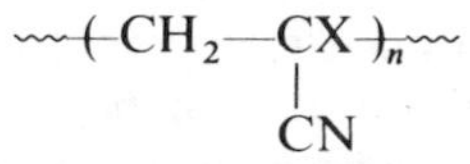

A wide range of other acrylic ester polymers have also been investigated, including alkoxyacrylates, cyanoacrylates, glycol dimethacrylates, α-chloroacrylates, hydroxyethylacrylate and methacrylate. Technologically, acrylic polymers are useful in a wide range of applications, individual members or their copolymers being important as plastics (especially polymethylmethacrylate), fibres (polyacrylonitrile), rubbers (acrylic elastomers) and adhesives (anaerobic and cyanoacrylate) and as thermoset coatings (usually complex copolymers). Acrylic monomers enter readily into copolymerisation with each other and with other monomers, so a wide range of copolymers has been investigated. The monomers are readily polymerised by free radical polymerisation, although they are also susceptible to anionic polymerisation.

ACRYLITE   Tradename for polymethylmethacrylate.

ACRYLOID Early tradename for polymethylacrylate and other acrylates and methacrylates in solution for use as adhesives and coatings. Later also used for higher methacrylate polymers and copolymers useful as lubricating oil viscosity modifiers.

ACRYLONITRILE (AN) CH_2=CHCN. B.p. 77·3°C. Produced largely by ammonoxidation of propylene:

$$2\,CH_2{=}CHCH_3 + 3\,O_2 + 2\,NH_3 \rightarrow$$
$$2\,CH_2{=}CHCN + 6\,H_2O$$

but also by oxidation of propylene with nitric oxide and by dehydration of ethylene cyanohydrin, itself obtained from ethylene oxide. It may also be produced by the addition of hydrogen cyanide to acetylene. Most monomer is used for the production of acrylic and modacrylic fibres. Other major uses are for the production of nitrile rubbers, acrylonitrile–butadiene–styrene and styrene–acrylonitrile copolymers. It is also converted to adiponitrile by hydrodimerisation, to acrylamide and to acrylic esters. The monomer is polymerised to polyacrylonitrile by free radical polymerisation or by anionic mechanisms. In the latter case stereoregular polymer may be obtained, but usually, like the free radical polymers, atactic polymer is produced. Unusually, the polymer is insoluble in the monomer so that the polymerisation is heterogeneous unless conducted in a suitable solvent. The monomer is highly volatile, flammable and toxic.

ACRYLONITRILE–BUTADIENE–STYRENE COPOLYMER (ABS) Tradenames: Abson, Bexan, Blendex, Cycolac, Dylel, Extir, Formid, Hi-Blen, Kostiline, Kralac, Kralastic, Kralon, Lacqran, Lorkaril, Lustran, Lustropak, Magnum, Marbon, Nivionplast, Novodur, Ravikal, Restiran, Ronfalin, Royalite, Sicoflex, Sternite, Terluran, Tybrene, Ugikral, Urtal, Uscolite. A terpolymer of the three monomers, often considered as a modified polystyrene since its properties resemble this polymer, except that its impact strength is much higher. A very wide variety of commercial products are available, varying not only in composition but also in morphology, due to the different procedures used in the copolymer preparation.

Early products were mechanical mixtures of styrene–acrylonitrile copolymer and nitrile rubber or they were obtained by co-coagulating latices of these two polymers. Present products are almost entirely produced by the copolymerisation of styrene and acrylonitrile in the presence of polybutadiene latex—the so-called latex grafting method. This results in considerable grafting and the products have higher impact strengths.

The morphology is similar to that of high impact polystyrene, but the dispersed rubber particles are much smaller (1–10 μm) and contain styrene–acrylonitrile copolymer inclusions. Other rubbers which are sometimes used are ethylene–propylene–diene monomer rubber and solution polybutadiene. A typical ABS contains about 20% rubber, about 25% acrylonitrile and about 55% styrene, having a T_g value of about 105°C, a tensile modulus of 2·5 GPa and an impact strength of about 4 J

$(12\cdot7\,\text{mm})^{-1}$ on an Izod test. However, owing to the many variations possible, the properties can vary considerably, e.g. 'super' ABS can have a notched Izod impact strength of up to about 8 J $(12\cdot7\,\text{mm})^{-1}$. Furthermore, some or all of one of the comonomers may be replaced, e.g. methylmethacrylate may replace acrylonitrile, as in MBS and MABS, with improvement in transparency, or a saturated rubber may replace the polybutadiene, as in ASA and ACS, with an improvement in oxidation resistance. ABS is mainly used as an injection moulding material where good appearance and strength, together with reasonable stiffness and softening point are required.

ACRYLONITRILE–CHLORINATED POLYETHYLENE–STYRENE COPOLYMER (ACS) Tradename ACS Resin. A terpolymer, obtained by the copolymerisation of acrylonitrile and styrene on the presence of chlorinated polyethylene, similar in properties to acrylonitrile–butadiene–styrene copolymer, except that it is more resistant to embrittlement due to oxidative degradation, e.g. during weathering, and has better fire resistance.

ACRYLONITRILE–ETHYLENE/PROPYLENE RUBBER–STYRENE COPOLYMER (AES) Tradenames Hostyren XS, Novodur AES, Rovel. A terpolymer obtained by grafting styrene–acrylonitrile copolymer onto ethylene–propylene or ethylene–propylene–diene monomer rubber. Similar to ABS but with better weathering resistance.

ACRYLONITRILE–STYRENE–ACRYLATE COPOLYMER (ASA) Tradenames Geloy, Luran S. A terpolymer, similar to acrylonitrile–butadiene–styrene copolymer, but with a saturated acrylate rubber replacing the unsaturated polybutadiene. This improves the resistance of the polymer to oxidative degradation, e.g. during weathering.

ACRYSOL Tradename for polyacrylic acid and its salts.

ACS Abbreviation for acrylonitrile-chlorinated polyethylene–styrene copolymer.

ACS RESIN Tradename for acrylonitrile-chlorinated polyethylene–styrene copolymer.

ACSZ Abbreviation for aryl cyclic sulphonium zwitterion.

ACTH Abbreviation for adrenocorticotropin hormone.

ACTIGUM Tradename for xanthan.

ACTIN A major, about 20%, contractile protein found in the striated muscles of higher animals. Occurs in the thin filaments of the striated myofibrils, which themselves make up the muscle fibres. Extracted from dried muscle, after the other major component myosin, with cold, slightly alkaline water. As extracted the actin consists of globular molecules (G-actin) of molecular weight 46 000 and diameter about 5·5 nm, with a fairly high, about 5%, proline content and cysteine, about 1%, as well as the unusual amino acid 3-methylhistidine. In the presence of potassium chloride and copper ions, when bound to ATP, G-actin polymerises to fibrous form (F-actin):

$$n(\text{G-actin—ATP}) \rightarrow (\text{G-actin—ADP})_n + n\,\text{phosphate}$$
$$\text{F-actin}$$

The F-actin consists of two helically entwined chains of spheres of G-actin monomers. In the thin filaments of the myofibrils tropomyosin and troponin are also associated with the F-actin chain.

ACTINIC DEGRADATION   Degradation induced by exposure to ultraviolet light (photodegradation) or to high energy radiation.

ACTIVATED MONOMER   In chain polymerisation, the product of the reaction of the primary radical, usually derived from the initiator, with a monomer molecule: $I^* + M \rightarrow I\text{—}M^*$. Thus in free radical polymerisation of a vinyl monomer:

$$R\cdot + CH_2{=}CHX \rightarrow R\text{—}CH_2\text{—}CHX\cdot$$

Subsequently the activated monomer $I\text{—}M^*$ rapidly grows by many successive propagation steps to form a polymer molecule.

ACTIVATION GRAFTING   A method of making a graft copolymer by producing, by irradiation, free radical active sites on one polymer in the presence of the monomer for a second polymer. Some selectivity of grafting sites may be achieved by ultraviolet light irradiation by photolysis of carbonyl groups or halogen atoms in a side group. On the other hand, for high energy irradiation, the production of grafting sites is much more indiscriminate.

ACTIVATOR   (1) A compound used to increase the vulcanisation rate of a rubber by activating the accelerator. Most vulcanisation systems contain two activator components—a metal oxide (usually zinc oxide) at about 3–5 phr and a fatty acid (usually stearic acid) at about 0·5–2 phr. The activator is thought to work by forming zinc stearate which itself forms rubber soluble complexes with the accelerator.

(2) In certain chemical reactions, a subsidiary chemical species capable of activating the main species which brings about the chemical change. The term is used, particularly, when metal oxides and organic acids, e.g. zinc oxide and stearic acid, activate rubber vulcanisation accelerators, when transition metal ions, e.g. Fe^{2+}, activate redox polymerisation catalysts and when reducing agents activate supported metal oxide, e.g. CrO_3 and MoO_3, polymerisation catalysts.

ACTIVE CENTRE   (1) In chain polymerisation, the atom, or reactive bond between two atoms, at which addition of successive monomer molecules occurs during polymerisation propagation through interaction with the π-bond of the monomer. In free radical polymerisation the active centre is a free radical, e.g. $\sim\!CH_2CHX\cdot$ in a vinyl polymerisation. In cationic polymerisation it is usually a carbenium ion, e.g. $\sim\!CH_2CHX^+$, or sometimes an oxonium ion, e.g. $\sim\!(CH_2)_x\text{—}O^+$ in cyclic ether polymerisation. In anionic polymerisation it is a carbanion, e.g. $\sim\!CH_2CHX^-$. In coordination polymerisation, e.g. Ziegler–Natta polymerisation, the monomer is often considered to add on by inserting itself in the highly polar 'bond' between the growing chain end and the catalyst component to which it is coordinated, when this bond is considered to be the active centre.

(2) Alternative name for active site.

ACTIVE ESTER   An ester whose carbonyl group is particularly active towards nucleophilic attack due to withdrawal of electrons by the R and R′ groups of the ester RCOOR′. Of especial importance in peptide synthesis by reaction of a free amino group of an amino acid or peptide with an active ester of a second amino acid or peptide. Here aromatic R′ groups are used, containing electron-withdrawing groups as in p-nitrophenyl and trichlorophenyl esters or N-hydroxysuccinimidyl.

ACTIVE SITE   (Active centre) The region in an enzyme which contains the binding site for the substrate and catalytic site at which the enzyme-catalysed transformation takes place. Only a small part of the overall volume occupied by the folded enzyme molecule and usually situated near the surface for ready access by the substrate. In some cases it consists of a cleft or pocket into which the substrate or part of it neatly fits. Owing to polypeptide chain folding, amino acid residues widely separated along the chain length may be brought close together at the active site. Binding is usually through specific amino acid residues but usually only involves non-covalent forces. Enzymes may be classified according to the essential amino acid present at the binding site, as with a serine enzyme, such as the serine protease, trypsin. However, amino acid residues not involved in binding or catalysis often contribute to the specificity of the enzyme as a result of their steric and polar effects.

ACYL ENZYME   The enzyme–substrate intermediate formed by reaction of a specific amino acid residue at the enzyme active site with a specific acyl group on the substrate, e.g. reaction of a serine hydroxyl of a serine enzyme with an acyl group of an ester.

ACYL PEROXIDE   Alternative name for diacyl peroxide.

ADAMANTANE-POLYBENZOXAZOLE   (AD–PBO) A high temperature resistant polymer with potential use as an electrolytic cell membrane material. Produced by reaction of adamantane dicarboxylic acid chloride (obtained from acenaphthene or methylcyclopentadiene dimer) with 4,4'-diamino-3,3'-dimethyloxydiphenyl or *o*-dianisidine:

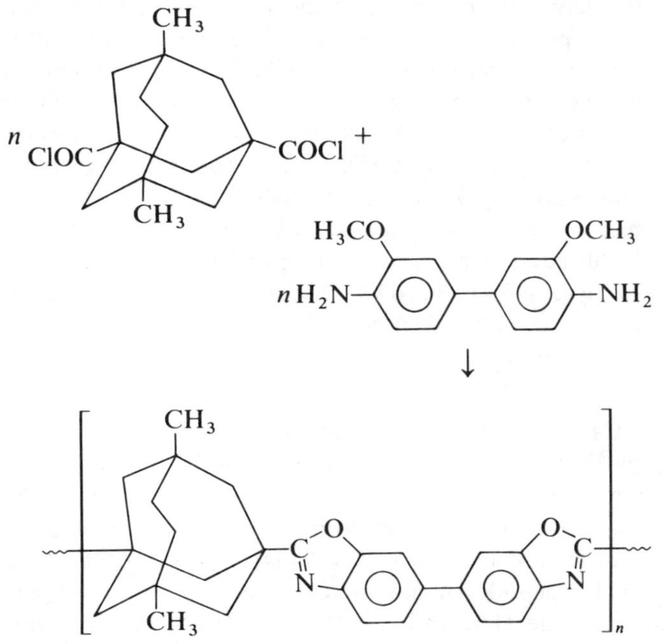

1,2-ADDITION Occurs in diene polymerisation, with monomers of the type $CH_2{=}CX{-}CH{=}CH_2$, when addition of monomer to the active centre is through reaction of the 1,2-double bond, giving the polymer

$$\pm CH_2{-}CX\pm_n$$
$$\quad\quad\;\; |$$
$$\quad\quad CH{=}CH_2$$

Very few 1,2-structures are formed in free radical polymerisation, but polymers with about 50% 1,2-structures may be formed using metal alkyl initiators. Certain Ziegler–Natta polymerisations produce polymers with up to 99% 1,2-structures in either the iso- or syndiotactic forms.

1,4-ADDITION Occurs in diene polymerisation, with monomers of the type $CH_2{=}CX{-}CH{=}CH_2$, when addition of monomer to the active centre involves both of its double bonds. It therefore appears to have added across carbons 1 and 4 by opening of the conjugated double bond, best represented as

$$\overset{1}{C}H_2{=\!=}\overset{2}{C}X{=\!=}\overset{3}{C}H{=\!=}\overset{4}{C}H_2$$

It may also be viewed as the result of addition of monomer to the resonance isomer (**II**) of the active centre (**I**), formed normally by a 1,2- or 3,4-addition:

$$\sim\!\!CH_2{-}\dot{C}X{-}CH{=}CH_2 \leftrightarrow$$
$$\qquad\qquad\quad \mathbf{I}$$

$$\sim\!\!CH_2{-}CX{=}CH{-}\dot{C}H_2 \xrightarrow{n\ CH_2{=}CX{-}CH{=}CH_2}$$
$$\qquad\quad\; \mathbf{II}$$

$$\pm CH_2{-}CX{=}CH{-}CH_2\pm_n$$

The 1,4-units in the polymer may be formed as either the *cis* or *trans* isomers, i.e. exhibit *cis–trans* isomerism. 1,4-addition predominates in free radical polymerisation, usually *trans* isomeric units being in the majority.

3,4-ADDITION Occurs in diene polymerisation, with monomers of the type $CH_2{=}CX{-}CH{=}CH_2$, when addition of monomer to the active centre is by reaction of the 3,4-double bond, giving the polymer

$$\pm CH_2{-}CH\pm_n$$
$$\quad\quad\;\; |$$
$$\quad\quad CX{=}CH_2$$

With a symmetrical diene, e.g. butadiene, 3,4-addition is indistinguishable from 1,2-addition. High 3,4-addition is rare, but examples occur in the anionic polymerisation of isoprene with sodium *n*-butyl and in Ziegler–Natta polymerisations with vanadium acetylacetonate/aluminium triethyl.

ADDITION POLYMER Alternative name for chain polymer.

ADDITION POLYMERISATION Alternative name for chain polymerisation.

ADDITIVE A material used in conjunction with a polymer to produce a polymer compound or composite in order to modify the polymer properties in a desired direction for processing or end use. Usually the term is restricted to materials mixed reasonably intimately with the polymer. Thus, for instance, in composite laminates containing fibrous or sheet reinforcements, where the mixing is not on a small scale, the reinforcing filler is not considered to be an additive since it consists of relatively large 'particles'. The amount of additive used is usually expressed in terms of phr—parts per hundred (by weight) of resin (i.e. polymer). An additive may be used in very small amounts, e.g. less than 0·01 phr for a crystallisation nucleating agent, or in exceedingly large amounts, e.g. several hundred phr with some fillers. In this latter case the compound will have properties considerably different from that of the polymer. However, additives are more commonly used in the range of a few to a few tens of phr, when the properties are generally those of the base polymer, suitably modified. In individual cases it may be necessary to modify processing properties, e.g. by the use of lubricants, heat stabilisers or peptisers, or to modify end-use properties, e.g. by the use of fillers, plasticisers, blowing agents, impact modifiers, colourants, flame retardants, antioxidants, etc.

ADHESION The attraction between a solid surface and a liquid or other solid surface. The adhesion between a liquid which may subsequently solidify (the adhesive) and two solid surfaces (the adherends) gives such an adhesive joint its strength. Most adhesives are based on polymers. Adhesion results from the intermolecular forces of attraction and also possibly from chemical bonding of the liquid to the solids. However, in an adhesive joint the strength of adhesion, as measured by some destructive test, is much less than that calculated for intermolecular forces. This is due to several factors. Defects in the joined surfaces may result in stress concentrations. In testing, the surfaces will not remain exactly parallel so peeling will occur and the molecular interactions will be disrupted sequentially, not all simultaneously. Additional stress may be present due to shrinkage during setting of the adhesive or differential expansion/contraction between adhesive and adherends. Often uniform stress transfer between adherend and adhesive does not occur, so local stresses may be many times the mean stress. Furthermore, defects may be present at the interface, where the adhesive has not covered the adherend surface completely, owing to its inability to sufficiently wet it. These also cause stress concentrations. Thus joint strength increases with decreasing contact angle and with roughness, as given by Wenzel's equation. Spreading will also depend on the critical surface tension of the solid surface. Such an effect is expressed in the de-Bruyne rule.

ADHESION SHEAR STRENGTH Alternative name for interlaminar shear strength.

ADIPIC ACID $HOOC(CH_2)_4COOH$. M.p. 153°C. Usually synthesised by air oxidation of cyclohexane, initially to a mixture of cyclohexanone and cyclohexanol, which is then further oxidised with nitric acid to adipic acid. A monomer for the formation of nylon 66 (and other polyadipamides) via formation of the nylon 66 salt by reaction with hexamethylenediamine. The latter is also produced from adipic acid, as are adipate polyesters and adipate ester plasticisers. Sometimes used in place of phthalic anhydride for the formation of unsaturated polyester resins and alkyd resins, giving cured products with greater flexibility.

ADIPRENE C Tradename for a millable polyurethane elastomer based on poly-(1,4-oxybutylene glycols) containing glycerylmonoethyl ether and tolylene diisocyanate.

ADIPRENE L Tradename for a cast polyurethane elastomer system based on a prepolymer process whereby a storage stable polytetramethylene glycol polyether polyol, terminated with tolylene diisocyanate, is chain extended and crosslinked by heating at about 100°C with 3,3'-dichloro-4,4'-diaminodiphenylmethane. Sometimes a diol or diol/triol mixture is used for softer products.

ADJACENT RE-ENTRY The return of a polymer molecular chain at an adjacent site on the fold surface of a lamellar crystal, after emerging from the lamellar surface and undergoing chain folding. This implies a fairly tight fold, since loose folding (giving a loose loop) would more likely result in non-adjacent re-entry into the crystal, as in the switchboard model. Adjacent re-entry is thought to occur frequently in polymer single crystals grown from dilute solution.

AD–PBO Abbreviation for adamantane–polybenzoxazole.

ADRENOCORTICOTROPIN (Corticotropin) (ACTH) A polypeptide hormone secreted by the anterior pituitary gland which acts on the adrenal cortex to stimulate the production of certain steroids. Consists of a single polypeptide chain of 39 amino acids (of molecular weight 4700) whose sequence was one of the first of any protein to be fully determined. It appears that only the 24 residues at the N-terminal are necessary for activity.

ADSORPTION CHROMATOGRAPHY The separation of different molecular species in a solute by adsorption at the top of a column packed with a solid stationary phase, followed by elution with a liquid phase which moves down the column with respect to the stationary phase in a countercurrent manner. Thus as the eluate solvent front moves, the solute in it undergoes a continuous series of distributions between the stationary phase and eluent and very efficient separations can be achieved. The method has occasionally been applied to the fractionation of synthetic polymers according to molecular weight, the lower molecular weight species being more powerfully adsorbed, but it is not very efficient. It is also occasionally used with proteins, but the related technique of ion-exchange chromatography is more successful. Hydroxy apatite (a calcium phosphate mineral) columns are most frequently used with proteins, from which the proteins are eluted with a phosphate buffer. Oligosaccharides are often separated on charcoal columns. Better separated fractions on elution may be obtained by continuously increasing the polymer solvent power of the eluting liquid (gradient-elution chromatography). For preparative separations the stationary phase is usually packed in a column (column chromatography), but for analytical work it is often spread as a thin layer on an inert glass or plastic sheet—a type of thin layer chromatography. Even for biopolymers, the main area of application, the technique has been largely superceded by methods giving better resolution, e.g. ion-exchange chromatography.

ADSORPTION POLYMERISATION A polymerisation brought about by adsorption of the monomer onto a catalytic surface followed by a 'zipping-up' type of polymerisation of the adsorbed monomer. An example is the polymerisation of acetaldehyde by γ-alumina, chromium oxide and other metal oxide surfaces. Here adsorption causes polymerisation of the carbonyl group by attack by an independent ionic species. This contrasts with coordination polymerisation where the initiating species is not a separate species but is part of the adsorption rate.

AERODUX Tradename for a resorcinol–formaldehyde adhesive.

AEROLITE Tradename for a urea–formaldehyde adhesive.

AEROSIL Tradename for pyrogenic silica.

AES Abbreviation for acrylonitrile–ethylene/propylene–rubber–styrene copolymer.

AFCODUR Tradename for polyvinyl chloride.

AFCOLENE Tradename for polystyrene.

AFCOPLAST Tradename for polyvinyl chloride.

AFCOVYL Tradename for polyvinyl chloride.

AFFINE DEFORMATION A deformation in which the changes in the dimensions of a small part of a body are in the same proportion as the overall macroscopic changes in the whole body. In particular, in the statistical theory of rubber elasticity, the small part is the vector length of each polymer chain between the junction points of the crosslinked network.

AFFINITY CHROMATOGRAPHY A chromatographic technique by which some biopolymers, especially proteins, may be isolated from even complex mixtures in a pure homogeneous form in a single step. The mixture is passed through a column of an inert support (often agarose) to which are covalently attached molecules or groups (ligands) which bind specifically to the biopolymer it is desired to isolate. All other components are eluted from the column whilst the desired polymer is tightly bound. The required biopolymer may be subsequently released either by elution with a solution containing the free ligand to which it tightly binds or by use of a solution with a very different pH or ionic strength. Thus an enzyme may be purified by use of its specific coenzyme as ligand. For example, heparin may be used to bind blood clotting proteins or an antibody may be used to bind its antigen and vice versa. Most commonly, ligands based on nucleotide or nucleoside representing derivatives of substrates or cofactors are used. These may be used for the separation of groups of polymers, e.g. kinase or dehydrogenase enzymes, and are then called general ligands. Examples are Blue Dextran and Cibacron Blue F3GA. Despite the difficulties in finding a suitable ligand and in attaching it to the matrix, the method is increasingly being used, especially for protein purification, due to the high purification achieved in a single step.

AFFINITY-ELUTION CHROMATOGRAPHY A variation of affinity chromatography in which the polymer mixture is bound to an ion-exchange resin packed in a column and the specific polymer to be separated is eluted by use of a solution of its own specific ligand. Especially useful for the isolation of enzymes.

AFFINITY LABELLING Reaction of an enzyme with a synthetic compound which resembles a normal (biological) substrate and binds at the active site in a similar way. In addition, it contains a functional group which reacts with, and therefore covalently binds to, a specific amino acid residue near the active centre. A well-known example is TPCK reaction with chymotrypsin.

AFLAS Tradename for tetrafluoroethylene–propylene copolymer.

AFLON Tradename for tetrafluoroethylene–ethylene copolymer.

AFT-2000 Tradename for an aromatic polyquinazolinedione—amide copolymer developed as a fibre with useful high temperature and fire resistant properties.

AFTER EFFECT That part of the polymerisation that occurs in free radical polymerisation following an abrupt decrease in the rate of initiation, such as occurs at the end of a radiation period in a photo- or radiation-induced polymerisation. During the after effect the radical concentration decreases so that the steady state assumption is no longer valid. If ΔM_{aft} is the excess reaction that occurs by the time the new (lower) rate of initiation ($\mathscr{I}_1$) is reached, over that which would have occurred had the new stationary state been reached instantaneously, then $\Delta M_{aft} = k_p[M]\ln[(\theta+1)/2]/k_t$, where k_p and k_t are the propagation and termination rate constants respectively, $[M]$ is the monomer concentration and

$$\theta = [(\mathscr{I}_1 + \mathscr{I}_2)/\mathscr{I}_1]^{1/2}$$

with $(\mathscr{I}_1 + \mathscr{I}_2)$ being the initial rate of initiation.

AGAR A mixture of polysaccharides, containing 1–5% sulphate ester groups, obtained from many species of the red seaweed *Gelidium*, widely used as a gel medium for microbiological cultures. Even 1–2% aqueous solutions set to a firm gel at room temperature. It consists of two main components, separated by fractionation of their acetates—agarose and agaropectin.

AGAROPECTIN The minor (about 30%) polysaccharide component of agar, separated from the agarose by the insolubility of its acetate in chloroform. More complex in structure than agarose, containing in addition to D-galactose and 3,6-anhydro-L-galactose units, D-glucuronic acid, pyruvic acid and a much higher proportion of sulphate ester groups.

AGAROSE Tradename Sepharose. The major (about 70%) polysaccharide component of agar, separated from the agaropectin by the solubility of its acetate in chloroform. A linear alternating polymer of equimolar proportions of 1,3-linked D-galactose and 1,4-linked 3,6-anhydro-L-galactose units, with very few hydroxyls being sulphated. Useful as a material for gel permeation chromatography in aqueous medium, for separating very large polymer molecules (of molecular weight greater than about one million) due to its open pore structure. Also used for forming agarose–protein complexes for affinity chromatography. Further used as the gel support medium for zone electrophoresis.

AGEING The long-term deleterious change in the properties of a polymer composition during its service life. Usually due to changes in the chemical structure of the polymer (i.e. degradation) initiated by any of the usual degradation agencies—heat, ultraviolet light, chemical attack, etc., or a combination of agencies.

AGERITE Tradename for a range of antioxidants. Agerite powder is N-phenyl-β-naphthylamine, Agerite DPPD is NN'-diphenyl-p-phenylenediamine, Agerite white is N,N'-dinaphthyl-p-phenylenediamine, Agerite Superlite is butylated 4,4'-isopropylidenediphenol.

A-GLASS A soda-lime glass whose oxide weight per cent composition is: SiO_2, 72·0; Al_2O_3, 0·6; CaO, 10·0; MgO, 2·5; Na_2O, 14·2; SO_3, 0·7. Fibres have a tensile strength of about 3·0 GPa, Young's modulus of 72·5 GPa; a density of $2·50 \, g \, cm^{-3}$ and a refractive index of 1·512. Owing to its poor thermal properties and low chemical resistance, especially to water, it is only rarely used as a polymer reinforcement, except in low cost/performance applications.

AH ANTIOXIDANT A chain-breaking antioxidant containing an active hydrogen atom capable of being abstracted by the propagating free radicals (Ṙ) of the oxidation chain reaction:

$$AH + \dot{R} \rightarrow \dot{A} + RH$$

Inhibition of oxidation results if Å is of lower activity than Ṙ. Further, Å may pair its spin with another free radical and so act as a radical trap. Secondary aryl amines, e.g. N-phenyl-β-naphthylamine, and hindered phenols are the most important types.

AH SALT Alternative name for nylon salt.

AI Tradename for a range of polyamide–imide copolymers based on trimellitic anhydride.

AIBN Abbreviation for azobisisobutyronitrile.

AIR DRYING The hardening of a drying or semi-drying oil or of an oil modified alkyd resin caused by atmospheric oxygen reacting with labile sites allyl to the double bond of the unsaturated group of the fatty acid components of the oil. This produces peroxy and other free radicals which combine to join the oil molecules together, which in the case of alkyds crosslinks the polymer chains. Reaction is limited to the surface layers, but when the oil is spread as a film, the whole film will harden. In alkyd resins used as coatings, the drying process is accelerated by the use of metal soaps as driers.

AKABORI METHOD Alternative name for hydrazinolysis.

AKULON Tradename for nylon 6 and nylon 66.

ALA Abbreviation for α-alanine.

α-ALANINE (Ala) (A) $H_2NCH(CH_3)COOH$. An α-amino acid found in nearly all proteins, occurring in especially large amounts in fibroin. Has pK' values of 2·34 and 9·69, with the isoelectric point at 6·00. Often associated with helical conformations in proteins. Synthesised by the Strecker synthesis from acetaldehyde, ammonia and potassium cyanide, via an aminonitrile.

ALATHON Tradename for low density polyethylene and ethylene–ethyl acetate copolymer.

ALBERIT Tradename for phenol–formaldehyde polymer.

ALBIGEN A Tradename for poly-(N-vinylpyrrolidone).

ALBUMIN A globular protein, originally defined as being soluble in both pure water and dilute salt solutions, whereas the closely related globulins are only soluble in dilute salt solutions. Albumins may be further distinguished from globulins on the basis of their electrophoretic mobility, since they move faster towards the positive pole at a pH of about 8·5. However, some proteins that behave electrophoretically as globulins are water soluble so that the strict distinction between the two is not possible. Albumins have a relatively low molecular weight, 20 000 to 70 000. They occur widely, important examples being ovalbumin from eggs, serum albumins of blood and plant seed albumins, e.g. barley albumins.

ALCOTEX Tradename for polyvinyl alcohol.

ALCRYN Tradename for a plasticised chlorinated polyethylene, useful as a thermoplastic elastomer with good oil, heat and weather resistance. Used for hose, seals, sheeting and mechanical goods.

ALDOHEXOSE An aldose containing six carbon atoms. There are sixteen possible configurational isomers existing as eight enantiomorphic pairs. However, only the D- and L-forms of glucose, mannose and galactose are commonly found in polysaccharides.

ALDONIC ACID A sugar acid in which the carbonyl group of a monosaccharide has been replaced by a carboxyl group. D-gluconic acid is the most important example.

ALDOPENTOSE An aldose containing five carbon atoms. There are eight possible configurational isomers existing as four enantiomorphic pairs. They are D- and L-xylose, ribose, arabinose and lyxose. Only the last member does not occur naturally.

ALDOSE A monosaccharide whose simplest structural representation is as a polyhydroxyaldehyde in open-chain form:

$$OHC(CH_2OH)_nCH_2OH$$

where $n = 1\text{--}4$, e.g. glucose. A five-carbon aldose is an aldopentose and a six-carbon aldose is an aldohexose.

ALEURITIC ACID

$$HO(CH_2)_6\overset{|}{C}H\overset{|}{C}H(CH_2)_7COOH$$
$$\quad\quad\quad OH\,OH \quad\quad\quad M.p.\ 102°C.$$

One of the major, 30–40%, components of the complex esters that comprise shellac.

ALFIN CATALYST An early stereoregular polymerisation catalyst useful for the polymerisation of dienes. It is capable of giving predominantly *trans*-1,4-polymer but not with the high stereoregularity of the later developed metal alkyl and Ziegler–Natta catalysts. The catalyst is typically prepared by formation of amyl sodium from amyl chloride and metallic sodium:

$$C_5H_{11}Cl + 2Na \rightarrow C_5H_{11}Na + NaCl$$

which is then reacted with isopropyl ether to form sodium isopropoxide and propylene:

$$C_5H_{11}Na + [(CH_3)_2CH]_2O \rightarrow$$
$$Na\overset{+}{O}CH(CH_3)_2 + CH_3CH{=}CH_2 + C_5H_{12}$$

The propylene in turn reacts with more amyl sodium to yield allyl sodium:

$$CH_3CH{=}CH_2 + C_5H_{11}Na \rightarrow$$
$$NaCH_2CH{=}CH_2 + C_5H_{12}$$

The catalyst thus typically consists of sodium isopropoxide and allyl sodium deposited on sodium chloride crystals. The mechanism of polymerisation may be a free radical mechanism but ionic mechanisms have also been suggested.

ALFREY APPROXIMATION Alternative name for Alfrey's Rule.

ALFREY–PRICE Q–e SCHEME (Q–e scheme) A method of quantitatively expressing monomer and radical activities in copolymerisation. The scheme assumes that the rate constant for addition of monomer B to radical $\sim\!\!\sim$A (k_{AB}) is given by

$$k_{AB} = P_AQ_B\exp(-e_Ae_B)$$

where P_A is a measure of radical $\sim\!\!\sim$A* reactivity and Q_B is a measure of monomer B reactivity and e_A and e_B are measures of the polarity of the radical and monomer respectively. Then, assuming that the e values are the same in both radical and monomer, the reactivity ratios of A and B (r_A and r_B) are given by:

$$r_A = (Q_A/Q_B)\exp[-e_A(e_A - e_B)]$$

and

$$r_B = (Q_B/Q_A)\exp[-e_B(e_B - e_A)]$$

Although the theoretical basis for the method may not be very sound, it nevertheless provides a useful basis for the prediction of reactivity ratios of pairs of monomers that have not been previously studied. However, such predictions are really only semi-quantitative since Q and e may not have unique values for each monomer, and their precision relies on previously determined reactivity ratios, which themselves are often inaccurate.

ALFREY'S RULE (Alfrey Approximation) An approximation that enables viscoelastic functions to be more easily interrelated than with exact relationships. It enables functions to be calculated from experimental data over a wide range of frequencies. In particular, if it is assumed that the function $e^{-t/\tau} = 0$ up to time $t = \tau$ and $e^{-t/\tau} = 1$ from $\tau > t$, then since

$$G(t) = G_r + \int_{\ln t}^{\infty} H(\tau)\,d\ln\tau$$

where τ is the relaxation time and $G(t)$ the relaxation modulus, G_r being the relaxed modulus, then the relaxation spectrum ($H(\tau)$) is given by

$$H(\tau) = [-dG(t)/d\ln t]_{t=\tau}$$

ALGAL POLYSACCHARIDE A polysaccharide which occurs naturally in algae, such as brown seaweeds, e.g. laminaran, alginic acid and fucoidin, green seaweeds, e.g. xylan and complex sulphated polysaccharides and red seaweeds, e.g. xylan, porphyran, agarose and carrageenan. Often extractable by hot water or dilute acid or alkali. It is hygroscopic, mucilageneous and serves as either a food reserve polysaccharide, e.g. laminaran, or a cell-wall polysaccharide, e.g. sulphated polysaccharides.

ALGINATE Generic name for a fibre based on metallic salts of alginic acid. The fibres are produced by wet spinning of sodium alginate into a solution of the metal salt, usually calcium chloride, thus producing a water insoluble metal alginate fibre. The fibres are fire resistant but are swollen, or dissolve, in alkaline solutions.

ALGINIC ACID A glycuronan in which the uronic acid units are β-D-mannouronic and α-D-guluronic acids 1,4′-linked and arranged as blocks of either type of unit, or as an alternating arrangement of each type, within the same molecule. Found in the cell walls of brown algal seaweeds from which it is extracted with dilute sodium carbonate solution. As the sodium salt or propylene glycol ester, it is widely used as a thickening and emulsifying agent, especially in foodstuffs. Sodium alginate forms a viscous aqueous solution, but treatment with divalent ions, e.g. Ca^{2+}, leads to gelling.

ALGOFLON Tradename for polytetrafluoroethylene.

ALIPHATIC RESIN A petroleum resin consisting of the low molecular weight polymer obtained by cationic polymerisation of mixed C_4–C_6 olefin fractions obtained from the cracking of petroleum. The resins are widely used as tackifying agents in adhesives and rubber compositions.

ALKALI CELLULOSE The 'complexes' obtained by treating cellulose with aqueous sodium hydroxide, consisting mostly of soda cellulose II. Used as an intermediate in the preparation of cellulose xanthate and various cellulose ethers.

ALKATHENE   Tradename for low density polyethylene.

ALKON   Tradename for polyoxymethylene copolymer.

ALKOXYACRYLATE   An acrylic monomer of the type CH_2=CHCOOROR′, useful in replacing some of the alkyl acrylate monomer in acrylic elastomers in order to achieve better low temperature flexibility with good oil resistance. Usually ethoxyethylacrylate or methoxyethylacrylate is used.

ALKOXYMETHYLACRYLAMIDE   An acrylic monomer of the general type

$$CH_2=CHCONHCH_2OR$$

where OR is an alkoxy group, usually butoxy, useful as a crosslinking site comonomer in thermosetting acrylic stoving enamels. It participates in crosslinking by reaction with diepoxy compounds.

ALKYL-2-CYANOACRYLATE   Alternative name for cyanoacrylate.

ALKYD POLYESTER   Alternative name for polyester alkyd.

ALKYD RESIN   A saturated polyester resin prepared by esterification of a polyfunctional alcohol, commonly glycerol, with a difunctional acid, frequently phthalic anhydride. The polymer formed is therefore branched at low conversion, and forms a network at higher conversions. Usually the term is reserved for such polyesters as modified by the incorporation of a drying oil (oil modified alkyd) when used as the film former in many surface coating materials. Incorporation of the oil enhances solubility, gives tougher films and enables crosslinking to occur by air drying.

The oil used is usually a naturally occurring plant triglyceride, e.g. linseed oil, tung oil, safflower oil, soya bean oil, castor oil, tall oil, cotton seed oil or coconut oil. In these oils the acid groups are long chain fatty acids containing unsaturated links through which crosslinking can occur by atmospheric oxygen—air drying. This is promoted by the use of a drier. Oils, and the alkyds based on them, are said to be drying, semi-drying or non-drying according to the rate of hardening of a coating obtained using the uncrosslinked alkyd. Other fatty acids, e.g. perlargonic and isooctanoic acids, or fatty acid containing natural products, such as rosin, are sometimes used as modifying acids. Polyols other than glycerol sometimes used are pentaerythritol, trimethylolpropane and sorbitol. Sometimes a diol, e.g. ethylene or propylene glycol, is also used to reduce crosslink density. Sometimes phthalic anhydride is replaced by maleic anhydride, isophthalic acid, adipic acid or sebacic acid.

The resins are made either by first hydrolysing the oil to free the fatty acids, which are then heated at 200–240°C with the polyol or anhydride, or by first subjecting the oil to alcoholysis with the polyol by heating to about 240°C. With an oil:glycerol ratio of 1:2, this largely yields a monoglyceride:

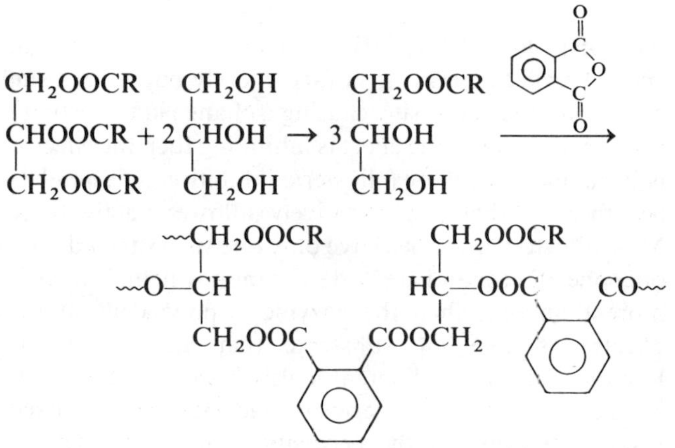

The resin is then formed by adding anhydride, or diacid, and heating further. This is the monoglyceride process.

The amount of oil incorporated is denoted by the oil length, and resins may be short, medium or long oil resins. As coatings the liquid alkyd is applied to the substrate and subsequently dries, i.e. hardens, by diffusion of atmospheric oxygen, causing formation of hydroperoxides α to the double bond of the fatty acid. These subsequently break down to free radicals which combine to form crosslinks. Usually this process is speeded up by the use of a metal soap drier.

Alkyd resins are often modified by the incorporation of a second polymer, e.g. cellulose nitrate, a phenolic, amino or silicone resin, to modify physical properties, or by the use of a polymerisable monomer, especially styrene, as in styrenated alkyd resins.

ALLENE LADDER POLYMER   A ladder polymer of the general structural type

$$\equiv\!\!\left[R\right]_n\!\!\equiv$$

where R is a ring structure, e.g.

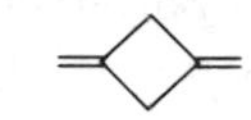

Polyindigo is one of the few polymers of this type.

ALLOMER   Alternative name for polyallomer.

ALLOPHANATE   A compound of the type

$$\begin{array}{c} R\!-\!N\!-\!O\!-\!CO\!-\!R' \\ | \\ CO\!-\!NH\!-\!R'' \end{array}$$

where R, R′ and R″ are alkyl groups. Formed by reaction of a urethane group with an isocyanate at temperatures of 120–140°C. When this reaction occurs in polyurethanes, branching or crosslinking results:

$$\sim\!\!NH\!-\!CO\!-\!O\!\sim + NCO\!\sim \rightarrow \sim\!\!N\!-\!CO\!-\!O\!\sim$$
$$\quad\quad\quad\quad\quad\quad\quad\quad\quad\quad\quad | $$
$$\quad\quad\quad\quad\quad\quad\quad\quad\quad\quad CO\!-\!NH\!\sim$$

ALLOPRENE   Tradename for chlorinated rubber.

ALLOSTERIC ACTIVATION The enhancement of enzyme catalytic activity by the binding of an effector at a site different from the active site on an allosteric enzyme. An example of positive cooperativity.

ALLOSTERIC ENZYME An enzyme at which an effector molecule may be non-covalently bound at a site other than the active site, causing a change in the activity at the active site. If the effect is inhibiting then the effector is behaving negatively (allosteric inhibition), if stimulating, then it is behaving positively (allosteric activation). Most allosteric enzymes have only one such site and bind only one effector and are said to be monovalent. If there is more than one, then the enzyme is polyvalent. If the effector is also the substrate then the enzyme is homotropic, if it is a different molecule then the enzyme is heterotropic. Most allosteric enzymes behave in a mixed fashion. Because of the necessity of providing several different sites, allosteric enzymes are usually quite complex, usually being oligomeric. Their catalytic behaviour is often unconventional, e.g. the rate of a catalysed reaction may vary sigmoidally with substrate concentration. The inhibiting effects are also often non-classical.

ALLOSTERIC INHIBITION The reduction of enzyme catalytic activity by the binding of an effector at a site different from the active site on an allosteric enzyme. One of the commonest types is feed-back inhibition. An example of negative cooperativity.

ALL PURPOSE FURNACE BLACK (APF) A type of furnace carbon black with reasonably high structure and hence having good processing characteristics. Useful for extruded products.

ALLYL GLYCIDYL ETHER

$$CH_2\text{—}CH\text{—}CH_2OCH_2CH=CH_2 \quad \text{B.p. } 154°C.$$

A useful comonomer for polyether (propylene oxide and epichlorhydrin) rubbers, providing crosslinking sites.

ALLYLIC ABSTRACTION Abstraction by a free radical (R·) of a hydrogen atom α to a double bond (an allylic hydrogen):

$$R· + CH_2=CH\text{—}CH_2X \rightarrow RH + CH_2=CH\text{—}\dot{C}HX$$
$$\quad\quad\quad\quad\quad\quad\text{I} \quad\quad\quad\quad\quad\quad\quad\text{II}$$

The resultant radical (**II**) is resonance stabilised, so that in the attempted free radical polymerisation of monomer (**I**), the reaction competes heavily with propagation, thus inhibiting polymerisation (autoinhibition). The radical (**II**) may be so stable that it does not propagate but instead terminates by combination with another radical (degradative chain transfer).

ALLYLIC POLYMERISATION Polymerisation of an allyl monomer, i.e. one of the type $CH_2=CHCH_2X$. Such polymerisations are abnormally slow and yield only low molecular weight polymer. Not only are these monomers unreactive, due to a lack of resonance stabilisation in the transition state, but they also readily participate in a monomer transfer reaction:

$$\sim\!\!CH_2\dot{C}H + CH_2=CHCH_2X \rightarrow$$
$$\quad\quad | $$
$$\quad CH_2X$$
$$\quad\quad\quad\quad\quad \sim\!\!CH_2CH_2CH_2X + CH_2=CH=\dot{C}HX$$
$$\quad\quad\quad\quad\quad\quad\quad\quad\quad\quad\quad\quad\quad\quad\quad \text{I}$$

This yields such a highly stable allyl radical (**I**) that instead of adding monomer it disappears by bimolecular combination, i.e. degradative chain transfer occurs giving an effective termination by autoinhibition.

ALLYLMER 39 Tradename for poly-(diethylene-glycol-bis-(allyl carbonate)).

ALLYL METHACRYLATE

$$\quad\quad\quad CH_3$$
$$\quad\quad\quad |$$
$$CH_2=CCOOCH_2CH=CH_2 \quad \text{B.p. } 43°C/15\,mm.$$

A methacrylate monomer sometimes used as a co-monomer for increasing the hardness and softening point of polymethylmethacrylate. More specifically the surface hardness may be increased by polymerising the monomer on the surface.

ALLYL MONOMER A monomer of structure

$$CH_2=CHCH_2X$$

e.g. where $X = \text{—}OOCCH_3$ as in allyl acetate. Such monomers only polymerise to very low molecular weight polymers since they are powerful chain transfer agents, owing to the formation of a resonance stabilised free radical:

$$\sim\!\!CH_2\dot{C}HCH_2X + CH_2=CHCH_2X \rightarrow$$
$$\quad\quad\quad\quad \sim\!\!CH_2CH_2CH_2X + CH_2=CH\dot{C}HX$$
$$\quad\quad\quad\quad\quad\quad\quad\quad\quad\quad\quad\quad\quad\quad \updownarrow$$
$$\quad\quad\quad\quad\quad\quad\quad\quad\quad\quad\quad\quad \dot{C}H_2CH=CHX$$

The resultant radical is too stable to re-initiate a polymer kinetic chain, so the monomer acts as its own inhibitor (autoinhibition) and degradative chain transfer is said to occur.

ALLYL RESIN A polymer formed by polymerisation of an allyl compound, i.e. a derivative of allyl alcohol $CH_2=CHCH_2OH$. Examples include the polymers of diallylphthalate, diallyl isophthalate, diethyleneglycol bis(allyl carbonate) and diallyl carbonate.

ALMOST ELASTIC MATERIAL A material in which the stress depends strongly on the current strain, but only slightly on the strain history. An almost constant deformation is supported by an almost constant stress.

ALON Tradename for a cellulose acetate fibre produced by acetylation of a viscose rayon fibre. This method has the advantage of using an aqueous rather than an acetone spinning solution. It has a tenacity of about 2·8 g denier^{-1} and an elongation at break of 23%, twice as high as cellulose acetate rayon, but otherwise it is similar to this fibre.

ALPHA HELIX   The helical conformation adapted by many natural and synthetic polypeptides. It contains eighteen peptide groups for every five turns of the helix. Each peptide group is involved in an intramolecular hydrogen bond through its N—H and C=O groups, the hydrogen bonds lying almost parallel to the helix axis. There are 3·60 peptide groups per turn and the residue repeat distance, i.e. distance moved along the helix axis on passing from one peptide to the next, is 1·50 Å.

ALPHA TRANSITION   The highest temperature amorphous transition of a polymer, the temperature of which (T_α) is higher than that of other more minor transitions (β-, γ-, etc.) which are also often observed. Usually it is the main transition and is designated the glass transition. As T_α is reached the onset of gross molecular motions of segments of polymer molecules several repeat units long becomes possible.

ALPOLIT   Tradename for an unsaturated polyester resin and a vinyl ester resin.

ALRESIN   Tradename for phenol–formaldehyde polymer.

ALTERNATING COPOLYMER   A copolymer in which the two different comonomer units (A and B) alternate down the polymer chain, i.e. the polymer has the structure

$$\sim\!\!ABABABABABAB\!\sim\!\sim$$

In chain copolymerisation such a copolymer results when neither monomer will add on to an active growing chain end terminated by a unit of the same type, i.e. the reactions

$$\sim\!\!A^* + A \rightarrow \sim\!\!AA^*$$

and

$$\sim\!\!B^* + B \rightarrow \sim\!\!BB^*$$

do not occur, but only the cross-propagation reactions

$$\sim\!\!A^* + B \rightarrow \sim\!\!AB^*$$

and

$$\sim\!\!B^* + A \rightarrow \sim\!\!BA^*$$

occur. Neither monomer will homopolymerise and both reactivity ratios are therefore zero. Very few monomer pairs exhibit this behaviour exactly, but with many pairs essentially alternating copolymers are formed due to at least one of the monomers having an extremely low reactivity ratio together with a strong tendency towards cross-propagation. Such behaviour is found in the copolymerisation of maleic and fumaric acids and their esters and anhydrides with a wide variety of comonomers.

In step-growth polymers, dyadic polymers formed by reaction of an AA with a BB monomer, can strictly be considered as alternating copolymers of AA and BB units or of BA and AB units. However, by convention these polymers are normally considered as homopolymers of AABB units, since their essential characteristics are determined by the $\sim\!\!A$—$B\!\sim$ link, as with the dyadic nylons. Many alternating chain polymers are thought to be formed by homopolymerisation of a 1:1 complex formed between the two comonomers. The complex is often considered to be a charge-transfer complex, leading to charge-transfer polymerisation. The use of a Lewis acid to promote alternating copolymerisation of a donor–acceptor comonomer pair has been extensively investigated. Alternating copolymers are often named by linking the names of the monomers by the term -alt-, as in poly-(styrene-alt-maleic anhydride).

ALTERNATING COPOLYMERISATION   A binary copolymerisation in which both monomer reactivity ratios are zero, or nearly so. Thus each of the two types of propagating species $\sim\!\!A^*$ and $\sim\!\!B^*$ show a strong preference for adding on the other monomer. Hence an alternating copolymer is produced. Many free radical polymerisations show a tendency towards alternation. If both monomer reactivity ratios are less than unity, then at a certain monomer feed composition, azeotropic copolymerisation will occur. Alternation tendency increases as the difference in polarity between the two monomers increases. Although such polar effects may result from an interaction of an electron donor radical propagating centre with an electron acceptor monomer (or vice versa), in many cases it is thought that alternation results from prior formation of a 1:1 complex between the two monomers. Such a complex may be a charge-transfer complex of an electron donor monomer, such as an olefin or vinyl ether, and an electron acceptor, such as maleic anhydride, carbon dioxide or sulphur dioxide. Complex formation, and hence alternation, may be enhanced by addition of a Lewis acid such as $EtAlCl_2$.

ALVANOL   Tradename for a novolac phenol–formaldehyde polymer.

ALVAR   Tradename for polyvinyl acetal.

AMANIM   Tradename for an aromatic polyamide–imide film material.

AMBER   A fossil resin from pine tree species now extinct. It consists of a complex mixture of acidic substances that softens at about 150°C and may be moulded as a thermoplastic. A hard material but not so brittle as other natural resins.

AMERIPOL   Tradename for styrene–butadiene rubber and high density polyethylene.

AMERIPOL CB   Tradename for high or medium *cis*-polybutadiene.

AMERIPOL SN Tradename for synthetic *cis*-1,4-polyisoprene.

AMEROID Tradename for a casein plastic material.

AMIDE INTERCHANGE Alternative name for trans-amidation.

AMIDEL Tradename for a transparent polyamide produced from isophthalic acid, lauryllactam and the diamine:

AMILAN Tradename for a nylon 6 fibre.

AMINO ACID An organic acid containing both an amino group (usually a primary amino group NH_2) and a carboxylic acid group, i.e. in general $H_2N—R—COOH$ (**I**). The monomer from which polyamides may be considered to be formed by loss of water:

$$n\,\mathbf{I} \rightarrow \text{─[NH—R—CO]}_n + n\,H_2O$$

When both functional groups are terminal, i.e. an α,ω-amino acid, and $R = \text{─[CH}_2\text{]}_{\overline{x}}$, a monadic nylon

$$\text{─[HN─[CH}_2\text{]}_{\overline{x}}CO]}_n$$

results. However, more important are the α-amino acids,

$$\begin{array}{c} H_2N—CH—COOH \\ | \\ R \end{array}$$

which are the monomers from which protein and other polypeptides may be considered to be formed.

α-AMINO ACID An amino acid substituted on the α-carbon atoms by the substituent R, i.e.

$$\begin{array}{c} H_2N—CH—COOH \\ | \\ R \end{array}$$

The monomers from which proteins and other polypeptides may be considered to be derived by the loss of water. In these polymers the amide repeat unit

$$\begin{array}{c} \text{─[HN—CH—CO]─} \\ | \\ R \end{array}$$

is referred to as the amino acid residue, being joined to its neighbours by peptide bonds.

About 30 different α-amino acid residues are found in proteins, although only 20 (the standard amino acids) are commonly found. Most proteins contain all 20 different monomers. Many other non-protein amino acids also occur naturally. Apart from glycine ($R = H$), the protein amino acids contain an asymmetric carbon atom at C^α, but only the L-configurational isomer is found in proteins. Since this isomer is of overwhelming importance,

compared with the D-isomer, the prefix L- is sometimes omitted.

Each amino acid has a trivial name, which is often abbreviated to its first three letters, e.g. Gly for glycine, Ala for alanine, etc., or even to a single letter, e.g. G for glycine, A for alanine, to simplify the description of the long sequences of amino acid residues found in proteins and peptides. The amino acids are classified according to the nature of the substituent R. Thus, for the standard amino acids, the non-polar acids with hydrophobic R groups are alanine, valine, leucine, isoleucine, proline, phenylalanine, tryptophane and methionine. Polar R groups are considered to be present in glycine, serine, cysteine, threonine, tyrosine, asparagine and glutamine, whereas polar charged groups (at least near neutral pH) occur in aspartic and glutamic acids (the acidic amino acids) and in lycine, arginine and histidine (the basic amino acids). Several other amino acids, e.g. 4-hydroxy-proline, hydroxylysine, desmosine and isodesmosine, are also occasionally found in proteins. Cystine is considered to be an oxidation product of cysteine. Strictly, proline and 4-hydroxyproline are imino acids. The amino acids have high melting points with decomposition, high water solubility, but low solubility in non-polar solvents, owing to their dipolar (zwitterion) nature, existing as $H_3\overset{+}{N}CHRCOO^-$ in the crystalline state and in neutral aqueous solutions.

They have both acidic and basic properties, i.e. they are amphoteric, the state of ionisation being pH dependent, i.e.

$$H_3\overset{+}{N}CHRCOOH \underset{H^+}{\rightleftharpoons}$$
$$H_3\overset{+}{N}CHRCOO^-\ (\mathbf{I}) \underset{H^+}{\rightleftharpoons} H_2NCHRCOO^-$$

The pH for electrical neutrality (form **I**) is the isoelectric point (pH_I) at which the amino acid has minimum aqueous solubility and rate of diffusion in electrophoresis. The dissociation behaviour is expressed by the pK values, which are typically 1·5–2·5 for ionisation of the —COOH group (pK_1) and 8·5–10·5 for ionisation of the —NH_3 group (pK_2). A further pK_R characterises ionisation of the substituent R when this contains an ionisable group. Free amino acids may be quantitatively determined by formol titration. Various reactions, especially the ninhydrin reaction, are used in the identification and characterisation of amino acids. Reactions of the amino group, e.g. with fluorodinitrobenzene, dansyl chloride, fluorescamine and phenylisothiocyanate, are useful in protein analysis for N-terminal groups.

The amino acid composition of a protein is determined first by hydrolysis with 6N hydrochloric acid at 110°C to give a complex mixture of amino acids; these are then reacted with ninhydrin, separated chromatographically (by partition, paper, thin layer or ion-exchange chromatography) and determined by their ultraviolet light absorption. The latter method is the most frequently employed, in an automated form using an amino acid analyser.

Reactions of amino acids are used in peptide and homopolypeptide synthesis. In the former, the stepwise synthetic methods used require all but one of the functional groups to be blocked by use of a protecting group. In the latter, the amino acid is usually converted to its N-carboxyanhydride which is then polymerised.

AMINO ACID ANALYSER An apparatus for the determination of the amino acid composition of proteins. The hydrolysate at pH 3, from 6N hydrochloric acid hydrolysis at 100–200°C, is passed down an ion-exchange column, often sulphonated polystyrene, which separates the individual amino acids on elution with increasing pH and sodium chloride concentration. Each amino acid has a characteristic retention time. The eluate is collected in fractions, the fractions are separately reacted with ninhydrin reagent, fluorsceine or o-phthalaldehyde, and are then examined for their light absorption at 570 nm and 440 nm to determine the amino acid concentrations. The whole analysis is automated and takes about 2–4 h, requiring only about 0·1 mg of sample, being sensitive for quantitative determinations down to 10^{-10} mol.

6-AMINOCAPROIC ACID (ε-Aminocaproic acid) (6-Aminohexanoic acid) $H_2N(CH_2)_5COOH$. M.p. 200–206°C. The amino acid from which nylon 6 can be considered to be derived. May be polymerised to nylon 6 by high temperature melt polymerisation. However, commercially, nylon 6 is made from ε-caprolactam, which is more readily available in higher purity.

ε-AMINOCAPROIC ACID Alternative name for 6-aminocaproic acid.

2-AMINO-2-DEOXY-D-GALACTOSE Alternative name for D-galactosamine.

2-AMINO-2-DEOXY-D-GLUCOSE Alternative name for D-glucosamine.

ξ-AMINOENANTHIC ACID Alternative name for 7-aminoheptanoic acid.

ω-AMINOENANTHIC ACID Alternative name for 7-aminoheptanoic acid.

N-β-(AMINOETHYL)-γ-AMINOPROPYLTRIMETH-OXYSILANE

$$H_2NCH_2CH_2NH(CH_2)_3Si(OCH_3)_3$$

A coupling agent for fibre, especially glass, reinforced epoxy resin, phenolic, melamine and thermoplastic composites.

N-AMINOETHYLPIPERAZINE

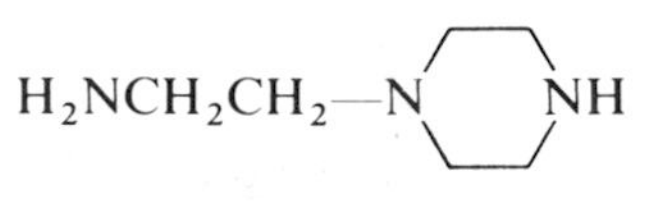

$$H_2NCH_2CH_2—N\bigcirc NH \quad \text{B.p. } 222°C.$$

Produced by reaction of 1,2-dichloroethane with ammonia. Useful as a curing agent for epoxy resins, giving cured castings with a higher impact strength than is

obtained using diethylenetriamine and triethylenetetramine.

7-AMINOHEPTANOIC ACID (ξ-Aminoenanthic acid) (ω-Aminoenanthic acid)

$$H_2N(CH_2)_6COOH \quad \text{M.p. } 187–195°C.$$

The monomer for nylon 7. Synthesised from the trimer obtained by the telomerisation of ethylene in the presence of carbon tetrachloride. The ethyl ester, also used as a monomer for nylon 7, is obtained from caprolactone, which is converted to 6-chlorohexanoic acid by treatment with a $ZnCl_2/HCl$ mixture. This is esterified, the chlorine replaced by nitrile, which is then reduced to give the required ethyl ester.

6-AMINOHEXANOIC ACID Alternative name for 6-aminocaproic acid.

9-AMINONONANOIC ACID Alternative name for 9-aminopelargonic acid.

9-AMINOPELARGONIC ACID (9-Aminononanoic acid)

$$H_2N(CH_2)_8COOH \quad \text{M.p. } 185–188°C.$$

The monomer for nylon 9. Synthesised either from 1,1,1,9-tetrachlorononane (obtained by telomerisation of ethylene with carbon tetrachloride) or by reductive aminolysis of methyl azelealdehyde (MAZ, $OHC(CH_2)_7COOCH_3$), itself obtained by ozonolysis of soya bean oil hydrolysis products. Converted to nylon 9 by heating at 225–260°C.

AMINOPLAST Alternative name for aminoresin.

AMINOPLASTIC Alternative name for aminoresin.

AMINOPOLYSACCHARIDE (Glycosoaminoglycan) A polysaccharide which contains aminomonosaccharide (aminosugar) units in the polymer molecule. Mainly of animal origin, and indeed, apart from glycogen, most animal polysaccharides are aminopolysaccharides. The aminosugars are often D-glucosamine or D-galactosamine. Chitin is the most important example and is the only homopolysaccharide of this type. The acidic mucopolysaccharides are the largest group, but these and the related blood-group substances are often formally glycoproteins, containing covalent links between monosaccharide and amino acid units.

γ-AMINOPROPYLTRIETHOXYSILANE $H_2N(CH_2)_3Si(OC_2H_5)_3$. A commonly used coupling agent for fibre, and other filler, reinforced epoxy resin, phenolic, melamine and thermoplastic composites, especially with glass fibres.

AMINORESIN (Aminoplast) (Aminoplastic) A polymer formed by reaction of an amide or amino compound with formaldehyde, either as a linear low molecular weight prepolymer or in a subsequent crosslinked network form. Aminoresins are important commercial polymers, being widely used as thermosetting plastic moulding and laminating resins, in coatings and varnishes and as textile (especially cotton textile) finishes. The most important types are urea–formaldehyde and melamine–formaldehyde polymers. Of lesser importance are benzoguanamine—formaldehyde, aniline–formaldehyde and other urea–formaldehyde resins such as thiourea, cyclic ureas, e.g. ethyleneurea, and dicyanamide–formaldehyde resins. The initially formed polymers are often condensed methylol derivatives of the amines or amides with methylene bridges (in urea–formaldehydes) and/or ether methylene bridges, as in melamine–formaldehydes. The prepolymer, after fabrication, may then be crosslinked to form a network thermoset plastic product.

AMINOSILANE A silane coupling agent containing amino groups. Examples include γ-aminopropyltrimethoxysilane and N-β-(aminoethyl)-γ-aminopropyltrimethoxysilane.

AMINOSUGAR A monosaccharide (or sugar) or monosaccharide unit in which a hydroxyl group has been replaced by an amino group or by an N-acetyl-amino group ($—NHCOCH_3$). In their nomenclature the aminosugars are formally regarded as being formed via the deoxysugar. Thus common naturally occurring examples (mostly in animals and micro-organisms) are D-glucosamine (or 2-deoxy-2-amino-D-glucose), D-galactosamine (or 2-deoxy-2-amino-D-galactose) and its acetylated derivative sialic acid.

11-(or ω-)AMINOUNDECANOIC ACID Alternative name for 11-aminoundecylenic acid.

11-(or ω-)AMINOUNDECYLENIC ACID (11-(or ω-) Aminoundecanoic acid)

$$H_2N(CH_2)_{10}COOH \quad \text{M.p. } 187–191°C.$$

The monomer for nylon 11. Obtained from castor oil, by methanolysis to methyl ricinoleate, which is pyrolysed to methyl undecylenate; the hydrolysed ester is converted to 11-bromoundecylenic acid, which is finally reacted with ammonia to give the desired monomer. Polymerised to nylon 11 (together with about 0·5% lactam in equilibrium) by heating at about 215°C.

AMMONIA CELLULOSE Alternative name for cellulose III.

AMORPHOUS DEFECT A disordered region either within, or on, the surface of a crystal; of no specified structure, but intermediate between the well-defined line, e.g. dislocation, or point defects and the amorphous phase structure. Postulated in order to account for the large loss in crystallinity on introduction of small amounts of comonomer (or other structural impurity) into the polymer chains.

AMORPHOUS ORIENTATION The component of the overall orientation due to the amorphous regions in a polymer. There will similarly be an amorphous component to each anisotropic property, e.g. birefringence. The amorphous orientation function may be determined by infrared dichroism, laser-Raman scattering or dye-doped polarised fluorescence.

AMORPHOUS POLYMER A polymer in which the molecular chains exist in the random coil conformation; since there is no regularity of structure, there is no crystallinity. Some polymers which are nominally amorphous may have some short-range order. Use of the term often implies that the polymer is amorphous in the solid state, since polymers are usually amorphous in solution or melt. An irregular conformation is adopted if the molecular structure of the polymer is irregular. Thus atactic polymers, random copolymers and thermoset polymers cannot crystallise due to molecular irregularity, and hence are amorphous. Even regular polymers, which normally crystallise, may often be quenched from the melt state to the amorphous state. Amorphous polymers exhibit a strong T_g, often with additional lower temperature, but weaker, transitions. If non-crosslinked, they are more readily soluble than crystalline polymers. They are normally isotropic (unless oriented) and homogeneous. Since they do not contain crystals to scatter light they are also transparent.

AMORPHOUS REGION A region in a crystalline polymer sample which has not crystallised and therefore in which the polymer chains exist in the random coil conformation, i.e. where the polymer is amorphous. Since crystallisation is limited in a crystalline polymer, amorphous regions are always present, typically accounting for 10–70% of the material. Thus the whole sample behaves as a 'composite' of amorphous and crystalline polymer. Both regions contribute their characteristic properties to the overall behaviour, with the amorphous regions exhibiting a T_g.

AMORPHOUS SCATTER The X-ray scattering produced by an amorphous polymer or region, consisting of a few diffuse halos. Although no short- or long-range order of a crystalline kind exists, a short-range order of the most probable distances between neighbouring atoms does exist. This is often expressed in terms of the atomic radial distribution function, obtained from the experimental scattering curve.

AMOSITE A fibrous, amphibole asbestos of structure $MgFe_6[(OH)Si_4O_{11}]_2$ (similar to crocidolite), containing a higher proportion of iron than anthophyllite. Forms fibres with a tensile strength of about 2 GPa and a tensile modulus of about 150 GPa, typically 60–100 nm wide and with good acid resistance. Sometimes used as a filler in polypropylene.

AMPHIBOLE A naturally occurring crystalline silicate with a ladder polymer structure consisting of two linked chains of alternating SiO_4^{4-} tetrahedra. The double chains are bonded to each other through planes of counterions (Mg, Fe and Na) whose size determines the stacking of the ladders. Hydroxyl groups may also be present. Not only are there numerous possibilities of cation replacement, but in some amphiboles some silicon may be replaced by aluminium. Some important amphiboles (anthophyllite, amosite, crocidolite and tremolite) are fibrous and, together with chrysotile, comprise the varieties of asbestos.

AMPHIPHILIC POLYMER   Alternative name for polyampholyte.

AMYL ACETATE   A mixture of the acetates of various pentanols, sometimes containing other higher alcohols, depending on the source of the alcohols used in its preparation. Petroleum based materials are acetates of five of the possible isomers. Fusel oil based materials contain isobutyl, n-hexyl and n-heptylalcohols. Typical boiling range is 120–145°C (fusel oil based), or 130–155°C, the purest grades boiling at 138–142°C (petroleum based). A solvent for cellulose nitrate, many natural resins and polyvinyl acetate. Swells polyethylene, polyvinyl chloride and polymethylmethacrylate.

AMYL ALCOHOL   Any of the eight isomeric pentanols. Commercial amyl alcohol is a mixture of 3-methylbutan-1-ol, $(CH_3)_2CHCH_2CH_2OH$ (isoamyl alcohol, b.p. 131°C) and 2-methylbutan-1-ol,

$$CH_3CH_2CH(CH_3)CH_2OH$$

(active amyl alcohol, b.p. 128°C) when obtained from fusel oil by fermentation, or, when obtained from petroleum, a mixture of five different isomers—pentan-1-ol (n-amyl alcohol, b.p. 137·5°C), pentan-2-ol (b.p. 119°C) and pentan-3-ol (b.p. 115·7°C) in addition to the two previous isomers. All form azeotropes with 20–40% water and dissolve a few per cent of water. Amyl alcohols are solvents for several natural resins, the fermentation alcohol also dissolving low molecular weight silicones, ethyl cellulose and urea–formaldehyde resins.

α-AMYLASE   (Liquefying amylase) An enzyme capable of degrading starch to maltose, in which the reducing groups set free during amylolysis retain the α-configuration of the amylose or amylopectin undergoing degradation. Since α-amylase attacks internal 1,4′-linkages in the polymer, rapid molecular weight loss occurs with a decrease in solution viscosity, forming initially, dextrins of low molecular weight.

β-AMYLASE   An enzyme capable of degrading starch by acting on the amylose and amylopectin from the non-reducing end, liberating maltose, and hence causing a rapid increase in reducing power of the starch solution. The action does not proceed beyond branch points, and hence with amylopectin, high molecular weight products (the β-amylase limit dextrins) are formed which resist further enzyme action.

AMYLOID   A water soluble cell-wall polysaccharide found in many seeds which, like amylose, gives a blue colour with iodine. An example is Tamarind seed mucilage—a 1,4′-linked β-D-glucopyranose, like cellulose. but with α-D-xylopyranose and 2-O-β-D-galactopyranosyl-α-D-xylopyranose side units.

AMYLOPECTIN   A branched α-D-glucan, the glucose sugar units being 1,4′-linked as well as 1,6′-linked at the branch points. The major component of starch, usually accounting for 70–85% of the starch. The branches consist of 20–25 α-D-1,4′-linked glucose units; how these are joined is not clear. Laminated, tree-like (herringbone) and ramified structures have all been proposed. Enzymatic degradation and amylolysis limit studies favour the tree-like structure. Total molecular weight is very high and hard to determine. The location of the branch points has been determined by the classical methylation and hydrolysis technique and by the isolation of isomaltose and trehalose as hydrolysis products.

AMYLOSE   A linear glucan in which the glucose sugar units are 1,4′-linked, i.e. poly(1,4′-α-D glucopyranose):

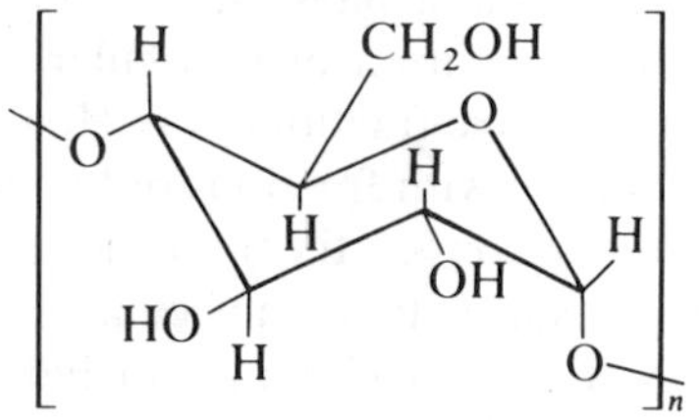

One of the two components of starch usually comprising 15–30% of the starch. Separated from the amylopectin by precipitation from aqueous starch dispersion by addition of a polar organic solvent. It has a DP of a few thousand, but is degraded by the enzymes α- and β-amylase to maltose. In solution it forms helical conformations in the presence of complexing agents with which it forms complexes, e.g. the blue complex with iodine. Concentrated solutions may undergo retrogradation. The X-ray diffraction pattern of both native starch and retrograded amylose demonstrates crystallinity. This may be of the A-form (cereal starches and amylose retrograded at > 50°C), the B-form (tuber starches and retrograded amylose at room temperature) or the U-form (the helical complexes).

A-AMYLOSE   A crystalline form of amylose occurring in cereal whole starches and in amylose retrograded at > 50°C.

B-AMYLOSE   A crystalline form of amylose found in whole tuber starches and in amylose retrograded at ambient temperature.

U-AMYLOSE   A crystalline form of amylose occurring in the complexes with polar molecules, e.g. iodine, butanol. It has a helical conformation with about six amylose units per turn and with the complexing molecule occupying the interior of the helix.

AN   Abbreviation for acrylonitrile.

-AN Suffix added to the end of the stem of a monosaccharide name to indicate a polysaccharide of that monosaccharide, e.g. glucan—a polymer of glucose.

ANABARIC CRYSTALLISATION Polymer crystallisation carried out at high pressure. This frequently leads to formation of extended chain crystals with thick lamellae and little chain folding.

ANAEROBIC ADHESIVE An adhesive, usually based on a dimethacrylate monomer, which remains fluid in the presence of air (which inhibits polymerisation), but which cures through polymerisation when applied to a substrate surface, especially steel (the iron of which catalyses polymerisation), and formed into a closed joint excluding air. A typical monomer used is tetramethyleneglycol dimethacrylate, which might contain a peroxide as initiator together with an amine promotor.

ANALYTICAL FRACTIONATION Fractionation of whole polymer into a series of fractions which are not isolated. The quantity of each fraction is determined by measurement of a suitable physical property. Much more rapid for the determination of molecular weight distribution than preparative fractionation. However, for this purpose a calibration with fractions of known molecular weight is required to convert the frequency distribution of the property measured to a molecular size frequency distribution. Analytical techniques include gel permeation chromatography, turbidimetric titration, sedimentation in an ultracentrifuge, fractional precipitation, gel volume and Brownian diffusion methods.

ANALYTICAL ULTRACENTRIFUGE An ultracentrifuge in which an optical system is used to monitor the solute concentration variation in the cell as a function of distance from the centre of rotation.

ANATASE TiO_2. A particular crystalline form of titanium dioxide useful as a pigment in polymers when exceptional whiteness is required. Its use is restricted, compared with the other crystalline form, rutile, due to its action as a photosensitiser in many polymers.

ANDRADE CREEP LAW An expression for the creep of metals of the form, $\varepsilon(t) = \varepsilon_0 + \beta t$, where $\varepsilon(t)$ is the creep strain at time t, ε_0 is a function of stress and β is a constant. The creep behaviour of some plastics reduces to this behaviour at low stresses.

ANDRADE EQUATION A relationship between the viscosity of a fluid (η) and temperature (T) of the form, $\ln \eta = k/T$, where k is a constant. An alternative form of the equation is in an Arrhenius form: $\eta = k' e^{-E/RT}$, where E is an activation energy, R is the universal gas constant and k' is another constant. Often the equation does describe the behaviour of low molecular weight fluids and polymer melts at temperatures above $(T_g + 100)°C$.

ANELASTIC Mechanical behaviour in which the stress and strain are not single-valued functions of each other.

This occurs particularly when a periodic stress is applied, due to internal friction in a viscoelastic material. Thus the dynamic mechanical behaviour of viscoelastic polymers can be said to be anelastic, but usually the term viscoelastic is used. Similarly a relaxation spectrum, at least when referring to a mechanical relaxation, is sometimes referred to as an anelastic spectrum.

ANELASTIC SPECTRUM Alternative name for relaxation spectrum when this is derived from a dynamic mechanical experiment, i.e. a dynamic mechanical spectrum, or when it is concerned with mechanical relaxation.

ANGLE OF ROTATION Alternative name for torsional angle.

ANGLE OF SHEAR The decrease in the angle between two lines originally in the directions of the x and y axes, after a pure shear deformation.

ANHYDRO- Prefix used in conjunction with the name of a monosaccharide to denote a monosaccharide unit in a polysaccharide, since, strictly, on incorporation into the polymer, a molecule of water has been lost by the monosaccharide. Thus, for example, the units in a glucan can be described as anhydroglucose units, and the polymer as a polyanhydroglucose.

3,6-ANHYDRO-L-GALACTOSE

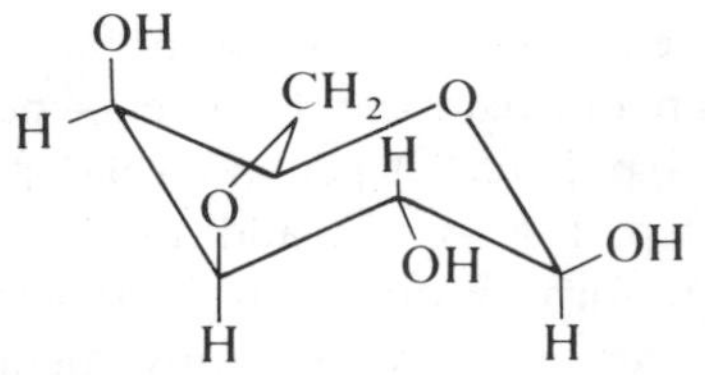

An anhydrosugar, which occurs as a monosaccharide component in agarose, agaropectin and carrageenan.

ANHYDROSUGAR A monosaccharide (or sugar) or monosaccharide unit from which a molecule of water has been eliminated between two of the hydroxyl groups, thereby forming a bicyclic compound, e.g. 3,6-anhydrogalactose.

ANIDEX Generic name for a fibre composed of a polymer in which at least 50% by weight of the repeating units are esters of acrylic acid.

ANILINE–FORMALDEHYDE POLYMER Tradename Cibanite. A polymer formed by reaction of aniline hydrochloride with formaldehyde in aqueous solution. It is probably formed via an intermediate p-aminobenzyl alcohol:

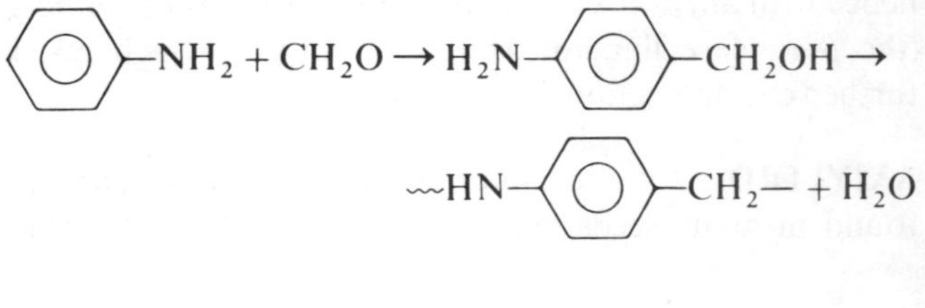

An excess of formaldehyde, with an aniline: formaldehyde ratio of about 1:1·2, is used and some *ortho* substitution and crosslinking also occurs. This makes the moulding of the material rather difficult, so that its commercial use has been rather restricted, despite the good electrical properties of the polymer.

ANIONIC-COORDINATION POLYMERISATION
A coordination polymerisation in which the active centre is anionic. Most Ziegler–Natta polymerisations are of this type.

ANIONIC GRAFTING   Formation of a graft copolymer by producing anionic sites on a polymer chain capable of polymerising the monomer for a second polymer. For example, poly-(*p*-chlorostyrene) reacted with sodium naphthalene produces polystyrene anions which will initiate polymerisation of acrylonitrile (AN):

ANIONIC POLYMERISATION   Ionic chain polymerisation in which the active centre is an anion, usually a carbanion. Like cationic polymerisation the counter-ion (Y^+) may remain in close association with the active centre as a tight or solvent-separated ion pair, or the ions may be free ions, depending on the solvent polarity. Generally, vinyl monomers carrying electron-withdrawing substituents (X) are prone to anionic polymerisation, i.e. monomers of the type $CH_2{=}CHX$, where X is —CN, —COOR, —COR, —Aryl or —CH=CH$_2$ (diene monomers). In addition, oxirane and thiirane ring compounds, such as ethylene and propylene oxides and sulphides, other oxygen ring compounds, such as β-propiolactone, δ-valerolactone and *N*-carboxy-anhydrides, and nitrogen ring compounds such as lactams, will also undergo anionic polymerisation.

Initiation is often by alkali metal alkyls such as *n*-butyl lithium:

$$LiC_4H_9 + CH_2{=}CHX \rightarrow C_4H_9CH_2CHX^- \cdots Li^+$$

In hydrocarbon solvents the initiator largely exists as a hexamer in equilibrium with the active polymerisation centres. Alkali metals themselves will initiate polymerisation forming radical anions:

$$M + CH_2{=}CHX \rightarrow \dot{C}H_2CHX^- \cdots M^+$$

The latter may dimerise to dianions and propagate from both ends. The radical anion from an alkali metal and an aromatic hydrocarbon, e.g. sodium naphthalene, can also act as an initiator: $Ar + Na \rightarrow \dot{A}r^- \cdots Na^+$.

As a result of the presence of the counterion, propagation often involves the addition of monomer to ion pairs, which requires the insertion of the monomer between the active centre and its counterion. This can have a stereoregulating effect on the polymerisation. In particular, anionic polymerisation is very effective at stereoregulation in diene polymerisation. The rate and stereospecificity of an anionic polymerisation is highly dependent on the solvent polarity (even small amounts of an added polar cosolvent to a non-polar solvent can have a large effect) and the size of the counter-ion. Often initiation is extremely rapid compared with propagation, so that all chains start to grow at the same time. Furthermore, termination of growing chains can be absent in highly purified systems, so that all chains can grow to the same length, producing polymer with a very narrow distribution of molecular sizes—sometimes essentially monodisperse polymer.

If termination is absent then a living polymer is produced. Such living polymers are more readily prepared by anionic polymerisation than by any other means. Subsequent addition of a second monomer to a living polymer leads to block copolymer formation. Addition of an appropriate reagent to kill the living ends can be used to form end groups of a known type, e.g.

$$\sim\!\!\sim\!CHX^-Y^+ + ROH \rightarrow \sim\!\!\sim\!CH_2X + ROY$$

Chain growth will similarly be terminated by reaction of the active centre with impurities, especially water and oxygen. Hence, these must be rigorously excluded in order to obtain high molecular weight polymer.

ANISOMETRY   The difference in the magnitude of the dimensions of a particle or body in different directions. Thus a sphere has a minimum anisometry—it is isometric. A plate-like particle is more anisometric and a long, thin cylinder (or fibre) is even more so. Frequently used in the characterisation of reinforcing fillers, especially carbon black, where the high structure blacks have high anisometry. Generally, the higher the anisometry the higher the modulus enhancement. A quantitative measure of anisometry is given by the ratio of the major to minor axis lengths of the ellipse constructed such that the particle or particle aggregate (or a planar projection of it) has the same two moments of inertia around the two axes. For long, thin particles, especially for fibres, and needle-like particles, the length to diameter ratio, i.e. the aspect ratio, is the measure of anisometry. Polymer molecules when extended have extreme anisometry, being typically about 1 nm in diameter and 10^3 nm in length.

ANISOTROPY   The dependence of the properties of a material on the direction in which they are being observed. In polymers anisotropy results when the polymer molecules are oriented or when anisotropically shaped and oriented fillers (e.g. fibres) are present. Mechanical and optical anisotropy are most important. The material properties are most conveniently referred to a coordinate system which coincides with any of the axes of symmetry (the principal axes) that may be present.

ANNEALING The improvement of crystallinity by heating to temperatures below the melting point. This may result from the growth of crystalline regions, e.g. lamellar thickening, the increase in crystal perfection by reduction of defects, or from a change to a more stable crystal structure in polymorphic polymers. Usually annealing has a beneficial effect on properties—increasing modulus and impact strength and reducing any tendency to crazing and cracking on excessive stressing. Amorphous polymers are also said to be annealed when heated to remove internal frozen-in stresses.

ANOMER One of the pair of configurational (or stereo) isomers of the ring form of a sugar that differ only in the configuration of the hemi-acetal carbon, i.e. carbon 1- (called the anomeric carbon) designated by the prefixes α- or β-, e.g. for D-glucose:

β-D-glucose   α-D-glucose

for a mixture of anomers

ANTHOPHYLLITE A fibrous, amphibole asbestos, of empirical composition $(Mg, Fe)_7Si_8O_{22}(OH)_2$, but having rather weak fibres compared with crocidolite. Sometimes used as a filler in polypropylene.

ANTIBODY Alternative name for immunoglobulin.

ANTICRYSTALLISING RUBBER (AC rubber) Natural rubber whose regular cis-1,4-polyisoprene structure has been disrupted by isomerisation of some of its cis units to trans units. This results in a reduced tendency to crystallise at low temperatures, e.g. below 0°C, and hence in a better retention of high elasticity compared with unmodified rubber. Isomerisation is achieved either by reaction with an organic thiol at 40–50°C or with butadiene sulphone or cyclohexylazocarbonitrile at 170–180°C.

ANTIDEGRADANT Alternative name for stabiliser.

ANTIFREEZE PROTEIN A blood plasma glycoprotein of Antarctic fishes. It consists of recurring sequences of Ala–Ala–Thr, with galactosyl-N-acetylgalactosamine disaccharide units attached to every threonine residue. These proteins have molecular weights of 10 000–25 000. Their function is to keep the blood from freezing, possibly due to their expanded structures, preventing the formation of ice crystals.

ANTIOX Tradename for a range of antioxidants, including Antiox 116 (N-phenyl-β-naphthylamine), Antiox 123 (N,N'-diphenyl-p-phenylenediamine, Antiox 2246 (2,2'-methylene-bis-(4-methyl-6-t-butyl phenol)).

ANTIOXIDANT An additive used to protect a polymer against oxidation by atmospheric oxygen. Most antioxidants are only effective against thermal oxidation and require the use of a photostabiliser to resist the effects of photo-oxidation. Early antioxidants were developed for use in rubbers and were aromatic amines, e.g. diphenyl-p-phenylenediamine, which discolour during use (staining antioxidant). Hindered phenols are mostly used in plastics materials since they are non-staining, e.g. 2,6-di-t-butyl-4-methylphenol. For high temperature use materials of low volatility (and therefore higher molecular weight), e.g. 1,1,3-tris(2-methyl-4-hydroxy-5-t-butylphenyl)-butane, are used. Amines and phenols are chain-breaking antioxidants of the AH type, where H is an active hydrogen. Preventive antioxidants are often peroxide decomposers and include many organosulphur compounds, e.g. thio-bisphenols. Often only about 0·1% is adequate and rarely more than 1% is needed. Frequently a synergistic mixture of an AH and a preventive antioxidant is used.

ANTIOZONANT An additive used to protect a polymer against the effects of ozone-induced degradation and hence used mainly in diene rubbers. Works either by providing a physical barrier to ozone penetration by forming a thin surface film of an ozone-resisting wax or by chemically reacting with ozone or polymer ozonolysis products, as do aromatic diamines such as p-phenylenediamine derivatives.

ANTIPARALLEL PLEATED SHEET The beta sheet conformation with adjacent polymer molecules running in opposite directions.

ANTIPLASTICISER A plasticiser which causes an increase in the stiffness of a polymer when used in small amounts but which, when used at higher concentrations, acts normally as a plasticiser. This happens particularly in polyvinyl chloride for concentrations of plasticiser of up to about 15 phr. Consequently, in this polymer plasticisers are generally not used below about 30 phr. Antiplasticisation has also been observed in nylons, polycarbonates and in polymethylmethacrylate. Thus it is seen to occur in polymers with polar side groups and with relatively rigid chains. The mechanism is not fully understood, but obviously involves an increase in intermolecular forces between polymer chains via polymer/plasticiser interactions, possibly by increased ordering of the chains.

ANTITHIXOTROPY (Negative thixotropy) (Rheopexy) Time-dependent fluid behaviour in which the apparent viscosity increases with time of shear and recovers when shearing, and hence flow, ceases. It occasionally occurs with polymer systems, such as with aqueous solutions of polyelectrolytes.

ANTRON Tradename for a nylon 66 fibre.

APEC Tradename for an aromatic polyester copolymer produced by reaction of bisphenol A with iso- or terephthaloyl chloride and some phosgene. The copolymer therefore contains both polycarbonate and polyarylate units.

APF Abbreviation for all purpose furnace carbon black.

APO Abbreviation for tris-(aziridinyl)phosphine oxide.

APOENZYME An enzyme protein without its cofactor and therefore catalytically inactive.

a-PP Abbreviation for atactic polypropylene.

APPARENT MOLECULAR WEIGHT The value of the molecular weight obtained from the sedimentation equilibrium method of ultracentrifugation, which is not the true value (M), owing to its concentration dependence: $M_{app} = M(1 + kMc + \cdots)$, where c is the concentration and k is a constant. M must be obtained by extrapolation to infinite dilution.

APPARENT VISCOSITY Symbol η or η' or η'_a. In rheology, the ratio of the shear stress to the shear rate when the latter is not constant, i.e. when the fluid is non-Newtonian. Thus η may vary with shear rate or with time. Such variability may be described by a suitable constitutive equation, that most commonly used for polymer systems being the power law equation.

APPROACH TO EQUILIBRIUM METHOD Alternative name for Archibald method.

A-PROTEIN An aggregate of protein subunits of the tobacco mosaic virus, which is frequently formed in mildly alkaline medium especially at low temperatures and low ionic strength. It consists mostly of trimers, with sedimentation values of 4–5 S. It is the usual form of the native virus protein.

ARABAN Alternative name for arabinan.

ARABIC ACID The acid whose calcium or magnesium neutralised salt is gum arabic, from which it is obtained by precipitation of the acidified solution with ethanol or acetone.

ARABINAN (Araban) A polymer of arabinose, found associated with pectic substances from which it is extracted with boiling 70% ethanol or hot lime water. These pectic arabinans are of uncertain structure but probably consist of 1,5'-linked α-L-arabinofuranose main chains with single α-L-arabinofuranose side groups 1,3'-

linked to the main chain. However, they may be artifacts of isolation procedures, being derived from more complex heteropolysaccharides.

ARABINOGALACTAN A polysaccharide containing both L-arabinose and D-galactose units, occurring frequently in coniferous woods, especially larches (which may consist of up to 25% arabinogalactan). Often separated into A and B fractions of differing molecular weight. The arabinose to galactose ratio is typically 1:6, the main chain being 1,3'-linked β-D-galactopyranose units with side groups of dimers and trimers of 1,6'-linked β-D-galactopyranose units and single or dimer units of β-L-arabinose groups. Some polymers also contain glucuronic acid units. Similar polymers are also found in exudate gums, e.g. from *Acacia*. Arabic acid, the main component of gum arabic, is a complex branched galactan whose main chain is 1,3'-linked β-D-galactopyranose.

L-ARABINOSE Fischer projection formula:

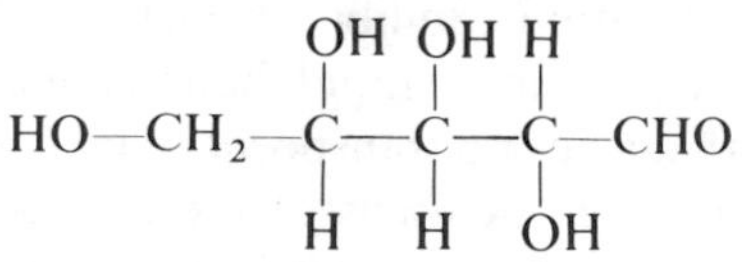

$$HO-CH_2-\overset{\overset{\displaystyle OH}{|}}{\underset{\underset{\displaystyle H}{|}}{C}}-\overset{\overset{\displaystyle OH}{|}}{\underset{\underset{\displaystyle H}{|}}{C}}-\overset{\overset{\displaystyle H}{|}}{\underset{\underset{\displaystyle OH}{|}}{C}}-CHO$$

Ring form (β-L-arabinofuranose):

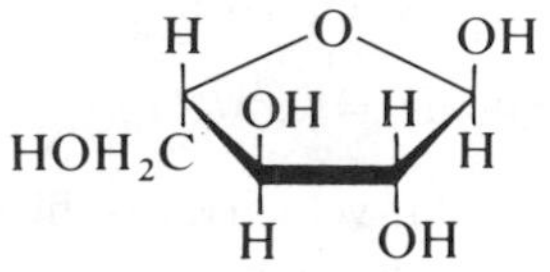

M.p. 160°C. $\alpha_D^{20,H_2O} + 105°$.

An aldopentose monosaccharide, found widely in many plant polysaccharides, including bacterial polysaccharides; examples include the heteropolysaccharides arabinogalactan, arabinoxylan, pectic substances and gum arabic. Often prepared by controlled hydrolysis of cherry gum.

ARABINOXYLAN A polysaccharide containing both xylose and arabinose units. Many of the so-called plant stem xylans are really of this type, consisting of 1,4'-linked β-D-xylose main chains with side groups of 1,3-sugar units containing α-L-arabinose units.

ARACAST Tradename for hydantoin-containing epoxy resins of structure.

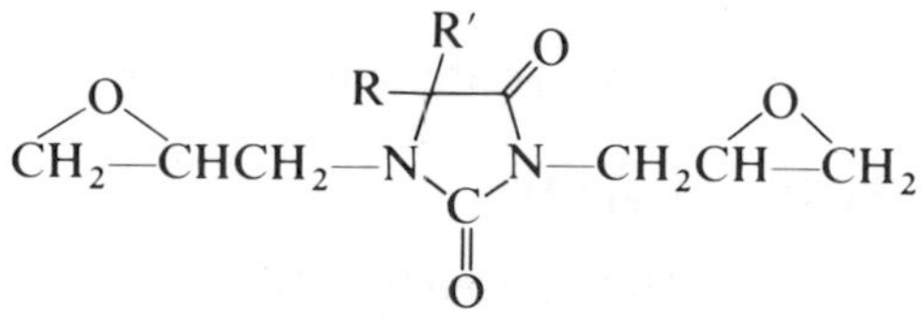

where R and R' are alkyl groups. The resins are of lower viscosity than conventional epoxy resins and may be cured using the usual range of epoxy resin hardeners. Cured resins have better heat resistance, dielectric strength, ultraviolet and electric arcing and tracking resistance than conventionally cured products.

ARACHIN A crystallisable protein globulin found in peanuts. It contains all the essential amino acids in reasonable amounts and is therefore nutritionally valuable. It is particularly rich in arginine and has been used for synthetic fibre production (Ardil), the fibres being hardened by reaction with formaldehyde.

ARALAC Tradename for a man-made protein fibre from casein.

ARALDITE Tradename for epoxy resin.

ARAMID Alternative name for those wholly aromatic polyamides which have at least 85% of the amide links attached directly to two aromatic rings. Poly(m-phenylene-isophthalamide) poly-(p-benzamide) and poly-(p-phenyleneterephthalamide) are the best known examples.

ARCHIBALD METHOD (Approach to equilibrium method, Pseudoequilibrium method) A method of the determination of the weight average molecular weight ($\bar{M}_w$) of a polymer in dilute solution by a modification of the sedimentation equilibrium method of ultracentrifugation, and much more rapid than that method. It relies on the fact that since no transport can take place through the top or bottom meniscus of the cell, an equilibrium concentration is rapidly established at these points. This, or the concentration gradient (dc/dr), is determined at one of these points and is related to $\bar{M}_w$ by:

$$(1/c)(dc/dr) = (\omega^2 r/RT)(1 - \bar{v}_2\rho)M$$

where ω is the angular velocity, r is the distance from the centre of rotation, $\bar{v}_2$ is the partial specific volume of the polymer, ρ is the solvent density, R is the universal gas constant and T is the temperature (K). Measurements are made at various time intervals and from these the molecular weight can be obtained by extrapolation to zero time.

ARDEL Tradename for a polyarylate which is a copolymer produced from tere- and isophthalic acids and bisphenol A.

ARDIL Tradename for a man-made protein fibre from groundnut protein, which consists largely of arachin.

ARENKA Tradename for an aramid fibre similar to Kevlar.

ARG Abbreviation for arginine.

ARGININE (Arg) (R)

H₂NCH((CH₂)₃NHC—NH₂)COOH
‖
NH

M.p. 238°C (deomposes) (loses water at 105°C).

A basic α-amino acid found widely in proteins, but in especially large amounts in fish proteins, e.g. salmine is 90% arginine. Its pK′ values are 2·17, 9·04 and 12·48 with the isoelectric point at 10·8. In the hydrochloric acid hydrolysis stage of amino acid analysis of proteins, arginine may be partially converted to ornithine. Arginine residues in proteins are preferentially cleaved by trypsin, but not by carboxypeptidase.

ARNEL Tradename for a cellulose triacetate fibre.

ARNITE Tradename for polyethylene terephthalate and polybutylene terephthalate moulding materials.

ARNITE A Tradename for a polyethylene terephthalate moulding material.

ARNITEL E Tradename for the thermoplastic elastomer based on the polyether/ester block copolymer, poly-(tetramethyleneterephthalate-b-polyoxytetramethyleneterephthalate).

ARNOX Tradename for poly-(2,6-dimethylphenylene oxide) modified by the incorporation of polystyrene or a modified polystyrene.

AROCHLOR Tradename for chlorinated diphenyl.

AROLAC Tradename for a casein fibre.

AROMATIC NYLON Alternative name for aromatic polyamide.

AROMATIC OIL A rubber oil containing a high proportion (about 50%) of the carbon atoms in aromatic ring structures. Such aromatics occur naturally in petroleum and are separated from the paraffinic components by solvent extraction and distillation. Good quality oils contain very little wax but do contain up to about 20% polar heterocyclic components. A typical aromatic oil will have a viscosity gravity constant of about 0·98, a refractivity intercept of 1·07, giving a carbon type analysis of about 45% aromatic, 37% paraffinic and 18% naphthenic carbon atoms. In terms of molecules, a typical composition is about 80% aromatic compounds (with saturated side chains or rings attached) 8% polar heterocyclic compounds and 12% saturated molecules. Aromatic oils are highly compatible with hydrocarbon rubbers and in vulcanisates give the highest tensile strengths and tear strengths due to their ability to aid carbon black dispersion. However, they have relatively poor oxidation stability due to the presence of the heterocyclics.

AROMATIC POLYAMIDE (Aromatic nylon) (Aramid) A polymer containing both aromatic rings and amide groups in the polymer chain. Often subdivided into partially aromatic polyamides (which also contain aliphatic chain carbon atoms, e.g. nylon 6T) and wholly aromatic polyamides, e.g. poly-(m-phenylene-isophthalamide). The polymers have much higher glass transition and melting temperatures than their aliphatic counterparts, typically with a $T_m \sim 350°C$ when the rings are *para*-linked, but may be below 200°C if *ortho*-linked. They are difficult to prepare by melt polymerisation owing to their high T_m values and to the low reactivity of aromatic diamines. Therefore low temperature solution or interfacial polymerisation is used with reactive monomers, e.g. with diacid chlorides. The polymers are difficult to process; fibres and films must be spun or cast from solution. They are difficult to dissolve and generally highly polar solvents are required, e.g. dimethylacetamide or hexamethylphosphoramide. The fibres are very stiff and are useful in high temperature and fire resistant applications and as reinforcements. Commercial examples are poly-(*p*-phenyleneterephthalamide), poly-(*m*-phenyleneisophthalamide), poly-(*p*-benzamide) and nylon 6T. The thermal stability of partially aromatic polyamides (e.g. nylon 6T) and cycloaliphatic polyamides (e.g. piperazine polyamides) are lower, being limited by the aliphatic portion of the structure. Tractability of the *para*-linked wholly aromatic polyamides may be improved by *meta*- or *ortho*-linking of the rings or by synthesising alternating ordered copolyamides.

AROMATIC POLYAMINE (Polyphenyleneamine) A polymer of structure $-\text{[}-\text{NHArNHAr}'\text{NH}-\text{]}_n$ where Ar and Ar' are aromatic rings, one of which contains a linking group, frequently

$$-CH_2-\langle \bigcirc \rangle-CH_2-$$

Synthesised by reaction of an aromatic diamine with an aromatic dihalide, e.g. a xylene dichloride, or by heating an aromatic diamine with a diphenol. May be crosslinked through the secondary amine groups by anhydrides.

AROMATIC POLYCARBONATE A polymer of the type $-\text{[}-\text{ArOCOO}-\text{]}_n$ where Ar is an aromatic ring structure. An ester of carbonic acid and a dihydric phenol. The most important example is the polymer derived from bisphenol A. Synthetic methods used are phosgenation either in the presence of pyridine or interfacially:

$$n\,\text{HOArOH} + n\,\text{COCl}_2 \rightarrow -\text{[}-\text{OArOCO}-\text{]}_n + 2n\,\text{HCl}$$

or by transesterification between the phenol and diphenyl carbonate:

$$n\,\text{HOArOH} + n\,\text{PhOCOOPh} \rightarrow$$
$$2n\,\text{PhOH} + -\text{[}-\text{OArOCO}-\text{]}_n$$

Only a few have melting points $>300°C$ whilst thermal stability is lower than in the corresponding aromatic polyamides. Bisphenol A polycarbonate, however, does have a useful balance of properties.

AROMATIC POLYESTER A polyester containing aromatic rings in the polymer chain. The polymer may be partially aromatic, e.g. PETP and PTMTP, or wholly aromatic, e.g. polyarylates and polyhydroxybenzoic acid. The polymers may by synthesised by self-condensation of the hydroxy-acid, by reaction of a diacid with a diol or, often better, by ester interchange between the diester of a diacid and a diol. Low temperature solution or interfacial methods, e.g. utilising a diol and diacid chloride, may also be used. The presence of aromatic rings increases thermal stability and raises T_m, especially when the rings are *para*-linked. Thus the polymer from hydroquinone and terephthalic acid has a T_m value of $>500°C$, compared with that of PETP of 265°C. Aromatic polyester block and other copolymers can show thermotropic liquid crystalline behaviour, as with the commercial moulding materials Vectra, Ultrax and Xydar.

AROMATIC POLYETHER A polymer of the type $-\text{[}-\text{ArOR}-\text{]}_n$, i.e. containing an ether link directly to an aromatic ring (Ar) with R as a further linking group (e.g. $-\text{SO}_2-$ as in polyethersulphones). Often R is absent, as in the polyphenylene oxides.

AROMATIC POLYHYDRAZIDE A polymer of the type

$$\left[\langle \bigcirc \rangle - \text{CONHNHCO} - \langle \bigcirc \rangle \right]_n$$

Best synthesised by low temperature solution polymerisation between hydrazine (or a dihydrazide) and a diacid chloride in a basic solvent such as NMA or HMPA. The polymers have good thermal stability, high T_m values, but are fairly soluble, e.g. in cold DMSO. They form polychelates with many metal salts and at $\sim250°C$ undergo intramolecular cyclodehydration to poly-(1,3,4-oxadiazoles). They also form polyphenyltriazoles when heated with aniline. They yield useful high temperature resistant fibres, e.g. X-500, PABM-T.

AROMATIC POLYMER A polymer containing aromatic ring structures, usually benzene rings, in the polymer chain. In the simplest example, polyphenylene, the polymer consists entirely of benzene rings, but any of a great variety of linking groups or atoms ($-\text{X}-$) are usually present giving polymers of the general structural type

$$\left[\langle \bigcirc \rangle - \text{X} \right]_n$$

These may be either of a simple atom, e.g. $-\text{O}-$ or $-\text{S}-$ in polyphenylene oxide and polyphenylene sulphide respectively, or a simple group, e.g. $-\text{SO}_2-$ in polysulphones and $-\text{NHCO}-$ in aromatic polyamides, or an ester group as in PETP and bisphenol A polycarbonate.

The presence of aromatic rings stiffens the polymer chain, raising both glass and melting temperatures. The stiffest chains have the rings *para*-linked thereby having the most symmetrical structures with the best possibility of crystallisation. Many aromatic polymers are also good high temperature resistant polymers. Heterocyclic polymers are also sometimes classified as aromatic polymers, as are polymers with aromatic side groups, e.g. polystyrene and related polymers. Polymers in which the linking

X group contains in-chain aliphatic carbons are often referred to as partially aromatic polymers. If no such carbons are present then the polymer may be called a wholly aromatic polymer.

AROMATIC POLYSILOXANE A polysiloxane containing aromatic rings as well as siloxane links in the polymer chain, in order to increase the thermal stability of the polymer. Synthesised, for example, by reaction of a bisphenol with a difunctional silane:

$$HO—Ar—OH + R_2Si(OR')_2 \rightarrow$$
$$+SiR_2—O—Ar—O+_n + 2n\ R'OH$$

AROMATIC POLYSULPHONAMIDE A polymer of the type

$$+NHRNHSO_2NHArSO_2+_n$$

where Ar is an aromatic ring. Synthesised by solution or interfacial polymerisation of an aliphatic diamine and an aromatic disulphonyl chloride. Wholly aromatic polymers are difficult to prepare due to the low reactivity of aromatic diamines. The polymers are soluble in polar solvents, e.g. m-cresol, and in aqueous alkali, have T_m values in the range 150–300°C (lower than the corresponding polyamides) and may be readily spun to fibres, which may be made highly oriented and crystalline by drawing.

AROMATIC POLYSULPHONATE A polymer of the type $+ArOSO_2Ar'SO_2O+_n$, where Ar and Ar' are aromatic rings. Synthesised by solution or interfacial reaction between an aromatic diol or bisphenol and an aromatic disulphonyl chloride. The polymers are often amorphous, slowly crystallising only from solution, not from the melt. They have lower T_g and T_m values than the corresponding phenyl esters and show a rapid weight loss on heating to 300–400°C.

AROMATIC POLYSULPHONE A polymer of the type $+ArSO_2+_n$, where Ar is an aromatic ring. Synthesised by oxidation of an aromatic sulphide, condensation of an aromatic disulphonyl chloride with a reactive dinuclear aromatic compound or by self-condensation of an aromatic dinuclear disulphonyl chloride. The polymers have high T_m values, e.g. poly-p-phenylenesulphone (520°C), and are rather intractable. The introduction of flexibilising ether groups as in the polyethersulphones, lowers the softening point and several useful high temperature resistant polymers of this type, which may be melt processed, have been developed.

AROMATIC POLYUREA A polymer of the type $+ArNHCONHAr'NHCONH+_n$, where Ar and Ar' are aromatic rings. Usually synthesised by reaction of a diamine with an aromatic diisocyanate:

$$n\ H_2NArNH_2 + n\ OCNAr'CNO \rightarrow$$
$$+NHArNHCONHAr'NHCONH+_n$$

usually by solution polymerisation at low temperature, since aromatic diisocyanates are very reactive. Aromatic diamines are relatively unreactive compared with aliphatic diamines, but wholly aromatic polyureas may be synthesised. They have T_m values in the range 150–300°C. However, the urea link is thermally unstable above ~200°C.

AROMATIC POLYURETHANE A polymer of structure $+RNHCOOR'OCONH+_n$ where R and/or R' are aromatic groups. However, the term is usually restricted to polymers in which both R and R' are aromatic. Thus polymers from the commonly used isocyanate, toluene diisocyanate, but with R' aliphatic, are not considered as aromatic polyurethanes. The wholly aromatic polymers may be synthesised by any of the usual polyurethane solution polymerisation methods, e.g. reaction between diisocyanate and diol. Although they have high T_m values, thermal stability is lower than in the corresponding polyamides and the polymers are of only limited interest.

AROMATIC RESIN A petroleum resin consisting of the low molecular weight polymer obtained by cationic polymerisation of mixed C_8–C_{10} unsaturated aromatic hydrocarbon fractions from petroleum cracking, containing styrene, α-methylstyrene, indene and vinyltoluene. They are widely used in adhesives, coatings, rubbers, sealants and caulks to improve tack and other mechanical properties.

ARTIFICIAL SILK A name given to early man-made fibres, which were produced with the specific aim of resembling silk. They did so to the extent that they were produced as long continuous fibres. However, many present day man-made fibres are not continuous and are not intended to resemble silk.

ARYL CYCLIC SULPHONIUM ZWITTERION (ACSZ) A compound of the type

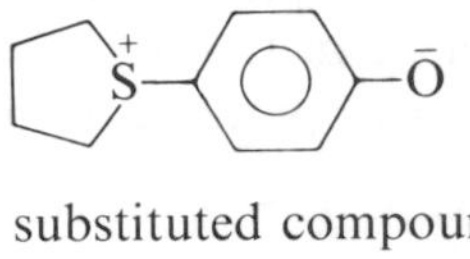

or a similar ring substituted compound, used as monomers for the synthesis of poly-(thio-1,4-phenylenetetramethylenes). Synthesised by reaction of a phenol with tetrahydrothiophene-1-oxide:

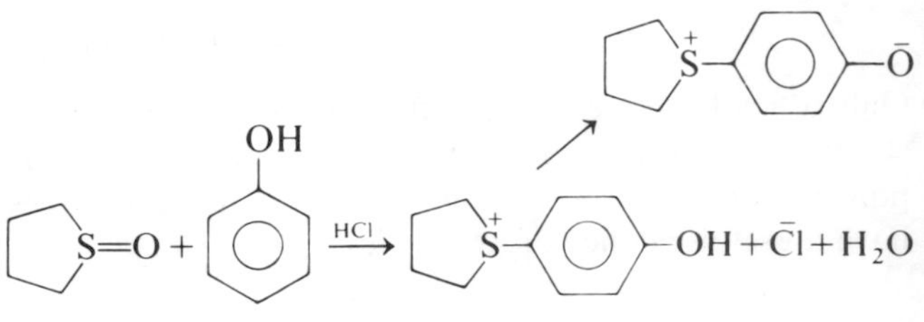

ARYLEF Tradename for a polyarylate which is a copolymer produced from tere- and isophthalic acids and bisphenol A.

ARYLENE POLYMER Alternative name for a poly-(arylene...).

ARYLON Tradename for a polyarylate which has a lower T_g (softening point about 155°C) but is cheaper than other commercial polyarylates such as Ardel and Arylef.

ARYLON T Tradename for a polysulphone/ABS blend, the ABS being based on an α-methylstyrene copolymer.

ASA Abbreviation for acrylonitrile–styrene–acrylate copolymer.

ASBESTOS A generic name for a group of naturally occurring silicate minerals, often found in fibrous form. Useful as fillers in polymer composites, especially with phenolic resins, either as a reinforcement or to improve thermal properties. Two classes of asbestos are found, a serpentine form, chrysotile, which is by far the most widely used, and an amphibole form, which is found in several different varieties—crocidolite, amosite and anthophyllite being the best known. Asbestos in most forms, but especially as crocidolite, can present severe toxic hazards, and as a consequence its use as a reinforcement is much less than formerly.

ASN Abbreviation for asparagine.

ASP Abbreviation for aspartic acid.

ASPARAGINE (Asn) (N)

$$H_2NCH(CH_2CONH_2)COOH.$$

M.P. 263°C.

(decomposes) (hydrate). A polar α-amino acid, occurring widely in proteins. Its pK' values are 2·02 and 8·80 with the isoelectric point at 5·41. During the hydrochloric acid hydrolysis stage of amino acid analysis of proteins it is converted to aspartic acid and so its content in proteins is often counted with this acid.

ASPARTIC ACID (Asp) (D)

$$H_2NCH(CH_2COOH)COOH.$$

M.p. 269–271°C.

An acidic α-amino acid occurring widely in proteins. Its pK' values are 1·88 and 3·65 with the isoelectric point at 2·77. It is often associated with hairpin loops in protein conformations.

ASPECT RATIO The ratio of the major to the minor dimension of a particle. In particular, for a fibre or rod-like particle, it is the length to diameter (l/d) ratio. For an elliptical particle it is the ratio of the major to minor axis lengths. It is important in determining the effect of dispersed particles on the viscosity of a fluid and on the mechanical properties of a filled solid.

ASPHALT (1) A natural or artificial mixture of a bitumen with particulate mineral matter.
(2) Alternative name for bitumen.

ASPLIT Tradename for phenol–formaldehyde polymer.

A-STAGE POLYMER or RESIN The first intermediate stage in the formation of a network by step-growth polymerisation in which a linear or branched low molecular weight prepolymer is formed which is soluble and fusible. The term is specifically used to refer to phenol–formaldehyde prepolymers, such as resoles. Subsequent reaction converts the A-stage to the B-stage and, finally, to the C-stage crosslinked network.

ASTERITE Earlier, a tradename for polymethyl-methacrylate modified by copolymerisation with butyl acrylate. Later, a tradename for polymethylmethacrylate filled with a mineral powder.

ASTREL 360 Tradename for an aromatic polyether-sulphone containing both

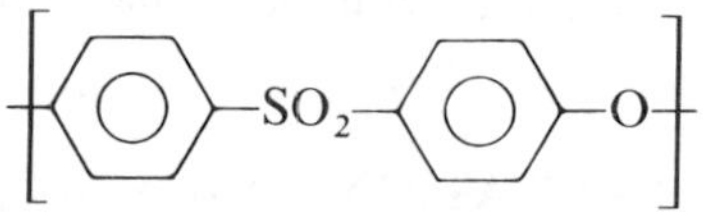

$$\left[\langle O \rangle - SO_2 - \langle O \rangle - O \right]$$

and

$$\left[\langle O \rangle - \langle O \rangle - SO_2 \right]$$

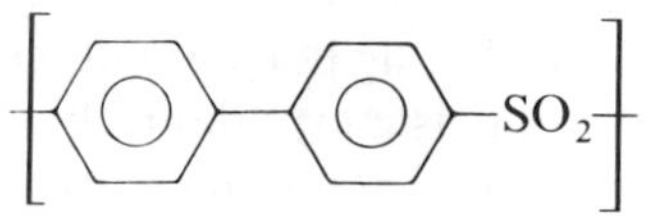

units in which the latter predominates. Synthesised by electrophilic attack of an arylsulphonyl chloride on an aromatic hydrocarbon, probably involving the monomers

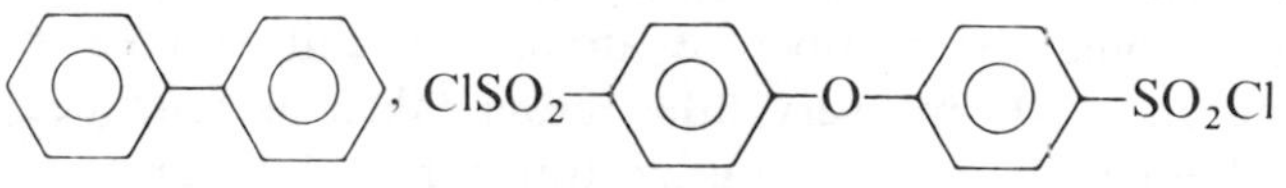

$$\langle O \rangle - \langle O \rangle, \quad ClSO_2 - \langle O \rangle - O - \langle O \rangle - SO_2Cl$$

and

$$\langle O \rangle - \langle O \rangle - SO_2Cl$$

ASx Abbreviation for asparagine and/or aspartic acid in a protein.

ATACTIC POLYMER A polymer in which at least one chain atom of the repeat unit can exhibit stereochemical configurational isomerism, but in which there is no preference for one particular configurational isomer, so that a random distribution of isomers exists. Polymers produced by free radical polymerisation are usually essentially atactic, with perhaps a slight tendency to syndiotacticity. Since atactic polymers have irregular structures, they do not usually crystallise and hence are amorphous. Vinyl polymers of the type $+CH_2CHX+_n$ are the most common type of polymer for which stereochemical isomerism is possible, and since they are normally free radically produced, vinyl polymers are usually amorphous.

ATACTIC POLYPROPYLENE (a-PP) Polypropylene in which there is a random distribution of configuration in the repeat units. Since the structure is irregular, the polymer does not crystallise and is a weak, rubbery material compared with isotactic polypropylene, which is hard and stiff. It is produced as a by-product during isotactic polypropylene manufacture and is separated by its solubility in, e.g., hexane. It has found some uses as a carpet backing and road surfacing material.

A-TELL Tradename for poly-(ethyleneoxybenzoate) in fibre form.

ATLAC Tradename for an unsaturated polyester resin.

ATOM FORM FACTOR (Atom scattering factor) Symbol f. A parameter describing the angular dependence of the scattering of X-rays by the electrons of an atom in a coherent manner, i.e. without change of wavelength or phase relationship. By convention f_0 (at scattering angle $2\theta = 0°$) is set equal to the number of electrons present in the atom.

ATOMIC POLARISATION A dielectric polarisation in which an applied electric field causes displacements of the atoms in the molecules of a dielectric. Typically, in organic molecules including polymers, its contribution to the total polarisation is only about 10% that of the electronic polarisation. The motions involved are more sluggish, so resonance frequencies are at about 10^{12} Hz, i.e. in the infrared region.

ATOMIC RADIAL DISTRIBUTION FUNCTION (Radial distribution function) Symbol ρ_r. A parameter describing the number of atoms per unit volume at distance r from every other atom taken as reference. Obtained from the Fourier transform of amorphous X-ray scattering data of amorphous polymers. It is frequently used to express the short-range order in an amorphous polymer.

ATOM SCATTERING FACTOR Alternative name for atom form factor.

ATR Abbreviation for attenuated total reflection spectroscopy.

ATTENUATED TOTAL REFLECTION SPECTROSCOPY (ATR) (Internal reflection spectroscopy) A technique for exposing a sample to an infrared beam in spectroscopy, which is often more useful than the conventional transmission technique using a thin film of sample. The sample is held directly in contact with the reflecting surface of a prism of a high refractive index substance such as thallium iodide or KRS-5. If the angle of incidence at the prism/sample interface is greater than the critical angle, then total internal reflection takes place. During the reflection the beam penetrates a few micrometres into the sample and is therefore 'attenuated' by sample absorption during this short path length penetration. The spectrum recorded is thus similar to the transmission spectrum of the surface layers of the sample. Usually in order to increase sample absorption and hence spectrum intensity, a trapezoidal prism is used in which

multiple, often 25, reflections take place (multiple internal reflection). For polymer work, ATR is particularly useful since the sample in the form of an opaque film, a coating or paint film, an adhesive layer, bonded film, fibre bundle, etc., may be directly examined.

AUTOACCELERATION (Gel effect) (Tromsdorff effect) (Tromsdorff–Norrish effect) An increase in the rate of a free radical polymerisation with an increase in conversion. It is due to the increasing viscosity of the monomer/polymer mixture and/or occlusion of growing active centres, causing a decrease in the mobility of the growing active centres, which in turn slows termination. The effect is most pronounced in concentrated systems, such as in mass polymerisation, especially if the monomer is a poor solvent for the polymer, and in solution polymerisation in poor solvents.

AUTOHESION Alternative name for tack.

AUTOINHIBITION Inhibition of free radical polymerisation when the monomer itself is the inhibitor. This results in a low polymerisation rate and the formation of only low molecular weight polymer. It occurs with allylic monomers (of the type $CH_2{=}CH{-}\overset{\alpha}{C}H_2X$) as a result of degradative chain transfer since the C—H bond α to the double bond (the allylic hydrogen) is weak. In addition, the allyl radical produced is highly resonance stabilised. The reaction is thus an allylic abstraction reaction with chain transfer to monomer:

$$\sim CH_2{-}\overset{\cdot}{C}H + CH_2{=}CH \quad \rightarrow$$
$$\qquad | \qquad\qquad |$$
$$\quad CH_2X \qquad\quad CH_2X$$

$$\sim CH_2{-}CH_2 + CH_2{=}CH \quad \leftrightarrow \overset{\cdot}{C}H_2{-}CH{=}CHX$$
$$\qquad | \qquad\qquad\quad |$$
$$\quad CH_2X \qquad\qquad \overset{\cdot}{C}HX$$

The allyl radical produced is too stable to re-initiate polymerisation, so inhibition has occurred.

AUTOORIENTATION MECHANISM A mechanism which explains the spherulitic mode of crystallisation, whereby a crystallite is formed and then induces formation of a neighbouring crystallite oriented perpendicularly to it. The process repeats itself to produce a herringbone-like array of crystallites, eventually growing into a spherulite.

AUTOXIDATION Free radical oxidation by atmospheric oxygen in which the products catalyse further oxidation and therefore cause an acceleration in the oxidation rate. Usually hydroperoxides are responsible since they dissociate either thermally or on ultraviolet light irradiation to free radicals which initiate new oxidation chains. The progress of autoxidation is measured, for example, by oxygen uptake and shows an induction period followed by the autoacceleration stage.

AV Abbreviation for acid value.

AVERAGE FUNCTIONALITY   Symbol f_{av}. In a step-growth polymerisation when a mixture of monomers of different functionality is used, the average functionality is the average number of functional groups per monomer molecule, for all types of monomer molecules, i.e. $f_{av} = \sum N_i f_i / \sum N_i$, where N_i is the number, or number of moles, of monomer with functionality f_i. f_{av} is used in the Carothers equation for non-linear polymerisation in predicting the critical conversion for gelation.

AVERAGE MOLECULAR WEIGHT   Alternative name for molecular weight average.

AVIMID   Tradename for a high temperature resistant polyimide.

AVLIN   Tradename for a viscose rayon staple fibre.

AVRAMI EQUATION   A relationship expressing the dependence of the amount of a crystallising polymer that has crystallised with time, i.e. a crystallisation rate equation. Its most general form is

$$W_L / W_0 = \exp(-Kt^n)$$

where W_L and W_0 are the masses of melt at time $t = t$ and $t = 0$ respectively, K is a rate constant and n is the Avrami exponent. Ideally n has an integer value which is the sum of the number of dimensions in which growth has taken place (three for spherulites, two for discs and one for rods or fibrils) and of the order of the time dependence of the nucleation process. This latter is either unity for sporadic nucleation (where the number of nuclei increases linearly with time) or zero (instantaneous nucleation). Thus for the frequently observed spherulite growth from sporadic nucleation, $n = 3 + 1 = 4$. In practice experimental data often give non-integral values of n due to the simultaneous operation of different growth processes.

AVRIL   Tradename for a crosslinked viscose rayon with a high tenacity of $3 \cdot 2$ g denier^{-1} (dry), $2 \cdot 2$ g denier^{-1} (wet) and about 10% elongation at break.

AVRON   Tradename for a medium tenacity viscose rayon.

AXIAL FLOW   A pressure flow which occurs in the direction of the axis of a pipe or channel, as in a circular bore pipe, or in the annular space between two concentric tubes.

AXIALITE   A multilayer lamellar crystalline aggregate similar to a hedrite, but grown from solution and having greater symmetry.

AXIAL ORIENTATION   Alternative name for uniaxial orientation.

AZDN   Abbreviation for azobisisobutyronitrile, which is the commonest example of an azodinitrile, hence the abbreviation.

AZELAIC ACID $HOOC(CH_2)_7COOH$. Produced by ozonolysis of oleic acid,

$$CH_3(CH_2)_7CH{=}CH(CH_2)_3COOH.$$

Useful mostly for the production of azelate ester plasticizers, but also a monomer for nylon 69. Also converted to 1,9-diaminononane.

AZEOTROPIC COPOLYMERISATION   A copolymerisation in which the copolymer being formed at any instant has the same composition as that of the monomer feed. This occurs if both monomer reactivity ratios are unity no matter what the feed composition, or if both the reactivity ratios are less than unity at one particular feed composition. In any other situations, drift in monomer, and therefore also in copolymer, composition occurs due to the more reactive monomer being consumed more rapidly. When both reactivity ratios (r_A and r_B) are < 1, then a plot of mole fraction of monomer A (F_A) in the copolymer versus mole fraction of monomer A in the feed (f_A) will cross the azeotropic line (the line $F_A = f_A$) at the cross-over point. The value of f_A at this point is $(1 - r_B)/(2 - r_A - r_B)$. Thus if the initial feed composition is less rich in monomer A, then for $r_A < r_B$, the copolymer formed will get richer in A until the azeotropic composition is reached, after which point all subsequent copolymer formed will have the azeotropic composition.

AZETIDIN-2-ONE   Alternative name for 2-azetidinone.

2-AZETIDINONE   (Azetidin-2-one) (β-Propiolactam) (β-Lactam) A lactam of structure

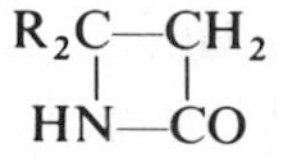

$$\begin{array}{cc} R_2C{-}CH_2 \\ | \quad\quad | \\ HN{-}CO \end{array}$$

conveniently synthesised, except when R is H, by addition of N-carbonylsulphamoyl chloride to olefins. Polymerises to yield substituted nylon 3 products.

AZIDE METHOD   A method of peptide bond formation useful in peptide and protein synthesis since the risk of racemisation of the L-amino acid involved is very low using this method. An acylazide is formed by reaction of a peptide or amino acid ester with hydrazine followed by treatment with nitrous acid. The acylazide is then reacted with a free amino group of a second amino acid or peptide:

$$\sim\!\!\text{NHCHRCOOR}' + N_2H_4 \rightarrow$$

$$\sim\!\!\text{NHCHRCONHNH}_2 + R'OH$$

$$\sim\!\!\text{NHCHRCON}_3 \xleftarrow{\text{HNO}_2}$$

$$\downarrow \text{NH}_2\text{CHR}''\text{CO}\sim$$

$$\sim\!\!\text{NHCHRCONHCHR}''\text{CO}\sim$$

AZIRIDINE   Alternative name for ethyleneimine.

AZLON Generic name for regenerated protein fibres. Examples include Ardil, Aralac, Fibrolane, Lanitol, Merinova and Vicara.

AZOBISCYCLOHEXYLNITRILE

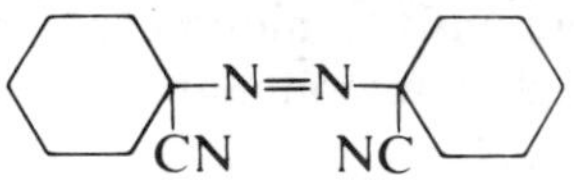

An azo initiator which decomposes at higher temperatures than does azobisisobutyronitrile, having a half-life time of 32 h/80°C, and therefore useful as a free radical polymerisation initiator at higher temperatures, especially for styrene and for unsaturated polyester crosslinking.

2,2'-AZOBIS-2,4-DIMETHYLVALERONITRILE

$$\left[(CH_3)_2CHCH_2\overset{\underset{CN}{|}}{C}(CH_3)N \!\!=\!\! \right]_2$$

An azo initiator of higher activity than azobisisobutyronitrile and therefore useful as a free radical polymerisation initiator in the lower temperature range of 35–60°C, especially for vinyl chloride.

AZOBISFORMAMIDE Alternative name for azodicarbonamide.

AZOBISISOBUTYRONITRILE (2,2'-Azobisisobutyronitrile) (ABIBN) (AIBN) (AZDN)

$$(CH_3)_2\overset{\underset{CN}{|}}{C} \!\!-\!\! N\!\!=\!\!N \!\!-\!\! \overset{\underset{CN}{|}}{C}(CH_3)_2$$

By far the most commonly used azo initiator for free radical polymerisation. Decomposes thermally or by ultraviolet light irradiation in a particularly simple manner to produce free radicals:

$$(CH_3)_2\overset{\underset{CN}{|}}{C} \!\!-\!\! N\!\!=\!\!N \!\!-\!\! \overset{\underset{CN}{|}}{C}(CH_3)_2 \rightarrow 2CH_3\overset{\underset{CN}{|}}{\dot{C}}CH_3 + N_2$$

Unlike many peroxide initiators it is free of induced decomposition and the rate of radical formation is relatively insensitive to choice of solvent. Therefore it has been widely used in polymerisation kinetics studies. It has moderate activity, having half-life times of 83 h/50°C, 13 h/60°C, 2·4 h/75°C and 0·5 h/85°C, and is therefore useful in the range 50–90°C, usually at 0·01–1% concentration. Its use is a little restricted by the formation of certain toxic products, especially tetramethylsuccinylnitrile, by geminate recombination.

2,2'-AZOBISISOBUTYRONITRILE Alternative name for azobisisobutyronitrile.

AZODICARBONAMIDE (Azobisformamide) (AC) (ADC) $H_2NOC\!-\!N\!\!=\!\!N\!-\!CONH_2$. Decomposition temperature 190–230°C. One of the most widely used chemical blowing agents for the production of cellular polymers. It decomposes on heating to produce large quantities of gas (nitrogen, carbon dioxide and carbon monoxide). The high decomposition temperature may be reduced by the use of any of several metal salts as so-called 'kickers'.

AZO INITIATOR A compound of general structure $R\!-\!N\!\!=\!\!N\!-\!R'$, where very often $R = R'$. Decomposes thermally or in ultraviolet light to yield free radicals:

$$R\!-\!N\!\!=\!\!N\!-\!R' \rightarrow \dot{R} + \dot{R}' + N_2$$

$\dot{R}$ and $\dot{R}'$ can then initiate polymerisation. Usually R and R' are alkylnitrile groups since in these compounds decomposition occurs at convenient rates in the usual polymerisation temperature range of 40–100°C. Azobisisobutyronitrile is by far the commonest example, others include azobiscyclohexylnitrile and 2,2'-azobis-2,4-dimethylvaleronitrile.

AZOMETHINE POLYMER Alternative name for polyazomethine.

AZO POLYMER Alternative name for polyazobenzene.

AZOTON Tradename for cyanoethylated cotton.

B

B Symbol for asparagine and/or aspartic acid in a protein, when it is not known which is present.

BABINET COMPENSATOR A device for the determination of the birefringence by nullifying the retardation of a birefringent sample by placing a calibrated, wedge-shaped crystal in the light path. Movement of the crystal in or out of the light path, using a calibrated micrometer, introduces varying thickness and therefore retardation to the light.

BABINET'S RECIPROCITY LAW An optical principle which states that it is not possible to distinguish the radiation scattering by a system of particles in space from the scattering by a complementary system of microvoids in a solid. In small angle X-ray scattering the scattering pattern does not indicate whether the scattering elements are particles or voids.

BACK BITING (Self transfer) A chain transfer reaction in which a polymer free radical abstracts an atom (usually hydrogen) from its own polymer chain. This may occur in polymerisation, for example in the free radical polymerisation of ethylene, resulting in the formation of butyl branches:

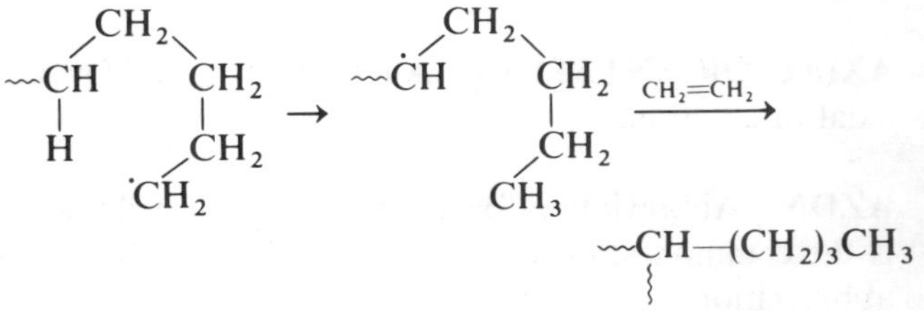

It can also occur in depolymerisation, for example causing the formation of 'dimers', 'trimers', etc., in the pyrolysis of polystyrene.

BACK REFLECTION A correction applied to light scattering data to subtract the light reflected back by the cell. A fraction $(n-1)^2/(n+1)^2$ of the light scattered at the complementary angle is subtracted from the total scattering at angle θ, where n is the refractive index of the glass.

BACTERIAL CELLULOSE Cellulose produced by certain bacteria of the *Acetobacter* type, especially *Acetobacter xylinum*. In most respects identical to cotton cellulose.

BACTERIOPHAGE (Phage) A virus which infects a bacterium. By far the most widely studied are the coliphages, i.e. those infecting the bacterium *Eschericia coli*. The phage infection involves injection of the nucleic acid, which may be DNA or RNA, into the host cell in which it is reproduced. The protein content may assist in infection but is not directly involved in replication. Simple phages, such as ϕX174, are polyhedral, with a diameter of about 15 nm and virion molecular weight of a few million, whose nucleic acid has about 5000 nucleotides. More complex phages, such as the T-even phages (T_2, T_4, T_6), have a tadpole shape with a 'head' and a 'tail', being about 20 nm long and containing about 40–60% DNA which consists of several thousand nucleotides. The virion molecular weight may be up to a few hundred million. The MS2 and R17 phages contain RNA not DNA. The nucleic acid occurs in two chains as a double helix. The nucleic acid sometimes contains several unusual bases, e.g. 5-hydroxymethylcytosine instead of cytosine in the T-even phages and 5-hydroxymethyluracil in place of thymine in a phage of *Bacillus subtilis*. Owing to their rapid replication and ease of isolation in a pure form, the phages have been very widely used in studies of genes and gene action. The ϕX174 and λ phages have been particularly widely used. ϕX174 DNA was the first DNA to be synthesised artificially.

BAGLEY END CORRECTION (End correction) (Entrance effect correction) (Head effect) A correction applied to capillary rheometer data, to allow for viscous and elastic effects at the entrance to the die. These have the effect of increasing the apparent length of the capillary by an amount eR, where R is the die radius; e is found experimentally by obtaining pressure difference (ΔP)–volume flow rate data, i.e. determining the flow curve, for dies of different length to diameter ratios (at constant R) and extrapolating to zero ΔP, when $e = L/R$, where L is the die length.

BAGLEY END-EFFECT CORRECTION Alternative name for end-effect correction.

BAKELITE Tradename for phenol–formaldehyde, phenoxy resin, epoxy resin, acrylic resin or ketone resin polymers.

BAKER–WILLIAMS FRACTIONATION (Chromatographic fractionation) (Gradient elution fractionation) (Precipitation chromatography) (Temperature gradient fractionation) (Thermal gradient elution) A method of fractionation in which a small amount of polymer, dispersed on a support, is placed on the top of a column of an inert support material. The polymer is eluted from the column by extraction with solvents of increasing solvent power, as in column extraction, but, in addition, a temperature gradient is imposed along the length of the column. In this way, as individual polymer species of different molecular weight pass down the column, they undergo a continuous series of precipitations and extractions.

BALATA The solid obtained by coagulation of the latex of the tree *Mimusops balata*. Like gutta-percha, it consists of trans-1,4-polyisoprene, but contains more resin (35–50%), and is therefore inferior in its useful properties.

BAMBOOING A manifestation of melt fracture in extrusion, in which the extrudate consists of smooth sections interrupted by a periodic distortion of the surface, having the appearance of bamboo.

BAND BROADENING A broadening of the retention volume of a gel permeation chromatogram due to instrumental and operational factors, owing to the dispersing nature of the solvent flow in the tubing, cell and, most of all, in the columns of the chromatograph. It has the effect of spreading a rectangular pulse input on injection to a Gaussian output in the detector. It increases with diameter of the tubing and of the column, solvent flow rate and narrowness of the polymer molecular weight distribution.

BANDED HACKLE A regular banded surface feature of a fracture surface, in which the crack has propagated through bundles of crazes formed ahead of it, by jumping through one set and forming a new set. A rough region at the start of a band forms as a result of the crack jumping from one craze to another. This is followed by a smooth region in which one craze predominates.

BANDED SPHERULITE Alternative name for ringed spherulite.

BAND SEDIMENTATION (Zone sedimentation) A variation of the sedimentation velocity method of ultracentrifugation in which a thin film of polymer solution is deposited on top of a solvent layer in the cell. On centrifugation the faster molecules separate from the slower molecules, forming one or more discrete bands. The sedimentation coefficient can then be found from the velocity with which the band migrates through the cell.

BARCOL HARDNESS An arbitrary number expressing the hardness of a polymer material, usually of the harder plastics, as measured by a standard impressor type indentation test. The scale is in 100 divisions, each corresponding to 0·0076 mm penetration.

BAREX Tradename for a copolymer of about 75% acrylonitrile with about 25% methyl acrylate, and about 8% of a nitrile rubber as a blend to improve impact strength, which has very good barrier properties as a packaging material, i.e. it is a barrier polymer.

BARRIER POLYMER (Barrier resin) A polymer which has a low permeability coefficient for a particular substance or group of substances and is therefore useful in packaging materials, particularly films and bottles, to provide a barrier to prevent loss of the contents of the package, such as moisture or flavours, or to prevent ingress of undesirable environmental substances. Polyvinylidene chloride and ethylene–vinyl alcohol copolymer are widely used as barrier polymers, often as a layer in a laminate with other layers which provide the main mechanical properties. Earlier, high acrylonitrile copolymers were used, but there is a possible toxic hazard due to residual acrylonitrile monomer.

BARRIER RESIN Alternative name for barrier polymer.

BARRY DEGRADATION Treatment of a periodate oxidised polysaccharide with phenylhydrazine, which results in the removal of the peroxidised-cleaved sugar units as phenylosazones of two, three and four carbon fragments. Therefore useful in polysaccharide structural analysis, but largely superceded by the cleaner Smith degradation method.

BARUS EFFECT Alternative name for die swell.

BASE FERROPORPHYRIN Alternative name for haemochromagen.

BASIC LEAD CARBONATE (White lead) $2PbCO_3 . Pb(OH)_2$. A cheap and effective thermal stabiliser for PVC of the basic lead type. However, it is subject to 'gassing' (formation of CO_2), and hence porosity in the polymer at high processing temperatures, and also to sulphur staining. Its dusty nature presents a toxic hazard in its use.

BASIC LEAD STABILISER (Lead stabiliser) Any of a group of basic lead compounds widely used as thermal stabilisers in PVC. Cheap and effective in preventing discolouration when used at levels of 2–8 phr, but their use involves a toxic hazard and renders the material opaque. Examples include basic lead carbonate, tribasic lead sulphate, dibasic lead phthalate and dibasic lead phosphite.

BAST FIBRE A natural fibre obtained from the inner bark (the bast) of certain plants. Flax, hemp, sunn, jute and ramie are all bast fibres of commercial significance.

BAUSCHINGER EFFECT A reduction in the elastic modulus of a viscoelastic material, in a hysteresis experiment, when the stress is reversed after having produced the initial plastic deformation. Alternatively, the non-equivalence of the tensile and compressive yield stresses. Such behaviour is quite common in polymers and direct evidence for it comes from yield stress measurement in oriented polymers.

BAYBLEND Tradename for an acrylonitrile–butadiene–styrene copolymer blended with bisphenol A polycarbonate.

BAYLON Tradename for low density polyethylene.

BAYPREN Tradename for polychloroprene.

BAYTOWN Tradename for styrene–butadiene rubber.

BBB Abbreviation for poly-(bis-benzimidazobenzophenanthroline).

BBL Abbreviation for poly-(benzimidazoimidazophenanthroline) ladder, a poly-(bis-benzimidazobenzophenanthroline) of a completely ladder structure.

BBP Abbreviation for butylbenzyl phthalate.

BD-CELLULOSE Abbreviation for benzoylated diethylaminoethylcellulose.

BDS K-RESIN Alternative name for the tradename K-resin.

BEAD-CHAIN POLYMER A partial ladder polymer consisting of a series of polycyclic block-like structures connected linearly through single bonds, found for example in some siloxane polymers. An example is:

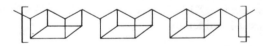

BEAD POLYMERISATION Alternative name for suspension polymerisation.

BEAD-SPRING MODEL A model for a polymer molecule which represents the chain as a series of beads connected by springs. The springs are of equal length and behave as freely jointed chains. The springs are considered long enough for the separations of the chain ends to behave as they do in Gaussian chains. The model forms the basis for the Rouse–Bueche–Zimm theory of viscoelastic behaviour.

BECKACITE Tradename for phenol–formaldehyde polymer.

BECKAMID Tradename for fatty polyamides.

BECKAMIN Tradename for urea–formaldehyde polymer.

BECKOPOX Tradename for an epoxy resin.

BEETLE Tradename for urea–formaldehyde or thiourea–formaldehyde moulding polymers, urea– and melamine–formaldehyde textile treatment polymers, melamine–formaldehyde paper treatment polymers, melamine– and urea–formaldehyde laminating resins and unsaturated polyester resins.

BEETLE NYLON Tradename for nylon 6.

BEETLE PET Tradename for a polyethylene terephthalate moulding material.

BEMBERG Tradename for cuprammonium rayon yarn.

BEMBERG PROCESS Alternative name for cuprammonium process.

BENCE–JONES PROTEIN A protein excreted in large amounts in the urine of patients suffering certain cancerous conditions. No two patients produce identical protein and each patient only produces a single such protein which is therefore much more readily available as a molecularly homogeneous species than is usual in other immunoglobulins. It consists of a dimer of the light polypeptide chain of the major immunoglobulin produced in the patient. The carboxyl terminal half has a very similar sequence of about 214 amino acid residues in all patients, but in the amino terminal half the sequence varies considerably among different patients, although some sequence homology exists.

BENZENE

B.p. 80·1°C. M.p. 5·5°C.

Although a solvent for many polymers, including those useful as varnishes and lacquers, its commercial use as a solvent is limited by its toxicity. However, it is also important as an intermediate in the manufacture of ethylbenzene (and hence styrene), phenol, cyclohexane and maleic anhydride. Mostly it is obtained by careful distillation from petroleum, when it is usually of high purity, although it may contain toluene, xylene, organo-sulphur (e.g. thiophene) and nitrogen (e.g. pyridine) compounds. A solvent for polystyrene and its co-polymers, many rubbers, some acrylics, alkyd and natural resins, polyisobutene, polyvinyl acetate and fluid silicones. It somewhat swells polyethylene, polyvinyl chloride and polymethylmethacrylate but not nylons.

BENZENE-1,3-DICARBOXYLIC ACID Alternative name for isophthalic acid.

BENZENESULPHONYLHYDRAZIDE (BSH)

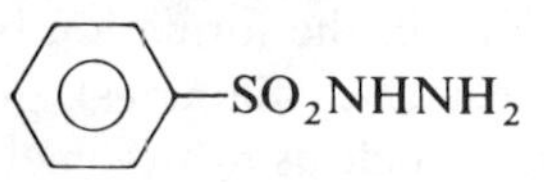

SO_2NHNH_2

Decomposition temperature 140°C. The decomposition temperature is lowered in the presence of bases. A chemical blowing agent, which decomposes to yield nitrogen as the blowing gas. Useful in the production of cellular polymers, especially microcellular rubber.

BENZOGUANAMINE (1,3-Diamino-5-phenyl-2,4,6-triazine)

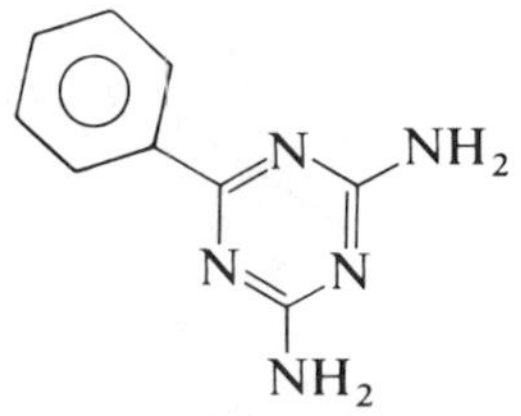

M.p. 227°C.

Prepared by heating dicyandiamide with benzonitrile in the presence of sodium or sodium hydroxide. Useful in the preparation of benzoguanamine–formaldehyde polymer.

BENZOGUANAMINE–FORMALDEHYDE POLYMER (Guanamine–formaldehyde polymer) A polymer formed by reaction of benzoguanamine with formaldehyde in aqueous solution, with a benzoguanamine:formaldehyde ratio of more than 1:2, at above 85°C. Initially methylolbenzoguanamines are formed in a similar way to melamine–formaldehyde reactions. These also similarly condense on further heating, especially in the presence of acid catalysts, to give network thermoset polymers. Butylated benzoguanamines have been used in stoving enamels.

BENZOPHENONE

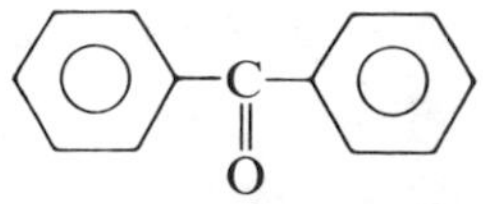

M.p. 49°C. B.p. 306°C.

A frequently used photosensitiser in organic chemistry; can also promote photodegradation in some polymers. In contrast, 2-hydroxybenzophenones are one of the most important groups of ultraviolet absorber type photo-stabilisers.

BENZOPHENONETETRACARBOXYLIC ACID DIANHYDRIDE (BTDA)

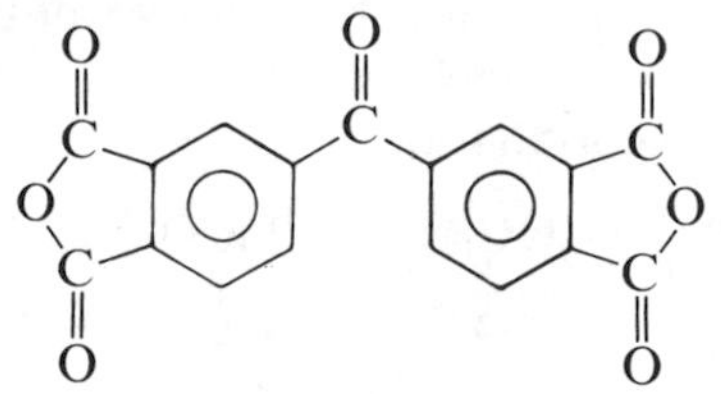

M.p. 236°C.

Synthesised by oxidation of

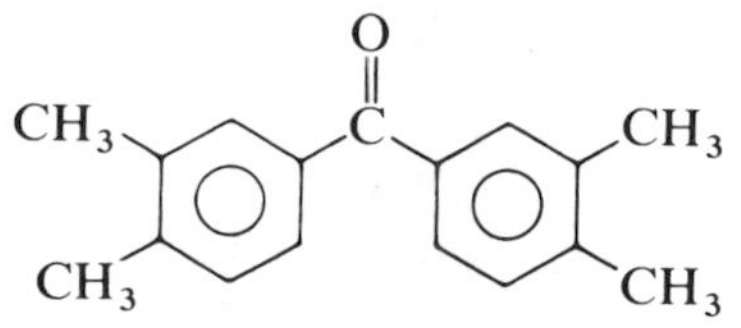

itself derived from *o*-xylene. Useful as a tetrafunctional monomer, especially in the formation of polyimides (by reaction with aromatic diamines), which are more tractable than those made using pyromellitic dianhydride.

p-BENZOQUINONE   (Quinone)

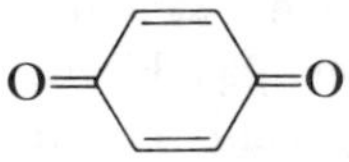

$$O= \bigcirc =O$$

M.p. 115·7°C.

A powerful free radical polymerisation inhibitor, capable of acting as a radical scavenger, possibly by such reactions as:

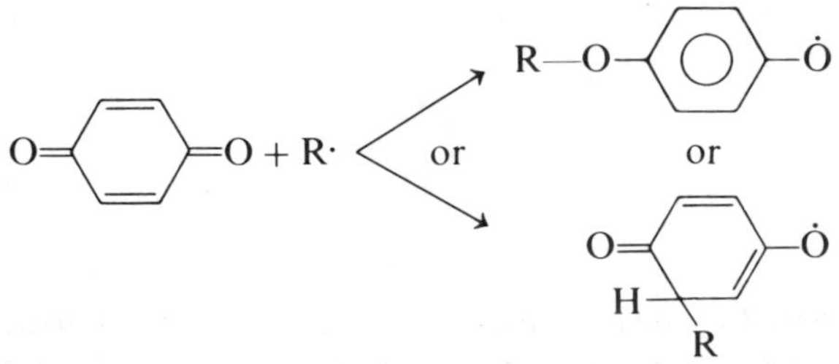

The fate of the resultant relatively stable free radicals determines the radical/inhibitor stoichiometry. The radicals may dimerise, disproportionate or react with further radicals R·.

BENZOTHIAZOLE   Alternative name for thiazole.

BENZOTRIAZOLE   Alternative name for 2-hydroxy-benzotriazole.

BENZOYLATED DIETHYLAMINOETHYL-CELLULOSE   (BD-cellulose) Cellulose in which some of the hydroxyl groups have been replaced by diethyl-aminoethyl and benzoyl groups. Such a polymer has a greater affinity for aromatic compounds than diethyl-aminoethylcellulose and is sometimes used as an ion-exchange resin in ion-exchange chromatography of ribonucleic acids.

BENZOYL PEROXIDE

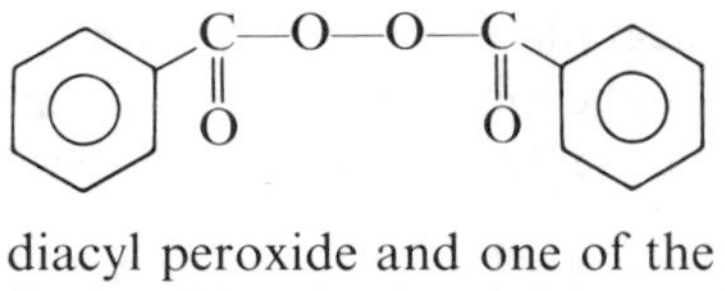

M.p. 103–105°C.

A diacyl peroxide and one of the most widely used free radical polymerisation initiators. Decomposes on heating, with half-life times of 220 h/50°C, 13 h/70°C and 0·36 h/100°C and temperatures for half-lives of 10 h/72°C, 1 h/92°C and 1 min/136°C, or on irradiation with ultraviolet light according to:

$$PhCO{-}OCPh \rightarrow 2\,PhCO\cdot$$

$$PhCO\cdot \rightarrow Ph\cdot + CO_2$$

Either

$$PhCO\cdot$$

or Ph· may initiate polymerisation, although it is usually the former unless the monomer has low activity. In the pure, dry form it is an explosive hazard and is therefore handled damped with water, as a paste with plasticiser or dispersed on an inert solid filler. Useful as a poly-merisation initiator in the range 70–100°C at concentrations of 0·1–0·5%, especially for polystyrene, acrylic polymers and for crosslinking polyester resins at temperatures of 10–70°C, when combined with an accelerator such as a tertiary amine, e.g. dimethylaniline. Also used for the crosslinking of many elastomers, especially silicone rubbers at about 125°C.

BENZYL ALCOHOL

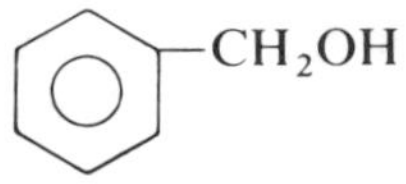

$$\bigcirc{-}CH_2OH$$

B.p. 205°C.

A high boiling point solvent which will dissolve cellulose acetate (when it can act as a plasticiser), cellulose ethers, polystyrene and natural rubber. When hot it will also dissolve many nylons. Dissolves about 8% water at 20°C.

BENZYL BENZOATE

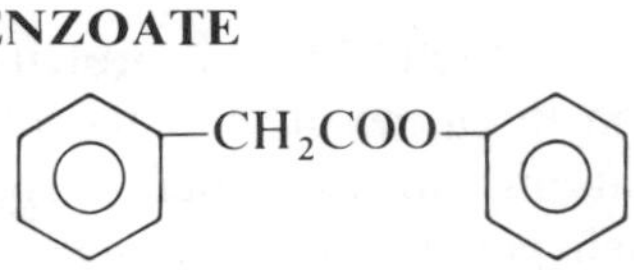

$$\bigcirc{-}CH_2COO{-}\bigcirc$$

M.p. 19°C. B.p. 327°C/760 mm, 170°C/11 mm.

A plasticiser for many natural resins such as rosin, copal and kauri.

BENZYL CELLULOSE   A cellulose ether in which some cellulose hydroxyl groups have been replaced by benzyloxy

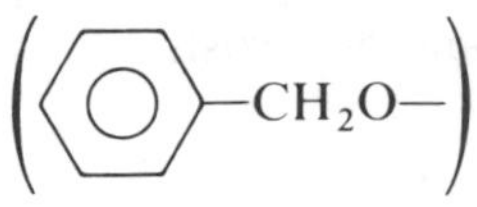

$$\left(\bigcirc{-}CH_2O{-}\right)$$

groups. It has had minor use as a plastic moulding material, at a DS of about 2, but due to its low softening point and poor thermal and light stability it is no longer used.

BENZYLCHLOROCARBONATE   Alternative name for benzylchloroformate.

BENZYLCHLOROFORMATE

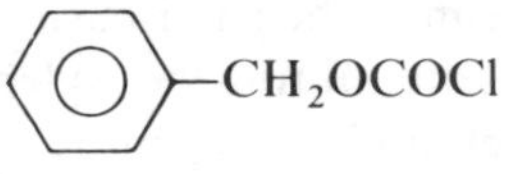

$$\bigcirc{-}CH_2OCOCl$$

B.p. 103°C/20 mm.

(Benzyloxycarbonyl chloride) (Benzylchlorocarbonate) The reagent used to form benzyloxycarbonyl protecting group derivatives of amino groups in peptide synthesis. It is readily synthesised by reaction of benzyl alcohol and phosgene.

BENZYLOXYCARBONYL CHLORIDE   Alternative name for benzyl chloroformate.

BENZYLOXYCARBONYL GROUP (Carbobenzyloxy group) (Z group) The group

$$\langle\bigcirc\rangle\text{—CH}_2\text{OCO—}$$

One of the most widely used protecting groups for the amino groups of amino acids and peptides in peptide and protein synthesis. It is introduced by the reaction of the amino group with benzyl chloroformate:

$$\langle\bigcirc\rangle\text{—CH}_2\text{OCOCl} + \text{H}_2\text{N—R—COOH} \rightarrow$$

$$\langle\bigcirc\rangle\text{—CH}_2\text{OCONH—R—COOH} + \text{HCl}$$
$$\text{I}$$

The protecting group may subsequently be removed by hydrogenolysis over a palladium catalyst:

$$\text{I} \xrightarrow{\text{H}_2/\text{Pd}} \text{H}_2\text{N—R—COOH} + \langle\bigcirc\rangle\text{—CH}_3 + \text{CO}_2$$

the products usefully being inert and readily removed. Alternatively, the group may be removed by sodium and liquid ammonia, which is a less selective method, or by hydrobromic acid in acetic acid.

BERNOUILLIAN PROCESS (Random chain process) (Zero order Markov chain process) A stochastic process defined by a series of independent events, i.e. one in which past events have no influence on future events, and which is therefore a non-memory process. Nevertheless, the events do have inherent probabilities of occurring. Some polymerisation events, e.g. random copolymerisation with no terminal or penultimate effects, or some stereoregular polymerisations, may be Bernouillian and produce polymers with sequences with Bernouillian statistical distributions. In the former case, knowledge of the comonomer sequence dyads is sufficient to test for Bernouillian statistics, whereas in the latter, triad sequence data is required. Simple polymerisation/termination/transfer relationships are also governed by random events and are also therefore Bernouillian.

BERNSTEIN, KEARSLEY AND ZAPPAS THEORY (BKZ theory) A theory of viscoelastic behaviour in which a strain energy function (U) is postulated and defines two relaxation functions (μ_1 and μ_2) given by: $\mu_1 = -2\partial U/\partial II_{C^{-1}}$ and $\mu_2 = 2\partial U/\partial I_{C^{-1}}$, where $I_{C^{-1}}$ and $II_{C^{-1}}$ are the first and second invariants of the inverse Cauchy–Green tensor ($\mathbf{C}^{-1}$), i.e. the finger strain tensor—a deformational tensor. The relaxation function is dependent on the elapsed time ($t - t'$) and on the rate of deformation, given by $II_{C^{-1}}$. The theory gives the shear stress τ as

$$\tau = \int_{-\infty}^{t} \{\mu_1(t - t')\mathbf{C} + \mu_2(t - t')\mathbf{C}^{-1}\}\,dt'$$

The theory has been quite successful in predicting response to one type of deformation of polymers, using measurements of μ_1 and μ_2 obtained from a different type of deformation.

BETA SCISSION Scission of a carbon–hydrogen or carbon–carbon bond adjacent to a carbon or oxygen free radical. Thus if the bond is part of a polymer chain, scission results. The bond dissociation energy of the relevant bond is reduced by the presence of the free radical due to the lowered energy of the product double bond. This provides a driving force for the reaction:

$$\sim\!\!\text{CHR—CH}_2\text{—CHR—}\dot{\text{C}}\text{H—CHR}\!\!\sim \rightarrow$$
$$\sim\!\!\text{CHR—}\dot{\text{C}}\text{H}_2 + \text{CHR}\!\!=\!\!\text{CH—CHR}\!\!\sim$$

Thus chain scission is a frequent result of the formation of polymer free radicals during degradation.

BETA TRANSITION A subsidiary glass transition occurring at a temperature (T_β) lower than that of the main α-transition. It involves small scale molecular movement, e.g. partial freedom of rotation of a few atoms or groups of atoms within the frozen glass.

BEXAN Tradename for styrene–acrylonitrile copolymer and acrylonitrile–butadiene–styrene copolymer.

BEXLOY Tradename for an amorphous nylon.

BEXTRENE Tradename for polystyrene.

BHT Abbreviation for butylated hydroxytoluene.

BIAXIAL BIREFRINGENCE Birefringence resulting from biaxial orientation. Two independent birefringences, measured perpendicularly to each other, must be determined to fully characterise the birefringence.

BIAXIAL ORIENTATION Orientation produced by stretching in two directions at right angles, so that there are two perpendicular characteristic axes (principal axes). Thus there are two maxima in the distribution of the oriented elements, e.g. polymer chains. It is most frequently encountered in films drawn in two mutually perpendicular directions. Biaxial orientation improves mechanical properties—modulus, impact strength and tensile strength in all directions.

BIAXIAL STRAIN Alternative name for plane strain.

BIAXIAL STRESS Alternative name for plane stress.

BICOMPONENT FIBRE (Conjugate fibre) A fibre composed of two constituent polymers lying either side-by-side (bilateral arrangement) or in a core-shell arrangement. When exposed to moisture or heat a crimp develops. Viscose, rayon and nylon fibres are all produced in bicomponent forms.

BICONSTITUENT FIBRE A fibre composed of two constituent polymers, one dispersed as short fibrils in a matrix of the second polymer. The fibres are combined by blending the two polymers before spinning, as opposed to the more usual blending of spun yarns.

BICYCLO-(2,2,1)-HEPTENE-2　Alternative name for norbornene.

BICYCLOPENTADIENE　(Dicyclopentadiene) (Endo-4,7-methylene-4,7,8,9-tetrahydroindene)

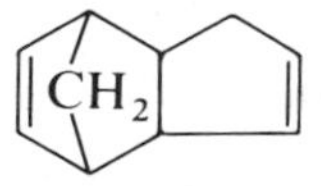

M.p. 32°C. B.p. 170°C.

The dimer product of attempted self-Diels–Alder polymerisation of cyclopentadiene at low temperature. Cannot be polymerised further due to the double bonds in the dimer being much less reactive than in the monomer. Reverts to monomer on heating to about 150°C. Occasionally used as the termonomer in ethylene–propylene–diene monomer rubber.

BIIR　Abbreviation for brominated butyl rubber, i.e. brominated isobutene–isoprene copolymer.

BIMETALLIC MECHANISM　A mechanism of propagation in Ziegler–Natta polymerisation in which both the transition metal and the metal of the metal alkyl are involved directly. Thus for the typical system $AlR_3/TiCl_3$, the bimetallic mechanism may be depicted as involving a bridged structure:

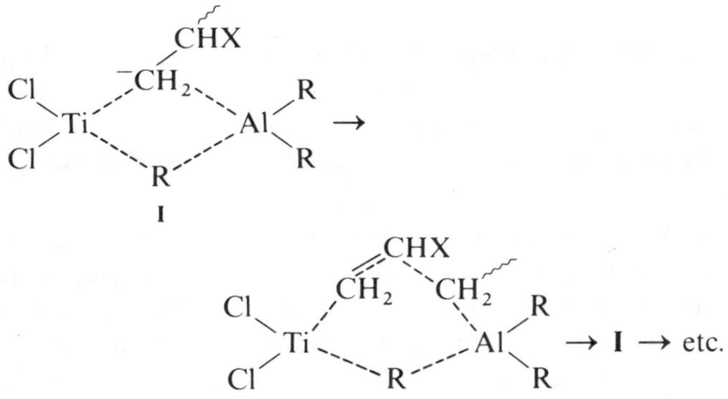

The monomer is usually also coordinated with the titanium before inserting itself into the titanium chain-end bond. It is likely that bimetallic mechanisms operate mostly in homogeneous catalysts, whilst monometallic mechanisms operate in heterogeneous catalysts.

BIMODAL DISTRIBUTION　(Binodal distribution) A molecular weight distribution in which the differential weight distribution function, $W(r)$, has two maxima. Often found in polymer fractions obtained by fractionation.

BINARY COPOLYMER　(Bipolymer) A copolymer which contains only two types of repeating unit. The most common type of copolymer.

BINGHAM BODY　(Bingham model) (Bingham plastic) (Bingham solid) (Ideal plastic) (Saint Venant solid) A rheological model in which flow only occurs above a certain shear stress—the yield value (τ_y). Above the yield value the flow is Newtonian. This behaviour occurs in some suspensions which are considered to have some internal structure which must first be broken down for flow to occur. Thus $(\tau - \tau_y)$ is given by $(\tau - \tau_y) = \mu\dot{\gamma}$, where τ is the shear stress, $\dot{\gamma}$ is the shear rate and μ is the coefficient of viscosity. Sometimes in this case μ is called the plastic viscosity and a Bingham material is referred to as an ideal plastic, despite the fact that polymer melts do not show Bingham model behaviour. However, such behaviour may be considered as a limiting case for pseudoplastic behaviour.

BINGHAM MODEL　Alternative name for Bingham body.

BINGHAM PLASTIC　Alternative name for Bingham body.

BINGHAM SOLID　Alternative name for Bingham body.

BINODAL　The line on a temperature versus composition phase diagram for a mixture of two components, which separates the single phase region from the two phase region and hence points on this line represent the limit of stability of a two phase system, e.g. of a polymer blend. If miscibility increases with temperature, the binodal exhibits a maximum, the upper critical solution temperature, above which temperature the two components are miscible in all proportions. On cooling a mixture from a point above the binodal to a point below it, phase separation occurs by a process of nucleation and growth. However, for polymer/polymer mixtures phase separation may be delayed and may not occur until after the spinodal has been crossed (spinodal decomposition). The reverse is true for a system where miscibility decreases with increasing temperature.

BINODAL DISTRIBUTION　Alternative name for bimodal distribution.

BINOX M　Tradename for 4,4′-methylene-bis-(2,6-di-t-butylphenol).

BIODEGRADATION　A deterioration due to the action of natural organisms such as bacteria and fungi. Usually caused by chemical attack by the enzymes and other products of the organism, leading to chain scission of the polymer. Most synthetic polymers of molecular weight greater than about 1000 are resistant. Notable exceptions are high polymers containing ester groups in the main chain, e.g. polycaprolactone and polyester based polyurethanes. Many natural polymers (e.g. cellulose and its derivatives) are biodegradable, as are some additives, e.g. straight chain plasticisers.

BIOGEL-P　Tradename for lightly crosslinked polyacrylamide gels suitable for gel permeation chromatography of biopolymers in aqueous systems.

BIOPOL Tradename for polyhydroxybutyrate.

BIOPOLYMER (Natural polymer) A polymer which is produced by biosynthesis in nature, as opposed to a human controlled polymerised synthetic polymer. Biopolymers may be classified into three main groups—polysaccharides, proteins and nucleic acids. Certain less important polymers are also produced in nature, such as natural rubber, lignin and other exudates of plants and trees. Biopolymers fulfil many different functional roles in nature: as structural supports, e.g. cellulose, as catalysts for biochemical reactions, e.g. enzymes, and as informational macromolecules (nucleic acids) to ensure inheritance of genetic characteristics. Different types of biopolymers may be combined chemically to form a kind of hybrid such as nucleoprotein (nucleic acid and protein) or glycoprotein (polysaccharide and protein). Some biopolymers are homopolymers, such as natural rubber and some polysaccharides (e.g. cellulose), but in general they are copolymers. Polysaccharides contain up to five different repeat units, proteins usually have 20 different amino acid residues as their repeat units and nucleic acids usually have four different repeat units. In the last two types, however, the different repeat units are arranged in a precise sequence (which is the same in all molecules of the sample) which is therefore molecularly homogeneous and monodisperse with respect to structure and molecular weight.

BIPLANAR FRACTURE Fracture in which crack propagation has taken place on two parallel planes. Often the planes contain steps running parallel to the crack propagation which are the secondary fracture front parabola markings. Each of the pair of fracture surfaces thus carries complementary surface markings.

BIPOLYMER A polymer containing two types of repeat unit, i.e. a binary copolymer, often frequently referred to simply as a copolymer, since copolymers with three types of repeat unit are called terpolymers and synthetic copolymers with more than three types are very rare.

BIREFRINGENCE (Double refraction) (Optical anisotropy) The difference between the refractive indices along two perpendicular directions as measured with polarised light along these directions. It results from molecular orientation, and the measurement of birefringence is the most common method of characterising polymer orientation. It is determined by measurement of the retardation by either a compensation or a transmission method. Positive birefringence results when the principal optic axis lies along the chain; negative birefringence when transverse to the chain. In cartesian coordinates there are three birefringences, two being independent. Thus $\Delta xy = n_x - n_y$, the differences in refractive indices along the x and y axes. Uniaxial orientation only requires one of these to describe the orientation.

BIS-(4-AMINOCYCLOHEXYL)METHANE (PACM)

$$H_2N-\langle\ \rangle-CH_2-\langle\ \rangle-NH_2$$

M.p. 64–65°C (*trans–trans* isomer), 36–37°C (*cis–trans* isomer), 60·5–62°C (*cis–cis* isomer).

A diamine for certain polyamide syntheses, including poly-(bis-(4-cyclohexyl)methane dodecanoamide) and some copolymers. Produced by hydrogenation of bis-(4-aminophenyl)methane(4,4'-methylenedianiline). It can exist in three isomeric forms—*trans–trans*, *cis–trans* and *cis–cis*. The usual commercial product is a mixture, whose composition depends on the conditions of synthesis.

BIS-(4-AMINOPHENYL)METHANE Alternative name for 4,4'-diaminodiphenylmethane.

BISAZODICARBOXYLATE (Bisazoester)

$$(CH_3OOCN{=}NCOO{-})_2(CH_2)_{10}$$
(I)

A vulcanisation agent for diene rubbers. Although not used commercially for vulcanisation it has been extensively used in experimental studies due to crosslinking occurring quantitatively according to:

$$2\,{\sim}CH_2{-}CX{=}CH{-}CH_2{\sim} + I \rightarrow$$

$${\sim}CH{=}CX{-}CH{-}CH_2{\sim}\quad {\sim}CH_2{-}CH{-}CH{=}CH_2{\sim}$$
$$CH_3OOC{-}N{-}NHCOO{-}$$
$$\qquad\qquad {-}(CH_2)_{10}{-}OOCNH{-}NCOOCH_3$$

Furthermore the number of crosslinks may be determined by nitrogen analysis. Such vulcanisates, with accurately known crosslink structure, are useful for checking the predictions of theoretical relationships between structure and mechanical behaviour.

BISAZOESTER Alternative name for bisazodicarboxylate.

BISCHLOROFORMATE A compound of the type ClOCOROCOCl, formed by reaction of phosgene with a diol, capable of forming polyurethanes by reaction with a diamine:

$$n\,ClOCOROCOCl + n\,H_2NR'NH_2 \rightarrow$$
$$\text{-(-COOROCONHR'NH-)}_n + 2n\,HCl$$

by either interfacial or solution polymerisation. Also reacts with diols to form alternating polycarbonates:

$$n\,ClOCOROCOCl + n\,HOR'OH \rightarrow$$
$$\text{-(-COOROCOR'O-)}_n + 2n\,HCl$$

3,3-BIS-(CHLOROMETHYL)OXACYCLOBUTANE
(3,3-Bis(chloromethyl)oxetane)

$$(ClCH_2)_2C{-}{-}CH_2$$
$$\qquad\quad |\qquad\ |$$
$$\qquad\ CH_2{-}O$$
I   M.p. 19°C.

Prepared from pentaerythritol by conversion to its tetraacetate, treatment with hydrochloric acid to give the trichloroacetate, followed by ring closure by treatment with sodium hydroxide:

$$C(CH_2OH)_4 \rightarrow C(CH_2OOCCH_3)_4 \rightarrow \underset{\underset{OOCCH_3}{|}}{C(CH_2Cl)_3} \rightarrow I$$

The monomer for poly-(3,3-bis(chloromethyl)oxacyclobutane).

3,3-BIS(CHLOROMETHYL)OXETANE
Alternative name for 3,3-bis(chloromethyl)oxacyclobutane.

BISCYCLOPENTADIENE
A monomer of type

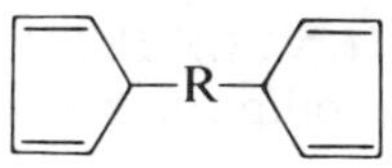

where R is usually a hydrocarbon group. Synthesised by reaction of a sodium salt of a cyclopentadiene with a dihalide (RX$_2$). Often spontaneously polymerises by a self-Diels–Alder polymerisation or by reaction with an activated dienophile, e.g. a quinone.

BISCYCLOPENTADIENONE
A compound of the type

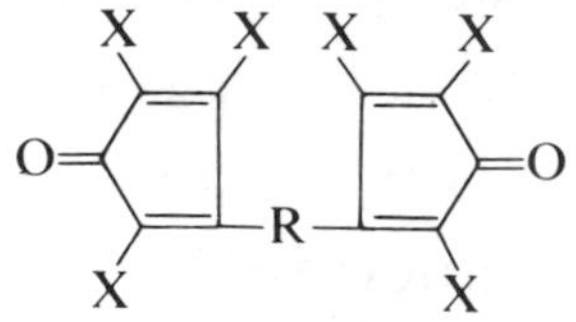

where R is a hydrocarbon group and X is a substituent. Phenylated biscyclopentadienones (X = phenyl) are copolymerised with diacetylenes by an inverse Diels–Alder polymerisation to produce phenylated polyphenylenes. The tetraphenyl substituted compounds (tetracyclones) are of most interest for polymer formation since unsubstituted and disubstituted compounds often form stable dimers. The phenyl derivatives are easiest to synthesise and purify.

BIS-(o-DIAMINE)
A compound of the type

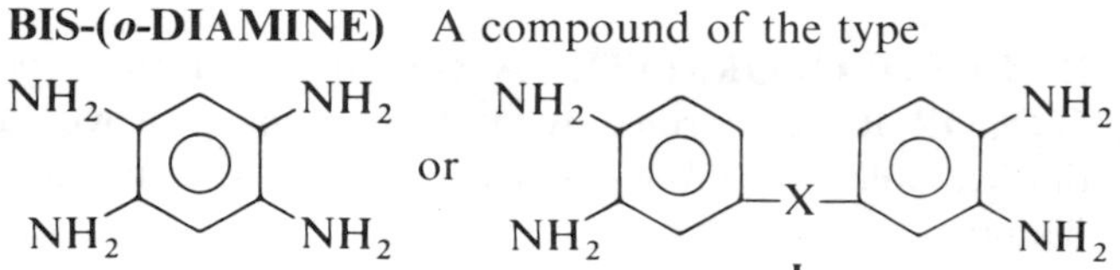

Useful as a tetrafunctional monomer for the synthesis of a variety of heterocyclic polymers by cyclopolymerisation. The best known example is 3,3'-diaminobenzidine (I with X absent). Polymers synthesised using bis-(o-diamines) are polybenzimidazoles, polyquinoxalines and the ladder polymers BBB and BBL.

BIS-(3,5-DI-t-BUTYL-4-HYDROXYPHENYL)-METHANE
Alternative name for 4,4'-methylene-bis-(2,6-di-t-butylphenol).

BIS-(DIETHYLENE GLYCOL MONOETHYL-ETHER) PHTHALATE
Alternative name for dicarbitol phthalate.

BIS-(β-DIKETONE) POLYMER
A chelate polymer containing repeating units of the type:

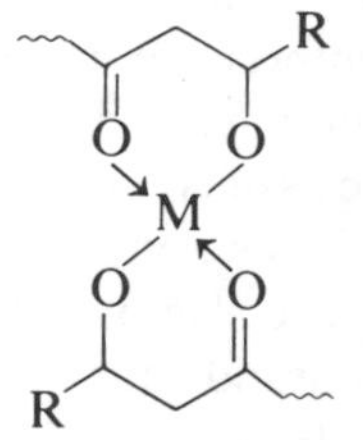

Formed by ligand exchange between a bis-β-diketone and a metal acetonylacetate or by addition of metal ion to an alkaline solution of the bis-β-diketone. Such polymers have been investigated as potential high temperature resistant polymers owing to their structural similarity to metal acetylacetonates, which are well known to be thermally stable. These polymers are one type of polyacetylacetonate.

BIS-(β-N,N-DIMETHYLAMINOETHYL) ETHER
(DMEA) (N,N-Dimethylethanolamine) (Dimethylamino ethanol)

$$[(CH_3)_2NCH_2CH_2\!\!-\!]_2O \qquad \text{B.p. } 135°C.$$

A widely used catalyst component for polyether polol based flexible polyurethane foam. Usually used in combination with 1,4-diazabicyclo-2,2,2-octane and stannous octoate.

BIS-(DIMETHYLBENZYL) CARBONATE

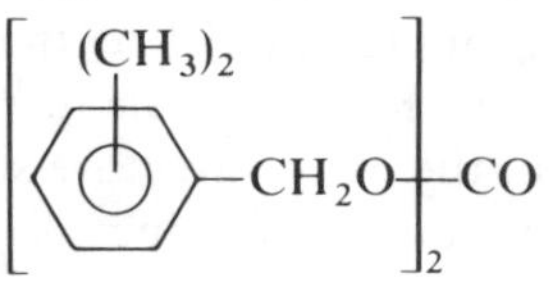

B.p. 219°C/4 mm.

A plasticiser for styrene–butadiene rubber and for polyvinyl chloride, of high permanence and with good flame retardancy and heat stability. Also compatible with polystyrene, polyvinyl butyral, but not with cellulose acetate.

BIS-(DIMETHYLBENZYL) ETHER

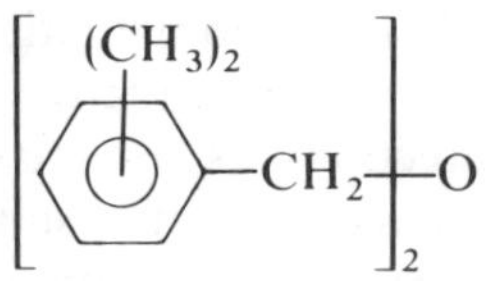

B.p. 184°C/4 mm.

A plasticiser for polyvinyl chloride and its copolymers, ethyl cellulose and for many synthetic rubbers, with good hydrolytic resistance, good heat stability and electrical properties.

BIS-(ETHYLENE GLYCOL MONOBUTYLETHER) ADIPATE
Alternative name for dibutoxyethyl adipate.

BIS-(2-HYDROXY-3-t-BUTYL-4-PHENYL)-METHANE
Alternative name for 2,2'-methylene-bis-(4-methyl-6-t-butylphenol).

BIS-(2-HYDROXYETHYL) TEREPHTHALATE

$$HOCH_2CH_2OOC-\underset{\text{I}}{\bigcirc}-COOCH_2CH_2OH$$

The main product, together with oligomers, of the first stage in the synthesis of polyethylene terephthalate by ester interchange of dimethyl terephthalate with ethylene glycol. By subsequent heating at 270–285°C, it is converted by ester interchange into the polymer. Alternatively, the polymer can be synthesised by separately preparing **I**, e.g. by reaction of terephthalic acid with ethylene oxide, followed by its polymerisation.

2,2-BIS-(HYDROXYMETHYL)-1,3-PROPANEDIOL
Alternative name for pentaerythritol.

1,1-BIS-(4-HYDROXYPHENYL)-CYCLOHEXANE
Alternative name for bisphenol Z.

2,2-BIS-(4-HYDROXYPHENYL)-PROPANE   Alternative name for bisphenol A.

BISMALEIMIDE   A compound of structure

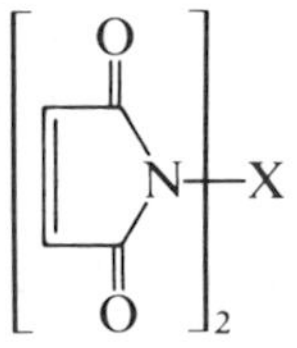

where X is often a polymethylene group or an aromatic ring, e.g. *m*- or *p*-phenylene. Useful as the dienophile monomer in Diels–Alder polymerisation and for the synthesis of polyaminobismaleimides by reaction with a diamine.

N,N'-BIS-(METHOXYMETHYL)URON

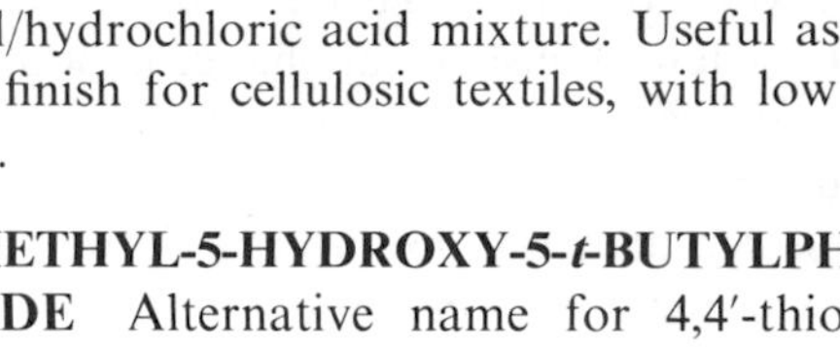

$$CH_3OCH_2-N\underset{H_2C}{\overset{CO}{\big|}}N-CH_2OCH_3$$

M.p. 29°C. B.p. 82°C/0·1 mm.

Formed by reaction of urea with formaldehyde, in a 1:4 ratio, followed by removal of water and treatment with a methanol/hydrochloric acid mixture. Useful as a crease resistant finish for cellulosic textiles, with low chlorine retention.

BIS-(2-METHYL-5-HYDROXY-5-t-BUTYLPHENYL)-SULPHIDE   Alternative name for 4,4'-thio-bis-(6-t-butyl-*m*-cresol).

BIS-MORPHOLINEDISULPHIDE   (N,N'-Dithio-bismorpholine)

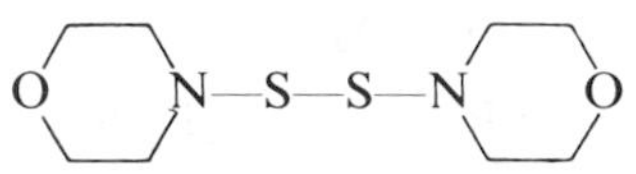

M.p. 122°C.

A vulcanising agent for the sulphurless vulcanisation of diene rubbers.

BISPHENOL A   (BPA) (Diphenylolpropane) (2,2,-bis-(4-hydroxyphenyl)propane) (4,4'-dihydroxy-2,2-diphenyl-propane)

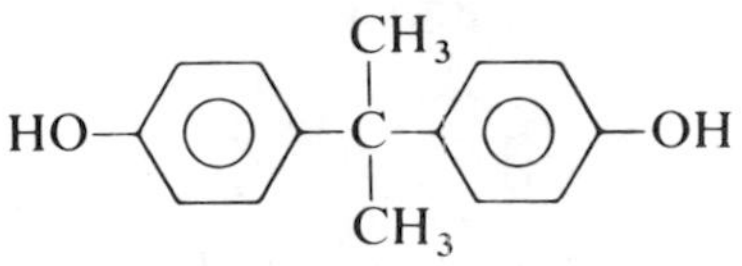

M.p. 155°C.

Prepared by reaction of acetone with phenol in the presence of a strong acid, e.g. 70% sulphuric acid:

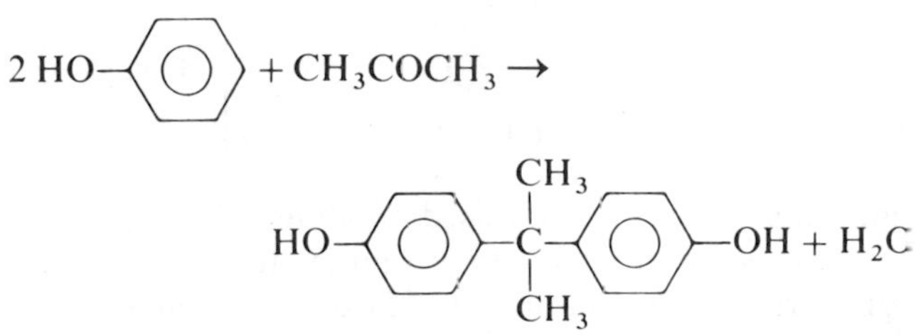

$$2\,HO-\bigcirc + CH_3COCH_3 \rightarrow$$

$$HO-\bigcirc-\underset{CH_3}{\overset{CH_3}{\underset{|}{\overset{|}{C}}}}-\bigcirc-OH + H_2C$$

Usually >95% of the *p–p'*-isomer is produced, the remainder being *o–p'*- and *o–o'*-isomers. Widely used as a diol for the formation of a variety of polymers. In particular, the reaction with epichlorhydrin, producing the diglycidylether of bisphenol A, forms the basis of most epoxy resin manufacture. Commercial polycarbonate is also based on bisphenol A, for which very high purity material is required. Other useful polymers produced using bisphenol A are polyethersulphones and phenoxy resin.

BISPHENOL A POLYCARBONATE   (Poly-(2,2-bis-(4-phenylene)propane carbonate)) (Poly-(4,4'-isopropylidenediphenylene carbonate)) (BPA polycarbonate) (PC) Tradenames Lexan, Makrolon, Merlon, Orgalan, Sinvet.

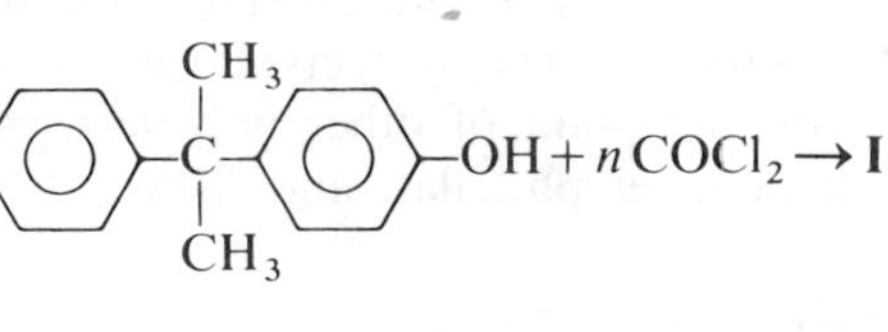

The only polycarbonate of commercial significance, being important as an engineering thermoplastic. Prepared by reaction of phosgene with bisphenol A:

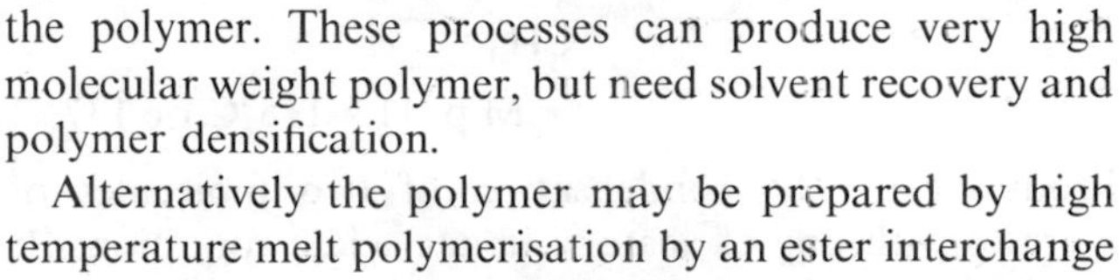

$$n\,HO-\bigcirc-\underset{CH_3}{\overset{CH_3}{\underset{|}{\overset{|}{C}}}}-\bigcirc-OH + n\,COCl_2 \rightarrow I + 2n\,HCl$$

either in pyridine (or other hydrochloric acid accepting solvent) or in a chlorohydrocarbon solvent, or by an interfacial process consisting of bisphenol A dissolved in aqueous sodium hydroxide with an organic solvent for the polymer. These processes can produce very high molecular weight polymer, but need solvent recovery and polymer densification.

Alternatively the polymer may be prepared by high temperature melt polymerisation by an ester interchange between an excess of diphenyl carbonate and bisphenol A.

In the first stage, carried out at about 200°C, phenol is eliminated and is lost by volatilisation:

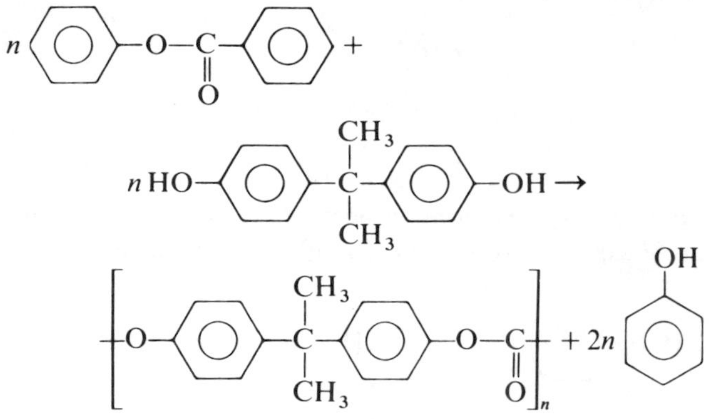

The temperature is then raised, to about 300°C, and the pressure is lowered, to about 1 mm, to complete the polymerisation by elimination of excess diphenyl carbonate. In this process the molecular weight is limited to about 30 000 due to the high melt viscosity, but otherwise the process is more convenient than the phosgene process.

The bisphenol A used must be of high purity otherwise the colour and mechanical properties of the polymer are impaired. The polymer can be crystallised, with a T_m value of 220–230°C, but is normally largely amorphous as cooled from the melt, as in moulding, and is therefore transparent with a T_g value of about 145°C. The polymer is stiff (tensile modulus about 2400 MPa, tensile strength about 65 MPa) and has a remarkably high impact strength (8–10 J $(12 \cdot 7 \, \text{mm})^{-1}$). The latter is possibly associated with a large free volume, as indicated by a broad transition extending from 0°C down to about -200°C. However, the impact strength is very notch sensitive, so that although products are strong in smooth sheet-like form, mouldings with sharp corners are much less strong. Its mechanical properties are remarkably constant over the temperature range -140 to $+140$°C.

The polymer has good electrical insulation properties and is self-extinguishing on burning. However, it does suffer from a susceptibility to crazing at strains of above about 0·75% and to limited chemical, solvent and ultraviolet light resistance. It also suffers from environmental stress cracking with some liquids.

The polymer is considered to be an engineering plastic and finds use for machine and appliance housings, for glazing, sterilisable medical products and also in blends to raise the softening point of other polymers, especially with polybutylene terephthalate and ABS.

BISPHENOL C

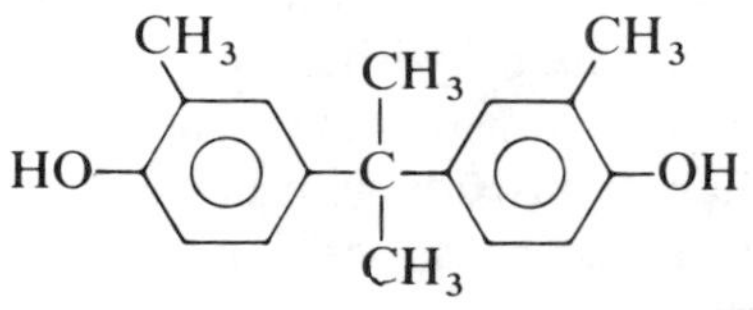

M.p. 113–115°C or 137°C.

Produced by the condensation of two molecules of o-cresol with one of acetone under acidic conditions. It yields a polycarbonate with a T_m value of 160–170°C, a T_g

value of about 110°C, and with outstanding resistance to hydrolysis.

BISPHENOL F   (4,4′-Dihydroxydiphenylmethane)

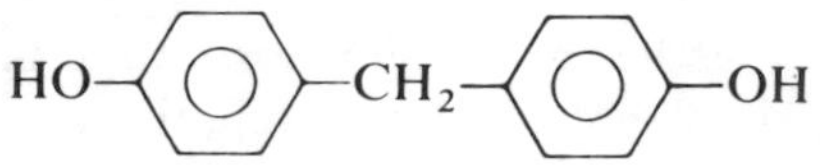

M.p. 163°C.

Produced by the condensation of two molecules of phenol with one of formaldehyde under acidic conditions. It yields a polycarbonate by reaction with phosgene with a T_m value of 300°C.

BISPHENOL Z   (1,1-Bis-(4-hydroxyphenyl)cyclohexane)

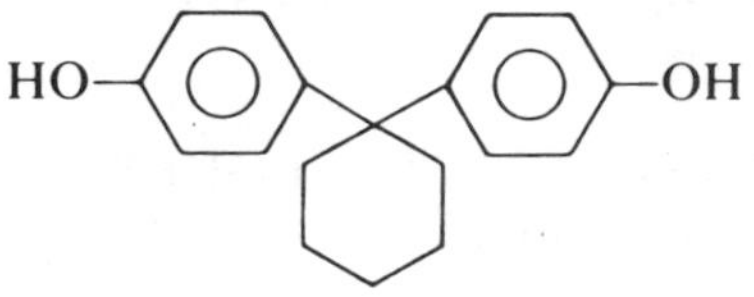

M.p. 184°C.

Prepared by the condensation of two molecules of phenol with one of cyclohexanone under acidic conditions. It forms a polycarbonate with a T_m value of 250–260°C and a T_g value of about 170°C. It is also used to make a commercial copolymer polycarbonate with bisphenol A which has been used in surface coatings and as film for electrical insulation.

BIS-(2,2,6,6-TETRAMETHYL-4-PIPERIDYL) SEBACATE

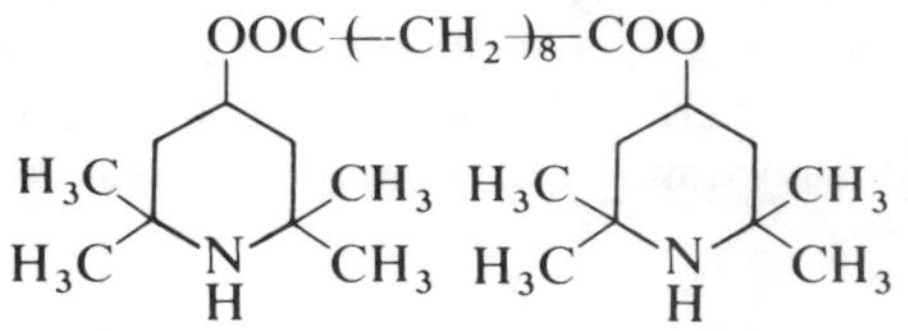

One of the most important hindered amine light stabilisers, whose mode of action is as a free radical trapping agent through abstraction of the hydrogen atoms attached to the nitrogens. Particularly useful in polypropylene.

BITUMEN A black resinous solid, or semi-solid, fusible but non-crystalline substance consisting of a complex mixture of high molecular weight (from a few hundred to a few thousand) aliphatic, naphthenic and aromatic hydrocarbons, occurring either naturally or as the residue of the distillation of crude oil. Natural bitumens soften at about 130–160°C. Although mostly used in road making, some use as plastics moulded products has occurred. Natural and artificial mixtures with particulate minerals are termed asphalts in the UK but are termed bitumen in the USA.

BIURET A compound of the type

$$\text{R—NH—CO—NR′—CO—NH—R″}$$

where R, R′ and R″ are alkyl groups. Formed by reaction of an isocyanate group with a urea group. Biuret groups are produced during polyurethane formation when the reaction temperature is above about 100°C. Like allophanate formation, biuret formation contributes to crosslinking. The urea groups arise from a prior reaction of amine groups with isocyanate groups. The amine groups may be present either in a diamine used as a chain extender or from reaction of water (often present as a blowing agent) with isocyanate.

BIURET REACTION The reaction of a peptide or protein, or of any compound containing two amide or peptide bonds linked through a single carbon atom, with alkali and Cu^{2+} ions, but not given by free amino acids. It yields a purple Cu^{2+} complex, which can be determined spectrophotometrically to measure the concentration of peptide or protein.

BKZ THEORY Abbreviation for Bernstein, Kearsley and Zappas theory.

BLACK ORLON A cyclised, fibrous form of poly-acrylonitrile, obtained by pyrolysis of polyacrylonitrile at 100–300°C in a controlled oxygen-containing atmosphere. Cyclisation produces a highly conjugated structure:

The conjugated sequences are occasionally interrupted by non-conjugated units. Higher pyrolysis temperatures yield hard brittle fibres of very high modulus and tensile strength (carbon fibres). The fibres have outstanding short-term resistance to high temperatures, even up to 700°C.

BLENDEX Tradename for acrylonitrile–butadiene–styrene copolymer.

BLOCK COPOLYMER (Block polymer) A copolymer in which the different repeating units, A, B, etc...., occur as long sequences or blocks joined together linearly. Thus a diblock copolymer has two blocks and may be represented as

$$\sim\!\!\!\sim\!AAA\!-\!(A)_x\!-\!AABBB\!-\!(B)_y\!-\!BBB\!\sim\!\!\!\sim$$

A triblock polymer has three sequences, etc. Usually the different types of blocks are incompatible with each other and separately aggregate to form domains, each domain containing only the A or the B sequences. If the B sequences (say) are rubbery (the soft sequences) and those of A are glassy or crystalline (the hard sequences) then the A sequences may act as thermally reversible crosslinks for the rubbery poly-B, especially in triblock ABA polymers. Such copolymers form the basis of many thermoplastic elastomers.

Block copolymer synthesis is achieved if active sites can be formed at the end or ends of one polymer (providing one sequence), which are then capable of polymerising the monomer for the second polymer when this is added, i.e.

$$\sim\!\!\!\sim\!AAAA^* + n\,B \rightarrow \sim\!\!\!\sim\!AAAABBBB\!-\!(B)_x\!-\!BBBB\!\sim\!\!\!\sim$$

or by coupling the active site with a second polymer. The most usual method for producing di- or triblock polymers is to use a preformed living polymer, prepared by anionic polymerisation, since any number of blocks of predetermined length may be built on. Multiblock polymers are best synthesised by step-growth polymerisation methods, as with polyether/esters, segmented polyurethanes and polysiloxane block copolymers. Use of free radical methods generally yields copolymers with varying and indeterminate block lengths, together with some homopolymer.

In most block copolymers each block consists of a sequence of only one type of repeat unit, but occasionally one or more of the blocks may themselves be random copolymer. If the only difference between the units in each block is stereochemical, then the polymer is a stereoblock copolymer. If there is a gradual change in composition along the molecule at the block junction, then the polymer is a gradient block copolymer. The existence of long blocks means that the polymer retains the properties of both types of sequences, as found in their corresponding homopolymers. Owing to incompatibility the long blocks aggregate into separate domains. The limited length of the blocks produces smaller phase separated domains than is found in normal polymer blends (microphase separation). Regular block lengths lead to regularity in the phase morphology, which may consist of spheres or cylinders of A blocks dispersed in a matrix of B blocks (or the inverse) or it may be lamellar. Morphology depends on block length and on the method of sample preparation. Such regular structures have been likened to large single crystals and lead to several unusual physical properties, including unusually high modulus and tensile strength and low gas permeability. Block copolymers may be named as poly-$(M_1$-b-$M_2)$, where M_1 stands for monomer 1, b stands for block and M_2 for monomer 2. In addition the number and sequence of the blocks may be indicated. Thus a diblock polymer may be named AB poly-$(M_1$-b-$M_2)$, where A is the first block of M_1 units and B is the second block of M_2 units. A triblock polymer is most commonly ABA poly-$(M_1$-b-$M_2)$, e.g. ABA poly-(styrene-b-butadiene).

BLOCKED ISOCYANATE An adduct of an isocyanate with a compound containing an active hydrogen atom, such as a phenol, which reacts with a polyol to produce a polyurethane only at elevated temperatures. Thus a blocked isocyanate may be mixed and stored with a polyol at ambient temperature and used in a one-component (one-shot) system for polyurethane production. However, as a result of urethane formation, the original active hydrogen component is liberated and must be eliminated from the product. This has restricted the use of blocked isocyanates to coatings applications where the undesirable liberated product may be lost by volatilisation.

BLOCKING GROUP Alternative name for protecting group.

BLOCK POLYETHERESTER Alternative name for polyetherester block copolymer.

BLOCK POLYMER Alternative name for block copolymer.

BLOOD-GROUP POLYSACCHARIDE Alternative name for blood-group substance.

BLOOD-GROUP SUBSTANCE (Blood-group polysaccharide) One of a group of structurally similar glycoproteins found in red blood cells. Each particular blood group type substance (A, B, H, etc.) has a specificity determined by the sequences and types of linkage of the monosaccharide units of the oligosaccharides. These latter consist largely of L-fucose (usually as a terminal group) D-galactose, *N*-acetyl-D-glucosamine and *N*-acetyl-D-galactosamine units and are thought to be O-glycosydically linked to either serine or threonine residues of the protein. Sometimes referred to as the neutral mucopolysaccharides.

BLOWING AGENT A gas, or material capable of producing a gas, which is incorporated into a polymer melt where it becomes trapped. The polymer is then allowed to cool and solidify, thus producing a cellular polymer, sometimes referred to specifically as an expanded polymer. The gas, often nitrogen, may be directly injected under pressure into the melt or it may be produced by the evaporation of a low boiling point liquid as in the production of expanded polystyrene. These types are said to use a physical blowing agent. Alternatively, and more commonly, a chemical blowing agent is used. Here the gas is produced *in situ* either by the chemical decomposition, often by heat, of the blowing agent, or by gas production from another type of chemical reaction, as in the reaction of water with isocyanates to produce carbon dioxide in polyurethane foam formation.

The use of a heat sensitive chemical blowing agent is the most versatile method of forming a cellular polymer, especially for expanded plastics where thermal decomposition occurs at the melt processing temperature to produce a large volume of gas under pressure. On emerging from the machine die (in extrusion) or into the mould (in injection moulding) the gas expands producing the expanded plastic product. A wide variety of chemical blowing agents, decomposing at the different melt processing temperatures used, is available. Examples are azodicarbonamide, dinitrosopentamethylenetetramine, 4,4'-oxy-bis-(benzenesulphonylhydrazide), trihydrazinotriazine and *N,N'*-dimethyl-*N,N'*-dinitrosoterephthalamide.

BLUE ASBESTOS Alternative name for crocidolite.

BLUE C Tradename for a spandex polyurethane elastomeric fibre, based on polyester polyurethane prepolymer, chain extended with a diamine and wet spun.

BLUE C NYLON Tradename for nylon 66.

BOAT CONFORMATION One of the two possible types of conformation of six-membered organic ring compounds, which is usually a higher energy form, and therefore less preferred than the alternative chair form. It may be represented, for the important case of a cyclohexane ring, as:

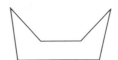

The possibility of boat and chair conformations also arise in monosaccharides and saccharide repeat units of polysaccharides. Here again the boat conformation is the higher energy form. As an example the boat form of glucose, in its usual α-D-glucopyranose form is:

$$
\begin{array}{c}
\text{H} \qquad\qquad \text{H} \quad \text{OH} \\
\text{HO} \qquad \text{CH}_2\text{OH} \\
\qquad\qquad \text{O} \\
\text{H} \qquad \text{H} \\
\text{HO} \quad \text{H} \qquad\qquad \text{OH}
\end{array}
$$

BOC Abbreviation for *t*-butyloxycarbonyl group.

BODY FORCE The force induced in the interior of a body by an applied external load, as distinct from the induced surface forces. Examples include gravitational, magnetic and inertia forces.

BOHR EFFECT The change in the oxygen affinity of haemoglobin with change of pH. The higher the pH the greater the saturation of haemoglobin (Hb) with oxygen, due to the release of a proton on oxygenation:

$$\overset{+}{\text{H}}\text{HHb} + \text{O}_2 \rightleftharpoons \text{HbO}_2 + \text{H}^+$$

Thus an increase in H^+ causes a shift in equilibrium to the left and hence less oxygen binding.

BOLTZMANN SUPERPOSITION PRINCIPLE
(Superposition Principle) In the theory of linear viscoelasticity, the assumption that the deformation $e(t)$ is dependent on the entire loading history and that the final deformation due to multiple loading steps ($\sigma_1, \sigma_2, \sigma_3, \ldots$ etc., which are the incremental stresses at time $\tau_1, \tau_2, \tau_3, \ldots$ etc.) is simply the sum of the various contributions of each loading step, which are independent of each other. Thus in creep:

$$e(t) = \sigma_1 J(t - \tau_1) + \sigma_2 J(t - \tau_2) + \cdots, \text{etc.}$$

where $J(t-\tau)$ is the creep compliance function. Thus if the stress alters continuously:

$$e(t) = \int_{-\infty}^{t} J(t-\tau)\,d\sigma(\tau)$$

or more correctly

$$e(t) = (\sigma/G_u) + \int_{-\infty}^{t} J(t-\tau)(d\sigma(\tau)/d\tau)\,d\tau$$

where the first term is the immediate elastic deformation with G_u being the unrelaxed modulus. Such a relationship is the integral representation of viscoelasticity, the integral being a Duhamel integral. A similar relationship exists for stress relaxation:

$$\sigma(t) = G_r e + \int_{-\infty}^{\infty} G(t-\tau)(de(t)/d\tau)\,d\tau$$

where G_r is the relaxed modulus and $G(t-\tau)$ is the stress relaxation modulus. It follows from the principle that creep and recovery responses are identical. Although the behaviour of some polymer systems, e.g. rubbers and amorphous polymers in tensile creep, are described by the principle, the behaviour of other polymers, notably crystalline polymers, is not.

BORATE GLASS (Polyborate) An amorphous network polymer of boric oxide (B_2O_3), obtained by the high temperature dehydration of boric acid, or of boric oxide and metal oxides, formed by melting mixtures of oxides with boric acid. The structure is:

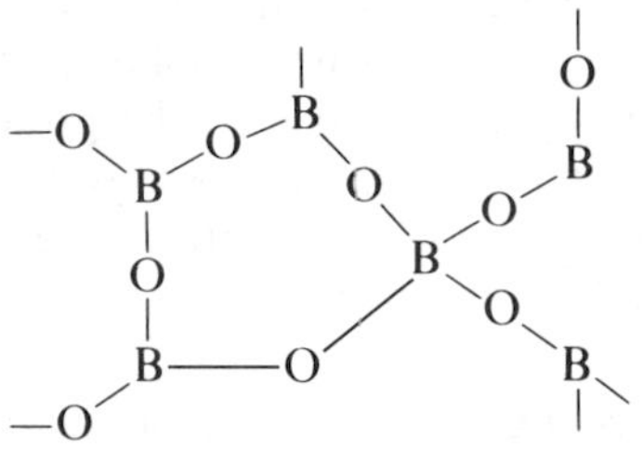

Metals used include the alkali or alkali earth metals. The glasses have relatively low softening points (boric oxide itself has a T_g value of about 250°C), the incorporation of metal oxide raising the T_g value up to about 20% metal content, then levelling off. Other properties change remarkably at about this metal level.

BORAZEN POLYMER A polymer of structure

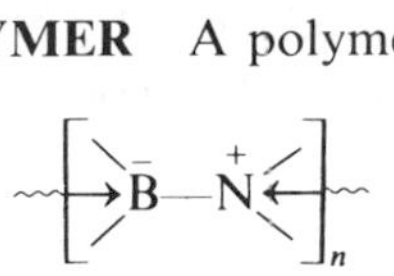

with each repeat unit being joined to the next by coordinate links. A polymer of this type is obtained by heating dimethylaminoborane to 150°C at 3000 atm.

BORAZIN POLYMER A polymer containing structures of the type

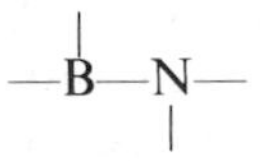

Boron nitride may be considered to be a borazin, with the units being joined to form fused rings.

BORAZON Alternative name for the cubic, diamond-like form of boron nitride.

BORIC ESTER POLYMER A polymer containing boric acid ester groups,

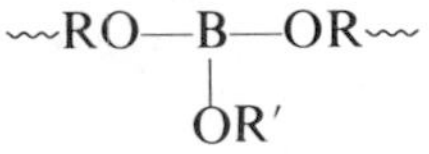

Thus polyborates may be formed by reaction of boric acid (H_3BO_3) with, for example, ethylene glycol or glycerol at about 150°C. Such polymers have been used as adhesives, plasticisers, etc., but are very sensitive to hydrolysis.

BORON Elemental boron is polymeric, being either amorphous and reactive, or crystalline (melting at about 2300°C), almost as hard as diamond and very inert, depending on the method of its preparation.

BORONAMIDE POLYMER Alternative name for polyboroamide.

BORON FIBRE A continuous filament material with exceptionally high tensile strength (3·4 GPa) and Young's modulus (310 GPa), yet having a relatively low density of 2·6 g cm^{-3}. It is therefore useful as a reinforcing fibre in high performance composites, mostly for aerospace applications, owing to the high cost of the fibres especially with epoxy resins. Produced by deposition from the vapour phase onto a tungsten or carbon fibre substrate.

BORONIC ESTER POLYMER A polymer containing the group $\sim R-B-(OR')_2$ in its repeat unit. Like the boric ester polymers, they have poor resistance to hydrolysis, except when formed from aromatic acids, as in, for example, the p-phenylene diboronic ester polymers:

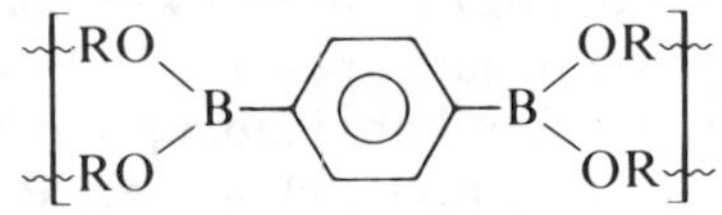

which have been widely studied.

BORON NITRIDE A network polymer of empirical formula BN. A sheet-like layered structure, like graphite, formed by heating boric acid with urea under a nitrogen atmosphere to 600°C, followed by sintering under ammonia at 1000°C. The material is relatively soft and machinable. It may be used in air to about 800°C and melts at about 3000°C. This planar hexagonal form is converted on heating at 1500–2000°C at 50–80 kbar pressure to a three-dimensional cubic crystalline form (borazon) similar in structure and properties to diamond, with which it is isomorphous.

BOROPHOSPHATE GLASS A glassy polymer formed by melting together, at 600–700°C phosphoric acid (or ammonium dihydrogen phosphate), boric oxide and 20–50 mole % of an alkali metal carbonate or alkaline earth oxide or carbonate. The polymer contains 3-coordinate or 4-coordinate boron units:

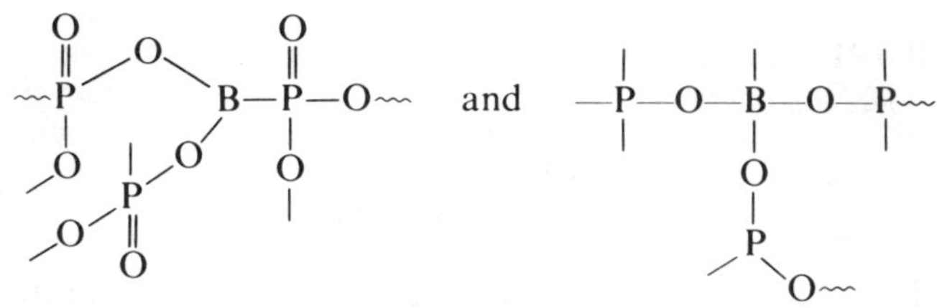

Boron phosphate polymer itself is crystalline. Most of the boron atoms are 4-coordinate (as opposed to 3-coordinate in boron oxide), i.e. as BO_4^- groups. Polymer T_g values are relatively low, varying from 250–450°C for a potassium polyphosphate–boric oxide glass as the B_2O_3 content increases from 0–50 mole%, since each 3-coordinate boron contributes one crosslink and each 4-coordinate boron contributes 3 crosslinks, with one crosslink at each phosphorus. The polymers are more resistant to hydrolysis than the ultraphosphate glasses. They also resist surface contamination and are non-misting in humid environments.

BOROSILICATE GLASS A range of glasses based on boric oxide, silica and a metal oxide, that have much lower thermal coefficients of expansion than soda-lime glass (with consequent better thermal shock resistance), greater chemical resistance (especially to alkalis) and good melt workability. Best known are the Pyrex glasses which consist of about 80% SiO_2, 12% B_2O_3 and about 12% metal oxide (chiefly sodium oxide). In another type, Vycor, with 55–70% SiO_2 and 20–40% B_2O_3, one component can be leached thus producing a porous glass.

BOROSILICONE A polyorganosiloxane containing main chain —Si—O—B—O—Si— links, usually based on polydimethylsiloxane. Prepared either by reacting a silanol terminated siloxane polymer with boric acid or by reacting ethoxy-terminated polymer with boron triacetate. Polymers with one boron atom per 200–500 silicon atoms are the basis of fusible elastomers. Polymers with higher boron contents are the basis of bouncing putty. In these polymers the unusual properties are thought to be due to the electron donor–acceptor complexes between polymer chains.

BOROSILOXANE POLYMER Alternative name for polyorganoborosiloxane.

BOUNCING PUTTY A polymer composition based on a borosilicone containing three —B—O—Si— links for approximately every 100 silicon atoms. Prepared by heating polydimethylsiloxane with boric oxide and ferric chloride. It shows very unusual extreme viscoelastic properties. It flows as a viscous liquid merely on storage; it can be shaped by kneading and can be drawn to fine threads by use of even low tension, yet shows high rubber-like resilience at the high strain rates involved on bouncing, but shatters on application of a sharp blow.

BOUNDARY LAYER The region of a flowing fluid close to its rigid boundary, where viscous forces cannot be ignored, as they are in the Euler equation. This must be so since the fluid velocity at the boundary must be zero but a velocity gradient must exist and the Navier–Stokes equation applies.

BOUNDARY SEDIMENTATION Alternative name for sedimentation velocity method.

BOUND ION-RADICAL MECHANISM A chain polymerisation mechanism in which the monomer is firmly adsorbed onto the solid surface of the polymerisation catalyst and is polymerised in this adsorbed state. The most important example is the polymerisation of ethylene by the use of activated metal oxide catalysts, as in the Phillips process, involving the use of partially reduced chromium oxides on an inert support. In this and other such cases it is a little uncertain whether the active centres are free radicals or ions.

BOUND RUBBER That part of a rubber in a rubber compound containing a fine particle reinforcing filler (especially carbon black), which becomes attached to the filler particles during compounding and is no longer soluble in the usual rubber solvents. The amount of bound rubber increases with time after compounding. It appears to form a network structure with the black and consists of a swollen gel in a rubber solvent. The amount of bound rubber increases with increase in surface activity of the black, being highest in the highly reinforcing blacks. Binding is thought to be due to the production of free radicals from the rubber (or possibly from the black), which react with the black or rubber respectively, forming covalent bonds between the two.

BOYER–BEAMAN RULE A relationship between the glass transition temperature (T_g) and the melting temperature (T_m) of a polymer, that states that the T_g value is between one-half and one-third of the value of T_m, expressed in K, having the particular values of $T_m/2$ for a symmetrical and $T_m/3$ for an unsymmetrical polymer. In general, this relationship is followed, indeed some such relationship would be expected since many of the factors affecting T_g similarly influence T_m. Notable exceptions are polydimethylsiloxane and bisphenol A polycarbonate.

BPA Abbreviation for bisphenol A.

BPA POLYCARBONATE Alternative name for bisphenol A polycarbonate.

BR Abbreviation for butadiene rubber.

BRACE COMPENSATOR A device consisting of a thin mica plate used to introduce a compensating birefringence into a light beam, and a half-shade plate covering half the field of view, to enable the birefringence of a polymer solution to be determined.

BRAGG EQUATION (Bragg Law) A relationship describing the angular dependence of diffraction of radiation (usually X-rays):

$$n\lambda = \lambda d \sin\theta$$

where λ = wavelength of radiation and n = an integer. The rays are considered to have been 'reflected' by sets of crystallographic (hkl) planes of separation d, 2θ being the angle of reflection, causing differences in path lengths of the reflected rays as the reflected waves must be in phase.

BRAGG LAW   Alternative name for Bragg equation.

BRANCHED CRYSTAL   (Crystal branching) A crystal which is divided into separate but connected lamella due to growth occurring on several fronts during crystallisation. The branching may be narrow angle as in multilayer crystals, wide angle as in dendritic crystals or fibril branching in spherulites.

BRANCHED POLYMER (Non-linear polymer) A polymer in which the molecules consist of a linear main chain to which are attached, usually randomly, other secondary chains, of molecular structure similar to that of the main chain. In long chain branching the secondary chain is of the same order of length as the main chain. In short chain branching the secondary chains are of a size up to that of only a few repeat units. Polymers in which side group chains, of different structure to the main chain, are part of the regular repeating unit are not considered as branched polymers, but rather as linear polymers. Thus in polyethylene, short chain branching as:

$$\sim\!CH_2CH\!\!-\!\!(CH_2CH_2\!-\!)_x CHCH_2CH_2\!\sim$$
$$\quad |\qquad\qquad\qquad\qquad |$$
$$(CH_2)_3CH_3 \qquad\quad CH_2CH_3$$

and long chain branching as:

$$\sim\!CH_2CH\!\!-\!\!(CH_2CH_2\!-\!)_y CH\!\!-\!\!(CH_2CH_2\!-\!)_z CH_2CH_2\!\sim$$
$$\quad |\qquad\qquad\qquad\qquad\quad |$$
$$(CH_2CH_2\!-\!)_x \qquad\quad (CH_2CH_2\!-\!)_z$$

where x, y and z are large, may be present, but poly-(octene-1)

$$\left[\begin{array}{l} CH_2CH\!\!-\!\!-\!\!-\!\!-\\ \quad |\\ (CH_2)_5CH_3 \end{array}\right]_n$$

and poly-(octylacrylate)

$$\left[\begin{array}{l} CH_2CH\!\!-\!\!-\!\!-\!\!-\\ \quad |\\ COOC_8H_{18} \end{array}\right]_n$$

for example, are considered as linear polymers.

The branches may be randomly distributed and be themselves branched, as in polymers formed by non-linear step-growth polymerisation before the stage of complete network formation has been reached. The branches may arise from a single atom or group (a star polymer) or several branches may be attached more or less regularly to a main chain (a comb polymer). In practice many polymers, which although nominally linear, may contain some branch points due to secondary reactions occurring during chain polymerisation. Branches may also be formed by post polymerisation reactions, as in graft copolymers. Polymer properties, especially mechanical and rheological properties, depend on the amount and distribution of branching. Thus a branched polymer of similar molecular weight to a linear polymer has a lower tensile strength and modulus, as well as lower melt and solution viscosities. Branched polymers are also more non-Newtonian. Indeed rheological measurements often provide the best methods of assessing the extent of branching, which is rather difficult to quantify. Branching can also sometimes be determined by end-group analysis or by infrared or nuclear magnetic resonance spectroscopy.

BRANCHING COEFFICIENT   (1) (α-Branching coefficient) (Flory α-branching coefficient) Symbol α. In the statistical theory of non-linear step-growth polymerisation, the probability that a given functional group attached to a branching unit (i.e. one derived from a monomer of functionality >2) at the end of the polymer chain, leads to another branching unit. Thus in a polymerisation of monomers

$$A\!\!-\!\!A + B\!\!-\!\!B + A\underset{A}{\overset{A}{\diagup\!\!\!\diagdown}}$$

the probability of the formation of structure

$$A\underset{A}{\diagdown}A\!\!-\!\![B\!\!-\!\!B\!\!-\!\!A\!\!-\!\!A\!]_n B\!\!-\!\!B\!\!-\!\!A\underset{A}{\diagup}A$$

is the branching coefficient. The critical branching coefficient may be deduced, at which an infinite network is formed and hence gelation is predicted to occur. The value of α is given by:

$$\alpha = p_B^2 \rho / [r - p_B^2 (1 - \rho)]$$

where p_B is the conversion of B groups, assuming A groups are in excess, r is the ratio of B to A groups and ρ is the ratio of A groups in branching units to all A groups.

(2) Alternative name for branching density.

α-BRANCHING COEFFICIENT   Alternative name for branching coefficient.

BRANCHING DENSITY (Density of branching) (Branching coefficient) The fraction of repeat units in a polymer which contain a branch point, given by $\lambda = \alpha(b/n)$, where b is the number of branch points, n is the number of repeat units, i.e. the degree of polymerisation, and α is a constant dependent on the functionality of the branch points. Depending on the theoretical model chosen and the topology of the branching (the branch length distribution and the functionality), λ may be related to the branching parameter, which may be determined by a combined gel permeation chromatography/viscosity method for long chain branching. In chain polymerisation where branching arises from transfer to polymer, ρ is given by

$$\rho = -C_p[1 - (1/p)\ln(1-p)]$$

where C_p is the chain transfer constant and p is the fractional conversion.

BRANCHING FACTOR Symbol G. A measure of the amount of branching in a polymer, as shown by the lower limiting viscosity number ($[\eta]_{Br}$) compared with that of the linear polymer ($[\eta]_{Lin}$) of the same molecular weight. G is given by $[\eta]_{Br}/[\eta]_{Lin}$, and is related to g, the ratio of the mean square radii of branched and linear molecules of the same molecular weight, by $G = g^a$, where a is a constant, frequently equal to 0·5. g is sometimes called the branching parameter and may be related theoretically to the number of branches per polymer molecule, which enables the degree of branching to be determined from a gel permeation chromatogram.

BRANCHING PARAMETER Symbol g. A parameter expressing the reduction in the radius of gyration (s) of a long chain branched polymer ($\langle s^2 \rangle_{Br}$) compared with that of a linear polymer of the same molecular weight ($\langle s^2 \rangle_{Lin}$), where $\langle s^2 \rangle$ is the mean square radius of gyration. It is defined as $g = \langle s_0^2 \rangle_{Br}/\langle s_0^2 \rangle_{Lin}$ and is always less than unity. The subscript '0' indicates that measurements are made in a theta solvent. The particular value of g depends on the functionality (f) of the branch points, the length (n_λ) and length distribution of the branches and their distribution along the chain. In general if the branching density is λ then:

$$g = \sum_\lambda [(3n^2/n^2) - (2n_\lambda^3/n^3)]^2$$

for randomly distributed branches where n is the degree of polymerisation. For star polymers: $g = (3/f) - (2/f^2)$. With variable length of branches the value of g increases. Comb polymers have a higher value of g than randomly branched polymers. g may be related to the branching factor G by $G = g^a$, but different theoretical models give various values for a from 0·5 to 1·5. Assuming a model and a particular branch topology, g and hence λ, may be determined by a combined viscosity/gel permeation chromatography method.

BRASSYLIC ACID

$$HOOC(CH_2)_{11}COOH \qquad M.p. \ 114°C.$$

Obtained by ozonolysis of erucic acid, itself obtained from crambe seed oil. A monomer for nylon 1313. Also converted to the corresponding diamine, 1,13-diaminotridecane, by reaction with ammonia followed by hydrogenation.

BREAKDOWN STRENGTH Alternative name for dielectric strength.

BREAKING LENGTH A measure of the breaking strength of a fibre. The length of a fibre which will just break under its own weight. Typically very high values of tens of km are found.

BREON Tradename for polyvinyl chloride and nitrile rubber.

BRICK DUST POLYMER A polymer, usually aromatic, of such high softening point and lack of solubility that it cannot be processed or fabricated into useful shapes or forms. Usually also coloured and resembling powdered bricks, hence the name.

BRILLOUIN SCATTERING (Rayleigh–Brillouin scattering) The inelastic scattering of light, usually observed with laser light and solid samples or solutions, which, with isotropic molecules, produces two pairs of scattered lines symmetrically placed with respect to the incident light frequency (the Rayleigh line), one shifted to higher and one to lower frequencies. With anisotropic molecules three pairs of lines are observed. The lines result from the interaction of the light with phonons (the thermal acoustic waves travelling with hypersonic velocity) in the material. This velocity may be determined from the frequency shift and related to the elastic constants. The lines are also broadened due to attenuation of the acoustic waves, and the attenuation coefficient and Landau–Placzek ratio ($I_{Rayleigh}/2I_{Brillouin}$) can also be measured. These vary with temperature, especially in the region of the transitions. Experimentally the light scattered at 90° to the laser beam is analysed using a Fabry–Perot interferometer and photomultiplier. The technique is particularly useful for studying diffusion and relaxation processes.

BRINELL HARDNESS A measure of the hardness of a material, originally designed for metals, in which the dimensions of the impression made by pressing a steel ball into a sheet sample of the material are measured. It is defined as:

Brinell Hardness (HB) $= 2F/(\pi D^2[1 - \{1 - (d/D)^2\}^{1/2}])$

where F is the load (kg), D is the diameter of the ball and d is the diameter of the impression made during 15 s application of the load. The HB value is quoted together with the F/D^2 ratio employed, which is usually 5 or 1 for plastics. One difficulty with plastics is their recovery on removal of the load, but methods are available, such as use of a carbon paper impression on the plastic, of overcoming this.

BRITTLE–DUCTILE TRANSITION (Tough–brittle transition) (Brittle point) (Brittle temperature) The temperature at which the mode of fracture changes from brittle to ductile fracture. Sometimes the term is used more generally to mean a similar change brought about by some other change in experimental or material conditions, such as rate of stressing, the presence of notches, crosslinking or plasticisation. According to the Orowan hypothesis, the transition occurs when the yield stress exceeds a certain critical value (the brittle stress). The transition may be related to the glass transition temperature or often to a lower temperature transition, e.g. a β-transition. However, since brittle behaviour is often controlled by the presence of flaws, such relationships often do not apply. Crosslinking (or raising T_g) and increasing the strain rate raises the transition, favouring brittle behaviour. Conversely, plasticisation (or lowering T_g) lowers the transition and favours ductile behaviour. Materials with high notch sensitivity also have a raised transition when notched.

BRITTLE FRACTURE   Fracture which, unlike ductile fracture, occurs without significant plastic deformation of the material at the crack tip. In the tensile mode the stress–strain curve is nearly linear up to fracture, which occurs at a maximum load at a strain of less than a few per cent. This type of fracture is frequently found in many amorphous and crystalline polymers well below their glass transition temperature, such as polystyrene and polymethylmethacrylate, which have been especially widely studied. As the temperature is raised to above the brittle–ductile transition, the mode of fracture changes to ductile fracture.

According to the Orowan hypothesis, brittle fracture occurs when the yield stress exceeds a certain value. A distinction between brittle and ductile fracture may also be made on the basis of the energy to fracture, although brittle fracture does not necessarily imply a low fracture energy as, for example, in fibre reinforced composites. The appearance of the fracture surfaces are also characteristic of the type of fracture, brittle fracture often resulting in a smooth surface with characteristic fracture markings, such as conic markings, mackerel and hackle.

An understanding of brittle fracture has been achieved through the development of fracture mechanics, often based on the Griffith theory. A sharp distinction cannot be made between brittle and ductile fracture since even in some very glassy materials some deformation takes place. Further, a given material will fail in a brittle manner under some conditions and a ductile manner under other conditions. Thus brittle fracture is favoured, the lower the temperature, the faster the loading and the more the state of stress approaches a uniform, i.e. triaxial or dilatational, state. In terms of molecular structure, any structural feature which raises T_g, such as bulky side groups or crosslinking, promotes brittle fracture. The role of crystallinity and the presence of reinforcing fillers is varied.

BRITTLENESS TEMPERATURE   Symbol T_B. The temperature at which 50% of test pieces of a material would fail on impact when tested at a low temperature. In the test, small rectangular bars are clamped horizontally and are tested using a vertically travelling striker. Temperatures are found at which all and none of ten samples fail, the temperature difference being ΔT. The test is then repeated ten times at intermediate temperatures. T_B is then given by:

$$T_B = T_h + \Delta T[(S/100) - \tfrac{1}{2}]$$

where T_h is the highest temperature at which failure of all

samples occurs and S is the sum of % breaks at each temperature.

BRITTLE POINT   Alternative name for brittle–ductile transition.

BRITTLE STRENGTH   The tensile strength of a material when brittle fracture occurs.

BRITTLE TEMPERATURE   Alternative name for brittle point.

BROAD-LINE NMR SPECTROSCOPY   Alternative name for wide-line NMR spectroscopy.

BROMINATED BUTYL RUBBER   (BIIR) (Bromobutyl rubber) Tradename Bromobutyl. Butyl rubber which has been brominated, preferably in solution rather than in the solid state, to give a homogeneously brominated product. Commercial products contain about 2% bromine, probably as

$$-CH_2CCHCH_2-$$
$$\overset{\displaystyle Br}{\underset{\displaystyle CH_2}{|}}$$

units, since bromination occurs on the isoprene units of the butyl rubber by an ionic reaction. Compared with normal butyl rubber, the brominated rubber vulcanises more rapidly and shows better adhesion. It may be covulcanised with natural rubber or SBR by sulphur vulcanisation. Otherwise its properties are similar to butyl rubber.

BROMOBUTYL   Tradename for brominated butyl rubber.

BROMOBUTYL RUBBER   Alternative name for brominated butyl rubber.

BRONSTED–SCHULTZ DISTRIBUTION FUNCTION   An expression relating the volume fractions of the different polymer molecular weight species of degree of polymerisation x present in the solution and swollen precipitate (coacervate) phases (ϕ_x and ϕ'_x respectively) of the two phase system resulting from partial precipitation (or solution) of the polymer. The expression is: $\phi'_x/\phi_x = e^{\sigma x}$, where σ is a parameter which is a complex function of the Flory–Huggins polymer–solvent interaction parameter, the molecular weight distribution and the concentration. Since the higher molecular weight species are more concentrated in the precipitated phase, the equation provides the basis for fractionation by the solubility methods of fractional precipitation and fractional solution.

BROOKFIELD VISCOMETER   A simple rotating cylinder viscometer, which consists of a rotatable cylinder which is immersed in the fluid under investigation. Readings are made of the torque at various speeds of rotation. However, although such readings are rapidly obtained, conversion to shear rate and shear stress is not simple.

BROWNIAN DIFFUSION FRACTIONATION

(Isothermal diffusion fractionation) An analytical fractionation method in which the polymer solution is layered onto a solvent layer in a cell, after which the development of a concentration gradient in the 'interfacial' region is followed with time by a special optical technique. The diffusion constants are thus determined and may be related to the molecular weight distribution.

BSH Abbreviation for benzenesulphonylhydrazide.

B-STAGE POLYMER or RESIN The second intermediate stage in the formation of a network polymer by step-growth polymerisation in which the polymer is of reasonably high molecular weight and may be somewhat branched and slightly crosslinked. However, network formation is not so advanced that the material is completely insoluble and infusible as in the subsequent C-stage, rather it is swollen by solvent and is rubbery. The term is used specifically to refer to phenol–formaldehyde polymers, which when prepared to form an initial resole, are called resitol in the B-stage.

BTDA Abbreviation for benzophenonetetracarboxylic acid dianhydride.

BUDENE Tradename for high *cis*-polybutadiene.

BUECHE THEORY A molecular theory of the non-Newtonian behaviour of polymer fluid which uses as a model the free-draining coil, which is considered to be caused to rotate in a simple shear field. The actual polymer chains are replaced by jointed subunits, which behave mathematically as a set of masses connected by springs. This enables the viscosity to be calculated since each unit experiences a sinusoidally varying tension and compression. For a monodisperse polymer in solution, the reduced viscosity is given by:

$$\frac{\eta - \eta_s}{\eta_0 - \eta_s} = 1 - \frac{6}{\pi^2} \sum_{n=1}^{N} \left[\frac{(\dot{\gamma}\lambda_1)^2}{n^2(n^4 + \dot{\gamma}^2\lambda_1^2)} \right] \left[\frac{2 - \lambda_1^2\dot{\gamma}^2}{n^4 + \lambda_1^2\dot{\gamma}^2} \right]$$

where η is the apparent viscosity, η_s that of the solvent, and η_0 that at zero shear rate (for polymer melts η_s is set equal to zero and c (the concentration) is replaced by ρ (the polymer density)), $\dot{\gamma}$ is the shear rate and λ_1 is a relaxation time, given by

$$\lambda_1 = 12(\eta_0 - \eta_s)M/\pi^2 cT$$

where M is the polymer molecular weight, N is the number of subunits per polymer molecule and T is the absolute temperature. Thus the viscosity is seen to be a function only of λ_1, as is often found in practice. For polydisperse polymers a molecular weight average may be taken. At high shear rates the theory gives $\eta/\eta_0 \sim (\lambda\dot{\gamma})^{1/2}$, i.e. n of the power law equation is $\frac{1}{2}$.

BULK COMPLIANCE Symbol B. The ratio of an elastic volume strain, i.e. change in volume, to the isotropic stress, i.e. hydrostatic pressure. In other words, the compressibility. The reciprocal of the bulk modulus for an elastic material.

BULK CREEP COMPLIANCE Symbol $B(t)$. The creep compliance for three-dimensional, i.e. volumetric, deformations usually being of more interest in compression rather than in hydrostatic tension (dilatation). Thus if a constant pressure P is suddenly applied to a sample and the change in volume $V(t)$ is measured with time, then $B(t)$ is given by: $B(t) = -V(t)/P$. Unlike tensile and shear behaviour, compression does not involve any change in long-range molecular conformations, so the behaviour is largely independent of molecular weight and crosslinking and is similar to that of low molecular weight liquids. Although like other bulk viscoelastic properties, the creep compliance can be measured directly, it can also be calculated from a knowledge of the shear ($J(t)$) and tensile ($D(t)$) compliances via the relation: $D(t) = J(t)/3 + B(t)/9$, or with more precision from bulk longitudinal creep measurements.

BULK LONGITUDINAL DEFORMATION Extension in which changes in dimensions occur only in one direction, being constrained in the two mutually perpendicular directions. This can be achieved by using a thin flat sample whose faces are bonded to two rigid plates, being subject to tension or compression in the 'thin' direction or by axial compression of a cylinder confined by a cylindrical container. Often characterised by the bulk longitudinal dynamic modulus (M^*) by measurements of the propagation of a longitudinal elastic wave.

BULK LOSS COMPLIANCE Symbol B''. The loss compliance for bulk deformation.

BULK LOSS MODULUS Symbol K''. The loss modulus for bulk deformation.

BULK MODULUS Symbol K. An elastic constant which is the ratio of an applied pressure (P) to the dilatation (Δ) in hydrostatic stressing, i.e. when $\sigma_{xx} = \sigma_{yy} = \sigma_{zz} = P$, where σ_{xx}, σ_{yy} and σ_{zz} are the normal stress components. Thus $K = P/\Delta$. Related to the other common isotropic elastic constants by $K = E/3(1 - 2v)$, $K = 2G(1 + v)/3(1 - 2v)$ and $3/E = 1/G + 1/3K$ (relation (1)) where E is Young's modulus, G is the shear modulus and v is Poisson's ratio. Hence for an incompressible material, for which $v = 0.5$, $E = 3G$ and K is infinite. This is approximately true for rubbers. Since K depends only on the dilatational part of the stress tensor it is a fundamental constant, whereas E depends on both the dilatational and deviatoric part and is therefore not a fundamental constant. K is usually determined experimentally using relation (1) above. Values of K for most polymers are approximately 10^6 MPa, decrease with increasing temperature and increase with hydrostatic pressure. K is independent of molecular weight and molecular weight distribution.

BULK POLYMERISATION (Mass polymerisation) Polymerisation in which only monomer and polymerisation initiator or catalyst are involved, i.e. carried out in the absence of solvent or other dispersion medium. The minimum of polymerisation additives are present, so the purest polymer is likely to be formed. Many step-growth polymerisations, e.g. formation of polyesters and polyamides, are conducted in the bulk, usually as melt polymerisations. In chain polymerisation, especially in vinyl polymerisation, the main disadvantage for large scale polymer production is the difficulty of dissipating the heat of polymerisation, especially at high conversion when the viscosity is high and stirring is difficult. Therefore bulk polymerisation is frequently taken only to about 50% conversion or is conducted in two stages, firstly to about 20% conversion, followed by a second stage in which the heat is dissipated by having a large surface area to the polymerisation reactor. This may be achieved by using a thin tube or by running the polymerising mixture as a thin layer down a tower. Bulk polymerisation directly in a mould is monomer casting. If the polymer is insoluble in its monomer the polymerisation becomes a precipitation polymerisation. At high conversions chain transfer to monomer becomes more likely leading to formation of branched polymer and as conversion increases so will autoacceleration.

BULK STORAGE COMPLIANCE (Dynamic bulk compliance) Symbol B'. The storage compliance for bulk deformation.

BULK STORAGE MODULUS (Dynamic bulk modulus) Symbol K'. The storage modulus for bulk deformation.

BULK STRAIN   Alternative name for dilatation.

BUNA   Abbreviation for an early polybutadiene produced by polymerisation of butadiene with metallic sodium.

BUNA AP   Tradename for ethylene–propylene, or ethylene–propylene–diene monomer, rubber.

BUNA CB   Tradename for high or medium *cis*-polybutadiene.

BUNA HULS   Tradename for emulsion styrene–butadiene rubber.

BUNA M   Tradename for butadiene–methylmethacrylate rubbers containing about 70% butadiene, but not significantly better than the similar and cheaper butadiene–styrene rubbers.

BUNA N   Tradename for nitrile rubber.

BUNA S   Tradename for an early emulsion styrene–butadiene rubber.

BUOYANCY FACTOR   The quantity $(1 - \bar{v}_2)\rho$ of a dilute polymer solution where $\bar{v}_2$ is the partial specific volume of the polymer and ρ is the solvent density. In the sedimentation of the solute by ultracentrifugation, the sign of this factor determines the direction of sedimentation of the polymer. If positive, sedimentation will be away from the centre of rotation and if negative, polymer molecules will move in the opposite direction, i.e. flotation will occur.

BURGER'S DISLOCATION   Alternative name for screw dislocation.

BURGER'S VECTOR   A measure of the size of a dislocation. A closed circuit (Burger's circuit) in the ideal crystal is made by joining lattice points and is placed onto the defect crystal around the dislocation. The circuit is now open, the Burger's vector points from the beginning to the end of the now open circuit in the defect crystal.

BUTACITE   Tradename for polyvinyl butyral.

BUTACLOR   Tradename for polychloroprene.

BUTACRIL   Tradename for nitrile rubber.

BUTADIENE

$$CH_2{=}CH{-}CH{=}CH_2 \quad \text{B.p.} \; -2{\cdot}6°C.$$

Obtained by cracking the C_4 stream from petroleum naphtha crackers or by dehydrogenation of butanes or butenes. Previously produced from ethanol or acetylene. The monomer for the production of several important synthetic rubbers, including polybutadiene, styrene–butadiene rubber and nitrile rubber. For polymerisation purposes highly purified butadiene is required. This is obtained either by complex formation between butadiene and cuprous ammonium acetate or by extractive distillation with a solvent which preferentially absorbs butadiene, such as acetonitrile, dimethylformamide or N-methylpyrrolidone.

BUTADIENE–ACRYLONITRILE COPOLYMER (Poly-(butadiene-co-acrylonitrile)) A copolymer containing acrylonitrile and butadiene units. Free radical emulsion polymerisation gives random copolymers which are important as nitrile rubbers. Ziegler–Natta polymerisation can produce highly alternating copolymers.

BUTADIENE RUBBER   (BR) A polybutadiene which is useful as a rubber. Three types of polybutadiene are important: high *cis*-, medium *cis*- and low *cis*-polybutadiene.

BUTADIENE–STYRENE COPOLYMER   Alternative name for styrene–butadiene copolymer.

BUTADIENE–VINYL PYRIDINE COPOLYMER

A copolymer containing butadiene and vinyl pyridine

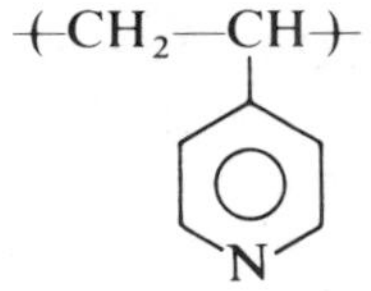

$$+CH_2—CH+$$

units. At one time widely investigated as a potentially useful vinyl pyridine rubber.

BUTAKON Tradename for emulsion styrene–butadiene rubber.

BUTAKON A Tradename for nitrile rubber.

BUTAN-1,3-DIOL Alternative name for butylene glycol.

1,4-BUTANEDIOL

$$HO(CH_2)_4OH \qquad B.p. \ 218°C.$$

A useful monomer for the production of polyesters, e.g. poly-(tetramethylene adipate), as prepolymers for polyurethane formation and for the chain extension of isocyanate prepolymers in the formation of some cast polyurethane elastomers.

n-BUTANOL (Butan-1-ol) (n-Butyl alcohol)

$$CH_3(CH_2)_3OH \qquad B.p. \ 117·7°C.$$

A good solvent for natural resins, low molecular weight silicones, urea and melamine–formaldehyde resins and polyvinyl acetate. Miscible with hydrocarbons, dissolves 20% water at 20°C and forms an azeotrope with 38% water.

BUTAN-1-OL Alternative name for n-butanol.

BUTAN-2-OL Alternative name for secondary butyl alcohol.

BUTAN-2-ONE Alternative name for methylethyl ketone.

BUTAPRENE Tradename for nitrile rubber.

BUTAREZ CTL Tradename for carboxy-terminated polybutadiene.

BUTAREZ ETS Tradename for hydroxy-terminated polybutadiene.

BUTENE-1 (1-Butene)

$$CH_2＝CHCH_2CH_3 \qquad B.p. \ -6·3°C.$$

Produced by thermal or catalytic cracking of petroleum fractions, by dehydrogenation of butane or by dimerisation of ethylene. The monomer for polybutene-1 by Ziegler–Natta polymerisation and also the most important comonomer, with ethylene, for the production of linear low density polyethylene.

1-BUTENE Alternative name for butene-1.

3-BUTEN-2-ONE Alternative name for methylvinyl ketone.

BUTON Tradename for a styrene–butadiene copolymer which is a thermosetting plastic.

2-(2-BUTOXYETHOXY)ETHANOL Alternative name for diethylene glycol mono-n-butylether.

BUTOXYETHYL STEARATE (Butylcellosolve stearate)

$$CH_3(CH_2)_{16}COOCH_2CH_2O(CH_2)_3CH_3$$

$$B.p. \ 245–250°C/4 \ mm.$$

A plasticiser for ethylcellulose, cellulose nitrate and some natural resins. A secondary plasticiser for polyvinyl chloride. It also has a lubricating action.

n-BUTOXYMETHYLACRYLAMIDE

$$CH_2＝CHCONHCH_2OC_4H_9$$

An alkoxymethylacrylamide monomer useful as a comonomer in the production of acrylic thermosetting stoving enamels.

BUTUF Tradename for polybutene-1.

BUTVAR Tradename for polyvinyl butyral.

n-BUTYL ACETATE

$$CH_3COO(CH_2)_3CH_3 \qquad B.p. \ 128°C.$$

Widely used as a solvent for cellulose nitrate in lacquers. Also a solvent for polystyrene, polyvinyl acetate, polyvinyl chloride and many natural resins.

BUTYLACETYL RICINOLEATE

$$CH_3(CH_2)_5CH(OOCCH_3)CH_2—$$
$$—CH＝CH(CH_2)_7COO(CH_2)_3CH_3$$

A plasticiser for cellulose esters and ethers, polyvinyl chloride and its copolymers.

n-BUTYLACRYLATE

$$CH_2＝CHCOOC_4H_9 \qquad B.p. \ 147°C.$$

Prepared by the same methods used for acrylic acid or methyl acrylate except that n-butanol is used in the final step. It is also prepared by transesterification of ethyl acrylate with n-butanol. The monomer for the formation of poly-n-butylacrylate by emulsion polymerisation using, e.g. persulphate initiation.

n-BUTYL ALCOHOL Alternative name for n-butanol.

BUTYLATED AMINORESIN An aminoresin in pre-polymer form in which the methylol groups have been converted to butyl ether groups by reaction with butanol, usually *n*-butanol. Urea–formaldehyde, melamine–formaldehyde and benzoguanamine–formaldehyde prepolymers are all used. The butyl ethers have higher solubility in organic solvents than the parent methylols and are useful as lacquers and stoving enamels, often being used in conjunction with oil modified alkyd resins. On heating at about 120–160°C crosslinking takes place:

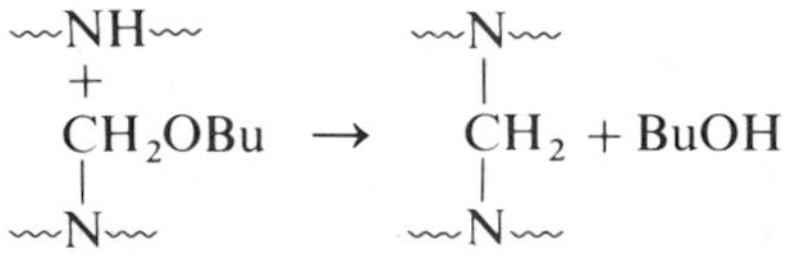

$$\begin{array}{ccc} \text{\textasciitilde\textasciitilde NH\textasciitilde\textasciitilde} & & \text{\textasciitilde\textasciitilde N\textasciitilde\textasciitilde} \\ + & & | \\ \text{CH}_2\text{OBu} & \rightarrow & \text{CH}_2 + \text{BuOH} \\ | & & | \\ \text{\textasciitilde\textasciitilde N\textasciitilde\textasciitilde} & & \text{\textasciitilde\textasciitilde N\textasciitilde\textasciitilde} \end{array}$$

probably together with condensation between the ether groups and the free hydroxyl groups on the alkyd.

BUTYLATED HYDROXYTOLUENE (BHT) Alternative name for 2,6-di-*t*-butyl-4-methylphenol.

BUTYLATED 4,4'-ISOPROPYLIDENE-DIPHENOL
Tradenames: Agerite Superlite, Stablite White.

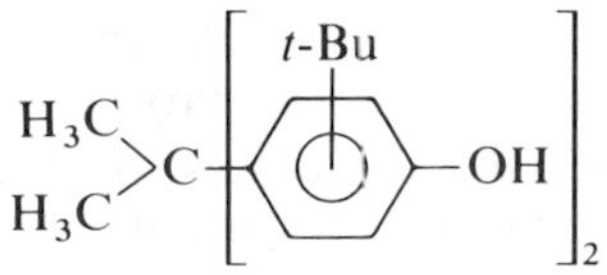

A non-staining, chain-breaking antioxidant widely used in rubbers and plastics.

BUTYLBENZYL PHTHALATE (BBP)

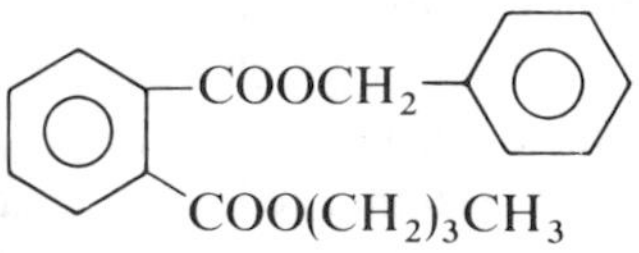

B.p. 370°C.

A plasticiser for polyvinyl chloride and its copolymers, which has good solvent action at processing temperatures and imparts good processing properties. Also useful for plasticisation of cellulose esters and ethers, polyvinyl formal and polyvinyl butyral.

BUTYL CARBITOL Tradename for diethylene glycol mono-*n*-butyl ether.

p-t-**BUTYL CATECHOL**

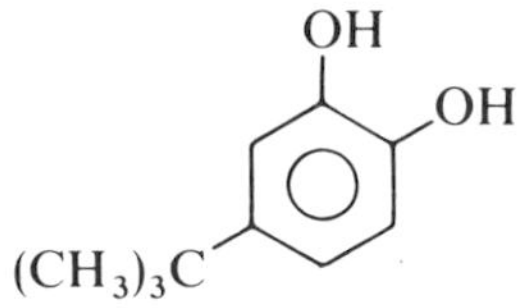

M.p. 57°C. B.p. 285°C.

A free radical polymerisation inhibitor, effective only in the presence of oxygen, when it is oxidised to a quinone. Widely used to prevent premature polymerisation during storage of monomers.

BUTYL CELLOSOLVE Alternative name for ethylene glycol mono-*n*-butyl ether.

BUTYL CELLOSOLVE STEARATE Alternative name for butoxyethyl stearate.

t-**BUTYLCHLOROFORMATE** (*t*-Butyloxycarbonyl chloride) (CH$_3$)$_3$OCOCl. A reagent useful for the introduction of *t*-butyloxycarbonyl protecting groups in peptide synthesis. Readily synthesised by reaction of *t*-butanol and phosgene.

n-**BUTYL-*n*-DECYL PHTHALATE**

B.p. 220°C/5 mm.

A plasticiser for polyvinyl chloride and its copolymers and for synthetic rubbers, being especially useful for latices, plastisols and organosols.

BUTYL DIOXITOL Tradename for diethylene glycol mono-*n*-butyl ether.

BUTYLENE GLYCOL (Butan-1,3-diol)

$$CH_3CH(OH)CH_2CH_2OH \qquad \text{B.p. } 207°C.$$

A useful solvent for some natural resins.

n-**BUTYLGLYCIDYL ETHER**

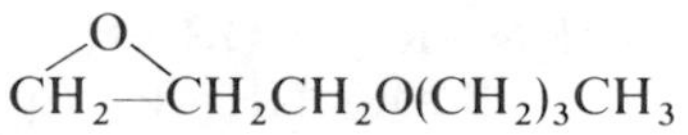

A reactive diluent for epoxy resins.

BUTYL HT Tradename for chlorinated butyl rubber.

t-**BUTYL HYDROPEROXIDE**

$$(CH_3)_3COOH$$

B.p. 34°C/20 mm, 80–90°C (decomposes).

A hydroperoxide free radical polymerisation initiator for use at >100°C, having half-life times of 170 h/100°C, 12 h/120°C and 0·44 h/150°C and half-life temperatures of 122°C/10 h, 140°C/1 h and 179°C/1 min. These values are for the usual commercial product which contains di-*t*-butyl peroxide. Useful for the crosslinking of unsaturated polyester and melamine resins at 100–150°C, but can be accelerated with cobalt for use at ∼40°C. Also used in the formation of styrenated alkyds.

4,4'-BUTYLIDENE-BIS-(6-*t*-BUTYL-*m*-CRESOL)
Tradename: Santowhite Powder.

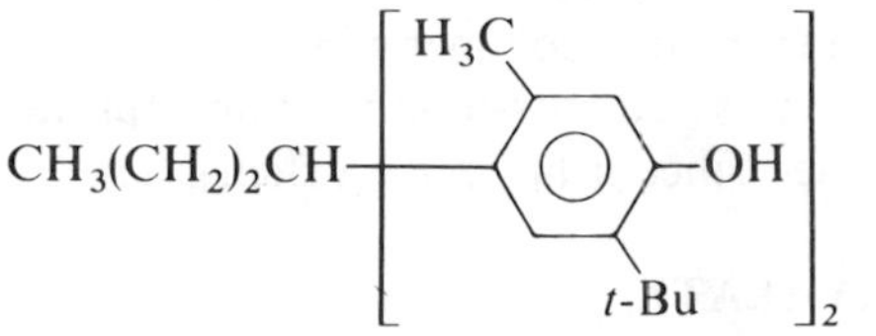

M.p. 209°C.

A non-staining, chain-breaking antioxidant widely used in rubbers and plastics.

n-BUTYLMETHACRYLATE

$$CH_2{=}\overset{\overset{\displaystyle CH_3}{|}}{C}COOC_4H_9 \quad \text{B.p. } 165{-}168°C.$$

Produced by transesterification of methyl methacrylate with *n*-butanol. The monomer for poly-*n*-butyl-methacrylate.

n-BUTYL OLEATE

$$CH_3(CH_2)_7CH{=}CH(CH_2)_7COO(CH_2)_3CH_3$$

B.p. 215°C/5 mm.

A plasticiser for cellulose esters and ethers, alkyd resins, phenol–formaldehyde resins, polyvinyl acetate, polyvinyl chloride and for cellulose acetate in particular.

BUTYL OXITOL

Tradename for ethylene glycol mono-*n*-butyl ether.

t-BUTYLOXYCARBONYL CHLORIDE

Alternative name for *t*-butylchloroformate.

t-BUTYLOXYCARBONYL GROUP

(Boc) One of the most widely used protecting groups for the amino groups of amino acids and peptides in peptide and protein synthesis. Introduced by the reaction of the amino group with *t*-butylchloroformate:

$$t\text{-BuOCOCl} + H_2N{-}R{-}COOH \rightarrow$$
$$t\text{-BuOCONH}{-}R{-}COOH + HCl$$

However, the chloroformate is of limited stability, so sometimes use of *t*-butylazidoformate (t-BuOCON$_3$) is preferred. The protecting group may be readily removed by treatment with acid, e.g. hydrochloric acid with dry ether or trifluoroacetic acid:

$$(CH_3)_3COCONH{-}R{-}COOH \rightarrow$$
$$(CH_3)_2C{=}CH_2 + CO_2 + H_2N{-}R{-}COOH$$

The products, being volatile, are easily removed. Unlike the similar benzyloxycarbonyl derivative, the protecting group is not removed by hydrogenolysis or by treatment with sodium and liquid ammonia.

t-BUTYL PERBENZOATE

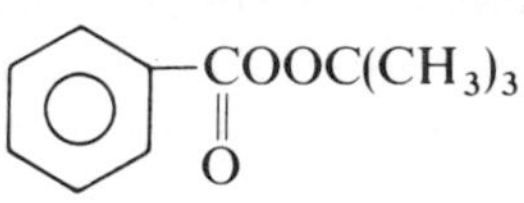

$$\text{(phenyl)}{-}\overset{\overset{\textstyle }{\|}}{C}{-}OOC(CH_3)_3$$

M.p. 8°C. B.p. 124°C.

A perester high temperature free radical polymerisation initiator for use at 80–120°C having half-life times of 17 h/100°C, 1·6 h/120°C and 0·5 h/150°C and half-life temperatures of 104°C/10 h, 121°C/1 h, and 164°C/1 min. Useful for high temperature polymerisation of styrene, vulcanisation of rubbers and crosslinking of unsaturated polyester compounds, since it has a long shelf-life.

t-BUTYL PERPIVALATE

$$(CH_3)_3C\overset{\overset{\textstyle }{\|}}{C}O{-}OC(CH_3)_3$$

A particularly active perester, having half-life times of 23 h/50°C, 1·9 h/70°C and 0·06 h/100°C and half-life temperatures of 56°C/10 h, 74°C/1 h and 113°C/1 min. Useful as a free radical polymerisation initiator for temperatures lower than is possible with benzoyl peroxide, especially for vinyl chloride and for crosslinking of unsaturated polyester resins at 70–90°C.

BUTYLPHTHALYLBUTYL GLYCOLLATE

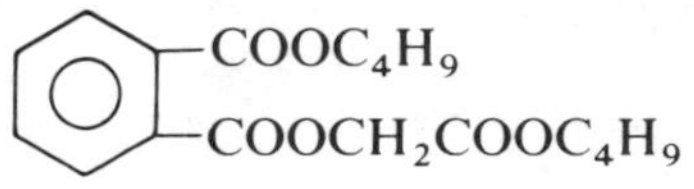

$$\text{(phenyl)}\begin{cases} {-}COOC_4H_9 \\ {-}COOCH_2COOC_4H_9 \end{cases} \quad \text{B.p. } 219°C/4\,mm.$$

A plasticiser of low volatility and good light stability, compatible with many polymers. Useful with alkyd resins, polyvinyl acetate, polyvinyl chloride, polystyrene, poly-methylmethacrylate and cellulose derivatives.

BUTYL RUBBER

(IIR) (GR-I) Tradenames: Esso Butyl, Polysar Butyl, Total Butyl. A polyisobutene copolymer containing 0·5–3% isoprene units to provide unsaturated sites for vulcanisation. Like the homo-polymer, it is produced commercially by low temperature cationic polymerisation. The presence of the 1,1-dimethyl group results in high steric hindrance and low resilience, but also in low gas permeability. Owing to its low unsaturation the polymer has good ozone and weathering resistance. It crystallises on stretching and hence shows high pure gum tensile strength which is improved by heating with filler, especially in the presence of an aromatic nitroso compound. Owing to its low un-saturation, vulcanisation is slow. Thiuram and dithio-carbamate accelerators are preferred. Alternatively vulcanisation occurs with *p*-quinonedioxime plus lead oxide which form *p*-dinitrosobenzene. Such vulcanisates have better oxidation and heat resistance but are subject to scorch. Other vulcanisation systems are low molecular weight phenolic resins and peroxides such as dicumyl peroxide. Use of the latter, however, causes much degradation as well, but this can be reduced by using a terpolymer containing some divinylbenzene units. Vulcanisation takes place via pendant vinyl groups, the vulcanisates having good ozone and heat resistance and fast cure rates. Butyl rubber may be chlorinated or brominated to chlorinated and brominated butyl rubbers respectively. These vulcanise more rapidly and show better adhesive properties.

n-BUTYL STEARATE

$$CH_3(CH_2)_{16}COO(CH_2)_3CH_3$$

B.p. 355–368°C/760 mm. M.p. 20°C.

A secondary plasticiser for cellulose derivatives, lacquers and rubbers, being most useful for its lubricating properties for processing purposes.

t-BUTYLSTYRENE

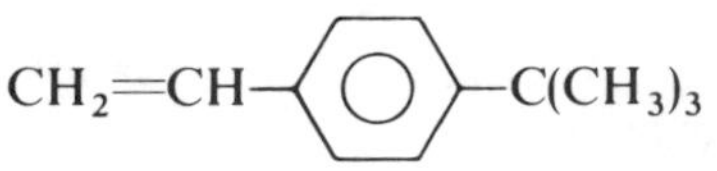

$$CH_2{=}CH{-}\text{(phenyl)}{-}C(CH_3)_3$$

B.p. 97°C/13 mm.

The commercial product contains about 5% of the *meta* isomer. A potentially useful monomer for the crosslinking of unsaturated polyesters due to its low volatility and low shrinkage on polymerisation.

γ-BUTYROLACTAM Alternative name for 2-pyrrolidone.

γ-BUTYROLACTONE

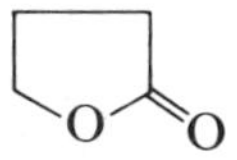

I   B.p. 204°C.

A viscosity reducing reactive diluent for epoxy resins, which reacts with amines during curing:

$$I + RNH_2 \rightarrow HO(CH_2)_3CONHR$$

producing amides which then enter into crosslinking through reaction of the hydroxyl group. A useful solvent for polyurethanes and, when hot, for polyoxymethylene.

BYION A small ion carrying the same charge sign as the macroion of a polyelectrolyte. It is introduced when an added electrolyte is present in a polyelectrolyte solution.

C

C Symbol for cysteine.

CA Abbreviation for cellulose acetate.

CAB Abbreviation for cellulose acetate butyrate.

CABANNES FACTOR The factor by which the Rayleigh ratio in light scattering is reduced due to anisotropy of the scattering particles. Given by: $(6 + 6\rho_u)/(6 - 7\rho_u)$, where ρ_u is the scattering depolarisation ratio for unpolarised light. For most synthetic polymers in solution the correction for anisotropy amounts to a few per cent at most and in any but the most accurate work may be ignored.

CABOSIL Tradename for pyrogenic silica.

CADON (1) Tradename for styrene–maleic anhydride copolymer and ABS copolymers based on styrene–maleic anhydride copolymer.
(2) Tradename for a nylon 66 fibre.

CADOXENE An aqueous solution of a complex of a cadmium salt and ethylenediamine (en), [Cd(en)₃OH₂], obtained by dissolving cadmium oxide in aqueous ethylenediamine, useful as a solvent for cellulose. It has the advantage over cuen and cuoxam that the solutions are colourless and the cellulose is only slowly degraded.

CAGE EFFECT Alternative name for geminate recombination.

CALCIUM CARBONATE CaCO₃. Occurs naturally as limestone and chalk in two crystalline forms—calcite and aragonite. One of the most widely used fillers in plastics composites and in rubbers, although it has little, or zero, reinforcing effect due to the trigonal shape of the crystals. However, many grades of material are very white, it is of low cost and the particle size is controllable. Coated forms, e.g. with a metal stearate coating, can give good melt flow properties and an attractive finish. Ground limestone, typically with particles of 0·5–30 μm is the least pure type, whilst ground chalk, usually called whiting, is purer, with up to 99% calcite, and particles mostly of about 1·5 μm. Synthetic precipitated calcium carbonates are even purer and can have exceedingly fine particle sizes, down to about 0·05 μm, consisting of mixtures of calcite and aragonite. The filler is best known as an extender for PVC compositions and GRP but is also used more widely.

CALLOSE A 1,3′-linked β-D-glucan found in the tissues and pollen of certain higher plants.

CAMPHOR

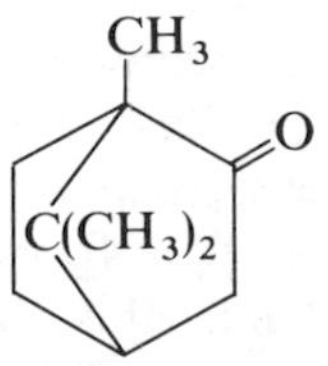

B.p. 204°C. M.p. 178°C (sublimes).

Obtained originally from the natural oil camphor oil, but now mostly manufactured from the α-pinene component of turpentine oil. An important plasticiser for cellulose nitrate, but rather volatile.

CANAL POLYMERISATION (Clathrate polymerisation) Polymerisation of a monomer locked into the large cavities in the crystals of another organic compound. Such complexes are often referred to as clathrates. Well studied examples include 2,3-dimethyl- and 2,3-dichlorobutadiene in urea and thiourea. The polymer produced, usually by high energy radiation, is highly *trans*-1,4-, since the regular organisation of the monomer in the clathrate results in a topotactic polymerisation.

CANNON FENSKE VISCOMETER A capillary viscometer of the Ostwald type in which the two arms are bent from the vertical so that the centres of the bulbs are in the same vertical axis. This arrangement eliminates errors caused by non-vertical mounting of the viscometer.

CANTRECE Tradename for a nylon 66 bicomponent fibre.

CAO Tradename for a range of antioxidants, including: CAO-1 (2,6-di-*t*-butyl-4-methylphenol), CAO-3 (2,6-di-*t*-butyl-4-methylphenol) (Food grade of CAO-1), CAO-4 and CAO-6 (2,2′-thio-bis-(6-*t*-butyl-4-methylphenol)), CAO-5 (2,2′-methylene-bis-(4-methyl-6-*t*-butylphenol)).

CAOUTCHOUC　　Alternative name for natural rubber.

CAPILLARY RHEOMETER　　One of the most popular types of rheometer for studying polymer melts in which the melt, contained in a reservoir, is forced by a piston or by pressure on the reservoir, through a capillary of known dimensions. The flow rate of the polymer is determined from the weight of polymer emerging from the capillary die in unit time. The instrument is fairly simple to use, temperature and shear rates are readily varied and, to a limited degree, elastic effects may be assessed as well as flow effects. However, the rate of shear varies across the capillary and several corrections have to be made to the data to obtain accurate apparent viscosity values. These corrections include the Bagley end correction and the Rabinowitsch correction. In addition since it is usually required to obtain the flow rate (Q) dependence on the pressure drop (ΔP) across the die, the latter must be corrected for the barrel head effect and for piston friction losses. To overcome the latter, the pressure should preferably be measured with a pressure transducer placed just above the die, rather than from the force on the piston. On the assumptions that the fluid velocity at the wall is zero, i.e. that there is no slip, and that the fluid is time independent and incompressible, the wall shear stress (τ_w) is given by $\tau_w = R\Delta P/2L$, where R and L are the radius and length of the capillary respectively and the wall shear rate ($\dot{\gamma}_w$) is given by $\dot{\gamma}_w = 4Q/\pi R^3$. However, strictly, for non-Newtonian fluids, this shear rate is only an apparent shear rate, the true wall shear rate being given by the Rabinowitsch equation. Thus a wall shear rate/shear stress curve may be drawn, i.e. a flow curve obtained, and the constants of the power law equation determined.

CAPILLARY RHEOMETRY　　Alternative name for capillary viscometry often used when the technique is applied to fluids of high viscosity, especially polymer melts, rather than dilute solutions.

CAPILLARY VISCOMETER　　(1) (Viscometer) An apparatus for the determination of the coefficient of viscosity of a fluid of relatively low viscosity (10^{-2}–1 P). Widely used in polymer work for the determination of the viscosity ratio of dilute polymer solutions in the solution viscosity method of polymer molecular weight estimation. Consists of a glass U-tube, part of one arm of which is a capillary about 10 cm long and surmounted by a bulb of about 10 cm^3 capacity. A further similar bulb is situated at a level below the capillary in the other arm.

In use, the viscometer is mounted vertically and the time is measured (the flow time) for a given volume of liquid to flow through the capillary under simple gravity flow. The flow time is the time required for the liquid level to fall between two inscribed marks, one just above and one just below the upper bulb. This simplest type is the Ostwald viscometer. Modifications of this type are also widely used, especially the Ubbelohde and Cannon Fenske viscometers. The theory of capillary viscometry is based on the Poisseuille equation.

(2) A capillary rheometer being used solely to obtain viscosity data rather than to obtain additional data on the elastic properties of a fluid.

CAPILLARY VISCOMETRY　　A technique for the determination of the coefficient of viscosity of a fluid by measurement of the rate of flow of the fluid through a capillary tube. Flow is induced by a pressure difference between the ends of the tube. This may be by either an applied pressure or by simple gravity flow. The theory of the method is based on the Poisseuille equation.

In polymer work, the dilute solution viscosity is used for solutions of up to about 1% in concentration in order to estimate polymer molecular weight. Such solutions have viscosities of 10^{-2}–1 P. The method uses a capillary viscometer in which a glass capillary about 10 cm long with gravity flow is employed. For polymer melts, of viscosity up to about 10^6 P, an extrusion rheometer is used and the technique is often referred to as capillary rheometry. A rheometer has a short (about 5 mm) capillary and the pressure difference is applied with a loaded piston.

CAPPED POLYOL　　Alternative name for tipped polyol.

CAPPING　　Alternative name for end-capping.

CAPRILACTAM　　Alternative name for caprinolactam.

CAPRINOLACTAM　　(Caprilactam)

$$(CH_2)_9\!-\!CO$$
$$\rule{0pt}{0pt}\qquad\quad\overline{}\!-\!NH \qquad \text{M.p. 128–135°C.}$$

A monomer for nylon 10. Synthesised by oxidation of decalin to cyclodecanone, followed by oximation and Beckman rearrangement, as in the synthesis of ε-caprolactam.

ε-CAPROLACTAM　　(6-Hexanolactam)

$$(CH_2)_5\!-\!CO$$
$$\rule{0pt}{0pt}\qquad\quad\overline{}\!-\!NH \qquad \text{M.p. 68–69°C.}$$

The monomer from which nylon 6 is produced. Synthesised by one of several routes, the traditional one being conversion of cyclohexane to cyclohexanol, which is dehydrogenated to cyclohexanone. This is converted to the oxime and rearranged (Beckman) to the lactam. Alternatively the oxime may be obtained by reaction of cyclohexane with nitrosyl chloride or by hydrogenation of nitrocyclohexane. Other routes include oxidation of cyclohexane with peracetic acid to caprolactone, followed by conversion to the lactam and nitration of cyclohexane to 2-nitrocyclohexane and subsequent hydrogenation and hydrolysis to ω-aminocaproic acid, which is then cyclised to the lactam. Polymerisation may be by a high temperature, water-catalysed reaction or by an anionic chain reaction at a lower temperature. The latter method may be carried out in the mass in a mould (monomer casting) and is also the basis for nylon reaction injection moulding processes, e.g. Nyrim.

ε-CAPROLACTONE

$$(CH_2)_5 — O$$
$$\quad\quad\quad\quad | \quad\quad\text{M.p. } -2°C. \text{ B.p. } 95°C/5\,mm.$$
$$\quad\quad\quad —CO$$

A cyclic lactone produced commercially by the oxidation of cyclohexanone with peracetic acid. The monomer for the production of poly-(ε-caprolactone).

CAPROLAN
Tradename for a polyurethane block copolymer based on a polyester or polyether polyol/MDI prepolymer and chain extended, and for a nylon 6 fibre.

CAPRON
Tradename for nylon 6.

η-CAPRYLLACTAM

$$(CH_2)_7 — CO$$
$$\quad\quad\quad\quad | \quad\quad\text{M.p. } 77°C.$$
$$\quad\quad\quad —NH$$

Can be obtained from acetylene or butadiene by cyclo-oligomerisation to 1,5-cyclooctatetraene, followed by reduction to cyclooctane, then oxidation to cyclo-octanone, oximation and Beckmann rearrangement to the lactam, as in the synthesis of ε-caprolactam. Although a potential monomer for nylon 8, the polymer is not produced commercially. Sometimes used as comonomer in the polymerisation of ε-caprolactam to nylon 6.

CAPSULAR POLYSACCHARIDE
An extra-cellular polysaccharide which often encapsulates a micro-organism, typically of the *Pneumococcus* type, in a jelly-like outer layer. A complex acidic polysaccharide, usually largely of 1,4'- or 1,3'-linked β-D-glucuronic acid and 1,4'-linked β-D-glucose units, often alternating. Sometimes some β-D-glucuronic acid is replaced by β-D-galactose. L-Fucose or L-rhamnose may also be present, as well as branches.

CARBITOL
Tradename for diethylene glycol monoethyl ether.

CARBITOL ACETATE
Tradename for diethylene glycol monoethyl ether acetate.

CARBOBENZYLOXY GROUP
Alternative name for benzyloxycarbonyl group.

CARBOCATIONIC POLYMERISATION
Cationic polymerisation that proceeds via a carbocation as the active centre. Such an ion is a carbenium (also called a carbonium) ion, which is a trigonal, trivalent ion of the type $\sim C^+R_2$. In current nomenclature a carbonium ion is a pentavalent ion of the type $\overset{+}{C}R_5$, which is not known to be involved in cationic polymerisation.

CARBOCHAIN POLYMER
A homochain polymer in which all the chain atoms are carbon atoms. The vast majority of homochain polymers are of this type since carbon is outstanding in its ability for self-enchainment. Vinyl polymers and polydienes are two important classes of polymers of this type.

CARBODIIMIDE
A compound of the type R—N=C=N—R', where R and R' are alkyl groups, formed by elimination of carbon dioxide from an isocyanate when it is heated with certain catalysts, e.g. phospholene oxides:

$$2\,RNCO \rightarrow R—N{=}C{=}N—R + CO_2$$

Similarly polycarbodiimides may be obtained from diisocyanates but which, when sterically hindered, only yield low molecular weight polymers. These are useful in forming polyester–urethanes with enhanced hydrolytic stability. The carbodiimide is thought to eliminate the carboxylic acid groups formed on hydrolysis, which catalyse further hydrolysis, by reacting with them and forming stable acylureas:

$$R—N{=}C{=}N—R + R'COOH \rightarrow$$
$$\quad\quad\quad\quad\quad\quad R—NH—CO—NR—COR'$$

Carbodiimides, especially N,N'-dicyclohexylcarbo-diimide, are useful in peptide synthesis, promoting the formation of peptide bonds.

CARBOHYDRATE
Originally a compound, often of simple empirical formula (CH_2O) and therefore thought to be a hydrate of carbon, i.e. $C_n(H_2O)_n$, n usually having a value of 5 or 6. Nowadays certain similar nitrogen and sulphur containing compounds, somewhat deficient in oxygen, are also considered to be carbohydrates. One of the most widespread and abundant naturally occurring groups of organic compounds. Subdivided into mono-saccharides, oligosaccharides and polysaccharides, the first two groups also being referred to as sugars. Carbohydrates serve naturally as sources of energy in sugars or as stores of energy, e.g. in starch and glycogen, or as structural supporting material in plants, e.g. cellulose, and some animals, e.g. chitin. They can also be found in bacterial cell walls, blood-group substances, connective tissues and in nucleic acids.

CARBON BLACK
A colloidal form of carbon, often referred to as being amorphous, but better described as polycrystalline, having a two-dimensional order of carbon atoms as graphitic structures lying parallel to each other and of 1–3 nm in length; bundles of layers are randomly oriented overall. It consists of more than 95% carbon (often >99%) with 0·1–4% oxygen present as surface carboxylic and phenolic hydroxyl groups and quinonoid structures, together with 0·5–0·6% hydrogen. Carbon blacks are primarily classified according to their method of manufacture, as furnace (the most common), channel, thermal, acetylene and lampblack types. Further classification schemes are confusing, the usual scheme consisting of various letters representing different qual-ities of the black, such as abrasion resistance of a filled rubber (A) as in HAF (high abrasion furnace), SAF and ISAF, structure (S) as in HAF-HS (high structure) and -LS (low structure), particle size as in FF (fine furnace),

electrical conductivity as in SCF (super conductive furnace), processability (P) as in HPC (hard processing channel), colour (also C) as in LCC (low colour channel). In general terms blacks are made by incomplete combustion and/or thermal decomposition of a hydrocarbon (oil or gas) feedstock. The particle size and size distribution are very important in determining black properties. Size may vary from 20–30 nm in channel and furnace blacks to 300 nm in thermal blacks. However, frequently the particles are fused together into clusters or chains forming the so-called primary structure. The primary structures themselves may form loose aggregates—the secondary structure. Carbon black is by far the most important filler in rubbers, having a pronounced reinforcing effect on many mechanical properties such as tensile and tear strengths, modulus and abrasion resistance. Generally the finer the particles the greater the reinforcement. Structure mostly affects processability, the higher the structure the lower the nerve and the stiffer the unvulcanised compound. Carbon black is also useful in plastics as a pigment, UV stabiliser and antioxidant.

CARBON DISULPHIDE CS_2. B.p. 46·3°C. A solvent for hydrocarbon rubbers and cellulose ethers. Swells polyethyene, polyvinyl chloride and polymethylmethacrylate. Highly inflammable due to its low auto-ignition temperature of about 125°C.

CARBON FIBRE (Graphite fibre) Tradenames: Celion, Fortafil, Grafil, Hitco, Hi-tex, Hyfil, Kureha, Magnamite, Modmor, Panex, Thornel, Torayca. A crystalline graphitic and fibrous form of carbon often with a high degree of structural perfection, thus providing fibres with exceptionally high modulus and tensile strength which are useful as reinforcing fibres in high performance composites—carbon fibre reinforced plastic (CFRP). Produced by one of several processes involving pyrolysis of an organic fibre precursor at temperatures up to about 400°C, followed by graphitisation at a much higher temperature. Both cellulose (as rayon) and pitch fibres have been used. Stretching during graphitisation considerably increases the modulus of the resultant carbon fibre. However, higher performance fibres are obtained by the pyrolysis of polyacrylonitrile (often as a copolymer) fibres of 1–3 denier in which during an initial low temperature stage (120–250°C), which is partially oxidative, the fibres are held under tension. This produces orientation of the crystallites. This is followed by a carbonisation stage at temperatures of up to 1500°C, which gives high strength fibres (type I or HS fibres). Further graphitisation by heat treatment of up to 2800°C produces high modulus fibres (type II or HM fibres) at some sacrifice to tensile strength. Fibres are often produced from yarns or tows of about 10 000 filaments. Typically, HS fibres have a tensile strength of 3·2 GPa, a modulus of 230 GPa and a density of 1·76 g cm^{-3}, whilst HM fibres have a tensile strength of 2·4 GPa, a modulus of 330 GPa and a density of 1·87 g cm^{-3}. The fibres are widely used in high performance composites, especially with epoxy resins as the matrix. Adhesion between fibre and matrix is improved by chemical treatment (oxidation) of the fibres. This considerably improves the otherwise low interlaminar shear strength. The fibre is often used as an epoxy resin prepreg to improve fibre alignment in the composite. Short carbon fibres are also used for the reinforcement of engineering thermoplastics.

CARBON FIBRE REINFORCED PLASTIC (CFRP) A composite material in which a plastic matrix is reinforced by carbon fibres. The main advantage of such reinforcement is the combination of high stiffness with reasonable strength possible in the composite, deriving from the high stiffness of the fibres. Early uses of carbon fibres were in thermoset polymers with long fibres, particularly for aerospace applications since the composites, mostly with epoxy resins, have specific stiffnesses and strengths that can exceed even the most sophisticated titanium based alloys. More recently short fibres have been used with thermoplastic engineering polymer matrices, such as nylons, especially for sporting goods products. Advanced plastics composites are now sometimes based on long carbon fibre reinforced thermoplastics matrices, such as with polyetherether ketone, and can have exceptional combinations of stiffness, strength and high temperature resistance.

CARBON-13 NMR SPECTROSCOPY Nuclear magnetic resonance spectroscopy in which the resonating nuclei are ^{13}C atoms. This increasingly used technique has the advantage over conventional proton magnetic resonance spectroscopy that ^{13}C nuclei have a much larger range of chemical shifts than protons; about 300 ppm compared with only about 10 ppm for protons. This gives resonant frequency lines chemically shifted about seven times more than for protons. Hence the technique is much more discriminating. However, it is much less sensitive both due to the lower sensitivity of the ^{13}C nucleus itself and to the low natural abundance (only about 1%) of the ^{13}C atoms, the remaining 99% being ^{12}C atoms. Nevertheless in the more recently developed spectrometers, sensitivity has been considerably enhanced by ^{13}C spin–spin decoupling, often also involving the nuclear Overhauser effect, and also by operation as Fourier transform NMR with spectrum accumulation. The technique is particularly valuable for the determination of long chain branching, especially in polyethylene and other polyolefins, for the analysis of polymer stereochemistry and for the study of individual ^{13}C nuclei molecular motions, which is possible due to their rarity, and hence isolation, in both solution and the solid states.

CARBON TETRACHLORIDE CCl_4. B.p. 76·6°C. A highly toxic solvent of low flammability. Tends to form hydrochloric acid on contact with water. A solvent for some cellulose esters, polystyrene, hydrocarbon rubbers and natural resins.

CARBONYL POLYMERISATION Polymerisation of a carbonyl containing monomer in which chain polymerisation occurs by opening of the carbonyl $>C=O$ double bond. However, many such compounds do not readily polymerise due to their low ceiling temperature. An exception is formaldehyde which does readily polymerise by anionic polymerisation to polyoxymethylene. Other aldehydes will similarly polymerise but only at low temperatures and with strong bases, such as alkali metal alkyls and alkoxides. Ketones are generally unreactive. Reactivity is increased by the presence of electron-withdrawing substituents, e.g. trichloroacetaldehyde, thiocarbonyl fluoride and hexafluorothioacetone. In general, free radical mechanisms do not operate because polarisation in a carbonyl group inhibits attack by a free radical and also because free radicals are difficult to generate at the very low temperatures required.

CARBORANE POLYMER (Polycarborane) A polymer containing carborane ring structures in the polymer chain. Carborane is a multi-ring cage-like structure of carbon and boron atoms. The most common types consist of the $C_2B_{10}H_{12}$ system (decacarborane), the atoms forming an icosahedron, and the pentagonal bipyramidal $C_2B_5H_2$ system. The two hydrogens joined to the carbons are acidic and polymers may be made by replacing the hydrogens with lithium, to form

$$Li—C(B_{10}H_{10})C—Li \ (I)$$

or other functional groups and linking the carboranes together by further reaction with an appropriate difunctional monomer. The decacarborane with carbons on non-adjacent *meta* positions has been the most studied:

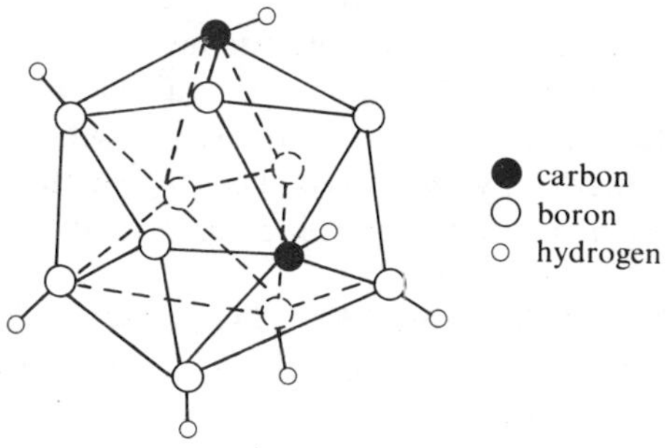

● carbon
○ boron
○ hydrogen

In particular, polycarborane–siloxane polymers, obtained by reaction of **I** with dichlorosilanes have been widely investigated as useful high temperature resistant polymers.

In a similar way aromatic fluoro-linked *m*-carborane polymers,

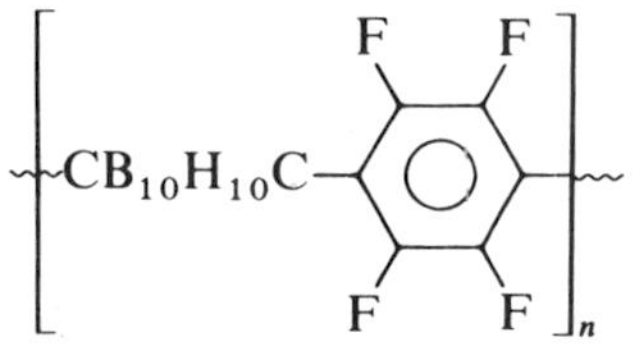

and amide-linked polymers have been prepared as well as metal-linked polymers, e.g. $\{-CB_{10}H_{10}C—Sn(CH_3)_2-\}_n$. These polymers show better thermal stability than most organic polymers, e.g. to 500°C, but oxidative stability is limited to about 300°C due to the hydrogen atoms attached to boron.

CARBOWAX Tradename for low molecular weight polyethylene oxide, which is useful as a lubricant and for cosmetics and as a pharmaceutical base.

N-CARBOXY-α-AMINO ACID ANHYDRIDE
(*N*-carboxy-anhydride)
(1,3-Oxazolidine-2,5-dione)
(Leuchs anhydride) (NCA)

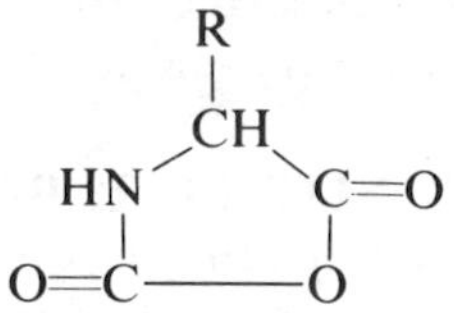

I

An unsymmetrical acid anhydride which can be considered to be derived by dehydration of an *N*-carboxy-α-amino acid. Polymerisation of NCAs is the most important method of synthesis of poly-α-amino acids and therefore has been intensively investigated. NCAs are best prepared by treatment of the amino acid with phosgene in an inert solvent, e.g. dioxane or toluene:

$$H_2NCHRCOOH + COCl_2 \rightarrow ClOCNHCRCOOH$$
$$\rightarrow I + HCl$$

Any side group function group (NH_2 or $COOH$) must be blocked. In these cases preparation via formation of the benzyloxycarbonyl derivative by reaction of the amino acid with benzylchloroformate may be preferred. NCAs are highly reactive and readily polymerised to poly-α-amino acids. Thermal polymerisation in bulk yields only low molecular weight polymers, whereas solution polymerisation in inert solvents, e.g. dioxane, dimethylformamide, chloroform or even water, readily yield high molecular weight polymers when initiated by base (B). Primary and secondary amines produce lower molecular weight polymers than tertiary amines or strong bases such as sodium methoxide. The kinetics and mechanisms are complex, a simplified mechanism is:

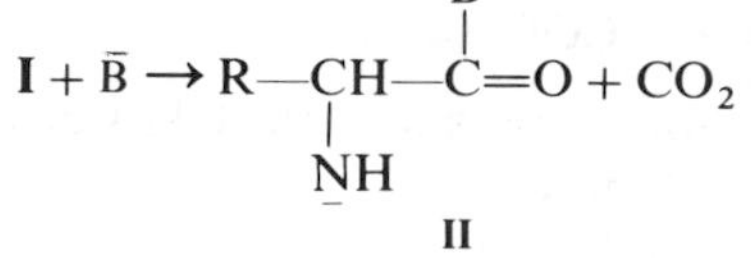

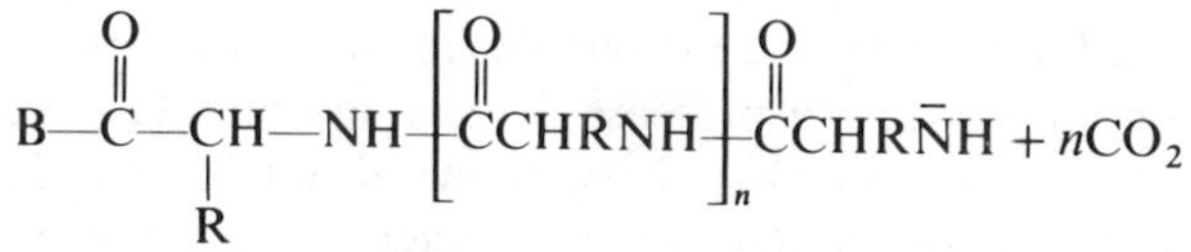

However, the molecular weight distribution is often very broad and the reaction autocatalytic. Homopolypeptides (e.g. polyalanine, polyglycine, polylycine, polyproline) of all the twenty standard α-amino acids, many random and block (but no sequential) copolymers, other polyamino acids (such as polysarcosine) and derived polymers (such as poly-γ-benzyl-L-glutamate) have all been prepared in high molecular weight form.

N-CARBOXY-ANHYDRIDE Alternative name for *N*-carboxy-α-amino acid anhydride.

CARBOXYLATED ELASTOMER An elastomer containing a proportion of a carboxyl comonomer, usually acrylic or methacrylic acid. Most commercial products are nitrile or SBR based polymers. The polymers may be vulcanised with polyvalent metal oxides, e.g. zinc or magnesium oxide, to give vulcanisates of higher strength than reinforced sulphur vulcanised non-carboxylated rubbers. Commercially the elastomers are mostly used in latex form as non-woven textile and paper binders and as adhesives, especially for porous substrates.

CARBOXYLATED NITRILE RUBBER Nitrile rubber containing a few per cent of acrylic or methacrylic acid units which may be crosslinked by metal oxides. The vulcanisates have better strength properties than ordinary nitrile rubber.

CARBOXYMETHYLCELLULOSE (CMC) (CM cellulose) Cellulose in which some hydroxyls have been replaced by —OCH_2COOH groups, which are negatively charged at neutral pH. Useful as cation exchange resins in ion-exchange chromatography, especially of proteins.

S-CARBOXYMETHYLCYSTEINEKERATEINE (SCMK) A kerateine in which the three thiol groups have been protected from oxidation by alkylation by reaction with iodoacetic acid. Often two fractions are obtained, one SCMKA of molecular weight 62–70 000 with low proline content and with about 50% α-helix structure and the other, SCMKB, of molecular weight about 25 000 being high in proline but with no α-helix structure.

CARBOXY-MODIFIED NITRILE RUBBER Alternative name for carboxy-terminated nitrile rubber.

CARBOXYNITROSORUBBER (CNR) Tradename PCR. A nitrosorubber which contains about 1% 4-nitroperfluorobutyric acid units

$$\sim\!\!+\!NO\!\!-\!\!\!\!\underset{(CF_2)_3COOH}{\overset{}{\big\downarrow}}\!\!\!\!\!\!\!\Big)_n$$

as crosslinking sites for vulcanisation using metal oxides or, better, chromium trifluoroacetate.

CARBOXYPEPTIDASE A An enzyme usually isolated from the pancreas, but also found in the kidney and spleen, which is a protease. It is an exopeptidase, being capable of removing the C-terminal amino acid fairly indiscriminately, but with a preference for non-polar terminal residues. Thus, it will not remove lysine, arginine or proline. It is therefore very useful for studying the C-terminal acids of a protein polypeptide, for which it can yield partial sequences by studying the rate at which the terminal amino acids are produced. It contains 307 amino acid residues and a zinc atom, which is essential for its enzyme activity. The active site is a shallow depression on the surface near which is a largely hydrophobic pocket which binds a non-polar side group of the substrate. The residues Arg 145, Tyr 248 and Glu 270 are involved at the active site in the catalysis, as is the zinc atom. This atom is tetrahedrally coordinated to His 69 and 196, the oxygen of Glu 72 and the carbonyl oxygen of the substrate peptide bond to be hydrolysed. Carboxypeptidase A is produced biologically initially as its zymogen, procoboxypeptidase A, by the action of trypsin.

CARBOXYPEPTIDASE B An intestinal enzyme produced in a similar manner to carboxypeptidase A, but showing a preference for removing C-terminal amino acids with basic side groups.

CARBOXYPOLYBUTADIENE A copolymer of butadiene with a few per cent of acrylic or methacrylic acid units. The acid units improve adhesion of the polymer to tyre cords when used in tyre construction.

CARBOXY-TERMINATED NITRILE RUBBER (CTBN) (Carboxy-modified nitrile rubber) Tradenames: Hycar CTBN, Butarez CTL. A low molecular weight (about 5000) nitrile rubber containing carboxyl end groups, which may be reacted with appropriate functional groups to chain extend or crosslink the polymer. The functional group may be attached to another polymer, especially an epoxy resin, when incorporation of CTBN gives a pronounced toughening effect of the resin.

CARBOXY-TERMINATED POLYBUTADIENE Tradenames: Butarez CTL, Hycar CTB, Nisso PBC. A low molecular weight telechelic polybutadiene with carboxyl end groups. Produced by use of a difunctional metal alkyl catalyst, by anionic polymerisation to a living polymer, whose living ends are reacted with carbon dioxide to form the carboxy end groups. Useful as a liquid rubber which may be crosslinked by reaction with a polyepoxy or polyethyleneimine compound.

CARBOXY-TERMINATED POLYISOBUTYLENE Prepared by the ozonolysis, in the presence of pyridine, of a high molecular weight copolymer of isobutylene containing 2–4 mol% of piperylene units:

$$\sim\!\!+\!\![CH_2C(CH_3)]_n CH_2CH\!\!=\!\!CHCH(CH_3)\!\!+\!\!\!\sim\!\xrightarrow{O_3}$$

$$HOOCCH(CH_3)[CH_2C(CH_3)_2]_nCH_2(COOH)$$

Liquid polymers of molecular weight 2000–4000 are produced and may be crosslinked with epoxides or aziridine.

CARDANOL

where $R = \!\!+\!\!(CH_2)_7CH\!\!=\!\!CHR'$
with

$$R' = CH_2CH\!\!=\!\!CH(CH_2)_2CH_3$$

or $+\!\!+\!\!CH_2)_5CH_3$. The mixed monohydric phenolic component of cashew nut shell liquid. It reacts very readily with formaldehyde forming phenol–formaldehyde polymers with good chemical and solvent resistance.

CARDOL

where $R = +\!\!+\!\!CH_2)_7CH\!\!=\!\!CHCH_2CH\!\!=\!\!CH(CH_2)_2CH_3$. The dihydric phenolic component of cashew nut shell liquid. It reacts readily with formaldehyde forming phenol–formaldehyde polymers with good chemical and solvent resistance.

CARIFLEX BR Tradename for high *cis*-poly-butadiene.

CARIFLEX IR Tradename for *cis*-1,4-polyisoprene.

CARIFLEX S Tradename for emulsion styrene–butadiene rubber.

CARIFLEX TR Tradename for styrene–buta-diene–styrene block copolymer.

CARIFLEX TR1107 Tradename for styrene–isoprene–styrene block copolymer.

CARINA Tradename for polyvinyl chloride.

CARINEX Tradename for polystyrene.

CARLONA Tradename for low density and high density polyethylene.

CARLONA P Tradename for polypropylene.

CAROTHERS EQUATION An equation relating the fractional conversion (p) in a step-growth polymerisation to the number average degree of polymerisation of the polymer ($\bar{x}_n$). In a simple linear polymerisation with equal moles of A and B reactive groups, the equation is: $\bar{x}_n = 1/(1-p)$. Thus formation of high molecular weight polymer requires a very high conversion, since, for example, to form a polymer with $\bar{x}_n$ of only 100 requires p to be 0·99, i.e. 99% conversion. When non-stoichiometric amounts of A and B groups are used, with, say, B in excess, the equation is: $\bar{x}_n = (1+r)/(1-r-2rp)$, where r is the ratio of moles of A to moles of B groups. For non-linear polymerisation, with equivalent amounts of A and B groups, the equation is: $\bar{x}_n = 2/(2-pf_{av})$ where f_{av} is the average functionality of all the monomers involved. In this case if the gel point is assumed to be reached when $\bar{x}_n \to \infty$, then the critical conversion for gelation (p_c) is given by $p_c = 2/f_{av}$. However, experimentally observed gel points occur at lower conversions than the predicted p_c.

CARRAGEENAN A sulphated galactan found as a cell-wall polysaccharide in several seaweeds, especially *Chondrus crispus* (of which it may account for 80% of the dry weight), and from which it is extracted with boiling water. It consists of a closely related mixture of sulphated polymers of alternating 1,3′- and 1,4′-linked D-galacto-pyranose units. The components differ in the degree and sites of sulphation and in the extent to which the units are present as 3,6-anhydride. κ-carrageenan is precipitated by potassium ions leaving λ-carrageenan in solution. μ-carrageenan is regarded as a contaminant of λ-carrageenan. It shows a gelling tendency, but as the degree of sulphation increases from κ to μ to λ, so this tendency decreases. Widely used in pharmaceutical and food preparations, as a suspension and emulsion stabiliser and binding agent.

i-CARRAGEENAN Alternative name for μ-carrageenan.

κ-CARRAGEENAN A component of carrageenan precipitated from its solution by addition of potassium ions. It consists of linear chains of alternating 1,3′-linked β-D-galacto-pyranose-4-sulphate and 1,4′-linked 3,6-anhydro-α-D-galactopyranose units.

λ-CARRAGEENAN A component of carrageenan, isolated from the mother liquor after precipitation of κ-carrageenan. It consists largely of linear chains of alternating 1,3′-linked β-D-galactopyranose-2-sulphate and 1,4′-linked α-D-galactopyranose-2,6-disulphate units, although variation in the patterns of sulphation are found.

μ-CARRAGEENAN (*i*-Carrageenan) A contaminant of λ-carrageenan from which it has not been isolated, but whose presence is inferred from methylation studies and conversion to 3,6-anhydro-galactose. It consists of linear chains of alternating 1,3′-linked β-D-galactopyranose-4-sulphate and 1,4′-linked α-D-galactopyranose-6-sulphate units with variable amounts of 3,6-anhydro groups.

CASEIN A phosphoprotein obtained from skim milk, which is about 80% of the total protein content of the milk. The casein occurs as a colloidal calcium phosphate complex in micelles. Solid casein is obtained by coagulation, either by acidifying to pH 4·6 (the isoelectric point) giving acid casein, or by treatment with a small quantity of the enzyme rennin, when the product is sometimes called paracasein.

Electrophoresis at pH 8·6 separates whole casein into three major fractions. In order of increasing mobility these are: α-casein (70%), β-casein (25%) and γ-casein (5%). Similar fractions can also be obtained by careful precipitation of dilute solutions. α-Casein consists of a component sensitive to precipitation by calcium (α_s-casein, 55%) and a component that is insensitive (κ-casein). κ-Casein is responsible for the stabilisation of casein micelles and is the component destroyed by the action of rennin when it coagulates casein. Each of these fractions is probably not homogeneous and varies in composition. The phosphorus contents (as phosphate) are approximately: α, 1%; β, 0·6%; κ, 0·03%; γ, 0·1%. The phosphorus is bound largely to serine residues. Overall, casein is rich in glutamine, proline, leucine, aline and lysine. In addition, a polysaccharide component is present in κ-casein. The polypeptide molecular weights are about 30 000.

Acid casein is useful in adhesives and has been used for the production of casein fibre (tradenames: Aralac, Fibrolan and Lanital) by regeneration by wet spinning from an alkaline solution. The spun fibres are hardened by treatment with formaldehyde. They form soft, warm textiles but have rather poor wet strength and are relatively weak. Rennet casein is useful for the production of casein plastics and contains about 5% calcium. The polymer is formalised to harden it, to reduce its water sensitivity and to reduce its tendency to putrify. Formalisation involves reaction with formalin, which crosslinks the polypeptide chains, probably via reaction with both amine groups and lysine residues. Many casein plastics have been produced (tradenames: Galalith, Erinoid, Ameroid) being important in the early days of the plastics industry, but they are now little used.

α-CASEIN The fastest moving component of casein on electrophoresis, it may also be separated from whole casein by its solubility behaviour in calcium and urea solutions using column chromatography. It comprises about 70% of whole casein. It can be separated into two further fractions, α_s-casein (about 55%) which is sensitive to calcium precipitation and κ-casein (about 15%) which is calcium soluble. It has a molecular weight of about 28 000 and a phosphorus content of about 1%.

β-CASEIN A component of casein, which follows α-casein on electrophoresis. It may be separated from whole casein by its insolubility in 1·7M urea, and solubility in 3·3M urea. It comprises about 25% of whole casein and has a phosphorus content of about 0·6% and a molecular weight of about 25 000.

γ-CASEIN A minor component of casein (about 5%), being the slowest moving electrophoretically. It has a very low phosphorus content, about 0·1%, and a molecular weight of about 25 000.

κ-CASEIN The main glycoprotein of milk, isolated from α-casein. A component of casein (about 15%) which may be isolated by its solubility in calcium salt solution. It is responsible for the stabilisation of whole casein micelles in milk and is the component destroyed by the action of rennin during the rennin coagulation of casein. It contains about 5% of the bound polysaccharide component. Bovine κ-casein carbohydrate contains D-galactose, 2-amino-2-deoxy-D-galactose and sialic acid units and has a molecular weight of about 28 000. Human κ-casein has, in addition, L-fucose and 2-amino-2-deoxy-D-glucose. In both, the carbohydrate is present as more than a single group, joined glycosidically to the protein.

CASHEW NUT SHELL LIQUID (CNSL) A mixture of mono- and dihydric phenols containing cardanol and cardol, which have long alkyl unsaturated chain substituents. These monomers react readily with formaldehyde forming phenol–formaldehyde polymers which crosslink by normal phenol–formaldehyde curing reactions as well as by chain polymerisation of the side groups. The products have very high resistance to attack by solvents and to chemical attack.

CASTOMER Tradename for a cast polyurethane elastomer, based on a polycaprolactone polyester/tolylene diisocyanate prepolymer and cured with 3,3'-dichloro-4,4'-diaminodiphenylmethane.

CASTOR OIL Obtained from the seeds of *Ricinus communis* (castor beans), it consists mostly of the glyceryl esters of ricinoleic acid, i.e. it is about 90% glyceryl ricinoleate

$$CH_2OR$$
$$|$$
$$CHOR$$
$$|$$
$$CH_2OR$$

where R is

$$—OC(CH_2)_7CH{=}CHCH_2CH(OH)(CH_2)_5CH_3$$

but with some glycerol hydroxyls esterified with palmitic and stearic acids (1–2·5%), oleic acid (0·1–7%) and linoleic acid (3–5%). Useful as a source of ricinoleic acid. Occasionally used as a plasticiser extender in cellulose derivatives. It is sometimes used as a non-drying oil in alkyd resins and its drying power may be increased by dehydration.

CAST POLYURETHANE ELASTOMER (Reaction moulded polyurethane elastomer) A polyurethane elastomer formed by chain extension of a liquid prepolymer (made from a polyester or polyether polyol and a diisocyanate) with a diol or diamine. The ingredients are batch or continuously mixed at 80–100°C, poured into moulds and after a short time the mouldings are removed and cured in an oven. The products have good wear properties, high strength and excellent oil and solvent resistance. Typical products include printing and wear resistant rollers, solid tyres and cold cure applications, as in cable jointing and potting and tyre filling. Most products are segmented polyurethanes which separate into hard and soft segment domains, the former acting as physical crosslinks. However, some chemical, i.e. covalent bond, crosslinks are also often present through reaction of excess diisocyanate to form allophanate and biuret links. A few products are one phase materials and only contain chemical crosslinks. Many commercial products have been developed. The earliest was Vulkollan, which required the use of a non-storage stable isocyanate prepolymer, but stable prepolymer systems have also been developed. The prepolymer process is the most frequently used, in which a polyester or polyether prepolymer is chain extended with a diisocyanate to yield a prepolymer, which is subsequently further extended and crosslinked by heating at about 100°C for typically 10–20 h with a diamine, usually 3,3'-dichloro-4,4'-diaminodiphenyl-methane, or a diol, often 1,4-butanediol. Examples include Baygal, Baymidur, Castomer, Formrez, Dorlastan, Duthane, Adiprene L, Cyanoprene, Multrathane, Prescollan, Solithane and Vibrathane. A one-shot process is also sometimes used in which the polyol is mixed with the chain extender and the diisocyanate is added later, catalysts also being needed to achieve the right balance of reaction rates.

CATALAC (1) Tradename for a phenol–formaldehyde polymer suitable for casting.
(2) Tradename for polyvinyl acetate.

CATALIN Tradename for a phenol–formaldehyde polymer in cast form.

CATARAX Tradename for coumarone–indene resin.

CATENANE POLYMER (Polycatenane) (Catenated ladder polymer) A polymer consisting of a series of rings joined linearly with each ring being threaded through its neighbour. Schematically represented by:

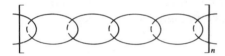

CATENATED LADDER POLYMER Alternative name for catenated polymer.

CATENATION The formation of chains of atoms. Use of the term is usually restricted to the formation of covalently linked homoatomic chains. The only element capable of this to any considerable extent is carbon, thus providing the basis for the extremely large number of structures found in organic compounds. To a much lesser extent sulphur is capable of catenation, as in polymeric sulphur, as are silicon (e.g. in the highly unstable $\{SiH_2\}_n$) and germanium. Boron and phosphorus are also capable of homoatomic bonding, but only to a very limited extent.

CATIONIC-COORDINATION POLYMERISATION Coordination polymerisation in which the active centre is a cation, usually a carbenium ion. Most studied are vinylalkyl ether polymerisations using catalysts such as $SnCl_4/Cl_3CCOOH$, $AlEtCl_2$ and boron trifluoride etherate. Often isotactic polymers are produced. Polymerisation of isobutylvinyl ether by boron trifluoride etherate at $-70°C$ was the first reported case of a stereospecific polymerisation by a coordination catalyst.

CATIONIC GRAFTING Formation of a graft copolymer by production of cationic sites on a polymer chain, capable of polymerising the monomer for a second polymer. The method works particularly well with polymers containing pendant chlorine atoms. In the

presence of a Friedel–Crafts catalyst these atoms can produce carbenium ions as initiation sites for the monomer. An example is the grafting of styrene onto polyvinylchloride:

$$\sim CH_2CH\sim + AlCl_3 \rightarrow$$

with Cl substituent, leading to

$$\sim CH_2\overset{+}{C}H\sim \quad \xrightarrow{\text{styrene(S)}} \quad \sim CH_2CH\sim$$
$$AlCl_4^-$$

CATIONIC POLYMERISATION Ionic chain polymerisation in which the active centre is a cation. Like other chain polymerisations the active centre is usually located on a carbon atom. In other words it is a carbocationic polymerisation with the active centre of the type (for a vinyl polymerisation) $\sim C^+HR$, i.e. it is a carbenium ion (previously called a carbonium ion, a term which now refers to a 'non-classical' pentavalent carbocation of the type $\sim C^+R_4$). However, many cationic polymerisations of non-vinyl monomers proceed via heteroatom cationic active centres, e.g. oxonium ions $\sim O^+R_2$, sulphonium ions $\sim S^+R_2$ or ammonium ions $\sim N^+R_3$. Electrical neutrality is maintained by the presence of a counter-ion, often associated with the active centre, e.g. as a tight or solvent separated ion pair. Alternatively the ion pair may be completely dissociated to free ions. The closeness of the association is highly dependent on the solvent polarity, which therefore influences the rate and the stereospecificity of the polymerisation since these also depend on the binding of the counter-ion.

Cationic polymerisation has not proved as useful a mechanism as free radical or anionic polymerisation for the formation of high molecular weight polymers, largely because chain breaking transfer reactions often interfere as do rearrangements of the active centre carbocations. Nevertheless many different monomers are susceptible to cationic polymerisation. These include vinyl monomers with electron-releasing substituents, notably alkyl groups (of which isobutene is the outstanding example) and alkoxy groups, i.e. the vinyl ethers (as in methylvinyl ether). Many cyclic monomers, especially cyclic ethers, are susceptible to ring-opening cationic polymerisation. They include ethylene and propylene oxides, ethylenimine, tri- and tetramethylene oxide, 1,3-dioxalone and trioxane.

Many different initiators, usually referred to as catalysts, are used. They frequently require the presence of a small quantity of a co-initiator (or cocatalyst). Frequently the initiation reaction involves a pre-initiation equilibrium in which the actual charged initiating species is formed. This may be either by a bimolecular reaction, for example between a Lewis acid and water, e.g.

$$BF_3 + H_2O \rightarrow H^+\bar{B}F_3OH$$

or by dissociation of an initiator molecule, e.g.

$$I_2 \rightarrow I^- + I^+$$

Such pre-equilibria considerably complicate the study of cationic polymerisation, especially since the small quantity of water present as an impurity is frequently involved as the cocatalyst. Initiators may be Lewis acids, such as BF_3, $AlCl_3$, AlR_3, $SnCl_4$, $TiCl_4$ and $SbCl_5$, often acting in conjunction with a co-initiator, protonic acids such as H_2SO_4, $HClO_4$ and CF_3SO_3H, electron acceptors such as tetracyanoethylene and maleic anhydride (which act by charge-transfer initiation) and stable carbocationic salts such as $Ph_3C^+\bar{S}bF_6$. The latter have proved particularly useful in fundamental studies of propagation since they do not involve the complications of pre-initiation steps. Additionally, electrochemical and radiation-induced initiation may be used.

Propagation can be very rapid, occurring at different rates with free ions and with ion pair active centres. Carbocations are prone to isomerise (thus leading to different repeat unit structures from those expected from the monomer used (isomerisation polymerisation)) either by bond or electron rearrangements, e.g. as in trans-annular polymerisation and in the formation of terpene resins, or, more commonly, by hydride ion shifts, e.g. as in the polymerisation of 3-methyl-1-butene, where 1,3-addition occurs.

Polymer molecular weights are frequently very low due to the important chain transfer reactions that occur. These include transfer of a proton to monomer:

$$\sim CH_2\overset{+}{C}HX\bar{Y} + CH_2{=}CHX \rightarrow$$

$$\sim CH{=}CHX + CH_3\overset{+}{C}HX\bar{Y}$$

or transfer to catalyst or cocatalyst (especially water):

$$\sim CH_2\overset{+}{C}HX\bar{Y} + H_2O \rightarrow \sim CH_2CHXOH + \overset{+}{H}\bar{Y}$$

Generally transfer has a higher activation energy than propagation, so that polymerisation at low temperatures, e.g. $-100°C$, is needed to form high molecular weight polymer. Unusually, cationic polymerisation can sometimes occur with a negative activation energy so that the rate decreases and polymer molecular weight increases as polymerisation temperature is increased.

CAUCHY STRAIN Alternative name for extension.

CAUCHY STRESS EQUATION An alternative name for the momentum equations in Eulerian form, which may be written using the summation convention as:

$$\rho(\partial u_i/\partial t + u_j \partial u_i/\partial x_j) = \partial \sigma_{ij}/\partial x_j + \rho g \qquad i = 1,2,3$$

where u_i and u_j are the velocity components, σ_{ij} are the total stress components, ρg is the body force (due to gravity) and ρ is the density of the fluid.

CBS Abbreviation for N-cyclohexylbenzothiazylsulphenamide.

CD Abbreviation for circular dichroism.

CED Abbreviation for cohesive energy density.

CEILING TEMPERATURE Symbol T_c. The temperature at which polymerising monomer of some specified concentration (M_c), often taken as $1\,mol\,dm^{-3}$, is in equilibrium with its polymer. Also often used to mean the temperature above which monomer cannot be successfully polymerised to high polymer by a chain-growth mechanism. Polymerisations are only successful up to a temperature of about $20°C$ below that at which $M_c = 1\,mol\,dm^{-3}$. Existence of a ceiling temperature is due to the tendency of many chain-growth polymers to depolymerise at elevated temperatures.

CELAFIBRE Tradename for a cellulose acetate staple fibre.

CELANESE Tradename for cellulose acetate rayon.

CELANEX Tradename for polybutylene terephthalate.

CELCON Tradename for polyoxymethylene co-polymer.

CELION Tradename for a polyacrylonitrile based carbon fibre.

CELIOX Tradename for a heat resistant fibre obtained by heating polyacrylonitrile, or acrylonitrile copolymer, to $200–300°C$ in air, which produces an oxidised, cyclised polymer, containing structures like:

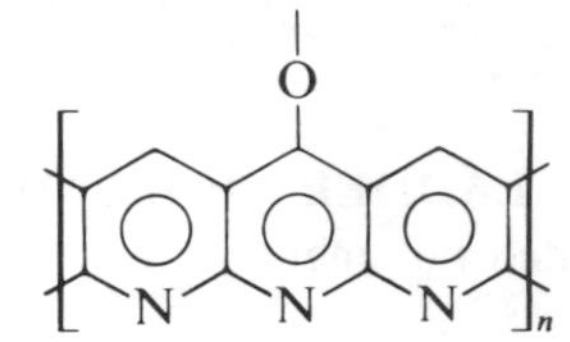

This is the first stage towards the formation of carbon fibres. The fibres are useful as an asbestos replacement in protective clothing.

CELITE Tradename for kieselguhr.

CELLIDOR A Tradename for cellulose acetate.

CELLIDOR B Tradename for cellulose acetate butyrate.

CELLIDOR S Tradename for cellulose acetate.

CELL NUCLEATING AGENT An additive used in the formation of cellular polymers to control cell size. Examples are citric acid and sodium bicarbonate.

CELLOBIOSE (β-D-glucopyranosyl-1,4′-D-glucose)

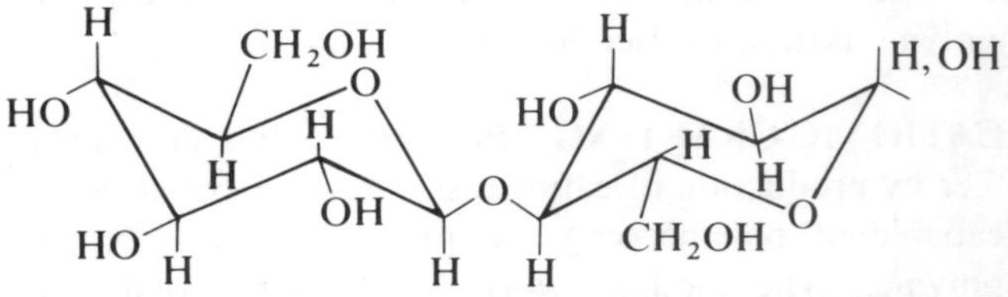

M.p. $225°C$ (decomposes). $\alpha_D^{20}\ 34{\cdot}6°$.

Isolated in low yields from the partial hydrolysis of cellulose. The octa-acetate may be isolated in about 50% yield by hydrolysis in a mixture of acetic anhydride and sulphuric acid, in which medium it is insoluble. Determination of the structure, together with those of other cellulose oligomers, e.g. cellotriose, confirmed that the cellulose glycoside links are of the β-1,4′-type, since cellobiose is 4-O-β-D-glucopyranosyl-D-glucopyranose.

CELLOBOND Tradename for phenol–formaldehyde, urea–formaldehyde, melamine–formaldehyde polymers used as adhesives, laminating and foundry resins and for unsaturated polyester resins.

CELLOPHANE A regenerated cellulose film, produced by extrusion of viscose through a slit die into a coagulating bath. Usually plasticised with ethylene glycol or glycerol and often coated, e.g. with polyvinylidene chloride or cellulose nitrate, to reduce moisture permeability.

CELLOSOLVE Alternative name for ethylene glycol monoethyl ether.

CELLOTRIOSE

Isolated in low yields from the partial hydrolysis of cellulose. Like cellobiose it is more readily isolated as its acetate and determination of its structure confirmed the β-1,4′-linked glucan structure of cellulose.

CELL STABILISER An additive used in flexible and rigid polyurethane foam formulations to stabilise the rising foam and assist in the dispersion of ingredients. Frequently a silicone–polyether block copolymer is used.

CELLULAR POLYMER A polymer 'composite' material which consists of a polymer containing occluded gas in bubbles or cells. The cells may be open and interconnecting with the polymer formed into a series of struts or they may be closed cells. In the former case the gas will be air, whereas in the latter case the gas is totally enclosed and so may be other than air. Cellular polymers can have very low densities, down to below $0.1\,\mathrm{g\,cm^{-3}}$, due to the high gas volume fraction, and hence can have exceptionally low thermal conductivities. They are therefore often used for thermal insulation purposes.

Structural foams have a solid outer skin and can be more efficient in material utilisation to achieve a specified stiffness in a moulded product, compared with a solid product. Cellular polymers are usually produced by introducing a gas into the polymer melt using a blowing agent, followed by solidification. The product is then often referred to as an expanded polymer. Alternatively, the cellular product may be produced from the polymer in a low viscosity form, e.g. as a latex, by mechanical whipping in the gas or by producing gas by a chemical reaction. Such products are often referred to as foamed polymers. The most important cellular polymers are expanded polystyrene and polyurethane foam.

CELLULAR RUBBER A general name for a material which consists of a rubber matrix containing a mass of gas (usually air) filled cells. The cells may or may not be interconnecting. Different varieties of cellular rubber are foam, sponge and expanded rubber.

CELLULOID Tradename Xylonite. Originally a tradename, now generally used to describe camphor plasticised cellulose nitrate (of nitrogen content about 11%), once a major plastic material, but now little used. Also sometimes used to mean other camphor plasticised cellulose derivatives, e.g. cellulose acetate, benzyl and ethyl cellulose.

CELLULOSAN The fraction of a hemicellulose that consists of the polysaccharides other than cellulose and the polyuronides, i.e. mainly mannans and xylan.

CELLULOSE The most abundant natural polymer, being the major cell-wall material in land plants and their main structural component. Native cellulose is found in both hard and soft woods (about 50% cellulose) and cotton (about 95%), which are the major sources for conversion to a wide variety of useful products including fibres, textiles, paper and various derivatives. These latter form useful plastics, fibres, films, emulsifiers and thickening agents. Other sources include the bast fibres and bacterial cellulose. Cellulose is isolated commercially from wood by pulping using the sulphite or Kraft process, which essentially involves solubilisation of the non-cellulose components (lignin and hemicelluloses). In the laboratory several special methods of isolating relatively pure cellulose have been developed. They yield initially a holocellulose, from which the α-, β- and γ-cellulose fractions are isolated by alkaline and acid treatments.

Cellulose is a linear D-glucan, as is shown by its ready hydrolysis to high yields of D-glucose (or better by methanolysis of cellulose acetate to give only methyl glucosides). Hydrolysis of methylated cellulose to over 90% yields of 2,3,6-tri-O-methylglucose, shows the 1,4′-linking of the glucose units. Partial hydrolysis to cellobiose and cellotriose (or better, formation of their acetates by treatment with acetic anhydride) shows the β-anomeric linkage. Thus cellulose is poly-(1,4′-β-D-glucose). Its degree of polymerisation is typically about 15 000 and it is partially crystalline (crystallinity 30–70%), the chains in the crystallites being intermolecularly hydrogen bonded. Native cellulose has a crystal structure of cellulose I, regenerated and mercerised cellulose that of cellulose II. Cellulose III and IV are further crystalline forms. In native cellulose fibres, the molecules lie along the length of the fibre in bundles of fibrillar aggregates (microfibrils) about 100 Å wide. The microfibrils may consist of protofibrils about 35 Å wide which may be individual cellulose molecules chain folded into ribbons.

Cellulose is insoluble in water, although quite hygroscopic. It will dissolve in aqueous solutions of complexing agents, e.g. cuam, cuene and cadoxene, which are useful for viscosity measurements. Strong acids dissolve cellulose but cause hydrolysis, eventually to D-glucose. Mild hydrolysis in dilute acid yields hydrocellulose. Hydrolysis often proceeds to the levelling off degree of polymerisation. Hydrolysis at elevated temperatures in air causes oxidation at the end groups and peeling. Oxidised cellulose is formed by oxidation with any of several reagents. Treatment with alkali causes swelling of cellulose fibres and is the basis of mercerisation. When treated with aqueous sodium hydroxide of $>12\%$ concentration, the alkali cellulose produced is a complex of soda cellulose I. With $<12\%$, soda cellulose II is formed. Many of the chemical derivatives of cellulose are much more readily soluble and have lower softening points than cellulose itself and can therefore be processed to useful products. Being partially crystalline, the chemical reactivity of cellulose is often limited by accessibility to the amorphous regions. Many substitution reactions are possible on one or more of the three hydroxyl groups, the amount of substitution being given by the degree of substitution. Useful commercial products are the cellulose esters (cellulose acetate, triacetate, nitrate, acetate-butyrate, acetate-propionate, acetate-phthalate) and the cellulose ethers (methylcellulose, hydroxyethylmethylcellulose, ethyl cellulose, benzylcellulose and cyanoethylcellulose). Cellulose xanthate is a soluble intermediate in the production of regenerated cellulose fibres (rayon) and film (cellophane). Regenerated cellulose is also produced by the Bemberg process using the cellulose–cuprammonium complex.

α-CELLULOSE   The insoluble high molecular weight cellulose remaining after plant tissues have been delignified and the hemicelluloses removed by extraction with 10% aqueous alkali.

β-CELLULOSE   The low molecular weight cellulose precipitated when the 10% aqueous alkaline extract of delignified plant tissue (obtained after separation of the insoluble α-cellulose) is acidified. It is thought to arise from degradation of the much higher molecular weight α-cellulose.

γ-CELLULOSE   The polysaccharides remaining in solution after acidification of the 10% aqueous alkaline extract of delignified plant tissue and removal of the insoluble α- and β-celluloses. It consists of a mixture of hemicelluloses. Together with the β-cellulose fraction, it

constitutes what, today, is understood by the term hemicellulose.

CELLULOSE I   The usual crystalline form of native cellulose, in which the anhydroglucose chains are arranged parallel to each other but slightly bent in places, with chains in alternate planes being anti-parallel. This arrangement permits the maximum amount of inter- and intramolecular hydrogen bonding. The unit cell has dimensions of $a = 8.35$ Å, $b = 10.3$ Å (fibre repeat distance corresponding to a cellobiose unit) $c = 7.9$ Å, $\beta = 84°$.

CELLULOSE II   (Hydrate cellulose) A crystalline modification of cellulose found in both regenerated and mercerised cellulose. The unit cell dimensions are the same as in cellulose I, except the angle $\beta = 63°$.

CELLULOSE III   (Ammonia cellulose) A crystalline modification of cellulose whose unit cell structure is similar to cellulose II. Obtained by swelling cellulose I or II in either liquid ammonia or in an aliphatic monoamine.

CELLULOSE IV   (Cellulose T) (High temperature cellulose) A high temperature polymorph of cellulose, obtained by heating cellulose I or II in water or glycerol to 280°C. The unit cell has dimensions similar to those of cellulose I, but with the angle $\beta = 90°$.

CELLULOSE ACETATE   (CA) Tradenames: Alon, Dexel, Forticel, Rhodialite, Tenite, Trolit (plastics materials), Celafibre, Celanese, Dicel, Fortinese, Lanese, Seraceta, Tohalon (fibres). Cellulose in which some or all of the hydroxyl groups have been acetylated. Industrially this is performed with a mixed acetic acid, acetic anhydride, sulphuric acid reagent to yield a solution (dope) of cellulose triacetate. The triacetate with a DS of >2.6 (also called primary acetate) is an important fibre. The dope may be hydrolysed by addition of water to a polymer DS of about 2—secondary cellulose acetate, also simply called acetate and also a major fibre forming polymer. Cellulose acetate with a DS of 2.2–2.5 was earlier an important plastic material. It is also useful as a photographic film and recording tape material. The solubility varies considerably with DS. At a DS >2.6 the polymer is only soluble in a few solvents, e.g. ethylene dichloride, at a DS of 2.6 it is soluble in acetone, but not at DS 1.88. Between DS 0.85 and 1.35 the polymer is water soluble.

CELLULOSE ACETATE BUTYRATE   (CAB) Tradenames: Tenite Butyrate, Cellidor B. A mixed cellulose ester formed by reaction of cellulose with a mixture of butyric and acetic anhydrides and sulphuric acid catalyst. Typically it contains about 15% combined acetic and about 40% combined butyric acids. A useful plastic material, being optically transparent and of good impact strength and appearance. Applications include sheet and tool handles. It is commonly plasticised with sulphonamides.

CELLULOSE ACETATE PHTHALATE A mixed cellulose ester, in which one carboxyl of the phthalic acid is esterified by reaction with the hydroxyl groups of cellulose. In acid form it is soluble in organic solvents and as its sodium salt it is soluble in water. It is useful therefore as an antihalation photographic film backing, which dissolves on development and as an enteric coating for medicinal tablets.

CELLULOSE ACETATE PROPIONATE Tradename: Tenite Propionate. A mixed cellulose ester, prepared similarly to cellulose acetate butyrate and having properties similar to cellulose propionate. Plastic grade material contains 2–7 wt% acetyl and about 40 wt% propionyl groups.

CELLULOSE ACETATE RAYON (Acetate) (Acetate rayon) Tradenames: Celanese, Estron, Aceta, Seraceta. The fibre produced by dry spinning from secondary cellulose acetate (of DS about 2·3) solution in acetone (dope). More water repellent and less hygroscopic than viscose rayon, but tends to swell in organic solvents or even dissolve, e.g. in acetone, methylethyl ketone and dioxan. It softens on heating and melts at 230°C. It has a specific gravity of 1·32, a typical tenacity of about 1·4 g denier^{-1} and an elongation at break of 25%, being elastic to about 5% elongation. Fabrics from the fibre are particularly warm and drape well, but can be difficult to dye. It is largely used for lingerie and dress fabrics. Staple fibre is often blended with cotton, wool or viscose rayon to improve handle and drape.

CELLULOSE BUTYRATE A cellulose ester in which some of the cellulose hydroxyl groups have been replaced with butyrate groups. Both the *n*- and iso-esters may be prepared by reaction of cellulose with butyric anhydride and sulphuric acid. Commercial polymers are alcohol soluble and are used as lacquers.

CELLULOSE ETHER A cellulose derivative in which some, or all, of the cellulose hydroxyl groups have been replaced by ether groups. A range of commercial products, having a DS generally of 1·5–2·5 are available. They are prepared by reaction of alkali cellulose with either an alkyl chloride for alkyl celluloses, e.g. methyl and ethyl cellulose, a chloroacid for carboxyalkyl celluloses, e.g. sodium carboxymethyl cellulose, an olefin oxide for a hydroxyalkyl cellulose, e.g. hydroxyethyl cellulose, or acrylonitrile for cyanoethyl cellulose. Mixed products are also made by using a combination of two of the above reagents, e.g. hydroxypropylmethyl cellulose, hydroxyethylmethyl cellulose, ethylhydroxyethyl cellulose, ethylmethyl cellulose and sodium carboxymethylhydroxyethyl cellulose. Their water solubility is made use of in most applications.

CELLULOSE NITRATE (CN) (Nitrocellulose) (Celluloid) Tradenames: Parkesine, Tenite, Trolit F, Xylonite. A cellulose ester in which some of the cellulose hydroxyl groups have been converted to nitrate groups. Nitration of cellulose is readily achieved, without degradation, using a mixture of nitric and phosphoric acids containing P_2O_5, or nitric and acetic acids containing acetic anhydride. Industrially, a mixture of nitric and sulphuric acids containing water is used:

$$Cell—OH + HNO_3 \rightleftharpoons Cell—O—NO_2 + H_2O$$

As the water content increases, so the DS achieved decreases. Complete nitration (DS = 3·0, nitrogen content 14·1%) may be achieved, but commercial products have nitrogen contents of 10·5–12·5% (for plastics and lacquers) or 12·0–13·5% (for explosives). The products are unstable unless boiled in water to remove traces of nitrating acids. The polymer is soluble in many organic solvents, e.g. ether, acetone. Collodion is a solution in a mixed ether/alcohol solvent of considerable historical interest since the first successful artificial fibre (Chardonnet silk) was spun from its solution and the first plastic material (Parkesine) was a plasticised cellulose nitrate. Celluloid was a major cellulose nitrate plastic, plasticised with camphor, but is little used now. It is also plasticised by the usual phthalate and phosphate plasticisers and by triacetin.

CELLULOSE PROPIONATE Tradenames: Cellidor CP, Forticell, Tenite. A cellulose ester, prepared in a similar manner to cellulose acetate, the commercial product typically having a DS of about 2·6. A plastic material of similar properties to cellulose acetate butyrate.

CELLULOSE T Alternative name for cellulose IV.

CELLULOSE TRIACETATE (Primary acetate) (Triacetate) Tradenames: Arnel, Courpleta, Tricel. A cellulose acetate with a DS > 2·6. A major fibre-forming polymer, relatively recently (for a cellulose-based material) developed, by dry spinning from a dichloroethane solution. It has a higher softening point (m.p. 290–300°C) than secondary cellulose acetate and may be heat set. It has a low moisture regain (4·5% at 65% RH) and relatively low strength (tenacity of 1·2 g denier^{-1} when dry, 0·8 g denier^{-1} when wet).

CELLULOSE X A crystalline modification of cellulose in which the unit cell dimensions are close to those of cellulose IV but with a distinctive X-ray diffraction pattern. Obtained by treating cotton or wood pulp with strong hydrochloric or phosphoric acids.

CELLULOSE XANTHATE Alternative name for sodium cellulose xanthate.

CELLURONIC ACID An oxycellulose, produced by treatment of cellulose with dinitrogen tetroxide, in which the C-6 carbon atom of each anhydroglucose unit is oxidised to a carboxyl group.

CELON Tradename for a nylon 6 fibre.

CENTIPOISE Abbreviation cP. A unit of viscosity equal to 10^{-2} poise (P). Typical organic liquids have viscosities of about 1 cP.

CEREAL GRAIN POLYSACCHARIDE (Cereal gum) One of a group of water soluble β-D-glucans found in the grains of oats and barley, with 1,3'- and 1,4'-links between the β-D-glucopyranose units in the approximate ratio 1:2·5. It often also contains protein, as glycoprotein, as well.

CEREAL GUM Alternative name for cereal grain polysaccharide.

CEREAL GUM POLYSACCHARIDE The polysaccharide obtained by extraction of the endosperm of cereal grains with water, followed by precipitation with alcohol. Although the extract is largely polysaccharide (often a β-D-glucan), it may also contain a glycoprotein component.

CERECLOR Tradename for chlorinated paraffin.

CERULOPLASMIN An enzyme glycoprotein present in blood plasma occurring in the α_2-globulin fraction. The protein part is of molecular weight about 130 000 and contains eight copper atoms per mole. The carbohydrate, comprising about 7·5%, consists of oligosaccharides of D-mannose, D-galactose, N-acetyl-hexose and sialic acid units, there being probably one carbohydrate group of molecular weight about 1150 for each protein subunit. The enzyme functions as an oxidase for polyamines, polyphenols and ascorbic acid.

CF Abbreviation for conductive furnace carbon black.

CFM An abbreviation for a fluoroelastomer based on vinylidene fluoride–chlorotrifluoroethylene copolymers.

CFM 11 Alternative name for trichlorofluoromethane.

CFM 12 Alternative name for dichlorodifluoromethane.

CFRP Abbreviation for carbon fibre reinforced plastic.

C-GLASS A soda-lime/borosilicate glass whose oxide weight % composition is: SiO_2, 65·0; Al_2O_3, 4·0; Fe_2O_3, 0·2; CaO, 14·0; MgO, 3·0; B_2O_3, 6·0; Na_2O, 8·0; SO_3, 0·1. It has excellent chemical stability and is therefore sometimes used as a fibre reinforcement in polymer composites when chemical resistance is required. Its fibres have a tensile strength of 3·1 GPa, a Young's modulus of 74 GPa, a density of 2·49 g cm^{-3} and a refractive index of 1·54.

CHAIN-BREAKING ANTIOXIDANT An antioxidant capable of interrupting an oxidative chain reaction by reacting with the propagating free radicals to form inactive products or products of reduced activity. Several types, operating by different mechanisms, exist—free radical scavengers, electron donors and, most importantly, the hydrogen donors (or AH antioxidants) which contain active hydrogen atoms, readily abstracted by the propagating free radicals ($\dot{R}$):

$$A—H + \dot{R} \rightarrow \dot{A} + R—H$$

where $\dot{A}$ has a lower activity than $\dot{R}$ and/or is capable of combining with another free radical.

CHAIN CLEAVAGE Alternative name for chain scission.

CHAIN COPOLYMERISATION Copolymerisation which proceeds by a chain polymerisation mechanism.

CHAIN DEFECT (Chain disorder) A point defect in a polymer crystal, which encompasses kinks, jogs, chain torsions, chain ends and isolated folds and thus causes only local disturbance to the perfect order of the crystal. For example, in polyethylene the simplest kink is a 2 gl kink, when the extended all-*trans* chain has two successive *gauche* conformations:

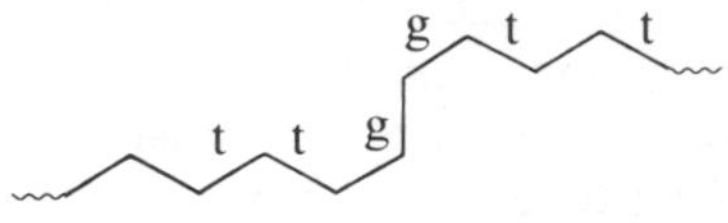

CHAIN DISORDER Alternative name for chain defect.

CHAIN FOLDING (Folded chain structure) The folding back on themselves of polymer molecular chains in polymer crystals, to give a regular array of polymer chain segments (the stems) as in, for example, polyethylene (the most widely studied polymer) where the stems are extended chains. The stems have been shown by selected area electron diffraction to be perpendicular to the small dimension, i.e. the thickness, of the lamellar crystals, and run from surface to surface. It is at the surface that chain folding occurs, where the folds join the stems together, so that a single molecule may consist of many stems and folds. The fold may extend over a few (4–5) (tight folding) or many (loose folding) repeat units. Chain folding has been studied mostly in polymer single crystals obtained by crystallisation from dilute solution but is also believed to occur in melt crystallised polymers. The fold length and lamellar thickness are larger, the higher the crystallisation temperature, and may be increased by annealing.

CHAIN INITIATION Alternative name for initiation.

CHAIN LENGTH DISTRIBUTION (Degree of polymerisation distribution) A molecular weight distribution in which the distribution of molecular sizes (i) is described in terms of the distribution of degrees of polymerisation rather than of the more usual molecular weights.

CHAIN POLYMER (Addition polymer) A polymer produced by chain polymerisation. Vinyl polymers are the commonest type, but diene polymers and polymers produced by ring-opening polymerisation are also chain polymers. In the last case, since some polymers, such as nylon 6, may be formed by either a chain or a step-growth mechanism (or indeed the mechanism may not be clear), the polymers may be considered as either chain or step-growth polymers.

CHAIN POLYMERISATION (Addition polymerisation) One of the two main mechanistically distinct types of polymerisation, in which the monomer is converted to polymer via a chain reaction. This involves the formation of activated monomer molecules which grow very rapidly to high polymer molecules by successive addition of many other monomer molecules to the active centre situated at the end of the growing polymer chain. The active centre (designated by an asterisk*) may be a free radical (free radical polymerisation) or an ion (ionic polymerisation). The ion may be a cation (cationic polymerisation) or an anion (anionic polymerisation). In anionic polymerisation the active centre may be coordinated with a metallic centre (coordination polymerisation). The activated monomer ($I—M^*$) is usually formed by reaction of a monomer molecule (M) with a suitable free radical or ionic initiating species (I^*) (initiation) which is (or in the case of free radicals is derived from) the initiator. Initiation may also be by direct formation of activated monomer by heating (thermal polymerisation), irradiation (radiation polymerisation) or electrolysis (electrolytic polymerisation) of the monomer:

$$I^* + M \rightarrow I—M^*$$

The rapid growth of active centres to form long polymer molecules (propagation) proceeds by successive addition of many monomer molecules to the active centres:

$$I—M^* + M \rightarrow I—M—M^* \xrightarrow{M} I—M—M—M^* \xrightarrow{M} \text{etc.}$$

In general:

$$I—M_n—M^* + M \rightarrow I—M_{\overline{n+1}}—M^*$$

Thus a single act of initiation produces a chain reaction with many propagation steps. Propagation of an individual chain stops either by termination to form dead polymer:

$$\text{\small\leavevmode}M_n—M^* \rightarrow \text{\small\leavevmode}M_{n+1}$$

or by a chain transfer reaction with another molecule X:

$$\text{\small\leavevmode}M_n—M^* + X \rightarrow \text{\small\leavevmode}M_{n+1} + X^*$$

to give an activated molecule X* which may initiate formation of another polymer chain. Propagation may also cease simply as a result of depletion of monomer as in the formation of living polymer.

The degree of polymerisation of the polymer formed depends on the rate constants of the individual steps of initiation, propagation, termination and transfer, and on the concentrations of the species involved. Unless transfer predominates, high molecular weight polymer is often readily obtained. In contrast to step-growth polymerisation, monomer is only slowly consumed, high molecular weight polymer is formed right from the start and polymer molecular weight does not usually vary with conversion. The instantaneous concentration of active species is always very low ($\sim 10^{-6}$M) and its constant value allows the steady state assumption to be made, so enabling solution of the kinetic equations of polymerisation.

The commonest monomers which are chain polymerised are unsaturated molecules, usually with carbon–carbon double bonds, i.e. vinyl monomers ($CH_2=CXY$) (vinyl polymerisation) or dienes (usually $CH_2=CX—CY=CH_2$) (diene polymerisation). Ring-opening polymerisation of saturated cyclic monomers can also be a chain polymerisation. Since polymerisation of unsaturated monomer involves conversion of a π-bond to a σ-bond, chain polymerisations are usually highly exothermic, having a high negative heat of polymerisation. In practical polymerisation systems (mass, suspension, solution and emulsion methods are used) the exotherm must be adequately controlled.

CHAIN PROPAGATION Alternative name for propagation.

CHAIN SCISSION (Chain cleavage) Breaking of an in-chain chemical bond in a polymer molecule, producing two new molecules of lower molecular weight. The most important chemical reaction in many polymer degradations, causing impairment of mechanical properties, e.g. embrittlement, and reduction of modulus, by molecular weight reduction. Random scission results when chain scission is equally probable at any repeat unit. Beta scission often results from the production of polymer free radicals.

CHAIN TERMINATION Alternative name for termination.

CHAIN TILT A tilting of the polymer chains in a chain folded lamellar crystal away from the normal angle to the lamellar surface, to some lower angle. This may occur on stressing and deforming the lamellae as, for instance, during the drawing of a fibre.

CHAIN TRANSFER (Transfer) The transfer of the active centre (chain carrier) in a chain polymerisation from a growing polymer molecule ($\text{\small\leavevmode}M_n^*$) to another molecule (X—Y) by abstraction of an atom (Y) from X—Y:

$$\text{\small\leavevmode}M_n^* + X—Y \xrightarrow{k_{tr}} \text{\small\leavevmode}M_n^*—X + Y^*$$

where k_{tr} is the rate constant for the transfer reaction. This terminates growth of the polymer molecule, but not the kinetic chain, since the new active centre (Y^*) may re-initiate polymerisation:

$$Y^* + M \xrightarrow{k_Y} Y—M^* \xrightarrow{M} Y—M—M^* \rightarrow \text{etc.}$$

where k_Y is the rate constant for initiation by Y. Transfer thus reduces the polymer molecular weight, according to the Mayo equation. In ionic polymerisation, when kinetic chain termination is absent, chain transfer is the main mode of polymer chain growth termination. The reactivity of X—Y is expressed by its chain transfer constant C, defined as the ratio k_{tr}/k_p, where k_p is the rate constant for propagation. The rate of polymerisation (R_p) will be unaffected if Y* re-initiates rapidly.

When transfer predominates ($k_p < k_{tr}$) and much X—Y is present, only very low molecular weight polymer will be formed (telomerisation). If $k_Y < k_p$, then R_p is reduced (retardation) and if, in addition, $k_p \ll k_{tr}$, degradative chain transfer results. X—Y may be monomer (transfer to monomer), initiator (transfer to initiator or induced decomposition), solvent (transfer to solvent), polymer (transfer to polymer) or a deliberately added chain transfer agent or an adventitious impurity.

CHAIN TRANSFER AGENT (Transfer agent) A substance capable of causing a chain transfer reaction in a chain polymerisation. It may be solvent, impurity or a material deliberately added to limit polymer molecular weight (modifier). Its effect on molecular weight is given by the Mayo equation.

In free radical polymerisation, hydrocarbons, acids, ethers, alcohols, carbonyl compounds and simple halides have low transfer constants (C_S). However, carbon tetrachloride and tetrabromide and thiols have high transfer constants due to the weak carbon–halogen or sulphur–hydrogen bond. The value of C_S does depend on the monomer being polymerised, generally increasing with increasing activity of the propagating radical.

In ionic and coordination polymerisation, due to the high reactivity of the active centres, any of many impurities, especially those with active hydrogen atoms, e.g. water and alcohols, can often severely limit polymer molecular weight unless special precautions are taken to exclude them. Thus in cationic polymerisation, transfer of A^- from an active hydrogen containing species H—A can be an important mode of termination:

$$\sim\!\!M_n^+ X^- + H\!\!-\!\!A \rightarrow \sim\!\!M_n\!\!-\!\!A + \overset{+}{H}\overset{-}{X}$$

In anionic polymerisation, termination of growth of polymer chains often occurs by a proton transfer:

$$\sim\!\!M_n^- X^+ + H\!\!-\!\!A \rightarrow \sim\!\!M_n\!\!-\!\!H + A^- X^+$$

CHAIN TRANSFER CONSTANT (Transfer constant) Symbol C. In chain polymerisation, the ratio of the rate constant of a transfer step (k_{tr}) to that of a propagation step (k_p). A measure of the reactivity of the species responsible for transfer and its effect on reducing polymer molecular weight, as given by the Mayo equation. The different constants are designated C_M (transfer to monomer), C_I (transfer to initiator or induced decomposition), C_S (transfer to solvent or transfer agent) and C_P (transfer to polymer).

CHAIR CONFORMATION Normally the most stable and therefore preferred conformation of six-membered organic ring compounds, particularly relevant to cyclohexane rings, where it may be represented as

It is also the preferred conformation of monosaccharides and saccharide units in polysaccharides, when these are in the pyranoside ring form. Thus, for example, α-D-glucose may be represented as:

$$\begin{array}{c} \text{H} \quad\quad \text{CH}_2\text{OH} \\ \text{HO} \quad\quad\quad \text{O} \\ \quad \text{H} \quad \text{H} \\ \text{H} \\ \text{HO} \quad\quad \text{OH} \quad\quad \text{H} \\ \quad \text{H} \quad\quad\quad \text{OH} \end{array}$$

The alternative boat conformation is usually a higher energy form so is not so commonly found.

CHALCOGENIDE GLASS A glassy, amorphous, crosslinked polymer of one of the chalcogenide elements (sulphur, selenium or tellurium) and one or more of the elements As, Sb, Bi, Cd, Ga, Ge, Pb, Mg, P, Si or Sn. The polymers are prepared by fusing the elements concerned at 500–1000°C such that volatilisation and oxidation do not occur. Typical binary polymers are the glasses of As, Sb or Ge with S, Se or Te, of which some specific compositions crystallise, e.g. As_2S_3. Many multi-component glasses have also been investigated, such as Si/P/Te, Ge/P/Se and As/Ge/Si/Te polymers.

The glasses are coloured but transparent to infrared radiation and have been used for infrared windows. Some polymers are semiconducting and have other unusual electrical properties, such as the conductivity varying with voltage. However, they have not been developed commercially for memory or switching devices due to the variability in their performance. The polymers can have low softening points and in this respect are more like organic polymers than inorganic polymers.

CHANNEL BLACK A type of carbon black produced by impinging a flame of a burning hydrocarbon fuel on a cool surface and scraping off the black particles deposited. This is an inefficient process and channel blacks have largely been replaced by furnace blacks. The process can produce very fine particles (10–20 nm) which are especially useful for pigmentation. The black is also useful in cables due to its electrical conductivity.

CHARACTERISTIC RATIO Symbol C. A measure of the expansion of a polymer chain in dilute solution due to steric interactions (given by the steric hindrance parameter σ) and valence bond angle (θ) restrictions. It is defined as:

$$C = \langle r^2 \rangle_0 / Nl^2 = [(1 - \cos\theta)/(1 + \cos\theta)]\sigma^2$$

where $\langle r^2 \rangle_0$ is the unperturbed mean square end-to-end distance and N is the number of links of length l in the chain.

CHARDONNET SILK A regenerated cellulose fibre, produced by denitration (with ammonium hydrosulphide or sulphocarbonate) of cellulose nitrate fibres. The latter are produced by solution spinning from collodion (a solution of cellulose nitrate in 40% ether and 60% alcohol). The first artificial silk to be produced in quantity, but now of historical interest only.

CHARGE MOSAIC MEMBRANE A membrane of an ion impermeable matrix material, e.g. a silicone polymer, containing dispersed particles of separate cationic and anionic exchange resin beads, such that each bead spans the thickness of the membrane. Such dispersions form a regular array, or 'mosaic', of particles and can form the basis of negative reverse osmosis membranes.

CHARGE-TRANSFER INITIATION Initiation of cationic polymerisation by a powerful electron acceptor such as tetracyanoethylene or maleic anhydride, with an electron-donating monomer such as N-vinylcarbazole. Partial electron transfer occurs from the donor monomer to the acceptor initiator to form a charge-transfer complex and a monomer radical cation which propagates:

$$\text{monomer} + \text{initiator} \rightarrow [\text{charge-transfer complex}]$$
$$\rightarrow \text{monomer}^{+\cdot} + \text{acceptor}^{-\cdot}$$

However, in some systems thought originally to polymerise in this way, subsequently an acidic impurity or water was found to be involved in initiation. Polymerisations initiated in this way are one type of charge-transfer polymerisation.

CHARGE-TRANSFER POLYMER A polymer which is capable of forming a charge-transfer complex by the complete or partial transfer of electron from a donor group (D) of low ionisation potential to an acceptor group (A) of high electron affinity. Such materials show electrical conductivity due to the mobility of electrons along the stacks of ~DADADADA~ complexes. Both D and A may be attached to the polymer chain or one of them may be an added small molecule, as in a doped polymer. The most common type is that of a polyamine donor with a monomeric acceptor, e.g. polyvinylpyridine-iodine (conductivity 10^{-3} S cm^{-1}), poly-(N-vinylcarbazole)-iodine (conductivity 10^{-3} S cm^{-1}) and the 2,4,5,7-tetranitrofluorenone complexes (conductivity 10^{3} S cm^{-1}). The radical-ion polymers are extreme types of charge-transfer polymer in which complete charge transfer has occurred.

CHARGE-TRANSFER POLYMERISATION Chain polymerisation which involves the formation of a charge-transfer complex between an electron donor and an electron acceptor molecule. It may be the result of charge-transfer initiation, where a monomer is the donor and an initiator is the acceptor, but more commonly it occurs in copolymerisation where the two monomers are the donor and acceptor respectively. This results in the complex homopolymerising to give an alternating 1:1 copolymer as, for example, in the copolymerisation of maleic anhydride with sulphur dioxide. Even a non-polar/polar monomer pair may polymerise in this way in the presence of certain metal halides (MX_n) such as $AlEtCl_2$ or $ZnCl_2$, which complex with the polar monomer (usually an ester or nitrile) to form the acceptor complex (A), the non-polar monomer (such as styrene, a diene or alkene) being the donor (D): $A + MX_n \rightleftharpoons A:MX_n$

$$yD + yA:MX_n \rightleftharpoons yD \rightarrow A:MX_n \rightarrow (D\text{—}A)_y + yMX_n$$

Charge-transfer formation can also make photoinitiation easier by lowering the energy for ionisation, thus leading to the initiation of cationic polymerisation on irradiation, e.g. with N-vinylcarbazole/acrylonitrile and styrene/tetracyanobenzene.

CHARLESBY–PINNER EQUATION A mathematical relationship between the sol content (expressed as the sol fraction, s) and the crosslinking index (γ) of a polymer: $s + \sqrt{s} = 1/\gamma$ ($= 2/\delta$, if the initial MWD is random), where δ is the crosslinking coefficient. Useful in the analysis of polymer samples which have been subject to chain crosslinking without chain scission, e.g. as a result of high energy irradiation.

CHARPY IMPACT STRENGTH The impact strength measured according to the Charpy method. In this test a falling weighted pendulum strikes centrally a rectangular bar test piece mounted horizontally on anvils but not clamped. The sample may be notched or unnotched. The notch is a 45°, v-shaped notch and the notch tip radius is often 0·25 mm. The impact strength is recorded as the energy expended in breaking the sample by recording the height to which the pendulum follows through. Results are quoted either as energy per unit length of notch (J m^{-1}) or as energy per unit area fractured (J m^{-2}). The test has the same shortcomings as the similar Izod test.

CHELATED PTO Tradename: Enkatherm. A metal chelate polymer of structure:

where M is a metal atom, e.g. Ca, Sr or Zn. Formed by treatment of fibres or woven fabrics of PTO with an ammoniacal solution of any of many metal hydroxides. The polymers are coloured, the colour depending on the metal. They do not burn below 1000°C and do not shrink or melt. They therefore form potentially useful fire resistant fabrics.

CHELATE POLYMER (Metal chelate polymer) (Polychelate) (Coordination polymer) A polymer which contains metal atoms bonded to organic functional groups by coordinate bonds, i.e. chelated to the polymer chain. Sometimes the term coordination polymer is used when the coordinate links are part of the polymer chain, when the term chelate polymer is reserved for polymers in which the chelated metal is not part of the polymer chain. Many such polymers have been investigated especially as potential high temperature resistant polymers, but with limited success in this respect. Several methods have been used for synthesis. Firstly, polydentate ligands may be joined through metal ions by reaction with metal salts or by ligand exchange with coordination complexes, as in the formation of phthalocyanine polymers, bis-(β-diketone) polymers and poly-(acetonylacetate). Alternatively, a metal chelate, if containing a suitable functional group, may be polymerised. Thus Schiff bases containing hydroxy groups can be converted to polycarbonates by reaction with phosgene. Finally, the coordinate bonds may be formed by complexation of a preformed polymer containing appropriate ligands. Preparation of Schiff base polymers has been widely studied by this last method.

CHEMICAL BONDING THEORY A mechanism for the action of coupling agents, usually with glass fibres, in reinforced composites, whereby stress transfer between matrix and fibre is aided by the existence of chemical bonds across the glass/coupling agent/matrix interface. Coupling agents usually contain one functional group which will react or form hydrogen bonds with silanol groups on the glass surface and a second functional group which reacts with the polymer during curing.

CHEMICAL DEGRADATION Degradation induced by chemical attack on a polymer. Hydrolytic degradation is the most common form and frequently occurs in polyesters and polyamides or at the glycoside links of cellulose and its derivatives. Ozone attack is a further example and occurs at unsaturated groups, e.g. ozone degradation of diene rubbers. In these examples attack occurs at any repeat unit and therefore results in random scission.

CHEMICAL GRAFTING Formation of a graft copolymer by the formation of free radicals on one polymer chain in the presence of the monomer for a second polymer. This can occur by a chemical reaction at suitable groups attached to the first polymer chains, e.g. by the reaction of pendant hydroxyl groups, such as in cellulose, with ceric ions:

$$\text{OH} + \text{Ce}^{4+} \rightarrow \text{O·} + \text{Ce}^{3+} + \text{H}^+ \underset{\text{M}}{\rightarrow}$$

$$\text{OMMMMMMM}\sim$$

CHEMICAL PLASTICISER Alternative name for peptiser.

CHEMICAL PROBE A reagent capable of reacting with a specific type of sulphur structure in a particular way, such as a monosulphide, disulphide or polysulphide in a rubber vulcanisate and therefore useful in characterising the vulcanisate network structure. Examples are methyl iodide, triphenyl phosphite and the thiol-amine system propane-2-thiol/piperidine.

CHEMICAL STRESS RELAXATION (Network breakdown) Stress relaxation of a rubber network occurring at high temperatures due to cleavage of either main-chain bonds or crosslinks. If N_t is the number of network chains present at time t and N_0 their number

initially, then $N_t/N_0 = e^{-kt}$, where k is the rate constant for cleavage.

CHEMICRYSTALLISATION A method of increasing the crystallinity by selective chemical reaction (normally involving chain scission) of the amorphous regions, permitting their molecules to reorganise themselves into crystallites. Ideally the polymer molecules should recombine subsequent to achieving the increased crystal perfection.

CHEMIGUM N Tradename for nitrile rubber.

CHEMIGUM SL Tradename for hydroxy-terminated polyester polyurethane prepolymers which may be cured by adding further diisocyanate or latent diisocyanate.

CHICLE A gummy polymeric product of the tree *Achras sapota*, composed of a mixture of *cis*- and *trans*-1,4-polyisoprene plus large amounts of resins.

CHINA CLAY Alternative name for kaolin.

CHINON Tradename for a fibre consisting of casein protein grafted with polyacrylonitrile.

CHITIN A highly insoluble aminopolysaccharide occurring widely in the external skeleton of many insects and crustaceans. A polymer of N-acetyl-2-amino-2-deoxy-D-glucose units joined by β-1,4'-links. The structure is therefore that of cellulose in which the hydroxyl groups on carbon-2 are replaced by —NHCOCH$_3$ groups. The usual crystalline form, α-chitin, has a unit cell similar to that of cellulose. The polymer may be deacetylated to chitosan. A major naturally occurring polymer for which little commercial use has yet been found.

α-CHITIN The usual crystalline form of chitin, whose unit cell is similar to that of cellulose and is orthorhombic.

CHITOSAN Tradename: Kytex. Deacetylated chitin, prepared by treatment of chitin with hot concentrated alkali. Partial hydrolysis of chitosan to oligosaccharides (chitobiose, etc.) establishes the structure of chitin. Like chitin it is insoluble in water, concentrated alkalis and organic solvents, but is soluble in dilute acids.

CHLORENDIC ACID (Hexachlororndomethylene tetrahydrophthalic acid (abbreviation HET acid) (1,4,5,6,7,7-Hexachlorobicyclo-(2,2,1)-hept-5-ene-2,3-dicarboxylic acid). Decomposes to anhydride on heating.

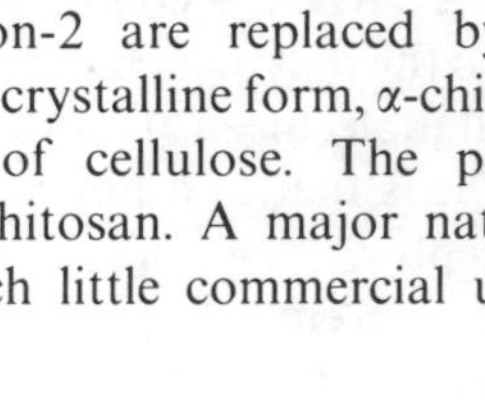

The Diels–Alder adduct of hexachlorocyclopentadiene and maleic acid. Frequently used as a reactive fire retardant comonomer in the formation of unsaturated polyesters, where the high chlorine content is responsible for imparting fire retardancy. However, by its use the polymer becomes susceptible to both thermal and ultraviolet degradation.

CHLORENDIC ANHYDRIDE (Hexachloroendomethylenetetrahydrophthalic anhydride (abbreviation HET anhydride)) (1,4,5,6,7,7-Hexachlorobicyclo-(2,2,1)-hept-5-ene-2,3-carboxylic anhydride).

M.p. 239–240°C.

The Diels–Alder adduct of hexachlorocyclopentadiene and maleic anhydride. Used as a reactive fire retardant comonomer in a similar way and with similar effects to chlorendic acid in unsaturated polyesters and epoxy resins.

CHLORINATED BUTYL RUBBER (Chlorobutyl rubber) (CIIR) Tradenames: Butyl HT, Chlorobutyl. Butyl rubber which contains about 1% chlorine. Obtained by chlorination of butyl rubber in hexane solution. Chlorination occurs adjacent to the isoprene units to form allylic chlorines. Through these the rubber may be vulcanised more readily than normal butyl rubber. As well as sulphur vulcanisation, curing may be brought about by zinc oxide alone or combined with thiourea or tetramethylthiuram disulphide, or by phenolic resins or diamines.

CHLORINATED DIPHENYL Tradename: Arochlor. A mixture of chlorinated diphenyl and terphenyl compounds, varying from mobile liquid to hard solid. Used as plasticisers in polyvinyl chloride, cellulose esters and rubbers for imparting fire resistance. Can contain from 20 to 60% chlorine.

CHLORINATED PARAFFIN Tradenames: Cereclor, Clorofin. A chlorinated paraffin wax which may contain 40–70% chlorine. The products are liquid except at high (e.g. 70%) chlorine content but decompose at about 250°C before they boil. Typically they contain molecules of about 18 to about 24 carbon atoms. Act as plasticisers for polyvinyl chloride, imparting fire resistance, with compatibility increasing with increasing chlorine content.

CHLORINATED POLYETHER Alternative name for poly-(3,3-bis-(chloromethyl)oxacyclobutane).

CHLORINATED POLYETHYLENE (CPE) (CM) Tradenames: Alcryn, Halothene, Hostapren, Kelrinal, Tyrin. Polyethylene which has been post-chlorinated by reaction with chlorine either in solution or in suspension. Typically it contains about 25% chlorine. Chlorination destroys structural regularity of the polyethylene and

hence crystallinity, so that by about 25% chlorine content the material is rubbery. However, the crosslinkable and closely related chlorosulphonated polyethylene is better known as a rubber. It is useful as a fire retardant polymeric additive and also as a compatibilising agent in polymer blends. At higher than 45% chlorine contents the material resembles polyvinylchloride. Rubbers, known as CM, have good ozone, weathering, oil and heat resistance and are vulcanised by peroxides, e.g. dicumyl peroxide.

CHLORINATED POLYVINYL CHLORIDE (CPVC) Tradenames: Genclor S, Lucalor, PeCe, Rhenoflex, Sovitherm. Produced by free radical chlorination of swollen polyvinyl chloride particles at about 50°C or below. The resultant polymer has chlorinated repeating units largely of the type $\sim CHCl-CHCl \sim$. This material has a usefully higher T_g value and therefore softening temperature than polyvinyl chloride. An earlier product, obtained by chlorination of polyvinyl chloride dissolved in a chlorinated solvent at 50–100°C, had increased solubility and was useful for the production of fibres and lacquers.

CHLORINATED RUBBER (Rubber chloride) Tradenames: Alloprene, Duraprene, Paravar, Parlon, Pergut, Rayolin, Tegofan. Natural rubber which has been chlorinated, typically to about 65% in commercial products, either by reaction with chlorine at about 100°C in an inert solvent or as a latex. The product contains some substituted as well as added chlorine and is thought to be somewhat cyclised. Useful in paints and adhesives due to its good chemical and fire resistance. Synthetic cis-1,4-polyisoprene is also available in chlorinated form (tradename: Pliochlor).

CHLORINE RETENTION The formation of chloramine groups in the crease resistant finishes of melamine–formaldehyde and urea–formaldehyde polymers, due to the reaction of polymer imino groups with hypochlorite bleaches:

$$\sim NH \sim + HOCl \rightarrow \underset{\underset{Cl}{|}}{\sim N \sim} + H_2O$$

On subsequent ironing, for urea–formaldehyde chloramines, breakdown occurs giving hydrochloric acid which tenders the fabric:

$$\sim NCl \sim + H_2O \rightarrow HCl + \tfrac{1}{2}O_2 + \sim NH \sim$$

Melamine–formaldehyde chloramines are more stable and do not break down but they do cause yellowing of the fabric. The problem can be alleviated by the use of methylol derivatives of cyclic ureas which have no free $\sim NH \sim$ imino groups. Examples of such compounds are dimethylolethyleneurea and propyleneurea, the methyl ether of hexamethylolmelamine and the urons.

CHLORITE HOLOCELLULOSE The holocellulose produced by removal of the lignin from land plant tissue by repetitive treatment with sodium chlorite.

2-CHLORO-1,3-BUTADIENE Alternative name for chloroprene.

CHLOROBUTYL Tradename for chlorinated butyl rubber.

CHLOROBUTYL RUBBER Alternative name for chlorinated butyl rubber.

CHLOROETHENE Alternative name for vinyl chloride.

2-CHLOROETHYLVINYL ETHER

$$CH_2{=}CHCH_2CH_2Cl \qquad B.p.\ 108°C.$$

Useful as a comonomer in the production of acrylic elastomers to act as a crosslinking site.

CHLOROFIBRE Generic name for a man-made fibre containing at least 50% by weight of polyvinyl chloride or polyvinylidene chloride.

CHLOROFORM $CHCl_3$. B.p. 61·2°C. A highly narcotic and toxic solvent, but of high solvent power, e.g. for cellulose esters, hydrocarbon rubbers, polymethylmethacrylate and natural resins. In the presence of light and moisture it forms acidic products but this can be prevented by the addition of small amounts of ethanol.

CHLOROMALEIC ACID

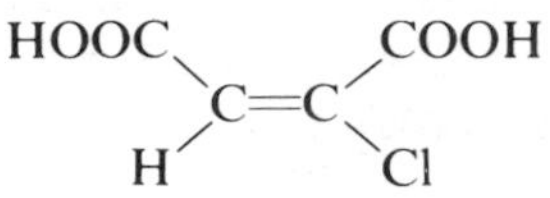

$$\begin{array}{ccc} HOOC & & COOH \\ & C{=}C & \\ H & & Cl \end{array} \qquad M.p.\ 108°C.$$

Occasionally used instead of maleic anhydride in the synthesis of unsaturated polyester resins with improved fire resistance.

CHLOROPRENE (2-Chloro-1,3-butadiene)

$$CH_2{=}CClCH{=}CH_2 \qquad B.p.\ 59·4°C.$$

Originally synthesised from acetylene by its dimerisation to monovinylacetylene, followed by hydrochlorination. Alternatively, it is obtained by the chlorination of butadiene to 1,2-dichlorobutadiene, followed by dehydrochlorination. Readily polymerised by free radical initiators to polychloroprene and also by Ziegler–Natta and cationic catalysts.

CHLOROPRENE RUBBER Alternative name for polychloroprene.

CHLOROSILANE A compound of general structure R_xSnCl_{4-x} where $x = 1, 2$ or 3 and R is an organic group or hydrogen, usually being an alkyl group. The organochlorosilanes form the most important monomers for the synthesis of polyorganosiloxanes. Chlorosilanes can be synthesised by several different routes, the Grignard and direct processes being used the most frequently. The former process is the most versatile, whilst the latter is the most economic. Both processes produce a mixture of chlorosilanes which must be separated by fractional distillation. However, for the most commonly employed methylchlorosilanes, the boiling points are very close together so that complete separation is difficult to achieve. An olefin addition process can be used to introduce an organofunctional group and yields a clean product. By-product chlorosilanes may be converted to useful monomers by the use of redistribution.

On hydrolysis with water chlorosilanes yield silanols but these are unstable and normally spontaneously decompose with the elimination of water to yield siloxanes. Monochlorosilanes, such as trimethylchlorosilane, dimerise to disiloxanes:

$$2\,R_3SiCl \rightarrow 2\,R_3SiOH \rightarrow R_3SiOSiR_3$$

Dichlorosilanes polymerise to a mixture of polyorganosiloxanes

$$2\,R_2SiCl_2 \rightarrow 2\,R_2Si(OH)_2 \rightarrow \{R_2SiO\}_n$$

and oligomeric cyclic organosiloxanes. Trichlorosilanes, such as methyltrichlorosilane, being trifunctional, produce crosslinked network siloxanes. These condensations are catalysed by both acids (including the hydrochloric acid produced in the reaction) and by bases. They are also catalysed by many metal compounds, which is useful in RTV elastomers.

CHLOROSTYRENE

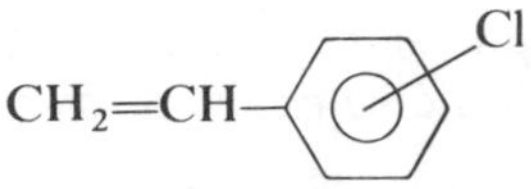

$$CH_2{=}CH-\!\!\!\bigcirc\!\!\!-Cl$$

The term may refer to any of the three isomers—*ortho-* (b.p. 186·5°C), *meta-* (b.p. 63°C/96 mm) or *para-* (b.p. 189·4°C) or, more commonly, a mixture of isomers. Commercial materials are usually *ortho-/para-*mixtures in the approximate ratio of 2:1 and boiling at 177–185°C/760 mm or 44–50°C/5 mm. Chlorostyrenes polymerise under similar conditions to styrene but at faster rates. Produced by chlorination of ethylbenzene to a mixture of *ortho-* and *para-*chloroethylbenzenes, which is then air oxidised to the alcohol and then dehydrated. The monomer is a useful replacement for styrene for crosslinking polyester resins due to its ready polymerisability, the high softening points of the products and their lower flammability.

CHLOROSULPHONATED POLYETHYLENE (CSM) Tradename Hypalon. Polyethylene which has been reacted with a mixture of chlorine and sulphur dioxide under ultraviolet light irradiation. A commercial rubber which contains 1–1·5% sulphur (as sulphonyl chloride groups) and 25–43% chlorine. The substitution destroys the polyethylene crystallinity. The raw polymer may be vulcanised in any of several ways: with litharge (PbO) or magnesia (MgO) in combination with mercaptobenzothiazole, with a disulphide or with a polyfunctional alcohol (such as penterythritol), or with epoxides.

The vulcanisates are characterised by their good ozone, heat, oxygen and weathering resistance together with relatively low flammability and reasonable oil resistance.

CHLOROTRIFLUOROETHYLENE

$$CF_2=CFCl \qquad \text{B.p. } -27°C.$$

Prepared by dechlorination of trichlorotrifluoroethane, itself prepared by reaction of hexachloroethane with hydrogen fluoride:

$$CCl_3—CCl_3 \xrightarrow{HF} CClF_2—CH_2F$$
$$\rightarrow CF_2=CFCl + Cl_2$$

The monomer for the preparation of polychlorotrifluoroethylene by free radical polymerisation in aqueous systems.

CHLOROTRIFLUOROETHYLENE–ETHYLENE COPOLYMER

(Poly-(chlorotrifluoroethylene-co-ethylene)) Tradename: Halar. A 1:1 alternating copolymer having the good chemical resistance and electrical properties typical of a fluoropolymer but with low creep and good impact strength.

CHOLESTERIC PHASE A liquid crystalline meso-phase of a modified nematic type, which occurs when the mesogen is chiral (having a mirror image). In addition to the long-range orientational order of the nematic phase, there is a helical distortion of the orientation direction progressing from one layer of molecules to another in a direction perpendicular to the layers.

CHONDROITIN A mucopolysaccharide found in animal connective tissue consisting of alternating units of 1,3'-linked β-D-N-acetyl-D-galactosamine and 1,4'-linked β-D-glucuronic acid units, with the former being partially sulphated.

CHONDROITIN SULPHATE A Alternative name for chondroitin-4-sulphate.

CHONDROITIN SULPHATE B Alternative name for dermatan sulphate.

CHONDROITIN SULPHATE C Alternative name for chondroitin-6-sulphate.

CHONDROITIN-4-SULPHATE (Chondroitin sulphate A)

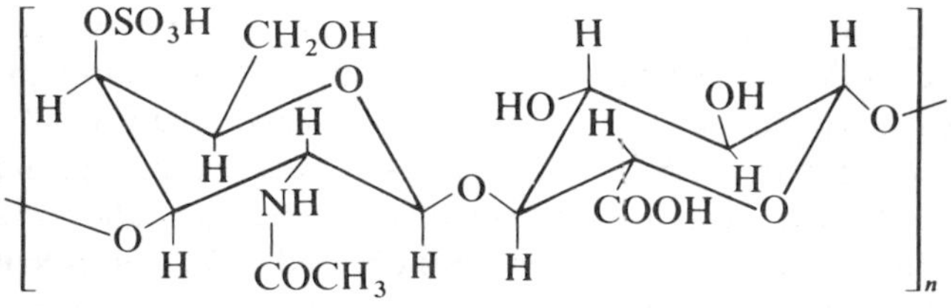

A mucopolysaccharide found in animal connective tissue especially cartilage. It consists of alternating units of 1,3'-linked β-D-N-acetyl-D-galactosamine and 1,4'-linked β-D-glucuronic acid, the galactosamine being sulphated at C-4.

CHONDROITIN-6-SULPHATE (Chondroitin sulphate C)

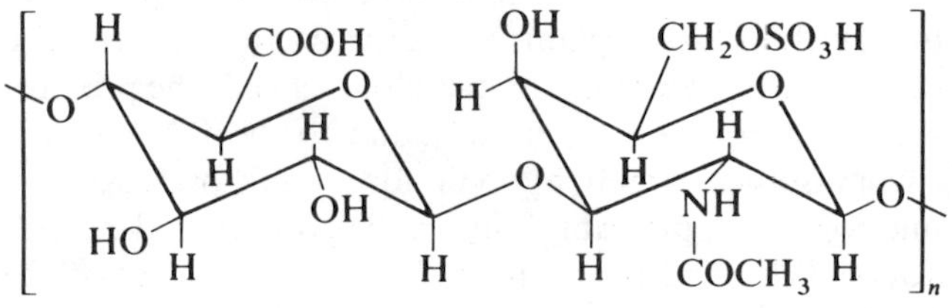

A mucopolysaccharide found in animal connective tissues, especially cartilage. It consists of alternating 1,3'-linked β-D-N-acetyl-D-galactosamine and 1,4'-linked β-D-glucuronic acid units, the former being sulphated at C-6.

CHOPPED FIBRE COMPOSITE Alternative name for short fibre composite.

CHOPPED STRAND MAT (CSM) A sheet-like, non-woven material, especially of glass fibres, formed by depositing chopped strands of fibres (about 50 mm long) on a belt in a non-random manner as a layer and applying a binder. Typical weights are $0.3–1.0\,kg\,m^{-2}$. The binder may be a polyvinyl acetate emulsion for hand lay-up moulding, an unsaturated polyester for roofing sheet, or a crosslinked polyester powder binder for press moulding. A very commonly used form of glass fibre reinforcement for polyester composites when the highest strengths are not required.

CHROMATOGRAPHIC FRACTIONATION Alternative name for Baker–Williams fractionation.

CHROMATOGRAPHY A method of separation of different molecular species even if they are very similar, based on the differential distribution of the species in a mixture, between a thin stationary phase and a bulk phase which is moving with respect to the stationary phase in a counter-current manner. This results in repeated transfers of the components of the mixture many times between the two phases and hence efficient separations.

The stationary phase may be the surface of a solid support packed into a column (adsorption chromatography) or it may be a liquid held in a solid support, as in partition chromatography. The moving phase is a liquid or a gas, as in gas chromatography. In polymer work, chromatographic methods are mostly used for fractionation, but pyrolysis/gas–liquid chromatography is also useful. Other polymer chromatographic methods are gel permeation chromatography, precipitation chromatography, ion-exchange chromatography, high pressure liquid chromatography and thin layer chromatography.

CHROME FINISH Alternative name for methacrylatochromic chloride.

CHROMOPROTEIN (1) A conjugated protein in which the prosthetic group is coloured. Examples include the haem proteins and the flavoproteins.

(2) Alternative name for chromosomal protein.

CHROMOSOMAL PROTEIN (Chromoprotein) The protein content of the chromosomes present in the nuclei of all cells of higher plants and animals. In fish sperm the proteins are protamines, but in all other cells the proteins are histones. The protein is bound to the DNA of the chromosomes, largely by ionic forces (salt bridges), as a nucleoprotein, probably lying in the grooves of the DNA double helix and neutralising the negative charge of the DNA phosphates.

CHRYSOTILE (White asbestos) A single-sided, crystalline layer silicate mineral, which occurs as silky fibres, forming, together with the fibrous amphiboles, the asbestos minerals. Its approximate composition is $Mg_2Si_2O_5(OH)_4$, made up of a silicate sheet of SiO_4 tetrahedra joined to a brucite $(Al(OH)_3)$ sheet in which two of the three hydroxyls are replaced by oxygen from the SiO_4 tetrahedra. By far the most widely used type of asbestos as a reinforcing filler in polymer composites. The fibre has a density of $2.5\,g\,cm^{-3}$, with a fibril diameter of about 20 nm, the fibres existing as bundles of 5–500 μm in diameter. Fibre length can be up to 10 cm, but the fibres are usually broken down during extraction, being typically about 1 mm in length. The fibrils have a tubular structure composed of an inner silica $(Si_2O_5^{2-})_n$ sheet to which is attached an $Mg(OH)_2$ (brucite) layer, the sheets being rolled up to form the tube. This gives the fibrils some flexibility. Fibre bundles have a tensile strength typically of about 4000 MPa and a Young's modulus of 145 GPa. Thus they are stiffer than E-glass fibres and produce stiffer polymer composites. With parallel fibre alignment, composites with tensile strengths of about 700 MPa and Young's moduli of about 85 GPa can be prepared. Chrysotile is not very resistant to acids but has high thermal stability, only about 4% water loss occurring to 450°C. Since strength and stiffness are retained to this temperature it is a very suitable reinforcement for high temperature composites including ablative materials. However, increased concern over the toxic hazards involved in the use of asbestos has considerably curtailed its use as a reinforcement.

CHYLOMICRON A blood serum lipoprotein fraction with density less than $0.94\,g\,cm^{-3}$ obtained by density gradient ultracentrifugation of serum. Contains only 1–2% protein, which is insufficient to cover the surface of the lipid, which must obtain some of its hydrophilicity from polar phospholipid components as well as from the protein. It is likely that all serum lipid is really lipoprotein even if only having a very small protein content as in chylomicron.

CHYMOTRYPSIN An enzyme found in the digestive tract where it acts as a protease, catalysing hydrolysis of peptide bonds on the carbonyl side of amino acid residues with large hydrophobic side chains. It is also useful in protein sequencing structural studies for causing preferential cleavage at these residues producing peptides whose sequences are then determined. Its active site has been widely investigated, e.g. by affinity labelling using TPCK. It consists of a hydrophobic pocket for binding the substrate and a nearby catalytic portion as a shallow depression on the surface, at which serine 195, histidine 57 and aspartic acid 102 residues are situated. These residues form a hydrogen bonded structure making the serine 95 strongly electrophilic, so that it reacts with a substrate acyl group (forming an acyl enzyme) as an intermediate step in the hydrolysis. Thus the enzyme is a serine enzyme. The enzyme is produced as an inactive zymogen, chymostrypsinogen which is converted to chymostrypsin by the action of trypsin by hydrolysis between residues 15 and 16.

CHYMOTRYPSINOGEN The zymogen precursor for chymotrypsin. It consists of a single polypeptide chain of 245 amino acid residues and five disulphide bridges. It is converted to chymotrypsin by the sequential action of trypsin and chymotrypsin itself, which cleaves the chain in three places with loss of two dipeptides. This leaves three chains held together by only two interchain disulphide and three intrachain disulphide bridges.

CIBANITE Tradename for aniline–formaldehyde polymer.

CIBANOID Tradename for urea– or urea/thiourea–formaldehyde polymer.

CIIR Abbreviation for chlorinated butyl rubber.

CILIUM The free chain end of a polymer molecule present in the amorphous surface layer of a crystallite.

CIRCULAR BIREFRINGENCE Birefringence due to the different velocities of propagation of light polarised circularly in the clockwise and anticlockwise directions in an optically active material. Plane polarised light, which is composed of these two circularly polarised components suffers a rotation of the plane of polarisation.

CIRCULAR DICHROISM (CD) The occurrence of a difference in the light absorption for right and left circularly polarised light in an optically active material. Quantitatively it is given by the difference in the extinction coefficients $(\Delta\varepsilon)$ for right and left circularly polarised light (ε_L and ε_R) as: $(\Delta\varepsilon)_\lambda = \varepsilon_L - \varepsilon_R = \log_{10}(I_R/I_L)/[M]l$ where I_R and I_L are the intensities of the right and left circularly polarised light (of wavelength λ) transmitted by a sample of path length l and molar concentration $[M]$. Sometimes ellipticity is used as a measure of circular dichroism, since CD arises from the emerging light being elliptically rather than linearly polarised. CD is dependent on wavelength (just as is optical activity where it gives rise to optical rotatory dispersion) and a peak in the CD spectrum of $\Delta\varepsilon$ versus λ is called a Cotton band. The CD spectrum is a miniature of the absorption spectrum and can only be observed in its vicinity. CD is widely used in conformational studies, e.g. of α-helix content and of helix–coil transitions of synthetic polypeptides and proteins.

CIS-9,CIS-12,CIS-15-OCTADECATRIENOIC ACID   Alternative name for linolenic acid.

CIS-9,CIS-12-OCTADECADIENOIC ACID
Alternative name for linoleic acid.

CIS-CONFORMATION   (1) The particular geometric isomeric form of a repeat unit of a polydiene or other double bond or saturated cyclic ring-containing (e.g. cyclohexane) repeat unit, in which the attachment of the polymer chain residues is cis to the double bond or ring, i.e.

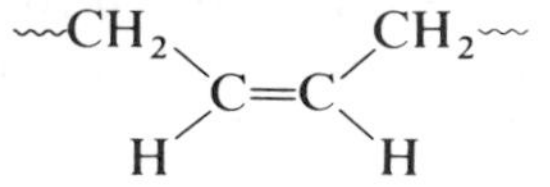

or

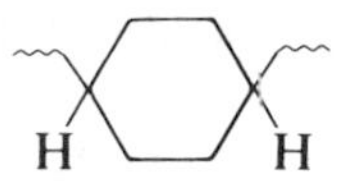

Polymers with high cis isomeric content have lower T_g and T_m values than the corresponding high trans polymers.

(2) Alternative name for gauche conformation.

CISDENE   Tradename for styrene–butadiene rubber.

CIS-9-OCTADECENOIC ACID   Alternative name for oleic acid.

CIS-1,4-POLYBUTADIENE

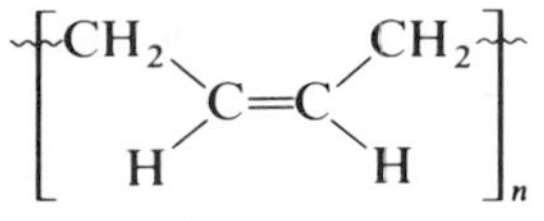

One of the stereoisomeric forms of polybutadiene. In butadiene rubbers the cis-1,4- structure usually predominates, as in high cis-polybutadiene (about 97%), medium cis-polybutadiene (about 92%) and low cis-polybutadiene (about 40%). Highly stereoregular cis-1,4- polymer has a T_g value of $-106°C$ and a T_m value of about 3°C and is produced by Ziegler–Natta polymerisation.

CIS-1,4-POLYISOPRENE   Tradenames: Cariflex IR, Coral, Natsyn, Ameripol SN.

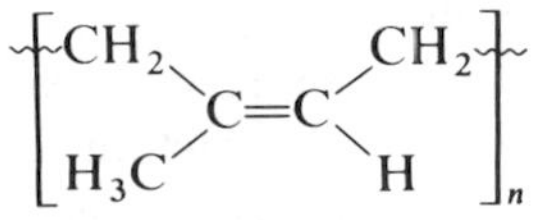

A polyisoprene in which the isoprene units are joined in the 1,4- manner and in which they occur in the cis isomeric form. The most important isoprene polymer since natural rubber is composed of polyisoprene in this isomeric form.

Similar synthetic polymer is also produced as a commercial product. Natural rubber polymer has at least 97% and probably 100%, of the cis-1,4-structure. The regularity of the structure enables the polymer to crystallise when highly strained, so that the modulus at high elongations is enhanced and pure gum vulcanisates have high tensile strengths. Classically, the 1,4- structure was identified by characterisation of the ozonisation products as laevulinic acid and laevulinic aldehyde:

$$\sim CH_2C(CH_3)\!\!=\!\!CHCH_2CH_2C(CH_3)\!\!=\!\!CHCH_2\sim \rightarrow$$
$$\sim CH_2C(CH_3)\!\!=\!\!O + O\!\!=\!\!CHCH_2CH_2C(CH_3)\!\!=\!\!O$$
$$\downarrow \qquad\qquad + O\!\!=\!\!CHCH_2\sim$$
$$HOOCCH_2CH_2C(CH_3)\!\!=\!\!O$$

Infrared and NMR methods are now more useful in the structural identification of polyisoprenes.

Such a pure cis-1,4- polymer as is found in natural rubber cannot be synthesised, but synthetic polymers with very high cis-1,4- contents are made by stereoregular polymerisation. Commercial polymers have cis-1,4- contents of greater than 90%, most being made by Ziegler–Natta polymerisation and having a cis-1,4- content of about 96%. The so-called 'low' cis-1,4- polymer is made by anionic solution polymerisation with an alkyl, usually butyl, lithium catalyst and has about a 92% cis-1,4- content, the remaining units being 3,4-polyisoprene structures. The high cis synthetic polymers naturally have properties closely resembling those of natural rubber. However, there are significant differences, e.g. slightly reduced tendency to crystallise. They also vulcanise a little more slowly than natural rubber whose protein content accelerates vulcanisation. Both natural and synthetic polymers have a T_g value of about $-73°C$ and a T_m value of about 20–30°C, so they do not normally crystallise unless highly strained. The synthetic polymer is used in a similar way to natural rubber, including the method of vulcanisation. Considerable efforts have been made to modify the synthetic polymer, e.g. by introducing anhydride groups, to improve its green strength.

CIS–TRANS ISOMERISM   Alternative name for geometric isomerism.

CIS-9,TRANS-11,TRANS-13-OCTADECA-TRIENOIC ACID   Alternative name for eleostearic acid.

CLARENE   Tradename for an ethylene–vinyl alcohol copolymer and for polyvinyl alcohol.

CLARITY   (See-through clarity) The ability of a transparent material to transmit a clear image of an object when viewed through it, without any aberration. It is caused by forward scattering of the light at very small angles to the undeviated beam. It is assessed by viewing charts through the film under test. The charts consist of sets of parallel lines perpendicular to each other and with different separations.

CLASSICAL LADDER POLYMER   Alternative name for a double strand polymer.

CLATHRATE POLYMERISATION Alternative name for canal polymerisation.

CLIMB (Dislocation climb) The movement of an edge dislocation out of its slip plane. This can occur only by removal or introduction of a row of motifs, by diffusion of interstitial motifs or vacancies and is thus a slow process.

CLORENE Tradename for a vinylidene chloride copolymer fibre.

CLOROFIN Tradename for chlorinated paraffin.

CLOUD POINT CURVE The curve on a phase diagram, e.g. of temperature versus composition, separating the region of miscibility from the two phase region, where phase separation occurs. For polymer solutions or blends, the curve shifts to higher temperatures as molecular weight increases, consistent with the Flory–Huggins theory, which also predicts a maximum in the curve at the upper critical solution temperature. However in practice, in some cases, a minimum in the curve occurs at a lower critical solution temperature. Cloud point curves may be determined experimentally by light scattering measurements.

CLUPEINE One of the most widely studied protamines, found in herring sperm. It contains about 90% arginine, as short sequences, separated by single or by two neutral amino acid residues of only seven different types. It contains several different, but closely related, polypeptide chains of 31 or 32 residues, N-terminated by proline or alanine and C-terminated by arginine. The sequences of three separated fractions have been determined. The total molecular weight is about 5000.

CM Abbreviation for chlorinated polyethylene rubber.

CM-1 Tradename for hexafluoroisobutylene–vinylidene fluoride copolymer.

CM-1 FLUOROPOLYMER Tradename for hexafluoroisobutene–vinylidene fluoride copolymer.

CMC Abbreviation for carboxymethylcellulose or sodium carboxymethylcellulose.

CM-CELLULOSE Abbreviation for carboxymethylcellulose.

CN Abbreviation for cellulose nitrate.

CNR Abbreviation for carboxynitrosorubber.

CNSL Abbreviation for cashew nut shell liquid.

COACERVATE The polymer-rich liquid phase which separates from a polymer solution when the solvent power is reduced by, for example, addition of a non-solvent. Coacervation extraction is a method of fractional solution.

COACERVATION EXTRACTION A method of fractional solution in which the polymer is first dissolved in a solvent to which a non-solvent is then added so that the polymer nearly completely separates as a liquid coacervate. The coacervate is then extracted with solvents of increasing solvent power to yield fractions of increasing molecular weight. In this way equilibrium between polymer and solvent during extraction is more rapidly achieved than in direct extraction of solid polymer.

COAGENT A vulcanisation agent used in the vulcanisation of ethylene–propylene rubbers to minimise chain scission.

COBALT NAPHTHENATE The most commonly used accelerator for the room temperature curing of unsaturated polyester resins when used with cyclohexanone or methylethyl ketone peroxide initiator. The cobalt salt of naphthenic acid, which is a complex mixture of mainly cycloaliphatic carboxylic acids obtained from crude petroleum.

COCATALYST Alternative name for co-initiator.

COCONUT OIL A non-drying oil whose triglycerides contain large amounts of lauric and myristic acid residues. Sometimes used in alkyd resins for good colour retention.

CODEFORMATIONAL DERIVATIVE Alternative name for Oldroyd derivative.

COEFFICIENT OF VISCOSITY Symbol μ or η. (Viscosity) (Viscosity coefficient) (Newtonian viscosity) (Dynamic viscosity) The quotient of the shear stress divided by the shear rate for steady flow of a Newtonian fluid. It is therefore independent of shear rate and is a constant. Non-polymeric fluids, which usually show Newtonian behaviour, have coefficients of viscosity in the range 10^{-3}–10^{-1} N s m^{-2} (i.e. Pa s) which corresponds to 10^{-2}–1 P. Most polymeric fluids, solutions or melts, are non-Newtonian so do not have a coefficient of viscosity which is constant with rate of shear. Their behaviour is therefore described by an apparent viscosity.

COEFFICIENT OF VISCOUS TRACTION Symbol η_t or τ. The quotient of a tensile stress to the rate of elongation for a Newtonian fluid undergoing elongational flow and therefore for which the coefficient is a constant, being independent of rate of elongation. This contrasts with the elongational viscosity of non-Newtonian materials, which is not constant with rate of elongation. For an incompressible Newtonian material the coefficient of viscous traction is three times the coefficient of viscosity. The term is also sometimes used as an alternative for elongational viscosity.

COEFFICIENT OF VULCANISATION The amount of sulphur combined with a vulcanised rubber network, defined as the parts by weight of sulphur combined per one hundred parts of rubber.

COENZYME An organic molecule which acts as the cofactor to an enzyme protein. Coenzymes function as carriers of specific groups, atoms or electrons in the enzymically catalysed reaction. The coenzyme has a vitamin as part of its overall structure. When the coenzyme is strongly bonded to the protein it is often referred to as a prosthetic group. On the other hand the bonding may be so loose that the coenzyme is acting more in the nature of a specific substrate. Examples of coenzymes are nicotinamide, adenine dinucleotide and flavin mononucleotide (involved in electron transfer) and coenzyme A (involved in aldehyde transfer).

COFACTOR The non-protein component of an enzyme which is necessary for enzyme activity. The protein–cofactor complex is the holoenzyme. The cofactor may be a metal ion (giving a metallo enzyme), e.g. Zn^{2+} in carboxypeptidase, Mg^{2+} in phosphohydrolases and Fe^{2+} or Fe^{3+} in cytochrome, or an organic molecule (when it is called a coenzyme). The cofactor may act as a primary catalytic centre (as with some metal ions) or as a bridging group to bind substrate and enzyme together. Alternatively, it may act to stabilise the enzyme protein in the particular conformation which is biologically active. Coenzymes act mostly as carriers of specific groups, atoms or electrons in the enzymically catalysed reaction.

COHEN AND TURNBULL EQUATION An equation derived from a free volume theory of viscoelastic behaviour, which states that for a material whose viscosity (η) follows the Doolittle equation, η is given by $\eta = \partial \exp(B'/(T - T_2))$ and the relaxation time (τ) is given by $\tau = \tau_0 - \exp(B'/(T - T_2))$ where a and B' are constants, T is the temperature and T_2 is the temperature (similar to the T_2 of the Gibbs–De Marzio equation) above which a redistribution of free volume can occur without a change in energy. T_2 is about 50 K below T_g. τ_0 is the relaxation time at T_g.

COHESIVE ENERGY DENSITY (CED) The quotient of the molar cohesive energy and the molar volume (V) of a substance. The former is equal to the molar heat of evaporation (ΔE) since on evaporation all intermolecular forces must be overcome. Thus $CED = \Delta E/V$. In the Hildebrand theory of solutions, the heat of mixing is given by:

$$\Delta H_m = V\phi_1\phi_2[(\Delta E_1/V_1)^{1/2} - (\Delta E_2/V_2)^{1/2}]^2$$

where ϕ_1 and ϕ_2 are the volume fractions of components 1 and 2. Thus it is more convenient to consider the solubility parameter (δ), where $\delta = (CED)^{1/2}$ when considering enthalpy of mixing. From the above equation it is apparent that the theory only considers cases of a positive heat of mixing.

CO-INITIATOR (Cocatalyst) One of the two components of a commonly used type of initiator for cationic polymerisation. Such initiators consist of mixtures of a Lewis acid, e.g. BF_3, $SnCl_4$, $TiCl_4$ and a Brönsted acid, such as water or HCl. The Brönsted acid is usually referred to as the co-initiator, and the Lewis acid is known as the initiator, since it can often initiate when used alone. The study of cationic polymerisation is often complicated by the presence of small amounts of impurities, especially water, which act as co-initiators. The term cocatalyst is sometimes used instead, but is often not preferred since the 'cocatalyst' is consumed during the reaction.

COIR A seed hair fibre from the husk surrounding the shell of the fruit of the coconut palm. The fibres are 0·1–1·5 mm in diameter and 10–35 cm in length and consist mainly of cellulose (about 40%) and lignin (about 40%). It is useful for matting, sacking and upholstery.

COLD BEND TEMPERATURE The lowest temperature at which a strip of polymer of specified dimensions does not crack or fracture when wound onto a standard mandrel.

COLD CURE (Cold vulcanisation) A method of vulcanisation of natural rubber at ambient temperature using sulphur chloride as the vulcanisation agent. This method is now replaced by the use of ultraaccelerators such as dithiocarbamates.

COLD DRAWING The plastic deformation that occurs after a neck has formed on tensile stressing beyond the yield point. A stable neck is formed and elongation of the sample occurs by movement of the neck along the sample, the cross-section thereby growing in length at the expense of the undrawn material. It follows the sharp drop in stress resulting from the formation of the neck.

COLD FLEX TEMPERATURE The temperature at which a rectangular strip of polymer of specified dimensions, when twisted using a specified torque at low temperatures, is twisted through 200°.

COLD FLOW Alternative name for creep occurring at ambient temperature.

COLD-HARDENING The hardening and loss of elasticity of an elastomer on storage at low temperature. It is frequently due to crystallisation, as with several polyester based polyurethane elastomers, the polyester segments of which, having melting points of about 50°C (as for example with poly-(ethylene adipate)), crystallise on storage even at room temperature.

COLD RUBBER A rubber, usually styrene–butadiene rubber, produced by emulsion polymerisation at about 5°C by use of a redox initiator.

COLD VULCANISATION Alternative name for cold cure.

COLE–COLE EQUATION An empirical equation, which is a modification of the Debye dispersion equation, relating the dielectric properties of a dielectric to the frequency (ω). It is: $\varepsilon^* = \varepsilon_\infty + (\varepsilon_s - \varepsilon_\infty)/(1 + (i\omega\tau)^\alpha]$ where ε^* is the complex relative, ε_s is the static and ε_∞ is the high frequency relative permittivity, α is a parameter, of value between zero and unity, which corresponds to introducing a range of relaxation times symmetrically distributed about the most probable relaxation time τ, rather than the single distribution time of the Debye equation. This leads to a broader relaxation, as is usually found with polymers. This has the effect of depressing the centre of the semi-circle on a Cole–Cole plot below the abscissa.

COLE–COLE PLOT A plot of the dielectric loss factor (ε'') against the relative permittivity (ε') of a dielectric for different values of the frequency. The Debye relaxation equations may be rearranged to give:

$$[\varepsilon' - (\varepsilon_s + \varepsilon_\infty)/2]^2 + \varepsilon''^2 = [(\varepsilon_s + \varepsilon_\infty)/2]^2$$

where ε_s is the static relative permittivity and ε_∞ is the high frequency value. This equation is the equation of a circle with centre at $((\varepsilon_s + \varepsilon_\infty)/2, 0)$, i.e. on the abscissa (the ε' axis). Thus a plot of ε'' against ε' should be a semi-circle. This is so for many polar liquids, but for polymers relaxations are broader and the points on the Cole–Cole plot fall outside the circle. The Cole–Cole equation can represent such behaviour and leads to a Cole–Cole plot in which the centre of the circle is depressed below the abscissa.

COLLAGEN A major animal structural protein, in higher animals, often the most abundant protein accounting for up to about 35% of the total protein. Found as a major component of skin, tendon, cartilage, bone and teeth. Various structural levels may be identified. The individual polypeptide chains exist as entwined triple helices (the tropocollagen 'molecules'). The tropocollagen molecules are aggregated into fibrils which pack parallel to each other to form fibres. The fibres are arranged differently in different tissues, being packed parallel in tendon giving high tensile strength but little elasticity, whereas in skin they are packed in a felt-like way. In bones and teeth network structures containing calcium salts give high rigidity.

Typically, collagen is composed of about 35% glycine, 10% alanine, 12% proline and 10% hydroxyproline with smaller amounts of other amino acids, including the unusual acid hydroxylysine. Variable small amounts of carbohydrate may also be covalently attached to the hydroxylysine residues. Each chain has about 1000 amino acid residues and, unusually for a protein, contains sequences of similarly repeating residues of the type Gly–X–Pro, Gly–Pro–X and Gly–X–Hypro. Thus glycine occupies every third position throughout the length of the chain, except at the chain ends, which contain less glycine and are highly polar and less regular. Polar and non-polar regions alternate along the chain, the latter containing a high proportion of Gly–Pro–Hypro tripeptide sequences.

The chains adopt an unusual helical conformation with three residues per turn. Three helices are entwined to form a tropocollagen 'molecule'. Two polypeptides (the α_1-chains) of the tropocollagen are usually identical whilst the third (α_2-chain) is different. The helix is held together by both hydrogen bonds and crosslinks. The tropocollagen molecules are packed in a parallel fashion and head-to-tail to give fibrils, but with a gap between the end-to-end arrays. The heads are staggered such that polar and non-polar portions of different tropocollagen molecules are brought together. This accounts for the 60/70 nm spacing of the bands seen on the electron microscope. The strength of the fibrils is enhanced by formation of dehydrolysine–norleucine crosslinks between tropocollagen molecules. In the tanning of hides to produce leather additional crosslinks are formed by reaction with aldehydes. Collagen is biosynthesised from its precursor procollagen. The boiling of collagen in water denatures the collagen and produces gelatine. However, when the gelatine solution is cooled, partial renaturation occurs to give a gel.

COLLIGATIVE PROPERTY Symbol P. One of several properties of a dilute solution of a solute which is dependent on $\ln a_1$, where a_1 is the activity of the solvent. For a very dilute solution which behaves ideally, $\ln a_1 = -x_2$, where x_2 is the mole fraction of the solute. The colligative property P is then directly proportional to the number of solute molecules (N_i) for a polydisperse solute containing many different species (i) per unit volume (V), i.e. $P = kN_i/V$, where k is a constant. If the weight concentration of solute is c, which equals $w_i/V = N_iM_i/VN_A$ where M_i is the molecular weight of species i and N_A is Avogadro's number, then $P/c = kN_AN_i/N_iM_i = kN_A/\bar{M}_n$ (eqn. (1)). Therefore measurement of P yields the number average molecular weight $\bar{M}_n$.

Osmotic pressure is the most useful colligative property for determination of polymer molecular weight values, measured by membrane osmometry, although vapour pressure lowering (using vapour pressure osmometry), ebulliometry and cryoscopy are sometimes useful. For dilute non-ideal solutions, the P–$\bar{M}_n$ relationship is a virial equation: $P/c = RT/\bar{M}_n + Bc + Cc^2 + Dc^3 + \cdots$ or $P/c = RT(1/\bar{M}_n + A_2c + A_3c^2 + A_4c^3 + \cdots)$ or

$$P/c = (P/c)_0(1 + \Gamma_2c + \Gamma_3c^2 + \cdots)$$

where $(P/c)_0$ is the value of P/c extrapolated to infinite dilution, when the equations simplify to eqn (1). B, C, D, ... (or $A_2, A_3, A_4, \ldots$) (or $\Gamma_2, \Gamma_3, \Gamma_4, \ldots$) are the second, third, fourth, ... etc. virial coefficients and $B = RTA_2 = RT\Gamma_2/\bar{M}_n$, $C = RTA_3 = RT\Gamma_3/\bar{M}_n$, ... etc. In dilute solution, less than about 1% w/v, terms in c^3 and above are negligibly small and the c^2 term may also often be ignored. The second virial coefficient (B_2 or A_2 or Γ_2) is related to the excluded volume and is dependent on the nature of polymer–solvent interaction, according to the Flory–Huggins equation. Membrane osmometry is limited generally to polymers with $\bar{M}_n > 20\,000$, whereas vapour pressure osmometry (and the lesser used techniques) are generally only applicable if $\bar{M}_n$ is $< 20\,000$.

COLLODION A solution of cellulose nitrate in a mixed ether/alcohol solvent.

COLLOID A system in which at least one of the components exists in a fine particle size, in the range 1–1000 nm. The term can also refer to the particles themselves. Originally colloids were designated on the basis of their transport properties, i.e. the slow diffusion and lack of permeation through a semi-permeable membrane. Colloids may be classified as one of three types. (1) Colloidal dispersions, in which the particles are dispersed in a continuous dispersing medium in which both the disperse phase and the medium may be either solid, liquid or gas. In polymer work, liquid/liquid dispersions (emulsions) and solid/gas dispersions (foams) are of most interest. The disperse phase may have little interaction with the medium (a lyophobic colloid) or it may show considerable interaction (a lyophilic colloid). (2) Most lyophilic colloids comprise the second type of colloid—the polymer solutions where the 'particles' are now individual molecules rather than aggregates of molecules. (3) A third type of colloid is the association colloid, in which aggregates of molecules, parts of which are lyophobic (usually hydrocarbon chains) and parts of which are lyophilic (usually ionic), form in the dispersion medium, which is usually water. Such colloids are said to contain micelles.

Because of the small size of the particles involved in colloids, and hence the large surface areas, interfacial effects are very important. In most colloids (especially in those first to be investigated) the surfaces carry electrical charges, which help to stabilise the dispersion. This is true for many early studied polymer systems, which were proteins. However, in most polymer solutions the molecular 'particles' do not carry charges nor do they have a defined 'surface'. For these reasons polymer scientists now prefer to regard polymer solutions as true solutions, like any other solutions, rather than as being one type of the very large group of colloidal systems.

COLLOIDAL SILICA Alternative name for precipitated silica.

COLOPHONY Alternative name for rosin.

COLUMN CHROMATOGRAPHY In general any chromatographic technique in which the stationary phase is packed into a column. However, the term is usually restricted to adsorption chromatography carried out in a column, as opposed to thin layer chromatography.

COLUMN ELUTION Alternative name for column extraction.

COLUMN EXTRACTION (Column elution) (Solvent gradient chromatography) (Column fractionation) A method of fractional solution in which the polymer is deposited as a thin layer on a support material which is usually inert, e.g. sand or glass beads. The coated support is then packed in a column and fractions of increasing average molecular weight are successively eluted with solvents of increasing solvent power. In this way more rapid equilibrium is established between polymer and solution than in simple extraction. A variation is gradient elution chromatography and several other fractionation methods have a superficial resemblance to this method.

COLUMN FRACTIONATION Alternative name for column extraction.

COMBINATION (Coupling) The commonest mode of termination in free radical polymerisation, in which two free radical active centres combine to form a covalent bond between the two growing polymer molecules. Thus in a vinyl polymerisation:

$$\sim\!\!CH_2\dot{C}HX + \dot{C}HXCH_2\!\!\sim \rightarrow$$
$$\sim\!\!CH_2CHX\!\!-\!\!CHXCH_2\!\!\sim$$

a head-to-head link is formed. The resultant dead polymer molecule has a degree of polymerisation equal to the sum of those of the two combining molecules. The number average degree of polymerisation is twice the kinetic chain length. Most growing radicals of the type $\sim\!\!CH_2\!\!-\!\!\dot{C}HX$, e.g. styrene and acrylonitrile, terminate in this way, whereas those with an α-methyl group terminate by disproportionation. The molecular weight distribution is of the Schultz type with a degree of coupling of two and hence a polydispersity index of 1·5.

COMBINATORIAL ENTROPY Alternative name for configurational entropy.

COMB POLYMER A branched polymer in which the branches, often reasonably evenly spaced, are attached to a definitely identifiable main chain. Such polymers are frequently graft copolymers in which active sites have been formed on a preformed main chain. These act as initiation sites for the growth of side chains (the branches) by polymerisation of an added monomer.

COMONOMER A monomer which is mixed with other monomers or monomer and polymerised, thus producing a copolymer. A binary copolymer is formed from two comonomers, a ternary copolymer from three comonomers, etc.

COMPARATIVE TRACKING INDEX (CTI) A measure of the resistance of an electrical insulating material to tracking. It is determined by a standard test in which drops of aqueous 0·1% ammonium chloride solution are allowed to fall between electrodes pressed onto the surface of the insulator. The CTI is the voltage for which the insulator will just withstand 50 drops before failing.

COMPATIBILITY (1) The ability of one material to exist in the presence of another without any undesirable results. In particular, the resistance of a polymer (especially as a coating) to attack (solvation) by a solvent.

(2) In polymer blends, the term has, in effect, the opposite meaning to that in (1) since incompatibility is used to describe miscibility of one polymer in another.

COMPATIBILITY CONDITION The condition that the compatibility equations are satisfied.

COMPATIBILITY EQUATION One of a set of mathematical relationships between the strain components of a stressed body. Such relationships must exist since there are six strain components defined in terms of only three displacements, and they may be simply derived from these definitions. In the most general case, there are six equations, and these, together with the three relationships obtained from the equilibrium condition of forces and the six constitutive equations, allow solutions of problems in elasticity (involving six stress components, six strain components and three displacements) to be obtained.

COMPENSATION Nullification of the birefringence of a material by use of a second compensatory birefringent material in a technique utilising a Babinet or other type of compensator for determination of polymer birefringence.

COMPENSATOR METHOD A method for the determination of birefringence which uses a known retardation to nullify or compensate for the retardation of a birefringent sample by placing a birefringent crystal in the light path, as, for example, in a Babinet compensator. Such a method is highly sensitive and also determines the sign of the birefringence, but is not suitable when rapid changes in birefringence occur due to rapid sample deformation.

COMPLEX BULK COMPLIANCE Symbol $B^*(\omega)$. The complex compliance for bulk deformation.

COMPLEX BULK MODULUS Symbol $K^*(\omega)$. The complex modulus for a sinusoidally varying deformation in bulk compression.

COMPLEX COMPLIANCE (Complex dynamic compliance) Symbol $C^*(\omega)$ in general (although in elasticity the symbol C sometimes refers to the modulus, i.e. the inverse of the compliance), $J^*(\omega)$ in shear, $D^*(\omega)$ in tension and $B^*(\omega)$ in bulk. A mathematical representation of both the storage and loss compliances, $C'(\omega)$ and $C''(\omega)$ respectively, of a viscoelastic material, as obtained from dynamic mechanical measurements. Defined by the usual complex number notation as $C^*(\omega) = C'(\omega) - iC''(\omega)$, where $i = (-1)^{1/2}$. For a sinusoidally varying stress (of maximum amplitude σ_0) and strain (of maximum amplitude e_0) of angular frequency ω radians s^{-1} ($= 2\pi \times$ frequency (Hz)),

$$C^*(\omega) = e^*/\sigma^* = (e_0/\sigma_0)(1/e^{i\delta})$$

where δ is the phase angle between stress and strain. Thus $C^*(\omega)$ is time invariant and is the inverse of the complex modulus $M^*(\omega)$, although the storage and loss moduli are not simply inversely related in this way. Often $C''(\omega)$ is small compared with $C'(\omega)$ so that $C^*(\omega) \sim C'(\omega)$.

COMPLEX DIELECTRIC CONSTANT Alternative name for complex relative permittivity.

COMPLEX DYNAMIC COMPLIANCE Alternative name for complex compliance.

COMPLEX DYNAMIC MODULUS Alternative name for complex modulus.

COMPLEXITY DISTRIBUTION An empirical molecular weight distribution for the branched polymer obtained by polymerising a monomer containing one A-type group and $(f - 1)$ B-type groups, when A reacts with B to produce a polymer by step-growth polymerisation and f is the functionality with $f > 2$.

COMPLEX MATERIAL Alternative name for composite material.

COMPLEX MODULUS (Complex dynamic modulus) (Dynamic modulus) Symbols $G^*(\omega)$ in shear, $E^*(\omega)$ in tension, $K^*(\omega)$ in bulk and in general, $M^*(\omega)$. A mathematical representation of both the storage and loss moduli $M'(\omega)$ and $M''(\omega)$ respectively, of a viscoelastic material. By standard complex number notation $M^*(\omega) = M'(\omega) + iM''(\omega)$ where $i = (-1)^{1/2}$. Thus in a dynamic mechanical experiment with a sinusoidally varying strain (e^*) of angular frequency ω and a resultant out of phase sinusoidally varying stress σ^*,

$$M^*(\omega) = \sigma^*/e^* = (\sigma_0/e_0)e^{i\delta} = \sigma_0/e_0 (\cos\delta + i\sin\delta)$$

where σ_0 and e_0 are the maximum amplitudes of the stress and strain respectively and δ is the angular phase lag (or phase angle) in radians. Thus $M^*(\omega)$ is time invariant. Often $M''(\omega)$ is small compared with $M'(\omega)$ so that $M^*(\omega) \approx M'(\omega)$. Typically $M'(\omega)$ is about 10^9 Pa and $M''(\omega)$ about 10^7 Pa. Although $M^*(\omega) = 1/C^*(\omega)$, where $C^*(\omega)$ is the corresponding complex compliance, their components $M'(\omega)$ and $C'(\omega)$ and $M''(\omega)$ and $C''(\omega)$ are not so simply inversely related. The absolute value of the complex modulus $|M(\omega)| = [M'(\omega)^2 + M''(\omega)^2]^{1/2}$ is sometimes called the absolute modulus.

COMPLEX PERMITTIVITY Alternative name for complex relative permittivity.

COMPLEX RELATIVE PERMITTIVITY (Complex dielectric constant) (Complex permittivity) Symbol ε^*. A mathematically convenient way of combining the two components of the relative permittivity of a dielectric, the real part ε' and the loss factor ε'', when the dielectric is subject to a sinusoidally alternating electric field. Thus in the usual notation for a complex quantity, ε^* is given by: $\varepsilon^* = \varepsilon' - i\varepsilon''$, where i is $(-1)^{1/2}$. The frequency dependence of ε^*, ε' and ε'', is given by the Debye dispersion equation. The real part, also called the relative permittivity, is responsible for the capacitative (storage) component of the current flowing across a capacitor filled with the dielectric and leads the voltage by 90°. The imaginary part, the loss factor, is responsible for the resistive component of the current in phase with the voltage. ε^* is analogous to the complex compliance involved in the dynamic mechanical behaviour of polymers.

COMPLEX SHEAR COMPLIANCE Symbol $J^*(\omega)$. The complex compliance for a sinusoidally varying deformation in shear.

COMPLEX SHEAR MODULUS Symbol $G^*(\omega)$. The complex modulus for a sinusoidally varying deformation in shear.

COMPLEX TENSILE COMPLIANCE Symbol $D^*(\omega)$. The complex compliance for a sinusoidally varying deformation in tension.

COMPLEX TENSILE MODULUS Symbol $E^*(\omega)$. The complex modulus for a sinusoidally varying deformation in tension.

COMPLEX VISCOSITY Symbol η^*. A mathematical representation of the overall time-dependent viscosity of a fluid undergoing oscillatory flow; in complex number form it is given by: $\eta^* = \eta' - i\eta''$, where η' is the dynamic viscosity and is the real component representing the viscous component, and η'' is the imaginary component, which is related to the shear storage modulus (G') by $G' = \omega\eta''$, where ω is the frequency. The imaginary component η'' is not in phase with the real component η' and represents the elastic response of the fluid. $|\eta''|$ equals $\tau_m/\dot{\gamma}_m$, where τ_m and $\dot{\gamma}_m$ are the maximum values of the shear stress and shear rate respectively.

COMPLIANCE (Elastic compliance) Symbol S. An elastic constant which is the ratio of a strain or strain component to a stress or stress component. For a perfectly elastic material it is the reciprocal of the elastic modulus. For a viscoelastic material the modulus and compliance are not reciprocally related due to their different time dependencies. In tensor notation the relationship $\varepsilon_{ij} = S_{ijkl}\sigma_{kl}$, where σ_{kl} and ε_{ij} are the generalised stress and strain components, defines 81 possible different compliances S_{ijkl} for a material, since i and j can have any of three values. However, the condition of static equilibrium reduces these to 36. Furthermore, the existence of strain energy function (giving $S_{ijkl} = S_{klij}$) reduces the number further to 21. Material symmetry reduces the number even more and for an isotropic material there are only two independent compliances. The common experimentally measured compliances are the tensile, shear, bulk and creep compliances.

COMPOSITE A shaped product made of a composite material, such as a moulding, laminate or extrudate. Sometimes used to refer to the composite material itself.

COMPOSITE MATERIAL (Complex material) A solid material which consists of a combination of two or more simple (or monolithic) materials and in which the individual components retain their separate identities. A composite material has properties different from those of its component simple materials; use of the term composite often implies that the physical properties are improved since the main interest technologically is in obtaining materials with superior physical (usually mechanical) properties to those of the composite's component materials. A composite material also has a heterogeneous structure containing two or more phases arising from its components. The phases may all be continuous phases or one or more may be dispersed phases within a continuous matrix. In the latter case it is necessary to state some lower limit to the size of the dispersed phase particles below which the material is considered to be monolithic. The size is commonly taken to be of the order of 10^{-8} m, since this is approximately the lowest size limit obtainable in particle manufacture. Furthermore, the range 10^{-9}–10^{-8} m is commonly regarded as the dividing line between true solutions and colloidal dispersions. Composites of many types are encountered in polymer materials technology. They may consist of polymer–polymer combinations (polymer blends) or polymer–gas combinations (expanded, cellular or foamed polymers), but the two most common types are the polymer–stiff filler combinations of polymer–fibre and polymer–particulate composites. In these latter two types, the aim is usually to obtain some enhancement of one or more of the mechanical properties, i.e. reinforcement, although cheap fillers are sometimes used merely to act as volume-filling, cost-reducing agents, i.e. as extenders. A full description of a composite material requires a knowledge not only of the composition of the phases but also the geometry (particle shape, size, size distribution and orientation) and concentration of any dispersed phases. Concentration is usually expressed as volume fraction of each phase. In addition, a distinct interfacial region may exist, whose nature can considerably influence the properties of the material, especially by the use of coupling agents. The shorter term, composite, usually refers to some particular shaped product made from a composite material.

COMPRESSION SET The set remaining after a predetermined time (often 30 min) after removal of a compressive deforming force. Usually refers to rubbers tested as circular discs.

COMPRESSIVE ELASTIC MODULUS The ratio of the stress to the strain in uniaxial compression. Often measured as the slope of the initial linear portion of the stress–strain curve. For isotropic materials the value is approximately the same as that of the Young's modulus.

COMPRESSIVE STRENGTH The maximum stress that a rigid material will withstand when under uniaxial compression. This is not necessarily to the point of rupture, since some polymers undergo cold flow. Typical values for polymers are about 100 MPa for rigid plastics and 100–500 MPa for polymer composites.

COMPTON SCATTERING Alternative name for incoherent scattering.

CONALBUMIN An egg-white albumin, accounting for about 14% of the protein content, capable of binding metal atoms, e.g. ferric iron, to give a red complex. Its molecular weight is about 80 000. Unlike ovalbumin it is not denatured by shaking.

CONDENSATION POLYMER A step-growth polymer produced by a polymerisation reaction in which the elimination of a small molecule, often water, has occurred, i.e. produced by a condensation polymerisation. Thus most step-growth polymers are also condensation polymers, important examples being the polyesters, polyamides and phenol–, urea– and melamine–formaldehyde polymers. However, some polymer types, notably the polyesters and polyamides, can also be formed by a non-condensation ring-opening polymerisation. Some other important step-growth polymers, polyurethanes being the outstanding example, are not condensation polymers.

CONDENSATION POLYMERISATION (Polycondensation) A step-growth polymerisation which involves the elimination of a small molecule, frequently water (as in polyesterification and polyamidation) but which may also be, for example, carbon dioxide, ammonia or hydrochloric acid, during the polymer forming reaction between the functional groups. Since the classic step-growth reactions (extensively studied in earlier work which established the principles of step-growth polymerisation) were condensations (mostly polyesterification and polyamidation) all step-growth polymerisations have been referred to as condensation polymerisations. However, in many such polymerisations, notably polyurethane formation, no small molecule is eliminated, 'condensed' out. Other examples of condensation polymerisation are phenol–, urea– and melamine–formaldehyde formation (elimination of water), epoxy resin formation, amidation and esterification with acid chlorides (hydrochloric acid elimination) and the formation of many heterocyclic polymers, often via the formation of a more tractable precursor polymer. The polymers so formed are often referred to as condensation polymers.

CONDUCTING POLYMER A polymer which exhibits higher than usual electrical conductivity, i.e. greater than about $10^{-10}\,S\,cm^{-1}$. Since nearly all such polymers have conductivities of $< 10^2\,S\,cm^{-1}$ they should be referred to as semiconducting polymers. Only a comparatively few polymers, such as poly-(sulphur nitride) fibres in the fibre direction, exhibit true, i.e. metallic, conductivity of $> 10^2\,S\,cm^{-1}$.

CONDUCTIVE FURNACE BLACK (CF) A type of furnace carbon black that is electrically conducting. Useful for the production of electrically conducting polymer composites.

CONDUCTIVE RUBBER A rubber with a much lower electrical resistivity than normal. This is achieved by high loadings (e.g. 50 phr) of one of the more highly conducting grades of carbon black, e.g. super conducting furnace or acetylene black. A typical resistivity is about $10^{15}\,\Omega\,cm$ compared with about $10^{15}\,\Omega\,cm$ in a normal rubber.

CONE AND PLATE RHEOMETER A particular design of rheometer in which the sample chamber consists of a lower, fixed flat horizontal plate and a wide angle cone, with apex angle about $178°$. The cone is suspended so that its apex is just out of contact with the plate. The gap between the cone and plate is filled with the fluid under investigation. On rotating the cone, the viscous drag is measured by the force required to maintain cone rotation at constant speed. The advantage of this arrangement is that the shear rate is constant throughout the fluid, i.e. it is independent of distance from the axis of rotation. This simplifies analysis of the data. The instrument may be modified to measure normal forces, as in the Weissenberg rheogoniometer, but is limited to relatively low rates of shear.

CONEX Tradename for poly-(m-phenylene-isophthalamide).

CONFIGURATION In structural organic chemistry, the arrangement in three-dimensional space of the atoms or groups attached to a central atom, so that one configurational isomer can only be converted to another by the breaking of chemical bonds. Stereochemical isomerism, as occurring in stereoregular polymers, geometric isomerism, as found in polydienes, and positional isomerism are all types of configurational isomerism. However, physical chemists use the term as an alternative to the organic chemists' term conformation, to describe the overall shape of a polymer molecule as a result of the rotations possible about the polymer chain chemical bonds.

CONFIGURATIONAL ENTROPY (Combinatorial entropy) Symbol S_{conf}. The entropy gain obtained on mixing different molecules due to the number of possible ways of arranging the two different types of molecule, in binary mixtures, by interchanging them. It is often considered as the sole contribution to the entropy of mixing (ΔS_m), being given by Boltzmann's equation as $S_{conf} = \Delta S_m = k \ln \Omega$, where k is Boltzmann's constant and Ω is the number of possible arrangements. In mixtures of small molecules this leads to:

$$\Delta S_m = - R(n_1 \ln x_1 + n_2 \ln x_2)$$

where R is the universal gas constant, n_1 and n_2 are the numbers of moles and x_1 and x_2 are the mole fractions of the two components 1 and 2. Polymer solutions show large deviations from this ideal solution behaviour due to their much lower entropies of mixing. These arise from the large number of conformations that a polymer molecule can adopt in solution compared with the pure polymer. The Flory–Huggins theory of polymer solutions leads to a similar equation to the above for ΔS_m, but with volume fractions replacing mole fractions.

CONFORMATION (Constellation) The overall three-dimensional shape adopted by a molecule by rotations about the single bonds present in the molecule. Each particular conformational state (sometimes referred to as a conformer or rotamer) has a certain potential energy due to energetic interatomic interactions. The more energetically favoured conformers are the more highly populated conformational states. Even in the case of fairly simple small organic molecules, the distribution of conformational states can be quite wide. For rotations about carbon–carbon single bonds, as for example in the simplest case of ethane, the eclipsed conformation

is less likely than the staggered conformation

due to hydrogen–hydrogen atom repulsion. In slightly more complex molecules, such as butane, considering only the staggered conformations, the *trans* (t) conformation will be of lower energy than the *gauche* (g):

Hence in carbon–carbon atom polymer chains the number of conformations is almost infinite and the overall shape adopted (the macroconformation) will be determined by the large number of individual bond rotations (or microconformations) possible.

In general, *trans* conformations are favoured in simple aliphatic hydrocarbon chains and the all-*trans* (or extended chain) conformation is found in the crystals of some crystalline polymers, such as polyethylene and nylons. When side groups are present in the chain, other conformations may be adopted. Thus in regular, e.g. isotactic, polymer chains, helical conformations such as the 3_1-helix (with an alternating tgtgtgtg conformational sequence) and the α-helix (tggtggtgg sequences) may be found in the crystals. In these conformational sequence descriptions, a bond conformer is called t or g even though deviations of up to about 30° from the precise t or g rotational position may be involved.

In chain folded crystals the chain folding is due to a succession of *gauche* conformers in the chain. Lattice defects in a crystal are due to the presence of irregular conformational states in an otherwise regular chain. This can give rise to the defects of kinks, gaps and Reneker defects.

When neighbouring substituents carry atoms which have electron pairs on electronegative substituents, then interatomic attractions may result in adoption of the conformation which permits the maximum number of *gauche* interactions between the atoms involved (the *gauche* effect). This is the case in polymers such as polyoxymethylene, $\text{+CH}_2\text{—O+}_n$ and $\text{+CH}_2\text{—CO+}_n$,

where compared with polyethylene, $\text{+CH}_2\text{—CH}_2\text{+}_n$, the potential energy barriers to rotation are less; the chains are therefore more flexible, resulting in lower T_g and melt viscosity values. A similar effect is found in polysiloxanes due to the longer bond lengths of the chain bonds.

In solution, polymers usually adopt a random coil conformation, except that sometimes the α-helix may be the most stable conformation, as with some polypeptides. It has always been assumed that the random coil is also adopted by polymers of irregular structure in the solid state, giving an amorphous material. This has only recently been confirmed by small angle neutron scattering measurements. Similarly in the amorphous regions of partially crystalline polymers, the polymer molecules are randomly coiled. The dimensions of a random coil are characterised by its end-to-end distance or by its radius of gyration. Several theoretical models for the random coil conformation have been analysed in order to relate the dimensions of the coil to the fundamental parameters of bond lengths and angles. Typical models are the freely jointed chain and freely rotating chain models. Analysis of models is often simplified by replacing actual bonds and chain atoms by statistically equivalent chain elements. The rotational isomeric state model pays particular attention to non-bonded interatomic forces so that the chain is no longer freely rotating.

β-CONFORMATION   The extended chain, zig-zag, conformation of a polypeptide chain. Neighbouring chains in the β-conformation are aggregated to form sheets of molecules packed parallel to each other. However, the term parallel β-conformation refers to the chains all running in the same direction, i.e. N-terminal to C-terminal. When alternate chains run in opposite directions the anti-parallel β-structure results. The angles of rotation are $\phi = -120°$ and $\psi = +115°$ for the parallel arrangement and $\phi = -140°$ and $\psi = +135°$ for the anti-parallel arrangement, so the chains are not fully extended. The chains are held together by interchain hydrogen bonding, in which all peptide groups participate, thus providing the driving force for sheet formation. Alternate α-carbons are up and down giving a pleated sheet. Some fibrous proteins, e.g. β-keratin and fibroin, largely exist in this conformation. Portions of the conformation of some globular proteins are also of this type. Owing to the extensive intermolecular hydrogen bonding a polymer in this conformation is usually highly insoluble.

CONGO COPAL   A copal natural resin obtained from the tree *copaifera* in central Africa. A hard insoluble type of copal and normally collected as a fossil resin. It softens at about 100°C and liquefies at about 150°C. It consists of about 80% monobasic hydroxy acid and a dibasic acid with about 20% resene. It is very widely used in varnishes and gloss paint making after being 'run', which chemically alters it by heating to about 300°C and increases its solubility.

CONIC MARKING   A fracture surface produced by the interaction of the main fracture front with a nearly coplanar secondary, circularly spreading fracture. The main fracture becomes diverted to the secondary fracture plane through a step, the locus of whose points is therefore a conic, which appears on the fracture surface. The conic may be a parabola (parabola marking) or a hyperbola. The secondary fracture arises from stress concentrations around the main fracture due to the presence of inhomogeneities in the material. It is often found in the fracture of nominally homogeneous or of filled brittle polymers.

CONIFERYL ALCOHOL

$$HOCH_2CH{=}CH{-}\underset{}{\bigcirc}\!\!-OH$$ with OCH$_3$ substituent.

M.p. 74°C.

One of the main 'monomers' from which lignin is built up by oxidative polymerisation.

CONJUGATED POLYMER
A polymer containing sequences of alternating double bonds (conjugated double bonds), usually involving carbon or carbon and nitrogen atoms, in the polymer chain. If delocalisation of the double bond π-electrons is extensive (eka-conjugated polymer) the polymers have enhanced electrical conductivity (some are semiconducting polymers) and often also high thermal stability. Often delocalisation is limited by the presence of structural defects, giving a rubiconjugated polymer. The sequences may be part of a simple single strand polymer chain, e.g. polyacetylene, polydiacetylenes, polyphenylene and dehydrochlorinated PVC, or they may be part of the fused ring structures of a ladder polymer produced by a cyclisation reaction, e.g. cyclised PAN. Often, however, and especially in the latter case, the sequences are of limited length, so the enhanced properties (especially conductivity) are much lower in value than expected for completely conjugated polymer chains.

CONJUGATED PROTEIN
A protein consisting of both a polypeptide and a non-polypeptide component, the latter being the prosthetic group. The two components may or may not be joined by covalent bonds. They are usually classified according to the nature of the prosthetic group as the nucleoproteins, lipoproteins, haemoproteins, phosphoproteins, flavoproteins and metalloproteins.

CONJUGATE FIBRE
Alternative name for bicomponent fibre.

CONSIDÈRE CONSTRUCTION
The construction of the tangent to a tensile true stress (σ)–strain (e) curve from the point -1 on the strain, i.e. elongation, axis. Since the maximum load is carried by the specimen when $d\sigma/de = \sigma/(1+e) = \sigma/R$, where R is the draw ratio, the tangential point gives the maximum load and hence the yield point. The construction leads to a useful criterion of whether the sample will neck or neck and cold draw. If $d\sigma/dR$ is always $> \sigma/R$, i.e. no tangent can be drawn, then the specimen will be uniformly extended on stressing, and necking does not occur. If $d\sigma/dR = \sigma/R$ at a single point, i.e. only one tangent can be drawn, then the specimen will neck, the neck will thin and, finally, the specimen fractures. If $d\sigma/dR = \sigma/R$ at two points, i.e. two tangents can be drawn, then the specimen will neck and cold draw.

CONSISTENCY
A general term expressing the resistance of a material to a permanent change in shape. Thus qualitatively it describes the overall fluid-like nature of a material, whether it is a gas, a mobile liquid, syruplike, toffee-like or a rigid solid. Quantitatively it can only be expressed by the overall force–flow behaviour which may be very complex. Sometimes the term refers specifically to the slope of the shear stress–rate of shear (τ–$\dot{\gamma}$) curve for a non-Newtonian fluid at a particular rate of shear, i.e. the consistency (η_c) $= d\tau/d\dot{\gamma}$.

CONSISTENCY CURVE
In rheological measurements, using a rotary torque rheometer, it is the plot of the rate of rotation, which is proportional to shear rate, against torque. Conversion of this curve to a flow curve is often not simple due to the shear rate varying with the distance from the axis of rotation, as in a coaxial cylinder rheometer. However, in a cone and plate rheometer the shear rate does not vary with this distance and the conversion is simple.

CONSISTENCY INDEX
Symbol k. One of the two parameters (the power law indices) of the power law equation: $\tau = k(\dot{\gamma})^n$, where τ is the shear stress and $\dot{\gamma}$ is the shear rate. It becomes equal to the viscosity for a Newtonian fluid, since n then equals unity. Sometimes the term refers to a related parameter k' in the relation: $\tau_w = k'(\dot{\gamma}_w)^n$, where τ_w and $\dot{\gamma}_w$ are the shear stress and shear rates at the wall respectively, for flow down a pipe with no slip at the wall. For a power law fluid $k' = k[(3n+1)/4n]^n$.

CONSTANT STRAIN RATE MODULUS
Symbol $F(\dot{e})$. A modulus characterising the behaviour of rubbers at large strains, derived from an extension to the linear theory of viscoelasticity. For rubbers extended at constant strain rate, $F(\dot{e})$ is a function of time only, being given by $F(\dot{e}) = g(e)[\sigma(e, t)]/e$, where σ is the strain- and time-dependent stress, e is the strain and $g(e)$ is some function of strain. Thus plots of log (stress) versus log (time) for different strain rates give a series of parallel straight lines.

CONSTANT VISCOSITY NATURAL RUBBER
(CVNR) (Viscosity stabilised natural rubber) Raw natural rubber in which the normal formation of crosslinks, which occurs after coagulation from the latex with a resultant increase in viscosity, has been eliminated. This is achieved by treatment of the latex either with hydroxylamine hydrochloride (about 0·15%) giving a stable viscosity of about 55–65 Mooney units or with hydrazine hydrate. Stabilisation at this low viscosity level can eliminate the need for subsequent mastication. The

stabiliser reacts with aldehyde groups present in the polymer at about 5–30 groups per million in molecular weight, which are thought to be the cause of crosslinking. If not eliminated, the crosslinking causes an increase in the Mooney viscosity of up to 20–30 units.

CONSTELLATION Alternative name for conformation.

CONSTITUTIONAL UNIT Alternative name for repeat unit.

CONSTITUTIVE EQUATION (Rheological equation of state) Any equation relating stress, strain, stress rate and strain rate. In ideal elastic solids the equation takes the form of a generalised Hooke's law involving only stress and strain in a linear relationship. However, most polymers are more complex in their behaviour since they are viscoelastic, so that stress rate and strain rate are involved. Further complications are that behaviour may be non-linear, strain may not be fully recoverable on removal of stress and the relationship may only be valid for small deformations. In addition, account may have to be taken of material anisotropy. Thus for most polymer materials simple models such as the Maxwell and Voight models do not represent the true behaviour of the material. However, they are still frequently used owing to their relative ease of mathematical handling of real flow problems. Solution of such problems requires that the constitutive equation defining the type of fluid (Newtonian, non-Newtonian, viscoelastic etc.), is used in conjunction with the dynamic equations, the continuity equation and the boundary conditions. Constitutive relationships are generalised for a particular material, independent of the size and shape of the sample.

CONTINUITY EQUATION (Equation of continuity) A mathematical statement of the principle of conservation of matter used in deformation and flow problems. Used with the momentum equation and the appropriate constitutive equation, it provides the powerful continuum mechanics approach to the solution of rheological problems. It may be written in its most general form, i.e. independent of choice of coordinates, in vector–tensor notation as $\partial\rho/\partial t = -(\nabla.\rho v)$, where ρ is the density, t is the time, ∇ is the nabla operator and v is a velocity component. This is the Eulerian form. If the coordinates move with the fluid element under consideration, then the Lagrangian form results: $D\rho/Dt = -\rho(\nabla.v)$, where D/Dt is the substantial time derivative. If the density is constant, i.e. for an incompressible material, then the equation is $(\nabla.v) = 0$. In expanded form, for rectangular cartesian coordinates, the equation is:

$$\partial\rho/\partial t = -[v_x\partial\rho/\partial x + v_y\partial\rho/\partial y + v_z\partial\rho/\partial z]$$
$$-[\rho\partial v_x/\partial x + \rho\partial v_y/\partial y + \rho\partial v_z/\partial z]$$

CONTINUOUS STRAND MAT (Swirl mat) A sheet-like, non-woven form of glass fibre in which strands, about 0·5 m long, have been distributed as a layer on a belt and then a binder applied. Particularly useful in the matched metal die moulding process for production of glass reinforced plastics.

CONTINUOUS ZONE ELECTROPHORESIS Alternative name for curtain electrophoresis.

CONTINUUM MECHANICS The analysis of flow and deformation on the assumption that the material under consideration is a continuum. This means that the properties of interest, although they may vary throughout the material, do so continuously, so that even very small elements are representative of the material as a whole. Obviously as the scale of scrutiny reaches molecular proportions, such an assumption is not valid; however, such a fine scale of scrutiny is not necessary for most problems. A particular deformational or flow problem can be analysed by equations representing the physical laws. One type is applicable to all materials and represents the various principles of conservation. Equations of this type are the balance equations for the conservation of mass (the continuity equation), the conservation of momentum (the momentum equation) and the conservation of energy (the energy equation), i.e. the first law of thermodynamics. Other equations represent the behaviour of individual classes of material. These are the thermodynamic equations of state (relating density, pressure and temperature), the constitutive equation (relating stress to kinematics), the heat transfer equation (relating heat flow to temperature distribution) and the energetic equation of state (relating internal energy to temperature, density and deformation). If the density may be considered constant, as is usually assumed in rheological problems, then the thermodynamic equation of state is $\rho = $ constant and problems are, in principle, soluble by use of the continuity, momentum and constitutive equations alone, together with the appropriate boundary conditions. In practice, however, especially for other than the simplest geometries, solution of the equations may be an impossibly complex task.

CONTOUR LENGTH (Displacement length) The length of a fully stretched out model for a polymer chain, as in the freely jointed and freely rotating models. If the chain consists of n links each of length l with valence angle of $180°$ between the links, then the contour length is nl. This compares with a value of $nl \sin(\theta/2)$ if the valence angle is θ.

CONTRACTED NOTATION An abbreviated nomenclature for representing stress and strain components, moduli and compliances. The double suffix notation is replaced by a single suffix notation. Thus the stress components σ_{xx}, σ_{yy} and σ_{zz} (or alternatively σ_{11}, σ_{22} and σ_{33} respectively) are written as σ_1, σ_2 and σ_3 respectively, whilst σ_{yz}, σ_{zx} and σ_{xy} (or alternatively σ_{23}, σ_{31} and σ_{12} respectively) become σ_4, σ_5 and σ_6 respectively. For the strain components, ε_{11}, ε_{22} and ε_{33} become e_1, e_2 and e_3 respectively, but $2\varepsilon_{23}$, $2\varepsilon_{31}$ and $2\varepsilon_{12}$ become e_4, e_5 and e_6 respectively. The reason for the factor 2 is that engineering shear strain (e) is defined as twice tensor shear strain (ε). The compliance (S) and stiffness (C) constants are given by the generalised Hooke's law by $\sigma_p = C_{pq}e_q$ and $e_p = S_{pq}\sigma_q$, in which for C_{pq}, suffix 1 is substituted for 11, etc., as above for the stress notation. For S_{pq}, $S_{ijkl} = S_{pq}$ when p and q are 1, 2 or 3, $2S_{ijkl} = S_{pq}$ when either p or q are 4, 5, or 6 and $4S_{ijkl} = S_{pq}$ when both p and q are 4, 5, or 6.

CONTRACTILE PROTEIN A protein which is involved in the contractile or motile systems of organisms. Myosin and actin are the best known examples.

CONVECTED COORDINATE A coordinate system which is embedded in a material and which therefore deforms as the material deforms, when acted upon by a system of forces. This is the Eulerian measure of strain, the coordinates are convected with the strain. In rheological problems, where constitutive equations are often written in convected coordinates, it is necessary to transfer these to fixed coordinates under which experimental measurements are made, without loss of material objectivity. This may be done, for example, by use of the Oldroyd or Jaumann derivatives.

CONVERSION Symbol p. In polymerisation, the fraction, or per cent of the monomer converted to polymer. In step-growth polymerisation involving non-stoichiometric amounts of the two types of reactive groups A and B, unless otherwise specified, conversion refers to the type of reactive group not in excess.

CONVOLUTION INTEGRAL A formal exact relationship between the creep compliance $J(\tau)$ and the stress relaxation modulus $G(\tau)$. It is:

$$t = \int_0^t G(\tau)J(t - \tau)\,d\tau$$

COORDINATION POLYMER Alternative name for chelate polymer. Sometimes use of the term is restricted to those polymers in which the coordinate bonds are part of the polymer chain.

COORDINATION POLYMERISATION Chain polymerisation in which the growing active centre, and often the monomer as well, is coordinated to a metallic centre of the polymerisation catalyst. Such coordination often plays an important role in stereoregulation of the polymerisation. Ziegler–Natta polymerisations are the most important type of coordination polymerisation, but other ionic polymerisation catalysts may involve coordination. Examples are butyl lithium and Grignard reagents in anionic polymerisation and boron trifluoride etherate in cationic polymerisation. In most Ziegler–Natta polymerisations the active centre is also anionic, so the polymerisation may be described as an anionic-coordination polymerisation.

COPAL One of a group of natural resins varying in hardness and solubility and including Congo copal, manila copal and kauri copal. Like other natural resins, chemically, copal resins consist mostly of a complex mixture of fused ring aromatic carboxylic acids.

COPO Tradename for styrene–butadiene rubber.

COPOLYETHERESTER Alternative name for poly-etherester block copolymer.

COPOLYMER A polymer which contains more than one type of repeat unit. In the simplest, and for synthetic polymers, the most common case, only two types of repeat unit (usually designated A and B or M_1 and M_2) are present (a binary copolymer). Often the term copolymer refers specifically to this type. More rarely, three (terpolymer) or more different types may be present in a synthetic polymer. In biopolymers very often many different repeat units may be present. This is especially true in proteins and nucleic acids.

Copolymers are classified according to the way in which the repeat units are arranged in the polymer molecular chains. For synthetic polymers, a random copolymer is the most common type, where the different repeat units are arranged in a random manner. In an alternating copolymer the units are arranged as ⎯ABABABABAB⎯. In a block copolymer long sequences of each type of repeat unit are present, whilst in a graft copolymer chains of one type of unit are attached to a chain of the second type. In the important biopolymers, proteins and nucleic acids, the many different types of repeat units are arranged in a precisely ordered sequence (hence the name sequential copolymer). Each molecule is identical and the precise ordering is essential for the correct biological function. In nucleic acids and some proteins, the precise sequences contain genetic information and hence such polymers are also called informational macromolecules.

Synthetic copolymers are prepared by copolymerisation. In the case of random copolymers this involves simultaneous polymerisation of a mixture of the comonomers, the structure of the copolymer being controlled by the amounts and reactivity ratios of the comonomers used. More specialised methods of preparation are required for the other types of copolymer.

COPOLYMER COMPOSITION EQUATION (Copolymer equation) (Copolymerisation equation) An equation relating the composition of the monomer mixture (the feed), for a mixture of two monomers (A and B) of mole fractions f_A and f_B, to the composition of the copolymer formed from the monomers, expressed in terms of the mole fractions of A and B in the copolymer (F_A and F_B) and the reactivity ratios r_A and r_B. Since f_A and f_B do not equal F_A and F_B respectively, except in the case of azeotropic copolymerisation, the composition of the feed changes during copolymerisation, thus also causing a change in copolymer composition known as drift. Hence the equation only describes the instantaneous copolymer composition. A copolymer formed over an appreciable range of conversion will show compositional variation. The initially formed polymer will be richer in the more reactive monomer (say A), the later formed polymer will correspondingly be poorer in A, as A becomes depleted in the feed. The equation is:

$$dA/dB = [A](r_A[A] + [B])/[B](r_B[B] + [A])$$

in terms of the concentrations [A] and [B] of monomers A and B, where dA/dB is the molar ratio of A to B entering the copolymer at any particular instant. A more useful form in terms of mole fractions is:

$$F_A/F_B = (r_A f_A/f_B + 1)/(r_B f_B/f_A + 1)$$

The equation was originally derived by making the steady state assumption for the concentrations of ⁓A* and ⁓B*, but later a statistical derivation was made without the necessity for this assumption. For multicomponent copolymerisations, similar but more complex equations may be developed. Thus in terpolymerisation nine propagation reactions are involved. The overall copolymer composition, for copolymer produced over a significant range of conversion, may be obtained by integration of the equation, as may the composition at any given conversion. The equation may be used for the determination of reactivity ratios provided the conversion is kept low (< 10%) and provided that the copolymer composition can be determined analytically.

COPOLYMER EQUATION Alternative name for copolymer composition equation.

COPOLYMERISATION A polymerisation in which more than one monomer (A, B, etc.) participates, thus producing a copolymer, which therefore contains more than one type of repeat unit. The term is sometimes restricted to polymerisations involving only two different monomers—binary copolymerisation, which is the commonest type. Polymerisations involving more than two monomers are multicomponent polymerisations. When three monomers are involved, the polymerisation is a terpolymerisation.

Chain copolymerisation, with simultaneous polymerisation of the monomers, usually results in an approximately random copolymer, although free radical copolymerisations show a distinct alternating tendency. In ionic chain copolymerisation this tendency is less. Monomer reactivity in copolymerisation is often very different from that in homopolymerisation. Although often cross-propagation rates are higher than for normal propagation, the overall rate of copolymerisation may be lower due to the cross-termination reactions being faster than in homopolymerisation. These effects are due to the polar effects, as expressed in the e value of the $Q-e$ scheme. Furthermore, the composition of the copolymer formed is not the same as the composition of the monomer mixture (the feed), except in azeotropic copolymerisation. The composition is given by the copolymer composition equation in terms of the reactivity ratios of the monomers

and their mole ratios in the feed. Ionic copolymerisations are much more selective than free radical copolymerisations, so that it is often difficult to produce copolymers containing appreciable amounts of both comonomers.

The different ways in which the monomers are incorporated into the copolymer molecules gives rise to several different types of copolymerisation behaviour, dependent upon the value of the product $r_A r_B$, where r_A and r_B are the reactivity ratios of monomers A and B respectively. Ideal copolymerisation results when $r_A r_B = 1$. Some free radical and many ionic copolymerisations are of this type. In the special case of $r_A = r_B = 1$, the copolymer composition is always the same as the feed composition. When $r_A = r_B = 0$, or more generally $r_A r_B = 0$, an alternating copolymer results. Many free radical copolymerisations show a tendency to alternation, i.e. $r_A r_B < 1$. If both r_A and r_B are <1 then at a certain feed composition azeotropic copolymerisation occurs (at the cross-over point). If r_A and r_B are both >1, so $r_A r_B > 1$, then block copolymerisation occurs. However, this situation is rare and block copolymers usually have to be produced in other ways.

In most copolymerisations, other than azeotropic, the copolymer is richer in the more reactive monomer than is the feed, so that drift in composition occurs. Drift may be minimised either by stopping the copolymerisation at low conversion, or by continuous or incremental addition of the more reactive comonomer.

Step-growth copolymerisation usually gives a random copolymer since the principle of equal group reactivity applies. Often copolymers produced by step-growth reactions expected to give block copolymers, as for example by use of chain extension reactions, nevertheless give random copolymers due to the occurrence of interchange reactions. Step-growth polymerisations of AA + BB monomers, producing dyadic polymers, are not usually considered as copolymerisations since both monomers are essential for the polymerisation and in the polymer produced the repeat unit is considered to be —AB— rather than the polymer being considered as an alternating copolymer of —AA— and —BB— units.

COPOLYMERISATION EQUATION Alternative name for copolymer composition equation.

COPPER NUMBER A measure of the reducing power of the aldehyde (or potential aldehyde) groups of a carbohydrate, especially of cellulose. Determined by heating with an alkaline copper salt solution and measuring the amount of copper oxide formed.

CORAL Tradename for synthetic *cis*-1,4-polyisoprene.

CORDOLAN Tradename for polyvinyl chloride/polyvinyl alcohol fibre.

CORDURA Originally a tradename for a high tenacity rayon, now used for nylon 66 twine and cord.

CORE–SHELL MORPHOLOGY A model of emulsion polymerisation, in which polymerisation occurs in an outer shell of monomer surrounding a polymer particle. In a seeded polymerisation, the polymerising monomer does not completely swell the seed particles and polymerisation occurs in a monomer-rich shell, thus producing an overcoating of the seed polymer particles with a second polymer.

COREZYN Tradename for a vinyl ester resin.

COROTATIONAL DERIVATIVE Alternative name for Jaumann derivative.

CORRELATION CHARACTERISTIC Alternative name for correlation function.

CORRELATION FUNCTION (Correlation characteristic) A mathematical function describing the spatial distribution of electron density ($\gamma(r)$) in a two phase system, which may be obtained by Fourier inversion of the small angle X-ray scattering curve, in a similar manner to the use of the Patterson function in wide angle X-ray scattering. It is defined as: $\gamma(r) = \overline{(\Delta\rho_1)(\Delta\rho_2)}/(\Delta\bar{\rho})^2$ where $\Delta\rho_1$ and $\Delta\rho_2$ are the local deviations of the electron density from the average value $\bar{\rho}$ at two points separated by a distance r. For a completely random system $\gamma(r) = \exp(-r/\bar{l}_p)$, where $\bar{l}_p$ is the correlation distance, itself related to the mean correlation lengths $\bar{l}_1$ and $\bar{l}_2$ of the two phases by $1/\bar{l}_p = 1/\bar{l}_1 + 1/\bar{l}_2$.

CORRELATION LENGTH A measure of the size of the dispersed phase particles in a two phase system, obtained from small angle X-ray scattering. It is defined as either (1) the lengths of intersection that a vector, representing the X-ray beam, makes on passing through the particles and through the matrix respectively (l_2 and l_1), the mean values being obtained from all l values for all directions of the vector (this is also called the intersection length) or (2) the scattering mass radius of the equivalent spherical particle, for which the electron density is given by the correlation function.

CORROLITE Tradename for a vinyl ester resin.

CORROSION CRACKING Alternative name for environmental stress cracking.

CORRUGATED CRYSTAL A form of a polymer single crystal containing a series of sharp folds or corrugations. Often collapses to a flat form during the specimen preparation necessary for electron microscope observation. Corrugated (as opposed to the more usual hollow pyramidal) crystals are formed at higher supercoolings. Corrugation arises from the special requirements of chain folding in polymer single crystals.

CORTICOTROPIN Alternative name for adrenocorticotropin.

CORVAL Tradename for a crosslinked viscose rayon no longer manufactured. Of higher wet tenacity than viscose rayon and used in blends to improve softness and handle of synthetic fibres.

CORVIC Tradename for polyvinyl chloride.

COTTON A natural seed hair fibre from species of the genus *Gossypium*, grown in subtropical climates. One of the most important textile fibres. When picked it is about 94% cellulose, but in finished fabrics it is about 99% cellulose. It is a staple fibre, the fibre length being 1–5 cm. The fibres have a ribbon-like twist giving them a natural crimp and have a collapsed hollow central canal (the lumen). The fibres are about 15–20 μm in diameter. The typical tenacity is about 3.5 g denier^{-1}, with a low elongation of about 3%. Moisture regain is high at about 8·5%, giving cotton textiles good comfort properties. Cotton is frequently mercerised by swelling in alkali, causing the fibres to become rounder and to have fewer convolutions. This increases strength, absorbancy, dyeability and lustre.

COTTON BAND A peak or trough in the optical rotatory dispersion or circular dichroism spectrum. An idealised Cotton band shows zero intensity (at a point of inflection) in the ORD spectrum at the band centre (where a peak in the CD spectrum occurs), with a steep rise and fall in rotation on either side, and a measurable optical rotation on either side extending to much higher and lower wavelengths. A positive Cotton band shows positive rotation on the long wavelength side of the ORD spectrum and a positive peak in the CD spectrum. For a negative band the curves are inverted.

COTTON COUNT A measure of the coarseness of a yarn. The number of hanks of 840 yd which weigh 1 lb.

COTTON EFFECT Sometimes this refers merely to an optical rotatory dispersion effect or a dependence of circular dichroism on wavelength, but often it refers more specifically to the existence of a Cotton band.

COTTON LINTERS The short, less than 5 mm, fibres remaining on the cotton seeds after the longer cotton fibres have been removed for spinning into yarn. Useful as a source of cellulose for conversion to various cellulose derivatives.

COTTONSEED OIL A non-drying oil whose triglycerides contain mostly palmitic, oleic and 9,12-linoleic acid residues. Sometimes used in alkyd resins for good colour retention.

COUETTE CORRECTION Alternative name for end-effect correction.

COUETTE FLOW Alternative name for drag flow, sometimes specifically referring to flow in the annular region between two concentric cylinders, which rotate with respect to each other, i.e. in a couette.

COUETTE RHEOMETER Alternative name for a couette viscometer.

COUETTE VISCOMETER (Couette rheometer) A coaxial cylinder viscometer, in which a cylinder is rotated inside another cylinder, with the fluid under investigation filling the annular space between the cylinders. Either cylinder may be rotated, but it is usually the inner one, and the torque (M) exerted on or by the cylinder is measured. If the radii of the inner and outer cylinders are R and R_o respectively, and their relative angular velocity is Ω, then for rheometers with only a narrow gap, having the advantage of a nearly constant shear rate throughout the fluid, the apparent viscosity is given by:

$$\eta = (M/4\pi h\Omega)(1/R^2 - 1/R_o^2)$$

where h is the length of the inner cylinder immersed in the fluid.

COULOMB YIELD CRITERION (Mohr–Coulomb yield criterion) A yield criterion which states that the critical shear stress for yielding (τ) in any plane increases linearly with the compressive stress applied normal to the plane (σ_N), i.e. $\tau = \tau_c + \mu\sigma_N$, where τ_c is the 'cohesion' of the material and μ is the coefficient of friction. For uniaxial compression, yielding occurs on a plane at an angle θ to the compressive stress (σ_1) with

$$\sigma_1(\cos\theta\sin\theta - \mu\cos^2\theta) = \tau_c$$

For the smallest value of σ_1 this gives $\tan 2\theta = 1/\mu$. Thus the criterion gives the direction of yield as well as the conditions for yielding to occur. The initial yield stress ($\sigma_{y,c}$) is given by $\sigma_{y,c} = 2\tau_c/[(\mu^2 + 1)^{1/2} - \mu]$. For tension $\sigma_{y,t} = 2\tau_c/[(1 + \mu^2)^{1/2} + \mu]$ i.e. compressive strength is greater than tensile yield strength. This is to be expected for polymers.

COUMARONE

M.p. $-18°C$. B.p. $173°C$.

Obtained from the coal tar naphtha fraction, but usually not isolated from the indene which occurs with it. The mixture with indene is polymerised to the commercial coumarone–indene resins by treatment with concentrated sulphuric acid or a Friedel–Crafts catalyst at low temperature, e.g. $0°C$.

COUMARONE–INDENE RESIN (Indene–coumarone resin) (Indene resin) Tradenames: Catarax, Cumar, Epok C, Piccoumaron. The commercial copolymer produced by treatment of the mixture of coumarone and indene obtained from coal tar naphtha fractions, with concentrated sulphuric acid or a Friedel–Crafts catalyst at low temperature. The polymers are usually of low molecular weight, less than 5000, dark in colour and vary from sticky liquids to hard but brittle solids, depending on molecular weight and the coumarone–indene ratio. They are highly soluble in many organic solvents and compatible with a wide range of other polymers. They are useful as constituents of paints and varnishes and as softeners in rubbers and polyvinyl chloride.

COUNTER-CURRENT DISTRIBUTION A technique for the separation of mixtures, in which the mixture is shaken with two immiscible liquids. The components of the mixture distribute themselves between the liquids according to their partition coefficients. The liquids are separated and shaken again with the appropriate immiscible liquid. By further repetition of the process partial or complete separation may be achieved. Even with the use of automated apparatus, the process is time consuming and has been replaced largely by chromatographic methods. The technique has been widely used for the separation of biopolymer mixtures.

COUNTERION Alternative name for gegenion.

COUPLING (1) The joining together of two or more polymer molecules, which contain terminal chemically reactive groups, by reaction of these groups forming covalent bonds directly between polymer molecules, or by reaction with a third, usually small, molecule capable of reaction with the polymer functional groups which are normally the same. In this latter case the polymer molecules are frequently low molecular weight prepolymers. This method is used in the formation of several important polymers, notably the coupling of polymeric polyols via diisocyanates to form polyurethanes and in the coupling of diglycidyl ethers in the curing of epoxy resins. Coupling leads to the formation of block copolymers if the two polymer molecules are different.

(2) Alternative name for spin–spin coupling.

(3) Alternative name for combination.

COUPLING AGENT (Keying agent) A material applied as a thin layer to the surface of a reinforcing filler, usually as a fibre and often glass, to improve the adhesion between the filler and a polymer matrix in a filler–polymer composite. This has the advantage of increasing the tensile strength and retention of mechanical properties during long-term immersion in water, but often at some sacrifice of impact strength. Chemical bonding often occurs between coupling agent and both fibre and matrix, the coupling agent containing chemical groups which will react with groups present on the glass surface and groups which participate in the chemical reactions of polymer crosslinking, at least in the important cases of glass reinforced unsaturated polyester and epoxy resins. Other theories of adhesion promotion are the deformable layer and restrained layer theories. Increasingly, coupling agents are also being used with mineral and other fillers incorporated into thermoplastics matrices. The best known coupling agents are the silane materials, although chromium complexes, phosphorus esters and titanate esters are also useful.

COURLENE Tradename for a low density polyethylene fibre.

COURPLETA Tradename for cellulose triacetate filament and staple fibres.

COURTELLE Tradename for a polyacrylonitrile fibre.

COVALENT CHROMATOGRAPHY A chromatographic method in which the polymer to be isolated forms a covalent bond with the packing material of a column, whereas components from which it is to be separated do not. The polymer is subsequently eluted with a reagent which breaks the bonds formed. It is usually used for the separation of proteins which can form disulphide bridges with the packing. These are then cleaved by elution with, for example, 2-mercaptoethanol.

CPE Abbreviation for chlorinated polyethylene.

CPVC Abbreviation for chlorinated polyvinyl chloride.

CR Abbreviation for chloroprene rubber, i.e. polychloroprene.

CR-39 Tradename for poly-(diethyleneglycol-bis-(allyl carbonate)).

CRANKSHAFT MOTION Alternative name for Schatzki mechanism.

CRASTINE Tradename for polybutylene terephthalate.

CRAZING The formation of small crack-like cavities in a material of a few micrometres in length and running perpendicular to the direction of stressing when a glassy polymer is stressed. Spanning the crazes are fibrils of highly oriented polymer. Eventually the crazes may form the sites for subsequent fracture by formation of larger cracks. Typically crazing occurs at about half the yield stress. For crazing to occur the stress system must contain a hydrostatic component.

CREEP The progressively increasing strain over a period of time of a viscoelastic material when subject to a continuously applied stress. Together with stress relaxation and dynamic mechanical behaviour, the most frequently investigated phenomenon of viscoelasticity. Creep is important because of its significance in understanding viscoelastic behaviour in general, its ease of measurement and the importance of quantitative creep data in the design of components made from polymers, when these will be subject to long-term stresses.

Creep is composed of three components—the instantaneous (purely elastic) deformation when the stress is applied, the time-dependent and increasing deformation arising from delayed elasticity and the longer-term viscous flow. The two former deformations are fully recoverable after removal of the stress (creep recovery). Like elastic deformations, the creep behaviour of a material may be characterised by a creep modulus (shear, tensile or bulk), but is more commonly expressed by the creep compliance. Unlike elastic behaviour, the creep compliances are not the inverse of the corresponding moduli due to the differences in their time dependencies. Thus the shear creep modulus $G(t)$ is $\neq 1/J(t)$, where $J(t)$ is the shear creep compliance.

Creep tests are comparatively easy to carry out in tension, compression, simple shear and torsion except for very rigid materials where exceptional rigidity of the apparatus is required. In these cases flexural tests are often used. In general, shear creep compliances are more useful in relating deformation behaviour to molecular mechanisms, since there is no volume change in shear. However, for materials with a Poisson's ratio of about one-half and/or at small deformations, the tensile compliance $D(t)$ equals $J(t)/3$, so tensile tests can be used to obtain $J(t)$.

For linear viscoelasticity, creep may be modelled simply by the Voigt model, for which the creep strain is given by $e(t) = J\sigma_0(1 - e^{-t/\tau'})$, where τ' is the retardation time, σ_0 is the (constant) stress and J is the spring compliance. For recovery the strain is given by $e = e_0 \exp(1 - t/\tau')$, where e_0 is the initial strain. The creep compliance is given by $J(t) = J(1 - e^{-t/\tau'})$. Other more complex models give a similar dependency on τ'. However, many polymers exhibit non-linear viscoelasticity and in these cases no satisfactory model has been found. With polymers, creep results are often obtained over very long times, and are usefully expressed as isochronous or isometric creep curves using logarithmic time scales. Several empirical equations, such as the Nutting equation, have been proposed to represent non-linear behaviour.

CREEP COMPLIANCE (Creep compliance function) Symbols $J(t)$ in shear, $D(t)$ in tension and $B(t)$ in bulk. The time-dependent ratio of strain to stress during creep. One of the most widely measured and useful parameters characterising the viscoelastic behaviour of a material. Unlike the elastic compliance (for purely elastic behaviour), the creep compliance is not the inverse of the creep modulus since the two have different time dependencies.

The creep compliance is an exponentially decreasing function of time. It can be considered to consist of three components. Firstly, the unrelaxed compliance characterising the immediate elastic response and the limiting value of the overall creep compliance at short times, i.e. characteristic of the 'glassy' behaviour of the material. Typically this has a value of about $10^{-9}\,\mathrm{Pa}^{-1}$. Secondly, the compliance component characterising viscous flow at very long times, typically having a value of about $10^{-5}\,\mathrm{Pa}^{-1}$. Its limiting value is the equilibrium or relaxed compliance. Both these components are time independent. The characteristic viscoelastic behaviour at intermediate times provides the third component and is indicated by the compliance component for the retarded elastic response. In this transition region this component dominates behaviour.

The effects of a multiple loading programme can be represented by the Boltzmann Superposition Principle. Creep experiments are commonly performed in shear, yielding the shear creep compliance $J(t)$, or in tension giving the tensile (or Young's or extensional) creep compliance $D(t)$. A bulk (usually compressive) compliance $B(t)$ is also occasionally measured. The three compliances are related by: $D(t) = J(t)/3 + B(t)/9$, and if Poisson's ratio is one-half then $D(t) = J(t)/3$, since $B(t)$ is very small. Usually since creep behaviour is of interest over a long period of time, e.g. 10–15 decades, the time dependency is plotted as compliance versus log time.

CREEP COMPLIANCE FUNCTION Alternative name for creep compliance.

CREEP FRACTURE Alternative name for static fatigue.

CREEP LATERAL CONTRACTION RATIO The lateral contraction ratio in creep.

CREEP MODULUS Symbols $E(t)$ in tension, $G(t)$ in shear and $K(t)$ in bulk. The time-dependent ratio of the stress to the strain in creep. It may be considered to consist of three components. Firstly, the relaxed component characterising the immediate elastic response of the material which is completely recoverable. Typically, this has a value of about 1 GPa for polymers. Secondly, a component characterising viscous flow at very long times (the relaxed modulus) typically having a value of about 10^5 Pa. Thirdly, a component characterising the retarded elastic response, which dominates behaviour in the transition region. In simple design problems the creep modulus at the time of interest may be used as a pseudoelastic modulus with the classical (Hookean) elasticity equations. Unlike the elastic modulus, the creep modulus is not the inverse of the creep compliance due to its different time dependency. Analysis and description of creep behaviour is more frequently expressed in terms of creep compliance than creep modulus.

CREEP RECOVERY Alternative name for recovery for a viscoelastic material.

CREEP RUPTURE Alternative name for static fatigue.

CREMONA Tradename for polyvinyl alcohol fibre.

CREPE (1) (Crepe rubber) The raw gum natural rubber obtained from natural rubber latex by coagulation followed by air drying. It is often coloured, pale yellow or brown, due to the presence of β-carotene. The palest crepes are made by repeated washing and milling or are even bleached by chemical treatment, e.g. with xylyl mercaptan. One of the commoner forms of natural rubber, occasionally used, e.g. in shoe soleing, without vulcanisation.

(2) A raw rubber or rubber compound, usually in the form of a rough sheet, as obtained from compounding, e.g. on a two roll mill.

CREPE RUBBER Alternative name for crepe.

CRESLAN Tradename for a polyacrylonitrile fibre.

CRESOL (Hydroxytoluene) (Methylphenol) *Ortho*-cresol (*o*-cresol)

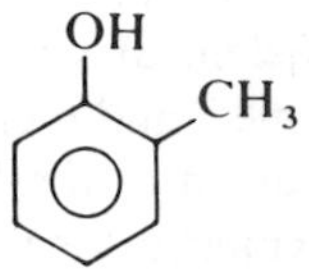

M.p. 30·9°C. B.p. 191·2°C.

Meta-cresol (*m*-cresol)

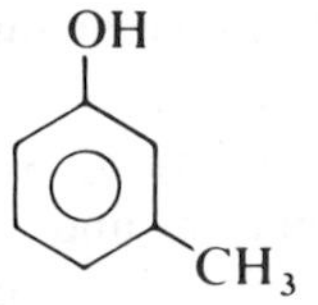

M.p. 12·2°C. B.p. 202°C.

Para-cresol (*p*-cresol)

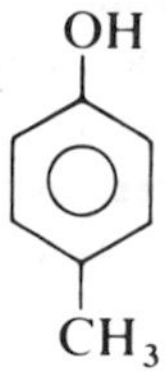

M.p. 34·7°C. B.p. 201·9°C.

Previously obtained from coal tar, now mostly produced from toluene or phenol. Cresols are obtained from toluene by sulphonation followed by alkali fusion. This gives largely *p*-cresol with some *o*-cresol. Alternatively, chlorination followed by hydrolysis gives a mixture with about 50% *m*-cresol and 25% each of the other isomers. In a further method alkylation of toluene with propylene gives cymene which on oxidation gives a *m*-/*p*-mixture of cresols. Phenol may be converted to a mixture of cresols and xylenols, mainly *o*-cresol and 2,6-xylenol, by reaction with methanol at about 350°C with aluminium oxide. Cresols are sometimes used instead of phenols in the formation of phenol–formaldehyde polymers with improved resistance to chemical attack. Only the *meta* isomer retains a functionality of three to reaction with formaldehyde and is therefore capable of network formation. *p*-Cresol is useful for conversion to various phenolic antioxidants, such as 2,6-di-*t*-butyl-4-methyl-phenol.

***m*-CRESOL**

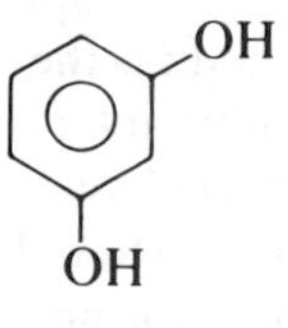

B.p. 202·2°C.

A solvent for polyamides, polyurethanes, polycarbonates and polyethylene terephthalate.

CRISTABOLITE A crystalline silica mineral which consists of sheets of linked SiO_4 tetrahedra to form fused six-membered rings joined to produce a three dimensional network similar to diamond, the Si atoms occurring at the positions of the carbon atoms in diamond with oxygen between each adjacent pair.

CRITICAL ASPECT RATIO The ratio of the critical fibre length to the diameter of a fibre in a fibre reinforced composite.

CRITICAL BRANCHING COEFFICIENT Symbol α_c. The value of the branching coefficient at which an infinite network is calculated to be formed, and therefore at which gelation is predicted to occur, in a non-linear step-growth polymerisation. For a polymerisation involving difunctional monomers A—A and B—B with multifunctional monomer A_f (where f is its functionality), then α_c is simply given by $1/(1-f)$. If more than one type of multifunctional monomer are polymerising then an average value of f is used. By combining α_c with the relationship between branching coefficient and conversion, the critical conversion at which gelation is predicted to occur may be calculated. This value is generally less than the experimentally observed value, since the statistical theory by which α_c is derived ignores the possibility of some wasted intramolecular reactions between A and B groups, whereby loops are formed, thus removing the possibility that the groups can contribute towards network formation.

CRITICAL CONVERSION Symbol p_c. The conversion during a non-linear step-growth polymerisation at which gelation occurs. p_c may be predicted theoretically either by a modified version of the Carothers equation or statistically by consideration of the branching coefficient. In practice the experimental value of p_c is approximately half-way between the two predicted values.

CRITICAL ELONGATION RATE The rate of elongation of a polymer melt above which the elongational viscosity decreases with increasing rate, i.e. tension thinning occurs.

CRITICAL FIBRE LENGTH Symbol l_c. The length of a fibre (l) in a fibre reinforced composite necessary so that a tensile stress applied to the composite, parallel to the fibres, may build up in the fibre just enough to reach the tensile strength of the fibre. l_c is proportional to the fibre diameter (d) and is of the order of $10d$ if the polymer matrix deforms elastically or about $100d$ if it deforms by plastic flow. Thus l_c is sometimes represented by the critical aspect ratio. For plastic flow $l_c = \hat{\sigma}_f d/2\hat{\tau}$, where $\hat{\sigma}_f$ is the tensile strength of the fibres and $\hat{\tau}$ is the shear strength of the matrix or of the matrix–fibre interface, whichever value is smaller. For elastic matrix behaviour, the value of l_c is more complex, but it is proportional to $(G_m/E_f)^{1/2}$ where G_m is the matrix shear modulus and E_f is the fibre tensile modulus.

CRITICAL FIBRE VOLUME FRACTION The minimum volume fraction of fibres in a fibre reinforced composite required to cause an increase in tensile strength over that of the polymer matrix alone. In these composites the fibres often break in tension, when stressed along their length, before the matrix and this can result in the composite having a lower strength than that of the matrix

alone due to the reduced cross-sectional area of the residual matrix. The effect only occurs at low fibre volume fractions, if at all.

CRITICAL SHEAR RATE Symbol $\dot{\gamma}_c$. The shear rate for a polymer melt above which melt fracture occurs. $\dot{\gamma}_c$ is dependent on melt temperature (increasing with increasing temperature), molecular weight and geometry of the flow channels.

CRITICAL SHEAR STRESS Symbol τ_c. The shear stress for a polymer melt above which melt fracture occurs. τ_c is dependent on polymer molecular weight, the product $\tau_c \bar{M}_w$ often being constant ($\bar{M}_w$ being the weight average molecular weight), molecular weight distribution and temperature.

CRITICAL STRAIN ENERGY RELEASE RATE (Work of fracture) Symbol G_c. The strain energy release rate required for the fracture of a material. It is related to the critical stress intensity factor (K_c) for an infinite sheet by $K_{1c}^2 = EG_{1c}$ for plane stress fracture, and by $K_{1c}^2 = (E/(1-v^2))G_{1c}$ for plane strain fracture, where E is the tensile modulus and v is the Poisson ratio. The subscript 1 denotes mode 1 fracture (the crack-opening mode), which is the only mode usually considered. It is a measure of the ease with which a crack can propagate in a material, since G is the energy absorbed per unit area of crack. Typical values for polymers are in the range $1–10\,kJ\,m^{-2}$, which compares with values of $100–1000\,kJ\,m^{-2}$ for metals and $0.01–0.1\,kJ\,m^{-2}$ for ceramics. Fracture will thus occur when $\sigma(\pi a)^{1/2}$ equals $(EG_c)^{1/2}$, where σ is the stress and $2a$ is the crack length. G_c may thus be calculated if E and K_c are known. However, E depends on the rate of strain, which is often uncertain for fracture. G_c may be determined experimentally by measurement of the dependence of the compliance (C) on crack length, since $G_c = (P^2/2t)(\partial C/\partial a)$, where P is the total load and t is the specimen thickness.

CRITICAL STRESS INTENSITY FACTOR (Fracture toughness) Symbol K_c. In fracture mechanics, the critical value of the stress intensity factor which, if exceeded, results in fracture. For an infinite plate, since $K = \sigma(\pi a)^{1/2}$, where σ is the stress and $2a$ is the crack length, it is related to the critical strain energy release rate (G_c) by $K_c^2 = FG_c$ for plane stress fracture and $K_c^2 = (E/(1-v^2))G_c$ for plane strain fracture, where v is the Poisson ratio. The usual units are $MN\,m^{-3/2}$, and values for polymers are typically about $1\,MN\,m^{-3/2}$, compared with about $100\,MN\,m^{-3/2}$ for metals. K_c can be determined by loading a specimen containing a crack of known length until fracture occurs. Usually only the crack-opening mode (mode 1) is considered, when the symbol used is K_{1c}. K_c is often more useful than G_c for design calculations since, unlike the latter, its use does not require a knowledge of E at the unknown rate of straining involved.

CRITICAL SURFACE TENSION Symbol γ_c. The maximum value of the surface tension of a liquid which will just allow it to spread over the surface of a particular solid. For liquids with a surface tension of higher value than γ_c for the solid of interest, the contact angle will be greater than $0°$ and spreading will not occur. For polymers, in general γ_c follows the value of the surface tension of the polymer (γ_s), i.e. the lower the intermolecular forces of attraction in the polymer, the lower the values of γ_s and γ_c.

CROCIDOLITE (Blue asbestos) An amphibole asbestos of chemical structure $Na_2MgFe_3[(OH)_4Si_4O_{11}]$, occasionally used as a filler in polymer composites. Unlike the commoner chrysotile asbestos, the fibres are lath-like and the molecules consist of ladder-like silica chains lying parallel to each other, being joined together by metal ions. This produces a stiff and brittle fibre which is much more difficult to separate from the fibre bundles than in chrysotile asbestos. Fibre diameter is about $100\,nm$ and bundle length about $20\,mm$. Mechanical properties are similar to chrysotile asbestos (tensile strength is $3·5\,GPa$ and tensile modulus is $175\,GPa$), although stiffness is somewhat higher and resistance to acids is much better. The toxic hazards associated with crocidolite are even more severe than those with chrysotile.

CROSS BREAKING STRENGTH Alternative name for flexural strength.

CROSS EQUATION An empirical equation for the shear rate dependence of the apparent viscosity (η) of a polymer fluid. It is: $\eta = \eta_\infty + (\eta_0 - \eta_\infty)/(1 + k\dot{\gamma}^m)$, where η_∞ and η_0 are the upper and lower Newtonian viscosities at very high and very low shear rates respectively and k and m are constants. m should have values between $0·66$ and 1 and is a function of molecular weight distribution. k is often taken to be unity. Such an equation gives a sigmoidal shape for an η versus $\dot{\gamma}$ plot, as is often found in practice.

CROSSLINK A covalent bond or relatively short sequence of chemical bonds joining two polymer chains together thus forming a crosslinked polymer. In polymers normally referred to as network polymers, produced by step-growth polymerisation of multifunctional monomers, a distinction cannot be made between the primary chains and the joins between the chains, so these latter cannot be identified as crosslinks. In contrast, crosslinks are of different structures from the primary chains that they join and are short relative to the primary chain segments between the crosslinks. They may only be single bonds, but usually consist of several bonds in length, e.g. in sulphur vulcanised rubber and styrene-crosslinked unsaturated polyester resins. Crosslinks may be formed either during chain polymerisation by use of a multifunctional monomer, e.g. divinylbenzene, or by subsequent chemical reaction on a preformed linear polymer, as in vulcanisation, radiation crosslinking, peroxide vulcanisation and the curing of unsaturated polyesters. The concentration of crosslinks can be expressed in several ways, e.g. as crosslink density, network parameter,

crosslinking index or crosslinking coefficient. Noncovalent bonding forces between chains may produce similar effects to crosslinking through 'virtual' crosslinks.

CROSSLINK DENSITY (Degree of crosslinking) (Density of crosslinking) Symbol q. The fraction of polymer chain units, i.e. normally the repeat units, that are crosslinked. If there are N such units, since each crosslink links two such units, the number of crosslinks will be $qN/2$. For high degrees of crosslinking, so that free ends may be ignored q is related to the network parameter (M_c) by $M_c = w/q$, where w is the molecular weight of the unit.

CROSSLINKED POLYMER A polymer in which linear polymer chains are joined together by covalent chemical bonds usually via crosslinking atoms or groups, although sometimes only a single covalent bond separates the chains. Effects similar to those found by covalent crosslinking may also be produced via the so-called 'virtual' crosslinks. For other than a very low extent of crosslinking, all molecules are joined to each other so the polymer may also be termed a network polymer. In general, this latter term tends to be used to describe polymers produced by step-growth polymerisation using multifunctional monomers, where the network is formed by joining together the branches. In this type of crosslinked polymer the individual polymer chains cannot be distinguished from the crosslinks.

The term crosslinked polymer tends to be used for polymers where the individual chains may be distinguished and where the crosslinks are short relative to the chain segments between crosslinks. Such structures may be formed either by crosslinking a preformed linear polymer, or during chain polymerisation using a proportion of a multifunctional monomer. Examples of the latter are divinylbenzene and the dimethacrylates.

There are many ways of crosslinking preformed linear polymers. In many cases a readily crosslinkable group is deliberately incorporated into the polymer to facilitate crosslinking, as in the important cases of the sulphur vulcanisation of polydiene rubbers and the crosslinking of unsaturated polyesters through linking the unsaturated groups via chain polymerisation of styrene. If free radicals can be generated on polymer chains then these may combine to produce crosslinks. Several methods can be used for this, e.g. by heating with peroxides, as in the peroxide vulcanisation of rubbers, by irradiation with high energy radiation (radiation crosslinking) or with ultraviolet or visible light (photocrosslinking).

The structural characterisation of crosslinked polymers is often difficult, especially due to their insolubility. Several measures of the amount of crosslinking have been used. These include the crosslink density, the crosslinking coefficient, the crosslinking index and the network parameter. The case of lightly crosslinked rubbers is one of the few in which the amount of crosslinking may be determined (as the network parameter). Crosslinking reduces solubility so that at other than very low crosslinked densities, polymers are insoluble being merely swollen even in good solvents. Crosslinking increases modulus, reduces elongation and reduces flow under stress. The main reason for vulcanising rubbers is to reduce cold flow when the rubber is stressed.

CROSSLINKING The process of forming the cross-links between linear polymer molecules. This may be achieved by vulcanisation of elastomers, by radiation crosslinking, by photocrosslinking, by copolymerisation of in-chain unsaturated groups with a vinyl monomer (as in the crosslinking of unsaturated polyester resins) or by reaction of unsaturated groups with atmospheric oxygen (as in the air-drying of alkyd resins).

CROSSLINKING COEFFICIENT Symbol δ. The number of crosslinked repeat units per weight average polymer molecule in a crosslinked polymer. Thus $\delta = q\bar{X}_w$ where q is the crosslink density and $\bar{X}_w$ is the weight average degree of polymerisation.

CROSSLINKING INDEX Symbol γ. The number of crosslinked repeat units per number average polymer molecule in a crosslinked polymer. Thus $\gamma = q\bar{X}_n$ where q is the crosslink density and $\bar{X}_n$ is the number average degree of polymerisation.

CROSS-OVER POINT In copolymerisation, the composition at which a plot of F_A versus f_A (or F_B versus f_B) crosses the azeotropic copolymerisation line ($F_A = f_A$ or $F_B = f_B$), where F_A and f_A are the mole fractions of monomer A in the copolymer and feed respectively and similarly F_B and f_B for monomer B.

CROSS-OVER REACTION Alternative name for cross-propagation.

CROSS-PROPAGATION (Cross-over reaction) In chain copolymerisation, the propagation step in which an active centre adds on a monomer of the opposite type rather than one of the same type (self-propagation). Thus in binary copolymerisation, the reactions, $\sim A^* + B \rightarrow \sim AB^*$ and $\sim B^* + A \rightarrow \sim BA^*$, are the cross-propagation reactions. The relative tendency of self- to cross-propagation is given by the monomer reactivity ratio. Cross-propagation of both types of active centre favours alternating copolymerisation.

CROSS-β-STRUCTURE A β-conformation sometimes found in synthetic and protein polypeptides in which the individual polypeptide chains are folded back on themselves to give an anti-parallel β-conformation, for which, in fibres, the chain axis is perpendicular to the fibre axis.

CROSS-TERMINATION In free radical copolymerisation, termination by reaction of two radicals terminated by monomer units of the opposite type, i.e. $\sim A^* + \sim B^* \rightarrow$ termination, by combination or disproportionation with rate constant k_{AB}. Cross-termination is often favoured over termination by reaction between two like radicals due to polar effects. This parallels the tendency towards cross-propagation (and alternating copolymerisation) in that it is favoured as the product of the two reactivity ratios approaches unity and as the difference between the polarities of the two monomers increases. The tendency to cross-termination is often expressed by the factor $\phi = k_{AB}/2(k_{AA}k_{BB})^{1/2}$, where k_{AA} and k_{BB} are the self-propagation rate constants, when cross-termination is favoured. However, termination may be controlled by diffusion effects, when it is expected that the value of k_{AB} will be dependent on copolymer composition.

CROWN ETHER POLYMER A polymer which contains a macrocyclic ether structure, which adopts a crown-shaped conformation. Vinyl crown ether polymers have been synthesised and chelate to alkali metal (M) ions more strongly than do monomeric crown ethers. A typical structure is:

CRUDE MDI Alternative name for polymeric MDI.

CRYLOR Tradename for a polyacrylonitrile fibre.

CRYOGLOBULIN A blood plasma globulin, isolated in some fractionation procedures with fibrinogen, which has low solubility in aqueous solutions at about $0°C$. Thus cooling a serum or solution can cause separation by precipitation, crystallisation or gelling. Cryoglobulins have similar molecular weights and other properties to normal γ-globulins. They are formed in relatively large quantities in plasma when individuals are suffering certain diseases, but may be present, in small amounts, in normal plasma.

CRYOLAC CIT Tradename for methylmethacrylate–acrylonitrile–butadiene–styrene or methylmethacrylate–butadiene–styrene copolymer.

CRYOLITE Tradename for methylmethacrylate–acrylonitrile–butadiene–styrene or methylmethacrylate–butadiene–styrene copolymer.

CRYOSCOPY A technique for the determination of the molecular weight of a solute by measurement of its ability to depress the freezing point of the solvent. Used occasionally for the determination of the number average molecular weight of polymers, but less frequently than the closely related colligative property method of ebulliometry. Particular difficulties restricting its use are precipitation of polymer close to the freezing point and supercooling.

CRYSTAL BRANCHING Alternative name for branched crystal.

CRYSTAL–CRYSTAL TRANSITION The transformation of one crystal structure of a polymer to a different structure. It usually results from a change in temperature, as a result of different structures being stable at different temperatures. Few examples are known in polymer crystals, but they include poly-(butene-1) and polytetrafluoroethylene.

CRYSTALLINE ORIENTATION The component of the overall orientation due to the crystalline regions of a polymer. Characterised by its orientation function, f_{cryst}, describing orientation of the crystallite axis with reference to some fixed direction, often the direction of deformation (machine direction). The angle between the two is the crystalline orientation angle, and since there are three crystallographic axes there are three such angles (α, β, ε), and therefore three distributions: $f_\alpha = (3\cos^2\alpha - 1)/2$ and similarly f_β and f_ε, any two of which may be independent. It is determined by intensity measurements of the Debye–Scherrer arcs in wide angle X-ray scattering. The separate reflections of each Miller hkl plane give a $\cos^2\phi_{hkl}$ term from which $\cos^2\alpha$, etc. may be calculated.

CRYSTALLINE RELAXATION A relaxation, with its accompanying transition, associated with the crystalline regions, i.e. the lamellae, of a polymer. The most important relaxation, the primary relaxation, is melting. Sometimes, in addition, a crystal–crystal relaxation and transition are observed, as with polyisoprene and polytetrafluoroethylene. Certain secondary transitions are also sometimes observed, such as premelting and even lower transitions. These may be identified with particular molecular processes, especially where considerable morphological information is available, as is the case with polyethylene. Such processes include short-range cooperative chain motions (similar to those found in amorphous regions), motions associated with defects and with chain folds and interlamellar shear.

CRYSTALLINITY The long-range (hundreds of ångströms) regular ordering of atoms or molecules in unit cells on a three-dimensional crystalline lattice. Non-polymeric solids are essentially 100% crystalline, whereas polymers only crystallise if the molecules have regular structures and then only do so to a limited extent. The extent of crystallinity is often called the degree of crystallinity (or merely crystallinity) and is typically 30–80%. The remaining material is randomly disordered (amorphous), the polymer chains being randomly coiled. The degree of crystallinity realisable decreases with increasing structural irregularity (e.g. atacticity, branching) of the polymer and varies with crystallisation conditions. Thus atactic polymers are usually completely amorphous, whereas 100% crystalline single crystals can be grown under certain special conditions. At the molecular level the crystal structure is characterised by wide angle X-ray diffraction. However, the polymer crystals (or crystallites) are very small and aggregate usually in an ordered fashion to produce different supermolecular morphologies, e.g. fibrils and spherulites. Crystallinity has a profound effect on polymer properties, especially the mechanical properties, since in the crystallites the polymer molecules are more tightly and evenly packed than in amorphous regions and hence intermolecular forces are higher. Thus as degree of crystallinity increases, so do the moduli, stiffness, yield and tensile strengths, hardness, density and softening points.

CRYSTALLISATION The process of formation of crystalline material from a disordered aggregate of molecules. This may result from cooling a melt to below the polymer crystalline melting temperature (T_m) (melt crystallisation) or from cooling or evaporating a solution (solution crystallisation). Polymer melt crystallisation requires considerable supercooling typically about 30 K. As the temperature is reduced the rate reaches a maximum then diminishes as the melt viscosity increases. Eventually the freezing of molecular motions (below the glass transition temperature) prevents any further crystallisation. Rapid quenching of a melt may prevent any crystallisation occurring. The degree of crystallinity achieved thus depends on the cooling conditions. Crystallisation may also be induced in the solid state by orientation or by annealing.

The progress of crystallisation may be followed by dilatometry, microscopic observation of spherulite growth, wide angle X-ray diffraction or differential scanning calorimetry. The rate–time plot is usually S-shaped. Both primary and secondary crystallisation stages may be observed. Analysis of the kinetics of crystallisation often follows an Avrami equation and yields information on the mechanism, especially nucleation. Nucleation is either heterogeneous due to the presence of impurities, of partially ordered polymer or of deliberately added nucleating agents, or it may be homogeneous, which arises from the random aggregation of polymer molecules to form an ordered region which subsequently grows.

CRYSTALLITE The name given to the crystals present in a crystalline polymer, which, in contrast to non-polymer crystals, are so small as to be observable only with an electron microscope. Typical sizes are 10^{-5} to 10^{-6} cm. Invariably they are of lamellar habit (lamellae) which aggregate such that each crystallite is related to its neighbour, e.g. linearly as in fibrils. The earlier fringed micelle model of a partially crystalline polymer was later replaced by the idea of highly disordered crystallites (as in the paracrystalline model). The demonstration of chain folding in polymer single crystals then led to the current view that crystalline polymers consist of aggregated lamellae, whose thickness of about 100 Å (measured by electron diffraction) is much less than the polymer molecular length (thousands of ångströms). The molecules must therefore lie across the thickness of the lamellae and be folded at the surface. The chain folds may also pass out of the crystallite forming loose loops and re-enter adjacently or non-adjacently (the switchboard model). Alternatively, the molecular chain may re-enter another crystallite forming a tie molecule.

CRYSTAL ORIENTATION ANGLE The angle between one of the crystallographic axes of the crystal unit cell and some reference direction, e.g. the fibre (or drawing) direction in a uniaxially oriented sample.

c-SHEAR (Intralamellar shear) A shearing process occurring in a polymer folded chain lamellar crystal, in which movement occurs parallel to the polymer chain axis. In many crystalline polymers this is the crystal lattice *c*-axis, for example in polyethylene, for which it is thought to be the cause of the high temperature crystal relaxations (the α_2 or α'' relaxation) at just below T_m.

CSM (1) Abbreviation for chopped strand mat.
(2) Abbreviation for chlorosulphonated polyethylene.

C-STAGE POLYMER or RESIN The final network polymer formed in a step-growth polymerisation, which has passed through the initial B- and A-stages. The crosslink density is high enough so that the material is completely insoluble and infusible. The term is specifically used to refer to phenol–formaldehyde polymers.

CTBN Abbreviation for carboxy-terminated nitrile rubber.

CTI Abbreviation for comparative tracking index.

C-TERMINAL END The end of a protein or other polypeptide chain which contains the free carboxyl group of the C-terminal amino acid residue.

C-TERMINAL RESIDUE The amino acid residue at the C-terminal end of a protein or other polypeptide chain.

CUAM Alternative name for cuoxam.

CUEN An aqueous solution of a complex of a cupric salt and ethylenediamine (a cupriethylenediamine hydroxide complex) useful as a solvent for cellulose, which is not as readily oxidatively degraded as in cuoxam.

CUMAR Tradename for coumarone–indene resin.

CUMENE HYDROPEROXIDE

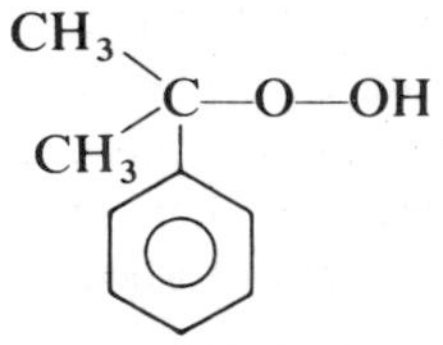

A peroxide initiator of very high thermal stability, having half-life times of 290 h/120°C, 10 h/150°C and half-life temperatures of 158°C/10 h, 180°C/1 h and 250°C/1 min.

Useful as a component of a redox initiation system in the low temperature emulsion polymerisation of styrene/butadiene (cold rubber production) and other diene monomers. Also useful as a crosslinking initiator component for unsaturated polyesters when prolonged curing times are required.

CUMULATIVE DISTRIBUTION FUNCTION A discontinuous distribution function giving the summation of the statistical weights, usually expressed as the weights (w_i) or numbers of molecules (moles) (n_i) of each polymer species of size i (and molecular weight M_i) in a polydisperse polymer sample, up to a certain value, as a function of i, i.e. the values of

$$\sum_{i=1}^{i} n_i(M_i) \quad \text{or} \quad \sum_{i=1}^{i} w_i(M_i)$$

When converted to the corresponding continuous function, an integral distribution results.

CUMULATIVE MASS FRACTION Alternative name for cumulative weight fraction.

CUMULATIVE WEIGHT FRACTION (Cumulative mass fraction) In a polydisperse polymer sample, the weight (or mass) fraction of all polymer molecular species of molecular weight less than and equal to the value of molecular weight of interest, i.e. it is

$$\sum_{i=1}^{i} w_i$$

where w_i is the weight fraction of each individual molecular species i.

CUMYL PEROXIDE Alternative name for dicumyl peroxide.

CUOXAM (Cuam) (Cuprammonium hydroxide) (Schweitzer's reagent) An aqueous solution of a complex of a cupric salt and ammonia, useful as a solvent for cellulose. Such solutions are subject to rapid oxidative degradation so that often cuen or cadoxen are preferred, especially for molecular weight determination. Regenerated cellulose is produced from a cuoxam solution in the cuprammonium process.

CUPRAMA Tradename for a cuprammonium rayon staple fibre, often blended with wool to improve softness.

CUPRAMMONIUM HYDROXIDE Alternative name for cuoxam.

CUPRAMMONIUM PROCESS (Bemberg process) The method of production of cuprammonium rayon by wet spinning of a cellulose solution in cuoxam into water.

CUPRAMMONIUM RAYON Tradenames: Bemberg, Cupresa, Cuprama. A regenerated cellulose fibre produced by wet spinning a cellulose solution (generally from cotton linters) in cuprammonium solution into water with some stretching. Very fine filaments (down to 0·4 denier) and yarn (down to 15 denier) may be made. The strength (tenacity, 2.3 g denier^{-1}, dry) is greater than viscose rayon, with an elongation at break of 15% (dry) and 25% (wet). The fine filament fabrics have good drape and can appear similar to silk.

CUPRESA Tradename for a cuprammonium rayon yarn.

CUPRO Generic name for regenerated cellulose fibres produced by the cuprammonium process.

CURE (1) The process of deliberately crosslinking a polymer to improve its properties, especially the mechanical properties such as stiffness. In plastics materials curing will convert linear or branched solid thermoplastic, or liquid low molecular weight precursor polymer, to a hard, stiff thermoset material. The term is particularly used to refer to the crosslinking of unsaturated polyester and epoxy resins. In rubbers the term is often used to mean vulcanisation.

(2) The development of useful properties as a result of crosslinking a polymer. Of most interest are the mechanical properties, especially important being the modulus increase with cure. Solubility and swellability decrease with increasing crosslinking (and cure). Sometimes the term, state of cure, is used in this way. In rubbers the course of the curing (vulcanisation) reaction is followed in a curemeter as the rate of cure.

CUREMETER An instrument for following continuously the progress of curing during vulcanisation of a rubber by measuring the stiffness of the rubber, contained in a heated chamber, as curing takes place. Typically the dynamic shear modulus is measured by measuring the force required to move a paddle oscillated whilst it is sandwiched between two rubber test specimens. In the oscillating disc curemeter, which is a kind of modified Mooney viscometer, the rotor is embedded in the sample and is oscillated through a small arc and the torque required to do this is measured. Thus a torque–time (cure–time) curve is generated from which the vulcanisation parameters, scorch time, cure time and final modulus, can be calculated. The cure curve is usually one of three types: (1) cure to a constant equilibrium torque, which is found with many synthetic rubbers, (2) cure to a maximum torque followed by reversion, often found with natural rubber and (3) cure with torque continuously increasing, found with some synthetic rubbers.

CURE TIME The time required during vulcanisation of a rubber for the rubber to reach a desired state of cure.

CURIE-POINT PYROLYSIS Pyrolysis of a thin layer of material coated onto a wire which is then heated to its Curie point by induction heating. A precise pyrolysis temperature is thus achieved, in contrast to other pyrolysis methods. The pyrolytic degradation products are usually determined by pyrolysing directly in the carrier gas stream of a gas–liquid chromatograph (pyrolysis gas chromatography).

CURTAIN ELECTROPHORESIS (Continuous zone electrophoresis) A technique of paper electrophoresis for the separation and purification of proteins on a larger scale than usual for paper electrophoresis. The sample is continuously applied to a point in the middle of the top edge of a vertically mounted sheet of filter paper. It is eluted by a continuous stream of eluent applied across the whole of the edge of the paper. Simultaneously an electric field is applied horizontally across the paper to separate the components of the mixture in this direction. The separated proteins are collected in a row of receivers along the bottom edge of the paper.

CVNR Abbreviation for constant viscosity natural rubber.

CYANACRYL Tradename for acrylic elastomers.

CYANAMIDE NH_2CN. M.p. 42°C. Its calcium salt is produced by reaction of calcium carbide with nitrogen and is converted to cyanamide in aqueous solution by treatment with dilute acid:

$$CaC_2 + N_2 \rightarrow CaNCN \xrightarrow[H_2SO_4]{} CaSO_4 + NH_2CH$$

In slightly alkaline aqueous conditions cyanamide is converted to dicyanamide.

CYANATE METHOD A method for the determination of the N-terminal amino acid of a peptide or protein. Reaction of the amino group with cyanate ion at pH 10 in 8 M urea yields an N-carbamyl derivative (**I**). On heating with acid this yields a hydantoin (**II**), which is separated from the amino acids produced by chromatography, and is identified and determined quantitatively:

$$NCO^- + H_3\overset{+}{N}CHRCONHCHR'CO\sim \rightarrow$$

$$H_2NCONHCHRCONHCHR'CO\sim \xrightarrow{HCl}$$

I

$$\text{R—HC——CO} + H_3\overset{+}{N}CHR'CO\sim$$

II

CYANOACRYLATE (Alkyl-2-cyanoacrylate)

$$\begin{array}{c} CN \\ | \\ CH_2{=}CCOOR \end{array}$$

where R is an alkyl group, normally an ethyl or methyl group. An acrylic monomer, which although it polymerises by a free radical mechanism, can also be polymerised anionically in the presence of a mild base—even water. This is made use of in cyanoacrylate adhesives. Methyl cyanoacrylate is produced by depolymerisation of the polymer formed by reaction of formaldehyde with methyl cyanoacetate:

$$NCCH_2COOCH_3 + HCHO \rightarrow$$

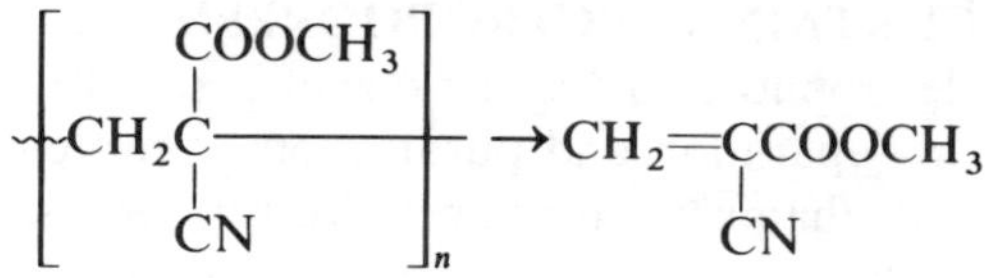

$$\left[CH_2C \underset{CN}{\overset{COOCH_3}{\mid}} \right]_n \rightarrow CH_2{=}CCOOCH_3 \underset{CN}{}$$

CYANOACRYLATE ADHESIVE Tradename: Eastman 910. An adhesive based on a cyanoacrylate monomer, which rapidly polymerises on exposure to moisture, especially that absorbed on substrate surfaces to be joined. Curing occurs especially rapidly when the joint is closed, thus excluding air which inhibits the polymerisation of acrylic monomers.

CYANOETHYL CELLULOSE Cellulose in which some of the cellulose hydroxyl groups have been replaced by —CH$_2$CH$_2$CN groups by reaction of alkali cellulose (derived from cotton) with acrylonitrile. The product containing about 3% nitrogen, i.e. with about one hydroxyl in six having reacted, has increased resistance to mildew, bacteria and heat, increased receptivity to dyestuffs and improved abrasion resistance. Further reaction results in too great a loss in tensile strength.

CYANOGEN BROMIDE BrCN. A reagent useful for the partial hydrolysis of proteins for the production of peptides in the sequencing of the protein. It cleaves the polypeptide chain specifically at a methionine residue which is converted to C-terminal homoserinelactone residue (**I**):

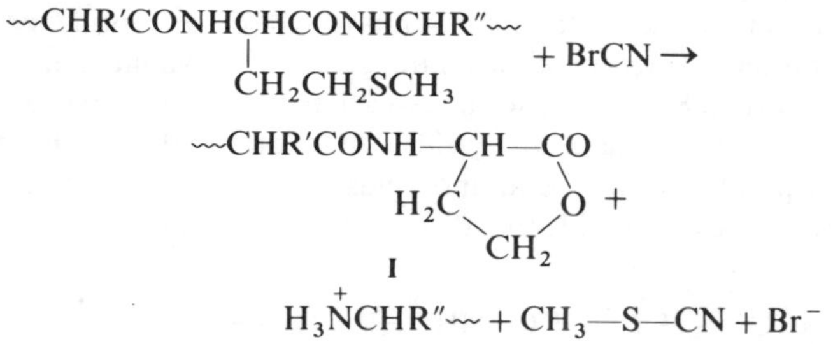

$$\sim\!\!\sim\!\!CHR'CONHCHCONHCHR''\!\!\sim\!\!\sim \underset{\begin{array}{c}\mid\\ CH_2CH_2SCH_3\end{array}}{} + BrCN \rightarrow$$

$$\sim\!\!\sim\!\!CHR'CONH\!-\!CH\!-\!CO$$
$$H_2C \qquad O +$$
$$CH_2$$
$$\mathbf{I}$$

$$\overset{+}{H_3}NCHR''\!\!\sim\!\!\sim + CH_3\!-\!S\!-\!CN + Br^-$$

CYANOPRENE Tradename for a cast polyurethane elastomer system based on a polyether polyol/diisocyanate prepolymer and cured with 3,3'dichloro-4,4'-diaminodiphenylmethane.

N-CYANOSULPHONAMIDE POLYMER (NCNS polymer) A polymer of general structure **I**, synthesised by reaction of a secondary with a primary biscyanamide, which on heating at about 150°C form crosslinks through triazine rings:

$$NCNRArNRCN + NCNHAr'NHCN \rightarrow$$

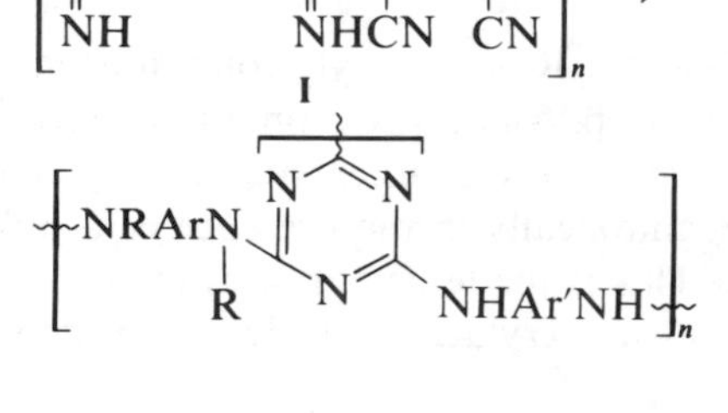

$$\left[\underset{NH}{\overset{\mid\mid}{CNRArNRC}}\!-\!\underset{NHCN\ CN}{\overset{\mid\ \ \ \mid}{NAr'NH}} \right]_n \rightarrow$$
$$\mathbf{I}$$

$$\left[NRArN \underset{R}{\overset{}{\mid}} \overset{N{=}N}{\underset{N}{\bigtriangleup}} NHAr'NH \right]_n$$

where Ar and Ar' are aromatic groups and R is an aryl sulphonyl group. The commercial polymers are useful as clear thermosetting moulding materials, for laminating with glass and other fabric reinforcements and in the manufacture of printed circuit boards.

CYANURAMIDE Alternative name for melamine.

CYCLIC AMIDE Alternative name for lactam.

CYCLIC AMINE (Imine) A compound of general structure:

$$\left[\underset{NH}{\overset{R}{\square}} \right]$$

where R is a hydrocarbon group. It may be polymerised to a polyamine by ring-opening polymerisation by acid catalysts, e.g. by Lewis acids or protonic acids, but not by bases since the anion $\sim\!\!\sim\!\!R\bar{N}H$ is unstable. Polymerisation is very fast due to the high strain in the ring and occurs via attack of monomer on the immonium ion:

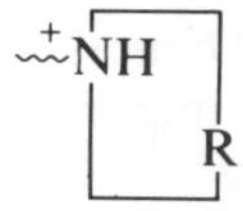

$$\sim\!\!\sim\!\!\overset{+}{N}H \underset{R}{\square}$$

CYCLIC ESTER Alternative name for lactone.

CYCLISED POLYACRYLONITRILE Polyacrylonitrile which has been converted to a partial ladder polymer by heating, usually with controlled oxidation, to temperatures in the range 100–300°C (for Black Orlon) or much higher (for Pluton).

CYCLISED RUBBER (Cyclorubber) (Isomerised rubber) Tradenames: Plioform, Pliolite. Natural rubber, or sometimes another diene rubber such as polybutadiene, which has been treated with sulphuric acid, a sulphonic acid or a Lewis acid, such as SnCl$_4$, AlCl$_3$ or BF$_3$. This causes a reduction of the unsaturation and formation of condensed ring structures, typically with two or three rings being fused together:

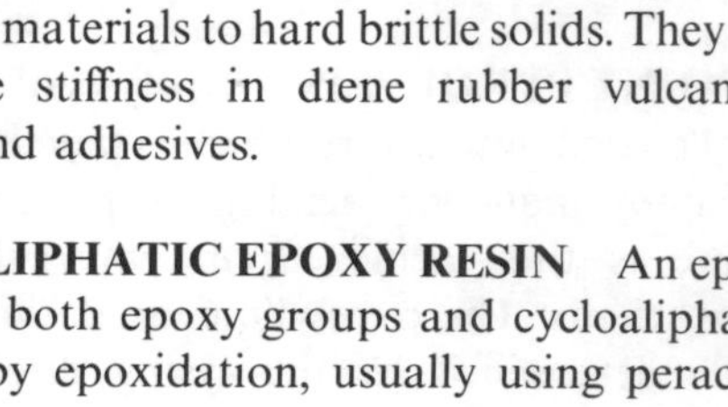

$$\left[\begin{array}{c} CH_2 \quad CH_3 \\ C \quad CH_2 \\ CH_2\ CH \quad CH_2 \\ CH_2\ C \quad C \\ CH_2\ CH_3 \quad CH_3 \\ CH_2 \end{array} \right]$$

Products, which are thermoplastic, can range from tough balata-like materials to hard brittle solids. They are useful to provide stiffness in diene rubber vulcanisates, in coatings and adhesives.

CYCLOALIPHATIC EPOXY RESIN An epoxy resin containing both epoxy groups and cycloaliphatic rings. Obtained by epoxidation, usually using peracetic acid, of unsaturated ring compounds. The latter are made by Diels–Alder reaction of butadiene and a dienophile such as crotonaldehyde or acrolein:

$$CH_2{=}CHCH{=}CH_2 + CH_3CH{=}CHCHO \rightarrow$$

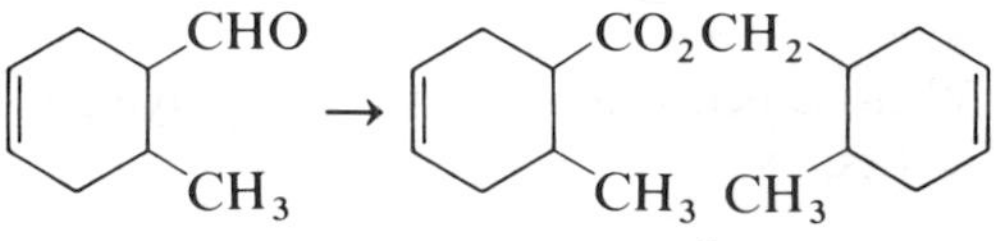

I

Examples are 3,4-epoxy-6-methylcyclohexylmethyl-3,4-epoxy-6-methylcyclohexanecarboxylate (epoxidised **I**), vinylcyclohex-3-ene dioxide

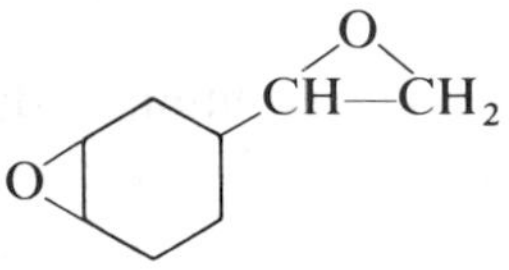

and dicyclopentadiene dioxide

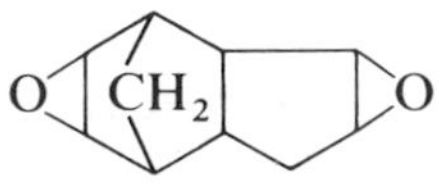

CYCLOHEXAAMYLOSE   (Schardinger α-dextrin) M.p. 200°C $\alpha_D^{20,H_2O} + 150°C$. A cyclic hexamer of D-glucose units which are α-1,4′-linked. A component of Schardinger dextrin.

CYCLOHEXANE

B.p. 80·7°C.

Manufactured by the hydrogenation of benzene, which it resembles in solvent power, but it is somewhat less toxic and therefore is preferred commercially as a solvent. It is useful as a solvent for hydrocarbon rubbers, silicone oils and natural resins. It is also useful as an intermediate for conversion by oxidation to cyclohexanone and cyclohexanol.

1,4-CYCLOHEXANEDIMETHANOL   Alternative name for 1,4-dimethylolcyclohexane.

CYCLOHEXANOL

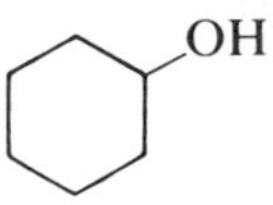

OH

B.p. 160°C. M.p. 25°C.

A solvent for cellulose ethers, low molecular weight silicones, phenol–formaldehyde resins and many natural resins. Also useful in epoxy resin coatings to give better finishes.

CYCLOHEXANONE

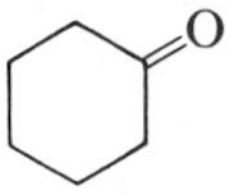

O

B.p. 156·5°C.

A solvent, more viscous than most organic solvents, for cellulose esters, many natural resins, natural rubber,

polyvinyl chloride (for which it is especially useful), polyvinyl acetate, polystyrene and phenol–formaldehyde resins. It swells polymethylmethacrylate and polyethylene, which it dissolves when hot. It has a slight mutual miscibility with water.

CYCLOHEXANONE HYDROPEROXIDE   The product formed by reaction of hydrogen peroxide with cyclohexanone. Several different peroxides and hydroperoxides are formed, but the main constituent of commercial products is

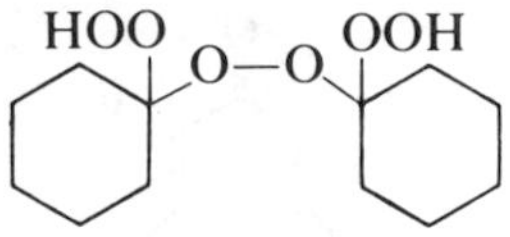

It has half-life times of 42 h/100°C, 0·3 h/120°C and 0·05 h/150°C and half-life temperatures of 115°C/1 h, 91°C/10 h and 166°C/1 min. Useful as an initiator in the crosslinking of unsaturated polyesters at ambient temperature, when a cobalt naphthenate or octoate accelerator is used. Sometimes used as an initiator component in emulsion polymerisation.

N-CYCLOHEXYLBENZOTHIAZYLSULPHEN-AMIDE   (CBS)

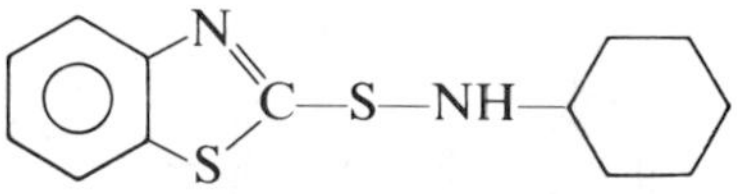

M.p. 79°C.

A delayed action accelerator for the sulphur vulcanisation of rubbers.

CYCLOHEXYL METHACRYLATE

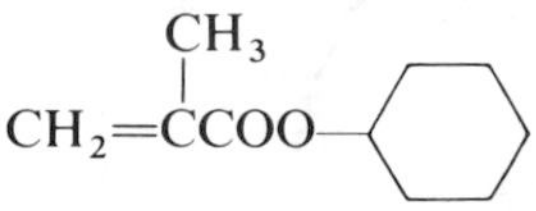

B.p. 72°C/6 mm.

The methacrylate monomer for the formation of polycyclohexylmethacrylate.

N-CYCLOHEXYLTHIOPHTHALIMIDE   Tradename: Santogard PVI.

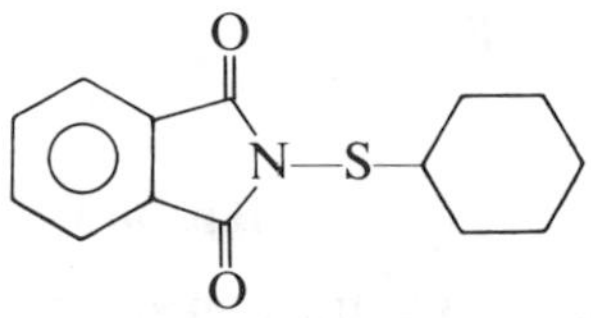

A prevulcanisation inhibitor, useful for preventing scorch during compounding and shaping, but not affecting subsequent crosslink formation during vulcanisation.

N-CYCLOHEXYL-p-TOLUENESULPHONAMIDE

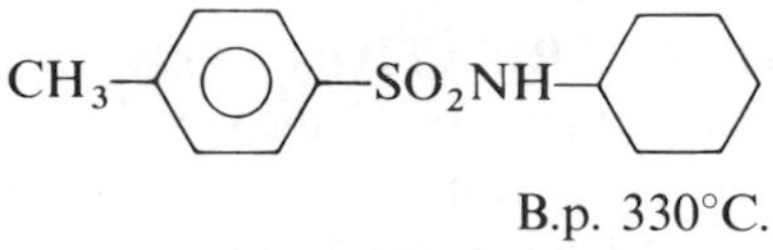

B.p. 330°C. M.p. 87°C.

A plasticiser for cellulose esters and natural resins.

CYCLOOCTENE

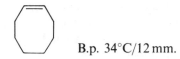

B.p. 34°C/12 mm.

The monomer for formation of polyoctenamer by ring-opening polymerisation. Obtained by hydrogenation of cyclooctadiene, itself obtained by cyclodimerisation of butadiene.

CYCLOPENTADIENONE A compound of the type

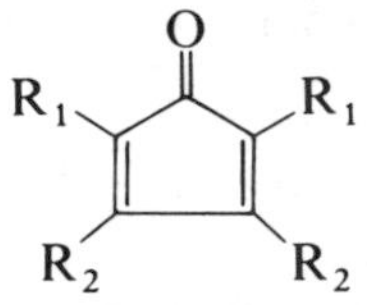

where R_1 is often methyl and R_2 is often phenyl. Acts as a pseudobisdiene monomer for Diels–Alder polymerisation. When tetrasubstituted it forms a stable monomer or reversible dimer, but when unsubstituted or mono- or disubstituted it forms an irreversible dimer. Polymers may be produced since the initial Diels–Alder adduct regenerates a new diene by loss of carbon monoxide to continue polymerisation to very high molecular weights. Thus with a bismaleimide:

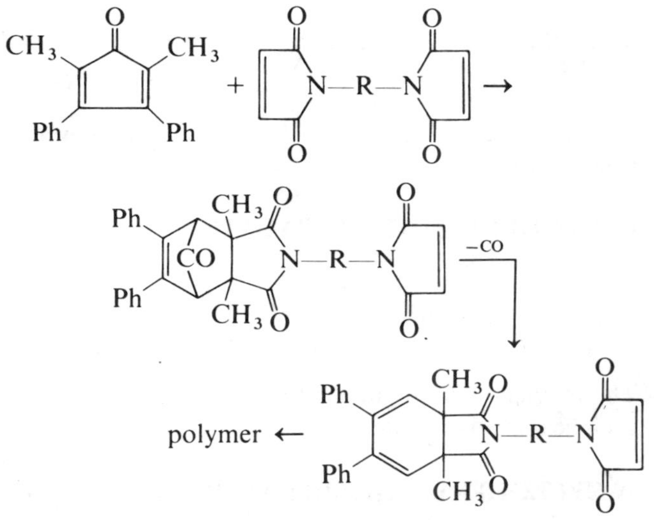

polymer ←

CYCLOPENTENE

B.p. 46°C.

Obtained from the C_5 fraction of cracked petroleum, where it occurs to the extent of only a few per cent. However, cyclopentadiene and dicyclopentadiene are major components of this fraction and may be hydrogenated to cyclopentene. The monomer for the formation of polypentenamer by ring-opening polymerisation.

CYCLOPOLYMERISATION Polymerisation in which ring structures are formed within the polymer chain. In one type, polymerisation of a divinyl monomer occurs, which instead of producing an insoluble cross-linked polymer network on polymerisation, produces a linear and soluble polymer containing cyclic units in the polymer chain:

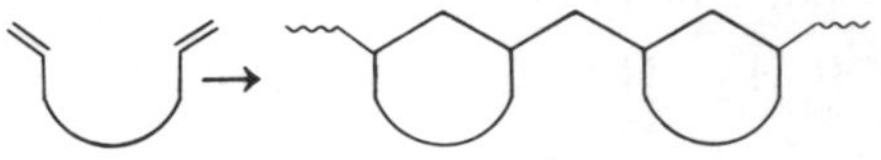

Well-studied examples are the polymerisations of 1,6-dienes:

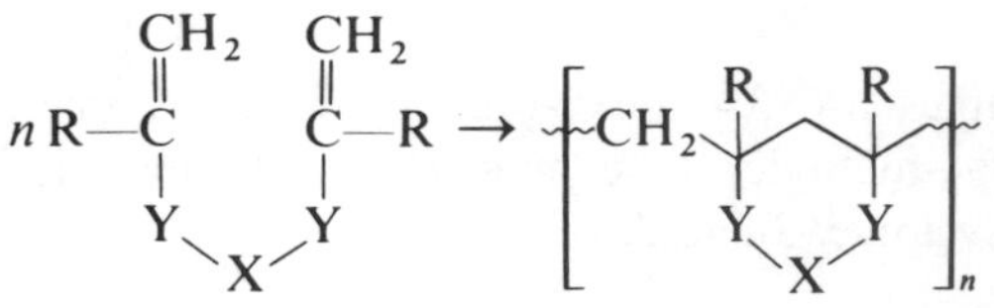

and also the cyclopolymerisation of divinyl ethers with maleic anhydride:

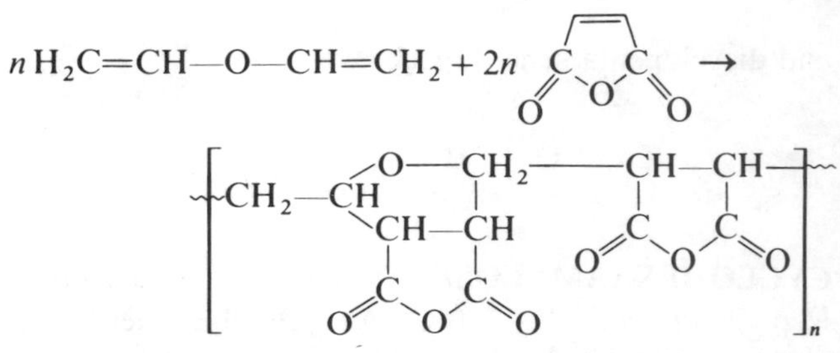

This copolymer, sometimes referred to as pyran polymer has anti-tumour biological activity. Certain diallylamines when cyclopolymerised produce commercially useful polymers for the dissipation of electrostatic charges in reprographic paper, e.g. poly-(dimethyldiallylammonium chloride).

In a second type, heterocyclic rings are formed, usually by step-growth polymerisation with elimination of a volatile (often water). Ring formation may be by a post-reaction on a precursor polymer (as in the conversion of a polyamic acid to a polyimide) or it may be simultaneous with chain formation and extension (as in the synthesis of polyphenylquinoxalines).

CYCLORUBBER Alternative name for cyclised rubber.

CYCLOSILOXANE A siloxane of cyclic structure,

$$\left[\!-\!Si(R_2)\!-\!O\!-\!\right]_n$$

Formed by the hydrolysis of the corresponding dichlorosilane together with linear polymer. The presence of an insoluble organic solvent together with the water and of a suitable catalyst with vigorous stirring, favours cyclisation. The cyclic isomers can be isolated by fractional distillation, although frequently the tetramer predominates. The cyclic tetramers, especially octamethylcyclotetrasiloxane, are useful monomers for the synthesis of very high molecular weight polyorganosiloxanes, which are required for silicone rubbers, since sufficiently pure dichlorosilanes are difficult to obtain. On equilibration with an alkaline catalyst at 150–200°C polymers with molecular weights of about 10^5 may be obtained.

CYCLSAFE Tradename for a nitrile barrier resin containing 70% acrylonitrile and 30% styrene.

CYCOLAC Tradename for acrylonitrile–butadiene–styrene copolymer.

CYCOLOY Tradename for a blend of acrylonitrile–butadiene–styrene copolymer with bisphenol A polycarbonate.

CYCOPAC Tradename for a high nitrile barrier polymer.

CYCOVYN Tradename for a blend of polyvinyl chloride with acrylonitrile–butadiene–styrene copolymer.

CYLINDRITE Cylindrically shaped spherulites present as the basic supermolecular element of crystalline aggregates in row-nucleated crystalline polymer samples.

CYMAX Tradename for polythiodiethanol.

CYMEL Tradename for melamine–formaldehyde polymer.

CYMENE

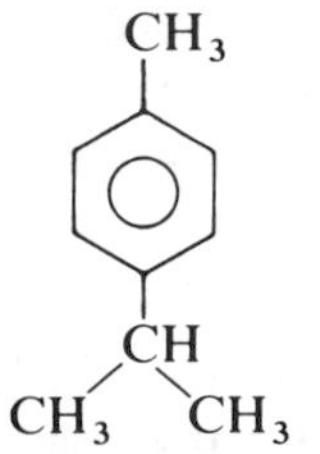

$$CH_3$$

$$CH$$
$$CH_3 \quad CH_3 \qquad B.p.\ 176 \cdot 5°C.$$

A by-product of the sulphite paper pulp process. A solvent for many natural resins, polystyrene and other hydrocarbon polymers. A high boiling point diluent for lacquers.

CYS Abbreviation for cysteine.

(CYS)₂ Abbreviation for cystine.

CYSTEIC ACID

$$H_2NCHCOOH$$
$$CH_2SO_3H$$

Cysteic acid residues are formed in protein polypeptide chains and peptides by oxidation of any disulphide cystine bridges. This arises when the disulphide bridges are treated with performic acid:

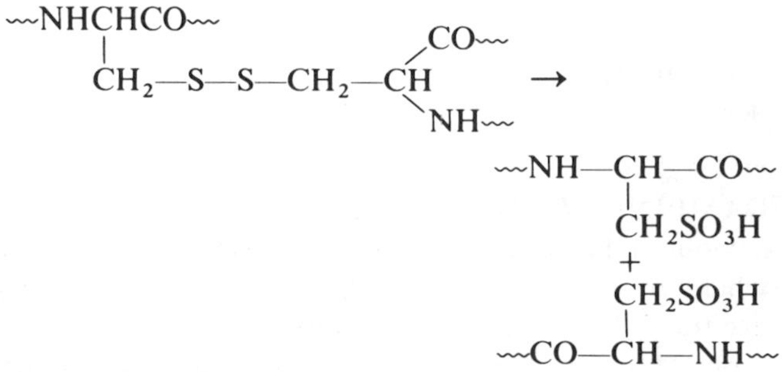

$$\sim\!\!\sim\!NHCHCO\!\sim\!\!\sim$$
$$CH_2\!-\!S\!-\!S\!-\!CH_2\!-\!CH \qquad\qquad CO\!\sim\!\!\sim$$
$$NH\!\sim\!\!\sim \qquad\to$$

$$\sim\!\!\sim\!NH\!-\!CH\!-\!CO\!\sim\!\!\sim$$
$$CH_2SO_3H$$
$$+$$
$$CH_2SO_3H$$
$$\sim\!\!\sim\!CO\!-\!CH\!-\!NH\!\sim\!\!\sim$$

Cysteine residues are similarly oxidised.

CYSTEINE (Cys) (C)

$$H_2NCHCOOH$$
$$CH_2SH \qquad M.p.\ 240°C\ (decomposes).$$

A polar α-amino acid widely distributed in proteins but usually present, except notably in some keratins, in only small amounts. Two cysteine residues in proteins are often found combined by oxidation to form the disulphide bridge of a cystine residue. If intermolecular, the bridge is a crosslink. Its pK′ values are 1·96 and 8·18 with the isoelectric point at 5·07. Cysteine may be destroyed during the hydrochloric acid hydrolysis stage of amino acid analysis of a protein, so it is often oxidised to the more stable cysteic acid with performic acid, prior to hydrolysis. Cysteine residues may be specifically determined in a protein by the use of Ellman's reagent. It is readily oxidised in neutral or alkaline solution to cystine, but may be protected from oxidation by reaction with iodoacetate. It may be determined separately from cystine by reaction with formaldehyde, yielding thiazoline-carboxylic acid

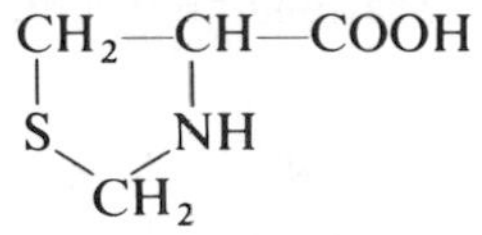

$$CH_2\!-\!CH\!-\!COOH$$
$$S \qquad NH$$
$$CH_2$$

CYSTEINE ENZYME Alternative name for thiol enzyme.

CYSTINE ((Cys)₂)

$$H_2NCHCH_2SSCH_2CHNH_2$$
$$COOH \qquad\qquad COOH$$

$$M.p.\ 258/261°C\ (decomposes).$$

A polar α-amino acid widely found in proteins but in only small amounts except, notably, in keratin. The disulphide bridge often arises from oxidative coupling of two cysteine units. If these are on different polypeptide chains then a crosslink is formed. Its pK′ values are 1·7 and 7·8 with the isoelectric point at 4·60. It is partially destroyed during the hydrochloric acid hydrolysis stage of amino acid analysis of a protein and therefore is often oxidised with performic acid to the more stable cysteic acid prior to hydrolysis.

CYTOCHROME A haemoprotein acting as the transporter of electrons in the oxidation of nutrients (fats and carbohydrates) with oxygen. This respiratory chain has been most intensively studied in the mitochondria of higher animals. There are at least five different cytochromes involved, of which cytochrome C is the most widely studied since it is the only one readily isolated in a homogeneous form.

CYTOCHROME C The most widely studied cytochrome. Samples from a wide variety of species have been studied and their amino acid sequences determined. It consists of a single polypeptide chain of 103–114 amino acid residues of molecular weight of about 12 000. Unlike the other common haemoproteins, haemoglobin and myoglobin, the iron(II) of the haem group undergoes oxidation to iron(III) and reduction back again as the cytochrome participates in the biological electron transfer process. The iron of the haem group is bound to the polypeptide by covalent links from two cysteine residues on the polypeptide to the porphyrin ring and by coordination of a side chain histidine nitrogen and a methionine sulphur to the iron. Since all six iron coordination positions are occupied cytochrome C cannot bind oxygen.

CYTOR Tradename for a polyurethane block copolymer based on a polyester or polyether polyol/MDI prepolymer and chain extended.

D

D (1) Symbol for a difunctional unit in a polyorganosiloxane, i.e. for a repeating unit in a linear polymer of the type

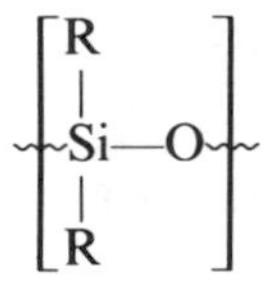

$$\left[\begin{array}{c} R \\ | \\ Si-O \\ | \\ R \end{array}\right]$$

Derived from the use of a dichloro (or dialkoxy) silane monomer. Often specifically refers to a dimethylsiloxane unit.
(2) Symbol for aspartic acid.

D- A prefix used in the naming of organic compounds containing an asymmetric (chiral) atom (usually carbon) to denote a particular enantiomer of the enantiomorphic pair. Used especially for carbohydrates, both in monosaccharides, e.g. D-glucose, and in polysaccharides, e.g. poly-(1,4'-β-D-mannose). Whether a molecule is D- or L- depends on the configuration of its parent. By convention the agreed parent configurations are those assigned to glyceraldehyde $OHCCH(OH)CH_2OH$. In the Fischer projection formula the convention is that with the carbonyl group uppermost, if the lowest asymmetric carbon has the hydroxyl to the right then the sugar is D-, and if to the left it is L-. The prefixes D- and L- are sometimes followed by a + or − sign to denote the direction of rotation of polarised light by the particular enantiomorph, e.g. as in D-(+)-glucose.

DAA Abbreviation for dialphanyl adipate.

DABCO Abbreviation for 1,4-diazabicyclo-2,2,2-octane.

DACON Tradename for polyvinyl chloride.

DACRON Tradename for polyethylene terephthalate fibre.

DAI-EL Tradename for fluoroelastomers based on vinylidene fluoride–hexafluoropropylene copolymers (sometimes with some tetrafluoroethylene termonomer units).

DAIFLON Tradename for polychlorotrifluoroethylene.

DAIP Abbreviation for diallyl isophthalate.

DALTOFLEX Tradename for a millable polyurethane elastomer, developed from the earlier Vulcaprene A, based on a polyesteramide and tolylene diisocyanate and cured by the addition of excess isocyanate or with formaldehyde.

DALTOMOLD Tradename for a partially crosslinked polyurethane elastomer, based on a polyester or polyether polyol/MDI prepolymer and then chain extended.

DAMAR A relatively soft type of natural resin soluble in organic solvents but not in ethanol. Unlike many natural resins it contains about 80% of inert resenes and only relatively low amounts of acidic compounds. Some of the resene is waxy and lowers the solubility. It is useful in certain varnishes and softens at about 70°C.

DAMPING The dissipation of mechanical energy as heat in a viscoelastic material, rather than the storage of energy as in a purely elastic deformation. Damping has been particularly widely studied in polymers as part of their dynamic mechanical behaviour, usually being characterised by the tangent of the loss angle. It can provide the most sensitive method for the detection of transitions, due to the onset of particular molecular motions in a polymer. Damping is both frequency and temperature dependent. A maximum in damping occurs in the transition region, which at low frequencies is close to T_g on the temperature scale. The observed transition temperature is raised by about 10°C for each decade increase in frequency. Molecularly, damping is associated with the onset of segmental motions with increasing temperature (or decreasing frequency). Thus below the main transition temperature molecular motion is frozen and the behaviour is elastic, whereas well above the transition rubber elasticity is observed.

DAMPING FACTOR Alternative name for tangent of the loss angle.

DAMPING INDEX In dynamic mechanical behaviour, a measure of the damping, being defined as $1/n$ where n is the number of oscillations between two arbitrarily fixed boundary conditions, e.g. amplitudes, in a series of waves. It is used particularly in torsional braid analysis.

DANSYL CHLORIDE (DNS) (1-Dimethylamino-naphthalene-5-sulphonyl chloride)

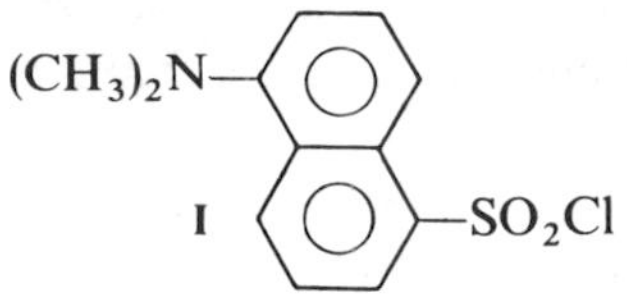

$$I \quad (CH_3)_2N-\text{[naphthalene]}-SO_2Cl$$

A reagent very widely used for identification and quantitative determination of N-terminal amino acids of peptides and proteins. The dansyl derivative is fluorescent, making the method about a hundred times more sensitive than the DNP method:

$$I + H_2NCHCO\text{---} \rightarrow$$

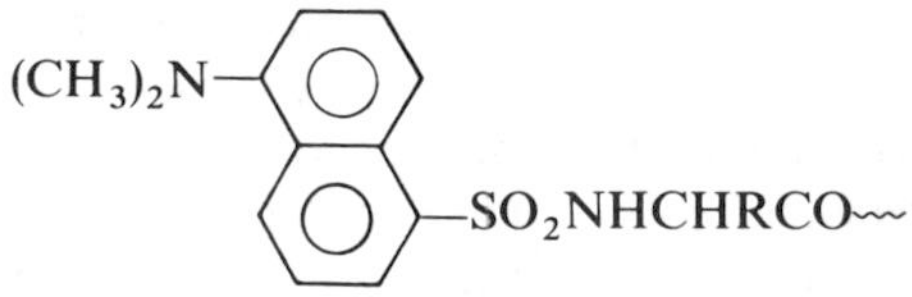

$$(CH_3)_2N-\text{[naphthalene]}-SO_2NHCHRCO\text{---} \quad + HCl$$

After hydrolysis of the dansylated material, the dansyl amino acid is separated by chromatography and determined quantitatively.

DANSYL DERIVATIVE The derivative formed by reaction of dansyl chloride with an N-terminal amino acid of a peptide or protein. Formation of this derivative followed by its separation from other amino acids by chromatography and quantitative determination provides a very sensitive method for the determination of the N-terminal amino acid.

DAP Abbreviation for diallyl phthalate or for dialphanyl phthalate.

DA$_{79}$P Abbreviation for heptylnonyl phthalate.

DAPLEN Tradename for low density polyethylene and polypropylene.

DAREX Tradename for styrene–butadiene rubber.

DARK FIELD MICROSCOPY Optical microscopy in which the light is incident upon the sample at some angle to the axis of the microscope. Only light reflected from the specimen enters the objective lens. This eliminates background light so the specimen appears bright against a dark background. It is particularly useful for observing polymer crystals, two phase systems, e.g. polymer blends, and other objects of low opacity.

DARLAN Tradename for a polyvinylidene cyanide/vinyl acetate copolymer fibre.

DARVAN Tradename for a vinylidene cyanide/vinyl acetate copolymer fibre.

DAWSONITE A synthetic microfibre material, which chemically is sodium aluminium hydroxycarbonate, $NaAl(OH)_2CO_3$, produced as acicular crystals about $0.5 \mu m$ long with an aspect ratio of 25–40. Useful as a reinforcing filler in thermoplastics with some fire retarding properties due to its endothermic decomposition at 300–350°C.

DBP Abbreviation for dibutyl phthalate.

DBPC Abbreviation for di-*t*-butyl-*p*-cresol.

DBS Abbreviation for dibutyl sebacate.

DCC Abbreviation for *N,N'*-dicyclohexylcarbodiimide.

DCCI Abbreviation for *N,N'*-dicyclohexylcarbodiimide.

DCHP Abbreviation for dicyclohexyl phthalate.

DCP Abbreviation for dicapryl phthalate.

DC STEP-RESPONSE METHOD Alternative name for time-domain method.

DDA Abbreviation for di-*n*-decyl adipate.

DDP Abbreviation for didecyl phthalate.

DDSA Abbreviation for dodecenylsuccinic anhydride.

DEAD-END POLYMERISATION A free radical polymerisation in which a very active initiator is used so that the rate of polymerisation continuously decreases as initiator is consumed and a limiting conversion of monomer to polymer occurs. By use of this method the rate constant for initiator decomposition (k_d) may be determined through the relationship

$$\ln[1 - \ln(1-p)/\ln(1-p_\infty)] = -k_d t/2$$

where p and p_∞ are the fractional conversions at any time t and at infinite time respectively.

DEAD POLYMER The polymer formed in a chain polymerisation which no longer contains active centres. Dead polymer is incapable of further growth by addition of monomer, except after a transfer to polymer step. In free radical polymerisation nearly all the polymer formed is dead polymer since the instantaneous concentration of active centres is always very low ($\sim 10^{-6}$M). In ionic, especially anionic, polymerisation and Ziegler polymerisation, termination reactions are often absent, so the active centres remain in the polymer, even after all the monomer has been consumed. This polymer, in contrast, is known as living polymer.

DEAE-CELLULOSE Abbreviation for diethylamino-ethylcellulose.

DEAE-SEPHADEX Abbreviation for diethylamino-ethyl-Sephadex.

DEATH CHARGE POLYMERISATION Polymerisation of an electrically charged monomer involving loss of charge, e.g. of an aryl cyclic sulphonium zwitterion to poly(thio-1,4-phenylenetetramethylene).

DEBORAH NUMBER Symbol De. The ratio of a characteristic time of a fluid, often taken as the relaxation time (itself given by the ratio of the viscosity to the modulus at steady flow), to the characteristic time of the flow process being considered. The latter is the time elapsed following a point in the material after which a non-negligible change in the flow conditions has occurred. The characteristic time is often taken as the ratio of a characteristic length to the mean velocity of the polymer. For a high value of De (say above about unity) steady-state conditions are unlikely to become established and relaxation phenomena, i.e. elastic or memory effects, will dominate, whereas at De < 1 deformation will be essentially that of a viscous flow, with little memory of previous flow history retained.

DE-BRUYNE RULE A strong adhesive joint cannot be made using a polar adhesive with a non-polar adherend or with a non-polar adhesive and polar adherend. The rule expresses the fact that polar adhesives will not have a surface tension below the critical surface tension of a non-polar adherend.

DEBYE DISPERSION EQUATION (Debye equation) A relationship giving the frequency dependence of the complex permittivity (ε^*) of a dielectric:

$$\varepsilon^* = \varepsilon_\infty + (\varepsilon_s - \varepsilon_\infty)/(1 + i\omega\tau^2)$$

where ε_∞ and ε_s are the very high frequency and the static values of the relative permittivity respectively, i is $(-1)^{1/2}$, ω is the angular frequency of the sinusoidally varying field (or voltage) and τ is the relaxation time. By equating the real and imaginary parts, the equation may be rewritten as

$$\varepsilon'(\omega) = \varepsilon_\infty + (\varepsilon_s - \varepsilon_\infty)/(1 + \omega^2\tau^2)$$

and

$$\varepsilon''(\omega) = [(\varepsilon_s - \varepsilon_\infty)/(1 + \omega^2\tau^2)]\omega\tau$$

where $\varepsilon'(\omega)$ is the relative permittivity and $\varepsilon''(\omega)$ is the loss factor. These equations correspond to a step in ε' and a peak in ε'' at $\omega = 1/\tau$. A relaxation characterised by these relations is a Debye relaxation. However, polymers exhibit broader relaxations which may be represented by empirical modifications to the Debye equation, e.g. as the Cole–Cole equation, incorporating a distribution of relaxation times.

DEBYE EQUATION (1) Alternative name for the Debye dispersion equation.

(2) A relationship for the relative permittivity (ε) of a dielectric material as a function of the electronic and atomic polarisability (α_a) and the orientational polarisability terms. The latter is given by $\mu^2/(3kT)$, where μ is the dipole moment of the molecule concerned, T is the temperature and k is Boltzmann's constant. It is

$$(\varepsilon - 1)/(\varepsilon + 2) = (4\pi N_1/3)(\alpha_a + \mu^2/3kT)$$

N_1 is the number of molecules present per cubic centimetre and equals $\rho N_A/M$, where N_A is Avogadro's number, ρ is the density and M is the molecular weight. The equation is derived by neglecting interactions between polar groups and therefore is only applicable to gases and dilute solutions, and, in the case of polymers, to copolymers containing a low concentration of polar comonomer groups. For most solids the Fröhlich and Onsager equations are more appropriate.

DEBYE RELAXATION A dielectric relaxation which obeys the Debye dispersion equation. Polymer relaxations are broader, with the loss factor having a lower loss maximum than the classical Debye case, suggesting a distribution of relaxation times.

DEBYE–SCHERRER DIAGRAM The wide angle X-ray scattering pattern obtained using a strip of photographic film surrounding the sample. For a randomly oriented collection of crystals, as in a powder or bulk (but unoriented) polymer, the pattern consists of a pattern of lines, which are the segments of the total scattering pattern of concentric rings. The line positions give the d-spacings between the hkl planes of the atoms.

DECAHYDRONAPHTHALENE Tradename: Decalin.

 B.p. 183–192°C.

Produced by the complete hydrogenation of naphthalene. Like tetrahydronaphthalene it is a solvent, but not a very powerful one, for natural resins, hydrocarbon rubbers and, at elevated temperatures, for polyethylene. It does not oxidise as readily as tetrahydronaphthalene.

DECALIN Tradename for decahydronaphthalene.

n-DECANE $CH_3(CH_2)_8CH_3$. B.p. 174°C. A solvent for hydrocarbon rubbers and, when hot, for polyethylene and polypropylene.

1,10-DECANEDICARBOXYLIC ACID (Dodecandioic acid) $HOOC(CH_2)_{10}COOH$. M.p. 129°C. Synthesised from cyclododecatriene (prepared by cyclotrimerisation of butadiene) via cyclododecane, which is air oxidised first, then nitric acid oxidised to the acid. A monomer for the synthesis of both nylon 612 and poly-(bis-(4-cyclohexyl)methanedodecanoamide).

DECITEX A measure of the coarseness of a fibre or yarn, similar to denier, but often used instead for synthetic fibres and yarns. The weight in grams of 100 m of the yarn or fibre.

DECOUPLING Alternative name for spin–spin decoupling.

DEENAX Tradename for 2,6-di-t-butyl-4-methylphenol.

DEFORMABLE LAYER THEORY A mechanism proposed to account for the action of coupling agents in improving the mechanical performance of fillers, especially glass fibres, in polymer composites. Originally the coupling agent was thought to provide a plastically deformable buffer zone at the matrix–filler interface. However, effective layers of coupling agent are not thick enough for this mechanism to operate. A later mechanism suggested that the agent acts by changing the chemical reactivity of the uncrosslinked polymer matrix so that the matrix near the interface is more flexible and ductile than the bulk material.

DEFORMATION The displacement of a point in a body relative to adjacent points. Strain is usually defined in terms of deformation.

DEFORMATION BAND Alternative name for shear band.

DEFORMATION BIREFRINGENCE Often used as an alternative name for orientation birefringence. The term also means the birefringence resulting from an axial compression or dilation of optically isotropic elements causing a change in the spacing of the elements in only one axial direction. Bond angle distortion can also result in a deformation birefringence without any orientation.

DEFORMATION RATIO (Extension ratio) (Stretch) (Unit elongation) A measure of strain. Defined as l/l_0, where l_0 is the initial length of an element in a body which attains a length l when a stress system is applied. Widely used in large strain analyses of elasticity problems, particularly in rubber elasticity, when it is given the symbol λ.

DEG Abbreviation for diethylene glycol.

DEGRADABLE POLYMER A polymer which has been deliberately sensitised to degradation. Usually applied to polymers used in packaging and containing photosensitive groups, so that on exposure to sunlight the polymer photodegrades and embrittles. The erosive action of the wind and rain then breaks down the material to small particles, which become incorporated into the soil where they may subsequently biodegrade. This constitutes one approach to alleviating the plastics litter problem. Photosensitisation is achieved either by incorporation of photosensitive comonomer units or by use of an additive (e.g. metal dithiocarbamate, benzophenone or ferrocene). A few homopolymers, e.g. 1,2-polybutadiene, are intrinsically sufficiently photosensitive to be regarded as degradable.

DEGRADATION In a technological sense, any undesirable change in polymer properties due to exposure to a degradation agency either during processing or service life of the polymer. In a strictly chemical sense, the breaking down of large molecules to smaller molecules. Generally therefore used to describe chemical changes occurring in a polymer (often chain scission) which cause a deterioration in useful properties, e.g. embrittlement or discoloration.

Degradations may be classified by their mechanisms or by the agencies causing degradation. The main mechanisms are depolymerisation, random scission and elimination. The agencies which initiate degradation are heat (thermal degradation), heat plus oxygen (thermo-oxidative degradation), ultraviolet light (photo-degradation), high energy radiation, mechanical stress (mechanochemical degradation), chemical attack, biological organisms (biodegradation). In practically important situations several agencies may be involved. Thus, in weathering, not only sunlight and air (causing photo-oxidation) but also water vapour, heat, ozone and pollutants may participate. Some degradations are, however, useful, e.g. pyrolysis for structure elucidation and the mastication of rubbers to improve processibility.

DEGRADATIVE CHAIN TRANSFER Chain transfer in which the chain transfer constant is high and the resultant free radical has such a low activity that it does not re-initiate a polymerisation chain. Best known in the autoinhibition of allyl monomer polymerisation, due to allylic abstraction followed by radical termination of the relatively stable allyl radical (**I**):

$$CH_2{=}CH\dot{C}HX + R\cdot \rightarrow CH_2{=}CHCHX$$
$$\phantom{CH_2{=}CH}\mathbf{I}\phantom{HX + R\cdot \rightarrow CH_2{=}CHC}\underset{\displaystyle R}{|}$$

DEGREE OF COUPLING Symbol k. The number of independent growing polymer molecules in a polymerisation required to react together to form a dead polymer molecule. Thus in free radical polymerisation with termination by disproportionation $k=1$ and by combination $k=2$.

DEGREE OF CROSSLINKING Alternative name for crosslink density.

DEGREE OF CRYSTALLINITY The amount of crystalline material in a partially crystalline polymer, usually expressed as a percentage. Values are commonly in the range 30–80% for polymers as a whole. For a particular polymer the value depends on the crystallisation conditions, especially the rate of cooling and the crystallisation temperature. The measured value is very dependent on the method used, which may be wide angle X-ray diffraction, infrared spectroscopy, differential scanning calorimetry, thermal analysis or density determination.

DEGREE OF DEGRADATION Symbol α. In random scission degradation the fraction of breakable bonds in a polymer molecule that have been broken. Analogous to the conversion in step-growth polymerisation.

DEGREE OF POLYMERISATION (DP) The number of repeat units (M) in an individual polymer molecule, i.e. the value of n in the generalised formula of a polymer molecule $-[M]_n$. The degree of polymerisation is simply related to the polymer molecular weight (MW) by $MW = DP \times M_0$, where M_0 is the molecular weight of the repeat unit. The molecular size may always be expressed by the DP as well as by the MW. A polymer sample usually contains a distribution of molecules of many different degrees of polymerisation, i.e. it is polydisperse, so some average value of DP must be used. The particular averages used are the same as those used for average molecular weight, namely the number, weight and z-averages. The degree of polymerisation is controlled by the type and conditions of polymerisation. In simple linear step-growth polymerisation it is simply related to the conversion p by $\overline{DP}_n = 1/(1 - p)$, where $\overline{DP}_n$ is the number average DP.

DEGREE OF POLYMERISATION DISTRIBUTION
Alternative name for chain length distribution.

DEGREE OF SUBSTITUTION (DS)
The average number of hydroxyl groups substituted, e.g. by ester or ether groups, per monosaccharide unit in a polysaccharide, usually referring to cellulose. The properties of any of the many useful cellulose derivatives depend not only on the particular derivative but also on the degree of substitution which may vary from zero to three. The maximum value obtainable can be limited by steric crowding around the pyranose ring or by accessibility.

DEHYDRATED CASTOR OIL
Obtained by dehydrating castor oil by heating to 250–300°C with an acid catalyst. Hence, it contains more unsaturation than castor oil, and is therefore faster drying. It is useful as a semi-drying oil in alkyd resins for its non-yellowing and water-resistant properties.

DEHYDROCHLORINATION
The loss of hydrogen chloride by elimination from a chlorine-containing molecule, resulting in the formation of carbon–carbon double bonds. Many chlorine-containing polymers dehydrochlorinate during thermal, high energy or photodegradation. The outstanding example is the thermal degradation of polyvinyl chloride, which occurs rapidly at processing temperatures. Here, as is often the case, dehydrochlorination is a sequential reaction by unzipping loss of hydrogen chloride, resulting in the formation of conjugated double bond sequences (polyenes) which absorb visible light. Hence the degraded polymer is coloured:

$$-(CH_2—CHCl)_x \rightarrow -(CH{=}CH)_x + x\,HCl$$

Here degradation, dehydrochlorination and discoloration are often used synonymously.

DEHYDROLYSINONORLEUCINE RESIDUE
A type of crosslink found in collagen formed by reaction between two lysine residues of adjacent tropocollagen 'molecules', thus forming intermolecular crosslinks:

$$
\begin{array}{ccc}
\sim\!\!NH—CH_4—CO\!\!\sim & & \sim\!\!NH—CH—CO\!\!\sim \\
| & & | \\
(CH_2)_3 & & (CH_2)_3 \\
| & & | \\
CHO & & CH \\
+ & \rightarrow & \| \\
NH_2 & & N \\
| & & | \\
(CH_2)_4 & & (CH_2)_4 \\
| & & | \\
\sim\!\!NH—CH—CO\!\!\sim & & \sim\!\!NH—CH—CO\!\!\sim
\end{array}
$$

i.e. a Schiff base is formed.

DELAYED ACTION ACCELERATOR
An accelerator for the sulphur vulcanisation of diene rubbers which is inactive at the temperatures used in the mixing and processing stages (about 120°C), but which accelerates the network formation at vulcanisation temperatures. Thus the accelerator shows little tendency to scorch. Sulphenamides are the best known examples of this type of accelerator.

DELAYED ELASTICITY
(Retarded elasticity) The time-dependent response to stress, or its removal, of a viscoelastic material. The development of the response is delayed since the molecular response, consisting of segmental rotation about polymer molecular chain bonds, is not instantaneous due to molecular entanglements and the time required to overcome the rotational barriers. Delayed elasticity is manifest as the recoverable part of the deformation observed in creep, i.e. as the recovery (strictly the strain recovery) and as stress relaxation.

DELRIN
Tradename for polyoxymethylene.

DE MATTIA MACHINE
An instrument for measuring the flex-cracking resistance of a rubber. A strip of rubber has a transverse groove moulded in it to act as a stress concentrator. In one mode of testing the development of cracks is noted at various stages of repeated flexing of the strip by bringing the ends of the strip together. In another mode, a cut is made in the groove and its growth is measured after various flexing cycles.

DEN
Tradename for an epoxy resin.

DENATURATION
The loss of the specific folded native conformation of a protein with consequent loss in biological activity. Thus changes occur in the tertiary and often also in the secondary structure with formation of a non-specific randomly coiled conformation. There is no change in the primary structure and whilst scission of covalent bonds is generally not believed to occur on denaturation, some alteration in disulphide bridges might be involved as one aspect of denaturation.

Denaturation can be brought about by several different agencies—heat, change in pH, change in solvent or ionic environment. High concentrations of aqueous salt solutions, e.g. urea, guanidine or of some detergents such as sodium dodecylsulphate, are frequently used as denaturing agents. In many cases the reduction of hydrophobic bonding is the most important influence on denaturation although hydrogen bonding changes are usually also involved. The term is sometimes used more widely to include the breakdown of oligomeric proteins or supermolecular complexes into their separate subunits. Often restoration of the original conditions necessary for existence of the native conformation results in renaturation, with restoration of biological activity.

DENATURING AGENT   An agency causing the denaturation of a protein in aqueous solution. Most frequently the term is restricted to the use of salts which, in concentrated aqueous solution, denature proteins, e.g. 8M urea or guanidine or concentrated solutions of detergents such as sodium dodecylsulphate.

DENDRITE   A branched multilayer lamellar crystal of tree-like morphology in which the overall shape of the single crystal is often retained. The branches diverge at wide angles to each other and are crystallographically related. Branching is more extensive with lower crystallisation temperatures. Formed by crystallisation from dilute solution at lower temperatures than used to grow single crystals.

DENDRITIC SPHERULITE   A spherulite whose growth is initiated by formation of a dendrite as a nucleus and whose large-scale branching, tree-like structure is still observable in the completely formed spherulite.

DENIER   A measure of the coarseness of a fibre or yarn, derived from its original use for silk and preferred for natural fibres. The weight in grams of a length of 9000 m of the fibre or yarn. The similar term tex (or decitex) is often used for synthetic fibres. Filament deniers vary from 1–7 for clothing to 15–25 for carpets. Yarn deniers may reach over 1000.

DENSITY GRADIENT CENTRIFUGATION (Zonal centrifugation) A method of preparative ultracentrifugation in which a density gradient is set up in a centrifuge tube by use of sucrose– or glycerol–water mixtures. Prior to centrifugation the polymer solution (which usually is a protein solution) is layered on top and on centrifugation the different species move differentially down the tube to form separate zones or bands depending on the size, shape and density of the species. Centrifugation is usually stopped before equilibrium is reached. The separated polymers are obtained from their bands either by piercing the bottom of the centrifuge tube, collecting small samples as they emerge and analysing them, or by pumping a dense solution carefully into the bottom and forcing successive bands out of the top of the tube. An alternative similar sedimentation procedure is density gradient sedimentation equilibrium centrifugation.

DENSITY GRADIENT COLUMN   An apparatus for the rapid determination of the density of small pieces of a solid polymer. A column of liquid is prepared by carefully mixing two immiscible liquids of slightly different densities and filling the column with the mixture in such a way that the density increases from top to bottom of the column. A sample when carefully placed in the column will sink to the level where its density is the same as that of the liquid in the column at that point. The column is calibrated using glass floats of known density.

DENSITY GRADIENT SEDIMENTATION EQUILIBRIUM CENTRIFUGATION (Isopycnic centrifugation) A variation of the sedimentation equilibrium method of ultracentrifugation, in which a mixed solvent of two components with quite different densities is used. On rotation, a solvent composition (and hence density) gradient is formed and dissolved polymer molecules collect at the position where solvent and polymer densities are the same. Owing to diffusion the polymer forms a band, whose width depends on polymer molecular weight. If the sample contains species of different densities then band broadening or separation of bands occurs. The method is of particular value in the analysis of proteins and nucleic acids, where density differences may be due to different chemical composition, conformation or isotropic content. Here aqueous solutions of inorganic salts, especially caesium chloride, are often used to establish the gradient. In synthetic polymers, copolymer composition and tacticity variation may similarly be studied. Polymer molecular weights may also be determined by this method, especially if a monodisperse band is obtained, as with nucleic acids.

DENSITY OF BRANCHING   Alternative name for branching density.

DENSITY OF CROSSLINKING   Alternative name for crosslink density.

DEOXY-   Prefix used in the naming of a deoxysugar.

2-DEOXY-2-AMINO-D-GALACTOSE   Alternative name for D-galactosamine.

2-DEOXY-2-AMINO-D-GLUCOSE   Alternative name for D-glucosamine.

6-DEOXY-L-GALACTOSE   Alternative name for L-fucose.

6-DEOXY-D-GLUCOSE   Fischer projection formula:

$$H_3C-\overset{\overset{\displaystyle H}{|}}{\underset{\underset{\displaystyle OH}{|}}{C}}-\overset{\overset{\displaystyle H}{|}}{\underset{\underset{\displaystyle OH}{|}}{C}}-\overset{\overset{\displaystyle OH}{|}}{\underset{\underset{\displaystyle H}{|}}{C}}-\overset{\overset{\displaystyle H}{|}}{\underset{\underset{\displaystyle OH}{|}}{C}}-CHO$$

Ring form 6-deoxy-α-D-glucopyranose:

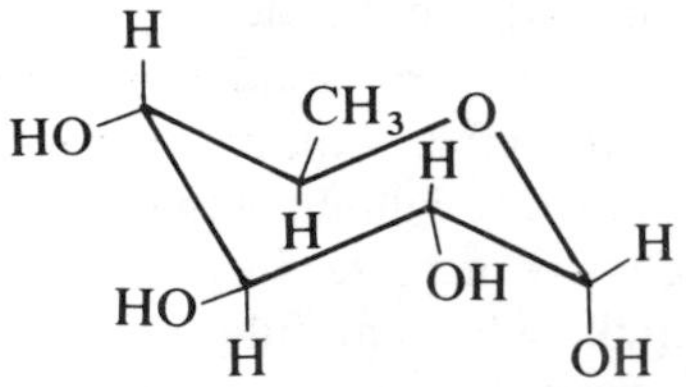

M.p. 140°C. $\alpha_D^{20,H_2O} + 29\cdot7°$.

Occurs in the bark of certain plants as a glycoside and in other natural glycosides.

6-DEOXYHEXOSE   A hexose in which the hydroxyl group on carbon-6 has been replaced by hydrogen. Examples are L-rhamnose and L-fucose.

DEOXYSUGAR   A monosaccharide (or sugar) or monosaccharide unit in a polysaccharide in which a hydroxyl group has been replaced by a hydrogen atom, e.g. L-rhamnose. Other sugar derivatives are also often regarded as being formed via the deoxysugar. Thus replacement of hydroxyl by amino group gives an amino-deoxysugar.

DEP   Abbreviation for diethyl phthalate.

DEPOLARISATION   A change in the direction of polarisation of light scattered by anisotropic particles in solution. The amount of light scattered by such particles at 90° is greater than that from isotropic particles by an amount given by the Cabbanes factor. For random polymer molecular coils of high molecular weight the effect of depolarisation on scattering intensity is negligible.

DEPOLYMERISATION   A degradation reaction in which chain scission occurs at neighbouring repeat units resulting in a sequential loss of monomer molecules from the end of an initiated polymer chain by a series of depropagation steps (unzipping). Essentially the reverse of chain polymerisation, having similar kinetics. Initiated by high temperature (usually above 250°C) ultraviolet radiation or by chemical attack. Only a few polymers completely depolymerise to give 100% yields of monomer (e.g. polymethylmethacrylate, polytetrafluoroethylene, poly-α-methylstyrene, polyoxymethylene) but others (e.g. polystyrene) show a tendency to depolymerise, giving appreciable yields of monomer.

DEPOLYMERISED RUBBER   (Liquid rubber) Tradename: Lorival. Natural rubber which has been heated to about 140°C. This produces chain scission degradation involving a drop in molecular weight to such an extent that the rubber becomes liquid. Products can then be made from the rubber simply by casting at ambient temperature followed by vulcanisation.

DEPROPAGATION   The reverse of propagation. The sequential chain scission step during depolymerisation responsible for the formation of monomer. Has a lower activation energy than propagation and hence is favoured at high temperatures. Thus many vinyl polymers, in the presence of an initiator, are formed at a low temperature but degrade by depolymerisation at a higher temperature. The temperature at which the rates of these two reactions are equal (i.e. at equilibrium between monomer and polymer) is the ceiling temperature.

DER   Tradename for an epoxy resin.

DERAKANE   Tradename for a vinyl ester resin.

DERMATAN SULPHATE   (Chondroitin sulphate B) (β-Heparin)

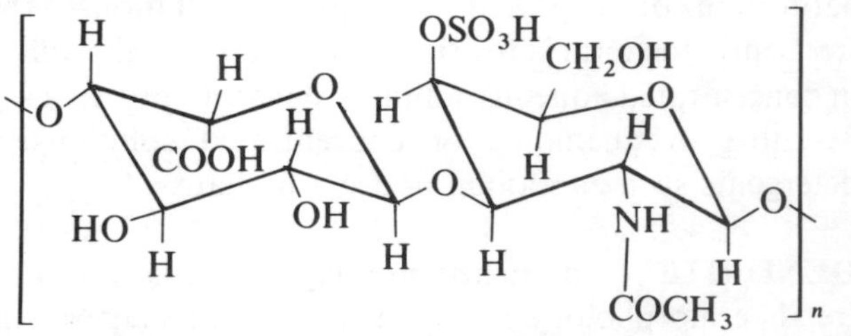

A mucopolysaccharide found in numerous animal tissues, notably skin and lung, often admixed with chondroitin sulphates. Of similar structure to chondroitin-4-sulphate but with β-L-iduronic acid units replacing β-D-glucuronic. It has a certain blood anticoagulant activity, but this is not as great as that of heparin.

DEROTON   Tradename for polybutylene terephthalate.

DESMEARING   Mathematical correction of small angle X-ray scattering intensity data when a slit collimation source is used. It corrects for a smearing of the diffraction pattern and a distortion of the peaks to lower angles, so that the pattern that would have been obtained had a point source been used, is calculated.

DESMOPAN   Tradename for a partially crosslinked thermoplastic polyurethane elastomer, based on a polyester or polyether polyol/MDI prepolymer and chain extended.

DESMOPHEN A   Tradename for a millable poly-urethane elastomer.

DESMOSINE

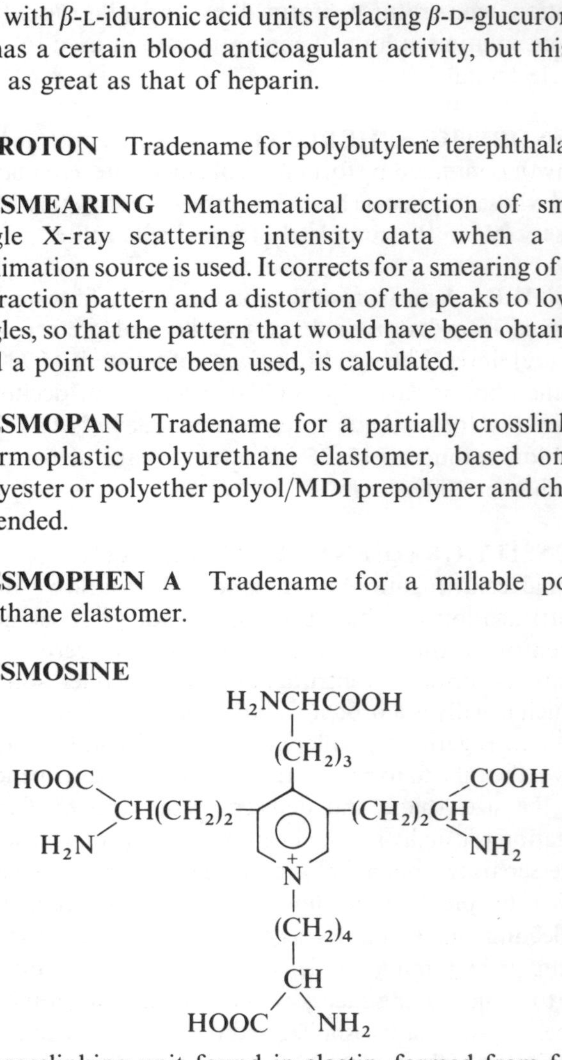

A crosslinking unit found in elastin, formed from four lysine residues coming into close proximity.

DET   Abbreviation for diethylenetriamine.

DEVIATORIC STRESS   (Stress deviator) (Extra stress tensor) The difference between a stress component and the hydrostatic stress ($P = (\sigma_{11} + \sigma_{22} + \sigma_{33})/3$). Thus, for example, the stress σ_{11} becomes the deviatoric stress

$$\sigma'_{11} = \sigma_{11} - P = \tfrac{2}{3}(\sigma_{11} - \tfrac{1}{2}(\sigma_{22} - \sigma_{33}))$$

Particularly useful for analysing plastic flow and yielding since these take place at constant volume. The deviatoric component alters the shape but not the volume, which is altered by the pressure (hydrostatic stress).

DEXEL   Tradename for cellulose acetate.

DEXON   Tradename for a polyglycollide fibre, useful as surgical sutures which disintegrate and are assimilated in the body.

DEXSIL   Tradename for polycarboranesiloxanes. The polymers may be designated SiB-2 or SiB-4, referring to the value of m in the repeat unit formula as 2 or 4:

$$\left\{ Si(CH_3)_2 - CB_{10}H_{10}C \left(Si(CH_3)_2O - \right)_m \right\}_n$$

DEXTRAN   One of a group of closely related α-D-glucans of molecular weight 10^5–10^7 produced by various bacteria when grown in sucrose solutions. The main chain consists of 1,6'-linked glucose units with some single unit glucose branches linked 1,3'- and 1,4'- and occasionally 1,2'-. Useful as a blood plasma extender (when partially hydrolysed) and as a packing material for gel filtration columns, when crosslinked.

DEXTRIN   An extensively degraded starch produced by the action of any of a variety of degradative agencies. Pyrodextrins result from the action of heat on dry starch, which causes loss of crystallinity and an increase in readiness to swell. Hydrolytic dextrins are known as white (produced at low temperatures and acidities) or yellow (produced at high temperatures or acidities) dextrins. British gums are produced by heating with little or no acid at high temperature for a long time. Dextrins are mostly used as adhesives. Limit dextrins are produced by the action of enzymes on starch or starch components, as in Schardinger dextrin.

DEXTROSE   Alternative name for D-glucose.

DFP   Abbreviation for diisopropylphosphofluoridate.

DGEBA   Abbreviation for diglycidyl ether of bisphenol A.

DIACETIN   Alternative name for glyceryl diacetate.

DIACETONE ALCOHOL
(4-Hydroxy-4-methylpentan-2-one)

$$\underset{\underset{CH_3}{|}}{\overset{\overset{OH}{|}}{CH_3COCH_2CCH_3}}$$ B.p. 167°C.

A solvent for cellulose nitrate and acetate (and hence useful in lacquers based on these polymers), phenol–formaldehyde resins, many natural resins, polyvinyl acetate and cellulose ethers. Produced by alkaline catalysed condensation of two molecules of acetone. Partly reverts to acetone in the presence of even traces of alkali, especially on heating or long storage.

DIACRYL 101   Tradename for an acrylic polymer laminating resin. A dimethacrylate of the addition product of ethylene oxide and bisphenol A, which on polymerisation crosslinks and produces a thermoset product.

DIACYL PEROXIDE   (Acyl peroxide) A peroxide of general structure

$$\underset{\underset{O}{\|}}{R}C\underset{}{} O - O C\underset{\underset{O}{\|}}{} R'$$

where R and R' are usually alkyl or aryl groups. Widely used as free radical polymerisation initiators, by thermal or ultraviolet light irradiation decomposition:

$$RCO - OCR \rightarrow 2\,RCO\cdot \rightarrow 2\,R\cdot + 2CO_2$$

Either

$$RCO\cdot$$

or R· free radicals may initiate. Benzoyl peroxide is the best known example; others include lauroyl peroxide, 2,4-dichlorobenzoyl peroxide and acetylcyclohexanesulphonyl peroxide.

DIAGONAL ELECTROPHORESIS   (Diagonal method) A paper electrophoresis technique used for the separation of cysteic acid peptides in the determination of the location of disulphide bonds in proteins. The protein is hydrolysed enzymically, keeping the disulphide bridges intact. The resultant peptide mixture is then subjected to paper electrophoresis to separate the peptides. The paper is then exposed to performic acid vapour which oxidises the disulphides, cleaving them and forming cysteic acid peptides. Electrophoresis is then continued at right angles to the first direction. Unmodified peptides appear on a diagonal line across the paper. The cysteic acid peptides are off-diagonal. They may then be eluted and analysed.

DIAGONAL METHOD   Alternative name for diagonal electrophoresis.

DIAKON   Tradename for polymethylmethacrylate moulding powder.

DIAKON MX   Tradename for a polymethylmethacrylate/acrylate rubber blend.

DIALKYL PEROXIDE   A compound of the general structure RO—OR' where R and R' are alkyl or aryl groups. One of the most stable groups of organic peroxy-compounds and therefore useful for high temperature ($>100°C$) production of free radicals for polymerisation or crosslinking: $RO—OR' \rightarrow RO· + R'O·$. Examples include dicumyl peroxide and di-*t*-butyl peroxide.

N,N'-DIALKYL-p-PHENYLENEDIAMINE A group of very effective antiozonants, particularly when used in conjunction with a wax. A common example is *N*-isopropyl-*N'*-phenyl-*p*-phenylenediamine.

DIALLYL ISOPHTHALATE (DAIP)

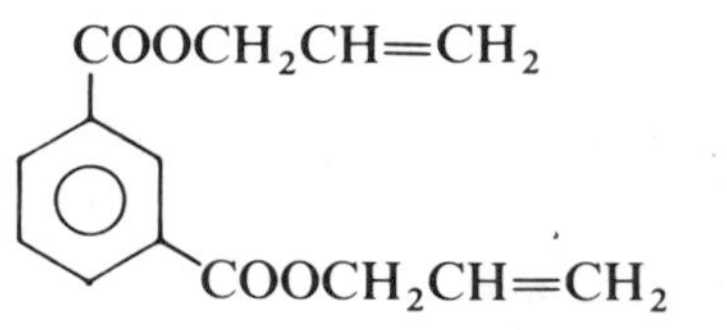

$$COOCH_2CH{=}CH_2$$

$$COOCH_2CH{=}CH_2 \quad \text{B.p. } 181°C/4\,mm.$$

Prepared by the esterification of isophthalic acid with allyl alcohol. It polymerises free radically, like diallyl phthalate, to crosslinked products but has even better thermal stability—up to 220°C for long periods.

DIALLYL PHTHALATE (DAP)

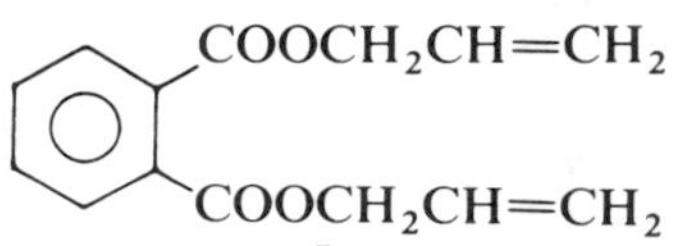

$$COOCH_2CH{=}CH_2$$
$$COOCH_2CH{=}CH_2$$
I $\qquad$ B.p. 160°C/4 mm.

Prepared by the esterification of phthalic anhydride with allyl alcohol. It may be polymerised with a free radical initiator, such as a peroxide, to a highly crosslinked polymer with good thermal stability and retention of electrical properties under conditions of wet and dry heat. For the production of electrical mouldings the monomer is prepolymerised to an essentially linear polymer (**II**) by heating to about 100°C. This is then compounded to give a thermosetting moulding powder, which is then further heated to give the crosslinked product (**III**):

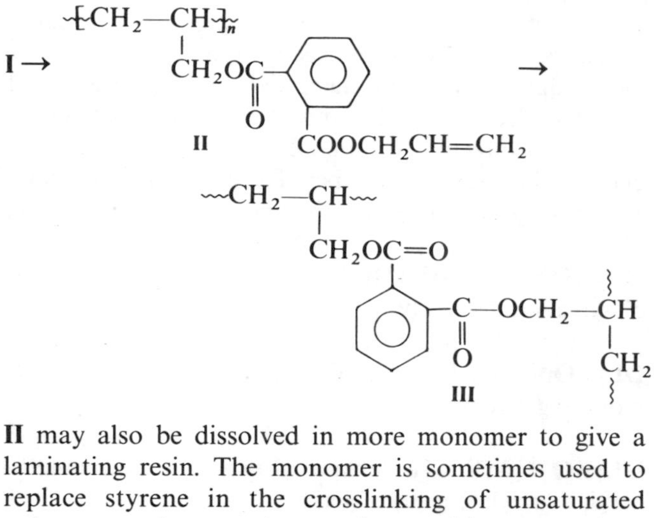

$$\text{I} \rightarrow$$

II

III

II may also be dissolved in more monomer to give a laminating resin. The monomer is sometimes used to replace styrene in the crosslinking of unsaturated polyester resins to give products with greater heat resistance. Also useful as a polymerisable plasticiser for cellulose esters and polyvinyl chloride.

DIALPHANOL ADIPATE Alternative name for heptylnonyl adipate.

DIALPHANOL PHTHALATE Alternative name for heptylnonyl phthalate.

DIALPHANYL ADIPATE Alternative name for heptylnonyl adipate.

DIALPHANYL PHTHALATE Alternative name for heptylnonyl phthalate.

DIALYSIS A technique for the removal of small molecules or ions from a polymer solution, which is usually aqueous, by enclosing the solution in a container with a semi-permeable membrane through which the small molecules may pass, but not the polymer. By placing the enclosed solution in a pure water medium the small molecules will continually permeate until their concentration in the polymer solution is very low. The technique is especially useful for desalting biopolymer solutions, particularly proteins, but such desalting may lead to a destruction of the tertiary structure and loss of biological activity. Usually regenerated cellulose membranes are used. A speeded-up version of dialysis is ultrafiltration.

DIAMIDRAZONE (Dihydrazide) A compound of the type

$$H_2NNHCRCNHNH_2$$
$$HN \quad NH$$

where R is usually an aryl group. Prepared by reaction of hydrazine with a dinitrile or its imido-ester:

$$R'OCRCOR'$$
$$HN \quad NH$$

Useful in the synthesis of a variety of heterocyclic polymers, including poly-(sym-triazines), poly-(1,2,4-triazoles), poly-(1,3,4-oxadiazoles) and poly-(as-triazoles).

3,3'-DIAMINOBENZIDINE

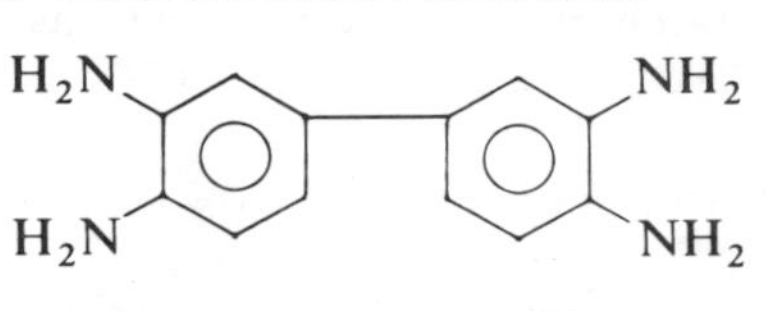

$$H_2N \qquad NH_2$$
$$H_2N \qquad NH_2 \quad \text{M.p. } 179{-}180°C.$$

The most commonly used bis-(*o*-dimine) monomer for the synthesis of a variety of heterocyclic polymers, e.g. polybenzimidazoles and polyquinoxalines.

DIAMINOBICYCLOOCTANE Alternative name for 1,4-diazabicyclo-2,2,2-octane.

1,4-DIAMINOBUTANE

$$H_2N(CH_2)_4NH_2$$
$$\text{M.p. } 27{-}28°C. \text{ B.p. } 158°C.$$

Produced by the addition of hydrogen cyanide to

acrylonitrile, to yield succinic nitrile, followed by hydrogenation:

$$CH_2{=}CHCN + HCN \rightarrow$$
$$NCCH_2CH_2CN \rightarrow H_2N(CH_2)_4NH_2$$

The diamine monomer for the production of nylon 46.

4,4′-DIAMINODIPHENYL ETHER

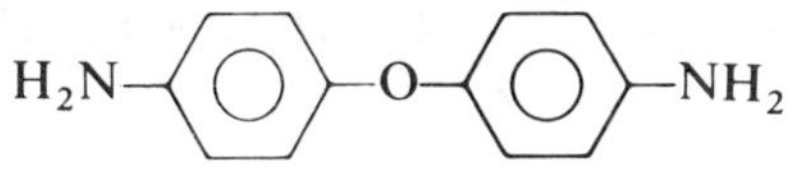

$$H_2N{-}\langle O \rangle{-}O{-}\langle O \rangle{-}NH_2$$

M.p. 93°C.

A useful monomer for the synthesis of several heterocyclic polymers, particularly by reaction with a dianhydride to form a polyimide.

4,4′-DIAMINODIPHENYLMETHANE

(4,4′-Methylenedianiline)(Bis-(4-aminophenyl)methane)

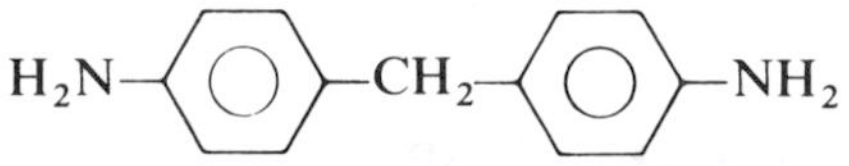

$$H_2N{-}\langle O \rangle{-}CH_2{-}\langle O \rangle{-}NH_2$$

M.p. 92–93°C.

A useful monomer for the synthesis of many heterocyclic polymers, including polyamides and polyimides. Produced by reaction between aniline and formaldehyde. A useful curing agent for epoxy resins, giving products similar to those that have been cured using *m*-phenylenediamine, but which does not stain the skin.

4,4′-DIAMINODIPHENYLSULPHONE

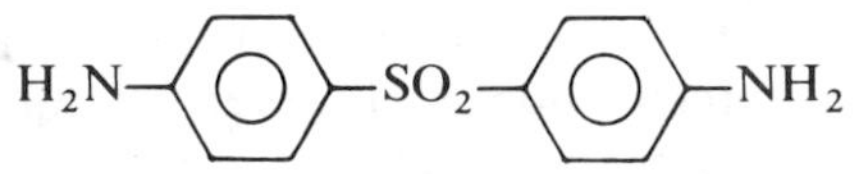

$$H_2N{-}\langle O \rangle{-}SO_2{-}\langle O \rangle{-}NH_2$$

M.p. 176°C.

A useful curing agent for epoxy resins, giving products with the highest heat distortion temperatures of all amine-cured resins.

1,6-DIAMINOHEXANE Alternative name for hexamethylenediamine.

1,3-DIAMINO-5-PHENYL-2,4,6-TRIAZINE Alternative name for benzoguanamine.

DIANHYDRIDE (Tetracarboxylic dianhydride) A compound of the type

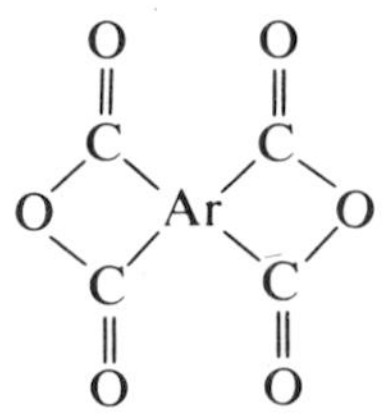

where Ar is usually an aromatic ring system, e.g.

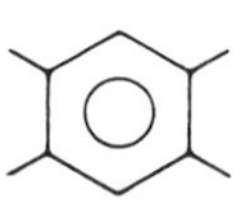

in pyromellitic dianhydride. Widely used tetrafunctional monomers in the synthesis of heterocyclic polymers, e.g. polyimides, BBB, and in the curing of epoxy resins.

1,4-DIAZABICYCLO-2,2,2-OCTANE
(DABCO) (Diaminobicyclooctane)(Triethylenediamine)

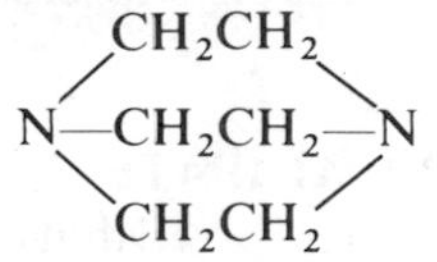

The most widely used amine catalyst component for flexible polyurethane foam formation based on polyether polyols. It is particularly active since the bicyclic structure exposes the nitrogen to reaction and steric hindrance is at a minimum.

DIBA Abbreviation for diisobutyl adipate.

DIBASIC LEAD PHOSPHITE

$$2PbO.PbHPO_3.\tfrac{1}{2}H_2O$$

A basic lead stabiliser for the thermal stabilisation of polyvinyl chloride. Useful for outdoor applications since it also imparts weathering resistance due to its action as an ultraviolet absorber.

DIBASIC LEAD PHTHALATE

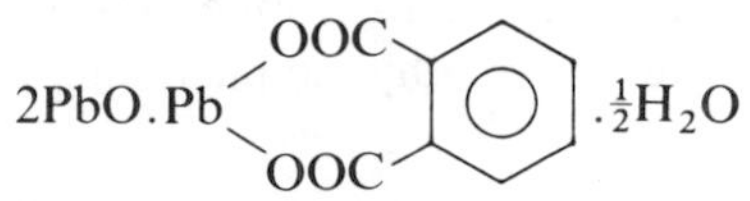

A basic lead stabiliser for the thermal stabilisation of polyvinyl chloride. Useful for high temperature applications, especially in electrical insulations, due to its low reactivity with plasticisers.

DIBENZOTHIAZYLDISULPHIDE (MBTS)

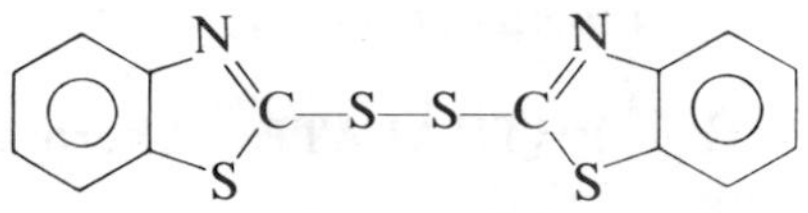

M.p. 170°C.

A useful accelerator for the sulphur vulcanisation of rubbers, with a lower tendency to scorch than mercaptobenzothiazole.

DIBENZYL SEBACATE

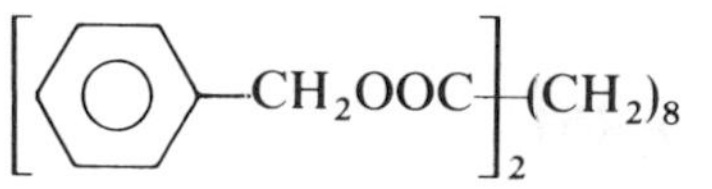

B.p. 265°C/4 mm.

A plasticiser for polyvinyl chloride and synthetic rubbers, imparting low temperature flexibility.

DIBLOCK POLYMER (AB block copolymer) A block copolymer consisting of two blocks, one of A repeating units and one of B repeating units. Thus its structure may be represented as

AAA......AAAABBBB......BBB

A well-known example is the styrene–butadiene block copolymer (or SB block copolymer). In this example, as with others in which one block is stiff and the other is rubbery, although the blocks segregate into domains, the stiff domains do not act as physical crosslinks as they do in triblock polymers.

DIBUTOXYETHYL ADIPATE

(Dibutylcellosolve adipate)
(Bis-(ethylene glycol monobutyl ether) adipate)

$$[CH_3(CH_2)_3OCH_2CH_2OOC(CH_2)_2]_2$$
B.p. 209°C/4 mm.

A useful plasticiser for polyvinyl chloride and its copolymers, synthetic rubbers and, in particular, polyvinyl butyral, especially for good low temperature flexibility and good ultraviolet light stability.

DIBUTOXYETHYL PHTHALATE

(Dibutylglycol phthalate)
(Dibutylcellosolve phthalate)

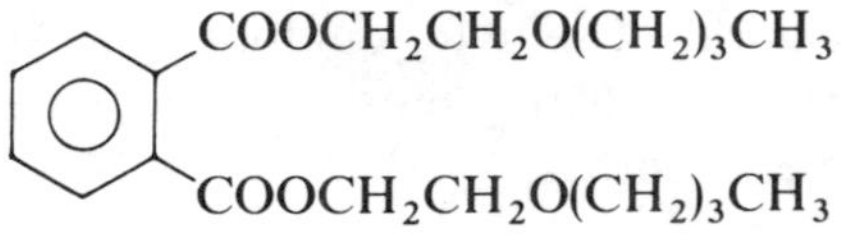

B.p. 370°C/760 mm (decomposes). 225°C/4 mm.

A plasticiser for polyvinyl chloride and its copolymers, cellulose nitrate, polymethylmethacrylate and synthetic rubbers.

DIBUTYLCELLOSOLVE ADIPATE Alternative name for dibutoxyethyl adipate.

DIBUTYLCELLOSOLVE PHTHALATE Alternative name for dibutoxyethyl phthalate.

2,6-DI-*t*-BUTYL-*p*-CRESOL (DBPC) Alternative name for 2,6-di-*t*-butyl-4-methylphenol.

DIBUTYLGLYCOL PHTHALATE Alternative name for dibutoxyethyl phthalate.

2,6-DI-*t*-BUTYL-4-METHYLPHENOL

(DBPC) (2,6-Di-*t*-butyl-*p*-cresol)
(Butylated hydroxytoluene (abbreviation BHT))
Tradenames: CAO-1, Ionol, Deenax, Nonox TBC.

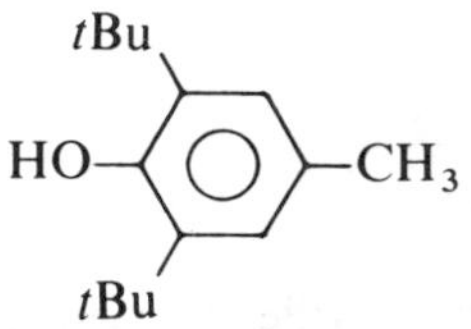

M.p. 70°C.

A chain-breaking antioxidant of the hindered phenol type. Non-staining but rather volatile for high temperature use. Widely used in both rubbers and plastics.

DI-*t*-BUTYL PEROXIDE

$$(CH_3)_3CO—OC(CH_3)_3$$
B.p. 109°C (decomposes), 12–13°C/20 mm.

A high temperature free radical polymerisation initiator for use at >100°C, having half-life times of 210 h/100°C,

20 h/120°C and 0·8 h/150°C and half-life temperatures of 126°C/10 h, 148°C/1 h and 193°C/1 min. Useful for high temperature polymerisations, e.g. of ethylene, and for the production of styrenated alkyds and oils.

DI-*t*-BUTYLPEROXYDIISOPROPYLBENZENE

Tradename: Vulcup.

An organic peroxide useful in the peroxide vulcanisation of diene and silicone rubbers.

2,5-DI-(*t*-BUTYLPEROXY)-2,5-DIMETHYLHEXANE

Tradename: Varox.

$$[(CH_3)_3COOC(CH_3)_2CH_2]_2$$

An organic peroxide useful in the peroxide vulcanisation of diene and silicone rubbers.

DI-*n*-BUTYL PHTHALATE (DBP)

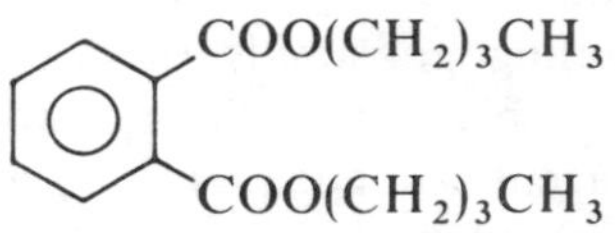

B.p. 340°C/760 mm, 182°C/5 mm.

A widely used plasticiser for cellulose nitrate and highly compatible with many other polymers, being particularly useful in polyvinyl butyral, polymethylmethacrylate, polystyrene and in synthetic rubbers. However, it is rather too volatile for many uses.

DIBUTYL SEBACATE (DBS)

$$[CH_3(CH_2)_3OOC(CH_2)_4]_2$$
B.p. 345°C/760 mm, 200°C/5 mm.

A plasticiser, imparting good low temperature properties, for polyvinyl acetals, polyvinyl chloride, and its copolymers, acrylics (especially for coatings), cellulose acetate-butyrate and nitrile rubber.

DIBUTYLTIN DILAURATE

$$(C_4H_9)_2Sn[OOC(CH_2)_{10}CH_3]_2$$
M.p. 22–24°C.

Sometimes used in place of stannous octoate as a catalyst component for the formation of flexible polyurethane foams based on polyether polyols, when greater resistance to hydrolysis is required, as in water containing formulations.

DIBUTYLTIN STABILISER A widely used type of organotin thermal stabiliser for chlorine containing polymers, especially polyvinyl chloride, of the general formula $(C_4H_9)_2Sn(X)_2$, where X is an organic functional group of any of several types—long chain carboxylate (e.g. laurate) or mercaptan (e.g. thioglycollate). Very efficient at preventing discoloration during thermal degradation and may be used for clear compositions. However, there is a significant toxic hazard in use, although this may be reduced by use of the similar dioctyltin stabilisers.

DICAPRYL PHTHALATE (DCP)
(Di-1-methylheptyl phthalate)
(Di-*sec*-octyl phthalate)

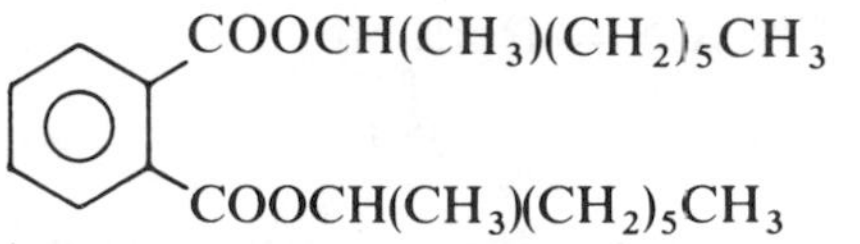

$$COOCH(CH_3)(CH_2)_5CH_3$$
$$COOCH(CH_3)(CH_2)_5CH_3 \quad \text{B.p. } 229°C/4\,mm.$$

A plasticiser for polyvinyl chloride and cellulose nitrate similar in performance to dioctyl phthalate, but with good weatherability.

DICARBITOL PHTHALATE
(Diethoxyethoxyethyl phthalate)
(Diethyldiglycol phthalate)
(Bis-(diethylene glycol monoethylether)phthalate)

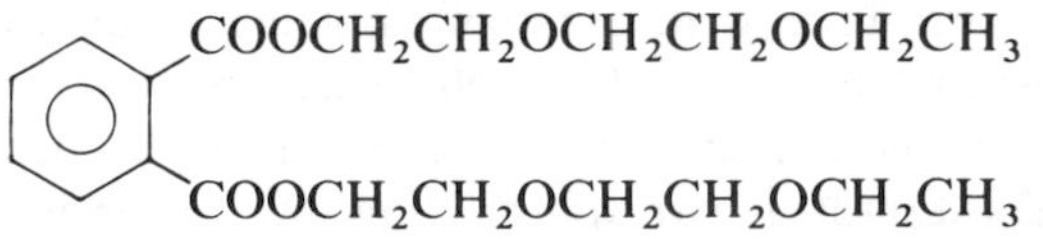

$$COOCH_2CH_2OCH_2CH_2OCH_2CH_3$$
$$COOCH_2CH_2OCH_2CH_2OCH_2CH_3$$
$$\text{B.p. } 220\text{–}260°C/4\,mm.$$

A plasticiser for cellulose ethers and esters, polyvinyl chloride and its copolymers and for synthetic rubbers.

DICEL Tradename for a cellulose acetate filament.

DICELLOSOLVE PHTHALATE Alternative name for diethoxyethyl phthalate.

o-DICHLOROBENZENE (ODCB)

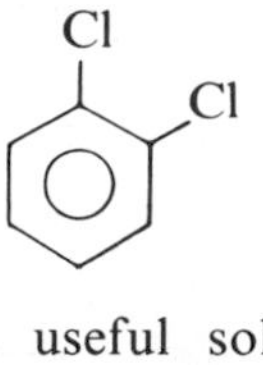

B.p. 180.5°C.

When hot, a useful solvent for polyethylene and polypropylene and the most frequently used solvent for the determination of the molecular weights of these polymers. It may contain some *p*-isomer, often at 15%, which is the eutectic composition.

2,4-DICHLOROBENZOYL PEROXIDE

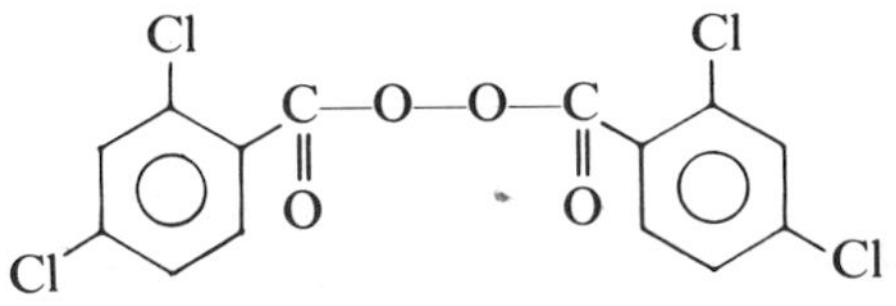

A diacyl peroxide useful as a free radical polymerisation initiator. Of greater activity than benzoyl peroxide, having half-life times of 18 h/50°C, 1.5 h/70°C and 0.05 h/ 100°C and half-life temperatures of 54°C/10 h, 72°C/1 h and 112°C/1 min. Especially useful for the polymerisation of acrylic syrups used in the production of polymethyl-methacrylate by monomer casting.

3,3'-DICHLORO-4,4'-DIAMINODIPHENYL-METHANE (MOCA) (4,4'-Methylene-bis-(*o*-chloro-aniline))

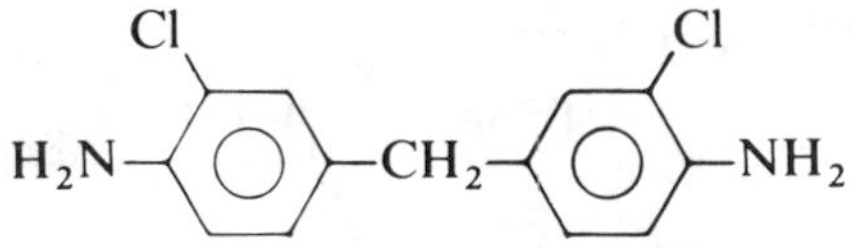

M.p. 110°C.

An aromatic diamine, used in the formation of polyurethane elastomers as a chain extender by reaction with isocyanate terminated prepolymers, yielding urea units. When excess free diisocyanate is present, chain extension and formation of urea sequences, which act as hard segments, results:

$$\sim\!NH_2 + OCN\!\sim \rightarrow \sim\!NH—CO—NH\!\sim$$

DICHLORODIFLUOROMETHANE (CFM 12)

$$CF_2Cl_2 \qquad \text{B.p. } -29.8°C.$$

Sometimes used as a physical blowing agent in polyurethane foams, often in conjunction with trichlorofluoromethane.

1,2-DICHLOROETHANE (Ethylene dichloride)

$$ClCH_2CH_2Cl \qquad \text{B.p. } 83.6°C.$$

A toxic, low flammability solvent, which tends to produce hydrochloric acid on contact with water. In general it is a powerful solvent, dissolving hydrocarbon rubbers, many natural resins, urea and phenol–formaldehyde prepolymers, polystyrene, polyvinyl acetate and cellulose ethers. It swells rayon, polyvinyl chloride and polymethylmethacrylate. Produced by the addition of chlorine to ethylene as an intermediate in the conversion of ethylene to vinyl chloride.

DI-(2-CHLOROETHYL)VINYL PHOSPHONATE

$$CH_2\!=\!CHPO(OCH_2CH_2Cl)_2$$
$$\text{B.p. } 132°C/3\,mm.$$

Prepared by dehydrochlorination of the 2-chloroethylphosphonate $ClCH_2CH_2PO(OCH_2CH_2Cl)_2$, itself obtained by rearrangement of tri-(2-chloroethyl)phosphite, $P(OCH_2CH_2Cl)_3$. It is useful as a comonomer for imparting fire retardancy to unsaturated polyesters and dyeability to polyacrylonitrile fibres.

DICHLOROMETHANE (Methylene dichloride)

$$CH_2Cl_2 \qquad \text{B.p. } 41°C.$$

Produced by the chlorination of methane. A powerful solvent for many polar and non-polar polymers, e.g. useful as a paint remover, but somewhat toxic.

DICHLOROSILANE The most important type of monomer for the production of linear polyorganosiloxanes. Of general structure $R^1R^2SiCl_2$, where R^1 and R^2 are usually methyl or phenyl groups. Produced by the direct process of alkylation of silicon or, especially for mixed silanes where $R^1 \neq R^2$, by the Grignard process. Hydrolysis yields unstable silanediols which condense to a mixture of linear and cyclic polyorganosiloxanes:

$$R^1R^2SiCl_2 \rightarrow R^1R^2Si(OH)_2 \rightarrow$$
$$\text{--}[R^1R^2SiO]_n + [R^1R^2SiO]_n$$

2,5-DICHLOROSTYRENE

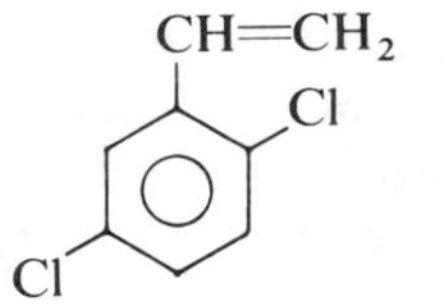

B.p. 74°C/3 mm.

Prepared by the side chain chlorination of 2,5-dichloro-ethylbenzene, followed by dehydrochlorination. It readily polymerises, by methods similar to those used for styrene, to poly-(2,5-dichlorostyrene).

DICHROIC RATIO The ratio of absorbances of polarised radiation, usually in the infrared region, by a sample with the light polarised in two mutually perpendicular directions. The usual measure of dichroism in an oriented polymer. For biaxial orientation three dichroic ratios need to be defined, for uniaxial only two. The ratios may be quantitatively related to certain orientation functions.

DICHROISM The dependence of the absorbance of polarised radiation on the direction of polarisation. In solid polymers the magnitude of the dichroism (expressed as the dichroic ratio) depends on the orientation of the radiation absorbing groups and hence of the polymer molecules. Dichroism in the infrared spectral region is therefore useful in characterising orientation in solid organic polymers which always absorb in this region. Occasionally ultraviolet or visible light dichroism of added dye molecules may be utilised since these tend to align themselves with polymer molecules in amorphous polymers or regions.

DICUMYL PEROXIDE (Cumyl peroxide) Tradename: Dicup.

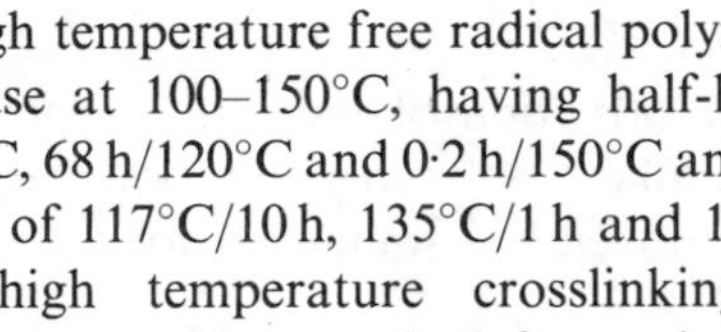

M.p. 42°C.

A high temperature free radical polymerisation initiator for use at 100–150°C, having half-life times of 100 h/100°C, 68 h/120°C and 0·2 h/150°C and half-life temperatures of 117°C/10 h, 135°C/1 h and 172°C/1 min. Useful for high temperature crosslinking of unsaturated polyesters, rubbers, polyolefins and silicones at 100°C.

DICUP Tradename for dicumyl peroxide.

DICY Abbreviation for dicyandiamide.

DICYANAMIDE (Dicy)

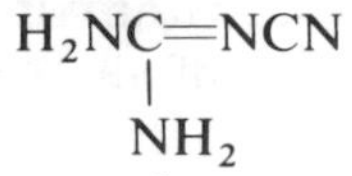

M.p. 209°C (decomposes).

Prepared by dimerisation of cyanamide. It exists in three tautomeric forms:

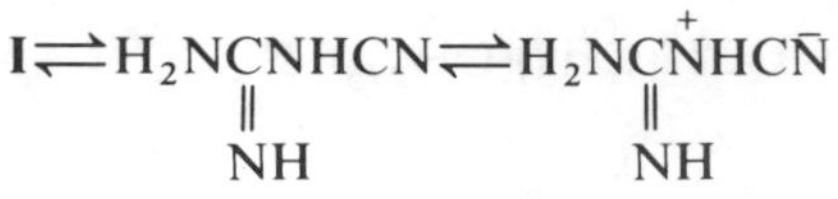

Useful as a latent curing agent for epoxy resins, giving a long pot life at ambient temperature and curing at 145–160°C. Useful as a source of melamine, which is formed when it is heated above its melting point. Occasionally it is used in the formation of dicyandiamide–formaldehyde resins.

DICYANAMIDE–FORMALDEHYDE POLYMER A polymer formed by reaction of dicyandiamide with formaldehyde in aqueous alkaline solution. Initially methylol-dicyanamides are formed, which, when the solution is made more alkaline, further react, as in urea–formaldehyde polymers, to form hydrophobic low molecular weight polymers. These are useful as cellulose textile finishes to improve dye fastness. Dicyandiamide/urea–formaldehyde polymers are useful in the wet strengthening of paper.

N,N'-DICYCLOHEXYLCARBODIIMIDE (DCC) (DCCI)

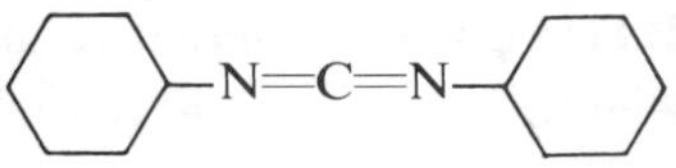

M.p. 35–36°C.

A widely used reagent for promoting the formation of a peptide bond by reaction of a carbonyl group of an amino acid or peptide with an amino group of another amino acid or peptide. Initially an O-acyl urea is formed:

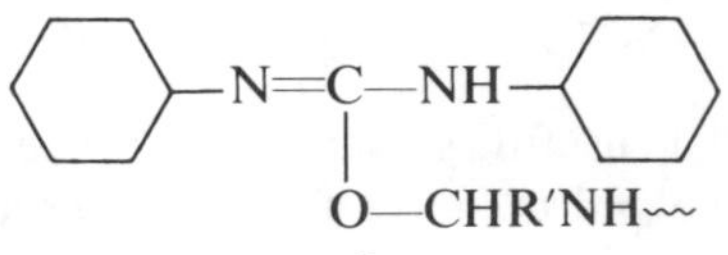

which reacts with amino group to form the peptide bond and N,N'-dicyclohexyl urea:

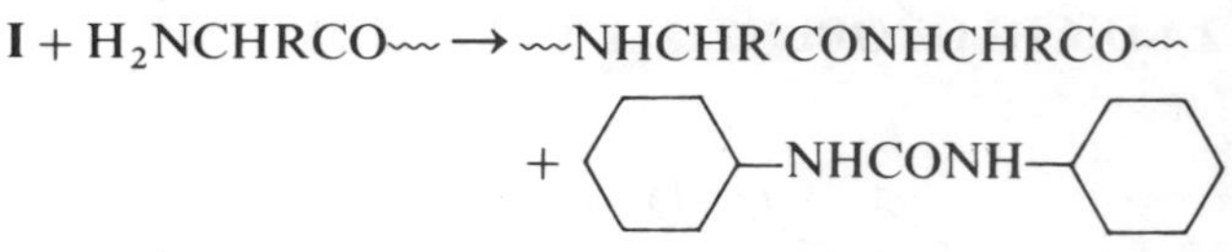

In the preferred solvents, methylene dichloride and tetrahydrofuran, the urea is insoluble and is readily removed.

4,4'-DICYCLOHEXYLMETHANE DIISOCYANATE ($H_{12}MDI$) (PICM) (methylene-bis-(4,4'-phenylisocyanate))

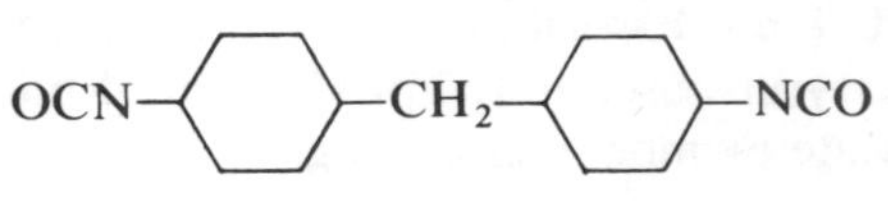

An alicyclic diisocyanate useful in the formation of polyurethanes by reaction with a polyol, with superior resistance to yellowing on thermal or photo-oxidation. Synthesised by reduction of 4,4'-methylene-bisaniline, itself a product of reaction of formaldehyde with aniline, then reaction with phosgene.

DICYCLOHEXYL PHTHALATE (DCHP)

B.p. 220–228°C/760 mm. M.p. 69°C.

A plasticiser for polyvinyl chloride and its copolymers, polystyrene, ethyl cellulose and cellulose nitrate, of low water solubility.

DICYCLOPENTADIENE Alternative name for bicyclopentadiene.

DICYCLOPENTADIENE RESIN A petroleum resin consisting of the low molecular weight polymer obtained by thermal polymerisation of fractions from petroleum cracking containing large amounts of dicyclopentadiene, but also some styrene and indene. The polymers are useful as plasticisers but can have high softening points (up to 175°C) when of higher molecular weight.

DIDA Abbreviation for diisodecyl phthalate.

DI-*n*-DECYL ADIPATE (DDA)

$$[CH_3(CH_2)_9OOC(CH_2)_2]_2$$

B.p. 244°C/5 mm.

A plasticiser for polyvinyl chloride and its copolymers for good low temperature flexibility and permanence.

DIDECYL PHTHALATE (DDP) Alternative name for diisodecyl phthalate.

DIDP Abbreviation for diisodecyl phthalate.

DIELECTRIC An alternative name for an electrically insulating material, which may be considered to be one with a resistivity of greater than about $10^{10}\,\Omega\,cm$. Most polymers, excepting those that are specially produced as conductive or semiconducting polymers, are thus dielectrics. From an electronic point of view, a dielectric has a filled valence band separated widely from the higher energy valence band, so that application of an electric field cannot elevate other than a few electrons to this level. However, application of an electric field causes polarisation and in alternating fields dielectric dispersion and dielectric loss will occur.

DIELECTRIC BREAKDOWN The permanent loss of the insulating properties of a dielectric material due to the application of an excessive electric field. At the high voltages involved, breakdown is usually sudden and catastrophic, with the material often becoming burnt out between the electrodes. The resistance to breakdown is measured as the dielectric strength. Although a dielectric has an intrinsic dielectric strength, under most conditions met with in practice breakdown occurs at much lower voltages. This may be due to thermal breakdown, breakdown by discharges, by tracking or by electrochemical breakdown.

DIELECTRIC CONSTANT Alternative name for relative permittivity.

DIELECTRIC DISPERSION The variation in the dielectric properties of a dielectric (particularly the relative permittivity and loss) with frequency or temperature. It is particularly marked in the region of a dielectric relaxation. The term sometimes refers in particular to the difference between the low frequency and the high frequency values of the relative permittivity (on either side of a relaxation), although this is more usually called the relaxation strength.

DIELECTRIC LOSS FACTOR (Loss factor) (Loss index) Symbol ε''. The imaginary part of the complex relative permittivity. It is equal to the product of the relative permittivity (ε') and the dielectric loss tangent, i.e. it is tan δ. ε'' is frequency dependent, the dependency being given by one of the Debye equations. In the region of a dielectric relaxation, ε'' passes through a maximum value. Determination of the frequency at which the maximum occurs (ω_{max}) is an important way of characterising polymer relaxation (by dielectric spectroscopy) and in determining the relaxation time (τ), since $\omega_{max} = 1/\tau$. ε'' may be determined, in the appropriate different frequency ranges, by the DC step-response, ultralow frequency or audio frequency, e.g. Schering, bridge, resonant current or wave transmission methods. Values for ε'' for polymers vary from about 0·0001 for non-polar polymers to about 1·0 for polar polymers in the region of a transition.

DIELECTRIC LOSS TANGENT (Loss tangent) (Dissipation factor) Symbol tan δ. The tangent of the loss angle when a dielectric is subject to a sinusoidally varying applied electric field. Since tan δ equals $\varepsilon''/\varepsilon'$, i.e. the ratio of the imaginary (loss) to the real (storage) relative permittivities, it is also the ratio of the energy dissipated to the energy stored for each cycle. It is the most commonly used parameter for expressing the 'lossiness' of a dielectric, when it is sometimes referred to as the power factor (sin δ). However, any confusion is, in practice, minimal, since for low loss dielectrics, e.g. with tan $\delta < 0·1$, the difference is $< 0·5\%$. Usually tan δ is determined experimentally at the same time as the relative permittivity using any of the methods available for the latter. Tan δ is highly frequency and temperature dependent, reaching a maximum value as a dielectric relaxation is passed. Values for non-polar polymers can be as low as

10^{-5}, when $\tan\delta$ approximately equals δ, with no loss peaks. Polar polymers may have values of up to about 0.5 at the maximum of their α-relaxation.

DIELECTRIC POLARISATION Alternative name for polarisation.

DIELECTRIC PROPERTIES Those electrical properties of a material relating to its behaviour as a dielectric. They comprise the relative permittivity and the dielectric loss properties. The latter are most commonly characterised by the loss factor.

DIELECTRIC RELAXATION A relaxation occurring as a result of a movement of dipoles or electric charges due to a changing electric field, which have been displaced from their equilibrium positions by the applied field, i.e. it is a time-dependent depolarisation process. It most commonly occurs when a polymer contains polar groups due to the dipolar orientation following the field. However, it can also arise from the migration of charged species to a boundary in a two-phase material (interfacial polarisation). At a molecular level, the time-dependent process of polarisation of a dielectric on application of an electric field. It is of greatest significance when alternating fields are involved, when relaxation is responsible for the occurrence of more or less sharp changes in the dielectric properties with frequency or temperature. In particular, a reduction in the relative permittivity and a peak in the loss factor occur over a relatively small frequency (or temperature) range, although for solid polymers this range is somewhat broader than usual. The rate of a polarisation, or depolarisation, process is characterised by the dielectric relaxation time. This is sufficiently short for electronic and atomic polarisations, so that these occur essentially instantaneously (having frequencies of about 10^{12} Hz). The frequencies of orientation polarisation of polar dielectrics occur in the practically important frequency range of 10^2–10^{10} Hz, where a dispersion of dielectric properties is often observed. Several different types of dipolar relaxations may be observed, corresponding to different motions of polar groups in a polymer dielectric. Thus if the loss factor is determined at a fixed frequency with increasing temperature, then thermal 'loosening' of the groups causes additional motions to occur with additional relaxations. These correspond to similar relaxations in dynamic mechanical behaviour, but the relative magnitudes of the transitions may not be the same. The highest temperature relaxation is the α-relaxation and corresponds to motions characteristic of the glass transition. Lower temperature (or higher frequency) relaxations are labelled the β-, γ-relaxations, etc., as the temperature is lowered and are generally less intense and broader. They are due to more local motions, such as those of side groups or of short chain segments, occurring in the polymer. Measurements may be made with an alternating field or as a result of a step function (as in the DC transient technique). The relaxation is non-exponential, i.e. it is non-ideal, due to time-dependent correlations between neighbouring chain dipoles. This causes a broadening of the relaxation over a wider frequency than in the ideal case. Non-ideal behaviour is often expressed by $\varepsilon^*/(\varepsilon_0 - \varepsilon_\infty) = (1 + i\omega\tau)^{-\beta}$, where ε^* is the complex permittivity, ε_0 and ε_∞ are the zero and infinite frequency permittivities, ω is the angular frequency, τ is the relaxation time and β is a distribution parameter having a value of unity for ideal behaviour but nearer zero for non-ideal behaviour. In the solid state, dielectric relaxation can characterise the transitions in polymers with their temperature and frequency dependencies. In dilute solution, dipole orientation frequencies are in the range 10^5–10^{10} Hz and result from either side group or main chain segmental motions.

DIELECTRIC RELAXATION TIME (Relaxation time) Symbol τ. A characteristic time constant for a dielectric relaxation such that, according to the Debye theory, maximum dielectric loss occurs at an angular frequency ω_{max}, where $\omega_{max} = 1/\tau$. Location of the peak loss, by dielectric spectroscopy, thus provides a way of determining τ. It is analogous to the retardation, not the relaxation, time of the dynamic mechanical behaviour of viscoelastic materials, since it refers to the change in polarisation (a strain) resulting from an imposed change in field strength (a stress). Nevertheless, the term relaxation time is used.

DIELECTRIC SPECTROSCOPY The determination of the dielectric properties, usually the loss factor (ε'') and the relative permittivity (ε') as a function of frequency (ω) at several fixed temperatures, by performing several isothermal scans. The data are often displayed as plots of ε'' and ε' against $\log\omega$—the dielectric spectra. The data may be replotted as ε'' and ε' against temperature to show the individual (α-, β-, etc.) relaxation processes. Alternatively a contour map or three-dimensional model may be constructed to show simultaneously the variation of ε'' and ε' with both frequency and temperature.

DIELECTRIC STRENGTH (Electric strength) (Breakdown strength) Symbol E. The voltage at which dielectric breakdown occurs divided by the sample thickness. The measured value is highly dependent on the method of test, the physical environment, sample history and purity. In the absence of spurious discharges and under carefully controlled conditions, an intrinsic dielectric strength may be measured. However, under most conditions of test the recorded strength is less than one-tenth of this value. It is usually measured under normal power conditions, i.e. 50 Hz alternating field, and the results obtained are thickness dependent.

DIELS–ALDER POLYMERISATION Polymerisation by a Diels–Alder reaction between a bisdiene and a bisdienophile as monomers. Although these monomers are often difficult to prepare, the method is of interest in the synthesis of aromatic polymers, e.g. phenylated polyphenylenes, and especially for ladder polymers with a high degree of structural perfection. The diene and dienophile groups may be present in the same monomer molecule, as in a biscyclopentadiene. Some monodiene monomers also appear to polymerise by a Diels–Alder reaction, thus butadiene and isoprene with EtAlCl$_2$ catalyst give the ladder polymer

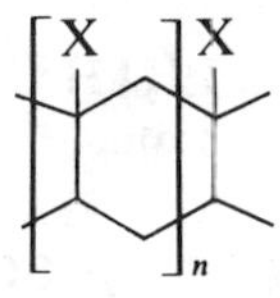

where X = H in butadiene and X = methyl in isoprene. Similarly diacetylene forms the polyacene

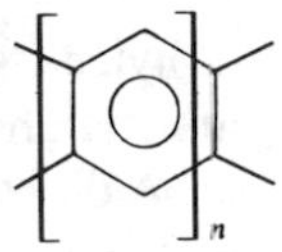

More frequently the diene and dienophile groups are in separate monomers, e.g.

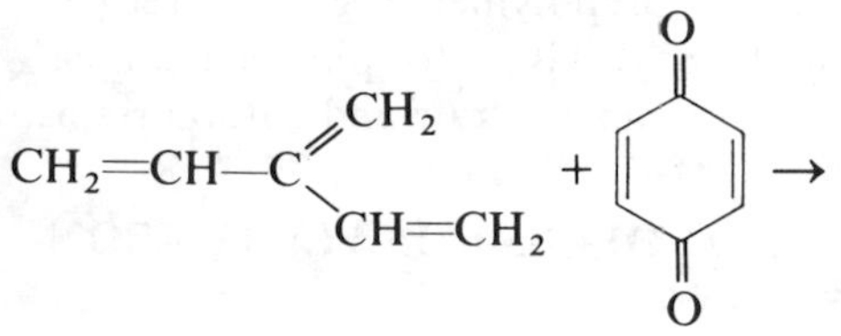

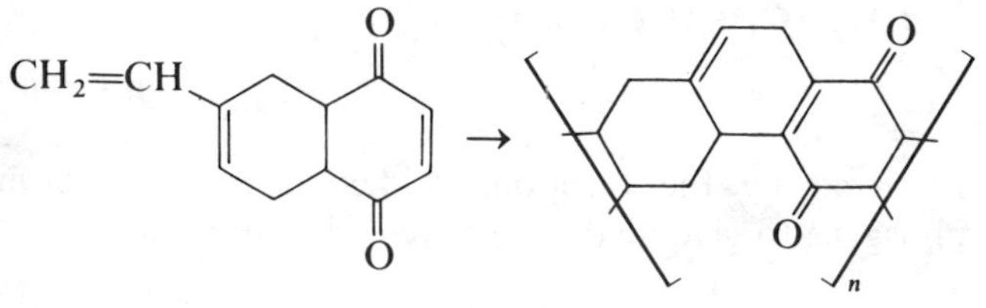

DIELS–ALDER REACTION (Diene synthesis) A reaction involving the addition of a 1,3-diene to another unsaturated molecule (the dienophile) to form a cyclic adduct. In general,

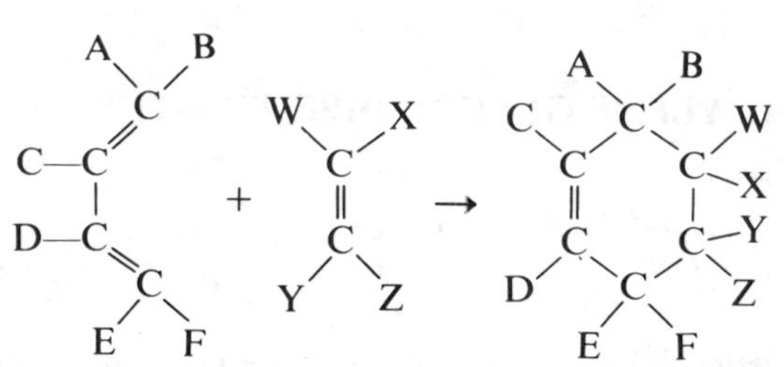

where substituents A, B, C, D, E and F are usually hydrogen or alkyl or other electron-donating groups, and W, X, Y and Z are electron-withdrawing groups such as —COOH, —COOR, —CN and —NO$_2$, which activate the dienophile. Polarity difference between diene and dienophile promotes reaction. In some cases polarities of the reactants is reversed—the inverted Diels–Alder reaction. The reaction has proved useful in the synthesis of many aromatic polymers by Diels–Alder polymerisation.

DIENE   Tradename for low *cis*-polybutadiene.

DIENE POLYMERISATION   Polymerisation of a conjugated 1,3-diene monomer, usually of the type

CH$_2$═CX—CH═CH$_2$, where X is either hydrogen (butadiene) or a substituent, e.g. CH$_3$ (isoprene) and Cl (chloroprene). Other diene polymerisations, such as those of diallyl phthalate and divinylbenzene, are usually referred to a divinyl polymerisations. Conjugated diene monomers may polymerise by 1,2-, 3,4- or 1,4-addition and often a mixture of such addition structures results. However, polymerisation conditions may be chosen to maximise one type of addition. When the diene is symmetrical, as in butadiene, 1,2- and 3,4-addition are identical.

In 1,4-addition both *cis* and *trans* isomers of the repeat unit may be formed:

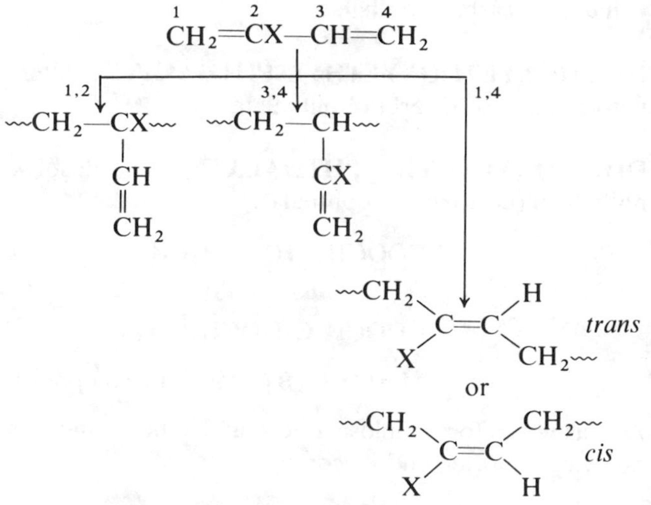

1,4-Addition predominates in free radical polymerisation, especially at low temperatures, with most repeat units being *trans* linked. High 1,4-addition also occurs with lithium or lithium alkyl initiation and either high *cis* or high *trans* polymers can be formed depending on polymerisation conditions. The highest specificity is obtained in Ziegler–Natta polymerisations, e.g. both 99% *cis*- and 99% *trans*-1,4-polymers of butadiene and isoprene may be synthesised. These polymerisations can also produce highly specific 1,2- or 3,4-polymers (both iso- and syndiotactic). Metal alkyl anionic initiators can only achieve ~50% 1,2- or 3,4-structures.

DIENE RUBBER   A rubber based on a polymer formed from a diene monomer or comonomer. Important diene homopolymers are polybutadiene, polychloroprene and polyisoprene, whilst several copolymers, mostly based on butadiene as the diene comonomer, such as poly-(acrylonitrile-co-butadiene) and poly-(styrene-co-butadiene), and also important. The diene units, as well as having low T_g values thus conferring rubbery properties, also provide sites for crosslinking by vulcanisation, usually by sulphur vulcanisation. The main disadvantage of diene units is their sensitivity to ozone and oxidative attack.

DIENE SYNTHESIS   Alternative name for Diels–Alder reaction.

DIE SWELL (Barus effect) The expansion of an extrudate on emerging from a die, so that its cross-section is larger than the die cross-section. Quantitatively it is given by the swelling ratio. It usually occurs only with elastic fluids and hence universally with polymer melts. It is a result of the normal forces that develop on shearing a polymer melt. Molecularly it may be explained by the recoiling, i.e. molecular relaxation, of the molecules on emerging from the die, from the orientation they have as a result of flow in the die. Die swell increases with shear rate up to a maximum (near the critical shear rate) then decreases. It increases with polymer molecular weight and any other factors which increase entanglements.

DIETHOXYDIGLYCOL PHTHALATE   Alternative name for dicarbitol phthalate.

DIETHOXYETHOXYETHYL PHTHALATE   Alternative name for dicarbitol phthalate.

DIETHOXYETHYL   PHTHALATE   (Dicellosolve phthalate) (Diethylglycol phthalate)

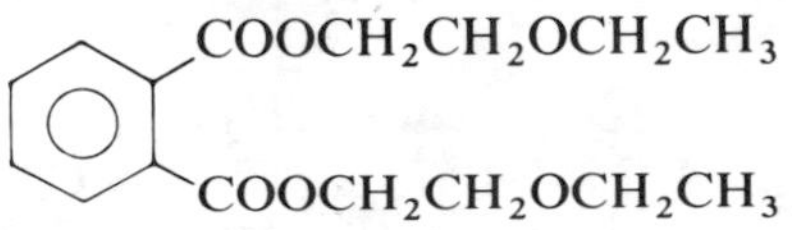

$$COOCH_2CH_2OCH_2CH_3$$
$$COOCH_2CH_2OCH_2CH_3$$

M.p. 31°C. B.p. 354°C (decomposes).

A plasticiser for cellulose esters and ethers, and for polyvinyl chloride and its copolymers.

DIETHYLAMINOETHYLCELLULOSE   (DEAE-cellulose) A cellulose derivative in which some of the hydroxyl hydrogens have been replaced by diethylaminoethyl groups

$$(-CH_2CH_2\overset{+}{N}H(C_2H_5)_2)$$

which are cationically charged at low pH. Such polymers act as anion-exchange resins and are widely used in ion-exchange chromatography of protein mixtures.

DIETHYLAMINOETHYL-SEPHADEX   (DEAE-Sephadex) Sephadex in which some of the dextran hydroxyl hydrogens have been replaced by

$$-CH_2CH_2\overset{+}{N}H(C_2H_5)_2$$

groups, carrying the positive charge at neutral or low pH. Useful as an ion-exchange resin and also exhibiting molecular size exclusion effects, in the ion-exchange chromatography of protein mixtures.

DIETHYLAMINOPROPYLAMINE

$$(C_2H_5)_2NCH_2CH_2CH_2NH_2$$
B.p. 169°C.

Useful as a curing agent for epoxy resins, being rather less reactive than diethylenetriamine.

DI-2-ETHYLBUTYL AZELATE

$$[CH_3CH_2CH(CH_2CH_3)CH_2OOC-]_2 (CH_2)_7$$
B.p. 230°C/5 mm.

A plasticiser for imparting good low temperature flexibility to polyvinyl chloride and its copolymers, cellulosic polymers, styrene–butadiene and nitrile rubbers.

DIETHYLDIGLYCOL PHTHALATE   Alternative name for dicarbitol phthalate.

DIETHYLENE GLYCOL   (DEG)

$$HOCH_2CH_2OCH_2CH_2OH$$
B.p. 245°C.

Formed, together with ethylene glycol, when ethylene oxide is reacted with water. High ethylene oxide:water ratios favour its formation. A diol sometimes used for the preparation of polyesters, e.g. unsaturated polyester resins with greater flexibility than those obtained using propylene glycol, and especially for poly-(diethylene glycol adipate) as a prepolymer for polyurethane formation. Also useful as a solvent for phenol and urea–formaldehyde resins, cellulose nitrate and natural resins.

DIETHYLENE GLYCOL-BIS-(ALLYL) CARBON-ATE)

$$(CH_2{=}CHCH_2OCOCH_2CH_2)_2O$$
$$\overset{\|}{O}$$
B.p. 160°C.

Prepared by the reaction of diethylene glycol with phosgene to give diethylene glycol chloroformate

$$(ClCOCH_2CH_2)_2O$$
$$\overset{\|}{O}$$

followed by reaction with allyl alcohol. It is polymerised to poly-(diethyleneglycol-bis-(allyl carbonate)) by free radical polymerisation using a peroxide initiator.

DIETHYLENE GLYCOL DIBENZOATE

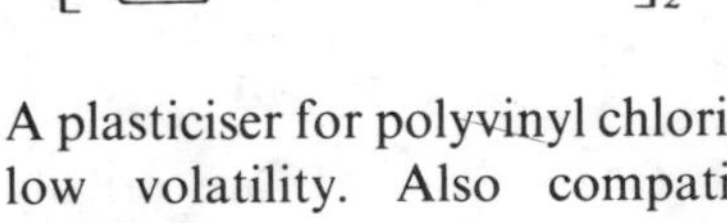

$$\left[{\bigcirc}{-}COOCH_2CH_2{-} \right]_2 O$$
B.p. 240°C/5 mm.

A plasticiser for polyvinyl chloride and its copolymers, of low volatility. Also compatible with many other polymers.

DIETHYLENE GLYCOL DINONANOATE   Alternative name for diethylene glycol dipelargonate.

DIETHYLENE GLYCOL DIPELARGONATE   (Diethylene glycol dinonanoate)

$$[CH_3(CH_2)_7COOCH_2]_2O$$
B.p. 229°C/5 mm.

A plasticiser for nitrile rubbers. Also used in combination with other plasticisers in polyvinyl chloride and its copolymers, cellulose derivatives and several synthetic rubbers.

DIETHYLENE GLYCOL MONO-*n*-BUTYL ETHER
(2-(2-Butoxyethoxy)ethanol) Tradenames: Butyl carbitol, Butyl dioxitol.

$$CH_3(CH_2)_3OCH_2CH_2OCH_2CH_2OH$$
B.p. 231°C.

A solvent for cellulose nitrate, polyvinyl acetate and many natural resins.

DIETHYLENE GLYCOL MONOETHYL ETHER
(2-(*β*-Ethoxyethoxy)ethanol) Tradenames: Carbitol, Dioxitol.

$$CH_3CH_2OCH_2CH_2OCH_2CH_2OH$$
B.p. 202°C.

A solvent for cellulose nitrate, polyvinyl acetate and for many natural resins.

DIETHYLENE GLYCOL MONOETHYL ETHER ACETATE   Tradename: Carbitol acetate.

$$CH_3CH_2OCH_2CH_2OCH_2CH_2OOCCH_3$$
B.p. 217·7°C.

A solvent for cellulose esters, polyvinyl acetate and some natural resins. Miscible with water in all proportions.

DIETHYLENETRIAMINE   (DET)

$$H_2NCH_2CH_2NHCH_2CH_2NH_2$$
B.p. 207°C.

A pungent liquid produced by reaction of ammonia with 1,2-dichloroethane. Useful as a curing agent for epoxy resins, giving a very fast reaction even at room temperature, by reaction with the epoxy groups via the five active amino hydrogens. It gives cured products with good all-round properties except at high temperatures.

DIETHYL ETHER

$$CH_3CH_2OCH_2CH_3 \qquad B.p.\ 34·6°C.$$

A hazardous solvent, being highly inflammable, narcotic and forming unstable peroxides. Of interest when mixed with a small amount of ethanol, as a solvent for cellulose nitrate and polyvinyl acetate.

DIETHYLGLYCOL PHTHALATE   Alternative name for diethoxyethyl phthalate.

DI-2-ETHYLHEXYL ADIPATE   (Dioctyl adipate) (DOA)

$$[CH_3(CH_2)_3CH(CH_2CH_3)CH_2OOCCH_2CH_2]_2$$
B.p. 224°C/10 mm.

A useful plasticiser, especially for retention of low temperature flexibility, for polyvinyl chloride and its copolymers ethyl cellulose, cellulose nitrate, polystyrene and many synthetic rubbers. It also has good compatibility with polymethylmethacrylate, but not with cellulose acetate, polyvinyl acetate or styrene–butadiene rubber.

DI-2-ETHYLHEXYL AZELATE   (Dioctyl azelate) (DOZ)

$$[CH_3(CH_2)_3CH(CH_2CH_3)CH_2OOC]_2(CH_2)_7$$
B.p. 237°C/5 mm. 376°C/760 mm.

A useful plasticiser for polyvinyl chloride, cellulose polymers and nitrile and styrene–butadiene rubber, for imparting good low temperature flexibility.

DI-2-ETHYLHEXYL HEXAHYDROPHTHALATE

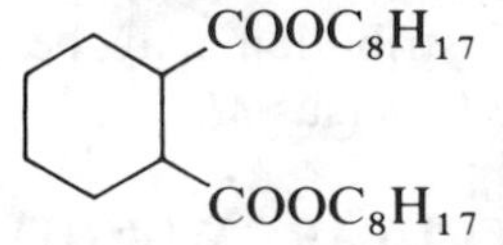

B.p. 216°C/5 mm.

A plasticiser for polyvinyl chloride and its copolymers, similar in action to dioctyl phthalate.

DI-2-ETHYLHEXYL ISOPHTHALATE   (Dioctyl isophthalate) (DOIP)

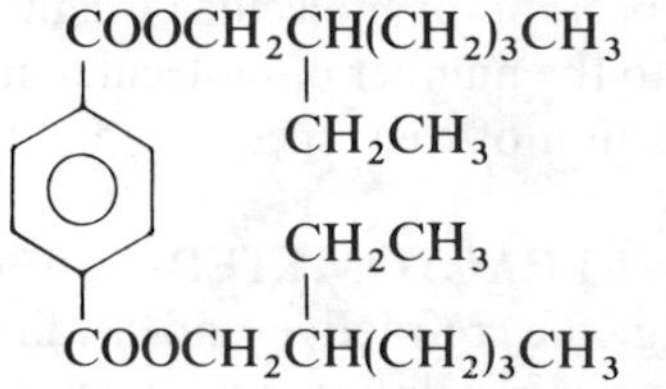

241°C/5 mm.

A plasticiser for polyvinyl chloride, especially useful for plastisols of low viscosity. Also compatible with polystyrene and some cellulose ethers and esters.

DI-2-ETHYLHEXYL PHTHALATE   (Dioctyl phthalate) (DOP)

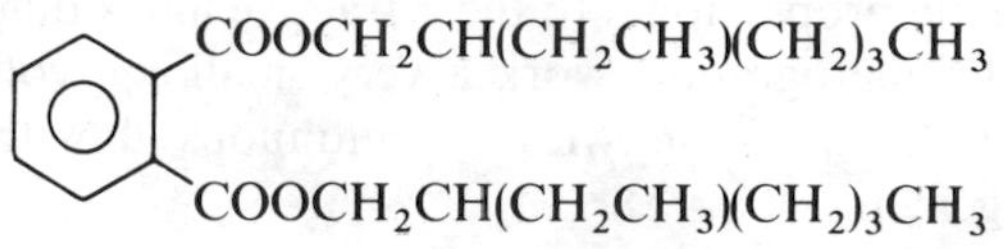

B.p. 370°C/760 mm, 230°C/5 mm.

One of the major general purpose plasticisers for polyvinyl chloride and its copolymers. Also useful for synthetic rubbers, especially polychloroprene, and natural resins.

DI-2-ETHYLHEXYL SEBACATE   (Dioctyl sebacate) (DOS)

$$(CH_2)_8[COOCH_2CH(C_2H_5)(CH_2)_3CH_3]_2$$
B.p. 256°C/5 mm.

A plasticiser for polyvinyl chloride and its copolymers, imparting good low temperature flexibility. Also compatible with a wide range of other polymers.

DIETHYL PHTHALATE   (DEP)

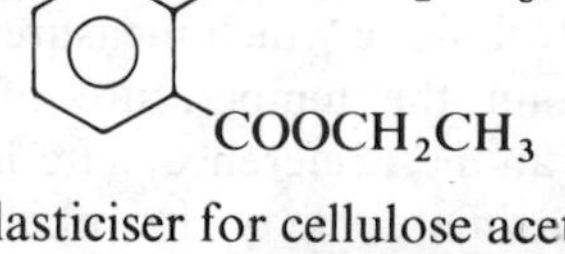

B.p. 296°C/760 mm.

A plasticiser for cellulose acetate (when used in conjunction with dimethyl phthalate) but too volatile for use in other polymers.

DIFFERENTIAL DISTRIBUTION (Differential molecular weight distribution) The continuous form of the frequency distribution, which describes the distribution of molecular sizes i (usually expressed as molecular weights, M_i) according to the amount of polymer, measured either as the weight (or mass) or number of molecules (or moles) of each species i, giving respectively the differential weight (or mass) and differential number distributions. Since the distribution is continuous it may be expressed mathematically by an analytic function $w(M_i)$ or $N(M_i)$ and the distribution may be plotted graphically as $w(M_i)$ or $N(M_i)$ versus M_i.

DIFFERENTIAL MOLECULAR WEIGHT DISTRIBUTION Alternative name for differential distribution.

DIFFERENTIAL NUMBER DISTRIBUTION A differential distribution in which the amount of each polymer molecular species i (of molecular weight M_i) is described according to the number of molecules, number of moles or mole fraction of each species present.

DIFFERENTIAL REFRACTOMETER An instrument for measuring the refractive index difference between a dilute polymer solution and pure solvent. Different designs are used for the measurement of refractive index increment and for the detection of polymer in the effluent from a gel permeation or other chromatographic column. A two-component cell is used, one component containing solvent, the other containing solution. The deflection of a light beam traversing the cell is directly proportional to the refractive index difference. For chromatographic work a very small cell volume is used in conjunction with a continuous flow-through arrangement.

DIFFERENTIAL REPRESENTATION OF VISCO-ELASTICITY Representation of the relationship between stress (σ) and strain (e) by a constitutive equation which is a differential equation involving differentials of stress and strain with respect to time. It has the general form

$$a_0\sigma + a_1 \, d\sigma/dt + a_2 \, d^2\sigma/dt^2 + \cdots, \text{ etc.}$$
$$= b_0 e + b_1 \, de/dt + b_2 \, d^2 e/dt^2 + \cdots, \text{ etc.}$$

Often only a few terms are required to represent actual behaviour. In the simplest mechanical models of linear viscoelasticity only the first-order differentials are used. Thus for a Maxwell model, $de/dt = (1/E)\,d\sigma/dt + \sigma/\eta$, where E and η are the spring modulus and the dashpot viscosity respectively of the model components.

DIFFERENTIAL SCANNING CALORIMETRY (DSC) A thermal analysis technique which measures the energy required to maintain the temperature of the sample the same as that of an inert reference material. A differential scanning calorimeter may be operated isothermally (the static mode) or dynamically (the scanning mode) at any of several fixed heating or cooling

rates. If the sample undergoes an exo- or endothermic change, the detector output is a direct measure of the enthalpy change involved. This contrasts with the related differential thermal analysis technique, where the output is merely a measure of the temperature difference between sample and reference. Applications of the technique to polymer science include detection and measurement of rates of chemical reactions (polymerisation, crosslinking or degradation), detection and measurement of transitions and studies of melting crystallisation and morphology.

DIFFERENTIAL THERMAL ANALYSIS (DTA) A thermal analysis technique which measures the temperature difference between the sample and an inert reference material. It therefore detects, and can quantitatively measure, chemical and physical changes which involve a change in the enthalpy of the sample. The technique is usually used dynamically, i.e. by heating or cooling the sample at a constant rate. Widely used in the study of polymerisation, curing and degradation reactions, melting and glass transitions and crystallisation.

DIFFERENTIAL WEIGHT DISTRIBUTION A differential distribution in which the amount of each polymer molecular species i (of molecular weight M_i) is described according to its weight (or mass) or weight fraction.

DIFFRACTOMETER A device for the determination of the angular dependence of the intensity of scattered X-rays by use of a Geiger scintillator or proportional counter, whose output may be recorded automatically.

DIFFUSE SCATTERING Diffuse small angle X-ray scattering from polymers, either solid or in solution. Usually at a maximum of $0°$ and decreases steadily to 1–$2°$. In the solid state, this is thought to be due to microvoids. In solutions of proteins scattering due to individual particles (polymer molecules) is observed and information about their shape and size (expressed as 'scattering mass ratio') may be obtained.

DIFFUSION COEFFICIENT (Diffusion constant) Symbol D. The constant of proportionality in Fick's laws of diffusion between the rate of diffusion and the concentration gradient (Fick's first law) and the variation of concentration with time and concentration gradient (Fick's second law). The diffusion coefficient may be determined experimentally by measurement of F, the amount of substance that has diffused through a sheet or membrane of thickness l under steady-state conditions (where the concentration remains constant with time at all points within the sheet) and the surface concentrations of the diffusing species (in the case of gases, the vapour pressure). Assuming D is constant, i.e. it does not depend on concentration, then D is related to the permeability coefficient (P) by $P = DS$, where S is the solubility of the diffusing substance in the solid. This relationship is only true if Henry's law holds. Often D is concentration dependent. The units of D are $L^2 T^{-1}$, e.g. $cm^2 s^{-1}$. D may also be determined by a time lag method in which the amount of diffusing substance passing through a sheet (Q_t) is measured with time and then the intercept of a plot of Q_t versus time on the time axis is given by $l^2/6D$, where l is the sheet thickness.

In ultracentrifugation by the sedimentation method, sedimentation is balanced by frictional forces, given by the frictional coefficient f. D is substituted for f, through the Einstein equation in the analysis of ultracentrifugation data by the Svedberg equation, in order to obtain the polymer molecular weight (M). The value of D is concentration dependent and the value at infinite dilution (D_0) is required.

For a polydisperse sample the particular average of D obtained depends on the method of measurement. In order to calculate the polymer weight average molecular weight a z-average D-value is required, together with a weight average sedimentation coefficient (S_0). When both S_0 and D_0 are weight averages, the polymer sedimentation–diffusion average molecular weight is obtained. However, the particular type of average of D used is usually not known, so the sedimentation velocity method is best suited to monodisperse polymers such as proteins and nucleic acids. D may be measured by the free boundary, porous plug, ultracentrifuge and intensity fluctuation spectroscopy methods.

DIFFUSION CONSTANT Alternative name for diffusion coefficient.

DIGLYCIDYL ETHER OF BISPHENOL A (DGEBA)

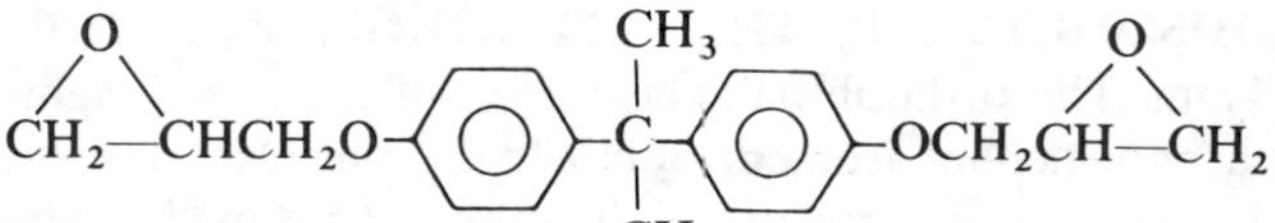

M.p. 43°C.

Prepared by reaction of a large excess of epichlorhydrin with bisphenol A at about 100°C in the presence of aqueous alkali, which acts as both a catalyst for the epoxy–hydroxyl group reaction and for dehydrochlorination of the initially formed chlorohydrin:

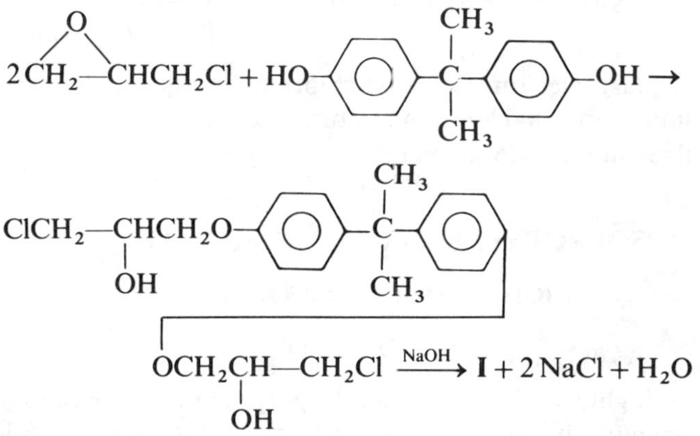

Many commercial epoxy resins, before curing, are essentially DGEBA, often having a molecular weight of about 380 (which compares with 340 as the molecular weight of DGEBA). Thus they contain small amounts of higher polymers of structure:

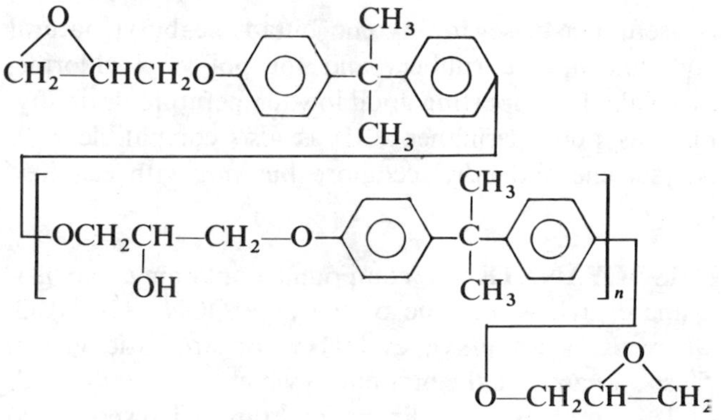

where $n > 0$. These materials are viscous fluids. Resins with higher molecular weights, which are mixtures with $n > 1$, are also produced by use of less epichlorhydrin and may be hard, brittle solids. Resins are characterised by their Durrans melting points and epoxide equivalent. The resins contain as many secondary hydroxyls as the value of n. For use in epoxy resin products, the resins are crosslinked by reaction with any of a variety of curing (or hardening) agents which contain active hydrogens which react with the epoxy groups. These include aliphatic and aromatic diamines and acid anhydrides. In the case of primary diamines the functionality is four so that network structures are formed. With anhydrides, network formation occurs since the effective functionality of the epoxy resin is increased to above two by reaction of the anhydride groups with the hydroxyls in the higher molecular weight resins and with the hydroxyls formed by reaction of the epoxy groups with anhydride.

DIHEDRAL ANGLE Alternative name for torsional angle, particularly when this refers to rotation about a carbon–carbon single bond.

DIHYDRAZIDE Alternative name for diamidrazone.

1,3-DIHYDROXYBENZENE Alternative name for resorcinol.

1,4-DIHYDROXYBENZENE Alternative name for hydroquinone.

2,2′-DIHYDROXY-3,3′-DI-α-METHYLCYCLO-HEXYL-5,5′-DIMETHYLPHENYLMETHANE Alternative name for 2,2′-methylene-bis-[6-(1-methylcyclohexyl)-p-cresol].

4,4′-DIHYDROXYDIPHENYLMETHANE Alternative name for bisphenol F.

4,4′-DIHYDROXY-2,2-DIPHENYLPROPANE Alternative name for bisphenol A.

2,2-DIHYDROXY-1,3-INDANDIONE Chemical name for ninhydrin.

DIISOBUTYL ADIPATE (DIBA)

$$[CH_3CH(CH_3)CH_2OOCCH_2CH_2]_2$$
B.p. 282°C/760 mm.

A useful plasticiser for styrene–butadiene, butyl, natural and chloroprene rubbers and for polyvinyl chloride, especially for imparting good low temperature flexibility, but has poor permanence. It is also compatible with polystyrene and ethyl cellulose but not with cellulose acetate.

DIISOCYANATE

A compound containing two isocyanate groups, i.e. one of the type OCN—R—NCO, where R is an alkyl, cycloalkyl or aromatic group. Diisocyanates are the monomers which, by reaction with diols or polyols, form linear (or branched respectively) polyurethanes by step-growth polymerisation:

$$n\,OCN—R—NCO + n\,HO—R'—OH \rightarrow$$
$$\text{-}[O—CONH—R—NHCO—O—R']_n$$

Branched and crosslinked polyurethanes result from the use of polyols with a functionality of more than two. Crosslinking also results from further reactions of isocyanate groups to form allophanate and biuret groups. Other reactions characteristic of isocyanate groups may also occur during polyurethane formation, e.g. reaction with water to give amino groups and carbon dioxide (useful in the blowing of foams). Subsequent reaction of the amine groups with isocyanate groups gives urea groups, which then in turn react with more isocyanate groups to form biuret groups.

Many of these reactions, including the main polyol/isocyanate polyurethane forming reaction, are catalysed by basic compounds and tertiary amines are frequently used as catalysts. Roughly, the more basic the catalyst, the greater is its catalytic effect. Examples include *N*-methylmorpholine and 1,4-diazabicyclo-2,2,2-octane. Certain metal salts, especially tin compounds such as dibutyltin dilaurate, are also effective catalysts.

Diisocyanates are prepared by reaction of phosgene with the appropriate diamine:

$$H_2N—R—NH_2 + 2COCl_2 \rightarrow OCN—R—NCO + 2HCl$$

Both aliphatic and aromatic diisocyanates are used in the formation of commercial polyurethane products. The commonest are tolylene diisocyanate (particularly for flexible foams, but also in cast and millable polyurethane elastomers), 4,4'-diphenylmethane diisocyanate (especially for rigid foams and thermoplastic elastomers), naphthalene diisocyanate and hexamethylene diisocyanate.

DIISODECYL ADIPATE (DIDA) B.p. 240°C/4 mm.

The adipic acid ester of mixed ten-carbon alcohol (i.e. decanol) isomers, such as 2,4,6-trimethylheptanol. A useful plasticiser for polyvinyl chloride and its copolymers, for good low temperature flexibility and permanence.

DIISODECYL PHTHALATE (DIDP) (Didecyl phthalate) B.p. 250–257°C/4 mm. The phthalate ester of

the mixed isomers of ten-carbon alcohols (i.e. decanols) obtained by the Oxo process. The isomers are usually highly branched as in

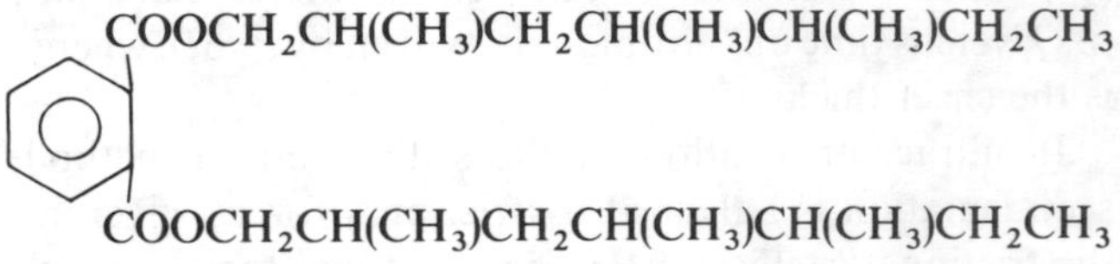

$$COOCH_2CH(CH_3)CH_2CH(CH_3)CH(CH_3)CH_2CH_3$$
$$COOCH_2CH(CH_3)CH_2CH(CH_3)CH(CH_3)CH_2CH_3$$

i.e. they are mostly trimethylheptanols. A low volatility plasticiser for polyvinyl chloride and its copolymers.

DIISONONYL PHTHALATE (DINP) Like dinonyl

phthalate, a phthalate ester of mixed isomeric nonanols obtained from the Oxo process, consisting mostly of 3,5-dimethylheptanol and 3-ethyl-5-methylhexanol. A plasticiser for polyvinyl chloride and its copolymers, similar in action to dinonyl phthalate.

DIISOOCTYL ADIPATE (DIOA) B.p. 218°C/4 mm.

The adipic acid esters of mixed isomeric octanols, e.g. 4,5-dimethyl-, 3,5-dimethyl- and 3,4-dimethylhexanols and 3- and 5-methylheptanols. Useful as a plasticiser in a similar way to di-2-ethylhexyl adipate.

DIISOOCTYL AZELATE (DIOZ) B.p. 235°C/4 mm.

The azelaic acid ester of the mixed octanol isomers obtained by the Oxo process, e.g. 4,5-dimethyl-, 3,5-dimethyl- and 3,4-dimethylhexanols with 3- and/or 5-methylheptanol. Similar in properties and uses as a plasticiser to di-2-ethylhexyl azelate.

DIISOOCTYL PHTHALATE (DIOP) B.p. 235°C/

4 mm. The orthophthalic acid ester of the mixed eight-carbon alcohol isomers obtained by the Oxo process. Typically a mixture of 4,5-dimethyl-, 3,5-dimethyl and 3,4-dimethylhexanols plus 3- and/or 5-methylheptanols. A widely used general purpose plasticiser for polyvinyl chloride and its copolymers, similar in performance to di-2-ethylhexyl phthalate.

DIISOOCTYL SEBACATE (DIOS)

$$(CH_2)_8[COO(CH_2)_3CH(CH_3)CH(CH_3)CH_3]_2$$
B.p. 256°C/5 mm.

A plasticiser for polyvinyl chloride and its copolymers, imparting good low temperature flexibility. Also compatible with a wide range of other polymers.

DIISOPROPYLPEROXYDICARBONATE

$$(CH_3)_2CHOCO—OCOCH(CH_3)_2$$
$$\overset{\|}{O} \qquad \overset{\|}{O}$$

A highly active free radical polymerisation initiator, having half-life times of 6 h/50°C and 0·25 h/70°C. Useful for initiation at 30–60°C, especially for vinyl chloride polymerisation.

DIISOPROPYLPHOSPHOFLUORIDATE (DFP)

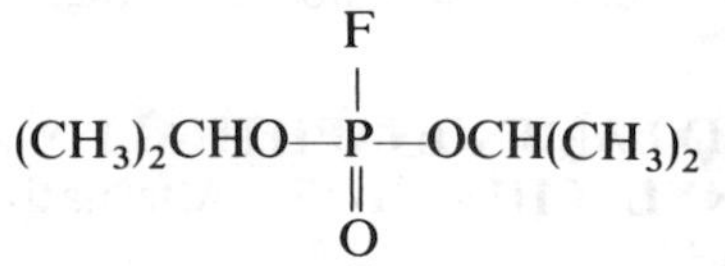

$$(CH_3)_2CHO-\overset{\overset{\displaystyle F}{|}}{\underset{\underset{\displaystyle O}{\|}}{P}}-OCH(CH_3)_2$$

I

A reagent useful in the identification of hydroxyl groups at the active site in enzymes. It reacts with these groups as

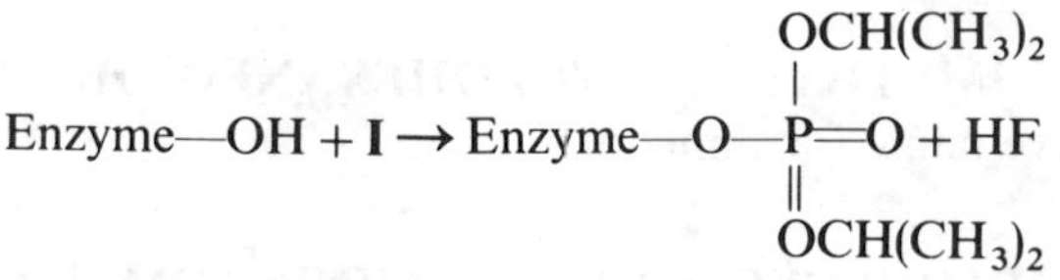

$$Enzyme-OH + I \rightarrow Enzyme-O-\overset{\overset{\displaystyle OCH(CH_3)_2}{|}}{\underset{\underset{\displaystyle OCH(CH_3)_2}{\|}}{P}}=O + HF$$

and inactivates the enzymes if these groups are involved. Thus active serine groups in serine enzymes such as chymotrypsin can be identified. It also inactivates acetylcholinesterase, important in nerve function, and therefore acts as a nerve poison.

DIKETOPIPERAZINE

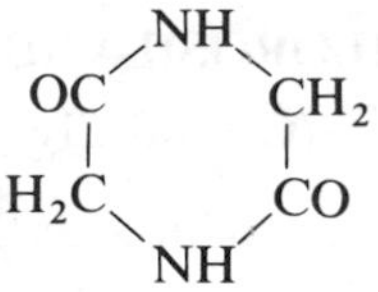

$$\begin{array}{ccc} & NH & \\ OC & & CH_2 \\ H_2C & & CO \\ & NH & \\ & I & \end{array}$$

M.p. 312°C (decomposes). Sublimes at 260°C.

The product of cyclodimerisation of the α-amino acid, glycine, formed when it is heated by elimination of water:

$$2\,H_2NCH_2COOH \rightarrow 2\,H_2O + I$$

Formation of a diketopiperazine from glycine or other α-amino acids is an important reason why these acids cannot be polymerised to poly-(α-amino acids) simply by heating.

DILATANCY   Alternative name for shear thickening.

DILATATION   (Bulk strain) (Dilatational strain) Symbol Δ. A deformation in which displacements occur in all three directions (x, y and z), i.e. a change (ΔV) in volume (V) occurs. Given by $\Delta = \Delta V/V$. ΔV is zero in incompressible materials for which the Poisson ratio is 0·5, as is approximately true for many rubbers. Produced by a change in pressure (P) to which it is related by the bulk modulus (K) by $K = P/\Delta$.

DILATATIONAL STRAIN   Alternative name for dilatation.

DILATATIONAL STRESS   Alternative name for hydrostatic stress, in which $\sigma_{11} = \sigma_{22} = \sigma_{33} = -P$, where P is the pressure.

DILATOMETRY   The measurement of the changes in the volume of a substance by filling a small containing vessel, usually a cylindrical glass bulb, with the substance and following the change in volume by following the change in level of the substance (if liquid) or of a liquid in which the substance is immersed (if solid), in an attached graduated capillary tube. The technique has several uses in polymer science, notably for the measurement of the rates of polymerisation and crystallisation by the shrinkage that occurs and for the measurement of the glass transition temperature by observation of the temperature at which the volume expansion coefficient changes.

DILAUROYLTHIODIPROPIONATE (DLTDP)

$$(C_{12}H_{25}OOCCH_2CH_2)_2\text{-}S$$

M.p. 37–40°C.

A preventive antioxidant capable of peroxide decomposition. Particularly effective in polyolefins when used in a synergistic mixture with a hindered phenol.

DILAURYL PHTHALATE

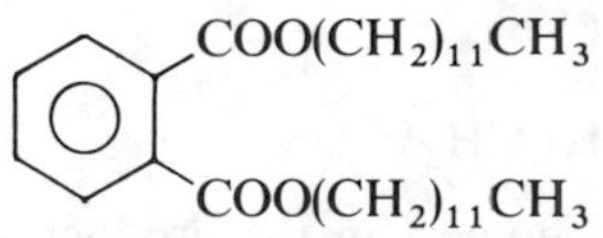

$$\underset{\displaystyle COO(CH_2)_{11}CH_3}{\overset{\displaystyle COO(CH_2)_{11}CH_3}{\bigcirc}}$$

B.p. 161°C/4 mm.

A plasticiser for cellulose esters, but not cellulose acetate, and cellulose ethers. Also plasticises polyvinyl chloride and its copolymers.

DILUTE SOLUTION VISCOMETRY   Alternative name for solution viscometry.

DILUTION VISCOMETER   Alternative name for Ubbelohde viscometer.

DIMEDONE   (5,5-Dimethyl-1,3-cyclohexanediol)

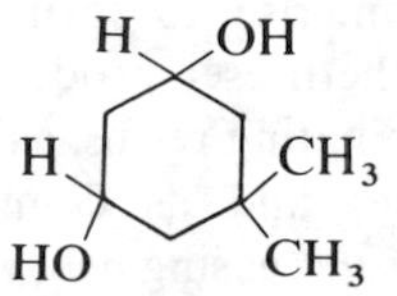

$$\begin{array}{ccc} H & OH & \\ H & & CH_3 \\ HO & & CH_3 \end{array}$$

M.p. 149°C.

A reagent which specifically reacts with aldehyde groups and therefore is useful in their determination. Used particularly in natural rubber where aldehydes are thought to be responsible for storage hardening. It can therefore be used to produce constant viscosity natural rubber, although in practice other reagents are used.

DIMER   A molecule formed by the joining together of two monomer molecules, i.e. the smallest possible oligomer. Dimers may be formed by the degradation of longer polymer chains by back biting during depolymerisation (as with polystyrene) or during the later stages of random scission degradation, as in the formation of disaccharides and dipeptides on extensive hydrolysis of polysaccharides and proteins respectively. Thus characterisation of the structure of the dimers so produced can give useful information about the structure of the original polymer. Cyclic dimers are frequently formed when a step-growth polymerisation is attempted with a monomer which can form a stable ring structure by cyclodimerisation, as with the formation of diketopiperazine when glycine is heated.

DIMER ACID (Dimer fatty acid) A dibasic acid formed by cyclodimerisation of the unsaturated fatty acid components of some fatty oils such as tall oil and castor oil fatty acids. This is done either by heating in steam at about 300°C or by heating with active clays at about 200°C. The product is complex, but may be fractionated to yield fractions containing, for example, dimers of oleic and linoleic acids:

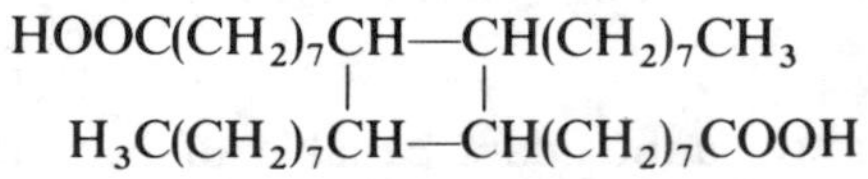

$$HOOC(CH_2)_7CH\!-\!CH(CH_2)_7CH_3$$
$$H_3C(CH_2)_7CH\!-\!CH(CH_2)_7COOH$$

and

$$HOOC(CH_2)_7CH\!-\!CHCH_2CH\!=\!CH(CH_2)_7COOH$$

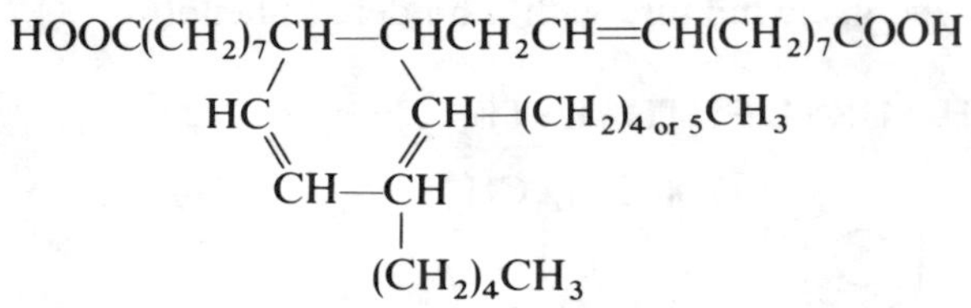

respectively. Useful as monomers in the production of fatty polyamides by reaction with ethylenediamine and other diamines.

DIMER FATTY ACID Alternative name for dimer acid.

DIMETHACRYLATE (Glycol dimethacrylate) An ester of methacrylic acid and a diol, having the structure

$$CH_2\!=\!CCOOROCOC\!=\!CH_2$$
$$CH_3 \qquad CH_3$$

Such monomers will polymerise to form a network polymer and therefore a thermoset product. They are useful as thermosetting laminating resins, as the basis of anaerobic acrylic adhesives and as comonomers in polymethylmethacrylate for increasing heat resistance. A common example is tetramethyleneglycol dimethacrylate.

DIMETHOXYETHYL PHTHALATE (Dimethylcellosolve phthalate) (Dimethylglycol phthalate)

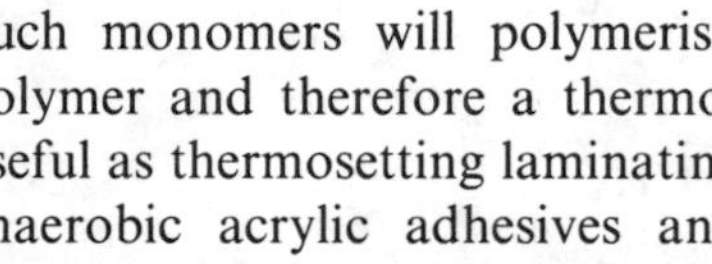

$$COOCH_2CH_2OCH_3$$
$$COOCH_2CH_2OCH_3 \qquad \text{B.p. } 350°C/760\,mm.$$

A plasticiser for cellulose acetate with low oil extractability. Also compatible with polyvinyl chloride and its copolymers, polyvinyl butyral and synthetic rubbers.

DIMETHYLACETAMIDE (DMAC)

$$CH_3CON(CH_3)_2 \qquad \text{B.p. } 166°C.$$

A useful dipolar aprotic solvent, especially as a solvent for carrying out the synthesis of, and for dissolving, polyimides and aromatic polyamides, and for spinning polybenzimidazole fibres.

DIMETHYLAMINOETHANOL Alternative name for bis-(β-N,N-dimethylaminoethyl) ether.

1-DIMETHYLAMINONAPHTHALENE-5-SULPHONYL CHLORIDE Alternative name for dansyl chloride.

DIMETHYLCELLOSOLVE PHTHALATE Alternative name for dimethoxyethyl phthalate.

5,5-DIMETHYL-1,3-CYCLOHEXANEDIOL Alternative name for dimedone.

DIMETHYLCYCLOHEXYLAMINE (DMCHA)

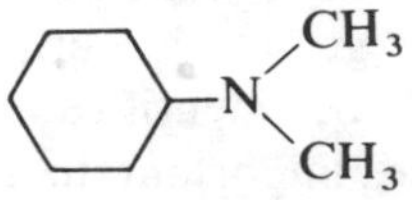

A widely used catalyst for the formation of flexible polyurethane foams based on polyester polyols.

DIMETHYLDICHLOROSILANE

$$(CH_3)_2SiCl_2 \qquad \text{B.p. } 70°C.$$

A chlorosilane usually produced as the major product by the direct process of alkylation of elemental silicon. The major monomer for the production of a variety of polyorganosiloxanes and important silicone products (silicone oils, greases, rubbers and resins). Hydrolysis with water yields the unstable silanediol which condenses to siloxanes. Both linear polydimethylsiloxane and dimethylcyclosiloxanes are formed. Use of moderately concentrated aqueous sulphuric acid yields largely linear polymer, whereas in the presence of water-immiscible solvents, such as toluene and diethyl ether, lower cyclosiloxanes, e.g. octamethylcyclotetramethylsiloxane, are favoured. Whilst hydrolysis can produce the lower molecular weight polymers useful as silicone fluids, higher molecular weight polymers required for silicone rubbers are difficult to obtain due to the need for extremely pure monomer free from mono- and trichlorosilanes.

N,N'-DIMETHYL-N,N'-DINITROSO-TEREPHTHALAMIDE

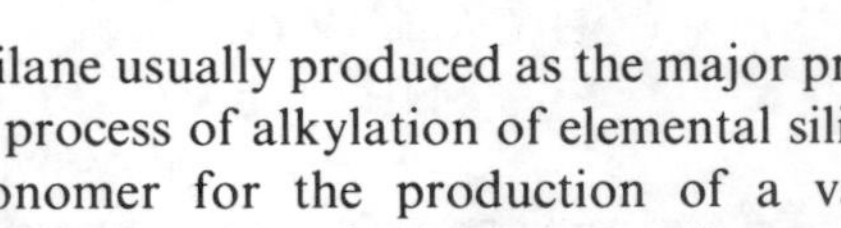

A chemical blowing agent with a low decomposition temperature (90–105°C) and producing a mixture of nitrogen, carbon dioxide and water.

3,6-DIMETHYL-1,4-DIOXAN-2,5-DIONE Alternative name for lactide.

N,N-DIMETHYLETHANOLAMINE Alternative name for bis-(β-N,N-dimethylaminoethyl) ether.

DIMETHYLFORMAMIDE (DMF)

$$HCON(CH_3)_2 \quad\quad \text{B.p. } 153°C.$$

A useful dipolar aprotic solvent, capable of dissolving many otherwise difficultly soluble polar polymers. A solvent for polyvinyl chloride, polyacrylonitrile, polyacrylic acid, polyvinyl acetate, polymethylmethacrylate and other acrylics, aliphatic, but not aromatic, polyamides, polyurethanes, polyimides and polycarbonates, polyoxymethylene (when hot), polar rubbers, alkyd resins and polystyrene.

DIMETHYLGLYCOL PHTHALATE Alternative name for dimethoxyethyl phthalate.

DI-1-METHYLHEPTYL PHTHALATE Alternative name for dicapryl phthalate.

DIMETHYL NYLON 3

$$\text{┤C(CH}_3)_2\text{CH}_2\text{NHCO┤}_n$$

Synthesised by ring-opening polymerisation of the lactam 4,4-dimethylazetidinone. The polymer is of interest as a potentially useful fibre, having a T_m value of about 325°C. It may be wet spun from aqueous salt solutions.

1,4-DIMETHYLOLCYCLOHEXANE (1,4-Cyclo-hexanedimethanol)

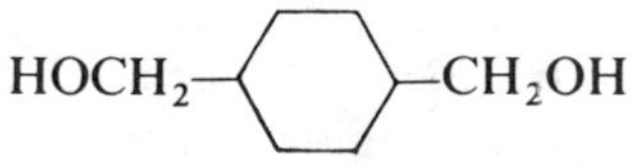

$$HOCH_2\text{—}\bigcirc\text{—}CH_2OH$$

M.p. 35°C.

Prepared by the reduction of dimethylcyclohexanedicarboxylate, itself prepared from terephthalic acid. The monomer for the formation of poly-(cyclohexane-1,4-dimethylene terephthalate) by ester interchange with dimethyl terephthalate.

DIMETHYLOLETHYLENEUREA (DMEU)

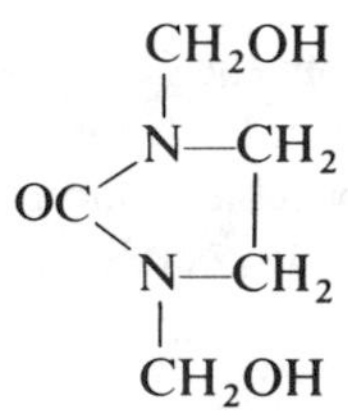

M.p. 133°C.

Formed by reaction of ethyleneurea with formaldehyde. Useful as a crease-resistant finishing agent for cellulosic textiles. On application to the fabric, with heating to about 130°C, it partially polymerises to a urea–formaldehyde type of polymer and crosslinks the cellulose chains through reaction with cellulose hydroxyl groups. Fabrics so treated only develop the chlorine retention problem after repeated washings.

DIMETHYLOLPROPYLENEUREA

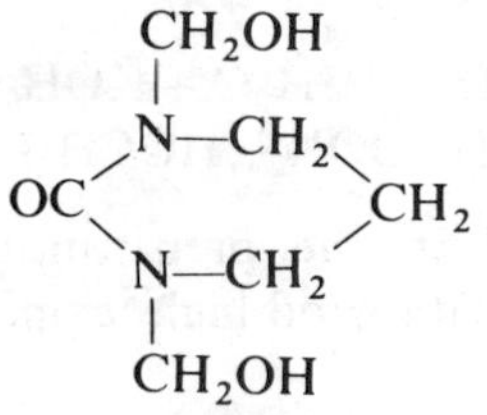

Useful as a crease-resistant textile finish in a similar way to dimethylolethyleneurea, but it is more durable to repeated washings.

3,3-DIMETHYLOXETAN-2-ONE Alternative name for pivalolactone.

DIMETHYLPHENOL Alternative name for xylenol.

2,6-DIMETHYLPHENOL (2,6-Xylenol)

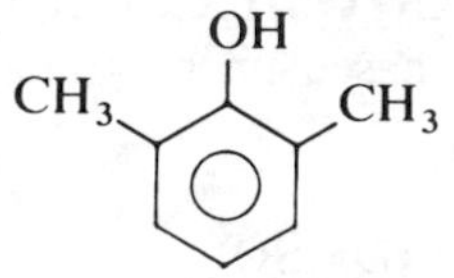

M.p. 49°C. B.p. 203°C.

The monomer for the synthesis of poly-(2,6-dimethyl-1,4-phenylene oxide) by oxidative coupling.

DIMETHYL PHTHALATE (DMP)

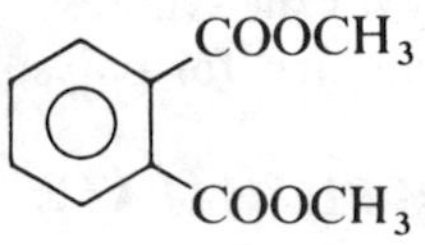

B.p. 284°C/760 mm. M.p. −1°C.

A plasticiser, especially for cellulose acetate, but highly compatible with many other polymers—cellulose esters and ethers, polystyrene, polyvinyl chloride, many rubbers and natural resins. However, its high volatility restricts its use largely to cellulose acetate, when it is used in conjunction with diethyl phthalate.

2,5-DIMETHYLPIPERAZINE

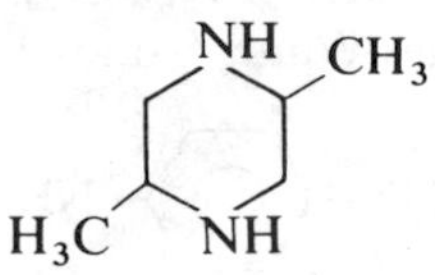

Synthesised by cycloamination of isopropanolamine, $CH_3CH(OH)CH_2NH_2$. This yields a mixture of cis (b.p. 162°C, m.p. 18°C) and trans (b.p. 160°C, m.p. 115°C) isomers. The latter is usually preferred for the preparation of piperazine polyamides.

2,2-DIMETHYLPROPANE-1,3-DIOL (Neopentylene glycol) (Neopentyl glycol)

$$HOCH_2C(CH_3)_2CH_2OH \quad \text{M.p. } 128°C.$$

Prepared by reaction of isobutyraldehyde with form-aldehyde:

$$(CH_3)_2CHCHO + 2 CH_2O + KOH \rightarrow$$
$$HCOOK + HOCH_2C(CH_3)_2CH_2OH$$

Sometimes used in the preparation of unsaturated polyester resins with good high temperature resistance.

α,α-DIMETHYL-β-PROPIOLACTONE   Alternative name for pivalolactone.

DIMETHYL SEBACATE

$$\text{--}[(CH_2)_4COOCH_3]_2 \qquad\qquad \text{B.p. } 294°C/760 \text{ mm.}$$

A plasticiser with a wide range of compatibility but too volatile for most practical uses.

DIMETHYLSILICONE   Alternative name for poly-dimethylsiloxane.

DIMETHYLSILICONE ELASTOMER   (MQ) The basic silicone elastomer, being based on polydimethyl-siloxane.

DIMETHYL SULPHOXIDE   (DMSO)

$$(CH_3)_2SO \qquad\qquad \text{B.p. } 189°C.$$

A useful dipolar aprotic solvent for many polar polymers, especially for polyacrylonitrile and polyvinyl acetate. A useful solvent for carrying out many polycondensation reactions, especially for the synthesis of aliphatic, but not aromatic, polyamides and for polysulphones.

DIMETHYL TEREPHTHALATE

$$CH_3OC\text{--}\bigcirc\text{--}COCH_3$$

M.p. 142°C.

Prepared by the esterification of terephthalic acid with methanol. Preferred to terephthalic acid as the monomer for the formation of polyethylene terephthalate by ester interchange with ethylene glycol, owing to its ease of purification, greater miscibility and greater reactivity.

N,N'-DINAPHTHYL-p-PHENYLENEDIAMINE
Tradenames: Agerite White, Santowhite CI, Nonox CI.

M.p. 235°C.

A chain-breaking antioxidant, widely used in rubbers. Staining, but has an inhibiting effect on copper, which, as an impurity, often catalyses oxidative degradation. A toxic hazard due to the presence of impurity β-naphthylamine and now little used.

2,4-DINITRO-1-FLUOROBENZENE   Alternative name for 1-fluoro-2,4-dinitrobenzene.

2,4-DINITROPHENYL DERIVATIVE   (DNP derivative)

The derivative formed by reaction of 1-fluoro-2,4-dinitrobenzene with a free α-amino group of amino acid, peptide or protein. Formation of DNP derivative is widely used in the determination of the N-terminal amino acid residue of peptides and proteins.

p-DINITROSOBENZENE

A vulcanisation agent for diene rubbers. Usually formed *in situ* by the oxidation of p-quinonedioxime.

DINITROSOPENTAMETHYLENETETRAMINE (DNO) (DNPT)

$$\begin{array}{c} CH_2\text{--}N\text{--}CH_2 \\ | \qquad\quad | \\ ON\text{--}N \quad H_2C \quad N\text{--}NO \\ | \qquad\quad | \\ CH_2\text{--}N\text{--}CH_2 \end{array}$$

Decomposition range 160–200°C. A chemical blowing agent decomposing to produce nitrogen, nitrous oxide, water and methylamine gases and widely used in the production of microcellular rubber.

DINONYL PHTHALATE   (DNP)   (Di-(3,5,5-tri-methylhexyl) phthalate) Largely

$$COOCH_2CH_2CH(CH_3)CH_2C(CH_3)_3$$
$$COOCH_2CH_2CH(CH_3)CH_2C(CH_3)_3$$

but contains other isomers from the Oxo alcohol mixture from which it is made. A general purpose plasticiser for polyvinyl chloride and its copolymers, similar in its action to di-2-ethylhexyl phthalate. Especially useful for low viscosity plastisols.

DINP   Abbreviation for diisononyl phthalate.

DIOA   Abbreviation for diisooctyl adipate.

DIOCTYL ADIPATE   (DOA) Alternative name for di-2-ethylhexyl adipate.

DIOCTYL AZELATE   (DOZ) Alternative name for di-2-ethylhexyl azelate.

DI-*n*-OCTYL-*n*-DECYL PHTHALATE (DNODP) A linear phthalate plasticiser, which is a mixture of the phthalate esters of the eight- and ten-carbon straight-chain alcohols, but often containing lower alcohol esters as well. Contains all possible combinations of mixed esters. A useful plasticiser for polyvinyl chloride and its copolymers of lower volatility than di-2-ethylhexyl phthalate.

DIOCTYL ISOPHTHALATE (DOIP) Alternative name for di-2-ethylhexyl isophthalate.

DIOCTYL PHTHALATE (DOP) Alternative name for di-2-ethylhexyl phthalate.

DI-*n*-OCTYL PHTHALATE (DNOP)

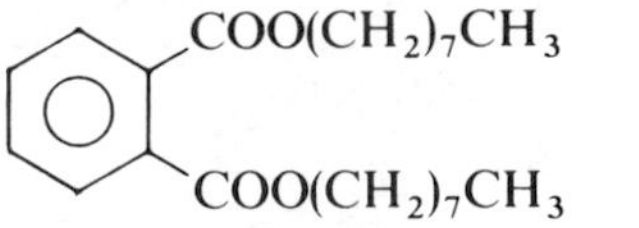

$COO(CH_2)_7CH_3$

$COO(CH_2)_7CH_3$ B.p. 340°C/760 mm.

A plasticiser for polyvinyl chloride and its copolymers, synthetic rubbers and cellulose esters.

DI-*sec*-OCTYL PHTHALATE Alternative name for dicapryl phthalate.

DIOCTYL SEBACATE (DOS) Alternative name for di-2-ethylhexyl sebacate.

DIOCTYLTIN STABILISER A widely used type of organotin stabiliser, of general formula $(C_8H_{17})_2SnX_2$, similar to dibutyltin stabilisers, but of lower toxicity.

DIOCTYL TEREPHTHALATE (DOTP)

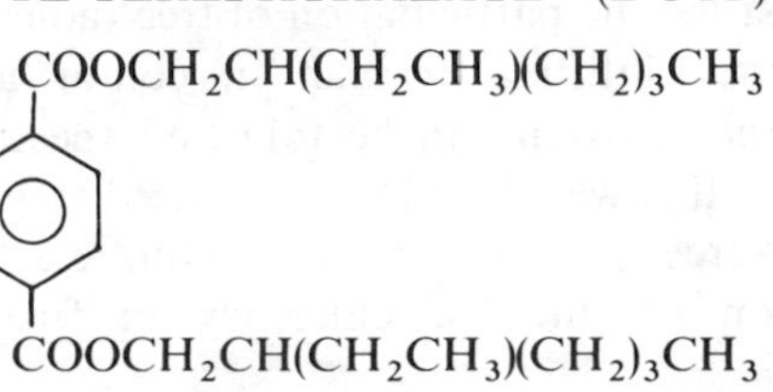

$COOCH_2CH(CH_2CH_3)(CH_2)_3CH_3$

$COOCH_2CH(CH_2CH_3)(CH_2)_3CH_3$

A plasticiser for polyvinyl chloride similar to di-2-ethylhexyl phthalate, but of lower volatility.

DIOLEN Tradename for a polyethylene terephthalate fibre.

DIOP Abbreviation for diisooctyl phthalate.

DIOPSIDE A linear, crystalline naturally occurring mineral silicate polymer of the pyroxene type, in which the counter-ions linking the chains are calcium and magnesium. The empirical formula is $MgCa(SiO_3)_2$. In addition, chromium is also frequently present.

DIOS Abbreviation for diisooctyl sebacate.

DIOXAN

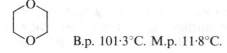

B.p. 101·3°C. M.p. 11·8°C.

A toxic solvent, miscible with water in all proportions (forming an azeotrope with 18% water) and also forming explosive peroxides. A solvent for polystyrene, polyvinyl acetate, polyvinyl chloride, natural rubber, natural resins, cellulose esters and ethers, polyesters and epoxy resins

1,4-DIOXAN-2,5-DIONE Alternative name for glycolide.

DIOXITOL Tradename for diethylene glycol mono-ethyl ether.

DIOZ Abbreviation for diisooctyl azelate.

DIPENTENE

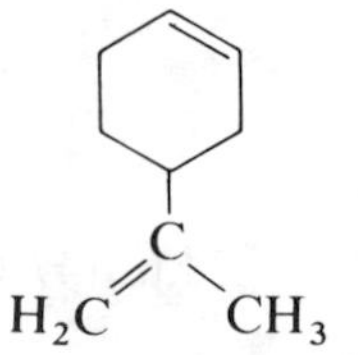

H_2C CH_3 B.p. 175–176°C.

Occurs widely in a variety of essential oils, including turpentine. Similar to turpentine as a solvent, but of much lower oxidation rate. A solvent for rubber, natural resins, phenol–formaldehyde resins and glyptals.

DIPENTENE RESIN A terpene resin obtained by the cationic polymerisation of dipentene, the reaction possibly proceeding as shown:

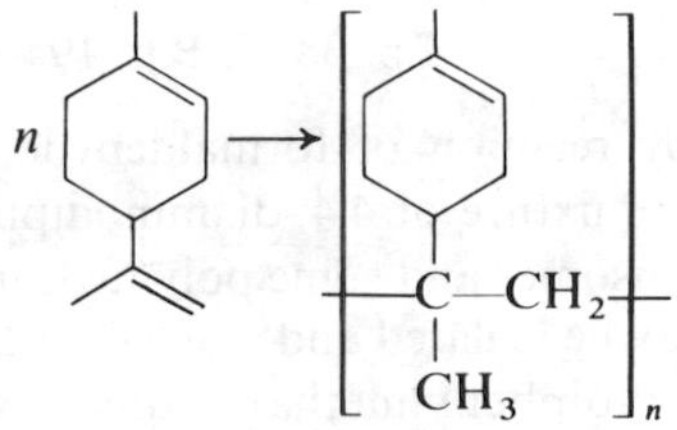

but probably also producing other types of repeat unit as well. Commercial polymers of low molecular weight (500–1000) are useful in hot-melt adhesives and coatings.

DIPEPTIDE The smallest possible peptide, containing only two α-amino acid derivatives and therefore only one peptide bond. In general its structure is

$$H_2NCHRCO—NHCHR'COOH$$

Occasionally dipeptides occur naturally, but they are also produced by severe, but not complete, hydrolysis of larger peptides and proteins.

DIPHENYL CARBONATE

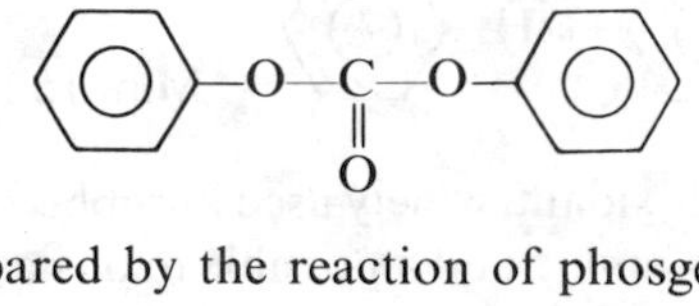

M.p. 78°C.

Prepared by the reaction of phosgene with an aqueous

alkaline solution of phenol, i.e. sodium phenoxide, in the presence of an inert solvent, such as dichloromethane:

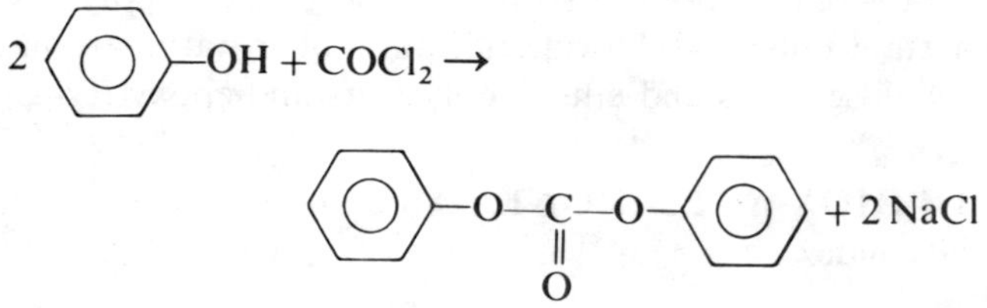

$$2 \langle \bigcirc \rangle - OH + COCl_2 \rightarrow$$

$$\langle \bigcirc \rangle - O - \underset{\underset{O}{\|}}{C} - O - \langle \bigcirc \rangle + 2\,NaCl$$

The monomer for the preparation of polycarbonates, especially bisphenol A polycarbonate, by ester interchange.

DIPHENYLGUANIDINE (DPG)

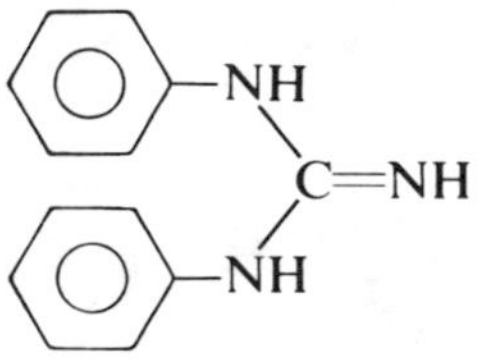

$$\langle \bigcirc \rangle - NH$$
$$\qquad\qquad C = NH$$
$$\langle \bigcirc \rangle - NH$$

M.p. 147°C.

An early accelerator for the sulphur vulcanisation of rubber, now mostly used as a secondary accelerator in combination with, for example, a thiazole.

4,4'-DIPHENYLMETHANE DIISOCYANATE
(MDI) (Methylene-bis-(4,4'-phenylisocyanate))

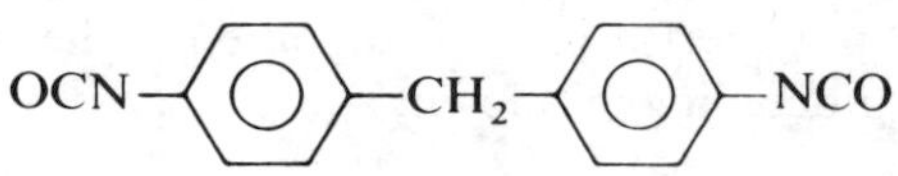

$$OCN - \langle \bigcirc \rangle - CH_2 - \langle \bigcirc \rangle - NCO$$

M.p. 38°C. B.p. 194–199°C/5 mm.

Synthesised by reaction of formaldehyde with aniline, which yields a mixture of 4,4'-diaminodiphenylmethane with some 2,4'-isomer and some polynuclear amines. The 4,4'-isomer may be isolated and reacted with phosgene to give pure 4,4'-diphenylmethane diisocyanate. More commonly the mixture is phosgenated directly to give polymeric MDI. This product has a low vapour pressure and therefore is less of a toxic hazard in use. Pure MDI is used in the manufacture of polyurethane elastomers, whereas polymeric MDI is the major isocyanate used in the manufacture of rigid polyurethane foams, where its higher functionality contributes to crosslinking.

DIPHENYLOLPROPANE
Alternative name for bisphenol A.

N,N'-DIPHENYL-p-PHENYLENEDIAMINE
(DPPD) Tradenames: Agerite DPPD, Antiox 123, Nonox DPPD.

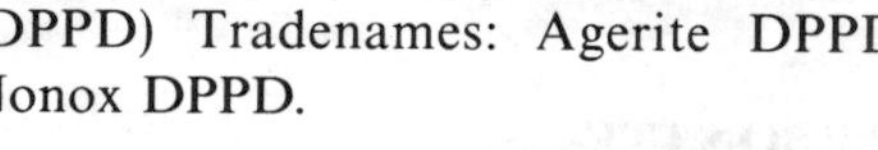

$$\langle \bigcirc \rangle - NH - \langle \bigcirc \rangle - NH - \langle \bigcirc \rangle$$

M.p. 152°C.

A chain-breaking antioxidant, widely used in rubbers, but rather incompatible, staining and a possible toxic hazard.

DIPHENYL PHTHALATE (DPP)

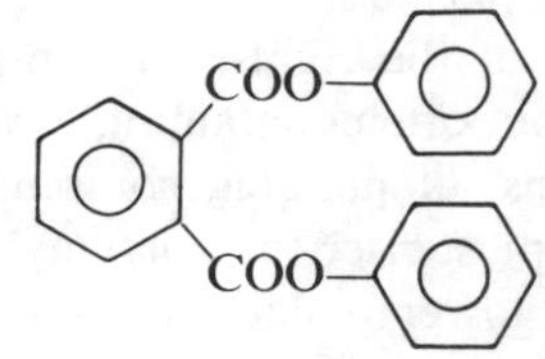

$$\langle \bigcirc \rangle \overset{COO - \langle \bigcirc \rangle}{\underset{COO - \langle \bigcirc \rangle}{}}$$

B.p. 405°C/760 mm. M.p. 69°C.

A plasticiser for polyvinyl chloride and its copolymers, polystyrene, ethyl cellulose and cellulose nitrate, of low water solubility.

DIPHENYLPICRYLHYDRAZYL (DPPH)

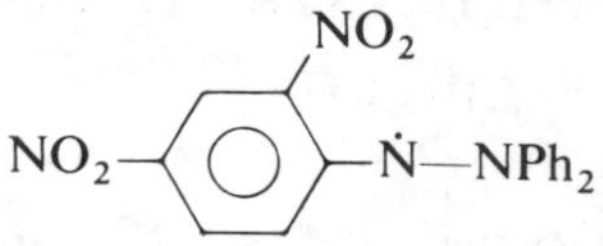

$$NO_2 - \langle \bigcirc \rangle - \overset{\cdot}{N} - NPh_2 \quad (NO_2 \text{ ortho})$$

A stable free radical that acts as a radical scavenger by combining with other free radicals (R·):

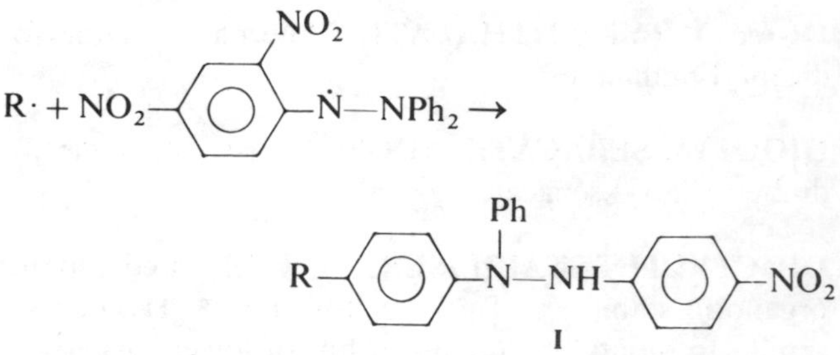

$$R\cdot + NO_2 - \langle \bigcirc \rangle - \overset{\cdot}{N} - NPh_2 \rightarrow$$

$$R - \langle \bigcirc \rangle - \overset{\overset{Ph}{|}}{N} - NH - \langle \bigcirc \rangle - NO_2$$
$$\mathbf{I}$$

Reduction of a reaction rate by addition of DPPH is often used as a test for the participation of free radicals in the reaction. Since DPPH is violet in colour and **I** is colourless, the reaction can be followed spectrophotometrically. In this way DPPH can be used to count the number of free radicals in a system, e.g. for the determination of initiator efficiency in free radical polymerisation.

1,3-DIPOLAR ADDITION
A cyclo-addition reaction between a 1,3-dipolar compound, e.g. a phenyl azide, and a dipolarophilic compound (containing a C=C, C=O, C≡C or C≡N group) with formation of a five-membered ring compound, e.g.

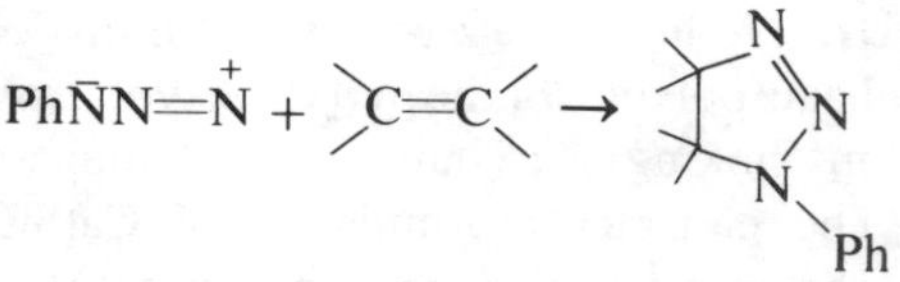

$$Ph\bar{N}N\overset{+}{=}N + \rangle C = C \langle \rightarrow$$

When the reactants each contain two such dipolar or dipolarophile groups, polymerisation may occur. The reaction is especially useful in the synthesis of many five-membered heterocyclic ring polymers. Typical dipolar monomers are nitrile oxides (e.g. terephthaloyl oxide) (**I**), sydnones, nitrilimines and nitrones. Diethynyl compounds, e.g. 1,4-diethynylbenzene (**II**), are common dipolarophilic monomers. A typical reaction is that of **I** and **II**, which yields a *p*-phenylene linked polyisoxazole:

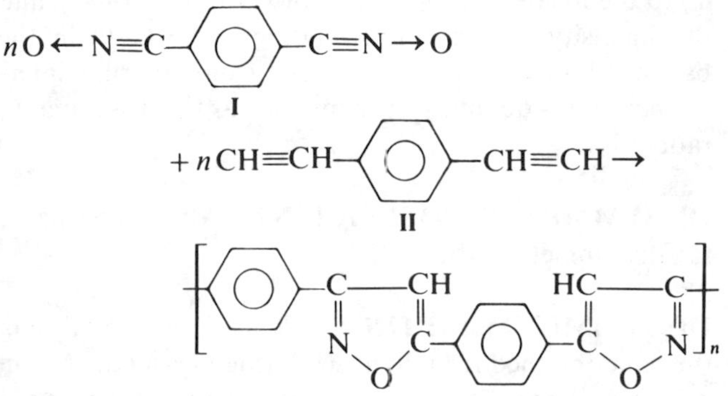

$$n O \leftarrow N \equiv C - \langle O \rangle - C \equiv N \rightarrow O$$
I

$$+ n\, CH \equiv CH - \langle O \rangle - CH \equiv CH \rightarrow$$
II

Similarly **II** with a disydnone yields a polyphenyl-pyrazole.

DIPOLAR POLARISATION Alternative name for orientation polarisation.

DIPROPYLENE GLYCOL DIBENZOATE

$$\langle O \rangle - COOCH_2CH(CH_3)OCH(CH_3)CH_2OOC - \langle O \rangle$$

B.p. 232°C/5 mm.

A plasticiser for polyvinyl chloride having high solvent power at high temperatures. Also useful in polyvinyl acetate and in cellulosics. Compatible with polystyrene and polymethylmethacrylate.

DIRECTION COSINE (1) The cosine of the angle between the axes of two sets of cartesian coordinate axes; there are nine such quantities, a_{11}, a_{12}, ..., etc.

(2) The cosine of the angles of inclination of the normal to the plane on which a stress is acting and the *x*, *y* and *z* directions (in cartesian notation), and given the symbols *l*, *m* and *n* respectively.

DIRECT PROCESS The direct conversion of elemental silicon to chlorosilanes by reaction with alkyl or aryl chlorides. Typically the commercially important methylchlorosilanes are prepared by heating silicon intimately mixed with a copper catalyst at 250–280°C. Most simply $2\,RCl + Si \rightarrow R_2SiCl_2$, but the mechanism is complex, and a mixture of products is obtained. A typical product composition is dimethyldichlorosilane (75%) (b.p. 70°C), trimethylchlorosilane (4%) (b.p. 58°C), methyltrichlorosilane (10%) (b.p. 68°C) plus other products. The closely boiling components are difficult to separate completely, even by careful fractional distillation. Although the product composition cannot be readily varied by variation in reaction conditions, higher yields of the more useful dimethyldichlorosilane may be achieved by redistribution. The process can be used also for synthesising phenylchlorosilanes, using chlorobenzene and a silver catalyst, but is not as versatile as the Grignard process. Nevertheless, it is more economic and is the most widely used process commercially.

DISACCHARIDE An oligosaccharide containing two monosaccharide units joined through a glycoside bond the particular carbons involved being indicated in the name, e.g. 1,4'-, by their numbers, i.e. a glycoside in which the aglycone is another monosaccharide. Naturally occurring examples include sucrose, which being 1,2'-linked (by reaction of both hemi-acetal hydroxyls) has lost the reducing power of its parent monomers (non-reducing disaccharide) and lactose, which is 1,4'-linked and is therefore a reducing disaccharide. Typical disaccharides from partial hydrolysis of polysaccharides are cellobiose and maltose derived from cellulose and starch respectively.

DISC ELECTROPHORESIS (Disc-gel electrophoresis) Gel electrophoresis in which the gel is divided into two sections. The protein migrates from a more porous gel into a zone of lower porosity which is also buffered at a different pH. Each individual protein component thus becomes concentrated into a sharp band giving much higher resolution than in simple gel electrophoresis.

DISC-GEL ELECTROPHORESIS Alternative name for disc electrophoresis.

DISCHARGE INCEPTION VOLTAGE The voltage at which internal discharges start in a dielectric as a result of the presence of voids, resulting in treeing and ultimately in dielectric breakdown.

DISCRETE SMALL ANGLE X-RAY SCATTERING X-ray scattering from a solid polymer that is not amorphous by wide angle X-ray scattering. Usually consists of a single maximum corresponding to a Bragg spacing of about 100 Å. Commonly observed in fibrils as the long period diffraction.

DISINCLINATION A disturbance in the perfectly regular molecular chain conformation in a polymer crystal such that a rotation about an axis will bring the disturbed part of the molecule into exact registry with the undisturbed part. If the rotation vector is parallel to the chain, the molecule contains a twist, if perpendicular it contains a bend. Two twists (as in a disinclination dipole) is equivalent to a kink.

DISINCLINATION DIPOLE Two twists, i.e. disinclinations with rotation vectors parallel to the molecular chain, at the same point in a polymer molecule with rotation vectors anti-parallel but of equal magnitude. Equivalent to a kink.

DISLOCATION A disturbance in the linear array of motifs in a crystal, whose influence only persists locally. In a polymer crystal the disturbance of the perfectly regular molecular conformation is such that a translation of one part of the molecule will bring it into exact registry with the second. If the translation vector is parallel to the molecular chain (and equal in pitch, if the conformation is helical) then the dislocation is equivalent to either a missing or to an extra link. If the vector is perpendicular to the chain then the molecule is kinked.

DISLOCATION CLIMB   Alternative name for climb.

DISPERSION POLYMERISATION   A suspension polymerisation using a much larger than usual quantity of suspension agent, thus producing a suspension of very fine particles as a creamy latex, but with greater latex stability than obtained with a conventional emulsion. Polyvinyl acetate and polyvinyl chloride are sometimes produced in dispersion, the former often as a dispersion in an organic solvent.

DISPERSITY   Alternative name for polydispersity.

DISPLACEMENT   The movement of a particle of a body when it is stressed. The displacement of one point in a body relative to adjacent points is the deformation. Deformation, rather than displacement, is the usual measure of strain. If the displacement of particles is uniform in the same direction then a rigid body displacement has occurred. The third type of displacement possible is a rotation.

DISPLACEMENT LENGTH   Alternative name for contour length.

DISPROPORTIONATION   A mode of termination in free radical polymerisation in which a hydrogen atom attached to a carbon adjacent to a growing free radical active centre is transferred to another free radical active centre. This usually occurs in the polymerisation of α-methyl substituted monomers, e.g. methyl methacrylate:

$$\sim\!\!CH_2\dot{C}CH_3 + \dot{C}CH_3CH_2\!\!\sim\ \rightarrow$$
$$\qquad | \qquad\quad |$$
$$\qquad X \qquad\quad X$$

$$\sim\!\!CH_2CHCH_3 + CH_2\!\!=\!\!CXCH_2\!\!\sim$$
$$\qquad\qquad\quad |$$
$$\qquad\qquad\quad X$$

In contrast, unsubstituted growing polymers (having $\sim\!\!CH_2\dot{C}HX$ active centres) usually terminate by combination. Two dead polymer molecules are formed, their degrees of polymerisation being equal to the kinetic chain length. The overall molecular weight distribution is the most probable one, the polydispersity index being equal to 2.

DISSIPATION FACTOR   Alternative name for tangent of the loss angle.

DISSYMMETRY   (1) Dissymmetry is present in a molecule that does not have mirror image symmetry but does have a translational or rotational axis of symmetry. In stereoregular polymers a syndiotactic sequence has translational symmetry, but it cannot be superimposed on its mirror image and is therefore dissymmetric. However, an isotactic sequence can be superimposed on its mirror image and so is not dissymmetric.

(2) In light scattering the difference between the intensity of light scattered by large particles at any angle (θ) to the incident beam in the forward direction (I_θ) and the intensity at the corresponding angle ($\pi-\theta$) in the backward direction ($I_{\pi-\theta}$). Arises from internal interference and is qualitatively expressed as the dissymmetry ratio.

DISSYMMETRY COEFFICIENT   Alternative name for dissymmetry ratio.

DISSYMMETRY METHOD   A method of analysing the data obtained from light scattering measurements on a dilute polymer solution, to obtain a value of the weight average molecular weight, $\bar{M}_w$. Simpler than the widely used Zimm plot method but requires an assumption to be made about the shape of the polymer molecules. The scattered light intensity is measured at 45°, 90° and 135°, and the dissymmetry ratio, z, is obtained at zero concentration by extrapolation of a plot of $1/(1-z)$ versus c, where c is the weight concentration. Assuming a particular shape, often a Gaussian coil, of the polymer molecule, a value of P_{90}, the particle scattering factor at 90°, is obtained from tables and the scattering data are thereby corrected for dissymmetry. Then the simple relationship between R_θ (the reduced scattering intensity) and $\bar{M}_w$ may be used: $Kc/R_\theta = 1/\bar{M}_w + 2Bc/RT$, where B is the second virial coefficient, c is the polymer concentration and K is an instrumental constant. The plot of Kc/R_{90} versus c has an intercept at $c=0$ of $1/\bar{M}_w$.

DISSYMMETRY RATIO   (Dissymmetry coefficient) Symbol z. The quantitative measure of the dissymmetry in light scattering, defined as $z_\theta = I_\theta/I_{\pi-\theta}$, where I_θ is the intensity of light scattered at an angle θ to the incident beam in the forward direction and $I_{\pi-\theta}$ is the intensity at the corresponding angle in the backward direction. Often measured at $\theta = 45°$.

DISTEARYLTHIODIPROPIONATE   (DSTDP)

$$(C_{18}H_{37}OOCCH_2CH_2)_2\!\!-\!\!S \qquad \text{M.p. } 58\text{--}62°C.$$

A preventive antioxidant capable of peroxide decomposition. Useful in polyolefins, especially in synergistic mixtures with a hindered phenol.

DISTORTION BIREFRINGENCE   Birefringence produced on stressing a glassy polymer in which the deformation involves distortion of bond angles and lengths, since appreciable molecular orientation is not possible. Nevertheless, stress-induced birefringence can result. The different mechanisms of production of birefringence are reflected in the stress optical coefficient values above and below T_g, which differ by about a factor of ten.

DISTRENE   Tradename for polystyrene.

DISTRIBUTED CIRCUIT METHOD   A method for the determination of the dielectric properties of a dielectric at frequencies above about 10^8 Hz. In this range the wavelength becomes comparable to the sample dimensions so that lumped circuit methods may not be employed. The sample provides a medium for the propagation of waves in a waveguide and the dielectric properties can be determined from the wavelength and attenuation characteristics of the standing wave using the wave transmission method. The basic electrical properties, capacitance, inductance and resistance, are then said to be distributed.

DISTRIBUTION FUNCTION   Alternative name for molecular weight distribution function.

DISULPHIDE BRIDGE   A structure of the type

$$\{\sim S—S\sim\}$$

being the most frequent type of intermolecular crosslink between polypeptide chains in proteins as cystine units:

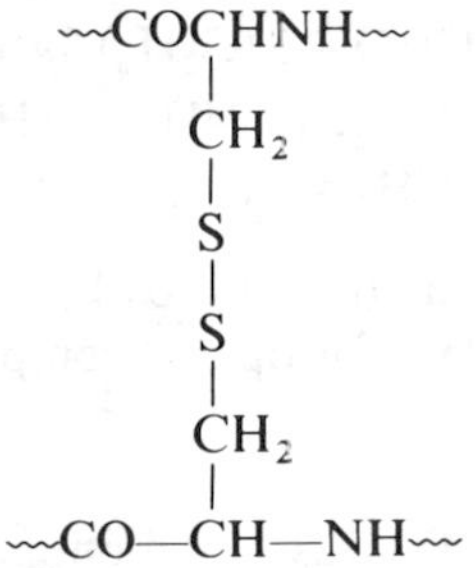

$$\sim COCHNH\sim$$
$$|$$
$$CH_2$$
$$|$$
$$S$$
$$|$$
$$S$$
$$|$$
$$CH_2$$
$$|$$
$$\sim CO—CH—NH\sim$$

Disulphide bridges may be considered to arise from oxidation of two cysteine residues. Intramolecular disulphide bridges may also be present. Such links play an important role in determining and stabilising the native conformation of the tertiary structure of proteins in which they occur. Cleavage of the disulphide bridges is a necessary preliminary to determination of primary structure in order to separate the polypeptide chains. Cleavage may be brought about either by oxidation to two cysteic acid residues, e.g. with performic acid, or by reduction with, e.g., β-mercaptoethanol, dithioerithrytol

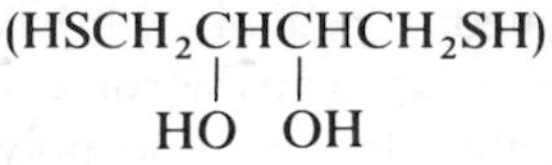

$$(HSCH_2CHCHCH_2SH)$$
$$|\quad|$$
$$HO\quad OH$$

dithiothreitol

$$(HSCH_2CHCH_2SH)$$
$$|$$
$$OH$$

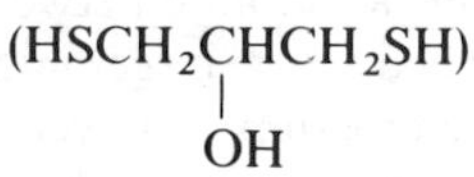

or tritolyl phosphine, to cysteine units. The reduced units may then be protected from becoming re-oxidised, by forming new disulphide bridges either by alkylation with iodoacetic acid or by reaction with acrylonitrile.

DISYNDIOTACTIC POLYMER   A ditactic polymer in which the configurations of each of the types of stereochemical centre alternate down the chain, i.e. both are syndiotactic. Both *erythro* and *threo* types may exist, but are identical except for a difference in the end group. Therefore in a high polymer they are essentially the same.

The structure may be represented by the Fischer projection formula:

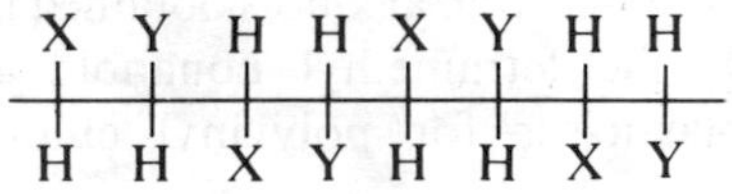

$$\begin{array}{ccccccccc} X & Y & H & H & X & Y & H & H \\ | & | & | & | & | & | & | & | \\ H & H & X & Y & H & H & X & Y \end{array}$$

or by

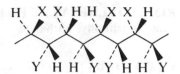

DITACTIC POLYMER   A polymer which possesses two stereoisomeric centres (labelled C* below) in each repeat unit. Ditactic polymers are mostly formed by polymerisation of 1,2-disubstituted ethylenes:

$$n\,CHX=CHY \rightarrow \ \{\!C^*HX—C^*HY\!\}_n$$

Theoretically many different configurations may exist since two configurations are possible for each repeat unit. However, in practice, the number of configurational isomers is limited due to the retention of configuration about the C*—C* bond during polymerisation. The polymers are thus either diiso- or disyndiotactic. The diisotactic polymer may be either erythrodiisotactic or threodiisotactic. Polymerisation of a *cis*-1,2-disubstituted ethylene gives a diisotactic polymer and polymerisation of a *trans* isomer gives a disyndiotactic polymer.

***N,N'*-DITHIOBISMORPHOLINE**   Alternative name for bismorpholinedisulphide.

5,5'-DITHIO-BIS-(2-NITROBENZOIC ACID) (DTNB) Chemical name of Ellman's reagent.

DITHIOCARBAMATE   A compound containing the structure

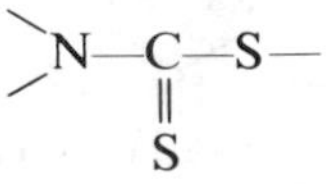

$$>\!\!N—C—S—$$
$$\|$$
$$S$$

Useful as ultra-accelerators in the sulphur vulcanisation of rubbers, especially as the metal salts such as zinc dithiocarbamate. Very fast, except at low temperatures, when they are useful in the vulcanisation of rubber latices.

DI-*o*-TOLYLGUANIDINE   (DOTG)

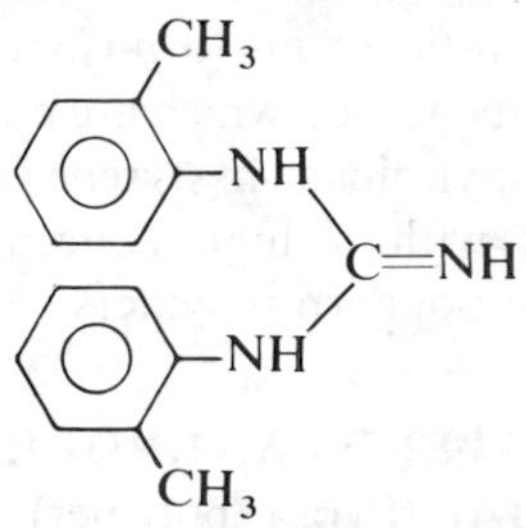

A medium speed accelerator for the sulphur vulcanisation of rubbers. Also used as a 'plasticiser' in polychloroprene.

DITRIDECYL PHTHALATE (DTDP) B.p. 285°C/3·5 mm. A mixed phthalate ester of the highly branched isomers of the 13-carbon alcohols obtained from the Oxo process, e.g. the tetramethyl nonanols. A very low volatility plasticiser for polyvinyl chloride and its copolymers.

DI-(3,5,5-TRIMETHYLHEXYL) PHTHALATE Alternative name for dinonyl phthalate.

DIUNDECYL PHTHALATE (DUP)

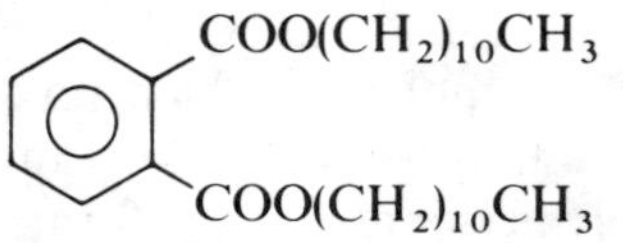

$$COO(CH_2)_{10}CH_3$$
$$COO(CH_2)_{10}CH_3 \quad \text{B.p. } 262°C/10 \text{ mm.}$$

A low volatility plasticiser for polyvinyl chloride and its copolymers.

DIVEMA Abbreviation for divinylether–maleic anhydride copolymer.

DIVIDED DIFFERENCE METHOD A method of handling colligative property data when the third virial coefficient is not negligible. Thus if (Π/c_1) and (Π/c_2) are the reduced osmotic pressures of two polymer solutions of concentrations c_1 and c_2, then the second and third virial coefficients, B and C, are given by

$$\Pi/c_2 - \Pi/c_1 = B(c_2 - c_1) + C(c_2^2 - c_1^2)$$

and a plot of the LHS/$(c_2 - c_1)$ versus $(c_2 + c_1)$ has an intercept of B and a slope of C.

DIVINYLBENZENE (DVB)

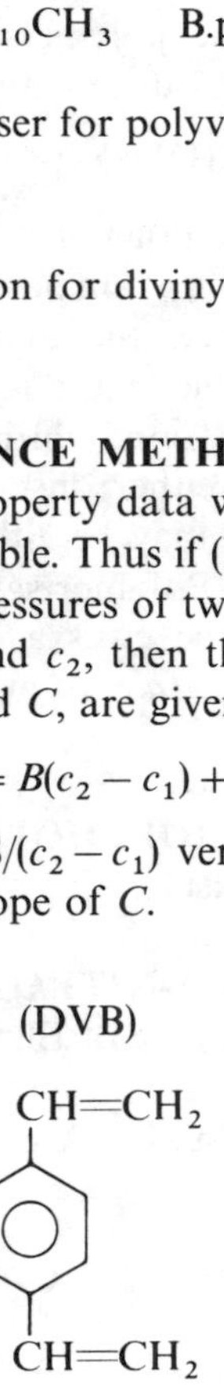

$$CH=CH_2$$
$$CH=CH_2$$

A monomer useful for the production of crosslinked polymers, especially when used as a comonomer in styrene copolymers, e.g. in ion-exchange resins and in polystyrene with a higher softening point. It is produced by the dehydrogenation of diethylbenzene in a manner similar to the production of styrene from ethylbenzene. Commercial products contain 25% or 55% of divinylbenzene (with a molar ratio of *m*- to *p*-isomers of 2:1) together with ethylvinylbenzenes, which are not desirable since they behave like vinyltoluene in styrene polymerisations. Polymerisations with a high concentration of divinylbenzene produce popcorn polymers.

DIVINYLETHER–MALEIC ANHYDRIDE CO-POLYMER (DIVEMA) (Pyran polymer) An alternating copolymer produced from divinylether and maleic anhydride by free radical polymerisation and possibly containing the units of type **I**, formed by a series of cyclisation steps:

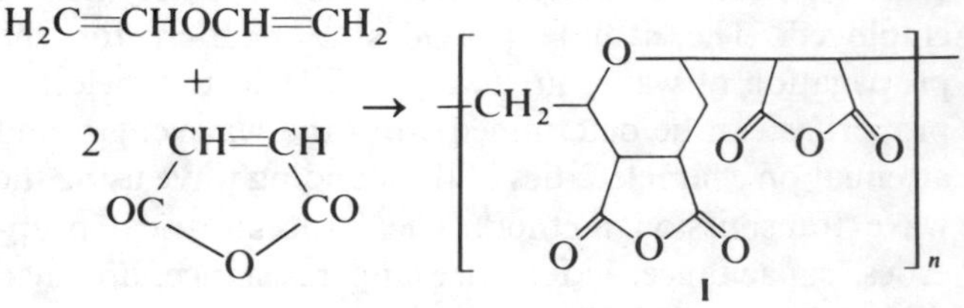

$$H_2C=CHOCH=CH_2$$
$$+$$
$$2 \quad CH=CH$$
$$OC \qquad CO$$
$$O$$

The hydrolysed, neutral polymer shows a wide range of biological activity of interest for a variety of potential medical applications.

DIVINYL KETONE

$$(CH_2=CH)_2CO \quad \text{B.p. } 44°C/75 \text{ mm.}$$

Prepared by the Mannich reaction of acetone or methyl vinyl ketone, with formaldehyde and diethylamine hydrochloride:

$$(CH_3)_2CO + 2 HCHO + 2(C_2H_5)_2NH \cdot HCl \rightarrow$$
$$OC(CH_2CH_2NH_2)_2 \cdot 2 HCl \rightarrow$$
$$(CH_2=CH)_2CO + 2(C_2H_5)_2NH \cdot HCl$$

It is occasionally used as a crosslinking agent, e.g. with methyl methacrylate for optical components and in ion-exchange resins.

DIVINYL MONOMER A monomer containing two vinyl groups. Examples are divinylbenzene and the dimethacrylates such as ethylene glycol dimethacrylate. These monomers are useful when used in small amounts as comonomers to produce crosslinked polymers which are insoluble or of increased hardness and softening point.

DIVINYL POLYMERISATION Polymerisation of a monomer which contains two vinyl groups. When the two double bonds are conjugated (as in 1,3-dienes) the polymerisation is called diene polymerisation. The term is usually restricted to monomers in which the double bonds are not conjugated, e.g. diallyl phthalate and divinylbenzene, and both are capable of entering into polymerisation separately. In these cases polymerisation at the second double bond forms a crosslink, so that at high enough conversions an infinite network will form as the gel point is reached. For large initial molecules this happens when one crosslink, on average, has formed per initial (primary) chain, i.e. the crosslinking index (γ) equals unity. Beyond this critical point for gelation, both sol and gel will be present.

DI-*p*-XYLYLENE Alternative name for 2,2'-paracyclophane.

DLTDP Abbreviation for dilauroylthiodipropionate.

DMA Abbreviation for dynamic mechanical analysis.

DMAC Abbreviation for dimethylacetamide.

DMCHA   Abbreviation for dimethylcyclohexylamine.

DMEA   Abbreviation for bis-(β-N,N-dimethylamino-ethyl) ether, which is alternatively known as dimethyl-ethanolamine.

DMEU   Abbreviation for dimethylolethyleneurea.

DMF   Abbreviation for dimethylformamide.

DMP   Abbreviation for dimethyl phthalate.

DMS   Abbreviation for dynamic mechanical spectroscopy.

DMSO   Abbreviation for dimethylsulphoxide.

DNFB   Abbreviation for 2,4-dinitro-1-fluorobenzene.

DNO   Abbreviation for dinitrosopentamethylene-tetramine.

DNODA   Abbreviation for (di)-n-octyl-n-decyl adipate.

DNODP   Abbreviation for di-n-octyl-n-decyl phthalate.

DNOP   Abbreviation for di-n-octyl phthalate.

DNP   Abbreviation for dinonyl phthalate.

DNP DERIVATIVE   Abbreviation for dinitrophenyl derivative.

DNPT   Abbreviation for dinitrosopentamethylene-tetramine.

DNS DERIVATIVE   The derivative formed by reaction of dansyl chloride with the free amino group of an amino acid, peptide or protein. Particularly useful in the determination of N-terminal amino acids.

DOA   Abbreviation for dioctyl adipate.

1-DODECANE THIOL   Alternative name for dodecyl mercaptan.

DODECANDIOIC ACID   Alternative name for 1,10-decanedicarboxylic acid.

DODECANOIC ACID   Alternative name for lauric acid.

DODECANOLACTAM   Alternative name for lauryl-lactam.

DODECENYL SUCCINIC ANHYDRIDE   (DDSA)

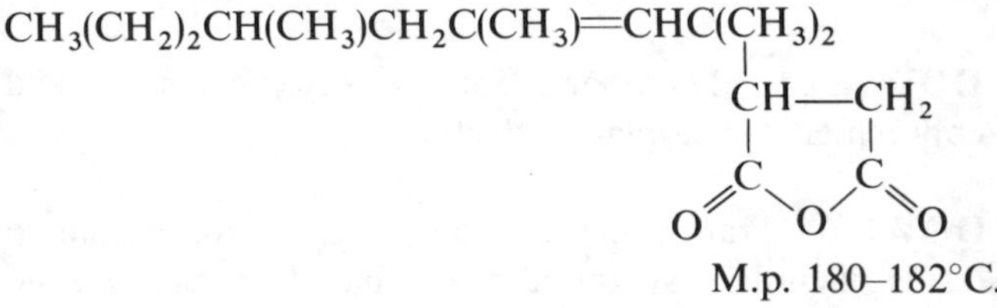

$$CH_3(CH_2)_2CH(CH_3)CH_2C(CH_3)\!\!=\!\!CHC(CH_3)_2$$

$$\begin{array}{c} CH\!\!-\!\!CH_2 \\ | \quad\quad | \\ C \quad\quad C \\ \diagup\;\diagdown\;\diagdown \\ O \quad O \quad O \end{array}$$

M.p. 180–182°C.

A low viscosity liquid obtained by reaction of maleic anhydride with tetrapropylene. Useful as a flexibilising curing agent for epoxy resins.

DODECYL MERCAPTAN   (Lauryl mercaptan) (1-Dodecane thiol)

$$CH_3(CH_2)_{11}SH$$
B.p. 165–169°C/39 mm, 111–112°C/3 mm.

A very effective and widely used chain transfer agent for free radical polymerisation. Often used as a modifier in the emulsion polymerisation of diene monomers in the production of synthetic rubbers. The value of its chain transfer constant can be as high as several million.

DOIP   Abbreviation for dioctylisophthalate.

DOLAN   Tradename for a polyacrylonitrile fibre.

DOMAIN   (Micelle) (Microphase) A region in a phase separated block copolymer which consists of aggregates of blocks of only one type, which are incompatible with blocks of the second type. Frequently, as in thermoplastic elastomers based on block copolymers, such as SBS, the domains contain blocks of glassy or crystalline material (the hard segments) as the dispersed phase in a rubber matrix consisting of blocks of the second type of unit. The domains are usually much smaller in size than the phase separated regions in other multiphase polymer systems, such as polymer blends. For example, in SBS block copolymer domain size is about 300 nm and in multiblock copolymers it may be down to 2–5 nm, as in segmented polyurethanes. The domains may exhibit different morphologies, varying from spheres to cylinders or may exist as lamellae. Domain morphology is frequently characterised by electron microscopy, often after staining any diene rubber phase by osmium staining. Domain size and morphology are dependent on block length and sample preparation conditions.

DOMINANT LAMELLA   A lamellar crystal which forms during the early stages of melt crystallisation and contributes to the skeleton of the resultant spherulite. Subsequently the cavities between the dominant lamellae are filled with subsidiary lamellae.

DONNAN DIALYSIS   Alternative name for ion-exchange dialysis.

DONNAN EFFECT   The unequal distribution of ions on either side of a membrane separating a polyelectrolyte solution from a simple electrolyte solution, resulting from the establishment of the Donnan equilibrium.

DONNAN EQUILIBRIUM   The equilibrium distribution of ions when a polyelectrolyte solution is separated from a small ion electrolyte solution by a semipermeable membrane. If the polyelectrolyte is separated from pure water then, although permeable to the small counter-ions, no permeation will take place due to necessary electrostatic charge separation. However, if an electrolyte is added, both its anions and cations may diffuse together and the presence of the macroions will cause an unequal distribution of small ions (the Donnan effect) on both sides of the membrane, there being more counter-ions and fewer byions on the polyelectrolyte side. This arises from the requirement of electroneutrality on both sides of the membrane and of the same chemical potential of the solute species in equilibrium on both sides. This diffusion contributes to the osmotic pressure of a polyelectrolyte solution. The effect is measured by increasing amounts of added electrolyte.

DOOLITTLE EQUATION An empirical equation, but derivable from the van der Waals equation, relating the viscosity of a fluid (η) to its free volume (V_f), where $V_f = V - V_0$, where V is the volume of the liquid and V_0 its volume on closest packing of its molecules at 0 K. It is $\eta = A \exp(BV_0/V_f)$, where A and B are constants. An analogy with this equation is the basis of the derivation of the Williams–Landel–Ferry equation and the constant B can be related to a_T of this equation.

DOP Abbreviation for dioctyl phthalate.

DOPE The solution of secondary cellulose acetate produced by hydrolysis of cellulose triacetate by heating with 95% aqueous acetic acid, or a solution of secondary cellulose acetate in acetone (25–35 wt%) used for spinning cellulose acetate rayon fibres.

DORLASTAN Tradename for a cast polyurethane elastomer, similar to Vulkollan.

DORYL Tradename for poly-(methylenediphenylene oxide), a phenol–aralkyl polymer.

DOS Abbreviation for dioctyl sebacate.

DOTG Abbreviation for di-o-tolylguanidine.

DOTP Abbreviation for dioctyl terephthalate.

DOUBLE ORIENTATION Orientation in which the specimen is both uniaxially and uniplanarly oriented. Obtained by rolling a fibre sample in a mill in the fibre direction.

DOUBLE REFRACTION Alternative name for birefringence.

DOUBLE RESONANCE TECHNIQUE In nuclear magnetic resonance spectroscopy, a technique in which one radio frequency source irradiates the sample at the resonant frequency of one nucleus, whilst a second source is used for scanning the spectrum. The first frequency causes spin–spin decoupling to occur.

DOUBLE STRAND POLYMER (Classical ladder polymer) The commonest type of ladder polymer, consisting of a series of rings each joined to the other by two covalent bonds from different, but usually adjacent, ring atoms, thus forming a series of fused rings, as in **I**:

Alternatively **I** may be considered to consist of two linear molecules (the two strands) regularly joined by crosslinks at each repeat unit.

Often synthesised by cyclisation of a single strand polymer through pendant functional groups (as in the cyclisation of polyacrylonitrile), by Diels–Alder polymerisation, by 1,3-dipolar addition or by reaction between tetrafunctional monomers, e.g. bis-(o-ring substituted) monomers. Examples include polyphenylsilsesquioxane, poly-(bis-benzimidazobenzophenanthroline), ladder polyquinoxalines and ladder pyrrone polymers.

The polymers have potentially a higher thermal stability than their single strand counterparts, since bond scission must occur in both strands at the same repeat unit, i.e. at points a and b in **I**, before polymer molecular weight is reduced. This is statistically unlikely. However, many polymers which are nominally ladders are incompletely cyclised and are in fact only partial ladders.

DOUBLE SUFFIX NOTATION A notation used in elasticity for denoting stress, strain, stress and strain components, and the elastic constants referred to cartesian axes, i.e. three mutually perpendicular axes lying in the x, y and z directions. Thus, for example, stresses or stress components acting in directions normal to the planes of a hypothetical small cube of material, with edges parallel to the coordinate axes, x, y and z, are designated σ_{xx}, σ_{yy} and σ_{zz} respectively. In a similar way strains and strain components may be represented as ε_{xx}, ε_{yy} and ε_{zz}. These are the normal components. Stresses or stress components acting within a face of such a cube, which contains the planes referred to above, are designated σ_{xy}, σ_{yz}, σ_{zx}, σ_{zy}, etc., where the first subscript refers to the plane defined as for the normal plane above and the second subscript refers to the axis in the direction in which the stress acts. These stresses are the shear stresses and may be alternatively denoted by τ_{xy}, τ_{xz}, τ_{yz}, etc. A similar notation applies to shear strains: ε_{xy}, ε_{xz}, ε_{yz}, etc. (also written γ_{xy}, γ_{xz}, γ_{yz}, etc.). An alternative way of denoting stresses and strains, or their components, is to replace x, y and z by 1, 2 and 3 respectively. Thus, for example, σ_{xx}, σ_{xy}, σ_{zx}, etc., become σ_{11}, σ_{12}, σ_{31}, etc. Sometimes the double suffix notation is abbreviated to the contracted notation.

DOW EAA Tradename for an ethylene–acrylic acid copolymer thermoplastic elastomer.

DOWEX Tradename for a range of ion-exchange resins, including synthetic resins based on polystyrene crosslinked with divinylbenzene. Thus a sulphonated polystyrene provides a cationic resin, whilst a chloromethylated polystyrene, which has been converted to a quaternary ammonium compound by reaction with a tertiary amine, provides an anionic exchange resin.

DOWLEX Tradename for a linear low density polyethylene containing octene-1 as the comonomer.

DOZ Abbreviation for dioctyl azelate.

DP Abbreviation for degree of polymerisation.

DPG Abbreviation for diphenylguanidine.

DPP Abbreviation for diphenyl phthalate.

DPPD Abbreviation for *N,N'*-diphenyl-*p*-phenylene-diamine.

DPPH Abbreviation for diphenylpicrylhydrazyl.

DRAG FLOW (Couette flow) Flow brought about by movement of the boundary, or boundaries, containing a fluid, which drags the fluid along. No pressure is imposed on the system, as occurs in pressure flow, but a pressure gradient may develop. The simplest type is plane couette flow, other types being axial, annular and circular couette flows. Drag flow is important in many polymer melt flow processes, e.g. extrusion (where the screw is the moving boundary) and wire coating (where the wire is the moving boundary).

DRALON Tradename for a polyacrylonitrile fibre.

DRAW DOWN The stretching of an extruded polymer melt often in the form of a film, usually in the direction of flow, causing it to be drawn and the extrudate thickness to become reduced. This is only possible for melts which can sustain a tensile stress, i.e. that are elastic, but nevertheless viscous behaviour must predominate for the necessary flow to occur. Drawn down ability increases with melt temperature and decreases with molecular weight and melt elasticity as measured, e.g., by die swell.

DRAW RATIO The ratio of the original length of a tensile specimen (especially fibres and films) to its deformed length after being subject to a tensile force. Usually refers to deformations such that the specimen has formed a neck and started to become drawn. Since cold drawing takes place at constant volume the draw ratio also equals the ratio of the original cross-sectional area to that of the drawn specimen.

DRAW RESONANCE A periodic variation in the cross-sectional area of an extrudate that is being drawn, which occurs above a critical draw ratio. Most notably it occurs in the spinning by extrusion of fibres (as a variation in fibre diameter) and in the film casting of polymer films. It may amount, in the case of films, to a ratio of maximum to minimum film thickness of up to about two and, in the case of fibres, to a ratio of fibre diameters of up to about four. It occurs particularly with tension thinning polymers, such as polypropylene.

DRESSING Alternative name for size.

DRIER A catalyst for the air drying of alkyd resins. Driers are the metal salts of commercial fatty acids, such as the octoates, naphthenates and linoleates, which are soluble in the organic medium used. The salts of cobalt, manganese and lead are primary driers and operate by the catalysis of hydroperoxide decomposition, or even by direct reaction with C—H bonds to produce free radicals:

$$R—H + Co^{3+} \rightarrow R\cdot + Co^{2+} + H^+$$

Salts of other metals, such as calcium, barium and zinc are promotor driers. They do not themselves accelerate drying but activate primary driers.

DRIFT The change in composition of the copolymer formed during copolymerisation. Drift always occurs, except during azeotropic copolymerisation, due to one of the monomers being more reactive and hence being consumed more rapidly. In order to obtain a copolymer of constant composition, either conversion must be limited to low values or the more reactive monomer must be continuously added to the feed to maintain constant feed composition.

DRUDE EQUATION An equation describing optical rotatory dispersion. A one-term equation often fits data for the visible and near ultraviolet regions and has the form $[M]_\lambda = a_c\lambda_c^2/(\lambda - \lambda_c^2)$, where $[M]_\lambda$ is the molar rotation at the wavelength λ and a_c and λ_c are adjustable constants. However, for α-helical polypeptides a two-term equation must often be used—frequently the Moffitt equation.

DRYING OIL An oil, such as linseed or tung oil, which air dries rapidly due to the high degree of unsaturation in the fatty acid residues of its triglycerides. Generally a film of oil will become dry in 2–6 days.

DRY SPINNING A process for the formation of a fibre in which a solution of a polymer is extruded through a multi-orifice die (a spinneret) into an atmosphere, e.g. of warm air, in which the solvent evaporates leaving a solid fibre. The technique is useful for the production of polyacrylonitrile, cellulose acetate and triacetate and polyvinyl chloride fibres.

DS Abbreviation for degree of substitution.

DSC Abbreviation for differential scanning calorimetry.

DSTDP Abbreviation for distearylthiodipropionate.

DTA Abbreviation for differential thermal analysis.

DTDP   Abbreviation for ditridecyl phthalate.

DTNB   Abbreviation for 5,5'-dithio-bis-(2-nitrobenzoic acid).

DUCTILE FRACTURE   (Tough fracture) Fracture in which significant plastic flow occurs before fracture. Strain at fracture is more than a few per cent, unlike brittle fracture, and may be several hundred per cent. However, a sharp distinction cannot be made between ductile and brittle fracture. Materials with low T_g values are more likely to suffer ductile fracture, but the mode of fracture also depends on the fracture conditions. Thus a ductile fracture is favoured at high temperature, at a low rate of load application and as the type of loading becomes uniaxial (with a large shear component leading to yielding) rather than triaxial.

DUHAMEL INTEGRAL   The type of integral found in the mathematical formulation of the Boltzmann Superposition Principle.

DUMMY INDEX   A repeated index sometimes introduced in the index notation of stresses and strains, to enable the summation convention to be used.

DUOLITE   Tradename for a range of ion-exchange resins, including synthetic resins based on crosslinked polystyrene, similar to Dowex resins.

DUP   Abbreviation for diundecyl phthalate.

DUPRENE   Early tradename for polychloroprene.

DURACON   Tradename for polyoxymethylene copolymer.

DURAFIL   Tradename for a high tenacity rayon staple fibre.

DURALOY   Tradename for a polyoxymethylene or polybutylene terephthalate/rubber toughened polymer blend.

DURANIT   Tradename for emulsion styrene–butadiene rubber.

DURAPHEN   Tradename for phenol–formaldehyde polymer.

DURAPRENE   Tradename for chlorinated rubber.

DURATEM   Tradename for a polyimide–polysulphone copolymer with a glass transition temperature of 273°C.

DUREL   Tradename for a polyarylate.

DURETHAN 740   Tradename for polyhexamethylene isophthalamide, which is a transparent polyamide.

DURETHAN BK   Tradename for nylon 6.

DURETHAN U   Tradename for polyurethane plastic moulding materials, often based on the polyurethane obtained by reaction of 1,6-hexamethylene diisocyanate with 1,4-butanediol. The earlier Igamid U polymer is similar. It is similar in properties to nylon 66, but with a lower T_m value of 183°C.

DURETTE   Tradename for a chlorinated poly-(m-phenylene isophthalamide).

DUREZ   Tradename for phenol–formaldehyde polymer.

DURIT   Tradename for a vinylidene chloride copolymer fibre.

DURITE   Tradename for phenol–formaldehyde polymer.

DUROMETER   A small, hand-held instrument for the measurement of the hardness of rubbers. Common types measure either the Shore hardness or International Rubber Hardness Degrees.

DURRANS MELTING POINT   A measure of the softening point of an uncured epoxy resin. A known weight of mercury is placed on the resin in a test tube and heated. The melting temperature is the temperature at which the molten resin rises to above the mercury.

DUTHANE   Tradename for a cast polyurethane elastomer similar to Vulkollan.

DUTRAL   Tradename for ethylene–propylene rubber.

DUTRALENE   Tradename for a thermoplastic polyolefin rubber.

DUTRAL TP   Tradename for a thermoplastic polyolefin rubber.

DVB   Abbreviation for divinylbenzene.

DYAD   A pair of adjacent repeat units in a polymer molecule. Some features of polymer microstructure may be analysed in terms of the dyads. Thus stereoregularity (configurational isomerism) can be discussed in terms of meso and racemic dyads. In copolymers, with monomer units M_1 and M_2, three types of dyad are possible, $—M_1M_1—$, $—M_1M_2—$ and $—M_2M_2—$. The copolymerisation reactivity ratios may be calculated if the dyad frequencies are known, possibly by use of nuclear magnetic resonance spectroscopic observation of the methylene resonances in vinyl polymers.

DYADIC NYLON   (AA–BB polyamide) (AA–BB nylon) A nylon synthesised by polyamidation of a dicarboxylic acid, HOOCR'COOH, or one of its derivatives, e.g. the diacid chloride, with a diamine, H_2NRNH_2, and therefore having the structure

$$-[NHRNHCOR'CO]_n$$

Common examples are nylon 66 and nylon 610.

DYFLOR   Tradename for polyvinylidene fluoride.

DYLAN   Tradename for low density polyethylene.

DYLARC   Tradename for styrene–maleic anhydride copolymer and ABS-type materials based on this copolymer.

DYLEL   Tradename for acrylonitrile–butadiene–styrene copolymer.

DYLENE   Tradename for polystyrene and high impact polystyrene.

DYLITE   Tradename for expanded polystyrene.

DYNAGEN XP-139   Tradename for propylene oxide rubber.

DYNAMIC BIREFRINGENCE   The time-dependent birefringence produced by an oscillating stress at constant frequency. There is usually a decrease in strain optical coefficient with increase in frequency, as orienting mechanisms with longer relaxation times have no time to occur.

DYNAMIC BULK COMPLIANCE   Alternative name for bulk storage compliance.

DYNAMIC BULK MODULUS   Alternative name for bulk storage modulus.

DYNAMIC EQUATION   Alternative name for momentum equation.

DYNAMIC FATIGUE   Fatigue which occurs under a load which is varying, usually periodically and often sinusoidally. The lower the loading stress, the more loading cycles that are needed for failure (the fatigue life). As the applied stress, or rather its amplitude when it is sinusoidally varying, is lowered further, a value is reached (the fatigue limit) below which the material does not fail even after an infinite number of cycles. The commonest mode of deformation is flexure and the input function, although usually sinusoidal, may be saw-tooth or square wave. A further variable is the frequency, whilst the excitation may be through the stress or strain. Hence there is great complexity in the modes of deformation that a sample may experience. As the temperature is increased, the time to failure decreases, and under the same conditions of temperature and frequency, when the polymer shows high loss of mechanical energy (high $\tan \delta$), high temperature rises in the material can occur. If these are significantly large, e.g. to above T_g, the material

becomes excessively permanently deformed and thermal failure is said to have occurred. However, failure usually results from the appearance of cracks, which may be initially present as microscopic flaws, which grow under repeated stressing until fracture of the whole specimen occurs.

DYNAMIC FLOW   A flow in which the external influence on the fluid causing the flow varies with time. If the time variation is carefully programmed, e.g. if it varies sinusoidally with known amplitude and frequency (oscillatory flow) and the response of the fluid is monitored, then, in principle, the flow characteristics of viscosity and normal stress coefficients may be determined. The particular advantage of dynamic measurements over steady flow conditions is that effects may be observed over a wider time scale, especially to include shorter times. Thus the applicability of any proposed constitutive equations may be tested over an extended range of flow conditions.

DYNAMIC MECHANICAL ANALYSIS (DMA)   Alternative name for dynamic mechanical spectroscopy.

DYNAMIC MECHANICAL BEHAVIOUR   The stress–strain behaviour of a material when subject to an applied sinusoidally varying stress or strain. For a perfectly elastic material the strain response is immediate and the stress and strain are in phase. For a viscous fluid, stress and strain are 90° out of phase. Thus for polymers which usually exhibit viscoelasticity, the strain response lags behind the stress by some phase angle (δ)—the loss angle—between 0° and 90°. The behaviour can be represented by a complex modulus (or compliance) consisting of the in-phase component (the storage modulus or compliance) and a 90° out-of-phase component, characterised by the loss modulus or compliance. The energy loss due to the viscous flow is given by the ratio of these moduli, which is the loss tangent (or $\tan \delta$). Values of the dynamic moduli and $\tan \delta$ are both frequency and temperature dependent.

Maximum loss, i.e. maximum damping, occurs in the transition region. Such changes are related to particular molecular motions in the polymer. Measurement of the dynamical mechanical behaviour, e.g. by torsion pendulum, vibrating reed, forced vibration non-resonance or elastic wave propagation methods, can give data over a very wide frequency range of 10^{-3}–10^6 Hz and frequently provides the most sensitive method of detecting transitions and associated molecular motions.

DYNAMIC MECHANICAL RELAXATION   A relaxation in which a specimen is subject to a periodic, usually sinusoidal, variation of the applied stress (or strain) and the response (as strain or stress respectively) is monitored either as a function of the frequency (at constant temperature) or as a function of temperature (at a constant frequency). From the data, the frequency (or temperature) dependence of the moduli, especially the shear moduli, may be obtained. The results are often displayed as the dynamic mechanical spectrum, also sometimes called the anelastic spectrum, and provide some of the clearest evidence for the existence of multiple transitions in both crystalline and amorphous polymers. Techniques used include the torsion pendulum, vibrating reed and sonic modulus methods. The frequency range involved is about 10^{-1}–10 Hz for the torsion pendulum and 10^2–10^4 Hz for forced vibration (notably the vibrating reed), although step-function excitation followed by stress relaxation or creep measurements can extend measurements to lower frequencies. The sonic modulus method is used at 10^3–10^7 Hz.

DYNAMIC MECHANICAL SPECTROSCOPY
(DMS) (Dynamic mechanical analysis) The determination of dynamic mechanical behaviour over a range of frequency or temperature. Most conveniently a variation of temperature is used at constant frequency. Temperature and frequency effects may be interrelated by the time–temperature superposition principle. Results may be expressed as a spectrum (the dynamic mechanical spectrum) of the storage and loss moduli or of tan δ against temperature or frequency. Often the plots are of log modulus against temperature or log frequency for convenience.

DYNAMIC MECHANICAL SPECTRUM (Anelastic spectrum) (Relaxation spectrum) A plot of the storage or loss moduli or of tan δ against temperature or frequency for a viscoelastic material. For convenience, given the wide range of values involved, the plots may be of log modulus or log frequency. The output of a dynamic mechanical spectroscopy experiment.

DYNAMIC MODULUS Alternative name for complex modulus.

DYNAMIC OSMOMETRY Membrane osmometry which is performed with a membrane which is permeable to at least part of the solute (the lower molecular weight species of a polymer sample). Thus the observed osmotic pressure is time dependent. The technique attempts to relate the variation of osmotic pressure with time to reach equilibrium. In osmodialysis, the study of the time dependency can give information on the molecular weight distribution.

DYNAMIC RHEOMETER (Oscillating rheometer) A rheometer in which the applied shear or tensile stress (or applied deformation rate) is varied periodically with time, usually sinusoidally. The rheometer may therefore be used to determine the dynamic and complex viscosities of fluids. In general, the sample is deformed by some oscillatory driver, often electromechanically, and the amplitude is measured by a strain transducer, such as a linear variable differential transformer or an optical device. The stress is measured by the small deformation of a relatively rigid spring or torsion bar, to which is attached a stress transducer. Such instruments can operate over a very wide frequency range, and therefore a correspondingly wide shear rate range, but only small deformations are possible. In addition to viscosity, elastic modulus may also be determined.

DYNAMIC SHEAR COMPLIANCE Alternative name for shear storage compliance.

DYNAMIC SHEAR MODULUS Alternative name for shear storage modulus.

DYNAMIC TENSILE COMPLIANCE Alternative name for tensile storage compliance.

DYNAMIC TENSILE MODULUS Alternative name for tensile storage modulus.

DYNAMIC VISCOSITY (1) Alternative name for coefficient of viscosity, used to distinguish it from the kinematic viscosity.
(2) Symbol η'. The real component of the complex viscosity. The component of the stress, in oscillatory flow, which is in phase with the rate of strain and 90° out of phase with the strain, divided by the rate of strain. It is given by $\eta' = \sigma_m \sin\delta/\omega\gamma_m$, where σ_m and γ_m are the maximum stress and strain variables respectively, ω is the angular velocity and δ is the phase angle. It is also related to the shear loss modulus (G'') by $G'' = \omega\eta'$ and is related to the steady-state shear viscosity, being very close to it in value at low frequencies. It represents the viscous response of a fluid to an oscillatory stress.

DYNAMICAL AMPLIFICATION FACTOR (Q-factor) Symbol Q. The reciprocal of the ratio f/f_r, where f is the half-power width of the resonance peak in a resonance forced vibration experiment and f_r is the resonance frequency. Thus Q is also the reciprocal of tan δ, the tangent of the loss angle.

DYNEL Tradename for a modacrylic fibre consisting of a copolymer of 40% acrylonitrile and 60% vinyl chloride repeating units.

DYNYL Tradename for a polyether block amide.

DYPHENE Tradename for phenol–formaldehyde polymer.

E

E Symbol for glutamic acid.

EAA Abreviation for ethylene–acrylic acid copolymer.

EASTMAN 910 Tradename for a cyanoacrylate adhesive.

EBONITE (Hard rubber) (Vulcanite) The product obtained by heating a highly unsaturated diene rubber such as natural rubber, synthetic polyisoprene, butadiene rubber, styrene–butadiene rubber or nitrile rubber, with a large quantity of sulphur. Typically it contains about one sulphur atom for each 4–8 chain carbon atoms, mostly combined as

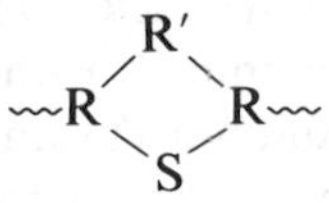

structures where R is a chain segment and R' is a short, e.g. two carbon, segment. Thus in natural rubber ebonite the structure

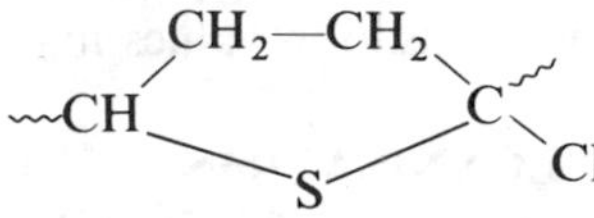

predominates. Some intermolecular units of the type

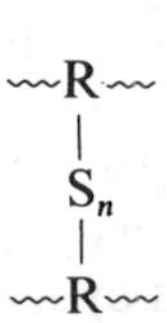

where n is mostly two are also present. Ebonite is a hard black substance, insoluble but swollen by aromatic hydrocarbons and other organic liquids. It has a high resistance to corrosive chemicals and good electrical properties. Above its T_g value, which varies from 35 to 90°C for sulphur contents of 25–50%, it softens and may be shaped, i.e. it is thermoplastic. Heat resistance is often expressed as the yield temperature, i.e. the temperature at which deformation begins to increase rapidly, which is close to T_g. This increases with sulphur content and time of vulcanisation. It is higher for most synthetic ebonites than for natural rubber ebonite. In use, the rubber and sulphur are often compounded with fillers and/or softeners and vulcanisation is performed at about 150°C for several hours. Useful as chemical resistant coatings, pipes, fittings, battery boxes and other mouldings.

EBULLIOMETER An apparatus for the determination of the boiling point elevation of a solvent due to the presence of a solute (ebulliometry). Occasionally used with polymer solutions for the determination of number average molecular weight.

EBULLIOMETRY (Ebullioscopy) A technique for the determination of the molecular weight of a solute, by measurement of its ability to raise the boiling point of the solvent (one of the colligative properties). Occasionally used for the determination of the number average molecular weight ($\bar{M}_n$) of a polymer solute, its use being restricted to polymers of $\bar{M}_n < 20\,000$. Although a typical boiling point elevation (ΔT) is only about 0·001°C, such temperature differences are easily measured by the use of a thermistor. However, the method is limited by the problems of superheating and foaming of the boiling solution.

For a dilute ideal solution: $\Delta T/c = RT^2 M_1/\Delta H \bar{M}_n = K/\bar{M}_n$, where c is the solution concentration, M_1 is the solvent molecular weight, ΔH is its heat of evaporation, R is the universal gas constant, T is the pure solvent boiling point (in kelvin) and K is the ebullioscopic constant.

EBULLIOSCOPIC CONSTANT A constant for a particular solvent which relates its boiling point elevation (ΔT) due to the presence of dissolved solute, to the molecular weight of the solute (M_2) and therefore used in ebulliometry. For polymer solutions, M_2 is the number average molecular weight. Equal to $RT^2 M_1/\Delta H$, where R is the universal gas constant, T is the pure solvent boiling point, M_1 is its molecular weight and ΔH is its molar heat of evaporation.

EBULLIOSCOPY Alternative name for ebulliometry.

EC Abbreviation for epichlorhydrin rubber.

ECH Abbreviation for epichlorhydrin.

ECO Abbreviation for epichlorhydrin/ethylene oxide rubber.

ECOLYTE Tradename for a styrene copolymer containing a small amount of methyl vinyl ketone as a comonomer. The comonomer is readily photodegradable and therefore the copolymer photodegrades and embrittles on exposure to sunlight. The polymer is useful as a packaging material which is degradable and therefore disappears when discarded, thus not producing a litter problem as with other plastic packaging materials.

EDGE DISLOCATION (Taylor–Orowan dislocation) A dislocation generated by the insertion of an extra plane of motifs into the crystal such that its dislocation line vector runs along the edge of the extra plane of motifs.

EDISTIR Tradename for polystyrene.

EDMAN DEGRADATION Alternative name for Edman method.

EDMAN METHOD (Edman degradation) A method for reacting, in a stepwise manner, the amino acid residues from the N-terminal end of a protein or peptide by reaction with isothiocyanate. It provides a method for the determination of the N-terminal residue and of the terminal sequences by repeated application. Under basic conditions (pH 9–10) a phenylthiocarbamyl (PTC) derivative (**I**) is formed:

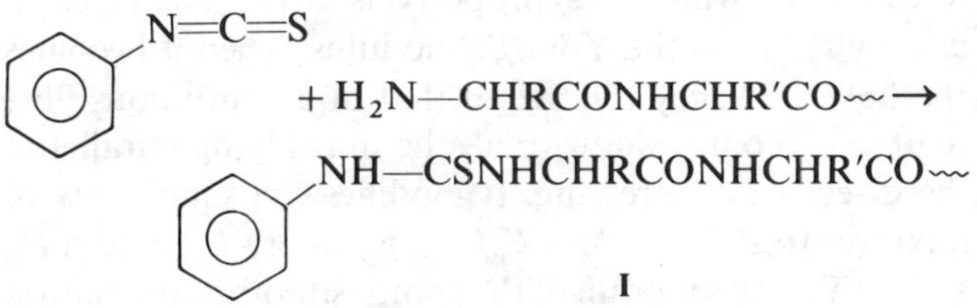

The PTC amino acid is cleaved in acid (anhydrous trifluoroacetic acid) to give the free N-terminal amino acid as a thiazolinone (**II**). In acid, **II** rearranges to a phenylthiohydantion (**III**):

The PTH can be separated by chromatography and determined. The process can be repeated on the remaining peptide/protein (**IV**) and so the amino acid sequence is determined in a stepwise manner. The whole process may be repeated many times in an automated apparatus (a protein sequenator). In some cases the peptide or protein can be bound to a solid support to facilitate washing it free of reagents, etc. at each stage.

EEA Abbreviation for ethylene–ethyl acrylate copolymer.

EFFECTIVE NETWORK CHAIN Symbol v_e. In a vulcanised elastomer the number of network chain segments between crosslinks per unit volume, which contribute to the vulcanisate elasticity. Chain ends do not do so and hence are not counted as effective network chains.

EFFECTOR (Modulator) (Modifier) (Regulator) A substance which modifies the activity of an allosteric enzyme by altering the conformation of the polypeptide chain by binding at a site other than the active site. The effector may also be the substrate (homotropic co-operativity), otherwise there is heterotropic cooperativity.

EFFICIENCY FACTOR FOR REINFORCEMENT Alternative name for efficiency of reinforcement.

EFFICIENCY OF REINFORCEMENT (Efficiency factor of reinforcement)(Fibre efficiency factor)(Krenckel's efficiency factor of reinforcement) Symbol η. The ratio of the amount of fibre contributing to reinforcement of the property under consideration to that not contributing. Thus η depends on the fibre orientation with respect to the direction in which the property is being measured. η usually refers to the Young's modulus, when it becomes the fibre efficiency relative to that of a continuous fibre reinforced composite with all the fibres lying parallel to the direction of stressing. It modifies the simple law of mixtures to $E_c = \eta E_f \phi_f + E_m(1 - \phi_f)$ where E_c, E_f and E_m are the Young's moduli of the composite, fibre and matrix

respectively. When lateral contraction is ignored, the values of η are: for all fibres lying parallel, $\eta = 1$ parallel to the fibres and $\eta = 0$ perpendicular to the fibres, for fibres randomly distributed in two dimensions, as in chopped strand mat, $\eta = 3/8$ and for fibres randomly distributed in three dimensions, $\eta = 1/5$.

EFFICIENCY PARAMETER The number of sulphur atoms chemically combined with a rubber for each crosslink formed by sulphur vulcanisation. A measure of the efficiency of use of the sulphur in forming short network supporting crosslinks, mainly mono- and disulphides, rather than polysulphides and cyclic structures. For non-accelerated vulcanisation, values of 40–55 are typical, for a typical accelerated system values are 15–20 and for efficient vulcanisation values may be 2–5.

EFFICIENT VULCANISATION (EV) Vulcanisation in which a high proportion of the sulphur is used in forming mono- and disulphide crosslinks and relatively little in forming polysulphide crosslinks and cyclic structures, which contribute little to improvement in vulcanisate properties. The efficiency is quantitatively expressed as the efficiency parameter (E). Efficient vulcanisation is achieved by the use of small amounts of sulphur (0·5–2 phr) and relatively large amounts of accelerator (3–6 phr). Values of E may be as low as 2–4, whereas values of 10–20 are typical with normal vulcanisation systems.

EGA Abbreviation for evolved gas analysis.

EGG ALBUMIN Alternative name for ovalbumin.

E-GLASS A borosilicate glass of oxide weight per cent composition: SiO_2, 54·3; Al_2O_3, 15·0; CaO, 17·3; MgO, 4·7; Na_2O/K_2O, 0·6; B_2O_3, 8·0. Similar to Pyrex glass in its composition. Originally developed for its good electrical properties, but now the almost universally used glass, as glass fibres, for glass reinforced polymer composites. Fibres are highly drawn during their manufacture so that their properties are very different from those of bulk glass. They are very sensitive to surface abrasion, which greatly reduces their strength, so they are treated with a lubricating size before being gathered into strands. Virgin fibres have a tensile strength of about 3·7 GPa, a Young's modulus of about 76 GPa and a density of 2·54 g cm^{-3}. Refractive index is 1·548. Among the commoner glasses the strength is only exceeded by S-glass and resistance of E-glass reinforced composites to hydrolysis is particularly good.

EHRENFEST EQUATIONS A set of relationships expressing the thermodynamic properties of a material existing at a second order transition. One example is: $\Delta\kappa\Delta C_p/TV(\Delta\alpha)^2 = 1$, or $dP/dT = \Delta C_p/VT\Delta\alpha = \Delta\alpha/\Delta K$, where $\Delta\kappa$, ΔC_p and $\Delta\alpha$ are the changes in compressibility, specific heat and expansion coefficient respectively at the transition and T, V and K are the temperature, volume and bulk modulus respectively. Adherence of polymer behaviour to these relations at T_g has been used as evidence that T_g is a true second order transition. However, in general the relations are only approximately true.

EILERS EQUATION   An empirical equation which gives the coefficient of viscosity of a filled material (η_c) in terms of the viscosity of the unfilled matrix (η_m) and the volume fraction of spherical filler particles (ϕ_f), when the filler is much more rigid than the matrix, as in mineral filled rubbers. It extends the Einstein equation to higher filler loadings and is: $\eta_c/\eta_m = [1 + (1\cdot25\phi_f)/(1 - S\phi_f)]^2$, where S is the hydrodynamic interaction parameter usually taken to be $1\cdot2$–$1\cdot3$, the theoretical value of close packed spheres being $1\cdot35$. The ratio η_c/η_m is often taken to be equal to the ratio of the Young's moduli or of the shear moduli, so the equation is useful for the prediction of moduli as well. It is similar in form to the Mooney equation.

EINSTEIN COEFFICIENT   Symbol k_E. A parameter relating the viscosity of a fluid, including a polymer, containing dispersed particles, such as filler, to that of the unfilled fluid, as expressed in the Einstein equation. Its value depends on the shape of the particles; for spheres it is $2\cdot50$, whereas as the particles become more elongated its value increases. Thus for rods of an aspect ratio of 10 the value is 6.

EINSTEIN EQUATION   A theoretical expression describing the effect of a dilute suspension of rigid spherical particles on the viscosity of a fluid. It is: $\eta/\eta_0 = 1 + k_E\phi_f$, where η and η_0 are the coefficients of viscosity of the suspension and pure fluid respectively, ϕ_f is the volume fraction of suspended particles (filler) and k_E is the Einstein coefficient, equal to $2\cdot50$ for spheres. Other equations, such as the Mooney and Eilers equations, are empirical modifications of the Einstein equation applicable to higher concentrations of particles than the 10% value to which the Einstein equation is restricted. The equation forms the basis of many theories of the elastic behaviour of composites, since the ratio η/η_0 may be related to the ratios of the Young's and shear moduli. It also forms the basis of the development of theories of the viscosity of dilute polymer solutions, as in the Simha and Flory–Fox equations.

EINSTEIN–GUTH–GOLD EQUATION   Alternative name for the Guth–Smallwood equation or the Guth–Gold equation, when these are formulated in terms of the Young's modulus.

EKA-CONJUGATED POLYMER   A conjugated polymer sufficiently free of structural defects (for example with the chain molecules locked in a planar conformation) so that extensive electron delocalisation can occur. Thus considerable electronic conductivity can occur, unlike a rubi-conjugated polymer. Examples include polyacene–quinone radical polymers, polyphthalocyanins, graphite pyropolymers and polydiacetylenes.

EKAVYL   Tradename for polyvinyl chloride.

EKKCEL   Tradename for copolymers of p-hydroxybenzoic acid with iso- or terephthalic acid and an aromatic dihydroxy compound, e.g. 4,4'-dihydroxydiphenylether. More easily processible than the p-hydroxybenzoic acid homopolymer poly-(p-oxybenzoyl).

EKONOL   Tradename for poly-(p-hydroxybenzoic acid).

EKTAR   Tradename for a polyester made from cyclohexanedimethylol and a second diol with terephthalic acid.

ELAPRIM   Tradename for nitrile rubber.

ELAPRIM AR   Tradename for acrylic elastomers.

ELASTANE   Alternative name for spandex.

ELASTIC COMPLIANCE   Alternative name for compliance.

ELASTIC CONSTANT   One of several material constants relating stress or stress components to the corresponding strain or strain components by a constitutive relationship. The most general such linear relationship is the generalised Hooke's Law, between the nine stress components and the nine strain components, given by the tensor equation, $\sigma_{ij} = C_{ijkl}\varepsilon_{kl}(i, j, k, l = 1, 2 \text{ or } 3)$ and its inverse $\varepsilon_{ij} = S_{ijkl}\sigma_{kl}$, where C_{ijkl} and S_{ijkl} are the elastic moduli (or stiffness constants) and elastic compliances respectively. Thus there are in principle 81 moduli and 81 compliances. However, for static equilibrium, the stress components σ_{ij} equal the components σ_{ji}, and by the definition of the strain tensor $\varepsilon_{ij} = \varepsilon_{ji}$. Thus the maximum number of independent C_{ijkl} and S_{ijkl} values is 36. In contracted notation, the generalised Hooke's Law is written $\sigma_i = C_{ij}\varepsilon_j$ $(i, j = 1, 2, \ldots 6)$ or in inverted form $\varepsilon_i = S_{ij}\sigma_j$. By a thermodynamic argument (the existence of a strain energy function) it can be shown that $C_{ijkl} = C_{klij}$ and $S_{ijkl} = S_{klij}$ and hence the maximum number of independent constants is reduced to 21 even for the most general case, no matter how complex the relationship between elastic properties and direction, i.e. for all anisotropic cases. Symmetry reduces the number of independent constants even more, e.g. orthorhombic (as in polyethylene crystals) to 9, hexagonal or transverse isotropic (as in uniaxially drawn fibres or film) to 5 and cubic to 3. The experimentally measured moduli can be related to the above compliance and stiffness constants. Thus for an isotropic material, which has only two independent constants, the Young's modulus $E = 1/S_{11}$, the shear modulus $G = 1/2(S_{11} - S_{12})$, the bulk modulus $K = 1/3(S_{11} + 2S_{12})$ and the Poisson's ratio $v = -S_{12}/S_{11}$. Hence E, G. K and v are interrelated, e.g. $G = E/2(1 + v)$ and $K = E/3(1 - 2v)$.

ELASTICITY   Mechanical behaviour in which a stress induces a deformation which is completely recovered on removal of the stress. If the deformation is completely recovered immediately then the material is perfectly elastic. Metals approximate to this behaviour at small strains as do some rubbers. Plastics generally show delayed elasticity and viscous flow effects and hence are viscoelastic. If the stress is directly proportional to the strain, not the case in most polymers, then the material exhibits linear elasticity. If not, then it exhibits non-linear elasticity. The relationships between stress and strain are determined by the elastic constants.

ELASTIC LIMIT   The greatest stress that a material can sustain without any permanent strain remaining on release of the stress. In many polymer materials this may be beyond the proportional limit.

ELASTIC LIQUID   A liquid, which unlike a purely viscous liquid, exhibits elastic effects as well as viscous flow. These may be either stress relaxation effects, where the stress does not become instantaneously isotropic or zero as soon as the liquid is held in a fixed shape, or they may be elastic recovery effects, where the shape does not remain constant as soon as the stress is made isotropic or zero. Many polymer fluids (melts and solutions) show such elastic effects.

ELASTIC MODULUS   (Modulus) (Stiffness constant) Symbol C. Use of this symbol is confusing since it can also mean compliance—the inverse of modulus. An elastic constant which is the constant of proportionality in the generalised Hooke's Law relationship between stress (σ_{ij}) and strain (ε_{kl}) which is, $\sigma_{ij} = C_{ijkl}\varepsilon_{kl}$, or $\sigma_i = C_{ij}\varepsilon_j$ in contracted notation. Since i and j can have any of three values there are 81 different possible elastic moduli for a material. However, for static equilibrium conditions, i.e. with $\sigma_{ij} = \sigma_{ji}$, these reduce to 36. Furthermore the existence of a strain energy function, giving $C_{ijkl} = C_{klij}$, reduces the number to 21. Material symmetry reduces the number even more. Thus for an isotropic material there are only two independent elastic moduli. Hence, of the commonly measured moduli, Young's modulus, shear modulus, bulk modulus and Poisson's ratio, since only two are independent, all are interrelated.

ELASTICOVISCOUS FLUID   Alternative name for viscoelastic fluid. Sometimes the term is used to refer only to those viscoelastic fluids in which the viscous flow effects predominate and only minor elastic effects are present.

ELASTIC–PLASTIC TRANSITION   The change from recoverable elastic behaviour to non-recoverable plastic strain, which occurs on stressing a material beyond the yield point.

ELASTIC SCATTERING   The scattering of radiation (light, X-rays or neutron beams) by a medium in which the scattered radiation has the same wavelength as the incident radiation. The type of scattering, as opposed to inelastic scattering, that occurs in Rayleigh scattering.

ELASTIC SOLID   A material which can exist in a unique equilibrium shape at zero stress and which when held at any other shape by stressing, the stress allows a non-isotropic equilibrium shape to exist. If both these conditions are attained instantaneously on changing the stress, then the material is perfectly (or ideally) elastic. If either of these conditions takes a finite time to be attained, then the material is non-ideally elastic or viscoelastic. In contrast a liquid can have any equilibrium shape.

ELASTIC TURBULENCE   Alternative name for melt fracture.

ELASTIN   An elastic protein occurring in blood vessels, lung tissue, ligaments and other tissues where relatively high elasticity is required. Although fibrous, the material is not crystalline, existing in random coils, with a little (less than 10%) α-helical conformation. When swollen with water, as in living tissues, it shows rubber-like elasticity due to its crosslinked network structure, which also imparts insolubility thus making structural characterisation difficult. Elastin contains over 90% non-polar amino acid residues (glycine, alanine and valine making up 50–70%), with many sequences of the type Lys–Ala–Ala–Lys and Lys–Ala–Ala–Ala–Lys in the crosslinking regions and Pro–Gly–Val–Gly–Val–Ala between crosslinks. The crosslinks are of two types formed by modification of lysine residues. One type is similar to the dehydrolysinonorleucine crosslinks of collagen, the other type involves desmosine and isodesmosine.

ELASTOLLAN   Tradename for a partially crosslinked thermoplastic polyurethane elastomer, based on a polyester or polyether polyol/MDI prepolymer and then chain extended.

ELASTOMER   Although often used as an alternative term for rubber, the term usually implies a vulcanised rubber, since it is only after crosslinking that the characteristic desirable rubber elasticity develops.

ELASTOMERIC POLYAMIDE   Alternative name for polyether block amide.

ELASTO-OSMOMETRY   A technique for the determination of the number average molecular weight of a polymer (I) by observing the change in a property of a swollen polymer (II) gel when it is immersed in a solution of polymer I, compared with pure solvent. The change results from a change in chemical potential on addition of polymer I solute causing a change in the properties of the gel. Usually the property measured is the elongation of the gel (as a fibre or strip) at constant load, or change in stress when held at constant elongation.

ELASTOTHANE   Tradename for a millable polyurethane elastomer, curable either by sulphur or peroxide vulcanisation.

ELASTRON   Tradename for polycarbamoylsulphonate.

ELECTRET A dielectric in a permanent state of polarisation and hence the electrical counterpart of a magnet. With polymers, electrets with field strengths of $30\,kV\,cm^{-1}$, stable for several years may be made. A suitable polymer is heated in an electric field to polarise it and with the field still applied, the temperature is reduced thus 'freezing-in' the polarisation when the field is removed. Electrets of several fluoropolymers, such as fluorinated ethylene–propylene and polyvinylidene fluoride, have been studied in particular. In addition such polymers often exhibit piezo- and pyroelectric effects and have been used in several useful devices such as electret microphones. Studies of depolarisation on reheating an electret, by the thermally stimulated discharge method, can provide information on polymer transitions.

ELECTRIC BIREFRINGENCE (Kerr effect) The birefringence produced by the application of an electric field causing orientation of molecular dipoles. By use of a sinusoidally or rectangularly alternating field, the relaxation times and hence rotary diffusion constants of many polymers in solution, especially proteins and polypeptides, have been investigated. Effects in solid polymers are only weak.

ELECTRIC STRENGTH Alternative name for dielectric strength.

ELECTROCHEMICAL POLYMERISATION (Electrolytic polymerisation) (Electro-initiated polymerisation) Chain polymerisation initiated by the electrolysis of a solution containing the monomer and an electrolyte capable of generating reactive radicals by discharge at the electrodes or by undergoing electron exchange to form a monomer radical anion or a monomer radical cation.

At cathode:

$$R^+ + e^- \rightarrow R\cdot$$
$$R\cdot + M \rightarrow \text{free radical polymerisation}$$

or

$$R^+ + e^- \rightarrow R\cdot$$
$$R\cdot + M \rightarrow R^+ + M^{\cdot-}$$
$$2M^{\cdot-} \rightarrow {}^-MM^-$$
$${}^-MM^- \rightarrow \text{anionic polymerisation}$$

At anode:

$$A^- \rightarrow A\cdot + e^-$$
$$A\cdot + M \rightarrow \text{free radical polymerisation}$$
$$\searrow$$
$$A^- + M^{\cdot+}$$
$$2M^{\cdot+} \rightarrow {}^+MM^+$$
$${}^+MM^+ \rightarrow \text{cationic polymerisation}$$

Examples include the free radical polymerisation of vinyl monomers using metal carboxylates (by the Kolbe reaction): $RCOO^- \rightarrow RCOO\cdot + e^- \rightarrow R\cdot + CO_2$ where $R\cdot$ then initiates, the anionic polymerisation of acrylonitrile, methyl methacrylate or styrene by electrolysis of tetra-alkyl ammonium or alkali metal salts and the cationic polymerisation by the electrolysis of $AgClO_4$ to polymerise styrene or N-vinylcarbazole.

ELECTRODIALYSIS A technique for the removal of dissolved salts from aqueous solution, especially useful for desalination of sea water. An electrical cell is separated into a series of subcells by alternate polyanion and polycation ion-exchange membranes. Under the influence of the applied electrical potential, the permselectivities of the membranes cause salt to be concentrated in alternate subcells and to be removed from the others.

ELECTROFOCUSSING Alternative name for isoelectric focussing.

ELECTRO-INITIATED POLYMERISATION Alternative name for electrochemical polymerisation.

ELECTROLYTIC POLYMERISATION Alternative name for electrochemical polymerisation.

ELECTROMECHANICAL BREAKDOWN Dielectric breakdown, occurring in soft polymers, caused by compression of the material by the Coulombic attractive forces of the electrodes. Thus the observed intrinsic dielectric strength will be lowered. Breakdown occurs at a maximum field strength (E_m) given by : $E_m = (Y/\varepsilon_0\varepsilon')^{1/2}$, where Y is Young's modulus, ε' is the relative permittivity and ε_0 is the relative permittivity of free space. It is often the cause of breakdown in rubbers and in plastics on heating.

ELECTRON DIFFRACTION The diffraction pattern obtained by Bragg reflections of an electron beam by a crystalline material. It is readily observed by use of an electron microscope by simple alteration of the electron optics. Small selected areas of a sample are readily examined. The pattern obtained depends on the orientation of the crystal relative to the incident beam and can be used to determine orientation of the crystal unit cells. In particular, use of the technique shows that in polymer lamellar crystals the polymer chains are oriented perpendicular to the large dimension of the lamellae which leads to the inevitable conclusion that the molecules must be chain folded.

ELECTRONICALLY EXCITED STATE Alternative name for excited state.

ELECTRONIC BREAKDOWN The most common type of intrinsic dielectric breakdown. At a high field, the small number of conduction electrons available are sufficiently accelerated that, on collision with the molecular structure, they cause ionisation and the production of an 'avalanche' of electrons, resulting in breakdown.

ELECTRONIC POLARISATION A polarisation resulting from the displacement of the electrons of the atoms of a material from their normal equilibrium positions with respect to their atomic nuclei when an electric field is applied. For non-polar dielectrics, such as hydrocarbon polymers, this is the major contribution to polarisation. Such polarisation occurs very rapidly and so follows the variation in the field even at the highest frequencies encountered. The resonance frequency is at about 10^{15} Hz, which is in the optical range, so that no power loss occurs in the normal electrical frequency range. Owing to electronic polarisation, the relative permittivity rarely falls below a value of two.

ELECTRON MICROSCOPY (EM) A microscopic technique analogous to optical microscopy but using a beam of electrons as the illuminant. Magnetic fields act as the lenses and the image is displayed on a phosphorescent screen. The instrument is usually used in transmission but can also be used for electron diffraction experiments. The main advantage is that very high resolution is possible, of the order of a few ångströms in the best instruments. Also the rapid decrease in intensity with specimen thickness produces good contrast between specimen and its background, but also requires the specimen to be very thin. Such thin specimens are usually produced by microtomy or ultramicrotomy. Contrast may be improved by shadowing the sample (usually mounted on a carbon grid) with heavy metal atoms, e.g. platinum, or by staining, e.g. with osmium or phosphotungstic acid. Etching is also useful. Sample surfaces are often examined after replication or by scanning electron microscopy. Electron microscopy is the main technique used in the examination of polymer single crystals, fracture surfaces and other micro-morphological features, e.g. polymer blend phase structures and multilayer crystals.

ELECTRON PARAMAGNETIC RESONANCE SPECTROSCOPY (EPR spectroscopy) Alternative name for electron spin resonance spectroscopy.

ELECTRON SPECTROSCOPY FOR CHEMICAL APPLICATION (ESCA) (X-ray photoelectron spectroscopy) A spectroscopic technique in which a sample is bombarded with soft X-rays, which cause the emission of core, i.e. non-valence, electrons from the atoms of the sample. The kinetic energies of the ejected electrons are determined as a spectrum of intensities versus kinetic energies. These energies are characteristic of the atom and its environment. The technique is suited to several types of polymer analysis and requires only minimum sample preparation. Film, solid and powder samples may all be used. It is uniquely capable of examining surface layers and probing to different depths. It has been applied to the study of surface modified (e.g. by surface post polymerisation or oxidation) polymers, detecting surface layers of additives or contaminants or surface degradation as in weathering.

ELECTRON SPIN RESONANCE SPECTROSCOPY (ESR spectroscopy) (Electron paramagnetic resonance spectroscopy) A technique for the characterisation of molecules containing unpaired electrons; for organic molecules this means molecules containing free radicals. A free radical sample, which may be solid, liquid or gas, is placed in a strong magnetic field, which produces two electron spin energy levels for the unpaired electron. On irradiation with microwave radiation, transitions occur between the two levels dependent on the applied field strength. A detector produces an ESR spectrum of absorbed microwave energy versus field strength. This is usually displayed and recorded as the first derivative spectrum.

The position of the absorption is denoted by the g-value, analogous to the chemical shift in nuclear magnetic resonance spectroscopy, and is dependent on the electronic environment of the unpaired electron and therefore on structure. Furthermore, splitting of the energy levels, and therefore of the spectral peaks (hyperfine splitting) occurs due to interaction of the electrons with the magnetic field of the neighbouring nuclei, e.g. protons. Thus the total spectrum can give useful information on the structure and concentration (from intensities) of the organic free radical. In polymer work, the technique is particularly applied to the study of free radicals in polymerisation, degradation and fracture processes. By the use of stable radicals, in the spin-labelling and spin-probe techniques, the molecular motions associated with transitions and relaxations may also be studied.

ELECTRON TRANSFER POLYMER Alternative name for oxidation–reduction polymer.

ELECTROPHORESIS (Ionophoresis) One of a range of related techniques widely used for the separation (analytical and in some cases preparative), purification and characterisation of individual proteins in mixtures of proteins. It utilises the fact that protein molecules carry electric charges due to ionisation of substituent acid (e.g. glutamic and aspartic) or basic (e.g. lysine and arginine) groups in aqueous solutions giving negative or positive charges respectively in the appropriate pH ranges. The molecule will carry a net charge which will depend on the pH. Like all charged molecules, such proteins will move through an aqueous medium across which an electric field is applied, except at the isoelectric pH when the molecules carry no net charge. The differing mobilities of different proteins in the electric field provide a method for their separation from their mixtures.

In the original electrophoretic method, moving boundary electrophoresis, each migrating protein forms a front where movement can be followed. However, this method has largely been replaced by the simpler and more sensitive zone electrophoresis methods carried out, not in aqueous solution, but with an aqueous solution immobilised on a solid support. If the support also separates on the basis of molecular size, as with polyacrylamide gel or starch, even higher resolution is possible and this separation may be scaled up to a preparative technique. SDS gel electrophoresis is universally used to confirm protein homogeneity as well as to determine protein molecular weights. Further variations capable of even greater resolution are disc electrophoresis and isoelectric focussing.

ELECTROSPRAY MASS SPECTROMETRY Mass spectrometry in which a polymer solution is sprayed through a fine jet to give a supersonic molecular beam. In this way it is possible to produce isolated and charged molecules which may be analysed according to their mass.

ELEMID Tradename for an ABS/nylon blend.

ELEOSTEARIC ACID (*cis*-9-, *trans*-11-, *trans*-13-octadecatrienoic acid)

$$CH_3CH_2CH_2CH_2CH = CHCH = CHCH$$
$$\overset{\|}{CH(CH_2)_7COOH}$$
M.p. 49°C.

The predominant (80%) acid residue of the triglyceride of tung oil and responsible for its rapid air drying properties.

ELEXAR Tradename for styrene–ethylene/butylene–styrene block copolymer.

ELLIPTICITY Symbol θ. A measure of circular dichroism. Quantitatively it is the angle (θ) whose tangent is the ratio of the major to the minor axis of the ellipse that is traced by the tip of the resultant electric vector of the propagating light. It is given by $\theta = \tan^{-1}[(A_1 - A_r)/(A_1 - A_r)]$, where A_1 and A_r are the electric vectors of the left and right circularly polarised components respectively. Since θ is dependent on both path length and solute concentration, a molar ellipticity $[\theta]_\lambda$ is defined in a manner analogous to molar rotation. It is approximately given by: $[\theta]_\lambda = 3300 \Delta\varepsilon$, where $\Delta\varepsilon$ is the circular dichroism.

ELLIS FLUID Alternative name for Ellis model.

ELLIS MODEL (Ellis fluid) A model describing the flow behaviour of a non-Newtonian fluid, giving a relationship, one form of which is:

$$1/\eta = 1/\eta_0(1 + (\tau/\tau_{1/2})^{(1-n)/n})$$

where η is the shear rate dependent apparent viscosity, n is a constant, η_0 is the limiting Newtonian viscosity at zero shear rate (at which the model predicts Newtonian behaviour) and $\tau_{1/2}$ is the shear stress (τ) at which the viscosity has fallen to $\eta/2$. At high shear rates the power law behaviour is approached (with flow index n). The equation often fits data over a wider range of shear rates than the power law, but is more complicated to use.

ELLMAN'S REAGENT (DTNB) (5,5′-Dithio-bis-(2-nitrobenzoic acid))

$$O_2N-\bigcirc-S-S-\bigcirc-NO_2$$
$$\overset{|}{COOH} \qquad\qquad \overset{|}{COOH}$$
I

Useful for the determination of the number of cysteine

residues in a polypeptide chain of a protein by the reaction:

$$I + HSCH_2-\overset{\overset{\displaystyle NH}{|}}{\underset{\underset{\displaystyle CO}{|}}{CH}} \rightarrow$$

$$O_2N-\bigcirc-S-S-CH_2-\overset{\overset{\displaystyle NH}{|}}{\underset{\underset{\displaystyle CO}{|}}{CH}} + HS-\bigcirc-NO_2$$
$$\overset{|}{COOH} \qquad\qquad\qquad\qquad\qquad\qquad \overset{|}{COOH}$$

The thionitrobenzoic acid concentration is then readily measured by its light absorption as it absorbs strongly at 412 nm.

ELMENDORF TEAR TEST The commonest test to assess the tear resistance of a film material, in which a specimen containing an edge slit is clamped with one moveable clamp attached to a pendulum (which is sometimes a sector of a wheel). On release of the pendulum from its raised position, it tears the specimen and the energy expended in doing this is recorded by a pointer reading on a scale on the dial of the pendulum.

ELONGATION The increase in length of a tensile test specimen, usually expressed as a percentage of the original length.

ELONGATIONAL FLOW (Extensional flow) (Stretching flow) (Tensile flow) Flow resulting from the stretching of a material, i.e. drawing it from a larger cross-sectional area to a smaller one. It is encountered in many polymer melt processing operations, e.g. when convergent flow occurs in a die or calender or in the free surface flows of fibre and film drawing, blow moulding or film blowing. It can give rise to processing defects such as parison sagging and draw resonance. It is characterised by the elongational viscosity. Naturally with fluid materials, the material cannot be 'pulled' in the same way as a solid, and the design of equipment to create simple elongational flow under controlled conditions is not easy.

ELONGATIONAL VISCOSITY (Coefficient of viscous traction) (Extensional viscosity) (Tensile viscosity) (Traction viscosity) (Trouton viscosity) Symbol λ. The tensile stress divided by the rate of elongation for elongational flow. Except for a Newtonian fluid, the elongational viscosity is not a constant, but is dependent on the rate of elongation. For polymers, it is often many times, up to several hundred, the apparent shear viscosity, and not simply three times as for a Newtonian fluid. It is important whenever a polymer melt is stretched, as in many polymer melt processing operations such as fibre spinning, blow moulding, injection moulding and calendering, when convergent flow occurs. For polymer melts λ may be three times η_0 at zero shear rates. However, unlike η, which for most polymer melts decreases with rate of shear, for different polymers, λ may increase, remain constant or decrease with increasing rate of shear. The rate of shear (which is usually low) at which such tension thinning or tension thickening occurs is the critical elongation rate. Typical values for λ for polymer melts are in the range 10^5–10^8 P.

ELONGATION AT BREAK (Extension at break) (Ultimate elongation) The maximum tensile strain to which a material can be subjected before it breaks. Often expressed as the percentage elongation. Typical values range from about 1–2% for thermosets and glassy thermoplastics, to 50–100% for tough plastics and up to several hundred per cent for unvulcanised and lightly vulcanised rubbers.

EL-REX Tradename for polypropylene.

EL-REY Tradename for low density polyethylene.

ELTEX Tradename for high density polyethylene.

ELVACET Tradename for polyvinyl acetate.

ELVANOL Tradename for polyvinyl alcohol.

ELVAX Tradename for ethylene–vinyl acetate copolymers with about 30% vinyl acetate, useful as adhesives and as wax additives.

EM Abbreviation for electron microscopy.

EMA Abbreviation for ethylene–maleic anhydride copolymer.

EMBRYO (Subcritical nucleus) A crystallisation nucleus smaller than the critical size required for crystal growth.

EMULSION POLYMERISATION Free radical polymerisation with the monomer dispersed in water by the use of an emulsifier (usually an anionic surfactant). In such a dispersion most monomer is initially present as relatively large droplets (typically about $100\,\mu m$ in diameter), some is present in surfactant micelles (typically about $1\,\mu m$ in diameter) and a little is in solution in the water. Polymerisation is initiated with a water soluble initiator, often a redox initiator, the free radicals being captured by the micelles where they initiate polymerisation. Only a few micelles become initiated and grow as more monomer diffuses from the droplets. After about 10% conversion (Interval I) no more micelles are initiated and uninitiated ones have disappeared. As the monomer/polymer particles grow they absorb more monomer and surfactant, until by about 60% conversion (Interval II) the monomer droplets have disappeared. Beyond this point (Interval III) all the monomer is present in the polymer particles. This is the Harkins model of emulsion polymerisation. In interval I the rate rises as the number of initiated micelles/particles increases. In interval II the rate remains steady. In interval III it decreases as the monomer concentration in the polymer particles decreases.

The Smith–Ewart theory, which provides a quantitative kinetic analysis of the polymerisation, shows that, in the case when the monomer is not very water soluble and when free radicals do not diffuse out of polymer particles, the rate of polymerisation (R_p) during interval II is given by $R_p = k_p[P][M]/2N$, where k_p is the rate constant for propagation, $[P]$ is the concentration of polymer particles, $[M]$ is the monomer concentration and N is Avogadro's number. $[P]$ depends on the initial number of micelles (which in turn depends on the surfactant concentration) and on the initial rate of initiation.

Thus in emulsion polymerisation, as opposed to other polymerisation techniques, the rate of polymerisation may be increased without the penalty of decreased polymer molecular weight. Other advantages are the ease of control of the exotherm because the water acts as a heat sink, the possibility of low temperature polymerisation (as in the production of 'cold' rubber) by the use of active redox initiators and the production of polymer in the form of latex, which may be directly usable, e.g. as a paint, adhesive or in polishes. In these applications latex stability and film forming properties as well as the rheological behaviour are important. These are affected by particle size and size distribution.

Other theories of emulsion polymerisation have been developed, especially the Medvedev theory, which assumes that polymerisation occurs in the emulsifier layer that surrounds the micelles. A core–shell theory has also been proposed in which polymerisation is assumed to take place in the outer layers of the particles.

EMULSION SBR Styrene–butadiene rubber produced by emulsion polymerisation.

EMULTEX Tradename for polyvinyl acetate.

ENANT Tradename for nylon 7 in fibre form.

ENANTHOLACTAM (7-Heptanolactam) (Oenantholactam)

$$(CH_2)_6-CO$$
$$\underline{\qquad\quad}-NH \qquad \text{M.p. } 65°C.$$

The lactam from which nylon 7 may be obtained by ring-opening polymerisation. Obtained by reaction of 7-chloroheptanoic acid (itself obtained from the telomerisation of ethylene with carbon tetrachloride) with ammonia.

END CAPPING (Capping) Chemical conversion of the polymer chain end groups responsible for initiating degradation, to more stable groups, thereby stabilising the polymer. Used especially when polyoxymethylene is stabilised by conversion of the thermally and hydrolytically labile hemi-acetal end groups (i.e. $HO-CH_2-O\sim$ groups) (which initiate depolymerisation), to stable acetate groups.

END CORRECTION Alternative name for Bagley end correction.

END-EFFECT CORRECTION (Couette correction) (Bagley end-effect correction) A correction applied in capillary viscometry to allow for the turbulent flow that occurs at the entrance to and exit from the capillary. An equivalent effect is considered to arise as if the capillary had an increased length but there was no turbulence. The increased length is $(L + cR)$, where L is the actual length, R is the capillary radius and c is a constant, evaluated to be in the range 0.5–0.9. Thus the Poisseuille equation should be modified to $\eta = R^4 Pt/8(L + cR)V$. For dilute polymer solution viscometry, the end-effect is usually negligible, but in the capillary rheometry of polymer melts it is taken into account by the Bagley end-effect correction method.

END GROUP The atom or group at the end of a polymer chain, which must necessarily be different from the regular repeat unit (—M—) due to its being bound to the polymer chain by only one chemical bond rather than two. Thus a linear polymer may be represented as $X\text{--}[\text{M}]_n\text{--}Y$, where X and Y are the end groups. Determination of the number of end groups, end group analysis, provides a method for the determination of polymer molecular weight. Branched polymers will contain more than two end groups and hence their determination provides a measure of the extent of branching. The nature of the end groups is determined by the nature of the chain growth and termination processes during polymerisation. In chain polymerisation one end group will be an initiator fragment, such as an ester group, RCO— (from the use of

$$\overset{\|}{O}$$

a diacyl peroxide initiator $(\text{RCO})_2\text{--O--}$); the other end

$$\overset{\|}{O}$$

group may be an unsaturated group arising from transfer to monomer or from termination by disproportionation or another group from a different transfer reaction. In step-growth polymerisation, the end groups will largely consist of the functional groups (A or B) present in excess in the original monomer mixture. If a stoichiometric balance of the two types of functional groups involved in the polymerisation is used, as in the polymerisation of an AB monomer, then one end group will be A and the other B.

END GROUP ANALYSIS (End group assay) The determination of the number of end groups in a polymer sample as the concentration of end groups (E) in moles per gram of polymer. If α is the number of end groups being determined per polymer molecule, then the polymer number average molecular weight $(\bar{M}_n)$ is α/E. One of the most frequently used methods of $\bar{M}_n$ determination of step-growth polymers, where the end groups are functional groups (e.g. hydroxyl and carboxyl in a polyester); usually determined by chemical analysis, especially titrimetry in a non-aqueous solvent. Modern methods allow $\bar{M}_n$ values up to about $50\,000$ to be determined.

If α for a particular type of end group is not known then all the end groups, of all types, must be determined by separate analysis for each type. Sometimes determination of only one type of end group is sufficient characterisation, e.g. acid value. End group analysis has also been widely performed on chain polymers, not for $\bar{M}_r$ determination, but to elucidate details of transfer and termination reactions, especially by the use of radioactively tagged initiator or transfer agent.

END GROUP ASSAY Alternative name for end group analysis.

***ENDO-CIS*-BICYCLO-[2,2,1]-HEPT-5-ENE-2,3-DICARBOXYLIC ANHYDRIDE** Alternative name for endomethylenetetrahydrophthalic anhydride.

***ENDO*-4,7-METHYLENE-4,7,8,9-TETRAHYDRO-INDENE** Alternative name for bicyclopentadiene.

ENDOMETHYLENETETRAHYDROPHTHALIC ANHYDRIDE (*Endo-cis*-bicyclo-[2,2,1]-hept-5-ene-2,3-dicarboxylic anhydride) (Nadic anhydride) M.p. 165°C.

Prepared by Diels–Alder addition of maleic anhydride to cyclopentadiene. Useful, in place of phthalic anhydride, in the formation of unsaturated polyester resins giving cured products with high softening temperatures.

ENDOPEPTIDASE (Proteinase) A protease which will cleave a peptide bond of a specific type anywhere in a polypeptide chain, as opposed to an exopeptidase, which only cleaves at the chain end. Most proteases are of this type.

END-PRODUCT INHIBITION Alternative name for feed-back inhibition.

END-TO-END DISTANCE Symbol r. The distance separating the two ends of a polymer chain. Its value may be computed for various theoretical models, such as the freely rotating and freely jointed chain models, of real polymer chains, from the chain parameters. Such computations involve a statistical averaging process and yield the mean square average end-to-end distance $\langle r^2 \rangle$. In the absence of external interactions, such as an applied force or with a solvent, the value is the unperturbed value, $\langle r^2 \rangle_0$. The end-to-end distance cannot be directly measured, but can be related to the radius of gyration which is measurable. On a freely jointed chain model, of n links each of length l, the distribution of end-to-end distances, $W(r)$, is given by $W(r) = (2\pi nl^2/3)^{3/2} \exp(-3r^2/2nl^2)\,dr$, which is Gaussian and spherically symmetrical, the most probable value being given by $(r^*)^2 = 2nl^2/3$. The above equations apply to a monodisperse polymer, or a single polymer chain, for which the averaging may also be denoted by a bar as $\bar{r}^2$. The further averaging required for a polydisperse polymer may then be denoted by $\langle \bar{r}^2 \rangle$.

ENDURANCE   Alternative name for fatigue life.

ENDURANCE LIMIT   Alternative name for fatigue limit.

ENERGY BALANCE CRITERION   The basic criterion in fracture methanics for fracture to occur, as, for example, used in the Griffith theory. The criterion for crack propagation is that the strain energy released as a result of crack propagation should be equal to, or exceed, the energy required for crack extension in order to form new surfaces. This is the surface energy plus the energy for plastic work done in forming the surfaces. Thus $\partial E / \partial A \geq T$, where E is the elastic energy stored, A is the area of the crack and T is the surface work. $\partial E / \partial A$ is the strain energy release rate.

ENGINEERING LAW   The relationship that shear stress (σ) equals shear strain (e or ε) multiplied by shear modulus (G), i.e. $\sigma_k = Ge_k$ ($k = 4, 5, 6$) in contracted notation, or $\sigma_{ij} = 2G\varepsilon_{ij}$ ($i \neq j$, $i, j = 1, 2, 3$) in tensor notation.

ENGINEERING POLYMER   A polymer that may be used in engineering applications. In the case of an engineering plastic, this means a polymer that has a combination of stiffness, toughness and low creep and the applications are mostly small engineering parts such as gears, cams and bearings. Many high temperature resistant polymers are also engineering polymers. The most widely used engineering plastics are nylons, polyacetal, bisphenol A polycarbonate, thermoplastic polyesters and polyphenylene oxide. Engineering uses of rubbers include bearing pads, e.g. for bridges and buildings, mounting blocks and rubber springs.

ENGINEERING STRAIN   (1) Alternative name for extension.
   (2) Alternative name for a strain or strain component, but which has twice the value of the corresponding tensor (or mathematical) strain when this is a shear strain.

ENGINEERING STRESS   Alternative name for nominal stress.

ENHANCEMENT FACTOR   The ratio of the modulus of a filled material to that of the unfilled material at low strain. As strain increases, its value decreases due to the rupture of filler–matrix bonds and to high localised strains. This effect may be minimised by the use of a coupling agent. Thus modulus measurement at low strain may give an optimistic impression of the reinforcement to be expected in service which may involve higher strains.

ENKALON   Tradename for nylon 6.

ENKATHERM   Tradename for chelated PTO polymers.

ENTANGLEMENT COUPLING   The localised twisting or looping of polymer chains about each other leading to the formation of a temporary network structure. The existence of such mobile networks is necessary to explain the two stage relaxation or creep behaviour of high molecular weight polymers, i.e. two sets of relaxation or retardation times and the larger than expected dependency of viscosity on molecular weight at high molecular weights. The exact nature of the coupling between chains is not understood, but is undoubtedly topological rather than being due to intermolecular forces of attraction. It is probable that the entanglement is in the form of a looping of chains around each other over quite a long range of their contours. The movement of such entanglements may be modelled by considering the escape of the chain by reptation from an imaginary tube surrounding it.

ENTANGLEMENT THEORY   Alternative name for Graessley theory.

ENTHALPY OF POLYMERISATION   Alternative name for heat of polymerisation.

ENTRANCE EFFECT CORRECTION   Alternative name for Bagley end correction.

ENVIRONMENTAL STRESS CRACKING (ESC) (Corrosion cracking) (Solvent cracking) (Stress corrosion cracking) The cracking produced in a stressed solid when placed in an aggressive liquid environment. The stress required to produce cracking, and sometimes fracture, is less than that required in the absence of the liquid. Residual frozen-in stress may be sufficient even in the absence of any additional applied stress to produce cracking. However, there is some minimum stress required to produce cracking. Sometimes the term specifically refers to the cracking of polyethylene or polypropylene when exposed to certain detergents or oils and triaxial but not uniaxial stress. If so then the term stress crazing is used for other cases, notably when cracking occurs when amorphous plastics, such as polymethylmethacrylate, polystyrene and polycarbonate, are exposed to certain organic solvents.

ENZYMATIC DEGRADATION   Degradation produced by enzyme attack, usually causing random scission. In biopolymers, enzymolysis often ultimately converts the polymer to its constituent monomer, monomers or dimers and is therefore useful in characterisation. Synthetic polymers may be exposed to enzyme producing natural organisms during their service life and be subject to biodegradation. This is usually undesirable (biodeterioration) but may be beneficial in deliberately designed biologically degradable polymers.

ENZYME   A protein which is a catalyst for a biochemical reaction. Most enzymes are capable of catalysing only one specific reaction or type of reaction, so any particular organism contains many hundreds of different enzymes. Enzymes are usually named according to the reaction they catalyse, by adding the suffix -ase to the name of the reaction, e.g. carboxypeptidase catalyses hydrolysis of a peptide from the C-terminal end of a protein. Enzymes require specific conditions, e.g. of temperature, pH, presence of co-enzymes, to be effective, but when the conditions are right enzymes can increase reaction rates by factors of 10^{10}–10^{20}. Thus most metabolic processes require the presence of enzymes for their continuation. Only small parts of the enzyme molecule are involved in interaction with the reactant (called the substrate) at the active site, although the rest of the molecule is essential to ensure that the active site peptide units (often widely spaced along the polypeptide chain) are held together in the correct close spatial relationship. Often the tertiary structure contains β-sheet regions. The active site often consists of a cleft or shallow surface depression in the tertiary structure, and its precise geometry determines which substrates will be activated. Earlier it was thought that the substrate fitted exactly into the 'hole' in the enzyme surface, but now an 'induced fit' mechanism is believed to be more correct, in which the binding of the substrate causes conformational changes at the active site. The precise mechanism of enzyme action is complex and has been widely investigated, particularly in trypsin, chymotrypsin, carboxypeptidase, lysozyme and ribonuclease.

Enzyme activity may be considerably altered by the presence of other small molecular species. Thus many enzymes require the presence of cofactors (metal ions or co-enzymes) to be effective. In contrast, some substances act as enzyme inhibitors, sometimes irreversibly, e.g. cyanide ion blocks the action of cytochrome oxidase which is involved in the utilisation of atmospheric oxygen in the body, and so leads to death. In the latter case, if the enzyme is inhibited (or activated) by binding an initiator at a site other than at the active site, then the enzyme is an allosteric enzyme. Such enzymes often have a complex structure, sometimes being oligomeric proteins.

ENZYME UNIT   (EU) (International Unit) A unit of enzyme activity. The amount of enzyme causing the transformation of 1 μmol of substrate per minute under specified conditions, often with the temperature at 25°C. Often replaced by the more recently introduced unit, the katal.

EPCAR   Tradename for ethylene–propylene rubber.

EPDM   Abbreviation for ethylene–propylene terpolymer, alternatively called ethylene–propylene–diene monomer rubber.

EPICHLORHYDRIN   (ECH)

$$CH_2\!\!-\!\!CHCH_2Cl \quad \text{B.p. } 116°C.$$

Produced by chlorination of propylene to allyl chloride which is converted to glycerol dichlorohydrin by reaction

with hypochlorous acid, followed by dehydrochlorination to epichlorhydrin:

$$CH_2\!\!=\!\!CHCH_3 \rightarrow CH_2\!\!=\!\!CHCH_2Cl \rightarrow$$
$$ClCH_2CH(OH)CH_2Cl \rightarrow CH_2\!\!-\!\!CHCH_2Cl$$

Useful for the formation of glycidyl compounds by reaction with a wide variety of substrates containing reactive hydrogen atoms:

$$CH_2\!\!-\!\!CHCH_2Cl + R\!\!-\!\!H \rightarrow R\!\!-\!\!CH_2CHCH_2Cl \rightarrow$$
$$R\!\!-\!\!CH_2CH\!\!-\!\!CH_2 + HCl$$

Thus reaction with carboxylic acids gives glycidyl esters, reaction with amine groups gives glycidyl amines, but, most importantly, reaction with a diol gives a diglycidyl ether. This last reaction is the basis for most epoxy resin production, particularly by reaction with bisphenol A. Epichlorhydrin may be ring-open polymerised to poly-epichlorhydrin by use of a Ziegler–Natta, aluminium alkyl or chelating agent catalyst.

EPICHLORHYDRIN RUBBER   Tradenames: Hydrin, Herclor, Epichloromer. A homopolymer of epichlorhydrin (abbreviation EC) or a copolymer with about 40% ethylene oxide (the equimolar amount) (abbreviation ECO). These polymers are useful rubbers and have excellent resistance to ozone, oils, heat and weathering. They also have very low gas permeability. The homopolymer has a brittle point of about $-15°C$, but this value is lowered to about $-40°C$ in the copolymer but with some loss of oil resistance. The polymers are vulcanised through the chlorine atoms by the use of diamines with a basic compound such as lead oxide. A terpolymer (ETE) containing some allyl glycidyl ether repeat units is peroxide or sulphur vulcanisable.

EPICHLOROMER   Tradename for epichlorhydrin rubber.

EPIDERMIN   A soluble pre-keratin preparation obtained by extraction of the skin of certain animals, e.g. cow's nose.

EPIKOTE   Tradename of an epoxy resin.

EPIMER   One of a pair of configurational (or stereo-) isomers of a sugar that differs from the other of its epimeric pair only in the configuration of the carbon atom adjacent to that carrying the carbonyl group. Thus D-glucose and D-mannose are epimers. The definition is sometimes extended to include any pair of stereoisomers that differ only in the configuration of a single asymmetric carbon atom. Conversion of one epimer to the other of a pair is epimerisation.

EPIMERISATION   The conversion of one of a pair of epimers to the other.

EPISULPHIDE (Thiirane) The sulphur analogue of epoxide, i.e. a compound or group of structure

$$\begin{bmatrix} S \\ R \end{bmatrix}$$

but as with the term epoxide, episulphide is often restricted to describe what are strictly 1,2-episulphides, i.e.

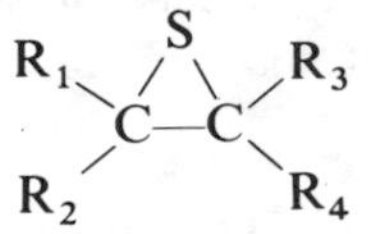

of which the two most important examples are ethylene sulphide and propylene sulphide. Episulphides may be ring-open polymerised to polysulphides by anionic or cationic catalysts and are in general more reactive than the corresponding epoxides.

EPITAXY The oriented growth of one crystalline substance on the surface of a crystalline substrate. In polymer crystallisation this frequently involves growth of a folded chain polymer on an extended chain crystal of the same polymer, e.g. in the formation of shish-kebabs. In a wider sense all polymer crystallisations (apart from single crystal formation) are epitaxial since they are multi-crystalline with one crystallite influencing the orientation of its neighbour. Non-polymer crystals also induce polymer crystal epitaxy, even if their surfaces show no orienting effect on the polymer crystal nucleus, as in transcrystallisation.

EPITHELIAL MUCIN A water soluble glycoprotein present in the oral and respiratory and other mucous secretions of higher animals and responsible for their properties. Usually it has a high L-fucose or sialic acid content and is closely related structurally to the blood-group substances. Certain mammalian submaxilliary mucins have been the most intensively studied. They usually contain 30–50% protein (largely of threonine, serine, proline, alanine and glycine units) as a core, linked to polysaccharide side chains (di- or oligosaccharides) through serine and threonine carboxylic acid groups.

EPM Abbreviation for ethylene–propylene rubber.

EPOCRYL Tradename for a vinyl ester resin.

EPOK Tradename for etherified urea–formaldehyde, phenol–formaldehyde (and rosin modified), melamine–formaldehyde, and benzoguanamine and butylated benzoguanamine polymers, alkyd resins and epoxy resins.

EPOK C Tradename for coumarone–indene resin.

EPOK V Tradename for polyvinyl acetate.

EPOLENE Tradename for low density polyethylene.

EPON Tradename for an epoxy resin.

EPOPHEN Tradename for an epoxy resin.

EPOREX Tradename for an epoxy resin.

EPOXIDE In general an alternative name for a cyclic ether, i.e. a compound of the general type

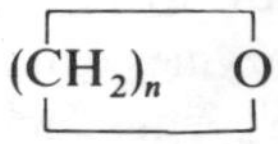

$$(CH_2)_n \quad O$$

or of a similar substituted derivative. Epoxides may be polymerised by ring-opening polymerisation to polyethers. However, in the context of polymer science, the term epoxide is reserved for 1,2-epoxides, i.e. compounds of the type

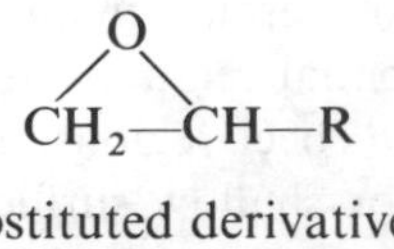

or other similar substituted derivatives. The best known are the simplest—ethylene and propylene oxides and the epoxy resins.

EPOXIDE EQUIVALENT The weight of an epoxy resin, in grams, which contains one equivalent of epoxide groups. Thus for a linear diepoxide the epoxide equivalent is half the number average molecular weight. It is determined either by infrared spectroscopy or by titration of the epoxy groups with hydrogen halide.

EPOXIDE RESIN Alternative name for epoxy resin.

3,4-EPOXY-6-METHYLCYCLOHEXYLMETHYL-3,4-EPOXY-6-METHYLCYCLOHEXANE CARBOXYLATE

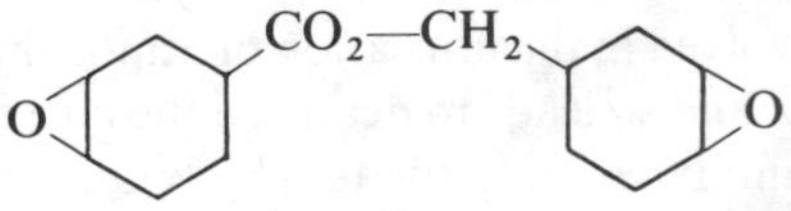

A liquid cycloaliphatic epoxy resin obtained by epoxidation of the corresponding cyclohexene derivative, itself obtained by Diels–Alder reaction of butadiene with crotonaldehyde, followed by the Tischenko reaction.

EPOXY–NOVOLAK An epoxy resin having, before curing, the generalised structure:

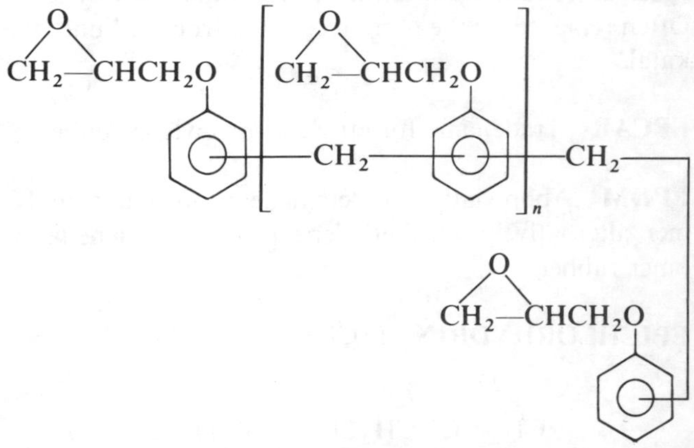

The most common commercial type of resin has $n = 1.6$ on average, corresponding to 3–6 epoxy groups per molecule. Thus on curing, due to the higher functionality than in normal resins, a higher crosslink density is produced in the cured product, which therefore has a higher heat distortion temperature. It is synthesised by reaction of a low molecular weight novolak with epichlorhydrin in the presence of sodium hydroxide.

EPOXY NUMBER   The gram equivalents of epoxide oxygen per 100 g of an epoxy resin. It is determined by titration of the epoxy groups with hydrochloric acid in pyridine solution.

EPOXY RESIN   (Epoxide resin) Tradenames: Araldite, Beckopox, DEN, DER, Epikote, Epon, Epophen, Eporex, Grilonit, Lopox, Uranox. A polymer, usually a mixture of low molecular weight oligomers, containing, on average, two or more epoxide groups per molecule. The term is also used to refer to the network polymers obtained by crosslinking the oligomeric prepolymers. Most commercial epoxy resins are very low molecular weight oligomers and form relatively tough products when crosslinked by reaction with an appropriate hardener (or curing agent). The most common type is a diglycidyl ether, particularly the diglycidyl ether of bisphenol A. Other glycidyl ethers, such as novolak–epoxy resins, and the glycidyl ethers of glycerol, polypropylene glycol or pentaerythritol, have a higher epoxy functionality than the difunctionality of the bisphenol A material. Much higher molecular weight polymers are also of interest (the phenoxy resins). Other types of epoxy resins are the glycidyl esters, glycidyl amines, epoxidised diene polymers and the cycloaliphatic epoxy resins.

Commercial epoxy resins, about 95% of which are DGEBA, are mixed oligomers characterised by their epoxide equivalent and have a number average molecular weight of up to about 500 and are viscous liquids. Above this molecular weight, the polymers are low melting point solids.

The linear epoxy resins are useful as surface coatings, adhesives, laminating plastics, castings and for encapsulation, for which they are crosslinked using any of a wide variety of crosslinking agents which are capable of reaction with the epoxide groups, by virtue of their having active hydrogen atoms:

$$R{-}H + \overset{O}{\underset{}{\diagup C{-}C \diagdown}} \rightarrow \overset{R \quad OH}{\underset{}{\diagup C{-}C \diagdown}}$$

Subsequently, the hydroxyl groups so formed may further participate in crosslinking. Since R—H may be di-, tri- or tetrafunctional, highly crosslinked products may be formed.

Aliphatic primary or secondary amines, such as diethylenetriamine, triethylenetetramine and diethylaminopropylene, give low viscosity mixtures with fast room temperature cures and are often used in stoichiometric amounts (based on the epoxide groups):

$$R{-}\overset{H}{\underset{H}{N}} + 2\,CH_2{-}CH\sim \rightarrow \sim CHCH_2{-}\underset{R}{N}{-}CH_2CH\sim$$

Cured resins have a heat distortion temperature of 70–110°C and a tensile modulus of about 3 GPa. Diethylaminopropylene, N-aminoethylpiperazine and dicyanamide are also useful curing agents. Fatty polyamides are also useful room temperature curing agents and give tough, even flexible, products. Various modified amines, such as ethylene oxide– and acrylonitrile–epoxy resin adducts and ketimines are also useful. Aromatic amines, such as m-phenylenediamine, 4,4′-diaminodiphenylmethane and diaminodiphenylsulphone give cured resins with higher heat distortion temperatures and higher chemical resistance but often need curing at elevated temperatures.

Acid anhydrides are widely used to produce cured resins with even higher heat distortion temperatures and thermal and chemical resistance. They are preferred to acids since less water is evolved in their use. Initially carbonyl groups are generated:

$$\overset{}{\underset{OH}{\mid}} + R \overset{C{=}O}{\underset{C{=}O}{\diagdown O \diagup}} \rightarrow \overset{O}{\underset{\underset{\mathbf{I}}{C}}{O}} R{-}COOH$$

followed by crosslinking mostly by:

$$\mathbf{I} + CH_2{-}CHCH_2\sim \rightarrow O \overset{R}{\underset{C \quad C}{}} {-}O{-}CH_2{-}\overset{OH}{\underset{}{CH}}\sim$$

and

$$\overset{}{\underset{OH}{\mid}} + CH_2{-}CH\sim \rightarrow \overset{OH}{\underset{CH_2CH\sim}{}}$$

Typically 0·85–1·0 mol of anhydride carbonyl to epoxy stoichiometry is used, with about 1% tertiary amine catalyst as well. The anhydrides used include phthalic, tetrahydrophthalic, hexahydrophthalic, maleic, pyromellitic, trimellitic, nadic methyl, dodecenylsuccinic and chlorendic anhydrides. Crosslinking may also be achieved by catalysing reaction between the epoxy resin molecules themselves by opening of the epoxy rings. Lewis acids, such as boron trifluoride–monoethylamine complex, and Lewis bases, such as o-(dimethylaminoethyl)phenol, tris-(dimethylaminomethyl)phenol and 2-ethyl-4-methylimidazole, are used.

Cured epoxy resins are characterised by their toughness, low shrinkage during cure, high adhesion to many substrates and good chemical resistance, although the properties depend very much on the particular curing system used.

EPR   Abbreviation for ethylene–propylene rubber.

EPR SPECTROSCOPY   Abbreviation for electron paramagnetic resonance spectroscopy.

EPS   Abbreviation for expanded polystyrene.

EQUATION OF CONTINUITY Alternative name for continuity equation.

EQUATION OF MOMENTUM Alternative name for momentum equation.

EQUILIBRATION The redistribution of molecular sizes in a polyorganosiloxane, brought about by siloxane interchange reactions by heating with acids and bases. The initially broad molecular weight range of linear and cyclic siloxanes formed by hydrolysis of chlorosilanes can be narrowed and the material made more molecularly homogeneous. If a chain stopper in the form of a trisubstituted silicon compound is also added to the equilibration mixture then a lowered molecular weight will result. In the preparation of silicone fluids the addition of known amounts of hexamethyldisiloxane results in a stable polymer of the desired molecular weight and viscosity.

EQUILIBRIUM CENTRIFUGATION Alternative name for sedimentation equilibrium method.

EQUILIBRIUM COMPLIANCE (Relaxed compliance) Symbols J_e (or J_r) in shear, D_e (or D_r) in tension and B_e (or B_r) in bulk. The compliance of a viscoelastic material at very long times when the material can be considered as a viscoelastic solid (rather than as a liquid) so that a final equilibrium strain is obtained, as modelled for example by the Maxwell model. In bulk compressive deformation an equilibrium always exists and B_r is the thermodynamic compressibility. For a linear polymer an equilibrium compliance may not exist since viscous flow may occur continuously. For a crosslinked polymer, however, an equilibrium is attained, which for rubbers is described by the theory of rubber elasticity.

EQUILIBRIUM DIALYSIS The equilibration of a polymer solution, usually a polyelectrolyte in water, with a solution of a third low molecular weight component, usually a salt, from which it is separated by a membrane permeable to the low molecular weight solute, but not to the polymer. Thus if a polymer solution is suspended in a sack of a permeable membrane in a solution of the third component, the Donnan equilibrium is established, the chemical potentials of the third component being the same in each solution.

EQUILIBRIUM DISTRIBUTION OF MOLECULAR WEIGHT The molecular weight distribution that results when a step-growth polymer undergoes interchange reactions until equilibrium is established between the different molecular species. Identical to the Schultz–Flory distribution, which is normally produced by straightforward step-growth polymerisation.

EQUILIBRIUM MODULUS (Relaxed modulus) Symbols G_e (or G_r) in shear, E_e (or E_r) in tension and K_e (or K_r) in bulk. The modulus of a viscoelastic material at very long times when the material can be considered as a viscoelastic solid (rather than as a liquid), so that a final, equilibrium, constant strain is obtained, as, for example, modelled by the Maxwell model. For an uncrosslinked polymer an equilibrium is usually not reached since viscous flow occurs. The equilibrium modulus is the inverse of the equilibrium compliance.

EQUILIBRIUM MONOMER CONCENTRATION Symbol $[M]_c$. The concentration of monomer in equilibrium with its polymer at a temperature at which depropagation (rate constant k_{dp}) competes with propagation (rate constant k_p), i.e. a higher than normal propagation temperature, so that the equilibrium,

$$\sim\!\!\sim\!\mathrm{M}_n^* + \mathrm{M} \underset{k_{dp}}{\overset{k_p}{\rightleftharpoons}} \sim\!\!\sim\!\mathrm{M}_{n+1}^*$$

is established. As the temperature is raised the equilibrium moves to the left, so $[M]_c$ increases. At high enough temperatures high polymer will not be formed at all. When the monomer concentration is at some relatively high specified value, e.g. 1 M, the temperature is the ceiling temperature.

EQUILIBRIUM POLYMERISATION (Reversible polymerisation) A polymerisation which proceeds to an equilibrium between monomer and polymer. Although many step-growth polymerisations proceed through equilibria, use of the term is normally restricted to chain polymerisation. This proceeds to an equilibrium with significant amounts of monomer when conducted near the ceiling temperature, i.e. at considerably above the normal polymerisation temperature. This is due to any tendency of the polymer to depolymerise. For polymers with low ceiling temperatures, e.g. poly-(α-methylstyrene), polymerisation proceeds to equilibrium even at ambient temperature. Several ring-opening polymerisations, e.g. of tetrahydrofuran, 1,3-dioxalone and trioxane, also behave in this way.

EQUIVALENCE PRINCIPLE The representation of complex, multiaxial stress–strain behaviour by a single relationship involving the use of equivalent stress and equivalent strain.

EQUIVALENT CHAIN Alternative name for statistically equivalent chain.

EQUIVALENT STRAIN Symbol $\bar{e}$. Defined as
$$\bar{e} = \sqrt{2}/3[(e_1 - e_2)^2 + (e_2 - e_3)^2 + (e_3 - e_1)^2]^{1/2}$$
where e_1, e_2 and e_3 are the principal strains. Useful in the analysis of stress–strain behaviour when combined, e.g. shear and tensile, stresses and strains are involved, using the equivalence principle.

EQUIVALENT STRESS Symbol $\bar{\sigma}$. Defined as
$$\bar{\sigma} = 1/2^{1/2}[(\sigma_1 - \sigma_2)^2 + (\sigma_2 - \sigma_3)^2 + (\sigma_3 - \sigma_1)^2]^{1/2}$$
where σ_1, σ_2 and σ_3 are the principal stresses. Useful in the analysis of stress–strain behaviour where combined, e.g. shear and tensile, stresses and strains are involved, using the equivalence principle.

ERACLENE Tradename for low density polyethylene.

ERACLENE HD Tradename for high density polyethylene.

ERINOID Tradename for polystyrene and for a caesin plastic.

ERYTHRODIISOTACTIC POLYMER A ditactic polymer in which the configuration of each type of stereoisomeric centre, such as the C* atoms in

$$-\!\!\left[C^*HX\!\!-\!\!C^*HY\right]_n$$

are both the same, both being isotactic. In the Fischer projection the substituents X and Y are on the same side of the chain:

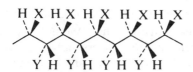

or

ERYTHROPOLYMER A ditactic polymer in which the configurations of the two types of stereoisomeric centre are the same.

ESC Abbreviation for environmental stress cracking.

ESCA Abbreviation for electron spectroscopy for chemical application.

ESCON Tradename for polypropylene.

ESCORENE Tradename for low density polyethylene and linear low density polyethylene.

ESCORENE ALPHA Tradename for linear low density polyethylene with hexene-1 as the comonomer.

ESR SPECTROSCOPY Abbreviation for electron spin resonance spectroscopy.

ESSENTIAL AMINO ACID An α-amino acid found in proteins, which forms an essential component of the human diet, since it is not synthesised in the human body. These α-amino acids comprise threonine, methionine, valine, leucine, isoleucine, lysine, phenylalanine and tryptophane. Arginine and histidine are also essential for many higher animals.

ESSO BUTYL Tradename for butyl rubber.

ESTANE Tradename for a linear thermoplastic polyurethane elastomer with no crosslinking, produced from poly-(tetramethylene adipate) of molecular weight about

1000, MDI and 1,4-butanediol with an exact equivalence of hydroxy to isocyanate groups, which gives the minimum of crosslinking.

ESTER GUM The ester produced by reaction of rosin with glycerol or other alcohols, e.g. pentaerythritol. It is usually mainly a mixture of glycerides of abietic acid. Esterification has the effect of increasing the hardness and softening point of the rosin; it is therefore useful in varnishes.

ESTER INTERCHANGE (Transesterification) An exchange of groups that occurs when an ester is heated with either a hydroxycompound (alcoholysis) or a carboxylic acid (acidolysis). For alcoholysis of a polyester with hydroxyl terminated polymer chains, the reaction may be represented as:

$$P_m\underset{\underset{O}{\|}}{O}CR\underset{\underset{O}{\|}}{C}OP_n + HOP_xOH \rightarrow P_m\underset{\underset{O}{\|}}{O}CR\underset{\underset{O}{\|}}{C}OP_x + P_nOH$$

where P_m, P_n and P_x represent polymer chains. Thus two different polyesters when melted together will form a simple random copolymer. Furthermore, if polymers of different number average degrees of polymerisation are mixed, the most probable molecular weight distribution will be produced, giving a change in $\bar{M}_w$ but not in $\bar{M}_n$, since the total number of molecules does not change. Ester interchange also occurs during polyesterification and enables non-stoichiometric amounts of monomers to be used, although excess glycol is usually preferred. Alternatively ester interchange may be used for polymer formation with advantage, by reaction of a diester with a diol:

$$n\,HOROH + n\,R''OOCR'COOR'' \rightarrow$$

$$-\!\!\left[OROCR'\underset{\underset{O}{\|}}{C}\right]_n + 2n\,R''OH$$

Provided R''OH is volatile (often R'' is CH_3) it may be removed and equilibrium forced towards polyester formation. Usually the reaction is catalysed by a protonic catalyst, e.g. p-toluene sulphonic acid, or weak base, e.g. a metal acetate. The method is particularly important for the synthesis of polyethylene and polybutylene terephthalates.

ESTRON Tradename for cellulose acetate rayon.

ETCHING A sample treatment helping to reveal the morphological features when viewed microscopically, especially by electron microscopy. In crystalline polymers sectioning and polishing of the sample surface, followed by etching with solvent vapour or by ion bombardment causes the crystal morphology to be thrown into relief. In polymer blends one of the two phases may be preferentially etched by a suitable solvent or chemical treatment.

ETE Tradename for epichlorhydrin/ethylene oxide/allyl glycidyl ether terpolymer rubber.

ETFE Abbreviation for ethylene–tetrafluoroethylene copolymer.

ETHANE-1,2-DIOL Alternative name for ethylene glycol.

ETHANOL (Ethyl alcohol) CH_3CH_2OH. B.p. 78·3°C. Like methanol, it is not a good solvent for most polymers, although it does dissolve polyvinyl acetate, low molecular weight silicones, some phenol–formaldehyde resins, ethyl cellulose and some natural resins. It is therefore useful in lacquers and varnishes. It forms an azeotrope with 4·4% water. Methylated spirits is ethanol diluted with 5% methanol and often contains up to 10% water. It is also called industrial methylated spirits (abbreviated to IMS). The water can considerably reduce its solvent power.

ETHENE Alternative name for ethylene.

ETHENESULPHONIC ACID Alternative name for vinylsulphonic acid.

2-ETHOXYETHANOL Alternative name for ethylene glycol monoethyl ether.

2-(β-ETHOXYETHOXY)ETHANOL Alternative name for diethylene glycol monoethyl ether.

ETHOXYETHYL ACRYLATE

$$CH_2=CHCOOCH_2CH_2OC_2H_5.$$
B.p. 174°C.

Useful as a comonomer in acrylic elastomers to replace some alkyl acrylate in order to improve low temperature properties and oil resistance.

ETHOXYTRIGLYCOL Alternative name for triethylene glycol monoethyl ether.

ETHYL ACETATE $CH_3COOC_2H_5$. B.p. 77·2°C. A good solvent for cellulose nitrate, cellulose butyrate (but not for cellulose acetate unless it contains 5–30% ethanol), ethyl cellulose, polyvinyl acetate, polystyrene and some natural resins. It swells polyvinyl chloride, polymethylmethacrylate and polyethylene.

ETHYL ACRYLATE $CH_2=CHCOOC_2H_5$. B.p. 99·5°C. The monomer for the preparation of polyethylacrylate by emulsion polymerisation. It has an unpleasant smell and is highly lachrymatory and irritating. It is prepared by the same methods as used for acrylic acid and methyl acrylate except that ethanol is used in the final step. Technologically it is the most important acrylate monomer and therefore it is also used for the conversion to other acrylate monomers by transesterification.

ETHYL ALCOHOL Alternative name for ethanol.

ETHYLBENZENE

B.p. 136°C.

An intermediate in the synthesis of styrene. Usually produced by the alkylation of benzene with ethylene using a Friedel–Crafts catalyst such as aluminium trichloride. Sometimes used in adhesives for polystyrene and its copolymers.

ETHYL CELLULOSE A cellulose ether in which some of the cellulose hydroxyl groups are replaced by ethoxy groups. Prepared by reaction of alkali cellulose with ethyl chloride. Commercial products have a DS of 2·2–2·6. At a DS of 0·8–1·3 the polymer is water soluble; above a DS of 1·4 solubility in non-polar solvents occurs until, at DS 2·8, the polymer is only soluble in non-polar solvents. It has the best water resistance and electrical insulation properties of all the cellulose polymers and has been used, e.g. as sheet in refrigerator parts, because of its good low temperature impact strength. It is also useful as a hot melt strippable coating and in other coating applications at a DS of about 2·5.

ETHYLENE (Ethene) $CH_2=CH_2$. B.p. −103·7°C. The largest tonnage industrial organic chemical, largely due to its important use as the monomer for the production of polyethylene and its conversion to vinyl chloride. It is produced by the pyrolysis of any of a range of hydrocarbon feedstocks, usually in a tubular reactor at 750–900°C with a residence time of about 1 s. Steam is frequently added to reduce hydrocarbon partial pressure and coke formation. Single component feedstocks, such as ethane or propane, are preferred, but mixed hydrocarbons, such as natural gas mixtures are also used. In Europe various fractions of crude oil distillation are often used, such as naphtha and gas oil. Even unrefined crude oil may be used. A product distribution of saturated and unsaturated hydrocarbons is formed, but yields of ethylene and other olefin monomers may be maximised. Typical yields are: for ethane feedstock—ethylene 50%, for propane feedstock—ethylene 40%, propylene 11%, methane 25% and for light naphtha feedstock—ethylene 32%, ethane 4%, propylene 15%, butadiene 5%, butenes 5% and methane 15%. In addition, ethylene may be obtained by the dehydration of ethanol, itself obtained by the fermentation of sugars. Most (about 50%) ethylene is used directly for polymerisation to polyethylene or ethylene copolymers, about 15% is converted to ethylene oxide, about 15% is converted to ethylene dichloride (itself for conversion to vinyl chloride) and about 10% is converted to ethylbenzene (for styrene). Other conversions include hydration to ethanol, oligomerisation to higher olefins and oxidation to linear primary alcohols.

ETHYLENE–ACRYLIC ACID COPOLYMER (EAA) Tradenames: Surlyn, Dow EAA, AC Polymer. Block (Surlyn) or random (Dow EAA and AC Polymer) copolymers of ethylene and acrylic acid, with up to 20% acid units, produced by free radical polymerisation. The polymers have high adhesion to metals and other substrates, combined with toughness, and are useful as packaging laminate component films.

ETHYLENE–ACRYLIC ELASTOMER Tradename: Vamac. A copolymer of ethylene, methyl acrylate and a crosslinking site monomer which contains a pendant carboxyl group. It is vulcanised by primary diamines or peroxides. Vulcanisates show a moderate oil resistance and good heat resistance, having a useful service temperature range of -30 to $+180°C$ and good ozone resistance. Applications include O-rings, seals and other under-bonnet automotive parts, as well as in cable insulation. It is also useful as a high temperature anti-vibration material since it shows good damping over a wide temperature range.

ETHYLENE–BUTENE-1 COPOLYMER
(Poly-(butene-1-co-ethylene)) A copolymer produced commercially by the Phillips process containing up to 5% butene-1. It is therefore similar to high density polyethylene but has a slightly lower density and better resistance to creep and environmental stress cracking. The polymer is therefore useful in bottle and pipe manufacture.

ETHYLENE DICHLORIDE Alternative name for 1,2-dichloroethane.

ETHYLENE–ETHYL ACRYLATE COPOLYMER
(EEA) Tradename: Zetafin. A range of copolymers produced commercially in a similar way to ethylene–vinyl acetate copolymer and with somewhat similar properties, but with better thermal stability. However, the mechanical and toxicological properties are not so attractive.

ETHYLENE GLYCOL (Ethane-1,2-diol)

$$HOCH_2CH_2OH. \qquad B.p. \ 197°C.$$

Prepared by the high temperature/high pressure hydration of ethylene oxide. A useful diol for the preparation of polyesters such as polyethylene terephthalate, polyester prepolymers for polyurethanes, e.g. polyethylene adipate, and polyester plasticisers. It is occasionally used for unsaturated polyesters, but the products have a tendency to crystallise. A plasticiser for casein and other protein plastics. A solvent for some phenol–formaldehyde resins and low molecular weight silicones. A plasticiser for regenerated cellulose and some cellulosic papers. Sometimes used to increase water tolerance in other solvent systems.

ETHYLENE GLYCOL DIMETHACRYLATE (Glycol dimethacrylate)

$$CH_2{=}\overset{\overset{\textstyle CH_3}{|}}{C}COOCH_2CH_2OCO\overset{\overset{\textstyle CH_3}{|}}{C}{=}CH_2$$

$$B.p. \ 97°C/4 \ mm.$$

A dimethacrylate which is useful as a comonomer in thermosetting acrylic materials.

ETHYLENE GLYCOL MONO-*n*-BUTYL ETHER
Tradenames: Butyl cellosolve, Butyl oxitol.

$$CH_3(CH_2)_3OCH_2CH_2OH. \qquad B.p. \ 171°C.$$

A solvent for cellulose nitrate and for many natural resins. Miscible with both water and hydrocarbons.

ETHYLENE GLYCOL MONOETHYL ETHER (2-Ethoxyethanol) (Cellosolve) (Oxitol)

$$CH_3CH_2OCH_2CH_2OH. \qquad B.p. \ 135°C.$$

A solvent for cellulose nitrate, many natural resins and for polyvinyl acetate.

ETHYLENE GLYCOL MONOMETHYL ETHER
(2-Methoxyethanol) (Methyl cellosolve)

$$CH_3OCH_2CH_2OH. \qquad B.p. \ 124·3°C.$$

A solvent for cellulose nitrate, cellulose acetate, polyvinyl acetate, polyvinyl acetal and many natural resins.

ETHYLENE GLYCOL MONOMETHYL ETHER ACETATE (Methoxyethyl acetate)

$$CH_3COOCH_2CH_2OCH_3. \quad B.p. \ 145·1°C.$$

A solvent for cellulose esters and ethers, polyvinyl acetate, epoxy resins, polystyrene, polyurethanes and some natural resins. Miscible with water in all proportions.

ETHYLENEIMINE (Aziridine)

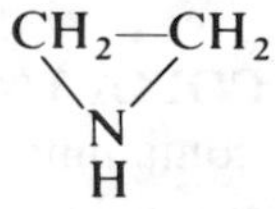

$$\overset{\textstyle CH_2{-}CH_2}{\underset{\underset{\textstyle H}{\textstyle N}}{\diagdown\diagup}} \qquad B.p. \ 55–56°C.$$

The monomer for polyethyleneimine, obtained by ring-opening polymerisation. It is prepared by reaction of 1,2-dichloroethane with ammonia via the intermediate β-chloroethylamine:

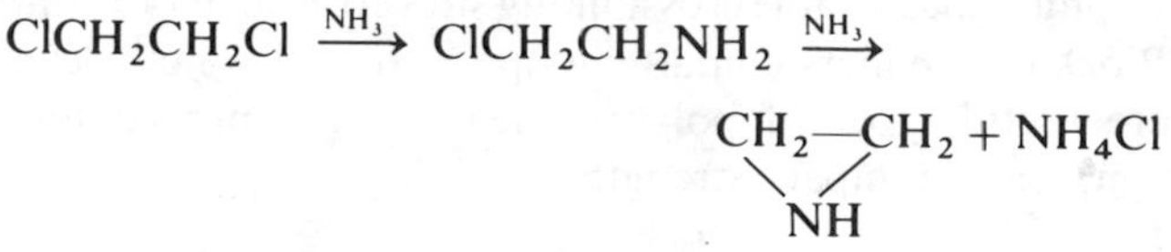

$$ClCH_2CH_2Cl \xrightarrow{NH_3} ClCH_2CH_2NH_2 \xrightarrow{NH_3}$$

$$\overset{\textstyle CH_2{-}CH_2}{\underset{\underset{\textstyle NH}{}}{\diagdown\diagup}} + NH_4Cl$$

Useful in the preparation of certain fire retardants, e.g. tris-(1-aziridinyl)phosphine oxide and sulphide, which are also sometimes simply called aziridine.

ETHYLENE–MALEIC ANHYDRIDE COPOLYMER
(EMA) An alternating copolymer of ethylene and maleic anhydride prepared by free radical polymerisation and having the structure

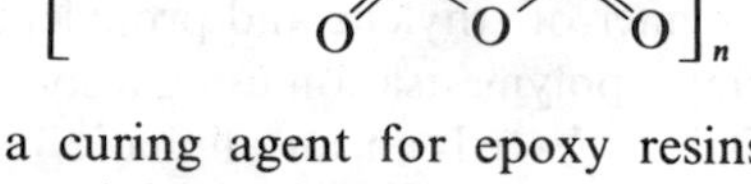

$$\left[CH_2{-}CH_2{-}\underset{\underset{O}{\overset{\textstyle |}{C}}}{CH}{-}\underset{\underset{O}{\overset{\textstyle |}{C}}}{CH} \right]_n$$

Useful as a curing agent for epoxy resins and other hydroxyl containing polymers.

ETHYLENE–METHACRYLIC ACID COPOLYMER
Tradename: Nucrel. A commercially produced copolymer, with 1–10% methacrylic acid, useful in the formation of ionomers.

ETHYLENE–METHYLACRYLATE COPOLYMER

Tradename: Vamac. A copolymer containing approximately equal amounts of the two comonomers together with a small amount of a carboxylic termonomer required for vulcanisation. A rubber which may be crosslinked by diamines or peroxides to give vulcanisates with good high temperature (to 175°C) and oil resistance.

ETHYLENE OXIDE

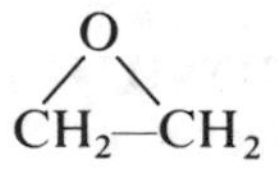

$$CH_2-CH_2 \quad \text{(with O bridging)}$$

B.p. 11°C.

Produced by direct oxidation of ethylene with oxygen at about 250°C and 20 atm pressure using a silver catalyst, or, in an older process from ethylene via the chlorohydrin. The monomer for the production of polyethylene oxide, but it is also useful for the production of acrylic acid, ethylene glycol, hydroxyethylcellulose and ethylenimine.

ETHYLENE–PROPYLENE BLOCK COPOLYMER

One of a range of copolymers containing up to about 15% ethylene, which considerably improves the low temperature impact properties compared with propylene homopolymers. Some early commercial types are referred to as polyallomers. Often the polymers are simply called propylene copolymers.

ETHYLENE–PROPYLENE COPOLYMER

One of a wide range of copolymers containing ethylene and propylene monomer units. Several different types are of importance commercially. Random copolymers containing relatively large amounts of both comonomers are useful as ethylene–propylene rubbers. Similar copolymers but also containing a third comonomer to provide sulphur vulcanisable crosslinking sites are also important. Block copolymers containing up to about 15% ethylene are useful types of polypropylene with improved low temperature impact strength.

ETHYLENE–PROPYLENE–DIENE MONOMER RUBBER

Alternative name for ethylene–propylene terpolymer.

ETHYLENE–PROPYLENE–ETHYLIDENE–NORBORNYL SULPHONATE

Alternative name for sulphonated EPDM.

ETHYLENE–PROPYLENE RUBBER (EPR) (EPM) (EPDM)

Tradenames: Buna AP, Dutral, Epcar, Intolan, Keltan, Nordel, Royalene, Vistalon. A rubber based on a random copolymer of ethylene and propylene. Prepared by Ziegler–Natta polymerisation using a soluble catalyst such as $AlEt_2Cl + VOCl_3$ in an aliphatic hydrocarbon solvent. A third diene monomer is incorporated to about 3–10% in the ethylene–propylene terpolymers, which being unsaturated may be vulcanised by conventional sulphur vulcanisation methods. Termonomers used are dicyclopentadiene, 1,4-hexadiene and ethylidene-norbornene. The latter is the most common. Commercial polymers contain 50–70% ethylene units and have a T_g value of about −50°C. On stretching the polymers crystallise, so the pure gum vulcanisates have high tensile strengths. EPM is vulcanised by peroxides, especially dicumyl peroxide. However, peroxides cause some chain scission as well. This may be reduced by the use of a co-agent which reacts with sterically hindered radicals which are reluctant to combine to produce crosslinks. EPDM may be vulcanised by methods similar to those used with butyl rubber, including sulphur vulcanisation with a powerful accelerator system such as mercaptobenzothiazole plus tetramethylthiurammonosulphide. Vulcanisates have very good ozone and oxygen resistance and high resilience. They may be oil extended and reinforced with fillers.

ETHYLENE–PROPYLENE TERPOLYMER

The terpolymer of ethylene (50–70%), propylene and a few per cent of a diene monomer. The latter enables the polymer to be vulcanised by conventional sulphur vulcanisation techniques.

ETHYLENE SULPHIDE

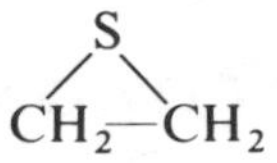

$$CH_2-CH_2 \quad \text{(with S bridging)}$$

B.p. 56°C.

Produced by the reaction of thiourea or potassium thiocyanate or carbonyl sulphide with ethylene oxide. It may be polymerised by ring-opening polymerisation to polyethylene sulphide by cationic or anionic polymerisation.

ETHYLENESULPHONIC ACID

Alternative name for vinylsulphonic acid.

ETHYLENE–TETRAFLUOROETHYLENE COPOLYMER

Alternative name for tetrafluoroethylene–ethylene copolymer.

ETHYLENETHIOUREA (ETU) (2-Mercaptoimidazoline)

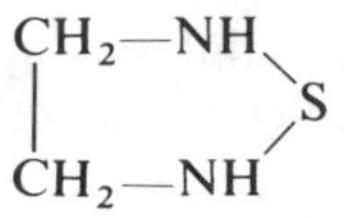

$$\begin{array}{c} CH_2-NH \\ | \quad\quad\quad\; \rangle S \\ CH_2-NH \end{array}$$

M.p. 185°C.

A vulcanisation agent widely used for polychloroprene. However, its use is now somewhat restricted due to its toxicity.

ETHYLENEUREA

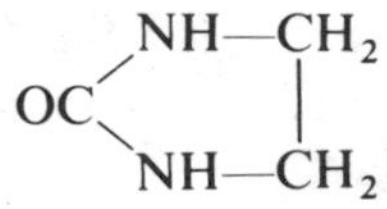

$$OC\begin{array}{c} NH-CH_2 \\ \quad\quad\quad | \\ NH-CH_2 \end{array}$$

M.p. 133·7°C.

Prepared by reaction of ethylenediamine with urea by refluxing a concentrated aqueous solution. Useful in the preparation of dimethylolethyleneurea. It is, however, under suspicion as a toxic material.

ETHYLENE–VINYL ACETATE COPOLYMER

(EVA) Tradenames: Alathon, Elvac, Elvax, Europrene EVA, Evatane, Lacqtene, Levapren, Montrathene, Ultrathene. Random copolymers with a wide range of vinyl acetate content which are produced commercially by the high pressure free radical ethylene polymerisation process. As the vinyl acetate content increases, so structural irregularity increases and hence crystallisation decreases. Copolymers with only a few per cent of vinyl acetate are like very low density polyethylene but with greater flexibility, tack and gloss and are useful as stretch and cling film. At 10–15% vinyl acetate content the copolymers resemble plasticised polyvinyl chloride in flexibility but without the necessity of having a plasticiser present. Copolymers with about 30% vinyl acetate are soluble, e.g. in toluene, and are useful as adhesives. Copolymers with 30–40% vinyl acetate are useful rubbers and may be vulcanised with peroxides.

ETHYLENE–VINYL ACETATE RUBBER (EVA

rubber) Tradename: Levapren. An ethylene–vinyl acetate random copolymer containing about 45% vinyl acetate units. It is vulcanised by peroxides, e.g. dicumyl peroxide and ditertiarybutyl peroxide. Its vulcanisates have good oxygen, ozone and heat resistance.

ETHYLENE–VINYL ALCOHOL COPOLYMER

(EVOH) Tradenames: Clarene, Eval, GL-Resin, Levasint, Selar OH. A copolymer containing both ethylene and vinyl alcohol repeat units. Copolymers are produced by hydrolysis of ethylene–vinyl acetate copolymers. Copolymers containing about 20–35 wt% of ethylene are crystalline and are useful as barrier polymers, having very low permeabilities to many vapours and odours, but not to water vapour. Indeed the polymers are very water sensitive and so are used only as laminates, sandwiched between two layers of the main packaging polymer, such as polyethylene.

2-ETHYLHEXYLDIPHENYL PHOSPHATE

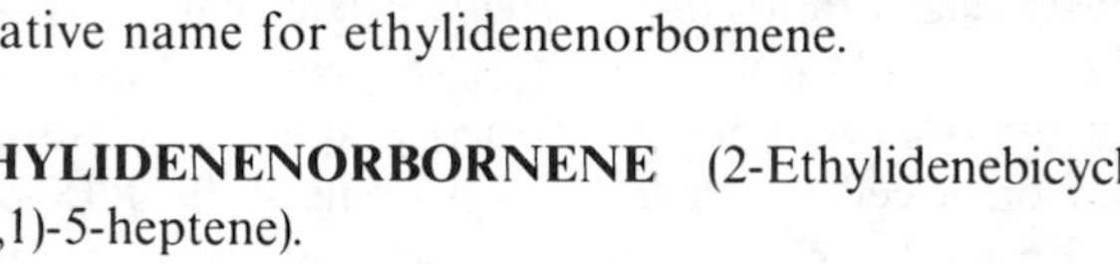

$$\left[\bigcirc -O \right]_2 POCH_2CH(CH_2CH_3)(CH_2)_3CH_3 \atop \quad\quad\quad\quad \| \atop \quad\quad\quad\quad O$$

B.p. 375°C/760 mm.

A flame retardant plasticiser, particularly for polyvinyl chloride and its copolymers and for nitrile rubbers.

2-ETHYLIDENEBICYCLO-(2,2,1)-5-HEPTENE Alternative name for ethylidenenorbornene.

ETHYLIDENENORBORNENE (2-Ethylidenebicyclo-(2,2,1)-5-heptene).

CHCH₃ / CH₂

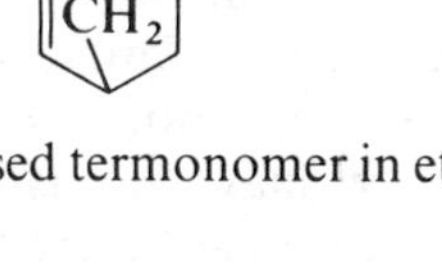

The most widely used termonomer in ethylene–propylene rubber.

ETHYL LACTATE $CH_3CH(OH)COOCH_2CH_3$. B.p 154·5°C. A good solvent for cellulose esters and ethers, many natural resins, urea–formaldehyde resins, polystyrene and polyvinyl acetate.

ETHYL METHACRYLATE

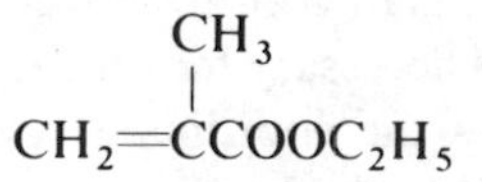

$$CH_2{=}\overset{\overset{\textstyle CH_3}{|}}{C}COOC_2H_5 \quad \text{B.p. } 117°C.$$

Produced by transesterification of methyl methacrylate with ethanol. The monomer for polyethylmethacrylate.

ETHYLPHTHALYLETHYL GLYCOLLATE

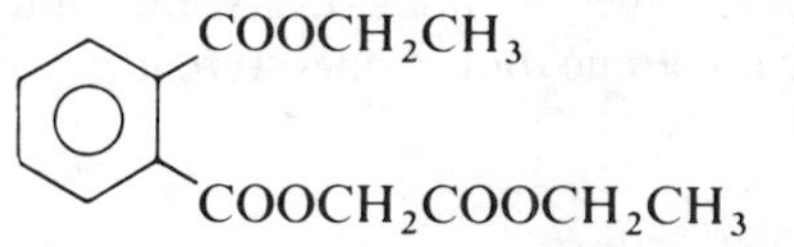

COOCH₂CH₃ / COOCH₂COOCH₂CH₃   B.p. 190°C/5 mm.

A plasticiser for cellulose esters and ethers, alkyd resins, phenol–formaldehyde resins, polyvinyl acetate, polyvinyl chloride and polystyrene.

N-ETHYL-p-TOLUENESULPHONAMIDE

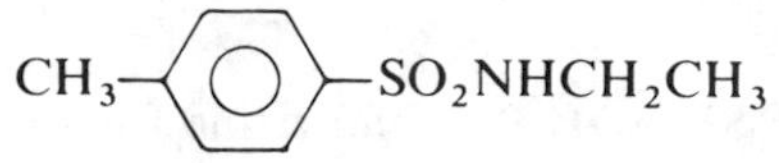

CH₃—◯—SO₂NHCH₂CH₃

B.p. 340°C/760 mm.
M.p. 59°C.

A plasticiser for amide polymers, such as proteins and nylons, and also for cellulose esters and ethers and polyvinyl acetate, especially for adhesive uses. Mixtures of o- and p-isomers are also frequently used.

ET POLYMER Tradename for a thermoplastic elastomer consisting of soft butyl rubber segments grafted onto segments of polyethylene.

ETU Abbreviation for ethylenethiourea.

EU Abbreviation for enzyme unit.

EULER EQUATION The form of the momentum equation for an ideal fluid, i.e. for which there are no shear components in the stress tensor, or more generally no deviatoric stresses. Thus since the total stress tensor is the isotropic pressure, it has the form:

$$\rho Dv/Dt = -\nabla P + \rho g$$

where ρ is the density, v is a velocity component, D/Dt is the substantial time derivative, P is the pressure, ∇ is the nabla operator and ρg is the body force.

EULERIAN ANGLE The angle between one of the set of three orthogonal reference axes in a crystal unit cell with respect to the corresponding axis in a set of reference axes. In a cartesian set of axes there are therefore three Eulerian angles which define successive rotations necessary to bring the two sets of axes into coincidence.

These angles are used in the description of crystalline orientation.

EULERIAN COORDINATES The description of deformation of a body in which the reference coordinates are those in the deformed body. For small deformations the Eulerian and Langrangian coordinates coincide, but the former are sometimes necessary for the description of large deformations. Thus the coordinate system is not fixed in space, as opposed to the Lagrangian system, and moves with the deformation. The continuity, momentum and energy equations described with the Eulerian frame of reference are thus said to be in their Eulerian form.

EUROPRENE Tradename for emulsion styrene–butadiene rubber and for a styrene–butadiene–styrene radial block copolymer.

EUROPRENE-*CIS* Tradename for medium *cis*-polybutadiene.

EUROPRENE N Tradename for nitrile rubber.

EUROPRENE SOL Tradename for solution styrene–butadiene rubber.

EUROPRENE SOL P Tradename for solution polybutadiene.

EUROPRENE SOL T Tradename for radial, star and diblock styrene–butadiene copolymers.

EV Abbreviation for efficient vulcanisation.

EVA Abbreviation for ethylene–vinyl acetate copolymer.

EVAL Abbreviation and tradename for ethylene–vinyl alcohol copolymer.

EVA RUBBER Abbreviation for ethylene–vinyl acetate rubber.

EVLAN Tradename for a viscose rayon staple fibre with a smaller cross-section, higher crimp and higher strength than usual.

EVOH Abbreviation for ethylene–vinyl alcohol copolymer.

EVOLVED GAS ANALYSIS (EGA) Alternative name for thermal volatilisation analysis.

EWNN Alternative name for FeTNa.

EXCITED STATE (Electronically excited state) An atom, molecule or group existing in an energy state with energy in excess of the normal (ground) state as a result of its absorption of a light quantum. The excited state is initially a singlet (S), with paired electron spins. Usually only the first excited singlet (S_1) state is involved in photochemical reactions. The S_1 state may be transformed to the lowest excited triplet state (T_1) having unpaired spins, by intersystem crossing. The excess energy may be lost by re-emission of radiation (by fluorescence or phosphorescence), by radiationless internal conversion back to the ground state as heat or by bond dissociation. In this latter case free radicals are produced which may initiate photodegradation. Alternatively the excess energy may be transferred to a second molecule (quenching).

EXCLUDED VOLUME The volume in a solution, in addition to the volume physically necessarily occupied by the solute molecules, from which other molecules are excluded due to the fact that the distance between two molecules cannot be less than the sum of their radii. Thus a certain volume is unavailable for occupancy by solute molecules. Such a concept has been widely used in considering polymer solutions, particularly in the Flory–Krigbaum theory, where the polymer segments exclude other polymer segments. This leads to long-range interactions between segments widely separated in a polymer chain, but which may approach each other due to molecular flexing. At the theta temperature, excluded volume effects are eliminated, whereas above this temperature polymer coil expansion takes place and the excluded volume increases as segments interact more with solvent.

EXCLUSION LIMIT In gel permeation chromatography, the lowest molecular weight of polymer of a given shape which does not penetrate the pores of the gel. Molecules of a size above this are therefore eluted from the column by a volume of solvent equal to the solvent volume surrounding the gel. This is the void volume.

EXOPEPTIDASE (Peptidase) A protease which will only cleave a terminal peptide link of a polypeptide chain, e.g. carboxypeptidase A.

EXPANDED POLYMER Alternative name for cellular polymer, but the term is often restricted to those cases where the material has been produced by allowing a gas to expand within a polymer melt and then cooling the melt, thus trapping the gas bubbles in the now solid polymer. The gas is produced either by the injection under pressure or by the chemical decomposition, usually induced by the high temperatures of melt processing, of a blowing agent.

EXPANDED POLYSTYRENE (EPS) (XPS) (Foamed polystyrene) Tradenames: Dylite, Fostafoam, Hostapor, Pelaspan, Polyzote, Resticell, Styrocell, Styrofoam, Styropor. A low density foamed polymer consisting of fused open cells, with thin polystyrene walls, filled with air. Formed from beads obtained by the suspension polymerisation of styrene containing a low boiling point liquid, often about 5% pentane. On heating, the beads soften and expand due to volatilisation of the solvent, which therefore acts as a blowing agent. When consolidated, foam density may be as low as $1\,lb\,ft^{-3}$ ($0.016\,Mg\,m^{-3}$). The material has a very low coefficient of thermal conductivity, being as low as $0.031\ W\,m^{-1}\,K^{-1}$ (or $0.22\ Btu\ in\ ft^{-2}\,h^{-1}\,°F^{-1}$). Expanded sheet typically has a density of about $0.05\ Mg\,m^{-3}$ and injection moulded foam a density of about $0.2\ Mg\,m^{-3}$. Higher density moulded foams are often termed structural foams.

EXPANDED RUBBER A cellular rubber made from solid starting materials and having a closed cell structure. To achieve this the expanding gas (air, nitrogen or the gas produced by use of a chemical blowing agent) must not diffuse too rapidly out of the cells and cause cell wall collapse. This is inhibited by using a rubber with a high viscosity.

EXPANSION FACTOR Alternative name for molecular expansion factor.

EXPONENTIAL DISTRIBUTION Alternative name for Schultz–Flory distribution.

EXTENDED CHAIN (Fully extended chain) (Zig-zag conformation) A polymer chain in its most extended conformation, consistent with the normal bond angles between the atoms of the chain. Thus for most chains, involving carbon, oxygen and nitrogen atoms, the most extended conformation is the all-*trans* conformation with a zig-zag shape. For a vinyl polymer this may be represented by:

$$\sim CH_2 \overset{CHX}{\diagup} CH_2 \overset{CHX}{\diagup} CH_2 \overset{CHX}{\diagup} CH_2 \overset{CHX}{\diagup} \sim$$

For certain simple chains, e.g. polyethylene, such a conformation is energetically the most favoured.

The conformation, being regular, is found as the conformation in some crystalline polymers, such as in polyethylene and polyamides, nylons and polypeptides having the β-pleated sheet structure for instance. However, in these cases the chains are not fully extended since chain folding occurs at the surface of the crystals, so that only certain lengths of the chains are extended. Truly fully extended chains may be found in extended chain crystals, often being produced by extensive drawing.

EXTENDED CHAIN CRYSTAL A polymer crystal or crystallite in which the polymer molecules adopt an extended chain conformation with little sign of chain folding. Often taken to be crystals in which the polymer molecules are at least $2000\ Å$ in the molecular axis direction. Such forms are favoured by slow crystallisation close to T_m, crystallisation under high pressure (anabaric crystallisation) or when highly oriented. They are also formed when certain very stiff molecules, e.g. poly-(p-phenyleneterephthalamide) are spun as fibres from solution.

EXTENDER An additive, used in relatively large amounts, to increase the bulk of a polymer based material at low cost and therefore to act as a cheapening agent. Often this results in the deterioration of some properties of the material. Fillers, when not used for the specific purpose of reinforcement are sometimes useful as extenders, e.g. cheap mineral fillers (especially calcium carbonate, as whiting, and clays), especially in polyvinyl chloride and in rubbers, However, their use often results in increased stiffness, reduced elongation at break and reduced impact strength, so that use is limited to less than about $50\ phr$. Liquid materials may also be used in flexible materials as with plasticiser extenders in polyvinyl chloride, where they are relatively incompatible with the polymer. On the other hand, when rubber oils are used as extenders, they are relatively compatible, so are acting more in the role of plasticisers.

EXTENSION (Cauchy strain) (Engineering strain) (Nominal strain) A measure of strain defined as $(l'-l)/l$, where l and l' are the distances between two points in the unstrained and strained material respectively. Thus the extension is the relative displacement of two points per unit length. In a tensile deformation it is the change in length divided by the original length. Useful for the description of small strains (when it equals the true strain) especially in linear elasticity theory.

EXTENSIONAL CREEP COMPLIANCE Alternative name for tensile creep compliance.

EXTENSIONAL FLOW Alternative name for elongational flow.

EXTENSIONAL MODULUS (Longitudinal modulus) The tensile modulus of a uniaxially drawn and oriented fibre or film in the direction of drawing. Alternatively the tensile modulus of a uniaxial, continuous fibre filled composite in the direction of the fibres. If the direction of orientation is the z-axis then it is $E_{33} = \sigma_{33}/e_{33}$.

EXTENSIONAL VISCOSITY Alternative name for elongational viscosity.

EXTENSION AT BREAK Alternative name for elongation at break.

EXTENSION RATIO Alternative name for deformation ratio.

EXTENSION SET The set of a material, especially a rubber, resulting from stretching a sample to a specified elongation or stress for a given time and allowing recovery to occur for a specified time. The set is the residual elongation. Extension set measurements are not so frequently made as compression set measurements but may be more reproducible.

EXTENSOMETER A device for following the extension of a sample which eliminates errors due to slipping of the sample in the jaws of the machine, by being directly attached to the sample. This also eliminates uncertainty as to the gauge length, since this is the distance between the two points of attachment. The strain is converted into the rotation of a beam of light or into an electrical signal by a displacement transducer, using a system of levers or arms.

EXTERNAL LUBRICANT A processing lubricant used for reducing friction, and hence sticking, between a polymer melt and processing machinery. Such a lubricant must therefore be relatively incompatible with the polymer so that it migrates to the surface. Use of excessive amounts can cause plate-out (transport of colourants and other additives to the surface along with the lubricant) especially under high shear conditions, and/or loss of clarity in transparent materials. External lubricants are always required in polyvinyl chloride compositions, but sometimes with other polymers as well. Examples include stearic acid (perhaps the most widely used) and its metal salts (especially calcium and lead) paraffin wax, low molecular weight polyethylene and some long chain esters such as ethyl palmitate. These are usually used at less than 1 phr.

EXTERNAL PLASTICISATION Plasticisation resulting from the use of a plasticiser as an additive, as opposed to internal plasticisation. The usual method of plasticisation, but can suffer from a lack of permanence due to a loss of plasticiser during use of a plasticised polymer by volatilisation or extraction of the plasticiser.

EXTIR Tradename for acrylonitrile–butadiene–styrene copolymer.

EXTRA STRESS TENSOR Alternative name for deviatoric stress.

EXUDATE GUM POLYSACCHARIDE Alternative name for gum polysaccharide.

F

F Symbol for phenylalanine.

FABRY–PEROT INTERFEROMETER A scanning interferometer used for measuring Brillouin spectra. It consists of two parallel glass plates with semi-reflecting/transparent surfaces on their inner sides. Multiple reflections occur between these surfaces producing interference fringes in the transmitted waves; about the first ten fringes are useful. The frequency range is scanned by varying the optical path between the plates, either by altering the air pressure (thereby altering the refractive index) or by varying the plate separation piezoelectrically.

FACTICE The soft, friable and elastic solid formed either by heating sulphur with unsaturated vegetable oils, such as linseed oil, or by treating them cold with S_2Cl_2. Sulphur crosslinks are formed between the oil molecules involving the double bonds of the unsaturated fatty acid chains, such as stearic and oleic acid, present in the oils. Useful as additives in rubbers to reduce nerve and enhance processing characteristics.

F-ACTIN The fibrous, 'polymeric' form of actin.

FAILURE ENVELOPE The line joining the co-ordinates of the points of fracture, e.g. those given by the tensile strength and elongation at break, on the stress–strain plots for a material, usually an elastomer, at different temperatures or different strain rates. In the latter case the temperature (T) is reduced to a reference temperature (T_0) by multiplying the fracture stress, e.g. the tensile strength, by the factor T/T_0. In the former case, similarly a reference strain rate is chosen. Sometimes the envelope is given by the plot of $\log \sigma_f/T$, where σ_f is the fracture stress. In either case the envelope is in effect a master curve eliminating the effects of strain rate and temperature and therefore is characteristic of the material itself.

FAR INFRARED SPECTROSCOPY Infrared spectroscopy in the region below $650\,cm^{-1}$, often considered specifically to cover the range $200–10\,cm^{-1}$, but sometimes referring to the conventional mid-range infrared region of $4000–200\,cm^{-1}$. A useful region for the study of aromatic, halogen and organometallic polymers, which absorb in this region and for the study of interchain vibrations.

FASIL Abbreviation for polyfluoroalkylarylenesiloxanylene.

FAST EXTRUSION FURNACE BLACK (FEF) A type of furnace carbon black which has good processing characteristics and is therefore useful for extruded products such as tyre carcass and sidewalls, inner tubes and hose.

FAST FRACTURE Fracture in which a crack grows catastrophically rapidly at up to the speed of sound, as distinct from slower fracture such as creep rupture or tearing. The type of fracture involved in impact failure which may be analysed by fracture mechanics treatments such as the Griffiths and Irwin theories.

FATIGUE In general the progressive weakening of a material component with increasing time under load, such that the load which would not produce failure at short times, does produce failure at long times. This general use of the term includes both static fatigue (creep rupture), where a steady constant load is experienced, and dynamic fatigue, where the load varies, usually in a periodic manner. The latter is also often simply called fatigue.

FATIGUE CURVE (S/N curve) A plot of the number of cycles (N) required to produce failure by fatigue at a particular stress amplitude (S) with stress being plotted linearly against $\log N$. On this curve the value of S at a particular number of cycles is the endurance limit. However, as S is decreased the curve flattens out so that there is a definite fatigue limit.

FATIGUE LIFE (Endurance) The number of loading cycles to produce failure, usually by fracture, in dynamic fatigue. The lower the stress, or more precisely the lower the amplitude of the loading cycle, the longer the fatigue life will be. Lowering the temperature has a similar effect. At a low enough load, failure does not occur even after an infinite number of cycles.

FATIGUE LIMIT (Endurance limit) (Fatigue strength) The applied stress in dynamic fatigue below which failure does not occur even after an infinite number of loading cycles.

FATIGUE STRENGTH Alternative name for fatigue limit.

FATTY POLYAMIDE Tradenames: Beckamid, Versamid. A polyamide formed by reaction of a dimer acid with an aliphatic diamine, usually ethylenediamine or diethylenetriamine. Usually of low molecular weight (2000–5000), largely amorphous and having amine end groups. Useful in epoxy resins as a combined curing agent and flexibiliser and as a component of thixotropic paints.

FDNB Abbreviation for 1-fluoro-2,4-dinitrobenzene.

FEED-BACK INHIBITION (End-product inhibition) (Retroinhibition) Inhibition of an enzymically catalysed reaction by the final product of the reaction acting as an inhibitor for the reaction, usually acting in the first stage (the committing reaction) by allosteric inhibition.

FEF Abbreviation for fast extrusion furnace black.

FELDSPAR The most abundant mineral in the earth's crust, which consists of a complex network alumino-silicate. It may be considered to be derived from silica in which either one-quarter or one-half of the Si-atoms have been replaced by aluminium, the resulting electrostatic charge being neutralised by Na, K or Ca. Thus the repeat unit ions are $AlSi_3O_8^-$ and $Al_2Si_2O_8^-$. The network consists of eight-membered rings of four $(AlSi)O_4$ tetrahedra linked through further tetrahedra into chains, which are themselves linked by further rings of four tetrahedra. This yields a very open framework, the interstices being occupied by the cations.

FEP Abbreviation for fluorinated ethylene–propylene copolymer, i.e. tetrafluoroethylene–hexafluoropropylene copolymer.

FERRITIN A conjugated protein in the spleen and liver, consisting of a polypeptide (apoferritin) of molecular weight of about 450 000 and a ferric hydroxide–phosphate complex of approximate composition $(Fe(O.OH)_8(FeO\!-\!OPO_3H_2)$ as the prosthetic group. Biologically, ferritin is involved in the absorption of iron and in making iron available for biosynthesis of haemoglobin.

FERROCENE POLYMER (Polyferrocene) A polymer containing ferrocene structures, i.e.

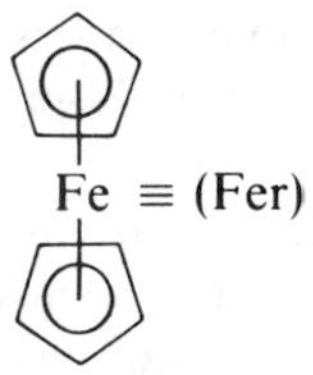

$$Fe \equiv (Fer)$$

either as pendant groups, as in poly-(vinylferrocene), or as part of the chain. Despite the good thermal stability of ferrocene itself, ferrocene polymers show a disappointingly low thermal stability. Typical syntheses are: $ClOC\!-\!Fe\!-\!COCl$ + diamine → ferrocene-amide polymers,

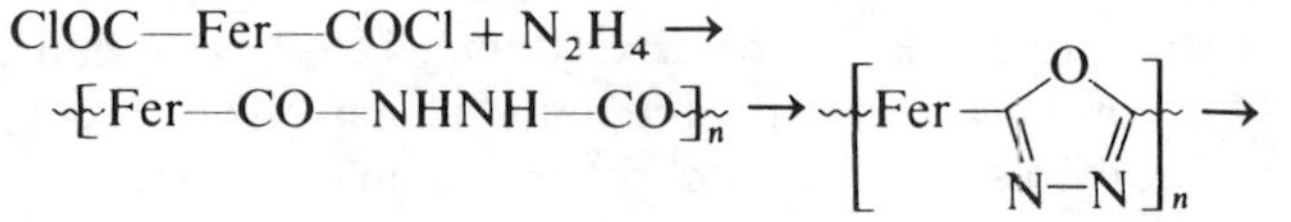

$$ClOC\!-\!Fer\!-\!COCl + N_2H_4 \rightarrow$$
$$\{Fer\!-\!CO\!-\!NHNH\!-\!CO\}_n \rightarrow \{Fer\overset{O}{\underset{N-N}{\diagdown\diagup}}\}_n \rightarrow$$

(ferrocene–oxadiazole polymer). Ferrocene–siloxane and ferrocene–pyrazole polymers have been similarly synthesised.

FERROPROTOPORPHYRIN Alternative name for haem.

FERTENE Tradename for low density polyethylene.

FeTNa (EWNN) A complex of iron, tartaric acid and sodium hydroxide, which in aqueous solution is a relatively powerful solvent for cellulose and in which the cellulose is remarkably free from oxidative degradation.

FETUIN A glycoprotein found in foetal calf serum, of molecular weight 4400, consisting of a single polypeptide chain, with three carbohydrate groups attached, containing sialic acid, D-galactose, D-mannose and N-acetylhexosamine in molar ratios 1:1·5:1·8:1·0 respectively.

FIBERITE Tradename for phenol–formaldehyde and melamine–formaldehyde polymers.

FIBER Y Tradename for fibres of poly-(bis-(4-cyclohexyl)methanedodecanoamide).

FIBRAVYL Tradename for a polyvinyl chloride fibre.

FIBRE A solid material in the form of a piece whose length is very much greater than its diameter and characterised by its fineness (typically having a diameter of 10–50 μm) and flexibility. Fibres are of the greatest importance due to their being capable of being woven or knitted into textiles.

Most fibre forming materials are organic polymers. Natural fibres include cotton, wool, silk and other less important animal, leaf and bast fibres. Man-made fibres may be regenerated fibres (from natural fibres), such as rayon, cellulose acetate and some protein fibres, or synthetic fibres, when the polymer used is a synthetic polymer. The three outstandingly important synthetic fibres are the nylons, polyethylene terephthalate (usually referred to as polyester) and polyacrylonitrile (usually referred to as acrylic), but many other synthetic fibres are used for textile and other purposes.

A very long continuous fibre is also called a filament. Usually fibres are used in textiles as yarns composed of either continuous bundles of smaller filaments or of shorter staple fibres which have been twisted together by spinning. All natural fibres, except silk, are staple fibres. Man-made fibres are produced initially as continuous filaments, which may be subsequently chopped into staple fibre, typically with fibres of a few centimetres in length. Fibre size is denoted by its denier or tex and its mechanical properties are often expressed per unit denier or tex, i.e. as the specific values. Fibres may be circular in cross-section (as with silk, nylons and polyester), or oval (as with wool and cotton), serrated, trilobal or of other shape. Fibres also often have crimp, i.e. waves, bends, twist or curl along their length, which either occurs naturally, as in wool and cotton, or may be imparted mechanically in man-made fibres.

Natural fibres, except silk, have three distinct parts—an outer skin (or cuticle), an inner area and a central core that may be hollow. Fibre properties depend on all these physical characteristics as well as on the polymer chemical structure. Mechanical behaviour is usually improved by the polymer molecules being highly oriented along the length of the fibre. High orientation is imparted to man-made fibres by drawing.

FIBRE B   Early tradename for poly-(p-phenyleneter-ephthalamide), later called Kevlar. Early versions were based on poly-(p-benzamide).

FIBRE DIAGRAM   The wide angle X-ray diffraction pattern produced by an oriented polymer in fibre form. It consists of a series of arcs, each of which results from scattering from a particular set of *hkl* planes with varying orientation. As orientation becomes more perfect the arcs become shorter eventually becoming spots, reminiscent of a single crystal pattern for a perfectly oriented specimen.

FIBRE EFFICIENCY FACTOR   Alternative name for efficiency of reinforcement.

FIBRE ORIENTATION   Alternative name for uni-axial orientation.

FIBRIL   Fibre like aggregates of crystalline subunits. They are present in spherulites in which the fibrils radiate from the centre. When rapidly grown, the fibrils have crystallites as subunits or when slowly grown the fibrils may themselves be aggregates of microfibrils. In highly crystalline samples the fibrils may be considered as defect crystals. Low angle branching during fibrillar growth causes the space-filling outward expansion of the spherulite. Regular twisting of the fibrils produces a ringed spherulite. They are also present in fibres as flat ribbon-like subunits of a few thousand ångströms thick and about 1 μm wide. However, sample preparation techniques required for observation may be responsible for their appearance and they may not be fundamental subunits. In general, the molecules lie perpendicular to the long axis, except in certain natural polymers such as cellulose, when they lie parallel.

FIBRILLAR CRYSTALLISATION   Crystallisation from solution in which the crystallites are fibrous and not lamellar, with the polymer chains parallel to the fibril direction.

FIBRILLAR STRUCTURE   Supermolecular structure composed of fibrils, in which the individual fibrils are fairly readily observable as in fibres. In spherulites, observation of the fibrils is often obscured by their dense packing.

FIBRIN   The protein released from fibrinogen in blood by the action of thrombin, or certain other enzymes, and largely responsible for the clotting of blood. The fibrin monomer so formed consists of fibrinogen chains less those fibrinogen peptides which have been cleaved off. The monomer 'polymerises' by aggregation through formation of hydrogen bonds and electrostatic attraction between chains. A hard clot is then formed, in the presence of calcium and an enzyme fibrininase, by a crosslinking reaction resulting from transamidation linking the γ-carboxyls of glutamic acid with ε-amino groups of lysine residues. The clot may be subsequently solubilised by the proteolytic action of the enzyme plasmin, present in blood plasma as its zymogen plasminogen.

FIBRINOGEN   A blood plasma protein which migrates electrophoretically between the β- and γ-globulins. It is separated from plasma by salting out or by precipitation in the cold with dilute ethanol. It has a molecular weight of about 340 000 and is a glycoprotein with about 3% carbohydrate. The molecules are highly elongated ellipsoids in shape, consisting of dimers each composed of three polypeptide chains linked through disulphide bridges. The carbohydrate consists of branched oligosaccharides containing D-mannose, D-galactose, 2-amino-2-deoxy-D-glucose and sialic acid units in the approximate ratios 3:3:2:2. The conversion of fibrinogen to insoluble fibrin is largely responsible for blood clotting. This is brought about by the action of thrombin and some other enzymes such as papain. The action of thrombin cleaves the polypeptide chain at arginyl residues forming a peptide of 18 (A) and one of 20 (B) amino acid residues and the fibrin 'monomer' terminated by four amino glycinyl acid residues.

FIBRO   Tradename for a viscose rayon staple fibre.

FIBROCETA   Tradename for a cellulose acetate staple fibre.

FIBROIN   The major component of silk, typically accounting for about 80% of the silk, whose threads consist of two fibroin filaments bound together by sericin. A protein fibre whose composition varies with origin, the most widely studied being that of *Bombyx mori*. It consists mostly of non-polar amino acid residues, e.g. about 45% glycine, 30% alanine and 12% serine with very low amounts of cystine, acidic and basic amino acids. Thus close packing of the chains is possible with extensive hydrogen bonding between chains but with little inter-chain covalent bonding. Thus fibroin will dissolve in hydrogen bonding solvents such as formic acid and aqueous lithium chloride solutions.

In *B. mori* fibroin every alternate residue is glycine with many sequences of the type

$$-Gly-Ala-Gly-Ala-Gly\rceil$$
$$\lfloor [Ser-Gly-(Ala-Gly)_2]_8\rceil$$
$$\lfloor Ser-Gly-Ala-Ala-Gly-Tyr$$

comprising about 60% of the molecule. These sequences are responsible for the crystalline (about 50% crystallinity) structure, whereas segments containing bulky ionic residues form amorphous regions. These are responsible for the limited elasticity of fibroin. The molecular weight is approximately 100 000. *B. mori* fibroin crystallises in the β-pleated sheet conformation with anti-parallel packing of chains.

Other fibroins may have quite different structures, e.g. some sawfly fibroin has many $\left(Ala-Gln \right)_n$ sequences, and adopt different conformations, e.g. the polyglycine II structure of Solomon seal sawfly which has about 66% glycine content. Many synthetic silk-like polypeptides have been studied, especially the polydipeptide poly-(Ala–Gly). These, however, often crystallise in the cross-β-structure.

FIBROLAN   Tradename for a casein fibre.

FIBROLANE A   Tradename for a man-made protein fibre from casein.

FIBROLANE BC   Tradename for a man-made protein fibre from casein, insolubilised by reaction with formaldehyde and stabilised by chrome tanning.

FIBROLANE BX   Tradename for a man-made protein fibre from casein, insolubilised by reaction with formaldehyde.

FIBROLANE C   Tradename for a man-made protein fibre from peanut protein.

FIBROUS CRYSTAL   A crystal whose shape (or habit) is long and thin or needle-like. In polymers the term is used in two ways. Firstly, to describe crystals composed of highly connected lamellae with the polymer molecular chain axis perpendicular to the long direction of the crystal. These are also described as laths or ribbons. Secondly, the term may refer to crystals in which the molecules have extended chain conformations, with the molecules parallel to the long (fibre) axis. These may be formed by crystallisation during polymerisation, deformation by drawing, stirring or flow of a solution (forming shish-kebabs) or by fibrillation of extended chain crystals on fracture or crystallisation.

FIBROUS PROTEIN   A protein whose polypeptide chains exist in elongated conformations (α-helix or β-conformation) parallel to a single axis, giving super-molecular fibrous or sheet-like structures, as opposed to the tightly folded conformations of globular proteins. Such structures have good mechanical properties and these proteins are the structural proteins and contractile proteins in organisms. Unlike globular proteins they are highly insoluble in water and aqueous salt solutions. Important examples of fibrous proteins are keratin, collagen, fibroin, elastin, actin and myosin.

FICK'S FIRST LAW OF DIFFUSION   The hypothesis that the rate of transfer (F) of a diffusing substance through unit area of a section of a solid is proportional to the concentration gradient normal to the section, i.e. $F = -D\,dC/dx$, where C is the concentration of the diffusing substance, x is the direction normal to the section and D is the diffusion coefficient.

FICK'S SECOND LAW OF DIFFUSION   The rate of change of the concentration of a diffusing substance diffusing only in a single direction, the x direction, is proportional to the rate of change in concentration gradient with x, i.e.

$$\partial C/\partial t = D\,\partial^2 C/\partial x^2$$

where D is the diffusion coefficient and C is the concentration. This relationship may be derived from Fick's first law of diffusion assuming that D is constant, i.e. that it does not depend upon C.

FIKENTSCHER *K*-VALUE   (Fikentscher number) (*K*-value) An empirical molecular weight parameter obtained by dilute solution viscometry measurements of a polymer solution. Used particularly for vinyl chloride homo- and copolymers, poly-(vinyl ethers), poly-(vinylpyrrolidone) and nylons. Defined as 10^3 times the K of the empirical Fikentscher equation:

$$(1/c)\log_{10}(\eta_{rel}) = (K + 75K^2)/(1 + 1\cdot 5Kc)$$

where c is the polymer concentration and η_{rel} is the viscosity ratio. Unfortunately the K-value is dependent on the solvent used, so this should be specified. The alternative ISO viscosity number is therefore often preferred. For commercial polyvinyl chloride polymers, K-values range from about 45–80, increasing as molecular weight increases.

FIKENTSCHER NUMBER   Alternative name for Fikentscher *K*-value.

FILAMENT   A very long fibre, as opposed to the short (few centimetre) fibres present in staple. Extremely long, continuous filaments are produced in the spinning of man-made fibres, although natural fibres are, apart from silk, all staple fibres. Filaments are usually used as bundles in yarns (filament yarns), although single filaments, especially if thick, may be used as such (monofilament). Several thousand loose filaments collected together as a rope or strand form a filament tow.

FILAMENT COUNT   The number of filaments in a yarn.

FILLER   An additive for a polymer composition, usually used in relatively large amounts (10 to several hundred phr, typically of the order of about 50 phr) to modify the polymer physical properties. Usually the term filler refers to a material which is stiffer than the polymer matrix and its use produces a polymer composite. Fillers are classified in several different ways. Since the most common use of a filler is to improve some aspect of mechanical behaviour, a filler may be reinforcing or non-reinforcing. In the latter case a filler may be used to increase thermal conductivity (as with silica or alumina), to raise or lower the electrical conductivity or merely to lower costs (use as an extender). Fillers are also classified according to their physical shape as fibrous, such as glass, carbon, cellulose or aramid, or as particulate, such as with many minerals (talc, calcium carbonate, silica, Woolastonite,..., etc.). Sometimes fillers are classified according to their chemical nature as organic or inorganic.

FILM EXTRACTION   A method of fractional solution in which the whole polymer is in the form of a thin film, usually as a coating on aluminium foil. In this way equilibrium is rapidly achieved between polymer and solution.

FILTER PAPER CHROMATOGRAPHY   Alternative name for paper chromatography.

FINDLEY EQUATION   A semi-empirical creep equation of the form $e = A \sinh B\sigma + Ct^n \sinh D\sigma$, where e is the creep strain at time t, σ is the stress and A, B, C and D are material constants. Often found to represent creep data well but for simple step loading only.

FINEMANN–ROSS METHOD   A method of copolymerisation data analysis for the estimation of the monomer reactivity ratios r_A and r_B of the monomers A and B respectively. The copolymer composition equation, when rearranged yields:

$$f_A(1 - 2F_A)/F_A(1 - f_A) = r_B + [f_A^2(F_A - 1)/F_A(1 - f_A)^2]r_A$$

where f_A and F_A are the mole fractions of monomer A in the feed and copolymer respectively, when copolymerisation is taken to low conversion. If the left hand side is plotted against the coefficient of r_A, then a straight line results with slope of r_A and intercept of r_B.

FINE THERMAL BLACK   (FT) A type of thermal carbon black which is only of low reinforcement, giving compounds with low modulus, hardness, hysteresis and tensile strength, but high elongation, tear strength and flex resistance. It is useful in mechanical goods, tubes and footwear uppers.

FINGERPRINT   Alternative name for peptide map.

FINISH   A treatment given to a glass fibre reinforcement to be used for a polymer composite in order to improve the composite properties. The finish is a coupling agent to improve mechanical performance, but it may also lubricate the glass surface. Before applying the finish from aqueous solution, the size must be removed since it is incompatible with the polymer matrix.

FINITE PLATE CORRECTION FACTOR   (Geometrical factor) Symbol Y. A factor used to correct the stress intensity factor for cases other than that of the simplest one of an infinitely thin sheet with an embedded crack, for which $Y = 1$. Its value depends on the shape and size of both the crack and specimen and location of the crack. Thus, for example, for the case of a finite width (w) plate with a central crack of length $2c$,

$$Y = [(w/\pi c)\tan(\pi c/w)]^{1/2}$$

and for a single edge crack in a plate of finite width,

$$Y = 1{\cdot}12 - 0{\cdot}23(c/w) + 10{\cdot}6(c/w)^2 - 21(c/w)^3 + 30{\cdot}4(c/w)^4$$

FINITE STRAIN ELASTICITY   Alternative name for large strain elasticity.

FIREMASTER TSA   Tradename for poly-(2,6-dibromophenylene oxide).

FIRE RETARDANT   Alternative name for flame retardant.

FIRST ORDER MARKOV CHAIN PROCESS   A Markov chain process in which the occurrence of an event is dependent on the last previous event. Copolymerisation is frequently such a process since the propagation step of the addition of one or the other of the monomer molecules depends on the terminal unit, i.e. on the previous propagation event. Analysis of the triad sequences then tests whether the terminal model holds. Similarly in considering stereoregulation in polymerisation, a racemic (r) or a *meso* (m) addition of monomer may be determined by the configuration of the previous unit. Here tetrad sequence data are needed to test for first order Markov statistics to fit a penultimate model.

FIRST ORDER TRANSITION   A transition which is characterised by a discontinuity in the first derivative of the Gibbs free energy with respect to temperature and by a discontinuity in the intensive properties, such as enthalpy, entropy and volume, of a material. This is in contrast to a second order transition where a discontinuity occurs in the derivatives with respect to temperature of the intensive properties. Melting and vaporisation are the most important first order transitions.

FISCHER PROJECTION FORMULA   A way of representing, in two dimensions, the three-dimensional configurational (or stereo-) isomers of monosaccharide (or sugar) molecules, by planar projection of the molecule in open-chain form. The projection is drawn with the carbonyl group uppermost, the convention then being that if the hydroxyl on the lowest asymmetric carbon is to the right the sugar has the D-configuration and if to the left the L-configuration. Thus in the hexoses, for example, there are 16 possible projections (or configurational isomers) consisting of 8 enantiomorphic pairs, e.g. D-glucose is (**I**). In the process of ring closure to form the ring structure of a sugar (depicted by the Hawarth formula, e.g. **III**), the Fischer projection does not show the straight chain distribution of atoms:

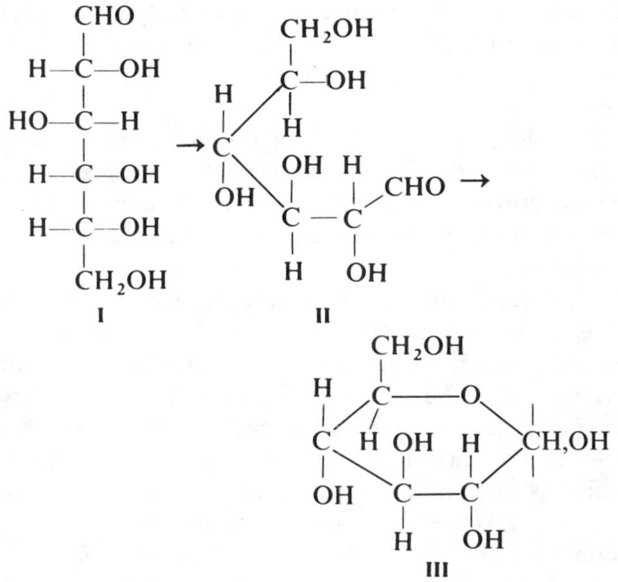

FIVE REGIONS OF VISCOELASTICITY (Regions of viscoelasticity) The five different types of viscoelastic behaviour exhibited by a viscoelastic material, typically an amorphous polymer, in five different temperature regions or, via the time–temperature superposition principle, when subject to deformation at different rates, e.g. at different frequencies in dynamic mechanical testing, at a fixed temperature. With increasing temperature, or decreasing frequency, the five regions of behaviour (or states) are the glassy, retarded elastic, rubbery, rubbery flow and viscous flow states. As temperature (or time of application of the deforming stress) is increased so more modes of molecular motion become possible, i.e. become 'unfrozen'. Hence different types of deformation are observed.

The transition from one state to another is not sharp and is best characterised by the changes that occur in a modulus or compliance with temperature or in a dynamic modulus or compliance with frequency. Since a modulus may change by about six orders of magnitude its value versus temperature is often plotted on a logarithmic scale. A very large range of frequency is needed to cover the whole range of viscoelastic behaviour (12–15 orders of magnitude), so changes in modulus with frequency are displayed on a log–log scale.

The above behaviour is most simply observed with a linear amorphous polymer, the precise behaviour being dependent on the molecular weight and on the presence of diluents such as solvent, plasticiser and filler. In particular, above a critical molecular weight, usually between 5000 and 30 000, entanglement coupling occurs which considerably reduces long-range molecular motions. Crystallinity and crosslinking also modify behaviour, in particular their presence eliminates the viscous flow region since under constant stress such polymers eventually reach an equilibrium deformation.

FIXED CHARGE The electrostatic charges carried by the polymer molecules of a polyelectrolyte, as opposed to the counter-ion charges, which, being small ions, are mobile charges.

FKM Abbreviation for fluoroelastomers based on vinylidene fluoride–hexafluoropropylene copolymer.

FLAGELLIN A protein found in bacterial flagella, of molecular weight about 40 000. It often contains residues of the unusual amino acid ε-N-methyllysine. On acidifying flagella, they dissociate into flagellin subunits which will reaggregate to identical structures under appropriate pH/salt conditions.

FLAME RETARDANT (Fire retardant) A material capable of increasing the fire resistance, especially by reducing the ease with which a polymer burns. Usually contains one of the elements phosphorus, boron, halogen or antimony. It may be incorporated either as a comonomer (a reactive flame retardant) e.g. chlorendic acid or anhydride, by chemical reaction with the polymer, e.g. APO, or as an additive. Of the many materials of this latter type, the system organohalogen plus antimony oxide is most frequently used, but unfortunately increases smoke formation. A flame retardant can function in many ways—by modification of condensed phase reactions to reduce the flow of combustible gases to the flame, e.g. by promoting char formation (phosphorus compounds), by encouraging endothermic reactions (phosphorus and boron compounds) or by interfering with flame reactions by reducing flame free radical reactivity (organohalogen compounds).

FLAVOENZYME Alternative name for flavoprotein.

FLAVOPROTEIN (Flavoenzyme) A conjugated protein in which the prosthetic group consists of a flavin nucleotide which is tightly, but usually not covalently, bound to the polypeptide chain. Examples of the prosthetic group are flavin mononucleotide (the vitamin riboflavin combined with phosphate) and flavin adenine dinucleotide (riboflavin combined through phosphate and sugar alcohol ribitol with adenine). Flavoproteins participate in various biological oxidative degradations and electron transport processes, where the flavin part can undergo reversible reduction. The oxidised form is highly coloured, red, yellow or green. Metalloflavoproteins contain, in addition, one or more metal atoms.

FLAX A bast fibre from the plant *Linum usitatissimum* grown in temperate climates. It contains 80–90% cellulose, 8–10% water and some lignin and wax. The fibres are 1–4 cm in length and 10–20 μm in diameter. Tenacity is about 6 g denier^{-1} with low elasticity (elongation at break of 2–5%). Moisture regain is about 12%.

FLEMION Tradename for a polytetrafluoroethylene

containing pendant perfluorocarboxylate groups, i.e. units of:

$$\sim CF_2-CF_2\sim$$
$$|$$
$$O$$
$$|$$
$$CF_2$$
$$|$$
$$CF-CF_3$$
$$|$$
$$O-(CF_2)_3COOH$$

Like the similar Nafion materials, it is useful as a membrane for use in electrolytic separation processes in aggressive chemical environments.

FLEXIBILISER Alternative name for a plasticiser, when, as is usually the case, it is used to increase the flexibility of the polymer. The term is often used in rubbers instead of the term plasticiser.

FLEXIBLE POLYURETHANE FOAM A resilient crosslinked polyurethane in cellular form, usually with open cells, produced by reaction of a polyester, or more commonly a polyether polyol, with a diisocyanate, usually the 80/20 tolylene diisocyanate or a modified tolylene diisocyanate, to yield a polyester or polyether foam respectively. Foaming is caused by the inclusion of water, which reacts with isocyanate groups to produce carbon dioxide and urea crosslinks. Additional foaming can be induced by the inclusion of a chlorofluorocarbon blowing agent.

One-shot processes are used with polyesters and polyethers and prepolymer and quasi-prepolymer processes with polyethers. The polyesters used are usually the adipates of diethylene glycol with a molecular weight of about 2000, being slightly branched by the incorporation of some trimethylolpropane or pentaerythritol units. These are reacted, exclusively by a one-shot process, with tolylene diisocyanate, usually the 80/20 isomeric mixture or the similar 65/35 isomeric mixture.

The polyethers used are usually polyoxypropylene triols, made by base catalysis with some trimethylolpropane or glycerol acting as the branching points. Polytetrahydrofuran was first used but was rather expensive and later polyoxypropylene diols were used in a prepolymer process. A triol (of molecular weight 3000–6000) is now used, tipped with ethylene oxide to introduce some reactive secondary hydroxyls, so that a one-shot process may be employed. Ethylene oxide units may also be incorporated into the chain to achieve satisfactory compatibility with other foam components. Quasi-prepolymer processes are also known.

Mostly 80/20 tolylene diisocyanate is used but also some 65/35 mixed isomers, with both polyester and polyether polyols. Sometimes the diisocyanate is modified with some polymeric MDI. Water is used as the blowing agent, but if a softer foam is needed then a chlorofluoromethane is employed, since increasing the water content also increases urea crosslinks.

Catalysts, often as complex mixtures, are used to obtain the right balance of chain extension, crosslinking and foaming reactions, which all occur simultaneously. Polyether foams typically use a combination of stannous octoate with dimethylethanolamine and 1,4-diazabicyclo-2,2,2-octane. Sometimes dibutyltin dilaurate is used since stannous octoate is sensitive to hydrolysis in water containing blends. For polyester foams usually only an amine catalyst is required, such as dimethylcyclohexylamine.

One-shot processes, both for polyester and polyether foams, use a silicone–polyether block copolymer (based on an ethylene oxide–propylene oxide copolymer) to stabilise the rising foam and to assist in the dispersion of ingredients.

Flexible polyurethane foams are characterised by a low density, typically $20-50 \, \mathrm{kg \, m^{-3}}$, with high resiliency and elasticity under dynamic loading and with little compression set. The polyester foams, although the type originally developed, are now much less important than polyether foams. The latter show a greater resiliency which is an important advantage for cushioning and upholstery. However, the better solvent and load-bearing properties (higher modulus and tensile strength) of polyester foams makes them more useful for textile lamination, footwear and luggage applications.

FLEXOMETER An instrument for assessing the fatigue resistance of a rubber. A block of rubber is subjected to an alternating shear or compression, or a combination of both, which is also sometimes superimposed on another fixed deformation, and the number of cycles to failure is measured.

FLEXTEN Tradename for an aromatic poly-(amide-hydrazide) fibre of structure

$$\left[NH-\bigcirc-CONHNHCO-\bigcirc-CO \right]_n$$

produced by reaction of p-aminobenzylhydrazide and terephthaloyl chloride. Useful as a high performance tyre cord reinforcement.

FLEXURAL MODULUS The ratio of the stress to the strain in a beam undergoing three point bending. Given by: $Fl^3/4bd^3e$, where F is the force at the mid-point, l is the length between outer loading points, b is the beam width, d is its thickness and e is the deflection at the mid-point.

FLEXURAL RIGIDITY The stiffness of a material in bending, defined as the bending moment (M) required to produce unit curvature ($1/R$) of a beam, where R is the radius of curvature. When a beam is deformed in this way, the convex side is extended and therefore experiences a tensile force, whereas the concave side experiences a compressive force. In a rectangular cross-section beam, an unstrained neutral plane (or axis in a circular cross-section beam) runs through the centre of the beam. M is equal to EI/R, where E is the Young's modulus and I is the second moment of area of cross-section with respect to the neutral axis. For a rectangular cross-section of width b and thickness t, $I = bt^3/12$ and $EI = Ebt^3/12$ and is equal to the flexural rigidity. Thus doubling the thickness gives eight times the rigidity. Flexural rigidity is often measured by a three point bending test, where for a beam loaded at mid-point, the deflection $d = PS^3/48EI$, where P is the load and S is the span; hence the flexural rigidity is $PS^3/48d$.

FLEXURAL STRENGTH (Cross breaking strength) The maximum stress (σ_m) which occurs at fracture for a brittle material loaded in flexure, usually as a beam. For a beam of rectangular cross-section of depth d and breadth b, under three point loading, $\sigma_m = 3W_mL/2bd$, where W_m is the maximum load and L is the length between the supports. For a circular cross-section rod, $\sigma_m = 8W_mL/\pi d^3$, where d is the rod diameter. These formulae assume that the beam bends to a circular shape which is only approximately true, but no significant error occurs if the ratio of L to d is above 16:1. The value of the flexural strength is higher than that of the tensile strength, the ratio of the two is called the rupture factor. The difference is at least partly due to the different way that the cracks grow in each case. Fibre reinforced polymers are often tested in flexure, often due to the clamping difficulties experienced for a tensile test. Frequently the strength is determined by the debonding of fibre to matrix, which is often tested in flexure using a short beam specimen, as the interlaminar shear strength $\sigma_s = 0.75W_m/bd$.

FLOOR TEMPERATURE The temperature below which high polymer cannot be formed due to its tendency to depolymerise. Analogous to the ceiling temperature which, in contrast, is an upper temperature limit. It arises when a polymerisation is endothermic and the entropy of polymerisation is positive. This rarely occurs but is found in the polymerisation of S_8 rings and in some solid state polymerisations, e.g. of ω-amino acids.

FLORY α-BRANCHING COEFFICIENT Alternative name for branching coefficient.

FLORY DILUTE SOLUTION THEORY Alternative name for Flory–Krigbaum theory.

FLORY DISTRIBUTION Alternative name for most probable distribution.

FLORY–FOX EQUATION A relationship between the limiting viscosity number ($[\eta]$) of a dilute polymer solution and polymer molecular weight (M). Derived as an extension of the Kirkwood–Riseman theory for the sedimentation of polymer molecules, using the pearl-necklace model of the polymer molecule. The general form of the equation is: $[\eta] = kM^{1/2}\alpha^3$, where k is a constant given by $k = \Phi(\langle r_0^2 \rangle/M)^{3/2}$, where $\langle r_0^2 \rangle$ is the unperturbed mean square end-to-end distance of the polymer molecule and α is the molecular expansion factor given by $\alpha^2 = \langle r^2 \rangle/\langle r_0^2 \rangle$, where $\langle r^2 \rangle$ is the mean square

end-to-end distance. In a theta solvent $\langle r^2 \rangle = \langle r_0^2 \rangle$ and $\alpha = 1$. Φ is a universal constant sometimes called the intrinsic viscosity parameter, calculated theoretically to be 2.6×10^{21} when $[\eta]$ and $\langle r^2 \rangle$ are expressed in dl g^{-1} and cm^2 respectively.

FLORY–HUGGINS INTERACTION PARAMETER (Polymer–solvent interaction parameter) Symbol χ. In the simple Flory–Huggins theory, χ is taken as $z\,\Delta e/kT$, where z is the lattice model coordination number, $\Delta \varepsilon$ is the enthalpy of interaction of a polymer segment with solvent, k is Boltzmann's constant and T is the absolute temperature. Thus χ may be considered as a measure of polymer–solvent interaction energy. However it is probably better considered as a semi-empirical parameter consisting of both entropic and enthalpic contributions as $\chi = \chi_S + \chi_H$. The enthalpic contribution is given by $V_1(\delta_1 - \delta_2)^2/RT$, where V_1 is the solvent molar volume and δ_1 and δ_2 are the solvent and polymer solubility parameters respectively.

The Flory–Huggins theory gives the partial molar energy of dilution as

$$\bar{G}_1 = RTV_1\phi_2/V_2 + RT(\tfrac{1}{2} - \chi)\phi_2^2 + \ldots, \text{etc.},$$

where V_2 and ϕ_2 are the molar volume and mole fraction of the polymer respectively. Since the first term is very small, and ignoring higher power terms, we have, that for solution to occur (since $\Delta \bar{G}_1$ must be negative) χ must be <0.5; the precise value, the critical value χ_c, depends on V_1/V_2, i.e. the polymer molecular weight. In fact, $\chi_c = \tfrac{1}{2}(1 + 1/x^{1/2})^2$, where x is the degree of polymerisation. The theory also leads to the conclusion that for solution, the solubility parameters of the solvent and polymer should lie within about $2.5\,(\text{MJ m}^{-3})^{1/2}$. Since χ is related to the second virial coefficient (A_2) through $A_2 = (\tfrac{1}{2} - \chi)/V_1\rho_2$, where ρ_2 is the polymer density, χ values may be determined experimentally by colligative property measurements, e.g. vapour pressure lowering for concentrated solutions or osmotic pressure for dilute solutions, or from equilibrium swelling measurements by use of the Flory–Rehner equation.

FLORY–HUGGINS THEORY A theory of the thermodynamic properties of polymer solutions which accounts for the large deviations from ideal solution behaviour found with these solutions. The theory calculates the number of ways that the polymer molecules can be placed on a lattice, assuming that the polymer can be represented by segments of similar size to the solvent molecules. This leads to an expression for the configurational entropy of mixing as

$$\Delta S_m = -R(n_1 \ln \phi_1 + n_2 \ln \phi_2)$$

where R is the universal gas constant, n_1 and n_2 are the number of moles of polymer and solvent respectively and ϕ_1 and ϕ_2 are their mole fractions. This is the same expression as that for the mixing of small molecules except that volume fractions have replaced mole fractions.

The enthalpy of mixing is given by: $\Delta H_m = z\,\Delta \varepsilon N_1 \phi_2$, where z is the lattice coordination number, $\Delta\varepsilon$ is the energy of formation of a polymer segment–solvent molecule contact and N_1 is the number of solvent molecules. Thus, applying the concept of a regular solution, the free energy of mixing is $\Delta G_m = RT[N_1 \ln \phi_1 N_2 \ln \phi_2 + \chi N_1 \phi_2]$ where χ is the Flory–Huggins interaction parameter and equals $z\Delta\varepsilon/kT$, where k is Boltzmann's constant. The partial molar free energies of dilution (or chemical potentials) of the solvent and polymer are:

$$(\partial G_m/\partial n_1) = \Delta \bar{G}_1 = RT[\ln \phi_1 + (1 - 1/x)\phi_2 + \chi\phi_2^2]$$
(eqn (1))

and

$$(\partial G_m/\partial n_2) = \Delta \bar{G}_2 = RT[\ln \phi_2 - (x - 1)\phi_1 + \chi x\phi_1^2]$$

where x is the number of segments in the polymer molecule, often taken as the degree of polymerisation (DP). For a polydisperse polymer the number average DP is used. Expansion of the logarithmic term in eqn (1) gives: $\Delta \bar{G}_1 = -RTV_1c_2(1/M_2 + A_2c_2 + \cdots,$ etc.) where M_2 is the polymer molecular weight, A_2 is the second virial coefficient and equals $(\rho_1/M_1\rho_2^2)(\frac{1}{2} - \chi)$ (with ρ_1 as the solvent density and M_1 the molecular weight), V_1 is the solvent molar volume and c_2 is the polymer concentration.

Since $\Delta \bar{G}_1 = RT \ln a_1$, where a_1 is the solvent activity, the osmotic pressure (π) of a polymer solution is given by $\pi V_1 = RT \ln a_1$ and hence it and the other colligative properties may be related to χ. The theory correctly describes osmotic behaviour for some systems, but not for others. In particular, since the theory assumes that a random distribution of polymer segments exists, it does not work very well for dilute solutions where the polymer molecules exist as individual coils, separated by regions of pure solvent. Here the Flory–Krigbaum theory is more successful. In addition, mixing is assumed to be random (which cannot be so unless $\Delta\varepsilon$ is zero), dipolar and hydrogen bonding forces are not allowed for and the theory predicts that χ is independent of c_2, which is often not so. Nevertheless, although deviations from the predicted values of $\bar{H}_1$ and $\bar{S}_1$ do occur, they often compensate for each other and the Flory–Huggins expression for $\Delta \bar{G}_1$ is often good. The theory has been successfully used for analysing many aspects of polymer solutions, particularly the phase equilibria involved.

FLORY–KRIGBAUM THEORY (Flory dilute solution theory) A theory of dilute polymer solutions which does not make the unrealistic assumption of the Flory–Huggins theory of the uniform distribution of polymer segments. In reality, in dilute solution the polymer molecules will exist as coils (of clusters of segments) separated by regions of pure solvent. A model for this structure is used which takes account of the excluded volume within the coils. Furthermore, in the model the coil does not have a uniform segment density and this is assumed to have a Gaussian distribution about the centre of mass. The theory leads to the excess partial molar enthalpy, entropy and free energy being given by $\Delta H^E = RT\kappa\phi_2^2$, $\Delta S^E = R\psi\phi_2^2$ and $\Delta G^E = RT(\kappa - \psi)\phi_2^2$ respectively, where R is the universal gas constant, T is the

absolute temperature, ϕ_2 is the polymer volume fraction and κ and ψ are enthalpy and entropy parameters, which by comparison with the Flory–Huggins theory are seen to be related by $\psi - \kappa = \frac{1}{2} - \chi$, where χ is the Flory–Huggins interaction parameter. For ideal behaviour, $\Delta H^E = T\Delta S^E$ and $\psi = \kappa$ and the theta point is reached, with Θ (the theta temperature) being given by $T\kappa/\psi$.

FLORY–REHNER EQUATION A relationship giving the maximum, i.e. equilibrium, swelling (given by v_{2m}, the ratio of unswollen to swollen volumes, or $1/v_{2m}(= q_m)$ the swelling ratio) of a lightly crosslinked polymer in terms of the crosslink density (as M_c, the average molecular weight between crosslinks, or the network parameter) and the quality of the solvent (as expressed by the polymer–solvent interaction parameter χ_1). The full form of the relationship is:

$$-[\ln(1 - v_{2m}) + v_{2m} + \chi_1 v_{2m}^2]$$
$$= (V_1/\bar{v}M_c)(1 - 2M_c/M)(v_{2m}^{1/3} - v_{2m}/2)$$

where $\bar{v}$ is the specific volume of the polymer, i.e. the reciprocal of the density, M is the primary molecular weight, which is infinite for a perfect network with no loose ends and V_1 is the solvent molar volume. If M_c is large, i.e. at low degrees of crosslinking, then v_{2m} is small, say <0.1, and $v_{2m}^{1/3}$ is much bigger than $v_{2m}/2$ so that the equation may be approximated by:

$$q_m^{5/3} = (\bar{v}M_c)(1 - 2M_c/M)^{-1}(\frac{1}{2} - \chi_1)/V$$

The equation is useful for the determination of M_c from swelling measurements.

FLORY–SCHULTZ DISTRIBUTION Alternative name for most probable distribution.

FLORY SOLVENT Alternative name for theta solvent.

FLORY TEMPERATURE Alternative name for theta temperature.

FLOW BEHAVIOUR INDEX Symbol n. One of the parameters of the power law equation, $\tau = k(\dot{\gamma})^n$, expressing the extent of deviation from Newtonian behaviour of a fluid; the more n differs from unity the greater is the deviation. For pseudoplastic fluids, as with most polymer melts, n is less than unity, typically having values in the range 0.5–1.0.

More generally, when n is not independent of shear rate $(\dot{\gamma})$, it is given the symbol n' ($n = n'$ for a true power law fluid) and is defined as

$$n' = d[\log(R\Delta P/2L)]/d[\log(4Q/\pi R^3)]$$

for laminar flow in a pipe of length L and radius R, with a volume flow rate of Q caused by a pressure difference ΔP, and an apparent wall shear rate of $4Q/\pi R^3$. Hence n' is constant for a fluid at different shear rates and is often more useful than n.

FLOW BIREFRINGENCE The formation of an optically anisotropic, and therefore birefringent, polymer fluid (solution or melt) caused by the shearing stresses of laminar flow orienting the polymer molecules by rotation and translation. Experimental determination of the birefringence is usually made by using a cylindrical couette type of fluid container, viewing the fluid along the direction of the rotating cylinder, using either a compensator or a photoelectric device. By studying the birefringence of a polymer solution, information may be obtained on the dimensions and hydrodynamic properties of the dissolved polymer molecules. The technique has been widely applied to both natural and synthetic polymers.

FLOW CURVE In rheology, a plot of shear stress against shear rate. For a Newtonian fluid this will be a straight line passing through the origin, with a slope equal to the Newtonian viscosity. For a non-Newtonian fluid the curve is not linear, although if plotted in the log–log form, as is often done, it will be linear for a power law fluid. However, this postulates a relationship between shear rate and shear stress which is not followed by many polymeric fluids, at least over a wide range of flow conditions. The flow curve is often a better representation of behaviour than a mathematical equation representing such a relationship.

FLUIDITY The reciprocal of coefficient of viscosity.

FLUON Tradename for polytetrafluoroethylene.

FLUON FEP Tradename for tetrafluoroethylene–hexafluoropropylene copolymer.

FLUOREL Tradename for vinylidene fluoride–hexafluoropropylene copolymer.

FLUORESCAMINE

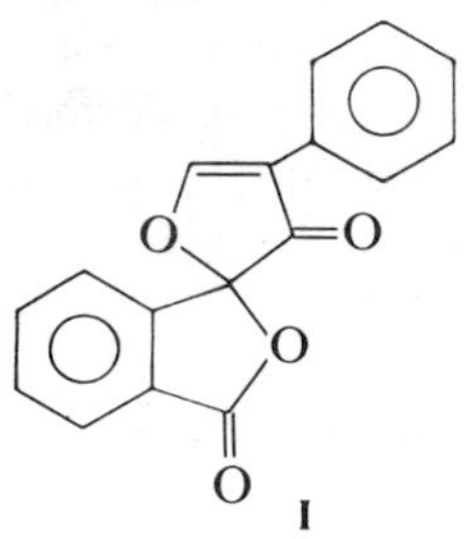

A reagent useful in the determination of the N-terminal amino acid of peptides and proteins, reacting with the free α-amino group to form a fluorescent derivative, which may be analysed in a similar way to the dansyl derivatives:

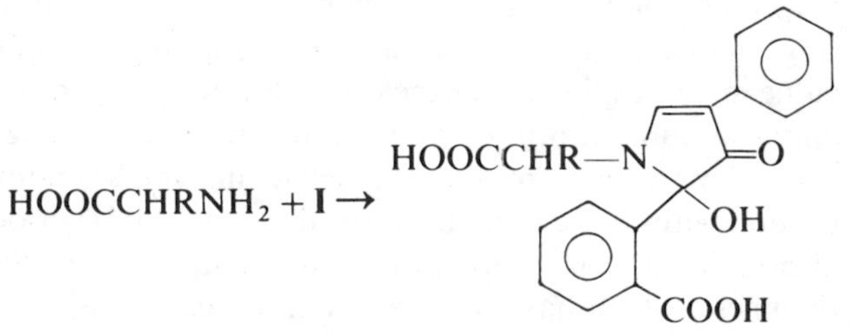

FLUORESCENCE The re-emission of radiation from a molecule after an initial absorption at a shorter wavelength. Occurs from the lowest excited singlet state (S_1) to the ground state (S_0). The fluorescence emission spectrum is often a mirror image of the absorption spectrum. The greatest interest is in fluorescence after irradiation by ultraviolet light; applications in polymer studies include fluorescence spectroscopy and the use of optical brightening agents.

FLUORESCENCE DEPOLARISATION Luminescence depolarisation in which the luminescent process is fluorescence.

FLUORESCENCE SCATTERING Alternative name for polarised fluorescence spectroscopy.

FLUORESCENCE SPECTROSCOPY A technique for the determination of the fluorescence intensity of a fluorescent molecule dispersed in a solid polymer or solution. Useful for studying relaxation processes, local viscosity and rotational diffusion constants. Polarised fluorescence spectroscopy is used in orientation studies.

FLUORINATED ETHYLENE–PROPYLENE CO-POLYMER Alternative name for tetrafluoroethylene–hexafluoropropylene copolymer.

FLUOROACRYLATE RUBBER Alternative name for polyfluoroacrylate.

1-FLUORO-2,4-DINITROBENZENE (FDNB) (2,4-Dinitro-1-fluorobenzene) (DNFB)

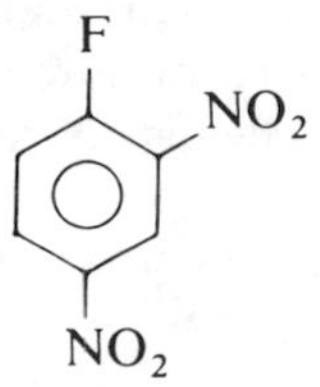

A reagent widely used to identify the N-terminal residue of peptides and proteins. It reacts with terminal α-amino acid groups at pH 8–9:

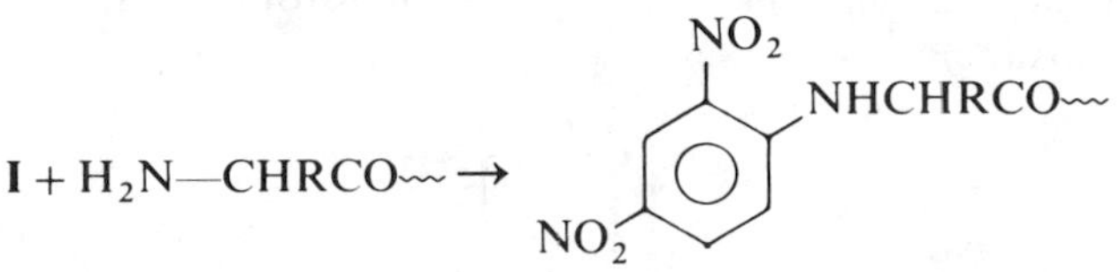

In the subsequent hydrolysis of the dinitrophenyl (DNP) peptide or protein with hydrochloric acid, a labelled terminal free amino acid is formed, which may be separated, e.g. by chromatography, identified and quantitatively determined spectroscopically. This method has been replaced to some extent by the use of the more sensitive dansyl chloride reagent.

FLUOROELASTOMER (Fluororubber) A fluoropolymer which has a low T_g value and does not crystallise, and which is therefore rubbery. As rubbers, the fluoroelastomers are outstanding for their thermal, thermo-oxidative and chemical resistance and their resistance to swelling by most solvents. Most commercial materials are copolymers, usually either of vinylidene fluoride, e.g. with hexafluoropropylene or chlorotrifluoroethylene, or of tetrafluoroethylene, e.g. with a fluoroalkylvinyl ether. The latter, being fully fluorinated, offers the best swelling and thermal resistance. Most polymers are difficult to vulcanise and need the use of special curing agents such as diamines or bisphenols.

FLUORONITROSORUBBER Alternative name for nitrosorubber.

FLUOROPHOSPHAZENE RUBBER Alternative name for phosphonitrilic fluoro-elastomer.

FLUOROPOLYMER A polymer whose repeating units contain fluorine. Such polymers often have outstanding thermal, thermo-oxidative and chemical resistance, both to chemical attack and to swelling by solvents. Although polytetrafluoroethylene is by far the most important fluoropolymer commercially as a plastic material, many other fluoropolymers have been developed as plastic materials, such as polyvinyl fluoride, polyvinylidene fluoride, polychlorotrifluoroethylene and polyfluoroalkylvinyl ether copolymer. Fluoropolymers have also been developed as high temperature and solvent resistant rubbers, such as vinylidene fluoride–hexafluoropropylene copolymers, vinylidene fluoride–chlorotrifluoroethylene copolymers, tetrafluoroethylene–perfluoromethylvinyl ether copolymers, nitrosofluoropolymer, fluorophosphazene and fluorosilicone polymers. Some fluoropolymers of lower molecular weight are useful as heat and chemically resistant oils and greases.

FLUOROPRENE Alternative name for poly-(2-fluoro-1,3-butadiene).

FLUORORUBBER Alternative name for fluoro-elastomer.

FLUOROSILICONE ELASTOMER (FVMQ) A silicone elastomer based on poly-(trifluoropropylmethyl-siloxane)

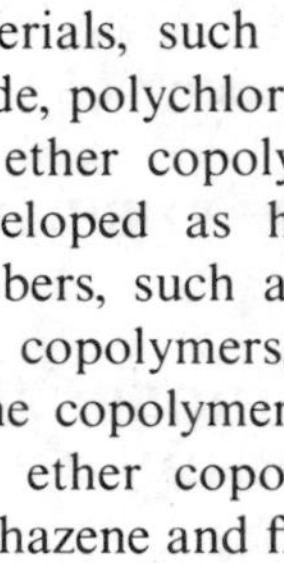

$$\left[\begin{array}{c} CH_3 \\ | \\ Si-O \\ | \\ CH_2 \end{array} \right]_n$$
$$\begin{array}{c} | \\ CH_2 \\ | \\ CF_3 \end{array}$$

with a small amount of vinylmethyl units for ease of peroxide vulcanisation. The vulcanisates have much improved solvent resistance, especially to fuels and other hydrocarbons, compared with other silicone rubbers, whilst retaining the low temperature flexibility and the good high temperature resistance of silicones and fluoropolymers. Useful despite its high cost for seals and O-rings, for fuel pumps and other aerospace applications.

FLUOROTHENE Tradename for polychlorotrifluoroethylene.

FLUOROTRIAZINE ELASTOMER (Perfluoroalkylenetriazine elastomer) A rubbery copolymer consisting of s-triazine rings linked through perfluoroalkylene groups, i.e.

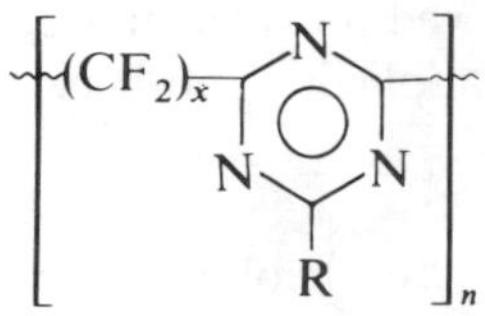

The polymers have outstanding thermo-oxidative stability but this is reduced by the use of curing agents and fillers and hydrolytic stability is not good.

FLUOSITE Tradename for phenol–formaldehyde polymer.

FOAMED POLYMER Alternative name for a cellular polymer whose use is often restricted to those cases where the material is produced by forming gas bubbles in a low viscosity liquid polymer system. Examples include the expanded polymers produced by the mechanical whipping of air into a rubber latex to give a foamed rubber and the formation of polyurethane foams from monomer/precursor polymer mixtures.

FOAMED POLYSTYRENE Alternative name for expanded polystyrene.

FOAM RUBBER A cellular rubber made from a liquid starting material. There are two types. Latex foam rubber is produced by mechanically whipping air into a rubber latex, gelling the latex and vulcanising it. An interconnecting cellular structure is formed. Foamed polyurethane rubber is produced from liquid monomer/prepolymer mixtures which gel by chemical reaction whilst gas is being formed.

FOLDED CHAIN STRUCTURE Alternative name for chain folding.

FOLD LENGTH Alternative name for fold period.

FOLD PERIOD (Fold length) The length of the folds in a chain folded crystalline polymer. In single crystals or single lamellae, it is the same as the lamellar thickness. Its value increases the higher the crystallisation temperature and also on annealing. However, the annealing of single crystal mats may produce lamellae in which folding occurs within the lamellae, so that the fold period, although increased, is now less than the lamellar thickness, which has increased even more.

FOLD PLANE The growth surface of a polymer crystal against which polymer molecules crystallise by regular chain folding. One of the lateral surfaces of the polymer crystal.

FOLD STEM Alternative name for stem.

FOOD DEPOSIT POLYSACCHARIDE Alternative name for reserve deposit polysaccharide.

FORAFLON Tradename for polyvinylidene fluoride.

FORCED VIBRATION METHOD A technique for the determination of the dynamic mechanical properties of a polymer in which a specimen is firmly attached at one end to a force transducer and at the other to a strain gauge. A sinusoidal tensile or shear displacement is applied by a powerful vibrator. The complex modulus is given by $E^* = FL/\Delta L$, where F and ΔL are the force and elongation amplitudes and L and A are the length and cross-sectional area of the sample. The phase angle difference (δ) between the force and displacement signals may be recorded electronically. Frequency ranges of $0.1–10^3$ Hz may be studied for materials with moduli in the range 0.1 MPa to 10 GPa. The method is especially useful for high loss materials for which other methods, e.g. torsion pendulum, may not be suitable.

FORMALDEHYDE

$$CH_2{=}O \qquad \text{B.p. } -21°C.$$

Produced by the dehydrogenation of methanol with air, $CH_3OH + \frac{1}{2}O_2 \rightarrow CH_2{=}O + H_2O$ at $250–400°C$ with an FeO/MoO or silver catalyst. It is usually handled as an aqueous solution (formalin), when it exists as a polymethylene glycol, $HO{+}CH_2O{)_n}H$ with $n = 1–10$. In the presence of acids, the hemi-acetal end groups further react, eliminating water to form acetals and giving a low molecular weight polymer. In the presence of bases, formic acid is formed by the Canizzaro reaction:

$$2\,CH_2{=}O + NaOH \rightarrow CH_3OH + CH_3COONa$$

Formaldehyde solutions thus always contain small amounts of formic acid. Aqueous solutions are normally stabilised with a little methanol. High molecular weight polymer may be produced by anionic polymerisation in an inert, e.g. hydrocarbon, solvent.

The major use (about 55%) is as 30–55% aqueous solutions for the production of phenol–formaldehyde, urea–formaldehyde and melamine–formaldehyde polymers. Smaller amounts are used for the production of polyoxymethylene, hexamethylenetetramine and pentaerithrytol. Sometimes polymers or other derivatives of formaldehyde are used as a source of formaldehyde in polymerisation or curing reactions. These include paraformaldehyde (prepared by the distillation of aqueous formaldehyde solutions), trioxane (a cyclic trimer prepared by heating a strong (about 60%) solution with 2% sulphuric acid) and hexamethylenetetramine (prepared by reaction of formaldehyde with ammonia).

FORM BIREFRINGENCE The contribution to the total birefringence in two phase materials, due to deformation of the electric field associated with a propagating ray of light at anisotropically shaped phase boundaries. The effect may also occur with isotropic particles in an isotropic medium if they are dispersed with a preferred orientation. The magnitude of the effect depends on the refractive index difference between the two phases and on the shape of the dispersed particles. In polymer systems the two phases may be crystalline and amorphous regions, polymer matrix and microvoids, or polymer and filler.

FORM FACTOR Symbol b. A geometrical factor which depends on the shape of a specimen and on the mode of its deformation, i.e. on apparatus geometry. The form factor relates the stress–strain ratio to the displacement—force ratio or angular displacement–torque ratio for specimens of specific shapes and deformation modes. In rheological measurements, for example, for cone and plate geometry, $\gamma_{21}/\sigma_{21} = b\alpha/S$, where γ_{21}/σ_{21} is the shear strain/shear stress ratio, α is the angular displacement, S is the torque and b is given by $2\pi R^3/3\theta$, where R is the radius of the cone and θ is the angle between cone and plate. In simple tension $e/\sigma_t = bx_1/f$, where x_1 is the linear displacement, σ_t is the stress, e is the strain, f is the force and $b = A/L$, where A is the cross-sectional area and L is the length of the specimen. For flexure, $e/\sigma_t = bx_2/f$, where e/σ_t is the stress–strain ratio, x_2 is the linear displacement, f is the force and $b = cd^3/4L^3$ for cantilever flexure of a beam of length L, width c and thickness d (in the x_2 direction). For three point loading with knife-edge support $b = cd^3/L^3$.

FORMIC ACID

$$HCOOH \qquad \text{B.p. } 100.6°C.$$

A solvent for polyamides, aliphatic polyesters, urea– and phenol–formaldehyde resins, often used as a 90% mixture with water. Often dissolves nylons and polyurethanes even in the cold. Slowly decomposes to water and carbon monoxide at room temperature.

FORMID Tradename for acrylonitrile–butadiene–styrene copolymer.

FORMOL TITRATION An analytical method for the determination of free amino groups, especially useful for following the formation of free amino acids during the hydrolysis of proteins. In the presence of excess formaldehyde unprotonated amino groups react with formaldehyde to give methylol derivatives with the release of a proton from the original protonated amino acid:

$$\overset{+}{H_3}NCHRCOO^- \rightleftharpoons H_2NCHRCOO^- + H^+$$

$$H_2NCHRCOO^- + 2\,HCHO \rightarrow$$
$$(HOCH_2)_2NCHRCOO^-$$

The released proton may then be titrated with alkali.

FORMREZ Tradename for a cast polyurethane system based on a tolylene diisocyanate terminated polyester prepolymer, chain extended and cured by heating with 3,3'dichloro-4,4'-diaminodiphenylmethane.

FORMVAR Tradename for polyvinyl formal.

FORTAFIL Tradename for a polyacrylonitrile based carbon fibre.

FORTICELL Tradename for cellulose propionate.

FORTIFLEX Tradename for high density polyethylene.

FORTILINE Tradename for high density polyethylene.

FORTINESE Tradename for a high tenacity cellulose acetate fibre.

FORTISAN Tradename for a high tenacity rayon fibre produced by stretching cellulose acetate fibre in steam, then saponifying with alkali to regenerate the cellulose. Tenacity is about 7 g denier^{-1} and elongation at break about 8%.

FORTREL Tradename for a polyethylene terephthalate fibre.

FORTRON Tradename for polyphenylene sulphide.

FOSSIL RESIN A natural resin dug from the ground, having been the exudation from trees long since dead and decayed. Congo and kauri copals and amber are examples.

FOSTACRYL Tradename for styrene–acrylonitrile copolymer.

FOSTAFOAM Tradename for expanded polystyrene.

FOSTARENE Tradename for polystyrene.

FOURIER NUMBER Symbol F_N. A dimensionless parameter defined as $F_N = Dt/(\text{section thickness})^2$, where D is the thermal diffusivity and t is the time, thus giving a measure of heat transfer through a section. For $F_N < 0.1$ the bulk of the material will be little affected by its environment. If F_N is 1.0 then the material has come to equilibrium with its surroundings. For intermediate values, a large temperature gradient will exist across the section.

FOURIER TRANSFORM INFRARED SPECTROS-COPY (FTIR) Infrared spectroscopy in which the incident beam is split into two waves and the spectral information of the sample is contained in the phase difference between the two waves, after they have been recombined. This is performed by the use of a Michelson interferometer. An interferogram is obtained from the detector which is digitised and transformed from the time domain by a mathematical Fourier transform operation and the resultant signals are converted to a conventional infrared spectrum. These operations are carried out by a computer. The advantages of FTIR over conventional IR are its abilities to detect weak signals and to obtain a complete spectrum in a few seconds.

FOURIER TRANSFORM NUCLEAR MAGNETIC RESONANCE SPECTROSCOPY (Pulsed Fourier transform nuclear magnetic resonance spectroscopy) Nuclear magnetic resonance spectroscopy in which wide frequency band radio frequency is applied as a pulse, which sets all the ^1H or ^{13}C nuclei resonating. Spectra in the time domain are detected as interferograms when the pulse is switched off and are added together (accumulated). The spectra are then Fourier transformed to the usual frequency domain. In this way a greater sensitivity is achieved, which is particularly valuable for ^{13}C NMR. Further, an entire spectrum can be scanned in a few seconds, so that even several hundred or thousand accumulations only require a few minutes. Further increase in sensitivity is possible with a nuclear Overhauser effect.

FOX–FLORY THEORY Alternative name for Flory–Fox theory.

FRACTIONAL ELUTION Alternative name for fractional solution.

FRACTIONAL EXTRACTION Alternative name for fractional solution.

FRACTIONAL PRECIPITATION Fractionation in which whole polymer is dissolved in a solvent and successive fractions of decreasing molecular weight are obtained as precipitates by progressively reducing the solvent power. This is usually done by addition of precipitant miscible with the solvent. Sometimes lowering of the solution temperature or evaporation of the most volatile solvent component in a mixed solvent system are used. The method is especially useful for high molecular weight polymers. The first few fractions obtained are of higher molecular weight than the last few obtained from fractional solution methods. The method is usually a preparative fractionation method but has been adapted to an analytical fractionation method in turbidimetric titration, summative fractionation and gel-volume fractionation. Occasionally reverse-order precipitation occurs and complicates analysis of the results. For biopolymers, separation of different polymers from mixtures in aqueous solution, e.g. for proteins, may be achieved by variation of pH (isoelectric precipitation) or ionic strength (salting-in and salting-out) or by use of organic solvents, often ethanol, at low temperatures to prevent denaturation.

FRACTIONAL RECOVERY A measure of the recovery from creep of a viscoelastic material. Defined as the ratio of strain recovered to maximum creep. Useful as a means of reducing complex recovery data to a 'master curve' of fractional recovery versus reduced time, for a variety of recovery conditions. However, such a simple plot does not properly represent data at long creep times or at high creep strains.

FRACTIONAL SOLUTION (Fractional extraction) (Fractional elution) Fractionation in which whole polymer is successively extracted with liquids of gradually increasing solvent power. The residue is removed at each stage and a series of fractions of increasing molecular weight are obtained from the solutions.

The method is less widely used than fractional precipitation due to the difficulties of establishing equilibrium between polymer and liquid (especially in the direct sequential extraction method) and in scaling-up. However, it is readily automated and can be performed with only about 1 g of polymer. Several techniques are used to speed up establishment of equilibrium, including film extraction, column extraction, gradient elution and coacervation methods.

FRACTIONATION The separation of a polymer sample consisting of different molecular species (whole polymer) into fractions which are more homogeneous, with respect to molecular structure, than the whole polymer. Although fractionation may be performed according to differences in configuration (tacticity) and chemical composition (especially with copolymers), the term usually refers to fractionation with respect to molecular size. Since most polymer samples are polydisperse, fractionation is important both as a means of obtaining samples with narrow molecular weight distributions and of characterising molecular weight distribution of the whole polymer. In preparative fractionation, samples are isolated, whereas in analytical fractionation, they are not. It is not possible to produce fractions which are monodisperse and lengthy repetitive fractionation is needed to give fractions of polydispersity approaching unity.

Fractionation methods mainly use solubility differences between molecular species, as in the techniques of fractional precipitation, fractional solution, turbidimetric titration and summative precipitation. In fractionation by chromatography, gel permeation chromatography (gel filtration) is outstandingly important, but absorption chromatography, partition chromatography, ionexchange chromatography and precipitation chromatography are also used. Fractionation by ultracentrifugation is especially valuable for biopolymers. Other techniques include thermal diffusion, ultrafiltration, zone melting and Brownian diffusion.

FRACTOGRAPHY The examination of fracture surfaces, usually microscopically. This not only gives information about the internal structure of the material, such as the nature of the dispersed phase in a composite, but also on the course of the fracture process itself by the appearance of fracture markings such as hackle and Wallner lines. Replication of the surface is often used followed by transmission electron microscopy, but scanning electron microscopy is much more direct.

FRACTURE The creation of new surfaces within an object which leads to its disintegration, usually as a result of an applied load. In the practical use of an object, fracture results in failure of the object, for example by impact loading. Fracture may be classified by its observed effects on the fractured body, as either brittle fracture or, if plastic flow has occurred, as ductile fracture, or according to the cause of fracture. Most commonly fracture is caused by direct loading, different systems of forces leading to, for example, tensile fracture, fracture in bending, crushing, etc. Such fractures are often catastrophic, but more controlled fracture occurs in tearing and also in carefully designed experimental setups which provide only slow propagation of the crack in order to study the fracture process. In direct loading the applied force is increasing very rapidly, e.g. under impact loading, but with a constant load applied for a long period of time, fracture may occur at a lower stress. This is creep fracture or static fatigue. In many types of fatigue, the load is cyclically varying and this can also cause fracture to occur at a lower stress than in fast fracture.

In addition to stress, other factors may be important, such as the environment. Both liquids and gases can give rise to environmental stress cracking. In abrasion and other wear processes, fracture is also important.

The fracture resistance of a polymer is a function both of the material mechanical properties (elasticity, plasticity and viscoelasticity) and of the microstructure, especially if the material is two phase. Many two phase materials are deliberately designed for high fracture resistance as with fibre-reinforced composites, rubber toughened plastics and fine particulate filled rubbers. Other types of polymer materials are naturally two phase, as with crystalline polymers, which also enhances their fracture resistance.

The process of fracture usually involves initiation of a crack, often at a flaw, e.g. a microcrack, in the material or other stress raiser, followed by crack propagation. Fracture mechanics analyses these processes quantitatively. The Griffith theory provides an energy criterion for crack propagation, as well as providing an explanation for the low strength of brittle materials, by postulating the existence of stress raising microcracks either on the surface or within the bulk of the material. Fracture may be accompanied by associated crazing around the crack, especially with brittle materials, or by shear yielding, especially with ductile materials.

FRACTURE MODE The particular type of fracture according to the way in which the specimen is stressed. There are three modes usually considered: the tensile crack-opening mode (mode I), the shearing mode (mode II) and the tearing mode (mode III). Mode I is the most widely analysed type.

FRACTURE TOUGHNESS In general, the ability of a material to resist fracture by the absorption of energy before breaking. Thus in a simple stress–strain plot, the area under the curve is an indication of the toughness: ductile materials which have large elongations at break will have high toughness, whereas brittle materials will not. In particular, in fracture mechanics, fracture toughness means the ability of a material to withstand propagation of a crack as measured by the critical stress intensity factor.

FRANK'S CAMERA A camera for small angle X-ray scattering measurements in which all the rays of the X-ray beam are focussed by total reflection from a horizontal and a vertical mirror. This requires very accurate construction and a point source of X-rays.

FRAME INDIFFERENCE A property of some physical laws, e.g. the conservation of mass, as expressed in the continuity equation, which means that the law can be expressed in an equation of the same form, independently of the frame of reference. A valid constitutive equation should also be frame indifferent. However, the momentum equations are not frame indifferent.

FREE-BOUNDARY ELECTROPHORESIS (Moving-boundary electrophoresis) The classical, and original, electrophoresis technique, now largely supplanted by the simpler and more sensitive zone electrophoresis methods. The buffered protein solution is placed in the bottom of a U-tube and buffer is carefully layered on top to cover the electrodes at the top of each arm. Usually protein solution pH is chosen so that all protein molecules carry the same charge and so move in the same direction. On passing a current the charged protein molecules move into the layered buffer and the fastest moving forms a front or boundary. Slower moving proteins do not separate completely but give separate sharp boundaries. The refractive index changes allow each boundary to be identified, e.g. by use of Schlieren optics, and the concentration of protein in each layer to be estimated. However, the equipment required is elaborate, requires large samples and is not very suited to preparative work. Thus zone electrophoresis methods, which also have higher resolving power, have largely supplanted this technique.

FREE DRAINING COIL A coiled polymer molecule in solution such that when the solution is subjected to a shear gradient due to flow, the solvent freely permeates through the coil. This results if there are no hydrodynamic interactions between the flow patterns of solvent around different segments of the coil. Probably only exists in coils of fairly low molecular weight ($< 30\,000$) polymer, since in larger molecules the solvent is usually considered to be immobilised in the interior of the coil, which rotates in the shear field.

FREELY JOINTED CHAIN (Random flight chain) A simple model for a polymer chain, which consists of identical infinitely thin segments each of length l, joined so that the angle between the segments can take any value. In addition, bond rotation is freely allowed. The mean square end-to-end distance ($\langle r^2 \rangle$) is then equal to nl^2, where n is the number of segments. Such a simple model hardly adequately represents a real polymer chain for which the characteristic ratio, $\langle r^2 \rangle/nl^2$, is greater than unity. However, a real chain is sometimes reasonably represented in this way by considering it to consist of n' statistically equivalent segments (each of length l') of a freely jointed chain provided that $\langle r^2 \rangle_0 = n'l'^2$, where $\langle r^2 \rangle_0$ is the unperturbed value and $r_{ex} = n'l'$, where r_{ex} is the fully extended value of length of the real chain.

FREELY ROTATING CHAIN A simple model for a polymer chain, similar to the freely jointed chain, but in which n chain segments each of length l are joined so that the angle between the segments now has a single fixed value θ. Now the unperturbed end-to-end distance is given by:

$$\langle r^2 \rangle_0 = [(1 - \cos \theta)/(1 + \cos \theta)]nl^2$$

FREE RADICAL POLYMERISATION (Radical polymerisation) Chain polymerisation in which the active centres are free radicals. The most common type of chain polymerisation due to the wide variety of monomers that may be polymerised by this mechanism. The monomers are usually either vinyl or diene monomers (vinyl or diene polymerisation) and only if the monomers have substituents which are powerfully electron withdrawing or electron donating are they not susceptible to free radical polymerisation. 1,2-Disubstituted monomers are not readily polymerised except when the substituent is small, e.g. a fluorine atom as in tetrafluoroethylene.

Initiation is usually by thermal decomposition of an added thermal free radical initiator (R—R), commonly an organic peroxide, hydroperoxide or azo compound or, in aqueous systems, by a redox initiator. However, any method which produces free radicals may be used, e.g. ultraviolet light (photopolymerisation), high energy radiation or thermal activation of the monomer (M) itself (thermal polymerisation): R—$R \rightarrow 2\,R\cdot$ or $M \rightarrow R\cdot$.

The primary radical $R\cdot$ initiates polymerisation by addition to monomer to form an activated monomer: $R\cdot + M \rightarrow R$—$M\cdot$. By rapid successive addition of further monomer molecules, a high molecular weight polymer molecule is rapidly formed (propagation),

$$R\text{—}M\cdot \xrightarrow{M} R\text{—}M\text{—}M\cdot \xrightarrow{M}, \text{ etc.}$$

or in general,

$$R\text{—}M_n\text{—}M\cdot + M \rightarrow R\text{—}M_{\overline{n+1}}M\cdot$$

When M is a vinyl monomer, CH_2=CHX (the most common case), we have

$$R\cdot + CH_2\text{=}CHX \rightarrow R\text{—}CH_2\text{—}CHX\cdot \quad \text{Initiation}$$
$$R\text{(}CH_2CHX\text{)}_n CH_2CHX\cdot + CH_2\text{=}CHX\cdot \rightarrow$$
$$R\text{(}CH_2CHX\text{)}_{n+1}\text{—}CH_2CHX\cdot \quad \text{Propagation}$$

and the active centre is the substituted carbon $\sim CHX\cdot$ not $\sim CH_2\cdot$, since X usually contributes to resonance stabilisation of the radical. For the same reason, and also for steric reasons, addition of further monomer molecules is predominantly head-to-tail and not head-to-head. In contrast to many ionic polymerisations, the polymer produced is essentially atactic, although there is a tendency to syndiotacticity as polymerisation temperature is lowered.

Propagation ceases due to either a termination or a transfer step. In the latter case we have:

$$\sim M\cdot + X{-}Y \rightarrow \sim\sim X + Y\cdot$$

The product free radical $Y\cdot$ will be of different activity from $M\cdot$. At one extreme it may be so stable that it cannot re-initiate polymerisation (inhibition by $X{-}Y$), or it may be capable of re-initiation but with a lower rate constant than that for propagation (retardation). Transfer usually results in a lower polymer molecular weight. $X{-}Y$ may be an impurity, it may be deliberately added (as an inhibitor or chain transfer agent), or it may be monomer (transfer to monomer) or even initiator (induced decomposition).

The kinetics of free radical polymerisation have been extensively analysed. The overall rate of polymerisation (R_p), synonymous with rate of monomer consumption ($-d[M]/dt$), may be derived in the simple case of the absence of transfer using the steady state assumption, as $R_p = k_p[M](R_i/2k_t)^{1/2}$, where R_i is the rate of initiation ($=2fk_d[I]$), k_d, k_p and k_t are the rate constants for initiator decomposition, propagation and termination respectively (k_t being defined as the rate constant for $\sim\sim\cdot + \sim\sim\cdot \rightarrow$ dead polymer), f is the initiator efficiency and [M] and [I] are the monomer and initiator concentrations respectively. The kinetic chain length (v) is given by, $v = k_p^2[M]^2/2k_tR_p$ and governs the number average degree of polymerisation ($\bar{x}_n$) since $\bar{x}_n = 2v$ for termination by combination and $\bar{x}_n = v$ for termination by disproportionation.

FREE RADICAL SCAVENGER   Alternative name for radical scavenger.

FREE RADICAL TRAP   Alternative name for radical scavenger.

FREE SURFACE FLOW   A flow occurring where the boundary of the fluid is not rigidly fixed by the wall of a container or conduit. Examples include the drawing of a fibre of film, and the expansion of a bubble as in blow moulding or film blowing. In such cases the flow is often an elongational flow rather than a shear flow.

FREE VIBRATION METHOD   A method for the determination of the dynamic mechanical properties of a polymer in which a test sample is clamped rigidly at one end and is free to move at the other end in an oscillatory manner, e.g. by attaching a weight (for longitudinal oscillations) or, more commonly, by attaching to a torsion wire as in a torsion pendulum.

FREE VOLUME   The volume of the vacant sites or 'holes' present in a liquid or amorphous solid, not occupied by the molecules of the material. This equals the total volume minus the occupied volume, which is itself the sum of the van der Waals radii of the atoms plus the volume associated with vibrations. An important concept in the interpretation of the viscoelastic properties of polymers. Thus viscous flow can be envisaged as occurring by segmental jumps of polymer molecular chain segments into vacant holes. The higher the free volume (V_f) the more readily these jumps can occur and the lower the viscosity—as given by the Doolittle equation. The concept is also important in the interpretation of many of the aspects of the T_g behaviour using free volume theory. As temperature increases, thermally activated molecular motions increase and V_f increases linearly with temperature as, $f = f_g + \alpha_f(T - T_g)$, where f is the total fractional V_f, f_g is its value at T_g and α_f is the coefficient of expansion of V_f. α_f may be identified with $\Delta\alpha = (\alpha_r - \alpha_g)$, the difference between the polymer thermal expansion coefficients above (α_r) and below (α_g) T_g. In this way α_f values may be obtained experimentally. These values are often found to be equal to those obtained from the coefficient C_2 of the Williams–Landel–Ferry equation since $C_2 = f_g/\alpha_f$. For many polymers f_g reaches the constant value of 0.025 as the temperature is lowered towards T_g and remains at this value as the temperature is lowered beyond T_g. This leads to the iso-free volume theory of T_g. It is believed that this value is simply an effective free volume relevant to certain relaxation experiments (and is sometimes called the WLF free volume), whereas the actual free volume is 10–15% less.

FREE VOLUME THEORY   A theory based on the concept of the free volume of a polymeric fluid or glass, which is very useful in quantitatively and qualitatively explaining many of the phenomena associated with the glass transition. On lowering the temperature of a polymer melt, a transition is considered to occur when the diminishing free volume has reached a fractional value (f_g) such that polymer chain segmental jumps are no longer possible. The fractional value (f) is given by $f = f_g + \alpha_f(T - T_g)$ where α_f is the coefficient of thermal expansion of the free volume and T is the temperature.

The theory is useful in predicting the effects of various parameters (pressure, presence of diluent and polymer molecular weight) on the value of T_g. Since pressure (P) increases cause a squeezing together of the molecules, the free volume is expected to decrease and T_g to increase. This effect is given by $dT_g/dP = \Delta\beta/\Delta\alpha$, where $\Delta\beta$ and $\Delta\alpha$ are the differences in compressibility and expansion coefficients above and below T_g. A better fit with experimental data is found if $\Delta\beta$ and $\Delta\alpha$ are replaced by β_f and α_f, the corresponding coefficients for the fractional free volume, since the given equation assumes that the occupied volume has the same values of β and α as those of the glassy state. The effect of diluent, e.g. plasticiser, is to increase the free volume and depress T_g, and is given quantitatively by,

$$T_g = [\alpha_P V_P T_{g,P} + \alpha_d(1 - V_P)T_{g,d}]/[\alpha_P V_P + \alpha_d(1 - V_P)]$$

where V_P, V_d, α_P, α_d, $T_{g,P}$, $T_{g,d}$ are the volume fraction, expansion coefficients and T_g values of the polymer and diluent respectively. The theory also correctly predicts that T_g decreases with decreasing polymer molecular weight, due to the increased number of chain ends generating more free volume, expressed by $T_g = T_g^\infty - K/M$, where T_g^∞ is the T_g value of the polymer with infinite molecular weight, K is a constant and M is the molecular weight which may be the number average value. A similar relationship, $1/T_g = (1/T_g^\infty) + (A/M)$, where A is a constant, does not give significantly different results except at low molecular weight.

FREEZE-DRYING A technique for obtaining solid polymer from solution, which is often useful for heat-sensitive materials or to produce the polymer in a finely divided, e.g. fluffy, form. The polymer solution is frozen and subjected to a sufficiently high vacuum so that the solid solvent sublimes. Frequently the heat of sublimation is sufficient to keep the solution frozen without additional cooling. Thus the volume of solid polymer produced is that of the original solution, and hence the polymer is not obtained (as it would be by simple evaporation of the solvent) as an inconvenient solid lump, but as an expanded mass. However, the sublimation process may take a long time if the solid solvent vapour pressure is low.

FREQUENCY DISTRIBUTION A discontinuous distribution giving the distribution of molecular sizes (i) (often expressed as molecular weights M_i) in a polydisperse polymer sample. Describes the variation in amount of polymer, measured usually as weight (or mass) (hence sometimes called the weight distribution) or as moles (hence sometimes called the number distribution), with change in molecular size per unit of i. When this stepwise function is converted to a corresponding continuous function, a differential distribution function results.

FREQUENCY DOMAIN METHOD A method for the determination of the dielectric properties of a material by application of a sinusoidally varying field and measurement of the response, as in the Schering and transformer ratio arm bridges, resonance and wave transmission methods. For low frequency ranges the time-domain methods are used.

FRICTIONAL COEFFICIENT (Frictional constant) Symbol f. In the sedimentation of polymer molecules in dilute solution during ultracentrifugation, it is the constant relating the frictional force F resisting sedimentation to the sedimentation velocity (dr/dt), i.e. $F = f(dr/dt)$. For spherical molecules of radius R, $f = 6\pi\eta R$, where η is the coefficient of viscosity of the medium. Both F and f are concentration dependent and for analysis of ultracentrifugation data the value at infinite dilution (f_0) is needed. This cannot be measured, but it may be related to, and replaced by, D_0 (the diffusion constant) by the Einstein relation $f_0 = kT/D_0$, where k is Boltzmann's constant and T is the absolute temperature. This substitution leads to the Svedberg equation.

FRICTIONAL CONSTANT Alternative name for frictional coefficient.

FRIEDEL–CRAFTS POLYMER In general, polymers synthesised by the Friedel–Crafts reaction. In particular, aromatic polymers synthesised by reaction between an aromatic chloromethyl compound or ether compound and an aromatic ring compound. The commercial Xylok resins are synthesised by a variation of the latter reaction.

FRIEDEL–CRAFTS POLYMERISATION A polymer forming reaction performed using a Friedel–Crafts reaction. The monomers used are a dihalide and an aromatic compound, frequently a diphenyl, diphenyl ether or diphenylmethane, most often the *para*-linked isomers. The polymers produced (Friedel–Crafts polymers) often contain mixed *ortho*, *meta* and *para*-linkages between the benzene rings and the molecular weight is often limited by the occurrence of side reactions.

FRIEDEL–CRAFTS REACTION An electrophilic substitution reaction on an aromatic ring resulting in alkylation or acylation of the ring. Reaction occurs between carbenium ions formed from the reagent (often an alkyl chloride) and the aromatic compound in the presence of a Lewis acid e.g. $AlCl_3$, $FeCl_3$ or BF_3:

$$RCl + H—Ar—H \rightarrow$$
$$R—Ar—H + HCl \xrightarrow{RCl} R—Ar—R + HCl$$

Substitution, e.g. by —OH, —OR, —Ph, of the aromatic compound often activates it towards alkylation at both *ortho* and *para* positions and therefore mixed isomeric products result. Di- or trisubstitution may occur since the alkyl groups activate for further substitution. The reaction is of some use in polymer syntheses, which are then described as Friedel–Crafts polymerisations.

FRINGED FIBRIL A modified fringed micelle model for describing fibre morphology, in which the polymer molecules are more nearly aligned (in the fibre direction) rather than being randomly oriented. In one version, the molecules diverge from alignment at the same position along the fibril, which therefore contains discrete crystallites. In another version, the crystallites are considered to be continuous fringed fibrils, in which the molecules diverge at different positions. The long period spacings are not completely accounted for by this model.

FRINGED MICELLE A model for a partially crystalline polymer widely accepted as explaining many of the properties of such a polymer. Used from the 1930s until the discovery of polymer single crystals. It consists of crystallites (the 'micelles') embedded in an amorphous matrix. It is based on the X-ray diffraction line broadening which indicates a crystallite size of about 100 Å and the appearance of an amorphous halo. Since polymer molecules are known to be much longer, they are supposed to meander from one crystallite through the amorphous phase to another. The discovery of chain folding in polymer single crystals led to the abandonment of this model, although some features are retained in the present understanding of polymer crystallite morphology.

FR-N   Tradename for nitrile rubber.

FRÖHLICH EQUATION   A relationship between the relative permittivity (ε) and the dipole moment (μ) of a dielectric. It is identical to the Onsager equation, except that it incorporates a correlation factor (g) to take account of the local ordering of the dipoles. g is given by, $g = (1 + z\overline{\cos\gamma})$, where γ is the angle of the reference molecule to a nearest neighbour and z is the number of nearest neighbours. It reduces to the Onsager equation when $g = 1$. It is,

$$[(\varepsilon - n^2)(2\varepsilon + n^2)/\varepsilon(n^2 + 2)^2]M/\rho = N_A g\mu^2/9\,\varepsilon_0 kT$$

where M is the molecular weight, n is the refractive index, N_A is the Avogadro number, ρ is the density, k is Boltzmann's constant, T is the temperature and ε_0 is the relative permittivity of free space. However, there is considerable difficulty in evaluating g. In a polymer, the dipoles are constrained from moving by being linked to a polymer chain so the g factor must take into account this strong intramolecular ordering.

FRONT FACTOR   A factor added to the basic equation of state of rubber elasticity giving the retractive force (f) of a stretched rubber in terms of the extension ratio (λ), $[f = nRT(\lambda - 1/\lambda^2)]$, where R is the universal gas constant and T is the absolute temperature. The right hand side is multiplied by the front factor X/X_0, which allows for energy differences between the *trans* and *gauche* conformations in the polymer chain. X/X_0 relates the chain dimensional characteristics to the *trans–gauche* energy differences and volume changes.

FROZEN-IN ORIENTATION   Orientation present in a melt cooled polymer produced by melt deformation, e.g. by flow through a narrow channel, and retained when the deforming forces are removed due to rapid cooling (quenching). This prevents relaxation of the orientation by freezing of molecular motions. The extent depends on the speed of cooling, relaxation time of the polymer and deformation forces. It usually results in an undesirable drop in impact strength.

FROZEN-IN STRESS   The stress present in a material or object, in the absence of an externally applied load, which arises from an earlier stressing whilst the material was in a more fluid state and which becomes 'locked-in' by cooling the material whilst still in the stressed state. This may occur during polymer melt processing operations, especially injection moulding, when the polymer melt is rapidly cooled to well below its glass transition temperature. Frozen-in stress causes a reduction in fracture resistance, especially by environmental stress cracking, but may be reduced by annealing.

FR-S   Tradename for styrene–butadiene rubber.

FRUCTAN   A polymer of fructose. Such polymers occur widely as reserve deposit carbohydrate polymers in plants, especially in the *Gramineae* and *Compositeae* (including many grasses), where they are synthesised from sucrose. Also produced by a number of micro-organisms. The fructans have β-D-fructofuranose units either 2,1'-joined (in the inulins) or 2,6'-joined (in the levans). Certain branched fructans contain both types of link. Plant fructans are of low molecular weight (about 8000) and are often terminated by D-glucose non-reducing end groups. They are very readily hydrolysed and must be isolated with care.

FRUCTOFURANOSE   A fructan in which the fructose units are in the furanose ring form, as in levan and inulin.

FRUCTOSE   (Laevulose) Fischer projection formula:

Ring form (β-D-fructopyranose):

M.p. 102–104°C. $\alpha_D^{20.H_2O}$ $-92\cdot4°$.

A ketohexose occurring free in many fruits and contained in oligosaccharides, often with D-glucose as in sucrose, but rarely in other glycosides. Homopolysaccharides of D-fructose (fructans) include levan and inulin and occur in many plants especially of the *Compositeae*. Heteropolysaccharides are rare. Usually prepared by inversion of sucrose after which the fructose is separated from the D-glucose either by crystallisation, oxidation of the latter or precipitation of calcium fructosate. It crystallises in the β-D-pyranose form, but when combined with other carbohydrates, it is in the furanose form.

FT   Abbreviation for fine thermal black.

FTIR   Abbreviation for Fourier transform infrared spectroscopy.

FUCAN SULPHATE   A sulphated fucan polysaccharide of which the naturally occurring fucoidan is the only known example.

FUCOIDAN   A fucan sulphate polysaccharide found, together with laminaran and alginic acid, in certain brown seaweeds especially the *Fucoseae*. Complete structural details are not known, but it is mainly composed of 1,2'-linked 4-fucose units sulphated at the 4-position to the extent of approximately one sulphate group per fucose unit.

L-FUCOSE (6-Deoxy-L-galactose) Fischer projection formula:

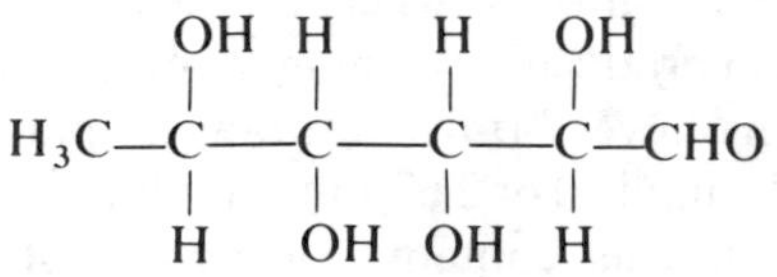

$$H_3C-\overset{\overset{\displaystyle OH}{|}}{\underset{\underset{\displaystyle H}{|}}{C}}-\overset{\overset{\displaystyle H}{|}}{\underset{\underset{\displaystyle OH}{|}}{C}}-\overset{\overset{\displaystyle H}{|}}{\underset{\underset{\displaystyle OH}{|}}{C}}-\overset{\overset{\displaystyle OH}{|}}{\underset{\underset{\displaystyle H}{|}}{C}}-CHO$$

Ring form (α-D-fucopyranose):

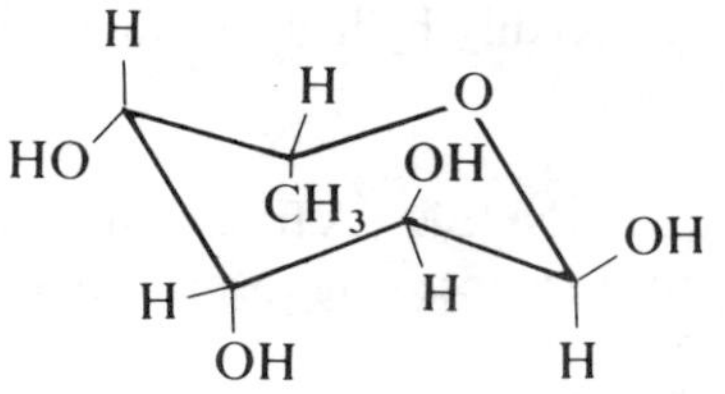

M.p. 145°C. $\alpha_D^{20.H_2O}$ −76°.

A deoxy-aldohexose occurring as a constituent of the heteropolysaccharide in gum tragacanth, seaweeds, e.g. fucoidan, and a common constituent of the glycoprotein in mucins, blood-group substances and in milk oligosaccharides. Obtained by hydrolysis of seaweed.

FULLY EXTENDED CHAIN Alternative name for extended chain.

FUMARIC ACID

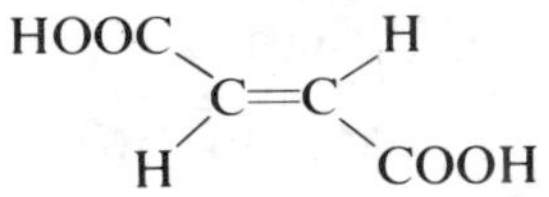

M.p. 284°C.

Prepared by *cis–trans* isomerisation of maleic acid by heating. It is sometimes preferred to maleic anhydride as a source of unsaturated units in unsaturated polyester resins, since it imparts a lighter colour and higher heat resistance to the product.

FUMED SILICA Alternative name for pyrogenic silica.

FUNCTIONALITY Symbol f. In step-growth polymerisation, the number of functional groups present per monomer molecule, which participate in polymer forming reactions. When the functionality of the monomer, as in an AB polymerisation, or monomers, as in an AABB polymerisation, is two, then linear polymerisation occurs. When tri- or higher functionality monomers are used, then non-linear polymerisation results. When monomers with different functionalities are used together, an average functionality may be calculated.

FUNGAL POLYSACCHARIDE A polysaccharide occurring naturally in a fungus. Examples include pachyman, yeast-β-D-glucan, D-mannans, galactocaralose and nigeran.

FUOSS–MEAD OSMOMETER A twin-capillary membrane osmometer which consists of two cylindrical metal blocks, on one face of each of which is a series of circular grooves, connected by a vertical groove. The grooves form the compartments of the cell, which is formed by clamping the two blocks together with the membrane sandwiched between the two faces. In this way the cell volume is kept at a minimum, whilst a large membrane area is maintained. A filling/draining tube and a measuring capillary is attached to each block.

FURANOSE The ring form of a monosaccharide or monosaccharide unit in a carbohydrate that contains five ring atoms and is therefore considered to be derived from furane.

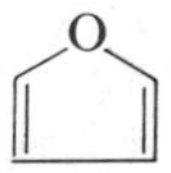

Few unsubstituted monosaccharides (although several substituted ones) exist in this form in the crystalline state but it is quite common in glycosides and oligosaccharides. In aqueous solution the furanose form is frequently present in equilibrium with the pyranose form. In naming a furanose monosaccharide or unit, the suffix-furanose is added to the name of the sugar, e.g. β-D-glucofuranose. A glycoside in furanose form is a furanoside.

-FURANOSIDE A suffix added to the name of a glycoside to show that it is in the furanose ring form.

FURAN RESIN Tradename: Quacorr. A polymer produced by the polymerisation of, usually, furfuryl alcohol, or sometimes furfural, or a copolymer obtained using these monomers. Polymerisation of the former occurs by acidic catalysis via dehydration, yielding initially linear polymers of the type:

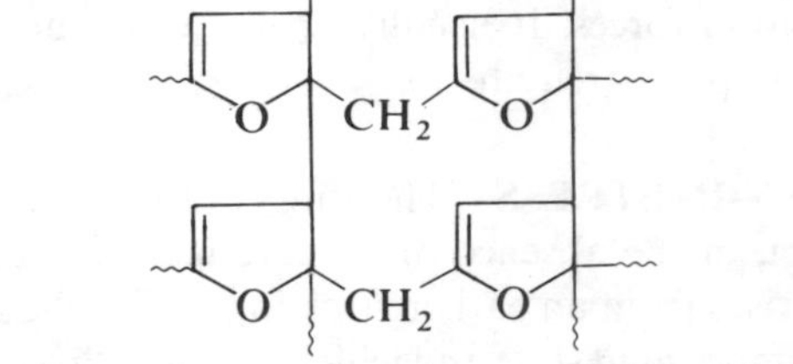

Subsequently the polymers become insoluble and crosslinked with loss of unsaturation possibly forming structures like:

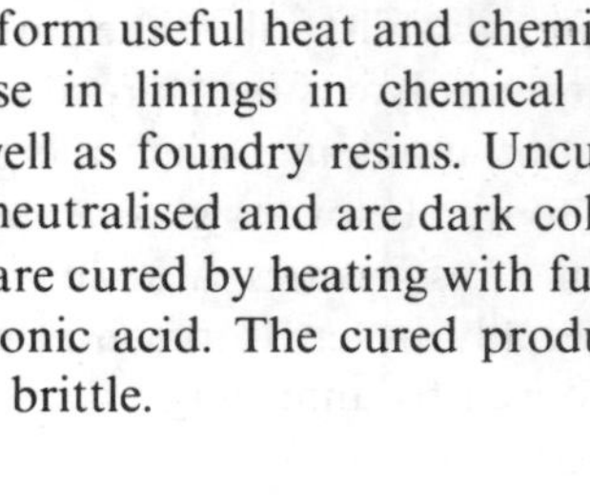

The polymers form useful heat and chemically resistant plastics for use in linings in chemical plant and as laminates as well as foundry resins. Uncured resins are prepared and neutralised and are dark coloured viscous liquids. These are cured by heating with further acid, e.g. *p*-toluenesulphonic acid. The cured products are black and are rather brittle.

FURFURAL

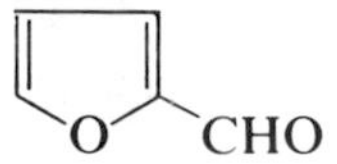

B.p. 162°C.

Produced by the degradation of naturally occurring hemicelluloses, e.g. by heating corncobs or oat hulls with sulphuric acid. A useful solvent for certain surface coating materials, e.g. rosin, for conversion to furfuryl alcohol and as a monomer in the production of furan resins and phenol–furfural resins. A reactive material which darkens in air and readily polymerises in the presence of acids.

FURFURYL ALCOHOL

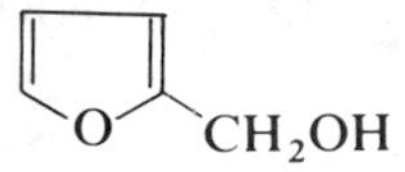

B.p. 170°C.

Produced by the catalytic hydrogenation of furfural. The main monomer for the production of furan resins. It readily polymerises when heated or exposed to oxygen, air or acids. A solvent for cellulose esters, natural resins and phenol–formaldehyde resins. It swells polyethylene.

FURNACE BLACK The most common type of carbon black produced by incomplete combustion of natural gas or of heavy aromatic oils in furnaces, the black being separated from the combustion products by bags or cyclones and then pelletised. A very wide variety of blacks with particle sizes 20–80 nm are made. Each type is designated by an abbreviation, such as SAF, ISAF, HAF, FF, FEF, GPF, HMF, SRF, CF, SCF. Typically these blacks have high structure and contain about 97% carbon. They are widely used for reinforcement in all rubber products. Highly reinforcing types are used in tyre treads and low reinforcing types in carcasses. Fine particle types are used for their good abrasion resistance in, for example, conveyor belts and shoe soles and coarser particle types are used in hose and cables.

FUSIBLE ELASTOMER An elastomeric gum based on a borosilicone with one boron atom to every 200–500 silicon atoms which is sticky after a precure and which may be welded to itself on application of slight pressure after final vulcanisation, i.e. it shows high self-tack and is therefore useful in self-adhesive electrical insulating tapes.

FVMQ Abbreviation for fluorosilicone elastomer.

FVSI Older alternative abbreviation for FVMQ.

FYBEX Tradename for a single crystal inorganic titanate fibre of chemical composition $K_2Ti_4O_9$, whose fibres are 40–70 μm long with an aspect ratio of about 10. Has been used as a reinforcing filler in thermoplastics composites.

G

G Symbol for glycine.

GABRITE Tradename for urea–formaldehyde polymer

G-ACTIN The globular, 'monomeric' form of actin.

GAFITE Tradename for poly-(methyl-α-chloroacrylate) and for polybutylene terephthalate.

GALACTAN A polymer of galactose, occurring widely in plants and animal sources. Both D- and (more rarely) L-enantiomorphic forms of the galactose units are found in both pyranose and furanose ring forms. Strictly homo-polymers are rare. Examples are the 1,4′-linked β-D-polymer in lupin seeds and a highly branched beef lung galactan containing both 1,3′- and 1,6′-links. The galactan from the Roman snail (also called galactogen) contains both D- and L-enantiomers in the ratio 6:1. Acidic galactans are found in compression wood and contain 1,4′- and/or 1,6′-links. Galactocarolose is a low molecular weight polymer containing about 10 galactofuranose units 1,5′-linked. A large number of arabinogalactans containing both D-galactose and L-arabinose units have been identified, especially in coniferous woods.

GALACTOCAROLOSE A galactan polymer formed, together with the analogous mannocarolose, in the culture medium *Penicillium charleseii* G. Smith grown in D-glucose. A low molecular weight polymer of about 10 units of β-D-galactofuranose, 1,5′-linked.

GALACTOGEN Alternative name for the galactan obtained from the Roman snail.

GALACTOGLUCOMANNAN A glucomannan polymer found in coniferous woods, having D-galactopyranose as non-reducing end groups probably as single unit side groups attached to C-6 of the main chain units.

GALACTOMANNAN One of a group of polysaccharides consisting of a backbone of 1,4′-linked β-D-mannopyranose units with single α-D-galactopyranose units as side groups. Guaron is the best known example, found in seeds. Similar polymers are also found in dermatophyte fungi, but in these the galactose side units are in the non-reducing furanose form, and are 1,2′-linked to the main chain, which itself is 1,6′-linked. They account for 5–15% of the sugar units.

β-D-GALACTOPYRANOSE The usual anomeric ring form of D-galactose units in carbohydrates which are therefore more fully described by this name.

D-GALACTOSAMINE (2-Deoxy-2-amino-D-galactose)

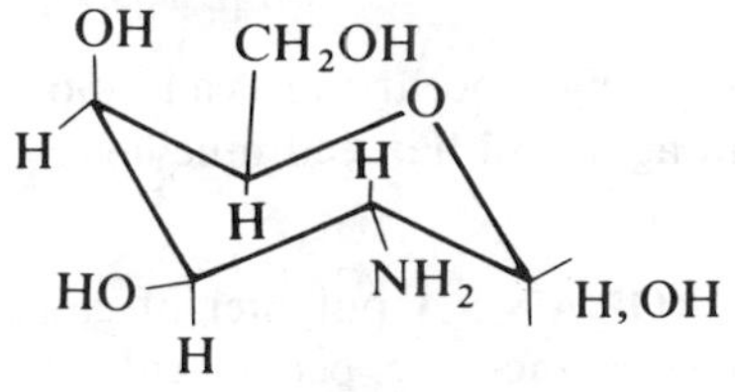

M.p. 187°C (hydrochloride). α_D^{20} +95° (hydrochloride).

An aminosugar monosaccharide found combined as a component of several aminopolysaccharides, usually in the *N*-acetylated form (*N*-acetyl-D-galactosamine), e.g. in chondroitin, or in a sulphated form, e.g. in the chondroitin sulphates.

D-GALACTOSE Fischer projection formula:

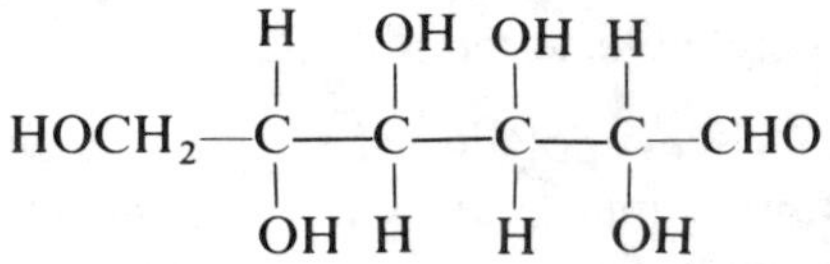

Ring form (α-D-galactopyranose):

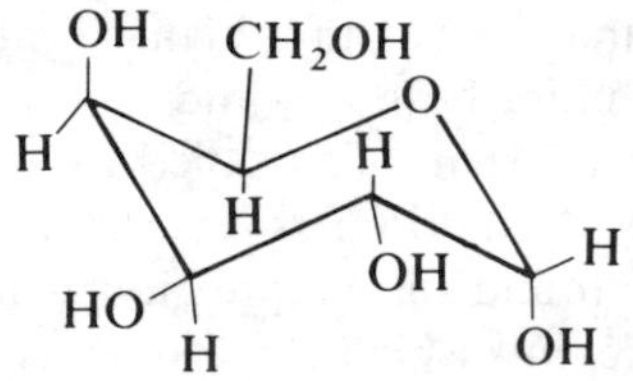

M.p. 167°C. α_D^{20,H_2O} +80.2°.

An aldohexose sugar occurring in oligosaccharides e.g. lactose, in homopolysaccharides (galactans) and in many heteropolysaccharides including guaran, galacto-glucomannan, arabinogalactan, pectic substances, gum arabic, agarose, agaropectin, carrageenan, keratan sulphate and glycoproteins. Usually obtained by acid hydrolysis of lactose, being separated from the D-glucose by fractional crystallisation. Also obtained from larch gums. Usually crystallises as α-D-galactopyranose, but shows some tendency to exist in the furanose form, especially when acetylated.

L-GALACTOSE Fischer projection formula:

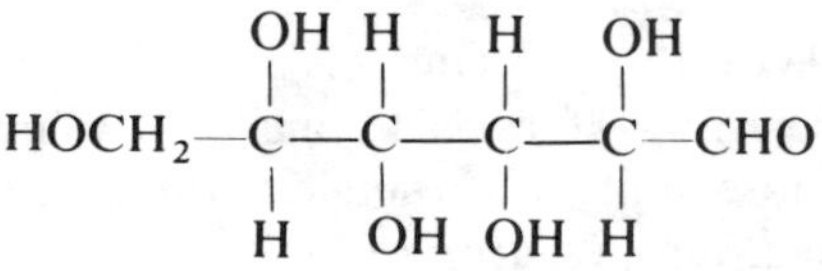

Ring form (β-L-galactopyranose):

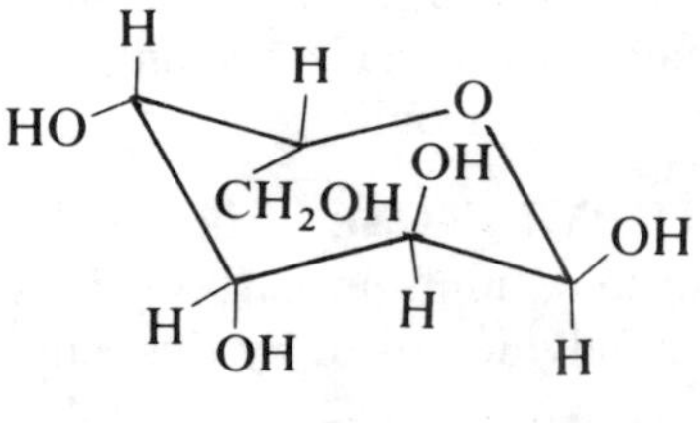

M.p. 165°C. α_D^{20,H_2O} −81°.

An aldohexose sugar occurring as a homopolysaccharide (galactan) in agar and flaxseed mucilage.

GALACTURONAN A polymer of galacturonic acid. Most pectic substances e.g. pectic acid, pectinic acid, are such polymers.

D-GALACTURONIC ACID

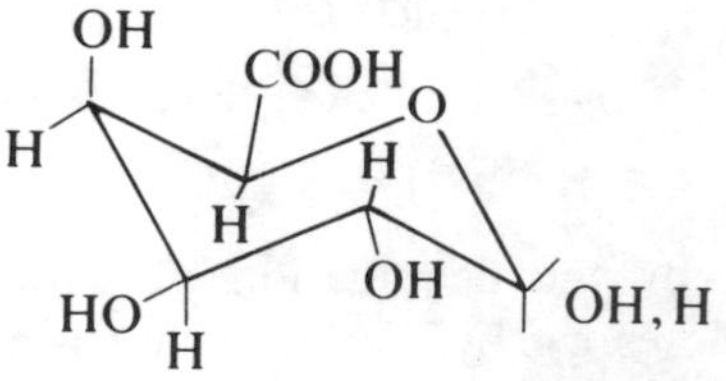

M.p. 159–160°C (decomposes). α_D^{20} +5.2°. A uronic acid, occurring as a monosaccharide component in pectic substances and some gums, e.g. gum tragacanth.

GALALITH Tradename for a casein plastic material.

GAMMA GLOBULIN Alternative name for γ-globulin.

GANTREZ AN Tradename for methylvinyl ether–maleic anhydride copolymer and derived polymers.

GANTREZ B Tradename for polyisobutylvinyl ether.

GANTREZ M Tradename for poly-(methylvinyl ether).

GANTREZ VC Tradename for isobutylvinyl ether–vinyl chloride copolymer.

GARAN Tradename for vinyltriethoxysilane.

GAS-PHASE POLYMERISATION Chain polymerisation in which the monomer is in the gaseous state, although as soon as monomer molecules are initiated and begin to grow, they must condense. Such polymerisations occur by photoinitiation of monomer producing a fog of solid polymer particles, so that most polymerisation actually occurs in the particles. Like emulsion polymerisation usually only one active centre can exist in each particle and fresh monomer diffuses from the vapour. Polymerisations in which the gaseous monomer is fed to a solid polymerisation catalyst, e.g. a Ziegler–Natta catalyst, often dispersed in a fluidised bed of inert supporting solid, are also termed gas-phase polymerisations, although here all the polymerisation must occur on the surface of the solid. Such a system has the advantage, especially in large scale commercial polymer production, that the handling of large quantities of solvent and the isolation of polymer from the solvent are avoided. Polyolefins are often produced commercially in this way.

***GAUCHE* CONFORMATION** (*Cis* conformation) Symbol g, g^+ or g^-. The conformation of a small molecule, or a small segment of a larger molecule, e.g. of a polymer chain, in which rotation about a pair of 1,2-disubstituted atoms is such that the two substituents are placed in a staggered *gauche* position to each other. This may be represented in a Newman projection for the simplest example, by *n*-butane, as:

(looking down the carbon–carbon bond), or alternatively by a saw-horse representation:

The *gauche* conformation is less energetically favoured than the *trans* conformation. In carbon–carbon polymer chains all-*gauche* conformations are not usually found. However polyoxymethylene crystallises in an all-*gauche* form and regular sequences of *trans* and *gauche* conformations, i.e. —tgtgtgtg— structures, comprise a 3_1-helix. *Gauche* conformations are sometimes especially favoured when neighbouring atoms or groups attract each other—the *gauche* effect.

GAUGE LENGTH The distance between two parallel marks on a tensile test sample and perpendicular to the sample length, between which elongation during test is measured.

GAUSSIAN CHAIN A chain, as a model for a polymer molecule, whose statistical distribution of chain end-to-end distances is a Gaussian distribution. This arises from the nature of the model which consists of a chain of n links each of length l in which the direction of any link in space is entirely random, not being influenced by the direction of any other link. It is thus a randomly jointed chain. The resultant probability $P(x, y, z)$ that one chain end will be found in a volume element $\mathrm{d}x\,\mathrm{d}y\,\mathrm{d}z$ at distances x, y and z from the origin where the other chain end is fixed, is

$$P(x, y, z)\,\mathrm{d}x\,\mathrm{d}y\,\mathrm{d}z$$
$$= (b^3/\pi^{3/2})\exp\left[-b^2(x^2 + y^2 + z^2)\right]\mathrm{d}x\,\mathrm{d}y\,\mathrm{d}z$$

where $b^2 = 3/2nl^2$. This is a probability density. The additional assumption made is that the end-to-end distance (r) is much less than nl. In terms of r, the probability density is $P(x, y, z) = (b^3/\pi^{3/2})\exp(-b^2/r^2)$. These relations form the basis of much of the statistical molecular theory of rubber elasticity by being extended to Gaussian networks, but are only valid for relatively low extensions for which $r \ll nl$. At high extensions non-Gaussian behaviour is observed since the condition $r \ll nl$ no longer holds.

GAUSSIAN DISTRIBUTION Alternative name for normal distribution.

GEDEVYL Tradename for polyvinyl chloride.

GEDEX Tradename for polystyrene.

GEGENION (Counter-ion) The ion of opposite charge to the ion under consideration. For an electrostatically charged polymer, e.g. a polyelectrolyte or the active centre in ionic polymerisation, it is the small ion present, of opposite charge to that of the polymer.

GEHMAN CREEP APPARATUS A simple apparatus for the rapid determination of the torsional creep of films. The film is clamped to an elastic wire and the assembly is twisted through 180°. On release the film exerts a torque on the wire, which prevents it from returning to zero angle. The angle of twist varies with time. Usually the angle is measured at 10s and the 10s modulus is calculated. A rapid assessment of the overall viscoelastic properties of the material may be obtained over a wide temperature range.

GEL That part of a polymer sample that is crosslinked, forming a network, and therefore is insoluble and is only swollen by contacting solvating liquids. The other part of the sample in linear or branched polymer and hence is soluble and is the sol. Thus on forming a network polymer by polymerisation using multifunctional monomers or by crosslinking of a linear polymer, both sol and gel are formed; the higher the conversion (in step-growth polymerisation) or the greater the extent of curing of linear polymer, the more gel there will be at the expense of sol.

GELATIN Partially denatured collagen, in which the intermolecular hydrogen bonds between the individual chains of the tropocollagen 'molecules' have been destroyed, e.g. by heat, but in which covalent crosslinks remain. On cooling a gelatin solution, partial renaturation occurs, causing gel formation in concentrated solutions.

GELATION In general any process in which a gel is formed. In particular, in a non-linear step-growth polymerisation, the rapid change that occurs during the polymerisation from a fluid, partially polymerised mixture to a swollen gel. This change is associated with the formation of a crosslinked, network structure in the polymer, at a critical conversion p_c, as the branched polymer molecules begin to join up. p_c may be predicted either by use of a modified version of the Carothers equation or statistically by consideration of the branching coefficient. The former method predicts gelation to occur at a higher p_c than it does in practice, whilst the latter predicts it to occur sooner. In practice gelation is observed by some method of detecting the loss of fluidity, most simply, for example, by the inability of bubbles to rise through the gel. The onset of gelation occurs at the gel point. In the formation of prepolymers, which are subsequently to be further reacted to a cross-linked product, it is important that conversion is not taken beyond the gel point, otherwise the fabrication required in the prepolymer stage is not possible.

GEL CELLOPHANE Regenerated cellulose which has never been allowed to dry out, but has always been stored swollen with water or other solvent. Thin films are frequently used as the semi-permeable membrane in membrane osmometry.

GEL CHROMATOGRAM The recorded output of a gel permeation chromatography experiment, which consists of a plot of detector response against time. The former is directly proportional to the weight concentration of polymer whilst the latter is directly proportional to the retention volume (V_R). By a suitable calibration, e.g. by the universal calibration method, V_R can be related to molecular weight of the polymer species in solution. Hence the molecular weight distribution of the polymer can be calculated from the chromatogram.

GEL EFFECT Alternative name for autoacceleration.

GEL ELECTROPHORESIS Zone electrophoresis in which the support material is a water swollen gel. This usually acts to separate the components by molecular size as well as by electrophoretic mobility. Thus higher resolution is obtained than with paper electrophoresis. The gels most frequently used are of starch, polyacrylamide or agarose. The gels may be used in columns or, more often, as flat slabs (block electrophoresis) or in thinner layers. Variations of the technique are disc electrophoresis and SDS-gel electrophoresis (frequently used for protein molecular weight measurement), isoelectric focusing and isotachophoresis.

GEL FILTRATION Alternative name for gel permeation chromatography, used particularly for aqueous systems, in which the polymer solute is frequently a biopolymer.

GELOY Tradename for an acrylonitrile–styrene–acrylate copolymer.

GEL PERMEATION CHROMATOGRAPHY (GPC) (Gel filtration) (Molecular sieve chromatography) (Size exclusion chromatography) A method of polymer fractionation, both preparative and analytical, in which a sample of the polymer in dilute solution is injected into a stream of flowing solvent at the top of a packed column. The packing usually consists of a porous gel having pore sizes such that, as the polymer solution passes through, the smaller polymer molecules can enter the pores so that their passage is retarded, whereas the large molecules are excluded from the gel and are eluted from the column more rapidly. Thus a series of fractions may be collected from the effluent, with gradually increasing average molecular weight. Alternatively, the effluent passes through a concentration detector, usually a differential refractometer, whose response is proportional to the concentration of polymer in the solution passing through it. The output is a plot (the gel chromatogram) of response, i.e. concentration, versus time, i.e. versus the retention volume (V_R) for constant flow rate. The molecular weight distribution (MWD) may be calculated from the chromatogram.

The technique has transformed the ease with which fractionated polymers and the MWD may be obtained, being much more rapid and reliable than classical fractionation methods. The usual gels are crosslinked polystyrene gels which swell in organic solvents. Crosslinked dextran gels are usually used for aqeuous solvents, when the technique is often called gel filtration. More rigid and high temperature stable porous glass and silica packings may also be used. The separation mechanism is probably that of steric exclusion, although diffusion may be important at high flow rates. A calibration curve is required to convert V_R to molecular weight. This is obtained using either monodisperse samples or well characterised narrow MWD fractions or whole polymer. The universal calibration method, enabling many different polymer types to be used with a single calibration, is the most widely used method. Calibration is not required if a low angle laser light scattering detector is used in series with the concentration detector, since it gives a direct continuous record of the molecular weight of the polymer as it is eluted from the column.

GEL POINT The stage in a non-linear step-growth polymerisation at which gelation occurs. The critical conversion at the gel point (p_c) may be predicted either by a modified version of the Carothers equation or statistically by consideration of the branching coefficient.

GELVA Tradename for polyvinyl acetate.

GELVATOL Tradename for polyvinyl alcohol.

GEMINATE PAIRING Alternative name for geminate recombination.

GEMINATE RECOMBINATION (Geminate pairing) (Cage effect) (Recombination) Recombination of the two free radicals formed by homolysis of a molecule, thus usually restoring the original molecule. In the liquid state, motion of the free radicals is restricted—the radicals are trapped in a solvent cage. In the solid state motions are even more restricted. Recombination may therefore occur before the two radicals have diffused apart. For example the thermal decomposition of benzoyl peroxide may be represented as:

$$\underset{\underset{O}{\parallel}}{PhCO}\!\!-\!\!\underset{\underset{O}{\parallel}}{OCPh} \underset{\text{recombination}}{\overset{\text{decomposition}}{\rightleftharpoons}} \left[2\underset{\underset{O}{\parallel}}{PhCO}\!\cdot \right] \rightarrow Products$$

GEMON Tradename for a polyaminobismaleimide similar to Kerimid.

GENAL Tradename for phenol–formaldehyde polymer for injection moulding.

GENCLOR S Tradename for chlorinated polyvinyl chloride.

GENERALISED EXPONENTIAL DISTRIBUTION A molecular weight distribution which is described by a three parameter equation for the differential weight distribution function:

$$W(r) = [my^{(k+1)/m}r^k][\exp(-yr^m)/\Gamma((k+1)/m)]$$

where $y = -\ln p$, k and m are adjustable parameters, Γ is the gamma function, p is the polymerisation conversion or the ratio of rates of propagation to termination and r is the size (e.g. degree of polymerisation or molecular weight) of the individual molecular species. When $m = 1$ it is the Schultz distribution and when $k = m - 1$ and $k > 0$ it is the Tung distribution.

GENERAL LINEAR SOLID Alternative name for standard linear solid.

GENERAL PURPOSE FURNACE BLACK (GPF) A type of furnace carbon black which is moderately reinforcing with good processability giving compounds of high modulus and hardness. Useful for tyre innerliners, carcasses, radial belts and sidewalls and in extruded goods such as hose.

GENTHANE A millable polyurethane elastomer based on a polyester copolymer from ethylene and propylene glycols with adipic acid and MDI, cured by heating with a peroxide, such as dicumyl peroxide.

GENTRO Tradename for styrene–butadiene rubber.

GEOLAST Tradename for a polypropylene/vulcanised nitrile rubber blend.

GEOMETRICAL FACTOR Alternative name for finite plate correction factor.

GEOMETRIC DISTRIBUTION Alternative name for Schultz–Flory distribution.

GEOMETRIC ISOMERISM (*Cis–trans* isomerism) A type of configurational isomerism which occurs in molecules, including polymers, which contain a carbon–carbon double bond. In particular in poly-(1,4-dienes), e.g. of repeat units —CH_2CX=$CHCH_2$—, two forms are possible: the *cis* conformation,

and the *trans* conformation,

The two configurational isomers are not generally interconvertible. 1,4-Polymerisation of the appropriate diene monomer with a stereospecific catalyst, such as a Ziegler–Natta catalyst, can lead to predominantly one type of isomeric placement. In general the *cis* polymers have lower T_g values and T_m values than the *trans* polymers, as illustrated by the polyisoprenes and polybutadienes.

GEON Tradename for polyvinyl chloride.

GIBBS–DIMARZIO THEORY A theory of the glass transition based on the postulate of the existence of a true thermodynamic transition at a temperature T_2 somewhat below the observed T_g value. At T_2, the conformational (configurational) entropy at equilibrium is zero and remains zero until absolute zero temperature is reached, thus resolving Kauzmann's paradox. The observed value of T_g depends on the rate of cooling; at an infinitely slow rate it equals T_2. In practice, however, the postulated lowest energy state conformation cannot be reached on cooling due to the 'freezing' of molecular motions.

A lattice model is used in combination with a hole formation energy and a polymer chain flexing energy to derive relationships between the thermodynamic properties by a statistical calculation of the configurational entropy. The theory predicts reasonable values for the energy differences between polymer chain rotational states and predicts that the unoccupied volume at T_2 is the product of T_g and $\Delta\alpha$, where $\Delta\alpha$ is the increase in the coefficient of thermal expansion at T_g. This has a value of 0·025, precisely the value predicted by free volume theory for fractional free volume at T_g. Extensions of the theory enable the effects of copolymerisation, plasticisation and molecular weight on T_g to be predicted.

g-INDEX (Polydispersity) A measure of the width of a molecular weight distribution, defined as,

$$g = [(\bar{M}_z/\bar{M}_w) - 1]^{1/2} = S_w/\bar{M}_w$$

where $\bar{M}_z$ and $\bar{M}_w$ are the z- and weight average molecular weights respectively and S_w is the standard deviation of the weight distribution of molecular sizes.

GLASS A solid material existing in a non-crystalline, i.e. amorphous, state. Glass formation is due either to such rapid cooling from the melt that crystallisation is prevented by quenching or to the lack of structural regularity in the molecules of the glass forming material. Such possibilities only occur with polymeric materials or occasionally with low molecular weight materials in which very strong intermolecular forces, such as hydrogen bonding, operate. Glass forming materials exhibit their typical hard, but brittle, glassy properties at temperatures considerably below the glass transition temperature.

The best known and traditional glassy materials are inorganic polymers, although amorphous organic polymers can also show glassy properties and are sometimes referred to as organic glasses. Most inorganic glasses are based on the acidic oxides: SiO_2 (giving silicate glasses), B_2O_3 (giving borate glasses) and P_2O_5 (giving phosphate glasses). Commercial glasses are silicate based, although the properties of silica itself when glassy (vitreous silica) are usually modified by the incorporation of additional oxides such as Na_2O, K_2O, MgO, CaO and Al_2O_3. Important types are soda-lime glass, borosilicate glass and aluminoborosilicate glass. In these glasses the oxides cause a reduction of softening point and so improve processability.

Glasses are hard and stiff materials but are of low strength due to a lack of molecular mobility in the network structures. Basic mechanical properties are rather independent of chemical structure with typical values of the tensile modulus of 50–90 GPa and a low tensile strength of 300–700 MPa. For many commercial applications such as glazing and containers, the most important property is the optical transparency found in most glasses. Glasses are electrical insulators with low relative permittivity and loss. Thermal conductivity is low due to the disordered structure, especially in vitreous silica and Pyrex borosilicates, which have good thermal shock resistance.

GLASS-CERAMIC   A polycrystalline silicate polymer, produced by the controlled crystallisation of a largely amorphous glass, often embedded in a minor proportion of a residual amorphous glass. Glass-ceramic objects are made by forming, e.g. by casting, the object in a suitable glass, then heating to above T_g to induce crystallisation. Crystallisation is enhanced to give a fine dispersion by the use of large amounts of nucleating agents. Final crystallite size is typically 100–500 nm. Suitable glasses are magnesium aluminosilicates, lithium–magnesium and lithium–zinc silicates. Glass-ceramics combine the resistance to high temperatures of crystalline silicates with isotropic mechanical properties due to the fineness of the crystalline dispersion.

GLASS–RUBBER TRANSITION   Alternative name for glass transition.

GLASS–RUBBER TRANSITION TEMPERATURE Alternative name for glass transition temperature.

GLASS TEMPERATURE   Alternative name for glass transition temperature.

GLASS TRANSITION (Glass–rubber transition) (Rubber–glass transition) The dominant transition in amorphous polymers, which also occurs at the highest temperature (except when a liquid–liquid transition also occurs), and is therefore labelled the α-transition. It also occurs in the amorphous regions of crystalline polymers but its importance then depends on the degree of crystallinity. The precise details of the molecular motions, i.e. the relaxation processes, responsible for the transition are usually uncertain. However, it is generally thought to involve several polymer chain bond rotations occurring in a cooperative manner such that segments of the chain, thought variously to involve 10–20 or up to 50 chain atoms, move simultaneously.

The temperature at which the transition occurs is the glass transition temperature (T_g). Although the transition may take place over several degrees, it is usually sharper than the lower temperature (β-, γ- and δ-) transitions. Amorphous polymers at temperatures well below the T_g value are hard, stiff, glassy materials, although they may not necessarily be brittle. On the other hand, at temperatures well above T_g polymers are rubbery. Hence the alternative name glass–rubber transition. The temperature region around T_g is often called the leathery region. There is thus a major change in the mechanical behaviour as the transition is traversed, for example the elastic modulus decreases by a factor of about 10^3. Thus T_g is of major importance in determining the upper use temperature for a polymer used as a plastic or fibre in load bearing applications.

For a crystalline polymer the changes in properties on traversing T_g are not so dramatic, but nevertheless it does contribute importantly to the overall softening point.

Many theories of the glass transition have been developed. These include the important free volume theory, the isoviscous, the isoelastic and iso-free volume theories. Considerable controversy surrounds the suggestion that the transition is a true second order thermodynamic transition, as is supposed in the Gibbs–DiMarzio theory, representing an equilibrium between the glassy and rubbery states. It fulfils some of the criteria, e.g. there is a change in the rate of change of the extensive thermodynamic properties (e.g. volume and specific heat) with temperature at T_g. However, true equilibrium has never been observed experimentally since the transition and the value of T_g are rate dependent and it can be regarded as a kinetic phenomenon. Furthermore, the values of the specific heat, thermal conductivity and expansion coefficients are smaller below T_g than above it, which is the reverse behaviour to that expected for a second order transition. Many other physical properties undergo a change at T_g and methods for determining T_g are based on these changes, for example, changes in the refractive index and the viscosity. Under dynamic conditions, as with periodically varying applied stress, strain or electric field, a maximum loss occurs in the region of the T_g value.

GLASS I TRANSITION   A transition occurring in a crystalline polymer which, although having its origin in the type of molecular motion responsible for a glass transition, is not accorded the status of a glass transition (T_g) due to its small magnitude. The relative insignificance of the transition is due, in highly crystalline polymers, to there being few chains between the crystals large enough to exhibit the required motions, involving 20–50 in-chain atoms.

GLASS II TRANSITION   A transition resulting from the onset of a Schatzki mechanism in polymers containing at least four —CH_2— groups in the main chain repeat unit, as in polyethylene and ethylene copolymers. It occurs at about $-105°C$. When the polymer is highly crystalline, the transition can become the dominant transition and sometimes takes on the status of the glass transition.

GLASS TRANSITION TEMPERATURE (Glass temperature) (Glass–rubber transition temperature) (Rubber–glass transition temperature) Symbol T_g.

The temperature at which the glass transition occurs. Since the transition occurs over a temperature range, T_g is more precisely the temperature at which the maximum rate of change of the property being observed, occurs. Since the transition is a manifestation of molecular motion, the value of T_g is dependent on the rate of testing. The lowest values result from a very slow rate of testing, as with dilatometry, which is often referred to as a static method and the value obtained is the usually quoted value of T_g. If the glass transition is regarded as a true thermodynamic transition, then the true equilibrium value of T_g is expected to be about 50°C below the observed static value. In general, as the frequency of testing is increased by a factor of 10, so the value of the observed T_g increases by 5–7 K.

T_g values for linear organic polymers range from about −100°C to above 300°C. Although some organic polymers may be expected to have T_g values considerably above 300°C, decomposition of the polymer occurs before T_g is reached so that its value cannot be observed. However, in inorganic polymers, very high T_g values may be observable, e.g. at about 1200°C for SiO_2.

An amorphous polymer existing at somewhat above its T_g value will be rubbery, whilst it will be glassy at temperatures well below its T_g value. Thus polymers commonly known as rubbers are amorphous with T_g values somewhat below ambient, say below about −20°C. Since the molecular origin of the glass transition is in cooperative chain motions involving main-chain bond rotations, the value of T_g is determined by those molecular structural features which influence these motions. The most important feature is the intrinsic flexibility of the chain bonds. Both C—C and C—O bonds are very flexible, thus polyethylene has an important transition at a T_g value of about −20°C and polyethylene oxide has a T_g value of −55°C. The incorporation of ring structures in the chain raises the T_g value, thus polyphenylene oxide has a T_g value of about 120°C and poly-p-xylylene has a T_g value of about 280°C. Surprisingly, the presence of in-chain double bonds may lower T_g, thus cis-1,4-polybutadiene has a T_g value of −108°C. However, in such polymers, cis–trans isomerism can considerably affect T_g, as shown by the T_g value of trans-1,4-polybutadiene which is −18°C.

Another major influence is that of bulky side groups; T_g increases with the size and number of substituents. Thus in vinyl polymers $\text{-[CH}_2\text{CHX]}_n\text{-}$ T_g values are −120°C (X = H), −10°C (X = CH_3), +100°C (X = phenyl) and 264°C when

$$X = \text{[indene/naphthalene-type ring structure]}$$

However, if the substituent itself has a chain structure, its flexibility can depress T_g. Thus in the methacrylates,

$$\text{-[CH}_2\text{C(CH}_3\text{)]}_n\text{-} \quad \begin{array}{c} | \\ \text{COOR} \end{array}$$

as R increases in size, so T_g decreases, the values being 105°C (R = CH_3), 65°C (R = ethyl), 35°C (R = n-propyl) and 20°C (R = n-butyl). Polarity in the side group also increases T_g due to increased interchain and intrachain attractions. Thus comparing vinyl polymers with substituents of approximately equal size but with increasing polarity, we have the values −20°C (R = CH_3), +80°C (R = Cl), 105°C (R = CN). Tacticity only seems to influence T_g in vinyl polymers when a second substituent, in addition to X, is present on the α-carbon. Thus in poly-(methylmethacrylate) T_g values are 45°C (isotactic), 115°C (syndiotactic) and 105°C (atactic). Of course the presence of the second α-substituent also increases T_g. Thus polystyrene has a T_g value of 100°C, but poly-(α-methylstyrene) has a T_g value of 172°C.

Not surprisingly, random copolymers exhibit a T_g value between the T_g values of the corresponding homopolymers. This effect is often given by the Gordon–Taylor equation. Block copolymers and polymer blends will exhibit two T_g values if the different blocks or polymers are relatively incompatible and two phases exist. However, some broadening of the two T_g values will occur if there is partial miscibility. When the different blocks or polymer chains are miscible at the molecular level then a single intermediate T_g value is observed.

The T_g value is also dependent on the polymer molecular weight, at least in the lower molecular weight range. The dependency is often expressed by, $T_g = T_g^\infty - c/M$, where T_g^∞ is the T_g value of a polymer of infinite molecular weight, c is a constant and M is the molecular weight, usually the number average. The T_g value may be considerably lowered by plasticisation, converting a material which is a stiff plastic at ambient temperature to a rubbery material. Thus for polyvinyl chloride the T_g value is 80°C (unplasticised) and −20°C when plasticised with 30 phr of dioctyl phthalate. Crosslinking can cause a considerable increase in T_g and, if extensive, T_g may be above the polymer decomposition temperature and be unobservable.

The value of T_g is best determined experimentally by a static, or quasi-static, method, such as dilatometry. Alternatively the change in specific heat, as determined by differential scanning calorimetry or differential thermal analysis, may be used. As previously mentioned, use of dynamic methods will give a higher value of T_g, dependent on the rate of testing. Such methods include dynamic mechanical analysis (e.g. using torsion pendulum or vibrating reed methods), dielectric relaxation and nuclear magnetic relaxation.

GLASSY COMPLIANCE Alternative name for unrelaxed compliance.

GLASSY MODULUS Alternative name for unrelaxed modulus.

GLASSY STATE The state of a polymer in which cooperative chain motions are 'frozen', so that only limited local motions, such as side-group rotations, are

possible. The material therefore behaves largely elastically since stress causes only limited bond angle deformations and stretching. Thus the material behaves as a hard, rigid and often brittle glass with a modulus of approximately 10^8–10^{10} Pa. Such behaviour is found in amorphous polymers existing approximately 20°C or more below their glass transition temperatures. Deformation is usually limited to about 1%, so the polymer is subject to brittle fracture and often has a relatively low impact strength. However, some local motion is often possible so that a limited plastic deformation may occur.

GLIADIN   The alcohol soluble protein, i.e. a prolamine, of wheat gluten. Like other cereal proteins it has a high content of glutamic acid and proline, but is deficient in lysine, histidine and arginine and therefore is relatively non-polar. Gliadin is a mixture of protein molecular components, which are frequently separated into five components by moving boundary electrophoresis at pH 3·1 in aluminium lactate, and into up to about 20 components by gel electrophoresis. These components all have a similar molecular weight of about 25 000 to 50 000.

GLIDE   Alternative name for glide dislocation.

GLIDE DISLOCATION   (Glide) Motion of a dislocation in a crystalline material on deformation, which is parallel to the Burgers' vector and requires only a small displacement of the rows of motifs. An edge dislocation can glide only in its slip plane, whereas a screw dislocation can glide in any plane parallel to the dislocation line.

GLIDE PLANE   The slip plane when slip is by glide.

GLN   Abbreviation for glutamine.

GLOBIN   The polypeptide component of haemoglobin from which it may be separated by treatment with acid or base. Globin will recombine with haem to form haemoglobin at pH values near neutrality.

GLOBULAR PROTEIN   A protein which exists in its native state in a folded, compact spherical or globular shape (the tertiary structure). The majority of proteins are of this type, as distinct from the smaller group of fibrous proteins. Unlike fibrous proteins they are usually soluble in aqueous solvents and solubility difference is a traditional way of separating individual components from mixtures of globular proteins. The main types are albumins, globulins, nucleoproteins, haemoproteins, hormones and most enzymes.

GLOBULIN   One of a large group of globular proteins characterised by water insolubility (in contrast to the similar, but water soluble, albumins) but soluble in dilute salt solutions. Sometimes they are further subdivided into euglobulins and pseudoglobulins. They are very widely distributed in both plants and animals. The more important groups are the blood serum globulins, the plant seed globulins, the ovoglobulins, β-lactoglobulin of milk and myosin of muscle.

They are readily denatured by heat and by denaturing agents and contain all the common amino acids, being relatively rich in aspartic and glutamic acids. Some globulins, such as the γ-globulins and lactoglobulins, have antibody activity and are called immune globulins.

The blood plasma proteins are the most widely studied, accounting for one-third to one-half of the plasma protein. They are separated into fractions and characterised according to their electrophoretic mobility as the α-, β- and γ-globulins. However, these fractions are themselves mixtures of different individual protein polypeptides. Further, or more complete separations may be achieved by density gradient centrifugation, gel electrophoresis or by column chromatography.

The plant seed globulins form another large group of globulins and include gliadin (wheat), zein (corn), legumin (pea) and arachin (peanut). The prolamines are one group of seed proteins which are soluble in 50–90% ethanol. Many of these plant globulins are oligomeric. The protein content of seeds varies from 10–15% in cereals to about 50% in soya beans. Most of this protein is globulin, and is readily extracted with dilute salt solutions. Other globulins occur in milk (lactoglobulin), egg white (ovoglobulin) and egg yolk (livetin).

α-GLOBULIN   A blood plasma protein fraction which migrates electrophoretically just slower than the serum albumin, so that it is the fastest moving globulin fraction. The fraction accounts for 10–20% of the total plasma globulins. Its major components are the α_1-globulins (which include orosomucoid and an α_1-glycoprotein), the α_1-lipoproteins, haptoglobin and the α_2-globulins which themselves are comprised of α_2-glycoproteins—sometimes subdivided to include a separate fraction termed macroglobulin, cerruloplasmin and prothrombin.

β-GLOBULIN   The blood plasma protein fraction that migrates electrophoretically at a speed between that of the α- and the γ-globulins. It comprises the β_1-lipoproteins, transferrin and plasminogen. It accounts for 20–30% of the total serum protein.

γ-GLOBULIN   A group of blood plasma proteins, accounting for about one-third of the total plasma globulin fraction; it makes up the slowest moving fraction on electrophoresis. γ-Globulins are glycoproteins containing a few per cent carbohydrate. They have been very widely studied since the important immunoglobulins comprise the largest part of the γ-globulin fraction. γ-globulins may be further divided on the basis of their ultracentrifugation behaviour into a 7S (Svedberg) fraction of molecular weight about 150 000 and a much smaller 19S fraction of molecular weight about one million (sometimes called the γ_1-macroglobulin fraction) and containing about 10% carbohydrate. The 7S fraction can be subdivided further according to carbohydrate content, amino acid sequences and immunological properties. However, each category comprises a mixture of many different protein molecules.

GLOSPAN Tradename for a spandex polyurethane elastomeric fibre, based on a polyester polyurethane chain extended with a diol.

GLOSS The degree to which a surface simulates a perfect mirror in its capacity to reflect incident light. It may be characterised by the ratio between the light flux scattered within a specified solid angle of the geometrical reflection direction (the specular direction) and the incident flux, for one or more angles of incidence. Often an angle of 45° is used.

GL-RESIN Tradename for ethylene–vinyl alcohol co-polymer.

GLU Abbreviation for glutamic acid.

GLUCAGON Hyperglycemic–glycogenolytic hormone. A small protein hormone of molecular weight 3500, consisting of a single polypeptide chain of 29 amino acid residues. Secreted by the pancreas, where it is synthesised as an inactive precursor (proglucagon) which contains eight more residues at the C-terminal end. It has the opposite effect on blood sugar to insulin, causing blood glucose to rise by breakdown of glycogen in the liver.

GLUCAN A polymer of D-glucose, which can formally be considered to be formed by step-growth polymerisation of D-glucose with elimination of water. Sometimes referred to as a polyanhydroglucose. The glycoside links so formed may be of the α-anomer (α-D-glucan) or β-anomer (β-D-glucan). Examples of linear polymers include α-D-1,4'-linked (amylose), β-D-1,4'-linked (cellulose), α-D-1,4'-linked and α-D-1,6'-linked (pullulan), β-D-1,3'-linked and β-D-1,4'-linked (lichenan). Important branched polymers include amylopectin and glycogen, both containing α-1,4'- and α-1,6'-links.

α-D-GLUCAN A glucan in which the glycoside links are of the α-type.

β-D-GLUCAN A glucan in which the glycoside links are of the β-type.

GLUCOMANNAN One of a group of linear polysaccharides consisting of 1,4'-linked β-D-mannopyranose and β-D-glucopyranose units in ratios varying from 1:1 to 1:4 and randomly arranged. Found in the bulbs of irises, orchids and lilies, as cell-wall constituents in coniferous woods and in small amounts in deciduous woods.

β-D-GLUCOPYRANOSE The normal anomeric ring form of D-glucose units in a carbohydrate, which are therefore more fully described by this name.

4-O-β-D-GLUCOPYRANOSYL-D-GLUCO-PYRANOSE The full chemical name of cellobiose.

D-GLUCOSAMINE (2-Deoxy-2-amino-D-glucose)

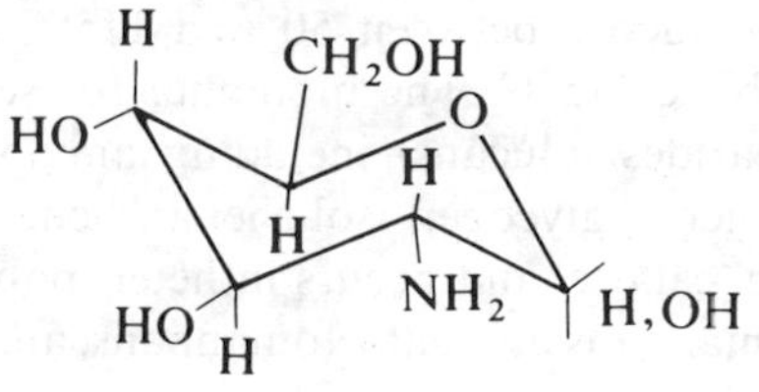

M.p. 88°C. α_D^{20} +47·5°.

An aminosugar monosaccharide, found combined as a component of several naturally occurring aminopolysaccharides, usually in the N-acetylated form (N-acetyl-D-glucosamine), e.g. chitin (a homopolysaccharide) and hyaluronic acid and, in the sulphated form, in heparin. Readily obtained by hydrolysis of chitin with concentrated hydrochloric acid.

D-GLUCOSE (Dextrose) Empirical formula $C_6H_{12}O_6$. Fischer projection formula:

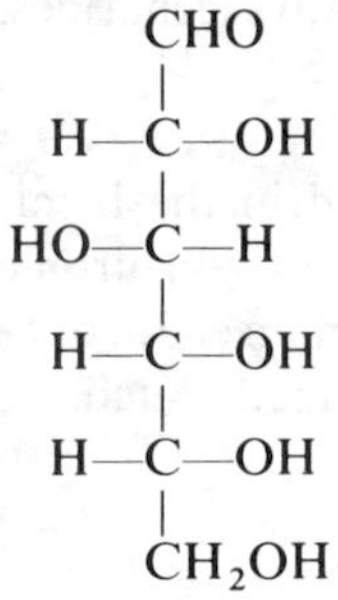

Hawarth formula:

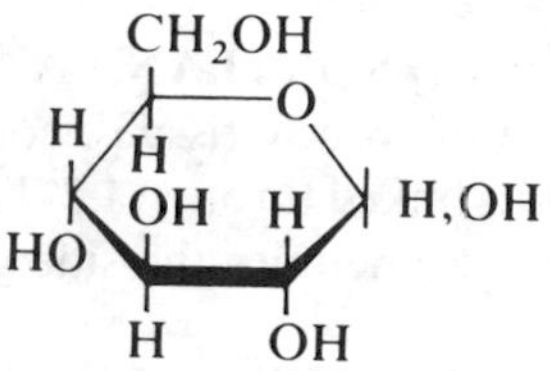

Conformational formula:

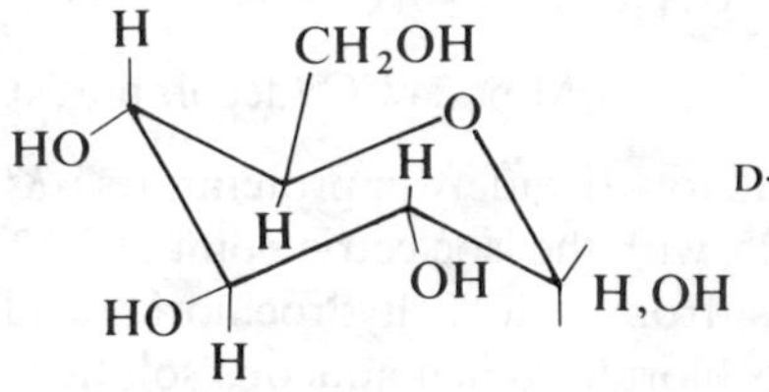

D-glucose

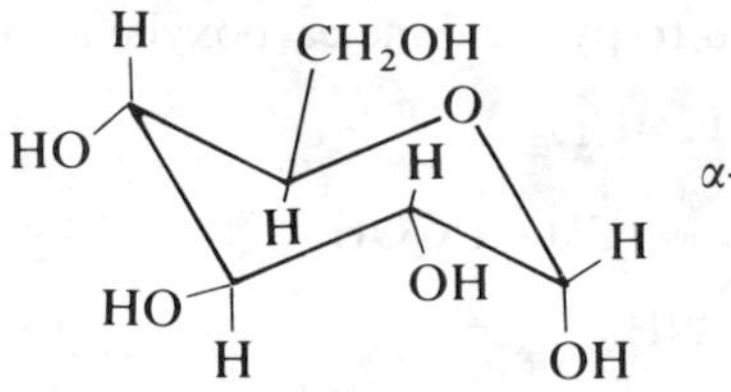

α-D-glucose

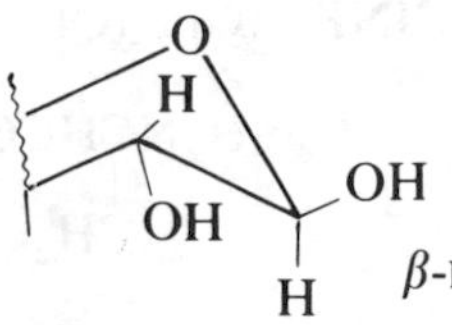

β-D-glucose

M.p. 146°C.

An aldohexose, which normally exists in the pyranose ring form (called glucopyranose) as shown above. The most commonly occurring monosaccharide, found free in many fruit juices and as the sugar 'monomer' in many polysaccharides and oligosaccharides where it occurs as

α-D-glucose (with α_D^{20,H_2O} of $+112°$, mutarotating to $+52\cdot7°$). Best obtained by hydrolysis of starch with dilute HCl. Readily soluble in water, from which it crystallises as the monohydrate (m.p. 83–86°C) below 50°C, as anhydrous α-D-glucose between 50 and 115°C and as β-D-glucose above 115°C. The monomer for several homopolysaccharides (glucans) including amylose, cellulose, dextrin, nigeran, glycogen, isolichenan, lichenan, pullulan, laminaran, callose, and occurs in heteropolysaccharides, e.g. glucomannans and galactomannans, and in oligosaccharides, e.g. sucrose.

D-GLUCURONIC ACID

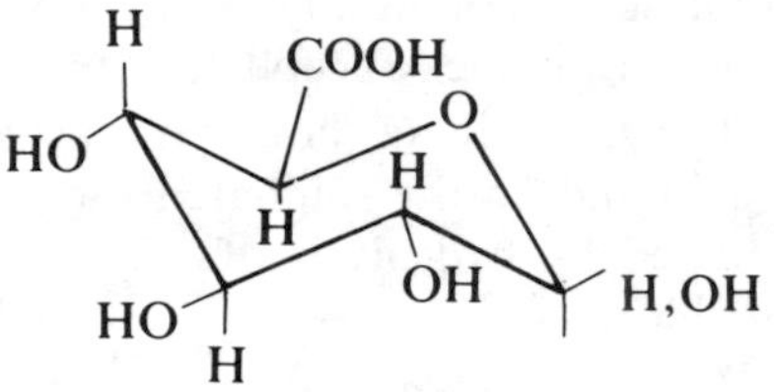

M.p. 165°C (decomposes). α_D^{20} $+36\cdot3°$.

A uronic acid, occurring as a monosaccharide component in glucuronoxylan, gum arabic, capsular polysaccharides, chondroitin and heparin.

GLUCURONOXYLAN A xylan found in the hemicellulose of woody tissues, consisting of α-D-glucuronic acid units linked through 1,2′-links to the main chain of β-D-xylose. Sometimes the side groups are methylated.

GLUTAMIC ACID (Glu) (E)

$$H_2NCHCOOH$$
$$|$$
$$CH_2CH_2COOH$$

M.p. 247°C (decomposes).

An acidic α-amino acid, found widely in proteins. Its pK' values are $2\cdot19$ and $4\cdot25$, with the isoelectric point at $3\cdot22$. It readily crystallises from dilute hydrochloric acid solutions as the hydrochloride. When aqueous solutions are heated it is converted to pyrrolidone carboxylic acid

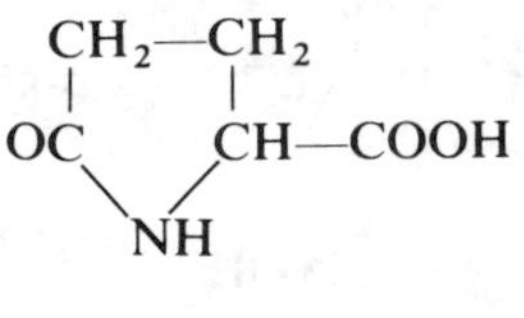

GLUTAMINE (Gln) (Q)

$$H_2NCHCOOH$$
$$|$$
$$CH_2CH_2CONH_2$$

M.p. 184–195°C (decomposes).

A polar-amino acid, widely occurring in proteins. Its pK' values are $2\cdot17$ and $9\cdot13$, with the isoelectric point at $5\cdot65$. In the hydrochloric acid hydrolysis stage of amino acid analysis of proteins it is converted to glutamic acid, so that its content in proteins is often determined combined with glutamic acid. Often associated with the α-helix conformation in proteins. On heating aqueous solutions it has a tendency to cyclise to ammonium pyrrolidone carboxylate

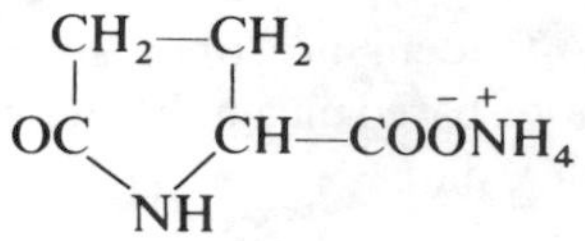

GLUTATHIONE A naturally occurring tripeptide found in the cells of higher animals. It is γ-glutamyl-cysteinylglycine, i.e. it has an unusual peptide linkage involving the γ-carboxyl rather than the α-carboxyl group with the formula:

$$HOOCCHCH_2CH_2CONHCHCONHCH_2COOH$$
$$|\qquad\qquad\qquad\qquad |$$
$$NH_2\qquad\qquad\qquad CH_2SH$$

GLUTEIN Alternative name for gluten.

GLUTELIN The protein fraction obtained from cereals which is insoluble in water, salt solutions and 70–80% ethanol. It is soluble in dilute acids and alkalis. Important glutelins are glutenin (from wheat and maize), oryzenin from rice and hordein from barley.

GLUTEN (Glutein) The residue obtained through the washing of wheat flour with water, which removes the starch. The dry material contains about 80% protein with 5–10% lipid, some starch and some water soluble globulin and albumin. The protein represents about 80% of the original protein and consists of a mixture of gliadin and glutenin, which may be separated by extraction of the gliadin, thus converting the elastic sticky gluten to an insoluble residue.

GLUTENIN The alcohol insoluble protein remaining after the extraction of gliadin from wheat gluten. Soluble in dilute acid and alkali. A glutelin of similar amino acid composition to gliadin. A single component by moving boundary electrophoresis, but heterogeneous with respect to molecular weight: the weight average molecular weight is about two million and the number average molecular weight is about 121 000. Unlike gliadin, the molecular weight is considerably reduced by reduction of disulphide bonds, i.e. glutenin is crosslinked through the disulphide bonds at least when extracted from dough.

GLX Abbreviation for glutamine and/or glutamic acid in a protein.

GLY Abbreviation for glycine.

GLYCAN Alternative name for polysaccharide, i.e. a polymer, named using the suffix -an, of glycosidically linked monosaccharide units.

GLYCEROL (Propane-1,2,3-triol) (1,2,3-Propane triol)

$$CH_2OH$$
$$|$$
$$CHOH$$
$$|$$
$$CH_2OH$$

M.p. 18·2°C. B.p. 290°C.

Occurs in many natural oils and fats as the triglyceride, from which it may be isolated by saponification with sodium hydroxide. Produced commercially from propylene either by a chlorination route via allyl chloride (and either epichlorhydrin or allyl alcohol), which is wasteful of chlorine, or by oxidation of propylene with steam and air at about 450°C to acrolein, conversion to allyl alcohol and oxidation to glycerol with hydrogen peroxide:

$$CH_3CH{=}CH_2 \rightarrow OHCCH{=}CH_2 \rightarrow$$

$$HOCH_2CH{=}CH_2 \rightarrow HOCH_2\overset{\overset{\displaystyle OH}{|}}{C}HCH_2OH$$

Glycerol is the predominantly used polyol for the preparation of medium and short oil alkyd resins. This is due to its high boiling point and to its imparting high solubility to the resin and good film properties in the cured resin. It is sometimes used in the formation of polyoxypropylene polyols, producing branches at the glycerol unit and for polyester polyols for use in flexible polyurethane foams. It is a solvent for casein, gelatin and other protein products.

GLYCERYL DIACETATE   (Diacetin)

$$\begin{array}{l} CH_2OOCCH_3 \\ | \\ CHOH \\ | \\ CH_2OOCCH_3 \end{array}$$

B.p. 140°C/10 mm, 259°C/760 mm.

Commercial samples also contain the other isomer together with some mono- and triacetate. A plasticiser for cellulose derivatives.

GLYCERYL MONOSTEARATE   (GMS)

$$\begin{array}{l} CH_2OOCC_{17}H_{35} \\ | \\ CHOH \\ | \\ CH_2OH \end{array}$$

M.p. 54–59°C.

A widely used lubricant for polymer processing, especially for polyvinyl chloride.

GLYCERYL TRIACETATE   (Triacetin)

$$\begin{array}{l} CH_2OOCCH_3 \\ | \\ CHOOCCH_3 \\ | \\ CH_2OOCCH_3 \end{array}$$

B.p. 285°C/760 mm, 152°C/20 mm.

A plasticiser for cellulose esters and ethers and for polyvinyl acetate. It suffers from high water solubility and high volatility.

GLYCERYL TRIBUTYRATE   (Tributyrin)

$$\begin{array}{l} CH_2OOCCH_2CH_2CH_3 \\ | \\ CHOOCCH_2CH_2CH_3 \\ | \\ CH_2OOCCH_2CH_2CH_3 \end{array}$$

B.p. 315°C/760 mm.

A plasticiser for cellulose acetate, acetate/propionate, and acetate/butyrate.

γ-GLYCIDOXYPROPYLTRIMETHOXYSILANE

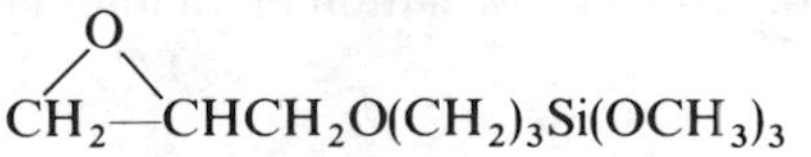

$$CH_2{-}CHCH_2O(CH_2)_3Si(OCH_3)_3$$

A coupling agent for reinforcements used in polymer composites. Used especially for glass fibres in epoxy resins.

GLYCIDYL   Prefix for the epoxy-containing group

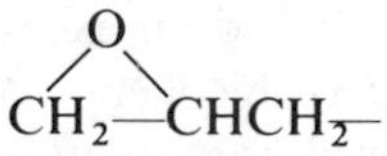

$$CH_2{-}CHCH_2{-}$$

This is the commonest epoxy group found in epoxy resins, as in the diglycidyl ether of bisphenol A.

GLYCIDYL ACRYLATE

$$CH_2{=}CHCOOCH_2CH{-}CH_2$$

B.p. 57°C/2 mm.

An acrylic monomer useful as a comonomer in the production of thermosetting acrylic stoving enamels. It provides crosslinking sites through reaction of the epoxide groups with polyfunctional amines.

GLYCIDYLAMINE   A compound of general structure

$$CH_2{-}CHCH_2NH{-}R$$

or, more usually, the diglycidyl compound

$$\left[CH_2{-}CHCH_2{-}\right]_2N{-}R$$

Prepared by reaction of epichlorhydrin with the appropriate amine. Sometimes useful in the formation of epoxy resins especially diglycidylaniline and triglycidylcyanurate.

GLYCIDYL ESTER   A compound of general structure

$$CH_2{-}CHCH_2OOCR$$

formed by reaction of epichlorhydrin with a carboxylic acid:

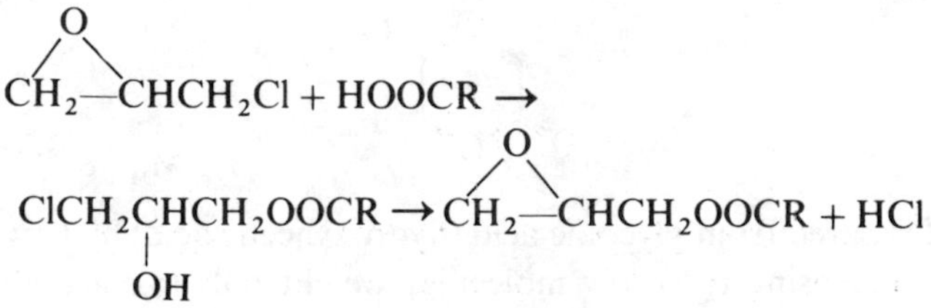

$$CH_2{-}CHCH_2Cl + HOOCR \rightarrow$$

$$ClCH_2\overset{\overset{\displaystyle OH}{|}}{C}HCH_2OOCR \rightarrow CH_2{-}CHCH_2OOCR + HCl$$

Some diglycidyl esters, e.g. diglycidyl phthalate, are occasionally used in the formation of epoxy resins.

GLYCIDYL ETHER A compound of general structure

$$CH_2 \overset{O}{\overset{\triangle}{—}} CHCH_2 — OR$$

formed by reaction of an alcohol with epichlorhydrin via an intermediate chlorohydrin. Diglycidyl ethers are the most common type and form the basis of most epoxy resins. They are prepared by reaction of a diol with epichlorhydrin, via formation of an intermediate chlorohydrin,

$$2\ CH_2 \overset{O}{\overset{\triangle}{—}} CHCH_2Cl + HO—R—OH \rightarrow$$

$$\underset{Cl}{HOCH_2CHCH_2} — ORO — \underset{Cl}{CH_2CHCH_2OH}$$

by heating with alkali and an excess of epichlorhydrin. The diglycidyl ether of bisphenol A is by far the most important, being used for about 95% of commercial epoxy resins. However, glycidyl ethers of novolak resins (epoxy–novolak), tetrabromobisphenol A, glycerol, polypropylene glycol and pentaerythritol are also useful for epoxy resins. Some monoglycidyl ethers, such as phenylglycidyl ether and n-butylglycidyl ether, are also useful in epoxy resins as reactive diluents.

GLYCINE (Gly) (G) H_2NCH_2COOH. The simplest α-amino acid and somewhat polar. Found in nearly all proteins, in particularly large amounts in fibroin and collagen. It is very soluble in water. It has pK' values of 2·34 and 9·60, with the isoelectric point at 5·97. It is often associated with hairpin loops in protein conformations. Its N-methyl derivative is sarcosine.

GLYCININ A soya bean protein globulin, which has been used for production of a textile fibre.

GLYCOGEN The principal reserve deposit polysaccharide of animals. Similar in structure to amylopectin, i.e. a branched α-D-1,4'-glucan, branched through 1,6'-links, except that the degree of branching is greater and the average branch degree of polymerisation is 10–14. Enzymatic degradation studies of the amylolysis limits favour the tree-like structure, as in amylopectin.

GLYCOL DIMETHACRYLATE Alternative name for dimethacrylate, often specifically used for ethyleneglycol dimethacrylate.

GLYCOLIDE (1,4-Dioxan-2,5-dione)

M.p. 87°C.

Prepared from glycollic acid (hydroxyacetic acid) by first condensing it to low molecular weight poly-(oxyacetyl)

then thermally decomposing it. The monomer for the synthesis of poly-(oxyacetyl) by anionic or cationic polymerisation.

GLYCOLIPID A lipid that contains a carbohydrate portion, occurring in the tissue lipoproteins of animals, plants and bacteria. The animal glycolipids are nearly always glycosides of ceramides (which are the N-fattyacyl derivatives of an unsaturated dihydroxyamine, e.g. sphingosine). They include the cerebrosides, gangliosides and sulphatides from brain and nerve tissues and have molecular weights of only 1000–2000. The high molecular weight glycolipids (the lipopolysaccharides) are found in certain micro-organisms and are fatty-acid esters of the carbohydrate polymer. The plant glycolipids are low molecular weight glycosides, e.g. of diacylglycerol and inositol.

GLYCOPROTEIN (Mucoprotein) In the most general sense, a protein to which a carbohydrate is attached. The term is usually restricted to polymers where the protein part is the major component, when the carbohydrate consists of mono- or oligosaccharide but not polysaccharide units. Sometimes the term is restricted to polymers with a carbohydrate content of less than 4%, in general as hexosamine. The term mucoprotein is sometimes used to mean glycoprotein or glycoprotein with more than 4% polysaccharide as hexosamine. When the carbohydrate is the dominant part, usually as polysaccharide, the term proteoglycan is frequently used.

A very widely occurring group of substances, found in blood plasma, e.g. fibrinogen, immune globulins, α_1- and α_2-glycoproteins and blood-group proteins, mucus secretions, e.g. submaxillary glycoproteins and egg whites (ovalbumin, ovomucoid), connective tissue, e.g. collagen, and enzymes in animals. Glycoproteins also occur in plants, e.g. in the cell walls, bacteria and viruses.

A glycoprotein, which may contain as little as less than 1% carbohydrate (as in ovalbumin) consists of a polypeptide chain to which are attached the linear or branched mono- or oligosaccharide units, with two to several tens of units, often terminated by N-acetylneuraminic acid (a sialic acid). The saccharide is covalently attached via an N-acetylhexosamine group to a specific peptide side group, e.g. as in the glycosylamine link between N-acetyl-D-glucosamine and the amide nitrogen of asparigine in immunoglobulins and ovalbumin, or in the glycoside link between N-acetyl-D-galactosamine and the hydroxyl of serine or threonine as in submaxillary glycoprotein.

α_1-**GLYCOPROTEIN** (α_1-Globulin) A blood plasma globular protein which migrates electrophoretically with α_1-globulins, accounting for 5–10% of this fraction. It has a molecular weight of about 55 000 containing about 15% carbohydrate. Sometimes the orosomucoid is also part of the α_1-glycoprotein.

α_2-**GLYCOPROTEIN** A blood plasma protein, which is a globulin, which migrates electrophoretically with the other α_2-globulins and contains about 10% carbohydrate having a molecular weight of up to about 800 000. Sometimes it is further subdivided into separate low molecular weight α_2-glycoprotein and higher molecular weight α_2-maenoglobulin fractions.

GLYCOSAN An internal glycoside, resulting from acetal formation by reaction between an aldose or ketose carbonyl group with a hydroxyl group in the same monosaccharide molecule.

GLYCOSIDE The cyclic mixed acetal formed by reaction of an aldose or ketose with an alcohol ROH. The R group is called the aglycone. It may exist in either the furanose ring (furanoside) or pyranose ring (pyranoside) forms. An internal glycoside is called a glycosan. It is readily hydrolysed by dilute acid to the parent monosaccharide.

GLYCOSIDE BOND (Glycoside link) The bond formed by reaction of an aldose or ketose in cyclic form, with an alcohol ROH to form a mixed acetal or glycoside:

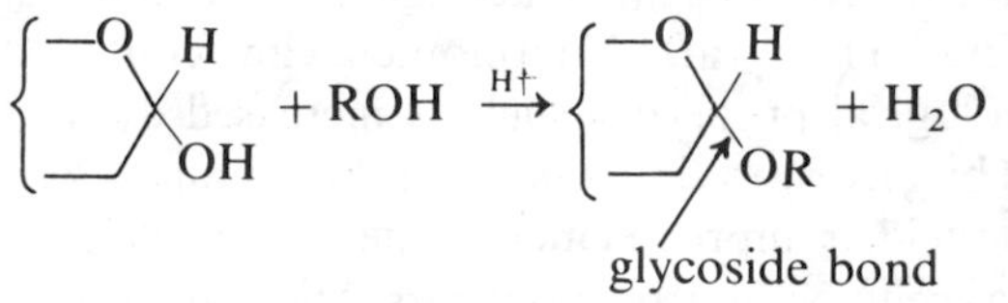

When the alcohol is itself the hydroxyl group of a monosaccharide, then the product is a disaccharide. This may similarly react further to form higher oligosaccharides and eventually a polysaccharide. Thus the monosaccharide units of a polysaccharide are linked through glycoside bonds. The bond may be formed by reaction of either the C-1 or C-2 hemi-acetal hydroxyl with any of the hydroxyls of the other monosaccharide. The bond may be formed in such a way that the anomeric carbon has either configuration. Conventionally, in a β-glycoside link the link points upwards and in an α-glycoside link, downwards.

GLYCOSIDE LINK Alternative name for glycoside bond.

GLYCOSOAMINOGLUCURANOGLYCAN The generic name for a polysaccharide containing alternating monosaccharide and uronic acid units, of which the acidic mucopolysaccharides form the most important group.

GLYCOSOAMINOGLYCAN The generic name for a polysaccharide containing aminomonosaccharide units. Alternative name for aminopolysaccharide. Those usually found in animal connective tissue are acidic glycosoaminoglycans or glycosoaminoglycuronans.

GLYCOSOAMINOGLYCURONAN A polysaccharide containing alternating sequences of hexosamine, e.g. D-glucosamine, and a hexuronic acid, e.g. a D-glucuronide.

GLYCURONAN (Polyuronide) A polysaccharide comprised mainly or entirely of hexuronic acid sugar units. Alginic acid and pectic acid are two important examples.

GLYPTAL A simple early investigated alkyd resin, obtained by reaction of glycerol with phthalic anhydride, modified by the incorporation of a fatty acid. Such polymers had limited commercial applications, requiring long curing times and not forming suitable films. The later oil modified alkyd resins of a similar type were developed into film formers of major importance.

GMS Abbreviation for glyceryl monostearate.

GOLD DECORATION The vacuum evaporation of gold onto a crystal surface, on which it preferentially nucleates in locations of increased surface roughness, e.g. on steps in stacked polymer crystal lamellae. This enables such surface features to be observed more clearly in electron microscopy.

GOLD DISTRIBUTION A molecular weight distribution resulting from a polymerisation with a fixed number of initiation sites (random growth with no termination) but in which $k_p \neq k_i$, where k_p and k_i are the propagation and initiation rate constants respectively. The distribution is thus wider than the similar Poisson distribution (which results from $k_p = k_i$) and results from some anionic polymerisations giving living polymers. The differential weight distribution function, $W(r)$, is a complex function of the molecular size r and of $\gamma (= k_p/k_i)$

GORDON–TAYLOR EQUATION An equation relating the glass transition temperature (T_g) of a random copolymer to the T_g values of the homopolymers of the constituent monomers A and B ($T_{g,A}$ and $T_{g,B}$) and to the weight fractions present in the copolymer (X_A and X_B). It is $(T_g - T_{g,A})X_A + K(T_g - T_{g,B})X_B = 0$, where K is given by $K = (\alpha_{1,B} - \alpha_{g,B})/(\alpha_{1,A} - \alpha_{g,A})$ ($\alpha_{1,A}, \alpha_{1,B}, \alpha_{g,A}$ and $\alpha_{g,B}$ are the volume coefficients of expansion of homopolymers of A and B in the liquid (i.e. rubber) and glassy states respectively). Derivation of the equation is based on the iso-free volume theory. Whilst experimental data fit the equation for some comonomer pairs, e.g. for styrene and butadiene, the fit is often not very good, especially when the chemical characteristics of the monomers are widely different. When block copolymers exhibit only a single T_g value, this is often correctly predicted by the equation.

GOUGH EFFECT Alternative name for Joule–Gough effect.

GPC Abbreviation for gel permeation chromatography.

GPF Abbreviation for general purpose furnace carbon black.

GPO Abbreviation for propylene oxide rubber.

GR-A Abbreviation for government rubber–acrylonitrile. The nitrile rubber produced under the U.S. government sponsored synthetic rubber programme during World War II.

GRADED BLOCK COPOLYMER Alternative name for gradient block copolymer.

GRADIENT BLOCK COPOLYMER (Graded block copolymer) (Tapered block copolymer) A block copolymer in which there is a gradual change in the composition at the junction(s) between the blocks from that of the repeat units of one type (A) to that of the other (B). Thus the structure at the junction may be represented as

∼∼∼AAABAABABAABBABBBABBABABBBABBBB∼∼∼

Such copolymers can be prepared by first copolymerising monomer A to give a living polymer poly-A, e.g. by anionic or Ziegler–Natta polymerisation, then adding monomer B before all the A has been polymerised. Some commercial propylene copolymers contain tapered ethylene blocks.

GRADIENT ELUTION CHROMATOGRAPHY Adsorption chromatography or ion-exchange chromatography where the mixture to be separated, e.g. a protein mixture, is adsorbed on a column of stationary phase and eluted with a progressively more powerful solvent. Thus the most weakly bound component is first eluted, followed progressively by the more strongly bound components.

GRADIENT ELUTION FRACTIONATION A fractionation method in which fractions are obtained by elution of polymer using a packed column. The term refers either to the simple packed column method, where the gradient is one of solvent power, or to the Baker–Williams fractionation method where, in addition, a temperature gradient is imposed on the column.

GRAESSLEY THEORY (Entanglement theory) A molecular theory of the non-Newtonian flow behaviour of polymer fluids, based on the entanglement and disentanglement of polymer molecules in a shear field. As with the Bueche theory the result is that the viscosity (η) is a function of $\lambda\dot{\gamma}$, where $\dot{\gamma}$ is the shear rate and λ is a relaxation time. At high shear rates $\eta/\eta_0 \approx (\dot{\gamma})^{-3/4}$, where η_0 is the viscosity at zero shear rate. The theory also predicts the existence of shear-dependent normal stresses, the normal stress coefficient ψ_{12} being given by $\psi_{12} = \psi_{12}^0(\eta/\eta_0)^2$.

GRAFIL Tradename for polyacrylonitrile based carbon fibres.

GRAFT COPOLYMER A copolymer which consists of a polymer chain (poly-A) to which are attached one or, usually, more polymer chains of another polymer (poly-B), i.e.

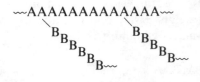

Block copolymers may be considered as graft copolymers in which the B chains are attached to the ends of the A chains. The copolymers are generally formed by forming active sites on poly-A in the presence of monomer B, which are capable of initiating polymerisation of monomer B. However, unlike block copolymers, it is usually difficult to attach the poly-B chains in a regular manner or to control their growth to chains of equal length. Thus the resultant copolymer is rather heterogeneous in structure. Furthermore, frequently homopolymerisation of monomer B also occurs and the homopolymer is difficult to separate from the copolymer. Grafting sites may be produced by generating free radicals on poly-A, by transfer grafting, activation grafting or by chemical grafting. Alternatively, ionic grafting sites may be introduced giving anionic or cationic grafting. The grafted monomer can confer useful technological properties such as increased dyeability in fibres by incorporating a polar grafted monomer such as acrylic acid or improved membranes, e.g. for desalination, such as cellulose graft copolymers. These are examples of surface grafting.

GRAFTING EFFICIENCY In graft copolymerisation, the ratio of the amount of a monomer grafted onto a polymer to the total amount of monomer polymerised as both grafts and homopolymer.

GRAFTING RATIO In graft copolymerisation, the ratio of the weight of the grafted polymer to that of the original polymer, when grafting results from a monomer/polymer mixture.

GRAPHITE A crystalline form of carbon, having a sheet-like polymer structure and consisting of layers of fused benzene rings at a separation of 3·40 Å. The intermolecular forces between the layers are low, hence they readily slide past each other and thus graphite has lubricating properties. Owing to the extensive electron delocalisation, the material has high electrical conductivity and is useful as a conducting and lubricating filler in polymers. The crystalline regions of carbon black and the crystallites of carbon fibres have a graphite structure.

GRAPHITE FIBRE Alternative name for carbon fibre.

GREEN–RIVLIN THEORY A theory of non-linear viscoelasticity in which the Boltzmann Superposition Principle is modified so that the strain $e(t)$ at time t is given by a multiple integral representation of the incremental stresses $\sigma_1, \sigma_2, \sigma_3, \ldots$, etc., applied at times $\tau_1, \tau_2, \tau_3, \ldots$, etc., of the form:

$$e(t) = \int_{-\infty}^{t} [J_1(t-\tau)\,d\sigma(\tau)/d\tau]\,d\tau$$
$$+ \int_{-\infty}^{t} \int_{-\infty}^{t} [J_2(t-\tau_1, t-\tau_2)\,d\sigma(\tau_1)/d\tau_1]$$
$$\times d\sigma(\tau_2)/d\tau_2(d\tau_1\,d\tau_2) + \cdots$$

$J_1(t)$ is the creep compliance function and $J_2, J_3, \ldots$, etc., are functions of the differences in times $t-\tau_1$, $t-\tau_2$, $t-\tau_3, \ldots$, etc. The first term is the linear Boltzmann term. Such an expression is sometimes called the multiple integral representation of viscoelasticity. This form relates to the one-dimensional case and may be inverted to describe stress relaxation. For the more general three-dimensional case such an inversion is not possible. The relationship is only useful if it can be restricted to a limited number of terms. For only a few polymers, notably polypropylene monofilament, have sufficient data been obtained to enable J_1, J_2 and J_3 to be determined. In some cases the experimental data fits an expression containing only the first two or three terms.

GREEN'S STRAIN (Kirchhoff strain) Symbol ε'. A measure of large strain deformation given by, $\varepsilon' = \frac{1}{2}[(l'/l)^2 - 1]$, where l' and l are the distances between two points in the strained and unstrained material respectively. At very low strain it is the same as natural strain.

GREEN STRENGTH The tensile strength of a raw unvulcanised rubber or rubber compound. Natural rubber has one of the highest values. A high green strength is desirable in those processing operations in which the integrity of a shaped piece of rubber needs to be maintained.

G-RESIN Tradename for a butene-1 based linear low density polyethylene.

GR-I Abbreviation for government rubber–isobutene, i.e. butyl rubber. The butyl rubber produced under the U.S. government synthetic rubber programme during the Second World War.

GRIFFITH THEORY A theory of brittle fracture based on the idea that the presence of cracks determines the brittle strength and crack propagation occurs, resulting in fracture, when the rate of decrease in elastically stored energy at least equals the rate of formation of fracture surface energy due to the creation of new surfaces. The theory results in expressions for the strength of a material in terms of crack length ($2c$) and fracture surface energy (γ). For example, the tensile strength for an infinitely thin sheet is given by $(2\gamma E/\pi c)^{1/2}$, where E is the tensile modulus. However, the surface energies found in the polymer fracture are several orders of magnitude higher than the values expected on the basis of the separation of planes of atoms. The high values are thought to be due to the plastic work performed near the crack tip during fracture.

GRIGNARD PROCESS A method for the synthesis of chlorosilanes by reaction of an ethereal solution of a Grignard reagent (RMgX) (where X is usually chlorine if R is alkyl or bromine if R is aryl) with a silicon compound, usually silicon tetrachloride. A mixture of silanes results, with R_2SiCl_2 predominating, but unlike the cheaper direct process, the product composition may be considerably varied by altering the reaction conditions. The sequence of reaction steps is:

$$RMgX + SiCl_4 \rightarrow RSiCl_3 + MgXCl$$
$$RMgX + RSiCl_3 \rightarrow R_2SiCl_2 \rightarrow R_3SiCl \rightarrow R_4Si$$

The process is very versatile being suitable for alkyl and aryl, or mixed alkyl/aryl, silanes, as well as for organo-hydrogen containing silanes and for organofunctional group containing silanes.

GRILAMID Tradename for nylon 12.

GRILAMID ELY 1256 Tradename for a polyetheramide thermoplastic elastomer based on C_{12} polyamide hard blocks and polyether soft blocks.

GRILAMID ELY 60 Tradename for a polyetheramide block copolymer, previously called Grilamid ELY 1256.

GRILAMID TR55 Tradename for a transparent polyamide produced from isophthalic acids, lauryllactam and

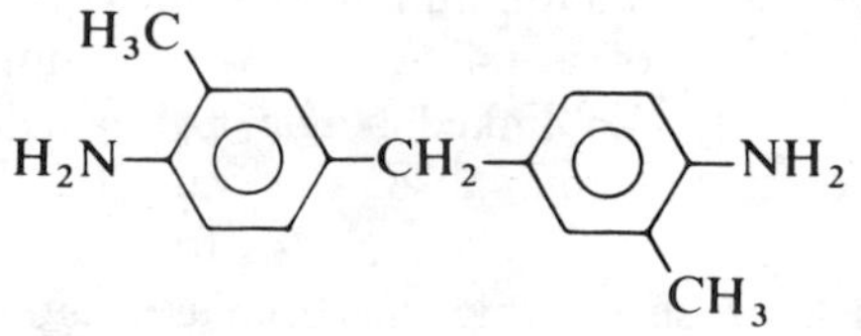

GRILENE Tradename for a polyethylene terephthalate fibre or a modified polyethylene terephthalate fibre produced from terephthalic acid, ethylene glycol and *p*-hydroxybenzoic acid.

GRILESTA Tradename for an unsaturated polyester resin.

GRILON Tradename for nylon 6 and nylon 66.

GRILONIT Tradename for an epoxy resin.

GR-M Abbreviation for government rubber–monovinylacetylene, the polychloroprene rubber produced under the U.S. government synthetic rubber programme during the Second World War. Monovinylacetylene refers to the precursor for the monomer chloroprene.

GROWTH HORMONE Alternative name for somatotropin.

GROWTH SPIRAL Alternative name for spiral growth.

GR-P Abbreviation for an early polysulphide rubber (government rubber–polysulphide) produced by a U.S. government sponsored programme during the Second World War.

GR-S Abbreviation for government rubber–styrene, the styrene–butadiene rubber produced under the U.S. government synthetic rubber programme during the Second World War, largely by the mutual recipe.

GUANAMINE–FORMALDEHYDE POLYMER
Alternative name for benzoguanamine-formaldehyde polymer.

GUANIDINE A compound of general structure $(ArNH)_2C=NR$, where Ar is an aromatic ring and R is an alkyl or phenyl group or hydrogen. An early type of accelerator for the sulphur vulcanisation of rubbers. Now mostly used as secondary accelerators. Examples include diphenylguanidine and di-o-tolylguanidine.

GUAR GUM A water soluble galactomannan polysaccharide found in the seed of the guar plant, consisting of linear chains of 1,4'-linked β-D-mannopyranosyl units with α-D-galactopyranosyl units attached by 1,6'-links in the ratio 2:1. Useful in mineral froth flotation and water treatment, as a binder and thickening agent in foods and pharmaceuticals and as a binder in paper.

GUARON A galactomannan obtained from guar seeds. It consists of a 1,4'-linked β-D-mannopyranose main chain with 1,6'-linked single α-D-galactose side groups, the galactose:mannose ratio being 1:2.

GUAYULE A shrub which has been used as a source of natural rubber.

GUINIER PLOT In small angle X-ray and small angle neutron scattering, a plot of $\ln I(\mathbf{h})$ versus $\mathbf{h}^2$ (or $(2\theta)^2$), where $I(\mathbf{h})$ is the scattered radiation intensity and $\mathbf{h}$ is the scattering vector (or θ is the scattering angle). According to the Guinier relation, $I(\mathbf{h}) = I_0 \exp(-\mathbf{h}^2 R^2/3)$, where R is the radius of gyration of the scattering particles. Thus the plot should be a straight line with a slope of $-R^2/3$. In addition, extrapolation to zero angle gives I_0, from which the molecular weight may be obtained.

L-GULURONIC ACID

M.p. 141·2°C (lactone). α_D^{22} +81·7°.

A uronic acid, occurring as a monosaccharide component of alginic acid.

GUM Strictly, a water soluble polysaccharide, yielding a viscous solution, but insoluble in organic solvents, for example the gum mucilages. The term is also used technologically as an alternative name for natural resin; however, in contrast, these are water insoluble, have some solubility in organic solvents and also melt on heating.

GUM ARABIC A complex gum polysaccharide, exuded by species of *Acacia*. A highly branched polymer, typically composed of L-arabinose, D-galactose, L-rhamnose and D-glucuronic acid units in the approximate molar ratio of 3:3:1:1. A partial structure consists of 1,3'-linked β-D-galacturonic acid units with 1,6'-linked side groups composed of 1,6'-glucuronic-1,6'-β-D-galactose units to which are attached 1,4'- and 1,3'-linked L-arabinose, L-rhamnose or α-D-galactose-1,3'-L-arabinose units at various points. Such structures are identified by controlled hydrolysis of the neutralised acid, arabic acid.

GUM GHATTI A water soluble polysaccharide exuded by the tree *Anageissus latifolia*. Composed of L-arabinose, D-galactose, D-mannose, D-xylose and D-glucuronic acid (as the calcium or magnesium salt) units in the ratio 10:6:2:1:2, i.e. it is mainly a galacto-arabinan. Useful as an emulsifying agent in foods and pharmaceuticals.

GUM KARAYA A partially acetylated polysaccharide exuded by the three *Stericolia urens*, consisting of D-galactan, L-rhamnose and D-galacturonic acid units and about 8% acetyl groups. It absorbs water very readily to form a viscous mucilage effective as a laxative. Also useful as a binder in paper and as an emulsion stabiliser in foodstuffs.

GUM POLYSACCHARIDE (Plant gum polysaccharide) (Exudate gum polysaccharide) A complex acidic polysaccharide formed spontaneously at the site of injury to certain plants as a liquid exudation, which dehydrates in air to give a hard clear material. Widely used commercially as thickening agents, binders, emulsion stabilisers and textile sizes. The polysaccharide molecules are highly branched, containing many D-glucuronic acid (or galacturonic acid) units together with two or more neutral sugar unit types. The acid units are usually found neutralised with calcium or magnesium. Examples include gum arabic, gum tragacanth, gum Ghatti and gum karaya.

GUM RUBBER (Pure gum) A raw unvulcanised rubber as obtained from polymerisation (for synthetic rubber) or, in case of natural rubber, after isolation from the natural latex. The term also sometimes includes unvulcanised rubber compounds, which are also referred to as the stock. Usually a gum rubber is soft, of low elastic recovery, low softening point, is readily swollen and dissolved by organic solvents and has little mechanical strength. Only after vulcanisation, producing a vulcanised rubber or vulcanisate, do useful strength, stiffness and high rubber elasticity result.

GUM TRAGACANTH   A plant gum exuded by the thorny shrubs *Astralagus*, consisting of a water soluble component, which is itself a mixture of polysaccharides but mainly of tragacanthic acid, and a highly branched L-arabino-D-galactan. The gum is useful as a thickening agent and suspension and emulsifying agent.

GUN COTTON   A cellulose nitrate in which the degree of substitution approaches three. It is useful as an explosive.

GUTH EQUATION   Alternative name for Guth–Gold equation when this is formulated in terms of Young's moduli. Sometimes also an alternative name for the Guth–Smallwood equation or the Einstein–Guth–Gold equation.

GUTH–GOLD EQUATION   (Guth equation) A relationship expressing the shear modulus (G_c) of a filled polymer composite as a function of filler volume fraction (ϕ_f). The equation is $G_c = G_0(1 + 2.5\phi_f + 14.1\phi_f^2)$, where G_0 is the shear modulus of the unfilled polymer. It is particularly applicable to rubber vulcanisates filled with spherical particles. A similar expression, sometimes called the Guth equation or the Einstein–Guth–Gold equation, is also used for the Young's modulus (E_c).

GUTH–SMALLWOOD EQUATION   An equation expressing the coefficient of viscosity (η_c) of a filled rubber as a function of the volume fraction of the filler (ϕ_f). It is based on the Einstein equation and is,

$$\eta_c = \eta_0(1 + 2.5\phi_f + 14.1\phi_f^2)$$

where η_0 is the coefficient of viscosity of the unfilled rubber. It only applies to non-reinforcing, spherical filler particles such as calcium carbonate and not to highly reinforcing fillers such as carbon black.

GUTTA   Alternative name for gutta percha.

GUTTA PERCHA   (Gutta) The solid isolated from the latex of the tree *Palquium oblongfolium*. Consists of *trans*-1,4-polyisoprene, but contains, in addition, large quantities of resin, mostly esters of phytosterols. Unlike natural rubber it crystallises even when unstretched and is therefore hard and stiff. Three crystalline forms exist. The β form is stable at above 70°C but is metastable at ambient temperatures, and hence is the usual form. Stretched samples exist in the α or β forms. It is useful, unvulcanised, as a cable insulation.

***G*-VALUE**   On the irradiation of a material, it is the number of molecules or groups reacted or formed for each 100 eV of energy absorbed.

H

H   Symbol for histidine.

HACKLE   The rough region of a fracture surface further from the origin of the fracture than the main mirror or mist regions. Formation of hackle is particularly favoured by fast fracture and is due to the fracture crack propagating on different levels over small regions of surface. Sometimes it has a banded structure (banded hackle) and sometimes the markings are elongated (river markings).

HAEM   (Haem group) (Protohaem) (Ferroprotoporphyrin) The prosthetic group of haemoproteins. It consists of a protoporphyrin ring surrounding and coordinated to a central iron(II) atom as a square planar complex through four nitrogen atoms. The porphyrin ring is flat, the π-electrons are extensively delocalised. In a haemoprotein the iron is octahedral with six ligands as a haemochromogen, the two remaining ligands are through a histidine residue of the protein and the oxygen molecule which binds to the haemoprotein (in haemoglobin and myoglobin) or to a methionine sulphur (in cytochrome C). The structure is:

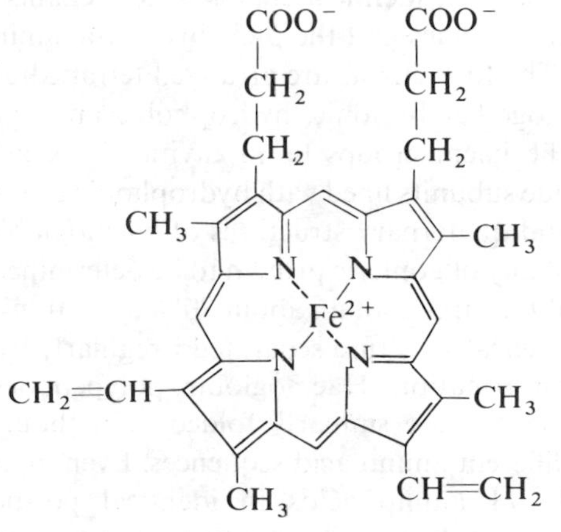

The haem group provides the red colour characteristic of blood, the oxygenated form being bright red, the deoxy form being bluish-red. Binding of the haem to the protein is augmented by hydrophobic bonding (the haem lies in a hydrophobic pocket of the folded polypeptide chain) and by salt bridges between the propionic acid units and basic amino acid residues of the protein. Oxidation of the iron(II) to iron(III) converts haem to haemin.

HAEM GROUP   Alternative name for haem.

HAEMIN   Haemin has the same structure as haem, except that the iron(II) has been oxidised to iron(III). Such oxidation readily occurs and is responsible for the role of cytochrome C as an electron carrier.

HAEMOCHROME   Alternative name for haemochromogen.

HAEMOCHROMOGEN (Haemochrome) (Base ferro-porphyrin) The six-coordinate octahedral complex formed from a haem which is square planar and four-coordinate, by coordination of two more ligands, e.g. with a histidine amino acid residue and an oxygen molecule as in oxy-haemoglobin or with a histidine and a methionine as in cytochrome C.

HAEMOCYANIN A haemoprotein acting as the respiratory protein in lower animals, snails, crustaceans, spiders, etc. Iron is absent, being replaced by copper (about 0.2%) which is chelated to the polypeptide chain. Measured molecular weights are very high (0.5 to 10 million) but are reduced as pH is increased, suggesting the dissociation of associated complexes into two or eight subunits which bind one oxygen molecule for every two copper atoms.

HAEMOGLOBIN A conjugated protein of a poly-peptide component (globin) with haem as the prosthetic group. Responsible for the transport of oxygen in the blood from the lungs to the body tissues by forming loose reversible complexes with oxygen. An oligomeric protein consisting of four polypeptide chains and four haem groups making a roughly spherical conjugate of molecular weight about $64\,000$.

Normal human haemoglobin (haemoglobin A) contains two pairs of identical chains—the α-chains of 141 amino acid residues and the β-chains of 146 amino acid residues. The four chains are arranged tetrahedrally and are held together by ionic, hydrophobic and hydrogen bonds. The haem groups lie in cavities in each of the polypeptide subunits lined with hydrophobic groups. The tertiary and quaternary structures of haemoglobin were the first of any oligomeric protein to be determined. Both the α- and β-chains contain about 70% of sections with a similar α-helical structure separated irregularly by bends in the conformation. Haemoglobin polypeptides from different species are similarly folded even though they contain different amino acid sequences. Even myoglobin with only 21 amino acids in identical positions to haemoglobin A has similar folding. However, replacement of even a single critical amino acid can have serious consequences. Thus replacement of a glutamic acid at position 6 in the β-chain (forming haemoglobin S) results in sickle cell anaemia.

Oxygen is bound to haemoglobin as the sixth ligand to the iron(II) atom of the haem group (the other ligands are the four porphyrin nitrogens and the imidazole ring of a histidine residue). The iron(II) is not oxidised to iron(III), being protected by its hydrophobic surroundings. The oxygen binding is favoured by higher pH values (Bohr effect) and increased oxygen partial pressure. Since both are lower in the tissues than in the lungs, the haemoglobin binds less oxygen and gives some up. Oxygen binding curves are sigmoidal in shape meaning that haemoglobin has relatively low affinity for the first one or two oxygen molecules bound, but binding of subsequent oxygen molecules is enhanced (allosteric effect). Such an effect is explained by cooperativity.

HAEMOGLOBIN A The normal form of human haemoglobin as distinct from haemoglobin S and other modifications where changes in the amino acid sequences have occurred.

HAEMOGLOBIN S A human haemoglobin in which the glutamic acid residue at position 6 of the β-chain is replaced by valine. This alters the conformation of the β-chain and is responsible for the disease sickle cell anaemia.

HAEMOPROTEIN A conjugated protein in which the prosthetic group is a haem group. This includes many of the transport proteins usually found in animals, such as haemoglobin, myoglobin, cytochrome C and haemocyanin of molluscs. The haem group is bound to the protein by a combination of coordination from, for example, an electron pair of a basic amino acid residue, to the iron, salt bridges and hydrophobic bonding. The protein provides a hydrophobic pocket in which the haem group resides, the iron thereby being protected from oxidation from iron(II) to iron(III). For both the simpler, e.g. myoglobin, and for the more complex oligomeric proteins, e.g. haemoglobin, the tertiary and quaternary structures were established quite early using mainly X-ray diffraction analysis.

HAF Abbreviation for high abrasion resistant furnace carbon black.

HAGENBACH CORRECTION Alternative name for kinetic energy correction.

HAIRPIN LOOP A looped conformation, frequently found in polypeptides, both synthetic and protein, in which the molecular chain is folded back on itself so that two segments of the chain are parallel. Single loops can be responsible for the overall folding or tertiary structure of globular proteins. The folding may be repeated many times, e.g. in the cross-β structure, when it is called the β-bend or β-turn. Hairpin loop formation is favoured by the presence of glycine, proline and aspartic acid residues. Such loops involving four amino acid residues per turn are called β-turns.

HALAR Tradename for chlorotrifluoroethylene–ethylene alternating copolymer.

HALATOPOLYMER (Polycarboxylate) A polymer which contains ionic links in the polymer chain, i.e. a polymer of the type:

$$\sim\!\!\sim + - \sim\!\!\sim + - \sim\!\!\sim + - \sim\!\!\sim + - \sim\!\!\sim$$

so displaying both polymeric and salt-like properties. Most studied are the divalent metal salts of dicarboxylic acids, e.g. calcium sebacate:

$$\sim\!\!\sim\text{COO}\,\text{Ca}^{2+}\,\bar{\text{O}}\text{OC}(\text{CH}_2)_3\text{COO}\,\text{Ca}^{2+}\,\bar{\text{O}}\text{OC}\!\sim\!\!\sim$$

HALF-LIFE (Half-life time) The time for half the quantity of a chemical species (often a free radical polymerisation initiator) to disappear due to a chemical reaction (thermal homolysis to free radicals in the case of initiators) at a specified temperature.

HALF-LIFE TEMPERATURE The temperature at which the half-life of a free radical initiator (or any species disappearing in a chemical reaction) is a specified length of time e.g. 1 min, 1 h or 10 h.

HALF-LIFE TIME Alternative name for half-life.

HALF-POWER WIDTH OF THE RESONANCE PEAK A measure of the damping during dynamic mechanical behaviour, obtained when the specimen is subject to the resonance forced vibration method, e.g. as a vibrating reed. At resonance, the amplitude–frequency curve peaks and the half-power width, Δf, is the width, in hertz, of this peak at $1/\sqrt{2}$ of the maximum amplitude. The ratio $\Delta f/f_r = \tan \delta$, where f_r is the resonance frequency and $\tan \delta$ is the tangent of the loss angle. The reciprocal of $\Delta f/f_r$ is the Q-factor or dynamic amplification factor.

HALF-WIDTH OF THE RESONANCE PEAK (Resonance peak half-width) A measure of the damping during dynamic mechanical behaviour, when the specimen is subject to the resonance forced vibration method, e.g. as a vibrating reed. A plot of amplitude versus frequency has a peak at resonance. The ratio of the difference between the frequencies at the points having half the maximum amplitude to the resonance frequency is approximately equal to E''/E' or $\tan \delta$, where E'' and E' are the loss and storage Young's moduli respectively and $\tan \delta$ is the tangent of the loss angle.

HALON Tradename for polychlorotrifluoroethylene.

HALOTHENE Tradename for chlorinated polyethylene.

HALPIN–TSAI EQUATION A modified and more general form of the Kerner equation for describing the elastic properties of a composite material. It is capable of allowing for various filler particle morphologies, packing modes and matrix Poisson's ratio. A widely used form is, $M_c/M_m = (1 + AB\phi_f)/(1 - B\phi_f)$ where M_c and M_m are the moduli (shear, Young's or bulk) of the composite and matrix respectively, A is a constant involving filler morphology and matrix Poisson's ratio (given by $A = k_e - 1$, where k_e is the filler Einstein coefficient (equals 2·5 for a rigid spherical filler in a matrix with Poisson's ratio equal to 0·5, for example)) and B is given by, $B = (M_f/M_m - 1)/(M_f/M_m + A)$, which is approximately equal to unity if M_f/M_m is very large.

HALS Abbreviation for hindered amine light stabiliser.

HAMON APPROXIMATION In the time-domain method for the measurement of the dielectric properties, the assumption that the current transient $I(t)$ is given by, $I(t) = At^{-n}$, where $0·5 < n < 1$ and A is a constant. It enables the dielectric loss in the frequency domain to be determined from the relation, $\varepsilon^*(\omega) = I(t)t/0·63C_0V_0$, where t is the time, $\omega t = \pi/5$, $\varepsilon^*(\omega)$ is the frequency-dependent complex permittivity, C_0 is the vacuum (or air) capacitance of the cell used and V_0 is the step voltage applied.

HAPTOGLOBIN A blood serum glycoprotein, which migrates electrophoretically with the α_2-globulins and comprises about one-quarter of the α-globulin fraction. It contains about 25% carbohydrate which consists of N-acetylneuraminic acid and galactose units. Three different genetic groups (haptoglobulins 1-1, 2-1 and 2-2) have been characterised. These contain two different polypeptide chains (α and β) which combine in different ways in the different types. The total molecular weight is about 80 000 in human haptoglobulin 1-1, comprising two α-chains of molecular weight 9 000 and one β-chain of molecular weight 65 000. Haptoglobulins are capable of binding to haemoglobin, the resulting complex having peroxidase activity.

HARD BLOCK Alternative name for hard segment.

HARD RUBBER Alternative name for ebonite. Sometimes used for rubbers hardened, not by vulcanisation but by the incorporation of a stiff polymer.

HARD SEGMENT (Hard block) A block in a block copolymer thermoplastic elastomer that has a T_g value (or T_m if the block is crystallisable) well above ambient temperature and so imparts stiffness to the material. It contains the thermoreversible crosslinks, so making the material melt processable as a thermoplastic. Examples include the styrene blocks (high T_g) in SBS, the tetramethyleneterephthalate blocks (high T_m) in polyester/polyether block copolymers and the polyurethane blocks in thermoplastic polyurethanes.

HARKINS MODEL A qualitative description of emulsion polymerisation in which free radicals, generated in the aqueous phase, are captured by monomer-swollen surfactant micelles, thus initiating polymerisation of the monomer in the micelles. A monomer/polymer particle is therefore nucleated which eventually grows by transport of further monomer from the monomer droplets through the aqueous phase to become a polymer latex particle in the final product. In some cases, however, homogeneous nucleation of particles may occur by initiation in the aqueous phase followed by precipitation of polymer particles. The Smith–Ewart theory provides a quantitative treatment of emulsion polymerisation on the basis of this model.

HARTSHORNE AND WARD METHOD A resonance circuit method for the determination of dielectric properties over the frequency range 10^5–10^8 Hz. An inductance is connected in parallel with a variable thickness air condenser which can also accept the sample dielectric in the form of a thin disc. The inductance is loosely coupled to a stable oscillator and a voltmeter and side capacitor are also connected in parallel. The frequency of the oscillator is tuned to resonance with the sample in the electrode, giving a maximum in the voltage V_m. It is then detuned with the capacitor, to the 'half-power' condition (with output $V_r/\sqrt{2}$), first on one side of the resonance peak and then on the other, the total change for this being ΔC_i. Resonance is now restored, the specimen removed, and with air between the electrodes, their gap is narrowed (by an amount Δx) to re-establish resonance. A similar detuning to $V_r/\sqrt{2}$ is again performed, requiring ΔC_0 to do this. Then, $\varepsilon' = t/(t - \Delta x)$, where t is the sample thickness and ε' is the relative permittivity and $\tan \delta = (\Delta C_i - \Delta C_0)/2 C_s'$, where C_s' is the capacitance of the sample. Measurements may be made over different frequency ranges by changing the inductance coil but in practice frequencies are limited to about 10^8 Hz.

HASHIN–SHTRIKMAN BOUNDS The theoretical bounds of the elastic moduli of a composite material, derived by the use of a variation method involving the elastic polarisation tensor. The bounds are narrower than those of the laws of mixtures, especially when the matrix and filler moduli differ little. However, when the moduli are widely different then the bounds are widely separated. For the shear moduli G_m (for the matrix) and G_c (for the composite), the lower bound is given by,

$$\frac{G_c}{G_m} = \frac{(G_f - G_m)\phi_f}{1 + (6\phi_m/5)[(G_f - G_m)(K_m - 2G_m)/(G_m(3K_m + 4G_m))]}$$

where G_f is the filler shear modulus and K_m is the matrix bulk modulus. The upper bound is given by interchanging the subscripts f and m.

HAWARTH FORMULA A way of representing the structure of monosaccharides, and saccharide repeat units in polysaccharides, in their usual cyclic form. The relative positions of the groups attached to the ring are shown as being either above or below the ring and the ring atoms are numbered as shown below. The full form of the formula is shown, for example, for D-glucose in the usual α-D-glucopyranose form, and for D-fructose in the usual α-D-fructofuranose form, as:

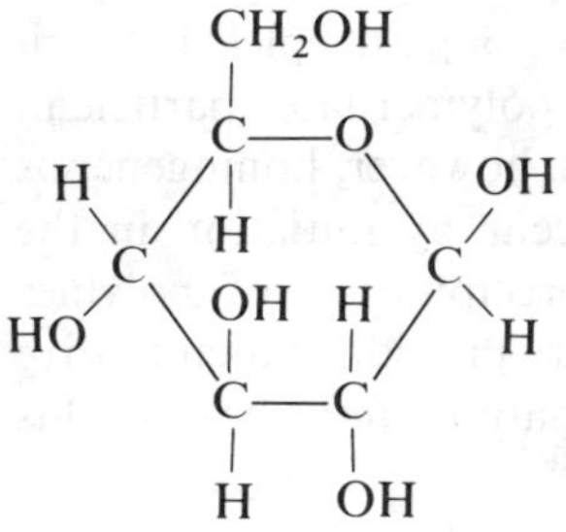

or in the more usual abbreviated form as:

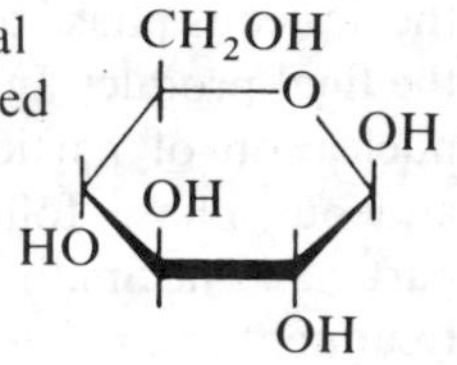

α-D-glucose

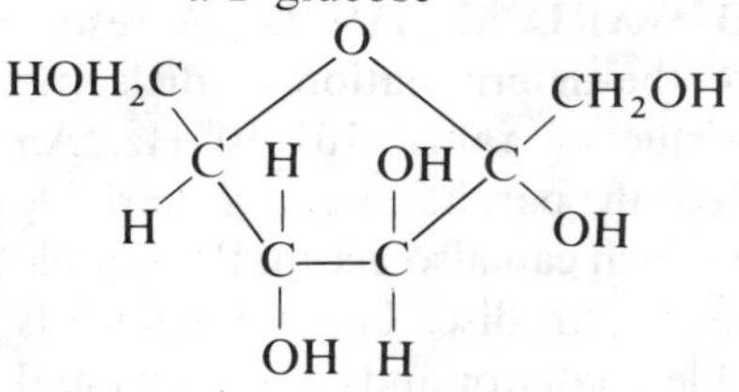

α-D-fructose

or in the more usual abbreviated form as:

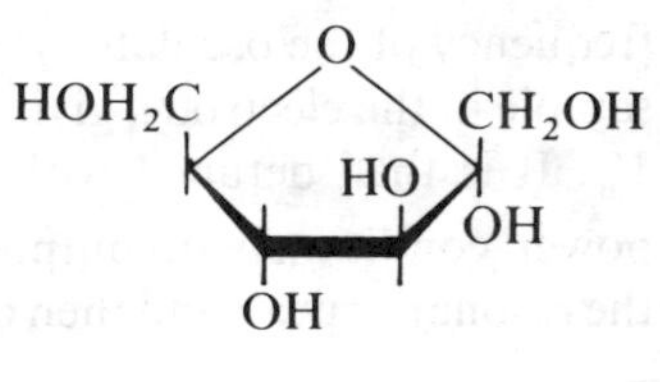

HAZE The reduction in contrast between light and dark parts of an object when viewed through a transparent material. It is caused by the scattering of light at refractive index discontinuities in a polymer, such as impurities, voids, crystallites, at relatively large angles. Light scattered at small angles does not contribute to haze but causes a loss of clarity.

HCC Abbreviation for high colour channel carbon black.

HDI Abbreviation for hexamethylene diisocyanate.

HDPE Abbreviation for high density polyethylene.

HDT Abbreviation for heat distortion temperature.

HEAD EFFECT Alternative name for Bagley end correction.

HEAD-TO-HEAD POLYPROPYLENE An alternating copolymer of ethylene and *cis*-2-butene. The copolymer has a T_m value of 66°C and a T_g value of -10°C.

HEAD-TO-HEAD STRUCTURE A structure of the type $\sim\!CH_2CHX\!-\!CHXCH_2\!\sim$ in a vinyl or related type of polymer, which is one of the two possible positional isomers, the other being a head-to-tail structure. During polymerisation, it results from a growing chain $\sim\!CH_2C^*HX$ adding on a monomer ($CH_2\!=\!CHX$) to give $-CH_2CHX\!-\!CHXCH^*_2$, the substituted carbon atom being designated the head. Normally only a few ($<1\%$) such structures are produced since head-to-tail links are favoured due to steric and resonance effects. Formation of a head-to-head link also results from a termination by combination step. These links are thought to provide the weak links in some polymers at which degradation reactions are initiated.

HEAD-TO-TAIL STRUCTURE A structure of the type $\sim\!CH_2CHX\!-\!CH_2CHX\!\sim$ in a vinyl or related type of polymer, which is one of the two possible positional isomers, the other being a head-to-head link. During polymerisation, it results from a growing chain $\sim\!CH_2C^*HX$ adding on a monomer ($CH_2\!=\!CHX$) with the unsubstituted carbon (considered the tail) adding on to the head to give: $\sim\!CH_2CHX\!-\!CH_2CH^*_2X$ (**I**). Normally this mode of addition predominates ($>99\%$) over the alternative head-to-head addition, since formation of **I** is favoured for steric and resonance reasons.

HEAT DISTORTION TEMPERATURE (HDT) The temperature at which a rectangular bar of a polymer deflects by a specified amount under three point bending when heated in a fluid, whilst being subject to a specified bending stress at the central point. The stress is usually either 1·82 or 0·45 MPa.

HEAT OF POLYMERISATION (Enthalpy of polymerisation) Symbol ΔH_p. The enthalpy difference between one mole of monomer and one mole of repeat units in its polymer. Chain polymerisations are highly exothermic (negative heat of polymerisation) due to the energy difference between the π-bond in the monomer and the σ-bond to which it is converted in the polymer. Therefore much heat is evolved in chain polymerisation (but not in ring-opening types) and the polymerisation method used must be capable of adequately dissipating the heat, otherwise the polymerisation cannot be controlled. Typical values are about 100 kJ mol⁻¹ (in ethylene), but lower values occur when the substituent in a vinyl monomer (CH_2=CHX) is capable of resonance stabilisation of the monomer, e.g. styrene (72 kJ mol⁻¹), acrylonitrile (78 kJ mol⁻¹), butadiene (35 kJ mol⁻¹) and methyl methacrylate (55 kJ mol⁻¹).

HEAVY MEROMYOSIN The fragment produced on trypsin cleavage of myosin, with a molecular weight of about 350 000 and with only partial α-helical structure. The remainder, of molecular weight about 115 000, is associated with the globular head of myosin.

HEDGEHOG DENDRITE An irregular dendrite in which, although branching is regular, there is no crystallographic relationship between branches.

HEDRITE A polyhedral multilayer lamellar crystal usually about 1 nm thick and about 10 μm in diameter. It is formed by crystallisation from the melt or concentrated solution. The lamellae are connected with each other by tie molecules or by fibrous tie crystals. Thus a hedrite is intermediate in complexity between a single crystal and a spherulite.

HELICAL FLOW A flow in which an axial pressure flow is combined with a circular drag flow, which results in the flow path of the fluid elements following a helical path. Thus helical flow results, for example, when a fluid is flowing along a circular cross-section pipe whilst at the same time an inner cylinder in the pipe is being rotated. For shear thinning fluids, as with many polymer fluids, the axial flow may be increased by the angular drag flow of the rotating cylinder. Such flows can also occur in extrusion.

3_1-HELIX A helical conformation in which each complete turn of the helix contains three repeat units of the polymer chain. Such a conformation is adopted by several isotactic vinyl polymers, e.g. polypropylene, polystyrene and polyvinyl allyl ethers, as well as by some polypeptides, as in the poly-L-glycine II and poly-L-proline II conformations, which form the basis of the triple helical structure of tropocollagen.

α-HELIX One of the two almost universally occurring regular conformations of polypeptide chains, of both synthetic and of protein origin. In globular proteins only segments of the polypeptide chain adopt an α-helical conformation, whereas in fibrous proteins the bulk of the chains may be present as α-helices, and hence the α-helix is

the secondary structure of many fibrous proteins. For L-amino acids (as found in proteins) a right-handed helix is favoured, for example, as in poly-(β-benzyl-L-aspartate).

The helix is non-integral and contains eighteen amino acid residues for each five turns of the helix, extending 0·54 nm for each turn in the axial direction, i.e. a residue rise of 0·15 nm. This agrees with the X-ray diffraction data, which originally suggested that a helical structure existed in some structural proteins, notably keratin. Such a helix was deduced from consideration of the double bond character (with consequent planarity) of the peptide group, the non-overlapping van der Waals radii and the intramolecular hydrogen bonding. In fact all the amide groups are hydrogen bonded, the bonds lying parallel to the axis. The R substituents of the peptides,

point away from the helix. The angles of rotation are $\phi = -58°$ and $\psi = -47°$ for a right-handed helix and $\phi = +58°$ and $\psi = +47°$ for a left-handed helix. Either right- or left-handed helices may be constructed for both D- and L- amino acid polypeptides, but for L-amino acids (as found exclusively in proteins) the right-handed helix is more stable and it is this form that is found. An α-helix cannot be constructed from mixed L- and D-amino acids.

The relationship between amino acid composition and sequence and α-helix forming ability has been widely studied using synthetic polypeptides, both homopolymers and sequential polydipeptides and polytripeptides. Both the state of charge and the bulkiness of the R group are important factors. Thus, for example, polyalanine (R is small and neutral) forms an α-helix, as does polylysine at pH 12 (but not at pH 7, when the R groups are positively charged) and polyglutamic acid at pH 2 (but not at pH 7, when the R groups are negatively charged). Furthermore, polyisoleucine (with a large R group next to the α-carbon) does not form an α-helix, whilst the extreme behaviour of polyproline or polyhydroxyproline is due to the ring nature of R and the inability to form intrachain hydrogen bonds. Thus proline (or hydroxyproline) units in a polypeptide chain cause a kink or bend, contributing to hairpin loops. In general alanine, leucine, phenylalanine, tryptophane, methionine, histidine, glutamic acid and valine favour α-helix formation, whereas tyrosine, cysteine, aspartic acid, serine, isoleucine, threonine, glutamine, asparagine, lysine, arginine and glycine all destabilise an α-helix.

The content of α-helix in an overall polypeptide conformation may be assessed by comparing its optical rotation before and after denaturation in solution. The α-helical structure contributes to optical rotation (it has a non-superimposable mirror image), the reduction in dextrorotatory power on denaturation is a measure of α-helical content. α-Helix formation may also be detected by a hyperchromic shift on the far ultraviolet absorption of the peptide group. Furthermore, bands characteristic of the α-helix may be observable in the CD or ORD spectrum. Also the frequencies of the N—H and C=O stretch and N—H deformation bands in the infrared region, and more especially their dichroism, are dependent on α- or β-conformation. Raman spectroscopy is also diagnostic of α- or β-conformation.

The α-helix to random coil transition has also been widely studied, especially by carbon-13 and proton NMR. In general, non-hydrogen bonding solvents (such as chloroform, trifluoroethanol, hexafluoropropylene) support α-helices, whereas hydrogen bonding solvents (such as trichloro- and trifluoroacetic acids) favour the random coil. Theoretical predictions of the allowed helix formation can be made by consideration of a Ramachandran plot.

γ-HELIX   A regular helical conformation proposed as a possible stable conformation for polypeptides, having a rise of only 0·99 Å per residue, a pitch of 5·03 Å and 5–14 peptides per turn. However, it is not known to occur in practice.

π-HELIX   A regular helical conformation for polypeptides with 4·40 peptides per turn and a rise of 1·15 Å per residue. However, it is not known to occur in practice.

ω-HELIX   A helical conformation found in a few synthetic polypeptides, notably poly-(β-benzyl-L-aspartate) annealed at about 160°C. It contains four peptide residues per turn (it is a 4_1 helix) and, unusually for a polypeptide based on an L-amino acid, it is a left-handed helix. The residue repeat (rise per peptide) is 1·32 Å, considerably smaller than the 1·50 Å of the α-helix. The angles of rotation are $\phi = +38°$ and $\psi = +78°$.

HELIX–COIL TRANSITION   (Helix–random coil transition) The change from an α-helix conformation to a random coil, observed in solutions of polypeptides. It has been widely studied for synthetic poly-(α-amino acids), especially for poly-(γ-benzyl-α-L-glutamate) and poly-(β-benzyl-L-aspartate) as well as for homopolymers of the standard amino acids, e.g. poly-(L-alanine). The transition may be brought about by a change in solvent polarity or hydrogen bonding ability, pH or temperature. The ease with which the transition occurs gives a useful indication of the stability of the α-helix. A widely studied system is the solvent system chloroform–dichloroacetic acid–trifluoroacetic acid. Deuterated forms of these solvents are particularly suitable for study by ORD or NMR.

Typical results show an order of stability of α-helices of poly-L-alanine > poly-L-leucine > poly-L-methionine > poly-γ-benzyl-L-glutamate > poly-γ-methyl-L-glutamate > poly-β-benzyl-L-aspartate. The effects of incorporation of a second L-residue, a D-residue or ionic residues have all been investigated. The transition is pH sensitive if the polypeptide contains acidic or basic amino acid residues. The effect of pH has been widely studied in poly-L-glutamic acid (a coil is found as pH is raised to pH 5 due to charged COO⁻ group repulsion) and in poly-α-L-lysine, where a coil to helix transition occurs at a pH of about 10 in water as the pH is raised. This is due to a proton being

lost from the NH_3^+ group. However, there is some doubt as to how random the coil is.

HELIX–RANDOM COIL TRANSITION   Alternative name for helix–coil transition.

HEMA   Abbreviation for 2-hydroxyethyl methacrylate.

HEMI-ACETAL   A compound or group, of general structure

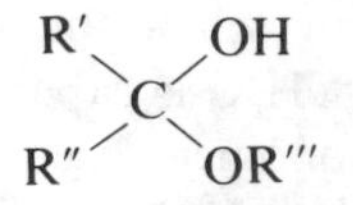

formed by reaction of an aldehyde or ketone

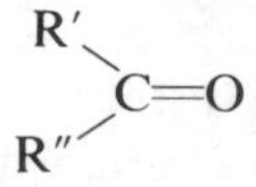

with an alcohol R‴OH. Of special importance in carbohydrates when, with both the hydroxyl and carbonyl groups in the same molecule, hemi-acetal formation occurs with ring closure to form a cyclic lactol:

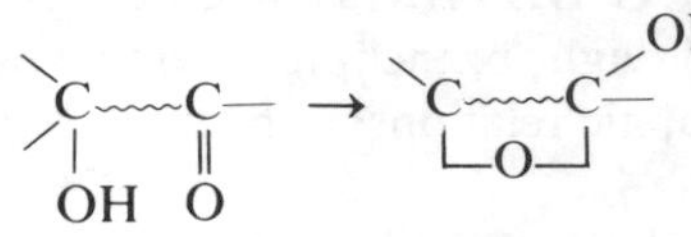

In this way hexoses and pentoses can exist in either open-chain or ring forms. Generally ring formation occurs to give either five- or six-member rings (furanose and pyranose respectively)

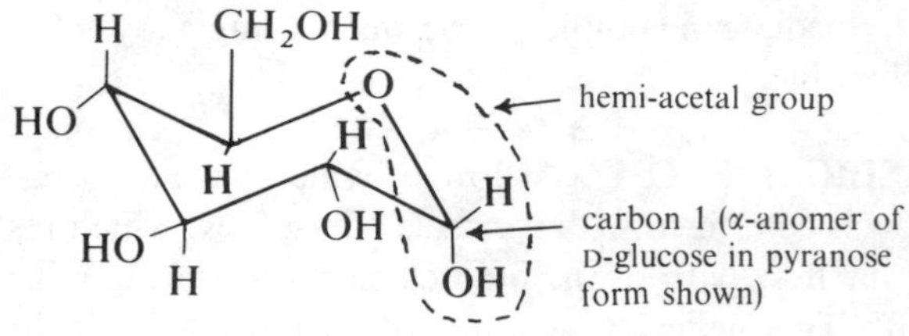

containing the hemi-acetal group, e.g. in D-glucose. Formation of the lactol results in the production of an asymmetric (chiral) carbon atom at carbon-1 (the anomeric carbon), thus giving rise to the possibility of α- or β-anomers.

HEMICELLULOSE   Originally those polysaccharides extractable from plants by aqueous alkali. These occur in close association with cellulose and were, incorrectly, thought to be cellulose precursors. The term is now used to describe land plant cell-wall polysaccharides, except cellulose and pectin. They are classified according to the sugar units present. Homopolysaccharides such as D-xylan, D-mannan and D-galactan occur only in small proportions, most hemicellulose being heteroglycans such as L-arabino-D-xylan, L-arabino-D-glucurano-D-xylan, D-gluco-D-mannan, D-galacto-D-gluco-D-mannan and L-arabino-D-galactan. They have branched structures of degree of polymerisation 50–200 and are sometimes O-methylated or acetylated. They are extracted from plant tissue after removal of the lignin (often with chlorous acid) by aqueous alkali.

HEMP A natural bast fibre from the plant *Cannabis sativa*. A dark, strong and durable fibre consisting of cells 1–2 cm in length, aggregated into strands of up to 200 cm long. It consists of about 80% cellulose with 10% water and some lignin, pectin and wax. It is widely used for ropes, twine and sacking.

HENCKY STRAIN Alternative name for natural strain.

HENEQUEN A leaf fibre from the plant *Agave fourcroydes*, very similar to sisal. It consists of strands about 150 cm long and is used, in particular, for agricultural twine.

HEPARIN A mucopolysaccharide found in several animal tissues, e.g. liver, lung, spleen, having a powerful blood anticoagulant activity. It contains alternating 1,4′-linked units of α-D-glucuronic acid and α-D-galactosamine, which are heavily sulphated at both oxygen and nitrogen. L-Iduronic acid units are also present. It occurs naturally as a glycoprotein, the polysaccharide amino acid link involving O-glycosides of serine.

β-HEPARIN Alternative name for dermatan sulphate.

HEPARIN MONOSULPHATE A type of N-acetyl-heparan sulphate.

HEPARITIN SULPHATE A type of N-acetyl-heparan sulphate.

n-HEPTANE

$$CH_3(CH_2)_5CH_3 \qquad \text{B.p. } 98.2°C.$$

A hydrocarbon solvent useful for the polymerisation of hydrocarbon monomers in solution.

7-HEPTANOLACTAM Alternative name for enantholactam.

HEPTYLNONYL ADIPATE (Dialphanyl adipate) (Dialphanol adipate) (DAA) B.p. 224°C/10 mm. The adipic acid esters of mixed isomers of seven- and nine-carbon alcohols, consisting mainly of linear heptylnonyl adipate.

$$CH_3(CH_2)_6OOC(CH_2)_4COO(CH_2)_8CH_3$$

but with diheptyl and some dinonyl adipates and with some branched isomers. A useful plasticiser for polyvinyl chloride and its copolymers, some cellulose esters and some synthetic rubbers, for good low temperature flexibility.

HEPTYLNONYL PHTHALATE (Dialphanyl phthalate) (Dialphanol phthalate) (DAP) (DA$_{79}$P) (Phthalate 79) B.p. 235°C/6 mm. Nominally

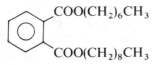

but commercial samples contain a mixture of linear and branched alcohols as well as containing diheptyl and dinonyl isomers. An important general purpose plasticiser for polyvinyl chloride and its copolymers, similar in action to di-2-ethylhexyl phthalate. Also used in polar rubbers such as nitrile and chloroprene rubbers.

HERCLOR Tradename for epichlorhydrin rubber.

HERCOLYN Tradename for methyl dihydroabietate.

HERCULON Tradename for polypropylene fibre.

HERMANN'S ORIENTATION FUNCTION Alternative name for orientation function.

HET ACID Abbreviation and tradename for hexachloroendomethylenetetrahydrophthalic acid. Alternative name for chlorendic acid.

HET ANHYDRIDE Abbreviation and tradename for hexachloroendomethylenetetrahydrophthalic anhydride. Alternative name for chlorendic anhydride.

HETEROATOMIC POLYMER Alternative name for heterochain polymer.

HETEROCHAIN POLYMER (Heteroatomic polymer) (Heteropolymer) A polymer which contains more than one type of atom in the polymer chain. The most common type is a polymer with carbon as the main atomic type in the chain, with oxygen (as in polyesters and polyethers), nitrogen (as in polyamides and polyurethanes) and sulphur (as in polysulphides and polysulphones) heteroatoms. However, a wide variety of inorganic polymers, such as the polysiloxanes (silicon and oxygen atoms), phosphonitrilic polymers (phosphorus and nitrogen atoms), polyborazoles (boron and nitrogen atoms) and sulphur–nitride polymers are all heterochain polymers.

HETEROCYCLIC POLYMER A polymer containing a heterocyclic ring in the polymer chain. The term therefore covers a very wide range of polymers—with nitrogen (e.g. polytriazoles, polytriazines, polybenzimidazoles, polyimides, polyquinoxalines, polypyrazoles), with nitrogen and oxygen (e.g. polyoxadiazoles, polybenzoxazoles, polybenzoxazinones), or with nitrogen and sulphur (e.g. polythiazoles, polythiadiazoles), as the heteroatoms. Often the rings are aromatic in character and the polymers have good high temperature resistance, having both high T_m values or softening points and good thermal stability. They are often synthesised by ring closure reactions such as cyclopolymerisation.

HETEROGENEOUS NUCLEATION Primary nucleation in the presence of a foreign surface which enhances nucleation by reducing the critical size needed for crystal growth. The surface may be an adventitious impurity, a localised surface, e.g. of the containing vessel, added nucleating agent or polymer crystals that have survived melting or dissolution (self-nucleation). Usually the number of nuclei increases with decreasing temperature as more potential nuclei are brought into play. Supercoolings required for crystallisation are less than for homogeneous nucleation, being 10–30 K.

HETEROGLYCAN Alternative name for heteropolysaccharide.

HETEROPOLYMER (1) A polymer containing more than one type of repeating unit, i.e. any polymer that is not a homopolymer. The term is therefore synonymous with copolymer, however use of this latter term is often restricted to synthetic polymers, the term heteropolymer being used instead for biopolymers.
(2) Alternative name for heterochain polymer.

HETEROPOLYMERISATION (1) Chain copolymerisation in which one of the monomers is a normal vinyl, or other doubly bonded monomer, whilst the other is of a non-vinyl type, such as SO_2, CO_2 or O_2, which will not homopolymerise itself but which will enter into copolymerisation. Sometimes the term also covers copolymerisations in which the second monomer is a carbon–carbon doubly bonded monomer but does not readily homopolymerise, such as maleic anhydride.
(2) A physically heterogeneous polymerisation system, such as suspension, emulsion, dispersion, or precipitation polymerisation.

HETEROPOLYSACCHARIDE A polysaccharide in which more than one type of monosaccharide (or sugar) unit is present.

HETEROTACTIC TRIAD A triad in a polymer exhibiting stereochemical configurational isomerism, in which the configuration of the central repeat unit is the same as one of its neighbours but opposite to that of the other neighbour. In a vinyl polymer, $\{CH_2CHX\}_n$, such a triad may be represented by a Fischer projection formula as:

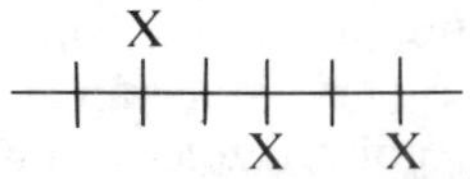

It is thus composed of one *meso* (m) dyad and one racemic (r) dyad, being either rm or mr.

HETEROTROPIC ENZYME An allosteric enzyme in which the effector substance is other than the substrate.

HETRON Tradename for an unsaturated polyester resin produced using some halogenated comonomer for fire resistance.

HEVEA BRASILIENSIS The tree, cultivated in tropical climates, whose latex, obtained by tapping the bark, is the sole useful source of natural rubber. Trees giving high yields of latex are propagated by bud grafting. The trees derived from the same mother tree are known as clones.

HEVEACRUMB Natural rubber in the form of a fine crumb, produced by incorporation of about 0·5% castor oil either into the latex before coagulation or into the wet coagulated rubber.

HEVEA LATEX Alternative name for natural rubber latex.

HEVEAPLUS MG Tradename for a natural rubber–methyl methacrylate graft copolymer, obtained by polymerisation of methyl methacrylate in the presence of a natural rubber latex. Typically contains 30–50% methyl methacrylate, which confers increased modulus and the ability to stick two unlike surfaces together.

HEVEAPLUS SG Tradename for a natural rubber–styrene graft copolymer.

HEVEA RUBBER Natural rubber obtained from the latex of the tree *Hevea Brasiliensis*, as is nearly all natural rubber.

HEX Abbreviation for hexachlorocyclopentadiene.

HEXA Alternative name for hexamethylenetetramine.

1,4,5,6,7,7-HEXACHLOROBICYCLO-(2,2,1)-HEPT-5-ENE-2,3-DICARBOXYLIC ACID Alternative name for chlorendic acid.

1,4,5,6,7,7-HEXACHLOROBICYCLO-(2,2,1)-HEPT-5-ENE-2,3-DICARBOXYLIC ANHYDRIDE Alternative name for chlorendic anhydride.

HEXACHLOROCYCLOPENTADIENE (Hex)

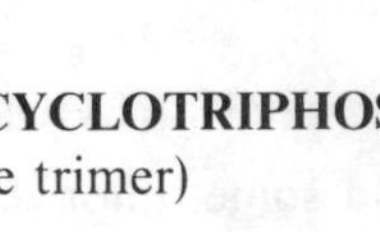

B.p. 234°C.

Synthesised by the chlorination of pentanes. Reacts by Diels–Alder addition to maleic anhydride to give the useful fire retardant monomers chlorendic anhydride and chlorendic acid.

HEXACHLOROCYCLOTRIPHOSPHAZENE (Phosphonitrilic chloride trimer)

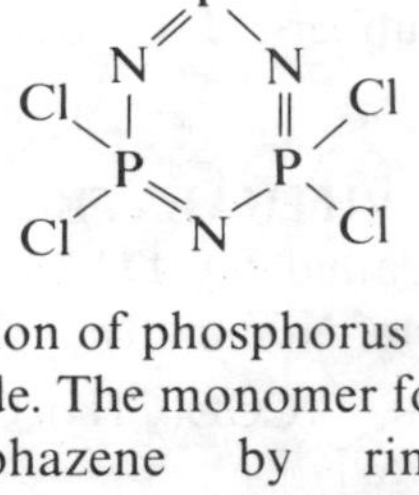

Produced by reaction of phosphorus pentachloride with ammonium chloride. The monomer for the formation of polydichlorophosphazene by ring-opening polymerisation on heating at about 300°C. In addition, crosslinking and formation of macrocycles occur.

HEXACHLOROENDOMETHYLENETETRAHYDROPHTHALIC ACID (HET acid) Alternative name for chlorendic acid.

HEXACHLOROENDOMETHYLENETETRAHYDROPHTHALIC ANHYDRIDE (HET anhydride) Alternative name for chlorendic anhydride.

HEXADECANOIC ACID Alternative name for palmitic acid.

1,4-HEXADIENE

$$CH_2=CHCH_2CH=CHCH_3$$

B.p. 72·5°C.

Sometimes used as the termonomer in ethylene–propylene rubber.

HEXAFLUOROISOBUTENE–VINYLIDENE FLUORIDE COPOLYMER Alternative name for hexafluoroisobutylene–vinylidene fluoride copolymer.

HEXAFLUOROISOBUTYLENE–VINYLIDENE FLUORIDE COPOLYMER Tradename: CM-1. An alternating copolymer of the two monomers, therefore having a structure

$$\left[CH_2CF_2CH_2\underset{\underset{CF_3}{|}}{\overset{\overset{CF_3}{|}}{C}}- \right]_n$$

which crystallises and is a thermoplastic rather similar to polytetrafluoroethylene. However, it has an even higher temperature of use and, apart from poor notched impact strength, it has better mechanical properties, especially creep resistance, than other fluoropolymer plastics. It requires very high melt processing temperatures (350–400°C) and is swollen by some organic solvents.

HEXAFLUOROPROPYLENE

$$CF_2=CFCF_3$$

B.p. −29·4°C.

Obtained as a by-product in the preparation of tetrafluoroethylene by pyrolysis of chlorodifluoroethane, or by passing tetrafluoroethylene over platinum at 1400°C. The monomer for the formation of copolymers, e.g. with vinylidene fluoride, as useful rubbers by free radical polymerisation.

HEXAFLUOROPROPYLENE–VINYLIDENE FLUORIDE COPOLYMER Alternative name for vinylidene fluoride–hexafluoropropylene copolymer.

HEXAHYDROPHTHALIC ANHYDRIDE

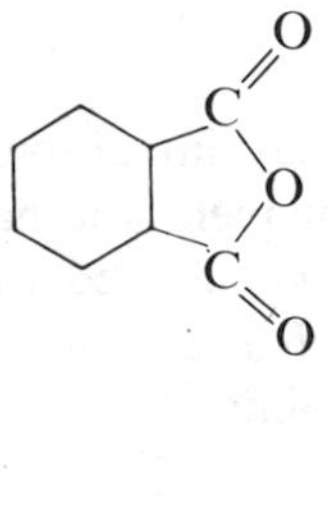

M.p. 35°C.

Useful as a curing agent for epoxy resins, giving light coloured products with good electrical properties to 150°C and high heat distortion temperatures (to 130°C) on curing, which usually also needs a catalyst.

HEXAMETHYLDISILOXANE

$$(CH_3)_3SiOSi(CH_3)_3$$ B.p. 100°C.

Produced by the hydrolysis of trimethylchlorosilane. Useful as a chain stopper for limiting the molecular weight of silane fluids during their production by the equilibration of the mixtures of linear and cyclic siloxanes formed by the hydrolysis of dichlorosilanes.

HEXAMETHYLENEDIAMINE (HMD) (1,6-Diaminohexane) (1,6-Hexanediamine)

$$H_2N(CH_2)_6NH_2$$

M.p. 42°C. B.p. 204–205°C, 90–92°C/14 mm.

Synthesised from adipic acid by reaction with ammonia to give adiponitrile, which is then hydrogenated. Alternatively adiponitrile may be obtained by electrolytic hydrodimerisation of acrylonitrile, by reaction of butadiene with hydrogen cyanide, or from reaction of 1,4-dichlorobutane with sodium cyanide. The dichlorobutane is obtained from 1,4-butane diol or from reaction of tetrahydrofuran with HCl. The monomer used for the production of several important dyadic nylons, including nylons 66, 69, 610, 612 and MXD6. These are often made by formation of an intermediate 1,1-nylon salt.

HEXAMETHYLENEDIAMMONIUM ADIPATE (Nylon 66 salt)

$$\overset{+}{H_3}N(CH_2)_6\overset{+}{N}H_3$$
$$^-OOC(CH_2)_4COO^-$$

M.p. 190–191°C.

A 1:1 adduct salt of hexamethylenediamine and adipic acid, formed by mixing solutions of these monomers at ambient temperature. It exists in equilibrium with a small amount of free amine and carboxylic acid groups. Used, instead of the separate diamine and dicarboxylic acid monomers, for the formation of nylon 66, since exact stoichiometry of the reactive groups is assured, as is necessary for high molecular weight polyamide formation.

HEXAMETHYLENEDIAMMONIUM SEBACATE (Nylon 610 salt)

$$\overset{+}{H_3}N(CH_2)_6\overset{+}{N}H_3$$
$$^-OOC(CH_2)_8COO^-$$

M.p. 170°C.

A 1:1 adduct salt of hexamethylenediamine and sebacic acid, formed and used for the production of nylon 610 in the same way as hexamethylenediammonium adipate is used for nylon 66.

HEXAMETHYLENE DIISOCYANATE (HDI)

$$OCN(CH_2)_6NCO$$

B.p. 140–142°C/21 mm.

One of the first isocyanates to be used in the formation of polyurethanes, e.g. as the fibre Perlon U. It is still used, but only occasionally due to its high volatility and associated toxic hazard. It is less reactive than tolylene diisocyanate and MDI and more toxic. A trimer is formed by reaction of three moles of HDI with one mole of water and is much less toxic due to its lower volatility; it is used in the formation of polyurethane coatings. It is sometimes also used when polyurethanes with good resistance to discolouration by oxidation are required. It is synthesised by reaction of hexamethylenediamine with phosgene.

HEXAMETHYLENETETRAMINE (HMTA) (HMT) (Hexa) (Hexamine)

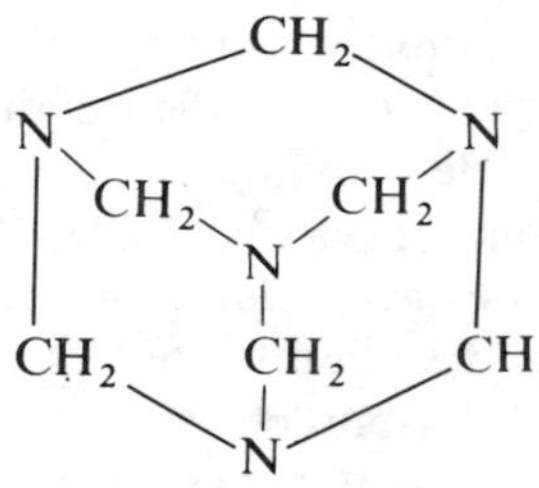

Prepared by the reaction of ammonia with aqueous formaldehyde solution at 20–30°C:

$$4NH_3 + 6CH_2O \rightarrow (CH_2)_6N_4 + 6H_2O$$

A crystalline solid which sublimes with decomposition at 260–270°C. Aqueous solutions are hydrolysed by the reverse of the above reaction at elevated temperatures. The usual curing agent in the crosslinking of novolacs, but also sometimes used to replace the ammonia catalyst in resole formation and as a secondary accelerator in rubber sulphur vulcanisation.

HEXAMETHYLETHER OF HEXAMETHYLOL-MELAMINE (HMEHMM)

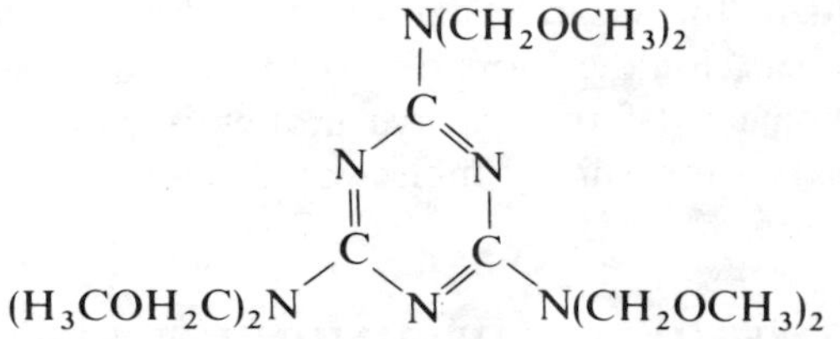

M.p. 55°C.

Formed by reaction of melamine with formaldehyde (in a 1:9 melamine to formaldehyde ratio) in aqueous solution, followed by reaction with acidified methanol. Commercial products may be liquid and contain some condensed molecules and some unreacted amino hydrogens. It is useful, in combination with dimethylolethyleneurea, in non-chlorine retentive cellulose textile finishes and, due to its solubility in a wide range of organic solvents, as a crosslinking agent in alkyd, epoxy and other resins.

HEXAMETHYLPHOSPHORAMIDE (Hexamethylphosphoric triamide) (HMP) (HMPA)

$$[(CH_3)_2N\text{---}]_3 P\!=\!O$$ 　　B.p. 232°C.

A powerful aprotic solvent, especially useful for polyamides, and for the preparation of these and other aromatic polymers.

HEXAMETHYLPHOSPHORIC TRIAMIDE
Alternative name for hexamethylphosphoramide.

HEXAMINE
Alternative name for hexamethylenetetramine.

n-HEXANE
$$CH_3(CH_2)_4CH_3$$ 　　B.p. 68·6°C.

A hydrocarbon solvent useful for the polymerisation of olefins and in certain adhesives.

1,6-HEXANEDIAMINE
Alternative name for hexamethylenediamine.

1,2,6-HEXANETRIOL

$$HO(CH_2)_4CH(OH)CH_2OH$$

M.p. 32·8°C. B.p. 178°C.

Sometimes used as a comonomer to provide branching points in polyester polyols used in polyurethane formation.

6-HEXANOLACTAM
Alternative name for ε-caprolactam.

HEXENE-1
$$CH_2\!=\!CH(CH_2)_3CH_3$$ 　　B.p. 63·5°C.

A comonomer sometimes used for copolymerisation with ethylene to produce one type of linear low density polyethylene. It is produced by oligomerisation of ethylene using a Ziegler or other catalyst.

HEXENE-1–4(and 5)-METHYL-1,4-HEXADIENE COPOLYMER
Tradename Hexsyn. Produced by Ziegler–Natta copolymerisation, e.g. with $AlEt_3/TiCl_3$ catalyst, of a mixture of hexene-1 and 2–30% of 4- and 5-methyl-1,4-hexadienes. The latter monomer mixture is obtained by reaction of ethylene with isoprene:

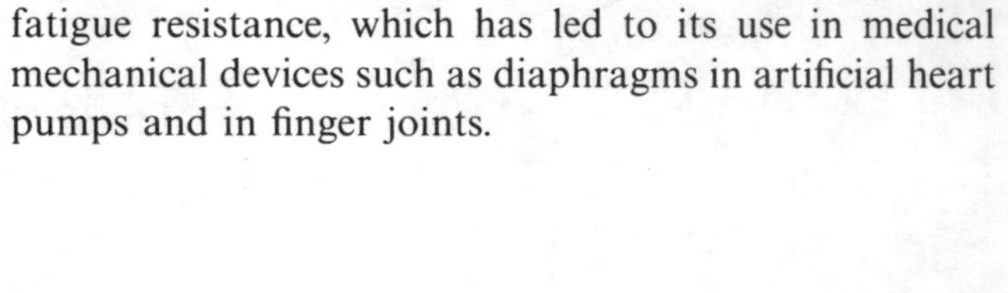

The polymer is somewhat similar to ethylene–propylene–diene monomer copolymer, but has a very high flex fatigue resistance, which has led to its use in medical mechanical devices such as diaphragms in artificial heart pumps and in finger joints.

HEXOSAN A polymer of a hexose. Thus glucans, mannans, galactans and fructans are all hexosans.

HEXOSE A monosaccharide containing six carbon atoms. Examples include the aldohexoses D-glucose, D-mannose, D- and L-galactose, the ketohexose D-fructose and the deoxy-hexoses L-fucose and L-rhamnose.

HEXSYN Tradename for hexene-1–4(and 5)-methyl-1,4-hexadiene copolymer.

HEXURONIC ACID A uronic acid containing six carbon atoms, i.e. $OHC(CH_2OH)_4COOH$.

H-FILM Alternative, older, tradename for the polyimide obtained by reaction of pyromellitic dianhydride and 4,4'-diaminodiphenyl ether, in film form. Also known as Kapton or Kapton H.

HI-BLEN Tradename for acrylonitrile–butadiene–styrene copolymer.

HIFAX Tradename for high density polyethylene.

HIGH ABRASION FURNACE BLACK (HAF) One of the most widely used types of carbon black of particle size about 30 nm, giving medium to high reinforcement with processability improving and modulus increasing in compounds with increasing structure. High structure black (HAF-HS) is a standard choice for tyre treads and other grades are useful for belt, sidewall and carcasses and for mechanical extruded goods.

HIGH CIS-POLYBUTADIENE A butadiene containing about 97% cis-1,4-butadiene structures. An important butadiene rubber, obtained by Ziegler–Natta polymerisation using, for example, $AlEt_2Cl/CoCl_2$ catalyst.

HIGH COLOUR CHANNEL BLACK (HCC) A type of channel carbon black of particle size about 10 nm and of the highest tinting power.

HIGH DENSITY LIPOPROTEIN The blood serum lipoprotein fraction obtained by density gradient ultracentrifugation having a density of $1.06–1.21 g cm^{-3}$ and containing about 50% of the protein. On electrophoresis this fraction migrates with the α-globulins and is therefore also an α_1-lipoprotein. It contains two different polypeptide chains of molecular weights 17 500 and 28 000. Like other blood lipoproteins the polypeptide chains cover the surface of the roughly spherical lipid particle, so conferring hydrophilicity and enabling lipid transport to occur.

HIGH DENSITY POLYETHYLENE (HDPE) Tradenames: Ameripol, Carolona, Daplen, Eltex, Eraclene HD, Fortiflex, Fortilene, Hifax, Hostalen, Ladene, Lacqtene HD, Lupolen, Marlex, Rexene, Rigidex, Rotene, Stamylan HD, Petrothene, Staflex, Super Dylan, Vestolen A. Polyethylene of density $0.940–0.960 g cm^{-3}$. It is produced by a low pressure process either by Ziegler–Natta polymerisation or by the Phillips or Standard Oil process and therefore is sometimes also called low pressure polyethylene. It is the most linear type and the most molecularly regular of all commercial polyethylenes and thus has the highest crystallinity, melting range and stiffness. Its major use is in blown bottles and other containers, but large quantities are also used for film, pipe and injection mouldings.

HIGH ELASTICITY (1) Elasticity in which very large strains are possible, as in rubber elasticity.
(2) Alternative name for delayed or retarded elasticity.

HIGH ENERGY RADIATION INDUCED DEGRADATION Alternative name for radiation induced degradation.

HIGHER NYLON A monadic nylon whose repeat unit is larger than that of the most common monadic nylon, nylon 6. Thus a higher nylon is one of the group of nylons, nylon 7, 8, 9, ..., etc. The commonest examples are nylons 11 and 12.

HIGH IMPACT POLYSTYRENE (HIPS) (Impact polystyrene) Tradenames: Dylene, Lustrex, Styron, Superflex. Polystyrene modified by the incorporation of rubber particles to improve the impact strength. Much polystyrene is produced in this form as a plastic material since unmodified polystyrene is rather brittle. It is usually produced by the mass polymerisation of styrene containing dissolved rubber. Products obtained merely by milling together rubber and polystyrene or by coagulation of mixed latices of these two polymers have inferior impact properties. As the rubber content increases so does the impact strength but correspondingly the softening point and stiffness decrease and the products are no longer transparent. Up to about 20% rubber may be used, at which level the impact strength is about $3.5 J (12.7 mm)^{-1}$ on the Izod test. The rubber should have a low T_g value, polybutadiene often being preferred, and it is usually somewhat crosslinked. Some grafting of the rubber to the polystyrene occurs during the polymerisation and this increases interfacial adhesion and hence impact strength. The rubber is dispersed as particles about $1–10 \mu m$ in diameter but containing inclusions of polystyrene particles within them. This effectively increases the rubber volume fraction. The study of particle morphology by osmium staining and transmission electron microscopy is a classical application of these techniques to synthetic polymers. Reinforcement by the rubber is largely due to the rubber particles promoting crazing or shear yielding in the polystyrene matrix on impact loading.

HIGH NITRILE POLYMER (Nitrile resin) Tradenames: Barex, Cyclsoft, Cycopac, Lopac, Soltan S. A styrene–acrylonitrile or other monomer–acrylonitrile copolymer with a high acrylonitrile content of about 70–80%. Such copolymers have a low permeability to gases, including carbon dioxide, and liquids and are therefore known as barrier polymers. Their potential usefulness as packaging polymers, especially for bottles for carbonated drinks, is limited by the toxic hazard associated with residual acrylonitrile monomer. To achieve higher softening points, the acrylonitrile may be partially or wholly replaced by methacrylonitrile, as in Lopac. Other commercial products are Barex and Cycopac.

HIGH-*ORTHO* NOVOLAC A novolac prepared in the pH range 4–6, using the salt of a divalent metal, especially zinc acetate, as a catalyst. This gives a novolac with a high proportion of methylene bridges between phenol nuclei, with linking through the *ortho* positions. The polymers have higher curing rates than conventional novolacs, with hexamethylenetetramine due to the free *para* position having a higher reactivity with formaldehyde.

HIGH POLYMER A polymer with a high degree of polymerisation (DP) and hence of high molecular weight. 'High' is often interpreted as meaning a sufficiently high DP so that the effects of end groups may be ignored. In addition, many of the physical properties, e.g. T_m, T_g and moduli (but not melt or solution viscosity or strength properties which continue to increase with DP even at very high DP), do not alter significantly with further increase in DP. Typically this means a polymer with a DP of more than about 100.

HIGH PRESSURE POLYETHYLENE Polyethylene produced by the original free radical polymerisation process conducted at very high pressures—hundreds to a few thousand atmospheres pressure. Normally referred to as low density polyethylene.

HIGH RESOLUTION NUCLEAR MAGNETIC RESONANCE SPECTROSCOPY Nuclear magnetic resonance spectroscopy in which the slight differences in the resonance frequencies (or fields) of the resonating nuclei (due to their different chemical bonding and electronic environments) are resolved. This is the most common type of NMR since it is of great value in the structural elucidation of organic molecules, including polymers. Classical NMR is proton NMR, but more recently ^{13}C NMR has proved to be of increasing value. ^{19}F NMR is useful with fluoropolymers.

The variation in resonance position is due to the differential shielding of the magnetic field by the electronic environment. Thus the protons of hydrocarbons resonate at higher fields than those near electronegative groups such as —OR and —COOR. These variations are the chemical shifts, whose position is related to the arbitrarily assigned zero reference position of the protons in tetramethylsilane. For all types of protons the total range of shifts is about 10 ppm, but for ^{13}C atoms it is about 600 ppm—hence the major advantage of ^{13}C NMR. The resolution is proportional to the field strength. With the increasing field strengths available, the corresponding radio frequencies employed give rise to 60 MHz, 100 MHz, 220 MHz and 300 MHz instruments.

Samples are examined as dilute (about 10%) solutions in deuterated solvents (for proton NMR). Spectra are complicated and made more difficult to interpret (but can also provide more information) due to the coupling of the spins of neighbouring nuclei (spin–spin coupling). This causes a splitting (spin–spin splitting) of the resonance peaks of increasing complexity (multiplicity) with an increase in the number of spin states in the coupled nuclei. The spacing between the split peaks and strength of coupling is given by the coupling constant (symbol J). Spin–spin decoupling, employing the double resonance technique, can be used to simplify the spectrum.

In polymer work the technique is uniquely valuable in the determination of tacticity by analysis of the triad or even longer configurational sequences, especially using ^{13}C NMR, and also in the determination of copolymer sequence lengths, especially in polymers of the type —CH_2CX_2—, which have no asymmetric centres. Geometric isomerism in diene polymers and chain branching in vinyl polymers have also been investigated in detail, again using ^{13}C NMR.

HIGH SPEED OSMOMETER A modern commercial osmometer which requires only minutes to about half an hour to enable an equilibrium osmotic pressure reading to be obtained. This contrasts with the period of hours to weeks which the traditional capillary osmometer requires and which this type has now largely replaced. These osmometers sometimes operate in a dynamic equilibrium mode by detecting electrically incipient solvent flow through the membrane. By means of a servomechanism the flow is countered by application of a hydrostatic head of pressure of solvent. Some models operate in a static mode, where solvent flow is balanced by the deflection of an attached diaphragm, which also acts as a sensor for pressure, e.g. by means of an attached strain gauge. In either case very little transport of solvent through the membrane is necessary to achieve equilibrium.

HIGH STYRENE RESIN A random copolymer of styrene and butadiene containing much more styrene (50–85%) than in the more common styrene–butadiene rubber copolymers. It is prepared in a similar way to this copolymer by emulsion polymerisation. The copolymers are somewhat rubbery plastic materials with a T_g value of about 50°C or below. They are useful for reinforcing rubbers by increasing the stiffness. The BDS-K resin materials are also of this type but are stiffer and have higher softening points.

HIGH TEMPERATURE CELLULOSE Alternative name for cellulose IV.

HIGH TEMPERATURE RESISTANT POLYMER A polymer which retains its desirable properties at high temperatures and which does not deteriorate in use at these temperatures. This usually implies a polymer with a high softening point (and therefore high T_g, if amorphous, or high T_m, if crystalline) and with good thermal, or more importantly, good thermo-oxidative stability. Temperatures at which a polymer is said to be high temperature resistant are widely interpreted. Often temperatures as low as 150°C (at least for use in air) are deemed high, whereas others regard only those temperatures above 250°C as high.

A wide variety of polymers have been synthesised and studied for their high temperature resistance, often for potential uses as fibres or electrical insulation materials, but also as plastics moulding materials and adhesives. Only a few high temperature resistant rubbers are known, e.g. silicones and phosphazenes. Among the different structural types, ladder, spiro, inorganic and organo-metallic polymers often have good high temperature resistance but have various defects. The aromatic and heterocyclic polymers have been the most successful and several are commercial materials (e.g. polyimides, polybenzimidazoles, aromatic polyamides, polysulphones).

HIGH TENACITY RAYON Tradenames: Cordura, Durafil, Fortisan, Rayflex, Tenasco. A highly oriented rayon produced by any of several modified rayon spinning processes. The modifications include the use of high zinc sulphate contents or of a quaternary ammonium compound to delay coagulation in the spinning bath and stretching the fibre in hot water or dilute acid during spinning. The fibres, typically, have a tenacity of about 3.5 g denier^{-1}, low elongation at break of about 10% and low water absorption compared with normal viscose rayon. They are mainly used for tyre reinforcement. Later developments produced 'super' high tenacity rayons with the use of quaternary ammonium compounds or polyethylene oxide in the spinning bath. They have tenacities of about 4 g denier^{-1} (dry) and 2.8 g denier^{-1} (wet). Fortisan is a regenerated cellulose from cellulose acetate.

HINDERED AMINE LIGHT STABILISER (HALS) An efficient photostabiliser which is especially useful as a weathering stabiliser in polypropylene, and acts in a similar manner to an antioxidant, although the mechanisms of action are very complex. Many examples are based on hindered piperidine compounds, such as bis-(2,2,6,6-tetramethyl-4-piperidyl)sebacate.

HINDERED PHENOL A phenol containing bulky substituents (e.g. *t*-butyl groups) *ortho* to the hydroxy group. The most common type of phenolic antioxidant, the substituents preventing the phenol itself from being readily oxidised by steric hindrance. Typical examples are 2,6-di-*t*-butyl-4-methylphenol and the bisphenols, e.g. 2,2'-methylenebis-(4-methyl-6-*t*-butylphenol).

HIPS Abbreviation for high impact polystyrene.

HIS Abbreviation for histidine.

HISIL Tradename for precipitated silica.

HISTIDINE (His) (H)

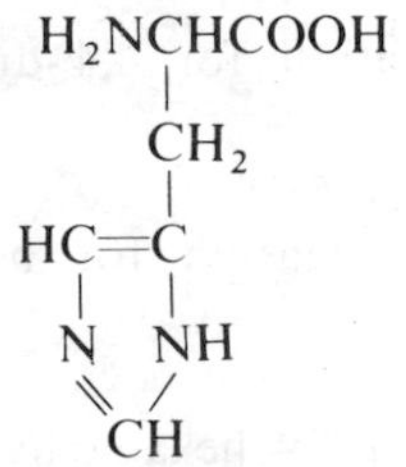

A basic amino acid found widely in proteins, in especially large amounts in haemoglobin. Its pK' values are 1.82, 6.00 and 9.17, with the isoelectric point at 7.59.

HISTONE A low molecular weight (10 000–20 000) basic protein, rich in arginine and lysine, occurring as a nucleoprotein associated with the DNA of the chromosomes of the nuclei of eukaryotic cells through ionic electrostatic forces, the histone neutralising the charge on the DNA phosphate groups. There are five main types of histone differing in their lysine and arginine contents. They are F1 (or HI) with 27% lysine, 15% glycine, 207 residues and a molecular weight of 21 000, f2a1 (H4) with 14% arginine and a molecular weight of 11 000, f2a2 (HIIb1 or H2A) of molecular weight 15 000 with a lysine content of 11% and an arginine content of 9%, f2b (HIIb2 or H2B) of molecular weight 13 800 and a lysine content of 16% and f3 (HIII or H3) of molecular weight 15 300, an arginine content of 15% and a lysine content of 10%. Each one contains all the common amino acids, but is low in sulphur and aromatic amino acids. Further, some residues may be modified, e.g. methylated or acetylated lysine or phosphorylated serine.

The sequences in many histones are known; generally basic sequences tend to be concentrated at one end, with non-polar and acidic sequences in other regions. Remarkably, for some histones the sequence structure is very similar for histones from widely different species. In the nucleoprotein, the histones are probably bound to the DNA regularly in the deep groove of the DNA helix. They may influence supercoiling of the DNA and are probably involved in controlling protein synthesis, e.g. as specific gene repressors of protein formation.

HITCO Tradename for carbon fibre.

HI-TEX Tradename for polyacrylonitrile based carbon fibre.

HM-50 Tradename for an aromatic polyetheramide high modulus fibre, similar to Kevlar, having a repeat unit structure of:

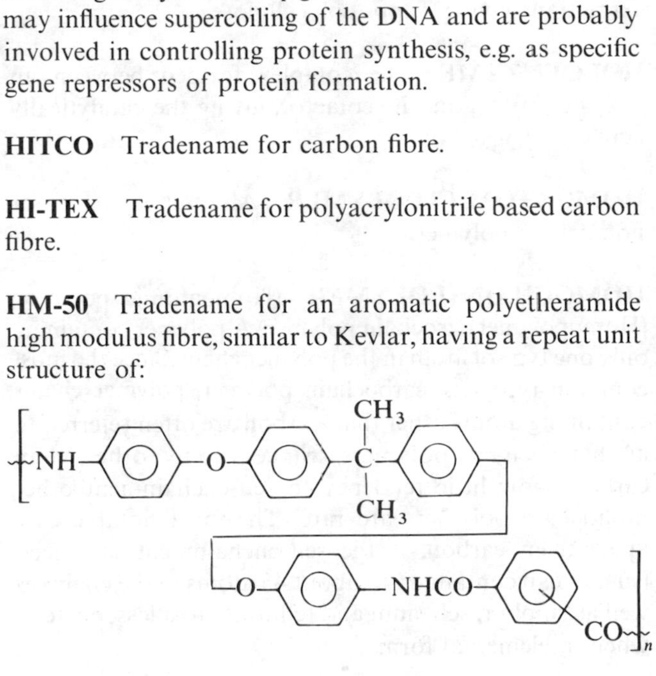

HMD Abbreviation for hexamethylenediamine.

H₁₂MDI Abbreviation for 4,4'-dicyclohexylmethane diisocyanate.

HMEHMM Abbreviation for hexamethylether of hexamethylolmelamine.

HMP Abbreviation for hexamethylphosphoramide.

HMPA Abbreviation for hexamethylphosphoramide.

HMT Abbreviation for hexamethylenetetramine.

HMTA Abbreviation for hexamethylenetetramine.

HOLE THEORY A theory which derives thermodynamic relationships for the glass transition temperatures of polymeric liquids and glasses, based on a lattice model, the lattice being partially occupied by chain segments and partially by holes. The theory therefore has some similarities to the Gibbs–DiMarzio theory. The T_g is considered as an isoconfigurational transition with the number of holes and chain conformations being frozen. This leads to a prediction for the pressure dependence of the T_g which is confirmed by some experiment measurements.

HOLLOW PYRAMIDAL CRYSTAL Alternative name for pyramidal crystal.

HOLOCELLULOSE The mixture of cellulose and hemicelluloses obtained after removal of the lignin from land plant tissues. The hemicelluloses may subsequently be removed by aqueous alkali extraction. Several methods of lignin removal (delignification) are used in the laboratory, including repeated treatment with sodium chlorite giving chlorite holocellulose. A standard method involves alternating chlorinations and methanolamine extractions giving Tappi holocellulose.

HOLOENZYME The complex formed between an enzyme protein and its cofactor, giving the catalytically active enzyme.

HOMOATOMIC POLYMER Alternative name for homochain polymer.

HOMOCHAIN POLYMER (Homoatomic polymer) (Homopolymer) (Isochain polymer) A polymer containing only one type of atom in the polymer chain. Since the most common type is a carbochain polymer, polymer chains containing atoms other than carbon are often referred to as heterochain polymers. Since atoms other than carbon show little tendency to self-enchainment other homochain polymers are rare. The most notable case apart from carbon, is the self-enchainment of silicon (silanes can contain up to about 45 atoms in the chain) as well as sulphur, selenium and tellurium to a lesser extent, when in elemental form.

HOMOGENEOUS MATERIAL A material whose properties, from the point of view of elasticity theory, are independent of position within the material.

HOMOGENEOUS NUCLEATION Primary nucleation with no preformed nuclei or crystalline surfaces present. It is a result of the spontaneous aggregation of polymer chains at supercoolings of about 50 K. The nucleus volume is about 10^4 Å^3 and thus involves either chain folding of part of a single polymer molecule or aggregation of smaller parts of several molecules in a fringed-micelle manner.

HOMOGENEOUS STRAIN A deformation in which the displacement of a point within a body is a linear function of the coordinates so that all the e_{ij} components in the strain tensor are constant.

HOMOGENEOUS STRESS A stress which has the same value and direction, with respect to a given plane, at any point within a body.

HOMOGLYCAN Alternative name for homopolysaccharide.

HOMOLOGOUS PROTEIN A protein whose biological function is the same or similar to that of the protein under consideration, but from a different species. Thus haemoglobins from man, horse and pig are homologous proteins. The homologous proteins of cytochrome C have been particularly intensively studied and relationships established between their sequence homology and taxonomy and position in the phylogenetic tree of the species concerned.

HOMOPOLYMER (1) (Unipolymer) A polymer containing only one type of repeat unit, as distinct from a copolymer. Its structure can therefore be most simply represented by $\text{-}[\text{M}]_n\text{-}$. Many important synthetic polymers useful as plastics materials are homopolymers of the vinyl type $\text{-}[\text{CH}_2\text{CHX}]_n\text{-}$. Some biopolymers, mostly polysaccharides, are also homopolymers (cellulose and starch are examples). Homopolymers are produced by polymerisation of a single monomer in chain polymerisation, or by an AB type polymerisation in step-growth polymerisation. Although AABB step-growth polymerisation uses two different monomers, the resultant polymers are considered to be homopolymers of repeat unit —AABB—, e.g.

$$\text{-}[\text{NH}(\text{CH}_2)_x\text{NHCO}(\text{CH}_2)_y\text{CO}]\text{-}$$

in a dyadic nylon, rather than as alternating copolymers of separate repeating units —AA— and —BB—, e.g.

$$\text{-}[\text{NH}(\text{CH}_2)_x\text{NH}]\text{-} \quad \text{and} \quad \text{-}[\text{CO}(\text{CH}_2)_y\text{CO}]\text{-}$$

in the above example.

(2) Alternative name for homoatomic polymer.

HOMOPOLYPEPTIDE

$$-\left[NHCHRCO\right]_n$$

A poly-α-amino acid with only one type of amino acid residue as the repeating unit.

HOMOPOLYSACCHARIDE (Homoglycan) A polysaccharide with only one type of monosaccharide (or sugar) unit present, e.g. a glucan, mannan, etc.

HOMOTROPIC ENZYME An allosteric enzyme in which the effector is also the substrate.

HOOKEAN ELASTICITY Elastic behaviour which obeys Hooke's law.

HOOKE'S LAW In its simplest form the law states that stress is linearly proportional to strain. Thus in the well-known case of simple extension, the law gives the Young's modulus as the constant of proportionality, i.e. the constitutive relation is: stress $= E \times$ strain. A more generalised form of the law states that any stress component is linearly related to all the strain components. The law is obeyed by many materials at relatively small strains (say up to a few per cent), such as by metals. However, only relatively few polymers obey the law and then only at very small strains, typically of less than 1%.

HOOP STRESS The tangential stress (σ_θ) at the wall of a thin walled pipe (diameter D and wall thickness h), resulting from an internal pressure (P) and given by the Barlow formula: $\sigma_\theta = PD/2h$.

HORDEIN The prolamine protein component of barley.

HORMONE A substance released by certain glands in vertebrates and carried by the blood to certain target organs where it regulates the metabolic processes occurring. Many hormones are polypeptides, some only of peptide size (e.g. oxytocin, vassopressin, bradykinin) whilst others, of molecular weight up to about 35 000, can be considered as proteins. The amino acid sequences of many hormones have been determined, notably that of insulin, which was the first protein whose sequence was resolved and is one of the most widely studied. Other well-studied protein hormones are adrenocorticotropin, glucagon, thyrotropin, somatotropin, prolactin and parathyroid hormone. The biological action depends on the tertiary structure, as often in proteins, but may only require a part of the chain sequence of amino acids.

HOSTADUR Tradename for a polyethylene terephthalate moulding material.

HOSTADUR B Tradename for polybutyleneterephthalate.

HOSTADUR E Tradename for polyethyleneterephthalate.

HOSTAFLON C Tradename for polychlorotrifluoroethylene.

HOSTAFLON ET Tradename for tetrafluoroethylene–ethylene copolymer.

HOSTAFLON TF Tradename for polytetrafluoroethylene.

HOSTAFLON TFA Tradename for tetrafluoroethylene—perfluoroalkylvinyl ether copolymer.

HOSTAFORM Tradename for polyoxymethylene copolymer.

HOSTALEN Tradename for high density polyethylene.

HOSTALEN PP Tradename for polypropylene.

HOSTALIT Tradename for polyvinyl chloride.

HOSTAMID Tradename for polynorbornanamide.

HOSTAMID LP700 Tradename for a transparent polyamide produced from terephthalic acid, the bisaminomethylnorbornene

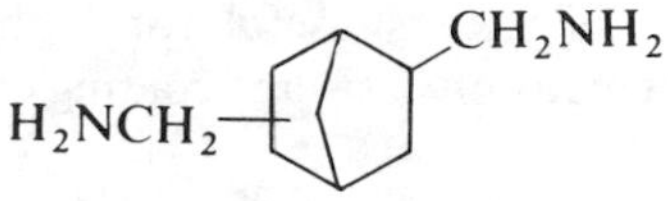

and caprolactam as comonomers.

HOSTAPHAN Tradename for a polyethylene terephthalate film material.

HOSTAPOR Tradename for expanded polystyrene.

HOSTAPREN Tradename for chlorinated polyethylene.

HOSTATEC Tradename for a polyetherketone.

HOSTYREN Tradename for polystyrene.

HOSTYREN XS Tradename for acrylonitrile–ethylene/propylene rubber–styrene copolymer.

HOT MELT ADHESIVE An adhesive, based on a thermoplastic polymer, which is applied to the adherend in the melt state when hot and subsequently solidifies on cooling. Such adhesives are attractive because of their ease of application, but as they are thermoplastic and are not crosslinked they are susceptible to creep and, in general, give low strength joints. Polyvinyl formal, polyvinyl butyral and polyamides are of this type.

HOT RADICAL A free radical which is thermally excited to a higher energy state than the ground state and has greater chemical activity than normal. Such radicals are formed in exothermic reactions, such as free radical chain polymerisation. Their presence may explain occasional unusual effects, e.g. that of an inhibitor (Q) becoming copolymerised:

$$\sim\!\!\sim\!\!M\cdot + Q \rightarrow \sim\!\!\sim\!\!MQ\cdot \,(\text{hot}) \xrightarrow{M} \sim\!\!\sim\!\!MQ\!-\!M\cdot \xrightarrow{M} \text{etc.}$$

This is thought to occur with benzoquinone in styrene polymerisations.

HOT RUBBER A rubber, usually styrene–butadiene rubber, produced by emulsion polymerisation at about 50°C using a peroxide or persulphate initiator.

HOT STAGE A thermostatted heated metal block mounted on a microscope stage so that a sample may be viewed microscopically whilst being heated or held at an elevated temperature. Widely used for observing polymer melting behaviour and spherulite development and growth rates.

HOT ZONE MECHANISM A mode of solid state polymerisation, when polymerisation is radiation induced, whereby the radiation excites zones of monomer molecules to short-lived active species, which can then polymerise without the necessity of an activation energy.

H-RESIN Tradename for crosslinkable, branched polyphenylene oligomers useful for high temperature resistant and corrosion resistant coatings.

HT-I Early tradename for poly-(m-phenylene-isophthalamide), later called Nomex.

HUGGINS CONSTANT Symbol k'. A constant of the Huggins equation having a value dependent on the particular polymer/solvent/temperature combination under consideration. Usually it has a value in the range 0·3–0·4, the value increasing with increasing solvent power.

HUGGINS EQUATION An empirical equation expressing the concentration (c) dependence of the viscosity number (η_{sp}/c) of a polymer in dilute solution. The relation is,

$$\eta_{sp}/c = [\eta] + k'[\eta]^2 c$$

where $[\eta]$ is the limiting viscosity number and k' is the Huggins constant for the particular polymer/solvent/temperature system being considered. Thus a plot of η_{sp}/c versus c is linear with intercept at $c = 0$ equal to $[\eta]$ and k' may be determined from the slope. This is often found experimentally, but sometimes an upward curvature occurs at high values of c, so that extrapolation to obtain $[\eta]$ can only be from the initial part of the curve.

HYALURONIC ACID A mucopolysaccharide occurring widely in connective tissue and synovial fluids of animals, probably as a glycoprotein with collagen, whose functions include lubrication and shock absorption in joints. A linear polymer of alternating 1,3′-linked β-D-N-acetyl-D-glucosamine and 1,4′-linked β-D-glucuronic acid units:

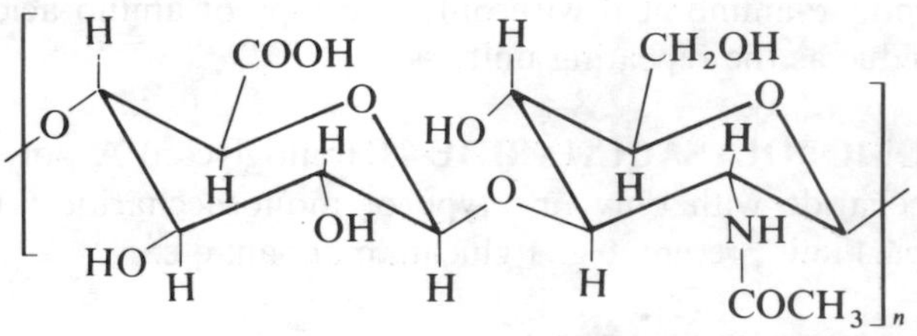

HYCAR Tradename for nitrile rubber and acrylic elastomers.

HYCAR CTB Tradename for carboxy-terminated polybutadiene.

HYCAR PA-21 Tradename for an early acrylic elastomer similar to Lactoprene EV.

HYDRATE CELLULOSE Alternative name for cellulose II produced by treatment of alkali cellulose with water.

HYDRAZINOLYSIS (Akabori method) A method for the determination of the C-terminal amino acid of a peptide or protein, by reaction with hydrazine. The C-terminal acid produces a free amino acid, whilst the other residues give hydrazides:

$$H_2NCHR'CO\cdots NHCHR''CO\cdots NHCHR'''COOH$$
$$+ H_2N\!-\!NH_2$$
$$\rightarrow H_2NCHR'CONHNH_2 + H_2NCHR''CONHNH_2$$
$$+ H_2NCR'''COOH$$

The free amino acid may be separated on a cation exchange resin and identified. Its determination thus provides a further method for the determination of the number of polypeptide chains in a protein.

HYDRIN Tradename for epichlorhydrin rubber.

HYDROCARBON RESIN One of a large group of low molecular weight (number average about 2000) polymeric hydrocarbons which range from viscous liquids to hard brittle solids. There are two types, the petroleum resins and the terpene resins. They are useful as tackifying agents in rubbers, adhesives and coatings.

HYDROCELLULOSE (Partially oxidised cellulose) The product of mild dilute acid hydrolysis of cellulose. Of similar structure to cellulose but of lower degree of polymerisation and still insoluble. The hemi-acetal groups formed on hydrolytic chain scission can react as aldehyde groups—hence the name partially oxidised. Their concentration may be determined as the copper number of the product.

HYDRODYNAMICALLY EQUIVALENT SPHERE A hypothetical spherical particle whose behaviour is equivalent to that of the non-spherical particles it represents, as far as the influence of the particles on the viscosity of a medium in which they are immersed, is concerned. Thus for a polymer in solution, existing as random coils, the coils may be replaced by the hydrodynamically equivalent spheres. This is necessary in the theoretical treatment of the viscosity of suspensions and solutions, e.g. in the Einstein equation, and is implicit in other theories of polymer solutions, e.g. in the derivation of the Flory–Fox equation.

HYDRODYNAMIC CHROMATOGRAPHY A technique for the determination of the particle size distribution in polymer latices. Separation according to particle size occurs on passage through a packed column. The latex particles must pass through the capillary-like paths between the packing particles. The larger latex particles travel with higher velocity since they are excluded from the capillary walls where axial velocities are small. Thus they are eluted first. The eluate is monitored by a detector and a plot is obtained of detector response versus retention time, as in gel permeation chromatography. This yields the particle size distribution after calibration. The carrier liquid is usually water under pressure and the packing is an ion-exchange resin.

HYDROGEL An insoluble slightly crosslinked polymer which is highly swollen by water of solvation. Hydrogels of poly-(2-hydroxyethylmethacrylate) are useful as soft contact lenses.

HYDROGEN BONDING INDEX Symbol γ. A measure of the ability of a material, usually a liquid, to hydrogen bond. It is based on the shift in the absorption peak of CH_3OD at $268\,cm^{-1}$ in the infrared spectrum, from its value in benzene to its value in the liquid concerned. It is defined as the shift in reciprocal centimetres per $10\,cm^{-1}$. Consideration of the value of γ, as well as that of the solubility parameter, can give a more reliable guide to the solubility of a polymer in the liquid, than consideration of the solubility parameter alone.

HYDROGEN TRANSFER POLYMERISATION A polymerisation which involves the migration of a hydrogen atom of the monomer during the formation of the polymer repeat unit. The best known example is during polyurethane formation by step-growth polymerisation, where a hydroxyl hydrogen becomes attached to nitrogen from the isocyanate group, thus forming the urethane group: $\sim\!OH + ONC\!\sim \rightarrow \sim\!O\!-\!CONH\!\sim$. Such a polymerisation is an example of a polyaddition reaction, rather than of a polycondensation, since unlike most step-growth polymerisations no small molecule is eliminated.

HYDROLYTIC DEGRADATION Degradation induced by the chemical attack of water on a polymer. Accelerated by acids, bases and elevated temperatures. Polymers containing hydrolysable groups in the main chain in each repeat unit suffer a rapid reduction in molecular weight on hydrolysis by a random scission process. Such polymers include polyesters, polyamides, polyurethanes and cellulose and its derivatives.

HYDROLYTIC POLYMERISATION Polymerisation initiated by a small amount of water. The best known example is the water initiated ring-opening polymerisation of lactams, especially of caprolactam.

HYDRON Tradename for poly-(2-hydroxyethylmethacrylate).

HYDROPHOBIC BOND (Hydrophobic interaction) The association of the non-polar (usually hydrocarbon) parts of an amphipathic molecule or collection of such molecules, to form non-polar regions, specifically excluding highly polar water. This exclusion is the main driving force for the formation of these interactions since it involves an increase in entropy. This is due to the excluded water forming an ordered structure around a hydrocarbon group, so that when the latter is transferred to an associated collection of such groups the excluded water is more disordered. Such interactions occur extensively in proteins (by association of the α-substituents) and in nucleic acids. In proteins hydrophobic bonds are thought to be the most important factor stabilising the tertiary structure of globular proteins in their native conformations. The action of denaturing agents, such as urea, is to reduce the hydrophobic interaction rather than to destroy hydrogen bonds.

HYDROPHOBIC CHROMATOGRAPHY Chromatography in which the interaction between the solid column matrix and the species to be separated is hydrophobic bonding. It is useful for the purification and separation of proteins, typically using an agarose matrix to which alkyl and aralkyl groups capable of hydrophobic bonding are attached. Elution of the bound species can be achieved by changing the ionic strength or hydrophobicity of the eluent.

HYDROPHOBIC INTERACTION Alternative name for hydrophobic bond.

HYDROQUINONE (Quinol) (1,4-Dihydroxybenzene)

$$HO\!-\!\langle O \rangle\!-\!OH$$

M.p. 170°C.

An inhibitor for free radical polymerisation, effective when oxygen is also present, because it becomes oxidised to *p*-benzoquinone.

p-HYDROXYBENZOIC ACID

$$HO\!-\!\langle O \rangle\!-\!COOH$$

M.p. 214°C.

The monomer for poly-(*p*-hydroxybenzoic acid). Decarboxylates on heating to 200°C, but can be polymerised by heating with trifluoroacetic anhydride at 75°C. Best polymerised by heating the diphenyl ester, when ester interchange, with elimination of phenol, occurs.

2-HYDROXYBENZOPHENONE A compound of the type

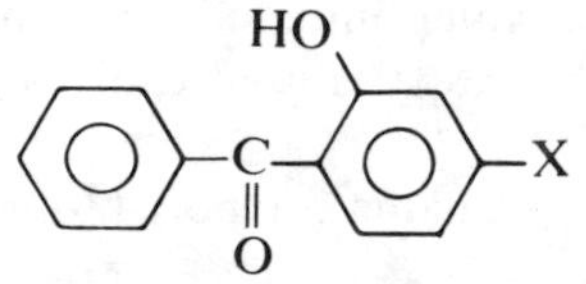

where the X substituent is frequently an alkyl or alkoxy group. Widely used as ultraviolet-absorbing-type ultraviolet stabilisers. Variation of X varies the wavelength of maximum absorption, which may be matched to the wavelength causing maximum damage to a particular polymer. The *ortho* placing of the hydroxy group causes it to be hydrogen bonded to the carbonyl oxygen. After absorption, the excited state excess energy is thought to be readily lost through rapid tautomeric shifts. The best known examples are the 2,4-dihydroxy-, 2-hydroxy-4-methoxy- and 2-hydroxy-4-*n*-octoxy-benzophenones.

2-HYDROXYBENZOTRIAZOLE Alternative name for 2-hydroxyphenylbenzotriazole.

12-HYDROXY-*CIS*-9-OCTADECENOIC ACID Alternative name for ricinoleic acid.

HYDROXYETHYLCELLULOSE A cellulose ether in which some cellulose hydroxyl groups are replaced by hydroxyethyl groups. Prepared by reaction of alkali cellulose with ethylene oxide, which can also form long polyoxyethylene chains by continued reaction with the hydroxyethyl groups:

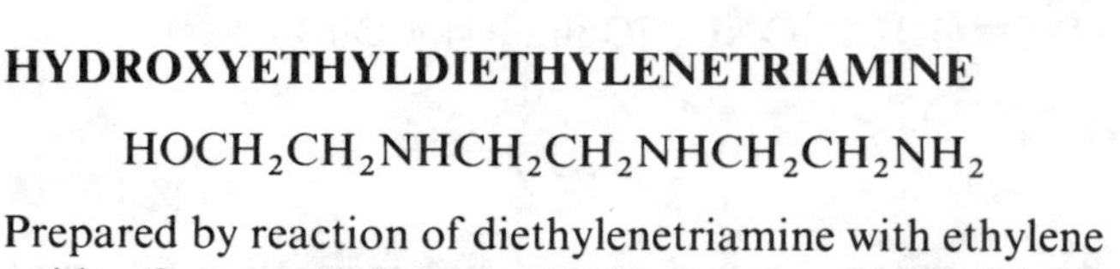

$$Cell-OH + CH_2-CH_2 \xrightarrow{NaOH}$$
$$O$$

$$Cell-OCH_2CH_2OH \rightarrow Cell-O{\left[CH_2CH_2O\right]}_xH$$

Water insoluble polymer of DS 0·2–0·5 is commercially produced for use as an alkali soluble textile and paper size, whereas polymers of DS about 1·5 are water soluble and are used as thickening agents and adhesives.

HYDROXYETHYLDIETHYLENETRIAMINE

$$HOCH_2CH_2NHCH_2CH_2NHCH_2CH_2NH_2$$

Prepared by reaction of diethylenetriamine with ethylene oxide. Commercial materials have about 15% of the dihydroxyethyl derivative. Useful for the fast room temperature curing of epoxy resins and less of an irritant than diethylenetriamine.

2-HYDROXYETHYL METHACRYLATE (HEMA)

$$CH_3$$
$$|$$
$$CH_2{=}CCOOCH_2CH_2OH$$

B.p. 95°C/10 mm.

Prepared by the addition of ethylene oxide to methacrylic acid. The monomer for the formation of poly-(2-hydroxyethylmethacrylate).

HYDROXYETHYLMETHYLCELLULOSE A cellulose ether in which some cellulose hydroxyl groups have been replaced by methoxyl and some by hydroxyethyl groups. Produced by reaction of alkali cellulose first with ethylene oxide and then with methyl chloride. The hydroxyethyl DS is usually 0·12–0·7 and the methoxyl DS is 1·5–2·0 in commercial products, which have better water solubility and dispersibility than methylcellulose and are useful thickening and binding agents.

HYDROXYL NUMBER The amount of potassium hydroxide, in milligrams, equivalent to the free hydroxyl groups, usually as end groups (as in a polyester) present in 1 g of a polymer. Determined by acetylation with acetic anhydride and titration of the excess anhydride with potassium hydroxide.

5-HYDROXYLYSINE (Hyl)

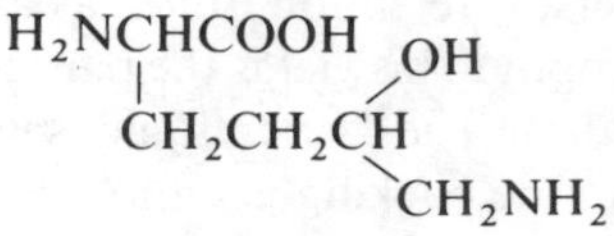

$$H_2NCHCOOH \quad OH$$
$$| \qquad\qquad |$$
$$CH_2CH_2CH$$
$$CH_2NH_2$$

A basic α-amino acid found in collagen.

4-HYDROXY-4-METHYLPENTAN-2-ONE Alternative name for diacetone alcohol.

HYDROXYMETHYLPHENOL Alternative name for methylolphenol.

o-HYDROXYMETHYLPHENOL Alternative name for saligenin.

2-(2-HYDROXY-4-METHYLPHENYL)-BENZO-TRIAZOLE Alternative name for Tinuvin P.

2-HYDROXYPHENYLBENZOTRIAZOLE (Benzotriazole) (2-Hydroxybenzotriazole) A compound of the type

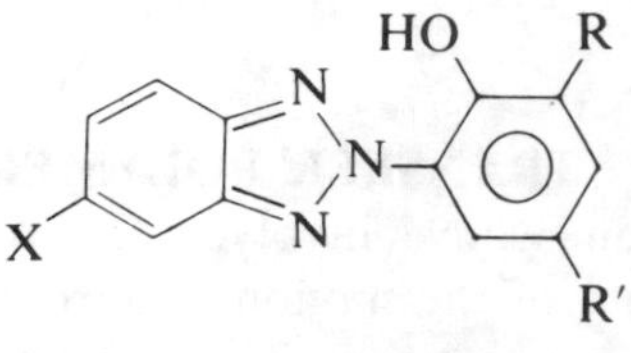

where X = H or Cl, R = H or alkyl and R′ = alkyl. Widely used as ultraviolet absorbers for ultraviolet stabilisation of polymers. Those with X = Cl absorb at longer wavelengths. Variation of R and R′ is used to achieve adequate compatibility with the polymer. A particular example is Tinuvin P where R = X = H and R′ = CH$_3$.

4-HYDROXYPROLINE (Hypro)

M.p. 273°C (decomposes).

A non-polar α-imino acid, although usually considered to be one of the α-amino acids, found in some proteins, notably collagen and elastin. Its pK' values are 1·92 and 9·73, with the isoelectric point at 5·83. Like proline it gives a yellow, rather than a purple, colour in the ninhydrin reaction.

HYDROXYPROPYLMETHYLCELLULOSE   A cellulose ether in which some cellulose hydroxyl groups have been replaced by hydroxypropyl and some by methyl groups. Several commercial products of DS 1·5–2·0 are made by reaction of alkali cellulose, first with propylene oxide and then with methyl chloride. The polymers are more easily dissolved in water than methylcellulose and are useful thickening and binding agents.

HYDROXY-TERMINATED POLYBUTADIENE
Tradenames: Butarez ETS, Hycar CTB, Nisso PBC. A low molecular weight telechelic polybutadiene with hydroxyl end groups. Prepared by anionic polymerisation of butadiene using difunctional initiator to produce a living polymer with two living ends. The chain ends are then reacted with ethylene oxide to give the hydroxyl end groups. Useful as a liquid rubber which may be crosslinked by use of a polyfunctional isocyanate.

HYDROXY-TERMINATED POLYISOBUTYLENE
Prepared from carboxy-terminated polyisobutylene, either by reduction or by reaction with propylene oxide:

$$\sim COOH + CH_2 \!-\! CHCH_3 \rightarrow \sim COCH_2CHOH$$

$$\overset{O}{\underset{}{}} \qquad \qquad \overset{\parallel}{O} \quad \overset{|}{CH_3}$$

The polymer may be crosslinked by reaction with isocyanates.

HYDROXYTOLUENE   Alternative name for cresol.

HYFIL   Tradename for polyacrylonitrile based carbon fibre.

HYL   Abbreviation for 5-hydroxylysine.

HYPALON   Tradename for chlorosulphonated polyethylene.

HYPERGLYCEMIC-GLYCOGENOLYTIC HORMONE   Alternative name for glucagon.

HYPRO   Abbreviation for 4-hydroxyproline.

HYSTERESIS   The dissipation of energy in a cyclic process. Thus in dynamic mechanical behaviour, the loss is a hysteresis effect, its magnitude being given by the loss modulus or compliance. In fatigue, repeated stressing cycles may cause excessive temperature rise due to hysteresis, resulting in failure either by thermal softening (in rigid thermoplastics) or by oxidative degradation (in rubbers). Hysteresis also results in many polymer fluids (especially thixotropic fluids)—for example, in a rotational viscometer experiment when the torque in plotted against shear rate at increasing shear rates, followed by decreasing shear rates, the two curves do not coincide but form a hysteresis loop.

HYSTYL   Tradename for a polybutadiene resin.

HYTREL   Tradename for a thermoplastic elastomer based on a polyether/ester block copolymer containing hard segments of crystallisable tetramethylene terephthalate and soft segments of polyoxytetramethylene terephthalate.

I

I   Symbol for isoleucine.

IDEAL COPOLYMER   Alternative name for random copolymer.

IDEAL COPOLYMERISATION   A copolymerisation in which, in the simple binary case, the product of the monomer reactivity ratios r_A and r_B for monomers A and B ($r_A r_B$) equals unity. Thus each type of growing centre $\sim\!\!A^*$ and $\sim\!\!B^*$ shows the same preference for adding one or the other monomers during propagation. This results in a completely random distribution of A and B units in the copolymer (a random or ideal copolymer). Most ionic copolymerisations are of this type, but many free radial copolymerisations show a tendency to alternating copolymerisation. For the special case of $r_A = r_B = 1$, the copolymer has the same composition as the monomer feed (azeotropic copolymerisation) at all feed compositions. When $r_A \neq r_B$, the copolymer will be richer in the more reactive monomer. As the difference between r_A and r_B increases, so it becomes more difficult to make copolymers containing appreciable amounts of both monomers.

IDEAL ELASTOMER   (Ideal rubber) An elastomer which on deformation at constant pressure, e.g. by stretching by application of a force f, satisfies the conditions: $(\partial H/\partial l)_{T,P} = 0$ and $f = -T(\partial S/\partial l)_{T,P}$, where l is the length, T is the temperature, P is the pressure, S is the entropy and H is the enthalpy. The name is given by analogy with the behaviour of an ideal gas, where $(\partial E/\partial V)_T = 0$ and $P = T(\partial S/\partial V)_T$, where E is the internal energy and V the volume. An assumption made in the development of the thermodynamic theory of rubber elasticity. Extension of an ideal elastomer thus merely involves a change in the entropy, i.e. the elasticity is purely entropic. In practice deviations from this ideal behaviour occur due to the energy which is expended in overcoming energy barriers to rotation on stretching, so that the internal energy depends on elongation and $f = (\partial E/\partial l)_{T,P} - T(\partial S/\partial l)_{T,P}$, i.e. the force depends on an energy as well as on an entropy term. However, this energy term is small except at very low elongations.

IDEAL FLUID   Alternative name for Newtonian fluid.

IDEAL PLASTIC   Alternative name for Bingham body.

IDEAL PLASTIC BEHAVIOUR   Alternative name for perfectly plastic behaviour.

IDEAL RELAXATION   A relaxation in which the rate of return to equilibrium on removal of the applied constraint is proportional to the displacement from equilibrium $(P - P_{eq})$. When the constraint is applied as a pulse, the decay is given by,

$$(P - P_{eq}) = (P - P_{eq})_0 \exp(-t/\tau)$$

where $(P - P_{eq})_0$ is the initial value of $(P - P_{eq})$, τ is the relaxation time and t is the time. For a sinusoidally varying constraint of frequency ω, $P^* = P' + iP''$, where $P' = P/(1 + \omega^2\tau^2)$ and $P'' = \omega\tau P/(1 + \omega^2\tau^2)$. Although some relaxation processes, e.g. low frequency dynamic mechanical relaxations, are ideal, non-ideal processes are also known. Examples are viscoelastic relaxation involving retardation, many dielectric relaxations and nuclear magnetic relaxation. The non-ideal behaviour is often formulated in terms of a distribution of relaxation times or non-exponential correlation functions.

IDEAL RUBBER   Alternative name for ideal elastomer.

IDEAL SOLID   Alternative name for Hookean solid.

IDEAL TEMPERATURE   Alternative name for theta temperature.

L-IDURONIC ACID

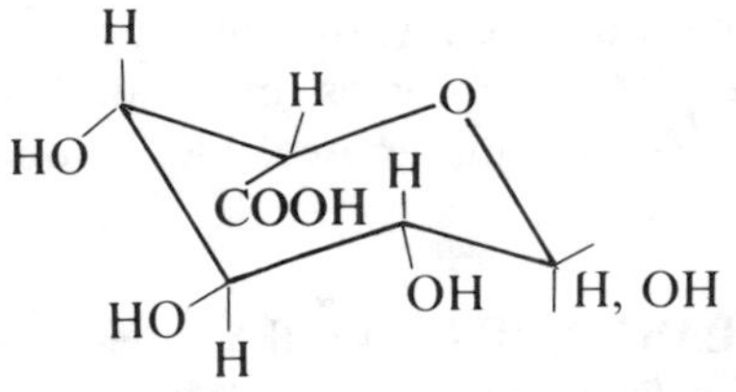

M.p. 131–132°C. α_D^{25} +33°.

A uronic acid, occurring as a monosaccharide component of dermatan sulphate.

IEN   Abbreviation for interpenetrating elastomeric network.

IGAMID U   Tradename for an early polyurethane plastic moulding material based on the reaction of 1,6-hexamethylene diisocyanate with 1,4-butanediol. Its properties are similar to nylon 66, but its T_m is lower at about 183°C. Later similar materials were designated Durethan U.

I-GUMMI   (I-Rubber) Tradename for an early polyurethane elastomer made by reaction of hydroxy-terminated polyester prepolymer and diisocyanates with a small amount of trimethylolpropane to give limited crosslinking.

IIR   Abbreviation for isobutene–isoprene rubber, i.e. butyl rubber.

ILE   Abbreviation for isoleucine.

IMIDEX   Tradename for an aromatic polyester–imide.

IMIDITE   Tradename for a polybenzimidazole based glass reinforced adhesive and laminating resin.

IMINE   Alternative name for cyclic amine.

α-IMINO ACID   An acid of general structure

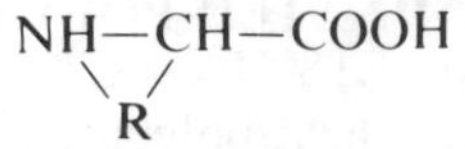

$$NH-CH-COOH$$
$$R$$

and therefore closely related to the α-amino acids. In effect an α-amino acid in which the R group is a substituent of both the α-carbon and the amino group. The two protein imino acids, proline and 4-hydroxyproline, are commonly referred to as α-amino acids. Other non-protein imino acids and their polymers, e.g. sarcosine and ornithine, are also of interest.

IMMUNE GLOBULIN   Alternative name for immunoglobulin.

IMMUNOELECTROPHORESIS   A technique for the separation of immunoglobulin proteins, as antibody–antigen mixtures, in which the mixture as a narrow band in agar gel is first subject to electrophoresis and then antiserum is placed in narrow channels cut in the gel parallel to the direction of electrophoretic migration. Diffusion of both antisera and antigen is then allowed to occur (double diffusion technique). This results in the formation of a series of bands, each representing a different antigen–antibody system.

IMMUNOGLOBULIN   (Antibody) (Immune globulin) A protein occurring in blood serum produced in response to the presence of a foreign body (an antigen) in order to render it harmless through formation of an antibody–antigen complex—the immune response. A glycoprotein with a few per cent carbohydrate content which migrates electrophoretically with the γ-globulins. Each antibody is specific for a particular antigen, containing complementary binding sites (usually two) and capable of forming a three-dimensional network (precipitin) with the antigen. Specificity is so great that even homologous protein antigens require different antibodies; however, when from closely related species a response may be obtained.

There are five classes of human immunoglobulins (IgG, IgA, IgM, IgD and IgE) IgG being the most important and the most widely studied. They have a molecular weight of about 150 000, containing four peptide chains, two heavy (or H chains) containing about 430 residues and two light (or L chains) with about 214 residues, linked through disulphide bridges into a Y-shaped structure. The carbohydrate (as oligosaccharide) is covalently bound to the H chain and consists of about 1% hexose (a varying ratio of galactose to mannose), hexosamine (1–1·5%), sialic acid (0·06–0·3%) and sometimes fucose (about 0·2%). In parts of each chain the sequences are the same in different antibodies, whereas differences occur nearer the N-terminal ends, which form the antigen binding sites. Since normal blood serum contains many different antibodies it is difficult to isolate a molecularly homogeneous sample. However, serum from certain cancerous patients produces large amounts of a single type of immunoglobulin and these have been widely studied. Such patients also produce excessive amounts of dimers of the L-chains, excreted in the urine—the Bence–Jones proteins.

IMPACT POLYSTYRENE   Alternative name for high impact polystyrene.

IMPACT STRENGTH   The ability of a material or object to withstand a sharp blow. It is usually expressed as the impact energy obtained from a particular impact test, i.e. as the energy absorbed by the object during fracture at a very high testing rate. The energy can be expressed in several forms: as the energy per unit cross-sectional area fractured, the energy per unit volume, the energy per unit width of object (or length of notch in a notched test specimen) or, most recently, energy per width of specimen (usually 12·7 mm).

Impact measurements are made by high speed tensile loading, of most interest in fundamental studies, by recording the fracture rate when a weight is allowed to fall on to a specimen, or by recording the energy loss of a weighted pendulum which strikes a specimen in the form of a bar. The latter is the most used test method, either in the Izod or Charpy configuration. However, although simple and rapid, the results are not suitable for design purposes, and really can only give a rough ranking of different materials or can be used for quality control purposes. This is because in general, but particularly in pendulum tests, the impact strength values obtained are sensitive to sample preparation method, sample dimensions and also to random variations from one specimen to another due to the presence of microcracks and other voids. Thus impact strength can hardly be described as a material property. In pendulum tests it is a combination of the energy needed to initiate fracture, the energy for crack propagation, the kinetic energy of the fragment flying off and possible energy losses to the testing machine. Frequently, notched samples are used in pendulum tests, which can reduce variability due to random flaws in samples.

The value of impact strength increases with increasing temperature, especially in the region of the glass transition temperature (T_g). Polymers with high impact strength below T_g have important secondary transitions. Orientation increases impact strength in the direction of orientation, but decreases it in the perpendicular direction. Since impact loading is multiaxial and samples break in the weakest direction, uniaxial orientation frequently lowers impact strength. Generally, crystallisation decreases impact strength, especially if large spherulites are formed. Impact strength may be considerably increased, especially in brittle polymers, by the addition of rubbery polymers as toughening agents. Typical values for the standard Izod impact strengths of polymers vary from about 0·15 J $(12·7 \, \text{mm})^{-1}$ for brittle polymers, such as polystyrene, to about 10 J $(12·7 \, \text{mm})^{-1}$ for really tough polymers, such as toughened PVC, polycarbonate and glass reinforced polyesters and epoxy resins.

IMPET   Tradename for polyethyleneterephthalate.

IMPLEX   Tradename for impact modified polymethylmethacrylate.

IMS   Abbreviation for industrial methylated spirits.

INCIDENT LIGHT POLARISATION MICROSCOPY   Optical microscopy in which the incident polarised light is split into two beams. One passes to the specimen mounted on a silvered surface, which reflects the light back. The other passes to the objective via an analyser set at 90° to the polariser. Thus the field appears dark. The specimen rotates the electric vector of the beam and hence some light passes the analyser. The specimen thus appears light against a dark background.

INCLUSION CELLULOSE   (Occlusion cellulose) Cellulose containing physically entrapped organic solvent rather like a clathrate complex; the solvent is not lost even on heating above its boiling point. Formed by swelling the cellulose in water and then drying by immersion in solvent and evaporation. Inclusion of solvent activates cellulose towards organic reagents especially for acetylation.

INCOHERENT SCATTERING   (Compton scattering) Scattering of radiation with no phase relationship between incident and scattered rays. The latter also consist of a band of wavelengths longer than the incident beam. It contributes to the diffuse background in X-ray diffraction patterns. It must be subtracted from the diffuse scattering from amorphous or highly disordered polymer, since only the coherent scattering gives useful structural information.

INDAPOL   Tradename for polybutene.

INDENE

   M.p. −2°C, B.p. 182°C.

Obtained from the coal tar naphtha fraction mixed with smaller amounts of coumarone. Such mixtures are polymerised by low temperature (about 0°C) treatment with concentrated sulphuric acid or a Friedel–Crafts catalyst, to coumarone–indene resins.

INDENE–COUMARONE RESIN   Alternative name for coumarone–indene resin.

INDENE RESIN   Alternative name for coumarone–indene resin.

INDEX NOTATION   (Suffix notation) A shorthand method of representing stress and strain components, coordinates of a point, etc., by using suffixes. Thus, for example, a stress designated σ_i, where i is the index which can have values 1, 2, or 3 (corresponding to the x, y and z coordinate directions) is a shorthand for σ_1, σ_2 and σ_3.

INDIAN HEMP   Alternative name for sunn.

INDIA RUBBER   Early name for natural rubber.

INDUCED DECOMPOSITION   (Transfer to initiator) Chain transfer in a free radical polymerisation by reaction of a growing active centre with an undissociated initiator molecule, especially with a peroxide:

$$\sim\!\!M_n^{\cdot} + RO\!\!-\!\!OR \rightarrow \sim\!\!M_n\!\!-\!\!OR + RO\cdot$$

where R is an alkyl or aryl group. This results in wastage of initiator, since a molecule of initiator is consumed without increasing the number of free radicals produced. However, this wastage is not taken account of in the definition of initiator efficiency. Generally azo initiators show little tendency to induced decomposition but in dialkyl and diacyl peroxides the effect is significant, especially at high initiator concentrations. Typical values of C_I, the initiator chain transfer constant, are 0·0001–0·1 (for peroxides with styrene); values for hydroperoxides are often even higher.

INDUCTION PERIOD   (Inhibition period) The time during which no observable change occurs in a chemical reaction or in a physical property. Subsequently followed by such a change. In polymer work, induction periods occur in degradation such that no apparent degradation occurs, e.g. oxygen absorption in an oxidative degradation, for a time even though the polymer is exposed to degradative conditions. Usually the result of incorporation of a stabiliser and frequently used as a measure of stabiliser effectiveness. In free radical polymerisation, the time during which no (or negligible) polymerisation occurs due to the presence of an inhibitor. Its magnitude is directly proportional to the amount of inhibitor present. By measurement of the variation in the length of the induction period, with inhibitor concentration, at constant initiator concentration, the rate of initiation of a polymerisation may be determined, provided that the number of radicals reacting with each inhibitor molecule is known.

INDUSTRIAL METHYLATED SPIRITS   (IMS) Ethanol which has been denatured with 5% methanol and also frequently contains up to 10% water.

INEFFECTIVE FIBRE LENGTH   The apparent length of a short fibre, or broken continuous fibre, in a fibre reinforced polymer composite over which no load transfer may be considered to occur near the end of the fibre. It arises from the fibre not carrying its maximum load over its transfer length.

INELASTIC FLUID   Alternative name for purely viscous fluid.

INELASTIC SCATTERING   The scattering of radiation by a medium in which the scattered radiation has a different wavelength to the incident radiation due to an energy interchange with the medium. When the interchange is due to the incident radiation causing molecular motions, these motions can be usefully studied using inelastic scattering. It is also the basis of Raman spectroscopy.

INFINITESSIMAL STRAIN ELASTICITY   Alternative name for small strain elasticity.

INFRARED DICHROISM   Dichroism (i.e. differential absorption in two directions at right angles) when using infrared radiation. This is the usual wavelength region for the determination of dichroism, since most polymers have suitable infrared absorption bands. Indeed dichroism is observed with nearly all infrared bands of an oriented polymer and is widely used for the study of chain conformation and morphology.

INHERENT VISCOSITY   Alternative name for logarithmic viscosity number.

INHIBITION   The prevention, or delay, of free radical polymerisation due to the presence of an inhibitor, which acts as a free radical scavenger. This results in a polymerisation induction period during which the inhibitor is consumed and after which polymerisation occurs, often at its normal rate. Inhibitor effectiveness is measured by the inhibition constant. Autoinhibition occurs when the monomer itself acts as the inhibitor.

INHIBITION CONSTANT   Symbol z. In a free radical polymerisation in the presence of an inhibitor, the ratio of the rate constant for reaction of inhibitor with a propagating free radical to that for simple propagation. A measure of the effectiveness of the inhibitor at suppressing polymerisation.

INHIBITION PERIOD   Alternative name for induction period.

INHIBITOR   A substance which reacts with initiating or propagating free radicals to form either non-radical products or radical products of low reactivity, and so prevents free radical polymerisation. It is an efficient radical scavenger which is consumed as it reacts during the polymerisation induction period, after which polymerisation occurs at the normal rate. A retarder is similar in kind but different in degree, merely slowing the polymerisation. Many products of reaction of inhibitors with radicals subsequently act as retarders, e.g. nitrobenzene. Inhibition may be caused by the presence of impurities, or an inhibitor may be deliberately added to prevent premature polymerisation. Many compounds act as inhibitors and/or retarders. Diphenylpicrylhydrazyl is very effective, forming non-radical products. However, usually an inhibitor forms radical products of low reactivity, e.g. the quinones such as p-benzoquinone. Dihydroxybenzenes such as hydroquinone and t-butylcatechol are also effective inhibitors in the presence of oxygen, due to their oxidation to quinones. Oxygen is also a powerful inhibitor, since in its reaction with free radical $R\cdot$, $R\cdot + O_2 \rightarrow ROO\cdot$, the resultant peroxy free radical is relatively stable, although it may re-initiate polymerisation.

INHOMOGENEITY Symbol U. A measure of the width of a molecular weight distribution, defined as

$$U = (\bar{M}_w / \bar{M}_n) - 1 = S_n^2 / \bar{M}_n^2$$

where $\bar{M}_w$ and $\bar{M}_n$ are the weight and number average molecular weights respectively and S_n is the standard deviation of the number distribution, given by

$$S_n = \left[\int_0^\infty (M - \bar{M}_n)^2 N(M) \, dM \right]^{1/2} = \bar{M}_w \bar{M}_n - \frac{1}{\bar{M}_n^2}$$

The parameter $\Delta = U^{1/2} = S_n / \bar{M}_n$ has also been used as an index of the width of a distribution.

INITIATING RADICAL Alternative name for primary radical.

INITIATION (Chain initiation) The first step in the formation of a growing polymer molecule in a chain polymerisation. Consists of the reaction between the active initiating species (I*) and a monomer molecule (M): $I^* + M \rightarrow I{-}M^*$. In free radical polymerisation I* is usually derived from an added initiator by its thermal homolysis, but may also arise from a redox reaction (redox initiation), by photochemical decomposition of initiator or monomer (photo-polymerisation), by high energy irradiation (radiation induced polymerisation) or heating of the monomer itself (thermal polymerisation). In ionic polymerisation the added initiator is often called the polymerisation catalyst, since it is often not consumed.

INITIATION FACTOR Alternative name for initiator efficiency.

INITIATOR A substance capable of causing the polymerisation of a monomer by a chain reaction mechanism (chain polymerisation). The initiator (I) or an active species (I*) produced from it by thermal or photo-activation or by chemical reaction, reacts with a monomer molecule (M) to produce an activated monomer: $I^* + M \rightarrow I{-}M^*$. This then rapidly adds on many other monomer molecules (propagation) through the active centres $\sim\!M^*$.

In ionic polymerisation I* is an ion and since the active centre often does not disappear, but is merely transferred to another molecule, the initiator is not consumed; it is often therefore simply called the polymerisation catalyst. The commonest initiators for cationic polymerisation are protons from strong acids, or the species produced by reaction of the polymerisation catalyst, often a Lewis acid (e.g. BF_3) and a co-catalyst (e.g. water):

$$BF_3 + H_2O \rightarrow H^+ BF_3 \bar{O}H$$

In anionic polymerisation typical initiators (or catalysts) are metal amides (e.g. $Na\overset{+}{N}\overset{-}{H}_2$), alkoxides ($M\overset{+}{O}\overset{-}{R}$), alkyls (e.g. lithium butyl), aryls (e.g. sodium naphthalene) and cyanides.

In free radical polymerisation a wide variety of initiators capable of forming free radicals (the primary radicals) are used. Thermal homolyses of diacyl peroxides (e.g. benzoyl peroxide), alkyl peroxides (e.g. di-t-butyl peroxide), hydroperoxides (e.g. cumene hydroperoxide), peresters (e.g. t-butylperbenzoate) and azo initiators are the most common. For example:

Diacyl peroxide:

$$\underset{\underset{O}{\|}}{RCO}{-}\underset{\underset{O}{\|}}{OCR} \rightarrow 2\,\underset{\underset{O}{\|}}{RCO}\cdot (\rightarrow 2\,R\cdot + 2\,CO_2)$$

Azo compound: $\quad R{-}N{=}N{-}R \rightarrow 2\,R\cdot + N_2$

These initiators may also be activated by ultraviolet light (photo-polymerisation). Redox reactions are also used, especially in aqueous systems for emulsion polymerisation, e.g. $H_2O_2 + Fe^{2+} \rightarrow \bar{O}H + HO\cdot + Fe^{3+}$. The activity of an initiator is important as it determines the temperature range over which it may be used. A useful guide is its half-life at the temperature of interest, or alternatively, the half-life temperature. Not all free radicals react with monomer, some are wasted, the wastage being given by the initiator efficiency.

INITIATOR EFFICIENCY (Initiation factor) Symbol f. The fraction of radicals produced from a free radical initiator which is successful in initiating a polymerisation chain. Some free radicals are wasted by recombination with each other rather than by reaction with monomer. When they arise from the same initiator molecule and are trapped within a solvent cage the reaction is geminate recombination (eqn. (1)). Reactions can also occur outside the cage, as in eqns. (2) and (3). Thus, for example, with benzoyl peroxide

$$\underset{\underset{O}{\|}}{PhCO}{-}\underset{\underset{O}{\|}}{OCPh} \rightarrow \left[2\,\underset{\underset{O}{\|}}{PhCO}\cdot \right]$$

$$\left[2\,\underset{\underset{O}{\|}}{PhCO}\cdot \right] \overset{\text{or}}{\rightarrow} \underset{\underset{O}{\|}}{PhCOPh} + CO_2 \qquad (1)$$

$$2\,\underset{\underset{O}{\|}}{PhCO}\cdot - \text{diffusion outside the cage}$$

$$\longrightarrow Ph\cdot + CO_2$$

$$Ph\cdot + \underset{\underset{O}{\|}}{PhCO}\cdot \rightarrow \underset{\underset{O}{\|}}{PhCOPh} \qquad (2)$$

$$2\,Ph\cdot \rightarrow Ph{-}Ph \qquad (3)$$

Thus f is always < 1, usually having values in the range 0·3–0·8. However, the reduction in initiator efficiency by induced decomposition is not taken into account in f values.

INNOVEX Tradename for linear low density polyethylene.

INORGANIC POLYMER A polymer in which all, or a very high proportion, of the chain atoms are not carbon atoms. Polymers which contain chain atoms as part of what are normally considered as organic groups, such as oxygen in polyesters and polyethers, nitrogen in polyamides and polyamines and sulphur in polysulphides and polysulphones, are considered as organic, not inorganic, polymers. Elements other than carbon do not show much tendency to catenation, so that although in principle a much greater variety of chain structures should be available in inorganic polymers, in reality the range is rather restricted. The only elements which are capable of forming homochain polymers are sulphur, selenium and tellurium, all of which can exist as polymer chains in elemental form. In a very broad sense most inorganic compounds could be considered to be polymeric, at least in the solid state, due to the very strong ionic lattice forces that operate between the atoms and molecules. However, only those materials in which the macromolecular bonding is covalent and therefore largely persists in solution and in the melt (as evidenced by high viscosity) are normally considered as polymeric.

Most inorganic polymers, in particular the extremely abundant naturally occurring silicate minerals, are highly crosslinked and are therefore hard, strong but brittle materials, highly insoluble and with high softening points. In addition, most inorganic polymer structures have good heat stability. These good high temperature properties led to an extensive search for synthetic inorganic polymers as useful high temperature resistant materials, which, it was hoped, would show the easy processing of organic polymers. Such hopes have not been fulfilled since the materials investigated proved to be difficult to synthesise and fabricate and often had very poor hydrolytic stability. With the exception of the silicones, and to a lesser extent phosphonitrilic polymers, very few synthetic inorganic polymers have achieved any commercial significance.

The properties of many groups of inorganic polymers are dominated by considerations of crosslink density. Linear polymers are often soft and rubbery; examples include the polymetaphosphates, polyphosphazenes, polysiloxanes, sulphur nitride polymers and polycarboranes. Crosslinked polymers with trifunctional crosslinking centres include the chalcogenide glasses, ultraphosphate glasses and the graphite form of boron nitride. Borate and borophosphate glasses contain some tri- and some tetrafunctional centres.

The most important polymers are those based on silica, in which the silicon atom provides tetrafunctional crosslinking sites. Important silicate materials are the silicate glasses (vitreous silica itself and the borosilicate glasses) and the crystalline silicate minerals, which include quartz, the feldspars, and the zeolites.

Intermediate in structure and properties between the inorganic and organic polymers is the wide range of organometallic polymers. These include the chelate polymers, ferrocene polymers, polymetalloxanes, polysiloxanes, polyorganometalloxanes and polymetallosiloxanes.

INSULIN One of the most widely studied hormones. A small protein of molecular weight 6000 consisting of two polypeptide chains, one of 21 residues (A chain) and one of 30 residues (B chain), with one intrachain and two interchain disulphide bridges. These maintain the three-dimensional tertiary structure essential for biological activity. Insulin was the first protein for which the complete amino acid sequence was determined. It is secreted by the pancreas and has several roles including glycogen and fatty acid synthesis and uptake of glucose and amino acids from the blood. Biologically it is synthesised via a single polypeptide chain precursor of 104–109 amino acid residues (preproinsulin) from which 23 residues are hydrolysed to give proinsulin which then undergoes proteolytic cleavage at two points to give insulin.

INTEGRAL DISTRIBUTION (Integral molecular weight distribution) Symbol $I(M_i)$. The continuous form of a cumulative distribution which describes the distribution of molecular sizes (i—usually expressed as the molecular weight M_i) of a polydisperse polymer sample according to the sum of the amounts of all species of sizes up to i (or M_i) as a function of i (or M_i). The amounts may be expressed as either the weights (or masses) or as the numbers of molecules, or as their fractions of the whole, giving the integral weight (or mass) and the integral number distributions respectively. For example,

$$I(M_i) = \int_0^{M_i} W(M)\,dM$$

for the weight distribution. Since these are continuous distributions they may be expressed mathematically by an analytical expression or graphically as a plot of $I(M_i)$ versus M_i.

INTEGRAL MOLECULAR WEIGHT DISTRIBUTION Alternative name for integral distribution.

INTEGRAL NUMBER DISTRIBUTION An integral distribution in which the amount of each polymer molecular species i present is expressed as the number of molecules, or the number of moles, or mole fraction, of i.

INTEGRAL REPRESENTATION OF VISCOELASTICITY The mathematical statement of the Boltzmann Superposition Principle describing linear viscoelastic behaviour.

INTEGRAL WEIGHT DISTRIBUTION An integral distribution in which the amount of each polymer molecular species i (of molecular weight M_i) is described according to its weight (or mass) or weight fraction.

INTENE Tradename for low *cis*-polybutadiene rubber.

INTERCHANGE REACTION The reaction of a terminal functional group of a step-growth polymer with an interunit linking functional group of another polymer molecule to give two new polymer molecules of different lengths. The reaction most frequently occurs at the high temperatures required for many melt polymerisations used for the preparation of step-growth polymers. In general the reaction may be represented as,

$$P_mXRXP_n + P_x\text{—}Y \rightarrow P_mXRXP_x + P_n\text{—}Y$$

where Y is the chain end functional group, X is the interunit functional group and P_n, P_x and P_m are long segments of polymer molecules. Equilibration of polymer molecules of different size distributions by interchange reactions leads to the formation of the most probable distribution of molecular sizes. Since the number of molecules does not change, the number average molecular weight is unaltered, but the other averages will be altered. Interchange between two different homopolymers of the same type, e.g. two nylons, will initially give a block copolymer, but full equilibration will eventually lead to random copolymer formation. Interchange occurs most readily in polyesters (transesterification) where it is useful for preparing polyesters from diester monomers. Interchange is less rapid in polyamides and may also occur in other step-growth polymers.

INTERCRYSTALLINE DISLOCATION (Interface dislocation) A dislocation which lies in the plane separating the adjacent fold surfaces of a bilayer single crystal, having a Burgers' vector which is a translation vector of the two-dimensional lattice of the chain folds. This model assumes that regular tight folding occurs at the crystal layer surfaces.

INTERFACE DISLOCATION Alternative name for intercrystalline dislocation.

INTERFACIAL POLARISATION (Space-charge polarisation) A dielectric polarisation which occurs in heterogeneous materials when the ratio of the relative permittivity to the DC conductivity varies throughout the material. It arises as a result of the trapped charge carriers at the interfaces distorting the applied electric field. Such processes are slow and have even longer relaxation times than orientation polarisations. Interfacial polarisation can arise owing to the presence of crystalline/amorphous boundaries in crystalline polymers, as a result of polarisation occurring at the electrodes or from the presence of conductive impurities such as water or metals. In this last case the effects are often termed Maxwell–Wagner effects.

INTERFACIAL POLYMERISATION A step-growth polymerisation of the AA—BB type, where polymerisation occurs at the interface between an aqueous solution of one monomer and an immiscible organic phase containing the second monomer. To be successful the polycondensation reaction chosen must be extremely rapid under mild, e.g. ambient temperature, conditions. Most of these polymerisations utilise the Schotten–Baumann reaction of a highly reactive diacid chloride, in the organic phase, with a monomer containing active hydrogen atoms. Thus polyamides result from reaction of diacid chlorides with diamines, polyesters from reaction of diacid chlorides with diols, polyurethanes from reaction of bischloroformates with diamines, polysulphonamides from reaction of disulphonyl chlorides with diamines and polycarbonates from reaction of phosgene with diamines. Polymerisation usually occurs just within the organic layer and the aqueous layer contains a base to neutralise the hydrochloric acid produced. The polymerisation is diffusion controlled and incoming monomer reacts so rapidly with polymer chain ends that the two different monomers are unlikely to diffuse all the way through the interface and so react with each other. Thus few new polymer chains are produced and very high molecular weight polymer is formed. There is no need for exact stoichiometric balance of the two monomers and monomer purity is not critical. However, for high molecular weight polymer the organic solvent must be chosen so that, preferably, only high molecular weight polymer precipitates; there must also be a favourable partition coefficient between the liquid phases of the water soluble monomer and high insolubility of water in the organic phase.

INTERFERENCE MICROSCOPY Optical microscopy in the commonest arrangement of which (Nomarski interference microscopy) the incident beam is split into two non-parallel oppositely polarised beams. The beams are reflected back from the specimen and its substrate respectively and recombined. The sample beam follows a shorter path length and the beams recombine destructively producing a dark image.

INTERFERENCE OPTICS An optical system used in an ultracentrifuge for the determination of the solute concentration at any point in the cell. A slit beam of light passes through a double sector cell, one compartment containing solvent, the other the polymer solution. Owing to the difference in optical path length between the two cells, an interference pattern is produced. This is photographed and the concentration is derived from the number of fringes over which the pattern has shifted. The method is more sensitive than the Schlieren method, enabling solutions of about 0·05% concentration to be used.

INTERLAMINAR SHEAR The shear produced by bending a beam of a composite material. In an interlaminar shear test the length to diameter ratio of the beam is kept short so that failure on bending is due to the surface shear stress exceeding the shear strength of the material. Thus the test is supposed to measure the interlaminar shear strength.

INTERLAMINAR SHEAR STRENGTH (Adhesion shear strength) The flexural strength of a composite measured by bending a short beam such that the length to depth ratio is small enough so that the tensile stress at half depth, i.e. in the neutral plane, is less than the tensile or compressive strength of the material, whilst the shear stress at the surface (τ_0) is greater than the shear strength of the material. Since τ_0 is independent of the length to diameter ratio, if it has any value smaller than a certain critical value, then shear failure will occur. However, there is considerable doubt as to whether the interlaminar shear strength test does measure only shear failure, since as soon as the beam begins to fail, it becomes in effect two beams each with a length to diameter ratio twice the original. With these new values the tensile strength may be exceeded, so that the next failure event may be tensile failure on the newly created beam surface. The interlaminar shear strength is drastically reduced by even very low (e.g. 0·5%) void contents.

INTERMEDIATE SUPER ABRASION FURNACE BLACK (ISAF) One of the most widely used types of carbon black of particle size about 25 nm giving high reinforcement and high tear resistance in compounds, with good processing characteristics. Useful in tyre treads of most types.

INTERNAL CONVERSION A radiationless process converting one electronic excited state to another of the same multiplicity, i.e. a singlet–singlet state ($S_1 \rightarrow S_0$) or triplet–triplet ($T_1 \rightarrow T_0$) state transition. Very rapid, having a rate constant in the range $11^{11}-10^{14}\,s^{-1}$, except for the $S_1 \rightarrow S_0$ transition which has a rate constant of $10^6-10^{12}\,s^{-1}$. Thus a means of dissipation of excess energy as heat by 'cascading' down through the upper rotational and vibrational states to the ground state.

INTERNAL INTERFERENCE The interference of the light scattered from a particle of size greater than about one-twentieth of the wavelength of the light, due to the fact that the scattering occurring from the different parts of the particle is out of phase at the point of observation. Thus the particle cannot be considered as a point scatterer. Hence the scattered intensity is attenuated, except at zero angle, such that it is less at any angle in the backward direction than at the same angle in the forward direction. This difference in intensities is the dissymmetry.

INTERNAL LUBRICANT A lubricant for reducing friction within a polymer melt during processing. Hence the lubricant must be relatively compatible with the polymer, otherwise it will act as an external lubricant. Internal lubricants, which are used particularly in processing unplasticised polyvinyl chloride, include polar

waxes, such as amide waxes and montan waxes, and glyceryl monoesters, e.g. glyceryl monostearate.

INTERNAL PLASTICISATION Plasticisation resulting from a modification to the polymer molecules themselves, rather than using a plasticiser as an additive, which is external plasticisation. It is usually achieved by copolymerisation, e.g. by the use of vinyl acetate in polyvinyl chloride, or sometimes by chemical modification of the preformed polymer. Thus the conversion of cellulose to cellulose acetate, to make it melt processable, may be considered an internal plasticisation. It has the advantage over external plasticisation that it is more permanent, since there is no possibility of plasticiser loss by volatilisation or extraction.

INTERNAL REFLECTION SPECTROSCOPY Alternative name for attenuated total reflection spectroscopy.

INTERNAL STABILISATION Stabilisation of a polymer against degradation by an alteration to the polymer molecular structure rather than by using a stabiliser additive. This may be achieved either during the polymerisation by the incorporation of a stabilising comonomer (as in the incorporation of ethylene oxide units to stabilise polyoxymethylene) or by subsequent chemical reaction on the already formed polymer (as in the end-capping method of stabilisation of polyoxymethylene).

INTERNATIONAL UNIT Alternative name for enzyme unit.

INTERPENETRATING ELASTOMERIC NETWORK (IEN) A type of polymer blend in which two elastomeric polymers have been intimately mixed, often as latices, and then vulcanised together (covulcanised), thus forming a rubber network with crosslinks between both types of elastomer.

INTERPENETRATING POLYMER NETWORK (IPN) A polymer blend in which a crosslinked polymer has been swollen with a second monomer, together with crosslinking agents, followed by its polymerisation to yield a mixture containing two networks which are interpenetrating at least if the polymers are reasonably compatible. Usually phase separation occurs due to polymer incompatibility, but this is limited to a fine scale owing to the interlocking of the networks, often having a cellular type of morphology.

INTERPOLYMER The complex copolymer formed when mixtures of polymers are mechanically sheared or masticated, e.g. by extrusion or milling, or when a mixture of a polymer and a monomer is similarly treated, causing polymerisation of the monomer. Frequently the product is a mixture of block and graft copolymers as well as containing the homopolymers. Such mixtures are usually now referred to as polymer blends.

INTERPOLYMERISATION The formation of an interpolymer either by the mechanical mixing of two polymers or by polymerising a monomer in the presence of a preformed polymer by mechanical action.

INTERSECTING LINES METHOD (Mayo and Lewis method) A method of copolymerisation data analysis for the estimation of the monomer reactivity ratios (r_A and r_B) of the monomers A and B. The data is obtained as a series of monomer feed concentrations ([A] and [B]) with the corresponding instantaneous copolymer compositions (d[A] and d[B]). The copolymer composition equation may be rearranged to:

$$r_B = ([A]/[B])\{(d[B]/d[A])(1 + r_A[A]/[B]) - 1\}$$

If r_B is then plotted as a function of various assumed values of r_A, then each pair of data points yields a straight line. The intersections of these lines for different feeds gives the best values of r_B and r_A. The spread of the various intersection points is a measure of the experimental errors.

INTERSECTION LENGTH Alternative name for correlation length.

INTERSYSTEM CROSSING A radiationless process converting one electronic excited state to another with spin inversion, i.e. between singlet and triplet states and vice versa. The $S_1 \rightarrow T_1$ transition has a rate constant of 10^4–$10^{12}\,s^{-1}$ and the $T_1 \rightarrow S_0$ transition has a rate constant of 10^{-1}–$10^5\,s^{-1}$.

INTERUNIT LINK The covalent bond joining the repeat units in a polymer molecule. In step-growth polymers the interunit link is frequently part of a chemical group capable of undergoing reaction leading to scission of the polymer chain. Thus in polyesters the interunit link is an ester link which is susceptible to hydrolysis, as is the glycoside interunit link in polysaccharides.

INTEX Tradename for emulsion styrene–butadiene rubber.

INTOL Tradename for emulsion styrene–butadiene rubber.

INTOLAN Tradename for ethylene–propylene rubber.

INTRALAMINAR SHEAR Alternative name for c-shear.

INTRINSIC BIREFRINGENCE Symbol Δ° or Δ_n°. The maximum possible orientation birefringence due to perfect alignment of the birefringence elements.

INTRINSIC DIELECTRIC STRENGTH (Intrinsic electric strength) The dielectric strength of a dielectric, which is an intrinsic material property, measured in the absence of spurious discharges and in the absence of secondary breakdown mechanisms such as thermal breakdown. In relatively stiff polymers, such as plastics, values are often 500–$1000\,MV\,m^{-1}$, electronic breakdown causing failure. In softer materials, such as rubbers, electromechanical breakdown can occur, giving lower values.

INTRINSIC ELECTRIC STRENGTH Alternative name for intrinsic dielectric strength.

INTRINSIC VISCOSITY Alternative name for limiting viscosity number.

INTRINSIC VISCOSITY PARAMETER Symbol Φ. The universal constant of the Flory–Fox equation. Calculated theoretically to have a value of 2.6×10^{21} when the limiting viscosity number and the mean square end-to-end distance are expressed in $dl\,g^{-1}$ and cm^2 respectively.

INTUMESCENCE The formation of an inert expanded foam layer on the surface of a material, especially a coating, during burning. This layer acts as a material transport and thermal insulation barrier and is hence fire retardant. Produced either by the incorporation of a polyol, e.g. penta-erithrytol, or a dehydrating agent, e.g. ammonium phosphate, and a char expanding agent, e.g. dicyanamide.

INULIN A β-D-fructan with the fructofuranose units 2,1'-linked. Occurs as a reserve deposit polysaccharide in many plants, especially the *Compositeae*, e.g. Dahlia tubers, and *Gramineae*, e.g. grasses and cereals. The main chain contains about 35 fructose units terminated by an α-D-glucose unit.

INVARIANT (1) A quantity that remains constant with a change in coordinate axes direction. In elasticity the strain invariants are often used.

(2) Symbol Q. In small angle X-ray scattering, an integral value of the scattered ray intensity, over the whole angular range, defined as

$$Q = \int_0^\infty I(s)s^2\,ds$$

for pinhole optics, where s is $2\sin\theta/\lambda$ (which is approximately $2\theta/\lambda$ at low θ). It is proportional to the scattering power and inversely proportional to the surface area per unit volume between the phases of a two phase scattering system. It may therefore be used to calculate the surface area of a dispersed phase, such as a filler in a polymer.

INVERSE EMULSION POLYMERISATION Emulsion polymerisation in which a water soluble monomer, such as acrylic acid, is dispersed as an aqueous solution in a continuous water-immiscible organic phase by means of a water-in-oil emulsifier.

INVERSE GAS CHROMATOGRAPHY (Molecular probe technique) Gas chromatography in which the material of interest is the stationary phase rather than the moving phase as is usual in gas chromatography. Thus if a polymer is the material of interest, it may be coated on an inert support and used as the packing in a gas chromatographic column. The gas phase moving through the column contains an injected vapour which becomes partitioned between the gas and the solid polymer phase by sorption/desorption. The vapour, which is usually organic, may thus be used to probe the structure of the polymer through its interactions with it. As with normal gas chromatography, the specific retention volume (V_g) of the probe is measured. Often plots of V_g versus reciprocal temperature are made (a retention diagram), from which glass and melting transitions can be identified. In addition, polymer crystallinity and surface area may be estimated. A wide range of polymer/probe interactions may be investigated giving diffusion constants, solubilities, Flory–Huggins interaction and solubility parameters.

INVERSE LANGEVIN APPROXIMATION An extension of the Gaussian theory of rubber elasticity (i.e. to non-Gaussian theory) to cases where the chain (consisting of n links each of length l) end-to-end distances (r) cannot be considered to be much less than the extended chain length. The probability distribution for chain ends $P(r)$ is then given by,

$$\ln [P(r)] = \text{constant} - n[(r/nl)\beta + \ln (\beta/\sinh \beta)]$$

where $r/nl = (\coth \beta - 1/\beta) = \mathscr{L}(\beta)$ and $\mathscr{L}$ is the Langevin function; correspondingly $\beta = \mathscr{L}^{-1}(r/nl)$ where $\mathscr{L}^{-1}$ is the inverse Langevin function.

INVERSE LAW OF MIXTURES A mathematical relationship, relating the value of a property P of a mixture, to the values of that property for the individual components (P_1 and P_2, for the usual case of a binary mixture), weighted according to their volume fractions ϕ_1 and ϕ_2. It has the form: $1/P = \phi_1/P_1 + \phi_2/P_2$. Commonly P is a modulus or other thermoelastic property. The relationship applies to some composite materials when the composite can be considered to behave as a series arrangement of components (the Reuss or isostress model). In this case the relationship gives the lowest bound value for the modulus.

INVERSE POLE FIGURE A pole figure in which the orientation distribution is plotted as a distribution of reference axis orientation with respect to a fixed position of the orientated polymer, usually a crystallographic axis. This requires the determination of pole figures for several *hkl* planes but leads to an accurate description of molecular orientation.

INVERTED COMPOSITE A composite in which the matrix is a high modulus material and the dispersed phase is a low modulus material, as compared with a normal composite where the reverse is true. Thus many polymer blends may be considered as inverted composites, especially the rubber toughened plastics, such as high impact polystyrene, where a rubber is dispersed in a rigid plastic matrix. Foams and many block copolymers may also be considered as inverted composites.

IODOACETATE $I—CH_2CO_2^-$. An alkylating agent which is useful in identifying the nature of the amino acid residues involved in the active site of enzymes. For example alkylation of histidine 119 in ribonuclease

$$\text{Enzyme} \sim CH_2 - \left\langle \begin{array}{c} N \\ \diagdown \end{array} \right\rangle NH + I—CH_2CO_2^- \rightarrow$$

$$\text{Enzyme} \sim CH_2 - \left\langle \begin{array}{c} N \\ \diagdown \end{array} \right\rangle N—CH_2CO_2^- + HI$$

inactivates the enzyme and identifies the histidine as being active. It also alkylates cysteine residues, e.g. cystine 25 in papain

$$\text{Enzyme} \sim SH + I—CH_2CO_2^- \rightarrow$$
$$\text{Enzyme} \sim SCH_2CO_2^- + HI$$

IONENE POLYMER A cationic polyelectrolyte containing quaternary ammonium groups in the polymer chain. Synthesised by reaction of a dihalide with a ditertiary amine, e.g. 1,4-diaza-2,2,2-bicyclooctane

$$n \left\langle \begin{array}{c} N \quad N \end{array} \right\rangle + n \; Br—R—Br \rightarrow \sim \left[\overset{+}{N}B\bar{r} \; N^+ B\bar{r}—R \right]_n$$
$$Cl$$

or with a primary amine

$$RNH_2 + Cl—R'—Cl \rightarrow \left[R'—\overset{+}{\underset{R}{N}}H \right]_n$$

ION-EXCHANGE CHROMATOGRAPHY A technique of partition chromatography useful for the separation of electrically charged molecules. Most frequently it is used for amino acids, peptides and proteins whose charge in aqueous solution depends on the prevailing pH. Such species carry charges due to their acidic and basic amino acid residues. When a solution of a mixture in a buffer of controlled ionic strength is to be separated it is passed down a column of an ion-exchange resin; the cation species will then bind to a cation-exchange resin or any anionic species to an anion-exchange resin. The most weakly bound species will be eluted first and the most strongly, the last. The pH, and/or the ionic strength, of the eluent can be changed to help progressive elution. A continuous change in eluent can be used to give gradient elution chromatography.

With cation exchangers, the resin is usually used in the Na^+ or H^+ form. Amino acids and peptides are usually separated with such resins, such as sulphonated polystyrene, at low pH to maintain the species being separated in cationic form. The eluate may be collected as separate fractions and the protein content analysed quantitatively using ninhydrin. These processes are automated in the amino acid analyser. Proteins may be similarly separated or, at higher pH, they may be separated on anion-exchange resins such as diethylaminoethylcellulose. Other cellulose derivatives used are carboxymethylcellulose and phosphocellulose which are cation exchangers. Polysaccharides are sometimes separated on anion-exchange resins, often by making them negatively charged by forming borate complexes. Certain ribonucleic acids may be separated on benzoylated diethylaminoethylcellulose. Sometimes dextrans, such as DEAE-Sephadex, are used and provide a separation mechanism according to molecular size as well as by electric charge.

Analytical separations and small scale preparations may also be performed on thin layers (one form of thin layer chromatography) as well as in columns. More rapid separations may be achieved with very short columns operating under pressure. Only a fraction of a milligram of sample may be required.

ION-EXCHANGE DIALYSIS (Donnan dialysis) An ion separation process in which two compartments of a container are separated by an ion-exchange membrane, say a polyanion, which is therefore permselective, allowing cations to diffuse through it but, since anions cannot diffuse, the amount of electrolyte must remain the same on each side of the membrane. If one compartment contains a low concentration of a valuable cation (M_1^+) then on placing a large amount of a second cation (M_2^+) in another compartment, migration of M_1^+ and M_2^+ will occur in opposite directions, to concentrate M_1^+ in the second compartment.

ION-EXCHANGE MEMBRANE A membrane formed from a polyion often with a charge density intermediate between that of an ionomer and a polyelectrolyte. When exposed to an electrolyte it will allow passage of counter-ions but not of byions and it is said to be permselective. Under the influence of an electric potential, a flux of byions will be set up. This principle is exploited when using the membranes in batteries, fuel cells, ion-selective electrodes and in electrodialysis.

ION-EXCHANGE RESIN An insoluble network polymer which carries ionic groups. The resin is capable of exchanging its mobile counter-ions with other ions of the same charge and may thus be used to exchange its counter-ions with other ions present in a surrounding solution. In practice resins are produced in the form of small beads of 0·1 to 0·5 mm diameter, which swell in the liquid medium, usually water. The amount of swelling is controlled by the degree of crosslinking. If the resin is packed into a column, and if an aqueous solution containing ions whose removal is desired is allowed to permeate through the column, the eluate will contain the exchanged counter-ions from the polymer in place of the undesired ions. Examples of cationic-exchange resins are the strongly acidic resins, sulphonated polystyrene (crosslinked with divinylbenzene), and the weakly acidic crosslinked acrylic acid resins. Examples of anionic-exchange resins are crosslinked polystyrene which has been chloromethylated to introduce —CH_2Cl groups, which have then been quaternised by reaction with

tertiary amine to give the quaternary ammonium groups —$CH_2\overset{+}{N}R_3\overset{-}{Cl}$.

IONIC POLYMER (Polyion) A polymer chain or network which carries electrostatic charges. These are usually of the same type, being neutralised by counter-ions. Linear polymers carrying a high charge density and which contain mobile counter-ions are sufficiently hydrophilic to be considered as polyelectrolytes. Other important types of ionic polymers are the network minerals (sheet and network silicates), the ion-exchange resins, the linear polyphosphates, the ionomers, the carboxylated rubbers, polyacrylate and methacrylate salts and the ionenes. Polyampholytes carry both positive and negative charges.

IONIC POLYMERISATION Chain polymerisation in which the growing active centres are ions. Usually, therefore, it is brought about by using an ionic initiator, although this is often referred to as the catalyst since, unlike free radical polymerisation, it is often not consumed during the polymerisation. Other methods of initiation may generate ions, as in radiation-induced polymerisation and electrochemical polymerisation. An ionic polymerisation may be either a cationic polymerisation or an anionic polymerisation. The active centres are extremely reactive to many species, so that it is much more important than in free radical polymerisation that highly purified monomers and solvents are used, usually with great care especially to exclude moisture.

Although ionic polymerisations involve the same basic steps as free radical polymerisation, i.e. initiation, propagation and termination, there are important differences. The active centre concentration is usually much higher, which is an important factor in making these polymerisations extremely rapid. Unlike free radical polymerisation, true termination by the reaction of two growing chain ends is impossible since their similar electrostatic charges repel each other. Termination of growing chains takes place rather by a transfer step, which may, however, also result in true termination (disappearance of active centres) since the product may be incapable of further initiation and growth. An important feature is that a counter-ion, of opposite electrostatic charge to that of the active centre, is often closely associated with the active centre either as a tight ion pair or as a solvent separated ion pair. Sometimes the active centre and its counter-ion may be completely dissociated as free ions. The freer the ions the more rapidly the active centre propagates. Furthermore, the closeness of the association of the counter-ion means that addition of incoming monomer must take place by insertion between the active centre and its counter-ion. This can often take place only by the monomer approaching in some preferred orientation, which leads to the formation of stereoregular polymer. Since dissociation of ion pairs is highly dependent on solvent polarity, the choice of solvent can strongly influence both the rate of polymerisation and the stereoregularity of the polymer produced. Stereospecificity is further enhanced if either the active centre or the monomer is coordinated to a metal catalyst centre (coordination polymerisation), as with Ziegler–Natta catalysts. Here, although the active centres are ionic in character, the coordination effects are so powerful that these polymerisations are usually considered as a separate type.

IONOL Tradename for 2,6-di-*t*-butyl-4-methylphenol.

IONOMER Tradename Surlyn. A copolymer of ethylene with 1–10% methacrylic acid, which has been converted to methacrylate salt, often the sodium, magnesium or zinc salt, by neutralisation with the appropriate base. The resultant ionic groups tend to aggregate to form domains which act as physical crosslinks for the polyethylene. However, the domains break down on heating, so the material may be melt processed as other thermoplastics. The copolymers are produced by the high pressure ethylene polymerisation process and so are similar to low density polyethylene. The comonomer decreases crystallinity but consequent loss of stiffness is restored by the physical crosslinks. The material is more transparent than LDPE and shows better adhesion, which makes it useful as a layer in laminated coextruded packaging films.

IONOPHORESIS Alternative name for electrophoresis.

IP 380 Tradename for a polyimide of structure:

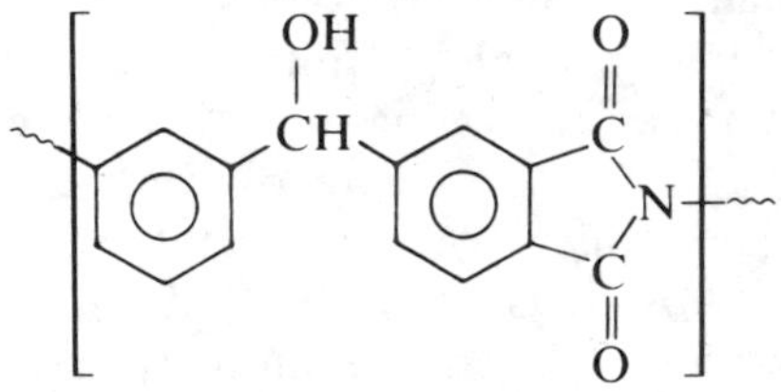

produced by self-condensation of

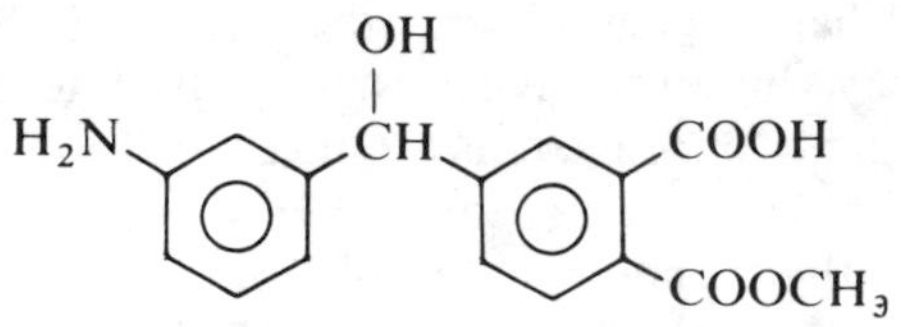

In contrast to the polypyromellitimides, produced from a tetracarboxylic acid or anhydride, the intermediate polyamic acid is reasonably hydrolytically stable.

IPA Abbreviation for isophthalic acid.

IPDI Abbreviation for isophorone diisocyanate.

IPN Abbreviation for interpenetrating polymer network.

IPORKA Tradename for urea–formaldehyde foamed polymer.

i-PP Abbreviation for isotactic polypropylene.

I-RUBBER Alternative name for I-Gummi.

IRWIN THEORY A theory of brittle fracture which assumes that fracture occurs when the decrease in strain energy per unit increase in crack length (the strain energy release rate, G), reaches a critical value, G_c, equal to 2γ, where γ is the fracture surface energy of the Griffith theory. The theory is formulated in terms of the stress field near a crack of length $2c$ and characterised by the stress intensity factor, K. For an infinite sheet and uniform stress σ, $K = \sigma(\pi c)^{1/2}$ and the tensile strength is given by $(G_c E/\pi c)^{1/2}$, where E is the tensile modulus. This result is equivalent to that of the Griffith theory. For thin sheets $EG = K^2$ under plane stress, and for thick sheets $K^2 = EG/(1 - v)^2$, where v is the Poisson ratio.

ISAF Abbreviation for intermediate super abrasion furnace carbon black.

ISOBUTANOL (Isobutyl alcohol) (2-Methylpropan-1-ol)

$$(CH_3)_2CHCH_2OH \qquad \text{B.p. } 108°C.$$

A solvent, similar to *n*-butanol in solvent power, and useful as a diluent in surface coatings.

ISOBUTENE

$$CH_2\!\!=\!\!C(CH_3)_2 \qquad \text{B.p. } -69°C.$$

The monomer for polyisobutene and butyl rubber. Isolated from the C_4 fraction from petroleum cracking by absorption in 65% sulphuric acid, followed by liberation with steam.

ISOBUTENE–ISOPRENE COPOLYMER Alternative name for poly-(isobutene-co-isoprene).

ISOBUTYL ALCOHOL Alternative name for isobutanol.

ISOBUTYLVINYL ETHER

$$CH_2\!\!=\!\!CHOCH_2CH(CH_3)_2$$
$$\text{B.p. } 83°C.$$

Prepared by the reaction of acetylene with isobutanol, in a similar way to vinylethyl ether. It may be polymerised cationically to poly-(isobutylvinyl ether). Copolymers with vinyl chloride are useful as coatings materials.

ISOCHAIN POLYMER Alternative name for homopolymer.

ISOCHRONOUS CREEP CURVE Alternative name for isochronous stress–strain curve.

ISOCHRONOUS STRESS–STRAIN CURVE (Isochronous creep curve) A method of presenting creep data for materials exhibiting non-linear viscoelasticity, where the response is not specified by a specific stress level.

Thus a family of creep curves at different stress levels is needed. The isochronous curve is the curve obtained by taking a cross-section of the family of curves at a particular time. It is thus a plot of stress against creep strain at a given time after application of the stress.

ISOCYANATE A compound containing the group $-N{=}C{=}O$, usually abbreviated to $-NCO$. A very reactive chemical group, especially with compounds containing active hydrogen atoms, such as in amines, alcohols, carboxylic acids and water, the products being a urea (**I**), a urethane (**II**), an amide (**III**) and a carbamic acid (**IV**) respectively

$$RNCO + R'NH_2 \rightarrow RNHCO{-}NHR' \qquad (1)$$
$$\text{I}$$
$$RNCO + R'OH \rightarrow RNHCO{-}OR' \qquad (2)$$
$$\text{II}$$
$$RNCO + R'COOH \rightarrow RNHCO{-}OCOR'$$
$$\rightarrow RNHCOR' + CO_2 \qquad (3)$$
$$\text{III}$$
$$RNCO + H_2O \rightarrow RNHCOOH$$
$$\text{IV}$$
$$\rightarrow RNH_2 + CO_2$$
$$+ R'NCO \searrow$$
$$\qquad RNHCONHR'$$
$$\text{—urea formation}$$
$$(4)$$

These reactions are of great importance in polyurethene formation which utilises reaction (2) between a di-isocyanate and a diol or polyol to form a linear or crosslinked polyurethane respectively. However, reactions (3) and (4) are also of interest since the gaseous carbon dioxide evolved can be utilised to form a polyurethane foam, in which case amine groups formed in (4) can further react as in reaction (1). Further secondary reactions can also occur by reaction of further isocyanate with the products of the above reactions—with a urethane to form an allophanate (reaction (5)), with a urea to give a biuret (reaction (6)) and with an amide to give an acyl urea (reaction (7)):

$$RNCO + {\sim}NHCO{-}O{\sim} \rightarrow {\sim}NCO{-}O{\sim} \qquad (5)$$
$$\qquad\qquad\qquad\qquad\qquad OCNHR$$
$$RNCO + {\sim}NHCONH{\sim} \rightarrow {\sim}NCONH{\sim} \qquad (6)$$
$$\qquad\qquad\qquad\qquad\qquad OCNHR$$
$$RNCO + {\sim}NHCO{\sim} \rightarrow {\sim}NCO{\sim} \qquad (7)$$
$$\qquad\qquad\qquad\qquad\qquad OCNHR$$

Reactions (5) and (6) especially, lead to the formation of crosslinks in polyurethanes. These reactions can occur at appreciable rates at ambient temperature, but may be catalysed by tertiary amines or metal salts, such as organotin salts. Isocyanates can also dimerise to uretidine diones and trimerise to isocyanurates. They can also be polymerised, by strong base at low temperatures, to substituted nylon 1 products:

$$n\,RNCO \rightarrow {+}NRCO{+}_n$$

Several other reactions of isocyanates can yield other useful polymers.

ISOCYANATE DIMER (Uretidine dione) Formed by reaction of an isocyanate with itself, but it is not formed with aliphatic diisocyanates:

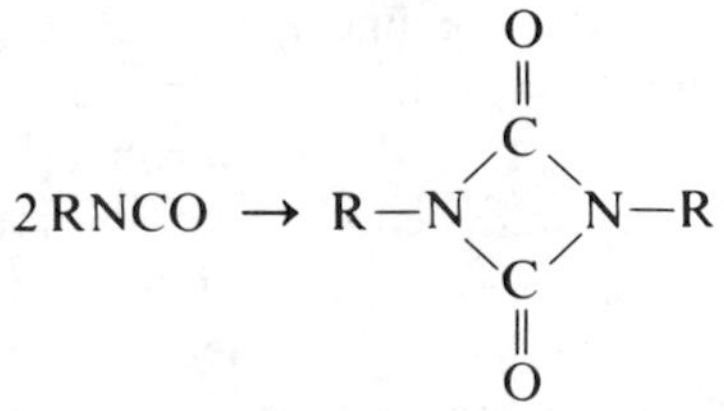

$$2\,RNCO \rightarrow$$

The reaction is catalysed by trialkylphosphines and trialkylamines, but can occur merely on storage, notably with 4,4'-diphenylmethane diisocyanate. The reaction results in the loss of isocyanate content, and hence activity in polyurethane formation, since the dimers are relatively inert to reaction with hydroxyl groups. However, the dimer of tolylene diisocyanate is used as a curing agent in certain millable polyurethane elastomers.

ISOCYANATE TRIMER Alternative name for isocyanurate.

ISOCYANURATE (Isocyanate trimer) A cyclic trimer of an isocyanate:

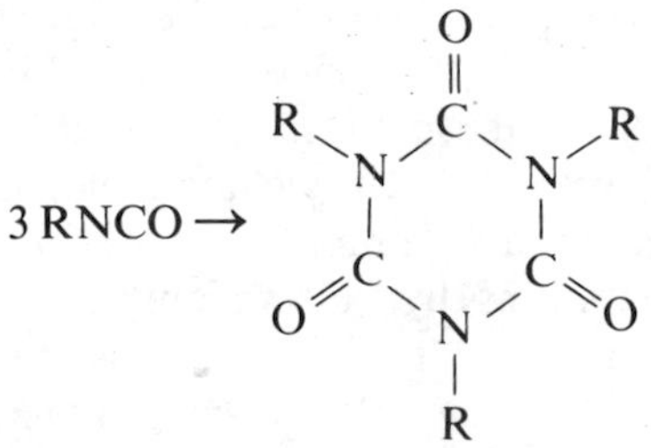

$$3\,RNCO \rightarrow$$

The reaction is catalysed by amines and certain metal salts. Trimer formation possibly provides a branching and crosslinking mechanism in polyurethane formation.

ISODESMOSINE

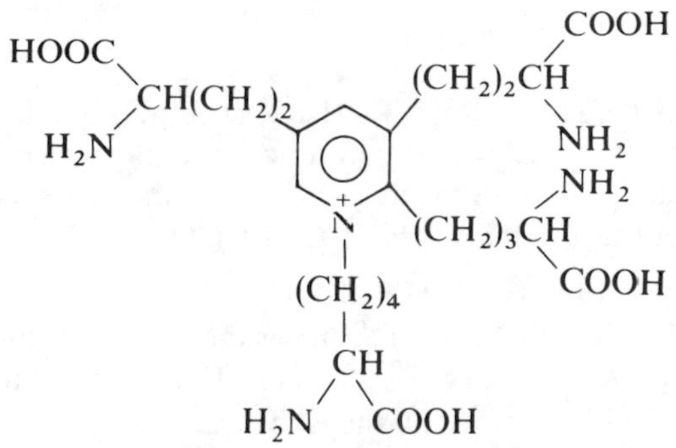

A crosslinking unit present in elastin, formed from four lysine residues coming into close proximity.

ISODIMORPHISM A type of repeat unit isomorphism in copolymers for which the homopolymers of the comonomers have different crystal structures. It occurs in many 1-alkene copolymer and amide copolymer pairs.

ISOELASTIC THEORY A theory of the glass transition in which it is supposed that at the glass transition temperature the rate of uncoiling of polymer molecules, due to an applied stress, becomes equal for all polymers. The rate is given by the reciprocal of the mean relaxation time (τ_m) for all molecules, defined as the time for the deformation resulting from the stress to reach $(1 - 1/e)$ of its final equilibrium value. As a consequence the deformation depends on the time (t) of the deformation. If $t \ll \tau_m$ then there is not much deformation (glassy behaviour) but if $t \gg \tau_m$ then there is a rubbery response. Since τ_m is temperature dependent, the observed T_g will also depend on time (or speed of testing) or frequency (in a dynamic experiment), as is in fact observed. This argument leads to the time–temperature superposition principle and to the kinetic theory of T_g.

ISOELECTRIC FOCUSSING (Electrofocussing) A variation of gel electrophoresis, usually used for polysaccharides and proteins, in which a pH gradient is imposed on a column of gel. The gradient is set up by passing an electric current through a column containing a mixture of amphoteric substances with a continuous range of isoelectric points, e.g. polymers containing pendant carboxyl and amino groups. On applying the electric field the protein molecules migrate until they reach the point where the pH is the isoelectric point for the protein and hence where they have zero mobility. Thus they concentrate there, i.e. they are focussed, forming a narrow band. Since different individual proteins have different isoelectric points, mixtures are separated into sharp bands. The technique can be used for analysis of very small samples or for small scale preparative separations by cutting out the individual bands and eluting the protein. Separations may also be performed in solutions; here the pH gradient is stabilised by providing, in addition, a density gradient using glycerol or sucrose solutions whose concentration increases down the column.

ISOELECTRIC pH Alternative name for isoelectric point.

ISOELECTRIC POINT (Isoelectric pH) Symbol pH_I. The pH at which an amphoteric molecule is electrically neutral. In the α-amino acids at pH_I the molecules exist as the neutral zwitterions $H_3^+NCHRCOO^-$, whilst at lower pH they form $H_3\overset{+}{N}CHRCOOH$ and at higher pH they form $H_2NCHRCOO^-$. The pH_I is the arithmetic mean of the two pK values of the amino acid. For amino acids with no ionisable groups in the substituent R, its value is usually at pH 5–6. At pH_I the amino acid has minimum solubility in aqueous solutions and will not move in an electric field. These properties are of help in separating and identifying amino acids by isoelectric precipitation and electrophoresis respectively.

ISOELECTRIC PRECIPITATION A method for the separation of mixtures of polymers, mostly proteins, whose solubility in aqueous solutions is dependent on pH. Proteins are least soluble at their isoelectric point since the molecules have no net electrostatic charge and tend to aggregate. Since each protein has a characteristic isoelectric point, if the pH of a protein mixture is adjusted to this pH, much or all of the protein in question will precipitate.

ISO-FREE VOLUME THEORY A theory of the glass transition which states that the free volume of all polymers becomes equal at the glass transition temperature (T_g). The fractional value of the total volume is often taken as 0·025 and is so small that segmental jumps become impossible below T_g.

ISOLENE Tradename for polyisobutene.

ISOLEUCINE (Ile) (I)

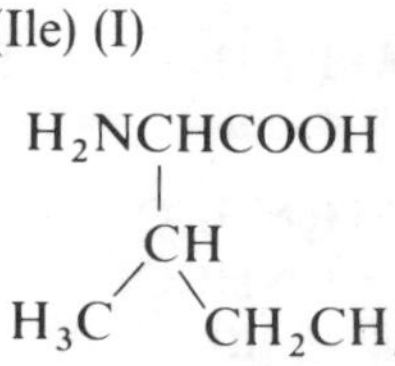

$$H_2NCHCOOH$$
$$|$$
$$CH$$
$$H_3C \quad CH_2CH_3$$

M.p. 285°C (decomposes).

A non-polar α-amino acid found widely in proteins. Its pK' values are 2·36 and 9·68, with the isoelectric point at 6·02. It is often associated with a β-conformation in proteins. In the hydrochloric acid hydrolysis stage of amino acid analysis of proteins, isoleucine is only slowly released, therefore the isoleucine content is determined by extrapolation to infinite hydrolysis time.

ISOLICHENIN An α-D-glucan occurring in Iceland moss lichen, having both 1,3'- and 1,4'-links between the glucose sugar units, but less regularly distributed than in the similar polymer nigeran.

ISOMALTOSE (α-D-glucopyranosyl-1,6'-D-glucopyranose)

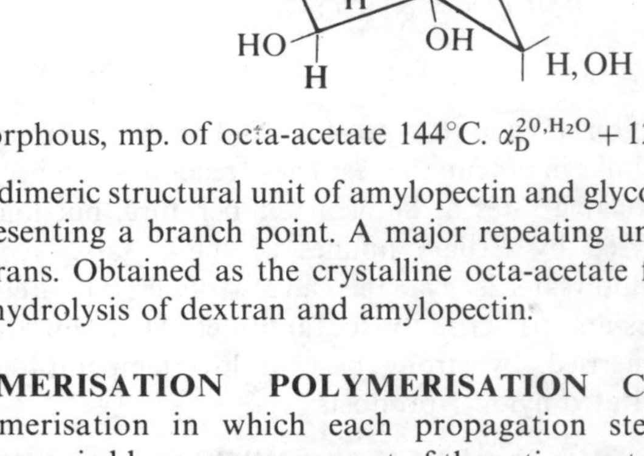

Amorphous, mp. of octa-acetate 144°C. $\alpha_D^{20,H_2O} + 122°$.

The dimeric structural unit of amylopectin and glycogen, representing a branch point. A major repeating unit in dextrans. Obtained as the crystalline octa-acetate from the hydrolysis of dextran and amylopectin.

ISOMERISATION POLYMERISATION Chain polymerisation in which each propagation step is accompanied by a rearrangement of the active centre, so that the polymer repeat unit is not that expected from consideration of the monomer structure. A similar result may also be produced by isomerisation of the monomer, especially by Ziegler–Natta catalysts, but this is not termed isomerisation polymerisation. Most frequently isomerisation is found in cationic polymerisation, because carbocations, in particular, tend to rearrange to thermodynamically more stable species. This may be a bond or an electron rearrangement, e.g. for an α,ω-diene:

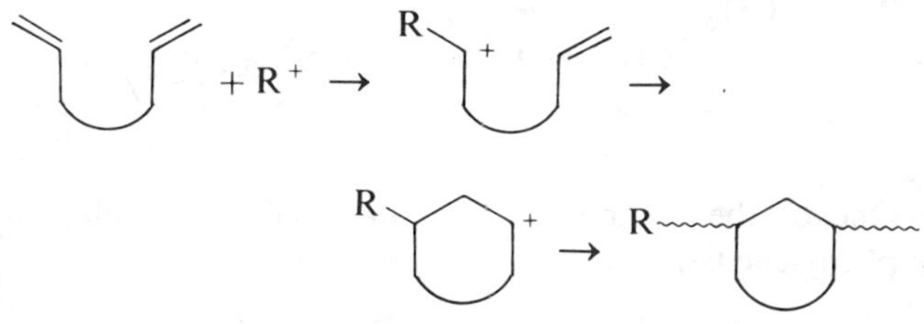

It may result in a transannular polymerisation, e.g. of 2,5,-norbornadiene to polynortricyclene:

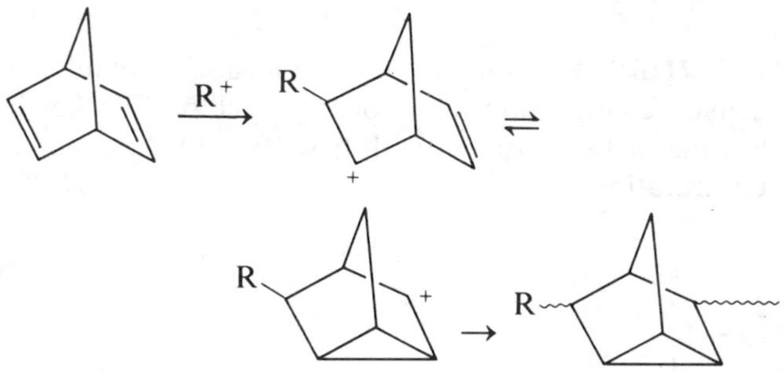

Alternatively, and more commonly, a hydride shift occurs, as in the polymerisation of 3-methylbutene-1 with Friedel–Crafts catalysts to give a 1,3-polymer at $-130°C$:

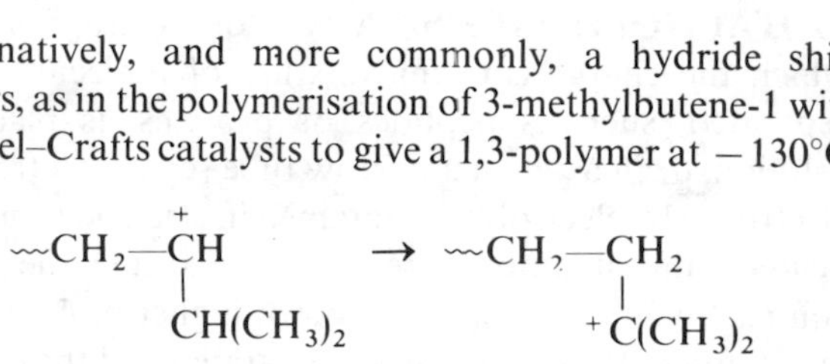

Above $-100°C$ both 1,2- and 1,3-units are formed. Similarly, 4-methylpentene-1 gives 1,4-, 1,3-, and 1,2-units.

ISOMERISED RUBBER   Alternative name for cyclised rubber.

ISOMETRIC CREEP CURVE   Alternative name for isometric stress–strain curve.

ISOMETRIC STRESS–STRAIN CURVE   (Isometric creep curve) A method of presenting creep data for non-linear viscoelastic behaviour, where the response is not specified by a given stress level. Thus a family of creep curves is needed, each curve corresponding to a certain stress. The isometric curve is the curve obtained by taking a cross-section of these curves at a given strain. It is thus a curve of stress versus time (or log time) for a given strain.

ISOMID   Tradename for a polyesterimide.

ISOMORPHISM   The crystallisation of different materials with a common crystal lattice. Co-crystallisation can lead to the formation of mixed crystals. In polymers, in addition to the usual requirements of chemical similarity, nearly equal bond lengths and atomic sizes, similar conformations should exist in the two polymers. Both chain isomorphism, where whole chains can replace each other, and, more commonly, repeat unit isomorphism exist. In the latter, both homopolymers may have similar crystal structures (e.g. polyvinyl alcohol and fluoride, polyhexamethylene adipamide and terephthalamide), the homopolymers may have different crystal structures (isodimorphism) or only one homopolymer may be crystalline. In this last case copolymerisation will reduce crystallinity, but less so than if isomorphism were absent.

ISONAMID   Tradename for a transparent polyamide produced by reaction of a mixture of adipic acid and azelaic acid with MDI:

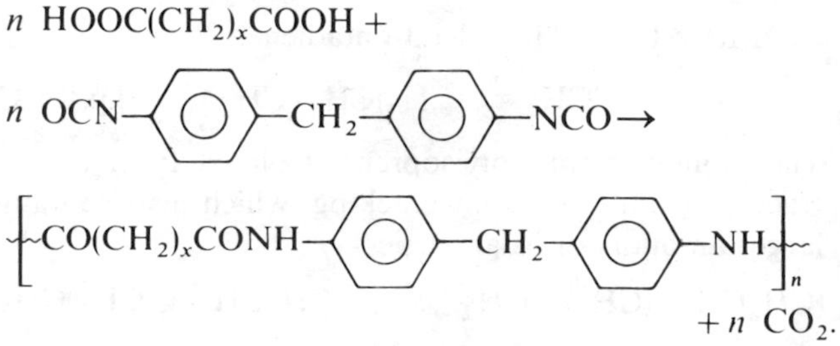

$$n\ HOOC(CH_2)_xCOOH +$$

ISOOCTANOIC ACID

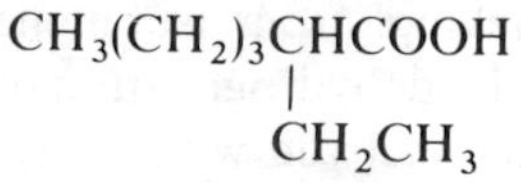

B.p. 227°C/760 mm.

Obtained by the aldol condensation of n-butyraldehyde to its dimer followed by its oxidation. Sometimes used as an acid modifier in non-drying alkyd resins.

ISOPHORONE DIISOCYANATE   (IPDI)

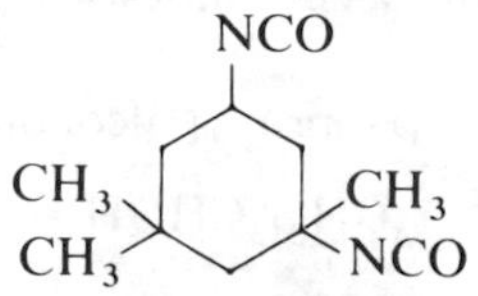

A diisocyanate useful in the production of polyurethanes, especially coatings, with a good resistance to thermal and oxidative yellowing.

ISOPHTHALIC   ACID   (Benzene-1,3-dicarboxylic acid) (IPA)

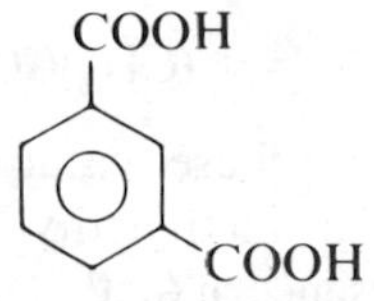

M.p. 347°C.

Prepared by the oxidation of m-xylene. Sometimes used in place of phthalic anhydride for the preparation of unsaturated polyester resins, giving cured products with higher stiffness, heat distortion temperature and chemical resistance. Also sometimes used in alkyd resins.

ISOPIESTIC METHOD A method, based on vapour pressure lowering, for the determination of the number average molecular weight ($\bar{M}_n$) of polymers whose $\bar{M}_n$ values are less than about 20 000. Separate solutions of polymer and of an involatile solute of known molecular weight are equilibrated under isothermal conditions in a closed vessel. Isothermal distillation takes place from one solution to the other, until at equilibrium the vapour pressures of the two solutions are equal. From the equilibrium concentrations of the two solutes, $\bar{M}_n$ of the polymer may be calculated. The method requires a long time to reach equilibrium and is rarely used. The time required may be shortened by measurement of the rate of isothermal distillation and extrapolation to equilibrium conditions.

ISOPRENE (2-Methyl-1,3-butadiene)

$$CH_2{=}C(CH_3)CH{=}CH_2 \qquad \text{B.p. } 35°C.$$

The monomer for polyisoprene. Isolated from the C_5 fraction from petroleum cracking, which also contains larger amounts of isoamylenes

$$(CH_3CH_2C(CH_3){=}CH_2 \quad \text{and} \quad CH_3CH{=}C(CH_3)CH_3)$$

which may be dehydrogenated to isoprene. Also obtained from the dimerisation of propylene, followed by isomerisation and cracking, from isobutene by reaction with formaldehyde, by dehydrogenation of isopentane and by reaction of acetaldehyde with acetylene to 3-methyl-1-butyn-3-ol, which is hydrogenated then dehydrated.

Isoprene may be polymerised to polyisoprene, most interest being in *cis*-1,4-polyisoprene since this is the isomeric form of natural rubber. It may be prepared by anionic polymerisation with butyl lithium to about 92% *cis* polymer or with a Ziegler–Natta catalyst to a 96–97% *cis* polymer. With other Ziegler–Natta catalysts *trans*-1,4- and 3,4-polymers may be synthesised.

ISOPROPANOL (Isopropyl alcohol) (Propan-2-ol)

$$(CH_3)_2CHOH \qquad \text{B.p. } 82·4°C.$$

A widely used solvent, often preferred to ethanol, since it is often anhydrous (although the 13% water content azeotrope is used) and need not be denatured. A solvent for many natural resins and useful in surface coatings.

ISOPROPENYLMETHYL KETONE (Methylisopropenyl ketone)

$$CH_2{=}C(CH_3)COCH_3 \qquad \text{B.p. } 98°C.$$

Prepared either by base catalysed condensation of formaldehyde with methylethyl ketone, or by their reaction in the presence of diethylamine hydrochloride by the Mannich reaction to the Mannich base which is then pyrolysed at about 200°C. It is readily polymerised to poly-(isopropenylmethyl ketone) by free radical, cationic or anionic polymerisation. The polymer is similar to polymethylmethacrylate in its physical properties with a T_g value of about 80°C, but it has poor thermal and photochemical stability.

ISOPROPYL ALCOHOL Alternative name for isopropanol.

N-ISOPROPYL-N'-PHENYL-p-PHENYLENE-DIAMINE

M.p. 70°C.

One of the commonest and most effective dialkyl *p*-phenylenediamine antiozonants.

ISOPYCNIC CENTRIFUGATION Alternative name for density gradient sedimentation equilibrium centrifugation.

ISO-RUBBER A modified natural rubber obtained by dehydrochlorination of rubber hydrochloride either by heating or by treatment with base. Part of the resultant unsaturation is of the type:

$$-CH_2CCH_2CH_2- \\ \quad \| \\ \quad CH_2$$

ISOTACHOPHORESIS A type of electrophoresis in which the charge carrying mixture of molecules to be separated, such as peptides or proteins, is placed in solution, or in a gel, in a narrow tube separated from the electrodes by electrolytes contained in electrode compartments. Thus if anions are being separated the anode compartment contains an anion of higher mobility than any anion in the mixture to be separated and the cathode contains an anion of lower mobility than any in the mixture. On passing a current the different anions separate into sharp zones whose positions depend on their mobility. Eventually a steady state is reached in which all zones move with the same speed. The zones are self-sharpening since if a very mobile ion from a very early zone falls back into the next slower zone, the higher potential gradient in the slower zone speeds it up so that it regains its original faster zone. The technique is a sensitive analytical tool, especially for protein fragments, i.e. peptides. When conducted in a gel, 'spacer' ions (low molecular weight ampholytes) are added to give better separation of the zones.

ISOTACTIC DYAD Alternative name for *meso* dyad.

ISOTACTIC INDEX The percentage of a polypropylene sample which is insoluble in boiling *n*-hexane. It is a measure of the amount of isotactic polypropylene present since atactic polymer is soluble in this solvent.

ISOTACTIC POLYMER A stereoregular polymer in which at least one in-chain atom of the repeat unit is capable of exhibiting stereochemical configurational isomerism and which has the same configuration in all repeat units. Thus in a vinyl polymer, $-[CH_2CHX]_n$, when the polymer is in its extended chain conformation, all the substituents X point away from the chain in the same direction:

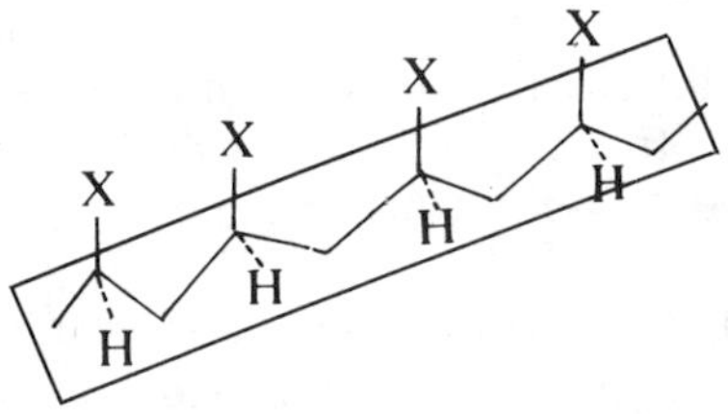

In a repeat unit with three chain atoms, such as polypropylene oxide, or poly-(α-amino acids), the substituent directions will alternate:

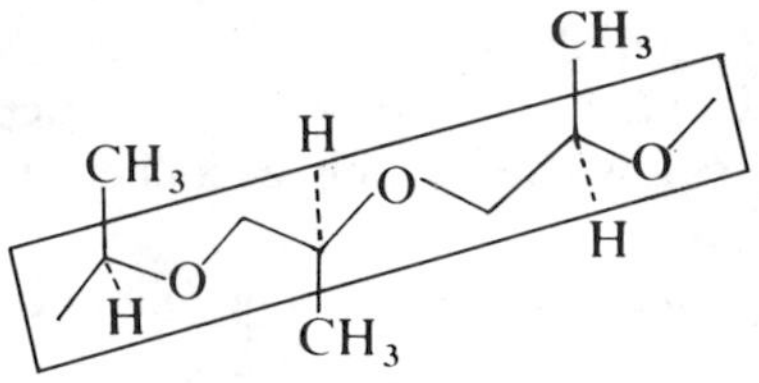

Isotacticity is the most common form of stereoregularity observed, especially in polyolefins produced by Ziegler–Natta polymerisation. Diisotacticity may be observed in ditactic polymers. Steric interactions frequently occur between the substituent R groups, forcing the chain to adopt a helical conformation in order to space out the R groups at larger distances.

ISOTACTIC POLYPROPYLENE (i-PP)

Polypropylene in which each repeat unit has the same configuration, i.e. the polymer is highly isotactic. It is produced by stereoregular polymerisation using Ziegler–Natta catalysts. Commercial polypropylene is highly isotactic, having an isotactic index of 90–95%, much of the atactic material having been extracted during the manufacturing process. Being regular in structure, the polymer crystallises readily and is thus hard and of high softening point compared with atactic polypropylene, which is a weak, gummy, rubbery material.

ISOTACTIC TRIAD

A triad in a polymer exhibiting stereochemical configurational isomerism in which the configurations of the asymmetric, or rather pseudo-asymmetric, centres are the same. It is thus composed of two *meso* dyads (mm). In a Fischer projection formula for a vinyl polymer, $-[CH_2CHX]_n$, it may be represented as

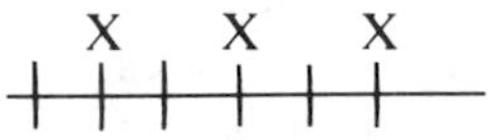

The central substituent X will have a different nuclear magnetic resonance peak compared with a heterotactic or a syndiotactic triad.

ISOTHERMAL DIFFUSION FRACTIONATION

Alternative name for Brownian diffusion fractionation.

ISOTHERMAL DISTILLATION

The distillation of solvent from a sample of solvent to a solution in that solvent held at the same temperature. It results from the lowering of the vapour pressure of the solvent by the solute. The occurrence of isothermal distillation is the principle on which the isopiestic and vapour pressure osmometry methods of polymer molecular weight determination rely.

ISO VISCOSITY NUMBER

The viscosity number of a polyvinyl chloride solution of concentration c g cm^{-3} (usually measured at $c = 0.005$ g cm^{-3}) in cyclohexanone at 25°C. It is therefore defined as $(t - t_0)/t_0c$, where t is the flow time of the solution and t_0 that of the solvent. It is commonly used as a measure of the molecular size of a polyvinyl chloride sample and has a value of 65–140 for commercial polymers.

ISOVISCOUS THEORY

A theory of the glass transition in which the glass transition temperature is thought to be the temperature at which, on cooling from the melt, the viscosity of all materials becomes equal, having a value of about 10^{13} P, and is so high that molecular motions are frozen. Whilst this might be true for low molecular weight polymers and inorganic glasses, it is not true for high molecular weight polymers, since melt viscosity is well known to be highly dependent on polymer molecular weight.

ISOVYL

Tradename for a polyvinyl chloride staple fibre.

ISOZYME

One of several multiple forms of the same enzyme occurring within the same species. Different isozymes differ only slightly in their amino acid sequence which allows differential control of the same metabolic function in different parts of the organism. The best studied example is lactate dehydrogenase.

ITACONIC ACID

$$\begin{array}{c} COOH \\ | \\ CH_2{=}CCH_2COOH \end{array}$$

M.p. 167–168°C.

Produced either by pyrolysis of citric acid or by fermentation of carbohydrates (e.g. in molasses) by, for example, *Aspergillus terreous*, both methods causing decarboxylation:

$$\begin{array}{c} OH \\ | \\ HOOCCH_2{-}C{-}CH_2COOH \end{array} \rightarrow$$
$$\begin{array}{c} | \\ COOH \end{array}$$

$$\begin{array}{c} CH_2{=}C{-}CH_2COOH + CO_2. + H_2O \\ | \\ COOH \end{array}$$

Dialkylesters may be made by esterification. Both the dimethyl and di-*n*-butyl esters are commercially useful. Owing to its high melting point the acid may only be polymerised in emulsion or in solution by free radicals, e.g. peroxide or persulphate initiators, to polyitaconic acid. It is also useful as a comonomer (1–5%) in various latices (styrene–butadiene, acrylic and vinylidene chloride) to improve rheological properties and adhesion when the latex is used as a coating. Acrylamide copolymers are useful as textile sizes. Sometimes used in the production of unsaturated polyester resins, but the uncured resins must be carefully inhibited due to the tendency of the itaconate units to homopolymerise.

IXAN Tradename for polyvinylidene chloride and copolymers.

IXEF Tradename for poly-(*m*-xylyleneadipamide).

IZOD IMPACT STRENGTH The impact strength measured according to the Izod method. In an Izod test a falling weighted pendulum strikes a rectangular bar specimen that is mounted vertically by being clamped at the lower end. The specimen may be notched or unnotched. If the former then the notch dimensions must be specified. Commonly a standard V-shaped notch of depth 2·5 mm and tip radius 0·25 mm is used. Other commonly used standard conditions are bar dimensions of 12·5 mm in width and 3·2 mm thick, temperature of 23°C and a pendulum travelling at 3·4 m s^{-1} and striking the specimen on the notched side 22 mm above the notch. The impact strength is the loss in energy on breaking the specimen, obtained by recording the height to which the pendulum follows through. The results may be expressed as the energy per unit width of specimen, i.e. per unit width of notch, usually in units of J m^{-1} or ft-lb inch^{-1}. However, this gives the false impression that the value is independent of sample width, so recent work has favoured units of energy per standard width of specimen (usually J (12·7 mm)$^{-1}$). Alternatively the results may be given as energy per unit cross-sectional area fractured, in units of J m^{-2}, which similarly falsely suggest that the value is independent of such surface area. The values are hardly of any use for design calculations for plastics parts since, apart from the limitations mentioned, the test specimens are much thicker than most plastics products and they contain notches which are usually deliberately avoided in products. Nevertheless the test results are very widely used in property descriptions of polymer materials because the tests are simple and quick to carry out. The values are of some use in giving a rough ranking of materials and for quality control purposes.

J

JAUMANN DERIVATIVE (Corotational derivative) Symbol $\mathscr{D}/\mathscr{D}t$. A time derivative operator for the transformation of convected to fixed coordinates in connection with rheological problems, whilst obeying the principle of material objectivity. Defined as,

$$\mathscr{D}/\mathscr{D}t\,\tau_{ij} = \partial\tau_{ij}/\partial t + u_k\,\partial\tau_{ij}/\partial x_k - \tfrac{1}{2}\omega_{jm}\tau_{mi} - \tfrac{1}{2}\omega_{im}\tau_{mj}$$

where τ_{ij} are the shear strain components, using the summation convention, u_k are the velocity components and ω_{jm} are the components of the vorticity tensor, given by $\omega_{ij} = (\partial u_i/\partial x_j - \partial u_j/\partial x_i)$. The Jaumann derivative, like the Oldroyd derivative, thus gives the strain components in fixed coordinates of the time derivative, as observed in a coordinate system which translates with the flow field, but which in addition also rotates with the local rotation as given by ω_{ij}. It is therefore also called the corotational derivative.

JECTOTHANE Tradename for a partially crosslinked thermoplastic polyurethane elastomer, based on a polyester or polyether polyol/MDI prepolymer and chain extended.

JELUTONG A mixture of polyisoprene and resin obtained from the latex of the tree *Dyera costulata*, mainly in Borneo. It has been used as a substitute for natural rubber, but is now used mainly as a chewing gum base.

JETRON Tradename for styrene–butadiene rubber.

JOG A polymer chain defect in a crystalline structure, in which part of the chain is displaced perpendicular to its long axis by the occurrence of a 'false' conformational isomer, e.g. a —tttg$^-$ttg$^-$ttt— sequence in an all-*trans* conformation giving:

If the displacement is less than the interchain distance, then the jog is a kink.

JOHNSTON–OGSTON EFFECT An effect in the sedimentation velocity determination of polymer solutions containing components which sediment at different rates (as in paucidisperse polymer), when the sedimentation coefficients are concentration dependent. The concentration in the leading concentration peak (due to the faster moving component) is higher than expected owing to the presence of some of the slower moving component. Similarly the second peak will be depleted in the slower component.

JOULE–GOUGH EFFECT (Gough effect) One of the anomalous thermoelastic effects found in rubbers. The contraction that occurs when a stretched rubber is heated, the increase in temperature that occurs when a rubber is rapidly adiabatically stretched or, conversely, the fall in temperature that occurs when a stretched rubber is allowed to retract. If high extensions are involved then the process is not exactly reversible owing to the heat evolved due to crystallisation. On removal of stress the crystallisation is not immediately destroyed.

JSR 1,2-PBD Tradename for 1,2-polybutadiene.

JSR RB Tradename for a 1,2-polybutadiene thermoplastic elastomer with amorphous soft blocks and crystalline hard blocks.

JUTE A natural bast fibre from plants of the genus *Corchorus*. The most widely used vegetable fibre after cotton. The brown fibre bundles are up to about 250 cm in length, individual fibres being 0·1–0·5 cm long. It consists of about 65% cellulose, about 10% water and 10–20% lignin. Typical tenacity is about 3–5 g dtex^{-1} with a moisture regain of 13% and an elongation at break of only 1·7%. The fibre is widely used for sacking, cordage and carpet backing.

K

K Symbol for lysine.

KADEL Tradename for an aromatic polyetherketone.

KALREZ Tradename for tetrafluoroethylene–perfluoromethylvinyl ether copolymer.

KANEBIYAN Tradename for polyvinyl alcohol fibre.

KAOLIN (China clay) The most important type of kaolinite, which is the major component of clays. The chemical unit structure is that of the aluminosilicate, $Al_2Si_2O_5(OH)_4$, consisting of silica sheets of $Si_2O_5^{2-}$ units alternately layered with sheets of gibbsite, $Al(OH)_3$. Cornish china clay is about 95% kaolin. Particles are mostly 1–2 μm in size and plate-like. The finest white grades are used as fillers in polymers, especially in polyvinyl chloride. Calcined clay is produced by heating kaolin to about 600°C when it loses some water, and is useful for its electrical insulation properties. Kaolin is also useful in controlling the flow properties of polyester and epoxy resins.

KAOLINITE A single-sided crystalline layer silicate mineral which is the main constituent of clays, china clay being pure kaolinite. Its composition is $Al_4Si_4O_{10}(OH)_4$, consisting of sheets of silicate tetrahedra linked as hexagonal rings and joined to a layer of gibbsite $(Al(OH)_3)$ on one side so that one of each of three hydroxyls is replaced with an oxygen from the SiO_4 tetrahedra.

KAPOK A seed hair fibre from the plant *Ceiba pentandra*, consisting of about 65% cellulose, 15% lignin, 12% water and some wax and protein. The fibres are about 30 μm in diameter and about 2 cm long and are tubular. It is useful for mattress and cushion stuffing and for thermal and acoustic insulation.

KAPRON Tradename for a nylon 6 fibre.

KAPTON (Kapton H) Tradename for the polypyromellitimide polyimide obtained from pyromellitic

dianhydride and 4,4′-diaminodiphenyl ether in film form. One of the best high temperature resistant polymers, especially for electrical insulation applications. Will withstand temperatures of up to 275°C in air for a year without deterioration. Owing to the intractability of the polyimide structure the material is processed as the soluble polyamic acid precursor polymer and then cyclised to the imide structure by heating at 300°C.

KAPTON H Alternative name for Kapton.

KAT Abbreviation for Katal.

KATAL (Kat) A unit of enzyme activity. The amount of enzyme that transforms one mole of substrate per second.

KAURI COPAL A fossil copal natural resin found in New Zealand. It is a hard resin which softens at about 120°C and is oil soluble. It finds wide applications in surface coatings.

KAURIT Tradename for urea–formaldehyde polymer adhesive.

KAUZMANN'S PARADOX The existence of an apparently negative value for ΔS, the entropy difference between the disordered supercooled liquid state and the crystalline state of a polymer. Such an impossible negative value is obtained if a plot of ΔS against temperature is extrapolated to below T_g. However, in practice, the straight line plot obtained above T_g in fact curves as T_g is approached, so avoiding a negative value for ΔS on extrapolation. Such a curvature provides evidence that T_g is a true second order transition.

KEDZIE DEFECT A combined dislocation and disinclination in which the former has a displacement (translation) vector parallel to the molecular chain and the latter has a rotation vector with the molecular axis as axis of rotation. The magnitudes of these vectors is such that the formation of the defect in a helical molecule untwists part of it by one extra repeat unit.

KELBURON Tradename for a polypropylene/ethylene–propylene–diene monomer blend thermoplastic polyolefin rubber.

KEL-F Tradename for polychlorotrifluoroethylene and chlorotrifluoroethylene–vinylidene copolymer.

KELLEY–BUECHE THEORY A theory based on the free volume theory and utilising the Williams–Landel–Ferry equation and the Cohen and Turnbull equation, to give an equation for the coefficient of viscosity (η) of a concentrated polymer solution, of the form:

$$\eta/\eta_p = \phi_p^4 \exp\left[1/(\phi_p f_p + \phi_L f_L) - 1/f_p\right]$$

where η_p is the polymer viscosity, ϕ_p is its volume fraction and f_p and f_L are the free volumes of polymer and solvent respectively. f_p is given approximately by $f_p = 0.025 + 4.8 \times 10^{-4} (T - T_{g,p})$ and f_L is given by $f_L = 0.025 + \alpha_L(T - T_{g,L})$, where T is the temperature, $T_{g,p}$ and $T_{g,L}$ are the polymer and solvent T_g values respectively and α_L is the solvent coefficient of cubical expansion. η is predicted to be very sensitive to f_p and f_L and to the addition of small quantities of solvent, especially close to T_g.

KELTAN Tradename for ethylene–propylene–diene monomer rubber.

KELTAN TP Tradename for a thermoplastic polyolefin rubber.

KELTRON Tradename for xanthan.

KELVIN ELEMENT Alternative name for Voigt element.

KELVIN MODEL Alternative name for Voigt model.

KEMATAL Tradename for polyoxymethylene copolymer.

KERATAN SULPHATE (Keratosulphate)

A mucopolysaccharide consisting mainly of alternating D-galactose and N-acetylated-D-galactosamine units, 1,3′-linked and 1,4′-linked respectively. The latter units are also sulphated at C-6, and some galactose units may also be sulphated. Isolated from bovine cornea and skeletal tissues.

KERATEINE A soluble derivative of keratin, obtained by reduction of the disulphide bridges, e.g. by reaction with a mercaptan. The resultant thiol groups are often protected from re-oxidation by alkylation, e.g. by reaction with iodoacetic acid, to give S-carboxymethylcysteine-kerateines.

KERATIN A group of closely related structural proteins, which are also fibrous, and which comprise the main component of the outermost layers of the epidermis and associated structures of higher animals, i.e. hair, fur, wool, horn, feathers, hooves, nails, claws and beaks. Keratins, coming from such divergent sources, may thus have quite a range of amino acid composition. However, they are characterised by a high sulphur (cysteine) and proline content and polar residues as well as containing all the other standard amino acids. Unlike the collagens and silk there do not appear to be repeating sequences. Most keratins are highly crosslinked, through disulphide bridges, which together with polar interactions provide elasticity (in hair and wool) and hardness in horn, etc. In addition, crosslinking renders keratins highly insoluble and hence makes structural characterisation difficult. The most studied keratins are those of wool.

For physical characterisation, soluble preparations may be produced in the form of keratohyalin granules, pre-keratin, keratose or kerateine. Most keratins are partially crystalline and give characteristic X-ray diffraction patterns. Two types of pattern were identified early—the α-pattern, now associated with α-helical conformations (in α-keratins), and the β-pattern, now associated with the β-pleated sheet (in β-keratins). Most keratins, e.g. wool and hair, are normally α-keratins but are often converted to give the β-form on stretching in steam. Feather keratins are β-keratins, and are the more crystalline (about 50%), whereas α-keratins have much lower crystallinities.

The X-ray pattern of α-keratin suggests the conformation is a coiled coil. It is thought that three polypeptide chains form protofibrils by winding around each other to form a triple stranded supercoil held together by disulphide bridges. These protofibrils may further aggregate to give microfibrils of about 70 Å in diameter and about 1 μm in length. Their substructure may consist of 11 protofibrils made up of two central protofibrils and nine outer ones (the 9 + 2 model).

Even keratin from a single source usually consists of different polypeptide chains. Thus wool fibres contain low sulphur, high sulphur and high glycine polypeptides, the low sulphur forming the α-helical microfibrils, the high sulphur (which contain up to about 30% cystine) and the high glycine (up to 50% glycine plus aromatic residues) forming the amorphous matrix. Feather keratins give a good β-X-ray pattern and are more highly crystalline. They have a similar amino acid composition to the α-keratins, but have lower cysteine and high proline and serine contents. Soluble feather keratin derivatives suggest the basic polypeptides have a molecular weight of about 10 000. Various models, e.g. a β-helix, a superfolded structure, have been proposed for the β-sheet structure.

α-KERATIN A keratin in which the polypeptide chains have a partial α-helical content, resulting in partial crystallinity (although this is still low, typically about 20%) and giving a wide angle α-X-ray diffraction pattern with a fibre axis repeat of 5·1 Å and a helix diameter of 9·8 Å. A coiled coil structure is also indicated. Narrow angle X-ray pattern and electron microscopy indicate a packing of microfibrils of diameter of about 70 Å and about 100 Å spacing. α-Keratin structures are present in many keratins, especially in hair and wool.

β-KERATIN A keratin in which the polypeptide chains partially adopt a β-pleated sheet structure, giving a crystalline structure which produces a β-X-ray diffraction pattern. It is found especially in feather keratins, but is also found on stretching α-keratins, e.g. wool, in steam.

KERATOHYALIN GRANULE A soluble α-keratin, usually obtained by extraction of rat-skin, with no sulphydryl or disulphide bridges or filaments. Useful for structural characterisation of keratin.

KERATOSE A soluble derivative of keratin, obtained by oxidation of the disulphide bridges, often with performic acid. α-Keratose gives an α-X-ray diffraction pattern and has a molecular weight of 40 000–80 000. β-Keratose is an insoluble fraction probably arising from the membranes of keratinised cells. γ-Keratose is a very soluble, low molecular weight (10 000–30 000) fraction, rich in cysteic acid, and does not give an X-ray pattern.

KERATOSULPHATE Alternative name for keratan sulphate.

KERIMID Tradename for polyaminobismaleimide laminating resin.

KERMEL Tradename for polyamide–imide copolymer fibres sythesised by reaction between trimellitic anhydride and a diisocyanate or diurethane:

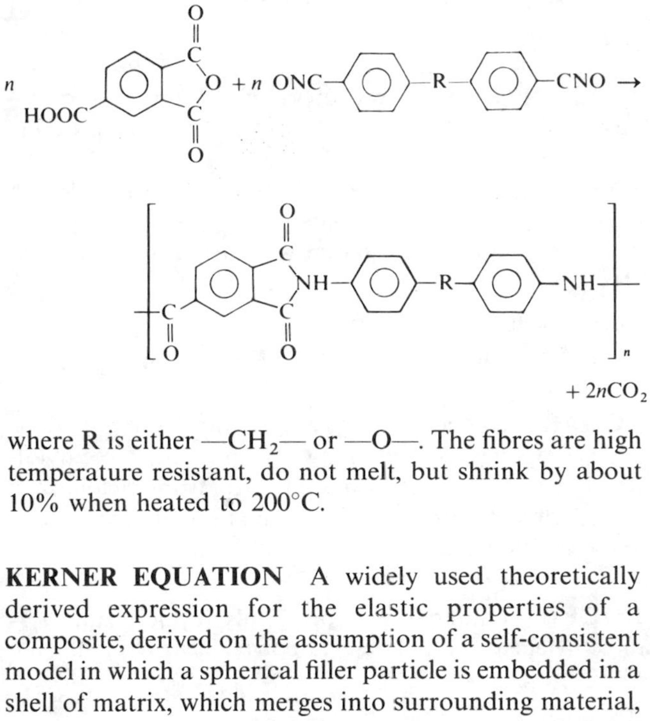

where R is either —CH$_2$— or —O—. The fibres are high temperature resistant, do not melt, but shrink by about 10% when heated to 200°C.

KERNER EQUATION A widely used theoretically derived expression for the elastic properties of a composite, derived on the assumption of a self-consistent model in which a spherical filler particle is embedded in a shell of matrix, which merges into surrounding material, which has the properties of the composite as a whole. A simple form of the equation, for fillers which are much stiffer than the matrix, which applies up to moderate filler volume fractions (ϕ_f), is:

$$G_c/G_m = 1 + [15(1 - v_m)/(8 - 10v_m)](\phi_f/\phi_m)$$

where G_c and G_m are the shear moduli of the composite and matrix respectively and v_m and ϕ_m are Poisson's ratio of the matrix and the volume fraction respectively.

KERR EFFECT Alternative name for electric birefringence.

KETOHEXOSE A ketose containing six carbon atoms.

There are eight possible configurational isomers existing as four enantiomorphic pairs. The only common keto-hexose in polysaccharides is D-fructose.

KETOPENTOSE A ketose containing five carbon atoms. There are four possible configurational isomers existing as two enantiomorphic pairs, D- and L-.

KETOSE A monosaccharide whose simplest structural representation is an open-chain polyhydroxyketose, e.g.

$$HOCH_2CO(CH_2OH)_nCH_2OH$$

where $n = 1$–3. A five-carbon ketose is a ketopentose (better represented in the cyclic furanose form) and a six-carbon ketose is a ketohexose (better represented in the cyclic pyranose or furanose form).

KEVLAR Tradename for poly-(p-phenyleneterephthalamide), earlier called Fibre B. Kevlar 49 (earlier called PRD-49) is a high modulus fibre which has been hot drawn at 250–550°C and is useful as a reinforcement in polymer composites. Kevlar 29 is of lower modulus, but of higher elongation at break, and is useful as a tyre cord reinforcement and in ballistic protective clothing.

KEYING AGENT Alternative name for coupling agent.

KHLORIN Tradename for a polyvinyl chloride fibre.

KIESELGUHR (Diatomaceous earth) Tradename Celite. A form of silica which is the fossilised remains of ancient primitive plants. Consists of 70–90% hydrated silica. Occasionally used as a filler in polymers to improve heat or electrical resistance, when it is often used in a calcined form. It has a high porosity and oil absorption.

KINEL Tradename for polyaminobismaleimide filled with glass, carbon fibre reinforcement, or PTFE, which may be cured during moulding to thermally stable thermoset mouldings which are notably void free owing to the lack of volatiles formed during curing.

KINEMATICS The analysis and description of motion without reference to the forces involved, which is the province of dynamics.

KINEMATIC VISCOSITY The ratio of the coefficient of viscosity of a fluid to its density.

KINETIC CHAIN LENGTH Symbol v. The average number of monomer molecules polymerised in a chain polymerisation for each free radical initiating a polymerisation chain. Thus v equals the ratio of the rate of polymerisation to that of initiation and

$$v = k_p^2[M]/2 k_t R_p = k_p[M]/2(f k_d k_t[I])^{1/2}$$

where initiation is by thermal decomposition of initiator I, of concentration [I] and efficiency f, [M] is the monomer concentration and k_d, k_p and k_t are the rate constants for initiator decomposition, propagation and termination respectively. v is simply related to $\bar{x}_n$, the number average degree of polymerisation, by $x_n = 2v$ when termination is by combination, and by $\bar{x}_n = v$ when termination is by disproportionation. The above equations show that v (and hence $\bar{x}_n$ and polymer molecular weight) is inversely dependent on polymerisation rate (R_p) (or radical concentration) and at constant R_p is independent of mode of initiation.

KINETIC ENERGY CORRECTION (Hagenbach correction) A correction sometimes necessary in capillary viscometry to allow for the dissipation of energy due to the driving pressure P, as kinetic energy of the fluid emerging from the capillary. The pressure is modified to give an effective pressure (P_{eff}) given by,

$$P_{eff} = P - m\rho(V/t)^2/\pi^2 R^4$$

where V is the volume of fluid, of density ρ, flowing through the capillary of radius R in time t and m is a function of the velocity distribution in the capillary. Thus the Poisseuille equation is modified to, $\eta = \rho(At - B/t)$, where A is a constant, η is the coefficient of viscosity and B is a constant given by $B = mV/8\pi L$, where L is the length of the capillary. Viscometers are usually designed and used to keep the correction to less than 1%, which is often considered to be negligible.

KINETIC THEORY OF RUBBER ELASTICITY (Statistical molecular theory of rubber elasticity) A theory of rubber elasticity developed on the basis of the consideration of the entropy change of a single molecular chain on stressing, the chain being linked to other chains to form a network. The chains are capable of existing in any of many configurations (or conformations in organic chemical nomenclature) due to thermal vibration. Highly coiled configurations will be found since these have the maximum entropy, unless stressed, when they elongate to produce a strain. The theory evaluates the stress–strain relationships by calculation of the configurational entropy change as a function of strain. By assuming a freely jointed chain the probability distribution of finding a chain end within a certain volume element with respect to other chain ends is calculated. The radial distribution function is of Gaussian form (Gaussian chains) provided the number of chain links is large, so that the distance between chain ends is much greater than the extended chain length. To calculate the elasticity (as a strain energy function) of a molecular network, it is assumed that the deformation is affine and that the junction points are fixed in space.

KINK An irregularity in a regular conformation (extended chain or helical) of a polymer chain resulting from the occurrence of a local conformation different from that required for the generation of the pure regular conformation, thus producing a kink in the chain. This is most simply exemplified by the occurrence of two higher energy *gauche* (g) conformational isomers in a regular all-*trans* (t) extended chain of polyethylene, producing:

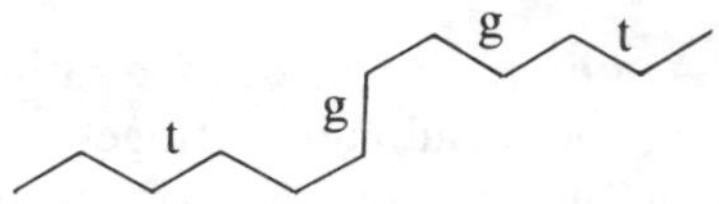

KINK BAND A type of microscopic deformation band found when a crystalline material is deformed by a compressive or shearing force acting normal to a slip plane. The process involves slip along a slip plane plus a rotation of the normal to the slip plane. Kink bands may also be formed as a result of crystallisation growth defects.

KINKING A deformation occurring when a crystalline material is stressed, usually in compression, giving a shear stress component, in which a portion of the crystal shears relative to material on either side of it. This results in the formation of a kink band. It occurs in both metals and crystalline polymers.

KIRCHHOFF STRAIN Alternative name for Green's strain.

KMEF PROTEIN A protein of the group keratin–myosin–epidermin–fibrinogen. A group of proteins which, in early studies, gave similar X-ray diffraction patterns (of the α-type) and were therefore classed together. The crystallinity responsible for this pattern is due to the α-helical structures present in these proteins.

KODAPAK Tradename for a polyethylene terephthalate moulding material.

KODAR Tradename for polyethylene terephthalate, for a copolymer containing cyclohexane dimethylol-terephthalate units and for a polyester from cyclohexane dimethylolterephthalate and a mixture of terephthalic and isophthalic acids.

KODEL Tradename for poly-(cyclohexane-1,4-dimethyleneterephthalate) and polyethylene terephthalate fibre.

KOLLIDON Tradename for poly-(N-vinylpyrrolidone).

KOSTIL Tradename for styrene–acrylonitrile copolymer.

KOSTILINE Tradename for acrylonitrile–butadiene–styrene copolymer.

KRAEMER CONSTANT Symbol k''. A constant of the Kraemer equation having a value dependent on the particular polymer/solvent/temperature combination under consideration. Usually has a value in the range 0.1–0.2, which is less than that of the similar Huggins constant (k'), since $k' + k'' = 0.5$.

KRAEMER EQUATION An empirical equation expressing the concentration (c) dependence of the relative viscosity (η_{rel}) of a polymer in dilute solution:

$$(\ln \eta_{rel})/c = [\eta] + k''[\eta]^2 c$$

where $[\eta]$ is the limiting viscosity number and k'' is a constant—the Kraemer constant. Thus a plot of $(\ln \eta_{rel})/c$ versus c should be a straight line with the intercept at $c = 0$ being $[\eta]$. This is often found experimentally. The equation is similar in form to the Huggins equation and $k' + k'' = 0.5$, where k' is the Huggins constant. Plots based on the equation are often preferred to those based on the Huggins equation since $k'' < k'$ so that the slope of the plot is less and the value of the intercept more certain.

KRAFT PROCESS (Sulphate process) A process for the production of cellulose pulp by heating wood with a mixture of sodium hydroxide and sulphide. The lignin is solubilised partly by formation of low molecular weight thiolignins. The pulp produced is strong but dark coloured.

KRALAC Tradename for acrylonitrile–butadiene–styrene copolymer.

KRALASTIC Tradename for acrylonitrile–butadiene–styrene copolymer.

KRALON Tradename for acrylonitrile–butadiene–styrene copolymer.

KRAMERS–KRÖNIG RELATIONS Relationships between the storage and loss parameters of dispersion behaviour, such as dielectric dispersion or dynamic mechanical behaviour of a viscoelastic material. Thus if P' is the storage parameter (such as dielectric constant or storage modulus) and P'' is the loss parameter (such as the power factor or loss modulus), the relations are:

$$P' - P_R = (2/\pi) \int_0^\infty (P''(\alpha)\omega^2 \, d\alpha)/\alpha(\omega^2 - \alpha^2)$$

and

$$P'' = (2/\pi) \int_0^\infty P'(\alpha) - P_R/(\alpha^2 - \omega^2) \, d\alpha$$

where P_R is the reduced value of the parameter, ω is the frequency and $\alpha = 2s_0 t$, where s_0 is any convenient value of s, where $s = 1/\tau$, τ being the relaxation time.

KRATKY U-BAR CAMERA An X-ray diffraction apparatus in which very good collimation of the incident beam is obtained by a system of extremely accurately machined metal blocks. It is particularly suitable for small angle X-ray diffraction experiments.

KRATON Tradename for styrene–butadiene–styrene block copolymer.

KRATON G Tradename for styrene–ethylene/butylene–styrene block copolymer.

KREHALON Tradename for a vinylidene chloride copolymer fibre.

KRENCKEL'S EFFICIENCY FACTOR OF REINFORCEMENT Alternative name for efficiency of reinforcement.

K-RESIN Tradename for a high styrene–butadiene radial block copolymer. Such commercial polymers are relatively rigid due to the high styrene content, with reasonable impact strength, but are transparent due to the small domain size.

KRONECKER DELTA Symbol δ_{ij}. A factor which has the value of unity when $i = j$ and zero when $i \neq j$. Useful in expressions in stress–strain analysis when the index notation is used.

KRYLENE Tradename for emulsion styrene–butadiene rubber.

KRYNAC Tradename for nitrile rubber and acrylic elastomers.

KRYNOL Tradename for emulsion styrene–butadiene rubber.

KRYTOX Tradename for polyhexafluoropropylene oxide.

KURALON Tradename for a polyvinyl alcohol fibre, crosslinked by reaction with formaldehyde to enhance water resistance.

KUREHA Tradename for carbon fibre.

K-VALUE Alternative name for Fikentscher K-value.

KYDEX Tradename for an impact resistant polyvinylchloride/acrylic polymer blend.

KYNAR Tradename for polyvinylidene fluoride.

KYNOL Tradename for phenol–formaldehyde polymer fibre.

KYTEX Tradename for chitosan.

L

L Symbol for leucine.

L- A prefix used in the naming of an organic molecule, especially of carbohydrates, to identify the configuration of an enantiomorph when the molecule contains an asymmetric (chiral) centre.

LAC The viscous secretion of the lac insect. Found in South Asia, from which the refined resin shellac is produced.

LACQRAN Tradename for acrylonitrile–butadiene–styrene copolymer.

LACQRENE Tradename for polystyrene.

LACQSAN Tradename for styrene–acrylonitrile copolymer.

LACQTENE Tradename for low density and high density polyethylenes.

LACQUER A coating composition consisting of a polymer solution which, after application as a liquid film on a substrate, dries rapidly to a solid film by evaporation of the solvent. The term often refers specifically, especially in the older literature, to cellulose nitrate or other cellulose based, film forming compositions.

LACQVYL Tradename for polyvinyl chloride.

α-LACTALBUMIN An albumin protein found in milk, whose function is to regulate lactose synthesis. Surprisingly, bovine lactalbumin has considerable homology with chicken egg lysozyme, with 47 of its 123 peptide residues being identical and in identical positions in the polypeptide chain. Furthermore, the four disulphide bridges are in corresponding positions in the two proteins.

LACTAM (Cyclic amide) A compound generally of structure

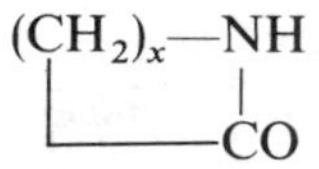

$$(CH_2)_x\!-\!NH \atop \qquad\quad\,|\atop \qquad\quad CO$$

often unsubstituted, but substituents may be present on the methylene groups, often α to the nitrogen. Four-membered ring lactams are β-lactams, e.g. 2-azetidinone, five-membered rings are γ-lactams, e.g. 2-pyrollidone, etc. The commonest method of synthesis is by Beckman rearrangement of the corresponding cyclic oxime, obtained by reaction of a cyclic ketone with hydroxylamine sulphate. The cyclic ketone is frequently obtained by cyclodimerisation or cyclotrimerisation of an olefin followed by hydrogenation to a saturated ring and oxidation.

Lactams are important monomers for polyamide formation by ring-opening polymerisation, especially for monadic nylons. The most important is ε-caprolactam, used for nylon 6 production. Polymerisation may be either by a high temperature, water-catalysed melt polymerisation, proceeding via an open-chain α,ω-amino acid, or by anionic chain polymerisation at lower temperatures using a strong base as an initiator. Substituted nylon 3 products (from azetidinones), nylon 4 (from pyrrolidone), nylon 5 (from piperidone), nylon 6 (from ε-caprolactam), nylon 8 (from capryllactam), nylon 10 (from caprinolactam) and nylon 12 (from lauryllactam) are all usually synthesised from the lactams indicated.

β-LACTAM Alternative name for a 2-azetidinone.

LACTIDE (3,6-Dimethyl-1,4-dioxan-2,5-dione)

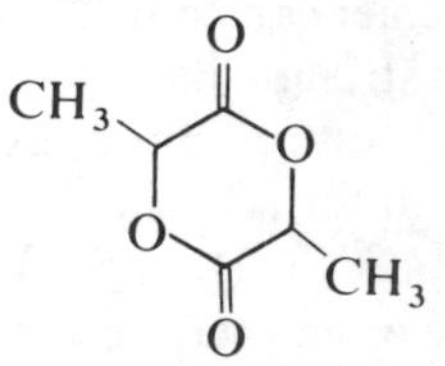

M.p. 128°C.

Polymerised by aluminium, magnesium or zinc alkyls or by stannic chloride to polylactide.

β-LACTOGLOBULIN A protein occurring in milk, of molecular weight 35 000 (for bovine β-lactoglobulin), containing two polypeptide chains, which contain all the amino acids, being especially rich in the essential amino acids. Therefore it functions well as food for growth in many animals. It is isolated from the milk whey (the supernatant liquid remaining after the casein has been precipitated by acidification of skimmed milk). It accounts for about 50% of the whey protein.

LACTONE (Cyclic ester) A compound most commonly of the simplest type

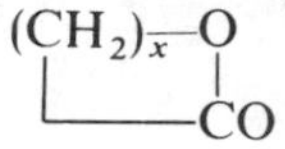

$$(CH_2)_{\overline{x}}\!-\!O \atop \qquad\quad\;\,|\atop \qquad\quad\,CO$$

e.g. β-propiolactone ($x=2$), γ-butyrolactone ($x=3$), δ-valerolactone ($x=4$), ε-caprolactone ($x=5$). The methylene groups may be substituted. Other types are cyclic diesters, e.g. of the type:

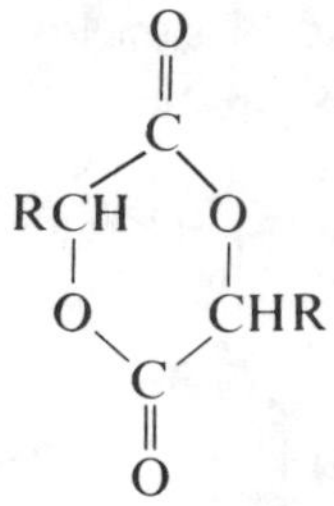

(glycolide when R = H, lactide when R = CH_3) or cyclic carbonates:

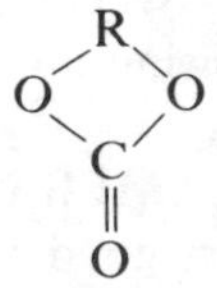

Lactones may often be polymerised to polyesters by ring-opening polymerisation:

$$n \; R\!-\!O \atop \quad\;\,|\atop \quad\;CO \;\;\to\; \text{—}[R\!-\!CO]_n\text{—} \atop \qquad\qquad\qquad\;\;\| \atop \qquad\qquad\qquad\;\,O$$

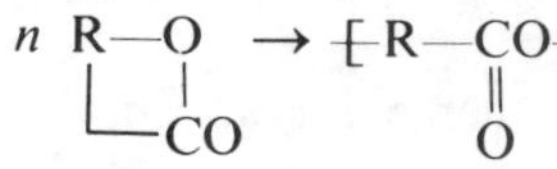

Polymerisability is dependent on both the ring size and the nature of the atoms in the ring. In general four-, seven- and eight-membered ring lactones polymerise readily but five- and six-membered rings do not. Anionic polymerisation may be initiated by organometallic complexes such as aluminium or zinc alkyls or cationic polymerisation may be initiated by Lewis acids. Polymerisation of ε-caprolactone has been widely studied since the polymer is a commercial product.

LACTOPRENE BN Tradename for an acrylic elastomer with good low temperature and oil resistance.

LACTOPRENE EV Tradename for an early acrylic elastomer based on polyethylacrylate containing about 5% 2-chloroethylvinyl ether comonomer for vulcanisation.

LADDER POLYMER A polymer in which the molecules consist of an uninterrupted sequence of rings interconnected by links in which steric restrictions prevent bond rotation between the rings, providing stiff molecules. The commonest type is the classical ladder or double strand polymer, where the rings are interconnected by two covalent bonds between each ring, thus forming a system of fused rings (**I**). Alternatively this structure may be viewed as consisting of two linear portions (or strands) regularly crosslinked. If the two bonds are between the same atoms (i.e. they form a double bond) an allene ladder (**II**) results. If the two bonds connect the same ring atom, but are not a double bond, the spiro-ladder (or spiro-polymer) (**III**) results. If structure **I** extends over two dimensions then a sheet or parquet polymer (**IV**) results.

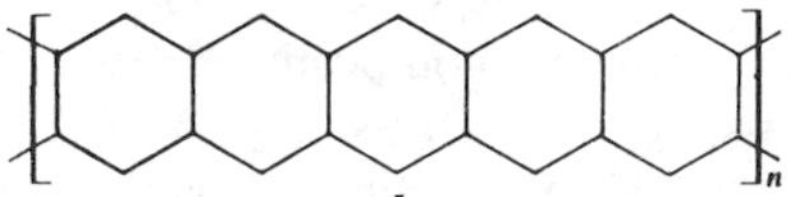

I

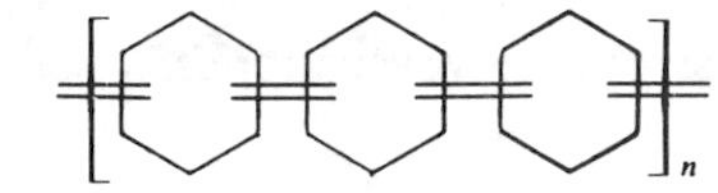

II

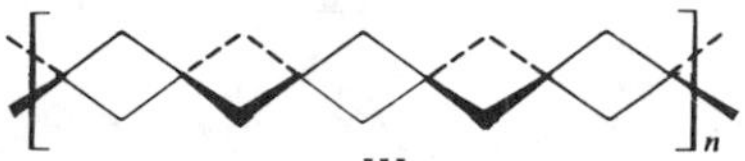

III

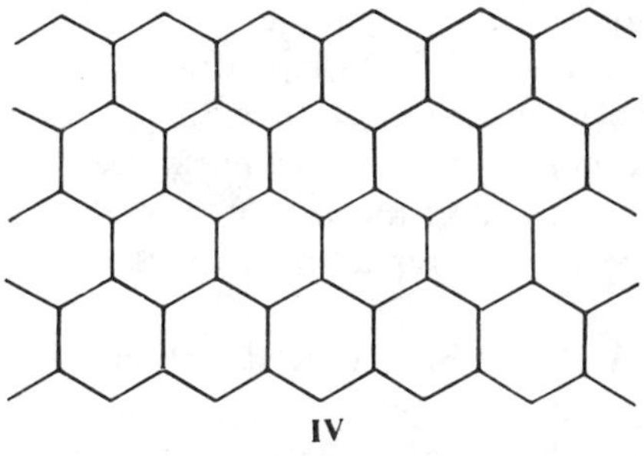

IV

LADDER POLYQUINOXALINE A polyquinoxaline with the quinoxaline rings joined through other fused rings. Synthesised by reaction between an aromatic 1,2,4,5-tetramine and a compound of type **I**,

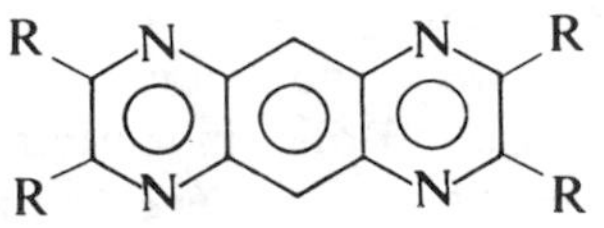

I

where R is —Cl, —OH or —OPh, by self-condensation of a 2,3-dihydroxy-7,8-diaminoquinoxaline (**II**) in PPA:

e.g. **II** →

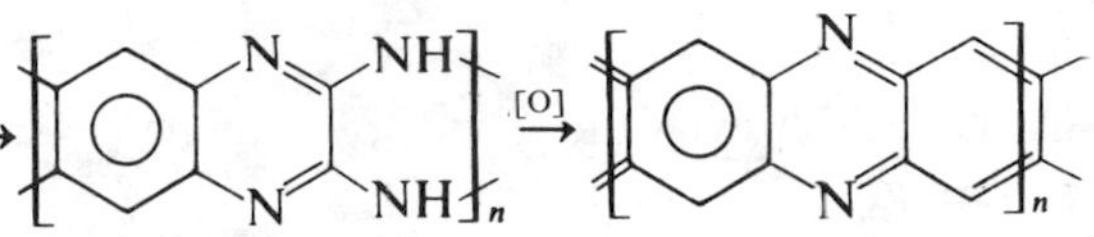

or by reaction of 1,2,6,7-tetraketopyrene with an aromatic 1,2,4,5-tetramine hydrochloride. The polymers can show good thermal stability (to 600°C in nitrogen) but this is often lower due to incomplete cyclisation resulting in partial ladder structure. Some polymers are useful high temperature resistant adhesives. Polymers with quinoxaline rings fused to thiazine and oxazine rings have also been synthesised.

LADDER PYRRONE Alternative name for polyimidazopyrrolone.

LADENE Tradename for polyvinyl chloride, polystyrene and low and linear low density polyethylene.

LAEVOPIMARIC ACID

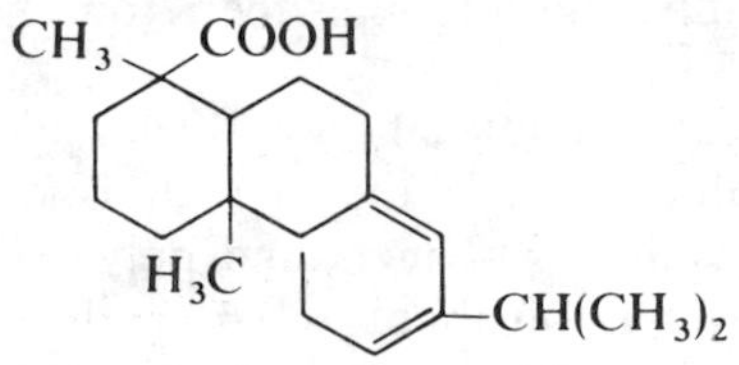

M.p. 153°C.

The major component of the oleoresin from certain pine trees. It is isomerised to abietic acid by the heat treatment used to obtain rosin from the oleoresin by distillation.

LAEVULINIC ALDEHYDE

$$O{=}CHCH_2CH_2C(CH_3){=}O$$

The only product, apart from a little laevulinic acid, of the hydrolysis of the ozonide of natural rubber, thus establishing the head-to-tail linking of the isoprene units in the polymer, since the product is formed by:

$$\sim CH_2C(CH_3){=}CHCH_2CH_2C(CH_3){=}CHCH_2\sim \xrightarrow{O_3}$$

$$\sim CH_2\underbrace{C(CH_3)CHCH_2}_{O_3}CH_2\underbrace{C(CH_3)CHCH_2}_{O_3}\sim \rightarrow$$

$$O{=}CHCH_2CH_2C(CH_3){=}O + H_2O_2.$$

LAEVULOSE Alternative name for D-fructose.

LAGRANGIAN COORDINATES The coordinate system used to describe the deformation of a body in which the reference coordinates are taken to be those in the undeformed body. The displacement vector **u** is also a function of the undeformed coordinates. The new position vector **r**′ is given by $\mathbf{r}' = \mathbf{r} + \mathbf{u}(\mathbf{r})$. The system is usually used in the description of infinitessimal deformation situations since the movement of axes within the body on deformation is negligible. The continuity, momentum and energy equations described with the Lagrangian frame of reference are thus said to be in their Lagrangian form.

LALLS Abbreviation for low angle laser light scattering.

LALS Abbreviation for large angle light scattering, often simply called light scattering. The alternative abbreviation WALS (wide angle light scattering) is sometimes used which reduces confusion with the abbreviation LALLS.

LAM Abbreviation for longitudinal acoustic mode.

LAMÉ CONSTANT Symbol λ. An elastic constant defined as, $\lambda = vE/(1 + v)(1 - 2v)$, where v is the Poisson ratio and E is the Young's modulus. In suffix notation this yields, for a stress component,

$$\sigma_{ij} = [E/(1 + v)]e_{ij} + \lambda\Delta \quad \text{for} \quad i = j$$

and $\sigma_{ij} = 2Ge_{ij} + \lambda\Delta\delta_{ij}$ for $i \neq j$, i.e. in shear, where Δ is the dilatation, G is the shear modulus, e_{ij} is the engineering strain component and δ_{ij} is the Kronecker delta.

LAMELLA (Lamellar crystal) The flat plate-like crystal or crystallite which is the characteristic crystal habit of most crystalline polymers and polymer single crystals. Typically it is of about 100 Å in thickness. Since the polymer chains have been shown by electron diffraction to lie across the thickness of the crystallite, they must be folded at the chain surface (chain folding) owing to their length of several thousand ångströms. In melt crystallised polymers the lamellae usually aggregate linearly to fibrils which in turn aggregate to spherulites. In single crystals the lamellae may be multilayer containing spiral growths, or may be aggregated to hedrites, axialites or dendrites.

LAMELLAR CRYSTAL Alternative name for lamella.

LAMELLAR FLOW Alternative name for laminar flow.

LAMELLAR THICKENING An increase in lamellar thickness brought about by heating the crystal (annealing) at a higher temperature than its original crystallisation temperature.

LAMELLAR THICKNESS The smallest dimensional size of a lamellar polymer crystal or crystallite. It is highly dependent on crystallisation conditions, especially the temperature. In folded chain crystals it has the same dimension as the polymer molecular length between folds. Generally it is in the range 50–500 Å. It may be determined by electron micrograph shadow length, small angle X-ray scattering or interference microscopy. It can be increased by annealing (lamellar thickening).

LAMELLAR TWISTING A regular twisting of the crystallites in the fibrils present in a spherulite giving rise to the ringed spherulite pattern.

LAMINARAN A 1,3'-linked β-D-glucan found in brown seaweeds, with some molecules being branched through 1,6'-links.

LAMINAR FLOW (Lamellar flow) (Streamline flow) A flow in which adjacent layers of fluid may be considered to move parallel to each other, i.e. flow without turbulence. It is assumed to occur in some simple flow models, such as plane couette flow. However, above a certain critical value of the Reynolds number (Re) associated with flow conditions (i.e. the nature of the fluid, the geometry of the containing channel), laminar flow becomes turbulent flow. Thus, for example, for a Newtonian fluid in a capillary, the initial value of Re is 2000. Laminar flows are only possible for long flow paths of constant cross-section. For non-Newtonian fluids they are only possible for a few geometries e.g. circular cross-section pipes or infinite slits.

LAMINATE (Laminated composite) (Layered composite) A composite in which the dispersed phase consists of layers of reinforcement bonded together. The laminate is built up by stacking individual layers of reinforcement impregnated with matrix polymer (the plies) and heat fusing or otherwise curing the matrix. Within each ply the fibres of the reinforcement are usually aligned so that whilst stiffness and strength is enhanced in the direction of alignment, the transverse properties are poor. Lamination therefore improves overall properties by aligning the plies in desired orientations so that properties may be maximised in the desired directions in the overall laminated composite. Typical laminating bases are paper, used especially with phenol– and melamine–formaldehyde resins, often as kraft or some tougher paper, woven fabrics, such as cotton, nylon, rayon and glass, or non-woven fabrics, which are cheaper, such as chopped strand mat.

LAMINATED COMPOSITE Alternative name for laminate.

LAMM'S SCALE METHOD An early technique for determination of the polymer concentration gradient in a cell as a result of ultracentrifugation. The cell is photographed in front of a uniform scale and the concentration gradient causes a refractive index gradient (dn/dc) which displaces the lines of the scale in proportion to dn/dc. Now often replaced by the Schlieren method.

LAMPBLACK (LB) A form of carbon black produced by burning a hydrocarbon fuel in pans and collecting the black from the smoke. Particles are 50–200 nm in diameter and therefore have a low surface area compared with other blacks. Usually lampblack has a high structure. It is used in high hardness and resilient rubber materials.

LANDAU–PLACZEK RATIO In Brillouin scattering, the ratio $I_R/2I_B$, where I_R and I_B are the intensities of the Rayleigh scattering and Brillouin scattering lines respectively. This ratio varies with temperature (markedly at T_g).

Measurement of the ratio in solution also provides a method for the determination of molecular weight.

LANESE Tradename for a cellulose acetate staple fibre.

LANITAL Tradename for a man-made protein fibre from casein.

LANITOL Tradename for a casein fibre.

LANON Tradename for a polyethylene terephthalate fibre.

LANSING–KRAEMER DISTRIBUTION A particular form of the generalised logarithmic normal distribution of molecular species, whose differential weight distribution function is,

$$W(r) = \exp\left[-(\ln r - \ln r_m)/2\,\sigma^2\right]/(2\pi)^{1/2}\exp(\tfrac{1}{2}\sigma^2)$$

where r is the size of each individual molecular species, e.g. degree of polymerisation, r_m is its mean value and σ is the standard deviation. An empirical distribution, not derived from any particular polymerisation mechanism.

LANTHIONINE

$$\underset{\substack{|\\NH_2}}{HOOCCHCH_2}SCH_2\underset{\substack{|\\NH_2}}{CHCOOH}$$

A thio-amino acid, similar to cysteine. Lanthionine residues are formed on the treatment of wool with alkali.

LANUSA Tradename for a viscose rayon staple fibre of circular cross-section and a wool-like handle. Coagulated slowly during spinning (by use of an initial treatment with water and dilute alkali) and stretched. A precursor for the later developed polynosic rayons.

LARGE ANGLE LIGHT SCATTERING (LALS) Alternative name for wide angle light scattering, commonly simply called light scattering.

LARGE ANGLE X-RAY SCATTERING (LAXS) Alternative name for wide angle X-ray scattering.

LARGE STRAIN ELASTICITY (Finite strain elasticity) Elastic behaviour in which the deformation is large enough for the small strain approximations not to be valid. Thus in the partial differential displacement equations giving the strain components, the second order terms, e.g. $(\partial u/\partial y)^2$, $(\partial u/\partial y \cdot \partial u/\partial z)$, where u, v and w are the displacements of x, y and z (the coordinates of a point) are not ignored. Furthermore, the stress components are different in the deformed compared with the undeformed material, so they must be specified with respect to one or the other; usually the former is chosen. Finite elasticity theory is applied to the elastic deformation of rubbers, which can sustain strains of several hundred per cent. Equations analogous to the generalised Hooke's law constitutive relations may be used. Materials which obey these relations are frequently called neo-Hookean.

LASER RAMAN SPECTROSCOPY Raman spectroscopy using a laser as the exciting light source. Most Raman spectroscopy is laser Raman spectroscopy.

LASTRILE Generic name for a fibre composed of a copolymer of acrylonitrile and butadiene containing between 10 and 50% acrylonitrile.

LATENE Tradename for low density polyethylene.

LATERAL CONTRACTION RATIO The ratio of the lateral contraction to the longitudinal extension in a tensile deformation. For an isotropic material behaving in a linear manner, lateral contraction ratio is identical to Poisson's ratio. Owing to the difficulties of measuring the small changes in dimension laterally, the lateral contraction ratio is difficult to determine experimentally with accuracy. For viscoelastic measurements in creep, changes in both lateral and axial dimensions occur with time in measuring the creep lateral contraction ratio, but for the relaxation lateral contraction ratio only the lateral dimension changes.

LATERAL MODULUS Alternative name for transverse modulus.

LATEX A polymer emulsion, i.e. a dispersion of small polymer particles (often approximately 1 μm spherical particles) in water. Produced synthetically by emulsion polymerisation, but polymer latices also occur naturally, natural rubber latex being particularly important. The latex is stabilised by the presence of an emulsifier (a soap or detergent) whose action depends on the correct pH or the presence of other salts. The solid polymer may be isolated from a latex by coagulation, i.e. aggregation of the latex particles, by destroying the stabilising action of the soap. This may be achieved by the addition of large quantities of ionic materials. Polymer latices are useful technologically in providing a polymer in a fluid form, from which products may be manufactured by liquid (rather than the more usual melt) technology, involving e.g. casting, dipping and coating processes.

LAURIC ACID (Dodecanoic acid)

$$CH_3(CH_2)_{10}COOH \qquad \text{M.p. } 44{\cdot}2°C.$$

Accounts for 44% of the acid residues of coconut oil. Metal salts of lauric acid are useful as lubricants and stabilisers for polyvinyl chloride.

LAUROLACTAM Alternative name for lauryllactam.

LAUROYL PEROXIDE

$$\underset{\substack{\|\\O}}{CH_3(CH_2)_{10}C}O-O\underset{\substack{\|\\O}}{C(CH_2)_{10}CH_3}$$

A diacyl peroxide widely used as a free radical polymerisation initiator, especially for styrene, vinyl chloride and acrylic monomers. Of somewhat greater activity than benzoyl peroxide, having half-life times of 60 h/50°C,

3 h/70°C and 0·1 h/100°C and half-life temperatures of 62°C/10 h, 80°C/1 h and 117°C/1 min.

LAURYLLACTAM (Laurolactam) (Dodecanolactam)

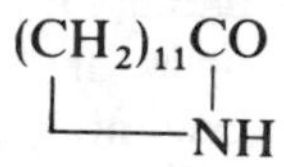

$$\begin{array}{c} (CH_2)_{11}CO \\ | \\ \underline{\qquad} NH \end{array}$$

M.p. 153–154°C.

The monomer used for the production of nylon 12. Synthesised from cyclododecatriene (obtained by cyclotrimerisation of butadiene), which is oxidised to cyclododecanone, oximated and Beckman rearranged to the lactam, as in the synthesis of ε-caprolactam. Polymerisation is slow due to ring stability and requires temperatures of 300–350°C and the use of acid catalysts.

LAURYL MERCAPTAN Alternative name for dodecyl mercaptan.

LAVSAN Tradename for a polyethylene terephthalate fibre.

LAW OF MIXTURES (Rule of mixtures) (Simple law of mixtures) A mathematical relationship relating the value of a property P of a mixture to the values of the property for the individual components P_1 and P_2 (for the usual case of a binary mixture) weighted according to their volume fractions ϕ_1 and ϕ_2. Usually of the form, $P = \phi_1 P_1 + \phi_2 P_2$, although the inverse law of mixtures is also sometimes found to hold. Very often P refers to a modulus, such as the Young's modulus E, and the relation is widely useful for composite materials. For example the relation $E_c = \phi_m E_m + \phi_f E_f$ describes the modulus of a composite (E_c) containing a matrix of modulus E_m and volume fraction ϕ_m and continuous fibres of modulus E_f and volume fraction ϕ_f, in the direction of the fibres, assuming that they are all parallel. In this case the composite can be considered to behave as a parallel arrangement of components (the isostrain or Voight model). For the isostress series arrangement of components (the Reuss model) the inverse law applies. In practice the simple form provides an upper bound to the modulus and the inverse form a lower bound. This lower bound value is approximately that found for the above composite when stressed perpendicular to the fibres. For other types of fibre orientation a fibre efficiency factor (η) is used to modify the equation. Thus for Young's modulus: $E_c = \eta \phi_f E_f + \phi_m E_m$. Thermal and electrical properties may also sometimes be represented by a law of mixtures relationship.

LAXS Abbreviation for large angle X-ray scattering.

LAYERED COMPOSITE Alternative name for laminate.

LAYER POLYMER Alternative name for parquet polymer.

LAYER SILICATE (Phyllosilicate) One of the two dimensional sheet-like naturally occurring crystalline mineral silicates, in which SiO_4 tetrahedra (as $Si_4O_{10}^{4-}$) are joined into infinite sheets and bound to either one or both sides of a layer of $Al(OH)_3$ (gibbsite) or $Mg(OH)_2$ (brucite). Any residual electrostatic charge is satisfied by additional cations. Many important minerals are of this type, such as chrysotile, kaolinite, vermiculite and talc.

LB Abbreviation for lampblack.

LCC Abbreviation for low colour channel carbon black.

LC POLYMER Abbreviation for liquid crystalline polymer.

LCST Abbreviation for lower critical solution temperature.

LDPE Abbreviation for low density polyethylene.

LEACRIL Tradename for a polyacrylonitrile fibre.

LEADERMAN THEORY A theory of non-linear viscoelasticity based on a modified Boltzmann Superposition Principle, in which it is assumed that creep and recovery for any load are a unique function of stress, $f(\sigma)$. The deformation is given by,

$$e(t) = \sigma/E + \int_{-\infty}^{t} [df(\sigma)/d\tau](t - \tau)\,d\tau$$

where the first term is the instantaneous elastic deformation (with modulus E) and τ is the time at which the load was applied. Some films, such as nylon and cellulosic films, behave in this way but others do not. The theory cannot take account of complex loading histories.

LEAD STABILISER Alternative name for basic lead stabiliser.

LEAF FIBRE A natural fibre obtained from the leaf stalks of certain plants. Manila hemp, sisal and henequen are examples.

LEATHERY REGION Alternative name for retarded elastic state.

LEATHERY STATE Alternative name for retarded elastic state.

LEAVIL Tradename for a polyvinyl chloride fibre.

LECITHIN Alternative name for lipovitellin.

LECTROFILM Tradename for poly-(2-vinyldibenzofuran).

LEFM Abbreviation for linear elastic fracture mechanics.

LEGUVAL Tradename for an unsaturated polyester resin.

LEU Abbreviation for leucine.

LEUCHS ANHYDRIDE Alternative name for *N*-carboxy-α-amino acid anhydride.

LEUCINE (Leu) (L)

$$H_2NCHCOOH$$
$$|$$
$$CH_2CH(CH_3)_2$$

M.p. 293°C (decomposes).

A non-polar α-amino acid occurring widely in proteins. Its pK' values are 2·36 and 9·60, with the isoelectric point at 5·98. It is often associated with the α-helix conformation in proteins. Leucine residues in proteins are preferentially cleaved by chymotrypsin. It is synthesised by the Strecker synthesis from isovaleraldehyde.

LEVAFLEX Tradename for a thermoplastic polyolefin elastomer.

LEVAN A β-D fructan, found in many grasses, and consisting of 20–30 fructofuranose units 2,6'-linked and terminated by an α-D-glucose unit. Similar bacterial levans have molecular weights of about 10^6, but are branched through 2,1'-links about every ninth fructofuranose unit.

LEVAPREN Tradename for ethylene–vinyl acetate copolymer with about 45% vinyl acetate content, which is a useful rubber and may be vulcanised with peroxides.

LEVASINT Tradename for ethylene–vinyl alcohol copolymer.

LEVELLING-OFF DEGREE OF POLYMERISATION (Levelling-off DP) The degree of polymerisation of the polymer remaining after the amorphous regions of a partially crystalline polymer have been removed by degradation. Applies particularly to hydrolytically degraded cellulose. The degradation reagent penetrates the polymer and preferentially reacts with the amorphous regions first. This causes an initial rapid drop in molecular weight and hence in the degree of polymerisation to the levelling-off value. This value may be related to the crystallite size, since the crystallites are subject to degradation at their fold surfaces. Further degradation within the crystallites is diffusion controlled and is very slow.

LEVELLING-OFF DP Alternative name for levelling-off degree of polymerisation.

LEVOFLEX Tradename for ethylene–propylene–diene monomer thermoplastic polyolefin rubber.

LÉVY MISES EQUATIONS The equations that result from the assumption that the magnitudes of the plastic strain increments that occur during yielding, due to the movements of the externally applied forces, can be described by a strain increment tensor whose principal components σ'_1, σ'_2 and σ'_3 are proportional to the principal components of the deviatoric strain tensor de_1, de_2 and de_3. Thus $de_1/\sigma'_1 = de_2/\sigma'_2 = de_3/\sigma'_3$. In terms of equivalent strain increments $d\bar{e}$, with the equivalent stress $\bar{\sigma}$, the equations may be written

$$de_1 = d\bar{e}/\bar{\sigma} = [\sigma_1 - \tfrac{1}{2}(\sigma_2 + \sigma_3)]$$

and similarly for de_2 and de_3. These are the basic equations describing plastic deformation. For problems involving constant stress and strain ratios, the equations may be integrated to give the Hencky plasticity equations.

LEXAN Tradename for bisphenol A polycarbonate.

LICHENAN A hot-water soluble β-D-glucan found in Iceland moss, containing both 1,3'- and 1,4'-links between the glucose units and with a degree of polymerisation of about 100.

LIFETIME OF A KINETIC CHAIN (Radical lifetime) Symbol τ_s. The average time that a growing polymer free radical active centre exists from its initiation to its termination in a free radical polymerisation. Used especially in the theory associated with the rotating sector method for the determination of the individual rate constants for propagation and termination. Typically τ_s is a few seconds.

LIGAND A molecule which will bind strongly to a protein but non-covalently, as with an enzyme and its specific co-enzyme. Binding to specific ligands is the principle of affinity chromatography.

LIGHT MEROMYOSIN The fragment produced on trypsin cleavage of myosin, of molecular weight of about 150 000 and of predominantly α-helical structure.

LIGHT SCATTERING (Large angle light scattering) A method for the determination of the weight average molecular weight ($\bar{M}_w$) of a polymer by determination of the angular dependence of the intensity of scattered light from a dilute solution of the polymer. Information on the virial coefficients, *z*-average molecular weight and radius of gyration may also be obtained.

For a simple solution in which the solute molecules act as point scatterers, i.e. for Rayleigh scattering, the intensity of scattered light, given by the reduced scattering factor (R_θ), is related to the molecular weight (M) by, $Kc/R_\theta = 1/M$, where K is a constant equal to $2\pi^2 n^2/\lambda^4 N_A$ $(dn/dc)^2$ (where c is the polymer concentration, n is the solvent refractive indexes, dn/dc is the refractive index increment, N_A is Avogadro's number and λ is the wavelength of the light). Thus the sensitivity increases as M increases, in contrast to the colligative properties, so the light scattering method is particularly suitable for the determination of M of high molecular weight polymers. The lower determinable limit of M depends on dn/dc, but is usually around 20 000, since solvent scattering becomes of similar magnitude below this limit. An upper determinable limit of M is around several million because of the difficulty in removing dust particles which scatter similarly to high molecular weight polymer.

However, R_θ is not independent of c and is better expressed by a virial equation, $Kc/R_\theta = 1/M + 2Bc/RT + \cdots$, where B is the second virial coefficient, T is the absolute temperature and R is the universal gas constant. In the limit as $c \rightarrow 0$, $Kc/R_\theta = 1/M$. For a polydisperse polymer the molecular weight M is $\bar{M}_w$.

If the polymer molecular weight is so large (usually greater than about 30 000) that the molecules cannot be considered as point scatterers, then internal interference occurs and scattering is greater in the forward direction than in the backward direction. This dissymmetry is characterised by the dissymmetry ratio. Internal interference effects can be eliminated by extrapolating measurements at low angles to zero angle. The attenuation of scattering is given by the particle scattering factor P_θ whose angular dependence is given by the particle scattering function $P(\theta)$, which is itself dependent on particle shape. At low angles for any particle shape,

$$\lim_{\substack{c \rightarrow 0 \\ \theta \rightarrow 0}} (Kc/R_\theta) = 1/\bar{M}_w P(\theta)$$
$$= 1/\bar{M}_w(1 + 16\pi^2 \langle s^2 \rangle \sin^2(\theta/2)/3\lambda^2)$$

where $\langle s^2 \rangle$ is the mean square radius of gyration. Thus a plot of Kc/R_θ versus $\sin^2\theta/2$, extrapolated to zero concentration, will approach asymptotically to a straight line at zero angle, whose intercept is $1/\bar{M}_w$ and whose slope yields $\langle s^2 \rangle$. This is the basis of the Zimm plot method of analysis of light scattering data. A simpler method is the dissymmetry method. Experimentally therefore, values of the scattered light intensity are required at various scattering angles for a series of polymer solutions of different concentrations. From these values the scattering due to solvent must be subtracted. The data is simply obtained using a light scattering photometer.

LIGHT SCATTERING PHOTOMETER (Photometer) An apparatus for measuring the intensity of light scattered by a polymer solution as a function of scattering angle (θ). The data obtained enable the weight average molecular weight of the polymer to be calculated using the light scattering method and a Zimm plot. The light source is either a mercury arc lamp or, better, a laser, giving a monochromatic beam incident upon a suitably shaped glass cell containing the scattering solution. The scattered light, of intensity I_θ, is detected by a rotatable photocell whose output is registered as a meter reading (M_θ) on a sensitive galvanometer. M_θ can be related to I_θ by a complex relationship involving various instrumental parameters.

LIGNIN The binding polymer for cellulose fibres in plants and trees. Typically a wood contains about 25%

lignin, which, together with the cellulose, acts as the structural component of the plant or tree, but apparently has no metabolic role. Lignin is a highly crosslinked network polymer, possibly existing as an infinite network, and therefore is insoluble. Thus structural studies have to be performed on degraded lignin. The very complex network is built up from aromatic alcohol-type monomers such as coniferyl alcohol and p-hydroxycumyl alcohol. The polymer is probably biosynthesised by oxidative polymerisation of these phenylpropane-type monomers forming carbon–carbon and ether links of many different types, forming a very complex network. In the pulping of wood the lignin is separated from the cellulose either by treatment with sulphur dioxide and calcium bisulphite, producing lignosulphonates, or with alkali, which solubilises it in water.

LIGNOSULPHONIC ACID The water soluble derivative of lignin produced by the bisulphite wood pulping process by treatment of wood with sulphur dioxide and calcium bisulphite.

LIGROIN Alternative name for petroleum ether.

LIMIT DEXTRIN The unhydrolysed polymer residue resulting from the enzyme catalysed hydrolysis of starch (or its components). Thus in the case of β-amylase hydrolysis of amylopectin, the resultant dextrin is produced through the inability of the β-amylase to hydrolyse beyond a branch point. Thus only the chain ends have been shortened.

LIMITING OXYGEN INDEX (LOI) (Oxygen index) The concentration of oxygen, expressed as volume per cent or volume fraction, in the oxygen–nitrogen mixture which will just sustain the burning of a material. For a solid polymer the LOI is determined by burning a stick of the polymer downwards like a candle (the candle test) under specified conditions. A material with LOI 21 would nominally burn in air. However, a value of 27 is taken as the self-extinguishing limit since candle-like burning neglects convective heating, which raises LOI.

LIMITING VISCOSITY NUMBER (LVN) (Intrinsic viscosity) (Staudinger Index) Symbol $[\eta]$. The value of the viscosity number of a dilute polymer solution extrapolated to infinite dilution, i.e. $(\eta_{sp}/c)_{c \rightarrow 0}$. It therefore gives a measure of the ability of an isolated polymer molecule to increase the viscosity of the solvent in the absence of intermolecular interactions between polymer molecules. Similar extrapolation of the logarithmic viscosity number, i.e. $(\ln \eta_{rel}/c)_{c \rightarrow 0}$, gives the same LVN.

This second extrapolation method is sometimes preferred since the slope of the plot of $\ln \eta_{rel}/c$ versus c is less than that of η_{sp}/c versus c. However, the data obtained by capillary viscometry are often plotted in both ways and the common intercept at $c = 0$ is taken as the LVN. The LVN may also be calculated from data obtained at only a single concentration by a single point method. Whilst this is much more rapid, it gives less reliable values.

Determination of the LVN is the most common method of measuring the molecular weight of a polymer, as the viscosity average molecular weight, using the Mark–Houwink equation. The usual units are $dl\,g^{-1}$ (decilitres per gram) since concentration is expressed as $g\,dl^{-1}$ in dilute solution viscometry.

LINEAR ELASTIC FRACTURE MECHANICS
(LEFM) An analysis of stress distribution near cracks in a stressed material, assuming that the material behaves linearly elastically up to fracture and that any yielding is confined to only a small region near the crack tip. This is reasonable for brittle polymers such as crosslinked thermosets and amorphous polymers well below T_g, but not for ductile materials.

LINEAR ELASTICITY
Elasticity in which stress is directly proportional to strain—the essence of Hooke's law and the basis of small strain elasticity. Applies only approximately to many polymers and then only at very small strains.

LINEAR LOW DENSITY POLYETHYLENE
(LLDPE) Tradenames Dowlex, Eraclere, Escorene, G-Resin, Innovex, Lacqtene, Ladene, Lotrex, Sclair, Stamylex. A low density polyethylene produced by a low pressure process, by Ziegler–Natta polymerisation or by use of the Phillips process. This is in contrast to conventional LDPE which is produced by the somewhat less convenient high pressure process. The low density is due to reduced crystallinity resulting from the use of up to about 10% of a 1-alkene comonomer such as butene-1 (the most common and the cheapest), hexene-1, octene-1 or 4-methylpentene-1. Thus, like conventional LDPE, the polymer contains short carbon chain side groups, which reduce structural regularity and hence crystallisation. Overall the polymer has similar properties to LDPE, but has somewhat better strength properties in films, and hence this is its major area of use. However, the polymers are less pseudoplastic than LDPE, so melt viscosities are higher during processing.

LINEAR PHTHALATE
A phthalate ester plasticiser made using a commercial mixture of, for example, six to ten carbon linear alcohols as opposed to the more traditional branched alcohols such as 2-ethylhexanol and isooctanol. The linear phthalates have better retention of low temperature flexibility and lower volatility than their branched counterparts.

LINEAR POLYMER
A polymer in which the molecules consist of single unbranched chains of atoms, such as a vinyl polymer

$$\sim CH_2-CHX-CH_2-CHX-CH_2$$
$$CHX-CH_2-CHX\sim \text{ or } +CH_2-CHX+_n.$$

Formed by simple step-growth polymerisation by reaction between difunctional monomer molecules or by simple chain polymerisation of singly unsaturated monomers or of conjugated dienes. If the monomer contains side groups, which themselves may be short chains, then these become side groups in the regular repeat units of the polymer and the polymer is still considered to be linear rather than branched. Linear polymers thus comprise all polymers that are not branched, crosslinked or of a network structure. Regular two-stranded polymers are also considered to be linear polymers.

LINEAR STEP-GROWTH POLYMERISATION
Step-growth polymerisation using difunctional monomers; therefore capable of forming only linear polymer molecules.

LINEAR VISCOELASTICITY
The simplest type of viscoelasticity, which can be described by a linear differential relationship between stress, strain and strain rate with constant coefficients. Thus there is direct proportionality between stress, strain and strain rate but no dependence of their ratios on the stress. If this latter were so, then the behaviour would be that of non-linear viscoelasticity. One simple example of linear behaviour is to assume that the stresses are additive, as in the Voight model, giving $\sigma = Ge + \eta\, de/dt$, where σ is the stress, e is the strain, G is the shear modulus and η is the coefficient of viscosity. Other more complex relations, e.g. based on the standard linear solid or arrays of Voight or Maxwell models, often more accurately represent real behaviour. However, for many polymer systems, except at small strains, non-linear behaviour is found.

LINOLEIC ACID   (*Cis*-9,*cis*-12-octadecadienoic acid)

$$CH_3(CH_2)_4CH{=}CHCH_2CH{=}CH(CH_2)_7COOH$$

M.p. $-5°C$. B.p. $202°C/1\cdot4$ mm.

Occurs in the triglycerides of most plant oils sometimes to a large extent, e.g. safflower oil (75% of the acid residues), soya bean oil (51% of the acid residues), cottonseed oil (40% of the acid residues). The double bonds are responsible for its air drying properties.

LINOLENIC ACID   (*Cis*-9,*cis*-12,*cis*-15-octadecatrienoic acid)

$$CH_3CH_2CH{=}CHCH_2CH$$
$$\|$$
$$CHCH_2CH{=}CH(CH_2)_7COOH$$

M.p. $-11°C$. B.p. $151–158°C/0\cdot001$ mm.

The major acid residue of the triglycerides of linseed oil (52%) and also occurs in other oils such as soya bean oil. It is responsible for the rapid air drying of linseed oil.

LINSEED OIL
An oil obtained from flax seeds, which is the glyceryl ester containing approximately 9% stearic, 24% oleic, 17% linoleic and 50% linolenic acid residues.

Thus it may be approximately represented by glyceryl oleate, dilinolenate:

$$CH_2OOCR$$
$$|$$
$$CHOOCR'$$
$$|$$
$$CH_2OOCR'$$

where

$R = -(CH_2)_7CH=CH(CH_2)_7CH_3$ and

$R' =$

$-(CH_2)_7CH=CHCH_2CH=CHCH_2CH=CHCH_2CH_3$

It is a drying oil and is widely used in alkyd resins for its good drying properties.

LIPOPOLYSACCHARIDE   A glycolipid in which the carbohydrate part is a polysaccharide. The most studied examples are the cell-envelope materials from certain Gram-negative bacteria, e.g. *Salmonella*. Consists of a core oligosaccharide, which consists of heptose, ethanolamine, phosphate, 3-deoxy-D-manno-octulosamine, D-galactose, D-glucose and 2-amino-2-deoxy-D-glucose units. The core is linked through 3-deoxy-D-manno-octulosamine to the lipid component (called lipid A) and through a hexose to a further polysaccharide containing specific sugar sequences responsible for the type or O-antigen specificity of these substances.

LIPOPROTEIN   A conjugated protein consisting of a complex of a protein with a lipid. Although covalent bonding between the two components may occur, frequently a lipoprotein is held together by hydrophobic bonding. There are three important types of lipoprotein. Firstly, those of the blood plasma, e.g. α_1-lipoprotein and β_1-lipoprotein, which frequently are classed as very low density, low density and high density lipoproteins according to their flotation rate. These can contain anything from about one to about 50% protein. They function in lipid transport from one organ to another. Secondly, the membrane lipoproteins of the cell membrane. These are usually composed of about 60% protein and consist of complex mixtures of polypeptides which seem not to be covalently bonded since they can be removed either by solvent extraction (extrinsic protein) or by more drastic methods, e.g. by use of SDS or other denaturing agents (the intrinsic proteins). Thirdly, there are the egg yolk lipoproteins such as lipovitellin, phosvitin and livitin.

α_1-LIPOPROTEIN   The blood plasma lipoprotein fraction that migrates with the α-globulins on electrophoresis of the plasma. It consists mostly of high density lipoprotein, containing about 50% lipid and having a molecular weight of $2-5 \times 10^5$. It consists of a mixture of polypeptides which can be further fractionated by density gradient centrifugation.

β_1-LIPOPROTEIN   The blood serum lipoprotein fraction that migrates with the β-globulins on electrophoresis of the serum. It consists mostly of low density (0·98–1·03 g cm^{-3}) lipoproteins containing 80–90% lipid, with a molecular weight from several million to 20 million, forming spherical particles. It can be further fractionated by density gradient centrifugation. Like the other serum lipoproteins it acts in the solubilisation of lipid for its transport in the blood.

LIPOVITELLIN   (Lecithin) The major lipoprotein fraction occurring in avian egg yolk. It is a conjugate of a phospholipid and vitellin with a very complex structure; possibly a complex of molecular weight 400 000 with smaller subunits of molecular weight down to about 30 000. It contains about 1·5% phosphorus, with 18% combined lipid as phosphatide.

LIPOVITELLININ   A lipoprotein occurring in avian egg yolk which floats when the yolk is centrifuged. It is a conjugate of a phospholipid and vitellinin. It contains about 1·7% phosphorus and about 40% phospholipid probably being a highly heterogeneous material.

LIQUEFYING AMYLASE Alternative name for α-amylase.

LIQUID CRYSTAL   A highly anisotropic fluid formed by certain substances (including some polymers—liquid crystalline polymers) with highly elongated or elliptically shaped, relatively rigid molecules. The molecules are oriented over a large distance, which imparts some solid-like properties, but the intermolecular forces are not sufficient to prevent flow. Liquid crystalline materials are thus highly anisotropic and have unusual optical and electrical properties due to their anisotropy. Liquid crystalline behaviour is also called mesomorphism.

LIQUID CRYSTALLINE MAIN CHAIN POLYMER A liquid crystalline polymer in which the mesogenic groups are part of the main polymer chain. The links between the mesogens may be rigid, giving a rigid backbone, as in aromatic polyesters, aromatic polyamides and poly-(p-phenylenebenzothiazole), or they may be flexible as in copolymers containing ethylene-terephthalate and p-oxybenzoate units. Polypeptides, which consist entirely of flexible main chain bonds, such as poly-γ-benzyl-L-glutamate, may also be liquid crystalline due to formation of the α-helix, which is rigid because of its powerful intramolecular hydrogen bonding.

LIQUID CRYSTALLINE POLYMER   (LC Polymer) (Mesomorphic polymer) A polymer which is capable of forming a liquid crystalline phase (a mesophase). The mesophase is so called since it consists of molecules (the mesogens) or rather (especially in the case of polymers) of parts of molecules (the mesogenic groups), which are ordered with a degree of order between the very regular three-dimensional orientational and positional order of a crystalline phase and the high disorder of a liquid. Liquid crystalline materials are characterised by long-range orientational order but by a lack of positional order in at least one dimension. Molecular ordering may be of different types—nematic, cholesteric or smectic. The mesogenic groups may be present either within a polymer chain (a liquid crystalline main chain polymer) or in side groups (a liquid crystalline side chain polymer). Often liquid crystallinity is enhanced by separating the mesogenic groups with spacer groups. Formation of a mesophase, and hence observation of liquid crystalline behaviour, may occur in a solution of a polymer (lyotropic polymer) or in the solid polymer within a certain temperature range (thermotropic polymer). Lyotropic behaviour was first observed in polymers with solutions of poly-(γ-benzyl-L-glutamate), and later with other polypeptides, when the α-helices act as the stiff mesogenic structures. Liquid crystallinity often results in enhanced mechanical performance, as in the high modulus aromatic polyamide fibres spun from liquid crystalline polymer solutions. Thermotropic behaviour is of interest, especially in some aromatic polyesters for use as plastics moulding materials, whose melts, being liquid crystalline, can have low viscosities due to orientation under the shear of moulding. The orientation persists in the solid material, enhancing modulus. Liquid crystalline materials also have interesting anisotropic electrical and optical properties which have applications in display devices.

LIQUID CRYSTALLINE SIDE CHAIN POLYMER A liquid crystalline polymer in which the mesogenic groups are part of the groups attached to the main chain. Ordering of the mesogens, and hence liquid crystalline behaviour, is enhanced by distancing them from the main chain with flexible groups. Many such polymers have been studied, such as

$$\left[\begin{array}{c} CH_2-CH- \\ | \\ (CH_2)_6OC_6H_4X \end{array} \right]_n$$

LIQUID–LIQUID TRANSITION Symbol $T_{l,l}$. A transition occasionally observed, occurring at a temperature above the main, or glass, transition, in certain amorphous polymers, such as in polystyrene at about 160°C. It is possibly due to the onset of motions of complete polymer chains. It is also thought to occur in polypropylene, polyisobutene and ethylene–propylene copolymers.

LIQUID RUBBER A liquid, usually a viscous low molecular weight polymer (typical molecular weight being a few thousand), which may be readily crosslinked to a product similar in behaviour to that of a normal rubber vulcanisate. The use of a liquid rubber offers the considerable technological advantage over conventional rubbers that expensive compounding and press vulcanisation procedures are not required. Some polyurethane elastomers, silicone rubbers and polysulphide rubbers may be processed as liquid rubbers. An early liquid rubber was depolymerised natural rubber and the term originally applied specifically to this material. More recently liquid

rubbers based on polybutadiene (of molecular weight about 10 000), which may be crosslinked through functional groups on the chain ends (telechelic polymer) have been developed, e.g. carboxy-terminated polybutadiene and hydroxy-terminated polybutadiene and similar nitrile and styrene–butadiene copolymers.

LIRELLE Tradename for a polyethylene terephthalate fibre.

LIVETIN The avian egg yolk phosphoprotein remaining soluble after separation of the vitellin and containing a lower phosphorus content (about 0·1%). Preparations consist of several components and can have enzymic activity.

LIVING POLYMER A polymer in which the active centres of the polymerisation are retained even though all the monomer has been consumed. This is not possible in a free radical polymerisation since free radicals mutually terminate. It can only be achieved occasionally in cationic polymerisation. Living polymers may however be frequently prepared by anionic, usually Ziegler–Natta, polymerisation provided care is taken to exclude all impurities that might 'kill' the active centres by chemical reaction with them. If a second monomer is added to a living polymer then this will continue the polymerisation forming a block copolymer. This process may be repeated to yield a multiblock polymer. Alternatively, the addition of an appropriate reagent which reacts with the active centre can yield a polymer with a desired predetermined end group,

$$\sim\overset{+}{CHXY} + ROH \rightarrow \sim CH_2X + ROY;$$
$$\sim\overset{+}{CHXY} + CO_2 \rightarrow \sim CHXCOO\overset{-}{Y}\overset{+}{} \overset{H^+}{\rightarrow}$$
$$\sim CHXCOOH + \overset{+}{Y};$$
$$\sim\overset{+}{CHXY} + CH_2\overset{}{-}CH_2 \rightarrow \sim CHXCH_2CH_2\overset{-}{O}\overset{+}{Y} \overset{H^+}{\rightarrow}$$
$$CHXCH_2CH_2OH + \overset{+}{Y}.$$

LLDPE Abbreviation for linear low density polyethylene.

LOCUST BEAN GUM A galactomannan polysaccharide obtained from the seed endosperm of the carob tree. It probably consists of 1,4'-linked β-D-mannopyranose units with lesser amounts of α-D-galactopyranose single side groups. Useful as a size and paper and food additive.

LOGARITHMIC DECREMENT Symbol Δ. The measure of damping of the oscillations in the torsion pendulum method for the characterisation of dynamic behaviour. Defined as

$$\Delta = \ln(A_1/A_2) = \ln(A_3/A_4) = \cdots \ln(A_n/A_{n+1})$$

where $A_1, A_2, A_3, \ldots$, etc. are the amplitudes of successive oscillations. Usually the reduction in amplitude between a number of oscillations (n) is measured, when

$$\Delta = (1/n)\ln(A_i/A_{i+n})$$

Δ is related to the tangent of the loss angle by

$$\tan \delta = (\Delta/\pi)/(1 - \Delta^2/4\pi^2) \approx \Delta/\pi$$

for low damping. For a circular cross-section rod of radius r and length l, the shear loss modulus is given by

$$G'' = (\omega^2 \Delta 2lM)/\pi^2 r^4$$

where M is the moment of inertia of the rod, or of a disc and ω is the frequency.

LOGARITHMIC LAW OF MIXTURES A mathematical relationship relating the value of a property P of a mixture, such as a composite material, to the values of the property for the individual components P_1 and P_2, for the usual case of a binary mixture. It has the form,

$$\log P = \phi_1 \log P_1 + \phi_2 \log P_2$$

where ϕ_1 and ϕ_2 are the volume fractions of the two components.

LOGARITHMIC NORMAL DISTRIBUTION A molecular weight distribution of similar mathematical form to the normal distribution, but with $\ln r$ replacing r, where r is the measure of the individual molecular sizes, e.g. degree of polymerisation. More useful than the normal distribution in describing broader distributions, especially those from which the lower molecular weight 'tail' has been removed. The differential weight distribution function is,

$$W(r) = \exp\left[-(\ln r - \ln r_m)^2/2\sigma^2\right]/r(2\pi)^{1/2}\sigma$$

where r_m is the mean value of the distribution and σ is the standard deviation, given by

$$\sigma^2 = \int_0^\infty (\ln r - \ln r_m)^2 \, W(r) \, dr$$

The number, weight and z-average degrees of polymerisation ($\bar{x}_n$, $\bar{x}_w$ and $\bar{x}_z$ respectively) are,

$$\bar{x}_n = r_m \exp(-\sigma^2/2) \qquad \bar{x}_w = r_m \exp(+\sigma^2/2)$$

and

$$\bar{x}_z = r_m \exp(+3\sigma^2/2)$$

hence

$$r_m = (\bar{x}_n \bar{x}_w)^{1/2} \text{ and } \bar{x}_w/\bar{x}_n = \bar{x}_z/\bar{x}_w = \cdots, \text{ etc.} = \exp \sigma^2$$

The maximun of $W(r)$ is at $\bar{x}_n^{5/2}/\bar{x}_w^{1/2}$ and the maximum of the number distribution, $N(r)$, is at $\bar{x}_n^{5/2}/\bar{x}_w^{3/2}$. The distribution has been generalised to

$$W(r) = r^A \exp\left[-(\ln r - \ln r_m)/2\sigma^2\right]/\sigma(2\pi)^{1/2} B r_m^{A+1}$$

where A is an adjustable parameter related to the molecular weight averages and $B = \exp\left[\frac{1}{2}(\sigma)^2(A+1)^2\right]$. When $A = 0$, $B = \exp(\frac{1}{2}\sigma^2)$, which gives the Lansing–Kraemer distribution. When $A = -1$, then $B = 1$, which gives the Wesslau distribution.

LOGARITHMIC STRAIN Alternative name for natural strain.

LOGARITHMIC VISCOSITY NUMBER (Inherent viscosity) Symbol $\ln(\eta_{rel})/c$. The ratio of the natural logarithm of the relative viscosity of a solution to the concentration of solute (c). In dilute polymer solutions c is usually expressed as gram per decilitre (dl) of solution. Like the viscosity number, it is concentration dependent even in dilute solutions, due to intermolecular polymer molecule interactions. To assess the effects of isolated polymer molecules, results must be extrapolated to infinite dilution, to obtain the limiting viscosity number,

$$[\ln(\eta_{rel})/c]_{c \to 0}.$$

LOI Abbreviation for limiting oxygen index.

LOMOD Tradename for a polyetherester block copolymer thermoplastic elastomer.

LONG CHAIN BRANCHING The occurrence in a polymer of branches or repeat units of the same order of length as the main chain. The designation of which chain is the main chain may be a mere formality, especially in the case of star polymers and when the branches are randomly distributed. In the case of comb polymers and graft copolymers, the branches are attached to a particular recognisable main chain. In the most common type, that of random branching, the branches can be formed by chain transfer to polymer in chain polymerisation. Many such polymers (low density polyethylene, polyvinyl acetate, polystyrene and polyvinyl chloride are examples) contain some long chain branches. Non-linear step-growth polymerisation necessarily results in branched polymer being formed. In this case the branches are terminated by unreacted functional groups, which, as conversion increases, will react further, thereby joining the branches together to form a network polymer.

The extent of branching may be expressed as the branching density. This is determined only with difficulty. The effect of the smaller hydrodynamic volume of branched molecules, compared with otherwise similar linear molecules, may be used to determine the branching parameter by a combined limiting viscosity number/gel permeation chromatography method. This parameter may be related to the branching density if the nature of the branch distribution is assumed. Ultracentrifugation, sedimentation and light scattering methods are also used to estimate branching.

LONGITUDINAL ACOUSTIC MODE (LAM) Alternative name for accordion mode.

LONGITUDINAL MODULUS Alternative name for extensional modulus.

LONGITUDINAL RELAXATION Alternative name for spin–lattice relaxation.

LONG OIL ALKYD RESIN An alkyd resin containing more than 70% of oil and therefore it is highly aliphatic and soluble in aliphatic solvents. It air dries rapidly and is used in household paints.

LONG PERIOD (Long spacing) (Long period spacing) The periodic spacing between lamellae in a single crystal mat and therefore closely related to lamellar thickness. It is determined by small angle X-ray diffraction which gives an average value of typically about 100 Å. A similar spacing may be observed in melt crystallised polymers by electron diffraction. Similar periodicities of structure, due to scattering units spaced at 50–1000 Å, are found in oriented synthetic fibres.

LONG PERIOD DIFFRACTION Discrete small angle X-ray diffraction due to a repetition of a structural feature about every 100 Å. Originally observed in oriented fibres and ascribed to regular alternating crystalline and amorphous regions. Now thought to be due to single crystal folded chain lamellae aggregated in close regularity. The diffraction measurements give the best measure of the fold period.

LONG PERIOD SPACING Alternative name for long period.

LONG SPACING Alternative name for long period.

LONZALIT Tradename for polyvinyl chloride.

LONZAVYL Tradename for polyvinyl chloride.

LOOSE END CORRECTION A correction applied in the theory of rubber elasticity to account for the fact that some chains are connected to the network at one end only and therefore make no contribution to the network elasticity. This leads to the derived value of the shear modulus, which is $(\rho RT/M_c)(1 - 2M_c/M)$, where ρ is the density, M_c is the molecular weight between crosslinks, M is the primary chain molecule molecular weight, R is the gas constant and T is the absolute temperature. The term $2M_c/M$ is the loose end correction.

LOPAC Tradename for a high nitrile polymer which is a copolymer of methacrylonitrile and styrene and which is useful as a barrier polymer.

LOPOX Tradename for an epoxy resin.

LORIVAL Tradename for depolymerised natural rubber.

LORKALENE Tradename for polystyrene.

LORKARIL Tradename for acrylonitrile–butadiene–styrene copolymer.

LOSS ANGLE Alternative name for phase angle.

LOSS COMPLIANCE Symbols C'' in general, J'' in shear, D'' in tension and B'' in bulk. In dynamic mechanical behaviour, the ratio of the strain 90° out of phase with the stress, to the stress. Mathematically, the imaginary component of the complex compliance C^*,

given by $C^* = C' + iC''$. C'' is frequency dependent; for a Voight model material the dependency is given by $C''(\omega) = C_v\omega\tau/(1 + \omega^2\tau^2)$, where ω is the frequency, τ is the relaxation time and C_v is the spring compliance. However, polymers do not behave so simply and a model with a retardation spectrum $L(\tau)$ is more realistic. For this model

$$C'' = \int_{-\infty}^{\infty} [L(\tau)\omega\tau/(1 + \omega^2\tau^2)] \, d\ln\tau + 1/\omega\eta_0$$

where η_0 is the steady state (zero frequency) viscosity.

LOSS FACTOR Alternative name for dielectric loss factor.

LOSS INDEX Alternative name for dielectric loss factor.

LOSS MODULUS Symbols M'' in general, G'' in shear, E'' in tension and K'' in bulk. For dynamic mechanical behaviour, the ratio of the stress 90° out of phase with the strain, to the strain. Mathematically, the imaginary component of the complex modulus $M^*(= M' + iM'')$ which represents the amount of energy dissipated, i.e. the loss, for each complete cycle $(\Delta\varepsilon)$, where $\Delta\varepsilon = \pi M'' e_0^2$ (e_0 is the maximum amplitude of the strain). Usually M'' is much less than the storage modulus M' unless damping is high. M'' is frequency dependent; at very low or very high frequencies it is very low, corresponding to the purely elastic behaviour of a rubber and a glass respectively. In the transition zone it passes through a maximum. For a Maxwell model material

$$M'' = (M_m\omega\tau)/(1 + \omega^2\tau^2)$$

where τ is the relaxation time, M_m is the spring modulus and ω is the frequency. For a more general model with a relaxation time spectrum of $H(\tau)$, which is more realistic of polymer behaviour,

$$M''(\omega) = \int_{-\infty}^{\infty} [H(\tau)\omega\tau/(1 + \omega^2\tau^2)] \, d\ln\tau$$

M'' is related to the loss compliance (C'') by,

$$M'' = (1/C'')/(1 + (\tan^2\delta)^{-1})$$

where $\tan\delta$ is the tangent of the loss angle. M'' may be determined experimentally by several different dynamic mechanical methods, such as the torsion pendulum method.

LOSS TANGENT Alternative name for tangent of the loss angle or dielectric loss tangent.

LOTRENE Tradename for low density polyethylene.

LOTREX Tradename for a butene-1 based linear low density polyethylene.

LOW ANGLE LASER LIGHT SCATTERING (LALLS) A technique for the determination of weight average molecular weight and molecular dimensions of a polymer in dilute solution. In contrast to conventional (wide angle) light scattering, the angles employed are in the range 2–10°. This avoids the need for the complex extrapolation of the conventional Zimm plot method and measurements at only one angle are required. In addition, only a small solution volume (about $0.1\,\mu l$) is needed, so that the technique is particularly useful for the continuous determination of the molecular weight of the polymer in a gel permeation chromatographic effluent. This application also relies on the fact that measurements only need to be made at a single concentration if the second virial coefficient (A_2) is known, since $Kc/R = (1/\bar{M}_w) + 2A_2c$, where K is a constant, c is the polymer concentration and R is the Rayleigh ratio.

LOW *CIS*-POLYBUTADIENE A polybutadiene prepared by anionic polymerisation using an alkyl lithium catalyst, with about 40% *cis*-1,4-butadiene units. An important commercial polybutadiene rubber. Sometimes referred to as medium *cis*-polybutadiene.

LOW COLOUR CHANNEL BLACK (LCC) A type of channel carbon black with particle size about 20 nm and therefore of lower tinting power than high colour channel black.

LOW DENSITY LIPOPROTEIN The blood serum lipoprotein fraction obtained by density gradient ultracentrifugation, having a density of 1.000–$1.003\,\mathrm{g\,cm^{-3}}$ and containing about 25% protein. It comprises most of the β_1-lipoprotein fraction from electrophoretic separation. An even lower density fraction (very low density lipoprotein) may also be obtained and contains even less (about 10%) protein, composed of four different polypeptide chains with characteristic amino acid sequences. Even lower density fractions, known as chylomicrons, can also be obtained.

LOW DENSITY POLYETHYLENE (LDPE) (Polythene) Tradenames: AC polyethylene, Alathon, Alkathene, Baylon, Carlona, Daplen, Dowlex, Dylan, El-Rey, Epolene, Eraclene, Escorene, Fertene, Lacqtene, Latene, Lotrene, Lupolen, Petrothene, Plastylene, Polyeth, Rexene, Riblene, Royalene, Sirtene, Stamylan LD and Tenite polyethylene, for plastic materials and Courlene for fibres. Polyethylene with a density in the range 0.915–$0.925\,\mathrm{g\,cm^{-3}}$. It is produced by the high pressure polymerisation process, and hence is sometimes called high pressure polyethylene, by free radical polymerisation of ethylene at exceedingly high pressures and at 150–200°C. The polymer is highly branched as a result of the extensive transfer to polymer reactions which occur during polymerisation. Therefore it does not crystallise to such a large extent as high density polyethylene. It is thus a softer, more flexible material with a lower melting range as well as lower density.

LOWER CRITICAL SOLUTION TEMPERATURE (LCST) The critical solution temperature below which a binary mixture exists as a homogeneous single phase solution at all compositions. Thus when such a mixture existing below the LCST is heated, phase separation occurs. Such behaviour is more unusual than upper critical solution temperature behaviour and is not accounted for by the simple theories of solution, e.g. the Flory–Huggins theory for polymer solutions. However, it does occur for several polymer/solvent pairs and not at all infrequently for polymer/polymer pairs, i.e. polymer blends. This behaviour implies an increase in χ (the Flory–Huggins interaction parameter) with temperature, and can be accounted for by the corresponding states theory, as being due to dissimilar thermal expansion coefficients between polymer and solvent. At high temperatures, the more highly expanded solvent must fit into a denser medium and mixing must involve a negative volume change and negative entropy of mixing.

LOW PRESSURE POLYETHYLENE Polyethylene produced by a low pressure (typically a few atmospheres) polymerisation process, either the Ziegler–Natta, Phillips or Standard Oil process. This usually produces linear and therefore high density polyethylene, but medium density polyethylene is also produced in this way. More recently linear low density polymers have been extensively produced using low pressure processes.

LOW SULPHUR VULCANISATION Sulphur vulcanisation using only a small amount (about 1 phr) of sulphur in combination with a relatively large quantity (3–4 phr) of a sulphur donor accelerator, especially a thiuram disulphide. Such vulcanisation systems often give efficient vulcanisation.

LOW TEMPERATURE TRANSITION A transition which occurs at a temperature lower than the α- (usually the glass) or the β-transition of a polymer. It is therefore usually specifically the γ-transition or, possibly, an even lower transition.

LUBRICANT An additive used in a polymer composition to reduce friction. It may be necessary for use of the final product, e.g. the use of graphite in engineering plastics employed as gears, cams and bearings, which are self-lubricating. More commonly lubricants are used to reduce friction in processing, either as external lubricants to reduce sticking to machinery, or as internal lubricants to reduce friction within a polymer melt and to aid melt flow and fusion. A particular processing lubricant may act partly internally and partly externally, depending on its compatibility with the polymer. Lubricants are used in relatively small amounts—typically at about the 1 phr level.

LUBRICATION APPROXIMATION The assumption that the flow in a channel or gap can be considered to be, locally, a parallel flow even though the surfaces involved may not be either flat or parallel. This is reasonable provided that the gap (H) is small compared with the length (L) in the direction (x) of flow along the boundaries over which the gap varies, i.e. $\partial H/\partial x \ll 1$. Thus, for example, in a tapered gap the angle of the taper should be $< 10°$. The approximation may be invalidated by large melt elasticity effects, but is usually valid when polymer melts are subjected to large steady rates of deformation over long times. Inertia effects may also invalidate the approximation. but these are low in polymer melts. The approximation often simplifies flow problems and an analytical solution may thus be obtained to the complex flows involved in, for example, extrusion, calendering and wire coating.

LUCABEN Tradename for an ethylene–acrylate copolymer.

LUCALOR Tradename for chlorinated polyvinyl chloride.

LUCITE Tradename for polymethylmethacrylate moulding material.

LUCOLENE Tradename for polyvinyl chloride.

LUCOREX Tradename for polyvinyl chloride.

LUCOVYL Tradename for polyvinyl chloride.

LUMINESCENCE DEPOLARISATION A technique for the observation of molecular movements, with measurement of their associated relaxation times, in which a luminescent group is attached to the polymer chain. By irradiation with polarised exciting radiation, only those molecules with a particular alignment are excited. Depolarisation of the emitted radiation occurs due to reorientation of the excited group as a result of rotational relaxation. In practice fluorescent groups, such as vinyl anthracene comonomer units, are used (giving fluorescence depolarisation) for which the rotational processes have relaxation times of $10^{-8}–10^{-10}$ s. These correspond to mobile side groups and dilute solution segmental motions.

LUMPED CIRCUIT METHOD A method for the determination of the dielectric properties of a dielectric in which the equivalent circuit of the dielectric in a capacitor is determined at a given frequency. The equivalent circuit is balanced in a bridge circuit against the dielectric by varying its inductances, resistances and capacitances, which may be said to be 'lumped' together. Schering and transformer ratio arm bridges are used. The techniques are useful over the frequency range 10 to about 10^6 Hz or over the range $10^5–10^8$ Hz by resonance methods such as the Hartshorne and Ward method. At higher frequencies distributed circuit methods using waveguides must be employed.

LUPAREN Tradename for polypropylene.

LUPHEN Tradename for phenol–formaldehyde polymer.

LUPKE PENDULUM An instrument for the determination of the rebound resilience of a material, usually used for rubbers. A metal rod, shaped to form an indentor at one end, is suspended from a wire at each end, so that it may swing whilst remaining horizontal. The 'pendulum' is released from a known height and strikes a vertically mounted test piece. The rebound resilience is the ratio of the rebound height to drop height.

LUPOLEN Tradename for low density or high density polyethylene.

LURAN Tradename for styrene–acrylonitrile copolymer.

LURAN S Tradename for acrylonitrile–styrene–acrylate rubber copolymer.

LURANYL Tradename for a poly-(2,6-dimethyl-1,4-phenylene oxide) blend with high impact polystyrene.

LURON 2 Tradename for a nylon 66/610 copolymer monofilament.

LUSTRAN Tradename for acrylonitrile–butadiene–styrene copolymer.

LUSTRAN A Tradename for styrene–acrylonitrile copolymer.

LUSTREX Tradename for polystyrene and high impact polystyrene.

LUSTROPAK Tradename for acrylonitrile–butadiene–styrene copolymer.

LUTONAL Tradename for poly-(vinylmethyl ether) or poly-(vinylethyl ether).

LUTONAL A Tradename for poly-(vinylethyl ether).

LUTONAL I Tradename for poly-(isobutylvinyl ether).

LUTONAL M Tradename for poly-(methylvinyl ether).

LUVICAN Tradename for poly-(N-vinylcarbazole).

LYCRA Tradename for a spandex polyurethane elastomeric fibre. Formed by reaction of a poly-(tetramethylene ether) glycol, such as polytetrahydrofuran or poly-(1,4- oxybutylene glycol) polyol, with excess tolylene diisocyanate to give an isocyanate terminated prepolymer. This is then chain extended with a diamine, such as hydrazine or ethylenediamine, in the solvent from which the polymer is to be spun to a fibre.

LYOTROPIC POLYMER A polymer which exhibits a transition from a glassy, crystalline or melt state to a liquid crystalline state only when in solution. The solvent may be necessary either for the formation of the mesogenic groups, (as in the formation of rigid α-helices in some polypeptides, such as poly-(γ-benzyl-D-glutamate)), or simply to lower the temperature to below the polymer decomposition temperature when molecular mobility permits orientation of the mesogens (as with aromatic polyamides such as polyphenyleneterephthalamide).

LYS Abbreviation for lysine.

LYSINE (Lys) (K)
$$H_2NCHCOOH$$
$$\mid$$
$$(CH_2)_4NH_2$$

M.p. 224°C (decomposes).

A basic α-amino acid found widely in animal proteins, but often lacking in plant proteins. Its pK' values are 2·18, 8·95 and 10·53, with the isoelectric point at 9·74. Lysine residues in a protein are readily cleaved with pepsin, but not with carboxypeptidase.

LYSOZYME An enzyme which catalyses the hydrolysis of glycosaminoglycans, e.g. in bacterial cell walls. A monomeric enzyme of molecular weight 14 600, consisting of 129 amino acid residues. The first enzyme whose tertiary structure was determined by X-ray diffraction. It contains a long cleft which accommodates six sugar rings of the substrate and also contains the active site. It is believed that the aspartate-52 and glutamic acid-35 amino acid residues of the enzyme are involved in the glycoside bond cleavage between N-acetylmuramic acid and an N-acetylglucosamine unit of the substrate (the substrate being an alternating copolymer of these units).

LYTRON Tradename for a partially esterified styrene–maleic anhydride copolymer.

M

M (1) Symbol for a monofunctional unit in a polyorganosiloxane, i.e. for the chain end unit R_3SiO—, often specifically referring to the most common type $(CH_3)_3SiO$—. Derived from the use of a monochlorosilane as a chain stopper.

(2) The letter used in silicone rubbers to denote a methyl substituted polysiloxane chain as in polydimethylsiloxane (MQ) for example.

(3) Symbol for methionine.

MABS Abbreviation for methylmethacrylate–acrylonitrile–butadiene–styrene copolymer.

MACKEREL A banded structure present on the surface of the outer mirror region of a fracture surface. It results from the crack, which propagates along the interface between the craze ahead of the crack and the matrix, jumping regularly from one interface to another.

The band spacing is usually very regular and the band shape corresponds to the shape of the advancing crack front.

MACROCONFORMATION The overall conformation, and therefore shape, of a polymer chain, as viewed on a larger scale (about 5–10 nm) than is the microconformation. Thus the extended chain, helical and folded chain conformations found in crystalline polymers can be said to be macroconformations, as can the tertiary structures of proteins and nucleic acids.

MACROCYCLIC POLYMER A polymer containing a large ring structure in the polymer chain as part of the repeating unit, as opposed to the irregularly sized large rings which may be considered to exist in network polymers. Several thermally stable polyers, e.g. phthalocyanine polymers, are of this type.

MACROFIBRIL Fibrils of a thickness of approximately 1 nm. Found in polyamide, polyester and regenerated cellulose fibres. Unlike many fibrils, composed not of microfibrillar subunits, but of lamellar structures such as concentric cylinders inserted inside each other.

MACROFORM ANISOTROPY The local anisotropy of optical polarisability at the repeat unit level of a polymer chain. It results from optical interaction of neighbouring elements with a polymer chain which causes a local anisotropy of the polarising field.

MACROION A polymer molecule carrying electrostatic charges due to the presence of ionisable groups on the polymer chain, i.e. the polymeric molecular component of an ionic polymer as distinct from the small ions neutralising the charge—the counter-ions.

MACROMECHANICS The analysis of the mechanical behaviour of composite materials in which the microstructure, i.e. the matrix and dispersed phase, is ignored, but in which the material is regarded as being homogeneous. However, allowance can be made for the anisotropic nature of the material. In fibrous composites the fibres are used as tapes, strands or sheets, and such groups of fibres can be used as elements for a macromechanics treatment. A simple example is the derivation of the simple law of mixtures for the modulus of a unidirectional fibre/matrix composite in the fibre direction. The properties of a composite structure may be calculated from the properties of the individual elements by assembling them in the appropriate way, including the possibility of arranging them at different orientations.

MACROMER Tradename for a styrene–butadiene star copolymer crosslinked with divinylbenzene.

MACROMOLECULAR HYPOTHESIS The idea, fundamental to polymer science, that polymer molecules are macromolecules, consisting of very long chains of repeating units, joined, often linearly, but sometimes as branched or network structures. At the time of the original suggestion of the macromolecular nature of certain natural and synthetic substances, such as natural rubber, cellulose and polystyrene, many scientists considered these substances to be either physical aggregates of small molecules (akin to colloids) or to be cyclic structures of only a few repeat units. However, careful measurements of molecular weights and other measurements refuted these suggestions and eventually the macromolecular nature of these and many other similar substances was accepted.

MACROMOLECULE A very large molecule. The term is often used synonymously with the term polymer. Thus the subject of polymer science could also be referred to as macromolecular science. The term polymer, however, implies that the molecules have many of the same type of repeat unit, whereas the term macromolecule does not. It is therefore favoured over the term polymer by many biochemists, since biological macromolecules (biopolymers) frequently consist of a complex arrangement of many different repeat units.

MACROPOROUS POLYMER Alternative name for macroreticular polymer.

MACRORADICAL A polymer molecule which is also a free radical. Thus a propagating active centre in free radical polymerisation is a macroradical. A preformed polymer molecule may be converted to a macroradical by abstraction of an atom (often hydrogen) from the polymer by a radical, by irradiation or simply by heating if it contains thermally labile bonds. Macroradicals may also be produced by mechanically shearing a polymer. Often the formation of macroradicals is the first step in a polymer degradation.

MACRORETICULAR POLYMER (Macroporous polymer) A crosslinked network polymer in which microvoids are present, giving a porous material which is more permeable to solvent than normal networks, but which is less swollen. Such networks are prepared by polymerisation in the presence of a polymer which is subsequently extracted, or by precipitating a polymer in the gel phase before complete polymerisation. Such polymers are useful in the preparation of ion-exchange resins and as gels for gel permeation chromatography. Such polymers contrast with microporous polymers which are single phase and owe their 'porosity' to their lightly crosslinked nature.

MAGNAMITE Tradename for carbon fibre.

MAKROBLEND Tradename for a bisphenol A polycarbonate/polybutylene terephthalate blend.

MAKROLON Tradename for bisphenol A polycarbonate.

MALEIC ACID

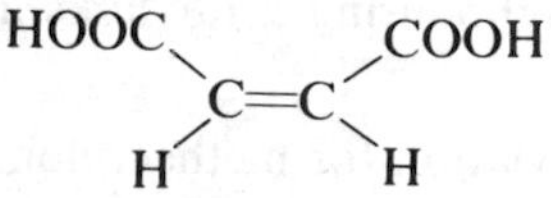

M.p. 130°C.

Prepared by the hydration of maleic anhydride. It may be used for the preparation of unsaturated polyester resins; however, maleic anhydride is usually preferred due to its lower melting point and lower water production on esterification.

MALEIC ANHYDRIDE

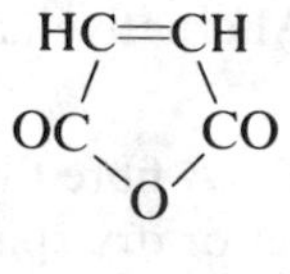

M.p. 52·6°C.

Prepared by the high temperature (about 450°C) air oxidation of benzene, catalysed by a vanadium compound. The preferred unsaturated monomer for the preparation of unsaturated polyester resins by polyesterification with phthalic anhydride and propylene glycol. It is preferred to maleic acid since it has a lower melting point, is more reactive and gives less water on esterification. It isomerises to the *trans* isomer (giving fumarate units) during esterification. It is also sometimes used in alkyd resins, when its double bond will polymerise with the unsaturated fatty acid residues, thus increasing viscosity.

MALTESE CROSS A dark cross-shaped 'shadow' appearing in a spherulite when observed between crossed polars in a polarising microscope. The extinction pattern is caused by crystallites whose optical indicatrix has one of its axes parallel to the polariser or analyser direction. Since the crystallites are arranged so that polymer molecules are tangential to the spherulite radius, extinction occurs in opposite quadrants.

MALTOSE (α-D-Glucopyranosyl-1,4′-D-glucopyranose)

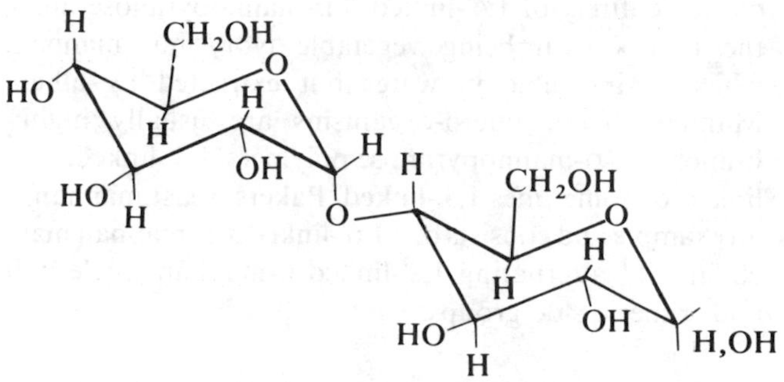

M.p. 102–103°C (hydrate). $\alpha_D^{20} + 130\cdot4°$.

A reducing sugar which is the dimeric structural unit of starch and glycogen. Prepared by the enzymatic hydrolysis of starch.

MALTOTRIOSE α-D-Gp-1,6′-α-D-Gp-1,6′-α-D-Gp. A glucose trimer isolated from the enzymatic hydrolysis of amylose, glycogen and isolichenan and demonstrating the presence of neighbouring α-1,6'-links in these polymers.

MAN Abbreviation for methacrylonitrile.

MANILLA COPAL A copal natural resin obtained from the tree *Agathis alba*. Unlike Congo copal the resins are soluble in ethanol and range in softness depending on the time after exudation from the tree. Some hard types, e.g. Pontianak, are fossil resins, used widely in spirit and oil varnishes. The copal softens at 80–90°C.

MANILA HEMP Alternative name for abaca.

MAN-MADE FIBRE A fibre that has been artificially produced either by wet or dry spinning from a polymer solution or by melt spinning a polymer. Usage of man-made fibres has grown at the expense of natural fibres, so that they now account for about 70% of textile use. However, the polymers used may be natural (usually cellulose, although proteins have been used in the past) to give a regenerated fibre, or the polymer may be a synthetic polymer to give a synthetic fibre. Synthetic fibres, especially nylons, polyethylene terephthalate and polyacrylonitrile, have become particularly important.

MAN-MADE POLYMER A polymer derived from a biopolymer by conversion to a chemical derivative, which is more useful as a material (often as a fibre) than the original biopolymer. Important examples are the various cellulose man-made fibre materials such as cellulose acetate and regenerated cellulose. The term, especially when used in connection with fibres, also sometimes encompasses synthetic polymers as well.

MANNAN A polymer of D-mannose. A frequent constituent of plants and micro-organisms either as a homopolysaccharide or with the main chain composed of mannose units but with galactose (galactomannan) or glucose (glucomannan) side branches. The plant mannans consist entirely of 1,4'-linked β-D-mannopyranose units, the best known being vegetable ivory nut mannan, which is insoluble in water but extracted by alkali. Mannans from micro-organisms are usually highly branched α-D-mannopyranose polymers, 1,2'-linked, 1,6'-linked or sometimes 1,3'-linked. Bakers' yeast mannan is an example and consists of a 1,6'-linked α-D-mannan main chain with alternating 1,2'-linked D-mannan single unit and trimeric side groups.

MANNOCAROLOSE A low molecular weight mannan of DP 8 produced in the culture medium of *Penicillium charleseii* G. Smith, in which the mannose units are 1,2'-linked, except for a branch at a C-6 atom of one unit.

β-D-MANNOPYRANOSE The usual anomeric ring form of D-mannose units in a carbohydrate which are therefore more fully described by this name.

D-MANNOSE Fischer projection formula:

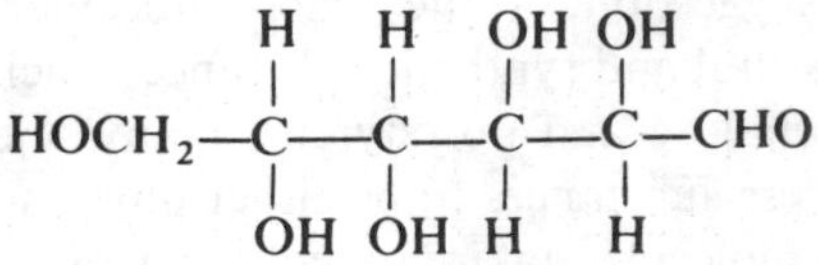

Hawarth formula of ring form (α-D-mannopyranose):

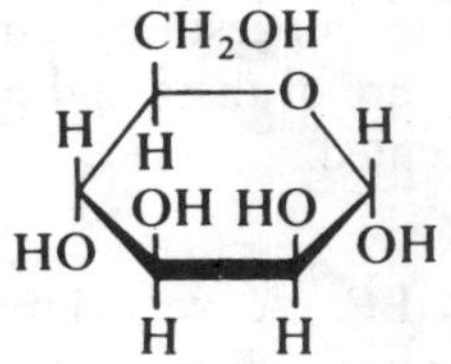

Chair conformation (α-D-mannopyranose):

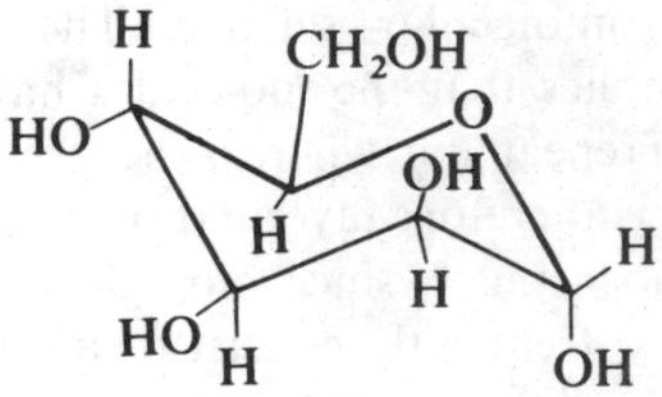

M.p. 132°C. $\alpha_D^{20,H_2O} + 14 \cdot 6°$.

An aldohexose monosaccharide occurring widely in many polysaccharides, either as a homopolysaccharide (mannan) or in heteropolysaccharides, e.g. guaran, glucomannan, galactoglucomannan.

D-MANNURONIC ACID

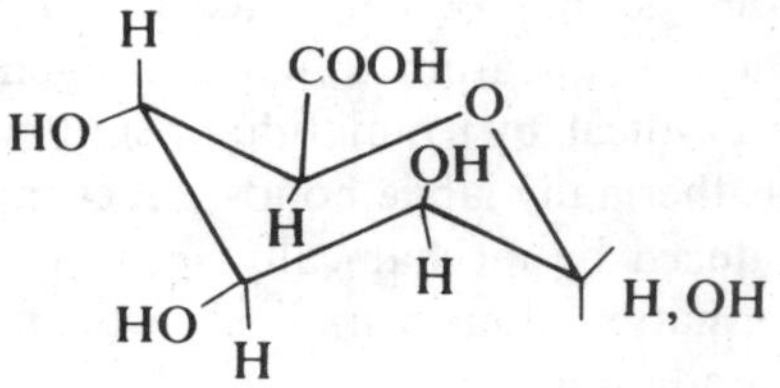

M.p. 110°C (sinters). $\alpha_D^{25} - 6 \cdot 05°$.

A uronic acid, occurring as a monosaccharide component of alginic acid.

MARANYL Tradename for nylon 6, 66 or 610.

MARBON Tradename for acrylonitrile–butadiene–styrene copolymer.

MARK–HOUWINK CONSTANT One of the constants, K or a, of the Mark–Houwink equation, $[\eta] = KM^a$, where $[\eta]$ is the limiting viscosity number of a polymer in solution and M is its molecular weight.

MARK–HOUWINK EQUATION An empirical relationship between the limiting viscosity number (LVN) $[\eta]$ of a polymer in solution and the polymer molecular weight (M). It is, $[\eta] = KM^a$, where K and a are constants (the Mark–Houwink constants) for a particular polymer, solvent and temperature. The constants are determined from a plot of log $[\eta]$ versus log M for a series of fractions (preferably nearly monodisperse) of the polymer. The basis of the dilute solution viscosity method for determination of polymer molecular weights. For a polydisperse polymer the viscosity average molecular weight $(\bar{M}_v)$ is obtained. In the determination of K and a, if monodisperse samples are not used, then the molecular weight distribution should be the same as the samples whose $\bar{M}_v$ is required.

Values of a vary from about 0·5 in poor solvents to about 0·8 in good solvents. These are in fact the limiting values predicted by the Flory–Fox equation. The values of K and a are not constant over the entire range of M values and should not be used outside the calibration range of M values, unless $[\eta]$ is determined under theta conditions, when $[\eta]_\theta = KM^{1/2}$. Occasionally a is greater than 0·8, but is still less than 1·0, when the polymer is either of very low molecular weight, or when the polymer molecules are very stiff. The value of K and its units are dependent on the units of polymer concentration used; conventionally these are grams per decilitre $(g\,dl^{-1})$, so that K has the units of decilitre per gram. Typical values are of the order of $10^{-4}\,dl\,g^{-1}$.

MARKOV CHAIN PROCESS A stochastic process in which the events involved are dependent on past events in a particularly simple way. If the event depends only on the last previous event, then the process is a first order Markov process, if on the previous two events it is a second order Markov process, etc. In copolymerisation the propagation steps usually depend on the terminal unit, and hence on the previous propagation event, hence copolymerisation is a first order Markov process. If propagation depends on the penultimate unit, then the copolymerisation is a second order Markov process. In considering stereochemical configurations in polymerisation, in a similar manner, first and higher order processes may be involved.

MARLEX Tradename for high density polyethylene.

MARTENS TEMPERATURE A measure of the softening temperature of a polymer. The temperature at which a standard test specimen, heated at a determined rate, bends to a certain extent under a defined load. A similar measurement to the heat distortion temperature, but giving a somewhat different result. For an amorphous polymer the value is about 20°C below T_g.

In the Martens test it is the temperature at which a bar of a polymer deflects by a specified amount under four point bending, when subjected to a specified bending stress. The bending is magnified by the loading arm attached to the upper end of the vertically mounted sample.

MARTIN–ROTH–STIEHLER EQUATION (MRS equation) An empirical stress–strain relationship for rubber elasticity of the form,

$$\sigma = E(\lambda^{-1} - \lambda^{-2})\exp[A(\lambda - \lambda^{-1})]$$

where σ is the stress, λ is the extension ratio for simple tensile deformation and A and E are constants.

MARTIN'S EQUATION An empirical relationship between the viscosity number (η_{sp}/c) and concentration (c) of a polymer in dilute solution, of the form, $\ln(\eta/c) = \ln([\eta]) + k[\eta]c$, where $[\eta]$ is the limiting viscosity number and k is a constant. Although often quite valid to high values of c, when the similar Huggins and Kraemer equations often fail, the linear plots of $\ln(\eta_{sp}/c)$ versus c can lead to serious errors in the determination of $[\eta]$ by extrapolation to zero c.

MASS POLYMERISATION Alternative name for bulk polymerisation.

MASS SPECTROMETRY (MS) A technique for the molecular structural analysis of organic compounds in which a small amount of the compound (a few micrograms) is ionised by electron impact, and volatilised ions (including the molecular parent ion but mostly fragment ions) are separated according to their mass/charge ratios (m/e ratio) often (as in a double focussing mass spectrometer) by passing them through electrostatic and/or magnetic fields. The resultant mass spectrum shows the variation in intensity of the different species and careful interpretation enables fine molecular structural information to be deduced.

For polymer analysis, the main difficulty is that, for other than low molecular weight species, polymers are not volatile. Thus, except when the electrospray technique is used, high polymers must be analysed by a prior degradation to lower molecular weight fragments. Indeed the main application of the technique to polymers is in the analysis of degradation products. In addition, other structural features such as stereoregularity, may also be examined indirectly.

MASTER CURVE The curve resulting from the application of the time–temperature superposition principle to viscoelastic data. This data is obtained at various temperatures (or frequencies) and consists of plots of log (relaxation modulus) or log (creep compliance) versus time (or frequency). It may be combined to give a single master curve for a chosen reference temperature, by use of the shift factors given by the WLF equation. Thus modulus or compliance values over an experimentally inaccessible range of frequencies (or temperatures) may be obtained.

MASTIC A natural resin obtained from the tree *Pistacia lentiscus* grown in Chios (Greece). It is alcohol soluble and of light colour and is used in spirit varnishes. It softens at about 55°C.

MASTICATION The mechanical shearing of a raw unvulcanised rubber (often natural rubber) on a mill or mixer in order to reduce its molecular weight, as measured, for example, by its Mooney viscosity, and hence to improve its processability. At low temperatures, molecular weight reduction is due to mechanochemical degradative chain scission, whereas above about 100°C oxidative degradation reduces the molecular weight. A minimum rate of molecular weight reduction occurs at about 100°C Mastication is often promoted using a peptiser. Mastication in the presence of a second monomer is one method of graft copolymerisation.

MATHEMATICAL STRAIN Alternative name for tensor strain.

MATRIX POLYMERISATION. Alternative name for template polymerisation.

MAXIMUM FIBRE STRESS In a three point bending test, the tensile stress on the upper surface of the specimen at the mid-point. For a rectangular specimen supported at the mid-point and the ends, it is $3Pl/2bd^2$, where P is the force at the mid-point, l is the distance between the end supports, b is the breadth and d is the depth of the specimen.

MAXIMUM PACKING FRACTION Symbol ϕ_{max}. The maximum volume fraction that a particulate solid may have as a result of packing difficulties arising from particle–particle contacts. It is the ratio of the true volume of the solid to its apparent bulk volume. It is important in determining the effect of fillers on the viscosity of fluids (as in the Mooney and Eilers equations) and on the moduli of filled polymer matrices (composites), as given by the Nielsen and Halpin–Tsai equations, for example.

Its value is highly dependent on the particle shape and state of agglomeration. For spheres, the maximum theoretical value is 0·74 for hexagonal close packing, but in practice values of 0·524 (simple cubic close packing) and 0·637 (random close packing) are used. Agglomeration and the use of non-spherical shapes reduces ϕ_{max}. Aligned rods or fibres can have high values, e.g. 0·907 for parallel random packing.

MAXWELL ELEMENT A Maxwell model as part of a more complex mechanical model for viscoelastic behaviour.

MAXWELL FLUID MODEL Alternative name for Maxwell model.

MAXWELL MODEL (Maxwell fluid model) A mechanical model of simple linear viscoelastic behaviour which consists of a spring of Young's modulus E in series with a dashpot of coefficient of viscosity η. Thus this is an isostress model (with stress σ), the strain (e) being the sum of the individual strains in the spring and dashpot. This leads to a differential representation of linear viscoelasticity as $de/dt = (1/E)\,d\sigma/dt + (\sigma/\eta)$. This model is useful for the representation of stress relaxation, since when $de/dt = 0$, integration gives $\sigma = \sigma_0 \exp(-t/\tau)$, where $\tau\,(= E/\eta)$ is the relaxation time. Such an exponential decay in stress is often observed in practice. However, the model may not be such a good representation since stress does not always decay to zero as $t \to \infty$ as predicted. Often more than one exponential term is required. Furthermore for creep, i.e. when $d\sigma/dt = 0$, we have $de/dt = \sigma/\eta$, i.e. Newtonian flow is predicted, which is generally not found for viscoelastic materials. The rheological equation for the model is $\tau = \eta\dot{\gamma} - (\eta/E)\dot{\tau}$, where τ is the shear stress and $\dot{\gamma}$ is the shear rate. Such a model is also often referred to as the Maxwell fluid model since, in contrast to the Voight model, the stress causes the strain to increase continuously, irrecoverably and indefinitely.

MAXWELL–WAGNER EFFECT An effect on dielectric properties caused by discontinuities or inclusions in the dielectric. This results in interfacial polarisation and the occurrence of a dielectric relaxation at low frequency. In some cases the relaxation may be at a frequency which is high enough to be attributed to a dipolar relaxation, so some care must be taken in the interpretation of observations. The heterogeneity may be a crack or void or a crystalline/amorphous boundary. However, a more serious case is one in which a highly conductive impurity is present in the insulating polymeric dielectric.

MAYO AND LEWIS METHOD Alternative name for intersecting lines method.

MAYO EQUATION. A relationship expressing the dependence of the number average degree of polymerisation ($\bar{x}_n$) of the polymer formed in a chain polymerisation on the transfer constant (C_s). It is, $1/\bar{x}_n = C_s[S]/[M] + 1/(\bar{x}_n)_0$, where $(\bar{x}_n)_0$ is the degree of polymerisation in the absence of transfer and [S] and [M] are the chain transfer agent and monomer concentrations respectively. A more complete expression allowing for transfer to monomer and initiator as well is,

$$1/\bar{x}_n = (k_t/k_p^2)(R_p/[M]^2) + C_M$$
$$+ C_s[S]/[M] + C_I k_t R_p^2 k_p^2 f k_d [M]^3$$

assuming termination is by combination and where R_p is the rate of polymerisation, k_d, k_p and k_t are the rate constants for initiator decomposition, propagation and termination respectively, f is the initiator efficiency and C_I and C_M are initiator and monomer chain transfer constants respectively.

MBS Abbreviation for methylmethacrylate–butadiene–styrene copolymer.

MBT Abbreviation for mercaptobenzothiazole.

MBTS Abbreviation for dibenzothiazyldisulphide.

MCC Abbreviation for medium colour channel carbon black.

MDI Abbreviation for 4,4′-diphenylmethane diisocyanate, which is alternatively called methylene-bis-(4,4′-phenylisocyanate).

MDPE Abbreviation for medium density polyethylene.

MEAN RESIDUE ROTATION. The way in which the molar rotation for copolymers is expressed, especially for proteins. The molar rotation per monomer unit.

MEAN SQUARE END-TO-END DISTANCE
Symbol $\langle r^2 \rangle$. The mean of the squares of the end-to-end distances of the statistically predicted values of a real, or statistically equivalent chain, or of the values of a collection of chains, i.e. $\langle r^2 \rangle = \sum r^2/n$, where there are n chains or n possible conformations of a single chain. Statistical analysis of the various models of polymer chains gives the results expressed as $\langle r^2 \rangle$ rather than as r due to the necessity of statistical averaging.

MECHANICAL ANISOTROPY The dependence of mechanical properties on their direction of observation, found in oriented or composite polymers. Usually greater in crystalline than amorphous polymers as a result of the greater orientations preset in the former. It is conveniently represented on a polar diagram. Mechanical anisotropy is more complex than optical anisotropy since it depends on both $\cos^2 \theta$ and $\cos^4 \theta$ terms, θ being the angle of orientation.

MECHANICAL RELAXATION A relaxation in which the applied constraint is mechanical stress or strain, the resulting strain or stress being followed with time or frequency. The relaxation may result either from a static experiment in which the decay in the response is followed after application of a pulse, as in classical stress relaxation, or, more commonly, it may be a dynamic mechanical relaxation. Ultrasonic relaxation is often considered to be a mechanical relaxation, since it involves the mechanical stressing of a material, resulting from the alternating compression and rarefaction from the passage of the ultrasonic wave.

MECHANOCHEMICAL DEGRADATION. A chain-scission type of degradation induced by mechanical shear. In contrast to non-polymer materials, shearing forces in polymers are not so easily relieved by molecular slippage and hence high shear rates can lead to local stresses high enough to cause polymer chain scission. Usually undesirable since it results in a lowering of molecular weight. Frequently occurs on mixing or extrusion of polymer melts, on ultrasonic irradiation and even on agitation of polymer solutions. Occasionally it is used beneficially to reduce undesirably high molecular weights by mastication.

MEDIUM *CIS*-POLYBUTADIENE A polybutadiene containing about 92% *cis*-butadiene structures. An important butadiene rubber, obtained by Ziegler–Natta polymerisation using, for example, an AlR_3/TiI_4 catalyst. Also sometimes referred to as high *cis*-polybutadiene or medium high *cis*-polybutadiene when the term medium *cis*-polybutadiene is used instead for low *cis*-polybutadiene.

MEDIUM COLOUR CHANNEL BLACK (MCC) A type of channel carbon black of particle size about 15 nm and of moderate tinting strength.

MEDIUM DENSITY POLYETHYLENE (MDPE) Polyethylene with a density in the range of 0·925–

0·940 g cm^{-3}. It may be produced either by a modified high pressure process or by the Ziegler–Natta or Standard Oil processes using a few per cent of a higher olefin to reduce crystallinity and hence density. As expected the polymer's mechanical behaviour is intermediate between low density and high density polyethylenes.

MEDIUM OIL ALKYD RESIN An alkyd resin containing 50–70% oil and soluble in aliphatic and aromatic solvents. Coatings may be cured either by air drying or by elevated temperatures. It is used in industrial finishes.

MEDIUM THERMAL BLACK (MT) A type of thermal carbon black of particle size about 500 nm giving compounds with low reinforcement, modulus, hardness, hysteresis and tensile strength but of high elongation. Can be used at high loadings. Useful in wire insulation, mechanical goods, footwear, belts, hose, gaskets, O-rings and mountings.

MEDIUM VINYL POLYBUTADIENE A polybutadiene containing about 35–55% 1,2-butadiene units in the polymer chain. Prepared by anionic polymerisation with lithium ethyl in a somewhat polar solvent such as a benzene/tetrahydrofuran mixture. Developed as a commercial rubber having similar properties to styrene–butadiene rubber.

MEDVEDEV THEORY A theory of emulsion polymerisation in which the free radical initiator reacts initially with an emulsifier molecule in the emulsifier layer surrounding a micelle where polymerisation takes place. This leads to different dependencies of the rate, number of particles and molecular weight on emulsifier and initiator concentrations, compared with the Smith–Ewart theory. The theory more satisfactorily accounts for the results obtained at high emulsifier or initiator concentrations.

MEK Abbreviation for methylethyl ketone.

MEKP Abbreviation for methyl ethyl ketone peroxide.

MELAMIN Tradename for melamine–formaldehyde polymer.

MELAMINE (1,3,5-Triamino-2,4,6-triazine) (Cyanuramide)

M.p. 354°C.

Traditionally produced by heating dicyanamide,

$$H_2NCNHCN \rightarrow I + ammonia,$$
$$\overset{\|}{\underset{NH}{}}$$

and deaminated products, such as ammeline (**I** with one —NH_2 group replaced with —OH). Alternatively it is produced by heating urea in the presence of ammonia at 250–300°C: $6 CO(NH_2)_2 \rightarrow I + 6 NH_3$. On heating, melamine sublimes below its melting point. Above its melting point, it decomposes with loss of ammonia. It is hydrolysed by boiling aqueous alkali to ammeline and by boiling aqueous acid to cyanuric acid (**I** with all the —NH_2 groups replaced by —OH groups). It has only a low water solubility (0·32 g per 100 g at 20°C and 5 g per 100 g at 100°C). It is useful as the monomer for the formation of melamine–formaldehyde polymers by reaction with aqueous formaldehyde via the formation of methylolmelamines. In neutral aqueous solution the amide form of the structure is correct, but in formaldehyde solutions the structure

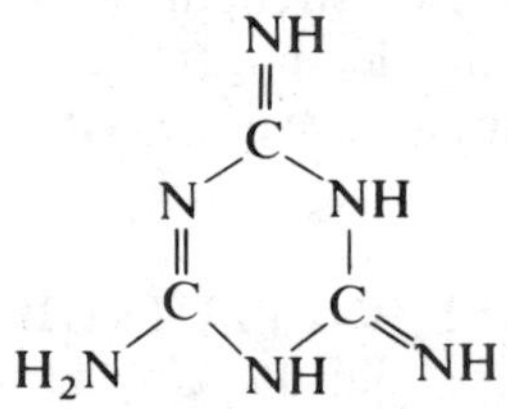

may be important.

MELAMINE–FORMALDEHYDE POLYMER (Melamine–formaldehyde resin) (MF) Tradenames: Cellobond, Cymel, Epok, Fiberite, Melamin, Melan, Melbrite, Melit, Melmex, Melmac, Melolam, Melsir, Melurac, Mouldrite, Plaskon, Plenco, Resamin, Resimine, Ultrapas. An amino resin formed by reaction of melamine (**I**) with formaldehyde in aqueous, usually slightly alkaline, solution. In crosslinked form it is useful as a thermoset plastic. In contrast to the somewhat similar thermoset phenol–formaldehyde polymers, it is colourless, and it has greater heat and chemical resistance to staining than the closely similar urea–formaldehyde polymers.

Initial reaction of the monomers, usually at above 80°C, forms methylolmelamines, the relatively insoluble melamine dissolving as methylolmelamines are formed. With a melamine:formaldehyde ratio of 1:2 to 1:3, dimethylolmelamine is formed. Hexamethylolmelamine is formed when the ratio is more than 1:8. In commercial polymerisations the ratio is 1:1·5 to 1:3. The reaction may be represented as:

$$I + 2HCHO \rightarrow$$

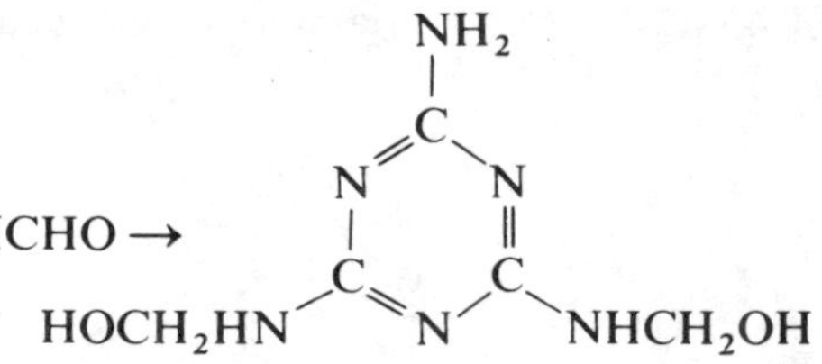

On further heating, water solubility diminishes and eventually a crosslinked gel is formed.

Commercial products are used as syrupy aqueous solutions which are condensed as far as the hydrophobic stage, but not as far as gelation. Alternatively the solid polymer may be isolated by evaporation or spray-drying. Subsequent further heating causes crosslinking to produce a hard thermoset product. Crosslink formation is usually by a combination of methylol–methylol condensations producing ether links,

$$\sim\!\!\sim\!NHCH_2OH + HOCH_2NH\!\sim\!\!\sim \rightarrow$$
$$\sim\!\!\sim\!NHCH_2OCH_2NH\!\sim\!\!\sim$$

or methylol–amine condensations producing methylene links,

$$\sim\!\!\sim\!NHCH_2OH + H_2N\!\sim\!\!\sim \rightarrow \sim\!\!\sim\!NHCH_2\!-\!NH\!\sim\!\!\sim + H_2O$$

Methylolmelamine ethers are formed by reaction with alcohols under acid conditions. The lower ethers have enhanced water solubility and are useful in textile finishing, especially the hexamethylether of hexamethylolmelamine. The higher ethers, especially butylated melamine–formaldehyde, have enhanced oil solubility and are useful in surface coatings. Moulding materials are usually filled with a cellulosic filler (woodflour or bleached woodpulp) and contain a crosslinking catalyst, called an accelerator, such as a zinc or ammonium salt. The salt catalyses by providing free acid at moulding temperature (150–170°C). Other applications are as laminating resins, adhesives and for the wet strengthening of paper.

MELAMINE–FORMALDEHYDE RESIN Alternative name for melamine–formaldehyde polymer.

MELAN Tradename for melamine–formaldehyde polymer.

MELBRITE Tradename for melamine–formaldehyde polymer.

MELINAR Tradename for polyethyleneterephthalate moulding material.

MELINEX Tradename for a polyethylene terephthalate film material.

MELIT Tradename for melamine–formaldehyde polymer.

MELMAC Tradename for melamine–formaldehyde polymer.

MELMEX Tradename for melamine–formaldehyde polymer moulding powders.

MELOLAM Tradename for melamine–formaldehyde polymer.

MELOPAS Tradename for melamine–formaldehyde polymer.

MELSIR Tradename for melamine–formaldehyde polymer.

MELT CRYSTALLISATION Crystallisation by cooling a melt to below its crystalline melting temperature (T_m). For polymers this requires supercooling (or

undercooling) of about 30 K. As the temperature is reduced the rate of crystallisation reaches a maximum, then diminishes as melt viscosity rises. Eventually, below the glass transition temperature, molecular motions become frozen preventing further crystallisation. Rapid quenching of a melt may prevent crystallisation completely. The rate of crystallisation and concentration of crystalline entities, e.g. spherulites, is also dependent on the preheat temperature of the melt and on the presence of nucleating agents and impurities, through their effect on nucleation. Melt crystallised polymers usually have spherulitic morphology. If the melt is stressed during crystallisation, e.g. by stirring or flow, then row nucleation can occur.

MELT FLOW INDEX (MFI) (Melt index) (Melt flow rate) The weight of a polymer extruded through a standard cylindrical die at a standard temperature in a laboratory rheometer carrying a standard piston and load. Thus MFI is a measure of the melt viscosity of a polymer and hence also of its molecular weight. It is frequently used for characterising polyolefins, especially polyethylene, when the standard conditions are: temperature, 190°C; die dimensions, 9·00 cm in length and, normally, 2·095 cm in diameter; load on the piston, 2·61 kg. For polypropylene a temperature of 220°C is usually used.

MELT FLOW RATE (MFR) Alternative name for melt flow index.

MELT FRACTURE (Elastic turbulence) Rupture of the streamlines during flow of a polymer melt, thus producing non-laminar flow. It is most commonly observed in extrusion, the extrudate having an uneven appearance, rather than being smooth. This may be observed as spiralling, rippling or bambooing, the unevenness being regular or irregular. It is the result of the very high tensile stresses that can occur at the entrance to an extrusion die due to the converging flow. If these stresses exceed the rupture stress then irregular flow can occur. The effect occurs above a critical flow (or shear) rate ($\dot{\gamma}_c$), corresponding to a shear stress of typically about 10^5–10^6 Pa, but is also dependent on temperature, $\dot{\gamma}_c$ increasing with temperature and with widening molecular weight distribution. The product $\tau_c \bar{M}_w$ (where $\bar{M}_w$ is the weight average molecular weight and τ_c is the critical shear stress) is often constant. The die geometry is also critical, for example $\dot{\gamma}_c$ is increased if the die L/D ratio is increased. Stick-slip at the die wall may also contribute to melt fracture.

MELT INDEX Alternative name for melt flow index.

MELTING POINT Sometimes used as an alternative name for melting temperature. Unlike well defined low molecular weight compounds, polymers do not melt at a precise temperature, but rather over a range of temperature. Thus the term is not frequently used for polymers.

MELTING TEMPERATURE Symbol T_m. The temperature at which the thermal energy in a solid material is sufficient to overcome the intermolecular forces of attraction in the crystalline lattice so that the lattice breaks down and the material becomes a liquid, i.e. it melts. Melting is a first order transition and occurs with an increase in entropy, volume and enthalpy. In low molecular weight materials melting takes place over a narrow, typically 0·5–2°C, temperature range, except in impure substances. However in polymers, due to imperfections in the crystallites, melting occurs over a much wider range of temperature, typically 10–20°C. Nevertheless, frequently a precise melting temperature is quoted. This temperature usually refers to the highest temperature of the melting range, i.e. when the last, and therefore the most perfectly ordered crystals, melt. Even this value of T_m is 10–20°C below the theoretical value quoted for perfect crystals. Sometimes T_m refers to the peak melting temperature, i.e. the temperature of maximum melting rate. Polymers may sometimes show a small premelting transition or may exhibit several T_m values, when capable of existing in different crystalline forms (polymorphism). T_m may be measured experimentally by direct microscopic observation (best using a polarising microscope), by differential scanning calorimetry or by differential thermal analysis.

MELT POLYMERISATION. A bulk polymerisation in which either, or both, the monomer and polymer are crystalline and which is carried out at a temperature above the melting temperature of both, i.e. in the molten state. This frequently applies to step-growth polymerisations carried out at the high temperatures needed in polyester and polyamide formation to ensure that the polymer remains molten. This is necessary so that chain ends remain accessible to each other for further reaction, and hence growth, and do not become buried in solid polymer.

MELT SPINNING A process for the formation of man-made fibres in which a polymer melt is extruded through a multi-orifice die (a spinneret) into a cool atmosphere which causes solidification of the polymer. The method is used for the commercial production of nylon, polyethylene terephthalate, polypropylene and polyvinylidene chloride fibres.

MELT STRENGTH The tensile stress at which a polymer melt fractures when subject to draw-down.

MELURAC Tradename for melamine–urea–formaldehyde polymer.

MEMBRANE FILTRATION Alternative name for ultrafiltration.

MEMBRANE FRACTIONATION Fractionation by ultrafiltration through a series of membranes of different porosities. The technique, although not widely used, produces fractions which are notably free of long low molecuar weight 'tails'.

MEMBRANE OSMOMETRY (Osmometry) A technique useful for the determination of number average molecular weight ($\bar{M}_n$) of a polymer by measurement of the osmotic pressure (one of the colligative properties) of its dilute solution. This is performed in an osmometer in which a cell containing the solution is separated from pure solvent by a semi-permeable membrane, which allows only solvent (but not solute) molecules to diffuse through it, by osmosis, to the solution side of the membrane. The excess pressure which must be applied to this side just to prevent osmosis is the osmotic pressure (π).

For an ideal solution, $\pi/c = RT/\bar{M}_n$, where c is the polymer concentration, R is the universal gas constant and T is the absolute temperature. For a non-ideal solution:

$$\pi/c = RT/\bar{M}_n + Bc + Cc^2 + \cdots$$

or

$$\pi/c = RT[1/\bar{M}_n + A_2c + A_3c^2 + \cdots]$$

where the c^2 and higher terms may usually be ignored and B (or A_2) and C (or A_3), etc. are the second, third, etc. virial coefficients. At the theta temperature, B, C, etc. are zero, otherwise a plot of reduced osmotic pressure (π/c) against c must be made. The slope gives B (or A_2) and the intercept gives $\bar{M}_n$. The plot is curved if C (or A_3) is not negligible, which may happen in very good solvents or with very high molecular weight polymer. A plot of $(\pi/c)^{1/2}$ against c is often linear in these cases. Alternatively, the divided difference method may be used. Osmometry is generally only reliable for polymers with $\bar{M}_n$ greater than about 20 000.

MEMBRANE POTENTIAL The electrical potential difference that exists across a semi-permeable membrane that separates a polyelectrolyte solution at equilibrium (Donnan equilibrium) with a mobile salt solution which can diffuse through the membrane. It arises from the fact that the membrane must be polarised to counteract the ion concentration gradients within the membrane, which themselves result from the membrane separating solutions with different salt concentrations.

MEMORY The ability of elastic materials to return to some equilibrium shape after deformation, when the deforming force is removed. The material may thus be said to remember this equilibrium shape. An elastic solid retains a memory indefinitely for its preferred shape, whereas at the other extreme, a purely viscous fluid has no memory at all. In the intermediate case of an elastic fluid, e.g. a polymer melt, memory effects are present, but do not last indefinitely, and hence they give rise to the concept of a 'fading' memory. Memory for the past shapes of melts leads to problems in extruded and moulded products, since the product wishes to return to its original shape. Examples of these problems are bananaing, weld line formation and post-extrusion swelling.

MENISCUS DEPLETION METHOD (Yphantis method) A variation of the sedimentation equilibrium method of ultracentrifugation, but requiring much less time for the establishment of equilibrium. Speeds of rotation are high enough so that the polymer concentration at the top of the cell (c_m) (i.e. at the meniscus) falls to zero. The method enables molecular weights to be more readily calculated than in the normal sedimentation equilibrium methods, since c_m is zero.

***p*-MENTHANE HYDROPEROXIDE** A mixture of hydroperoxides of structure:

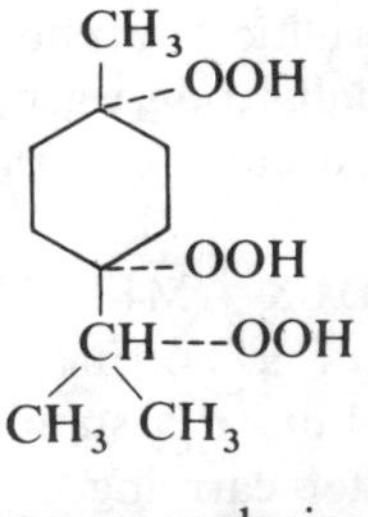

with hydroperoxide groups replacing hydrogen atoms at any of the positions shown. It has half-life times of 25 h/120°C and 2 h/150°C and half-life temperatures of 135°C/10 h, 110°C/1 h and 210°C/1 min. Sometimes used to replace cumene hydroperoxide as a redox initiator component in the production of cold rubber.

MER (Monomeric unit) The basic structural unit of a polymer, being derived from the monomer (or monomers) from which the polymer was made. In most cases it is identical with the repeat unit, although this may be smaller than the mer, as in polyethylene where the mer is $-CH_2-CH_2-$ but the repeat unit is $-CH_2-$, or larger than the mer, as in AABB step-growth polymers. However, often the terms mer and repeat unit are used interchangeably even in these cases.

MERAKLON Tradename for a polypropylene fibre.

MERCAPTOBENZOTHIAZOLE (MBT)

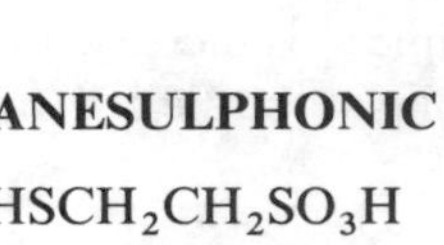

M.p. 176°C.

One of the most widely used accelerators for the sulphur vulcanisation of rubbers, However, it imparts a significant cure rate even at 100°C and therefore has a low scorch time.

MERCAPTOETHANESULPHONIC ACID

$$HSCH_2CH_2SO_3H$$

A sulphonic acid useful in place of hydrochloric acid for the hydrolysis of proteins to their constituent amino acids. Unlike hydrochloric acid it has the advantage that it does not destroy tryptophan.

2-MERCAPTOETHANOL $HSCH_2CH_2OH$. A reagent useful for the reduction of disulphide bridges in cystine groups in proteins to sulphydryl groups, forming cysteine groups, so separating the bridged polypeptide chains.

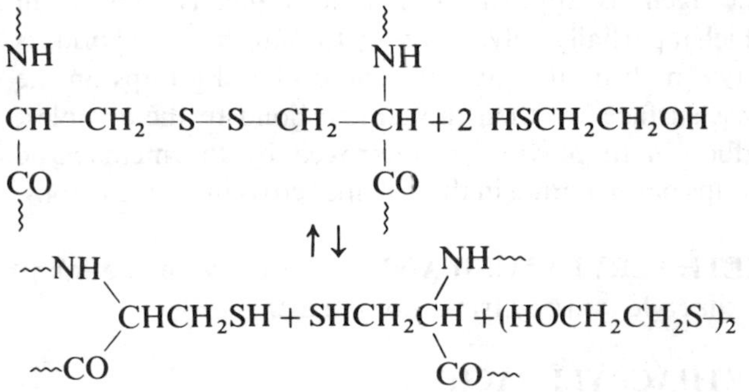

$$
\begin{array}{ccc}
\overset{\}{NH} & & \overset{\}{NH} \\
| & & | \\
CH-CH_2-S-S-CH_2-CH & + & 2\ HSCH_2CH_2OH \\
| & & | \\
CO & & CO \\
\underset{\}{} & \uparrow\downarrow & \underset{\}{}
\end{array}
$$

$$
\begin{array}{cc}
\sim NH & NH\sim \\
| & | \\
CHCH_2SH + SHCH_2CH & +(HOCH_2CH_2S\text{-})_2 \\
| & | \\
\sim CO & CO\sim
\end{array}
$$

The equilibrium proceeds to the right with excess 2-mercaptoethanol. The reaction is carried out at pH 8–10 in 6–10M urea to denature the protein.

2-MERCAPTOIMIDAZOLINE Alternative name for ethylenethiourea.

MERCERISATION The swelling of cellulose, usually as cotton yarn, usually under tension to prevent shrinkage, by immersion in concentrated alkali, which is subsequently washed out. The product, mercerised cotton, is in cellulose II form, has a lower density, higher extensibility, greater water absorption and higher lustre than untreated cotton, but has lower tensile strength.

MERINOVA Tradename for a man-made protein fibre from casein.

MERLON Tradename for bisphenol A polycarbonate.

MERLON T Tradename for a polycarbonate copolymer obtained from reaction of phosgene with a mixture of bisphenol A and 4,4'-dihydroxydiphenylsulphide.

MEROMYOSIN The fragments produced on enzymatic cleavage of myosin with trypsin, producing two fragments—heavy meromyosin and light meromyosin.

MERRIFIELD SYNTHESIS Alternative name for solid phase synthesis.

MESO DYAD (Isotactic dyad) Symbol m. A dyad in a polymer capable of exhibiting stereochemical configurational isomerism, in which the configurations of the two asymmetric, or rather pseudoasymmetric, centres, are the same. In the meso dyad of a vinyl polymer, represented by the Fischer projection formula:

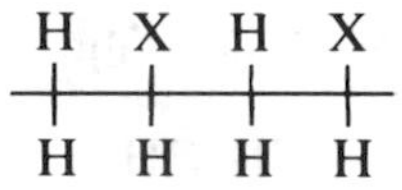

$$
\begin{array}{cccc}
H & X & H & X \\
| & | & | & | \\
H & H & H & H
\end{array}
$$

the two methylene protons are not equivalent and in their nuclear magnetic resonance spectrum their resonances appear as a characteristic quartet.

MESOGEN A molecule, or more frequently (and especially with polymers), part of a molecule which is responsible for the molecular ordering that occcurs in liquid crystalline polymers and other materials. Frequently mesogenic groups have elongated, roughly cylindrical, stiff molecular structures.

MESOMORPHIC POLYMER Alternative name for liquid crystalline polymer.

MESOMORPHISM Alternative name for liquid crystalline behaviour.

MESOPHASE The phase present in a liquid crystalline material which is intermediate in molecular ordering between the high order of a crystalline material and the disorder of a liquid. Different degrees of order may occur, as shown in the nematic, smectic and cholesteric type phases.

MET Abbreviation for methionine.

METACRYLENE Tradename for methylmethacrylate–acrylonitrile–butadiene–styrene or methylmethacrylate–butadiene–styrene copolymers.

β-METAL-BINDING GLOBULIN Alternative name for transferrin.

METAL CHELATE POLYMER Alternative name for chelate polymer.

METAL COMPLEX In polymer stabilisation, any of a large group of group II metal salts (especially cadmium, barium, zinc, calcium or magnesium) of ill-defined mixtures of short-chain organic acids such as phenates, akylated phenates and octoates. Useful as thermal stabilisers for polyvinyl chloride, being more compatible than the metal soaps. Less prone to separate out, especially under high shear processing conditions and may be used at higher concentrations (2–3%). Usually liquid and capable of being used in clear compositions.

METALLOENZYME An enzyme containing a metal ion as a cofactor. Several different metals may be involved, e.g. Zn^{2+} in carboxypeptidase, Mg^{2+} and Mn^{2+} in phosphotransferases, Fe^{2+} or Fe^{3+} in cytochromes and Cu^{2+} in cytochrome oxidase. Sometimes the metal itself shows catalytic activity, but this is greatly enhanced by combination with the enzyme protein. The metal ion may act in one of several ways in addition to being the primary catalytic centre. It may act as a bridge holding the enzyme–substrate complex together or by stabilisation of the particular protein conformation necessary for catalytic activity.

METALLOFLAVOPROTEIN A flavoprotein which contains one or more metal atoms (especially iron or molybdenum) as an enzyme cofactor.

METALLOPROTEIN A conjugated protein whose prosthetic group contains a metal atom or, occasionally, atoms. The haemoproteins are often considered as a separate group. Some members are coloured and are therefore also chromoproteins. Other types are the metalloenzymes, the metalloflavoproteins and certain individual examples, e.g. ferritin. One of the commonest metals is iron (e.g. the haem of haemoproteins and in the enzymes cytochrome and catalase), but other metals also occur especially in enzymes, e.g. zinc (Zn^{2+}) in alcohol dehydrogenase and Cu^{2+} in cytochromo-oxidase.

METALON Tradename for a polybenzoxazole produced by reaction of 3,3'-dimethoxy-4,4'-diaminodiphenylmethane and isophthalic acid and having the structure:

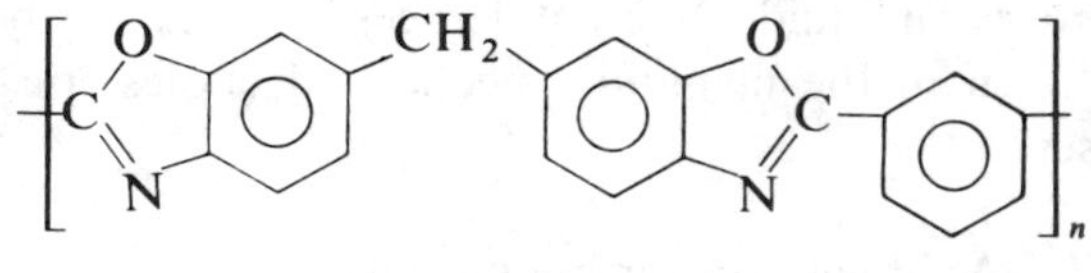

METAL SOAP Any of a wide group of group II metal (especially cadmium, barium, zinc, calcium or magnesium) salts of long chain fatty acids, especially stearic, lauric, naphthenic and ricinoleic. Useful as heat stabilisers for polyvinyl chloride. Frequently used in combination (e.g. mixed cadmium, barium and zinc stearates) when synergistic effects often occur. Can be very effective, giving clear compositions, at levels of about 1%. However, they are of limited compatibility with the polymer (indeed they have a lubricating action) and tend to separate out at levels above 1% or under high shear processing conditions.

METHACROLEIN (Methacrylaldehyde)

$$CH_2{=}C(CH_3)CHO \qquad B.p.\ 69°C.$$

Produced by catalysed vapour phase oxidation of isobutene or by condensation of propionaldehyde with formaldehyde. It readily polymerises by free radical, cationic or anionic mechanisms to polymethacrolein. It is also useful for the production of phenol–methacrolein and methacrolein–formaldehyde resins.

METHACRYLALDEHYDE Alternative name for methacrolein.

METHACRYLAMIDE

$$CH_2{=}CCONH_2$$
$$|$$
$$CH_3 \qquad M.p.\ 110°C.$$

Produced by partial hydrolysis of either methacrylonitrile or of acetone cyanohydrin, in a similar way to acrylamide. The monomer for polymethacrylamide.

METHACRYLATE An ester of methacrylic acid, i.e. a substance of chemical structure $CH_2{=}C(CH_3)COOR$, where R is an alkyl group.

METHACRYLATOCHROMIC CHLORIDE (Chrome finish) Tradename: Volan.

$$CH_2{=}C(CH_3)COOCr_2Cl_4(OH)_2$$

A coupling agent for glass fibre reinforced unsaturated polyester, epoxy resin and other polymer composites. The agent is applied from dilute aqueous solution in which it partially polymerises by forming hydroxyl bridges between chromium atoms. Acidic silanol groups on the glass surface are then thought to bond to the complex. Adhesion to polymer is improved by the methacrylic groups participating in the polymer crosslinking reactions.

METHACRYLATOSILANE Alternative name for γ-methacryloxypropyltrimethoxysilane.

METHACRYLIC ACID

$$CH_2{=}CCOOH$$
$$|$$
$$CH_3$$
$$B.p.\ 163°C.\ M.p.\ 14°C.$$

Prepared from acetone cyanohydrin, in a similar way to methyl methacrylate, or by oxidation of isobutene to hydroxybutyric acid, which is then dehydrated:

$$(CH_3)_2C{=}CH_2 \rightarrow (CH_3)_2C{-}COOH$$
$$|$$
$$OH$$
$$\rightarrow CH_2{=}C{-}COOH$$
$$|$$
$$CH_3$$

The monomer for polymethacrylic acid.

METHACRYLONITRILE (MAN)

$$CH_2{=}CCN$$
$$|$$
$$CH_3 \qquad B.p.\ 90·3°C.$$

Produced by ammonoxidation of isobutene:

$$2(CH_3)_2{=}CH_2 + 3O_2 + 2NH_3 \rightarrow CH_2{=}\overset{\displaystyle CH_3}{\underset{|}{C}}CN + 6H_2O$$

The monomer for the formation of polymethacrylonitrile and also for several copolymers useful as fibres and rubbers.

γ-METHACRYLOXYPROPYLTRIMETHOXYSILANE (Methacrylatosilane)

$$CH_2{=}CHCOOCH_2CH_2CH_2Si(OCH_3)_3$$

A very commonly used silane coupling agent for glass fibre reinforced unsaturated polyester, epoxy resin and thermoplastic composites.

METHANOL (Methyl alcohol)

$$CH_3OH \qquad B.p.\ 64·7°C.$$

Although a widely used organic solvent, it is more frequently used as precipitant for polymers, since many polymers are insoluble in methanol. However, it is miscible with other solvents, such as toluene, which are good polymer solvents. A solvent for polyvinyl acetate and ethyl cellulose.

METHIONINE   (Met) (M)

$$H_2NCHCOOH$$
$$|$$
$$CH_2CH_2SCH_3$$

M.p. 281°C (decomposes).

A non-polar α-amino acid, widely found in proteins but usually present in only relatively small amounts. Its pK' values are 2·28 and 9·21, with the isoelectric point at 5·74. In proteins, methionine residues are often associated with an α-helix conformation. Methionine residues are specifically cleaved by reaction with cyanogen bromide.

METHOLAN   Tradename for phenol–formaldehyde polymer.

2-METHOXYETHANOL   Alternative name for ethylene glycol monomethyl ether.

METHOXYETHYL ACETATE   Alternative name for ethylene glycol monomethyl ether acetate.

METHOXYETHYLACETYL RICINOLEATE

$$CH_3(CH_2)_5CH(OOCCH_3)CH_2-$$
$$-CH=CH(CH_2)_7COOCH_2CH_2OCH_3$$

B.p. 195°C/1 mm.

A plasticiser, mainly for polyvinyl butyral, but also used in polyvinyl chloride and its polymers, cellulose nitrate and in synthetic rubbers.

METHOXYETHYL ACRYLATE

$$CH_2=CHCOOCH_2CH_2OCH_3$$

Useful as a comonomer to replace some alkyl acrylate in acrylic elastomers in order to achieve good low temperature properties and oil resistance.

METHOXYETHYL OLEATE

$$CH_3(CH_2)_7CH=CH(CH_2)_7COOCH_2CH_2OCH_3$$
B.p. 360°C/760 mm, 190–224°C/4 mm.

A plasticiser for synthetic rubbers, ethyl cellulose, polyvinyl butyral and polyvinyl chloride and its copolymers (when used with a phthalate).

METHYL ABIETATE   Tradename: Abalyn.

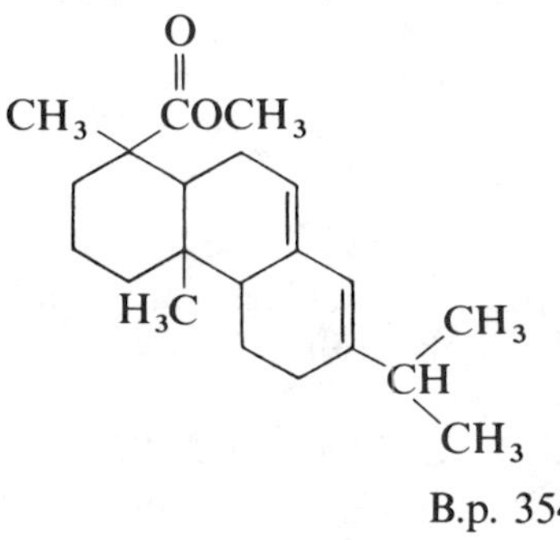

B.p. 354°C/760 mm.

The methyl ester of abietic acid, itself the principal component of rosin. A useful plasticiser in coatings and inks, especially because of its pigment wetting properties, and in films to improve adhesion. Compatible with a wide variety of polymers, e.g. polyvinyl chloride, natural rubber, polystyrene, ethyl cellulose, but not with mixed cellulose esters, cellulose acetate and polyvinyl acetate. Also used as a plasticiser for urea–formaldehyde resins.

N-METHYLACETAMIDE   (NMA)

$$CH_3CONHCH_3 \qquad B.p. 206°C.$$

A dipolar aprotic solvent useful for the synthesis of aromatic polymers, especially polyimides.

METHYL ACRYLATE

$$CH_2=CHCOOCH_3 \qquad B.p. 80·5°C.$$

The monomer for the formation of polymethylacrylate by solution or emulsion free radical polymerisation. Also useful as a comonomer for the production of a variety of acrylic copolymers for use as rubbers, coatings, textile sizes and adhesives. It is produced either by esterification of acrylic acid or by transesterification of ethyl acrylate with methanol. It may also be produced directly by dehydration and esterification of ethylene cyanohydrin (obtained from ethylene and hydrocyanic acid), from acetylene, methanol and carbon monoxide by a Reppe process, $C_2H_2 + CO + CH_3OH \rightarrow CH_2=CHCOOCH_3$ or from β-propriolactone (obtained from acetic acid via reaction of ketene with formaldehyde) and methanol, or from acrylonitrile via acrylamide sulphate.

METHYL ALCOHOL   Alternative name for methanol.

1-METHYL-1,3-BUTADIENE   Alternative name for piperylene.

2-METHYL-1,3-BUTADIENE   Alternative name for isoprene.

METHYL CELLOSOLVE   Alternative name for ethylene glycol monomethyl ether.

METHYLCELLULOSE   A cellulose ether in which some of the cellulose hydroxyl groups have been replaced by —OCH_3 groups. Prepared by reaction of alkali cellulose with methyl chloride. At a DS of 1·5 the polymer is soluble in cold, but not hot water. Commercial products of DS 1·6–2·0 are useful emulsion stabilisers, thickening agents, paper sizes and ceramic binding agents. At a DS approaching 3·0 the polymer is soluble in organic solvents.

METHYL-α-CHLOROACRYLATE

$$CH_2=CClCOOCH_3$$
B.p. 57–59°C/55 mm.

Prepared by dehydrochlorination of methyl-α,β-dichloropropionate (itself obtained by reaction of acrylonitrile, methanol, water and chlorine) or by chlorination of methyl acrylate. It may also be obtained by reaction of paraformaldehyde, trichloroethylene and methanol. An unpleasant and lachrymatory liquid which has been used to make poly-(methyl-α-chloroacrylate).

METHYLCHLOROSILANE
The most widely used type of chlorosilane for the formation of the polyorganosiloxanes of commercial importance, i.e. the methylsiloxanes. The best known compounds are trimethylchlorosilane, dimethyldichlorosilane and methyltrichlorosilane. Produced by the direct process of alkylation of silicon with methyl chloride. The resultant mixed product, mostly dimethyldichlorosilane, is separated by careful fractional distillation, although complete separation is difficult due to the similarity of the boiling points of the components.

METHYL-α-CYANOACRYLATE

$$CH_2{=}CCOOCH_3$$
$$|$$
$$CN$$

B.p. 48°C/2·5 mm.

The most important cyanoacrylate monomer used for cyanoacrylate adhesives. Produced by the depolymerisation of the polymer formed by reaction of formaldehyde with methyl cyanoacetate:

$$n \ NCCH_2COOCH_3 + n \ HCHO \rightarrow$$

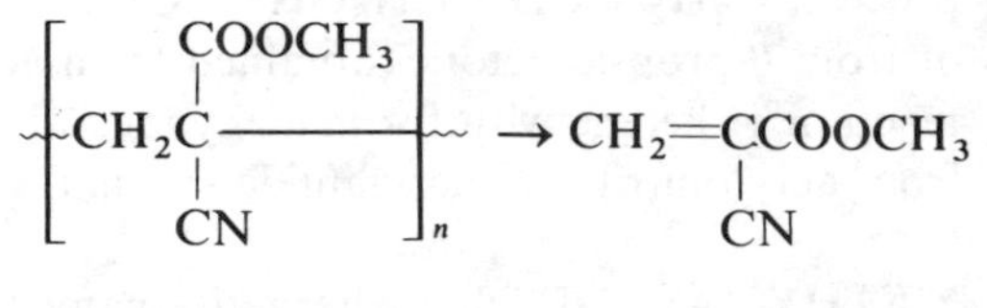

$$\left[CH_2C \begin{matrix} COOCH_3 \\ | \\ | \\ CN \end{matrix} \right]_n \rightarrow CH_2{=}CCOOCH_3$$
                                            $$|$$
                                            $$CN$$

METHYL DIHYDROABIETATE
Tradename: Hercolyn.

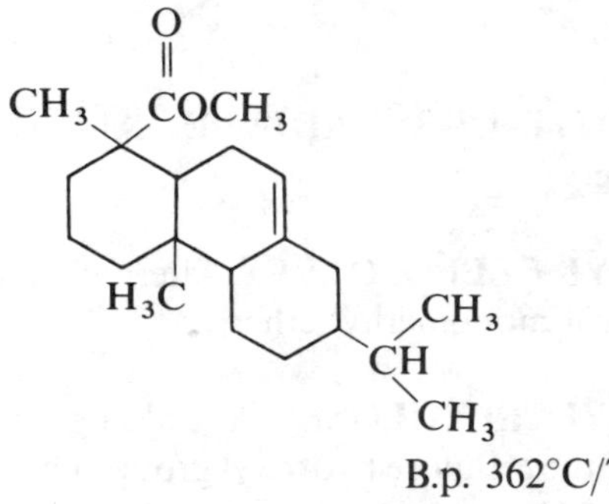

B.p. 362°C/760 mm.

The methyl ester of partly hydrogenated abietic acid. Commercial samples also contain the tetrahydro material. Useful as a plasticiser in a similar way to methyl abietate but has better ultraviolet light stability.

METHYLENDOMETHYLENETETRAHYDRO-PHTHALIC ANHYDRIDE
Alternative name for nadic methyl anhydride.

N,N'-METHYLENE-BIS-ACRYLAMIDE

$$(CH_2{=}CHCONH{)}_2CH_2$$

M.p. 185°C (decomposes).

Formed by the condensation of acrylonitrile with formaldehyde in the presence of 85% sulphuric acid. Useful as a crosslinking comonomer, e.g. with acrylamide for the formation of chemical grouts and in the formation of gels useful for chromatography and electrophoresis.

4,4'-METHYLENE-BIS-(2,6-DI-t-BUTYLPHENOL)
(Bis-(3,5-di-t-butyl-4-hydroxyphenyl)methane) Tradename: Binox M.

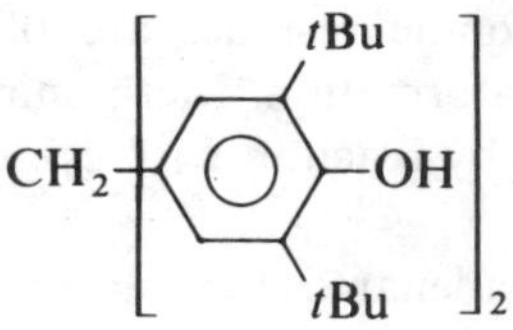

M.p. 154°C. B.p. 289°C/40 mm.

A chain-breaking antioxidant, non-staining and widely used in rubbers and plastics.

4,4'-METHYLENE-BIS-(o-CHLOROANILINE)
Alternative name for 3,3'-dichloro-4,4'-diaminodiphenylmethane.

4,4'-METHYLENE-BIS-(CYCLOHEXYLISO-CYANATE)
Alternative name for 4,4'-dicyclohexylmethane diisocyanate.

2,2'-METHYLENE-BIS-(4-METHYL-6-t-BUTYL-PHENOL)
(Bis-(2-hydroxy-3-t-butyl-4-methyl)methane) Tradenames: Antiox 2246, CAO-5.

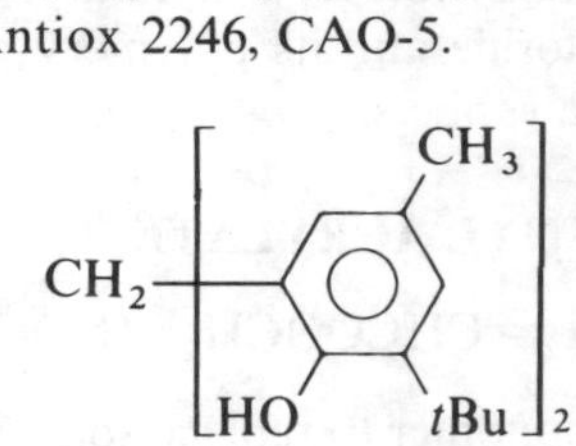

M.p. 125–133°C.

A chain-breaking, non-staining antioxidant for rubbers and polyolefins.

2,2'-METHYLENE-BIS-(6-(1-METHYLCYCLO-HEXYL)-p-CRESOL)
(2,2'-Dihydroxy-3,3'-di-(α-methylcyclohexyl)-5,5'-dimethyldiphenylmethane)

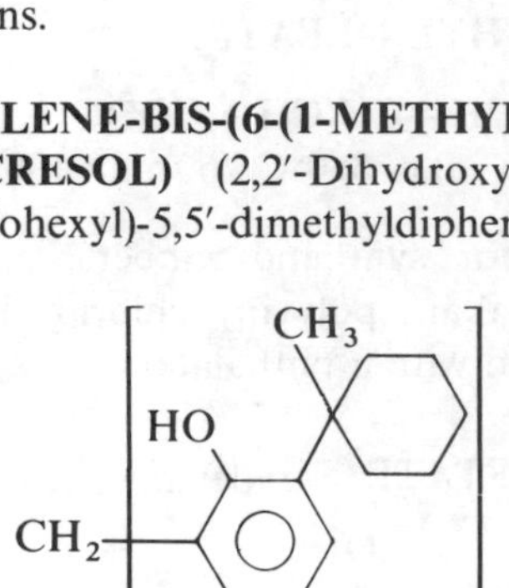

M.p. 130°C.

A powerful, chain-breaking, non-staining antioxidant, widely used in polyolefins.

METHYLENE-BIS-(4,4'-PHENYLISOCYANATE)
Alternative name for 4,4'-diphenylmethane diisocyanate.

4,4′-METHYLENEDIANILINE Alternative name for 4,4′-diaminodiphenylmethane.

METHYLENE DICHLORIDE Alternative name for dichloromethane.

METHYLETHYL KETONE (Butan-2-one) (MEK)

$CH_3CH_2COCH_3$ B.p. 79·6°C.

A solvent for cellulose nitrate, cellulose acetate/butyrate, some natural resins, polystyrene, some acrylic polymers, some vinyl copolymers, nitrile and chloroprene rubbers and useful in adhesives based on these polymers. Its miscibility with water varies greatly with temperature.

METHYL ETHYL KETONE PEROXIDE (MEKP) A complex peroxide produced by reaction of hydrogen peroxide with methyl ethyl ketone, containing both peroxy and hydroperoxy groups, including

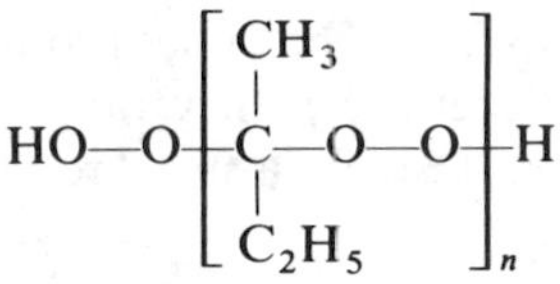

where $n = 1$–4 and the cyclic trimer

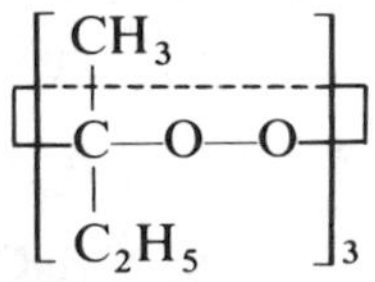

Highly unstable and handled as a solution in plasticiser or solvent. Useful as a free radical polymerisation initiator having half-life times of 162 h/100°C, 6 h/120°C and 0·2 h/150°C and half-life temperatures of 116°C/10 h, 134°C/1 h and 174°C/1 min. Also useful in the cross-linking of unsaturated polyesters at temperatures of 20–50°C when used with a cobalt naphthenate or octoate accelerator, or at above 50°C without accelerator.

3-METHYLHISTIDINE

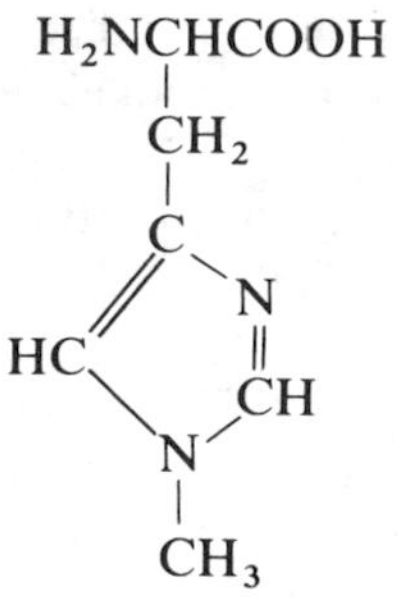

A rare amino acid found in small amounts in the contractile proteins myosin and actin.

METHYLISOBUTYL KETONE (4-Methylpentan-2-one) (MIBK)

$(CH_3)_2CHCH_2COCH_3$ B.p. 116°C.

A solvent for many natural resins, cellulose ethers, natural rubber, polyvinyl chloride, polyvinyl acetate and cellulose nitrate.

METHYLISOPROPENYL KETONE Alternative name for isopropenylmethyl ketone.

ε-*N*-METHYLLYSINE

$H_2NCH(COOH)(CH_2)_4NHCH_3$

A rare amino acid found in small amounts in myosin.

METHYL METHACRYLATE (MMA)

$CH_2{=}C(CH_3)COOCH_3$

B.p. 100·5°C.

The monomer for the production of polymethylmethacrylate, which is widely used as a plastic material. It is also widely used in fundamental studies of free radical polymerisation due to the relatively uncomplicated nature of its polymerisation. It is also useful as a comonomer, especially with other acrylic monomers, producing copolymers useful as coatings and adhesives and for the production of methyl methacrylate–butadiene–styrene and methyl methacrylate–acrylonitrile–butadiene–styrene copolymers. It is produced mostly from acetone by its reaction with hydrocyanic acid to acetone cyanohydrin, followed by reaction with sulphuric acid to methacrylamide sulphate, which is then reacted with methanol:

$$(CH_3)_2CO + HCN \rightarrow (CH_3)_2C(OH)CN \xrightarrow{H_2SO_4} CH_2{=}C(CH_3)CONH.H_2SO_4 \xrightarrow{CH_3OH} CH_2{=}C(CH_3)COOCH_3$$

It may also be produced from isobutene by the route:

$$(CH_3)_2{=}CH_2 \xrightarrow[\text{oxides}]{\text{nitrogen}} (CH_3)_2C(OH)COOH \rightarrow CH_2{=}C(CH_3)COOH \rightarrow CH_2{=}C(CH_3)COOCH_3$$

The monomer readily polymerises free radically (it is usually stored in the presence of an inhibitor) and also copolymerises with a wide variety of comonomers. It may also be polymerised anionically to give stereoregular polymethylmethacrylate.

METHYL METHACRYLATE–ACRYLONITRILE–BUTADIENE–STYRENE COPOLYMER (MABS) Tradenames: Cryolac CIT, Cryolite, Metacrylene, XT polymer. A copolymer of these four monomers similar to acrylonitrile–butadiene–styrene but with some of the acrylonitrile replaced by methyl methacrylate. Unlike ABS, with correct choice of comonomer composition, the copolymer may be transparent.

METHYL METHACRYLATE–BUTADIENE–STYRENE COPOLYMER (MBS) Tradenames: Cryolac CIT, Cryolite, Metacrylene, XT polymer. A terpolymer of these three monomers, similar to acrylonitrile–butadiene–styrene copolymer but with methyl methacrylate replacing acrylonitrile. By careful choice of comonomer composition, the copolymer may be transparent, unlike ABS. Sometimes the acrylonitrile is only partially replaced giving methyl methacrylate–acrylonitrile–butadiene–styrene copolymer

N-METHYLMETHOXYNYLON (Type 8 nylon) An N-substituted nylon containing the structure

$$\sim\!\!\text{CON}\!\!\sim$$
$$|$$
$$\text{CH}_2\text{OCH}_3$$

in the polymer chain. Formed by reaction of a precursor nylon with formaldehyde and methanol. Commercial products often have about 30% of the amide hydrogens substituted, are soluble in aliphatic alcohols and are useful as coating materials. They may be crosslinked by heating with an acid:

$$\sim\!\!\underset{\underset{\text{CH}_2\text{OCH}_3}{|}}{\text{N}}\!\!\sim \;+\; \sim\!\!\text{NH}\!\!\sim \;\xrightarrow{\text{H}^+}\; \sim\!\!\underset{\underset{\text{CH}_2\text{N}\sim}{|}}{\text{N}}\!\!\sim \;+\text{CH}_3\text{OH}$$

METHYL NADIC ANHYDRIDE Alternative name for nadic methyl anhydride.

N-METHYLOLACRYLAMIDE

$$\text{CH}_2\!\!=\!\!\text{CHCONHCH}_2\text{OH} \qquad \text{M.p. } 79°\text{C}.$$

Prepared by reaction of acrylamide with formaldehyde at about 50°C. Useful as a reactive, and therefore potentially crosslinking, acrylic comonomer, e.g. for durable press treatments of textile fabrics.

METHYLOLMELAMINE The initial product of reaction of melamine with formaldehyde in aqueous solution (usually slightly alkaline) in which some or all of the melamine amino hydrogen atoms have been replaced by methylol (—CH$_2$OH) groups. The amount of substitution depends on the melamine:formaldehyde ratio. If this is between 2 and 3 then dimethylolmelamine predominates. At high (> 8) ratios, all the amino hydrogens are replaced to form hexamethylolmelamine. On further heating, or by treatment with acids, the methylol groups react with themselves or with amino hydrogens, to produce linear and branched low molecular weight polymers, which eventually become crosslinked networks. The methylols may be converted to ethers by reaction with an alcohol, as in, for example, butylated amino-polymers (useful as lacquers due to their enhanced solubility in oils), and to the hexamethyl ether of hexamethylolmelamine.

METHYLOLPHENOL (Phenol alcohol) (Hydroxymethylphenol) A phenol containing methylol groups (—CH$_2$OH) substituted on the benzene ring. The initial product of reaction of a phenol, usually phenol itself, with formaldehyde, in the production of phenol–formaldehyde polymers. Substitution may be in either the *ortho* or *para* positions, so in phenol itself both mono-, di- and tri-methylols may be formed. Thus phenol is trifunctional and can form network polymers through reaction with formaldehyde. In resole formation, the methylolphenols are relatively stable, so di- and tri-methylols are formed. In novolacs the methylolphenols rapidly react with further phenol to form methylene bridges. With substituted phenols, the substituents may block the *ortho* or *para* positions, so the phenol may only be difunctional and have no potentiality for crosslinking, as with *ortho*- and *para*-cresols. Some substituted phenols retain all their *ortho* and *para* positions, as with *meta*-cresol, and so are still trifunctional.

METHYLOLUREA The initial product of reaction of urea with formaldehyde in neutral or slightly alkaline aqueous solution. Usually a mixture of monomethylolurea

$$\text{OC}\!\!\begin{array}{l} \diagup \text{NHCH}_2\text{OH} \\ \diagdown \text{NH}_2 \end{array} \qquad \text{(M.p. } 110°\text{C)}$$

and dimethylolurea

$$\text{OC}\!\!\begin{array}{l} \diagup \text{NHCH}_2\text{OH} \\ \diagdown \text{NHCH}_2\text{OH} \end{array} \qquad \text{(M.p. } 140°\text{C)}.$$

is formed. The latter is the major component if the urea:formaldehyde ratio is about 1:2. With this ratio at 1:4, trimethylolurea

$$\text{OC}\!\!\begin{array}{l} \diagup \text{N(CH}_2\text{OH)}_2 \\ \diagdown \text{NHCH}_2\text{OH} \end{array}$$

is also thought to be formed. On acidification, methylol groups condense with each other to form methylene-ether links and in acidic or alkaline solution they react with —NH— groups to form ether links. Thus, linear, branched and, finally, crosslinked network polymers are formed, as in cured urea–formaldehyde thermoset plastics.

METHYLOLUREA ETHER The reaction product of a methylolurea in acid solution, with an excess of an alcohol (ROH). For monomethylolurea the reaction is:

$$\text{H}_2\text{NCONHCH}_2\text{OH} + \text{ROH} \rightarrow$$
$$\text{H}_2\text{NCONHCH}_2\text{OR} + \text{H}_2\text{O}$$

Dimethylolurea usually forms a diether. The lower dialkyl ethers of dimethylolurea are water soluble and are useful (especially dimethylmethylolurea) as textile finishes. The higher ethers are less water soluble but are more soluble in organic solvents. Thus butylated urea–formaldehyde polymers are used with oil modified alkyd resins in stoving enamels. On heating above 100°C methylolurea ethers form crosslinked polymers with the evolution of some ROH, possibly by the sequence of reactions:

n ROCH$_2$NHCONHCH$_2$OR $\rightarrow$

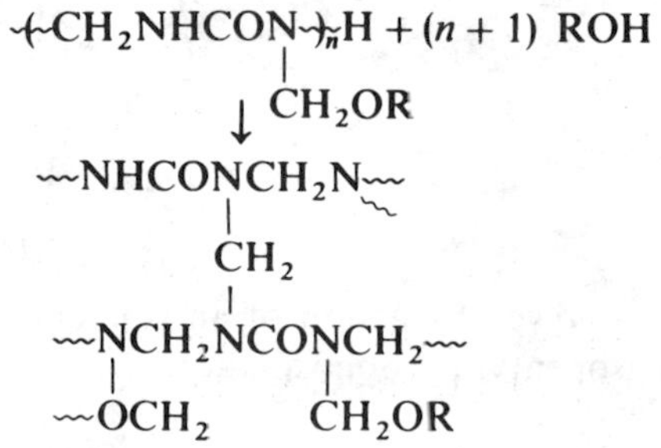

$$-(CH_2NHCON-)_nH + (n + 1)\ ROH$$
$$\downarrow CH_2OR$$
$$\sim\sim NHCONCH_2N\sim\sim$$
$$CH_2$$
$$\sim\sim NCH_2NCONCH_2\sim\sim$$
$$\sim\sim OCH_2 \qquad CH_2OR$$

4-METHYLPENTAN-2-ONE Alternative name for methylisobutyl ketone.

4-METHYLPENTENE-1

$$CH_2{=}CHCH_2CH(CH_3)_2 \qquad \text{B.p. } 53{\cdot}8°C.$$

Produced by dimerisation of propylene at 150–200°C using a sodium or potassium catalyst. The monomer for formation of poly-(4-methylpentene-1) by Ziegler–Natta polymerisation. Sometimes used as a comonomer with ethylene to produce one type of linear low density polyethylene.

METHYLPHENOL Alternative name for cresol.

METHYLPHENYLSILICONE ELASTOMER Alternative name for phenylsilicone elastomer.

METHYLPHTHALYLETHYL GLYCOLLATE

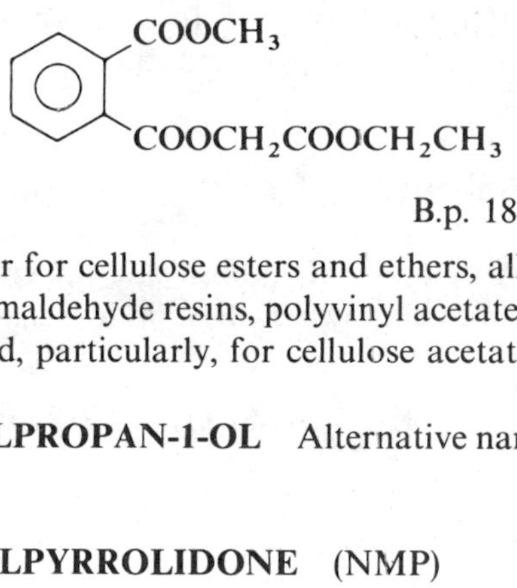

B.p. 189°C/2 mm.

A plasticiser for cellulose esters and ethers, alkyd resins, phenol–formaldehyde resins, polyvinyl acetate, polyvinyl chloride and, particularly, for cellulose acetate.

2-METHYLPROPAN-1-OL Alternative name for iso-butanol.

N-METHYLPYRROLIDONE (NMP)

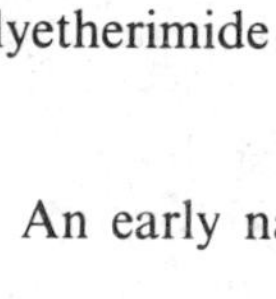

B.p. 202°C.

A useful solvent for aromatic polyamides, polyimides, polyamide–imides, polyetherimide and polyphenylene sulphide.

METHYL RUBBER An early name for poly-(2,3-dimethylbutadiene).

METHYLSILICONE Alternative name for polydmethylsiloxane.

o-, m-, and p-METHYLSTYRENE Alternative name for vinyltoluene.

p-METHYLSTYRENE (p-Vinyltoluene)

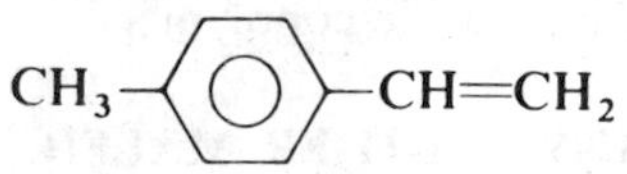

B.p. 170°C.

Prepared by the alkylation of toluene with ethylene to ethyltoluene, followed by dehydrogenation. However, this produces a mixture of *ortho* (14·2%), *meta* (19·3%) and *para* (11·9%) isomers of the ethyltoluenes. The *ortho* isomer is separated and the *meta/para* mixture is used as vinyltoluene. The monomer polymerises similarly to styrene and is useful for crosslinking unsaturated polyesters, as a comonomer instead of, or replacing some, styrene in styrene–acrylonitrile and acrylonitile–butadiene–styrene copolymers, to yield products with higher softening points.

α-METHYLSTYRENE

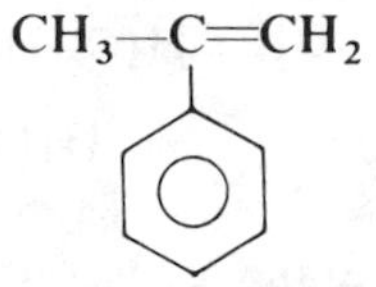

B.p. 165°C.

A vinyl monomer sometimes used as a replacement for styrene. It is produced either by the dehydrogenation of cumene (itself formed by the Friedel–Crafts alkyation of benzene with propylene) or as a by-product in the cumene process for the manufacture of phenol. It may be polymerised to poly-(α-methylstyrene).

METHYLTRICHLOROSILANE

$$CH_3SiCl_3 \qquad \text{B.p. } 66°C.$$

A chlorosilane normally produced by the direct process of reaction of silicon with methyl chloride. Useful as a monomer in the formation of polyorganosiloxanes, since on hydrolysis with water the intermediate trifunctional silanol, CH$_3$Si(OH)$_3$, eliminates water to produce a three-dimensional network polysiloxane:

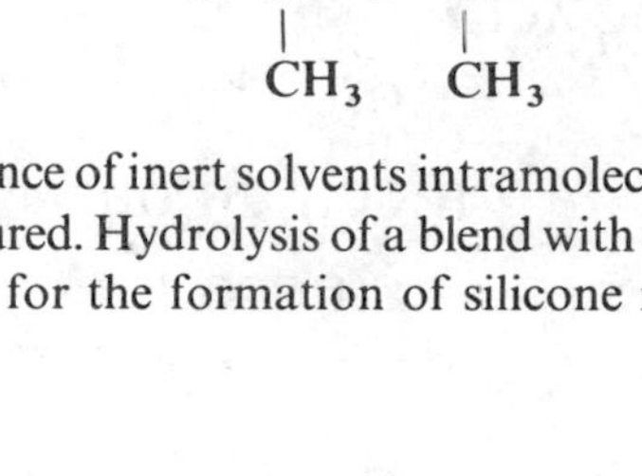

In the presence of inert solvents intramolecular condensation is favoured. Hydrolysis of a blend with dichlorosilanes is the basis for the formation of silicone resins.

METHYLVINYL ETHER (Vinylmethyl ether)

$$CH_2{=}CHOCH_3 \qquad \text{B.p. } 6°C.$$

Prepared by the vinylation of methanol with acetylene in the presence of potassium methoxide. It is polymerised by cationic polymerisation to poly-(methylvinyl ether) and is used as a comonomer for the preparation of methylvinyl ether–maleic anhydride copolymer.

METHYLVINYL ETHER–MALEIC ANHYDRIDE COPOLYMER (Poly-(vinylmethyl ether–co–maleic anhydride)) (Poly-(methylvinyl ether–co–maleic anhydride)).

Tradename: Gantrez AN. An alternating copolymer readily made by the free radical copolymerisation of the two monomers. It is commercially useful due to its water solubility (which causes hydrolysis of the anhydride groups) as a thickening, sizing and suspension agent. Various salts and half esters are also useful, including the ammonium salt of the half amide (**I**), prepared by bubbling ammonia into a solution of the polymer in benzene:

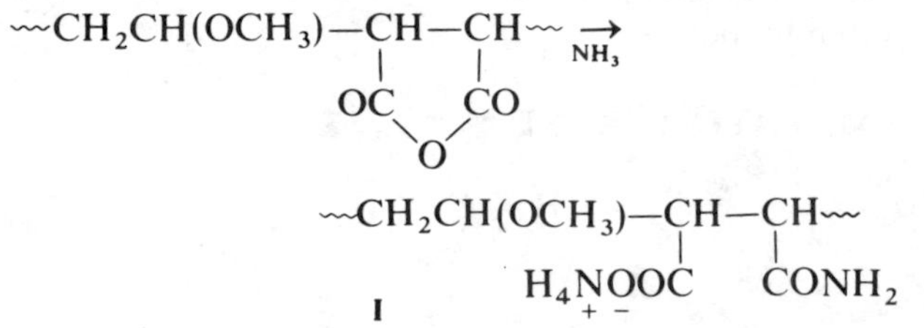

$$\sim\!CH_2CH(OCH_3)\!-\!CH\!-\!CH\!\sim \underset{NH_3}{\longrightarrow}$$
$$\underset{\displaystyle OC \qquad CO}{\qquad\qquad}$$
$$\underset{\displaystyle O}{\qquad}$$

$$\sim\!CH_2CH(OCH_3)\!-\!CH\!-\!CH\!\sim$$
$$\underset{\displaystyle \underset{+\ -}{H_4NOOC} \qquad CONH_2}{\qquad\qquad}$$

I

I is a polyelectrolyte and is useful as a thickener and emulsion stabiliser.

METHYLVINYL KETONE (3-Buten-2-one)

$$CH_3COCH{=}CH_2 \qquad \text{B.p. } 81.5°C.$$

Obtained by the catalytic hydration of vinylacetylene or by the base catalysed condensation of formaldehyde with acetone. It polymerises very readily by free radical, cationic or anionic polymerisation to poly-(methylvinyl ketone). Copolymers of ethylene, styrene, etc., with small amounts of methylvinyl ketone comonomer to provide ultraviolet light absorbing sites, are useful as photodegradable packaging materials, as in Ecolyte.

METHYLVINYLSILICONE ELASTOMER Alternative name for vinylsilicone elastomer.

METZNER FORM OF THE RABINOWITSCH EQUATION A form of the Rabinowitsch equation applying to a power law fluid, of the form, $\dot{\gamma}_w = [(3n' + 1)/4n'](2Q/\pi R^3)$, where $\dot{\gamma}_w$ is the wall shear rate, n' is the flow behaviour index, Q is the volume flow rate and R is the die radius.

MEVOLONIC ACID

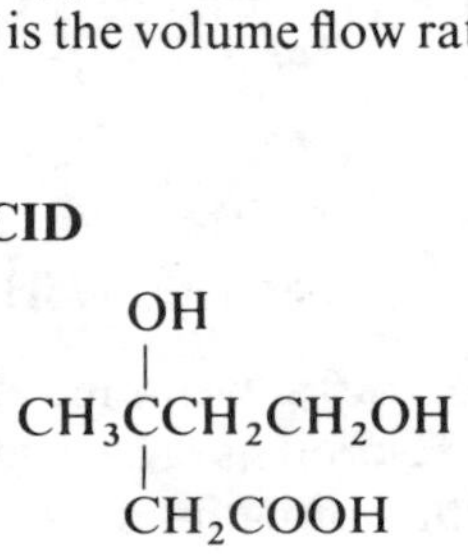

$$\underset{\displaystyle \underset{CH_2COOH}{CH_3CCH_2CH_2OH}}{\overset{\displaystyle \overset{OH}{|}}{}}$$

A key intermediate in the biosynthesis of polyisoprenoid compounds, including natural rubber. The steps involved are phosphorylation, decarboxylation and dehydration producing 3-methyl-3-butenyl-2-pyrophosphate,

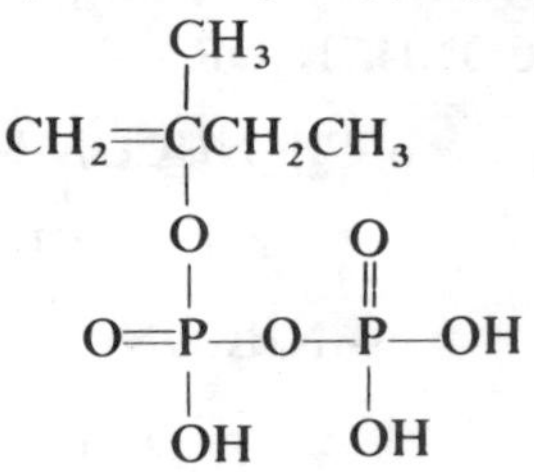

$$CH_2{=}\overset{\displaystyle \overset{CH_3}{|}}{C}CH_2CH_3$$
$$\underset{\displaystyle O{=}P{-}O{-}P{-}OH}{\overset{\displaystyle O \qquad O}{|}}$$
$$\underset{\displaystyle OH \qquad OH}{\qquad}$$

followed by action of an isomerase enzyme to produce isopentyl pyrophosphate

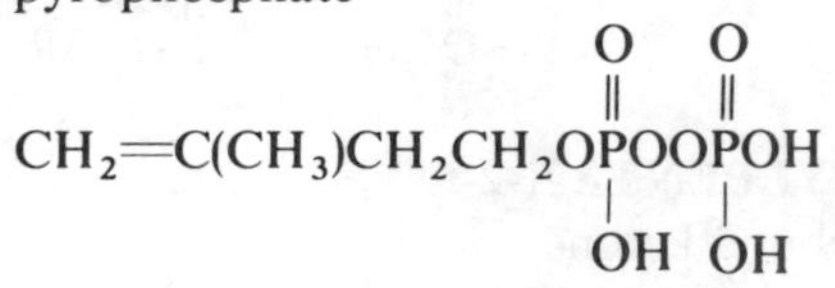

$$CH_2{=}C(CH_3)CH_2CH_2OPOOPOH$$
$$\overset{\displaystyle O \quad O}{\underset{\displaystyle OH \ \ OH}{}}$$

which is then polymerised to polyisoprene.

MEWLON Tradename for polyvinyl alcohol fibre.

MF Abbreviation for melamine–formaldehyde polymer.

MFI Abbreviation for melt flow index.

MFR Abbreviation for melt flow rate.

M-GLASS A high modulus glass whose oxide weight per cent composition is: SiO_2, 53.7; Fe_2O_3, 0.5; CaO, 12.9; MgO, 9.0; LiO, 3.0; BeO, 8.0; TiO_2, 7.9; ZrO_2, 2.0; CeO_2, 3.0. Occasionally used as a reinforcement in glass reinforced polymer composites when high rigidity is required. Its fibres have a high tensile strength of 3.5 GPa, a Young's modulus of 110 GPa, density of 2.89 g cm⁻³ and a refractive index of 1.635. Its use has been superseded by the more recently introduced stiffer carbon and boron fibres.

MIBK Abbreviation for methylisobutyl ketone.

MICA A complex aluminosilicate mineral sometimes used as a filler in polymers. Several forms are found, e.g. muscovite,

$$K_2Al_4(Al_2Si_6O_{20})(OH)_4$$

consisting of silica $(Si_2O_5)^{2-}$ layers sandwiched between which is gibbsite $(Al(OH)_3)$ and phlogopite,

$$K_2Mg_6(Al_2Si_6O_{20})(OH)_4$$

consisting of silica $(Si_2O_5)^{2-}$ layers sandwiched between which is brucite $(Mg(OH)_2)$. The three layer units are held together by potassium cations. The forces between layers are weak and the mineral is readily cleaved into sheets about 25 μm thick. Particles with high aspect ratios (about 100) provide useful plate-like reinforcing effects in polymers, although the major use of the filler is to improve electrical resistivity.

MICELLE Alternative name for domain.

MICROCELLULAR POLYURETHANE A cellular polyurethane with a density in the range $0.5–0.8 \, \mathrm{g \, cm^{-3}}$. Polyester based materials are usually formed by reaction of the polyester with MDI to form a quasi-prepolymer, which is reacted with a diol, usually 1,4-butanediol, and water. Polyether based systems are based on a liquid prepolymer formed from reaction of an excess of MDI with a glycol which is reacted with an ethylene oxide tipped polyoxypropylene, together with a fluorocarbon blowing agent.

MICROCELLULAR RUBBER A cellular rubber with small closed cells produced by moulding a mixture of a rubber with a blowing agent and high styrene resin. Expansion is allowed to occur only after removal from the mould.

MICROCONFORMATION. The conformation of only a small part of a molecule, e.g. a segment of a polymer chain, considering rotations about only a single or a few bonds. The *trans* and *gauche* conformations are microconformations which are often especially energetically favoured. In a large complex structure such as a polymer chain, the sum of the microconformations determines the overall macroconformation.

MICROCRYSTALLINE CELLULOSE Cellulose consisting of discrete, separated crystallites only, formed by hydrolysis of normal fibrous cellulose to its levelling-off degree of polymerisation (with removal of amorphous material) followed by mechanical beating. The microcrystals are about $1 \, \mu\mathrm{m}$ long and $50 \, \text{Å}$ in diameter. It is easily dispersed in water to give stable dispersions or gels.

MICROFIBRIL A fibrillar aggregate of crystallites of $50–1000 \, \text{Å}$ in thickness only observable by electron microscopy. Often found as subunits of fibres, or sometimes of spherulites. For observation, a mechanical disintegration treatment is required. Therefore there is some doubt as to whether microfibrils are true subunits or experimental artefacts.

MICROFORM ANISOTROPY The anisotropy of the optical polarisability of a polymer coil in solution due to interaction between chain elements far removed from each other. The anisotropy is dependent on the shape of the coil.

MICROGEL Small particles, of about $100 \, \mathrm{nm}$, of crosslinked and hence insoluble polymer, which may be considered to consist of single molecules. Sometimes microgel is present in rubbers, both natural and synthetic. In the latter case it is produced during emulsion polymerisation when small quantities of difunctional monomer are present; its size is then limited to the size of the emulsion particles. Microgel of molecular dimensions can dissolve in solvents to give true solutions, albeit with reduced solubility due to the presence of the crosslinks which also reduce the solution viscosity. Microgel can embrittle polymers and reduce tensile strength but may be

beneficial, e.g. in having a smoothing action in the extrusion of rubbers.

MICROMECHANICS The analysis of the mechanical behaviour of composite materials by considering the properties, concentration, geometry and packing of the individual components. This contrasts with macromechanics by recognising the inhomogeneous nature of the composite. By making various approximations of the packing geometry and stress fields within an element of the matrix, the average properties of the element may be calculated. These average properties may then be used in a subsequent macromechanics treatment of an assembly of such elements. The various methods that have been used differ with respect to the severity of the different assumptions made. They include the mechanics of materials, the self-consistent field and the variation methods. Numerical methods may also be applied to particular systems, especially when some symmetry exists in the phase geometry.

MICROMORPHOLOGY The shape of structural units whose dimensions are such that they can be observed by electron microscopy but not by optical microscopy.

MICROPHASE Alternative name for domain.

MICRORHEOLOGY The study of flow in relation to the microstructure of the material undergoing flow, as, for example, in the study of the effect of dispersed particle shape in a polymer blend or filled polymer on the flow properties.

MICROSTRUCTURE The structural features of a material at a fine scale of scrutiny. For polymers, from a chemical point of view, microstructure encompasses the molecular structural features of single polymer chains, often specifically referring to configurational isomerism (and particularly to tacticity). More broadly, the term also refers to a wider range of structural features, including geometric and positional isomerism, chain branching and structural irregularities. Copolymer microstructure refers to both composition and, more particularly, to monomer sequence length distributions. From a materials science viewpoint, microstructure of polymers refers to the grosser structural features of the arrangements of polymer chains in aggregates, as may be observed by optical and electron microscopic techniques. This encompasses the morphological aspects of the crystalline structure, of aggregates of crystals (such as spherulites) and of polymer blends and amorphous polymers, e.g. nodules, partially ordered regions and domains.

MICROTOMY The cutting of thin slices of a material with a glass, steel or diamond knife mounted in a microtome. The slices may then be examined by transmission optical or electron microscopy. Soft samples, e.g. rubber, are best cut by freezing at low temperature (cryomicrotomy) or by embedding in a rigid matrix, e.g. an epoxy resin.

MICROVOID A region in a polymer of lower electron density than its surroundings, of about 100 Å in size and amounting to about 1% of the volume. It is due to unfilled spaces around polymer segments, fibrils or lamellae. In fibres it is often the cause of the diffuse small angle X-ray scattering, the voids being regularly spaced.

MILLABLE POLYURETHANE ELASTOMER Tradenames: Adiprene C, Daltoflex, Desmophen A, Elastothane, Genthane, Millathane, Urepan 600, Urepan 640, Vibrathane 5000. A solid polyurethane elastomer which is capable of being compounded and processed by conventional rubber processing methods, i.e. compounded on a mill and processed by compression moulding. A prepolymer is made from a polyester or polyether polyol and a diisocyanate, with a slight excess of polyol, so that the polymer molecules are hydroxyl terminated. It is also of sufficiently high molecular weight, typically about 20 000, so that the material is gummy (rather than being liquid, as with the cast polyurethane elastomer prepolymers) and can be handled on a mill. The prepolymer can then be crosslinked by either subsequent addition of excess diisocyanate or by peroxide vulcanisation. If a polyester prepolymer is used which contains some unsaturated groups, then sulphur vulcanisation may be employed. The mechanical properties of the vulcanisates can approach those of cast polyurethane elastomers in the isocyanate cured materials, but the peroxide and sulphur cured rubbers do not achieve such good tensile and tear strengths, and are softer and often need to be reinforced. However, they do have lower compression set.

MINDEL Tradename for a polysulphone/ABS blend.

MINERAL SPIRIT Alternative name for white spirit.

MINLON Tradename for a mineral filled nylon 66.

MIPOLAM Tradename for polyvinyl chloride.

MIR Abbreviation for multiple internal reflection.

MIRROR (Mirror marking) The smooth part of a fracture surface, which reflects light specularly, that often surrounds the origin of fracture. It often passes into a region of mist, then of hackle, at larger distances from the origin. In this region a crack has propagated through a craze layer ahead of itself, thus leaving a craze layer on each fracture surface.

MIRROR MARKING Alternative name for mirror.

MIST The region of a fracture surface separating the mirror and hackle regions, which appears under the optical microscope as a dull matt surface. The region is rough, probably as a result of failure being due to cavitation.

MMA Abbreviation for methyl methacrylate.

MOBILITY The reciprocal of plastic viscosity.

MOCA Abbreviation for 3,3-dichloro-4,4'-diaminodiphenylmethane, which is alternatively known as 4,4'-methylene-bis-(o-chloroaniline).

MODACRYLIC Generic name for a fibre containing between 35 and 85 wt% acrylonitrile repeating units in the polymer chain. Examples are Dynel, Teklan and Verel.

MODAL Generic name for a regenerated cellulose fibre of high tenacity and high wet modulus.

MODEL COMPOUND A low molecular weight compound which is used as a model for a much larger polymer material. A compound is chosen which is close in structure to a repeating unit, or preferably to two, three or more repeating units, i.e. as model dimer, trimer, etc., of the polymer chain. Thus ethyl benzene

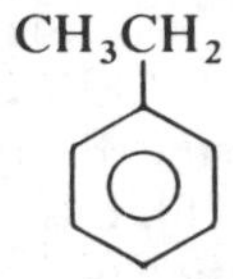

$$CH_3CH_2$$

is a model for a single repeat unit in polystyrene, but 2,4-diphenylpentane

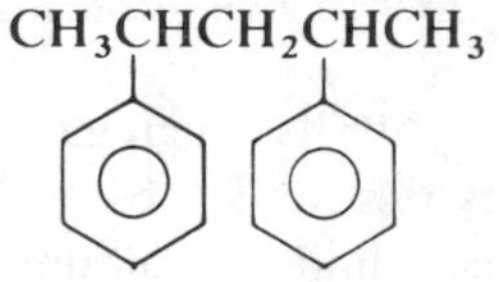

$$CH_3CHCH_2CHCH_3$$

is a better model for polystyrene since it models a dimer unit. Studies of the behaviour of the model compound may then be used either to predict or to explain the behaviour of the polymer. Alternatively, a comparison of the model and the polymer behaviour may be capable of identifying aspects of polymer behaviour which are characteristic of the macromolecular nature of the polymer. Model compound studies have been particularly useful in elucidating the characteristics of polymer chemical reactions, such as degradations, and in interpreting nuclear magnetic resonance and other spectroscopic results obtained with polymers.

MODIFIER (1) (Regulator) A chain transfer agent deliberately added to a polymerisation to limit the molecular weight of the polymer produced. Widely used in emulsion polymerisation, especially in diene polymerisation for rubber production. Compounds with very high chain transfer constants (C_S) are used, the commonest being the aliphatic thiols (mercaptans), especially those with 8–16 carbon atoms, e.g. dodecyl mercaptan. These have C_S values of up to several million.

(2) Alternative name for effector.

MODMOR Tradename for a polyacrylonitrile based carbon fibre.

MODULATOR Alternative name for effector.

MODULUS Alternative name for elastic modulus.

MODULUS OF ELASTICITY IN TENSION Alternative name for Young's modulus.

MODULUS OF RIGIDITY Alternative name for shear modulus.

MOFFITT EQUATION An equation originally derived theoretically to describe the optical rotatory dispersion behaviour of α-helical polypeptides, but, in practice, frequently used as an empirical equation with adjustable constants a_0, b_0 and λ_0 to describe the behaviour. It is of the form of a two-term Drude equation,

$$[M]_\lambda = (a_0\lambda_0^2)/(\lambda^2 - \lambda_0^2) + (b_0\lambda_0^4)/(\lambda^2 - \lambda_0^2)^2$$

where $[M]_\lambda$ is the molar rotation at wavelength λ. Estimates of the conformational helical content (the helicity) of a protein or synthetic polypeptide may be made from the values of the constants, obtained from suitable plots of the data.

MOHAIR The fibre from the coat of the Angora goat. The fibres are 10–25 cm in length and have no crimp, unlike sheep's wool. Tenacity is about $12\,\mathrm{g\,tex}^{-1}$, elongation at break is 30% and moisture regain is 13%.

MOHR–COULOMB YIELD CRITERION Alternative name for Coulomb yield criterion.

MOHR'S CIRCLE A graphical way of expressing the relationship between the shear and normal components of stress acting on a point in a body. It enables the plane stress components to be calculated provided the principal stresses (σ_1, σ_2) are known and vice versa. The equation of the circle is

$$[\sigma_x - (\sigma_1 - \sigma_2)/2]^2 + \tau_{xy} = (\sigma_1 - \sigma_2)^2/4$$

for the case of plane stress, with the z direction being perpendicular to the plane and in which σ and τ have their usual meanings.

MOIRÉ PATTERN A regular pattern of lines on an electron micrograph of two lamellae superposed but rotated by a small angle with respect to each other. Dislocations in the lamellae can be located as discontinuities in the lines and distortions by waviness in the lines.

MOISTURE REGAIN (Regain) The weight per cent of moisture present in a fibre or yarn, based on its oven-dry weight obtained by drying at 105–110°C, after being exposed to an atmosphere of 65% relative humidity at 21°C. Cellulosic fibres, such as cotton and rayon, have high regains of 8–14%, as have other natural fibres—wool

(16%), silk (11%). Synthetic fibres have much lower regains, e.g. polyacrylonitrile (1·5–2·5%), nylon 66 (4–5%) and polyethylene terephthalate (0·5–0·8%).

MOLAR ACTIVITY (Molecular activity) (Turnover number) The number of substrate molecules transformed per minute by a single enzyme molecular or active site. Expressed, for example, as katals per mole of enzyme.

MOLAR ATTRACTION CONSTANT (Small's constant) An energy term associated with an atom or group or bonding feature of an organic molecule, which contributes to the cohesive energy density. Thus the cohesive energy density (CED) of the material as a whole may be calculated by summing the molar attraction constants (F) as $\mathrm{CED} = \sum F/V$, where V is the molar volume. This provides a useful way of estimating the solubility parameters of polymers, which are often otherwise unavailable.

MOLAR MASS Alternative name for molecular weight when this is expressed in units of $\mathrm{g\,mol}^{-1}$.

MOLAR MASS AVERAGE Alternative name for molecular weight average.

MOLAR ROTATION Symbol $[M]_\lambda$. The optical rotation, and hence the optical activity, of an optically active material in solution at a concentration of $c\,\mathrm{g\,cm}^{-3}$ at wavelength λ. It is given by $[M]_\lambda = [\alpha]_\lambda M/100c$, where $[\alpha]_\lambda$ is the optical rotation (given by $[\alpha]_\lambda = (1\cdot8 \times 10^{10}/\lambda)(n_L - n_R)$, where n_L and n_R are the refractive indices for left and right circularly polarised light respectively) and M is the molecular weight. For polymers $[M]_\lambda$ is expressed as per mole of repeat units. For copolymers, especially for proteins, the mean residue rotation is used.

MOLECULAR ACTIVITY Alternative name for molar activity.

MOLECULAR COMPOSITE A composite consisting of highly oriented crystalline fibres of a polymer reinforcing a matrix of the same polymer.

MOLECULAR EXPANSION FACTOR Symbol α. A measure of the increase in size of a polymer molecular coil in solution as the solvent power increases from that of a theta solvent. Defined by $\alpha^2 = \langle r^2 \rangle / \langle r_0^2 \rangle$, where $\langle r_0^2 \rangle$ and $\langle r^2 \rangle$ are the mean square end-to-end distances in a theta solvent and in some other solvent respectively. α is related to the polymer molecular weight (M) by $\alpha^5 - \alpha^3 = cM^{1/2}$, where c is a thermodynamic term. At the theta temperature $\alpha = 1$, but in a good solvent $\alpha > 1$. For high molecular weight polymer, $\alpha^5 \gg \alpha^3$ and hence α is proportional to $M^{0\cdot10}$. This predicts, via the Flory–Fox equation, that the exponent a in the Mark–Houwink equation is 0·8 in a good solvent, which is confirmed experimentally.

MOLECULAR LEVEL ELECTRON MICROSCOPY

MOLECULAR LEVEL ELECTRON MICRO-SCOPY A technique for the observation of individual polymer molecules by electron microscopy. A very dilute polymer solution is sprayed on to a substrate, the solvent is evaporated and the specimen is shadowed. Individual molecules may then be observed, their dimensions can be measured, utilising the shadow length, and the molecular weight and distribution can be determined.

MOLECULAR NUCLEATION The process of incorporation of a new polymer molecule in a crystalline phase, either by the start of a new crystal, i.e. the earliest stage of primary nucleation, or of a new crystal layer, as in secondary crystallisation.

MOLECULAR PROBE TECHNIQUE Alternative name for inverse gas chromatography.

MOLECULAR SIEVE CHROMATOGRAPHY Alternative name for gel permeation chromatography.

MOLECULAR WEIGHT (Molar mass) (Relative molar mass) (MW). The ratio of the mass of an individual molecule of a substance to $\frac{1}{12}$ the mass of an atom of the ^{12}C isotope. On this basis MW is dimensionless. Alternatively, it is the mass of one mole of the substance in question, which gives the same numerical value but now has the units of $g\,mol^{-1}$ or $kg\,mol^{-1}$. For polymers, although each individual molecule has its own molecular weight, a macroscopic sample contains a distribution of molecules of different molecular weights—it has a molecular weight distribution. Thus an average of the molecular weight—the molecular weight average—must be used. The molecular weight of an individual polymer molecule is given by $M = M_0 \times DP$, where M_0 is the molecular weight of the repeat unit and DP is the degree of polymerisation.

MOLECULAR WEIGHT AVERAGE (Average molecular weight) (Molar mass average) Symbol $\bar{M}$. The molecular weight of a polymer sample that results from one of several possible methods of averaging of the different molecular sizes present in polymer samples that are not monodisperse. Since polymerisation is a random process, most polymer samples contain a range of molecular sizes (given by the molecular weight distribution), i.e. they are polydisperse, so some molecular weight average must be used. Several different averages, all ratios of the moments of the distribution, are used. They are given by

$$\bar{M} = \sum_i N_i M_i^{n+1} / \sum_i N_i M_i^n$$

where there are N_i molecules of molecular weight M_i for each molecular species i. In the number average molecular weight ($\bar{M}_n$), $n = 1$, in the weight average molecular weight ($\bar{M}_w$), $n = 2$, in the z-average molecular weight ($\bar{M}_z$), $n = 3$, in the $z + 1$-average molecular weight ($\bar{M}_{z+1}$), $n = 4$, etc., $\bar{M}_n$ is the most sensitive average to low M_i species and as we go to $\bar{M}_w$ to $\bar{M}_z$ to $\bar{M}_{z+1}$, etc., the average becomes more sensitive to molecules with high M_i. The viscosity average molecular weight and the sedimentation-diffusion average molecular weight are of a slightly different type. The average molecular weight is related to the average degree of polymerisation by $\bar{M} = M_0 \times \overline{DP}$, where M_0 is the molecular weight of the repeat unit and $\overline{DP}$ is the average degree of polymerisation.

The polymerisation conditions control the molecular weight of the polymer produced. In step-growth polymerisation high fractional conversions (p) are needed for high molecular weight, since $\bar{x}_n = 1/(1 - p)$ for simple linear polymerisation, where $\bar{x}_n$ is the number average degree of polymerisation. In chain polymerisation the polymer molecular weight decreases with increasing temperature, transfer and termination. Hence to achieve the desirable high molecular weights often needed, strict control over the reaction conditions is necessary. If too high a molecular weight results, then a modifier may be used.

A wide range of molecular weights is found in synthetic polymers; values of a few hundred (in oligomers), a few thousand (in prepolymers, which are often subsequently chain extended), several thousand to about 20 000 (common in synthetic fibres), 20 000 to 100 000 (common in plastics and some rubbers) and several million (in some other rubbers). Crosslinking greatly increases molecular weight and heavily crosslinked molecules (network polymers) have molecular weights approaching infinity. Biopolymers, which are often monodisperse, have molecular weights ranging from a few hundred (some simple proteins and polysaccharides) to hundreds of million (nucleic acids).

The value of the molecular weight average is frequently the most important structural feature, apart from the repeat unit structure, especially since the mechanical properties are molecular weight dependent. Generally in the range 10^3–10^4, a polymer is mechanically weak (soft or brittle solid), whereas above 10^5 little improvement in these properties occurs. Thus many polymers for use as plastics, rubbers, fibres, adhesives and coatings have molecular weights of 10^4–10^5. As molecular weight increases, so melt viscosity increases, and melt processing becomes more difficult.

Many techniques are used to determine molecular weight averages. Usually a particular method yields only one average. Thus colligative property methods (vapour pressure lowering, cryoscopy, ebulliometry, osmometry) yield $\bar{M}_n$, dilute solution viscometry yields $\bar{M}_v$, light scattering yields $\bar{M}_w$ and sedimentation can yield $\bar{M}_n$, $\bar{M}_w$ and $\bar{M}_z$. Values for all the averages can be obtained by gel permeation chromatography.

MOLECULAR WEIGHT DISTRIBUTION (MWD) The distribution of molecular sizes in a polydisperse polymer. Since nearly all synthetic polymers, and many biopolymers, are polydisperse, due to the random nature of the polymerisation reactions by which they are formed, a sample must contain a distribution of molecular sizes. The size distribution is described quantitatively by the molecular weight distribution function.

Many different molecular weight distribution functions have been discussed, some arising from an analysis of the polymerisation kinetics, others having merely an empirical mathematical form. The former include the Schultz–Flory, the Schultz, Poisson and Gold distributions and the latter include the Tung, Gaussian, logarithmic-normal, Lansing–Kraemer and Wesslau distributions. The molecular weight distribution is often characterised more simply by the polydispersity, which is easier to determine than the parameters of the distribution function.

The distribution may be determined either by determination of a sufficient number of the various molecular weight averages, or by fractionation. The first method only gives a limited description, whereas the second method was tedious and time-consuming until the development of gel permeation chromatography. This is much more reliable and rapid than traditional fractionation methods.

Classical linear step-growth polymerisation and simple chain polymerisation with termination by disproportionation both yield the Schultz–Flory distribution. In the latter case, if termination is by combination, the Schultz distribution with a degree of coupling of two results. However, frequently transfer and other reactions cause the distribution to be broader in chain polymerisation, with polydispersities of up to about five. Transfer to polymer particularly broadens the distribution (especially at high conversions or as a result of autoacceleration) to give polydispersities of up to about 100. Such very broad distributions also arise from Ziegler–Natta polymerisations and may be described by logarithmic-normal types of distribution function.

MOLECULAR WEIGHT DISTRIBUTION FUNCTION

(Distribution function) A mathematical expression describing the distribution of molecular sizes (i) (often expressed as molecular weight M_i) in a polydisperse polymer. Discontinuous functions (frequency functions) give the distribution of statistical weights of M_i, whereas cumulative distribution functions give the summation over all statistical weights up to M_i. The statistical weights may be the number (n_i), weight (w_i) or z (z_i) fractions (number, weight and z distributions respectively).

These stepwise distribution functions may be replaced by corresponding continuous functions, since the differences between the i values, in polymer size distributions, are unity, and hence are very small compared with the large size of the polymer molecules. Frequency distributions convert to differential distributions and cumulative distributions convert to integral distributions. The distribution can then be expressed as an analytic expression $f(M)$, where $f(M)\,dM$ is the number of moles of molecules in a sample with molecular weight between M and $M + dM$, and hence is a number density distribution. The functions may be derived theoretically by consideration of the kinetics of polymerisation (e.g. the Schultz–Flory distribution) or they may be merely empirical functions (e.g. the logarithmic-normal distribution).

MOLTOPREN

Tradename for an early flexible polyurethane foam produced by reaction of a lightly branched polyester polyol with 63/35 tolylene diisocyanate.

MOMENT OF DISTRIBUTION

Symbol v_r or μ_r. The moment of a distribution of values of a parameter X about a reference value X_0 (often conveniently taken as zero) is defined as

$$v = \sum_i g_i(X_i - X_0)^q \bigg/ \sum_i g_i$$

where X can have many values X_i and g_i is a statistical weighting factor for each value. q is a number, positive or negative, integral or fractional, giving the order of the moment, i.e. if $q = 1$ the moment is first order, if $q = 2$ the moment is second order, ..., etc. Moments of polymer molecular weight distributions are useful in describing the average molecular weights. Here $g_i = N_i$, the number of moles (or molecules) of each species i of molecular weight $M_i (\equiv X_i)$, so the qth moment is

$$\sum_i N_i M_i^q \bigg/ \sum_i N_i$$

An average of a distribution ($\bar{X}$) is defined in general as

$$\bar{X} = (v^{p+q-1}/v^{q-1})^{1/p} = \left(\sum_i g_i X_i^{p+q-1} \bigg/ \sum_i g_i X_i^{q-1} \right)^{1/p}$$

When $p = q = 1$ this is a simple one-moment average of the type

$$\bar{X} = \sum_i g_i X_i \bigg/ \sum_i g_i$$

The various polymer average molecular weights are thus given by the ratios of the moments (here usually given the symbol μ_r), as number average molecular weight, $\bar{M}_n = \mu_1/\mu_0 = \sum N_i M_i/\sum M_i$, weight average molecular weight, $\bar{M}_w = \mu_2/\mu_1 = \sum N_i M_i^2/\sum N_i M_i$, z-average molecular weight, $\bar{M}_z = \mu_3/\mu_2 = \sum N_i M_i^3/\sum N_i M_i^2$, $z + 1$-average molecular weight, $\bar{M}_{z+1} = \mu_4/\mu_3 = \sum N_i M_i^4/\sum N_i M_i^3, \ldots$, etc. These are all simple one-moment averages. However, if $q = 1$ and $p \neq q$, then a one-moment exponent average results, $\bar{X} = (\mu_r^p/\mu_r)^{1/p}$. In polymers, the viscosity average molecular weight ($\bar{M}_v$) is of this type, with $\bar{M}_v = (\sum N_i M_i^{1+a}/\sum N_i M_i)^{1/a}$ where a is a Mark–Houwink constant.

MOMENTUM EQUATION

(Equation of momentum) (Dynamic equation) (Dynamical equation) A mathematical statement of the principle of conservation of momentum in deformation and flow. When used with the continuity equation and the appropriate constitutive equation, it provides the powerful continuum mechanics approach to the solution of rheological problems. In Eulerian form, and using vector–tensor notation, the equation is, $-\nabla . (\rho vv) - \nabla P + \nabla . \tau' + \rho g = \partial(\sigma v)/\partial t$, where ∇ is the nabla operator, ρ is the density, v is a velocity component, P is the pressure, τ' is the corresponding deviatoric stress component, t is the time and ρg is the body force (per unit volume and usually due to gravity). In this Eulerian form the equations are also known as the Cauchy stress equations. Usually the equation is written in its Lagrangian form, i.e. considering the frame of reference to move with the element under consideration, as $\rho \, Dv/Dt = -\nabla P + \nabla . \tau' + \rho g$, where D/Dt is the substantial time derivative. The left-hand side is the inertial term, followed on the right-hand side by the sum of the surface and body forces. This represents three equations which, for rectangular cartesian coordinates, are

$$\rho \, Dv_x/Dt = -\partial P/\partial x + (\partial \tau_{xx}/\partial x + \partial \tau_{yx}/\partial y + \partial \tau_{zx}/\partial z) + \rho g_x$$

and similar equations for the y and z components. Written more fully, the equations are of the type

$$\rho(\partial v_x/\partial t + v_x \, \partial v_x/\partial x + v_y \, \partial v_x/\partial y + v_z \, \partial v_x/\partial z) =$$
$$-\partial P/\partial x + (\partial \tau_{xx}/\partial x + \partial \tau_{yx}/\partial y + \partial \tau_{zx}/\partial z) + \rho g_x$$

MONADIC NYLON (AB polyamide) (AB nylon) A nylon which may be considered to be derived from an ω-amino acid ($H_2NRCOOH$) and therefore having the structure $-\!\!\left[NHRCO\right]_n$ Well-known examples include nylons 6, 7, 8, 9, 11, 12 and 13.

MONOAXIAL ORIENTATION Alternative name for uniaxial orientation.

MONODISPERSE POLYMER A polymer sample in which all the molecules have the same degree of polymerisation and therefore the same molecular weight. Thus the polydispersity is unity and all the molecular weight averages have the same value. Many biopolymers, especially proteins, are monodisperse (or sometimes paucidisperse) due to the precisely controlled way in which they are formed. Synthetic polymers are always polydisperse, owing to the random way in which their growth is terminated during polymerisation. However, synthetic polymer samples approaching monodispersity may sometimes be synthesised by anionic polymerisation using the living polymer technique. Polystyrene is the best known example. Alternatively, samples with narrow distributions, but never monodisperse, may be obtained by careful preparative fractionation. Monodisperse, or nearly monodisperse, samples are useful for the calibration of non-absolute polymer molecular weight determination methods such as dilute solution viscometry and gel permeation chromatography.

MONOFIL Abbreviation for monofilament.

MONOFILAMENT (Monofil) A single filament, especially when the filament is being used singly rather than being used together with many other filaments as part of a yarn.

MONOGLYCERIDE PROCESS A method for the preparation of alkyd resins. First of all, the oil is heated with the polyol and alcoholysis occurs mainly producing a monoglyceride if an oil:glycerol ratio of 1:2 is used. The monoglyceride is then further heated with the anhydride or acid to produce the final alkyd resin.

MONOLITHIC MATERIAL A simple material consisting of only a single phase, even at a microscopic scale of scrutiny, as opposed to a complex or composite material.

MONOMER (1) A small molecule which is capable of undergoing polymerisation, in which many monomer molecules become covalently bonded to each other to produce a much smaller number of polymer molecules. Monomers capable of forming polymers by chain polymerisation do so either by the opening of a double bond, usually a carbon–carbon double bond, or by ring-opening polymerisation of a ring compound. The former type is the more common and involves such monomers as olefins and substituted olefins, especially the vinyl monomers. Conjugated diolefins (butadiene and its derivatives) may be similarly polymerised through either of the double bonds or by 1,4-polymerisation. Divinyl monomers, or other monomers containing two reactive double bonds, form crosslinked polymers. Sometimes carbonyl double bonds may be involved, as with formaldehyde ($CH_2{=}O$), but many carbonyl compounds cannot be polymerised, e.g. acetone ($(CH_3)_2C{=}O$).

Monomers capable of forming polymers by step-growth polymerisation contain functional groups (A and B) which are capable of reacting with each other in the polymerisation process. A monomer must have a functionality (f) of at least two to be capable of forming a polymer. If $f = 2$ then a linear polymer is formed, if f is greater than two then a branched or network polymer is produced. Monomer in which both types of functional group (A and B) are present in the same monomer molecule is an AB monomer and gives AB polymerisation, as in self-amidation of an amino acid to a polyamide. If the A and B groups are present in different monomers, then the monomer system is an AA + BB type and gives AABB polymerisation as in the formation of a polyester from polymerisation of a diacid with a diol.

(2) Alternative name for the subunit of an oligomeric protein.

MONOMER CASTING A bulk polymerisation performed in a mould so that shaped polymer products may be produced directly from monomer. In vinyl polymerisation this is possible only if the heat of polymerisation can be adequately dissipated, as, for example, in the casting of polymethylmethacrylate sheets from methylmethacrylate monomer. A further requirement is that there are no volatiles evolved, as in the casting of nylon 6 from ε-caprolactam by anionic polymerisation (in contrast to the usual nylon polymerisation methods where water is evolved). Very large mouldings may be cheaply made and the polymer can be of very high molecular weight, since no flow is required to shape the moulding.

MONOMERIC ENZYME An enzyme that consists of a single polypeptide chain. Only a few enzymes are of this type, most being oligomeric enzymes, but due to their simplicity they have been very widely studied structurally. They all catalyse hydrolytic reactions and are often proteases. Generally they have 100–300 amino acid residues and usually operate without a cofactor. Some well-known non-protease examples are ribonuclease and lysozyme. They are often synthesised as inactive zymogens, from which the active enzyme is formed by removal of peptide fragments.

MONOMERIC FRICTION COEFFICIENT Symbol ξ. A parameter of the Rouse–Bueche–Zimm theories of viscoelastic behaviour, which is a measure of the frictional resistance of a monomeric unit to translational motion. It is the average force required for a chain segment to move with unit velocity in its local surroundings. Values range from $1-10^{-9}$ dyne s cm^{-1}. Its value is specific to a particular polymer and is dependent on both temperature and molecular weight, reaching a limiting value at high molecular weights.

MONOMERIC UNIT Alternative name for mer.

MONOMER REACTIVITY RATIO Alternative name for reactivity ratio.

MONOMETALLIC MECHANISM A mechanism of propagation in Ziegler–Natta polymerisation in which only the transition metal atom (often titanium), and not the other metal, is involved directly in the chain propagation. The metal atom is thought to exist in an octahedral configuration with one vacant site ($\square$—, a d-orbital) at which the monomer can be complexed through its π-electrons. Bond making then takes place through a four-centred transition state. The added monomer unit now occupies a different position in the complex, which leads to the prediction of syndiotactic polymerisation. However, isotactic polymerisation is usually observed and this can be explained only by a rearrangement of the ligands surrounding the metal back to their original configuration:

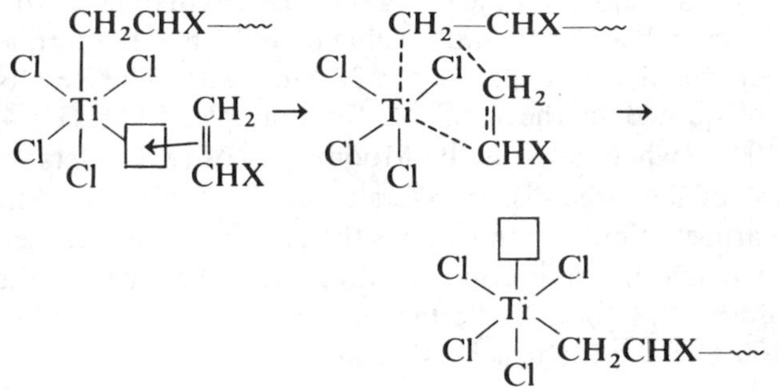

MONOSACCHARIDE A carbohydrate, usually of empirical formula $(CH_2O)_n$, which, together with the oligosaccharides, are called the sugars. The 'monomers' from which oligo- and polysaccharides may be considered to have been formed by step-growth polymerisation with elimination of water. Conversely an oligo- or polysaccharide may be hydrolysed back to its constituent monomers. Sometimes they occur naturally as the free monosaccharides, but most often they are combined as so-called monosaccharide (or sugar) units in the oligo- and polysaccharides. They are polyhydroxy compounds, which in their simplest representation (**I** below) also contain either an aldehyde or ketone group, when they are known as aldoses or ketoses respectively. Usually they contain five or six carbon atoms when they are specifically called aldo- (or keto-) pentose or hexose respectively. Threoses and tetroses are also known. Thus glucose, $C_6H_{12}O_6$, or

$$(\overset{6}{HOCH_2}.\overset{5}{CHOH}.\overset{4}{CHOH}.\overset{3}{CHOH}.\overset{2}{CHOH}.\overset{1}{CHO})$$

is an aldohexose, the corresponding ketose being fructose

$$(\overset{6}{HOCH_2}.\overset{5}{CHOH}.\overset{4}{CHOH}.\overset{3}{CHOH}.\overset{2}{CO}.\overset{1}{CH_2OH})$$

the carbon atoms are numbered as shown. The internal carbons carrying hydroxyls are asymmetric (chiral), each may exist as either configurational isomer of an enantiomorphic pair, distinguished by the letters D- or L- in the name, e.g. as in D-glucose. For a molecule with n asymmetric carbons, 2^n stereoisomers exist consisting of 2^{n-1} such pairs. Thus there are eight possible D-/L- pairs of isomeric aldohexoses and four such pairs of keto-hexoses. The isomers are optically active, although the direction (sometimes denoted by a + or − sign in the name, as in D-(+)-glucose) and the magnitude of rotation of polarised light varies widely among different sugars. The isomers may be represented by the open-chain Fischer projection formula (**I**) but many properties of monosaccharides suggest that they exist in the ring structures (**II**). The rings are obtained by forming a hemiacetal between carbon 1 and carbon 4 or 5 and are better represented by the Hawarth formula (**III**) or by **IV** which shows the precise chair conformation. Thus for D-glucose:

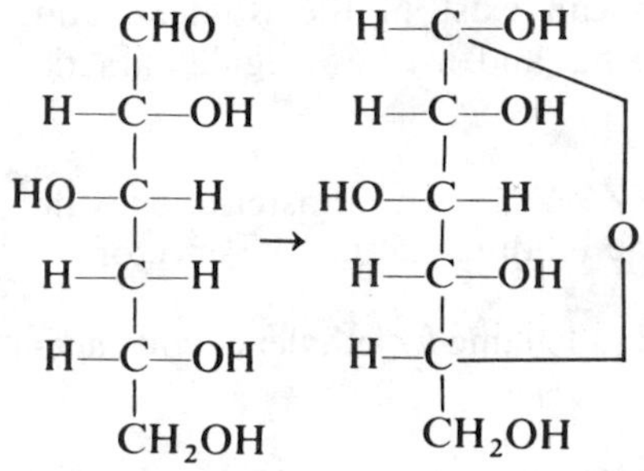

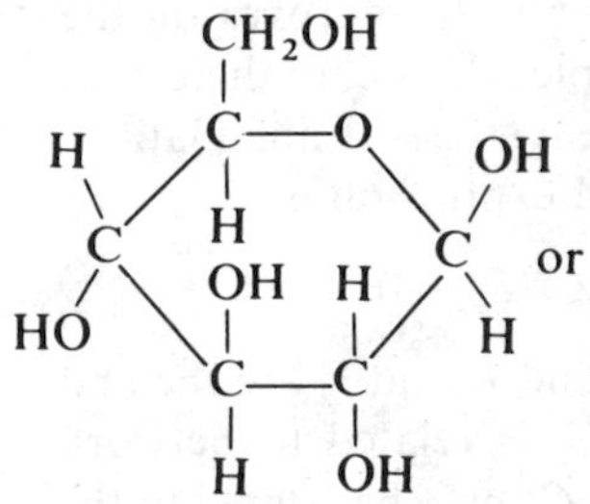

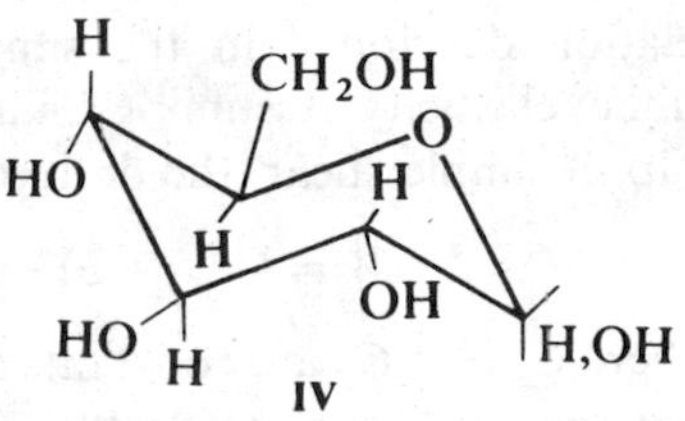

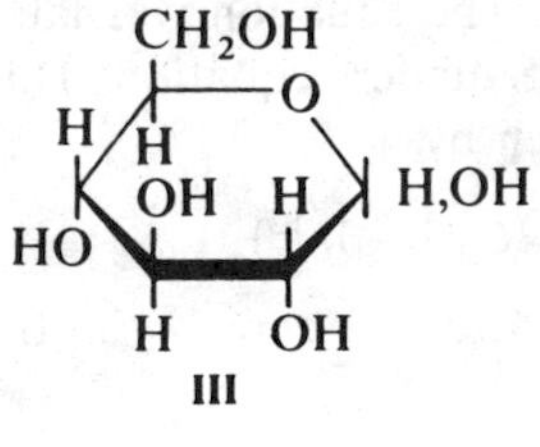

Five- and six-membered rings are considered to be related to the parent heterocyclic rings furan and pyran respectively, and the sugars are then named as being in the furanose or pyranose forms, e.g. D-glucopyranose. In forming these lactal rings a new asymmetric (chiral) centre is formed on carbon-1 (or carbon-2 with a ketose). The two stereoisomers (which are epimers) that result are called anomers and are distinguished by the prefixes α- or β-, e.g. α-D-glucose. A solution of either anomer rapidly partially reverts to the other to form an equilibrium mixture of intermediate optical rotation (mutarotation). Of the eight aldohexoses, glucose, galactose and mannose occur naturally in the D-form (galactose also occurs in the L-form), usually as pyranose rings. The only ketohexose found naturally is D-fructose, whilst of the aldopentoses, D-ribose, D-xylose and L-arabinose occur naturally, the first two as pyranoses, the latter as a furanose. A number of monosaccharide derivatives also exist, including the deoxysugars, the aminosugars, sugar alcohols, sugar acids and anhydrosugars. Glucose, mannose, galactose, xylose, galacturonic acid and N-acetylglucosamine occur as linear homoglycans, the first three, together with arabinose, fructose and fucose, also occur widely in heteroglycans.

MONOTACTIC POLYMER A polymer which contains only one atomic centre in each repeat unit which can exhibit stereochemical configurational isomerism. The commonest type of tactic polymer, most frequently found in vinyl polymers of the type $-[CH_2C^*HX]_n$, where C^* is the centre showing configurational isomerism. Other examples are polypropylene oxide and poly-α-amino acids. These polymers can exist in the isotactic and syndiotactic regular forms and in the irregular atactic form.

MONOVALENT ENZYME An allosteric enzyme which contains only one binding site for the effector.

MONTRATHENE Tradename for ethylene–vinyl acetate copolymer.

MOONEY EQUATION (1) An expression for the strain energy function (W) for finite strain deformations, e.g. as observed in rubber elasticity. It improves on the relation developed in the simple Gaussian theory of rubber elasticity. Assuming a linear stress–strain relationship in simple shear, the derived expression is

$$W = C_1(I_1 - 3) + C_2(I_2 - 3)$$

where C_1 and C_2 are constants and I_1 and I_2 are the first and second strain invariants. C_1 is related to network structure and hence to modulus. C_2 may be related to the effect of entanglements or to the different contributions of short and long network chains. The equation was later further modified to the Rivlin–Saunders equation. This leads to the stress (σ) being given by

$$\sigma = 2C_1(\lambda - 1/\lambda^2) + 2C_2(1 - 1/\lambda^3)$$

for a simple tensile deformation, where λ is the deformation ratio.

(2) An empirical equation, which extends the Einstein equation to higher filler loadings, for expressing the coefficient of viscosity of a composite (η_c) in terms of the viscosity of the matrix (η_m) and the volume fraction of spherical filler particles (ϕ_f) which are much more rigid than the matrix, as in mineral filled rubbers. It is $\eta_c/\eta_m = 2 \cdot 5\phi_f/(1 - S\phi_f)$, where S is a hydrodynamic interaction factor, often taken to be about $1 \cdot 4$. Since the ratio of the coefficients of viscosity is equal to the ratio of the shear moduli, the equation is also useful for predicting the moduli.

MOONEY–RIVLIN–SAUNDERS EQUATION (MRS equation) Alternative name for the Rivlin–Saunders equation.

MOONEY SCORCH TIME (Scorch time) A measure of the scorch resistance of a rubber compound. The compound is heated in a Mooney viscometer and the time measured for the Mooney viscosity to rise five units of viscosity above the minimum value recorded is taken as the scorch time.

MOONEY VISCOMETER (Shearing disc viscometer) A rheometer for measurement of the viscosity, as the Mooney viscosity, of an unvulcanised rubber. It is also useful for recording the changes in viscosity during vulcanisation and hence the course of curing. The rubber is enclosed in a small chamber with a serrated cavity within which a knurled disc rotates. The viscosity is recorded as the torque on the disc shaft, directly on a calibrated dial gauge.

MOONEY VISCOSITY The viscosity on an arbitrary scale for a raw rubber obtained using a Mooney viscometer. The scale is calibrated such that a value of zero is obtained on the dial gauge of the viscometer with the chamber empty and a value of 100 when the torque required to rotate the disc is $8 \cdot 30$ Nm ($73 \cdot 5$ lbf-in). Results are quoted in the form of, for example, 50 ML(1 + 4) (100C), where 50 M is the Mooney viscosity, L refers to use of the large disc (S refers to the small disc), 1 is the warm-up time in minutes, 4 is the time in minutes chosen to reach an equilibrium reading after the start of the motor and 100C is the temperature in degrees Celsius, 100°C being the usual test temperature.

MOPLEN Tradename for polypropylene and polypropylene/EPDM thermoplastic polyolefin elastomer.

MORPHOLOGY The physical structure of a polymer material on the microscopic or submicroscopic level, but not at the molecular level. Morphological features are most obvious in crystalline polymers at different levels of scrutiny from the small single crystals (the crystallites) through to the possible, very large (sometimes visible to the naked eye) crystalline aggregates, of which spherulites are the most common. Morphological features may also sometimes be observed in amorphous polymers, e.g. nodules. Sometimes morphology is also considered to embrace any orientational features found in a polymer material. Morphological structure in polymers is analogous to microstructure in other areas of materials science, but with polymers the use of the term microstructure in this context may cause confusion, since it also refers to aspects of the chemical structure of individual polymer molecules.

MOSAIC (Mosaic block) A description of the structure of a crystalline material in which the crystals consist of blocks of perfectly ordered atoms and molecules, which are slightly misaligned due to crystal imperfections, e.g. dislocations.

MOSAIC BLOCK   Alternative name for mosaic.

MOST PROBABLE DISTRIBUTION Alternative name for Schultz–Flory distribution.

MOTIF   The group of atoms which is repeated regularly within a crystal unit cell and throughout the lattice. In polymer crystals it may vary from a simple group, e.g. —CH_2— in polyethylene, to a collection of complex molecules, e.g. in tobacco necrosis virus where the whole virus is the motif. The crystal is characterised by the repetition scheme of the motif.

MOULDRITE   Tradename for urea–, melamine– or phenol–formaldehyde polymers.

MOVING-BOUNDARY ELECTROPHORESIS Alternative name for free-boundary electrophoresis.

MOWILITH   Tradename for polyvinyl acetate.

MOWIOL   Tradename for polyvinyl alcohol.

MP FIBRE   Tradename for a vinyl chloride/vinyl acetate copolymer fibre.

MQ   Abbreviation for a silicone elastomer based on polydimethylsiloxane homopolymer, i.e. dimethylsilicone elastomer.

MRS EQUATION   Abbreviation for the Mooney–Rivlin–Saunders equation or the Martin–Roth–Stiehler equation.

MS   Abbreviation for mass spectrometry.

MT   Abbreviation for medium thermal carbon black.

MUCILAGE   The plant kingdom counterpart of a mucin, consisting of a polysaccharide component forming a viscous mixture with water. Its natural function often relies upon its water-retention properties, e.g. to prevent dehydration. Examples include the seed and bark mucilages (D-galacto- and D-glucomannans, L-arabino-D-xylans and D-xylo-arabinans), the acidic mucilages (e.g. from slippery elm, containing D-galactose, 3-O-methyl-D-

galactose and D-galacturonic acid and L-rhamnose units) and the sulphated galactans agar, carrageenan and porphyran.

MUCIN   A glycoprotein consisting of a complex of a mucopolysaccharide with a protein. This produces a sticky, jelly-like or slippery substance whose biological function may be for joint lubrication. Sometimes also called mucoproteins.

MUCOID   Alternative name for mucoprotein when this is used to mean a glycoprotein with more than 4% carbohydrate content, measured as hexosamine, such as α_1-acid glycoprotein.

MUCOPOLYSACCHARIDE   A polysaccharide occurring naturally in mucins (viscous secretions of animal origin). Often restricted to the acidic mucopolysaccharides, but also sometimes grouped with the structurally related blood-group substances (neutral mucopolysaccharides).

MUCOPROTEIN   Sometimes this simply means a protein/carbohydrate complex or combination (through covalent bonding), i.e. glycoprotein or proteoglycan. Also, more restrictingly, the term can mean a glycoprotein with more than 4% polysaccharide as hexosamine, which is also sometimes referred to as a mucoid. In this sense the term encompasses such examples as ovomucoid, blood-group substances, submaxillary mucoid and the serum mucoproteins (orosomucoid, the haptoglobins, serum α- and β-glycoproteins). The term is also sometimes used as an alternative name for mucin.

MULLINS EFFECT   Alternative name for stress softening in filled rubbers.

MULTIBLOCK POLYMER   A block copolymer containing several, usually regarded as more than three, blocks of repeating units, frequently of the simple type ﹢$(AB)_n$﹢ where A and B represent the different types of block. Most block copolymers produced by step-growth polymerisation are of this type, such as the polyether/ester thermoplastic elastomers. In many cases one block type is rubbery (soft block or segment) and the other is stiff (hard block or segment). Owing to incompatibility between the different blocks, microphase separation usually takes place, when the domains of the hard blocks act as physical crosslinks for the matrix of the rubbery component. Such materials then behave as thermoplastic elastomers.

MULTICHAIN POLY-(α-AMINO ACID) (Multichain polypeptide) One of the best known types of multichain polymer, poly-D,L-alanyl-poly-L-lysine being particularly well known. They may be formed using a poly-(α-amino acid) containing functional side groups, e.g. the amino groups of lysine, as initiator for polymerisation of an N-carboxyanhydride. The initiating groups may also be attached to a protein, thus forming a polypeptidyl protein. They are of interest as possibly providing better models for globular proteins than linear poly-(α-amino acids) due to the proximity of the polypeptide chains and the similarity of hydrodynamic behaviour, i.e. low viscosities and high sedimentation coefficients.

MULTICHAIN POLYMER A polymer with more than two chains emanating from one centre or core, i.e. a star or comb polymer respectively. The star type may be formed by step-growth polymerisation of a mixture of an AB monomer with a multifunctional monomer, most simply of the type RA_f, where f is the functionality and is greater than two. Such a polymer will have a structure $R\text{--}[A(B\text{---}A)_x]_f$, where x is the number of repeating units in each chain and is not necessarily the same in each chain. The comb type may be formed by growing polymer chains from each of several functional groups present on a preformed polymer.

MULTICHAIN POLYMERISATION A step-growth polymerisation in which a multifunctional monomer A_f (where f is the functionality which is >2) is reacted with an AB monomer, so that although polymer molecules with f chains are produced, they cannot form a network. A similar type of polymer structure, with several chains growing from a single centre obtained by chain polymerisation, is termed a star polymer.

MULTICHAIN POLYPEPTIDE Alternative name for multichain poly-(α-amino acid).

MULTICOMPONENT COPOLYMER A copolymer containing more than two different repeat units, i.e. any copolymer other than a binary copolymer. Terpolymers are by far the commonest type of synthetic multi-component copolymer.

MULTICOMPONENT POLYMERISATION A polymerisation which in general involves the use of more than one comonomer, i.e. a copolymerisation, although the term is usually reserved for copolymerisations which involve three or more comonomers.

MULTILAYER CRYSTAL A stack of lamellar crystals, molecularly connected, which includes spiral growth, hedrite and axialite.

MULTIPLE INTEGRAL REPRESENTATION OF VISCOELASTICITY A representation of non-linear viscoelasticity by a constitutive relationship which is a modified Boltzmann superposition relationship in which, e.g. in creep, the deformation is due to joint contributions from different loadings applied to the polymer at different times. Such a treatment results in the deformation being given by an infinite series of multiple integrals, as in the Green–Rivlin theory.

MULTIPLE INTERNAL REFLECTION SPECTRO-SCOPY (MIR) The usual variety of attenuated total reflection spectroscopy, in which the prism is of trape-zoidal shape, such that many, often 25, reflections (and therefore sample absorptions) take place, thus intensify-ing the spectrum.

MULTRATHANE Tradename for a cast polyurethane elastomer system based on an MDI terminated polyester prepolymer or a tolylene diisocyanate terminated poly-(oxypropylene) glycol prepolymer, chain extended and cured by heating with 3,3'-dichloro-4,4'-diaminophenyl-methane.

MUREIN Alternative name for peptidoglycan.

MUTAROTATION The spontaneous partial conver-sion of one of the anomers of a monosaccharide to the other, in solution, to yield an equilibrium mixture of the α- and β-isomers. Since the two forms have different optical rotations the change is accompanied by a change in the optical rotation of the solution.

MUTUAL RECIPE A recipe for the production of hot styrene–butadiene rubber (GR-S) by emulsion poly-merisation used in the US Government-sponsored programme for synthetic rubber during World War II. It consists of (in parts by weight): butadiene (75), styrene (25), water (180), potassium persulphate (0·3), soap (5·0) and dodecyl mercaptan (0·5).

MW Abbreviation for molecular weight.

MWD Abbreviation for molecular weight distribution.

MYLAR Tradename for a polyethylene terephthalate film material.

MYOGLOBIN A haemoprotein found in skeletal muscles where its function is to receive oxygen from the blood haemoglobin and store it in the muscles. One of the first proteins whose three-dimensional tertiary structure was determined by X-ray diffraction, from sperm whale myoglobin. It consists of a single polypeptide chain of 153 amino acid residues and of molecular weight 17 000 bound to a single haem group. About 80% of the polypeptide exists in the α-helix conformation as eight helical sections, the helices being folded over to give a compact oblate spheroidal shape with the hydrophobic amino acid side chains inside and the polar amino acid groups on the outer surface. The folds occur where proline and other non-α-helix forming amino acid units, such as isoleucine and serine, occur. The haem group lies in a pocket between two long α-helical structures, such that a histidine residue is able to coordinate (by its nitrogen lone pair donation) to the iron. When oxygenated, a further histidine is coordinated indirectly to the iron through the oxygen molecule.

MYOSIN The major (approximately 50%) contractile protein found in the striated muscles of higher animals from which it may be extracted with cold alkaline 0·6M potassium chloride. It forms the thick filaments of the striated myofibrils, which themselves make up the fibres of the muscle. The molecules are about 1500 Å in length and of molecular weight about 500 000. There is a large content of polar amino acids (Glu, Asp, Lys acids) (about 38%) and some (about 2·5%) proline. The molecule exists as two long (heavy) polypeptide chains of molecular weight 100 000 each and of α-helical structure, wound around each other to give a rod-like, coiled-coil double stranded structure. This is attached to a double globular head consisting of the N-terminal region of the α-helical long chains plus four smaller (light) chains of molecular weight 15 000–20 000. The head region contains all the proline, plus some unusual amino acid residues, 3-methylhistidine, ε-N-methyllysine and ε-N-trimethyllysine. However, only four of the 170 lysine and one of the 35 histidine residues are methylated. The heads project outwards from the thick filaments, to bridge the other (thin) filaments of the myofibrils, which are largely composed of actin. In addition, the head catalyses hydrolysis of ATP, the energy provided being used to slide the thick and thin filaments past each other, producing muscle contraction. On cleavage with trypsin two meromyosin fragments (heavy and light) are produced.

MYRISTIC ACID   (Tetradecanoic acid)

$$CH_3(CH_2)_{12}COOH \qquad M.p.\ 53\cdot9°C.$$

Accounts for about 18% of the acid residues of coconut oil.

N

N   Symbol for asparagine.

NABLA OPERATOR   Symbol ∇. The mathematical operator defined by

$$\nabla = \sum_i \delta_i \partial/\partial x_i$$

for rectangular cartesian coordinates, where δ_i is a unit vector in one of the coordinate directions x_i. Thus

$$\nabla P = \sum_i \delta_i \partial P/\partial x_i$$

for a scalar P. For a vector **A**, the notation is

$$(\nabla \cdot \mathbf{A}) = \sum_i (\partial A_i/\partial x_i)$$

The operator is useful in vector–tensor notation in continuum mechanics, e.g. in the continuity and momentum equations.

NADIC ANHYDRIDE   Alternative name for endomethylene tetrahydrophthalic anhydride.

NADIC METHYL ANHYDRIDE   (Methyl nadic anhydride) (Methylendomethylenetetrahydrophthalic anhydride) (NMA)

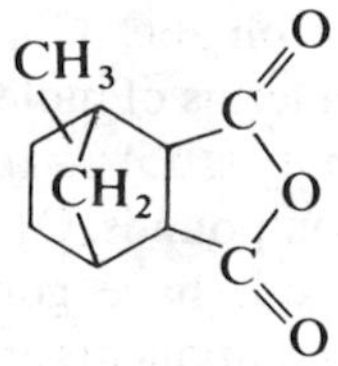

Commercial products are mixtures of isomers and are made liquid by the addition of about 0·1% phosphoric acid. Produced by Diels–Alder reaction of maleic anhydride with methylcyclopentadiene. A versatile curing agent for epoxy resins, capable of giving cured products with a high heat distortion temperature (to 200°C). Tougher products, with lower heat distortion temperatures may be formed using only low amounts of NMA.

NAFION   Tradename for tetrafluoroethylene–sulphonylfluoride vinyl ether copolymer in which the sulphonylfluoride groups have been hydrolysed to —SO₂H acidic groups.

NAIRIT   Tradename for polychloroprene.

NAPHTHA   (Solvent naphtha) A generic term for relatively high boiling point distillates (boiling over a range typically of about 30°C within the range 125–200°C) obtained from petroleum or coal tar sources. They consist of a complex mixture of aromatic, naphthenic and aliphatic hydrocarbons, in varying but controlled amounts in the different commercial products. Widely used as solvents for coatings and rubber materials. In general, solvents for natural resins, rubbers and polystyrene. Some grades, e.g. light solvent naphtha, are similar to xylene in solvency.

NAPHTHALENE-1,5-DIISOCYANATE   (NDI)

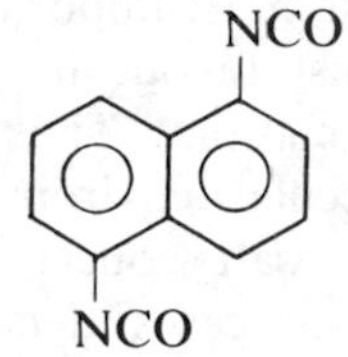

M.P. 128°C.

Synthesised by nitration of naphthalene followed by reduction to 1,5-diaminonaphthalene and then reaction with phosgene. Useful in the production of certain cast polyurethane elastomers.

NAPHTHENIC ACID   A complex mixture of cycloaliphatic acids extracted from the gas oil and kerosene fractions of petroleum. It consists mostly of highly substituted cyclopentane and cylcohexane carboxylic acids. Metal salts, of soap-like character, especially cobalt naphthenate, are useful polymerisation accelerators in the curing of unsaturated polyester resins and are also useful as stabilisers for polyvinyl chloride.

NAPHTHENIC OIL A rubber oil containing a high proportion of naphthene, i.e. cycloaliphatic, structures. A typical oil has a viscosity gravity constant of 0·885, a refractivity intercept of 1·050 and gives, on a carbon atom type analysis, 21% aromatic, 37% naphthenic and 40% paraffinic carbon. In terms of molecules this corresponds to about 42% aromatics, 55% saturated and about 3% polar heterocyclic compounds. Owing to the low content of the latter, these oils have good oxidative stability, whilst the paraffinic components are lubricating in rubber vulcanisates, thus imparting low heat build-up under dynamic stressing.

β-NAPHTHYLAMINE

M.p. 113°C.

An early rubber additive useful for both its antioxidant and vulcanisation effects. The *N*-phenyl substituted compound is now better known as an antioxidant.

NAPRYL Tradename for polypropylene.

NATIVE CELLULOSE Cellulose as obtained from its natural sources, as distinct from regenerated cellulose. A major difference between the two types is in their crystalline forms. Native cellulose crystallises as cellulose I, whilst regenerated cellulose crystallises as cellulose II.

NATIVE CONFORMATION (Native state) The conformation of the polypeptide chain of a protein in its normal biological condition and in which the protein has the required biological activity. It normally refers to globular proteins whose activity depends on the molecule adopting a precise three-dimensional shape (the tertiary structure). The destruction of this shape (denaturation) results in loss of biological activity. The adoption of the native conformation is usually the result of a fine balance of stabilising forces (hydrogen bonding, ionic forces, disulphide bridges and, most importantly, hydrophobic interactions) which just favour it over other slightly higher energy conformations. Thus even quite small changes in the molecular environment, e.g. heating, alteration of pH, or salt concentration, can cause denaturation. In many cases denaturation may be reversed (renaturation) with restoration of the native conformation and also biological activity.

NATIVE PROTEIN A protein existing in its native conformation, i.e. without having been subject to denaturation.

NATIVE STATE Alternative name for native conformation.

NATSYN Tradename for synthetic *cis*-1,4-polyisoprene.

NATURAL DRAW RATIO The maximum draw ratio that can be obtained. This limit is reached when cold drawing is complete and further extension would cause fracture. Cold drawing is thought to cause molecular conformational rearrangements between entanglements or crosslinks, resulting in orientation in the direction of drawing, and not molecular flow. Thus the existence of a natural draw ratio represents the limit of orientation possible and the observed natural ratio is dependent on the amount of pre-orientation.

NATURAL FIBRE A fibre which is produced naturally, as opposed to a man-made fibre. The traditional fibres from which useful products, notably textiles, are made. The natural fibres, cotton and wool, are still of importance although over the past 60 years man-made fibres have grown in importance to account for about 70% of textile use at present. Natural fibres may be of animal origin, e.g. wool and silk, vegetable origin, e.g. seed hair, leaf or bast fibres, or even mineral origin, e.g. asbestos.

NATURAL POLYMER Alternative name for biopolymer.

NATURAL RESIN A resin of natural origin, being obtained, except for shellac, from the secretions of certain trees, which exude the resin when the bark is wounded. The resins are hard, brittle, insoluble in water, but at least partially soluble in some organic solvents. They are non-crystalline, soften on heating and eventually melt to viscous liquids. The initial liquid exuding from a wounded tree is an oleoresin containing the natural resin plus some oils. The latter are often volatile and their loss, together with oxidation and polymerisation on exposure to the atmosphere, results in the formation of the hard resin. If the oils are non-volatile the exudation remains fluid, as in a balsam. Natural resins are widely used in the formulation of surface coating products and include rosin, copal, damar, sandarac and mastic. Some natural products, e.g. amber, Congo and kauri copals, are fossil resins, that is the exudations of trees long since dead and decayed. Most natural resins are mixtures of chemically inert resenes and complex acidic substances containing double bonds. Often the resins consist of polymerised substances.

NATURAL RUBBER (NR) (Caoutchouc) (India rubber) The rubber material obtained from the latex produced by certain plants and trees. Essentially *cis*-1,4-polyisoprene. The only rubber of any commercial significance until the 1930s, and still one of the most important. The commercial product is obtained almost exclusively from the tree *Hevea brasiliensis*, although the latices of many other plant species contain similar rubbers. The latex is obtained by tapping the bark of the tree. Some latex is used directly after concentrating by centrifuging (natural rubber latex) but mostly the solid rubber is obtained by coagulation by the addition of formic or acetic acid. Latex typically consists of about 35% rubber hydrocarbon (the *cis*-1,4-polyisoprene) and about 5% non-rubber products. The latter are retained to a greater or lesser extent in the solid rubber depending on the method of isolation. Typically, solid raw natural rubber contains about 95% cis-1,4-polyisoprene with 2·5% protein, and about 2·5% lipids, phenols, sugars and fatty acids. The non-rubber components play a significant part in determining rubber properties, which therefore differ somewhat from those of synthetic cis-1,4-polyisoprene. In particular, the protein content promotes vulcanisation but is susceptible to microbiological attack, whereas the phenols provide some antioxidant protection.

Many different grades of natural rubber are marketed, depending on the method used for coagulation and purification of the raw rubber. Commonest types are ribbed smoked sheet and pale crepe, but increasingly rubber is marketed according to technical performance rather than to colour, e.g. as technically classified rubber and Standard Malaysian Rubber. A variety of special grades are available which include constant viscosity natural rubber, heveacrumb, oil extended natural rubber, skim rubber and various modified types such as anti-crystallising rubber, grafted rubber and superior pro-cessing rubber. Natural rubber may also be converted to various rubber derivatives such as depolymerised rubber, cyclised rubber, chlorinated rubber and rubber hydro-chloride.

That the raw hydrocarbon is a polyisoprene was first indicated by its pyrolysis products (isoprene and di-pentene) and ozonolysis product (laevulinic aldehyde). X-ray diffraction of stretched crystalline rubber has shown that it is the cis rather than the trans form. Infrared and NMR spectroscopy have shown the absence of 1,2- and 3,4-isoprene structures, so that it is believed that the rubber hydrocarbon is more than 99% cis-1,4-poly-isoprene. In contrast synthetic polyisoprene is a less structurally pure cis-1,4-polyisoprene. The molecular weight of fresh raw rubber typically has a number average value of about one million and a weight average value of about 5 million. In addition, freshly tapped rubber contains crosslinked microgel. Furthermore, during storage raw rubber hardens forming macrogel. These changes are thought to be due to reactions of carbonyl groups in the rubber. In constant viscosity rubber such changes are prevented. For ease of compounding and processing the molecular weight of the raw rubber is reduced to about 130 000 (number average) by mastication.

Cis-1,4-polyisoprene is normally amorphous but crystallises on cooling to below 0°C, with a maximum rate at about −25°C, so that natural rubber hardens on storage at low temperatures. This may be arrested in anticrystallising rubber. The melting temperature is about +25°C. Crystallisation can also be induced by stretching and is responsible for the high tensile strength of natural rubber.

Linear raw rubber is a weak material at ambient temperature, becoming sticky at about 60°C and fluid at about 120°C. Like other diene rubbers, to produce a technologically useful material the molecules must be crosslinked by sulphur vulcanisation using an accelerator, e.g. mercaptobenzothiazole or a sulphenamide, and zinc oxide and stearic acid as accelerator activator. In addition, oils and plasticisers are sometimes used as softeners, as well as antioxidants and antiozonants. Frequently a reinforcing filler, usually carbon black, is used at 10–50 phr to increase modulus, tensile strength and abrasion resistance of the vulcanisate.

Despite competition from the many more recently developed synthetic rubbers, natural rubber is still of major importance due to its low cost, low hysteresis, high strength, high resilience, excellent dynamic properties and fatigue resistance. Its major uses are in tyre sidewalls, heavy tyres, belting hose, and in mechanical uses, such as shock absorbers, fenders and bridge bearings.

NATURAL RUBBER LATEX (Hevea latex) The latex obtained by tapping the bark of the Hevea brasiliensis tree, containing about 35% natural rubber hydrocarbon as particles about 1 μm in diameter, and about 5% non-rubber components consisting of protein, lipids, sugar and salts. Most latex is coagulated by the addition of acetic or formic acid to produce solid natural rubber. Some latex is used as the latex itself after concentration to about 60% rubber content by centrifugation or creaming. The stability of concentrated latex is preserved by the addition of about 1·5% ammonia. This releases fatty acids from the lipids which stabilise the latex after the protein has broken down by natural microbiological attack.

NATURAL STRAIN (Logarithmic strain) (True strain) (Hencky strain) The integral of the nominal strain, i.e.

$$\int_{l_0}^{l_1} dl/l$$

which equals $\ln(l_1/l_0)$, where l_0 is the original length and l_1 is the extended length in a tensile deformation. Identical to nominal strain at low strain, since then the strain is independent of the loading programme.

NAUGAPOL Tradename for styrene–butadiene rubber.

NAUGATEX Tradename for styrene–butadiene rubber.

NAVIER–STOKES EQUATION A form of the momentum equation for a Newtonian fluid, of coefficient of viscosity μ and density ρ, which is the starting point for classical Newtonian mechanics. It is,

$$\rho Dv/Dt = -\nabla P + \rho g + \mu \nabla^2 v$$

where v is a velocity component, P is the pressure, ∇ is the nabla operator, ρg is the body force, ∇^2 is the Laplacian operator and D/Dt is the substantial time derivative.

NBR Abbreviation for nitrile rubber.

NCA Abbreviation for N-carboxy-α-amino acid an-hydride.

NCNS Tradename for an s-triazine polymer produced by reaction between primary and secondary aromatic biscyanamides:

$$n \; NC-N-Ar-N-CN$$
$$\qquad\quad | \qquad\quad |$$
$$\qquad\quad R \qquad\quad R$$
$$+ \; n \; NC-NH-Ar'-NH-CN \rightarrow$$

$$\qquad\qquad\qquad\qquad NH$$
$$\qquad\qquad\qquad\qquad \|$$
$$-[C-N-Ar-N-C-N-Ar'-N-]\sim$$
$$\;\;\| \;\; | \qquad\quad | \qquad\quad | \qquad\quad |$$
$$\;\;NH \; R \qquad\quad R \qquad\; CN \qquad\; CN \;]_n$$

NCNS POLYMER Abbreviation for *N*-cyanosulphon-amide polymer.

NDI Abbreviation for naphthalene-1,5-diisocyanate.

NEAR INFRARED SPECTROSCOPY Infrared spectroscopy in the range $12\,500{-}4000\,cm^{-1}$. Most organic molecules, including polymers, absorb in this region, often due to overtone bands. Although much overlapping of bands may occur, bands due to C—H, N—H, O—H and C=O (first overtone) bonds have all proved useful. For solution studies, solvents free of absorptions in this region are carbon tetrachloride and carbon disulphide.

NECK-IN The reduction in width of an extruded film as it emerges from the extrusion die, due to a combination of swelling and surface tension effects. It decreases as melt elasticity increases and molecular weight distribution increases.

NECKING The rapid decrease in the cross-sectional area at a particular point along the length of a specimen when deformed in tension. This happens as the stress reaches the yield point, i.e. the necking accompanies yielding. Thus, due to the reduction in the cross-sectional area, although the true stress continues to rise, the nominal (or engineering) stress usually falls in a conventional test carried out at a constant rate of elongation. Subsequently the nominal stress may remain constant as the neck extends along the specimen and cold drawing occurs. The point at which necking occurs is given by the Considère construction as the point on the stress–strain curve at which

$$d\sigma/de = \sigma/(1 + e)$$

where σ is the true stress and e is the nominal strain. If the neck is to be stable and cold drawing is to occur then the above equality must hold at two points on the curve.

NEEDLE MAT A sheet-like material, usually of glass fibres, formed by needling chopped fibres about 50 mm in length through a backing material, usually of cotton.

NEGATIVE COOPERATIVITY The decrease in affinity for binding a molecule caused by the prior binding of another molecule. It occurs with allosteric enzymes when binding of one substrate or effector decreases affinity for a second substrate or effector.

NEGATIVE REVERSE OSMOSIS (Piezodialysis) The diffusion of a dissolved salt through a membrane (rather than the solvent, which is usually water) which can occur in certain two component ionic polymer materials, notably in charge mosaic membranes. The process works on application of an externally applied pressure and is preferred to conventional reverse osmosis for certain processes, especially desalination, since only the by-product salt (not the water) is transported through the membrane.

NEGATIVE SPHERULITE A spherulite in which the refractive index is greater along the molecular chains than across them. Owing to the tangential molecular orientation, the greater refractive index is also tangential.

NEGATIVE THIXOTROPY Alternative name for antithixotropy.

NEMATIC PHASE A liquid crystalline mesophase in which the mesogens show only long-range orientational order and no positional order, i.e. of their centres of gravity. If the mesogen is chiral then a cholesteric phase may be exhibited. Nematic phase behaviour is simpler than smectic phase behaviour, usually having only a single nematic transition (liquid crystal to isotropic melt) above T_g but below the normal T_m of a polymer.

NEOFLON Tradename for tetrafluoroethylene–hexafluoropropylene copolymer.

NEO-HOOKEAN Elastic behaviour found in inhomogeneous, isotropic and incompressible materials such as rubbers, which may be described by

$$f = \text{constant}\,(\lambda - 1/\lambda^2)$$

where f is the nominal stress and λ is the extension ratio. For simple extension the constant is $E/3$, where E is Young's modulus. This relationship may be derived from the statistical molecular theory of rubber elasticity as well as by purely phenomenological considerations. Such a material also has a strain energy function given by

$$W = E/6(\lambda_1^2 + \lambda_2^2 + \lambda_3^2 - 3)$$

with $E = 3NkT$, where N is Avogadro's number, k is Boltzmann's constant, T is the temperature and λ_1, λ_2 and λ_3 are the principal extension ratios.

NEOPENTYL ALCOHOL Alternative name for 2,2-dimethylpropane-1,3-diol.

NEOPENTYLENE GLYCOL Alternative name for 2,2-dimethylpropane-1,3-diol.

NEOPRENE Tradename for polychloroprene.

NERVE The elastic recovery from deformation of a raw unvulcanised rubber or rubber compound before vulcanisation. Excessive nerve must be avoided if a shaped piece of rubber is to retain its shape before vulcanisation. Nerve may be reduced by prior mastication, by the suitable choice of reinforcing fillers (especially useful is high structure carbon black) or by the addition of mineral rubber or factice. Addition of a proportion of crosslinked rubber, as in superior processing natural rubber, also controls nerve.

NESTORITE   Tradename for phenol– or urea–formaldehyde polymer.

NETWORK BREAKDOWN   Alternative name for chemical stress relaxation.

NETWORK PARAMETER   Symbol M_c. A measure of the crosslink density of a polymer. The molecular weight of the primary chain segment between the crosslinks. It is only readily determinable in lightly crosslinked elastomers, either by equilibrium swelling measurements through the use of the Flory–Rehner equation, or from measurements of the stress–strain behaviour by use of the equation, $M_c = RT\rho/G$, where R is the universal gas constant, T is the absolute temperature, ρ is the rubber density and G is the modulus factor (which is the slope of a plot of force against the function $(\lambda - \lambda^{-2})$, where λ is the extension ratio for a simple tensile test). For a lightly crosslinked rubber M_c may have values of several thousand.

NETWORK POLYMER   A crosslinked polymer where there is a high enough number of crosslinks for all the polymer molecules, or molecular segments, to be joined to each other, thus forming an infinite (molecularly speaking) network. Often the terms network polymer and crosslinked polymer are used synonymously, but the former is often preferred for polymers produced by step-growth polymerisation using multifunctional monomers, and the latter for polymers that have been formed by crosslinking preformed linear polymer molecules.

NEUTRAL MUCOPOLYSACCHARIDE   A glycosoaminoglycan which, unlike the acidic mucopolysaccharides, does not contain any uronic acid groups. The blood-group substances are the best known examples.

NEUTRON SCATTERING   The interaction of neutrons with a material such that they are deflected. Since neutrons are electrically neutral they are only scattered by collisions with the nuclei of the atoms. As a result the normal optical quantum rules do not apply and all molecular vibrational frequencies may be excited and observed. At the wavelengths normally used (about 5 Å), neutrons have velocities of about $10^3 \, \mathrm{m \, s^{-1}}$ and, hence, much lower kinetic energies than normal electromagnetic radiation for these wavelengths. Furthermore, owing to the large mass of the neutron, high momentum transfer occurs, giving a very wide range of scattering vectors. Inelastic scattering, i.e. with an exchange of energy, is useful for the study of molecular motions and vibrations in polymers. It may be analysed either by determining the velocity of the scattered neutrons by measuring their time of flight over a given distance, or by determining their wavelength by using a crystal monochromator.

Elastic scattering is widely used in small angle neutron scattering to investigate the molecular dimensions (e.g. as the radius of gyration) and conformation of polymer molecules both in solution and in the solid state. Scattering may be either coherent or incoherent. The ability of an atom to scatter is given by its scattering cross-section. Cross-sections vary randomly with increasing atomic number. Hydrogen has a very high incoherent cross-section, so that motions involving hydrogen are readily studied. Deuterium has a very different coherent cross-section compared with hydrogen and hence selective deuteration is widely used either to suppress incoherent background scattering, when coherent scattering is being studied, or, in incoherent scattering, to increase contrast.

NEWTONIAN FLUID   (Ideal fluid) A fluid for which the viscosity, here called the coefficient of viscosity and given the symbol μ, is independent of the rate of deformation, i.e. a fluid for which Newton's law of viscosity holds. In tensor notation, the shear stresses (τ_{ij}) (the only ones which are non-zero with respect to time and rate of deformation) are $\tau_{ij} = \mu\Delta_{ij}$, $i \neq j$, where Δ_{ij} are the components of the rate of deformation tensor. The more general equation of motion for a Newtonian fluid is the Navier–Stokes equation. Thus, for example, in plane couette flow we have $\tau = \mu(u/S)$ or $= \mu(\mathrm{d}u/\mathrm{d}S)$, where u is the relative velocity of the plates and S is their separation. Thus a plot of shear stress against shear rate will be a straight line of slope μ. Such is true of many ordinary liquids, but for most polymer systems, melts or solutions, such plots are non-linear, the apparent viscosity being shear rate dependent. Thus they are non-Newtonian, although they may show Newtonian behaviour at low rates of shear, say below $10^{-1} \, \mathrm{s^{-1}}$. Nevertheless, in the analysis of complex flows, as often found in polymer processing, solution of the problem may require the assumption of Newtonian behaviour for a result to be obtained.

NEWTONIAN VISCOSITY   Alternative name for coefficient of viscosity.

NEWTON'S LAW OF VISCOSITY   In a simple shearing flow, the shear stress components τ_{ij} are given by $\tau_{ij} = \mu\Delta_{ij}$, $i \neq j$, where Δ_{ij} are the components of the rate of deformation tensor and μ is the coefficient of viscosity. Specifically in plane couette flow $\tau_{ij} = u/S$, where u is the relative plate velocity and S is the plate separation.

NICKEL CHELATE   A nickel complex, usually of the square planar, four-coordinate type, useful as an ultraviolet stabiliser, especially for polypropylene. Thought to act largely by quenching of polymer excited state triplets. A typical example is:

NIELSEN EQUATION An extension of the Halpin–Tsai equation, giving the moduli (shear, bulk or Young's) of a composite (M_c) in terms of the moduli of the matrix and filler (M_m and M_f respectively). The equation is,

$$M_c/M_m = (1 + AB\phi_f)/(1 + \psi B\phi_f)$$

where ϕ_f is the filler volume fraction and A is a constant equal to $k_e - 1$, where k_e is the Einstein coefficient. B is a constant given by $B = (M_f/M_m - 1)/(M_f/M_m + A)$, which is about unity for $M_f \gg M_m$. ψ depends on the maximum packing fraction (ϕ_{max}) of the filler and is given by: $\psi = [1 + (1 - \phi_{max})/\phi_{max}^2]$.

NIGERAN An α-D-glucan produced by certain moulds and containing roughly equal proportions of 1,3'- and 1,4'-links between the glucose sugar units, the two types often alternating.

NINHYDRIN (Triketohydrindene hydrate) (2,2-Dihydroxy-1,3-indandione)

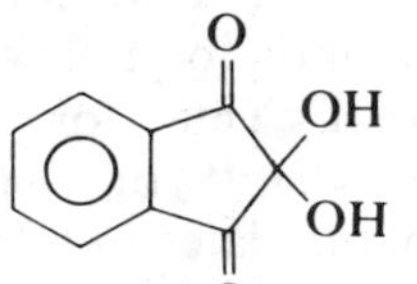

A very widely used reagent for the quantitative determination of amino acids. At pH 4 ninhydrin reacts with a free α-amino group of an amino acid or peptide formed by fragmentation of a protein polypeptide, to give a coloured product (Ruhemann's purple):

$$H_2NCHRCOOH + 2\ \mathbf{I} \rightarrow$$

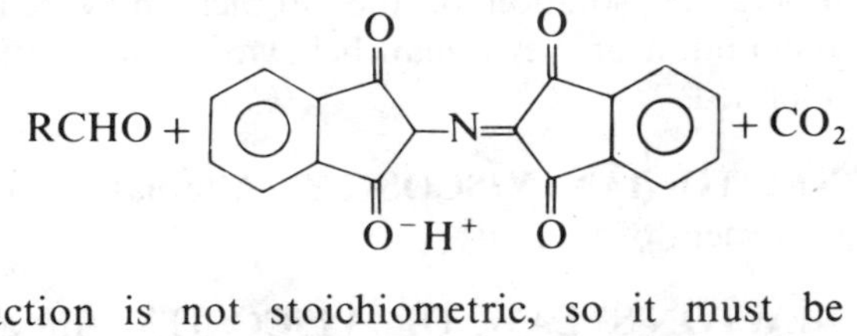

The reaction is not stoichiometric, so it must be standardised for each amino acid, but it is reproducible. Amino acids with no primary hydrogen, such as proline and hydroxyproline, give a yellow-brown product.

NISSO PB Tradename for a high vinyl polybutadiene resin.

NISSO PBC Tradename for carboxy-terminated polybutadiene.

NITRILE RESIN Alternative name for high nitrile polymer.

NITRILE RUBBER (NBR) Tradenames: Breon, Buna N, Butacril, Butakon A, Butaprene, Chemigum N, Elaprim, Europrene N, FR-N, GR-A, Hycar, Krynac, Paracril, Perbunan, Perbunan N. A butadiene–acrylonitrile copolymer diene rubber, important for its high oil and temperature resistance. Produced by emulsion polymerisation in a similar way to sytrene–butadiene

rubber processes, yielding both 'hot' and 'cold' rubbers. A wide range of commercial products with acrylonitrile contents of 25–50% is produced. As acrylonitrile content increases so hydrocarbon swelling resistance, T_g, hardness and tensile strength increase, but resilience and low temperature elastomeric properties diminish. Typical number average molecular weight is about 10^5 with a broad molecular weight distribution. Hot rubbers are branched and contain some crosslinked gel. Sometimes the more linear cold rubbers are slightly crosslinked by the incorporation of about 1% divinyl benzene to improve processability and compression set resistance. Typically the majority of the butadiene units are *trans*-1,4-linked with about 10% 1,2-linkages, but the values depend on polymerisation temperature. Ziegler–Natta produced solution rubbers are alternating in structure and can crystallise giving higher tensile strengths. Conventional general purpose nitrile rubber with an acrylonitrile content of about 35% has a T_g value of about $-35°C$.

Nitrile rubber is vulcanised by systems similar to natural rubber, i.e. normally by accelerated sulphur vulcanisation and is compounded with fillers, usually carbon black. Use of cadmium based vulcanisation systems improves the heat resistance of the vulcanisate. Nitrile rubber is frequently used as a blend with PVC to improve ozone, weathering, abrasion and oil resistance. Low molecular weight liquid nitrile rubbers, especially carboxy-terminated nitrile rubber, are useful toughening agents.

NITRILESILICONE ELASTOMER A silicone elastomer based on polydimethylsiloxane in which some methyl groups have been replaced by β-cyanoethyl ($CNCH_2CH_2$-) or γ-cyanopropyl ($CNCH_2CH_2CH_2$-) groups. This improves the solvent resistance but commercial polymers are not now produced.

***o*-NITROBIPHENYL** (ONB)

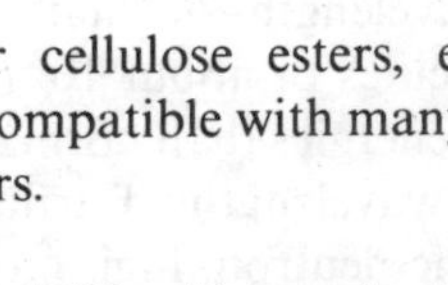

B.p. 330°C/760 mm. M.p. 35°C.

A plasticiser for cellulose esters, especially cellulose nitrate, but also compatible with many other natural and synthetic polymers.

NITROCELLULOSE Alternative name for cellulose nitrate.

NITRON Tradename for polyacrylonitrile fibre.

***N*-NITROSODIPHENYLAMINE**

M.p. 65°C.

A vulcanisation retarder, useful therefore in preventing scorch, but which also retards crosslink formation during vulcanisation.

NITROSOFLUORORUBBER Alternative name for nitrosorubber.

NITROSORUBBER (Fluoronitrosorubber) (Nitroso-fluororubber) An essentially alternating copolymer of tri-fluoronitrosomethane and tetrafluoroethylene, obtained by free radical polymerisation below 0°C. The polymer has the repeat unit structure

$$\left[\begin{array}{c} NO-CF_2CF_2 \\ | \\ CF_3 \end{array}\right]_n$$

The copolymer has some unusual properties being non-inflammable even in pure oxygen, having a T_g value of $-50°C$ (and hence being elastomeric), having outstanding chemical resistance, especially to N_2O_4 and ClF_3, and a very low solubility parameter. However, it has a poor thermal stability and stability to bases. It is difficult to vulcanise but this is overcome in carboxynitrosorubber.

NIVIONPLAST Tradename for nylon 66.

NMA (1) Abbreviation for N-methylacetamide.
(2) Abbreviation for nadic methyl anhydride.

NMP Abbreviation for N-methylpyrrolidone.

NMR SPECTROSCOPY Abbreviation for nuclear magnetic resonance spectroscopy.

NOBS Abbreviation for N-oxydiethylbenzothiazylsul-phenamide.

NODE A dislocation mark present in flax fibres. Several hundred are present in each fibre of length about 25 cm. It consists either of locally separated fibrils or of minute fissures. On stretching the nodes disappear but reform on removal of stress.

NOLIMID Tradename for a polyaminobismaleimide similar to Kerimid.

NOMARSKI INTERFERENCE MICROSCOPY The commonest type of interference microscopy.

NOMEX Tradename for poly-(m-phenyleneisophthal-amide). Previously called HT-I.

NOMINAL SHEAR RATE In most fluid flow situ-ations, the shear rate varies throughout the fluid. For non-Newtonian fluids it is important to have a measure of the shear rate and often a nominal shear rate is chosen, frequently as the maximum shear rate in the system, e.g. at the wall of a circular cross-section pipe in capillary flow.

NOMINAL STRAIN In a tensile deformation, the change in length divided by the original length. Natural strain is the integral of this and identical to it at small strains.

NOMINAL STRESS (Engineering stress) The force divided by the initial (undeformed) cross-sectional area of a body. Often more conveniently measured than the true stress and often taken to be equal to it to a good approximation at low (e.g. about 1%) strain. Especially useful when considering the properties of the body as a whole, such as tensile strength.

***n*-NONANE**

$$CH_3(CH_2)_7CH_3$$

B.p. 150·7°C.

A hydrocarbon solvent useful for the polymerisation of hydrocarbon monomers in solution and as a solvent for hydrocarbon rubbers.

NONANOIC ACID Alternative name for pelargonic acid.

NON-CHROMOSOMAL PROTEIN The protein con-tent of the nuclei of eukaryotic cells which is not associated with the DNA of the chromosomes. It can variously account for from a few per cent to 50% of the content of the cell nucleus.

NON-DRYING OIL An oil, such as castor oil, coconut oil or cottonseed oil, which contains so little unsaturation in the acid residues of its triglycerides, that a film is still fluid even after about 20 days, due to a lack of crosslinking.

NON-GAUSSIAN CHAIN A chain, as a model for a polymer molecule, where the Gaussian assumption that the chain end-to-end distance r is much less than the fully extended chain length is no longer valid. In the statistical theory of rubber elasticity, the results predicted by the Gaussian theory are thus no longer valid at high extensions since an appreciable proportion of the chains become highly extended. The non-Gaussian theory takes account of the extensibility of the chains. The Gaussian theory also becomes increasingly inadequate as chain length is reduced in networks as crosslink density is increased, since chain length between junction points is inversely proportional to the degree of crosslinking. The theory solves the problem of the chain end distribution by making use of the inverse Langevin function ($\mathscr{L}^{-1}$), giving the probability density $P(r)$ as

$$\ln P(r) = \text{constant} - n(r\beta/nl + \ln(\beta/\sinh\beta))$$

where $\beta = \mathscr{L}^{-1}(r/nl)$, n being the number of links of length l in the polymer chain.

NON-LINEAR ELASTICITY Elasticity in which the stress is not directly proportional to the strain. The usual behaviour shown by polymers at other than very small strains. Unlike metals, polymers may fully recover from strains beyond the proportional limit without any permanent deformation.

NON-LINEAR POLYMER Alternative name for branched polymer.

NON-LINEAR STEP-GROWTH POLYMERISATION Step-growth polymerisation involving at least one monomer with a functionality of > 2 and hence capable of forming branched polymer molecules, which if reacted to beyond the critical conversion will produce a network polymer, resulting in gelation.

NON-LINEAR VISCOELASTICITY Viscoelastic behaviour in which the relationships between stress, strain and time are not linear so that the ratios of stress to strain are dependent on the value of the stress. Furthermore, the Boltzmann Superposition Principle does not hold. Such behaviour is very common in polymer systems, non-linearity being found especially at high strains or in crystalline polymers. This complex behaviour makes prediction of deformations very difficult, e.g. for design purposes. The design engineering approach to characterising non-linear behaviour is to express experimental test results as isochronous or isometric stress–strain curves, or to make use of fractional recovery and reduced time concepts.

There is no satisfactory theory on non-linear behaviour based on a molecular approach. Several empirical equations, e.g. the Andrade creep law, have been proposed. Other theories based on the extension of the rheological theories of linear behaviour have been developed (e.g. Leadermann theory), especially by modification of the Boltzmann Superposition Principle to a multiple integral representation, as in the more generally applicable Green–Rivlin theory.

NON-NEWTONIAN FLUID A fluid which does not behave as a Newtonian fluid, i.e. one which does not obey the Navier–Stokes equation of motion. Deviations from this ideal Newtonian behaviour may be of several different types. Firstly, the viscosity, here called the apparent viscosity and given the symbol η, may not be independent of the rate of shear; it may increase with shear rate (shear thickening or dilatancy) or decrease with rate of shear (shear thinning or pseudoplasticity). The latter behaviour is commonly found with polymer melts and solutions. In general, such a dependency of shear stress on shear rate can be expressed as a power law. Secondly, the viscosity may be time dependent, as for fluids exhibiting thixotropy or rheopexy. Thirdly, again commonly found with polymer melts, the fluid may exhibit elastic effects, that is it is an elasticoviscous fluid.

NONOX Tradename for a range of antioxidants which include Nonox D (*N*-phenyl-β-naphthylamine), Nonox DPPD (*N,N'*-diphenyl-*p*-phenylenediamine), Nonox CI (*N,N'*-dinaphthyl-*p*-phenylenediamine), Nonox WSP

(butylated-4,4'-isopropylidenediphenol) and Nonox TBC (2,6-di-*t*-butyl-4-methylphenol).

NON-REDUCING SUGAR A carbohydrate which contains no potential (or actual) aldehyde groups and cannot therefore act as a reducing agent. Monosaccharides which are non-reducing are the ketoses. Non-reducing disaccharides are those in which the interunit glycoside bond is between the anomeric centres of both monosaccharide units, i.e. forming a 1,6'-link.

NON-STAINING ANTIOXIDANT An antioxidant which does not impart colour to the polymer which it is protecting, either initially or after degradation (due to its own degradation products being coloured). Most hindered and thio-phenols and peroxide destroyers are non-staining.

NORBORNENE (Bicyclo-(2,2,1)-heptene-2)

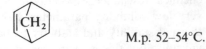

M.p. 52–54°C.

Produced by the Diels–Alder addition of ethylene to cyclopentadiene. The monomer for polynorbornene by ring opening polymerisation.

NORDEL Tradename for ethylene–propylene rubber.

NORDEL TP Tradename for a thermoplastic polyolefin rubber.

NORMAL DISTRIBUTION (Gaussian distribution) A bell-shaped distribution of values of a parameter, which is symmetrical about the mean value. Occasionally used to describe the molecular weight distribution of a narrow distribution polymer as an approximation of the Poisson distribution. The number distribution function is

$$N(r) = \exp\left[-(r - r_m)^2/2\sigma^2\right]/(2\sigma)^{1/2}$$

where σ is the half-width of the distribution at half the maximum height. The weight distribution is, $W(r) = (r/r_m)N(r)$, where r_m is the number average degree of polymerisation. The term has occasionally been used to mean the Schultz–Flory distribution.

NORMAL FORCE (Normal stress) A force, or stress, produced in a material at right angles to the direction of the applied force. This commonly occurs with polymer melts and other non-Newtonian fluids, although the mechanistic explanation is not clear. A normal force is primarily an elastic effect. The normal forces are the diagonal components of the stress tensor Δ_{ij}, i.e. when $i = j$, and are therefore σ_{11}, σ_{22} and σ_{33}. Experimentally, it is the normal stress differences, $(\sigma_{11} - \sigma_{22})$ (the first normal force difference) and $(\sigma_{22} - \sigma_{33})$ (the second normal force difference) that are measured. In a simple plane shear between parallel plates, the first normal stress difference tends to force the plates apart, whilst the second creates bulges at the edge of the plates parallel to the direction of the shearing forces. The first normal stress difference is normally positive and can exceed the shear stress at high shear rates. For polymer melts the second normal stress difference is only about a tenth of the first and is generally negative. The normal stress differences increase with the rate of shear ($\dot{\gamma}$) as, $\sigma_{11} - \sigma_{22} = \psi_1\dot{\gamma}^2$ and $\sigma_{22} - \sigma_{33} = \psi_2\dot{\gamma}^2$, where ψ_1 and ψ_2 are the first and second normal stress coefficients. Normal stress effects are manifest in many polymer melt flows, for example as die swell and the Weissenberg effect.

NORMALITY CONDITION (Normality rule) A condition of ideal plastic behaviour that the plastic strain increments are in the directions normal to the yield surface and hence also in the direction of the current yield stress. This follows from the Lévy–Mises equations.

NORMALITY RULE Alternative name for normality condition.

NORMAL STRESS (1) A stress, or stress component, at a point acting in a direction perpendicular to a plane through that point. Any stress acting at a point can be resolved into a normal stress and a shear stress. In suffix notation a normal stress would be designated σ_{xx}, σ_{yy} or σ_{zz} (or σ_{11}, σ_{22} or σ_{33}), i.e. the two suffixes are the same, since in each case the force is perpendicular to the plane on which it acts.

(2) Alternative name for normal force.

NORRISH MECHANISM The mechanism of fragmentation of organic carbonyl containing compounds, when subjected to ultraviolet radiation, i.e. to photolysis. Similar breakdown paths occur widely during photodegradation (especially photo-oxidation) of polymers. Both Norrish type I and type II mechanisms result in carbon–carbon bond dissociation of the excited state, producing chain scission in polymers, e.g.

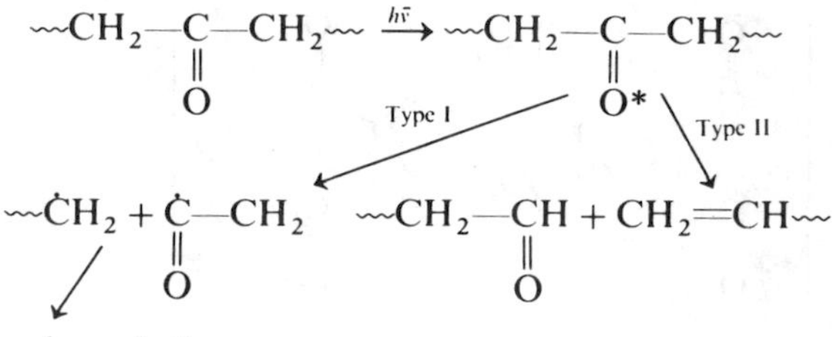

Further degradation

Type I scission results in the formation of free radicals which participate in further degradation reactions. Carbonyl groups are often present (e.g. in polyolefins) due to prior thermo- or photo-oxidation steps. In degradable polymers they are deliberately introduced by copolymerisation.

NORSODYNE Tradename for an unsaturated polyester resin.

NORSOPHEN Tradename for a phenol–formaldehyde polymer.

NORSOREX Tradename for polynorbornene.

NORVYL Tradename for polyvinyl chloride.

NORYL Tradename for a series of blends of poly-(2,6-dimethyl-1,4-phenylene oxide) with polystyrene or with high impact polystyrene.

NORYL GTX Tradename for a poly-(2,6-dimethyl-1,4-phenylene oxide) blend with a nylon.

NO-SLIP CONDITION The assumption that at the boundary between a rigid surface and a flowing fluid, there is no relative motion. This assumption is often made as a boundary condition for the solution of flow problems. It is often a reasonable assumption for polymer melts, at least below a certain critical shear stress.

NOTCH EMBRITTLEMENT A change in the mode of fracture from ductile to brittle fracture due to the presence of a notch in the sample. The change results from the change in the stress state at the notch, giving a more triaxial stress, which favours brittle fracture rather than yielding.

NOTCH SENSITIVITY The dependence of the fracture properties of a material on the presence of notches in a sample of the material. In particular the influence of the presence of a notch or crack on the tensile strength. It arises from a notch acting as a stress concentrator and thereby altering the state of stress near the notch. At the root of the notch a state of triaxial stress exists, different from that in the body of the material. Sometimes such a situation can cause fracture to be brittle fracture rather than ductile fracture (notch embrittlement). This is due to the triaxial stress field for a sharp notch increasing the yield stress by a factor of three, whereas, according to the Orowan hypothesis, the brittle stress is unaltered. Furthermore, since an apparent crack is already present in notched material, the energy to fracture, as measured in an impact test, emphasises crack propagation rather than crack initiation processes due to the presence of cracks or other flaws in an unnotched material. Therefore in impact testing, notched specimens are preferred.

NOVEX Tradename for low density polyethylene.

NOVODUR Tradename for acrylonitrile–butadiene–styrene copolymer.

NOVODUR AES Tradename for acrylonitrile–ethylene/propylene rubber–styrene copolymer.

NOVODUR W Tradename for styrene–acrylonitrile copolymer.

NOVOLAC (Novolak) A phenol–formaldehyde polymer prepared by reacting a phenol (usually phenol itself), in molar excess, with formaldehyde (e.g. with a phenol: formaldehyde molar ratio of 1·25:1) in aqueous solution under acidic conditions, commonly with oxalic acid at about 100°C. When the water is distilled after reaction, the polymer (after cooling) sets to a hard, brittle and soluble solid with a melting range of about 65–75°C. This is the A-stage polymer.

Reaction occurs by electrophilic attack on the phenol by protonated formaldehyde to give a mixture of o- and p-methylolphenols. These then react rapidly with further phenol to form dihydroxydiphenylmethanes:

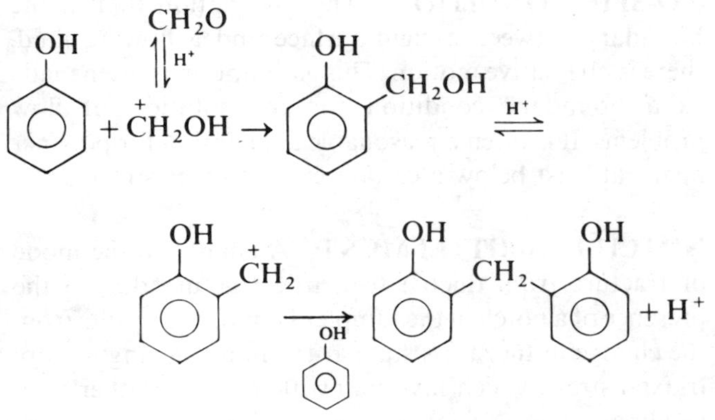

Further methylolation and condensation occurs so that a typical A-stage resin consists of a mixture of polynuclear phenols, linked through o- and p-methylenes and containing on average about six rings per molecule. This low molecular weight prepolymer is fairly linear since further substitution, beyond two groups per ring, is limited due to deactivation by the earlier substitution.

Unlike resoles, novolacs cannot be crosslinked and therefore cured, simply by further heating. A crosslinking agent, usually formaldehyde or a source of formaldehyde, must be used to increase the functionality. Hexamethylenetetramine (hexa) is the usual crosslinking agent, but the curing reactions are complex. The primary function of the crosslinking agent is to form further methylene bridges. These probably result from the decomposition, on heating at 180–190°C, of the secondary and tertiary benzylamines formed by the initial reaction of hexa with novolac prepolymer:

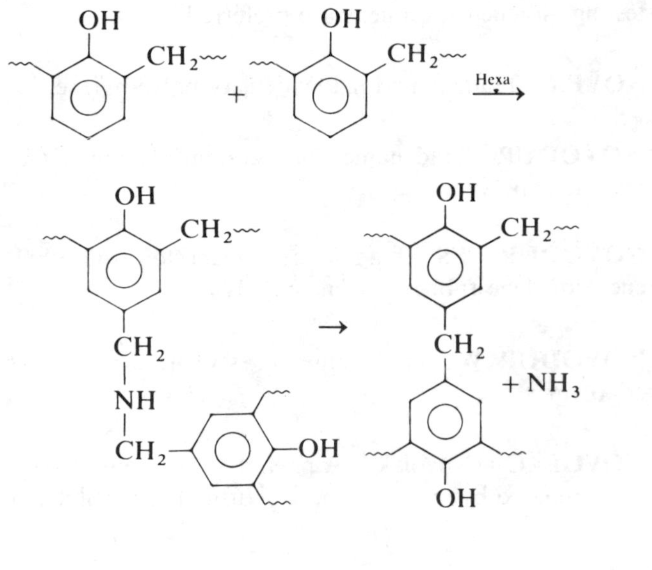

The azomethine groups (I) may be responsible for the brown colour of the cured products. In practical use a novolac is compounded with the curing agent (usually hexa), cellulosic filler (often woodflour) and lubricant for use as a moulding powder. Novolacs are also useful as adhesives.

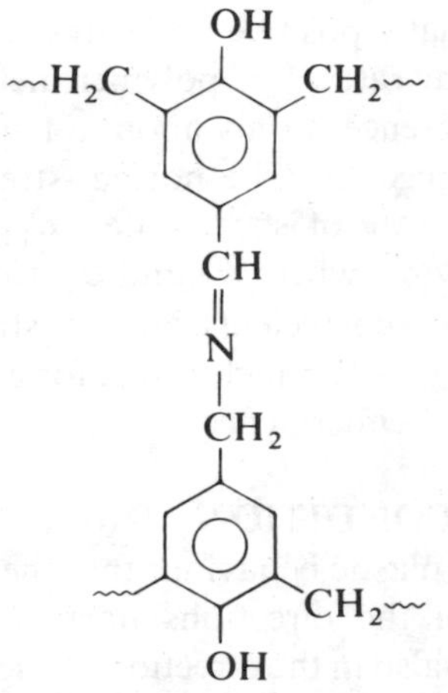

NOVOLAK   Alternative name for novolac.

NOVOLEN   Tradename for polypropylene.

NOVOLOID   Generic name for a fibre containing at least 85% of a crosslinked novolac. Kynol is an example.

NOVOPLAS   Tradename for a polysulphide rubber similar to Thiokol B.

NR   Abbreviation for natural rubber.

NR 150   Tradename for a thermoplastic polyimide containing hexafluoroisopropylidene groups for flexibilising the polymer chain, with repeat units of the type:

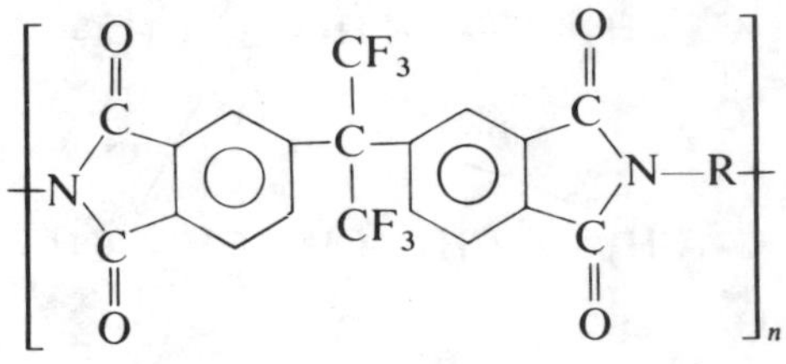

where R is ⟨⟩—O—⟨⟩ (polymer I)

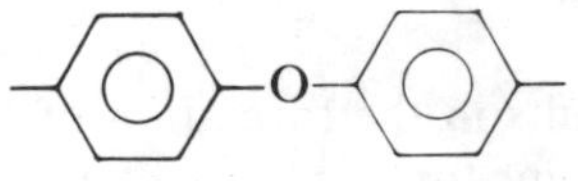

or 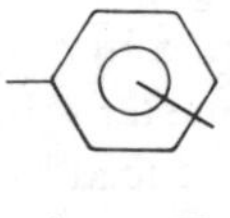 ⟨⟩ (polymer II)

The polymers have T_g values of 280–300°C (polymer I) or 350–371°C (polymer II) and are useful as high temperature resistant matrices for composites and adhesives.

N-TERMINAL AMINO ACID   The amino acid residue occurring at the end of a polypeptide or protein chain which has the free amino group.

N-TERMINAL RESIDUE The amino acid residue occurring at the N-terminal end of a polypeptide or protein molecular chain. The N-terminal residue may be determined by its reaction with either fluorodinitrobenzene or dansyl chloride to form a dinitrophenyl or dansyl derivative respectively. On hydrolysis of the modified protein the dinitrophenyl or dansyl amino acid is liberated and may be separated from the other amino acids and identified.

NUCLEAR MAGNETIC RELAXATION A relaxation of the magnetisation of the nuclear spins of the nuclei of atoms as a result of a change in the applied magnetic field. The field may be applied as a pulse or it may be periodically varying. The decay of magnetisation is characterised by two relaxation times. One for the relaxation in the direction of the applied field—the longitudinal or spin–lattice relaxation time (T_1) and one transverse to the field—the spin–spin or spin–phase relaxation time (T_2). The former corresponds to frequencies of 10^7–10^9 Hz, but when observations are made of relaxation in a rotating frame, the frequency is about 10^3 Hz. When T_1 is observed as a function of temperature, transitions may be observed. The nuclear magnetic method is of particular value in that the lower temperature transitions (the γ- and δ-transitions) are often readily observed whilst they are weak or absent in dielectric or mechanical measurements. Whilst T_1 and T_2 may be related by either the strong or weak collision theories to molecular correlation times, the precise molecular conformational changes involved are often not clear. Frequently the relaxation is non-ideal, involving either a distribution of correlation times or a non-exponential correlation function.

NUCLEAR MAGNETIC RESONANCE SPECTROSCOPY (NMR spectroscopy) A technique for the characterisation of organic molecules. When a molecule containing certain atoms, including hydrogen, ^{13}C and fluorine, is placed in a strong magnetic field and irradiated with radio frequency waves, transitions between different nuclear spin orientational states take place and energy is absorbed at specific frequencies (if the field is constant) or, more commonly, at specific field strengths (if the frequency is constant). NMR involving protons is proton magnetic spectroscopy (PMR). The exact field strength of energy absorption depends on the detailed atomic environment of the atom concerned and is expressed as the chemical shift. Thus NMR can give detailed information of molecular structure.

Two types of NMR have been applied to polymers. The usual high resolution NMR using dilute solutions is the best technique for studying tacticity and sequence distributions in copolymers. Wide-line NMR is used for studying molecular motions in solid polymers. Increasingly, ^{13}C NMR is proving of value for structural studies of fine details such as small amounts of branching. ^{19}F NMR has been widely applied to the study of fluoropolymers. NMR spectra can often be simplified by the double resonance technique. Fourier transform NMR is also used.

NUCLEAR OVERHAUSER EFFECT An enhancement in the population difference in the magnetic energy levels in NMR in nucleus A, when a nucleus of type B is being irradiated, as in spin–spin decoupling between A and B, with appropriate radio frequency waves. It occurs when ^{13}C NMR is carried out in the usual way, thus contributing to the high sensitivity of this type of NMR.

NUCLEATING AGENT An additive which provides nuclei for heterogeneous crystallisation, raising the crystallisation rate and temperature. More and smaller spherulites are consequently produced. Typical agents are metal salts of organic acids such as sodium benzoate, which are effective in parts per million amounts. Many inorganic fillers also promote nucleation.

NUCLEATION The formation of short-range ordered polymer aggregates in a melt or solution which act as growth centres for crystallisation. Primary nucleation results from fluctuations in local order/disorder and may be homogeneous or heterogeneous. Subsequently secondary and tertiary nucleation may follow.

NUCLEATION CAVITY A crack in a crystalline impurity (or additive) in which a small crystal of polymer may survive above the melting temperature, and so act as a self-seeding nucleus on subsequent cooling of the melt.

NUCLEATION DENSITY The number of nuclei at which crystal growth is initiated per unit volume. For polymer melts values in the range 10^3 cm^{-3} to $>10^{12}$ cm^{-3} have been observed. It varies with temperature (in either direction) and may be increased by addition of a nucleation agent.

NUCLEOPROTAMINE The nucleoprotein 'complex' occurring in the nuclei of fish sperm, consisting of an association of a protamine with the DNA of the chromosomes. Typically, the arginine (of the protamine) to phosphorus (of the DNA) ratio is about unity, thus protamine neutralises the charge on the phosphate groups. Probably the protamine lies in the narrow groove of the DNA.

NUCLEOPROTEIN A conjugated protein in which protein is associated with nucleic acid. The resultant supermolecular complex is held together by ionic, hydrogen bonding, hydrophobic bonding and van der Waals' forces, rather than by covalent bonding. Nevertheless owing to the many interactions the complex is very stable. There may be only one or several nucleic acid molecules usually associated with many (up to several hundred) individual protein molecules. There are four important types of nucleoprotein which have been widely studied. The ribonucleoproteins (which comprise the ribosomes of cells), the viruses and the chromosomal proteins, which contain both acidic and basic proteins, and comprise the histones and protamines.

NUMBER AVERAGE MOLECULAR WEIGHT

Symbol $\bar{M}_n$. A polymer molecular weight average defined as

$$\sum_{i=1}^{\infty} M_i N_i \Big/ \sum_{i=1}^{\infty} N_i$$

where N_i is the number of molecules and M_i is the molecular weight of each molecular size i. It is therefore the ratio of the total weight of the sample to the total number of molecules present. It is equal to both the ratio of the first to the zeroth moments of the distribution and to the mean of the number distribution. Another useful form of the definition is $\bar{M}_n = \sum w_i / \sum (w_i / M_i)$. The most easily measured of the molecular weight averages and therefore the most commonly used.

Compared with other averages the value of $\bar{M}_n$ is very sensitive to low molecular weight species, since they contribute more heavily to the number of molecules present than to the weight. Values for commercial polymers useful as plastics, rubbers, etc., are typically in the range 10^4–10^5.

The $\bar{M}_n$ value may be determined by any method which counts the number of molecules present in a dilute solution, such as colligative property methods (ebulliometry, cryoscopy, vapour pressure lowering). However, vapour pressure osmometry (for $\bar{M}_n$ values $< 20\,000$), membrane osmometry (for $\bar{M}_n$ values $> 20\,000$) and end-group analysis are the most widely used methods. $\bar{M}_n$ can also sometimes be determined by methods usually used for $\bar{M}_w$, e.g. light scattering and sedimentation.

NUMBER DISTRIBUTION

Alternative name sometimes used for frequency distribution, when this is describing the variation of molecular size i in a polymer sample as a function of the number of each species i present.

NUTTING EQUATION

An empirical power law relation between stress (σ), strain (e) and time (t), expressing non-linear viscoelastic behaviour, especially creep. It is $e = k\sigma^\beta t^n$, where k, β and n are constants for a particular temperature.

NYLENKA

Tradename for nylon 6.

NYLOFIL

Tradename for nylon 6, 66 or 610.

NYLON

A synthetic polyamide in which at least some of the carbon groups (R and/or R′) separating the amide groups in the polymer chain are aliphatic (usually methylene) groups. Thus wholly aromatic polyamides are not usually referred to as nylons, although aliphatic–aromatic polyamides such as nylon 6T are. Nylons are of two types. The monadic (or AB) nylons of general structure $-[RNHCO]_n-$, and the dyadic (or AA–BB) nylons of general structure $-[NHRNHCOR'CO]_n-$, derived from a diamine and a diacid. A nylon may be named either on the basis of the monomer(s) from which it is derived, e.g. polycaprolactam for nylon 6, or as a polyamide, e.g. poly-(caproamide) for nylon 6. Dyadic nylons are usually named using the latter method, e.g. poly-(hexamethyleneadipamide) for nylon 66. The abbreviated names of the types nylon x (e.g. nylon 6) or nylon xy (e.g. nylon 66) for monadic and dyadic nylons respectively, are very frequently used. The values of the numbers x and y are the number of carbon atoms in the monomer(s) from which the nylons are derived; in a nylon xy, x is the number in the diamine. This system applies to the most common nylons, for which R and R′ are methylene groups. For nylons derived from other common monomers, appropriate abbreviations for the monomers used, e.g. T for terephthalic acid, I for isophthalic acid, MXD for m-xylylenediamine, pip for piperidine, 2,5-diMepip for 2,5-dimethylpiperidine and TMD for trimethylhexamethylenediamine. Copolymers are designated by a solidus followed by the composition in parentheses. Thus nylon 66/610(60/40) is a copolymer made by polymerisation of hexamethylenediamine with adipic acid and sebacic acid in the molar ratio 60:40. Nylons, especially nylons 6, 66 and 610, are important commercial products. Those mentioned are major synthetic fibres, but are also useful as plastics. Nylons 11 and 12 are also well known as plastics, whilst nylons 7, 9, 13, 1313, 6T and TMDT have also been produced commercially.

One of the dominant characteristics of nylons is their intermolecular hydrogen bonding, which provides a driving force for crystallisation. The crystalline polymers often have high melting temperatures, e.g. 265°C for nylon 66, and are usually spherulitic. In the crystallites the nylon molecules exist in extended chain conformations, each chain being hydrogen bonded to its neighbour. The chains lie parallel to each other thus forming two-dimensional sheets of molecules (the β-sheet conformation). In most cases all amide groups are hydrogen bonded, although this sometimes necessitates a slight twisting of the chain. The sheets are stacked on top of each other to form a crystal lattice; different ways of stacking yielding different crystal structures, named the α-, β- and γ-structures. The spinning and drawing processes used in fibre production induce orientation in the fibre direction which enhances the stiffness and strength.

Nylon plastics are stiff and tough, with a high abrasion resistance and hence are 'engineering' plastics. The mechanical properties are considerably altered by water absorption from the atmosphere, which can amount to up to about 10%, depending on the nylon and the relative humidity. The absorbed water acts as a plasticiser and lowers the T_g value from about 50°C, typical for a dry nylon, to about 20°C, thus considerably enhancing impact strength.

Dyadic nylons are usually synthesised by melt polymerisation polyamidation of a dicarboxylic acid with a diamine, often via an intermediate nylon salt. More reactive derivatives of the acid, especially the chloride, may be used in a solution or interfacial polymerisation. Monadic nylons are synthesised either by self-amidation of an ω-amino acid

$$n\,H_2N(CH_2)_x COOH \rightarrow -[NH(CH_2)_x CO]_n- + n\,H_2O$$

or by ring-opening polymerisation of a lactam

$$n(CH_2)_x-CO \rightarrow \text{-}[NH(CH_2)_xCO]_n$$

with NH bridging.

In the latter either a high temperature water catalysed, or a low temperature anionic chain, polymerisation mechanism may be followed. In the formation of monadic nylons, the product polymer is in equilibrium with lactam, the equilibrium lactam content varying from about 10% in nylon 6 to about 1–2% in nylons 11 and 12. Monadic nylons lower than nylon 6 cannot be prepared by the above methods due to excessive ring formation. Other nylon synthetic methods include amidation of a dinitrile, NCRCN either with a diamine or by reaction with a dicarboxylic acid,

$$n \text{ NCRCN} + n \text{ HOOCR'COOH} \rightarrow$$

$$\text{-}[NHRNHCOR'CO]_n + 2n \text{ } CO_2$$

self-amidation of an aminonitrile or amidation with a carbonium ion forming reagent (Ritter reaction), especially formaldehyde.

NYLON 1 The unsubstituted polymer, $\text{-}[NHCO]_n$, cannot be made, but N-substituted polymers,

$$\text{-}[NCO]_n$$
$$\underset{R}{|}$$

can be synthesised by anionic chain polymerisation of isocyanates, RNCO, at low temperatures, where R is CH_3, C_2H_5, C_6H_5, etc. These polymers may be cast as tough films, but have a tendency to depolymerise. They are sometimes referred to as polyureas.

NYLON 2 Alternative name for polyglycine. Substituted nylon 2 products comprise the polypeptides and are named after the amino acid from which they are derived, e.g. when R is CH_3 in the general structure $\text{-}[CH(R)NHCO]_n$, the polymer is poly-(glutamic acid), rather than being named as a nylon.

NYLON 3 (Poly-(β-alanine)) $\text{-}[CH_2CH_2NHCO]_n$. Cannot be synthesised by self-amidation of β-alanine, since on heating this amino acid eliminates ammonia to give acrylic acid. However, high molecular weight polymers may be made by anionic chain polymerisation of acrylamide by treatment with a strong base. The reaction involves a Michael-type addition followed by a proton transfer. Nylon 3 has a melting temperature of about 325°C (with decomposition) which is too high for melt spinning fibres. Substituted nylon 3 polymers have been made by ring-opening polymerisation of the appropriate lactam,

$$R_2C-CH_2$$
$$|\qquad |$$
$$HN-CO$$

(an azetidinone), especially when R is CH_3, when dimethylnylon 3 results.

NYLON 4 (Polypyrrolidone) $\text{-}[(CH_2)_3NHCO]_n$. Synthesised by ring-opening polymerisation of pyrrolidone by anionic polymerisation. Cannot be prepared by heating its parent amino acid since it cyclises to the lactam. It has a melting temperature of 260–265°C and a relatively high water absorption—about 10% at 20°C and 65% relative humidity. These properties are suitable for a fibre-forming textile material.

NYLON 5 (Polypiperidone) $\text{-}[(CH_2)_4NHCO]_n$. Prepared by ring-opening polymerisation of piperidone. Melting temperature is about 260°C, but the polymer has attracted little attention. The corresponding amino acid, 5-aminovaleric acid, cannot be polymerised by self-amidation, since on heating it cyclises quantitatively to the lactam.

NYLON 6 (Polycaprolactam) (Poly-(6-aminocaproic acid)) (Poly-(ω-aminocaproamide)) (Poly-(ε-aminocaproic acid)) $\text{-}[(CH_2)_5CONH]_n$. Tradenames: Akulon, Amilan, Beetle nylon, Caprolan, Capron, Durethan BK, Enkalon, Grilon, Kapron, Maranyl, Nylenka, Nylofil, Nyrim, Organamid, Perlon L, Plaskon, Renyl, Sniamid, Technyl, Ultramid, Ultramid B. Melting temperature 215–225°C. One of the most important nylons commercially, being a major fibre-forming polymer, as well as a useful engineering plastic. Can be prepared by melt polymerisation of ε-aminocaproic acid, but synthesised commercially more conveniently from ε-caprolactam. This latter polymerisation may be either a high temperature (about 240°C) water-catalysed melt polymerisation or a lower temperature (about 100°C) mass polymerisation by an anionic mechanism. Large mouldings are readily made by the second method by the monomer casting technique. Nylon reaction injection moulding (RIM) is also based on anionic polymerisation of ε-caprolactam to nylon 6. Polymer is usually formed in equilibrium with about 10% lactam, which may be removed by extraction or volatilisation.

Nylon 6 absorbs atmospheric moisture to reach equilibrium water contents of about 3% at 50% relative humidity and 9·5% at 100% relative humidity and 20°C. It usually crystallises in the α-form having a fully extended zig-zag molecular conformation, the molecules lying parallel to each other in sheets, with adjacent chains running in opposite directions (the anti-parallel form). In this way all amide groups are hydrogen bonded to each other. The T_g (dry) value is about 70°C and is lowered to about 20°C at 50% relative humidity and −20°C at 100% relative humidity when containing the equilibrium water content. Typical bulk tensile properties at 23°C are tensile modulus, 2800 MPa (dry) and 1200 MPa (wet, 50% relative humidity), elongation at break (%), 100–200 (dry) and 300 (wet, 50% relative humidity). Izod impact strength is about 0·7 J $(12\cdot7 \text{ mm})^{-1}$ (dry) and 2·5 J $(12\cdot7 \text{ mm})^{-1}$ (wet, 50% relative humidity). The polymer is melt spun to a fibre, which is then drawn to 250–300% elongation to orient the polymer. This yields fibres with typical mechanical properties of: titer about 15 denier, tenacity about 5 g denier^{-1}, elongation about 40% and an elastic modulus of about 3000 MPa. Further hot drawing may improve these properties even more, e.g. tenacity up to about $10\,g\,denier^{-1}$.

NYLON 6T (Polyhexamethyleneterephthalamide)

$$-\!\!\!\left[NH(CH_2)_6NHCO-\!\!\left\langle\bigcirc\right\rangle\!-CO\right]_{\!n}$$

A partially aromatic polyamide synthesised by reaction between hexamethylenediamine and terephthalic acid. Has a very high T_m value (about 370°C) with mechanical properties similar to nylon 66, but with these properties retained to higher temperatures. The polymer also has lower creep. Fibres may be wet spun from concentrated sulphuric acid solution. The polymer is only soluble in strong acids.

NYLON 7 (Polyenantholactam) (Polyheptanoamide) $-\!\!\left[(CH_2)_6NHCO\right]_{\!n}$. Tradename: Enant. Synthesised either by self amidation of ζ-aminoenanthic acid or of ethyl-7-aminoheptanoate. The latter is obtained from caprolactone, by reaction with HCl and $ZnCl_2$ to give 6-chlorohexanoic acid, which is esterified, the chlorine replaced with nitrile and the nitrile reduced to give the desired monomer. The amino acid monomer is obtained from the telomerisation of ethylene, as is the alternative monomer, enantholactam. The polymer is formed in equilibrium with about 3% lactam. It has a T_m value of 233°C and a T_g value of 62°C, otherwise its properties are similar to nylon 6.

NYLON 8 (Polycaprylamide) (Polycapryllactam) (Poly-octanoamide) $-\!\!\left[(CH_2)_7NHCO\right]_{\!n}$. Synthesised either by self-amidation of 8-aminocaproic acid or by ring-opening polymerisation of capryllactam. The polymer has a T_m value of 200°C, a T_g value of 51°C, a tensile strength of about $4\,g\,denier^{-1}$ (dry) (fibre), an elongation at break of about 25% and water absorption of 1·8% at 65% relative humidity and 4% at 100% relative humidity and 20°C. Not a commercial product, owing to the relative inaccessibility of its monomers and to its properties being little different from other nylons.

NYLON 9 (Polypelargonamide) (Polynonanoamide) $-\!\!\left[(CH_2)_8NHCO\right]_{\!n}$. Synthesised by self-amidation at 225–260°C of 9-aminopelargonic acid. The polymer has a T_m value of 209°C, a T_g value of 51°C and a water absorption of 1·5% at 65% relative humidity and 3% at 100% relative humidity 20°C.

NYLON 10 (Polycapramide) (Polycaprinolactam) (Polydecanoamide) $-\!\!\left[(CH_2)_9NHCO\right]_{\!n}$. A nylon of little interest commercially due to the relative inaccessibility of its monomers, 10-aminocapric acid or caprinolactam. However, the latter may be made by a six stage synthesis from decalin. The polymer has a T_m value of 188°C and a T_g value of 43°C.

NYLON 11 (Polyundecanoamide) (Poly-(11-aminoundecanoic acid)) $-\!\!\left[(CH_2)_{10}NHCO\right]_{\!n}$. Tradenames: Rilsan, Rilsan B. Obtained by polymerisation at about 215°C of

11-aminoundecanoic acid, in equilibrium with about 0·5% lactam. Produced commercially as both a fibre and plastic material. Similar in properties to nylon 6, but having a lower T_m value of 190°C, a T_g value of 46°C and a lower water absorption of 1·2% at 65% relative humidity and 2% at 100% relative humidity and 23°C. Has a yield stress of 40 MPa, a tensile modulus of 1400 MPa, a tensile strength of 50 MPa, an Izod impact strength of 1·2 J $(12\cdot7\,mm)^{-1}$ and an elongation at break of 120–330% at 50% relative humidity. As a textile material it has a richer, drier handle than nylons 6 and 66 and is therefore favoured for underwear.

NYLON 12 (Polydodecanoamide) (Polylauryllactam) $-\!\!\left[(CH_2)_{11}NHCO\right]_{\!n}$. Tradenames: Grilamid, Rilsan A, Vestamid. A commercial nylon obtained by polymerisation of lauryllactam. Polymerisation is slow due to lactam ring stability and requires temperatures of 300–350°C and the use of acid catalysts. About 0·5% lactam remains in equilibrium with polymer. Has a T_m value of 179°C, a T_g value of 37°C, water absorption of 1·3% at 50% relative humidity and 2% at 100% relative humidity and 23°C. Polymer yield stress is 52 MPa(dry) and 41 MPa(wet, 50% relative humidity) at 23°C and elongation at break is about 250%(wet and dry). Tensile modulus is 1400 MPa (dry) and Izod impact strength is 1·4 J $(12\cdot7\,mm)^{-1}$.

NYLON 13 (Polytridecanoamide) $-\!\!\left[(CH_2)_{12}NHCO\right]_{\!n}$. Can be obtained by self-amidation of 12-aminotridecanoic acid, obtained from erucic acid

$$(CH_3(CH_2)_7CH\!=\!CH(CH_2)_{11}COOH)$$

which is found in crambe seed oil, by a route similar to that for 11-aminoundecanoic acid from ricinoleic acid. Has a T_m value of 1·80°C, a T_g value of 41°C, a yield stress of 33 MPa, a tensile modulus of 780 MPa and a water absorption of about 1% at 65% relative humidity.

NYLON 46 (Polytetramethyleneadipamide). $-\!\!\left[NH(CH_2)_4NHCO(CH_2)_4CO\right]_{\!n}$. Tradename: Stanyl. Produced by polycondensation between 1,4-diaminobutane and adipic acid. The polymer has an even higher melting temperature (295°C) than nylon 66 and thus can be used at somewhat higher temperatures.

NYLON 62 (Polyhexamethyleneoxamide) $-\!\!\left[NH(CH_2)_6NHCOCO\right]_{\!n}$. Prepared by reaction of hexamethylenediamine with oxaloyl chloride. Has a high T_m value (320°C) and melts with decomposition, but has a high stiffness and a low water sensitivity.

NYLON 66 (Polyhexamethyleneadipamide) $-\!\!\left[NH(CH_2)_6NHCO(CH_2)_4CO\right]_{\!n}$. Tradenames: Akulon, Antron, Blue C nylon, Cadon, Cantrece, Cordura, Grilon, Maranyl, Minlon, Nivionplast, Perlon T, Sniamid, Technyl, Trogamid, Ultramid A, Vydyne, Zytel. The most important nylon, being a major commercial fibre and the most important nylon plastic. Produced by high temperature melt polymerisation of the nylon 66 salt formed between adipic acid and hexamethylenediamine. Its popularity is due to its superior balance of properties and the relatively low cost of its monomers. The nylon salt is polymerised in a slurry in water, heated initially at 180–220°C, then at 280–290°C, steam being bled off.

The polymer has a T_m value of 260–265°C and a T_g value of about 50°C when dry. β- and γ-transitions also occur at about -80°C and -140°C respectively. Water absorption is about 2·5% at 50% relative humidity and 8% at 100% relative humidity and 23°C. Nylon 66 crystallises in the α-form with an extended zig-zag chain conformation, the chains lying parallel to each other in sheets and being intermolecularly hydrogen bonded. The β-crystal form is found in highly oriented fibres and consists of alternate sheets staggered up and down instead of always being displaced in the same direction as in the α-form. Moulded samples have a tensile modulus of about 3000 MPa, an elongation at break of 80–100%, a yield stress of about 80 MPa and an Izod impact strength of about 1·0 J $(12·7 \text{ mm})^{-1}$.

The polymer is melt spun to fibres which are then cold drawn to about 400% to improve their mechanical properties by orientation. Typical tensile modulus is about 50 g denier^{-1} (dry), falling to about 10 g denier^{-1} at 100% relative humidity. Tenacity is typically 6–10 g denier^{-1} and the fibres show almost complete recovery. The high abrasion resistance ensures long wear. These good properties make the fibre suitable for a wide range of textile applications. The low moisture regain and high wet strength result in easy laundering and rapid drying. Textiles do not need ironing. Nylon 66 is often blended with wool to improve the latter's wet strength, tensile strength and abrasion resistance and to reduce shrinkage. It also is used to improve the dimensional stability and abrasion and crease resistance of cotton and rayon in blends with these fibres.

NYLON 66 SALT Alternative name for hexamethylenediammonium adipate.

NYLON 69 (Polyhexamethylenenonamide) (Polyhexamethyleneazeleamide). $\{NH(CH_2)_6NHCO(CH_2)_7CO\}_n$. Synthesised by reaction between hexamethylenediamine and azeleic acid. It has a T_m value of 215°C and mechanical properties similar to nylon 610, and is therefore a possible replacement for this polymer as a plastic. Tensile strength is about 52 MPa, elongation at break is about 155% and the Izod impact strength is about 0·8 J $(12·7 \text{ mm})^{-1}$.

NYLON 610 (Polyhexamethylenesebacamide) $\{NH(CH_2)_6NHCO(CH_2)_8CO\}_n$. Tradenames: Maranyl, Nylofil, Perlon N, Ultramid S, Technyl, Zytel. An important commercial nylon produced by reaction of hexamethylenediamine with sebacic acid, initially to form a 1:1 nylon 610 salt, which is then melt polymerised at about 240°C. The polymer has a T_m value of 223°C and a lower water absorption (about 1% at 50% relative humidity and 3·5% at 100% relative humidity and 23°C) and therefore better dimensional stability and electrical

properties than nylons 6 and 66. It is often used in place of these as a plastic, although not as a fibre. Nylon 610 repeat units are often incorporated into copolymers, e.g. in nylons 66/610 and 6/66/610. Mechanical property values of the dry polymer are lower than for nylons 6 and 66, but are less affected by water absorption. Tensile strength is about 55 MPa, elongation at break is 100–150%, tensile modulus is about 2100 MPa and Izod impact strength is about 0·6 J $(12·7 \text{ mm})^{-1}$. Extruded monofilament is a major product.

NYLON 610 SALT Alternative name for hexamethylenediammonium sebacate.

NYLON 612 (Polyhexamethylenedodecanoamide) $\{NH(CH_2)_6NHCO(CH_2)_{10}CO\}_n$. Tradename Zytel. Produced by reaction of hexamethylenediamine with 1,10-decanedicarboxylic acid. Sometimes used to replace nylon 610, having an even lower water absorption, e.g. 0·4% at 50% relative humidity. The T_m value is 218°C, flexural modulus is about 2500 MPa, tensile strength is about 62 MPa, elongation at break is about 100% and Izod impact strength is about 0·7 J $(12·7 \text{ mm})^{-1}$.

NYLON 6/66 A copolymer containing nylon 6 and nylon 66 repeat units, which has a lower melt viscosity than the homopolymers and hence is easier to process to extruded plastic products. The copolymers also have lower T_m values, greater solubility and lower rigidity.

NYLON 6/66/610 A copolymer containing nylon 6, 66 and 610 repeat units, often in the ratio 30:40:30, having an even lower softening point and rigidity than the binary copolymers such as nylon 6/66.

NYLON 91 (Polynonamethyleneurea) $\{NH(CH_2)_9NHCO\}_n$. Tradename: Urylon. Synthesised by reaction of nonamethylenediamine (derived from azelaic acid) with urea. Has attracted commercial attention as a fibre. Similar in properties to nylon 11, with a T_m value of about 230°C. Fibres have a moisture regain of about 1·5% at 65% relative humidity and 25°C, a tensile strength of 4–5 g denier^{-1} and an elongation at break of about 20%.

NYLON 1010 (Polydecamethylenedecanoamide) $\{NH(CH_2)_{10}NHCO(CH_2)_8CO\}_n$. Synthesised by reaction of decamethylenediamine with sebacic acid, via the corresponding nylon salt. Both monomers are obtained from oleic acid. The polymer has a T_m value of 203°C.

NYLON 1313 (Polytridecamethylenetridecanoamide) $\{NH(CH_2)_{13}NHCO(CH_2)_{11}CO\}_n$. Synthesised by reaction of brassylic acid with 1,13-diaminotridecane. Has a T_m value of 174°C and properties generally similar to nylons 12 and 13, having a tensile strength of 39 MPa, a tensile modulus of 700 MPa and an elongation at break of 130%. Water absorption is low at 0·75% at 50% relative humidity.

NYLON FIBRE Generic name for a fibre composed of a synthetic polyamide polymer in which less than 85% of the amide groups are attached directly to two aromatic rings. If there are more than 85% of such links then the fibre is an aramid.

NYLON HPXD8   Alternative name for poly-(trans-hexahydro-*p*-xylylenesuberamide).

NYLON MXD6   Alternative name for poly-(*m*-xylyleneadipamide).

NYLON SALT   (AH salt) A 1:1 adduct of an aliphatic diamine and an aliphatic dicarboxylic acid of general structure

$$H_3 \overset{+}{N}(CH_2)_x \overset{+}{N}H_3$$
$$^{-}OOC(CH_2)_y COO^{-}$$

used as the intermediate monomer in the synthesis of dyadic nylons. Thus the nylon 66 or nylon 610 salts, often as a slurry in water, polymerise on heating to above 200°C to nylon 66 and 610 respectively. Reaction occurs by attack of free carboxylic acid groups on free amine groups present in very low concentration in equilibrium with the zwitterion nylon salt.

NYLOPLAST   Tradename for nylon 66.

NYMCRYLON   Tradename for a polyacrylonitrile fibre.

NYRIM   Tradename for a reaction injection moulding process using caprolactam as the monomer and hence producing mouldings in nylon 6.

NYTRIL   Generic name for a fibre consisting of a polymer with at least 85% vinylidene cyanide units.

O

OB   Abbreviation for 4,4'-oxybis-(benzenesulphonylhydrazine).

OCCLUSION   The isolation of a growing active centre in free radical polymerisation from other active centres when it becomes buried (occluded) within a polymer particle. Thus termination is reduced and the rate of polymerisation increases. This may occur in precipitation polymerisation, emulsion polymerisation, popcorn polymerisation or even in a homogeneous medium, especially if the medium is a poor solvent for the polymer, when it contributes to auto autoacceleration.

OCCLUSION CELLULOSE   Alternative name for inclusion cellulose.

OCTADECANOIC ACID   Alternative name for stearic acid.

OCTADECENE-1–MALEIC ANHYDRIDE COPOLYMER   Tradename: PA-18. An alternating copolymer of octadecene-1 (produced by the oligomerisation of ethylene) and maleic anhydride and thus having the structure:

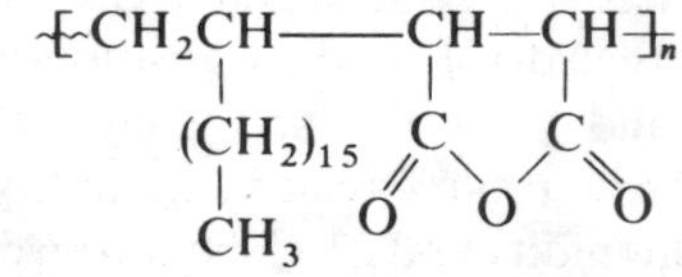

The polymer readily undergoes many chemical reactions, e.g. esterification, through its anhydride groups. It is a useful crosslinking agent for epoxy resins and hydroxyl containing polymers.

OCTAHEDRAL SHEAR STRESS   Symbol τ_{oct}. The stress acting on a plane having a normal that makes the same angle to all three principal directions. There are four such normals and a regular octahedron can be constructed from planes to these normals. Given by

$$\tau_{oct} = 1/3[(\sigma_{11} - \sigma_{22})^2 + (\sigma_{22} - \sigma_{33})^2 + (\sigma_{33} - \sigma_{11})^2]^{1/2}$$

where σ_{11}, σ_{22} and σ_{33} are the normal stresses. Sometimes used as a criterion of the modified von Mises type, for shear yielding.

OCTAMETHYLCYCLOTETRASILOXANE

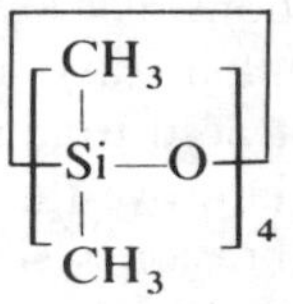

B.p. 175°C.

Formed by hydrolysis of dimethyldichlorosilane, together with linear polymer, e.g. by stirring an ethereal solution with a large volume of water when a 50% yield may be obtained. Readily separated and purified by distillation. The highly pure material is the monomer for the formation of high molecular weight polydimethylsiloxane for silicone rubber, by heating at 150–200°C with a trace of sodium hydroxide as catalyst.

***n*-OCTANE**

$$CH_3(CH_2)_6 CH_3$$

B.p. 125·6°C.

A hydrocarbon solvent, useful as a solvent for the polymerisation of hydrocarbon monomers in solution and as a solvent for hydrocarbon rubbers.

OCTENE-1

$$CH_2 = CH(CH_2)_5 CH_3$$

B.p. 121·3°C.

A comonomer, sometimes for copolymerisation with ethylene, to produce one type of linear low density polyethylene. Produced by oligomerisation of ethylene using a Ziegler or other catalyst.

n-OCTYL-n-DECYL ADIPATE (DNODA)
B.p. 220–250°C/4 mm. The adipic acid esters of the mixed isomers of eight and ten carbon alcohols. Useful as a plasticiser for imparting good low temperature flexibility, with reasonable permanence, in polyvinyl chloride and its copolymers, cellulose esters, polystyrene and polyvinyl acetate.

ODCB Abbreviation for o-dichlorobenzene.

OENANTHOLACTAM Alternative name for enantholactam.

OENR Abbreviation for oil extended natural rubber.

OGDEN'S THEORY A theory of rubber elasticity based on the assumption that the strain energy function (W) may be written in the very generalised form

$$W = \sum_n (\mu_n/\alpha_n)(\lambda_1^{\alpha_n} + \lambda_2^{\alpha_n} + \lambda_3^{\alpha_n} - 3)$$

where α_n may have any value, μ_n is a constant and λ_1, λ_2 and λ_3 are the principal extension ratios. This leads, for example, to the nominal stress (f) in simple extension being given as

$$f = \sum_n \mu_n(\lambda_1^{\alpha_n - 1} - \lambda_1^{(-\alpha_n/2) - 1})$$

Such results can provide good fits to the actual stress–strain behaviour (in pure shear and in equi-biaxial tension) with only a two term formula.

OI Abbreviation for oxygen index, which is an alternative name for limiting oxygen index.

OIL EXTENDED NATURAL RUBBER (OENR) Natural rubber containing an oil extender, such as a petroleum oil, up to about 50 phr to reduce the product cost and/or to soften it.

OIL EXTENDED RUBBER A rubber to which a compatible oil has been added. Up to about 50 phr of a naphthenic, paraffinic or aromatic oil may be used especially in natural rubber (to give oil extended natural rubber), styrene–butadiene rubber and butadiene rubber. The oil may be added to decrease cost, but also acts as a plasticiser, softening the rubber and therefore acting as a processing aid and reducing stiffness in the vulcanisate.

OIL LENGTH The amount of drying oil (drying, semi-drying or non-drying) present in an alkyd resin. Resins may be classified according to oil length as short oil, medium oil or long oil resins.

OIL MODIFIED ALKYD RESIN What is usually meant by the term alkyd resin, i.e. a branched polyester of a dibasic acid (frequently phthalic anhydride) plus a polyol (frequently glycerol) modified by the incorporation of the triglycerides of a natural plant oil. This improves solubility, enables air drying to occur and gives tougher films than with the unmodified resin.

OLDROYD DERIVATIVE (Codeformational derivative) Symbol d/dt. A time derivative operator for the transformation of convected to mixed coordinates in connection with rheological problems, whilst obeying the principle of objectivity. It is defined, in Cartesian coordinates, as,

$$d/dt\,\tau_{ij} = \partial\tau_{ij}/\partial t + u_i\partial u_i/\partial x_k - \tau_{ij}\partial u_i/\partial x_k - \tau_{ij}\partial u_i/\partial x_k$$

where τ_{ij} are the shear stress components (using the summation convention), u_i are the velocity components and x_k the coordinate directions.

α-OLEFIN A compound of structure

$$\begin{array}{c} CH_2{=}CH \\ | \\ R \end{array}$$

where R is an alkyl or cycloalkyl group, i.e. an olefin substituted on the α-carbon atom. α-Olefins may be polymerised to poly-(α-olefins), usually by Ziegler–Natta polymerisation to isotactic polymers. Examples include propylene (R = —CH_3), butene-1 (R = —CH_2CH_3), 4-methylpentene-1 (R = —$CH_2CH(CH_3)_2$), hexene-1 (R = —$(CH_2)_3CH_3$) and octene-1 (R = —$(CH_2)_5CH_3$).

OLEFIN FIBRE Generic name for a fibre composed of a polymer with at least 85wt% of ethylene, propylene or other olefin units, excepting amorphous rubbery polymers. Examples are Courlene, Fibralon, Herculon and Vectra.

OLEIC ACID (Cis-9-octadecenoic acid)

$$CH_3(CH_2)_7CH{=}CH(CH_2)_7COOH$$

M.p. 13·4°C (α-form), 16·3°C (β-form).

Occurs in the triglycerides of most plant oils to a significant extent (5–25% of the acid residues). Although the double bond enables it to react with atmospheric oxygen, unlike linoleic acid it does not air dry.

OLEORESIN The initial exudation from certain trees when the bark is wounded. An oleoresin consists of a resinous component together with an essential oil. The latter usually evaporates, leaving the hard resin. Best studied is pine oleoresin, from which the resin rosin and the essential oil turpentine are obtained.

OLIGOMER A polymer with only a few repeat units in each polymer molecule, i.e. having a degree of polymerisation of up to a value of about 10–20. Thus dimers, trimers, tetramers, etc. are oligomers. Oligomers are formed during the early stages of step-growth polymerisation or after extensive random scission degradation of a polymer. They are also formed by a chain polymerisation when extensive chain transfer occurs.

OLIGOMERIC ENZYME An enzyme consisting of two or more polypeptide chains (the subunits) usually associated through interactions other than covalent bonds. When the subunits are identical they are sometimes called protomers. Many enzymes are of this type. They are often also allosteric enzymes, where activity is regulated by feed-back inhibition. Oligomeric enzymes can have several biological advantages over the simpler monomeric enzymes. Thus they enable isozymes to exist with a fewer number of polypeptide chains, one subunit may be used to modify the behaviour of another, or the different subunits may have different functions.

OLIGOMERIC PROTEIN A protein consisting of more than one polypeptide chain, where the chains are not covalently bound to each other. The chains are nevertheless so tightly associated that the whole protein shows a considerable stability and behaves as a single molecule. Thus in dilute solution the molecular weight of the whole associated complex is determined. The separate chains are the subunits (or protomers or monomers). They may consist of identical or different polypeptide chains, and in the smaller oligomeric proteins there are 2–12 subunits (always an even number—except when sometimes three subunits are present). Many oligomeric proteins also contain a prosthetic group. The arrangement of the subunits in space is the quaternary structure of the protein. Denaturation destroys the quaternary structure. The most widely studied group is that of the haemoproteins. Supramolecular complexes, such as multienzyme complexes and virus particles are also sometimes considered to be oligomeric proteins.

OLIGOPEPTIDE A peptide consisting of only a few amino acid residues, e.g. a di-, tri- or tetrapeptide. The term is sometimes restricted to peptides containing only a single type of amino residue, i.e. a poly-(α-amino acid) oligomer.

OLIGOSACCHARIDE An oligomeric saccharide (or carbohydrate) containing from two to ten monosaccharide (or sugar) units linked through glycoside bonds, thus including the di-, tri-, tetrasaccharides, etc. They may be readily hydrolysed by dilute acid to the parent monosaccharide(s). They occur naturally or are formed by partial hydrolysis of polysaccharides. They may therefore be considered as low polymers of the constituent monosaccharide(s). Since the monosaccharide units may be in either pyranose or furanose ring forms, they may be glycosidically linked in either α- or β-anomeric forms through any of several hydroxyl groups (1,4'-, 1,6'-, etc.) and many structural variations are possible. However, in naturally occurring oligosaccharides 1,4'-linking is usually present.

ONB Abbreviation for o-nitrobiphenyl.

ONE-SHOT PROCESS A process used in the formation of polyurethane foams, both flexible and rigid, in which the polyol, diisocyanate, catalysts and other components are mixed and reacted in one operation. Earlier, the process had always been used with polyester polyols, where the primary hydroxyl groups are sufficiently reactive. However, the secondary hydroxyls of polyether polyols are less reactive, so a prepolymer process had to be used until the advent of sufficiently powerful catalysts, such as the combination of stannous octoate and 1,4-diazabicyclo-2,2,2-octane. Now the one-shot process is the more important, being more economical and, in flexible foam production, producing a foam with better cushioning properties.

ONE STAGE POLYMER Alternative name for one stage resin.

ONE STAGE RESIN (One stage polymer) A precursor linear polymer (the A-stage resin or polymer) which may be crosslinked to form a network polymer (the C-stage resin or polymer) without the use of crosslinking agents. The crosslinking reactions result solely from reactions of functional groups already present on the precursor polymer. The term is applied particularly to phenol–formaldehyde polymers of the resole type which, unlike the novolacs (typical two stage resins), may be crosslinked and cured merely by a change in pH and further heating.

ONSAGER EQUATION A relation between the relative permittivity (ε) and the polarity, as measured by the dipole moment (μ), of a polar material. It is derived on the basis of a more realistic model than that used for the Debye equation, since it allows for interactions between dipoles, and hence is more applicable to solids. The electronic and atomic polarisations are also incorporated through the optical refractive index (n) terms. It is

$$[(\varepsilon - n^2)(2\varepsilon + n^2)/\varepsilon(n^2 + 2)^2]M/\rho = N_A\mu^2/9\varepsilon_0kT$$

where M is the molecular weight, N_A is Avogadro's number, ρ is the density, ε_0 is the relative permittivity of free space, k is Boltzmann's constant and T is the temperature. A modified version of this equation is the Fröhlich equation.

OPALON Tradename for polyvinyl chloride.

OPPANOL Tradename for polyisobutene.

OPPANOL B Tradename for polyisobutene.

OPPANOL C Tradename for poly-(isobutylvinyl ether).

OPTICAL ANISOTROPY The dependence of the refractive index of a material on direction of observation, resulting in birefringence.

OPTICALLY ACTIVE POLYMER A polymer capable of rotating the plane of polarisation of polarised light due to the presence of asymmetric or dissymmetric centres (chiral atoms) usually of carbon. Isotactic and syndiotactic polymers contain dissymmetric centres but also contain planes of symmetry and hence are not optically active. Asymmetry and hence optical activity may arise in several ways. An asymmetric centre may be present in the monomer and may be preserved in the polymer, as in the polymerisation of propylene oxide:

$$n \; CH_3\overset{*}{C}H—CH_2 \rightarrow \text{+}\overset{*}{C}H—CH_2—O\text{+}_n$$
$$\underset{O}{} \quad\quad\quad CH_3$$

or asymmetric induction may occur during polymerisation, as in the polymerisation of 1,3-pentadiene using an optically active catalyst:

$$n \; CH_2\text{=}CH—CH\text{=}CH_2CH_3 \rightarrow$$

$$\text{+}CH_2—CH\text{=}CH—\overset{*}{C}H\text{+}_n$$
$$CH_3$$

or in the formation of alternating copolymers with maleic anhydride:

$$CH_2\text{=}C—CH_3 + CH\text{=}CH \rightarrow \text{+}CH_2\overset{*}{C}—CH—CH\text{+}_n$$

In these examples the asterisked carbon atoms (C*) are true asymmetric centres having four different groups attached to them.

OPTICAL ROTATORY DISPERSION (ORD) The variation of the optical rotation of plane polarised light by an optically active material, with wavelength. It is widely used for the study of the molecular conformation, especially the helical content and the helix–coil transition of synthetic polypeptides and proteins. The dispersion spectrum is often represented mathematically by a Drude equation. A peak or trough in the ORD spectrum is known as a Cotton band.

ORBITAL STEERING A possible mechanism for enzyme activity in which the enzyme active site induces such a precise orientation of the substrate with respect to the enzyme catalytic group that the relevant bonding orbitals of the atoms involved are so aligned that orbital overlap enables the transition state to be reached with high probability.

ORD Abbreviation for optical rotatory dispersion.

ORDERED COPOLYAMIDE An aromatic polyamide copolymer of a regular alternating structure of the type

$$\text{+}NHArCONHAr'NHCOAr''NHCO\text{+}_n$$

where Ar, Ar' and Ar'' are aromatic or heterocyclic rings. The polymers have high thermal stability and T_m values, especially when *para*-linked, but are rather more tractable than similar aromatic polyamide homopolymers. Their increased solubility means that solution casting of films or spinning of fibres can be performed.

ORDERED POLYMER A copolymer in which the repeating units are arranged in some regular sequence.

Both alternating and block copolymers are ordered polymers, as are polypeptides, where regular sequencing of the amino acid residues produces an ordered polymer.

ORGALAN Tradename for bisphenol A polycarbonate.

ORGANAMIDE Tradename for nylon 6.

ORGANOSOL A plastisol to which an organic solvent has been added to lower the viscosity. On gelation the solvent is lost by evaporation.

ORGANOTIN STABILISER An organometallic tin compound usually of the type $R_{4-n}SnX_n$, where $n = 1–3$. R is an alkyl group and X is an organofunctional group. Widely used as thermal stabilisers for polyvinyl chloride. Very effective even at low concentrations (e.g. $0.1–1\%$ or $0.1–1$ phr), especially for clear compositions, but rather expensive. Most commonly of the type R_2SnX_2 with R = butyl (dibutyltin stabiliser) or with R = octyl (dioctyltin stabiliser). X is usually a carboxylate group (e.g. maleate or laurate) or an organosulphur group (thiotin stabiliser) such as thioglycollate. Methyltin stabilisers (R = methyl and $n = 1$ or 2) and estertin stabilisers (R = esteralkyl group) have also been introduced.

ORGATER Tradename for polybutylene terephthalate.

ORIENTATION The alignment of the structural elements of a material. In polymers orientation at several different structural levels may be observed—polymer chains, segments of chains, crystallites or even additives, especially fibres in composite materials. Orientation causes anisotropy of properties. Drawn or spun fibres are deliberately oriented along their length to enhance strength and stiffness in this direction due to the uniaxial orientation. Films may be biaxially oriented. Orientation may be adventitious as in injection moulded objects, where it may be a disadvantage by causing mechanical weakness perpendicular to the direction of orientation. The precise description of orientation is given by the orientation distribution function. Orientation may be characterised by birefringence, wide angle X-ray diffraction, infrared dichroism, wide-line NMR, laser-Raman spectroscopy or polarised fluorescence measurements.

ORIENTATIONAL POLARISATION Alternative name for orientation polarisation.

ORIENTATION BIREFRINGENCE Birefringence resulting from the physical ordering of optically anisotropic elements, e.g. chemical bonds, along some preferred direction. In polymers this can occur by aligning polymer chains in crystalline and amorphous regions, i.e. molecular orientation, by stretching or drawing. The most common cause of birefringence.

ORIENTATION DISTRIBUTION The distribution of orientation angles of the oriented elements of an oriented polymer. Described by the orientation distribution function. The elements may be particular chemical bonds, whole polymer molecules or polymer crystallites. The orientation angle is some defined angle between a characteristic direction in the oriented element, e.g. a crystal axis, and some reference direction, e.g. the fibre axis.

ORIENTATION DISTRIBUTION FUNCTION Alternative name for orientation function.

ORIENTATION FACTOR Alternative name for orientation function.

ORIENTATION FUNCTION (Orientation factor) (Hermann's orientation function) (Orientation distribution function) Symbol f. A function expressing the orientation distribution of a polymer sample. One mathematical form, which is obtained by several of the techniques used for characterising orientation, is

$$f_\alpha = \frac{3\langle \cos^2 \alpha \rangle_{av}}{2} - 1$$

where α is the orientation angle, e.g. between a reference direction and a crystallographic direction. For random orientation $f_\alpha = 0$; for perfect alignment of all elements $\alpha = 0$ and $f_\alpha = 1$. For biaxial orientation the function must also contain a second characteristic angle.

ORIENTATION FUNCTION DIAGRAM A graphical representation of orientation function values (f_{iq}) with reference to several axes (q), where i is an axis of the oriented elements, e.g. a given crystal axis. This is done by plotting the values within an equilateral triangle such that the normals from the point plotted to the sides of the triangle have lengths proportional to the f_{iq} value.

ORIENTATION HARDENING Strain hardening in which orientation of the polymer molecules in the direction of deformation occurs and is responsible for the increased stiffness in this direction.

ORIENTATION POLARISATION (Orientational polarisation) (Dipolar polarisation) A dielectric polarisation which occurs when a polar, strictly a dipolar, material is placed in an electric field. The permanent dipoles tend to become aligned parallel to the field. The associated molecular or group motions are opposed by the viscous drag of the material, particularly if it is polymeric, so polarisation is much slower than electronic and atomic polarisations. It is also temperature dependent, decreasing with increasing temperature. At very long times, or at low frequencies, full orientation can occur. At very high frequencies, or short times, no contribution from dipolar orientation to polarisation occurs. Over a certain intermediate frequency range, of the order of about 10^6 Hz for polymers, in the region of the relaxation time (τ), a dispersion occurs with a step in the relative permittivity, when the polarisation cannot keep place with the changing field. Hence a considerable power loss occurs with a maximum in $\tan \delta$ (the loss tangent) at about $\omega_{max} = 1/\tau$, where ω_{max} is the frequency at which maximum loss occurs. This frequently occurs in the practically important audio/radio frequency range. The dispersion may be a Debye dispersion or, for polymers, it may be better characterised by a broader, modified Debye dispersion such as that of a Cole–Cole plot.

ORIENTED CRYSTALLISATION Crystallinity in which the crystallites have some definite spatial relationship to each other. Strictly speaking, crystalline polymers are always oriented, being polycrystalline, with their lamellar crystallites naturally tending to pack into stacks, e.g. in the fibrils of spherulites which themselves are randomly oriented. However, spherulites have no macroscopic orientation. More specialised types of crystalline orientation are single crystal mats, transcrystalline structures, row-nucleated material and fibrous crystals produced by drawing or rolling.

ORLON Tradename for a polyacrylonitrile fibre.

L-ORNITHINE

$$\begin{array}{l} H_2NCHCOOH \\ \quad | \\ (CH_2)_3NH_2 \end{array} \quad M.p.\ 226–227°C.$$

An α-amino acid which, although not found in proteins, can occur in protein hydrolysates from the decomposition of arginine. Its pK' values are 1·71, 8·69 and 10·76.

OROGLAS Tradename for polymethylmethacrylate sheet.

OROGLAS DR Tradename for polymethylmethacrylate/rubber blend.

OROSOMUCOID An acidic blood serum glycoprotein (sometimes called a mucoprotein) that migrates electrophoretically (in Veronal buffer) with the α_1-globulins. It has a molecular weight of 41 000. It contains about 40% carbohydrate, probably mostly through linking of N-acetylglucosamine of the carbohydrate with the β-amide of asparagine of the polypeptide, but also with some links via γ-carboxyl of the glutamic acid residues.

OROWAN HYPOTHESIS The assumption that brittle fracture occurs when the yield stress (σ_y) of a material exceeds a certain value, on the basis that brittle fracture and yielding (and plastic flow) are independent processes and that whichever process takes place at the lower stress will be that which occurs. Since yield stress is affected more than brittle stress (σ_B) by temperature, the brittle–ductile transition is observed. Furthermore, yield stress increases more with strain rate, so the transition is also strain rate dependent. In a notched sample, if $\sigma_B < 3\sigma_y$ then brittle failure occurs, whereas if $\sigma_B > 3\sigma_y$ the material is ductile even if notched.

ORTHOGONAL RHEOMETER A dynamic rheometer which consists of two flat parallel plates rotating in their own plane with different angular velocities, about two parallel but not coincident axes. Thus an eccentric oscillatory motion is set up.

ORTHOTROPIC MATERIAL A particular case of an anisotropic material which is symmetrical with respect to three mutually perpendicular planes. Such symmetry is often met in fibre reinforced composites, especially in laminates composed of polymer with fibres running in two directions perpendicular to each other or in which the fibres are randomly oriented within a plane. For such symmetry there are nine independent elastic moduli. If the fibres are oriented in only one direction, then the material is not only orthotropic, but is also transversely isotropic with the properties in the planes perpendicular to the fibre direction being independent of direction. In this case there are five independent elastic moduli.

OSCILLATING RHEOMETER Alternative name for dynamic rheometer.

OSCILLATORY FLOW Dynamic flow in which an external perturbation with known frequency and amplitude, and which is varying sinusoidally with time, is applied to the fluid concerned. Measurement of the time varying response of the fluid enables the components of the complex viscosity to be determined. A common experimental set-up is couette geometry with the fluid filling the annular space between two parallel concentric cylinders. The outer cylinder is subject to a sinusoidally varying velocity and the fluid response is measured by the displacement of the inner cylinder, as monitored by the torque of the torsion wire from which it is suspended. Alternatively, especially for polymer melts, a pair of oscillating discs may be embedded in the polymer and stress–strain transducers used. An advantage of dynamic measurements is that they give data on the elastic properties of the polymer using only very small deformations (<0·1%).

OSMODIALYSIS A dynamic osmometry technique performed with a leaky membrane, permeable to the lower molecular weight species of a polymer solution. From a study of the time dependence of the osmotic pressure, estimates of the molecular weight distribution may be made.

OSMOMETER An apparatus for the determination of osmotic pressure and used in the technique of membrane osmometry for the determination of polymer number average molecular weights. Consists of a cell containing a solution and a solvent compartment separated by a semipermeable membrane. In traditional osmometers a static equilibrium technique is used, whereby solvent is allowed to diffuse through the membrane by osmosis, to establish a hydrostatic (solvent) head of pressure on the solution. This is measured by attaching to each compartment a vertical capillary measuring tube. As osmosis occurs, the liquid level rises in the solution capillary and falls in the solvent capillary. When osmotic equilibrium is reached, the difference in liquid levels is the osmotic pressure in centimetres of solvent pressure. Widely used examples of this type are the Pinner–Stabin, Zimm–Meyerson and Fuoss–Mead osmometers. Unfortunately these require a long time to reach equilibrium (several hours to days or weeks) and have now been largely superseded by modern high speed osmometers.

OSMOMETRY Alternative name for membrane osmometry.

OSMOTIC PRESSURE Symbol Π. The excess hydrostatic pressure which must be applied to a solution, separated from pure solvent by a semi-permeable membrane, in order to just prevent osmosis, i.e. flow of solvent through the membrane from solvent to solution. Measurement of Π of a dilute polymer solution in an osmometer provides the most useful method of determination of its number average molecular weight.

OSTWALD–DE WAELE EQUATION Alternative name for power law equation.

OSTWALD VISCOMETER The simplest type of U-tube viscometer used for the determination of the viscosity ratio of dilute polymer solutions in the solution viscosity method of polymer molecular weight estimation. Since the determination of the limiting viscosity number requires measurements to be made at several concentrations, a modified form—the Ubbelohde viscometer—is more convenient.

OVALBUMIN (Egg albumin) An albumin protein occurring in egg whites and accounting for about 65% of the total protein content. It is a glycoprotein as well as being a phosphoprotein with a maximum of two phosphate groups per mole of serine base and contains a covalent link between an N-acetyl-D-galactosamine unit of an oligosaccharide side chain and the amide nitrogen of an asparagine residue. The carbohydrate accounts for less than 1% of the glycoprotein. Its molecular weight is about 45 000 with an isoelectric point of 4·6. It has a compact globular molecular shape and contains a high proportion of the essential amino acids, but its biological role is unknown. It is readily denatured irreversibly, e.g. by heating, treatment with organic solvents or even merely by shaking.

OVERCURE Curing that has proceeded for longer than the optimum cure time. Overcure can result in reversion, especially with natural rubber, when the modulus and strength decrease with increasing overcure. Alternatively, the rubber may continue to harden with continued curing.

OVERGROWTH The formation of smaller crystallites on the surface of larger crystals, due to a change in the crystallisation conditions. This can result from a change in temperature, e.g. overgrowths on polyethylene single crystals grown in dilute solution, or from lamellar overgrowth formation on shear-induced extended chain crystals as in shish-kebabs.

OVERSHOOT Alternative name for stress overshoot.

OVOMUCIN (Ovomucoid-β) An albumin occurring in egg whites and accounting for 1–2% of the total protein content. A glycoprotein which precipitates when egg white is diluted with water or dilute salt solution. It has a molecular weight of 210000 and contains about 13% carbohydrate.

OVOMUCOID An albumin occurring in avian egg whites and accounting for about 13% of the total protein content. It exhibits inhibition of proteolytic activity of trypsin. A mucoprotein containing about 25% polysaccharide. The carbohydrate composition varies from sample to sample but typically it consists of a central branched core of D-mannose and 2-acetamido-2-deoxy-D-glucose units to which are linked sialic acid and D-galactose units. The molecular weight is about 30000 and the protein may be a single peptide chain glycosidically linked through asparagine units. Unlike ovalbumin, it is not readily denatured by heating.

OVOMUCOID-β Alternative name for ovomucin.

1,3-OXAZOLIDINE-2,5-DIONE Alternative name for N-carboxy-α-amino acid anhydride.

OXAZOL-5-ONE A compound of structure

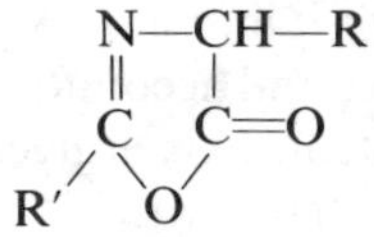

$$\begin{array}{c} N\!\!-\!\!CH\!\!-\!\!R \\ \| \qquad \quad | \\ C \quad\;\; C\!\!=\!\!O \\ \diagdown \quad \diagup \\ R' \quad O \\ \mathbf{(I)} \end{array}$$

Formation of oxazolone is responsible for racemisation during peptide synthesis using a protected amino acid, especially during alkaline conditions. Racemisation occurs due to the formation of the enol form of **I**:

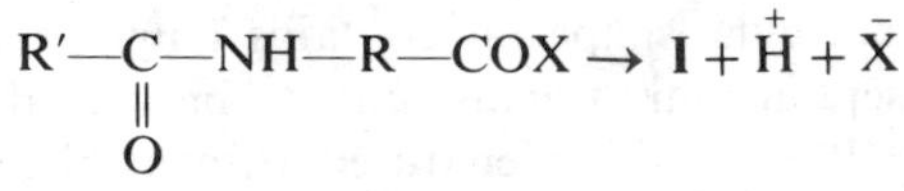

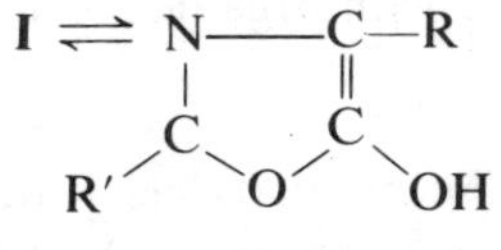

$$R'\!-\!\underset{\underset{O}{\|}}{C}\!-\!NH\!-\!R\!-\!COX \rightarrow I + \overset{+}{H} + \bar{X}$$

$$I \rightleftharpoons \begin{array}{c} N\!\!-\!\!\!-\!\!C\!\!-\!\!R \\ | \qquad\quad \| \\ C \qquad C \\ \diagup \quad \diagdown \; \diagdown \\ R' \quad O \quad OH \end{array}$$

Racemisation is minimised by using non-polar solvent and protecting groups such as benzoyloxycarbonyl and t-butyloxycarbonyl, by using DCCI or by the azide method of peptide bond formation. Racemisation is also minimised if the C-terminal residue is glycine or proline.

OXETANE POLYMER (Polyoxetane) Strictly a polymer obtained by the ring-opening polymerisation of an oxetane (oxacyclobutane), i.e. one of general structural type

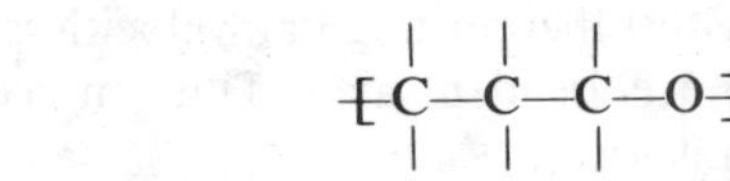

$$\begin{array}{c} | \quad\; | \quad\; | \\ \bm{-\!\!\!-}C\!\!-\!\!C\!\!-\!\!C\!\!-\!\!O\bm{\!\!-\!\!}_n \\ | \quad\; | \quad\; | \end{array}$$

However, the term is usually used as an alternative name for the only polymer of this type of commercial significance—poly-(3,3-bis-(chloromethyl)oxacyclobutane).

OXIDATION (Oxidative degradation) Degradation involving oxygen, often atmospheric oxygen in practice. Many polymer degradations proceed via free radicals ($\dot{R}$) which often result from polymer reaction with oxygen or on to which oxygen readily adds, since it is itself a free radical. The resultant peroxides or hydroperoxides can continue and extend degradative chain reactions:

$$\dot{R}_1 + O_2 \rightarrow R_1O\dot{O} \xrightarrow{R'_2-H} R_1OOH + \dot{R}_2$$
$$\downarrow$$
$$R_1\dot{O} + \dot{O}H$$

Thus oxygen accelerates degradation and alters the mechanism and the products. Thus when thermal and photodegradations are carried out in air they become thermo-oxidations and photo-oxidations respectively. Antioxidants are used as polymer additives to minimise the extent of oxidation.

OXIDATION–REDUCTION POLYMER (Redox polymer) (Electron transfer polymer) A polymer containing functional groups that can exchange electrons with other molecules or ions. Examples include polyvinyl hydroquinone:

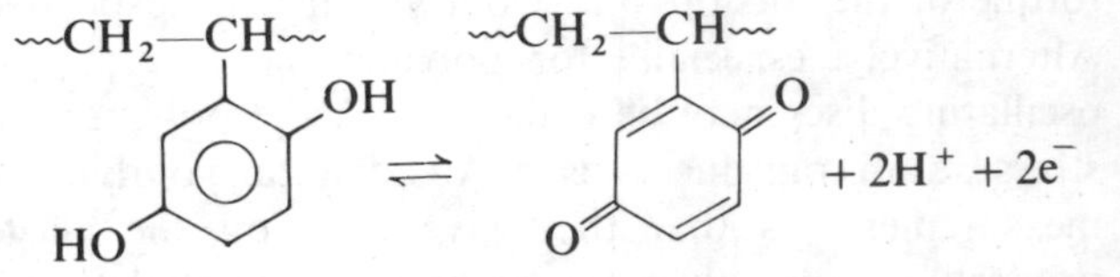

$$\sim CH_2-CH\sim \qquad \sim CH_2-CH\sim$$

$$\rightleftharpoons \qquad\qquad + 2H^+ + 2e^-$$

and mercaptan-containing polymers:

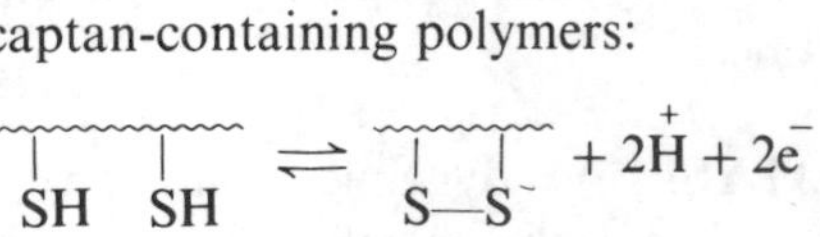

$$\underset{SH \quad SH}{|\quad\;|} \rightleftharpoons \underset{S\!-\!S}{|\quad\;|} + 2\overset{+}{H} + 2e^-$$

OXIDATIVE COUPLING (Oxidative polymerisation) Polymerisation in which monomer units are linked together via an oxidative reaction. The best known example is the polymerisation of 2,6-disubstituted phenols to disubstituted polyphenylene oxides:

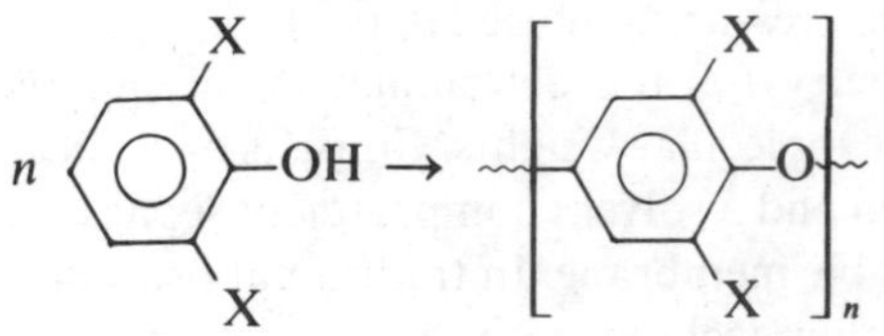

$$n \; \underset{X}{\overset{X}{\bigcirc}}\!\!-\!\!OH \rightarrow \sim\!\!\left[\underset{X}{\overset{X}{\bigcirc}}\!\!-\!\!O\right]_n$$

This may be brought about simply by bubbling oxygen into a solution of the phenol at moderate temperatures in the presence of a suitable catalyst such as CuCl/pyridine. Typically a chlorinated hydrocarbon solvent is used at about 60°C. The pyrolysis of *p*-xylene to produce poly-*p*-xylylene may also be considered to be an oxidative coupling.

OXIDATIVE DEGRADATION Alternative name for oxidation.

OXIDATIVE POLYMERISATION Alternative name for oxidative coupling.

OXIDISED CELLULOSE Alternative name for oxycellulose.

OXIDISED RUBBER Tradename: Rubbone. Natural rubber which has been heated in air with an oxidation catalyst, such as cobalt or manganese linoleate. A yellow-orange viscous fluid or resinous solid which contains 5–15% oxygen mainly as —OH groups but also with some —OOH, $>$C$=$O and —COOH groups. Contains 50–90% of the original *cis*-1,4-polyisoprene unsaturation and is of low molecular weight (about 3000). It has good resistance to thermal degradation and has been used in surface coatings.

OXIDISED STARCH A partially hydrolytically degraded starch produced by alkali oxidation with solutions of sodium hypochlorite. The product contains aldehyde, ketone and carboxyl groups which impede retrogradation and is used as textile and paper size.

OXITOL Alternative name for ethylene glycol monoethyl ether.

OXO PROCESS An industrial process for the production of alcohols, by the reaction of olefins (at high pressure and in the presence of a suitable catalyst, such as cobalt) with carbon monoxide and hydrogen. This yields aldehydes which hydrogenate to the corresponding alcohols. Usually a mixture of alcohols is produced, which nevertheless is frequently useful as such, e.g. for conversion to plasticisers. Thus, for example, nonyl alcohol contains a large amount of 3,5,5-trimethylhexan-1-ol, plus other isomers and octanols and is converted into dinonyl phthalates.

4,4'-OXYBIS-(BENZENESULPHONYL-HYDRAZIDE) (OB)

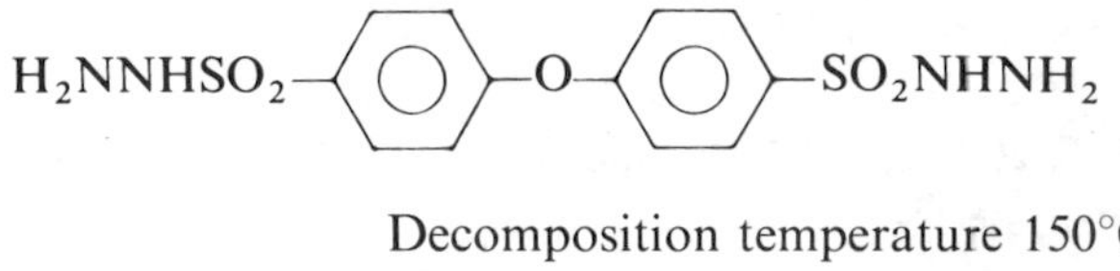

$$H_2NNHSO_2 - \bigcirc - O - \bigcirc - SO_2NHNH_2$$

Decomposition temperature 150°C.

A chemical blowing agent useful in the production of cellular polymers. Decomposes to produce nitrogen and water.

OXYCELLULOSE (Oxidised cellulose) A product of oxidation of cellulose in which carbonyl and/or carboxyl

groups have been introduced. Some reagents, e.g. oxygen, ozone, hydrogen peroxide and chlorine, are non-specific, but others, e.g. lead tetraacetate, periodic acid and dinitrogen tetroxide, cleave specific bonds in the anhydroglucose units. Such oxidations are frequently unwanted, e.g. in bleaching, dyeing and soda cellulose production since they cause a reduction in tensile strength of the fibre

N-OXYDIETHYLBENZOTHIAZYL-SULPHENAMIDE (NOBS)

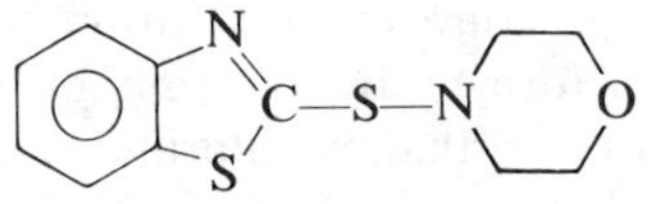

M.p. 79°C.

A delayed action accelerator for the sulphur vulcanisation of rubbers.

OXYGEN INDEX (OI) Alternative name for limiting oxygen index.

OZONE CRACKING Surface cracking or crazing in a diene rubber when exposed to low (even a few ppm) ozone concentrations. Results from the chain scission occurring during the ozone-induced degradation. Considerably accelerated when the rubber is stressed beyond the threshold stress.

OZONE-INDUCED DEGRADATION (Ozonolysis) The attack of ozone on in-chain unsaturated groups in a polymer, with the formation of a cyclic ozonide which subsequently breaks down causing polymer chain scission:

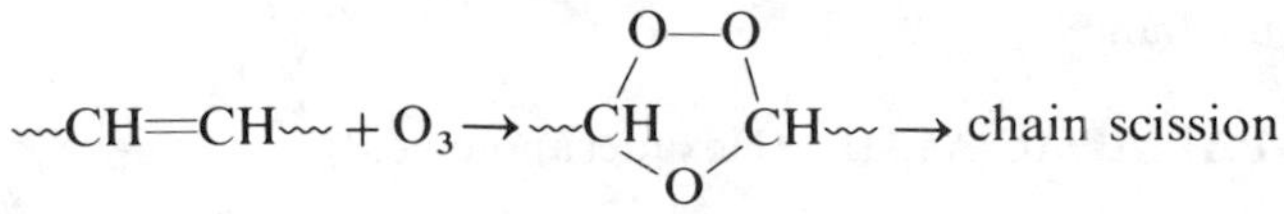

$$\sim CH=CH\sim + O_3 \rightarrow \sim CH \underset{O}{\overset{O-O}{\diagup \diagdown}} CH\sim \rightarrow \text{chain scission}$$

Thus diene rubbers are particularly sensitive, ozone cracking often being the result. Protection may be achieved using an antiozonant.

OZONOLYSIS Reaction of a carbon–carbon double bond to form, initially, a cyclic ozonide,

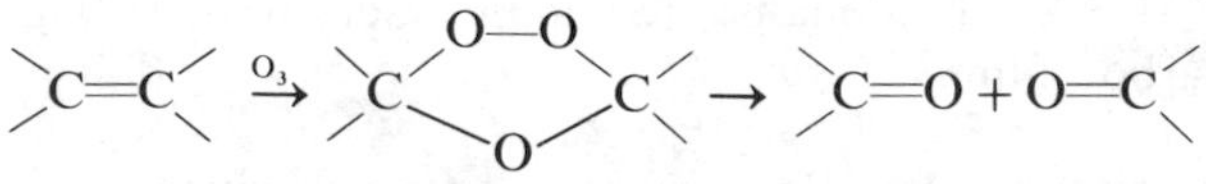

$$>C=C< \overset{O_3}{\rightarrow} >C \underset{O}{\overset{O-O}{\diagup \diagdown}} C< \rightarrow >C=O + O=C<$$

which is subsequently cleaved by further reaction giving carbonyl groups. Thus the presence of each double bond results in cleavage of the polymer chain. Such a reaction is useful in the structural elucidation of polymers containing double bonds in the main chain. Thus in diene rubbers ozonolysis gives fragments, e.g. of laevulinic aldehyde or acid, whose structure confirms the 1,4-joining of the units in most diene rubbers. In addition, small amounts of in-chain unsaturation may be determined by measuring the drop in the molecular weight on ozonolysis. Ozonolysis also causes ozone-induced degradation in diene rubbers.

P

P Symbol for proline.

PA Abbreviation for polyamide.

PA-18 Tradename for octadecene-1–maleic anhydride copolymer.

PABH-T Tradenames: X-500, Flexten. Abbreviation for a high temperature resistant aromatic polyamide-hydrazide fibre formed by reaction of *p*-aminobenz-hydrazide with terephthaloyl chloride:

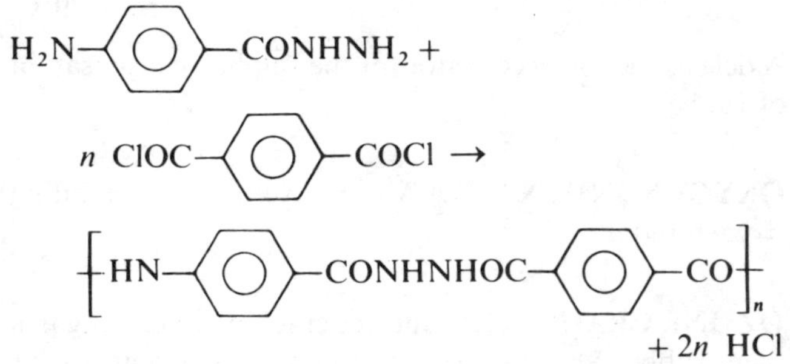

$$H_2N-\langle O \rangle-CONHNH_2 +$$

$$n \; ClOC-\langle O \rangle-COCl \rightarrow$$

$$\left[HN-\langle O \rangle-CONHNHOC-\langle O \rangle-CO \right]_n$$

$$+ 2n \; HCl$$

PABM Abbreviation for polyaminobismaleimide.

PACHYMAN A 1,3'-linked β-D-glucan found in certain fungi.

PACKING FRACTION Alternative name for maximum packing fraction.

PACM Alternative name for bis-(4-aminocyclohexyl)-methane.

PALMITIC ACID (Hexadecanoic acid)

$$CH_3(CH_2)_{14}COOH \qquad M.p. \; 63 \cdot 1°C.$$

Occurs in the triglycerides of many plant oils, especially in cottonseed oil, where it accounts for 29% of the acid residues.

PAN Abbreviation for polyacrylonitrile.

PANEX Tradename for a polyacrylonitrile based carbon fibre.

PANOSE An oligosaccharide trimer of glucose,

$$\alpha\text{-D-Gp1} \rightarrow 6'\text{-}\alpha\text{-D-Gp1} \rightarrow 4'\text{-}\alpha\text{-D-Gp}$$

isolated from hydrolysis of amylopectin, demonstrating the presence of both 1-4'- and 1-6'-links in the amylopectin.

PANOTEX Tradename for a heat resistant fibre, similar to Celiox, obtained by heating polyacrylonitrile to 200–300°C.

PAPA Abbreviation for polyazelaicpolyanhydride.

PAPAIN A plant enzyme, obtained from the fruit of the papaya, which is a protease. It has a molecular weight of about 21 000 and consists of a polypeptide chain with three disulphide bridges. It hydrolyses proteins at a wide range of different amino acid residues but not at acidic residues. A specific cysteine residue is involved at the active site in forming an acyl enzyme (a thiol ester) with the substrate, i.e. it is a thiol enzyme. It is sometimes useful in the cleavage of large peptides produced on initial reaction with, for example, trypsin, in sequence analysis.

PAPER CHROMATOGRAPHY (Filter paper chromatography) A type of partition chromatography in which a strip or sheet of paper, e.g. filter paper, is used as the support material. The stationary phase is thought to be the water adsorbed onto the cellulose from the moving phase (mixed organic liquid/water is normally used). It is very widely employed, both analytically and preparatively (on a small scale), for separations of the degradation products of hydrolysis of biopolymers in their structural elucidation, i.e. of peptides from proteins, of oligosaccharides from polysaccharides and of nucleosides from nucleic acids. In analytical work a drop of a solution of the mixture is applied as a spot near the end of the paper which is dipped into a trough of moving phase (solvent). The paper is mounted vertically and the solvent front rises up the paper by capillary action, separating the components as it progresses. The distance moved by a separated component is characteristic of it (and the nature of the stationary and moving solvent phases) and is given by its R_f value, defined as the ratio of the distance it has moved to the distance moved by the solvent front. The separated components are usually visualised by spraying with a solution of a reagent which forms coloured compounds with them, e.g. ninhydrin for peptides and amino acids. For preparative separations large thick sheets of paper are used and the sample is applied as a streak parallel to the solvent front.

PAPER ELECTROPHORESIS Zone electrophoresis in which the solid supporting material is a strip of paper or cellulose acetate. These materials are hydrophilic but do not absorb the proteins which are frequently separated by this method.

PAQR Abbreviation for poly-(acene quinone radical).

PARABOLA MARKING The most common type of conic marking observed in fracture, in which the locus of the step is a parabola.

PARACRIL Tradename for nitrile rubber and acrylic elastomers.

PARACRYSTALLINITY Crystallinity represented by a crystalline lattice that has been distorted in various characteristic ways. The degree of disorder may be estimated from the broadening and diffuseness of the reflections of the X-ray diffraction pattern. Distortions of the first kind are displacements of the structural elements—atoms, motifs or monomer units, from their equilibrium positions in the ideal lattice. Those of the second kind are of a longer range in which each lattice point varies its position with respect to its neighbour rather than with respect to the ideal lattice.

2,2′-PARACYCLOPHANE (Di-p-xylylene)

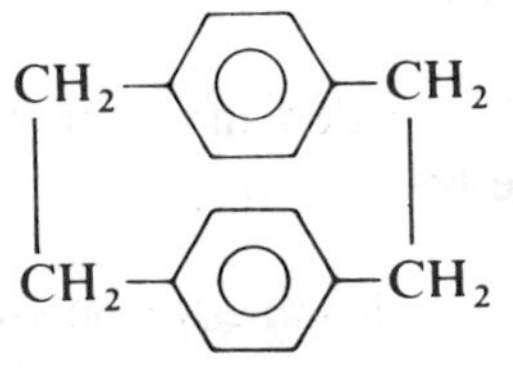

M.p. 280°C.

The monomer for the synthesis of poly-(p-xylylene) by pyrolysis at 600°C. Itself synthesised by pyrolysis of p-xylene at about 950°C.

PARAFFINIC OIL A rubber oil containing a high proportion of paraffinic, i.e. linear, saturated hydrocarbon, structures. A typical oil has a viscosity gravity constant of 0·81 and a refractivity index of 1·475, giving a carbon atom type analysis of 3·5% aromatic, 31% naphthene and 65·5% paraffinic carbon. In terms of molecules this corresponds to about 12% aromatic, 87·5% saturated and only about 0·5% polar heterocyclic compounds. The oil components occur naturally in the petroleum. They are separated from the aromatic compounds by extraction and further separated from the wax components. These oils impart low heat build-up in rubber compounds due to their lubricity. They have the best heat and light stability of all the rubber oils.

PARAFORMALDEHYDE

$$HO-(CH_2O)_n-H$$

A low molecular weight polymer of formaldehyde with a degree of polymerisation of 10–100 and low thermal stability, depolymerising to formaldehyde when heated to about 150°C. Prepared by the distillation of an approximately 35% aqueous solution of formaldehyde. A typical commercial product contains about 1–5% of free water and melts at 120–170°C. Sometimes used as a source of formaldehyde for the crosslinking of novolacs and resorcinol–formaldehyde polymers.

PARALAC Tradename for modified urea–formaldehyde polymer used for laminating and surface coatings.

PARALLEL PLATE VISCOMETER A viscometer in which a circular disc of viscous liquid is compressed between two parallel plates which are larger in diameter than the liquid disc. Such an instrument is only suitable for liquids of very high viscosity deformed at low rates of shear.

PARAMYLON A 1,3′-linked β-D-glucan found in certain algae.

PARAMYOSIN Alternative name for tropomyosin A.

PARA RUBBER Natural rubber, produced crudely, mainly in the early days of the rubber industry, by coagulation of a layer of natural rubber latex on a wooden paddle held in the smoke of a wood fire.

PARATHYROID HORMONE A protein hormone secreted by the parathyroid glands which regulates calcium and phosphorus metabolism. It consists of a single polypeptide chain of molecular weight 9500, containing 84 amino acid residues.

PARAVAR Tradename for chlorinated rubber.

PAREL 58 Tradename for propylene oxide rubber.

PARKESINE Tradename for cellulose nitrate, plasticised with camphor and often regarded as the first plastic material, later more successfully developed as celluloid.

PARLON Tradename for chlorinated rubber.

PARQUET POLYMER (Layer polymer) (Sheet polymer) A polymer consisting of infinite sheet-like molecules. Can be considered to be a crosslinked, multiple-stranded, ladder polymer. The classical example is graphite, consisting of planar networks of carbon atoms in parallel layers. Many silicates and talc are other examples. Synthetic examples are rare; the product of reaction of copper acetylacetonate with tetracyanoethylene is one example.

PARTIAL LADDER POLYMER (Step-ladder polymer) (Semi-ladder polymer) A double strand ladder polymer which does not consist completely of fused rings in the polymer chain, but rather of sequences of fused rings joined through single bonds. The case of single rings being linked through single bonds (e.g. a polyphenylene) is trivial, but may formally be regarded as a partial ladder polymer. However, the term is more frequently restricted to mean a polymer containing several fused rings so linked, e.g. polyisoindoloquinazolinedione. Many double strand polymers which nominally consist entirely of fused rings are in fact only partial ladder polymers. This arises from incomplete cyclisation during synthesis, especially when this involves zipping-up of pendant side groups.

PARTIALLY OXIDISED CELLULOSE Alternative name for hydrocellulose.

PARTIAL SPECIFIC VOLUME The volume increase resulting from the addition of 1 kg of solute (e.g. polymer) to a large (or infinite) volume of solvent. The reciprocal of the effective density of the polymer in solution. Occurs in the buoyant density term of the Svedberg equation relating sedimentation velocity to molecular weight. Determined by measuring the densities of pure solvent and polymer solutions of different concentrations using a density bottle or pyknometer.

PARTICLE SCATTERING FACTOR (Scattering factor) Symbol P_θ. The value of the particle scattering function at a particular value of the scattering angle θ.

PARTICLE SCATTERING FUNCTION Symbol $P(\theta)$. In light scattering with internal interference, the ratio of the scattered light intensity with interference to the intensity that would be observed in the absence of interference, measured at the angle θ to the incident beam, $P(\theta)$ being dependent on θ. At low angles $P(\theta) = 1 - \mu\langle s^2\rangle/3$, where $\mu = 4\pi/\lambda \sin(\theta/2)$ (λ is the wavelength of light) and $\langle s^2\rangle$ is the mean square radius of gyration of the scattering particle. At higher θ, $P(\theta)$ is also dependent on the shape of the scattering particles. For a Gaussian coil (such as many polymer molecules adopt in dilute solution) $P(\theta) = (2/n^2\mu^2)(e^{-na} - 1 + na)$, where n is the number of links in the polymer chain each of length a and μ is $(4\pi/\lambda)^2 \sin^2(\theta/2)(a/6)$. $P(\theta)$ may be determined experimentally by extrapolation of scattered intensities at low angles to zero angle to obtain P_0. The values of $P(\theta)$ can give useful information on scattering particle shapes.

PARTICLE WEIGHT An alternative name for molecular weight, usually preferred for oligomeric proteins, especially viruses, where the whole protein exists as a supermolecular complex, whose subunits are often not joined through covalent bonds. Thus the concept of individual molecules in the conventional sense is not appropriate.

PARTICULATE COMPOSITE A composite material in which the dispersed phase, often referred to as the filler, is composed of rigid particles whose dimensions are all of the same order of magnitude. Usually the function of the filler is to improve one or more of the physical properties of the material, especially by reinforcement to increase the tensile strength, stiffness, hardness and/or impact strength. Sometimes a cheap filler is used merely to lower the cost by acting as an extender. Common examples are carbon black and silica filled rubbers, and mineral powder filled, e.g. with calcium carbonate or talc, plastics. The filler may or may not be reinforcing depending on the interfacial interactions with the polymer matrix. Such interactions are particularly important in carbon black filled rubbers. In other systems interfacial adhesion may be improved by treating the filler with an appropriate coupling agent.

PARTITION CHROMATOGRAPHY A method of chromatography in which the separation of molecular species is by repeated distributions between two liquid phases, one mobile and the other held on a solid support. The support may be packed in bulk into a column (column chromatography) or it may be in a two-dimensional form as a strip of paper (paper chromatography) or as a spread layer (thin layer chromatography). It is mostly used for the separation of the hydrolytic degradation products of biopolymers (peptides from proteins, oligosaccharides from polysaccharides and nucleosides from nucleic acids) and for the separation of amino acid mixtures. Typical materials used in column or thin layer techniques are cellulose, starch and silica as supports onto which water is thought to be permanently adsorbed forming the stationary phase. The moving phase used for elution is usually an organic solvent/water mixture. Sometimes reversed-phase chromatography is performed in which the organic phase is stationary and the aqueous phase is moving.

PARYLENE C Tradename for poly-(p-xylylene) chlorinated in the benzene rings.

PARYLENE N Tradename for poly-(p-xylylene).

PASCAL-SECOND Symbol Pa s. The SI unit of viscosity. It has the dimensions of $N\,s\,m^{-2}$ or $kg\,m^{-1}\,s^{-1}$. One pascal-second equals $10\,P$ (the cgs unit). Polymer melts typically have viscosities in the range 10^2–$10^3\,Pa\,s$.

PAT Abbreviation for positron annihilation technique.

PATCH A fracture surface marking, often in the outer part of the mirror region, resulting from the propagating crack, propagating along the interface between the matrix and craze material, jumping irregularly from the interface on one side of the craze to the other.

PAUCIDISPERSE POLYMER A polymer sample which consists of only a few different molecular components differing in molecular weight. Often found in protein and other biopolymer samples. The various components may often be identified by their respective peaks on a sedimentation velocity concentration gradient curve if their sedimentation velocities are sufficiently different.

PB Abbreviation for polybutadiene.

PBA Abbreviation for polyether block amide.

PBD Abbreviation for polybutadiene.

PBG Abbreviation for racemic poly-(γ-benzyl-glutamate).

PBI Abbreviation for polybenzimidazole.

PBN Abbreviation for N-phenyl-β-napththylamine.

PBNA Abbreviation for N-phenyl-β-naphthylamine.

PBT Abbreviation for polybutylene terephthalate.

PC Abbreviation for polycarbonate, in particular for bisphenol A carbonate.

P-CELLULOSE Abbreviation for phosphocellulose.

PCR Tradename for carboxynitrosorubber.

PCTFE Abbreviation for polychlorotrifluoroethylene.

PCU Tradename for a polyvinyl chloride fibre.

PE Abbreviation for polyethylene.

PEARL NECKLACE MODEL A model for a polymer molecule consisting of a series of spherical particles connected by negligibly thin links. Useful in the development of the theory of viscosity of dilute polymer solutions in which the segments act as hydrodynamically equivalent spheres and where the distribution in space is Gaussian.

PEARL POLYMERISATION Alternative name for suspension polymerisation.

PEBAX Tradename for a polyether block amide.

PE CE Tradename for a chlorinated polyvinyl chloride fibre containing about 64% chlorine.

PECLET NUMBER Symbol Pe. A dimensionless parameter signifying the importance of thermal convection during flow. It is defined as $Pe = \rho C_p B U / \kappa$, where ρ is the fluid density, C_p is its specific heat, B is the width of the flow channel, U is the average velocity in the direction of flow and κ is the thermal conductivity.

PECTIC ACID A pectic substance in which few of the D-galacturonic acid units are esterified. In contrast to the pectinic acids it does not gel unless partially neutralised, e.g. with Ca^{2+}. They occur in the cell walls, probably as the calcium salt, from which they may be extracted by a complexing agent, e.g. sodium hexametaphosphate or EDTA.

PECTIC SUBSTANCE A plant polysaccharide in which D-galacturonic acid is the main constituent, although usually other sugar units are also present as either side groups or in the main chain. These include D-galactose, L-arabinose and L-rhamnose, sometimes as their 2-methyl esters. The esterified polymers are pectinic acids, whereas the ester free polymers are pectic acids. They occur widely in the cell walls of land plants, especially in the soft tissues, e.g. citrus fruit peel, apples. Often neutral polysaccharides, especially arabinans and galactans, occur with the pectic substance, but may be also chemically part of the polyuronide molecules. The polymers are linear with the α-D-galacturonic acid units 1,4'-linked. The precise linking of the neutral sugar units is not known. Possibly galactan blocks are present in the main chain in soya bean pectin and arabinan blocks in alfa-alfa pectin, whereas in lemon peel pectin 1,2'-L-rhamnose units are present in the chain. Side groups found include D-xylose, D-guluronic-1,6'-D-galactose and D-gluconic-1,4'-L-fucose units. If only free carboxylate groups are present, they repel, and the substance does not gel. Low concentrations of Ca^{2+} ions cause peeling, as does esterification to methyl esters, but not to ethyl esters.

PECTIN A pectic substance which is water soluble and readily forms gels. Sometimes used merely as an alternative name for pectic substance.

PECTINIC ACID A pectic substance in which a considerable proportion of the D-galacturonic units are methyl esters. They readily gel, in contrast to pectic acids, and occur in the intercellular layers of brown algal seaweeds from which they may be extracted with water or dilute acid.

PEEK Abbreviation for polyetheretherketone.

PEELING The removal of anhydroglucose units from the ends of cellulose molecules by exposure to the combined effects of alkali swelling, atmospheric oxygen and elevated temperature. Up to 50 units can be removed in this way before competing side reactions stop the reaction. Peeling can be minimised by any of several treatments, including reduction, glycosidation or oxidation.

PEG Abbreviation for polyethylene glycol.

PEK Abbreviation for polyetherketone.

PELARGONIC ACID (Nonanoic acid)

$$CH_3(CH_2)_7COOH$$

M.p. 12·5°C. B.p. 255·6°C.

Sometimes used as an acid modifier for non-drying alkyd resins.

PELASPAN Tradename for expanded polystyrene.

PELLETHANE Tradename for a polyurethane block copolymer based on a polyester or polyether polyol/MDI prepolymer and chain extended.

PELPRENE Tradename for a thermoplastic copolyester elastomer based on butylene glycol and terephthalic acid.

PEN-2,6 Abbreviation for poly-(ethylenenaphthalene-2,6-dicarboxylate).

PENTAD A sequence of five repeat units in a polymer molecule. Some aspects of polymer microstructure may be analysed in terms of pentads, in a similar way to the analysis of dyads and triads. In particular, configurational pentad sequences of vinyl polymers of the type

$$-[CH_2CHX]_n$$

can be determined by observing the resonances of the X substituent by its ^{13}C nuclear magnetic resonance spectrum or by using high field, e.g. 220 MHz, NMR techniques.

1,3-PENTADIENE Alternative name for piperylene.

PENTAERYTHRITOL (Tetramethylolmethane) (2,2-Bis-(hydroxymethyl)-1,3-propanediol)

$$C(CH_2OH)_4 \qquad \text{M.p. } 262°C.$$

Prepared by the reaction of acetaldehyde with formaldehyde under alkaline conditions. Used as a polyol for the production of long oil alkyd resins, for the preparation of 3,3-bis-(chloromethyl)-oxacyclobutane, and sometimes as a comonomer in polyethylene glycol adipates for use as a polyol prepolymer for polyurethane foams.

***n*-PENTANE**

$$CH_3(CH_2)_3CH_3 \qquad \text{B.p. } 36·0°C.$$

A hydrocarbon solvent, useful (due to its high volatility) as a physical blowing agent in the production of expanded polystyrene.

PENTON Tradename for poly-(3,3-bis-(chloromethyl)oxacyclobutane).

PENTOSAN A polymer of a pentose. Thus xylans and arabinans are pentosans. Since homopolysaccharide pentosans are rare, the term is often used to include those land plant polysaccharides in which arabinose or xylose units are important constituents in heteropolysaccharides.

PENTOSE A monosaccharide containing five carbon atoms. Examples include the aldopentoses L-arabinose, D-ribose and D-xylose, and the ketopentose D-fructose.

PENULTIMATE EFFECT An effect in copolymerisation in which the reactivity of the chain end propagating centre is affected by the penultimate repeat unit as well as by the end repeat unit carrying the active centre. It is thus a second order Markov effect. Thus one of the major assumptions of the kinetic theory, that reactivity depends only on the end unit, does not apply and the theory must be modified by the use of eight, rather than merely four, propagating rate constants. Thus instead of, for example, the reaction, $\sim\!\!A^* + B \rightarrow \sim\!\!AB^*$, there are now two possibilities, $\sim\!\!BA^* + B \rightarrow$ and $\sim\!\!AA^* + B$, and four rather than two reactivity ratios. The effect is often observed when the monomers contain very bulky or polar substituents, e.g. as found in styrene/acrylonitrile copolymerisation.

PEO Abbreviation for polyethylene oxide.

PEPSIN An enzyme found in the digestive tract, where it acts as a protease. It has a molecular weight of 35 000 and consists of a single polypeptide chain with three disulphide bridges and a phosphate diester bridge. It has a high hydroxy content from serine and threonine groups. It is one of the least selective proteases, causing hydrolysis of polypeptide chains at residues containing non-polar side groups on either side. It is therefore useful in the sequencing of the peptides formed from an initial cleavage of a protein polypeptide by a more specific protease, e.g. trypsin. Like other intestinal proteases, it is produced initially as its zymogen, pepsinogen, which is converted to pepsin by the action of pepsin itself (an example of autocatalysis) at the low pH in the stomach, by the loss of 44 amino acid residues as a mixture of peptides from the N-terminal end.

PEPSINOGEN The inactive zymogen of pepsin, to which it is converted by the action of free pepsin, losing 44 amino acid residues as a mixture of peptides from the N-terminal end.

PEPTIDASE Alternative name for exopeptidase.

PEPTIDE A molecule containing one or more peptide bonds, i.e. amide bonds, between α-amino acid residues. Formed, at least hypothetically, by reaction of an amino group and a carboxyl group of two, often different, α-amino acids. A peptide with two α-amino acid residues (and therefore one peptide bond) is a dipeptide, with three such residues—a tripeptide, etc. In general peptides with several residues are oligopeptides. Peptides with a large number of residues are polypeptides or proteins (if naturally occurring). The exact dividing line between a large peptide and a small protein is not clearly defined. It is often taken as 50 or 100 residues, i.e. a molecular weight of about 5000 to 10 000. Peptide structures are conventionally written from left to right starting with the N-terminal residue and ending with the C-terminal residue, using the usual abbreviations for the amino acid residues. For example the octapeptide bradykinin is Arg–Pro–Pro–Gly–Phe–Ser–Pro–Phe–Arg. Peptides are named from their parent amino acids beginning at the N-terminal residue, e.g. the dipeptide

$$\underset{\underset{CH_3}{|}}{H_2NCHCONH}\underset{\underset{CH_2OH}{|}}{CHCOOH}$$

is alanylserine. Naturally occurring peptides range in size from the tripeptide glutathione to the larger hormones, e.g. oxytocin, a decapeptide and adrenocorticotropic hormone (39 amino acid residues). The latter may be considered either as a large peptide or a small protein.

Peptides may be determined quantitatively spectrophotometrically after conversion to coloured products by the ninhydrin or biuret reactions. Peptides resulting from cleavage (usually partial hydrolysis) of proteins are important in the determination of protein primary structure. If the peptide fragments can be identified then it may be possible to build up a picture of the amino acid residue sequence in the protein. This is aided if hydrolysis can be performed at specific residues, e.g. by the use of trypsin or chymotrypsin. The mixtures of peptides obtained by partial protein hydrolysis may be separated by paper, column or gel permeation chromatography or by electrophoresis. Frequently two-dimensional chromatography is performed producing a peptide map or fingerprint. A similar useful procedure is diagonal electrophoresis. Formation of peptides requires protection of carboxyl and amino groups on the amino acid or precursor peptides involved, whose participation in the peptide bond formation is not wanted. Such protecting groups (X and Y) must be capable of removal after peptide bond formation. Thus a dipeptide may be synthesised by the reactions:

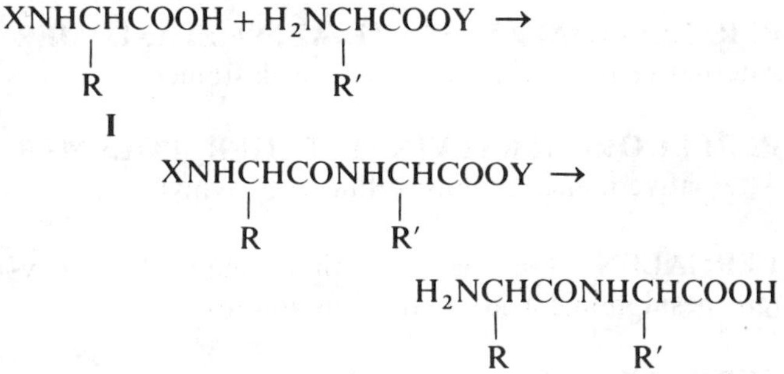

$$\underset{\underset{R}{|}}{\text{XNHCHCOOH}} + \underset{\underset{R'}{|}}{\text{H}_2\text{NCHCOOY}} \rightarrow$$

I

$$\underset{\underset{R}{|}\quad\underset{R'}{|}}{\text{XNHCHCONHCHCOOY}} \rightarrow$$

$$\underset{\underset{R}{|}\quad\underset{R'}{|}}{\text{H}_2\text{NCHCONHCHCOOH}}$$

The reaction is usually facilitated by prior conversion of **I** to a more reactive derivative, e.g. an acid chloride, azide or active ester, or by use of dicyclohexylcarbodiimide. Reaction conditions must minimise the risk of racemisation of the amino acid chiral centres.

Larger peptides are built up by a sequence of similar reaction steps, although the best choice of routes is difficult, e.g. an octapeptide could be made by sequential synthesis (mono → di → tri → ··· → octapeptide) or by linking two tetrapeptides. Thus synthesis of even small peptides is challenging and a lengthy process. Use of solid phase synthesis has considerably speeded up peptide synthesis.

The analysis of mixtures of peptides is frequently performed by use of column chromatography, especially ion-exchange chromatography, and can be performed automatically in an amino acid analyser. When combined with gel permeation chromatography it is especially effective. The structures of the separated individual short peptides are then determined by amino acid analysis, after complete hydrolysis, followed by sequencing by sequential Edman degradations. For longer peptides, the N-terminal and C-terminal residues are determined and then the peptide is hydrolysed to smaller identifiable fragments. If it is itself a hydrolysate, the second hydrolysis must be performed by a different method to the earlier hydrolysis.

PEPTIDE BOND   The amide bond in a polypeptide, i.e. the ⟿NH—CO⟿ bond in the amino acid residues (peptide units)

⟿NH—CHR—CO—NH—CHR'—CO⟿

Owing to delocalisation of an electron from the carbonyl double bond

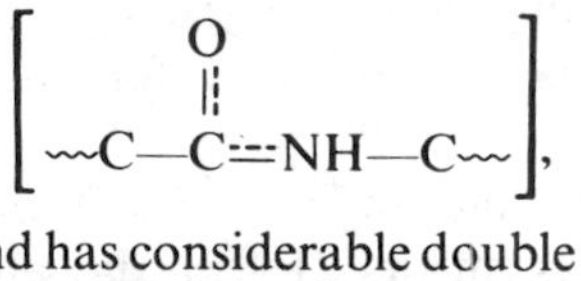

$$\left[\underset{}{\overset{\overset{\text{O}}{\|}}{\text{⟿C—C}\doteq\text{NH—C⟿}}} \right],$$

the peptide bond has considerable double bond character, having a length of about 1·32 Å (compared with the N—C peptide bond length of 1·47 Å and a typical C═N double bond length of 1·25 Å). Thus bond rotation about the peptide bond is very restricted and can be largely ignored, by convention this angle of rotation (ω) is assigned a value of $+180°$ for the *trans* configuration

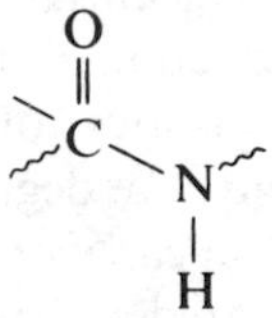

$$\underset{\underset{\text{H}}{|}}{\overset{\overset{\text{O}}{\|}}{\text{C—N}}}$$

which is always found in polypeptides, proteins and peptides. However, for an imino acid residue, e.g. proline, the *cis* and *trans* forms have similar energies and both may be found. All six atoms of the peptide group

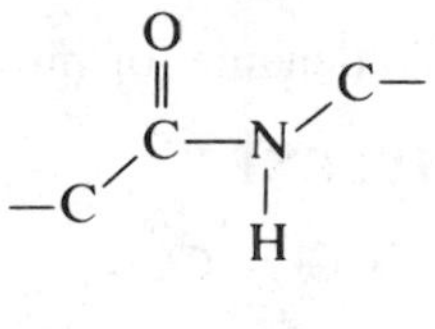

$$\underset{\underset{\text{H}}{|}}{\overset{\overset{\text{O}}{\|}}{\text{—C—C—N—C—}}}$$

are coplanar.

PEPTIDE FINGERPRINT Alternative name for peptide map.

PEPTIDE MAP (Peptide fingerprint) (Fingerprint) The distribution of peptides (as individual spots) on paper, resulting from their separation from a mixture by application of the two-dimensional method of paper chromatography and/or paper electrophoresis. The peptides may be visualised by spraying the paper with ninhydrin or other reagent. From an identical unstained map, the separate peptides may be obtained by cutting out the spots and eluting them. The production of peptide maps can reveal very small differences in the amino acid composition and sequence of proteins (from which the peptide mixtures are produced by partial hydrolysis) e.g. between homologous proteins of different species, and can locate the sites of amino acid replacements in mutant proteins such as in abnormal human haemoglobins.

PEPTIDOGLYCAN (Murein) A linear aminopolysaccharide, crosslinked by peptide bridges, occurring in bacterial cell walls and isolated by treatment with specific enzymes which induce solubilisation of the peptidoglycan. The main units usually present are *N*-acetyl-D-glucosamine and its lactyl ether (muramic acid), L-alanine, D-alanine, D-glutamic acid and L-lysine (or 2,6-diaminopimelic acid). Much structural evidence comes from studies of nucleotide precursors and by enzymatic degradation, e.g. with egg-white lysozyme which acts as a muramidase. The peptide–carbohydrate link in the most studied example from *Micrococcus lysodichleus*, is through L-alanine, the crosslink consisting of five or ten amino acid residues such that the polymer contains one pentapeptide per disaccharide unit. Certain parts of the carbohydrate chain are open, i.e. they contain no peptide side groups.

PEPTISED RUBBER Alternative name for softened rubber.

PEPTISER (Chemical plasticiser) An additive used (at about 0·5 phr) in raw rubber, usually natural rubber, prior to compounding and vulcanisation, in order to promote molecular weight reduction during mastication. Most effective at higher mastication temperatures where peptisers promote oxidative degradation. Some are also radical acceptors and so can work at lower temperatures by stabilising the sheared polymer free radicals which would otherwise recombine. Examples are di-(o-amidophenyl) disulphide, pentachlorothiophenol and its salts and thio-β-naphthol.

PERBUNAN Tradename for nitrile rubber.

PERBUNAN N Tradename for nitrile rubber.

PERCHLOROETHYLENE

$$CCl_2{=}CCl_2 \qquad \text{B.p. } 121°C.$$

A low flammability solvent for hydrocarbon rubbers, polystyrene, polyvinyl acetate and natural resins. Widely used as a textile dry cleaning solvent.

PERDUREN G Tradename for a polysulphide rubber similar to Thiokol B.

PERDUREN H Tradename for a polysulphide rubber similar to Thiokol ST.

PERFECT ELASTICITY Elasticity in which there is a unique relationship between stress and strain, independent of time scale and history of the sample loading, so that the deformation is completely and immediately recoverable. The strain which results from a stress system is only influenced by the current state of the stress system and is not influenced by how long it has been applied and in what manner. Most metals behave thus at low strains but most polymers do not, i.e. their behaviour is time dependent or viscoelastic. Some rubbers approximate to perfect elasticity.

PERFECTLY PLASTIC BEHAVIOUR (Ideal plastic behaviour) An idealised stress–strain behaviour in which either there is ideal elastic behaviour before yielding, after which the stress remains constant with increasing strain (elastic perfectly plastic behaviour) or there is no strain before yielding (rigid perfectly plastic behaviour). This is classical plasticity which may be characterised by the yield stress and the relationship between the magnitudes of the plastic strain increments and the movements of the applied loads.

PERFLUORINATED ELASTOMER (PFE) Alternative name for tetrafluoroethylene–perfluoromethylvinyl ether copolymer.

PERFLUOROALKOXY POLYMER (PFA) (Polyperfluoroalkylvinyl ether) A fluoropolymer containing repeating units of the type

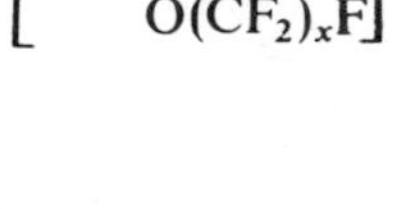

$$\left[\begin{array}{c} CF_2CF{-}{-}{-} \\ | \\ O(CF_2)_xF \end{array} \right]$$

Commercial copolymers, e.g. of tetrafluoroethylene and perfluoropropylvinyl ether (tradename Teflon PFA) or of other perfluoroalkylvinyl ethers (tradenames: Hostaflon TFA, Kalrez) are useful melt processable plastics similar to tetrafluoroethylene–hexafluoropropylene copolymer.

PERFLUOROALKYLENETRIAZINE ELASTOMER Alternative name for fluorotriazine elastomer.

PERFLUOROALKYLVINYL ETHER POLYMER Alternative name for polyperfluoroalkylvinyl ether.

PERGALEN Tradename for the sodium salt of poly-(ethenesulphonic acid) in aqueous solution.

PERGUT Tradename for chlorinated rubber.

PERISTON Tradename for aqueous solutions of poly-(N-vinylpyrrolidone).

PERLON Tradename for a nylon 6 fibre.

PERLON L Tradename for nylon 6.

PERLON N Tradename for nylon 610.

PERLON T Tradename for a nylon 66 fibre.

PERLON U Tradename for an early polyurethane fibre, based on the reaction of hexamethylene diisocyanate with 1,4-butanediol, i.e. the 6,4-polyurethane

$$-\!\!\left[OCONH(CH_2)_6NHCOO(CH_2)_4\right]_n$$

Its T_m value is about 183°C and it has similar mechanical properties to nylon 66, although textiles based on the polymer are rather harsh and stiff.

PERMALON Tradename for a vinylidene chloride copolymer fibre spun from Saran.

PERMANENT SET A deformation which shows no further change with time, after removal of the stress producing it.

PERMEABILITY COEFFICIENT (Permeability constant) Symbol P. The constant relating the rate of transfer of a diffusing substance (F) through unit area of a film or sheet of thickness l to the concentrations of the substance (C_1, C_2) on either side of the sheet, i.e. $F = P(C_1 - C_2)/l$. In the case of gases and vapours the concentrations may be replaced by vapour pressures provided that Henry's law is obeyed. Unfortunately many different units have been used for the permeability coefficient.

The SI units are mol m^{-1} Pa^{-1} s^{-1}, but, in practice, other units such as cm^3 mil (100 m^2 day atm)$^{-1}$ and cm^3 cm (cm^2 s cmHg)$^{-1}$ are used with conversion factors to SI units of $4·91 \times 10^{17}$ and $2·95 \times 10^5$ respectively. The values of P can vary widely depending on the particular gas/polymer being considered. Thus for oxygen, values (in mol $mN^{-1} s^{-1}$) vary from 1.3×10^{-18} for polyvinylidene chloride (a barrier polymer) to $205\,000 \times 10^{-18}$ for silicone rubber, and for water, values vary from 0.7×10^{-15} for polyvinylidene chloride to $14\,500 \times 10^{-15}$ for silicone rubber.

PERMEABILITY CONSTANT Alternative name for permeability coefficient.

PERMITTIVITY (1) Alternative name for absolute permittivity.
(2) Alternative name for relative permittivity.

PERMSELECTIVE MEMBRANE A membrane, such as an ion-exchange membrane, which will only allow passage of one type of ion, anion or cation, through the membrane.

PEROXIDE DECOMPOSER (Peroxide destroyer) A preventive antioxidant which decomposes peroxides and hydroperoxides to non-radical products and hence prevents them dissociating to free radicals and continuing oxidative chain reactions. Many organosulphur compounds, e.g. dilauroylthiodipropionate, are effective especially when combined with a phenolic antioxidant, forming a synergistic mixture.

PEROXIDE DESTROYER Alternative name for peroxide decomposer.

PEROXIDE INITIATOR An organic compound containing the peroxide link (—O—O—) which cleaves on heating or on ultraviolet light irradiation to produce free radicals capable of initiating free radical polymerisation. Free radicals may also be produced at or below ambient temperature by a redox reaction (redox initiation). The wide variety of peroxy compounds available with differing activities make them the most widely used group of initiators capable of giving convenient polymerisation rates at 30–150°C, depending on activity. They are also useful for promoting crosslinking reactions. The main groups are the diacyl peroxides, dialkyl peroxides, alkyl hydroperoxides (e.g. cumene, p-menthane, pinane, t-butyl and cyclohexanone hydroperoxides) and peresters (e.g. t-butylperbenzoate, t-butylperpivalate and diisopropyl-peroxydicarbonate).

PEROXIDE VULCANISATION Vulcanisation of a rubber by heating with a peroxide (ROOR). The peroxide decomposes to yield free radicals which abstract hydrogen from the rubber (R'—H) to give rubber free radicals which then combine to form a crosslink: ROOR → 2RO·; RO· + R'—H → ROH + R'·; 2R'· → R'—R'. With diene rubbers the vulcanisates may be fairly transparent and have low compression set but do not have such high tensile strengths as sulphur vulcanised products. Peroxide vulcanisation can also be used with saturated rubbers and is particularly useful for ethylene–propylene rubber and silicone rubber, especially when using benzoyl peroxide. High temperature decomposing peroxides such as dicumyl peroxide, 2,5-di-t-butyl-2,5-dimethylhexane and di-(t-butylperoxy)-diisopropylbenzene are used with diene rubbers.

PERSISTENCE LENGTH A parameter characterising the correlation of the direction of a polymer chain segment with the preceding segments. It is defined as the average projection of the end-to-end distance of an infinite chain in the direction of the first segment, and is given by $l_{pers} = 1/(1 + \cos\theta)$, where θ is the angle between the segments. It is particularly useful in the analysis of the worm-like chain model.

PERSPEX Tradename for polymethylmethacrylate sheet.

PERSULPHATE A salt of the type $M_2^+ S_2 O_8^{2-}$ where M is usually Na, K or NH_4. Useful as a component of a redox initiator, capable of producing free radicals at low temperatures (0–20°C), e.g. by

$$S_2 O_8^{2-} + Fe^{2+} \rightarrow Fe^{3+} + S\dot{O}_4^- + SO_4^{2-}$$

in aqueous systems, especially for emulsion polymerisation. At higher temperatures (40–70°C) the persulphate may be used alone:

$$S_2 O_8^{2-} \rightarrow 2 S\dot{O}_4^-$$

PES 200P Tradename for a polyethersulphone of structure:

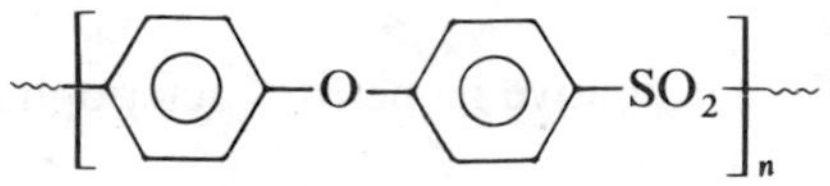

later called Victrex PES.

PES 720P Tradename for a polyethersulphone, similar to PES 200P but containing some

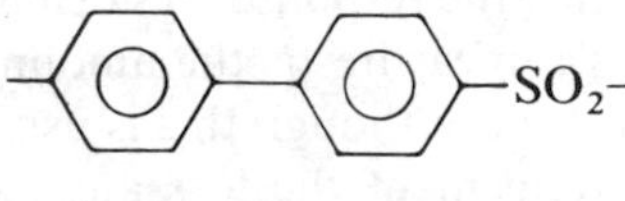

repeat units and therefore having a higher softening point.

PET Abbreviation for polyethylene terephthalate.

PETLON Tradename for a glass-filled polyethylene terephthalate for injection moulding.

PETP Abbreviation for polyethylene terephthalate.

PETROLEUM ETHER (Ligroin) The most volatile liquid fractions of petroleum hydrocarbon distillates, having fairly narrow boiling ranges of, e.g. 40–60°C, 60–70°C, 60–80°C, etc., from 30°C to up to about 120°C. They consist of mixtures of both linear and branched aliphatic hydrocarbons, with small amounts of naphthenes, olefins and aromatics. Used largely for extraction purposes.

PETROLEUM OIL A particular fraction of petroleum, or of cracked petroleum, with a certain boiling range. In a polymer context, the term often refers to the high boiling point fractions used as rubber oils.

PETROLEUM RESIN A low molecular weight polymer, obtained by cationic polymerisation, e.g. with $AlCl_3$ or BF_3 catalyst, of mixtures of olefins obtained from the cracking of petroleum fractions. Aliphatic resins are obtained from the C_4–C_6 fractions containing mixed olefins, isoprene and piperylene, whilst aromatic resins are obtained from higher boiling point (140–200°C) C_8–C_{10} fractions containing styrene, α-methylstyrene, indene, vinyltoluene and dicyclopentadiene. Dicyclopentadiene resins are obtained from fractions which largely contain only this latter monomer. The resins are widely used in adhesives, inks, coatings and rubber compositions, in which they have largely displaced the older coal tar based coumarone–indene resins.

PETROTHENE Tradename for low density and high density polyethylenes.

PEVALON Tradename for polyvinyl alcohol.

PF Abbreviation for phenol–formaldehyde polymer.

PFA Abbreviation for perfluoroalkylvinyl ether polymer.

PFE Abbreviation for perfluorinated elastomer.

PHAGE Alternative name for bacteriophage.

PHASE ANGLE (Loss angle) Symbol δ. In a sinusoidally varying disturbance of stress, strain or electric field, which causes a corresponding sinusoidally varying response of strain, stress or dielectric polarisation respectively, the phase angle is the amount, measured in radians, by which the response lags behind the disturbing influence. It is a measure of the amount of energy loss occurring per cycle, although this is usually expressed in terms of the tangent of the loss angle ($\tan \delta$). For low damping $\tan \delta \approx \delta$. If dynamic mechanical behaviour is represented vectorially, with both the stress and strains as rotating vectors, then for a viscoelastic material, the latter will lag behind the former by the angle δ. For dielectric behaviour, the current and voltage are out of phase by $\pi/2$ for a perfect dielectric with no loss, and the phase angle is the angle by which the phase difference between voltage and current differs from $\pi/2$.

PHASE CONTRAST MICROSCOPY An optical microscopy technique of enhancing the contrast in a specimen. The specimen is illuminated with a hollow cone of light and the transmitted light passes through a phase plate which phase shifts the light scattered by the specimen so that it interferes constructively or destructively with the unshifted directly transmitted beam. Thus a light or dark image of the scattering areas is produced. It is especially useful for examining polymer blends and other two phase samples. The phase contrast is greater, the greater the refractive index differences between the phases.

PHASE INVERSION The process by which, in a two phase system, the dispersed phase becomes the continuous phase (the matrix) and vice versa. This may occur as a result of a change in temperature or in chemical composition, as occurs in the formation of high impact polystyrene by polymerisation of styrene in the presence of a polybutadiene rubber.

PHB Abbreviation for polyhydroxybutyrate.

PHE Abbreviation for phenylalanine.

PHENODUR Tradename for phenol–formaldehyde polymer.

PHENOL

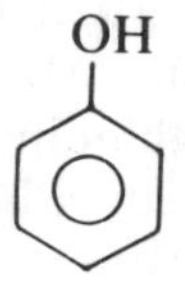

M.p. 40·9°C. B.p. 181·8°C.

Originally obtained from coal tar but now almost exclusively produced by various synthetic processes of which the cumene process is the most common. In this process benzene is alkylated with propylene to cumene, which is then air oxidised to cumene hydroperoxide, followed by its cleavage with sulphuric acid to produce phenol and acetone:

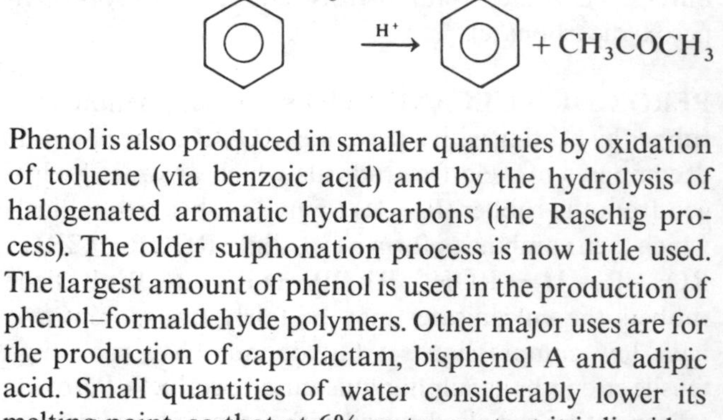

Phenol is also produced in smaller quantities by oxidation of toluene (via benzoic acid) and by the hydrolysis of halogenated aromatic hydrocarbons (the Raschig process). The older sulphonation process is now little used. The largest amount of phenol is used in the production of phenol–formaldehyde polymers. Other major uses are for the production of caprolactam, bisphenol A and adipic acid. Small quantities of water considerably lower its melting point, so that at 6% water content it is liquid at room temperature. It is readily soluble in many organic solvents but not in aliphatic hydrocarbons. When exposed to air a pink colour develops, the change being catalysed in the presence of iron and copper, when red colours can develop.

PHENOL ALCOHOL Alternative name for methylolphenol.

PHENOL–ARALKYL POLYMER A polymer produced by Friedel–Crafts reaction between an aralkylether and a phenol:

$$n \langle\bigcirc\rangle\!-\!OH + n \ CH_3OCH_2\!-\!\langle\bigcirc\rangle\!-\!CH_2OCH_3 \xrightarrow{SnCl_4}$$

$$\left[\langle\bigcirc\rangle\!-\!CH_2\!-\!\langle\bigcirc\rangle\!-\!CH_2\right]_n + 2n \ CH_3OH$$

The polymers may be cured with hexamine or epoxy resins to hard thermoset products, which may be considered as high temperature resistant types of phenol–formaldehyde polymers. The commercial Xylok resins are of this type.

PHENOL–FORMALDEHYDE POLYMER (PF) (Phenolic resin) (Phenoplast) Tradenames: Alberit, Alresin, Alvanol, Asplit, Bakelite, Beckacite, Catalac, Catalin, Cellobond, Duraphen, Durez, Durite, Dyphene, Epok, Fiberite, Fluosite, Genal, Kynol, Luphen, Metholan, Mouldrite, Nestorite, Norsophen, Phenodur, Plenco, Plyophen, Resinox, Rockite, Sirfen, Sternite, Tego, Trolitan, Varcum. The commonest type of phenolic resin. A condensation product of reaction of phenol and formaldehyde. The earliest synthetic polymer to be a manufactured as a commercial product and still a major product useful as a thermoset plastic, an adhesive and coating material. The structure of the final crosslinked product is highly complex and despite numerous studies the mechanism of its formation is still not fully understood.

The course of the reaction of phenol with formaldehyde is influenced by temperature, ratio of phenol to formaldehyde and, particularly, the pH. Under acid conditions in aqueous solution, electrophilic substitution on the phenol occurs to give transient o- and p-methylolphenols which react with further phenol to form dihydroxydiphenylmethanes. The reaction continues to form a complex mixture of linear and branched polynuclear compounds called novolacs (A-stage resin). On further heating with a substance capable of forming methylene bridges, usually hexamethylenetetramine, novolacs crosslink to form a network polymer (the C-stage resin).

Alternatively if the reaction of phenol and formaldehyde is carried out under alkaline conditions with an excess of formaldehyde, initially a soluble, fusible resole is formed as the A-stage resin. Again methylolphenols are first formed, but they are much more stable under the basic conditions so that polyhydroxy compounds are produced by further methylolation. Further condensation to polynuclear phenols occurs, largely by methylene bridge formation, to produce the resole which typically contains 1–5 rings in its A-stage molecules. After

neutralisation the resole may be crosslinked simply by heating at up to about 150°C to give a resit as the C-stage resin. At slightly higher temperatures methylene bridge formation predominates giving dibenzyl ether structures in the cured product.

Above about 160°C quinonemethide and other structures are also formed and are responsible for the dark colour found in many cured products manufactured by moulding, which restricts the range of applications of the polymer. In the initial stages of crosslinking the polymer passes through an insoluble, but fusible, rubbery stage, sometimes termed the B-stage, which in the case of resole crosslinking is called a resitol.

A-stage polymers are either viscous fluids or amorphous brittle solids, whilst the fully crosslinked C-stage polymers are hard and quite brittle infusible solids. In colour they vary from pale yellow to dark brown, depending largely on the temperature of crosslinking. The novolacs tend to be lighter in colour. The cured products typically have a tensile strength of 50–70 MPa and an Izod impact strength of 0.12–$0.25 \ J \ (12.7 \ mm)^{-1}$. However, the products usually contain a filler, such as woodflour, cotton fibre or a fabric laminate, when the mechanical properties may be much improved. The polymers are very resistant to chemical attack, the cresol- and resorcinol–formaldehyde polymers having even better resistance than the simple phenol–formaldehyde polymer. The polymers have reasonably good thermal stability, at least for short term use. They are very widely used as moulding materials for electrical and mechanical products (usually reinforced with woodflour or chopped cotton) and in paper and cotton fabric resin impregnated laminates. Products modified with drying oils or rosin are used as oil soluble coating materials. Use of the polymers as adhesives is also important.

PHENOLIC RESIN Alternative name for phenol–formaldehyde polymer.

PHENOPLAST Alternative name for phenol–formaldehyde polymer.

PHENOXYETHYL OLEATE

$$CH_3(CH_2)_7CH\!=\!CH(CH_2)_7COOCH_2CH_2O\!-\!\langle\bigcirc\rangle$$

A primary plasticiser for cellulose acetate–butyrate and a secondary plasticiser for polyvinyl chloride.

PHENOXY RESIN A polyester formed by reaction of a dihydric phenol and epichlorhydrin, but linear and of much higher molecular weight than the similar epoxy resins. The range of commercial products is based largely on bisphenol A, the polymers having the structure.

$$\left[O\!-\!\langle\bigcirc\rangle\!-\!\overset{\overset{\textstyle CH_3}{|}}{\underset{\underset{\textstyle CH_3}{|}}{C}}\!-\!\langle\bigcirc\rangle\!-\!O\!-\!CH_2\overset{}{\underset{\underset{\textstyle OH}{|}}{C}}HCH_2\right]_n$$

A two-stage synthetic method is used. In the first stage, excess epichlorhydrin in MEK solution is used and a low molecular weight epoxy resin is formed. This is then reacted with an equimolar amount of bisphenol. Alternating copolymers can be produced using different bisphenols in the two stages. T_g values are in the range 80–180°C, the polymer from bisphenol A having a value of 100°C. The polymers may be modified, e.g. by esterification, or crosslinked through the —OH group. They are soluble in many solvents and resistant to attack by acids and alkalis. The linear polymer is rigid and has reasonable impact strength. The polymer is used for primer coatings, which have excellent adhesion to metals and good compatibility with top-coat film formers.

PHENYLALANINE   (Phe) (F)

$$H_2NCH(CH_2C_6H_5)COOH$$

M.p. 283°C (decomposes).

An aromatic non-polar α-amino acid widely found in proteins. Its pK' values are 1·83 and 9·13, with the isoelectric point at 5·48. Either combined in a protein or as a free acid it may be readily quantitatively determined by its strong ultraviolet light absorption. Phenylalanine residues are cleaved by chymotrypsin and are often associated with the β-conformation.

PHENYLATED POLYPHENYLENE   Alternative name for polyphenylphenylene.

PHENYLENE-   Prefix for the group

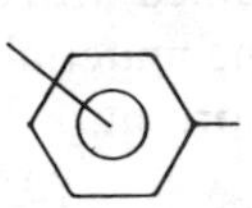

present in most aromatic polymers. Most frequently the rings are *para*-substituted.

m-PHENYLENEDIAMINE

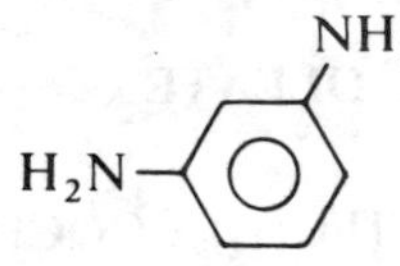

M.p. 62°C.

A useful curing agent for epoxy resins, giving products with good chemical resistance and high heat distortion temperatures. However, it oxidises with discolouration in air, and stains skin and clothing as well as being an irritant.

PHENYLENE POLYMER   In general, an alternative name for a polymer containing phenylene links in the polymer chain, i.e. an alternative name for poly-(phenylene...). In particular, an alternative name for polyphenylene itself.

PHENYLETHYLENE   Alternative name for styrene.

PHENYLGLYCIDYL ETHER

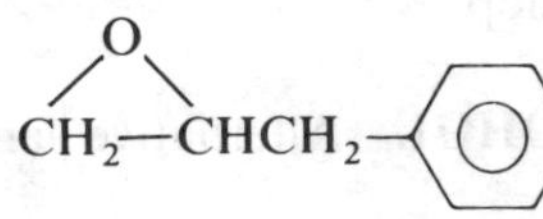

A reactive diluent for epoxy resins.

PHENYLISOTHIOCYANATE

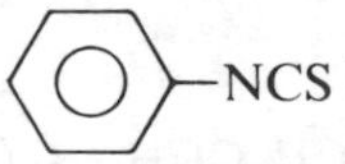

The reagent used for the reaction of the N-terminal amino acid of a peptide or protein in the Edman method for the determination of the N-terminal residue and also for sequence studies.

N-PHENYL-β-NAPHTHYLAMINE   (PBN) (PBNA)

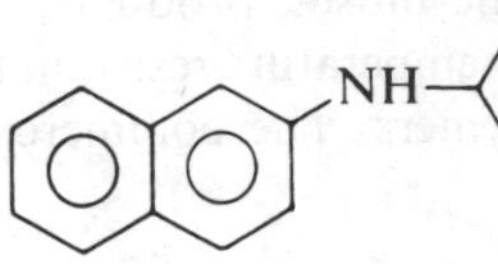

M.p. 108°C.

Tradenames: Agerite Powder, Nonox D, Antiox 116.
A chain-breaking amine antioxidant very widely used in diene rubbers. Cheap and effective but staining. Used much less than formerly due to its possible toxicity and the toxicity of β-naphthylamine sometimes present as an impurity.

PHENYLON   Tradename   for   poly-(m-phenylene-isophthalamide).

PHENYLPOLYETHYLENE GLYCOL PALMITATE

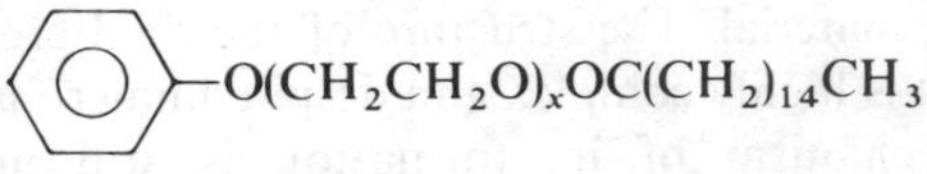

A plasticiser for cellulose acetate–butyrate, with good heat and light stability. Also used in synthetic rubbers and polyvinyl chloride.

PHENYL SALICYLATE

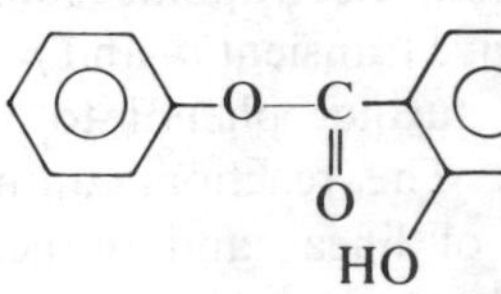

M.p. 43°C. B.p. 173°C/12 mm.

Tradename: Salol. This, and other alkyl phenyl salicylates, are useful ultraviolet absorbers for the ultraviolet stabilisation of polymers, especially cellulose esters. However, they themselves yellow on exposure to ultraviolet light due to the photo-Fries rearrangement:

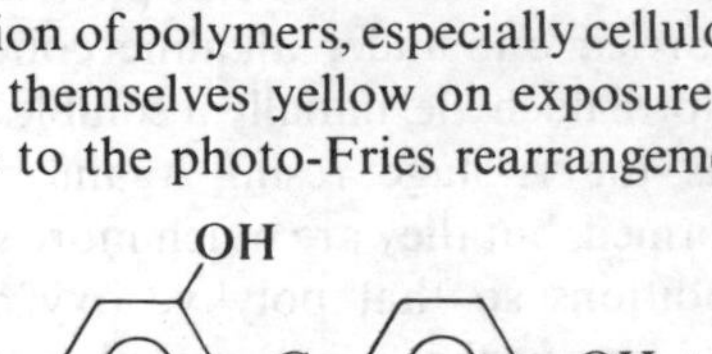

the 2,2'-dihydroxy isomer.

PHENYLSILICONE ELASTOMER (Methylphenylsilicone elastomer) (PMQ) A silicone elastomer based on polydimethylsiloxane but in which some of the methyl groups, 5–10%, have been replaced by phenyl groups. This reduces crystallisation of the dimethyl polymer, which occurs at about $-50°C$, and so lowers the low temperature brittle point, to a minimum of $-115°C$, at 7.5% phenyl content. The phenyl rich copolymers also have increased resistance to high energy radiation. Commercial products also usually contain vinylsiloxane units to help vulcanisation and are therefore designated PVMQ.

PHENYLTHIOCARBAMOYL AMINO ACID Alternative name for phenylthiocarbamoyl derivative.

PHENYLTHIOCARBAMOYL DERIVATIVE (Phenylthiocarbamoyl amino acid) (PTC derivative) The initial derivative of general formula

formed in the Edman method of analysis of peptides and proteins by reaction of phenylisothiocyanate with the N-terminal amino acid. It is subsequently converted to a thiazolinone and then to phenylthiohydantoin by treatment with acid.

PHENYLTHIOHYDANTOIN DERIVATIVE (PTH)

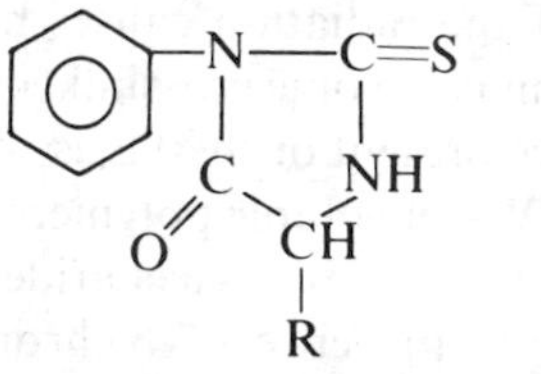

The final product of the Edman method of identifying the N-terminal amino acid of a peptide or protein by reaction with phenylisothiocyanate.

PHENYLVINYL KETONE

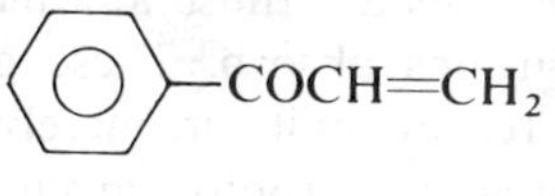

B.p. 116°C/18 mm.

Prepared by the Mannich reaction of acetophenone, formaldehyde and an amine hydrochloride:

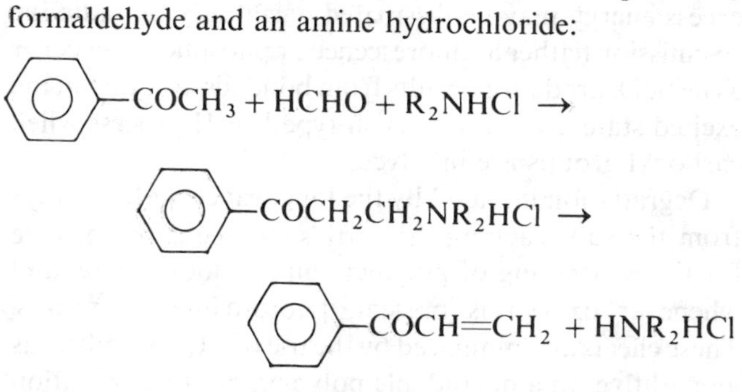

It is polymerised by free radical and anionic mechanisms to poly-(phenylvinyl ketone). Occasionally it is used as a comonomer in the preparation of certain lacquers and adhesives.

PHILLIPS PROCESS A low pressure polymerisation process for producing polyethylene. Typically ethylene is polymerised in solution in a hydrocarbon solvent at about 30 atm pressure and 130°C using a reduced chromium oxide catalyst supported on an inert material such as alumina. The process yields the most highly linear of all the commercial polyethylenes which therefore has the highest crystallinity and density (usually about 0.96 g cm^{-3}).

PHILPRENE Tradename for styrene–butadiene rubber.

PHOSGENE (Carbonyl chloride)

$$COCl_2 \qquad \text{B.p. } 8°C.$$

Prepared by the reaction of carbon monoxide with chlorine at about 200°C. A toxic gas, useful as the monomer for the formation of polycarbonates, especially bisphenol A polycarbonate, and in the production of diisocyanate monomers by reaction with diamines:

$$H_2NRNH_2 + COCl_2 \rightarrow OCNRNCO + 2HCl$$

PHOSPHAZENE POLYMER (Polyphosphazene) (Phosphonitrilic polymer) A polymer of repeat unit $\{PX_2=N\}_n$, where X may be halogen (if X = Cl, the polymer is polydichlorophosphazene—the most studied phosphazene polymer), alkyl, aryl, alkoxy or aryloxy. The latter types are frequently produced by appropriate substitution of the chlorines on polydichlorophosphazene. These types are the most useful since the dichloropolymer is highly susceptible to hydrolysis by atmospheric moisture. The perfluoroalkoxy polymers (X = OCH_2CF_3 or $OCH_2C_3F_7$) and especially copolymers containing both types of substituent, are useful rubbers—the fluorophosphazene rubbers. Other polymers, such as when X = CF_3, are not elastomeric and have high softening points. The polymers are synthesised by nucleophilic substitution reaction in solution, e.g. in tetrahydrofuran, on the carefully prepared linear and therefore soluble dichloropolymer, e.g. with alkoxide ion.

PHOSPHINATE POLYMER

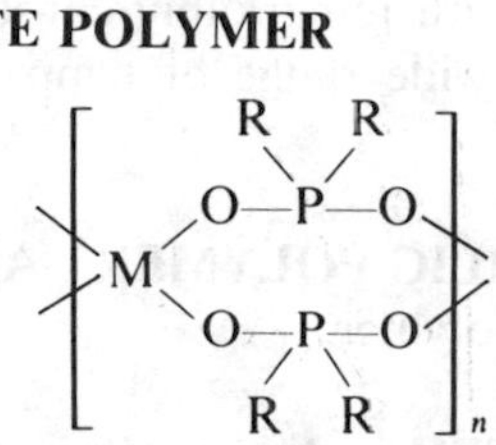

A coordination polymer of the general structural type shown, i.e. a spiro polymer containing double metal phosphinate bridges. Partly due to this spiro-ladder structure, such polymers have good thermal stability. In general they are prepared by reaction of dialkyl, diaryl or alkylaryl phosphinic acid with a metal salt, usually the acetate or acetylacetanoate. Metals investigated have included Zn, Co, Cr, Fe and Be (which gives polymers with the best thermal stability).

PHOSPHINOBORANE POLYMER A polymer of structure $\{PR_2^+ - \bar{B}R'_2\}_n$. It may be prepared by reaction of a borane with a phosphine, e.g.

$$PH_3 + (CH_3)_2BBr \xrightarrow{NEt_3} \{PH_2B(CH_3)_2\}_n + NEt_3HBr$$

The polymers have good thermal stability (to $> 220°C$).

PHOSPHOCELLULOSE (P-cellulose) Cellulose in which some hydroxyls have been replaced by phosphate groups. It is useful as a cation-exchange resin in ion-exchange chromatography, for example, of proteins.

PHOSPHOMANNAN A phosphorylated polysaccharide obtained from yeasts of the *Hansenula* genus consisting of D-mannose oligosaccharide units joined through phosphodiester links.

PHOSPHONITRILIC CHLORIDE TRIMER Alternative name for hexachlorocyclotriphosphazene.

PHOSPHONITRILIC FLUOROELASTOMER (Fluorophosphazene rubber) Tradenames: PNF, Eypel. A fluoroelastomer prepared by reaction of polydichlorophosphazene (I) with a mixture of sodium trifluoroethoxide and heptafluorobutoxide to give a polyfluoroalkoxyphosphazene:

$$\begin{bmatrix} Cl \\ | \\ P=N \\ | \\ Cl \end{bmatrix}_n + \begin{Bmatrix} n \; NaOCH_2CF_3 \\ n \; NaOCH_2(CF_2)_2CF_2H \end{Bmatrix} \rightarrow$$

I

$$\begin{bmatrix} OCH_2CF_3 \\ | \\ P=N \\ | \\ OCH_2(CF_2)_2CF_2H \end{bmatrix}_n + 2n \; NaCl$$

The rubber has a low T_g value of $-68°C$ and, unlike **I**, it has excellent hydrolytic stability. It also has excellent chemical, solvent and fire resistance. It may be vulcanised with peroxides or sulphur or by high energy radiation to give vulcanisates with remarkably constant mechanical properties over a wide range of temperature (-50 to $+200°C$).

PHOSPHONITRILIC POLYMER Alternative name for phosphazene polymer.

PHOSPHOPROTEIN A protein containing phosphorus, bound to the protein as phosphate ester through reaction with the hydroxyl group of the hydroxy amino acid residues, e.g. serine and sometimes threonine. Phosphoproteins occur mostly in milk, e.g. casein, and in eggs, e.g. lipovitellin, and usually contain about 1% phosphorus, which is readily cleaved with alkali or with enzymes. Phosphoproteins are often found as complexes with metals. Other phosphoproteins are found in other organs, e.g. the brain, membranes, cell nuclei, bones and blood.

PHOSVITIN A water soluble egg yolk phosphoprotein, with a high phosphorus content, about 10%. Preparations are electrophoretically homogeneous. It contains a large amount of serine (typically about 50%) with the phosphorus bound as phosphate and often as runs of phosphoserine residues. Its biological role is not clear.

PHOTOCHEMICAL POLYMERISATION Alternative name for photopolymerisation.

PHOTOCONDUCTIVE POLYMER A polymer which exhibits a relatively high electrical conductivity when irradiated with visible or ultraviolet light. Such polymers are of interest as forming the basis of electroimaging processes such as xerography, and in other photoelectric devices. The most widely studied polymer is poly-(*N*-vinylcarbazole) and its charge-transfer complexes, especially with 2,4,7-trinitrofluorenone.

PHOTOCROSSLINKING Crosslinking brought about by irradiation with ultraviolet or visible light. Polymer chains may be photocrosslinked if they contain pendant side groups which will dimerise under the action of light, e.g. cinnamic acid units or anthracene units. Alternatively, if the radiation causes hydrogen abstraction, the chains may become crosslinked. This can occur if stilbene units are present or by the use of a sensitiser, e.g. benzophenone. Water soluble polymers such as polyvinyl alcohol, proteins and polysaccharides, often become crosslinked in the presence of bichromate ion. Photocrosslinking is the basis for the production of photoresists.

PHOTODEGRADATION Degradation induced by exposure to ultraviolet (UV) radiation. The most damaging wavelengths are those at which the polymer UV absorbing groups absorb. These groups may be present in the repeat unit or merely as structural irregularities (often as carbonyl groups) at very low concentrations. UV absorption often involves electronic excitation in an unsatured group to an excited state. The excess energy may be dissipated harmlessly by radiative re-emission (either as fluorescence or phosphorescence) or as heat. Degradation results from bond dissociation in the excited state, e.g. by a Norrish type I or II process when carbonyl groups are involved.

Degradation initiated by the long wavelength UV rays from the sun reaching the earth's surface is responsible for the weathering of polymers out-of-doors. Here, and whenever oxygen is present, photo-oxidation results. These effects are minimised by the use of a UV stabiliser as an additive. In a degradable polymer, photodegradation is deliberately induced by incorporation of a photosensitiser.

PHOTOELASTICITY The property of a transparent material becoming birefringent when subject to stress. The effect is quantitatively described by the stress optical law. The birefringence arises from the material becoming anisotropic due to micro-orientation of the molecules on stressing. The photoelastic effect is especially useful in the technique of photoelastic stress analysis and also in the examination of residual stress and orientation in transparent plastic products by observing them in polarised light. If white light is used, then a series of coloured fringes is observed whose density depends on the amount of stress or orientation, due to interference between the two out of phase propagating light rays. For monochromatic light a series of light and dark fringes is observed.

PHOTOELASTIC STRESS ANALYSIS A technique for the determination of the stress components at any point in a stressed object by viewing a model of the object, constructed from a transparent plastic, with polarised light and observing the resultant interference fringes which arise from the photoelastic effect. Two variants of the method are used. In stress freezing, a three-dimensional scale model is first heated to anneal it and to remove any stresses present when not under external loading. Then the stress system it is desired to study is applied and the model is cooled whilst still being stressed in order to 'lock-in' the stresses and associated strains. Slices of material from the model are then examined in a polariscope. Alternatively, in the stress coating technique, the object itself is examined by coating it with a transparent plastic and subjecting it to the stress system of interest. The order (f) of each of the interference fringes observed in the polariscope, which arise from the birefringence (Δn) of the photoelasticity of the plastic, is determined. It is related to the birefringence through the relation $\Delta n = \lambda f/t$, where λ is the wavelength of the monochromatic light used, and t is the slice or coating thickness. The stress, or more precisely the difference in the principal stress components, may then be calculated using the stress optical law, provided the stress optical coefficient (C) of the plastic is known. The preferred plastics are epoxy resins and polycarbonate due to their high C values.

PHOTOINITIATED POLYMERISATION Alternative name for photo-polymerisation.

PHOTOINITIATION Initiation of free radical polymerisation by means of irradiation with ultraviolet light, causing photo-polymerisation.

PHOTOMETER Alternative name for light scattering photometer.

PHOTO-OXIDATION Photodegradation occurring in the presence of oxygen. Since photodegradation is frequently a free radical process, oxygen will also participate, accelerating the degradation processes (mainly chain scission) by promoting chain branching via hydroperoxide group formation and causing the incorporation of oxygen containing groups into the polymer. The main cause of the weathering of polymers. Alleviated by the use of an ultraviolet stabiliser.

PHOTO-POLYMERISATION (Photoinitiated polymerisation) (Photochemical polymerisation) Free radical or occasionally ionic, polymerisation initiated by the interaction of light, usually of ultraviolet wavelengths, with a photosensitive compound, producing free radicals. The compound may be the monomer itself—styrene has been particularly widely studied—which absorbs a photon to give an excited state, which itself then dissociates to free radicals: $M \rightarrow M^* \rightarrow R\cdot + R'$. The rate of initiation (R_i) is then $2\phi I_a$, where ϕ is the quantum yield for radical production and I_a is the absorbed light intensity which equals $\varepsilon I_0[M]$, where ε is the monomer extinction coefficient and $[M]$ its concentration. The rate of polymerisation is,

$$R_p = k_p[M](\phi I_a/k_t)^{1/2} = k_p[M][\phi I_0(1 - e^{-\varepsilon[M]b})/k_t]^{1/2}$$

where k_p and k_t are the propagation and termination rate constants respectively and b is the sample cell thickness.

The photosensitive compound may also be a free radical initiator (I) which is dissociated by absorption of a photon to produce free radicals at much lower temperatures than by thermal homolysis. R_p is then given by the same equation, but with $[I]$ replacing $[M]$ in the second term. Alternatively a photosensitiser (S) may be used (photosensitised polymerisation) which absorbs a photon and transfers its excess energy to monomer or initiator, which itself then becomes excited and dissociates to free radicals. Benzophenone and various dyes, e.g. eosin and fluorescein, have been used in this way. Here R_p is also as above but with $[S]$ replacing $[M]$ in the second term.

Photopolymerisation is of great use in the determination of the individual rate constants k_p and k_t, by comparison of rates in steady-state and non-steady-state conditions using the rotating sector method. It is also useful commercially in the fast curing of coatings and for the photocrosslinking of photoresists. Photopolymerisation should strictly be termed photoinitiated polymerisation, except when absorption of light is necessary for each propagation step. Such polymerisations are rare, but do occur with monomers of the type

$$R_1-CH=CH-R_2-CH=CH-R_1$$

which photo-cyclopolymerise in the solid state by topotactic polymerisation.

PHOTORESIST A photosensitive polymer system which, when applied as a coating to a substrate, after interaction with ultraviolet or visible light undergoes a change in solubility. This is usually an insolubilisation caused by photocrosslinking often by the use of a polymer with cinnamic acid or its derivatives as side groups. If the irradiation is performed through a suitable mask, only selected areas are insolubilised and the unexposed areas may be dissolved away leaving a raised image. Such a process is useful in lithography for reprographics and the production of integrated circuits.

PHOTOSENSITISED POLYMERISATION Photopolymerisation initiated by a photosensitiser.

PHOTOSENSITISER A compound or chemical group capable of promoting a photochemical reaction. In polymers, photo-polymerisation, photocrosslinking and photodegradation may be photosensitised. The sensitiser must absorb the radiation involved (ultraviolet or visible light) and be capable of transferring its excitation energy to the molecules, in which it is desired to initiate the free radical photo-reaction of interest. Many carbonyl containing organic molecules (e.g. benzophenone) and inorganic compounds (e.g. zinc oxide) are photosensitisers.

PHOTOSTABILISER Alternative name for ultraviolet stabiliser.

PHR Abbreviation for parts per hundred of rubber (or resin), i.e. of polymer, by weight. Commonly used as the way of expressing the composition of a rubber or plastic compound, when the amounts of the different additives used are given in phr. Thus, for example, a PVC composition containing 50 phr plasticiser and 20 phr filler would contain 50 g plasticiser and 20 g filler for every 100 g of polymer.

PHTHALATE An ester of one of the isomers of phthalic acid. If the particular isomer is not specified then it is assumed that the *ortho* isomer is implied, as in the naming of the very wide range of phthalate esters used as plasticisers.

PHTHALATE 79 Alternative name for heptylnonyl phthalate.

PHTHALIC ANHYDRIDE

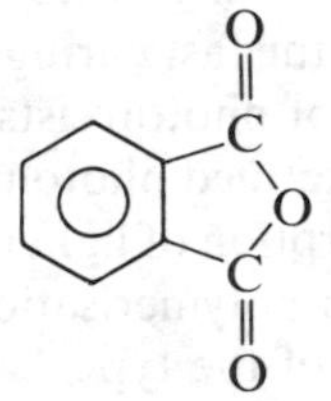

M.p. 131°C.

Produced by the catalytic oxidation of *o*-xylene or naphthalene, e.g. at about 500°C using a vanadium or molybdenum oxide catalyst. Widely used as the modifying acid in the formation of unsaturated polyester resins to space out the unsaturated groups and so reduce the crosslink density in the cured product. It gives polyesters which are highly compatible with styrene and hard, rigid cured products. Useful as a curing agent for epoxy resins, especially for the production of large electrical castings. It cures at 120–130°C, giving products with a relatively high heat distortion temperature (up to 110°C) and good electrical and mechanical properties.

PHTHALOCYANINE POLYMER Alternative name for polyphthalocyanine.

PHYLLOSILICATE Alternative name for layer silicate.

PHYSICAL CROSSLINK (Virtual crosslink) The existence of a restraining force between polymer chains, brought about by means other than covalent bonding, which is considered as conventional or chemical crosslinking, between the chains. Most simply it arises due to entanglements between chains. In thermoplastic elastomers several types of such crosslinks may occur. In block copolymers in which one block type separates into domains, the domains containing the hard segments act as physical crosslinks between the elastomeric blocks of the second type forming the matrix. The presence of a second crystalline polymer, simply dispersed as a polymer blend in an elastomeric matrix, can also produce a physical crosslinking effect. In all these cases the crosslinks are destroyed on heating and reform on cooling and are therefore thermoreversible crosslinks. The effects on polymer properties are similar to chemical crosslinking—higher stiffness, lower creep and greater elastic recovery.

PHYTOGLYCOGEN A plant polysaccharide having a structural similarity to glycogen and similar properties. An example is the phytoglycogen from sweetcorn, which is a water soluble glycan with an average unit chain length of 13, reflecting a high degree of branching.

PIB Abbreviation for polyisobutene.

PIBITER Tradename for polybutylene terephthalate.

PICCOUMARON Tradename for coumarone–indene resin.

PICM Abbreviation for 4,4'-dicyclohexylmethane diisocyanate.

PICS Abbreviation for pulse induced critical scattering.

PIEZODIALYSIS Alternative name for negative reverse osmosis.

PIEZOELECTRIC POLYMER A polymer whose polarisation changes under strain, due either to a change in dimensions or to electrostriction (strain dependence of the permittivity). Polymer electrets are piezoelectric. Polyvinylidene fluoride has been particularly widely studied and has been used in a variety of electrical devices such as microphones and ultrasonic transducers.

PINANE HYDROPEROXIDE A mixture of hydroperoxides of structure:

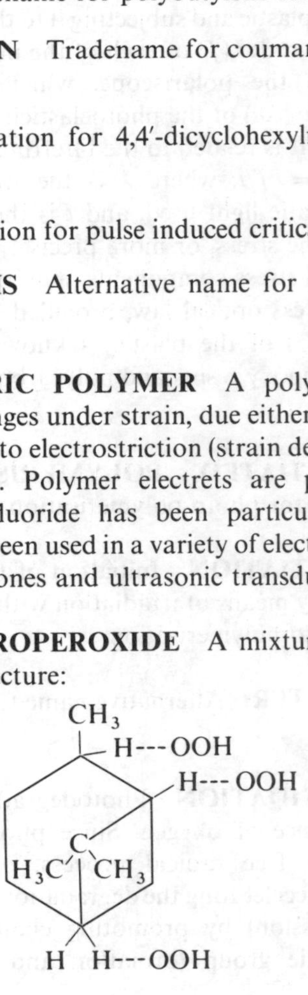

with hydroperoxide groups replacing hydrogen atoms at any of the positions shown. It has half-life times of 80 h/120°C and 5 h/150°C and half-life temperatures of 141°C/10 h, 169°C/1 h and 229°C/1 min. Sometimes used as a component of a redox initiator in emulsion polymerisation.

α-PINENE RESIN A terpene resin obtained by cationic polymerisation of α-pinene, the reaction possibly proceeding as:

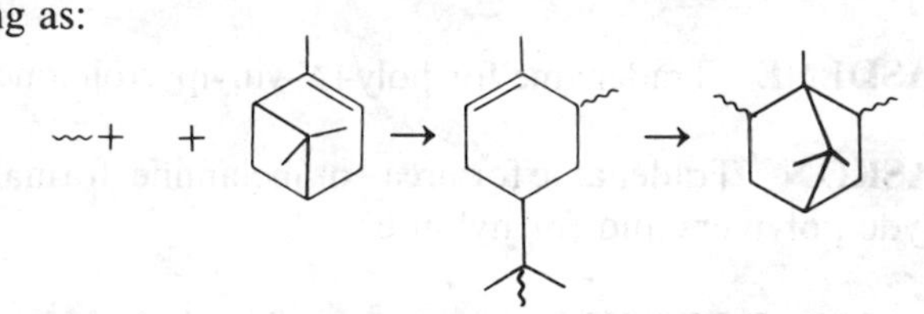

It is useful as a tackifier in styrene–butadiene and styrene–butadiene–styrene rubbers and in solvent based and hot melt adhesives. Typical molecular weight is about 800.

β-PINENE RESIN A terpene resin obtained by cationic polymerisation of commercial β-pinene, which also contains α-pinene and dipentene. The polymer consists mostly of 1,4-linked units formed via:

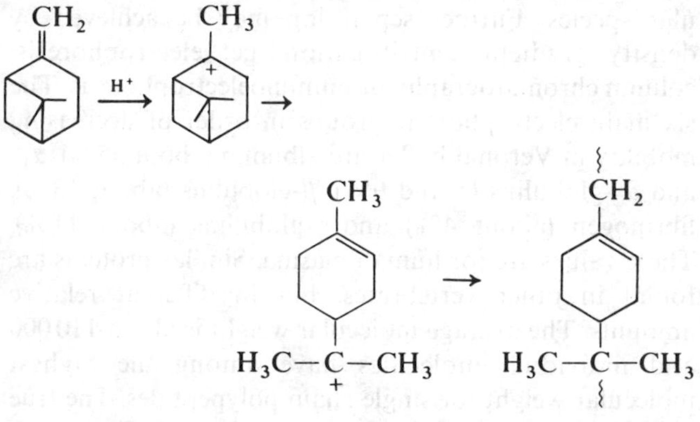

Typical commercial resins have a number average molecular weight of about 1000 and are mostly used in rubber solvent based adhesives.

PINNER–STABIN OSMOMETER (Stabin osmometer) A modified version of the Zimm–Meyerson osmometer with the filling tube extended to the bottom of the cell, enabling it to be filled and emptied in the vertical position. It is also completely enclosed in a glass jacket to reduce solvent evaporation when used at high temperature.

PIOLOFORM B Tradename for polyvinyl butyral.

PIOLOFORM F Tradename for polyvinyl formal.

PIPERAZIDINONE Alternative name for 2-piperidone.

PIPERAZINE

B.p. 145°C.

Synthesised by cyclisation of monoethanolamine, $HOCH_2CH_2NH_2$, in the presence of amine. Sometimes used as a monomer in the synthesis of nylons, i.e. for piperazine polyamides. 2,5-Dimethylpiperazine is most frequently used.

PIPERAZINE POLYAMIDE A polyamide formed by reaction of piperazine or a substituted piperazine (often trans-2,5-dimethylpiperazine) and a diacid, or more usually a diacid chloride, by a low temperature solution polymerisation:

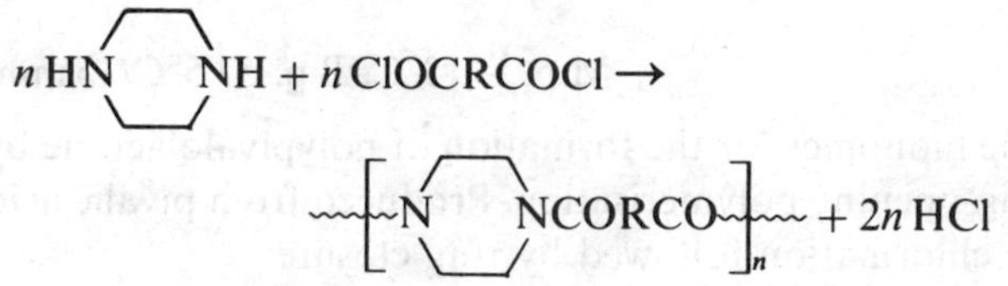

Widely studied examples are poly-(terephthaloyl-2,5-transdimethylpiperazine) and poly-(adipolypiperazine). The polymers have high T_m values (greater than 400°C and about 350°C for the above polymers respectively), are often soluble in formic acid, chloroform or dimethylformamide, but are more water sensitive than most polyamides.

PIPERAZINONE Alternative name for 2-piperidone.

PIPERIDINE

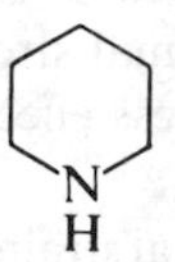

B.p. 106°C.

Useful as a curing agent for epoxy resins, especially for castings of large objects.

2-PIPERIDONE (α-Piperidone) (Piperazinone) (Piperazidinone) (δ-Valerolactam)

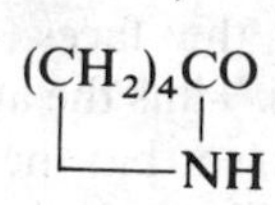

M.p. 39–40°C.

The monomer from which nylon 5 is formed by ring-opening polymerisation. Synthesised by performing the Beckman rearrangement on cyclopentanone oxime. Cyclopentanone is formed by reaction of butadiene and formaldehyde.

α-PIPERIDONE Alternative name for 2-piperidone.

PIPERYLENE (1-Methyl-1,3-butadiene) (1,3-Pentadiene)

$$CH_3CH{=}CHCH{=}CH_2 \qquad B.p. 42°C.$$

Exists in both the cis and trans forms. Obtained either from the C_5 fraction from cracking petroleum naphtha or as a by-product from the production of isoprene by dehydrogenation of isopentane. Both isomers can be polymerised to polypiperylene.

PIR Abbreviation for polyisocyanurate.

PITCH The black solid or semi-solid residue from the distillation of various organic substances such as coal tar, palm oil and wood tar. Similar in nature to bitumens but containing more aromatic hydrocarbons.

PIVALOLACTONE (α,α-Dimethyl-β-propiolactone) (3,3-Diethyloxetan-2-one)

$$\begin{array}{c} (CH_3)_2C\!\!-\!\!CO \\ |\qquad\ | \\ H_2C\!\!-\!\!O \end{array}$$

I

M.p. $-13°C$. B.p. $53.5°C/15\,mm$.

The monomer for the formation of polypivalolactone by ring-opening polymerisation. Produced from pivalic acid by chlorination followed by ring closure:

$$(CH_3)_3CCOOH \overset{Cl_2}{\to} (CH_3)_2CCOOH \overset{NaOH}{\longrightarrow}$$
$$\qquad\qquad\qquad |$$
$$\qquad\qquad\quad CH_2Cl$$

$$ClCH_2C(CH_3)_2COONa \to \mathbf{I}$$

PLANE COUETTE FLOW The simplest type of simple shear flow, which occurs when a fluid is sheared between two plane parallel plates moving relative to each other. For a Newtonian fluid the only non-zero components of the rate of deformation tensor are the diagonal components, so only the shear stress components are non-zero, i.e. there are no normal stress effects. However, in polymer fluids normal stress effects are usually present.

PLANE STRAIN (Biaxial strain) The state of strain in a body when one principal strain (ε) is zero, i.e. for which $\varepsilon_{xx} \neq 0$, $\varepsilon_{yy} \neq 0$ and $\varepsilon_{zz} = 0$.

PLANE STRAIN COMPRESSION A method of loading the central area of a sheet specimen by compressing it between the faces of two dies with rectangular cross-sections. Thus the area of the specimen being stressed is constant, so that the true, not nominal, stress is the most readily calculated. Furthermore, unlike uniaxial compression, the friction between dies and sample remains constant with increasing strain. It may be necessary to correct for edge effects.

PLANE STRAIN FRACTURE In fracture mechanics, a model for crack growth in which all the plastic strains in the yielded zone ahead of the crack tip occur in the plane defined by the directions of crack growth and the crack opening. Thus there is no strain component parallel to the crack front. Plane strain fracture is likely to occur during the initial stages of crack growth and when thick specimens are involved.

PLANE STRESS (Biaxial stress) A state of stress in which the normal stress to the plane under consideration is zero. Thus a thin sheet parallel to this plane has stress-free surfaces. If the z-direction is that of the normal to the plane, then it is a principal direction, because the shear stresses are zero.

PLANE STRESS FRACTURE In fracture mechanics, a model for crack growth in which there is a state of plane stress in the yielded zone ahead of the crack tip, such that the tensile stress parallel to the crack front is zero. This is most likely to apply in thin specimens and when the yielded zone is at least as large as the specimen thickness.

PLANT GUM POLYSACCHARIDE Alternative name for gum polysaccharide.

PLASDENE Tradename for poly-(N-vinylpyrrolidone).

PLASKON Tradename for urea– or melamine–formaldehyde polymers and for nylon 6.

PLASMA PROTEIN A protein found in blood plasma, i.e. the clear yellowish fluid remaining after removal of the cellular material by centrifugation. The plasma contains about 6–8 g per $100\,cm^3$ of dissolved protein, accounting for about 70% of all solutes. Traditionally the plasma proteins have been classified according to their moving boundary electrophoretic mobility, usually in Veronal buffer at pH 8.6, but such a separation largely yields fractions, not individual molecular species. Further separation may be achieved by density gradient centrifugation, gel electrophoresis, column chromatography or immunoelectrophoresis. The six main electrophoretic groups in order of decreasing mobility in Veronal buffer are albumin (about 55%), α_1- and α_2-globulins (5 and 9%), β-globulins (about 13%), fibrinogen (about 4%) and γ-globulins (about 11%). These values are for human plasma. Similar proteins are found in other vertebrates, but in different relative amounts. The average molecular weight is about 110 000 and individual molecules have among the highest molecular weight for single chain polypeptides. The true molecular weight is often obscured due to aggregation. The molecular shape is elongated (fibrinogen) ellipsoidal or spherical (lipoprotein).

PLASMIN An enzyme produced in the blood from its zymogen, plasminogen. A protease of molecular weight about 90 000 capable of solubilising (lysing) blood clots by acting on fibrin, thus removing intravascular blood clots. It will also cleave other proteins, e.g. casein and gelatine, in a similar way to trypsin, at arginyl links.

PLASMINOGEN A blood plasma protein found in the β-globulin fraction on electrophoresis. It has a molecular weight of 90 000. The zymogen for plasmin. It is converted to plasmin in the presence of many activators, e.g. trypsin, urokinase and other tissue enzymes, which cause cleavage of a single arginyl–valyl bond to convert the single polypeptide chain plasminogen to two chain plasmin molecules.

PLASTIC In the most general sense a plastic material is one that is capable of being shaped, through plastic flow, by the application of deforming forces. In a polymer context, when the term preferred is often plastics material, such a material is based on a high molecular weight polymer, usually organic, and may be distinguished from a rubber by its higher stiffness and lack of a large reversible elastic deformation (although no sharp division can be made between plastics and rubbers). Similarly the distinctions between plastics, fibres and coatings rests merely on the physical shape of the product being considered. Some polymers are useful in all three forms, as with some nylons.

PLASTIC DEFORMATION   (Plastic flow) The deformation that occurs after yielding. In classical plasticity, as applied to metals, it has been considered to be completely irreversible, as opposed to the completely reversible deformation that occurs elastically before yielding. It may thus be considered truly as plastic flow. However, for polymers, post yield deformation is wholly or partially recoverable. Nevertheless, in the glassy state polymers may approximate to ideal plastic behaviour, so the concepts of classical plasticity may sometimes be usefully applied.

PLASTIC FLOW   (1) Alternative name for plastic deformation in metals. For polymers the post yield deformation may be recoverable, but is nevertheless often referred to as plastic deformation. However, only the permanent, non-recoverable, deformation can be called plastic flow.

(2) In rubber technology an alternative name for plasticity.

PLASTICISATION   Usually refers to the softening and increase in flexibility of a polymer brought about by the incorporation of a plasticiser. In the context of polymer melt processing the term is also used to include softening of a polymer brought about simply by the action of heat or by mechanical working. Thus in general the term refers to an increase in deformability, whether permanent (i.e. as plasticity) or not. Occasionally, as with polyvinyl chloride, small amounts of plasticiser may act as antiplasticisers. Plasticisation is sometimes brought about by the use of a flexibilising comonomer, when it is referred to as internal plasticisation, rather than the more usual external plasticisation resulting from the use of a plasticiser as an additive.

PLASTICISER   A substance, usually a liquid, but occasionally a low melting or softening point solid, which solvates a polymer and therefore softens it, i.e. acts as a flexibiliser. To be practically useful a plasticiser must also exhibit permanence so that it must not be lost during use either by volatilisation or by extraction. Therefore practical plasticisers are normally high boiling point, and hence high molecular weight, organic liquids which are of similar solubility parameter to the polymer and may therefore be said to be compatible with the polymer being plasticised. As a result of its action a plasticiser also lowers the T_g value and the softening point of the polymer concerned and thus allows for easier processing. When small quantities, say less than about 20 phr, are used

specifically for this purpose, as is common for processing rubbers, plasticisers are referred to as process aids. Plasticisers are classified as primary (compatible over the whole composition range), secondary (of limited compatibility) or as plasticiser extenders (only compatible when used in combination with a primary plasticiser). The vast bulk of plasticisers are used in polyvinyl chloride, for which the commonest types are phthalates (usually dioctyl, diisooctyl or diheptylnonyl), the phosphates (such as tritolyl phosphate) and the aliphatic ester adipates and sebacates (such as dioctyl adipate or sebacate). Cellulose based polymers, such as cellulose acetate, ethyl cellulose and cellulose acetate–butyrate, are also plasticised by similar plasticisers but, in addition, several other types, e.g. triacetin and camphor, may sometimes be used. Rubbers such as natural rubber, styrene–butadiene rubber and nitrile rubber are often plasticised using rubber oils either as process aids or as softeners, although esters are also used. Plasticisers are frequently used in surface coatings, e.g. with polyvinyl acetate. Other polymers are much less commonly plasticised, especially crystalline polymers, in which it is very difficult to plasticise the crystalline regions. Plasticisers are typically used in the range of 10–50 phr, except in polyvinyl chloride, where the range is about 30–100 phr. The higher quantities can change the hard rigid unplasticised polyvinyl chloride to a soft rubbery material.

PLASTICISER 79A   Alternative name for heptylnonyl adipate.

PLASTICISER EXTENDER   (Extender) A substance, which although relatively incompatible (i.e. not at all or only slightly miscible) with the polymer concerned, may be used in conjunction with a primary plasticiser as a cheap diluent. It may be said to extend the material or fill out its bulk at low cost without causing a significant loss of flexibility as would occur if using a solid filler as an extender. Many plasticiser extenders are chlorinated paraffins and so also have a fire retardant effect. In these cases the compatibility is dependent on the chlorine content. Whether the material acts merely as an extender in a particular polymer, e.g. in polyvinyl chloride, or as a secondary or primary plasticiser depends on its chlorine content.

PLASTICITY   (1) The phenomenon of yielding and the subsequent plastic deformation that occurs with increasing stress. In very general terms, the ability of a material to be shaped by stressing and to retain its shape after the stress has been removed.

(2) (Plastic flow) A measure of the viscosity of a raw unvulcanised rubber or rubber compound, determined by measuring the thickness of a cylinder of rubber after it has been compressed by a specified force for a definite time, normally at elevated temperature. Plasticity is dependent on shear rate. Thus low shear rate plasticity tests may not give a reliable guide to processability when processing is at much higher shear rates. The measurement is made in a plastimeter and the result is often expressed as a

plasticity number. Although plasticity is inversely related to viscosity, the plasticity number increases as viscosity increases. The ability of a rubber, especially natural rubber, to retain its plasticity under degradative conditions, is often given by its plasticity retention index.

PLASTICITY NUMBER A measure of the plasticity of a raw rubber or rubber compound. The thickness of a sample of specified dimensions after being subject to a specified compressive force at a specified temperature for a specified time. The higher the plasticity number the lower the plasticity since less flow will have occurred. Thus plasticity number and viscosity are directly related.

PLASTICITY RETENTION INDEX (PRI) The plasticity number of a rubber, usually natural rubber, after heating for 30 min at 140°C as a percentage of the plasticity number before heating. A measure of the sensitivity of the rubber to atmospheric oxidation. Used particularly in the classification of natural rubber, good grades having a PRI of more than 70.

PLASTIC VISCOSITY For a Bingham model material, the ratio of the shear stress less the yield stress to the rate of shear.

PLASTIC WORK The energy expended in performing plastic deformation. By an analysis of plastic work, similar to that of the thermodynamic theory of rubber elasticity, the stress–strain relationships for plastic deformation (the Lévy–Mises equations) may be derived. However, unlike elasticity, the current state of strain in plasticity theory depends on the strain history, and the change in plastic work as a function of the change in strain must be considered. The increment of plastic work is a function of the strain invariants $I_1 = de_1 + de_2 + de_3$, $I_2 = de_1 de_2 + de_2 de_3 + de_3 de_1$ and $I_3 = de_1 de_2 de_3$, where de_1 de_2 and de_3 are the strain increments. If deformation occurs at constant volume, then $I_1 = 0$ and, ignoring terms higher than second order, the increment of plastic work is only a function of I_2.

PLASTIGEL A plastisol to which a thickening agent has been added to given a material of putty-like consistency.

PLASTIMETER (Plastometer) An instrument for the determination of the plasticity of a raw unvulcanised rubber or rubber compound. Usually, a cylinder of rubber is compressed under a known force for a known time and the thickness is then measured. In the Williams plastimeter large compression plates are used. In the rapid plastimeter the compression surfaces are of similar size to the sample. The result is often expressed as the plasticity number.

PLASTISOL A stable dispersion of fine particles (about 1 μm) of emulsion polyvinyl chloride in plasticiser, which is a viscous fluid. On heating, usually at 180–250°C

for a few minutes, the plasticiser is absorbed into the particles and solvates them so that they fuse together to produce a homogeneous mass. The fusion process is referred to as gelation. Plastisols, and the related organosols, plastigels and rigisols, provide a liquid form of polyvinyl chloride to which special processing techniques may be applied which are often more convenient for producing useful products than conventional melt processing methods. These techniques include dipping, spreading, low pressure injection moulding, rotational moulding and casting. Plastisol rheological behaviour is of great importance in these processes and can be very complex. Plastisols may be shear thinning or shear thickening, depending mostly on polyvinyl chloride particle size, size distribution and shape, but also on plasticiser type and other additives used.

PLASTOMETER Alternative name for plastimeter.

PLASTYLENE Tradename for low density polyethylene.

PLATEAU ZONE Alternative name for rubbery plateau. In particular, it refers to the rubbery behaviour in the time or frequency domains at temperatures other than those at which normal, i.e. long time, rubbery behaviour is observed, through the operation of the time–temperature superposition principle.

PLEATED SHEET (β-Pleated sheet) (β-Sheet) The two-dimensional sheet-like structure that forms in a collection of polymer chains (usually proteins, synthetic polypeptides or nylons) when the chains adopt a β-conformation. In polypeptides, alternate α-carbons are located up and down giving the pleated structure. All amides are intermolecularly hydrogen bonded giving the sheet great stability and hence the polymer has low solubility. Extensive pleated sheet structures are uncommon in solution. β-Sheet structures are also produced by folding of the polypeptide in some globular proteins, notably carboxypeptidase A. The chains may all run in the same direction (parallel pleated sheet), but usually they run alternately in opposite directions producing the anti-parallel arrangement.

β-Sheet structures are favoured by the presence of branched substituent groups at the α-carbon or other bulky groups close to the chain, e.g. as in valine, isoleucine, tyrosine and phenylalanine amino acid residues. Often a nominally β-conformation polymer may be crystallised in the cross-β form and frequently α-helical forming polymers may be crystallised in the β-form, e.g. by steam stretching, rolling or by use of low molecular weight polymer.

β-PLEATED SHEET Alternative name for pleated sheet.

PLENCO Tradename for phenol–, melamine– or phenol/melamine–formaldehyde polymer.

PLEXIDUR PLUS Tradename for a methyl methacrylate/acrylonitrile copolymer useful as a toughened polymethylmethacrylate plastic.

PLEXIGLAS Tradename for polymethylmethacrylate sheet.

PLEXIGUM Tradename for various acrylate and methacrylate polymers and copolymers useful as solutions or emulsions for adhesives, coatings and other finishes.

PLEXIGUM B Tradename for polyethylacrylate.

PLEXIGUM M Tradename for polymethylacrylate.

PLEXIGUM N Tradename for polyethylmethacrylate.

PLEXIGUM P Tradename for polybutylmethacrylate.

PLEXILEIM Tradename for polyacrylic acid and its salts.

PLIOCHLOR Tradename for chlorinated synthetic *cis*-1,4-polyisoprene.

PLIOFILM Tradename for rubber hydrochloride.

PLIOFLEX Tradename for styrene–butadiene rubber.

PLIOFORM Tradename for cyclised rubber.

PLIOLITE Tradename for cyclised rubber and for styrene–butadiene rubber.

PLUG FLOW Flow along a pipe in which a core of the fluid moves with zero velocity gradient, i.e. telescopic flow in which the fluid velocity is independent of the radius over some range of radius values, usually including the zero value, i.e. along the axis of the pipe.

PLURONIC Tradename for poly-(propylene oxide-b-ethylene oxide) block copolymers useful as non-ionic detergents.

PLUTON Alternative name and tradename for a carbonaceous form of cyclised polyacrylonitrile in which the polymer has been heated to a higher temperature than in the case of Black Orlon.

PLYOPHEN Tradename for phenol–formaldehyde polymer.

PMDA Abbreviation for pyromellitic dianhydride.

PMLG Abbreviation for poly-(γ-benzyl-α-L-glutamate).

PMMA Abbreviation for polymethylmethacrylate.

PM-POLYMER Tradename for thiol-terminated polybutadiene.

PMQ Abbreviation for phenylsilicone elastomer.

PMR Tradename for a norbornene or itaconic acid end-capped thermosetting polyimide prepolymer formed by polymerising the monomers (as a solution of 4,4′-diaminodiphenylmethane, the dimethylester of benzophenone tetracarboxylic acid and the monomethyl ester of 5-norbornene-2,3-dicarboxylic acid) *in situ* during the formation of the polyimide product (moulding, etc.), hence the term PMR (polymerisation of monomeric reactant). Useful for the formation of high temperature resistant composites and adhesives.

PMR SPECTROSCOPY Abbreviation for proton magnetic resonance spectroscopy.

P13N Tradename for a polyamic acid prepolymer of benzophenonetetracarboxylic dianhydride and 4,4′-diaminodiphenylmethane, end-capped with nadic anhydride. On heating (B-stage) a polyimide structure is formed and further heating during moulding brings about chain extension through the end groups.

PNF Tradename for phosphonitrilic fluoroelastomer.

PO Abbreviation for propylene oxide rubber.

POCAN Tradename for polybutylene terephthalate.

POINT MATCHING METHOD A technique of micromechanics useful for the calculation of elastic behaviour of composites, in which the exact solutions of the governing differential equations are used and the boundary conditions are satisfied at a finite number of discrete points.

POISE Symbol P. The cgs unit of viscosity. It has the dimensions of dyne s cm^{-2} or (g s cm)$^{-1}$ and equals 10^{-1} Pa s. Polymer melts typically have viscosities of 10^3–10^4 P, ordinary liquids having values of about 10^{-2} P.

POISEUILLE EQUATION (Poiseuille–Hagen equation) An equation relating the coefficient of viscosity (η) of a fluid to its rate of flow through a capillary of radius R and length L. It is, $\eta = \pi R^4 Pt/8LV$, where t is the time during which volume V of fluid has passed through the tube due to the pressure difference P across its ends. The basis for the determination of η by capillary viscometry. If the flow through a vertical tube due to gravity, as in an Ostwald or similar viscometer, is considered, then $P = \rho g \langle h \rangle$, where ρ is the fluid density and $\langle h \rangle$ is the average height of the fluid column. For a given viscometer and fixed fluid volume all quantities are constant, therefore $\eta = A\rho t$, where A is a constant. The equation is only valid for streamline flow (laminar flow) which occurs when the Reynolds number (N_R) is less than 2100.

In dilute solution viscometry of polymer solutions, the

viscometers used are designed to keep N_R less than 100. It is also assumed that the pressure is used solely to cause viscous flow. However, it may be partly used to impart a significant kinetic energy to the fluid as it emerges from the capillary. A kinetic energy correction is then applied which modifies the equation to $\eta = \rho(At - B/t)$, where B is a constant. For the determination of viscosity ratio (η_{rel}), as required in the solution viscosity method of polymer molecular weight estimation, η_{rel} simply equals t/t_0 in the absence of a kinetic energy correction, where t and t_0 are the flow times of solution and pure solvent respectively. An end-effect correction to L may also be necessary, especially in the capillary rheometry of polymer melts. When the assumption of Newtonian flow is not valid, then the Rabinowitz correction is also needed.

POISEUILLE FLOW An alternative name for pressure flow in general, but often refers specifically to axial streamline flow in a circular cross-section pipe for which the Poiseuille law holds for Newtonian fluids.

POISEUILLE–HAGEN EQUATION Alternative name for Poiseuille equation.

POISSON DISTRIBUTION A molecular weight distribution, whose differential distribution function is,

$$N(r) = [\exp(-v)]^{r-1} v^{r-1}/(r-1)!$$

where r is the size of the individual molecular species (such as degree of polymerisation) and $v = \bar{x}_n - 1$, where $\bar{x}_n$ is the number average degree of polymerisation. The corresponding weight function $W(r)$ is similar. The weight average degree of polymerisation $\bar{x}_w = (1 + 3v + v^2)/(1 + v) \approx 1 + \bar{x}_n$, hence this is a very narrow distribution ($\bar{x}_w/\bar{x}_n \rightarrow 1$ at high r). It results when a fixed number of polymerisation sites begin to grow simultaneously and growth is random with no termination. This often happens in the formation of living polymers by anionic polymerisation.

POISSON RATIO (Lateral contraction ratio) Symbol v. The ratio of the tensile strain produced by a tensile force, to the contraction perpendicular to the force. Thus if the force is along the z-axis, $v = -\varepsilon_{yy}/\varepsilon_{zz} = -\varepsilon_{xx}/\varepsilon_{zz}$ where ε_{xx}, ε_{yy} and ε_{zz} are the normal strains. One of the fundamental elastic constants of a material. Related to the shear modulus (G), Young's modulus (E) and the bulk modulus (K) by expressions such as $v = (E/2G) - 1$, and to the compliance constants by $v = -S_{12}/S_{11}$. For many rubbers v is approximately 0.5 (material is incompressible), whereas for many plastics $v = 0.30$–0.35.

POLAR DIAGRAM A method of graphical representation of the anisotropy of a property by plotting the value of the property radially against orientation angle.

POLARISATION (Dielectric polarisation) The displacement of the component charges of a dielectric by an applied electric field, causing the material to behave as an electric dipole. Quantitatively, the polarisation

P, a vector quantity, is the electric moment induced per unit volume and is proportional to the field strength (E) through the relation $P = (\varepsilon - 1)\varepsilon_0 E$, where ε is the relative permittivity of the material and ε_0 is the absolute permittivity of free space. At the molecular level, electronic polarisation, atomic polarisation, and orientation polarisation may all occur, the overall polarisation being given by $P = N_0 \alpha E_L$, where N_0 is the number of molecules per unit volume, α is the molecular polarisability and E_L is the local field strength. An additional interfacial polarisation may also occur. In an alternating field, as frequency increases, polarisation (and hence also ε) decreases, since the motions required for the various types of polarisation to occur, do not have time to occur. Progressively the interfacial, orientation, atomic and electronic components become inactive with increasing frequency.

POLARISED FLUORESCENCE SPECTROSCOPY The determination of the angular polarisation intensity of the fluorescence of fluorescent molecules dispersed in a polymer. It is useful for characterising orientation in amorphous polymers if it is assumed that the fluorescent molecules lie parallel to the polymer chains, since fluorescence polarisation depends on the direction of orientation of the fluorescent molecules. If the incident light is also polarised, then both the second and fourth moments of the orientation distribution may be obtained.

POLARISED RAMAN SCATTERING A technique for the measurement of the intensity and polarisation of Raman scattered light, useful for giving information on the second and fourth moments of the orientation distribution of an oriented polymer sample.

POLARISED RAMAN SPECTROSCOPY Raman spectroscopy using polarised light. Measurement of the depolarisation can yield the $\cos^2 \theta$ and $\cos^4 \theta$ orientation functions, often more readily than with other techniques. Variation of orientation across a sample, e.g. an injection moulding or flowing melt, can be readily investigated.

POLARISING MICROSCOPE An optical microscope in which the incident light passes through a polarising filter (the polariser) and the transmitted light through a further such filter (the analyser). If the directions of polarisation of the two filters are perpendicular no light reaches the eyepiece. However, a birefringent sample, e.g. a crystalline polymer, will cause light to be passed and will form an image. The technique is therefore very useful for the observation of spherulites, and crystallites and in indicating orientation.

POLECTON Tradename for poly-(N-vinylcarbazole).

POLE FIGURE A graphical representation of the distribution of crystalline orientation consisting of plots of contour lines of equal orientation against orientation angle. It can be obtained from analysis of the intensities of the diffraction arcs of a fibre diagram. Complete representation of orientation requires a three-dimensional figure, but two-dimensional stereographic projections are more convenient. Two are required for triaxial orientation, but uniaxial orientation only needs a plot of density of axis orientation against angle between the principal axis and the measuring direction (the polar angle).

POLE FIGURE ANALYSER A device for the simultaneous tilting and rotating of a sample in an X-ray diffractometer, required for the proper characterisation of crystalline orientation.

POLLOPAS Tradename for urea– or urea/thiourea–formaldehyde polymer.

POLYACENAPHTHYLENE

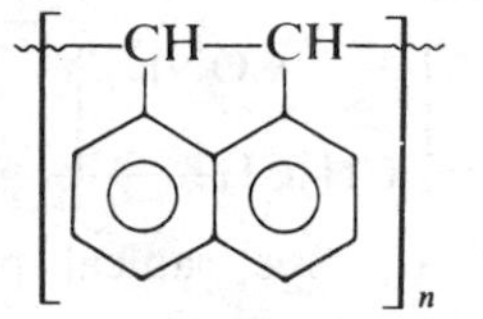

Obtained by free radical polymerisation of acenaphthylene. The polymer softens at around 250°C.

POLYACENE A ladder polymer of the type

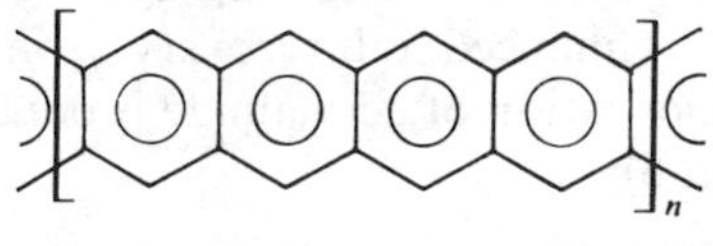

or a similar substituted polymer. Polymers of this type, with unusually high thermal stability are claimed to be formed by thermal polymerisation, probably by a Diels–Alder polymerisation, of a di-yne (a diacetylene).

POLY-(ACENE QUINONE RADICAL) (PAQR) A polymer of structural type:

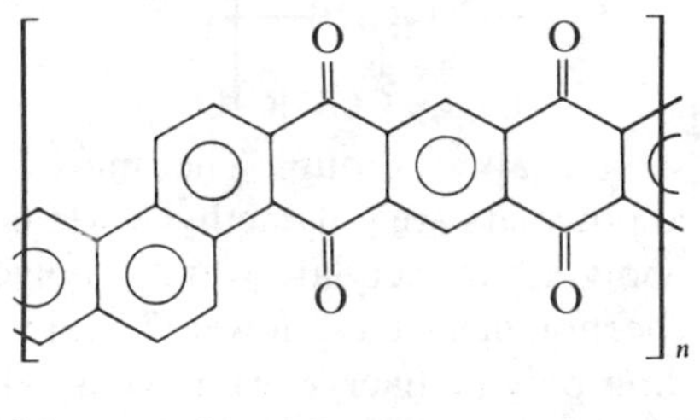

Synthesised by reaction of pyromellitic dianhydride with multinuclear aromatic compounds. A highly conjugated planar polymer, which therefore shows considerable electrical conductivity—up to 3×10^2 S cm^{-1}. It also has an extremely high relative permittivity of up to about 10^5. However, it is a very intractable material.

POLYACETAL In general a polymer containing repeat units of structure ─[O─CHR─O]─ where R is a hydrogen atom or an alkyl group. In particular the term often refers to the most important commercial polymer of this type where R = H, i.e. polyoxymethylene.

POLYACETALDEHYDE

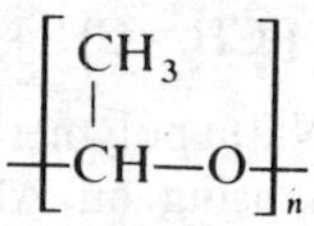

The polymer obtained by the polymerisation of acetaldehyde, which must be conducted at low temperatures due to the poor stability and low ceiling temperature (−40°C) of the polymer. A wide variety of cationic catalysts, e.g. protonic and Lewis acids, alumina, may be used, or even simple melting of frozen acetaldehyde (fusion polymerisation). Such polymers are amorphous, atactic and rubbery. Anionic polymerisation, e.g. with aluminium triethyl and water, can yield isotactic, crystalline polymers with a T_m value of about 165°C.

POLY-(1,4'-β-D-2-ACETAMIDO-2-DEOXY-GLUCOSE) Alternative full chemical name for chitin.

POLYACETONE

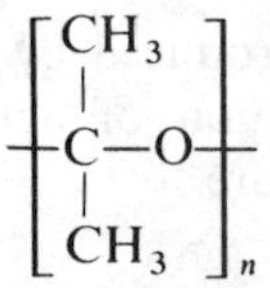

The polymer may be prepared by very low temperature polymerisation of acetone using a magnesium metal or Ziegler catalyst. The resultant polymer is very unstable even at room temperature.

POLY-(ACETONYLACETATE) A polymer containing acetonylacetate units chelated to a metal (M) as its repeat unit:

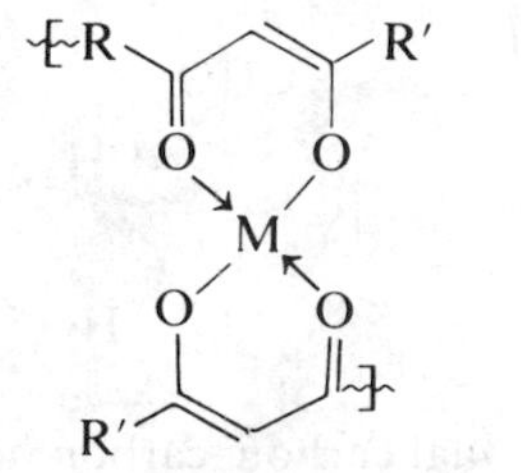

Monomeric acetonylacetates have good thermal stability, but the polymers are disappointing in this respect. Bis-(β-diketone) polymers containing similar structures have also been prepared.

POLY-(1,4'-β-N-ACETYL-2-AMINO-2-DEOXY-D-GLUCOSE

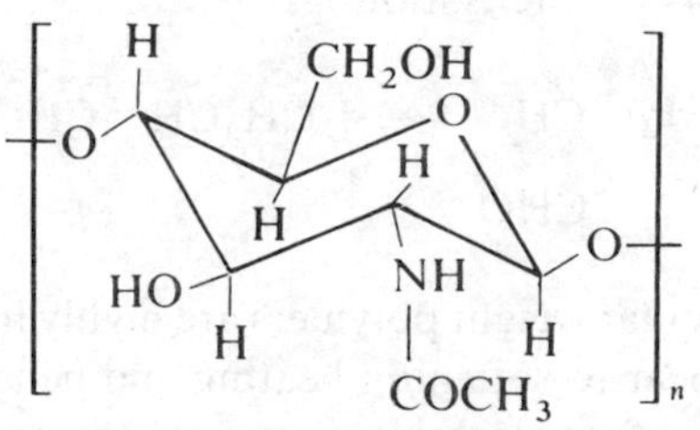

The full chemical name, describing the structure of chitin.

POLYACETYLENE

$$-\!\!\left[CH\!=\!CH\right]_n\!\!-$$

Prepared by Ziegler–Natta polymerisation of acetylene in an inert solvent, e.g. using an $Al(C_2H_5)_3/Ti(OC_4H_9)_3$ catalyst. The stereoregular, crystalline polymer is of interest because of its high electrical conductivity and is often conveniently prepared in film form for electrical studies by carrying out the polymerisation on a cast film layer. The polymer is slowly oxidised in air which causes a loss of conductivity and a darkening of its silvery lustrous appearance. By doping, with up to a few per cent dopant, the conductivity may be increased to the semiconducting or even metallic conducting range, i.e. up to about 10^4 S cm^{-1}. Dopants used include Br_2, I_2, AsF_5, $HClO_4$ and H_2SO_4 (dopants of the electron donating or n-type) and LiI or sodium naphthalide (electron attracting or p-type dopants). Unfortunately the films have poor mechanical properties as well as being unstable and the polymer is rather intractable.

POLYACROLEIN   Produced by free radical, anionic or cationic polymerisation of acrolein, but although nominally of the structure

$$\left[CH_2CH\!\!\sim\atop\ \ \ \ CHO\right]_n$$

usually it is of greater structural complexity. Free radical polymers typically contain only 10–20 mol% of aldehyde groups and are crosslinked. Most repeat units consist of cyclised tetrahydropyran in sequences of 2–4 units

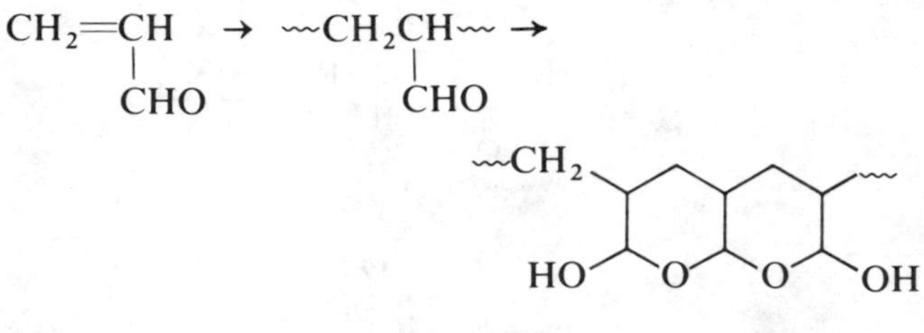

$$CH_2\!=\!CH \rightarrow \sim\!\!CH_2CH\!\!\sim \rightarrow$$
$$\quad\ \ CHO \qquad\qquad CHO$$

with some residual carbon–carbon double bonds as well as keto-carbonyl groups, acetal and hemiacetal links. Anionically produced polymers appear to result from carbonyl double bond polymerisation:

$$n\,CH_2\!=\!CH \rightarrow \left[CH\!-\!O\!-\!\atop CH\!=\!CH_2\right]_n$$
$$\qquad\ CHO$$

or from 1,4-polymerisation:

$$n\,CH_2\!=\!CH \rightarrow \left[CH_2CH\!=\!CH\!-\!O\right]_n$$
$$\qquad\ \ CHO$$

High molecular weight polymers are highly insoluble and do not appear to soften on heating and hence are rather intractable. More tractable polymers may be obtained by blocking the reactivity of the aldehyde groups in the monomer prior to polymerisation, e.g. by conversion to cyclic acetal

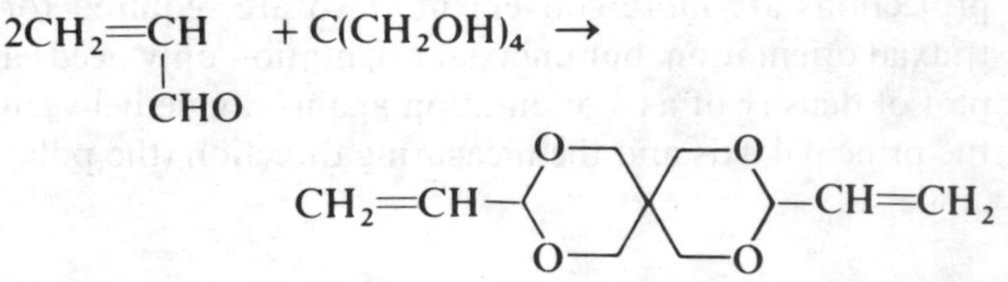

$$2CH_2\!=\!CH\ + C(CH_2OH)_4 \rightarrow$$
$$\qquad\ \ CHO$$

or to a diacetate or oxime. Alternatively the polymer may be chemically modified, e.g. by conversion to a water soluble bisulphite carbonyl adduct

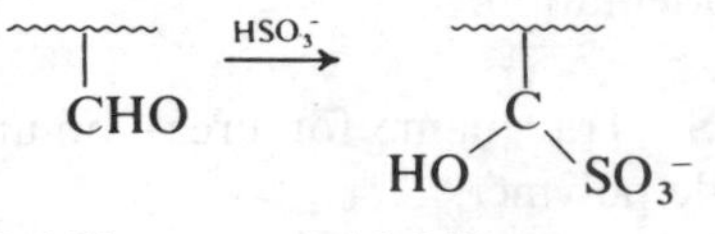

to impart solubility or mouldability.

POLYACRYLAMIDE

Readily prepared by free radical polymerisation of acrylamide, usually in aqueous solution with redox, peroxy, photo- or high energy radiation methods of initiation. The pH must be maintained at 3–5 otherwise the monomer hydrolyses and crosslinking occurs. The polymer has a T_g value of 165°C and is highly water soluble. It is useful as a flocculating agent, a thickening agent and for enhanced oil recovery from oil wells. Anionic polymerisation of acrylamide produces poly-(β-alanine) (nylon 3):

$$n\,CH_2\!=\!CHCONH_2 \rightarrow \left[NHCH_2CH_2CO\right]_n$$

not polyacrylamide. Polymerisation of acrylamide has been widely investigated in fundamental studies of solid-state polymerisation.

POLYACRYLATE   A polymer of an acrylic ester, i.e. one having a repeat unit structure of

$$\left[CH_2CH\!-\!\atop\ \ \ COOR\right]_n$$

where R is an alkyl group. The more thoroughly investigated polymers are polymethylacrylate, polyethyl-acrylate, poly-n-butylacrylate and polycyclohexyl-acrylate. The polymers have lower T_g values than the corresponding polymethacrylates, having values of 0°C for $R = CH_3$, and decreasing as R increases in size, until a minimum value is reached at $R = C_8H_{17}$ of about $-60°C$. Thus most members are essentially rubbery polymers best prepared by emulsion polymerisation.

POLYACRYLATE RUBBER   Alternative name for acrylic elastomer.

POLYACRYLIC ACID

$$\left[CH_2CH\!-\!\atop\ \ \ COOH\right]_n$$

Tradenames: Acrysol, Plexileim. Produced by free radical polymerisation of acrylic acid, usually in aqueous solution with, for example, a persulphate initiator. The polymer is readily soluble in water and in alcohols, in which it behaves as a polyelectrolyte. It becomes much more highly ionised on neutralisation, forming a polysalt. The salts in solution have highly expanded molecular coils and therefore also very high viscosities. Thus the polymer and its salts are useful as thickening agents for emulsions, textile finishes and paints, as well as being useful as suspension and flocculating agents. In cross-linked form the polymer is also useful as a cation-exchange resin.

POLYACRYLIC ELASTOMER   Alternative name for acrylic elastomer.

POLYACRYLONITRILE (PAN)

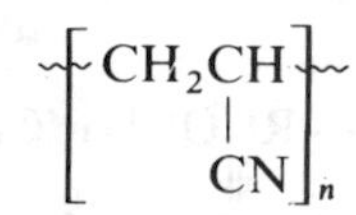

$$\left[CH_2CH \atop CN \right]_n$$

Tradenames: Acribel, Acrilan, Courtelle, Creslan, Crylor, Dolan, Dralon, Dynel, Leacril, Nitron, Nymcrylon, Orlon, Redon, Tacryl, Verel, Wolcrylon, Zefran. The basis of the important acrylic and modacrylic fibres. Produced by free radical polymerisation of acrylonitrile in solution or in suspension. Emulsion polymerisation is not particularly suitable due to the appreciable water solubility of the monomer. For fibre use solution polymerisation is particularly advantageous since the resultant polymer may be spun directly from the solution. Many commercial fibre polymers contain up to about 10% of a comonomer, such as 2-vinylpyridine or vinyl acetate, to improve dyeability. In bulk and suspension polymerisation the polymer precipitates early since it is highly insoluble in its monomer, so that the polymerisation is heterogeneous, containing polymer particles, and therefore shows early autoacceleration.

The polymer is highly polar with strong interchain interactions between nitrile groups. It therefore has a high softening point, with a T_g value if 97°C and a possible melting point of about 320°C. However, it decomposes before it softens sufficiently for flow to occur and therefore it cannot be melt processed as a plastic or melt spun to a fibre.

Fibres are spun from solution either in a highly polar solvent such as dimethylformamide, dimethylacetamide or dimethyl sulphoxide, or from strong aqueous solutions of certain inorganic salts such as calcium thiocyanate, zinc chloride or sodium perchlorate. It is highly resistant to swelling or solution by common organic solvents. On heating above 200°C the polymer darkens, eventually yielding a black solid, e.g. Black Orlon, of high thermal stability, in which pendant nitrile groups have cyclised to give a conjugated ladder polymer. Further heating to higher temperatures in a controlled oxygen-containing atmosphere causes graphetisation. This is the basis for the production of high modulus carbon fibres, when carried out on polyacrylonitrile fibres held under tension.

Acrylonitrile is also susceptible to anionic polymerisation with cyanide ion or Grignard reagents, such as phenyl magnesium bromide, or with other organometallic compounds. However, like the free radically produced polymers, the products are usually atactic.

POLY-(N-ACYLAMIDRAZONE) A polymer of structure

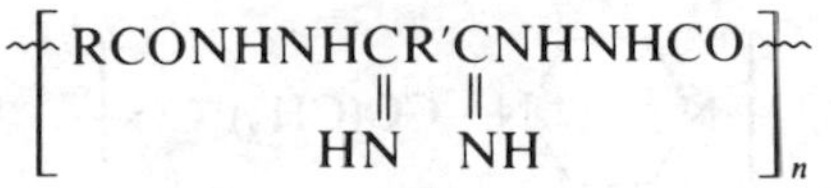

$$\left[RCONHNHCR'CNHNHCO \atop HN \quad NH \atop I \right]_n$$

where R and/or R′ are alkyl or aryl groups. Synthesised by reaction of a dihydrazide with a diimidate (diimido-ester):

$$n \; EtOCR'COEt + n \; H_2NNHCORCONHNH_2 \rightarrow I \atop HN \quad NH$$

or by reaction of a bisamidrazone with a diacid chloride, in solution or interfacially:

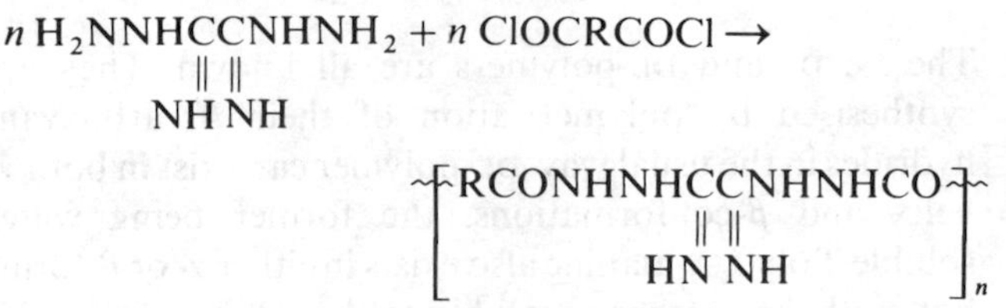

$$n \; H_2NNHCCNHNH_2 + n \; ClOCRCOCl \rightarrow \atop NH \; NH$$

$$\left[RCONHNHCCNHNHCO \atop HN \; NH \right]_n$$

Polyterephthaloyloxalamidrazone chelated to metals has been investigated as a commercial fibre. On heating to 300°C the polymers cyclodehydrate to poly-(1,2,4-triazoles), whereas on heating with a strong acid, e.g. dichloracetic, a poly-(1,3-oxadiazole) is formed.

POLYACYLHYDRAZONE   A polymer of structure

$$[NNHCORCONHN{=}CHR'CH]_n$$

where R and R′ are alkyl or aromatic groups. Synthesised by reaction of a bishydrazide with a dialdehyde or diketone. If R and R′ are aromatic then solubility is very limited and the polymers are coloured.

POLYACYLSEMICARBAZIDE   A polymer of structure

$$[NHCONHNHCORCONHNHCONHR']_n$$

Synthesised by reaction in a polar solvent of a diisocyanate with a bishydrazide. When R and/or R′ are aromatic T_m values are in the range 250–300°C, but thermal stability is poor.

POLYADDITION   A step-growth polymerisation in which no small molecule is eliminated, in contrast with polycondensation which is the much more common type of step-growth polymerisation. Polyurethane formation by hydrogen transfer polymerisation is the best known example.

POLYADIPAMIDE A nylon polymer formed by reaction of adipic acid with a diamine and having the general structure

$$\{NHRNHCO(CH_2)_4CO\}_n$$

Nylon 66 is the most important example.

POLYADIPOLYPIPERAZINE

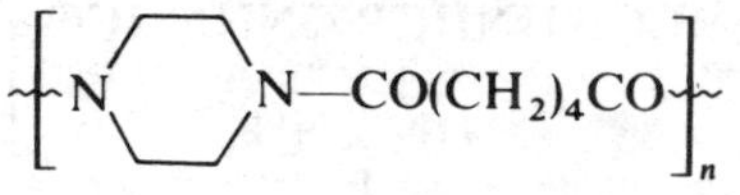

A piperazine polyamide formed by low temperature solution polymerisation by reaction of piperazine with adipoyl chloride in a hydrocarbon solvent. Has a T_m value of about 350°C and can be spun to tough fibres. However, the polymer is rather water sensitive.

POLY-(α-ALANINE)

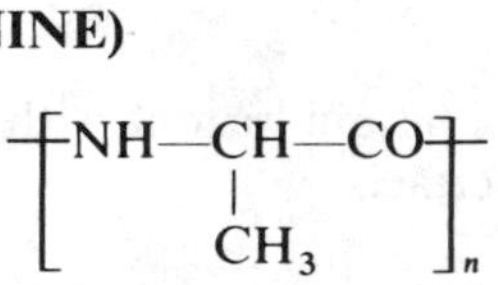

The L-, D- and DL-polymers are all known. They are synthesised by polymerisation of their N-carboxyanhydrides in the usual way. DL-polymer can exist in both α-helix and β-conformations, the former being water soluble. Poly-L-α-alanine also exists in either α- or β-forms but both are rather insoluble (although the former is soluble in, for example, dichloroacetic acid). Stretching of the α-form of the DL- or L-polymer produces the β-form.

POLY-(β-ALANINE) Alternative name for nylon 3.

POLY-DL-ALANYL-POLY-L-LYSINE A widely studied multichain poly-α-amino acid, consisting of a poly-L-lysine backbone with, typically, DL-alanyl side groups of five to twenty residues. Circular dichroism and infrared and X-ray studies indicate a random structure in both solution and solid state.

POLYALKENAMER A polymer of structure

$$\{(CH_2)_xCH{=}CH\}_n$$

or a similar substituted polymer. When $x = 2$ the polymers are the important polydiene rubbers—the unsubstituted polymer being polybutadiene. Although according to the strict definition of the term polyalkenamer such polymers are included, the term usually refers to polymers with $x > 2$. The polymers are formed by ring-opening polymerisation of a cyclo-olefin, i.e.

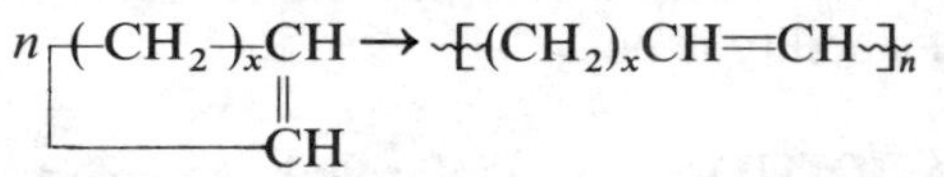

using a Ziegler–Natta catalyst such as a halide of acetylacetonate of tungsten, molybdenum or rhodium with $AlEt_3$ or $AlEtCl_2$ plus an activator. The polymers have a partially macrocyclic structure with low T_g values and are often crystallisable. They are potentially useful

rubbers and may be sulphur vulcanised and oil extended. Both polypentenamer and polyoctenamer are of commercial interest.

POLYALKYLENE CARBONATE A polymer of repeat unit structure

Such polymers may be produced from carbon dioxide, which reacts with zinc alkoxides to give zinc alkyl-carbonates:

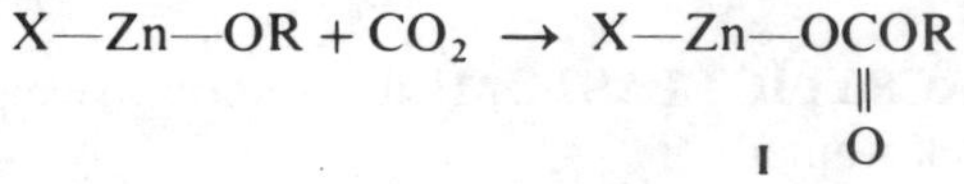

These are then further reacted with an epoxide to form the polycarbonate:

etc. The polymers have potential uses as biodegradable packaging and adhesives.

POLYALKYLENEPHENYLENE A polymer of the type

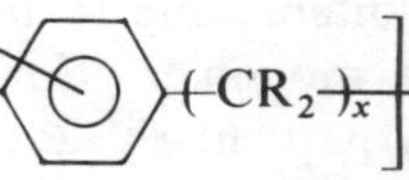

where R is hydrogen or an alkyl group. Polymethylene-phenylene has $R = H$ and $x = 1$. Polyxylylene has $R = H$ and $x = 2$. The *para*-linked isomers are of the most interest.

POLYALKYLENE SULPHIDE A polymer of structure $\{R{-}S\}_n$, where R is an alkylene or substituted alkylene group. Prepared by ring-opening polymerisation of episulphides, like the corresponding polyalkylene oxides. The sulphur polymers have higher melting temperatures and higher insolubility (especially in water) than their oxygen analogues. The two most important examples are polyethylene and polypropylene sulphides.

POLYALLOMER (Allomer) A name given to early ethylene–propylene block copolymers, containing up to about 15% ethylene to improve the low temperature impact strength of polypropylene. More frequently such polymers are called simply propylene copolymers.

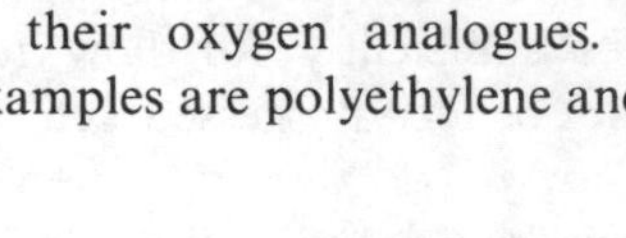

POLYALUMINOSILOXANE (Polyalumosiloxane) A polymer of the type formed by reaction of aluminium chloride or alkoxide with a difunctional silane, e.g. a silanol or acetoxysilane:

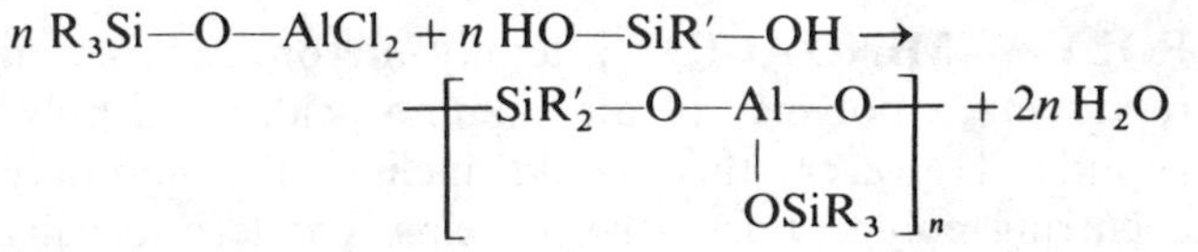

$$n\ R_3Si{-}O{-}AlCl_2 + n\ HO{-}SiR'{-}OH \rightarrow$$

$$\left[\begin{array}{c} SiR_2'{-}O{-}Al{-}O \\ | \\ OSiR_3 \end{array} \right]_n + 2n\ H_2O$$

The polymers are usually branched and crosslinked.

POLYALUMINOXANE A polymer with chains consisting of alternating aluminium and oxygen atoms, i.e.

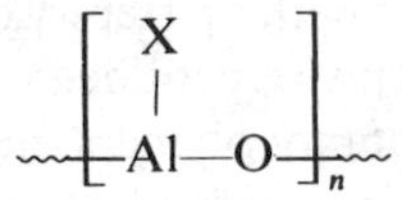

$$\left[\begin{array}{c} X \\ | \\ Al{-}O \end{array} \right]_n$$

Low molecular weight polymers may be prepared by hydrolysis condensation reactions such as:

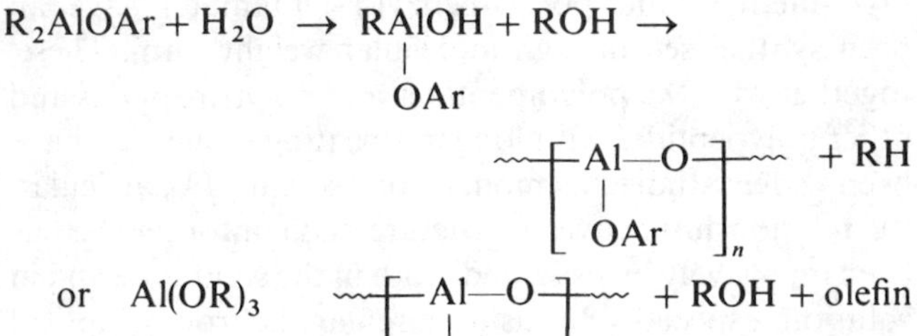

$$R_2AlOAr + H_2O \rightarrow RAlOH + ROH \rightarrow$$
$$\qquad\qquad\qquad\qquad\quad | \\ \qquad\qquad\qquad\qquad\ OAr$$

$$\sim\!\!\left[\begin{array}{c} Al{-}O \\ | \\ OAr \end{array} \right]_n\!\!\sim + RH$$

$$\text{or}\quad Al(OR)_3 \quad \sim\!\!\left[\begin{array}{c} Al{-}O \\ | \\ OAr \end{array} \right]_n\!\!\sim + ROH + \text{olefin}$$

POLYALUMOSILOXANE Alternative name for polyaluminosiloxane.

POLYAMIC ACID (Polyamide-acid) A polymer containing both amide and carboxylic acid groups. Aromatic polymers are the best known, as the precursor polymers for polyimides. They are formed by reaction of a dianhydride, e.g. pyromellitic dianhydride, and an aromatic diamine:

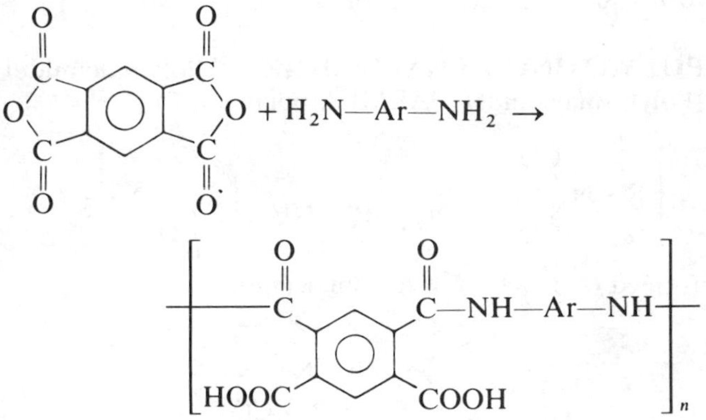

(structure of pyromellitic dianhydride) $+ H_2N{-}Ar{-}NH_2 \rightarrow$

(polyamic acid structure with HOOC and COOH groups, C=O, NH—Ar—NH)

On dehydration the polyamic acid yields the polyimide. Soluble in highly polar solvents such as DMF and *N*-methylpyrrolidone, whose solutions are used to fabricate fibres, films and enamels. Cyclisation to the polyimide structure is conducted on the fabricated article, e.g. Kapton film and Pyre ML wire enamel. Fabrication of the polyimide directly is not possible owing to its intractability.

POLYAMIDATION The reaction of carboxylic acid groups, or derived groups such as acid chlorides or esters, with amino groups, both types of groups being part of di- or polyfunctional monomer molecules, the reaction thus producing a polyamide. The amino and acid groups may be attached to the same monomer molecule (AB monomer) thus producing a monadic nylon, or may be on different monomer molecules (AA and BB monomers) thus producing a dyadic nylon. Although polyamidation, like polyesterification, proceeds to an equilibrium, e.g.

$$n\ HOOCR'COOH + n\ H_2NRNH_2 \rightleftharpoons$$
$$\qquad\qquad [COR'CONHRNH]_n + 2n\ H_2O$$

the position of the equilibrium is much more favourable to amide formation. Furthermore, transamidation is very slow. Amidation with carboxylic acid groups requires high temperatures ($> 150°C$) for reasonable rates. When a diacid is used, it is very convenient to proceed via a nylon salt. ω-Amino acids, $H_2N(CH_2)_xCOOH$, give mostly cyclic amide when $x < 5$, but yield polyamide in equilibrium with only a small amount of the cyclic lactam when $x > 4$. Polyamides produced by polyamidation have the Flory–Schultz distribution of molecular weights expected for a simple step-growth polymerisation.

POLYAMIDE (PA) A polymer containing the amide group, —NHCO—, in the repeat unit. The proteins are biopolymer polyamides of great importance. Proteins can be considered as polymers of α-amino acids with repeating units of general structure —CH(R)NHCO—, where R is any of about 30 different substituents. They are ordered copolymers and usually contain about 20 different amino-acid units in each molecule. Poly-amino acids are also known as polypeptides, although this term is usually restricted to synthetic polymers. However, the synthetic polyamides of greatest technological importance are the nylons, of general structure

$$[(CH_2)_xNHCO]_n$$

or

$$[NH(CH_2)_xNHCO(CH_2)_yCO]_n$$

Other synthetic polyamides are the fatty polyamides, the aliphatic–aromatic polyamides and the aromatic polyamides.

The properties of the polyamides are dominated by their ability to hydrogen bond through amide groups, either intramolecularly (as in proteins adopting the α-helix conformation) or intermolecularly (as in proteins and nylons adopting the β-sheet conformation). Hydrogen bonding stabilises these regular conformations and hence is a driving force for crystallisation. Thus polyamides of regular structure, as with proteins and most nylons, are partially crystalline in the solid state. Furthermore, proteins, nylons and aromatic polyamides may be oriented to produce strong, stiff fibres. Both natural protein fibres, such as wool and silk, and synthetic nylons are major fibres. In fibres, the most important application of nylons, orientation is induced during spinning and drawing.

Nylons are also useful as plastics and having high stiffness and toughness, they are often classed as engineering polymers. The powerful intermolecular hydrogen bonding forces contribute substantially to the strength of fibres and other products. Hydrogen bonds may also form between amide groups and other molecules, notably water, so polyamides generally exhibit water absorption, which can considerably alter their properties. Water absorption is usually an advantage, providing useful comfort properties in textiles and increasing toughness in plastics. Polyamides are generally highly insoluble due to the necessity for solvation to overcome the hydrogen bonding forces and crystallinity, so that the best solvents are themselves capable of hydrogen bonding.

The synthesis of protein polyamides is controlled in nature by ribonucleic acid. Synthetic polyamides may be synthesised by several methods. These include polyamidation by reaction of a diamine with a dicarboxylic acid or one of its derivatives, self-amidation of an amino acid and ring-opening polymerisation of a lactam. Polypeptides are also synthesised by several methods but especially by ring-opening of the N-carboxyanhydride of the appropriate α-amino acid.

POLYAMIDE-ACID   Alternative name for polyamic acid.

POLYAMIDE ELASTOMER   Alternative name for polyether block amide.

POLYAMIDE–HYDRAZIDE   A polymer containing both amide (—NHCO—) and hydrazide (—NHNH—) groups. Commercial examples are the high temperature resistant aromatic polyamide–hydrazide fibres X-500 and Flexten.

POLYAMIDE–IMIDE   A polymer containing both amide and imide groups in the polymer chain. Synthesised by reaction between a tricarboxylic acid, especially trimellitic acid, (or its anhydride or acid chloride), and a diamine or diisocyanate (preferably MDI), or by reaction of an amino-terminated amide and a dianhydride. Several commercial products, intermediate in properties between polyamides and polyimides, are produced. They have lower high temperature resistance than the polyimides but are more readily processable. Used as high temperature electrical insulation varnishes (AI, Rhodeftal), fibres (Kermel), moulding materials (Torlon) and film (Amanim).

POLYAMINE   A polymer containing amino groups in the polymer chain, not as side groups or as part of a heterocyclic ring. The most studied polymers are those containing piperazine units prepared, for example, by:

$$n \ H—N \bigcirc N—H$$

$$+ n \ CH_2{=}CHCO—N \bigcirc N—COCH{=}CH_2 \ \rightarrow$$

$$\left[N \bigcirc N—CH_2CH_2CO—N \bigcirc N—COCH_2CH_2 \right]_n$$

and polyethyleneimine prepared by ring-opening polymerisation.

POLY-α-AMINO ACID   Literally interpreted this term represents a polymer of an α-amino acid, i.e. a polypeptide. However, this would include the naturally occurring polypeptides—the proteins. The term usually refers to synthetic polypeptides, often being further restricted to homopolymers $\mathord{+}NH—CHR—CO\mathord{)}_n$, although random and block copolymers and poly-, di- and tripeptides are sometimes included. However, sequential copolymers built up stepwise from several amino acid residues by stepwise peptide synthesis reactions are not included. The homopolymers (homopolypeptides) of all 20 standard amino acids (e.g. poly-L-glycine, poly-L-alanine, etc.) as well as other amino acids (e.g. polysarcosine) and derivatives (e.g. poly-γ-benzyl-L-aspartate, ply-γ-methyl- and poly-γ-benzyl-L-glutamate) have all been synthesised in high molecular weight form. These, together with the polydipeptides and polytripeptides and other polypeptides with longer repeating sequences, have been widely studied as models for proteins. In particular, the relationship between structure and conformation has been intensively investigated, both in the solid state and in solution, especially by using circular dichroism, optical rotatory dispersion, nuclear magnetic resonance, ultra-violet and infrared spectroscopic techniques.

Much theoretical work involving calculations of preferred conformations as, for example, expressed on a Ramachandran plot, has been performed.

The homopolymers are best synthesised from their N-carboxyanhydrides, although the active esters of alanine and glycine may be polymerised and other peptide synthesis techniques are sometimes useful. Simple thermal polymerisation (150–200°C) of the parent amino acid is not successful due to decarboxylation, deamination and formation of diketopiperazines.

POLYAMINOBISMALEIMIDE   (Polymaleimide) (Polybismaleimide) (PABM) A polymer of the type

$$\left[R—N{\Big\langle}{\overset{CO}{\underset{CO}{}}} {—}NH{—}R'{—}NH{—} {\overset{CO}{\underset{CO}{}}}{\Big\rangle}N \right]_n$$

formed by reaction of a bismaleimide

$$ {\overset{CO}{\underset{CO}{}}}{\Big\rangle}N—R—N{\Big\langle}{\overset{CO}{\underset{CO}{}}} $$

with a diamine, $H_2NR'NH_2$. In commercial materials (Gemon, Kerimid, Kinel, Nolimid)

$$R = \bigcirc{—}CH_2{—}\bigcirc$$

(the bismaleimide of 4,4′-diaminodiphenylmethane) and the amine is 4,4′-diaminodiphenylmethane. Similar in structure to the aromatic polyimides, but as they are not aromatic, they do not have such good thermo-oxidative stability. However, unlike the polyimides, no volatiles are evolved during their formation, so void-free parts may be moulded.

POLY-(ω-AMINOCAPROAMIDE) Alternative name for nylon 6.

POLY-(6-AMINOCAPROIC ACID) Alternative name for nylon 6.

POLY-(ε-AMINOCAPROIC ACID) Alternative name for nylon 6.

POLY-(4-AMINO-1,2,4-TRIAZOLE) A polymer of structure

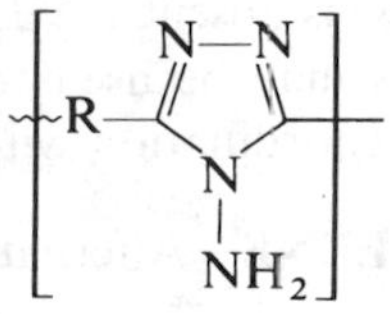

I

where R is an alkyl group. Synthesised by heating an excess of hydrazine with a dihydrazide at 200–300°C:

$$H_2NNHCORCONHNH_2 + H_2NNH_2 \rightarrow I$$

or by heating a diacid hydrazinium salt. If R is polymethylene (as is often the case) then fibres similar to nylons may be spun. They have T_m values of 250–350°C.

POLY-(11-AMINOUNDECANOIC ACID) Alternative name for nylon 11.

POLYAMPHOLYTE (Amphiphilic polymer) A polyelectrolyte which carries both positive and negative charges. A synthetic example would be a copolymer of acrylic acid, or its salt, with dimethylaminoethyl acrylate

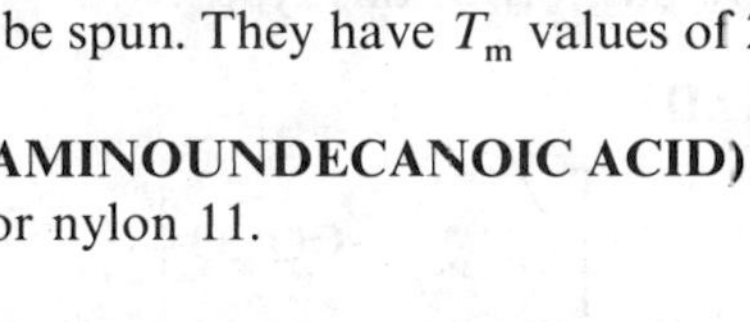

in which the polymer has a positive charge at low pH and a negative charge at high pH. Even more important is the amphoteric nature of proteins which carry both carboxyl groups (as in aspartic acid units) which are negatively charged and the positively charged units as in histidine, arginine, etc., the state of ionisation being pH dependent. The pH at which the net charge is zero is the isoelectric point, at which the polymer is tightly coiled due to attractions between opposite charges. Far from this pH, the coils will be expanded. Unlike normal polyelectrolytes an increase in ionic strength will increase expansion.

POLYANHYDRIDE A polymer of the type

$$-\!\!\!-\!\!\left[RCOOCO \right]_n\!\!\!-$$

where R is a hydrocarbon group. Best known are the aromatic polyanhydrides,

prepared by melt polymerisation of an aromatic dicarboxylic acid using acetic anhydride. The polymers have good hydrolytic stability, but with the exception of poly-(terephthalic anhydride) (T_m value of about 400°C), not exceptionally high T_m' values.

POLYANHYDROGLUCOSE Alternative name for glucan. Since such a polymer can be formally considered to be formed by step-growth polymerisation of glucose with loss of water, it can be said to consist of anhydroglucose units.

POLY-(ARYL...) Alternative name for poly-(arylene...).

POLYARYLATE Tradenames: Ardel, Arylef, Arylon, Durel. A wholly aromatic polyester derived from reaction of a dihydric phenol with an aromatic dicarboxylic acid. Best synthesised by reaction between a diacid chloride and the phenol either in the melt or in a high boiling point solvent. Characterised by their high T_m values (can be above 500°C), good thermal stability and solubility in chlorinated solvents. Several commercial polymers, which are copolymers of tere- and isophthalic acid and bisphenol A, have been developed as engineering thermoplastics intermediate in properties between bisphenol A polycarbonate and polyethersulphones. These commercial polymers are amorphous with T_g values of about 170°C and upper service life temperatures of 140–150°C. Arylon has somewhat lower softening behaviour but is somewhat cheaper.

POLY-(ARYLENE...) (Arylene polymer) (Poly-(aryl...)) Prefix for a polymer of the type

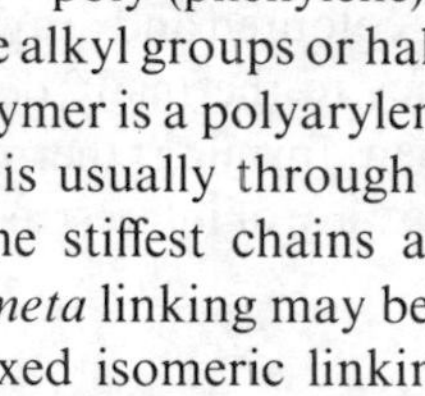

where X is a single atom or simple linking group and A, B, C and D are substituent atoms or groups (often A = B = C = D = hydrogen—poly-(phenylene), otherwise the substituents are simple alkyl groups or halogen atoms). When X is absent the polymer is a polyarylene. Linking between the benzene rings is usually through the *para* positions, since this gives the stiffest chains and the highest T_m values. However, *meta* linking may be desired to increase processability. Mixed isomeric linking is often present, arising from mixed substitution during synthesis, e.g. in a Friedel–Crafts polymerisation.

POLYARYLENESULPHONE Alternative name for polyethersulphone.

POLY-(ARYLENE TRIAZOLE) A polymer containing both arylene (usually phenylene) rings and triazole rings in the polymer chain. Synthesised by reaction of an aromatic polyhydrazide with aniline:

$$n \; \text{-}\!\!\left[\text{ArCONHNHCO}\right]_n \!\!\text{-} + n \; \text{PhNH}_2 \rightarrow$$

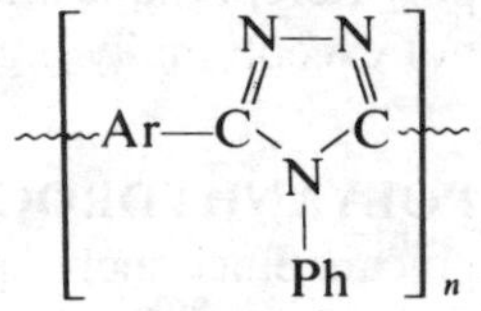

The polymers may be spun from DMF/LiCl solution to fibres or cast as films. They do not have such good thermal stability as the polyoxadiazoles or polybenzimidazoles.

POLY-(ARYL ESTER) A polymer containing both benzene rings and ester groups in the polymer chain. The polymers generally have T_m values in the range 130–350°C and good mechanical properties. Examples include poly-(ethylene terephthalate) and poly-(tetramethylene terephthalate).

POLY-(ARYL ETHER) A polyether containing benzene rings in the polymer chain as well as ether links. In particular, sometimes used as an alternative name for poly-(2,6-dimethyl-1,4-phenylene oxide).

POLYARYLETHERSULPHONE A polyethersulphone with benzene rings linking the ether and sulphone groups. All commercial materials are polyarylsulphones, but are usually referred to as polyethersulphones or simply as polysulphones.

POLY-(ARYL SULPHONE) Alternative name for an aromatic polyethersulphone, particularly Astrel 360.

POLYASPARTAMIDE Alternative name for poly-aminobismaleimide.

POLYAZELAIC POLYANHYDRIDE (PAPA)

$$\text{HO-}\!\!\left[\text{CO(CH}_2)_7\text{CO}_2\right]_n\!\!\text{-H}$$

Useful as a flexibilising curing agent for epoxy resins.

POLYAZINE A polymer of structure

$$\text{-}\!\!\left[\text{CH}\!=\!\text{N}\!-\!\text{N}\!=\!\text{CH}\!-\!\text{R}\right]_n$$

where R is usually an aromatic ring. Synthesised by reaction of hydrazine with a dialdehyde or diketone. The polymers are usually coloured and are difficult to dissolve. They show a tendency to thermally decompose with the evolution of nitrogen forming thermally stable polystilbenes and therefore are of interest as potential ablative polymers.

POLYAZOBENZENE (Polyazophenylene) (Azo polymer) A polymer of structure $\text{-}\!\!\left[\text{N}\!=\!\text{N}\!-\!\text{Ar}\right]_n$ where Ar is an aromatic ring. Synthesised by decomposition of aromatic bisdiazonium compounds with copper salts or by coupling with bisphenols or quinones. The polymers are usually coloured and of low molecular weight. They have reasonable thermal stability and show semiconductor properties. High molecular weight polymers may be synthesised by oxidative coupling of simple aromatic diamines, e.g. p-phenylenediamine, with oxygen or cuprous chloride, or by heating with t-butyl peroxide at 200°C.

POLYAZOMETHINE (Poly-(Schiff base)) (Schiff base polymer) (Azomethine polymer) (Polyimine) A polymer of structure

$$\text{=}\!\!\left[\text{N}\!-\!\text{R}\!-\!\text{N}\!=\!\text{CH}\!-\!\text{R}'\!-\!\text{CH}\right]_n$$

Synthesised by polycondensation between a diamine and a dialdehyde. Although reaction proceeds readily even at ambient temperature, high molecular weight polymers are difficult to obtain. The aromatic polymers have high T_m values, are coloured and have high thermal stability. Thus if R = Ph and R' is absent the T_m value is > 425°C. Substituted polymers may be used to form Schiff base chelate polymers by coordination with metal ions.

POLYAZOPHENYLENE Alternative name for polyazobenzene.

POLY-(p-BENZAMIDE) Tradenames: Fibre B, Terlon. Synthesised by solution self-polycondensation of p-aminobenzoyl chloride hydrochloride in tetramethylurea solution. Can be spun to high modulus, high temperature resistant fibres useful for reinforcement. Early versions of PRD-49 were possibly of this type.

POLYBENZIL

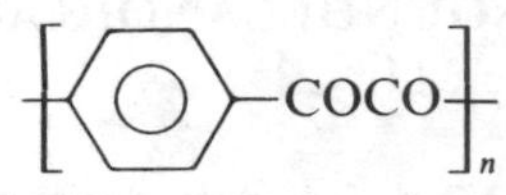

Synthesised by oxidation of polybenzoin with, for example, nitric acid. Melts with decomposition at about 350°C.

POLYBENZIMIDAZOLE (PBI) Tradename: Imidite. A polymer containing the structure

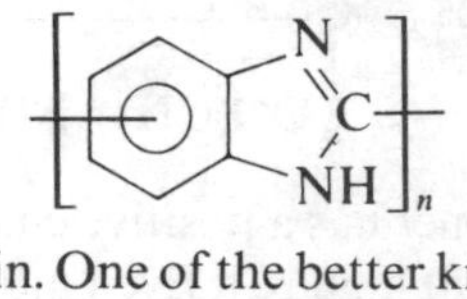

in the polymer chain. One of the better known of the really high temperature resistant polymers. Usually synthesised by reaction between an aromatic tetramine and a dicarboxylic acid derivative (preferably the diphenyl ester) or the acid itself in PPA solution. A widely studied PBI is made from 3,3'-diaminobenzidine:

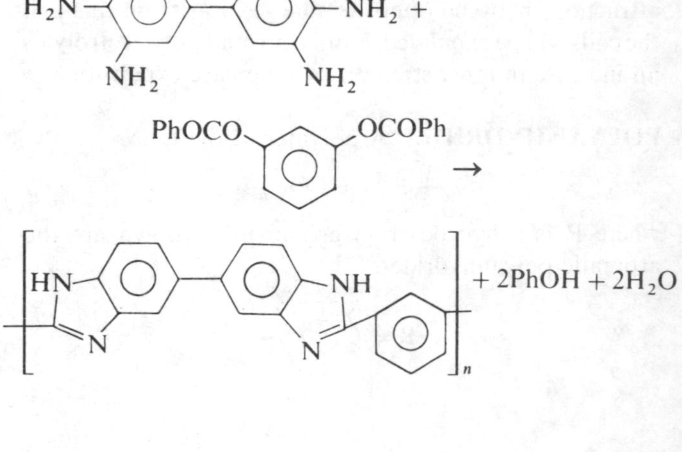

The conversion is often limited to low molecular weight, tractable polymers which can be melt or solution processed to fibres, films, etc. The fabricated product is then further cyclised by heating to 400°C. The resultant PBI is then only soluble in strong acids (e.g. formic acid) and is infusible ($T_m > 700°C$). The wholly aromatic polymers have outstanding resistance to basic hydrolysis and thermal stability in an inert atmosphere. However, thermo-oxidative stability is not as good as in the polyimides. Commercial products include laminating resins, adhesives (Imidite) and fibres.

POLYBENZIMIDAZOLINE A polyimidazoline with the imidazoline ring fused to a benzene ring.

POLYBENZIMIDAZOLONE A polymer of general structure:

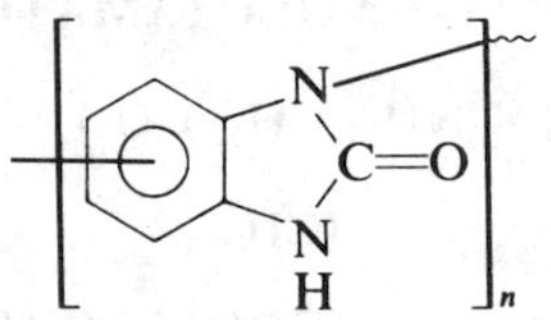

A commercial polymer of this type has been developed for use in reverse osmosis membranes.

POLYBENZIMIDAZOQUINAZOLONE A ladder polymer containing the heterocyclic ring system

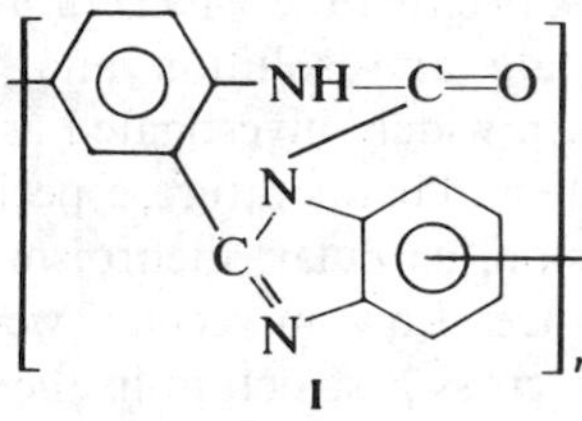

often of the type $-[(I)-(I)-X-]_n$. Synthesised by reaction between a bisbenzoxazinone and an aromatic bis-(o-diamine). A polybenzimidazoline precursor is first formed, alkoxycarboxylated, then cyclised.

POLYBENZIMIDAZOQUINOXALINE A polymer containing the heterocyclic ring system I, where R is a methyl or phenyl group. Synthesised by reaction between a bisbenzoxazinone (II) and a bis-(o-diamine) in PPA.

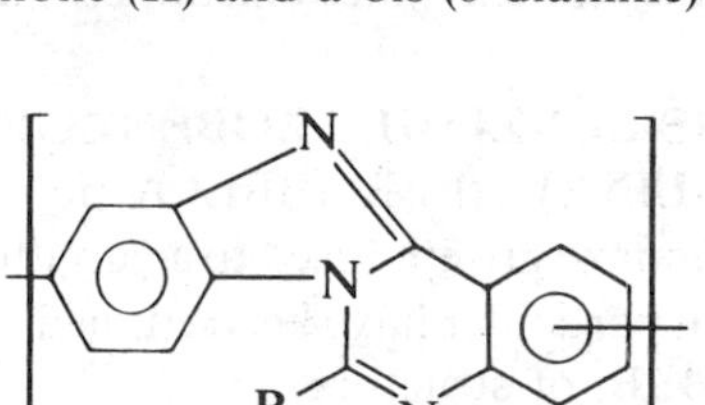

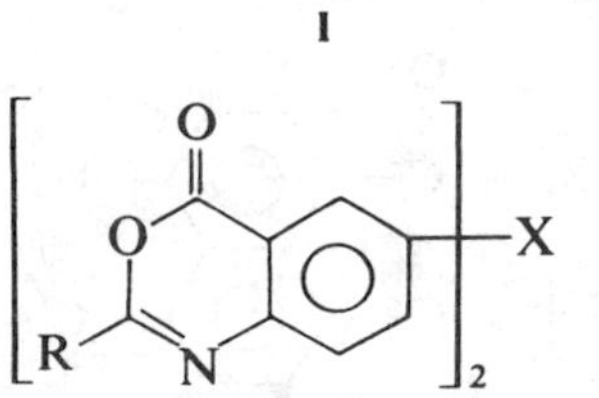

POLYBENZOIN

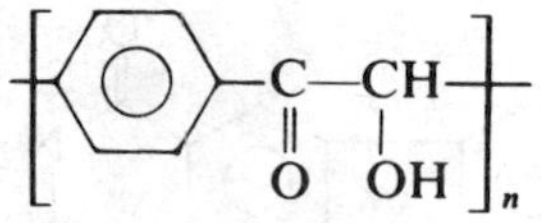

Synthesised, at least as a low molecular weight polymer, by carrying out the benzoin condensation on terephthalaldehyde. The polymer is highly insoluble, has a T_m value of about 180°C and is often coloured. It crosslinks on heating and is oxidised by nitric acid to polybenzil.

POLYBENZOPYRAZINE Alternative name for polyquinoxaline.

POLYBENZOTHIAZOLE A polymer containing the heterocyclic ring structure

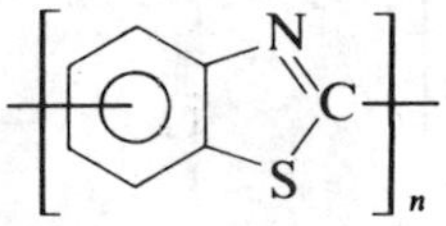

in the polymer chain. Synthesised by reaction between a bis-(o-mercaptoamine) (preferably as its hydrochloride, and commonly 3,3'-mercaptobenzidine) and a dicarboxylic acid derivative, especially the diphenyl ester, e.g.

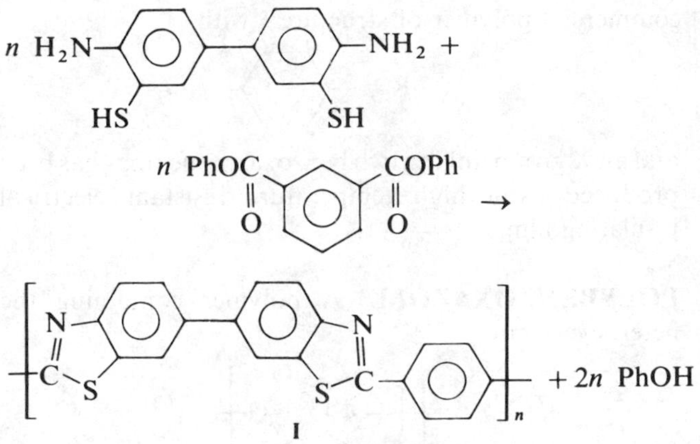

or by self-condensation of a mercaptoaminobenzoic acid in PPA solution. Also obtained simply by heating a toluidine with elemental sulphur at about 200°C. The polymers are only soluble in strong acids, e.g. sulphuric acid, and have excellent hydrolytic resistance to both acids and bases. The polymers have high T_g values (about 400°C for I) and high thermal and thermo-oxidative stability—up to about 400°C. Potential applications are for high temperature resistant coatings, adhesives and laminating resins.

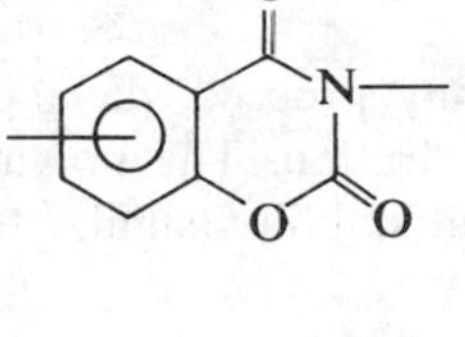

POLYBENZOXAZINDIONE A polymer containing the structure of the type

in the repeat unit. The polymer

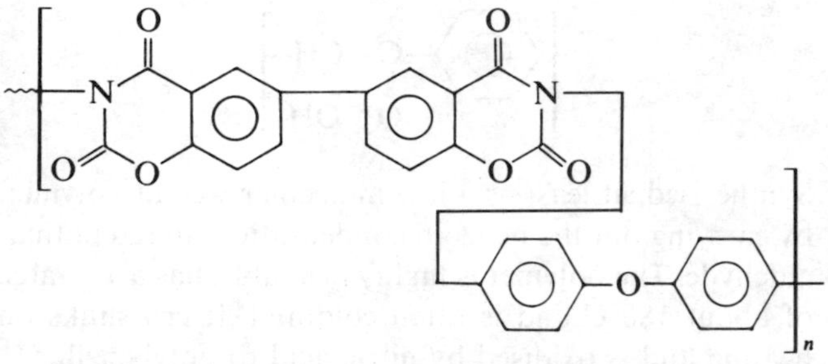

is produced commercially by reaction of an arylbis-(*o*-hydroxycarboxylic acid ester) with a diisocyanate.

POLYBENZOXAZINONE   A polymer of structure

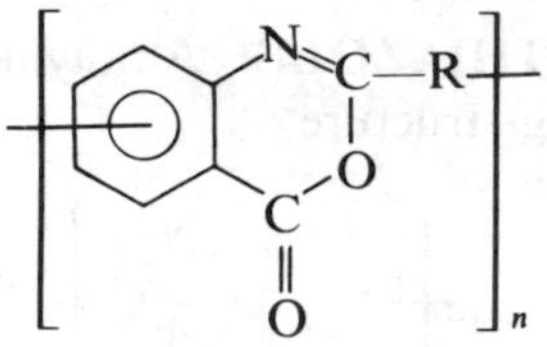

where R is usually an aromatic ring. Synthesised by polycondensation between a 4,4′-diaminodiphenyl-3,3′-dicarboxylic acid and a diacid halide, which occurs via a tractable polyamide–acid precursor polymer, which subsequently cyclodehydrates. The good thermal stability of the polymers is typical of a heterocyclic polymer. A commercial polymer of structure **I** with

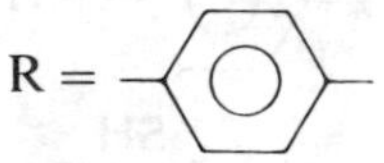

and an X group linking two benzoxazinone rings has been produced as a high temperature resistant electrical insulating film.

POLYBENZOXAZOLE   A polymer containing the heterocyclic ring

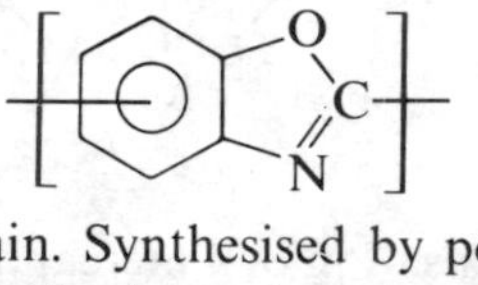

in the polymer chain. Synthesised by polycondensation between a bisaminohydroxyphenyl compound and a dicarboxylic acid or its derivative, or by self-condensation of an aminohydroxyphenylcarboxylic acid, e.g.

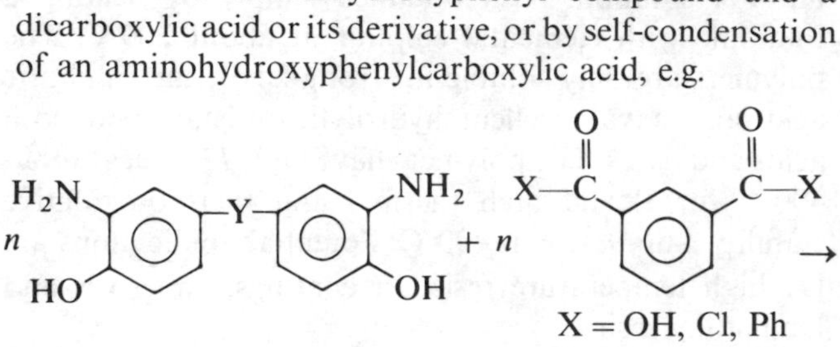

$$ + 2n\ HX + 2n\ H_2O $$

Polymerisation may proceed via a polyamide–phenol, followed by dehydration. The polymers are coloured, have good thermal stability (similar to **PBI** in nitrogen,

better in oxygen) and good hydrolytic stability. However, they are only soluble in strong acids (sulphuric and PPA) and are difficult to fabricate. Two commercial polymers are Metolon ($Y = CH_2$) and poly-(adamantylbenzoxazole).

POLYBENZYL   Alternative name for polymethylenephenylene.

POLY-β-BENZYL-α-L-ASPARTATE

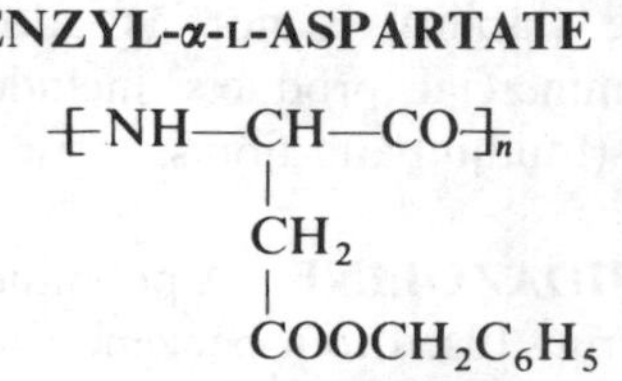

A synthetic polypeptide which can adopt the unusual left-handed α-helix in solution.

POLY-γ-BENZYL-α-L-GLUTAMATE   (PMLG)

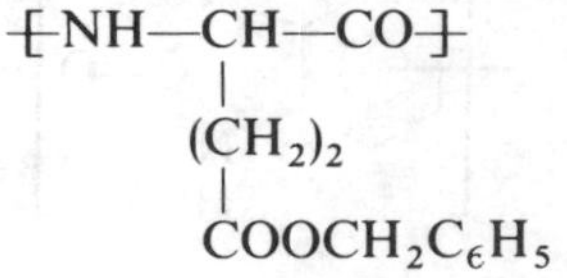

Synthesised by polymerisation of the *N*-carboxyanhydride of γ-benzyl-α-L-glutamic acid with strong base, e.g. $^-OCH_3$, in an inert solvent, e.g. dioxane. The conformation of the polymer has been extensively investigated in both the solid state and in solution. The polymer may exist in either the α-helix or β-conformation in both solid state and solution and the helix–coil transition has been widely investigated as a function of pH, solvent nature and temperature, especially by optical rotatory dispersion, circular dichroism and nuclear magnetic resonance. Low molecular weight polymers tend to form the cross-β structure in the solid state. In solution in, for example, chloroform, rigid rod-like helices exist, whereas in powerful hydrogen bonding solvents such as dichloroacetic acid the conformation is the random coil. The polymer provides the classical example of a lyotropic polymer, capable of liquid crystalline polymer behaviour in solution.

POLY-(BIS-(4-AMINOCYCLOHEXYLENE)-DECANEDICARBOXAMIDE)   Alternative name for poly-(bis-4-(cyclohexyl)methanedodecanoamide).

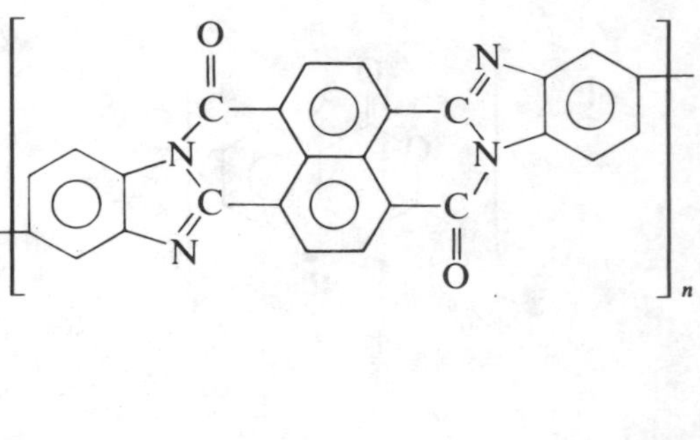

POLY-(BIS-BENZIMIDAZOBENZOPHEN-ANTHROLINE)   (BBB) (BBL) A polymer containing two benzimidazo groups fused to a naphthalene ring. Two polymers in particular have been studied. A partial ladder polymer (BBB) of structure has been synthesised by reaction between 1,4,5,8-naphthalenetetracarboxylic acid or anhydride and diaminobenzidine in PPA at about 200°C. This has outstanding thermal stability—no weight loss by TGA to 600°C in nitrogen or 500°C in air. BBL is the complete ladder polymer

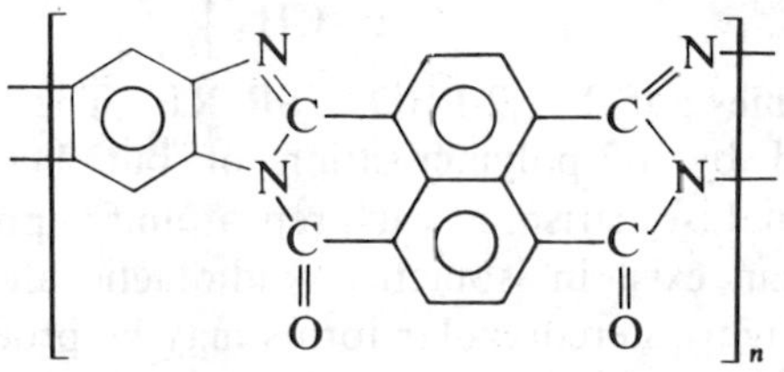

synthesised by reaction of 1,4,5,8-naphthalenetetra-carboxylic acid with 1,2,4,5-tetraminobenzene, but it does not show any significantly better thermal stability, probably because of incomplete cyclisation.

POLY-(3,3-BIS-(CHLOROMETHYL)OXACYCLO-BUTANE (Poly-(3,3-bis-(chloromethyl)oxetane)) (Oxetane polymer) (Chlorinated Polyether)

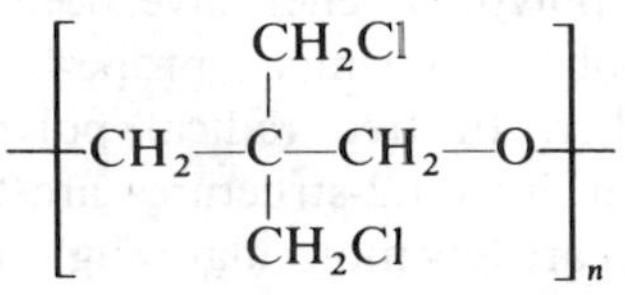

Tradename: Penton. Prepared by ring-opening polymerisation of 3,3-bis(chloromethyl)oxacyclobutane either by cationic polymerisation using a Friedel–Crafts catalyst such as BF_3 or $AlCl_3$ at low temperature (about $-50°C$), or by higher temperature polymerisation using an aluminium compound such as the alkoxide or hydride. The polymer crystallises in an α-form with a T_m value of 188°C from the melt, or in a β-form with a T_m value of 180°C by annealing the amorphous polymer above its T_g value (which is about 20°C). The polymer has only moderately good mechanical properties but has outstandingly good chemical and solvent resistance and has been used commercially as a protective coating material, e.g. in pipes and pumps.

POLY-(3,3-BIS(CHLOROMETHYL)OXETANE) Alternative name for poly-(3,3-bis-(chloromethyl)oxacyclobutane.

POLY-(BIS-(4-CYCLOHEXYL)METHANEDODE-CANOAMIDE) (Poly(bis-(4-aminocyclohexylene)de-canedicarboxamide))

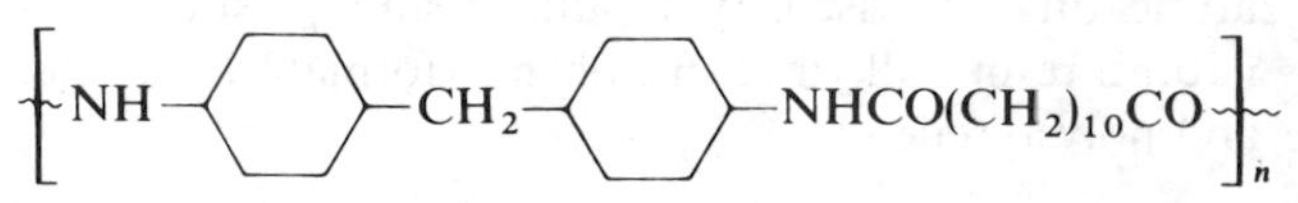

Tradenames: Qiana, Fibre Y. Prepared by reaction of bis-(p-aminocyclohexyl)methane with 1,10-decane-dicarboxylic acid. Useful as a fibre which provides fabrics of high aesthetic quality. Fibre tenacity is about 3 g denier^{-1}, elongation at break is about 30% and moisture regain is 2–2·5%. The polymer has a T_m value of 275°C.

POLYBISMALEIMIDE Alternative name for poly-aminobismaleimide.

POLY-(2,2-BIS-(4-PHENYLENE)-PROPANE CAR-BONATE) Alternative name for bisphenol A poly-carbonate.

POLYBIUREA (Polyurylene) A polymer of structure $-(NHCONHNHCONHR)_n$. Synthesised by reaction of a bisisocyanate with hydrazine hydrate or sulphate in a polar solvent. When R is aromatic, T_m values are in the range 250–400°C. The polymers are fibre forming with good ultraviolet light stability.

POLYBLEND Alternative name for polymer blend.

POLYBORATE Alternative name for borate glass.

POLYBOROAMIDE (Boronamide polymer) A polymer containing the repeat units

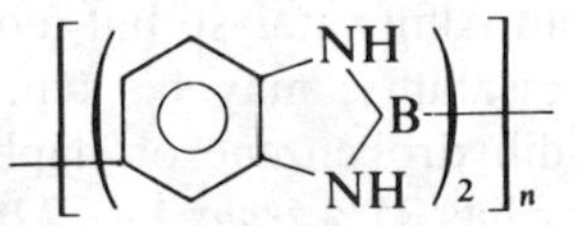

obtained by reaction of diboronic acid,

$$(HO)_2B—Ar—B(OH)_2$$

with a bis-(orthophenylenediamine). Aliphatic polymers have been similarly obtained and are water soluble. They have been suggested as useful textile sizes, coatings and adhesives.

POLYBOROSILOXANE Alternative name for poly-organoborosiloxane.

POLYBUTADIENE (BR) (PB) (PBD) Tradenames: Budene, Buna, Buna CB, Cariflex BR, Diene, Europrene Cis, Europrene Sol P, Intene, Solprene, Taktene. 1,4-Polybutadiene is

$$-(CH_2CH{=}CHCH_2)_n$$

and 1,2-polybutadiene is

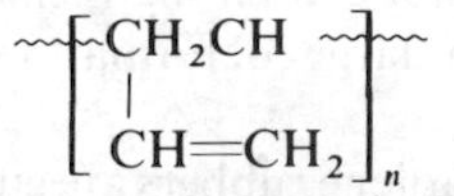

The polymer obtained by polymerisation of butadiene. Both 1,2- and 1,4-polymerisation can occur. In the former case both *cis* and *trans* isomers are possible. A major general purpose commercial rubber. Early rubber was produced by using sodium metal as the catalyst (Buna rubber). Such materials typically contained a mixture of isomers, including large amounts of the 1,2-isomer. They have a high gel content, wide molecular weight distribution (and hence poor processability) and high T_g values.

Lithium metal or lithium alkyls give high amounts of 1,4-polymer in non-polar media by anionic polymerisation. Commercially, n-butyl lithium is used, a typical product containing 40% *cis*-1,4-, 50% *trans*-1,4- and 10% 1,2-structures. These products are often called low *cis*-polybutadienes. They are linear and have molecular weights of $(1\cdot5-3\cdot0) \times 10^5$. Regularly branched star and comb shaped polymers may be prepared using either polyfunctional alkyl lithiums (CLi_4) or by terminating the living polymer chain ends using polyfunctional agents such as CH_3SiCl_3. Polymers with very high *cis*-1,4-contents may be synthesised by Ziegler–Natta polymerisation. Highest *cis*-1,4-contents (about 97%) (so-called high *cis*-polybutadiene) are obtained with an aluminium alkyl halide (e.g. $AlEt_2Cl$) and a cobalt compound (e.g. $CoCl_2$). Polymers with about 92% *cis*-1,4-content (medium *cis*-polybutadiene) are obtained with aluminium alkyl/TiI_4 combinations. Other Ziegler–Natta polymerisation catalysts can be used to produce *trans*-1,4- and 1,2-polymers. In the latter case syndiotactic polymer results from the use of $AlEt_3$ and the acetylacetonates of vanadium, molybdenum or cobalt, whereas isotactic polymer is produced with chromium acetylacetonate. *Trans*-1,4-polymer is also produced by the use of rhodium salts alone, due to their ability to form π-complexes with butadiene. Very high molecular weight polymer is produced using an Alfin catalyst, but molecular weight, and hence processability, may be kept lower by the addition of 1,4-dihydrobenzene or naphthalene. Typically, the polymer has 75% *trans*-1,4-, 22% 1,2- and 2% *cis*-1,4-structures. Free radical polymerisation of butadiene is not stereospecific and results in polymers typically having about 60% *trans*-1,4-, 20% *cis*-1,4- and 20% 1,2-structures. Such polymers, whilst not useful as rubbers as such, are useful for toughening of plastics as in HIPS and ABS.

1,4-Polybutadiene has one of the lowest T_g values of all polymers (about $-106°C$ for both *cis* and *trans* isomers), the value increasing with increasing 1,2-content, for example the isotactic 1,2-polymer has a T_g value of $-15°C$. Values for T_m are $3°C$ (*cis*-1,4-polymer), $145°C$ (*trans*-1,4-polymer), $128°C$ (isotactic 1,2-polymer) and $156°C$ (syndiotactic 1,2-polymer). Thus only the *cis*-1,4-polymer is a useful rubber, all the other stereo-regular isomers crystallising too readily. Only when the *cis*-1,4-polymer is very high in *cis* content will it crystallise and then only at very low temperatures. Thus polybutadiene rubbers do not exhibit high tensile strengths due to crystallisation on large deformation (as does natural rubber).

Useful polybutadiene rubbers are either the high (about 97%) *cis* or medium (about 92%) *cis* Ziegler–Natta polymers or the low (about 40%) *cis*-1,4-polymers produced with butyl lithium. They may be vulcanised by conventional sulphur vulcanisation to vulcanisates characterised by high resilience, high abrasion resistance and high ability to be oil extended. They are usually used in blends with natural or styrene–butadiene rubbers, e.g. in tyres where the mechanical hysteresis of polybutadiene when used alone would cause high heat build-up. Vulcanisate properties are not very dependent on initial raw polymer *cis*-1,4-content since considerable *cis*–*trans* isomerisation occurs on heating to vulcanisation temperatures. Equilibrium *trans* content is about 75%. Syndiotactic 1,2-polybutadiene has been used as a readily degradable plastic film material, whilst low molecular weight telechelic polymers of butadiene are useful as liquid rubbers.

1,2-POLYBUTADIENE (Vinylpolybutadiene)

$$\left[CH_2CH \atop \underset{\displaystyle CH=CH_2}{|} \right]_n$$

Tradenames: JSR 1,2-PBD, JSR-RB. The polymer produced by 1,2-polymerisation of butadiene. Configurational isomerism at each repeat unit is possible so that it can exist in isotactic, syndiotactic and atactic forms. Highly stereoregular forms may be produced by Ziegler–Natta polymerisation; for example, $AlEt_3$/vanadium acetylacetonate gives syndiotactic polymer. The polymers have T_g values of about $-15°C$ and T_m values of $128°C$ (isotactic) and $156°C$ (syndiotactic). Largely syndiotactic polymer (with 50–65% syndiotactic units) has been used as a degradable polymer for packaging, as it readily crosslinks and therefore embrittles on exposure to sunlight. Polymers containing 30–35% 1,2-content (the remainder being of *cis*-1,4-structure) (medium vinyl polybutadiene) have been developed as commercial rubbers of similar properties to styrene–butadiene rubber. In free radical polymerisation of butadiene about 20% 1,2-structures are formed. Polybutadiene resins are low molecular weight polymers with mostly 1,2-linkages.

POLY-(BUTADIENE-co-ACRYLONITRILE) Alternative name for butadiene–acrylonitrile copolymer.

POLY-(BUTADIENE-co-STYRENE) Alternative name for butadiene–styrene copolymer.

POLYBUTADIENE RESIN Tradenames: Buton, Nisso PB, Hystyl. A polybutadiene with mostly 1,2-linkages, i.e. a high vinyl polybutadiene. The commercial polymers are of low molecular weight, contain functional end groups for chain extension and are therefore crosslinkable, e.g. by peroxide vulcanisation.

POLYBUTENE Tradename: Indapol. Essentially low molecular weight polyisobutene, but obtained from a complex mixed C_4 petroleum fraction containing mainly isobutene but also butene-1 and *cis*- and *trans*-butene-2. It is produced by cationic polymerisation at low temperatures using a Friedel–Crafts catalyst, e.g. $AlCl_3$. Typical commercial products are viscous oils (of molecular weight from a few hundred to a few thousand) useful as lubricating oil viscosity modifiers and plasticisers for natural resins, alkyd resins, phenol–formaldehyde resins and polystyrene.

POLY-(1-BUTENE) Alternative name for polybutene-1.

POLYBUTENE-1

$$\left[\underset{\displaystyle CH_2CH}{} \atop CH_2CH_3 \right]_n$$

Tradenames: Butuf, Vestolen BT. Produced commercially in isotactic form by Ziegler–Natta polymerisation of butene-1. Unusually, isotactic polybutene-1 crystallises in three different polymorphic forms. On cooling from the melt form II crystallises initially, having a T_m value of 124°C and a density of 0·89 g cm^{-3}. This transforms spontaneously to form I at ambient temperatures having a T_m value of 135°C, a T_g value of -24°C and a density of 0·95 g cm^{-3}. Type III ($T_m = 106\cdot5$°C) is formed by crystallisation from solution. The polymer is a useful plastic, especially as a pipe material, since it has superior creep properties to both polyethylene and polypropylene, possibly due to the very high molecular weight of the commercial polymers. This enables pipes to be made with thinner walls. Otherwise its mechanical behaviour is between that of low density and high density polyethylene. Freshly extruded material (in crystalline form II) is mechanically weak and must be handled with care and then stored to allow the transformation to the stronger type I to occur.

POLY-(BUTENE-1-co-ETHYLENE)

Alternative name for ethylene–butene-1 copolymer.

POLY-(n-BUTYLACRYLATE)

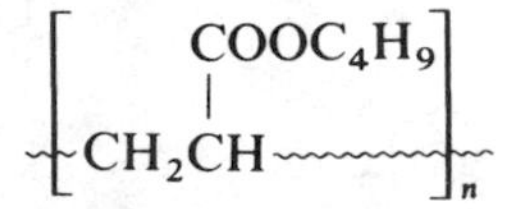

$$\left[\begin{array}{c} COOC_4H_9 \\ | \\ CH_2CH \end{array} \right]_n$$

Prepared by emulsion polymerisation of n-butyl acrylate. Its T_g value is about -55°C and it forms the basis of the acrylate rubbers (as copolymers with about 10% acrylonitrile) with better low temperature flexibility but poorer oil resistance than those based on polyethylacrylate. The rubbery polymers are also useful as impact modifiers in blends with polymethylmethacrylate and polyvinyl chloride.

POLYBUTYLENE TEREPHTHALATE (PBT)
(Polytetramethylene terephthalate)

$$\left[O{-}(CH_2)_4{-}OCO{-}\bigodot{-}CO \right]_n$$

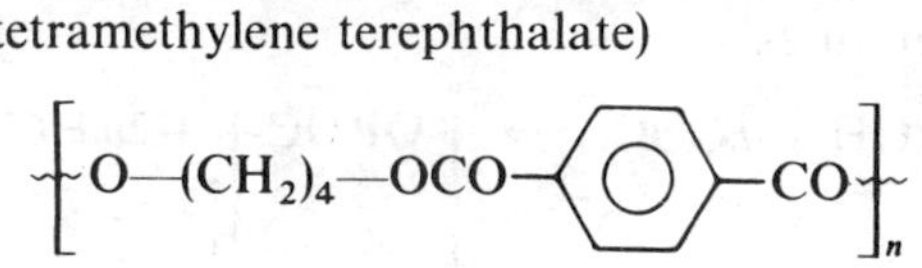

Tradenames: Arnite, Celanex, Crastine, Deroton, Gafite, Hostadur B, Orgater, Pibiter, Pocan, Techster T, Tenite, Ultradur, Versel, Vestodur. Prepared by ester interchange between butane-1,4-diol and dimethyl terephthalate in a similar manner to polyethylene terephthalate (PET). It has a lower T_m value (224°C) and T_g value (22°C when fully amorphous to 45°C when crystalline) than PET and due to the low T_g value it crystallises much more readily on moulding, giving mouldings without the defects of polyethylene terephthalate. Therefore it is used widely as an engineering plastic moulding material.

It has reasonably good mechanical properties with a tensile strength of about 55 MPa and a flexural modulus of about 230 MPa although these values are often higher in moulding materials due to the extensive use of glass fibre reinforcement. It has good dimensional stability,

especially in water, and high resistance to hydrocarbons. Together with polyethylene terephthalate it is often referred to as thermoplastic polyester.

POLY-(n-BUTYLMETHACRYLATE)

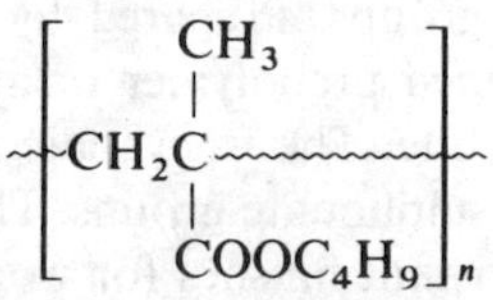

$$\left[\begin{array}{c} CH_3 \\ | \\ CH_2C \\ | \\ COOC_4H_9 \end{array} \right]_n$$

Tradename: Plexigum P. A somewhat rubbery polymer having a T_g value of about 20°C. Produced by free radical polymerisation of n-butyl methacrylate. It has found some use as a textile finish (as an emulsion) and as an adhesive.

POLY-(t-BUTYLSTYRENE)

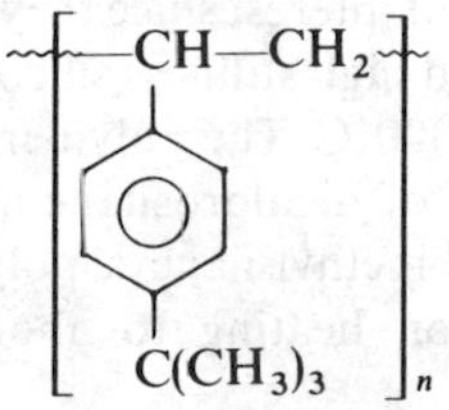

$$\left[\begin{array}{c} CH{-}CH_2 \\ \\ \bigodot \\ \\ C(CH_3)_3 \end{array} \right]_n$$

Produced by the thermal or free radical polymerisation of t-butylstyrene in a similar way to the preparation of polystyrene, but the monomer polymerises more readily. The polymer has a T_g value of 130–134°C.

POLYCAPRAMIDE
Alternative name for nylon 10.

POLYCAPRINOLACTAM
Alternative name for nylon 10.

POLY-(ω-CAPROAMIDE)
Alternative name for nylon 6.

POLYCAPROLACTAM
Alternative name for nylon 6.

POLYCAPROLACTONE

$$\sim\!\left[(CH_2)_5COO \right]_n$$

Tradename: Capa. A crystalline polyester with a T_m value of 62°C, usually prepared in a hydroxy-terminated form by polymerisation of caprolactone with a glycol as initiator to yield:

$$HO{-}\left[(CH_2)_5{-}COO \right]_n CH_2CH_2{-}OCO(CH_2)_5{-}OH$$

Useful as a polyester prepolymer for the preparation of polyurethane elastomers by reaction with a diisocyanate. It has a lower T_g value than poly-(ethylene adipate) but suffers from cold-hardening. It is also of interest for use as a modifying material in blends with other polymers, due to its unusually high compatibility. It is also unusual in that it is biodegradable.

POLYCAPRYLAMIDE
Alternative name for nylon 8.

POLYCAPRYLLACTAM
Alternative name for nylon 8.

POLYCARBAMATE Alternative name for polyurethane.

POLYCARBAMOYLSULPHONATE Tradenames: Elastron, Synthappret. A polymer containing —NHCOSO$_3^-$M$^+$ groups. Prepared by reaction of an isocyanate end-capped prepolymer or a polyisocyanate with sodium bisulphite. The isocyanate groups react to give the carbamoylsulphonate groups. The polymers are useful as crease resistant finishes for wool.

POLYCARBASILANE A polymer containing alternate silicon and carbon atoms in the polymer chain, i.e.

$$\left[Si - C \right]_n$$

Such polymers are of interest since they may be converted to high strength and high stiffness silicon carbide (SiC) on heating to about 1300°C. The polymers are produced by reacting a mixture of dichlorosilane and sodium metal, e.g. with dichlorodimethylsilane a polydimethylsilane is produced, which on heating to about 500°C gives a carbasilane:

$$CH_3 \atop SiCl_2 + Na \rightarrow \sim\!\!\left[\begin{array}{c} CH_3 \\ Si \\ CH_3 \end{array} \right]_n\!\!\sim + (2n\ NaCl) \rightarrow \sim\!\!\left[\begin{array}{c} CH_3 \\ Si \\ CH_2 \\ Si \\ CH_3 \end{array} \right]_n\!\!\sim$$

Similarly polysilastyrene is produced from dimethyldichlorosilane and phenylmethyldichlorosilane. This polymer, on heating to about 1400°C, forms β-SiC. Composites with SiC or Si$_3$N$_3$ (silicon nitride) or metal powders dispersed in a silicon carbide matrix can be formed by heating a dispersion of the required dispersed phase in a polycarbasilane matrix.

POLYCARBATHIANE A polymer whose chains consist of alternating carbon and sulphur atoms, i.e.

$$\sim\!\!\left[C - S \right]_n\!\!\sim$$

Examples include polymers of carbon disulphide and polyfluorothioacetone

$$\sim\!\!\left[CF_2 - S \right]_n\!\!\sim$$

POLYCARBAZANE A polymer whose chains consist of alternating carbon and nitrogen atoms linked through single bonds, i.e.

$$\sim\!\!\left[C - N \right]_n\!\!\sim$$

Polyisocyanates and polycarbodiimides are examples.

POLYCARBAZENE A polymer whose chains consist of alternating carbon and nitrogen atoms linked by double bonds, i.e.

$$\left[C = N \right]_n$$

Polynitriles, e.g. polyfumaronitrile, are examples.

POLYCARBODIIMIDE A polymer containing the group $-\!\!\left[N = C = N \right]\!\!-$ in the polymer chain. Synthesised by self-addition polymerisation of a diisocyanate using organophosphorous compounds, especially phosphoric oxides R$_3$PO:

$$n\ OCNRCNO \rightarrow \left[N = C = NR \right]_n + n\ CO_2$$

The polymers have properties rather similar to nylons. The CO$_2$ formation can yield useful foams with good sound and thermal insulation for use in buildings and with low flammability. Another type of polycarbodiimide is the polymer formed by anionic polymerisation of a carbodiimide, e.g.

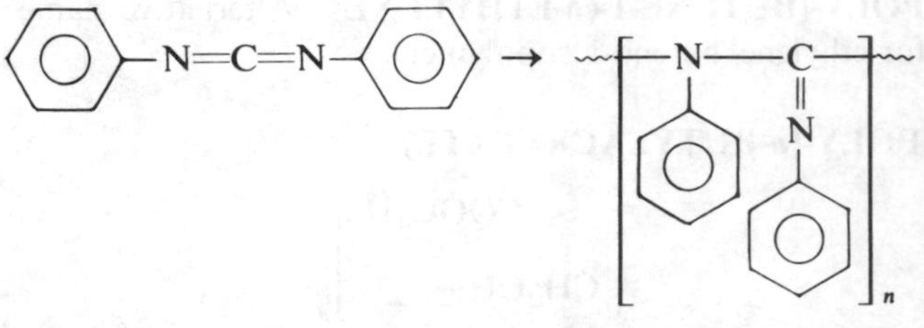

These polymers are highly unstable, readily reverting to monomer. An example of a polycarbazane.

POLYCARBONATE A polyester of carbonic acid, i.e. having repeating units of the type

$$\left[\begin{array}{c} ROCO \\ \| \\ O \end{array} \right]_n$$

Prepared either by the reaction of phosgene (the diacid chloride of the non-existent carbonic acid) with a diol at low temperatures

$$n\ HOROH + n\ COCl_2 \rightarrow \left[\begin{array}{c} OROC \\ \| \\ O \end{array} \right]_n + 2n\ HCl$$

or by ester interchange of a dicarbonate with a diol at elevated temperatures

$$n\ R'OCOR' + n\ HOROH \rightarrow \left[\begin{array}{c} OROC \\ \| \\ O \end{array} \right]_n + 2n\ R'OH$$

Aliphatic polycarbonates, although widely investigated, have melting temperatures and resistance to hydrolysis which are too low to be useful. However, polycarbonates of phenolic diols, especially bisphenols of the type

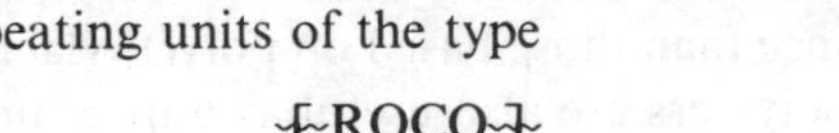

$$HO - \bigcirc - CR_2 - \bigcirc - OH$$

I

have much more useful properties. Since the only significant commercial product is the polycarbonate based on bisphenol A ($R = CH_3$ in **I**), the term poly-carbonate often refers to this particular material. This polymer is also frequently called bisphenol A poly-carbonate.

POLYCARBORANE Alternative name for carborane polymer.

POLYCARBORANESILOXANE (Polysiloxanecar-borane) (Si–B polymer) Tradenames: Dexsil, Ucarsil. A carborane polymer in which the carborane units are linked through siloxane bridges. In general the structure is,

$$\{CB_{10}H_{10}C(Si(CH_3)_2-O)_mSi(CH_3)_2\}_n$$
I

where m is 2 or 4 (Si-B2 and Si-B4 respectively). Sometimes some methyl groups are replaced by phenyl groups and the decaborane may be pentaborane ($-CB_5H_5C-$). Commercial high temperature resistant polymers contain vinyl borane end groups through which vulcanisation with peroxides may be achieved. T_g values are from $-60°C$ to $-30°C$, except when $m = 1$ when the T_g value is $25°C$ and the T_m value is $240°C$. The monomers are prepared from carborane by the sequence of reactions:

$$HCB_{10}H_{10}CH + 2Li \rightarrow LiCB_{10}H_{10}CLi \xrightarrow{(CH_3)_2SiCl_2}$$

$$ClSi(CH_3)_2CB_{10}H_{10}Si(CH_3)_2Cl \xrightarrow{CH_3OH}$$
II

$$CH_3OSi(CH_3)_2CB_{10}H_{10}CSi(CH_3)_2OCH_3$$
III

Self condensation of **II** with **III** with elimination of CH_3Cl gives **I** (with $n = 1$), whereas reaction of **III** with $(CH_3)_2SiCl_2$ gives **I** (with $n = 2$) and reaction with

$$ClSi(CH_3)_2-O-Si(CH_3)_2Cl$$

gives **I** (with $n = 3$). The rubbers may be reinforced with fume silica and are stable in air to about $260°C$; in an inert atmosphere they are stable to much higher temperatures than silicone rubbers.

POLY-(m-CARBORANESILOXANE) (SiB) Trade-name: Dexsil. A type of modified polyorganosiloxane whose repeat unit is typically of the structure:

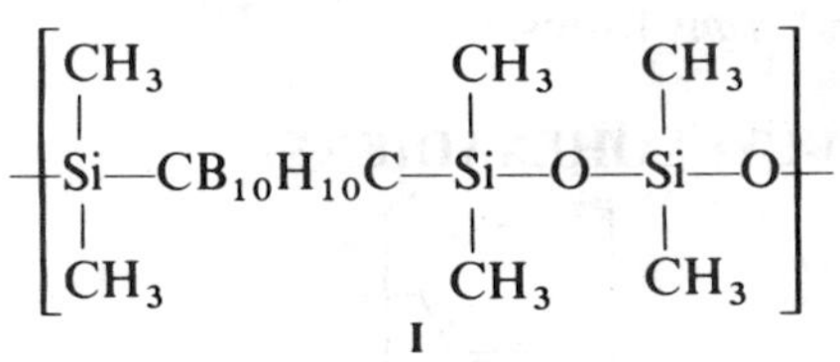

$$\begin{bmatrix} CH_3 & & CH_3 & CH_3 \\ | & & | & | \\ Si-CB_{10}H_{10}C-Si-O-Si-O \\ | & & | & | \\ CH_3 & & CH_3 & CH_3 \end{bmatrix}$$
I

where $-CB_{10}H_{10}C-$ is the m-decaborane structure. The above polymer would be designated SiB-2 since it has two siloxy groups, but other polymers with more siloxy groups between the carborane rings have been syn-thesised. The polymers are formed either by alkyl chloride elimination from equimolar mixtures of dichloro- and

dialkoxycarboranesiloxanes at $140–165°C$

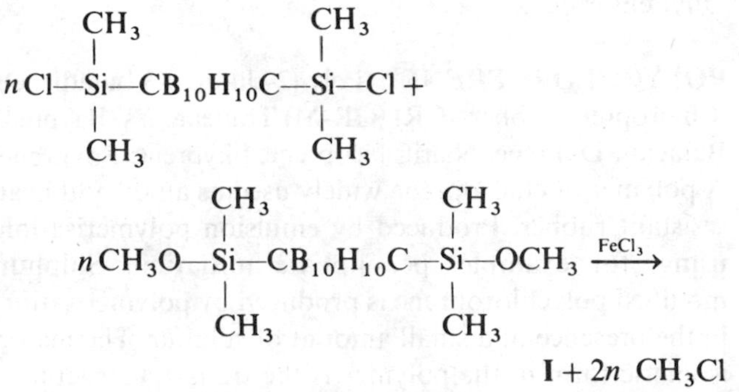

$$n \; Cl-\underset{\underset{CH_3}{|}}{\overset{\overset{CH_3}{|}}{Si}}-CB_{10}H_{10}C-\underset{\underset{CH_3}{|}}{\overset{\overset{CH_3}{|}}{Si}}-Cl +$$

$$n \; CH_3O-\underset{\underset{CH_3}{|}}{\overset{\overset{CH_3}{|}}{Si}}-CB_{10}H_{10}C-\underset{\underset{CH_3}{|}}{\overset{\overset{CH_3}{|}}{Si}}-OCH_3 \xrightarrow{FeCl_3}$$

$$\mathbf{I} + 2n \; CH_3Cl$$

or by condensing a carboranesilanol with a reactive silane

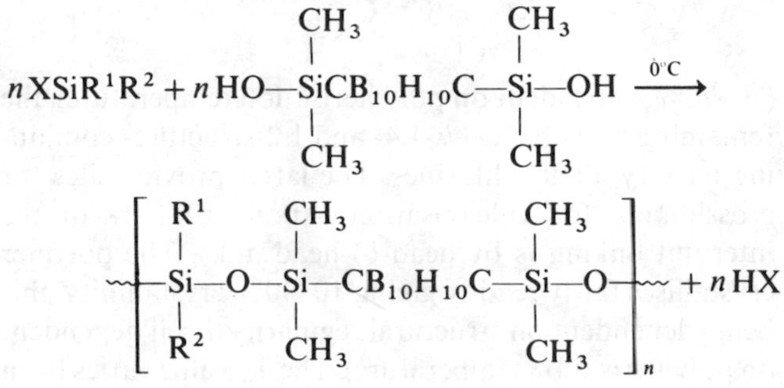

$$n \; XSiR^1R^2 + n \; HO-\underset{\underset{CH_3}{|}}{\overset{\overset{CH_3}{|}}{Si}}CB_{10}H_{10}C-\underset{\underset{CH_3}{|}}{\overset{\overset{CH_3}{|}}{Si}}-OH \xrightarrow{0°C}$$

$$\begin{bmatrix} R^1 & CH_3 & & CH_3 \\ | & | & & | \\ Si-O-Si-CB_{10}H_{10}C-Si-O \\ | & | & & | \\ R^2 & CH_3 & & CH_3 \end{bmatrix}_n + n \; HX$$

POLYCARBOXANE A polymer whose chains consist of alternating carbon and oxygen atoms, i.e.

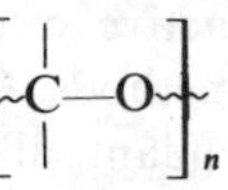

$$\begin{bmatrix} | \\ C-O \\ | \end{bmatrix}_n$$

The best known example is polyoxymethylene.

POLYCARBOXYLATE Alternative name for halato-polymer.

POLYCATENANE Alternative name for catenane polymer.

POLYCHELATE Alternative name for chelate polymer.

POLYCHLAL Tradename for polyvinyl chloride/polyvinyl alcohol fibre.

POLYCHLORAL

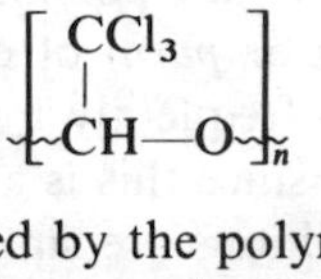

$$\begin{bmatrix} CCl_3 \\ | \\ CH-O \end{bmatrix}_n$$

The polymer obtained by the polymerisation of chloral (CCl_3CHO) using, for example, Lewis bases or organo-metallic catalysts at relatively low temperatures due to the low (about $58°C$) ceiling temperature for the polymer. With organometallic catalysts the polymer may be isotactic, crystallising with a T_m value of about $140°C$. The polymer has low flammability, good chemical and solvent resistance but is difficult to mould.

POLY-(2-CHLORO-1,3-BUTADIENE) Alternative name for polychloroprene.

POLYCHLOROETHENE Alternative name for poly-vinyl chloride.

POLYCHLOROPRENE (Poly-(2-chloro-1,3-butadiene)) (Chloroprene rubber) (CR) (GR-M) Tradenames: Baypren, Butaclor, Duprene, Nairit, Neoprene, Skyprene, Sovprene. A polymer of chloroprene widely used as an oil and heat resistant rubber. Produced by emulsion polymerisation using, for example, persulphate initiation. Sulphur modified polychloroprene is produced by polymerisation in the presence of a small amount of sulphur. The major isomeric form in the polymer is the trans-1,4-structure

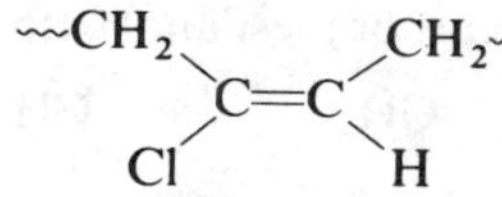

$$\sim CH_2 \qquad CH_2 \sim$$
$$C = C$$
$$Cl \qquad H$$

(70–90%, dependent on polymerisation temperature), the remaining units being *cis*-1,4- and 1,2-structures containing tertiary allylic chlorines. The latter provide sites for crosslinking for vulcanisation. About 10–15% of the interunit linking is by head-to-head links. The polymer crystallises fairly readily giving 10–40% crystallinity, this being dependent on structural regularity (itself dependent on polymerisation temperature). The T_m value varies from 45–75°C for polymerisation temperatures from -40°C to $+40$°C. The all-*trans* polymer has a T_g value of -45°C and a T_m value of $+45$°C.

Polychloroprene is inactive to conventional accelerated sulphur vulcanisation methods. Crosslinking is brought about through the pendant allylic chlorines on 1,2-chloroprene units, either by heating with a diamine, with a dihydric phenol or with a mixture of zinc and magnesium oxides or by reaction with ethylenethiourea. Sulphur modified polychloroprene can be vulcanised with zinc and magnesium oxides alone or combined with ethylene-thiourea. Polychloroprene is used for hose, belts, wire and cable covering, coatings, sheeting and adhesives where good oil resistance is required.

POLYCHLOROSTYRENE

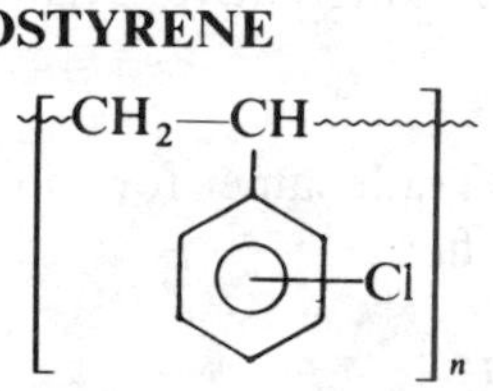

$$\left[CH_2 - CH \sim \right]_n$$

A polymer produced by the polymerisation of one of the isomers (*ortho*, *meta* or *para*) of chlorostyrene or of a mixture of isomers—frequently a mixture of *ortho* and *para* isomers is used since this is a commercial product. Polymerisation conditions are similar to those used for styrene, but polymerisation occurs more readily. The polymers have higher T_g values (about 110°C) than polystyrene (*para* polymer, 110°C; *meta* polymer, 90°C; *ortho* polymer 120°C) and have greater fire resistance.

POLYCHLOROTRIFLUOROETHYLENE (PCTFE)

$$\left[CF_2 CFCl \right]_n$$

Tradenames: Aclar, Daiflon, Fluorothene, Halon, Hostaflon C, Kel F, Voltalef. Like polytetrafluoroethylene it is prepared by free radical polymerisation in aqueous systems and is of similar properties. However, it does not crystallise so readily and has a lower T_m value (about 216°C). Its electrical and chemical resistances are also inferior. However, in thin films it is more transparent and has a higher tensile strength and hardness. It may be conveniently melt processed at 230–290°C and is useful for seals, gaskets and in certain electrical applications.

POLY-(CHLOROTRIFLUOROETHYLENE-co-ETHY-LENE) Alternative name for chlorotrifluoroethylene–ethylene copolymer.

POLYCLAR Tradename for poly-(*N*-vinylpyrrolidone).

POLYCONDENSATION Alternative name for condensation polymerisation.

POLYCYANOTEREPHTHALYLIDENE

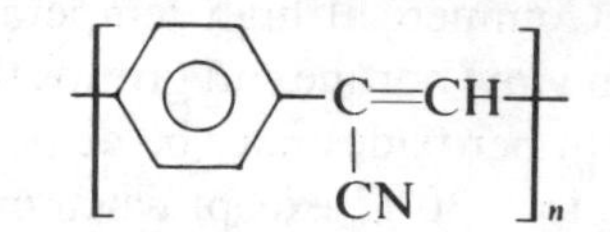

$$\left[\bigcirc - C = CH \right]_n$$
$$\qquad\qquad CN$$

Prepared by Knoevenagel condensation between terephthalaldehyde and α,α'-dicyanoxylene. Highly insoluble and coloured polymers result, which do not soften before decomposing at 450–500°C.

POLYCYCLAMIDE A polyamide containing aliphatic rings in the chain, e.g. poly-(1,4-cyclohexylenedi-methylenesuberamide). They have higher T_m values and softening points than linear aliphatic polyamides and are less sensitive to water absorption.

POLYCYCLOBUTENE

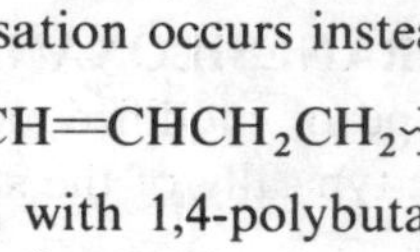

$$\left[\begin{array}{c} CH - CH \sim \\ | \qquad | \\ CH_2 - CH_2 \end{array} \right]_n$$

Prepared by Ziegler–Natta polymerisation of cyclobutene. The double bond opening occurs in a *cis* fashion. The rings may be on alternately opposite sides of the chain (erythrodisyndiotactic) or all on the same side of the chain (erythrodiisotactic). With some catalysts ring-opening polymerisation occurs instead to give

$$\left[CH = CHCH_2 CH_2 \right]_n \sim$$

which is identical with 1,4-polybutadiene and exists in both *cis* and *trans* forms.

POLY-(1,3-CYCLOHEXADIENE)

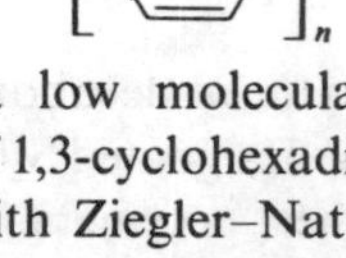

$$\left[\bigcirc \right]_n$$

Synthesised as a low molecular weight polymer by polymerisation of 1,3-cyclohexadiene with free radical or acid catalysts. With Ziegler–Natta or anionic catalysts, high molecular weight, amorphous polymers are obtained. The polymer may be dehydrogenated to poly-(*p*-phenylene) by heating with chloranil.

POLY-(CYCLOHEXANE-1,4-DIMETHYLENE TEREPHTHALATE) (Poly-(cyclohexylenedimethylene terephthalate))

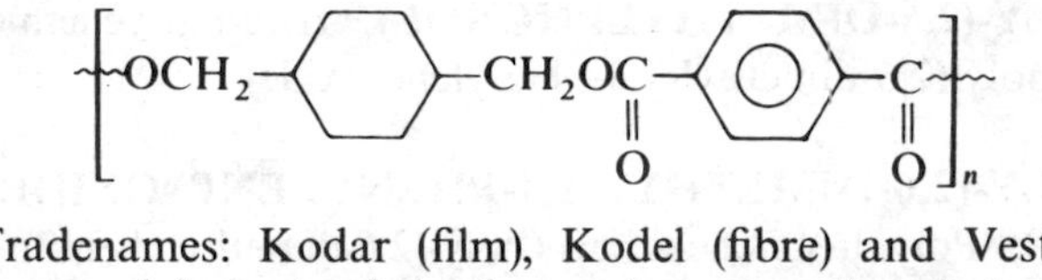

Tradenames: Kodar (film), Kodel (fibre) and Vestan. Prepared by ester interchange between dimethyl terephthalate and 1,4-dimethylolcyclohexane in a manner similar to that used for polyethylene terephthalate. It has a T_g value of about 130°C and a T_m value of about 290°C. Both the fibre and film form have higher water resistance and better weatherability than polyethylene terephthalate.

POLY-(1,4-CYCLOHEXYLENEDIMETHYLENE-SUBERAMIDE) Alternative name for poly-(transhexahydro-p-xylylenesuberamide).

POLY-(CYCLOHEXYLENE-1,4-DIMETHYLENE TEREPHTHALATE) Alternative name for poly-(cyclohexane-1,4-dimethylene terephthalate).

POLYCYCLOHEXYLMETHACRYLATE

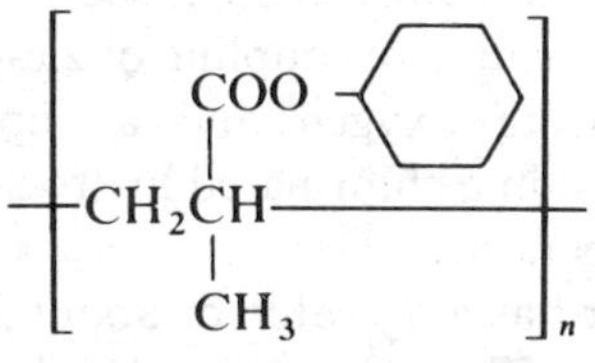

Produced by free radical polymerisation of cyclohexyl methacrylate. It is a clear, hard but brittle plastic which has a T_g value of 83°C. It finds some use as an optical lens material, having the advantage over polymethylmethacrylate of lower shrinkage during polymerisation, which is useful when casting large optical components.

POLYDECAMETHYLENEDECANOAMIDE Alternative name for nylon 1010.

POLYDECANOAMIDE Alternative name for nylon 10.

POLYDIACETYLENE A polymer of structural type

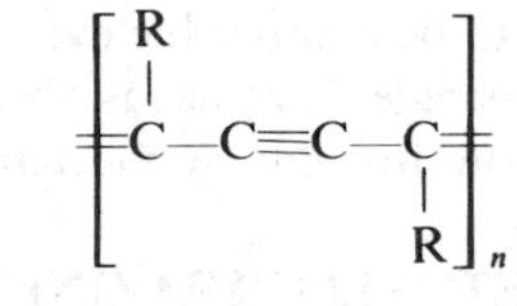

When prepared by photochemically induced topotactic 1,4-polymerisation of a crystalline diacetylene monomer, completely defect free, 100% crystalline, stereoregular single crystals can be produced. The molecules exist in the extended chain conformation. The crystals can be either lozenge shaped or fibrous and are of microscopic size. They have exceptional mechanical properties, for example a Young's modulus of about 60 GPa in the chain direction and a fracture strength of about 1·5 GPa. These values are close to those theoretically predicted for a perfect structure. Although the electrical conductivity is low (10^{-15}–10^{-10} S cm^{-1}), due to a large band gap (about 2 eV) between conductance and valence bands, electrons are promoted readily photochemically to give a photoconducting polymer.

POLY-(2,6-DIBROMOPHENYLENE OXIDE)

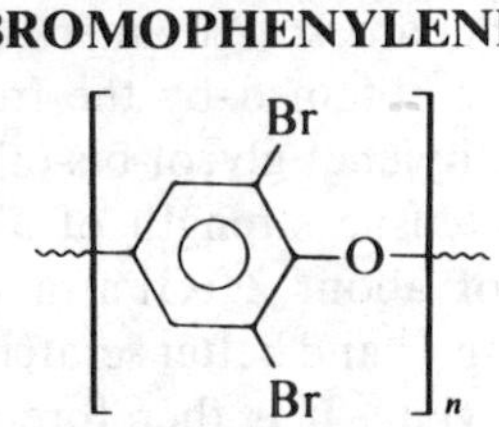

Tradename: Firemaster TSA. Produced by the oxidative coupling of 2,4,6-tribromophenol using alkali and $K_3[Fe(CN)_6]$. The polymer is useful as a non-migrating fire retardant for engineering polymers requiring high processing temperatures.

POLYDICHLOROPHOSPHAZENE (Polyphosphonitrilic chloride)

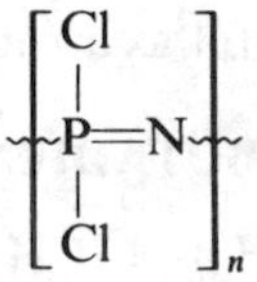

Prepared either by the reaction of phosphorus pentachloride with ammonium chloride in, for example, tetrachloroethane solution (which yields a mixture of low molecular weight polymer and cyclic oligomers), or by ring-opening polymerisation of hexachlorocyclotriphosphazene by heating at 250–300°C. Crosslinking occurs on prolonged heating producing an insoluble polymer. The polymer is rubbery with a T_g value of -64°C, but is very susceptible to hydrolysis by atmospheric moisture. However, it may be converted to other polyphosphazenes, which are hydrolytically resistant, by chemical reaction, e.g. with alkoxide ion in tetrahydrofuran solution replacing Cl with OR groups. Fluoroalkoxy substitution leads to the formation of the useful fluorophosphazene rubbers. Reaction with amines similarly gives N-substituted polymers.

POLY-2,5-DICHLOROSTYRENE

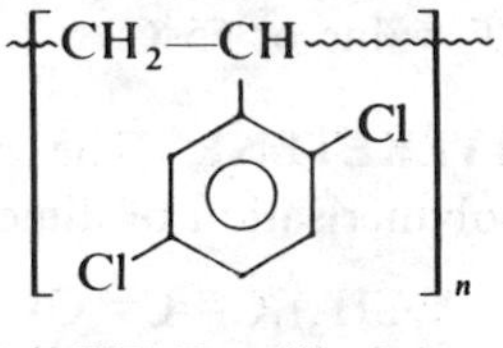

Tradename: Styramic HT. Readily formed by the thermal or free radical polymerisation of 2,5-dichlorostyrene. The polymer has a T_g value of about 120°C and has good fire resistance. At one time it was produced commercially.

POLY-(DIETHYLENEGLYCOL ADIPATE)

$$\{OOC(CH_2)_4COOCH_2CH_2\}_2O\}_n$$

A polyester synthesised by simple polycondensation between diethyleneglycol and adipic acid. When copolymerised with a small quantity of a polyol, such as trimethylolpropane or pentaerythritol, it forms the most common type of polyester polyol used for flexible polyurethane foam by reaction with a diisocyanate.

POLY-(DIETHYLENEGLYCOL-BIS-(ALLYL CARBONATE))

Tradenames: Allylmer-39, CR-39. The crosslinked polymer obtained by the free radical polymerisation of diethylene glycol-bis-(allyl carbonate). Typically it has a tensile strength of about 40 MPa, a flexural modulus of about 2·0 GPa, a notched impact strength of 10–20 J m^{-1} and better scratch resistance than polymethylmethacrylate. It is therefore useful for spectacle and optical lenses, safety shields and as an optical glass cement.

POLY-(1,1-DIHYDROPERFLUOROBUTYL ACRYLATE)   (Poly FBA) (Poly-1F4)

$$\left[\begin{array}{c} CH_2CH \\ | \\ COOCH_2CF_2CF_2CF_3 \end{array} \right]_n$$

A polyfluoroacrylate useful as a rubber.

POLY-(2,3-DIMETHYLBUTADIENE)   (Methyl rubber)

$$\left[CH_2C(CH_3){=}C(CH_3)CH \right]_n$$

A very early synthetic rubber, the monomer being made by conversion of acetone via pinacol to 2,3-dimethylbutadiene:

$$2\ (CH_3)_2CO \xrightarrow{\ H^+\ } \begin{array}{c} (CH_3)_2C-C(CH_3)_2 \\ |\quad\ | \\ OH\ OH \end{array} \rightarrow$$

$$CH_2{=}C(CH_3)C(CH_3){=}CH_2$$

The monomer was polymerised by heating for several months at 70°C or 30–35°C.

POLY-(5,5-DIMETHYLHEXENE-1)

$$\left[\begin{array}{c} CH_2CH \\ | \\ CH_2CH_2C(CH_3)_3 \end{array} \right]_n$$

Prepared by Ziegler–Natta polymerisation of 5,5-dimethylhexene-1 in isotactic form, which, similarly to poly-(4-methylpentene-1) is transparent although crystalline. It has a T_g value of 55°C.

POLYDIMETHYLKETENE

The polymer obtained by the anionic polymerisation of dimethylketene

$$((CH_3)_2C{=}C{=}O)$$

The reaction can lead to three different polymer repeat unit structures

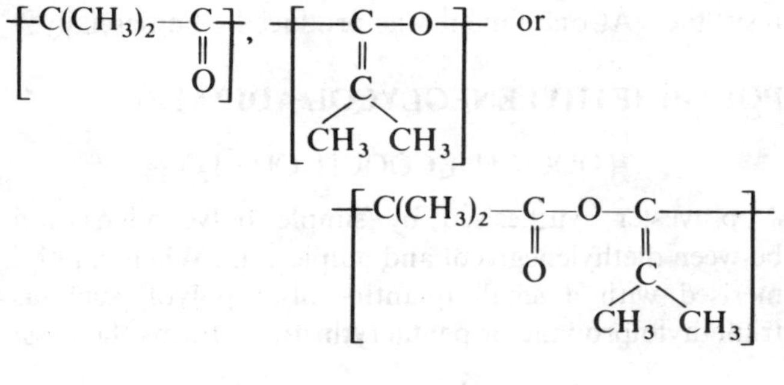

$$\left[\begin{array}{c} C(CH_3)_2-C \\ \| \\ O \end{array} \right], \quad \left[\begin{array}{c} C-O \\ \| \\ C \\ | \\ CH_3\ CH_3 \end{array} \right] \quad or$$

$$\left[\begin{array}{c} C(CH_3)_2-C-O-C \\ \|\qquad\quad \| \\ O\qquad\quad C \\ \qquad\quad | \\ CH_3\ CH_3 \end{array} \right]$$

Which one predominates depends on solvent polarity and the nature of the gegenion.

POLY-(2,6-DIMETHYLPHENOL)

Alternative name for poly-(2,6-dimethyl-1,4-phenylene oxide).

POLY-(2,6-DIMETHYL-1,4-PHENYLENE OXIDE)

(PPO) (Polyphenylene ether) (Poly-(2,6-dimethylphenol))

$$\left[\begin{array}{c} CH_3 \\ \\ \text{ring} - O \\ \\ CH_3 \end{array} \right]_n$$

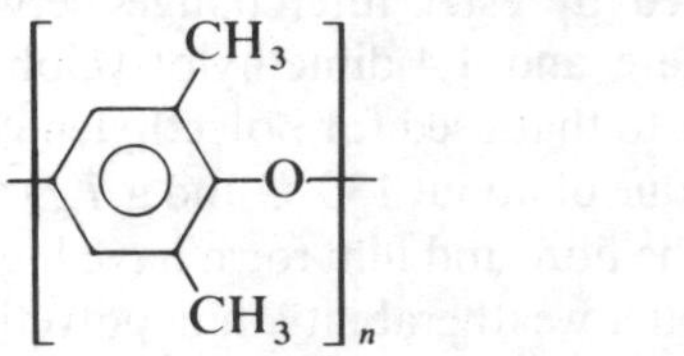

Tradenames: Luranyl, Noryl, Prevex, Vestoran (for blends with polystyrene and related polymers). The most important polyphenylene oxide polymer, often simply referred to as polyphenylene oxide. Incorporation of the methyl groups, by blocking the *ortho* positions during synthesis, ensures only 1,4-linking with the minimum of branching and crosslinking in the polymer. It also improves processability of the otherwise intractable unsubstituted polymer.

The polymer is readily synthesised in high molecular weight form by oxidative coupling of 2,6-dimethylphenol (2,6-xylenol) using oxygen and a cuprous salt plus pyridine catalyst in a chlorinated hydrocarbon solvent at moderate temperature.

The polymer has a T_g value of about 208°C and a T_m value of about 257°C. Owing to the close proximity of these transitions, melt cooled samples are amorphous due to quenching. Crystallisation can be induced by annealing or by solution crystallisation. The polymer therefore has a high softening point, good chemical resistance, electrical insulation and dimensional stability and has been used in a number of electrical, water pump, valve and meter applications. However, its relatively high price and the need for high processing temperatures has led to its main use as a blend with polystyrene in commercial materials. These retain many of the desirable properties of PPO at lower cost, but softening points are necessarily lower (polystyrene has a T_g value of about 90°C) to achieve easier processing. The blends are unusual in that the two polymers appear to be compatible even at the molecular level, exhibiting a single T_g value (as observed by DSC or dielectric relaxation, but not by mechanical relaxation).

POLY-(2,5-DIMETHYLPIPERAZINETEREPHTHALAMIDE)

A partially aromatic polyamide of structure

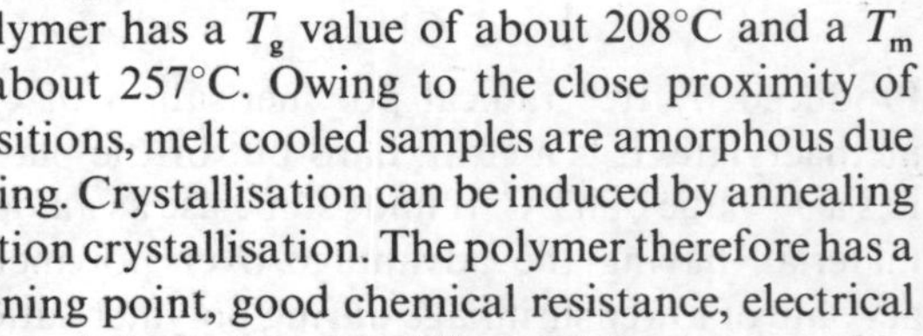

$$\left[\begin{array}{c} CH_3 \\ \\ HN \quad NHCO - ring - CO \\ \\ CH_3 \end{array} \right]_n$$

softening at 350°C and with an even higher T_m value. The T_g value is about 290°C. Fibres have a very high tenacity of 7·7 g denier^{-1} and a high (initial) modulus of 150 g denier^{-1}.

POLYDIMETHYLSILOXANE (Dimethylsilicone) (Methylsilicone)

$$\left[\begin{array}{c} CH_3 \\ | \\ Si-O \\ | \\ CH_3 \end{array}\right]_n$$

The most important polyorganosiloxane, being the basis of most technical silicone oils, greases, rubbers and resins. The polymer is formed by the hydrolysis of dimethyl-dichlorosilane with water often in the presence of an organic solvent. Hydrolysis proceeds via the unstable dimethylsilanol, which under the influence of the hydrochloric acid produced condenses to a mixture of cyclic oligomeric and linear polymeric siloxanes. This mixture may be homogenised with respect to molecular size by equilibration. Relatively low molecular weight polymers, with molecular weights from a few hundred to about 10 000, useful as silicone oils, are readily formed. However, the high molecular weight polymer, molecular weight about 10^5, required for silicone rubbers cannot be made this way owing to the difficulty of obtaining sufficiently pure dichlorodimethylsilane. Such polymers are made by ring-opening polymerisation of octamethyl-cyclotetrasiloxane. Several copolymer silicone rubbers are also of interest, including methylvinylsilicone, nitrilesilicone, fluorosilicone and phenylsilicone rubbers. The basis of crosslinked silicone resins is the copolymer formed by the hydrolysis of mixed di- and trichloro-silanes. As with all polyorganosiloxanes, polydimethyl-siloxane is characterised by high heat and chemical stability. It has an extremely low T_g value ($-123°C$), so retains its mechanical properties to low temperatures but does crystallise at about $-60°C$.

POLYDIPEPTIDE

A sequential polypeptide consisting of sequences of dipeptide units, i.e.

$$-[NHCHRCONHCHR'CO]_n$$

Synthesised by polymerisation of the dipeptide, e.g. as an active ester such as the N-thiophenylpeptide, the N-carboxyanhydride route not being available. The most widely studied polymers are those containing glycine, (especially poly-(gly–ala)), since glycine, being optically inactive, cannot racemise and furthermore, these are useful models for silk fibroin. Many of these polymers exist in the β- or cross-β conformations.

POLY-(2,6-DIPHENYL-1,4-PHENYLENE OXIDE)

Tradename: Tenax. An aromatic polyether of much higher thermal stability than the well-known poly-(2,6-dimethyl-1,4-phenylene oxide). Similarly synthesised by oxidative coupling of 2,6-diphenylphenol (obtained by coupling of cyclohexanone followed by dehydrogenation). Its high T_m (480°C) and T_g (235°C) values preclude melt processing Its main use is as a high voltage cable insulation in the form of a paper fabricated from dry-spun fibres.

POLYDISPERSE POLYMER

A polymer sample composed of individual molecules of different sizes, i.e. of different molecular weights, degrees of polymerisation and chain lengths. Such a sample is said to have a molecular weight distribution. Owing to the statistically random nature of the growth and termination processes in polymerisation, synthetic polymers are nearly always polydisperse. Almost monodisperse samples may sometimes be obtained by careful fractionation or by use of special polymerisation methods.

In contrast, many biopolymers, especially proteins and nucleic acids, are often monodisperse or paucidisperse. The full description of the spread of molecular sizes is given by the molecular weight distribution but this is difficult to measure. Simpler measures of the poly-dispersity, such as the ratios of the different molecular weight averages, are often used since they are more readily determined.

POLYDISPERSITY

(Dispersity) In general, the width of a molecular weight distribution, given quantitatively by any of several molecular weight distribution indices, including the polydispersity index ($\bar{M}_w/\bar{M}_n$), the inhomo-geneity and the g-index.

POLYDISPERSITY INDEX

A measure of the width of a molecular weight distribution, defined as the ratio of the weight to the number average molecular weights, $\bar{M}_w/\bar{M}_n$. Sometimes used to mean other indices of the molecular weight distribution width, such as g-index or inhomogeneity.

POLYDODECANOAMIDE

Alternative name for nylon 12.

POLYELECTROLYTE

An ionic polymer with a sufficient density of ionic groups to be highly hydrophilic and to act as an electrolyte. Most polyelectrolytes are water soluble. The charged groups attached to the polymer chains are the fixed ions, the counter-ions being mobile, except in polyelectrolyte complexes. Widely studied types include the salts of acrylic and methacrylic acids, natural polymeric acid salts such as alginates, polyphosphates, quaternary ammonium compounds and protonated amino containing polymers. Much interest in polyelectrolytes arises from the fact that proteins and nucleic acids carry ionisable groups. The high fixed charge density causes the chains to expand in solution resulting in high specific viscosities, which increase with dilution. These polyelectrolyte effects may be neutralised by the presence of a large amount of a low molecular weight electrolyte whose counter-ions reduce the electrostatic field surrounding the polymer chain.

POLYELECTROLYTE COMPLEX Alternative name for polysalt.

POLYENANTHOLACTAM Alternative name for nylon 7.

POLYEPICHLORHYDRIN

$$+CH_2CH—O\rightarrow_n$$
$$\qquad |$$
$$\qquad CH_2Cl$$

The polymer prepared by ring-opening polymerisation of epichlorhydrin

$$(CH_2—CH—CH_2Cl)$$
$$\qquad\backslash\quad/$$
$$\qquad O$$

using either Ziegler–Natta catalysts giving crystalline polymers, or an aluminium alkyl catalyst with water (probably operating by a cationic mechanism) or with a chelating agent, such as acetylacetone (possibly operating by a coordination mechanism). In the last two cases non-crystalline polymer results. The polymer forms the basis of the epichlorhydrin rubbers, the amorphous polymer being liquid or rubbery depending on molecular weight, or crystalline with a T_m value of about 119°C.

POLYESTER A polymer containing ester groups in the main chain, i.e. having repeating units of the type

$$—RCO—$$
$$\quad\ \|$$
$$\quad\ O$$

where R is a hydrocarbon group. Polymers with ester groups as side groups, e.g. polyvinyl esters and poly-acrylates and methacrylates, are not considered to be polyesters. A very wide variety of polyesters are known, many of them being of use commercially in a diverse range of important applications. Of the saturated polyesters, the linear polymers are structurally the simplest being of the types

$$+RCO\rightarrow_n \qquad or \qquad +ROCR'CO\rightarrow_n$$
$$\quad \| \qquad\qquad\qquad\qquad \| \quad \|$$
$$\quad O \qquad\qquad\qquad\qquad O \quad O$$

The aliphatic polymers are only useful in low molecular weight form as polyester plasticisers and as polyol prepolymers for polyurethanes, although high molecular weight poly-(ε-caprolactone) has found a limited use as a biodegradable plastic. Most of these polymers have T_m values and hence softening points which are too low for plastic and fibre use. On the other hand wholly aromatic polymers, such as poly-(hydroxybenzoic acid), are high temperature resistant plastics but are very difficult to melt process. The linear partially aromatic polymers are very important, especially polyethylene terephthalate (as a fibre), polybutylene terephthalate (as a plastic) and the polycarbonate of bisphenol A (as a plastic). Saturated branched polymers, often formed from glycerol, a diol and phthalic anhydride and known as alkyd resins, are important coating materials, especially as oil modified alkyd resins, which are crosslinkable. Other important polymers are the linear unsaturated polyester resins, often simply called polyester resins, which form the basis of traditional glass reinforced plastics (GRP) products. These last two groups are more or less complex copolymers and are made in a great variety of chemical structural forms by varying comonomer types and amounts. Copolymers which contain non-ester repeating groups are the polyester–amides, polyester–imides and polyester–ethers. The latter are produced as block copolymers and therefore are thermoplastic elastomers.

Polyesters are usually synthesised by polyesterification of a diol (or triol, etc.) with a diacid,

$$n\,HOROH + n\,HOOCR'COOH \rightarrow$$
$$+OROCR'C\rightarrow_n + 2n\,H_2O$$
$$\qquad\qquad \| \quad \|$$
$$\qquad\qquad O \ \ O$$

or with a derivative of a diacid (especially a diester) by ester interchange via alcoholysis,

$$n\,HOROH + n\,R''OOCR'COOR'' \rightarrow$$
$$+OROCR'C\rightarrow_n + 2n\,R''_2OH$$
$$\qquad\qquad \| \quad \|$$
$$\qquad\qquad O \ \ O$$

These methods require high temperatures and are frequently performed as melt polymerisations. However, diacid chlorides are much more reactive, usually being capable of forming high molecular weight polyester at ambient temperatures:

$$n\,HOROH + n\,ClOCR'COCl \rightarrow$$
$$+OROCR'C\rightarrow_n + 2n\,HCl$$
$$\qquad\qquad \| \quad \|$$
$$\qquad\qquad O \ \ O$$

This reaction is particularly useful for the preparation of polycarbonates and aromatic polyesters. Similarly an acid dianhydride may be used:

$$n\,HOROH + n\,R'\big\langle\!\!\begin{array}{c} O \\ \| \\ C \\ \ \ \ \ O \\ C \\ \| \\ O \end{array}\!\!\big\rangle \rightarrow +OROCR'C\rightarrow_n + n\,H_2O$$

This is especially useful for preparing alkyd resins using phthalic anhydride and unsaturated polyester resins using maleic anhydride.

Simple self-esterification of a hydroxyacid:

$$n\,HORCOOH \rightarrow +ORC\rightarrow_n$$
$$\qquad\qquad\qquad\quad \|$$
$$\qquad\qquad\qquad\quad O$$

is not used very often. Polymers of this type are usually prepared by ring-opening polymerisation of the corresponding lactone, as for poly-(ε-caprolactone) and poly-(γ-butyrolactone). Ester interchange between a diacid and a diester (acidolysis) is useful for preparing polyesters from dihydroxyacids and bisphenols.

Linear aliphatic polyesters including polycarbonates, crystallise but have low melting temperatures (from ambient to about 70°C) and low T_g values. Incorporation of a benzene ring into the polymer chain, as in polyethylene terephthalate, raises the T_m value, sometimes to above 300°C, and the T_g value. The unsaturated polyester resins are viscous liquids as low molecular weight prepolymers, which makes for easy application to a mould or as a coating, prior to crosslinking. When cured, the products are often hard but brittle materials. Linear saturated partially aromatic polyesters, such as polyethylene terephthalate and polybutylene terephthalate, are stiff and tough enough to be considered as engineering plastics. Orientation in the former is very important in fibres in developing high stiffness and strength. Some aromatic copolyesters show liquid crystalline behaviour and are useful high temperature plastic materials with low melt viscosities and hence easy processing. Polyesters are only slightly polar and have low water absorption and therefore good electrical properties. They are somewhat susceptible to hydrolysis especially under alkaline and high temperature conditions.

POLYESTER ALKYD (Alkyd polyester) An unsaturated polyester resin containing a partially polymerised diallyl phthalate as a crosslinking agent. It is used as a thermosetting moulding composition when compounded with filler, lubricant and peroxide curing agent, for electrical parts.

POLYESTER–AMIDE A copolymer containing both ester and amide links in the polymer chain. An example is the polyester–amide, obtained by reaction of ethylene glycol with monoethanolamine and adipic acid, used as a prepolymer in the formation of the polyurethane Vulcaprene.

POLYESTER–CARBONATE A copolymer polyester produced by reaction of bisphenol A with a mixture of phosgene and a dicarboxylic acid, especially terephthalic acid.

POLYESTER ELASTOMER Alternative name for polyetherester block copolymer.

POLYESTER FIBRE Generic name for a fibre composed of a polymer with at least 85 wt% of ester units of a dihydric alcohol and terephthalic acid. Examples are Dacron, Kodel, Terylene and Trevira.

POLYESTER FOAM A polyurethane foam based on the reaction of a polyester polyol with a diisocyanate. Both rigid and flexible foams are so made, but these have been largely replaced by polyether foams, except where foams of high load bearing capacity or good solvent resistance are required, or where semi-rigid foams are required.

POLYESTERIFICATION The process of formation of a polyester by formation of ester links by reaction between appropriate functional groups on suitable monomers. Most polyesters are made by polyesterification, which has been widely studied as a model reaction for step-growth polymerisation in general. Thus AB polymerisation occurs by self-condensation of a hydroxyacid:

$$n\,HORCOOH \rightarrow \text{-}(ORC\text{-})_n + n\,H_2O$$

However, few polyesters are made in this way. AABB polymerisation occurs between a diol and a diacid:

$$n\,HOROH + n\,HOOCR'COOH \rightarrow$$
$$\text{-}(OROCR'C\text{-})_n + 2n\,H_2O$$

However, esterification of a hydroxyl group with a carboxyl group requires temperatures of up to about 300°C, when the complication of ester interchange occurs. This may be promoted and taken advantage of by using an excess of diol, which increases the initial rate and favourably influences the equilibrium. The excess glycol is liberated near the end of the reaction by ester interchange and is removed by applying a vacuum or by an inert gas stream.

Alternatively a diester may be used instead of a diacid, so that polymer formation is solely by ester interchange. This has the advantages, especially in the formation of polyethylene and polybutylene terephthalates, that the diesters are easier to purify, they are more miscible with diols and they are more reactive. Nevertheless, as with diacids, high temperature melt, or sometimes solution methods may be used. Usually a proton donating, Lewis acid or weakly basic catalyst is used.

Other diacid derivatives may be used. Thus diacid chlorides react with diols at ambient temperatures,

$$n\,HOROH + n\,ClOCR'COCl \rightarrow$$
$$\text{-}(OROCR'C\text{-})_n + 2n\,HCl$$

especially when an alkali is present (Schotten–Baumann conditions) to react with the HCl formed and to activate the diol by conversion to alkoxide. The reaction may be conducted in an inert solvent or by interfacial polymerisation. It is useful for the preparation of high melting wholly aromatic polyesters and polycarbonates. Reaction of a diol with an acid anhydride is also important in the formation of alkyd resins and unsaturated polyester resins:

$$n\,HOROH + n\,R' \rightarrow \text{-}(OROCR'C\text{-})_n + n\,H_2O$$

Again only moderate temperatures are needed.

POLYESTER–IMIDE A polymer containing both ester and imide groups in the polymer chain. Several aromatic polymers of this type are commercially available (Cellatherm, Enamel Omega, Imidex, Isomid, Terabec). Synthesised by polycondensation between a dianhydride containing aromatic ester links and a diamine. The dianhydride is often pyromellitic dianhydride, BDTA or trimellitic anhydride. A typical diamine is 4,4'-diaminodiphenylmethane. The polymers are more tractable than the polyimides but have poorer high temperature resistance. Their main use is as high temperature resistant wire enamels.

POLYESTER PLASTICISER A polyester which is of sufficiently low molecular weight (500–10 000) to be a viscous liquid; it can be mixed with and is compatible with a polymer and therefore acts as a plasticiser. As with most plasticisers, polyvinyl chloride is the polymer with which they are mostly used. The advantage of using a polymeric plasticiser is that it has a very low volatility and low extractability, so that it shows much greater permanence than a similar low molecular weight plasticiser. Most polymeric plasticisers are polyesters, normally being esters of aliphatic diols (especially of ethylene, propylene, diethylene and triethylene glycols) and aliphatic diacids (especially adipic, azelaic and sebacic acids), although phthalates are sometimes used. Often a monocarboxylic acid, such as capric or lauric acid, is used to limit, in a controllable manner, the polymer molecular weight and to ensure that it has stable ester end groups.

POLYESTER POLYOL A low molecular weight polyester (typically of a few thousand) prepared by polyesterification using an excess of diol so that the polymer molecules are terminated by hydroxyl groups and have the general structure:

$$HO-[R-OOC-R'-COOR-O-]_nH$$

Useful in the formation of polyurethanes by reaction with a diisocyanate through the hydroxyl groups. It is used particularly for high strength polyurethane elastomers, although these do not have such good hydrolytic stability as those based on polyether polyols. Occasionally it is used for flexible polyurethane foams when a foam with low resiliency is required, as for packaging and fabric interlinings. It is also used for certain semi-rigid foams and for polyurethane coatings, when castor oil based polyester polyols are used. Polyester polyols are usually based on adipic (sometimes phthalic) acid and ethylene, diethylene, 1,2- and 1,3-propylene, 1,4- and 2,3-butane diols (occasionally branching is incorporated by the presence of glycerol), trimethylolpropane, 1,2,6-hexanetriol or pentaerythritol units. The polymers typically have molecular weights of 1000–4000.

POLYESTER–POLYURETHANE Alternative name for polyester–urethane.

POLYESTER–URETHANE (Polyester–polyurethane) A polyurethane formed by reaction of a polyester polyol with a diisocyanate.

POLYETH Tradename for low density polyethylene.

POLY-(ETHENESULPHONIC ACID) Alternative name for polyvinylsulphonic acid.

POLYETHER A polymer of general structure $-[R-O-]_n$, where R may be a simple alkylene group such as $-CH_2-$ in polyoxymethylene, $-CH_2CH_2-$ in polyethylene oxide,

$$-CH-CH_2-$$
$$\quad |$$
$$\quad CH_3$$

in polypropylene oxide and $-(CH_2)_4-$ in polytetramethylene oxide, or of more complex structure as in poly-(3,3-bis(chloromethyl)oxacyclobutane) and epichlorhydrin. The R group may also be aromatic as in polyphenylene oxide. The ether link in the polymer chain is flexible so the aliphatic polyethers have low T_g values; however, as they are often of simple symmetrical structure they often crystallise substantially. Polyoxymethylene has a high enough softening point, due to its high T_m value, to be a useful plastic, whereas low molecular weight hydroxyl-terminated polyethers are useful as polyol prepolymers for polyurethane formation. Other polyethers of high molecular weight are useful as rubbers, as with polypropylene oxide and polyepichlorhydrin.

POLYETHER–AMIDE A polymer containing both ether and amide links in the polymer chain. Several commercial polymers of this type are block copolymer elastomers and the term is an alternative name for polyether block amide. A further type is an aromatic polymer of which the commercial material, tradename HM-50, is an example.

POLYETHER BLOCK AMIDE (PBA) (Elastomeric polyamide) (Polyether amide) Tradenames: Dynyl, Ely 60, Ely 1256, Pebax, Vestamid X. A block copolymer obtained by coupling a hydroxy-terminated polyethylene, polypropylene or polytetramethylene glycol with a carboxy-terminated polyamide such as nylon 6 or 66. Thus a variety of materials may be produced, the higher the ether content the more flexible is the material. Commercial materials are generally stiff thermoplastic elastomers with good abrasion resistance and high tensile strength. Flexibility is comparatively unaffected by temperature over the range −40 to +80°C.

POLYETHER/ESTER BLOCK COPOLYMER (Block polyetherester) (Copolyetherester) (Polyester elastomer) (Thermoplastic polyetherester) Tradenames: Arnitel, Hytrel, Lomod, Pelprene. A block copolymer containing both polyether and ester blocks. The best known example is poly-(tetramethyleneterephthalate-b-polyoxytetramethyleneterephthalate).

POLYETHERETHER KETONE (PEEK)

Tradename: Victrex PEEK. A polyetherketone offering about the highest temperature resistance of any melt processable thermoplastic material. It has a T_g value of 143°C and crystallises with a T_m value of 334°C. The unreinforced polymer has a heat distortion temperature of 160°C but this increases to 315°C when 30% glass filled. Its chemical and hot water resistance are outstandingly good and it has a low flammability (oxygen index of 40) with very low smoke and toxic gas production on burning. It is useful in various electrical applications and in pumps and valves.

POLYETHER FOAM   A polyurethane foam based on the reaction of a polyether polyol with a diisocyanate. The most important type of both rigid and flexible foam, although polyester polyols are sometimes used.

POLYETHERIMIDE   Tradename: Ultem. An aromatic polymer containing both ether links and imide groups in the polymer chain. The commercial material has a repeat unit structure:

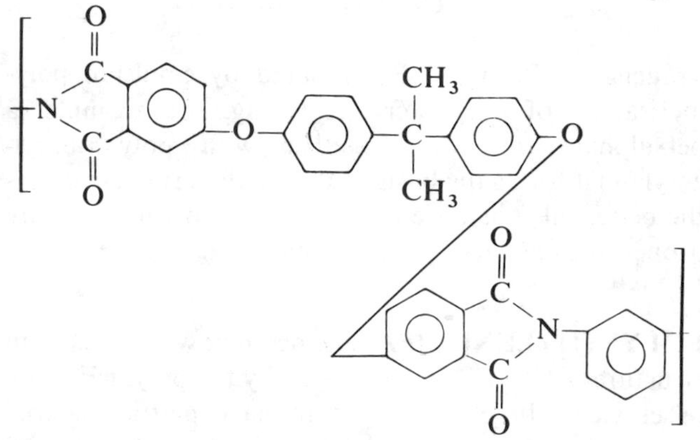

It has a heat distortion temperature of 200°C, with a continuous use temperature of 170°C and low flammability (oxygen index of 47). It is useful as an injection moulding material for heat resistant products such as in microwave ovens, circuit boards and under the car bonnet.

POLYETHERKETONE   (PEK) Tradenames: Hostatec, Kadel, Stilan, Ultrapek, Victrex PEK, Victrex PEK. An aromatic polymer containing both ketone and ether links between the benzene rings. Polymers may be prepared, but only in low molecular weight form, by self-condensation of aromatic acid chlorides using a Lewis acid catalyst, e.g. with aluminium chloride in dichloromethane solvent:

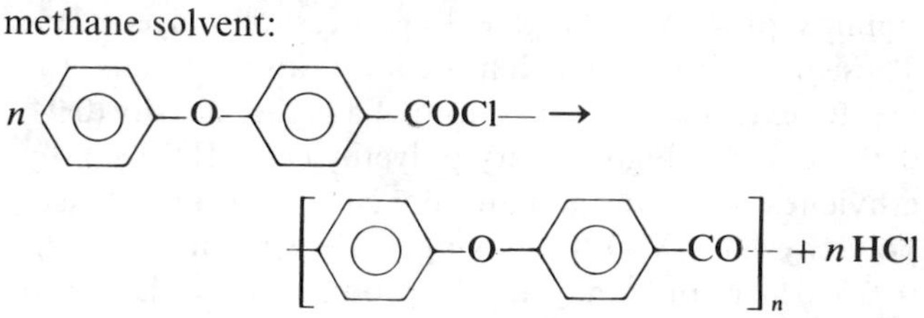

High molecular weight polymer is difficult to prepare because the polymers crystallise and therefore precipitate. An alternative synthesis involving reaction of a dihalide with a diphenoxide, or self-condensation of a phenoxide/halide monomer is:

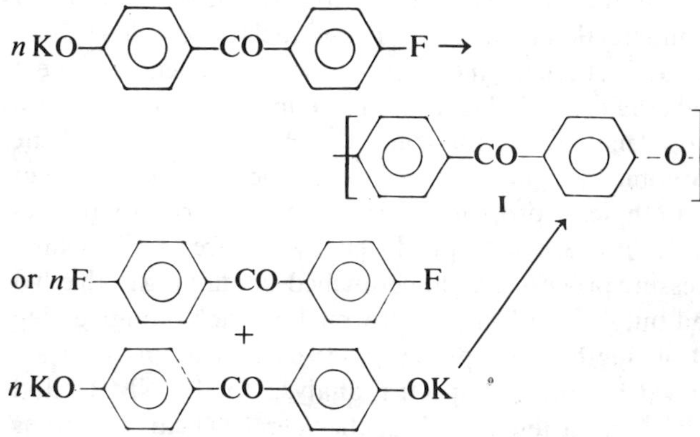

Solubility and hence formation of high molecular weight polymer is achieved using a high boiling point solvent such as an aromatic sulphone. The best established commercial polymer is polyetheretherketone, but recently several other polymers with somewhat higher melting temperatures and hence use temperatures, have become available. They may be polymers of structure **I** which has a T_g value of 154°C and a T_m value of 367°C. The polymers combine their high temperature resistance with good chemical (including hydrolytic) resistance and low flammability. They have the highest use temperatures of any melt processable thermoplastic, at least when reinforced with glass or carbon fibre.

POLYETHER POLYOL   A low molecular weight polyether, which may be linear or branched, with hydroxyl end groups. The most widely used type of prepolymer, usually with a molecular weight of a few thousand, for the formation of polyurethanes by reaction of the hydroxyl groups with a diisocyanate to form a polyether–urethane. Polyether polyols are particularly useful for flexible polyurethane foams, when polyoxypropylene triols are preferred. Such polyether foams have greater resiliency than polyester foams, which is useful in cushioning materials. The secondary hydroxyls are not very reactive and a prepolymer process has to be used. However, the use of a tipped polyol (containing more reactive primary hydroxyls from reaction of the end groups with ethylene oxide) or the use of a powerful catalyst system, enables a one-shot process to be used. Rigid foams are often made from a highly branched polyoxypropylene glycol polyol of molecular weight of a few hundred. Solid elastomers are made using polyethylene glycol or polytetrahydrofuran polyols as prepolymers. The latter gives products with better hydrolytic resistance than the former. Polyoxypropylene glycol is also used for polyurethane elastomers.

POLYETHER–POLYURETHANE   Alternative name for polyether–urethane.

POLYETHERSULPHONE   (Polyarylenesulphone) (Polysulphone) Tradenames: Astrel, Polyethersulphone 200P, Polyethersulphone 720P, Radel, Udel, Ultrason E, Ultrason P, Victrex PES, Victrex 720P. An aromatic polymer consisting of benzene rings linked by both sulphone ($-SO_2-$) groups and ether oxygen atoms. Several commercial products of this type have been developed as reasonably high temperature resistant engineering plastics. They may all be considered to be based on the *p*-phenylene polymer

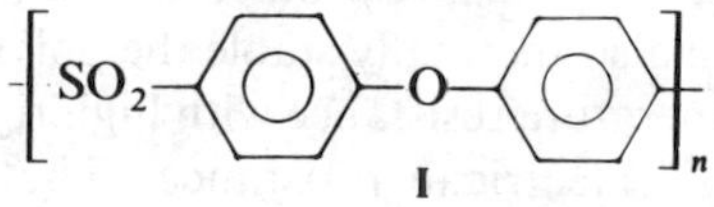

Two methods of synthesis, both solution polymerisations in a polar solvent, e.g. DMSO, are used. In the polyether synthesis, polycondensation occurs between a diphenate

salt (e.g. the sodium salt of bisphenol A) and a dihalosulphone:

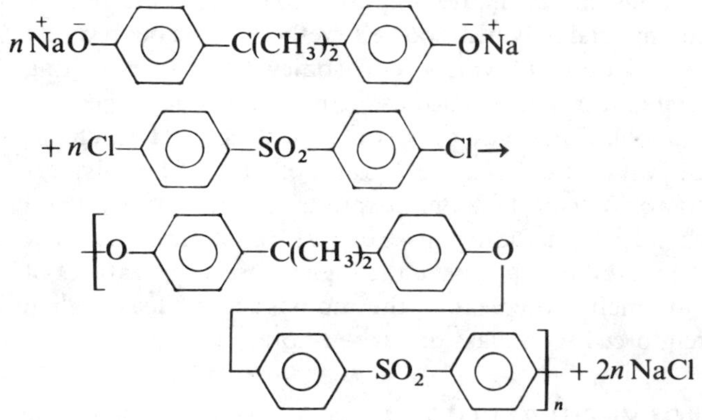

Polymer **II** is the commercial Udel (also simply called polysulphone). Alternatively self-condensation of

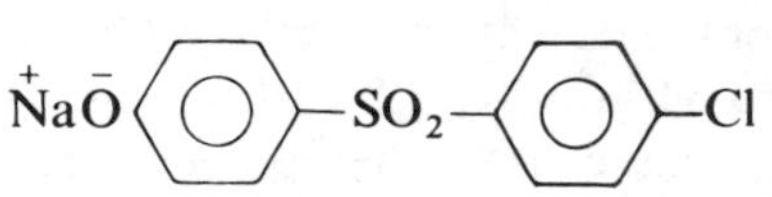

yields **I** (the commercial Victrex PES). In the sulphone polymerisation method, an aromatic hydrocarbon or diarylether is reacted with an aromatic disulphonyl chloride under Friedel–Crafts conditions, e.g. with $FeCl_3$ or $SbCl_5$ as catalyst

$$n \, HArH + n \, ClSO_2Ar'SO_2Cl \rightarrow$$
$$\text{\textlbrackdbl} ArSO_2Ar'SO_2 \text{\textrbrackdbl}_n + 2n \, HCl$$

where either or both Ar and Ar' are usually

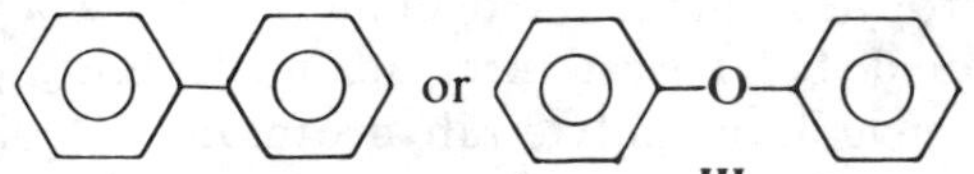

Polymer **I** resuts if Ar = Ar' = **III**, but contains both *ortho* and *para* linkages; therefore the earlier method is preferred for its synthesis, especially as the monomer is less expensive. Copolymers containing both type **I** ether units and

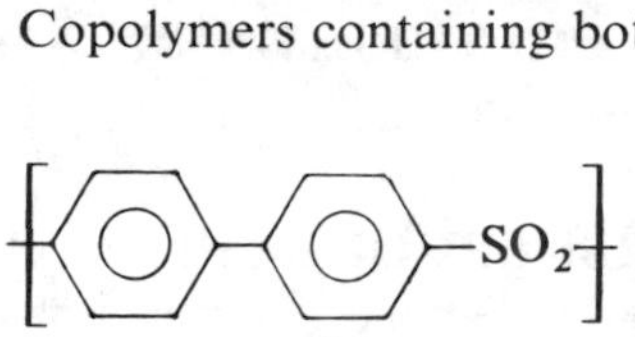

units are commercial products (Radel, Victrex 720P). Polyethersulphone (PES) 120P has more type **I** units and is preferably made by the polyether synthesis, whereas Astrel 360 has more type **IV** units and is produced by sulphone polymerisation.

The polymers are usually amorphous with high T_g values (Udel $\sim 190°C$, Victrex PES $\sim 230°C$, Victrex 720P $\sim 250°C$, Astrel 360 $\sim 285°C$). Since the sulphone and ether groups are also thermally stable the polymers have useful high temperature resistance with high rigidity, low creep and high electrical resistance. The are also transparent and self-extinguishing (Victrex PES has an LOI of 38). Typical applications include printed circuit boards, TV components, electric oven and heater components.

POLYETHER–URETHANE (Polyether–polyurethane) A polyurethane formed by reaction of a polyether polyol with a diisocyanate.

POLYETHYLACRYLATE

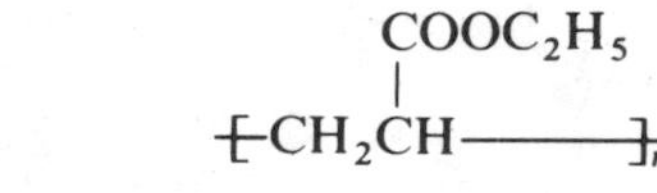

Tradename: Plexigum B. Prepared by emulsion polymerisation of ethyl acrylate, using, for example, a persulphate initiator. Together with poly-(n-butylacrylate) it forms the basis of the acrylate rubbers. It has the better oil resistance but the poorer low temperature properties, since its T_g value is higher at $-23°C$.

POLYETHYLENE (PE) The polymer with repeat unit structure $\text{\textlbrackdbl}CH_2CH_2\text{\textrbrackdbl}_n$ produced by the polymerisation of ethylene. It is the largest tonnage plastic material produced and is obtained in a variety of forms with various degrees of structural imperfection. The earliest laboratory polymers were polymethylenes, ethylene itself being a difficult monomer to polymerise. However, the earliest type of commercial polyethylene was produced by free radical polymerisation of ethylene at very high pressure and temperature. It is therefore sometimes referred to as high pressure polyethylene. This process produces a branched polyethylene which does not crystallise as readily as linear polyethylene and is therefore of lower density ($0.915–0.925 \, g \, cm^{-3}$)—low density polyethylene (LDPE). Ethylene may also be polymerised at normal pressures using very active catalysts, as in the Ziegler–Natta, Standard Oil and Phillips processes, to give low pressure polyethylene. These polymers are much more linear and crystallise to a greater extent and are therefore of a higher density ($0.94–0.96 \, g \, cm^{-3}$)—high density polyethylene (HDPE). Polyethylenes of intermediate density (medium density polyethylene—MDPE) having a density in the range $0.925–0.94 \, g \, cm^{-3}$ may also be produced by variations of the above processes.

More recently linear low density polyethylene (LLDPE), of similar properties to LDPE, has become an important commercial plastic. This is a copolymer of ethylene with a few per cent of a higher α-olefin comonomer. Several other ethylene copolymers are of commercial significance. These include ethylene–vinyl acetate, ethylene–methacrylic acid, ethylene–butene-1 and ethylene–ethylacrylate copolymers. Copolymers with propylene and terpolymers with propylene and a diene monomer are important as ethylene–propylene rubber and ethylene–propylene–diene monomer rubber respectively. Polyethylene produced by the free radical high pressure process is highly branched, containing both ethyl and butyl short branches (formed by back biting), giving 30–40 methyl groups per 1000 carbon atoms (characterised by infrared spectral analysis (or by NMR)) and containing a few long branches per 1000 carbon atoms formed by transfer to polymer. This irregularity of structure reduces the ability of the polymer to crystallise and hence reduces the density of the solid polymer. Thus LDPE is about 50% crystalline. Ziegler–Natta polymer is much more linear, but frequently has about five methyl groups per 1000 carbon atoms deliberately introduced by copolymerisation. It is more crystalline (about 70%) and of medium to high density. Phillips polyethylene is even more linear and crystalline (about 90%) and is of high density (about $0.96 \, g \, cm^{-3}$). Typical molecular weights of commercial polyethylenes are in the range 20 000–40 000 (number average) with a polydispersity of about 20–50 for LDPE and 10 000–50 000 with a polydispersity of 5–15 for HDPE.

The crystalline morphology of polyethylene has been very widely studied as a model for polymer crystallinity in general, particularly as single crystals may be readily grown from dilute solution. Crystallised from the melt, the polymer is spherulitic, having a very high nucleation rate and thus forming only small spherulites. Although it is impossible to quench a polyethylene melt to an amorphous solid, crystallisation, and hence density, may be somewhat reduced by quenching, especially if the polymer is of high molecular weight. The crystalline phase has a density of $1.00 \, g \, cm^{-3}$ and the amorphous phase a density of $0.85 \, g \, cm^{-3}$ at $25°C$.

The crystalline melting behaviour is dependent on density. Typically, LDPE melts over a wide range of about 80–110°C and HDPE over a range of about 120–135°C. Polyethylene exhibits several transitions apart from T_m. These are best investigated by dynamic mechanical spectroscopy. The best characterised are the α-, β- and γ-transitions. The α-transition is associated with motions in the crystalline phase and is at about $+50°C$, although the value is crystallinity dependent. The β-transition is at about $-20°C$ and is associated with motions involving branch points. The γ-transition is at about $-120°C$ and involves motions of sequences of a few $-CH_2-$ groups. Considerable controversy exists as to whether the β- or the γ-transition should be called T_g.

The mechanical behaviour of polyethylene is dependent on density. LDPE is a soft and flexible material, typically with a tensile modulus of $0.2 \, GPa$, a tensile strength of 10 MPa, an elongation at break of 800% and an Izod impact strength of $> 15 \, J \, (12.7 \, mm)^{-1}$. Its major use (about 70%) is as a film material especially for packaging. HDPE is a harder, stiffer material, typically having a tensile modulus of $1.0 \, GPa$, a tensile strength of 30 MPa, an elongation at break of 500% and an Izod impact strength of $2–8 \, J \, (12.7 \, mm)^{-1}$. Its main uses are as a pipe, container, film and injection moulding material. Although highly inert to swelling by all solvents, polyethylene may be susceptible to environmental stress cracking. It is also sensitive to thermal and photo-oxidation and is frequently protected with an antioxidant. Being completely non-polar it has a very high electrical resistivity and is an exceptionally low loss material, ideal for high frequency electrical insulation. Optically, polyethylene is opaque, or translucent in thin film, due to scattering of light by the spherulites.

POLYETHYLENE ADIPATE

$$-(CH_2CH_2OOC(CH_2)_4COO)_n-$$

A polyester (T_m, 47°C) frequently used when hydroxy-terminated, as the polyester polyol for the production of polyurethanes, especially for cast elastomers, by reaction with a diisocyanate. However, crystallisation of the polyester segments in the polyurethane can cause cold-hardening of the rubber. This is usually avoided by using polyester prepolymers of molecular weight of about 2000 or by using a copolymer with some ethylene glycol replaced by propylene glycol. It is synthesised by simple polycondensation between ethylene glycol and adipic acid.

POLYETHYLENE AZELATE

$$-(OCH_2CH_2OC(CH_2)_6C)_n-$$
$$\qquad\quad \| \qquad\qquad \|$$
$$\qquad\quad O \qquad\qquad O$$

A polyester formed by esterification of ethylene glycol with azelaic acid and having a melting point of 44°C.

POLYETHYLENE GLYCOL   (PEG)

$$HO-(CH_2CH_2O)_x-H \qquad\qquad M.p. \, 6°C.$$

(when $x = 8$) to 60°C (when $x = 150$). Viscous liquids with the degree of polymerisation (x) varying among different grades from about four (having a molecular weight of about 200, i.e. PEG 200) to about 200 (having a molecular weight of about 8000, i.e. PEG 8000). Useful as plasticisers for casein, gelatin, polyvinyl alcohol and in printing inks.

POLYETHYLENE GLYCOL DIBENZOATE

$$\langle O \rangle - COO(CH_2CH_2O)_x OC - \langle O \rangle$$

where $x = 6–12$. A plasticiser for phenol–formaldehyde resins and polyvinyl acetate. Also useful in adhesives.

POLYETHYLENE GLYCOL DI-2-ETHYLHEXOATE

$$CH_3(CH_2)_3CH(C_2H_5)COO$$
$$\qquad\qquad\quad |$$
$$-(CH_2CH_2O)_x-OCCH(C_2H_5)(CH_2)_3CH_3$$

A plasticiser for polyvinyl chloride and its copolymers and synthetic rubbers for good retention of low temperature flexibility.

POLYETHYLENE GLYCOL MONOPHENYL ETHER

$$\langle O \rangle - O(CH_2CH_2O)_x-H$$

A plasticiser for polyvinyl acetate. Also compatible with polyvinyl alcohol.

POLYETHYLENEIMINE

$$-(CH_2CH_2NH)_n-$$

Produced by cationic polymerisation of ethyleneimine, e.g. with protonic acid catalysts:

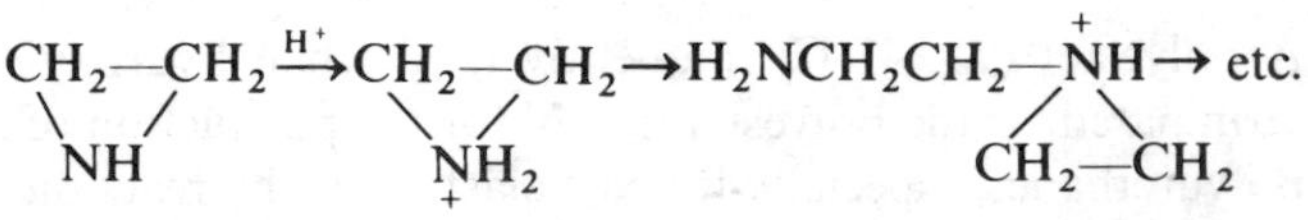

$$CH_2-CH_2 \xrightarrow{H^+} CH_2-CH_2 \rightarrow H_2NCH_2CH_2-\overset{+}{N}H \rightarrow \text{etc.}$$

Thus polymerisation occurs via a cyclic immonium ion. The polymers are frequently branched either through reaction of ⩗NH⩗ groups with end groups or by formation of immonium ions in the chain. The polymer is highly water soluble and hence finds use in paper and textile treatments.

POLY-(ETHYLENENAPHTHALENE-2,6-DICARB-OXYLATE)   (PEN-2,6)

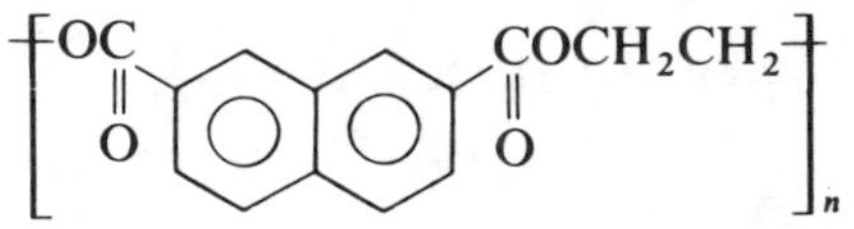

Synthesised by polyesterification of naphthalene-2,6-dicarboxylic acid with ethylene glycol. Shows much better high energy radiation stability than the similar PETP, having a tendency to slowly crosslink rather than undergo chain scission.

POLYETHYLENE OXIDE   (PEO)

$$\left[\!-CH_2CH_2-O-\right]_n$$

Tradenames: Carbowax, Polyox. Low molecular weight polymers may be prepared by passing ethylene oxide into ethylene glycol containing an alkaline catalyst, such as sodium hydroxide. Polymers of molecular weight 200–600 are viscous liquids useful as surfactants. Polymers of molecular weight 600–3000 are low melting waxy solids useful as pharmaceutical and cosmetic bases and as lubricants. Such polymers have terminal hydroxyl groups and are therefore referred to as polyethylene glycols. Low molecular weight polymers may also be prepared by cationic polymerisation of ethylene oxide using, for example, boron trifluoride/water, aluminium chloride or sulphuric acid as catalyst. High molecular weight polymers may be prepared using an alkaline earth (Ca, Ba or Sr) compound (e.g. the oxide, carbonate or alkoxide), but especially by using an amide/alkoxide such as,

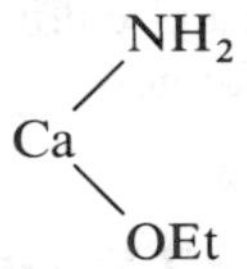

or an organometallic compound, such as an aluminium or zinc alkyl, often with a co-catalyst especially water. In either case a coordination polymerisation mechanism is probably involved. Polymers of high molecular weights $(10^5–10^7)$ are highly crystalline (with a T_m value of 66°C) and tough solids and may be formed by calendering, extrusion, etc. The polymers are highly water soluble as well as being soluble in a wide range of other liquids, including hydrocarbons, chlorinated hydrocarbons, ketones, alcohols. The solid polymers are useful as water-soluble packaging films and in solution as thickening agents, sizes, etc.

POLY-(ETHYLENE OXIDE) GLYCOL   Alternative name for poly-(oxyethylene glycol).

POLY-(ETHYLENEOXY BENZOATE)

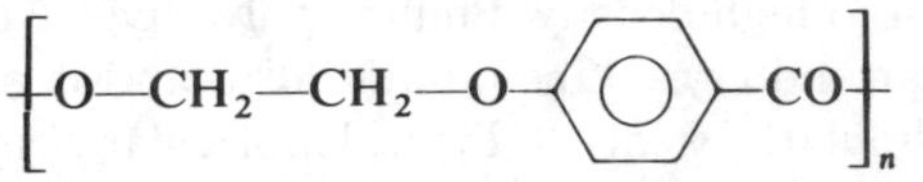

Tradename: A-Tell. A partially aromatic polyester, useful as a fibre, which crystallises with a T_m value of 223°C and has a T_g value of 65°C.

POLYETHYLENE SEBACATE

$$\left[\!-OCH_2CH_2OC(CH_2)_8C\!-\right]_n$$

A polyester formed by esterification of ethylene glycol with sebacic acid and having a melting point of 72°C.

POLYETHYLENE SULPHIDE

$$\left(\!-CH_2CH_2-S-\right)_n$$

Prepared by the ring-opening polymerisation of ethylene sulphide using ionic catalysts. Cationic catalysts such as sulphuric acid and boron trifluoride give only low molecular weight polymers with low stability due to the presence of acidic impurities. Anionic polymerisation with bases, alkali metals and alkoxides, can give high molecular weight polymers. However, organometallic catalysts are the most convenient in order to achieve high molecular weight, as with the use of zinc diethyl with water co-catalyst. The polymers crystallise with a T_m value of 208–212°C and have low solubility—they dissolve in few solvents and only above about 140°C.

POLY-(ETHYLENESULPHONIC ACID)   Alternative name for polyvinylsulphonic acid.

POLYETHYLENE   TEREPHTHALATE   (PET) (PETP)

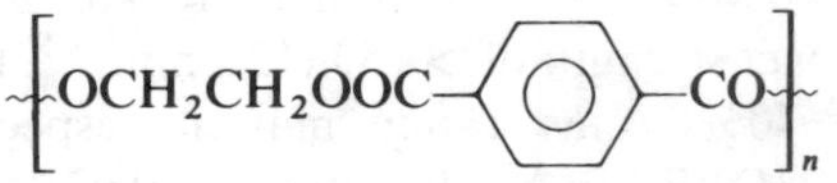

Tradenames: Dacron, Diolen, Fortrel, Grilene, Kodel, Lanon, Lavsan, Lirelle, Tergal, Terital, Terlenka, Terelene, Tetoron, Trevira, Vycron (fibres), Hostaphan, Impet, Melinex, Mylar (film), Arnite, Beetle PET, Hostadur E, Kodapak, Kodar, Melinar, Rynite, Petlon, Petlox, Techster E, Tenite Polyterephthalate (moulding materials). An important polyester known mainly as one of the major synthetic fibres, but also of interest as a plastic, for the manufacture of film, recording tapes and bottles.

It may be synthesised by simple polyesterification between ethylene glycol and terephthalic acid, but commercially it is manufactured by ester interchange melt polymerisation between dimethyl terephthalate and excess glycol. In the first stage, performed at 150–195°C, methanol distils, forming principally bis-(2-hydroxy ethyl terephthalate) ($x = 1$ in formula below) but with some other oligomers ($x > 1$):

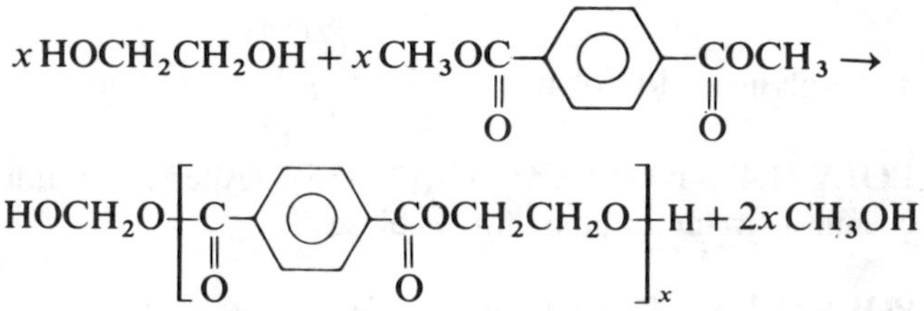

$$x\,HOCH_2CH_2OH + x\,CH_3OC\!-\!\!\big\langle O \big\rangle\!\!-\!COCH_3 \rightarrow$$

$$HOCH_2O\!\!\left[C\!-\!\!\big\langle O \big\rangle\!\!-\!COCH_2CH_2O\right]_x\!\!H + 2x\,CH_3OH$$

I

In a second stage, conducted at about 290°C the excess glycol, resulting from a second ester interchange, distils and polymerisation continues:

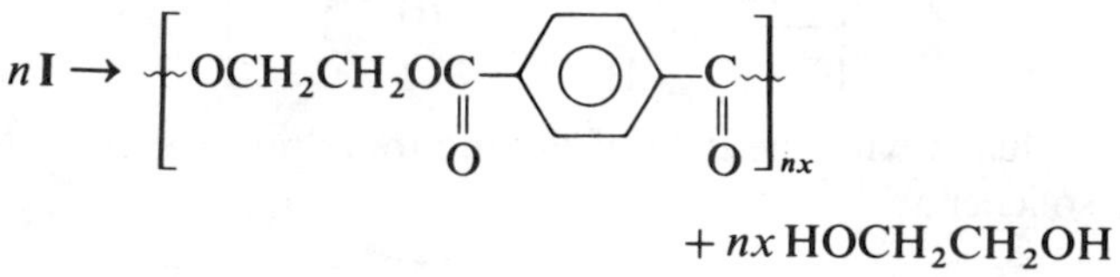

$$n\,I \rightarrow \left[OCH_2CH_2OC\!-\!\!\big\langle O \big\rangle\!\!-\!C\right]_{nx}$$

$$+ nx\,HOCH_2CH_2OH$$

By applying a vacuum, and thereby efficiently removing excess glycol, equilibrium is forced to the polymer direction. Generally a basic metal catalyst is also used.

Polymer for use as fibres and film has a $\bar{M}_n$ value of about 20 000, whereas moulding materials may have higher molecular weights. The polymer has a regular structure and will crystallise, with a T_m value 265°C, but material quenched from the melt, as extruded fibre, film or as a moulding, is largely amorphous and is mechanically weak. However, crystallisation is induced by drawing and hence orienting the polymer, above the T_g value, uniaxially in the case of fibre and biaxially in the case of film, followed by annealing at about 200°C. Crystallinity may then be up to about 50% and mechanical stiffness and strength are much improved. The T_g value is crystallinity dependent, ranging from about 80°C to about 120°C for crystalline polymer.

Fibres of the polymer are characterised by tenacities in the range 4·5–7 g denier^{-1} with corresponding elongations of 25–28%, depending on the amount of drawing after spinning. These properties are similar to those found in nylon fibres. However, in contrast, the fibres show only a low moisture regain of about 0·4%, so wet and dry properties are very similar. The fibres recover well from stretching, which imparts good crease and wrinkle resistance to textiles, as well as non-iron properties.

The biaxially oriented film is of sparkling clarity and high strength (a tensile strength of about 150 MPa) and good electrical properties. It is used for electrical insulation and in large amounts as an audio and video recording tape material. Only special grades, possibly of high molecular weight and containing crystallisation nucleating agents, may be successfully moulded. Normally, mouldings are mechanically weak and suffer from post moulding warping, especially when heated. Biaxially oriented moulded bottles may be successfully made by the stretch blow moulding process. The closely related polybutylene terephthalate is a much better moulding material.

POLYETHYLIDENE

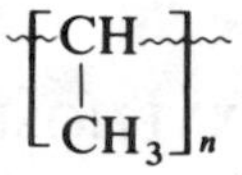

$$\left[\!\begin{array}{c} CH \\ | \\ CH_3 \end{array}\!\right]_n$$

Obtained by the polymerisation of diazoethane using a gold catalyst:

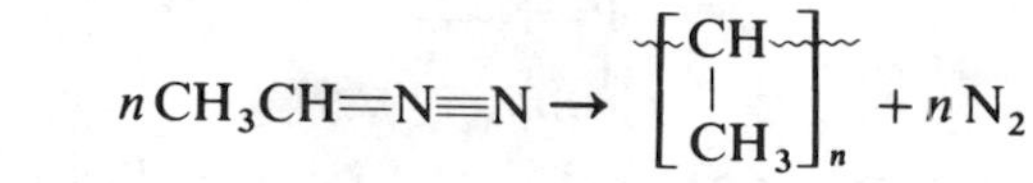

$$n\,CH_3CH\!=\!N\!\equiv\!N \rightarrow \left[\!\begin{array}{c} CH \\ | \\ CH_3 \end{array}\!\right]_n + n\,N_2$$

Isotactic polymer, which crystallises, or amorphous polymer may be produced.

POLYETHYLMETHACRYLATE

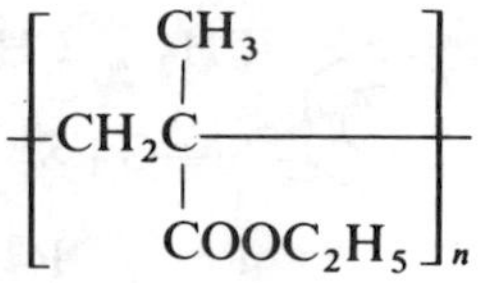

$$\left[\!\begin{array}{c} CH_3 \\ | \\ CH_2C\!\!-\!\!\!-\!\!\!-\!\!\!-\! \\ | \\ COOC_2H_5 \end{array}\!\right]_n$$

Tradename: Plexigum N. Produced by free radical polymerisation of ethyl methacrylate. It has a T_g value of about 65°C and has found limited use as an embedding medium and textile finishing material.

POLY-(p-ETHYNYLBENZENE) (Polypuff) (Polyxylylyne)

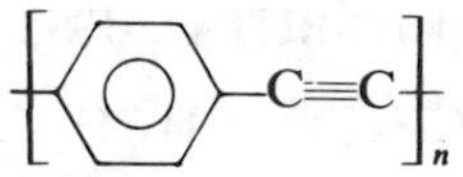

$$\left[\!\big\langle O \big\rangle\!-\!C\!\equiv\!C\right]_n$$

Polymers of this type have been synthesised by dehydropolycondensation of diethynylbenzenes, e.g. with cuprous chloride or pyridine. They are of interest due to the high carbon content (>96%), being readily converted to polymeric carbon, and due to their interesting electrical properties.

POLY-1F4 Abbreviation for poly-(1,1-dihydrofluorobutyl acrylate).

POLY-2F4 Abbreviation for poly-(perfluoromethoxy-1,1-dihydroperfluoropropyl acrylate).

POLY FBA Abbreviation for poly-(1,1-dihydrofluorobutyl acrylate).

POLYFERROCENE Alternative name for ferrocene polymer.

POLYFLON Tradename for polytetrafluoroethylene.

POLYFLUOROACRYLATE (Fluoroacrylate rubber) A fluoropolymer of structure

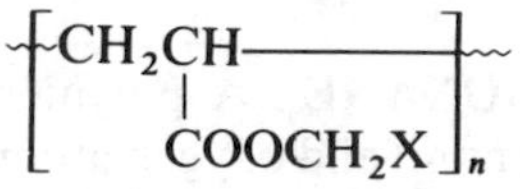

$$\left[\!\begin{array}{c} CH_2CH\!\!-\!\!\!-\!\!\!-\!\!\!-\! \\ | \\ COOCH_2X \end{array}\!\right]_n$$

where X is either —$CF_2CF_2CF_3$ (poly-(1,1-dihydroperfluorobutyl acrylate)) or —$CF_2CF_2OCF_3$ (poly-(perfluoromethoxy-1,1-dihydroperfluoropropyl acrylate)). These were early fluoroelastomers but are no longer commercially available.

POLYFLUOROALKOXYPHOSPHAZENE A polymer containing

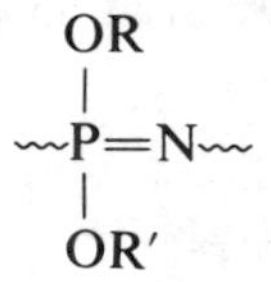

$$\text{OR} \atop \underset{\text{OR}'}{\overset{|}{\sim\sim P = N \sim\sim}}$$

groups, where R and R' are fluorinated alkyl groups. Such units are found in phosphonitrilic fluoroelastomer.

POLYFLUOROALKYLARYLENESILOXANYLENE
(Fasil)

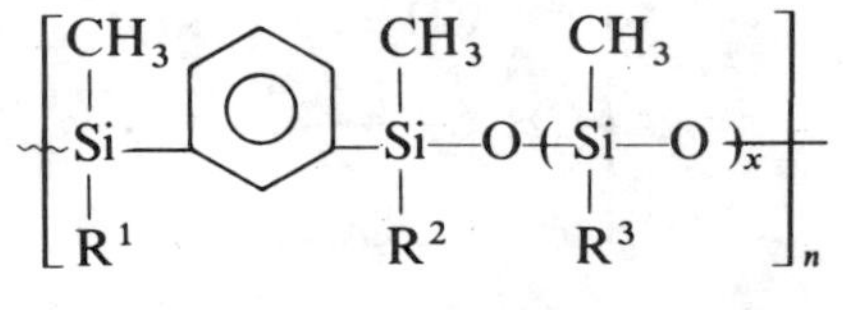

A polymer of the general structure shown above where $R^1 = R^2 = R^3 = CH_3$ or $CF_3CH_2CH_2$ and $x = 0$, 1 or 2. The polymer with $R^1 = R^2 = R^3 = CF_3CH_2CH_2$ and $x = 1$ (poly-(m-phenylene-1,3,5,7-tetrakis(3,3,3-trifluoropropyl)tetrasiloxylene) has been particularly intensively studied as a sealant for supersonic aircraft fuel tanks.

POLY-(2-FLUORO-1,3-BUTADIENE) (Fluoroprene)

$$+CH_2CF=CHCH_2+_n$$

The earliest fluororubber, similar in properties to polychloroprene, and therefore not outstanding as a rubber in the same way as later fluororubbers.

POLY FMFPA Abbreviation for poly-(perfluoromethoxy-1,1-dihydroperfluoropropyl acrylate).

POLYFORMALDEHYDE Alternative name for polyoxymethylene.

POLYFUMARONITRILE The polymer produced by free radical polymerisation of fumaronitrile:

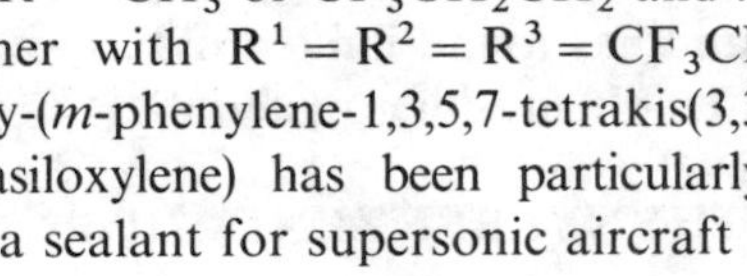

The polymer, being highly conjugated, is highly coloured and has good thermal stability, with a high concentration of unpaired electrons. It is an example of a polycarbazane.

POLYGARD Tradename for tris-(p-nonylphenyl)-phosphite containing some dinonyl substituted groups.

POLYGERMANOXANE A polymer with chains of alternating germanium and oxygen atoms. Polymers may be prepared by dehydration of dihydroxyorganogermanium compounds, e.g.

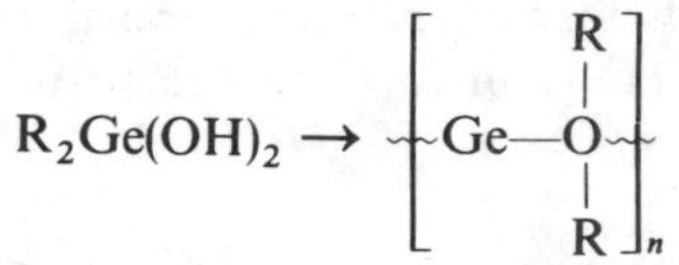

The polymers do not melt on heating but crosslink.

POLY-(1,4'-β-D-GLUCOPYRANOSE) Alternative full chemical structural name for cellulose.

POLY-(1,3'-α-D-GLUCOSE)

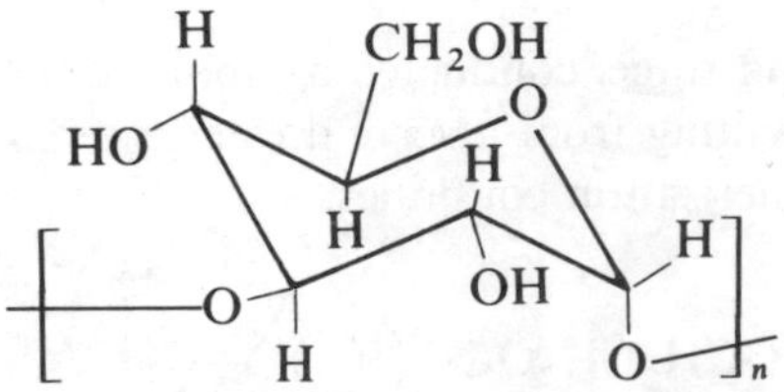

A glucan with the type of linking found in nigeran and isolichenan.

POLY-(1,3'-β-D-GLUCOSE)

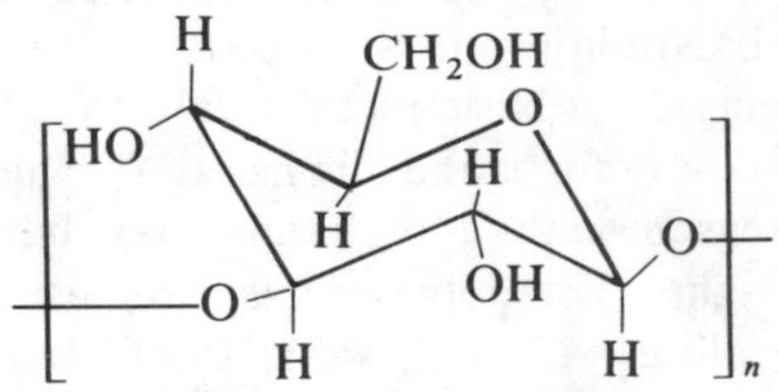

The structure of several polysaccharides found in fungi, algae and some higher plants. Laminaran and callose contain at least 98% 1,3'-linked β-D-glucopyranose units, whereas some algal β-D-glucans contain about 25% 1,6'-links and cereal gum polysaccharides have about 65% 1,4'-links as well. Lichenan has slightly less 1,4'-links.

POLY-(1,4'-α-D-GLUCOSE)

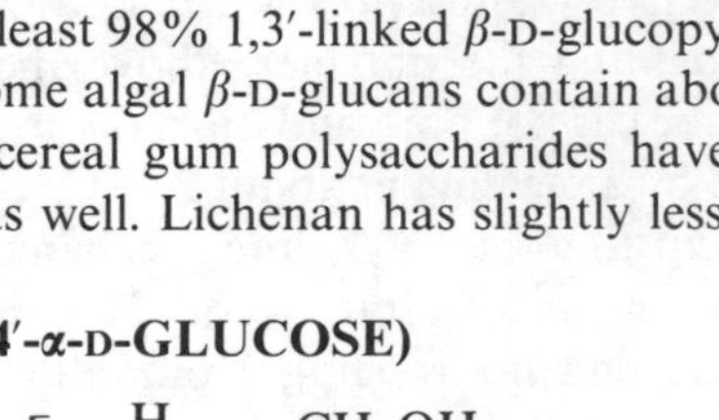

The glucan with the type of linking found in amylase. Also the stem structure of amylopectin, glycogen and pullulan, which are highly branched through 1,6'-links as well as 1,4'-links.

POLY-(1,4'-β-D-GLUCOSE) (Poly-(1,4'-β-D-glucopyranose))

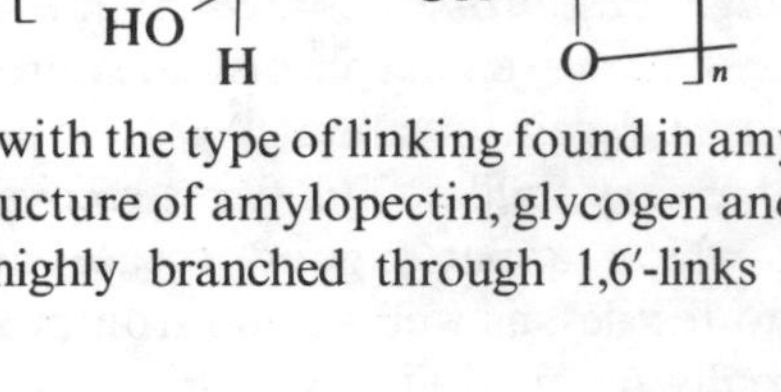

A glucan with the type of linking found in cellulose.

POLY-(1,6′-α-D-GLUCOSE)

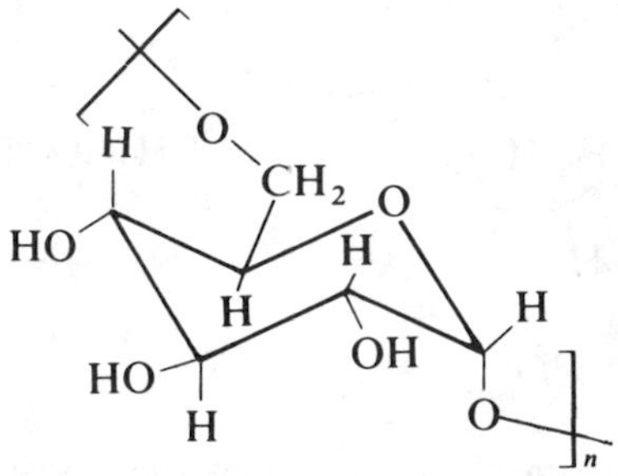

The main type of linking of glucose units found in dextran.

POLY-α-GLUTAMIC ACID

$$+NH—CH—CO\,]_n$$
$$|$$
$$(CH_2)_2$$
$$|$$
$$COOH$$

The optically pure L-isomer is best synthesised by polymerisation of γ-benzyl-α-L-glutamic acid N-carboxyanhydride in the usual way, e.g. by $\overline{O}CH_3$/dioxan, followed by removal of the γ-benzyl protecting group with hydrobromic acid. The acid, and especially its γ-benzyl and γ-methyl esters, have been widely investigated. In the solid state in films cast from the unionised acid, the polymer molecules are α-helical, but from ionised polyacid they are in the β-conformation. In solution it exists as a helix at low pH and as a random coil when ionised. In water the helix–coil transition occurs at pH 8·0 and may be followed by a far ultraviolet hyperchromic effect.

POLY-γ-GLUTAMIC ACID

$$+NH—CH—(CH_2)_2CO\,]_n$$
$$|$$
$$COOH$$

The D-isomer occurs naturally in the cell walls of certain *Bacillus* bacteria. Synthetic poly-γ-D-glutamic acid, identical in properties to the natural material, has been formed in several ways, e.g. from α,α′-dimethyl-γ-D-glutarylglutamate thiophenyl ester. It behaves as a typical polyelectrolyte and exists as a random coil in solution even when uncharged. Poly-γ-L-glutamic acid and alternating D-L- and α-, γ-copolymers have also been synthesised and studied.

POLYGLYCINE   (Nylon 2)

$$+NH—CH_2—CO\,]_n$$

The simplest poly-α-amino acid useful as a model for silk fibroin and collagen both of which have high glycine contents. It is synthesised by polymerisation of its N-carboxyanhydride in the usual way or even, at least in low molecular weight form, by use of a glycine ester. One of the least soluble polypeptides due to extensive intermolecular hydrogen bonding in the β-conformation (or polyglycine I) and polyglycine II structures which are found in the solid state. Sequential copolymers, e.g. the polydipeptide, poly-(gly–ala), and the polytripeptides, poly-(gly–gly–ala) and poly-(gly–pro–pro), have provided even better models especially for conformational studies.

POLYGLYCINE I CONFORMATION

The β-conformation of polyglycine (as opposed to the polyglycine II conformation) in which the sheets of intermolecularly bonded chains may be arranged as a rippled rather than as a pleated β-sheet structure, with intersheet distance being about 0·2 Å shorter and the angles of rotation being $\phi = -150°$ and $\psi = +46·5°$, rather than $\phi = -140°$ and $\psi = +135°$ for an anti-parallel pleated sheet.

POLYGLYCINE II CONFORMATION

A helical polypeptide conformation identical to the poly-L-proline II conformation, having a residue repeat of 3·12 Å, nearly that of a fully extended chain. Thus it is a 3_1 helix stabilised by interhelical hydrogen bonds. It is adopted by polyglycine when precipitated from solution by water.

POLYGLYCOLDIEPOXIDE

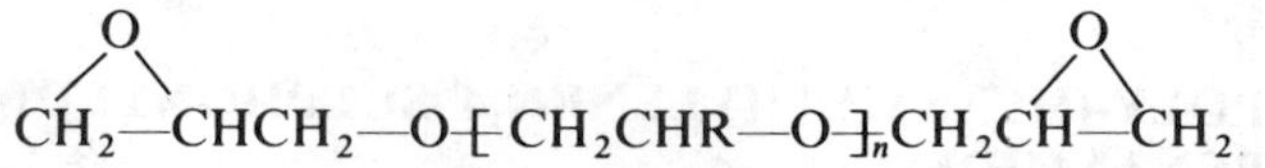

n having values of 2–7. Useful as a reactive flexibiliser for epoxy resins.

POLYGLYCOLIDE   (Polyoxyacetyl)

$$+OCH_2C\,]_n$$
$$\|$$
$$O$$

Tradename: Dexon. Prepared by anionic or cationic ring-opening polymerisation of glycolide, e.g. with SbF_5 or $SnCl_4$ as catalyst. The polymer is biodegradable and is therefore useful as a surgical suture material. It has a T_m value of 200°C.

POLYHEPTANOAMIDE   Alternative name for nylon 7.

POLY-(HEXAFLUOROiSOBUTENE-co-VINYLIDENE FLUORIDE)   Alternative name for hexafluoroisobutene–vinylidene fluoride copolymer.

POLYHEXAFLUOROPROPYLENE OXIDE   Tradename: Krytox.

$$+CFCF_2O\,]_n$$
$$|$$
$$CF_3$$

Produced in low molecular weight form (molecular weight 2000–7000) as heat and chemically resistant oils and greases.

POLYHEXAMETHYLENEADIPAMIDE   Alternative name for nylon 66.

POLYHEXAMETHYLENEAZELEAMIDE   Alternative name for nylon 69.

POLY-(HEXAMETHYLENE-1,3-BENZENEDISULPHONAMIDE)

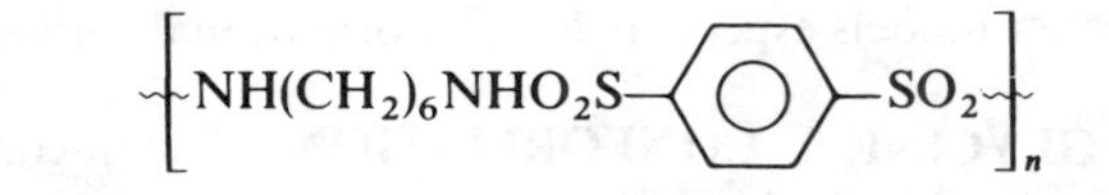

$$\left[NH(CH_2)_6NHO_2S-\underset{}{\bigcirc}-SO_2 \right]_n$$

A polysulphonamide prepared by low temperature solution (e.g. in tetramethylenesulphone) or interfacial polymerisation, by reaction of hexamethylenediamine with 1,3-benzenedisulphonylchloride in the presence of an acid acceptor such as sodium carbonate. Has a T_m value of 185–200°C and is readily melt spun to fibres.

POLYHEXAMETHYLENEDODECANOAMIDE Alternative name for nylon 612.

POLYHEXAMETHYLENENENONAMIDE Alternative name for nylon 69.

POLYHEXAMETHYLENEOXAMIDE Alternative name for nylon 62.

POLYHEXAMETHYLENESEBACAMIDE Alternative name for nylon 610.

POLY-(HEXAMETHYLENE-4,4'-SULPHONYLDIBENZAMIDE)

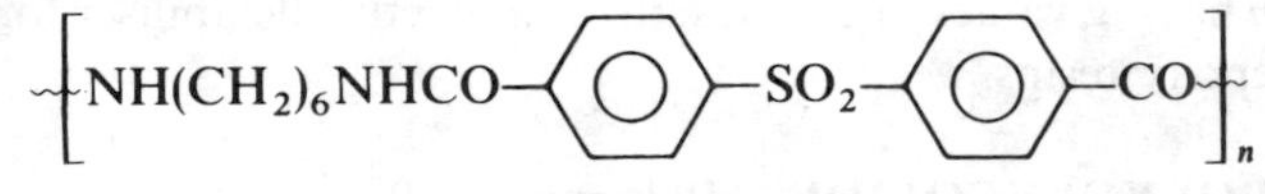

$$\left[NH(CH_2)_6NHCO-\underset{}{\bigcirc}-SO_2-\underset{}{\bigcirc}-CO \right]_n$$

A partially aromatic polyamide formed by reaction of hexamethylenediamine with 4,4'-sulphonyldibenzylchloride. Has a T_m value of 310°C and may be either melt spun or solution spun from trifluoroacetic acid to fibres which have, however, poor tensile properties, possibly due to poor drawability.

POLYHEXAMETHYLENETEREPHTHALAMIDE Alternative name for nylon 6T.

POLYHYDANTOIN (Polyimidazolidione) A heterocyclic polymer with the structure

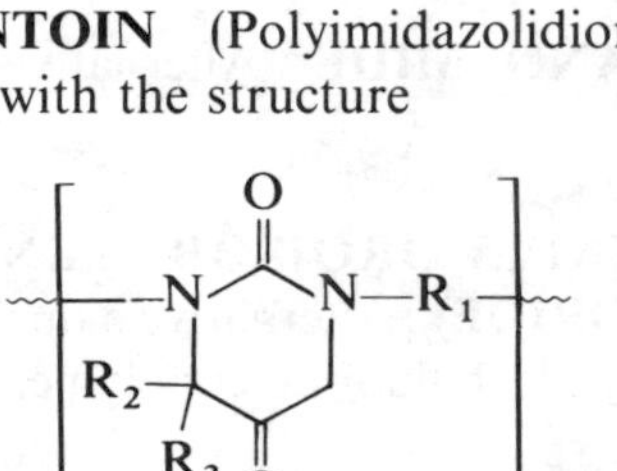

$$\sim\sim N \overset{O}{\underset{R_2 \quad R_3 \overset{}{\underset{O}{}}}{\bigcirc}} N-R_1 \sim\sim$$

in the polymer chain, where R_1 is an aliphatic or aromatic hydrocarbon group, and R_2 and R_3 are aliphatic groups or hydrogen. Synthesised by reaction of a diamine or N,N'-disubstituted diamine with a diisocyanate. Thus for a commercial material the amino monomer is produced by reaction of fumaric acid with a diamine:

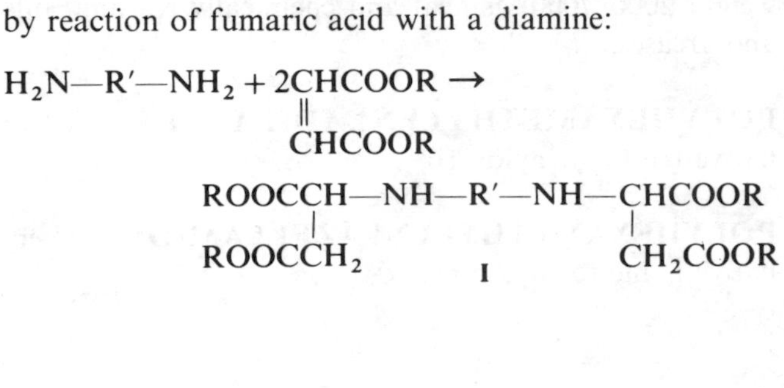

$$H_2N-R'-NH_2 + 2\underset{\overset{||}{CHCOOR}}{CHCOOR} \rightarrow$$

$$\underset{\overset{|}{ROOCCH_2}}{ROOCCH}-NH-R'-NH-\underset{\overset{|}{CH_2COOR}}{CHCOOR}$$
$$\qquad\qquad\qquad\qquad I$$

where R and R' are alkyl groups. Reaction with the diisocyanate

$$n\,I + n\,OCN-Ar-NCO \rightarrow$$

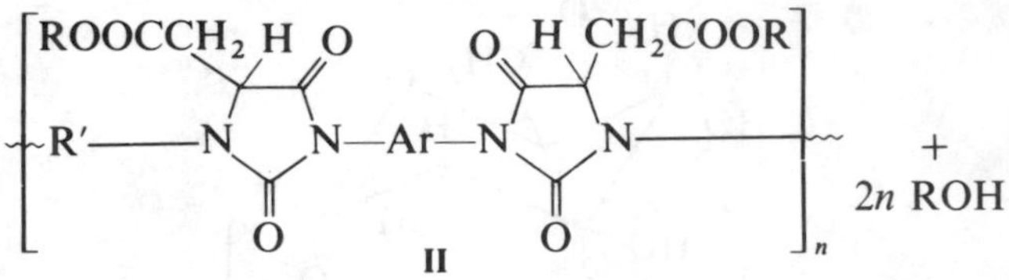

$$\left[\underset{R'}{\overset{ROOCCH_2 \; H \; O}{\underset{N}{\bigcirc}}} \underset{}{N-Ar-N} \overset{O \; H \; CH_2COOR}{\underset{N}{\bigcirc}} \right]_n + 2n\,ROH$$
$$\qquad\qquad\qquad\qquad II$$

(Ar is an aromatic group) gives the polyhydantoin. An alternative reactant to **I** is the reaction product of an aromatic diamine and a chloroester:

$$H_2N-Ar'-NH_2 + ClC(CH_3)_2COOC_2H_5 \rightarrow$$

$$[C_2H_5OOCC(CH_3)_2NH\!-\!]_2Ar'$$

which gives a similar product to **II** when reacted with a diisocyanate, but with R' as Ar' and the $-CH_2COOR$ and H ring atoms replaced by CH_3. Use of a blocked diisocyanate, e.g. a bisphenylurethane,

$$\underset{}{\bigcirc}-OOCNHArNHCOO-\underset{}{\bigcirc}$$

gives higher molecular weight polymers. The related polyiminoimidazolidiones are also of interest as precursors for polyparabanic acids. The polymers are useful as electrical film insulations and wire enamels.

POLYHYDRAZIDE A polymer of the structural type $\{RCONHNHCOR'\}_n$. The aromatic polyhydrazides (with R and R' as aromatic rings) are of most interest.

POLY-(p-HYDROXYBENZOIC ACID) (Poly-(p-oxybenzoyl)) (Poly-(p-hydroxybenzoate))

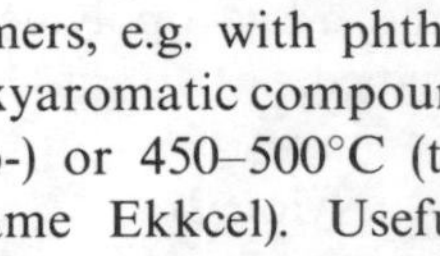

$$\left[\underset{\overset{||}{O}}{C}-\underset{}{\bigcirc}-O \right]_n$$

Tradename: Ekonol. Can be synthesised by heating p-hydroxybenzoic acid with trifluoroacetic acid, but best synthesised by ester exchange on heating diphenyl-p-hydroxybenzoate at 320–340°C. T_m is 550°C. Only moulded with difficulty or by special methods, e.g. sintering. Copolymers, e.g. with phthalic (iso- or tere-) acids and dihydroxyaromatic compounds have T_m values of 275–400°C (iso-) or 450–500°C (tere-) and may be moulded (tradename Ekkcel). Useful for their high temperature resistance and anti-wear properties. Can be used to 320°C continuously, only losing about 1 wt% at 400°C, by TGA.

POLYHYDROXYBUTYRATE (PHB)

$$\{CHCH_2COO\}_n$$
$$\;\;\;|$$
$$\;\;CH_3$$

Tradename: Biopol. A polyester produced by the bacterium *Alcaligenes eutrophus*, when grown on a solution of glucose. Commercial bioproduction has been developed since the polymer is a crystalline plastic material with a T_m value of 175°C and with a mechanical behaviour similar to polypropylene.

POLY-(2-HYDROXYETHYLMETHACRYLATE)

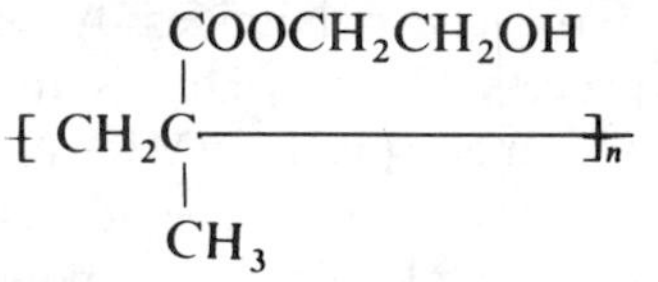

Tradename: Hydron. Produced by free radical polymerisation of 2-hydroxyethylmethacrylate. The polymer has a T_g value of 55°C or 86°C (conflicting data). A water soluble polymer that forms the basis of most soft contact lenses. These are hydrogels formed by copolymerisation in aqueous solution with a small amount of a dimethacrylate, e.g. ethyleneglycol dimethacrylate, as co-monomer, to obtain a crosslinked insoluble product.

POLY-γ-HYDROXYPROLINE

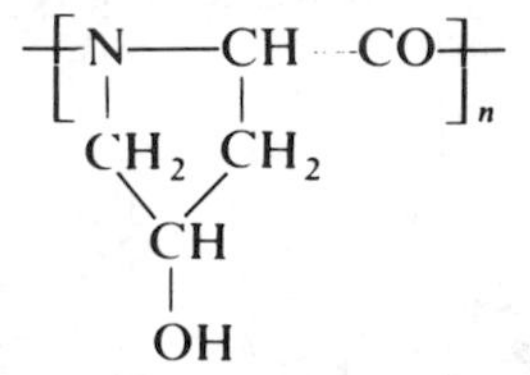

A poly-imino acid. The L-isomer is synthesised from L-hydroxyproline-N-carboxyanhydride with the hydroxyl group being protected with an acetyl group. The polymer and its acetyl derivative show a similar conformational behaviour to poly-L-proline. In the solid state hydrogen bonding occurs between the hydroxyl and carbonyl groups of adjacent chains.

POLYIMIDAZOLE

A polymer containing the heterocyclic ring

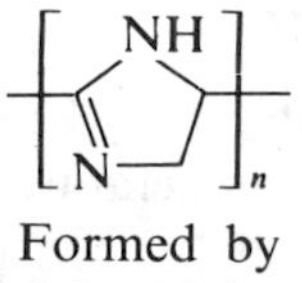

in the polymer chain. Formed by reaction between a dialdehyde and a 1,4-bis(phenylglyoxalyl)benzene:

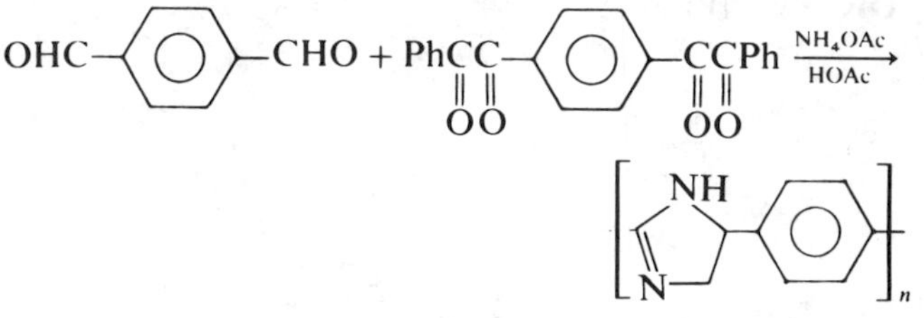

The polybenzimidazoles are much better known.

POLYIMIDAZOLIDIONE

Alternative name for polyhydantoin.

POLYIMIDAZOLINE

A polymer containing the heterocyclic ring

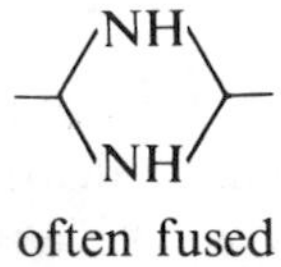

in the polymer chain, often fused to a benzene ring (a polybenzimidazoline). Synthesised by reaction between a bis-(o-diamine) and a diketone or dialdehyde:

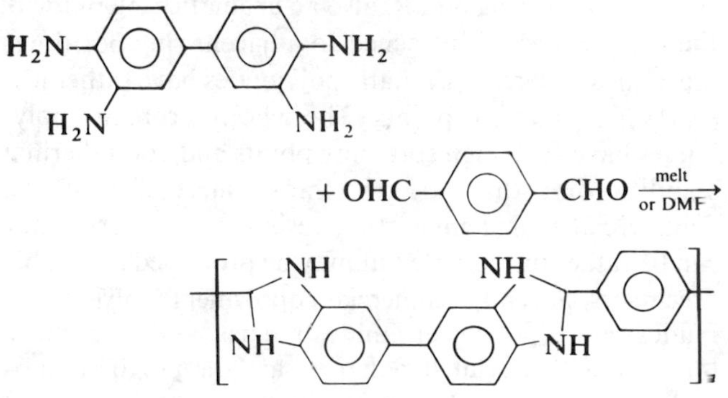

POLY-(1,3-IMIDAZOLINE-2,4,5-TRIONE)

Alternative name for polyparabanic acid.

POLYIMIDAZOLONE

A polymer containing the heterocyclic ring

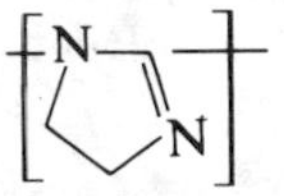

in the polymer chain, which may be fused to a benzene ring. Synthesised by reaction between an aromatic bis-(o-amine) and an aromatic diketone or dialdehyde, initially with formation of a poly(amino–amide) which subsequently cyclises on heating. Aliphatic polymers are formed by reaction between diethylenetriamine and a diacid. The polymers are of low molecular weight but are useful as textile anti-static finishes.

POLYIMIDAZOPYRROLONE

(Polypyrrolone) (Polypyrrone) (Pyrrone polymer) (Ladder pyrrone) A polymer of the structural type

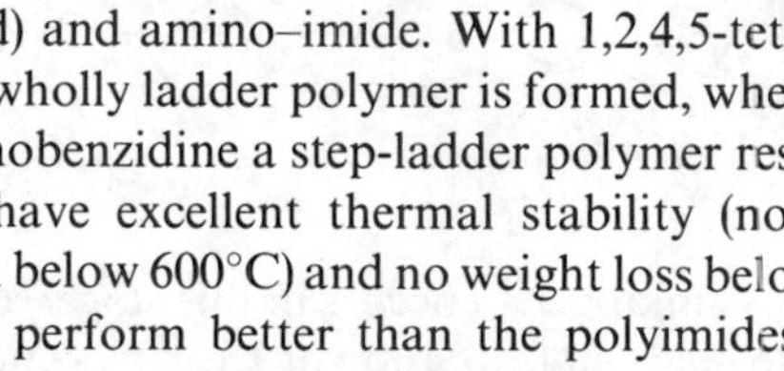

Synthesised by polycondensation between an aromatic bis-(o-diamine) and an aromatic dianhydride, e.g. PMDA. Polymerisation proceeds via an intermediate poly(amide–amino acid) and amino–imide. With 1,2,4,5-tetraaminobenzene a wholly ladder polymer is formed, whereas with 3,3'-diaminobenzidine a step-ladder polymer results. The polymers have excellent thermal stability (no exo- or endotherm below 600°C) and no weight loss below 550°C. Thus they perform better than the polyimides or PBI. They also have good resistance to high energy radiation. The polybisbenzimidazobenzophenanthrolines are closely related.

POLYIMIDE

A polymer of structure

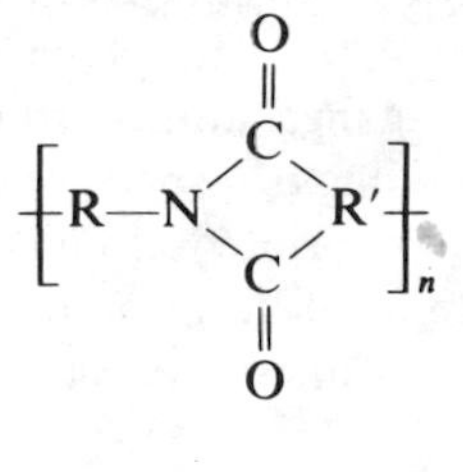

where R and/or R′ are aromatic or aliphatic hydrocarbons or other groups. The aromatic polyimides (with R an aromatic ring and R′ also an aromatic ring but with the imide carbonyls attached to adjacent ring positions) are of most interest. Aliphatic polyimides have rather low ($<150°C$) softening points. The wholly aromatic polyimides have such high softening points and good thermal stability that they are the most successful of the commercial high temperature resistant polymers. They can be rather intractable but may be processed as soluble precursors. Several commercial copolymers (poly(amide–imides) and poly(ester–imides)) exist and are more tractable. Other related polymers are the polybisaminomaleimides.

The best known polyimides are those synthesised by reaction of an aromatic dianhydride with an aromatic diamine. Thus the commercial Kapton H, Pyre ML and Vespel materials are condensation products of pyromellitic dianhydride and 4,4′-diaminodiphenylether, using a highly polar solvent such as DMF, or dimethylacetamide:

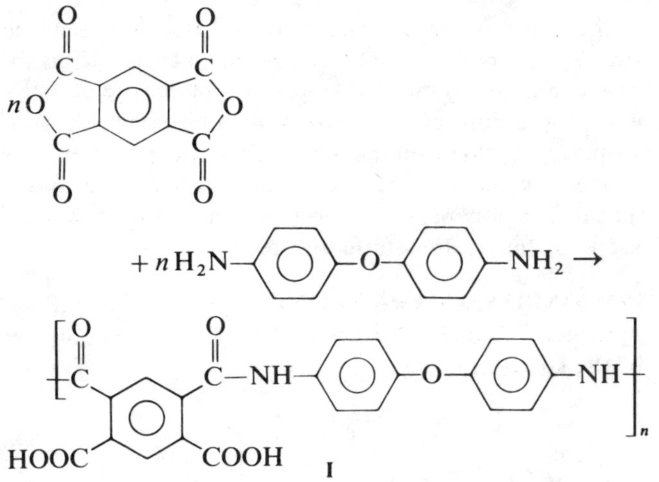

The intermediate poly(amic-acid) cyclises to the polyimide on heating at 250–300°C:

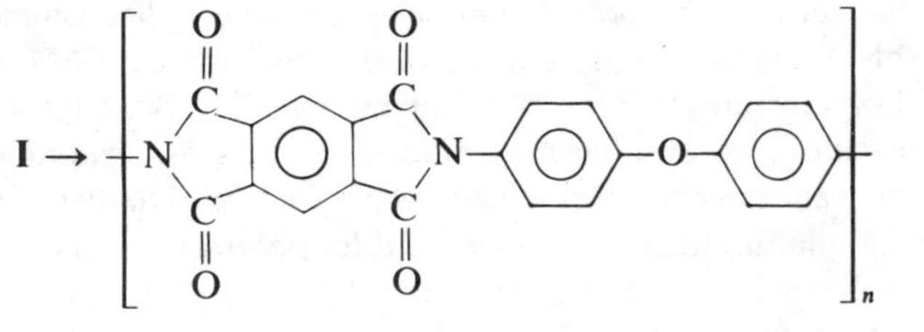

Some crosslinking also occurs in this stage, so processing must take place on formation of **I**, e.g. by solution casting or fibre spinning. Reaction of a diisocyanate with a 'capped' diamine also yields polyimides.

Many commercial products have been developed. Film (Kapton H) has similar room temperature properties to PETP, but much better high temperature resistance, being thermally stable to 420°C and having a T_g value of ~385°C and retaining mechanical properties indefinitely in air at 250°C. Mouldings (Vespel) are useful in bearings, electrical applications, printed circuits and, when glass and carbon fibre reinforced, in aerospace applications, e.g. as honeycomb. Adhesives (Skybond 700), foams

(Skybond), fibre (Kermel) and enamels (Pyre ML) are also produced. Since the early introduction of the commercial polypyromellitimides many other imide polymers have been developed with the main aim of achieving greater ease of processing. Many of these are imide copolymers, such as polyamide–imides, polyester–imides, polyether–imides and heterocyclic polyimides. A newer synthetic approach to processable polymers is to process from appropriate monomers (as in PMR) or prepolymers, which may already contain the imide groups, and to convert these to high molecular weight polymers either by chain or by condensation polymerisation. This may be achieved using appropriately end-capped monomers with reactive groups (as with Thermid M, P13N, Kerimid and Gemon).

POLYIMIDE 2080 Tradename for a thermoplastic polyimide produced from benzenetetracarboxylic dianhydride and tolylene diisocyanate by reaction in dimethyl sulphoxide solution, followed by reaction with 4,4′-diisocyanatodiphenylmethane, to give a block copolymer with structure:

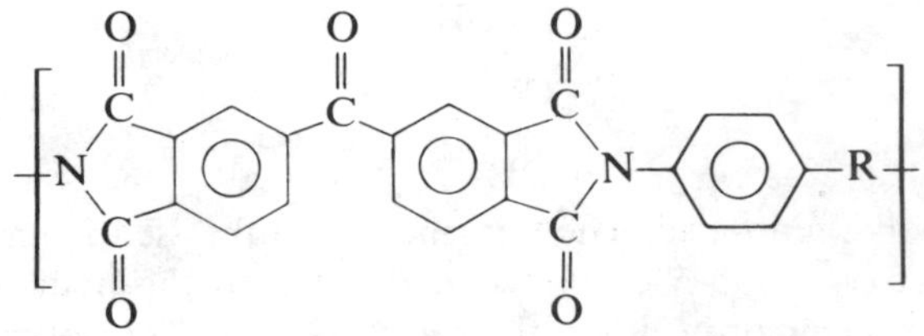

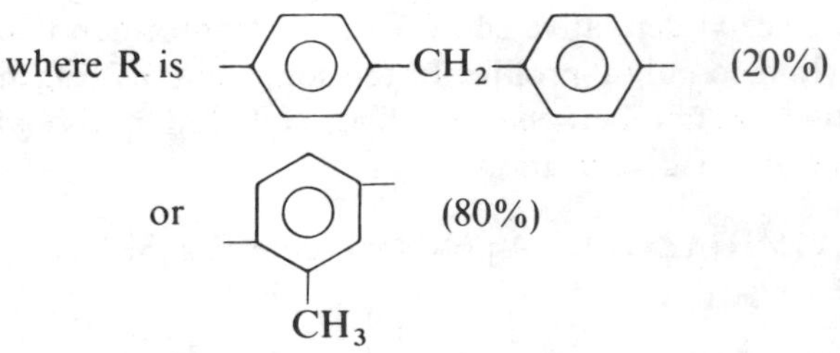

The polymer T_g value is 305°C and it is useful in the production of high temperature resistant laminates.

POLYIMIDE-co-ISOINDOLOQUINAZOLINE-DIONE (PIQ)

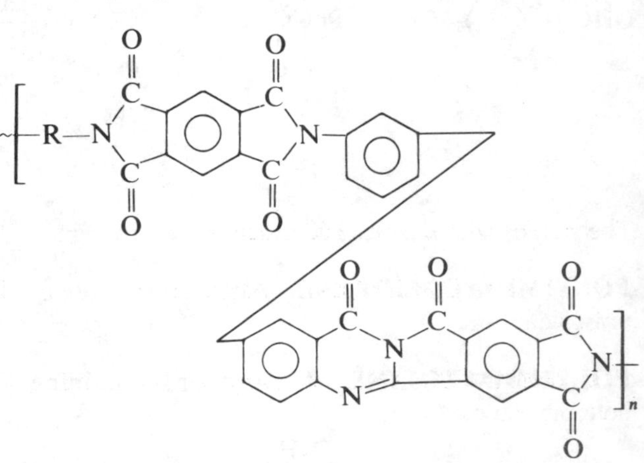

Prepared by reaction of 2 mol of an aromatic dianhydride and 1 mol of an aromatic diaminodicarbonamide. The commercial material is available as a polyamic acid

prepolymer in *N*-methylpyrrolidone solution for the manufacture of large scale integrated circuits.

POLYIMINE   Alternative name for polyazomethine.

POLYIMINOIMIDAZOLIDIONE   A polymer similar in structure to the polyhydantoins. Synthesised by reaction of a diisocyanate with a dicarbamoyl cyanide (obtained by reaction of a diisocyanate with HCN) in a polar solvent at 120°C:

OCNArNCO + NCCONHRNHCOCN →

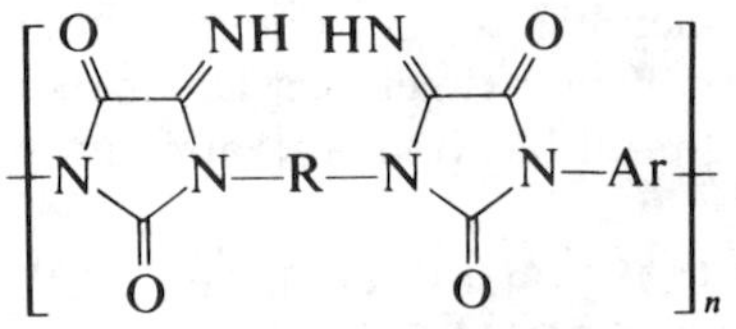

The polymers are precursors for polyparabanic acids.

POLYINDIGO   The allene ladder polymer

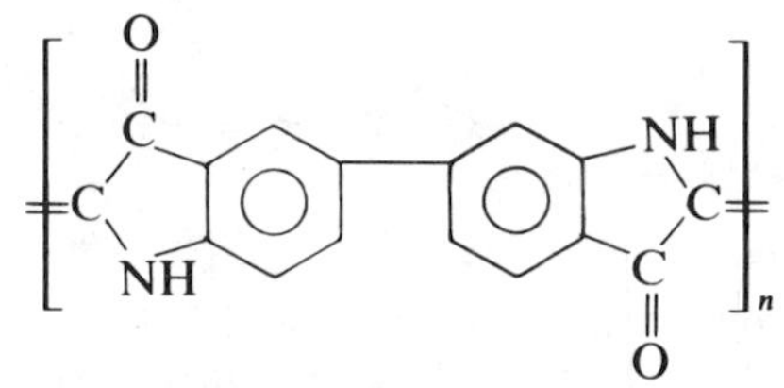

Synthesised by thermal elimination from a monomer of the type

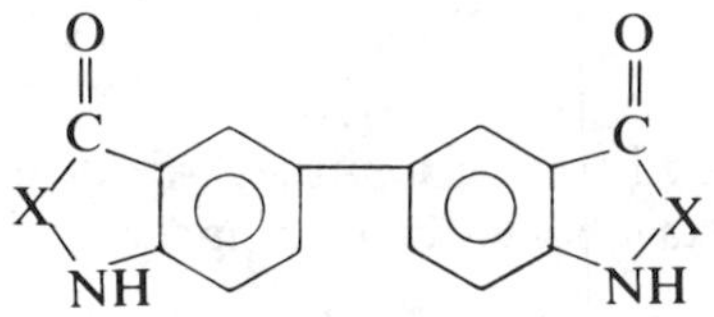

where X is —CH₂— or >CHCOOR. An example of a vat polymer.

POLYINDOLOQUINOXALINE   A polymer containing the heterocyclic ring

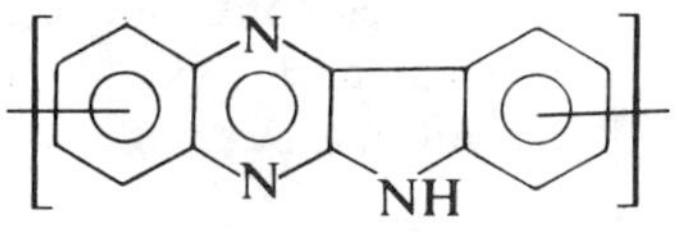

in the polymer chain. Synthesised by reaction of a bisisatin with an aromatic tetramine of the type

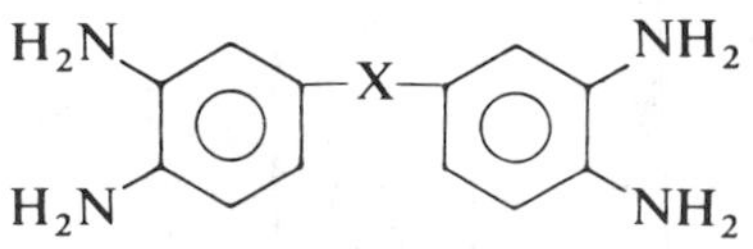

in PPA. The polymers have good thermal stability—stable to 400°C in air by TGA.

POLYION   Alternative name for ionic polymer.

POLYION COMPLEX   Alternative name for polysalt.

POLYISOBUTENE   (Polyisobutylene) (PIB)

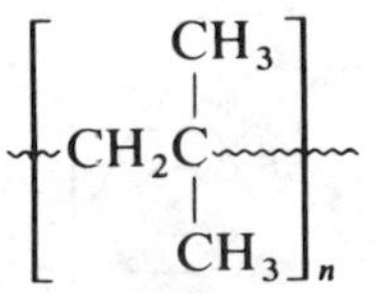

Tradenames: Isolene, Oppanol B, Vistanex. Produced by cationic polymerisation of isobutene, e.g. by use of a Friedel–Crafts catalyst such as AlCl₃ or BF₃, at low temperatures. It does not readily crystallise since no particular conformation is preferred. However, it may crystallise on stretching. The T_m value is 128°C and the T_g value is about −73°C. Commercial polymers may be viscous liquids if of low molecular weight or they may be rubbery solids. However, the polymer is most useful as a copolymer with about 2% isoprene (as butyl rubber) since it may then be sulphur vulcanised. The homopolymer finds uses in adhesives and as a viscosity modifier in motor oils. It has been used in polymer blends with LDPE to improve environmental stress cracking resistance.

POLY-(ISOBUTENE-co-ISOPRENE)   (Isobutene–isoprene copolymer) A copolymer of isoprene and isobutene units. Butyl rubber is such a copolymer with only a few per cent isoprene units.

POLYISOBUTYLENE   Alternative name for polyisobutene.

POLY-(ISOBUTYLVINYL ETHER)   (Poly-(vinylisobutyl ether))

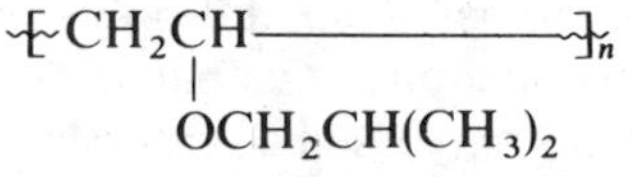

Tradenames: Lutonal I, Gantrez B, Oppanol C. Prepared by the cationic polymerisation of isobutylvinyl ether. It is a crystalline polymer that melts at about 170°C and has a T_g value of about −19°C. It is useful as a pressure sensitive adhesive and as a rubber tackifier. Isotactic polymer is produced by the BF₃ etherate catalysed polymerisation.

POLYISOCYANURATE   (PIR) A polymer containing isocyanurate rings (**I**) and commonly produced by trimerisation of an isocyanate:

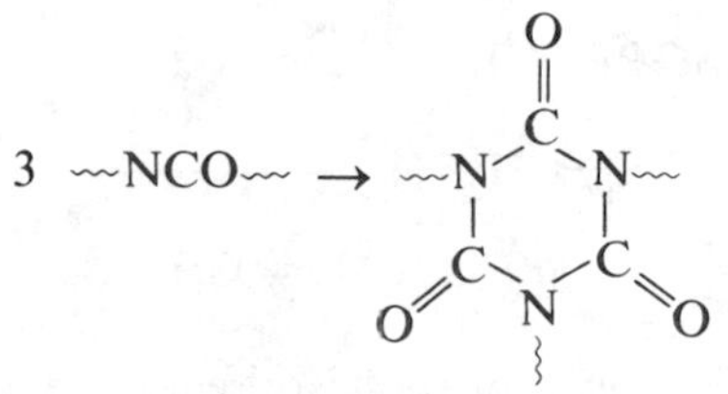

The reaction is catalysed by alkali metal phenolates, alcoholates and carboxylates. Commercially, polyisocyanurate foams are of interest as replacements for rigid polyurethane foams because of their much better fire resistance. They are produced by the use of a polymeric MDI,

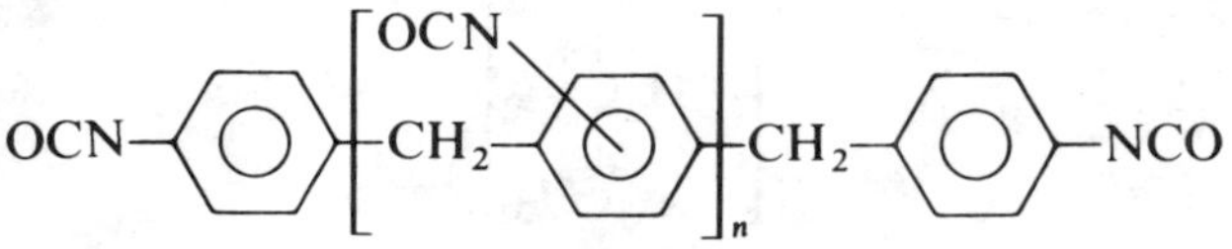

with n having values of 2–7, and a trichlorofluoromethane blowing agent. The products are necessarily highly crosslinked and somewhat brittle, so frequently poly-isocyanurate—polyurethane combinations are used. Isocyanurate ring containing polymers are also produced by a side reaction when 1,2-epoxides are reacted with diisocyanates to yield poly-(2-oxazolidines).

POLYISOINDOLOQUINAZOLINEDIONE A polymer of structure:

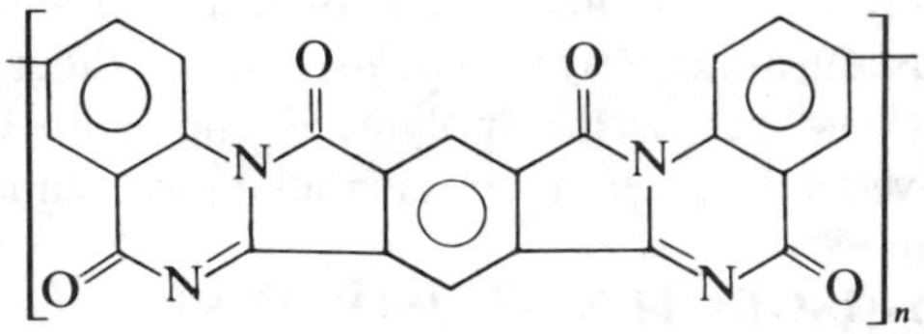

Synthesised by reaction between an aromatic diamine and an aromatic bis-(o-amino—amide) in PPA. The reaction proceeds via soluble uncyclised precursor polymers. The polymers have good thermal stability.

POLYISOPRENE (Poly-(2-methyl-1,3-butadiene)) A polymer of isoprene which can exist in any of several stereoisomeric forms. The commonest form is *cis*-1,4-polyisoprene which occurs in the latex of many trees and plants as natural rubber. *Trans*-1,4-polyisoprene can also be isolated from some plants as balata or gutta percha. Both of these forms may also be synthesised by the use of a stereospecific catalyst in anionic polymerisation. 3,4-polyisoprene may also be synthesised and can exist in both syndiotactic and isotactic forms. A further isomer is 1,2-polyisoprene. In addition to these regular structures, more than one different isomer may be present in the polymer, notably when it is prepared by free radical polymerisation. This generally results in poorer mechanical properties as found in the early synthetic polyisoprenes. Natural rubber is probably 100% *cis*-1,4-structure and although synthetic *cis*-1,4-polymers with *cis* contents of more than 90% are made, the difference in properties is significant.

3,4-POLYISOPRENE

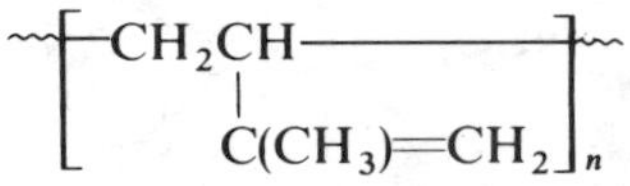

An isomeric form of polyisoprene of little interest compared with the 1,4-isomers. Produced, with about 90% 3,4-structures, by Ziegler–Natta polymerisation using a homogeneous catalyst system consisting of aluminium triethyl and a titanium alkoxide with an Al/Ti ratio of about 6. The polymer produced does not have

sufficient order at asymmetric centres to crystallise and is therefore amorphous.

POLY-(ISOPROPENYLMETHYL KETONE)

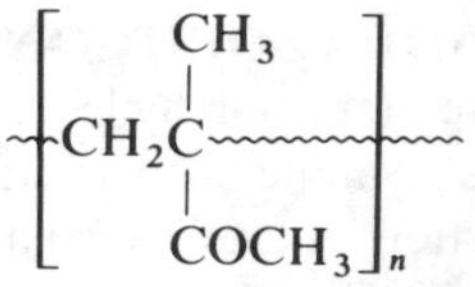

Readily produced by the free radical, cationic or anionic polymerisation of isopropenylmethyl ketone. The polymer has a T_g value of about 80°C and is similar to polymethylmethacrylate in its physical properties, except that it has poor thermal and photochemical stability.

POLY-(4,4′-ISOPROPYLIDENEDIPHENYLENE CARBONATE) Alternative name for bisphenol A polycarbonate.

POLYISOXAZOLE A polymer containing the heterocyclic ring

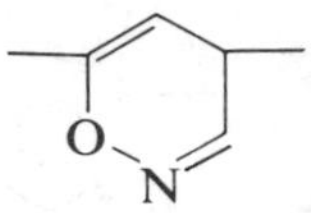

in the polymer chain. Formed by 1,3-dipolar addition between a dinitrile-N-oxide and a diacetylene. Closely related polyisoxazalines containing

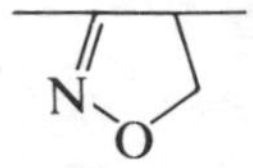

rings (obtained from a dinitrile-N-oxide and a diolefin) and polyisoxazolidines containing

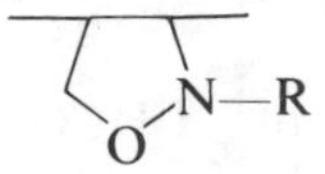

rings (obtained from 1,3-dipolar addition of a dinitrone to a diolefin) are also known, but only in low molecular weight form.

POLYKETAL A polymer containing ketal units, i.e. one of the general structural type $\{CR_1R_2{-}O\}_n$, where R_1 and R_2 are alkyl groups.

POLYLACTIC ACID Alternative name for polylactide.

POLYLACTIDE (Polylactic acid)

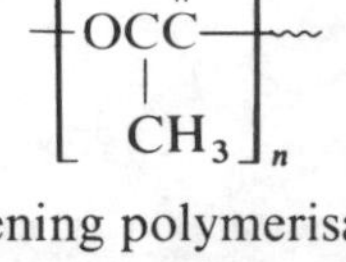

Prepared by ring-opening polymerisation of lactide with PbO, SbF_5 or Sb_2O_3 as catalyst, producing a polymer which softens at 100–130°C.

POLYLAURYLLACTAM Alternative name for nylon 12.

POLYLAURYLMETHACRYLATE

$$\left[CH_2C \begin{array}{c} COO(CH_2)_{11}CH_3 \\ | \\ | \\ CH_3 \end{array} \right]_n$$

A rubbery polymer, produced by free radical polymerisation of lauryl methacrylate, which is useful as a viscosity modifier in lubricating oils.

POLYLYSINE

$$\left[NH-CH-CO \begin{array}{c} | \\ (CH_2)_4 \\ | \\ NH_2 \end{array} \right]_n$$

Poly-L-lysine is synthesised from ε-N-carbobenzoxy protected L-lysine-N-carboxyanhydride in the usual way, followed by removal of the protecting group with hydrobromic acid/glacial acetic acid. The polymer is of interest as a model for basic proteins, itself being a water-soluble basic polyelectrolyte. At high pH it is uncharged and can assume a helical conformation. At lower pH it is a random coil polybase. In the solid state either an α-helix (hydrated) or β-conformation (dry) may be formed. In solution a β-conformation results on heating at high pH. Its interactions with a wide variety of polyanions, e.g. nucleic acids, acidic proteins and polypeptides, have been investigated. Like other basic polypeptides, polylysine and its copolymers have antibiotic properties of interest.

POLYMALEIMIDE Alternative name for poly-aminobismaleimide.

POLYMER
A substance whose molecules consist (as the term suggests from its Greek origins) of many (poly-) parts (Greek, meros) or units. The term refers to molecules with many units joined to each other through chemical covalent bonds, often in a repeating manner. The units are referred to as the mers or repeat units. When the units are all the same and are joined linearly (a linear polymer), the polymer is a homopolymer and its structure may be most simply represented as $\leftarrow M \rightarrow_n$, where M is the repeat unit and n is the number of repeat units or degree of polymerisation (DP). The bonds linking the units are the interunit links. When more than one type of repeat unit is involved the polymer is a copolymer.

Physical chemists sometimes use the term to refer to molecules consisting of only a few units, or maybe only aggregates of very small 'monomers'. However, in polymer science these are referred to as oligomers, the term polymer being reserved for molecules with many, usually above 10 or 20, repeat units. Usually polymer molecules of interest have much higher degrees of polymerisation (several hundred or thousand). To emphasise the long chain nature and high molecular weight of such molecules, sometimes the term high polymer is used. Another term emphasising the large size

of the molecules is the term macromolecule, although this term is normally applied to biopolymers.

Polymers are formed by the process of polymerisation of a monomer (for a homopolymer) or of more than one monomer (for a copolymer). Polymers produced synthetically are synthetic polymers, whereas those produced biologically in nature are natural polymers or biopolymers. Synthetic polymers produced by chain polymerisation are chain (or addition) polymers, of which the most common type are the vinyl polymers. Polymers produced by step-growth polymerisation are step-growth polymers, often also being condensation polymers.

Polymers may be named on the basis of the monomer or monomers from which they are derived by prefixing the name of the monomer (enclosed in parentheses) with the prefix poly-. Often the hyphen and parentheses are omitted when no ambiguity would arise. Examples of such names are poly-(propylene), or simply polypropylene, and poly-(vinyl acetate), or polyvinyl acetate. Alternatively a polymer may be named on the basis of its repeat unit structure. Thus the polymer obtained by polymerising ethylene oxide can be named as poly-(ethylene oxide) or as poly-(oxyethylene) or polyoxyethylene, using a structure based name. Sometimes, especially in the case of complex biopolymers, polymers are best referred to by some common trivial name. Examples are cellulose, which could be called a polyglucose in a monomer based name or poly-(1,4-β-D-glucopyranose) in a structure based name and nylon 66 (trivial name) which is also called polyhexamethylene adipamide (structure based name) or poly-(iminohexamethyleneiminoadipoyl) (IUPAC name).

Synthetic polymers and, now to only a limited extent, biopolymers, form the basis for plastics, rubber, fibre, adhesive and coating materials. Most monomers for such polymers are the products of the petrochemical industry. For such applications, and also for the structural function of some biopolymers in nature, adequate mechanical properties such as stiffness and strength are required. These are only achieved in polymers of high degrees of polymerisation (DP) typically above about 100. Indeed as DP increases so mechanical properties generally improve, but at the cost of ease of processing, at least when this is carried out by melt processing, which is most often the case.

Many important polymers, such as vinyl polymers and polydienes, contain only carbon atoms in the polymer chain (they are carbochain polymers), but other atoms are also found in the chains of heterochain polymers which may contain oxygen, nitrogen, sulphur, phosphorus or other atoms. Most biopolymers and synthetic commercial polymers are organic polymers. However, many organometallic and inorganic polymers are also known. It is often difficult to decide whether an inorganic polymer should be classed as a polymeric or non-polymeric material, since many naturally occurring inorganic minerals involve considerable ionic lattice bonding in building up their macromolecular structures. Normally the criterion of covalent bonding between the units is used in deciding if a material is polymeric. Ionic lattice and metallic bonded structures are not normally considered to be polymeric.

As well as existing as simple linear structures, polymers are often more complex. Polymers may be branched in structure and the branches may be joined to each other, or the linear chains may be crosslinked giving network polymer structures. Polymers used as plastics materials are classed as thermoplastic (when the polymer is linear) or thermoset (when the polymer is crosslinked). Most rubbers are crosslinked polymers.

Polymers may exhibit various types of isomerism, such as positional isomerism, configurational isomerism, and geometrical isomerism. In addition, various structural defects such as unsaturated groups (in otherwise saturated polymers), oxygenated structures and other chemically altered repeat units may be present. Thus although a simple linear polymer may be represented by the simple structure $Y\{M\}_nX$, where X and Y are the end groups, several other structural features may be present.

Owing to their long chain nature, polymer molecules may adopt an almost infinite number of conformations. However, when the molecular structure is regular, a regular conformation may be the most stable in the solid state and the polymer can potentially crystallise. Such polymers, however, are only ever partially crystalline and hence contain both crystalline and amorphous regions. Polymers which do not crystallise in the solid state are amorphous polymers and, if linear, the molecules adopt a random coil conformation. Such a conformation is also usually found for polymers in solution.

POLYMER ALLOY   In general, an alternative name for polymer blend. Sometimes the term is used specifically for a blend of two relatively rigid polymers, i.e. plastics, and sometimes for a blend of two amorphous polymers. Alternatively the term sometimes refers to a blend containing a crystallisable component.

POLYMER ANALOGOUS REACTION   A chemical reaction performed on a polymer for which the same, i.e. analogous, reaction may be performed on low molecular weight materials. The reaction may involve conversion of a functional group in each repeat unit to another group, e.g. the acetylation of hydroxyl groups in cellulose or the hydrolysis of polyvinyl acetate to polyvinyl alcohol. Complications exist in the polymer reactions in that reaction may be limited by coiling, insolubility or crystallinity in the polymer and unreacted material cannot be separated from reacted material since both types of groups are attached to the same polymer chains. If the reaction involves two functional groups reacting in pairs, as in the formation of polyvinyl acetal from polyvinyl alcohol, then reaction cannot go to completion due to single unreacted groups becoming isolated.

POLYMER BLEND   (Polymer alloy) (Polyblend) A physical mixture of two polymers; however, frequently some grafting of one polymer on to the other is also present as a result of the method of blend preparation, e.g. due to scission of polymer chains during melt mixing. For most pairs of polymers, molecular mixing is not favoured thermodynamically, since there is little entropy gain on mixing, compared with the mixing of small molecules. Thus most blends are two phase systems. In the relatively rare cases (e.g. polyphenylene oxide/polystyrene and nitrile rubber/polyvinyl chloride) of miscible (or compatible) polymers, the blend properties are intermediate (roughly the average) between those of the individual unblended polymers. For a dispersion of one polymer in another, the properties depend on the amount, size, shape and interfacial adhesion of the dispersed phase, but are primarily those of the continuous phase. An important type of blend is the toughened plastic (e.g. high impact polystyrene and ABS) in which rubber is dispersed in a relatively brittle and rigid plastic matrix. Other pairs of polymers are blended for many reasons, e.g. to increase stiffness (in the case of rubbers), to lower processing temperatures and cost or to raise softening point, e.g. polycarbonate/ABS. Blend compatibility decreases with increasing molecular weight and may (unlike the case with most small molecule mixtures) decrease with increasing temperature, i.e. exhibit a lower, rather than an upper, critical solution temperature.

POLYMER CEMENT   The hard network polymer produced by ionically crosslinking an aqueous solution of a carboxyl containing polymer, especially polyacrylic acid, with a metal oxide, especially zinc oxide. Such cements are useful as dental materials.

POLYMERIC MDI   (Crude MDI) The mixed isomeric product obtained by phosgenating the condensation product of formaldehyde and aniline, having the general structure

where $n = 0$–4, and containing about 35% diisocyanates and about 45% higher isocyanates. The major isocyanate used in the formation of rigid polyurethane foams by reaction with a polyol. Its relatively high molecular weight gives it a low volatility and therefore a low toxicity (for an isocyanate), whilst its high functionality contributes to crosslinking in the polyurethane.

POLYMERIC SULPHUR   Produced by heating low molecular weight rhombic sulphur (the normal form of sulphur) which consists of S_8 rings (i.e. it is octacyclosulphur) to above 150°C, when ring-opening polymerisation occurs with a large increase in melt viscosity. Above about 180°C, melt viscosity decreases again due to depolymerisation. On cooling, the polymer melt reverts to 'monomeric' octacyclosulphur. The polymeric melt may be quenched rapidly to a rubbery polymer, which has a T_g value of about −30°C; this is raised to about +80°C by the removal of the octacyclosulphur present, by extraction. The resultant so-called 'plastic sulphur' reverts to octacyclosulphur, but the rate of this change may be considerably reduced by incorporation of small amounts of phosphorus or arsenic.

POLYMERISATION The chemical reaction by which a monomer (M), or in the case of copolymerisation more than one monomer, is converted to polymer. Most generally the reaction may be represented as, $n\,M \rightarrow \text{---}[M]_n$, where n is the degree of polymerisation. Usually n has high values, i.e. 'high' polymer is formed, as low molecular weight polymers do not have the required mechanical properties for use as plastics, rubbers, fibres, coatings and adhesives. If a difunctional monomer is used, then a linear polymer is formed (as shown above), since each monomer molecule is only capable of joining to two other monomers. If monomer of higher functionality is used, then branched, and eventually network, polymer will be formed. Ring structures may be involved in polymerisation; thus cyclic monomers of the type $\text{---}M\text{---}$ may polymerise by ring-opening polymerisation. In contrast, ring structures may be formed as a result of polymerisation (cyclopolymerisation).

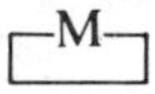

The molecular structure of a polymer is determined during its formation by polymerisation; therefore the conditions (temperature, time, monomer concentration, catalyst or initiator concentration etc.) must be chosen so that polymer with the desired structure is achieved. Naturally the repeat unit structure is determined by the choice of monomer, but the degree of polymerisation depends on polymerisation conditions, as do any structural irregularities formed by side reactions. In addition the polymer may be contaminated by unreacted monomer or other materials, especially solvent, required for the polymerisation.

Two main classes of polymerisation may be distinguished on the basis of polymerisation mechanism. In chain polymerisation (also called addition polymerisation) the double bond of an unsaturated, often vinyl, monomer is opened by reaction with an initiator (or catalyst) forming an activated monomer, to which subsequently many further monomer molecules may add on. The initiator or catalyst, and hence the active centres produced from its reaction with monomer, may be free radical or ionic, giving free radical or ionic polymerisation respectively. In step-growth polymerisation the monomer (or monomers) contain functional groups capable of reacting with each other, thus linking the monomer molecules together. Often a small molecule, e.g. water, is eliminated, when the step-growth reaction may be termed a condensation polymerisation.

As well as chemical variety, polymerisations may be conducted under a wide variety of physical conditions. Simplest is mass (or bulk) polymerisation, involving monomer and initiator or catalyst only. Many step-growth polymerisations are conducted in the mass at the high temperatures required to maintain monomer and polymer molten (melt polymerisation). Alternatively the monomer may be diluted with solvent (solution polymerisation). Occasionally the monomer is polymerised in

the solid state (solid state polymerisation). Frequently the monomer, if liquid, is dispersed in another liquid (usually water) as in suspension, emulsion and dispersion polymerisations. The polymer may be formed at the interface of a dispersed system (interfacial polymerisation) or occasionally from a monomer as a gas (gas phase polymerisation). Often the choice of polymerisation method is strongly influenced by the need to dissipate the heat of polymerisation, since many polymerisations, mostly chain polymerisations involving opening of a double bond, are highly exothermic.

POLYMEROGRAPHY Alternative name for resinography.

POLYMER POLYOL A graft copolymer of a polyether polyol, such as polyoxypropylene glycol, with a vinyl monomer, usually acrylonitrile, prepared by polymerising the vinyl monomer in the presence of the polyol. Useful in the preparation of flexible polyurethane foams with both enhanced modulus and high resilience.

POLYMER–SOLVENT INTERACTION PARAMETER Alternative name for Flory–Huggins interaction parameter.

POLYMETALLOSILOXANE A polysiloxane with some of the silicon atoms replaced by metal atoms, i.e. a copolymer containing both siloxane and metalloxane units. In some cases the polymer is an alternating copolymer of these two types of units, in others the metalloxane units are less frequent. Such polymers are of interest as high temperature polymers in which the performance of the siloxane units has been improved by the incorporation of the metal. This is to be anticipated since the metal will increase the polarity of the chain links, compared with silicon, and its higher coordination would be expected to increase intermolecular forces of attraction and so reduce the tendency of siloxane chains to degrade by cyclisation. When, as is usually the case, the metal atoms carry organic groups, the polymers are known as polyorganometallosiloxanes. Sometimes the metal atoms carry siloxane substituents, when the polymers are then called polyorganosiloxymetalloxanes. Although the main chain bonds may be more thermally stable than the bonds in a simple polysiloxane due to their greater polarity, the polymers are subject to hydrolytic degradation. Typical polymers investigated are those containing tin (polystannosiloxanes), aluminium (polyaluminosiloxanes) and titanium (polytitanosiloxanes). Polyborosiloxanes form the basis of the borosilicones.

POLYMETALLOXANE A polymer whose chains consist of alternating metal and oxygen atoms, although with metals of valency >2 branched or network structures may be formed. Examples include polyaluminoxanes, polytitanoxanes, polystannoxanes and polygermanoxanes. Many metal oxides, oxyacids and hydrated oxides are effectively polymetalloxanes. If organic groups are attached to the metal through metal–carbon bonds then the polymer is a polyorganometalloxane. An example is provided by the polymeric organotin oxides, such as

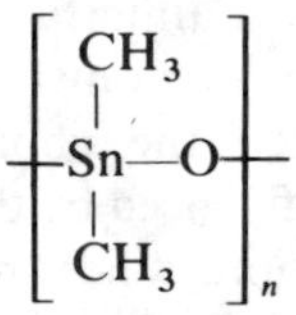

$$\left[\begin{array}{c} CH_3 \\ | \\ -Sn-O- \\ | \\ CH_3 \end{array}\right]_n$$

Many polymetalloxanes are known in which, in addition to the metal atoms, silicon atoms also separate the oxygens, as in the polymetallosiloxanes and the poly-organometallosiloxanes.

POLYMETAPHOSPHATE Alternative name for polyphosphate.

POLYMETHACROLEIN (Polymethacrylaldehyde) Produced by free radical, anionic or cationic polymerisation of methacrolein, and although nominally of structure

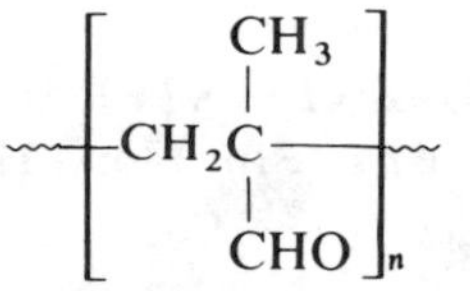

$$\left[\begin{array}{c} CH_3 \\ | \\ -CH_2C- \\ | \\ CHO \end{array}\right]_n$$

usually the structure is of much greater complexity, as in polyacrolein.

POLYMETHACRYLALDEHYDE Alternative name for polymethacrolein.

POLYMETHACRYLAMIDE

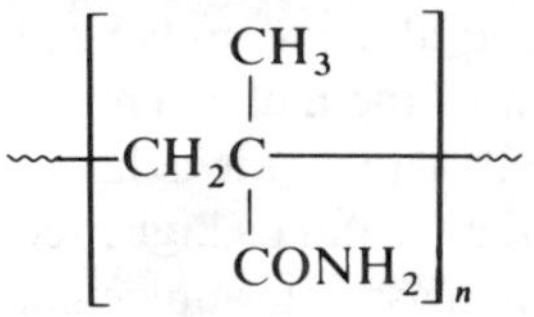

$$\left[\begin{array}{c} CH_3 \\ | \\ -CH_2C- \\ | \\ CONH_2 \end{array}\right]_n$$

Prepared by free radical polymerisation of methacrylamide, usually in aqueous solution, giving a highly water-soluble polymer of similar properties and applications to polyacrylamide.

POLYMETHACRYLATE A polymer of a methacrylate ester, i.e. one having a repeat unit structure

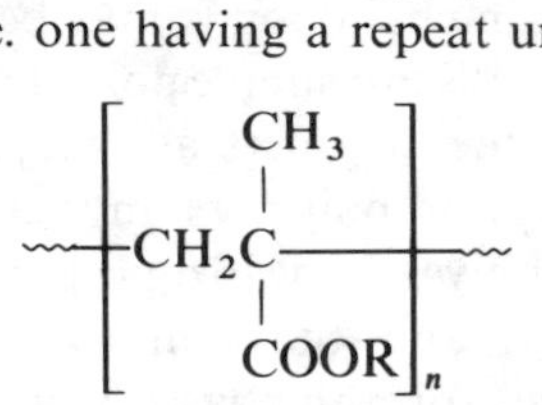

$$\left[\begin{array}{c} CH_3 \\ | \\ -CH_2C- \\ | \\ COOR \end{array}\right]_n$$

where R is an alkyl or substituted alkyl group. Polymethylmethacrylate (where $R = CH_3$) is by far the most important example, being the only acrylic polymer useful as a plastic since it has a relatively high T_g value (110°C). Other members have lower T_g values, the T_g value decreasing with increasing size of R, until $R = C_{12}H_{25}$ (when T_g is -30°C), after which T_g increases. Low molecular weight polymers of the higher methacrylates,

such as n-octyl and lauryl, have found minor uses as leather finishes and viscosity modifiers in lubricating oils.

POLYMETHACRYLIC ACID

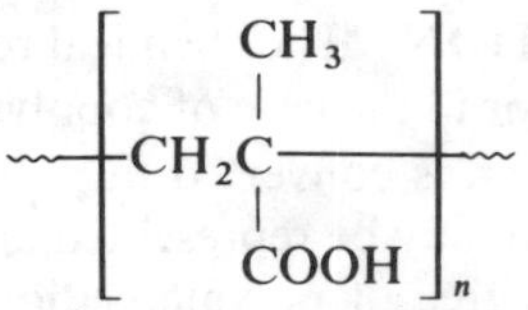

$$\left[\begin{array}{c} CH_3 \\ | \\ -CH_2C- \\ | \\ COOH \end{array}\right]_n$$

Produced by free radical polymerisation of methacrylic acid, frequently in aqeuous solution with a persulphate initiator. Similar in properties and applications to polyacrylic acid.

POLYMETHACRYLIMIDE Tradename: Rohacell. A polymer containing the cyclic repeat units:

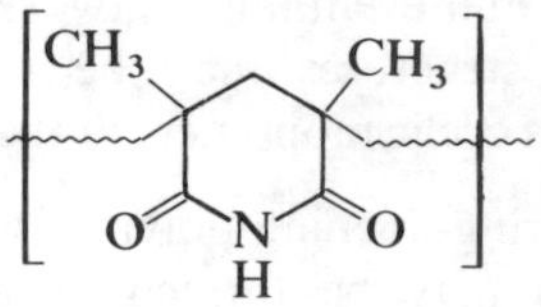

$$\left[\begin{array}{ccc} CH_3 & & CH_3 \\ | & & | \\ & & \\ O & N & O \\ & H & \end{array}\right]$$

Obtained from the copolymer of methacrylic acid and methacrylonitrile, i.e. containing the 'dimer' sequences:

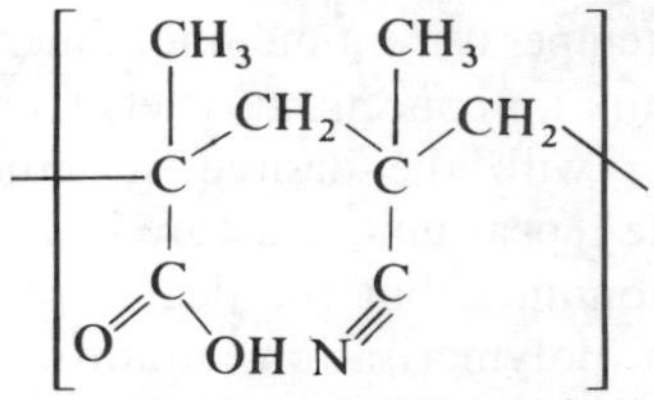

$$\left[\begin{array}{cccc} CH_3 & & CH_3 & \\ | & CH_2 & | & CH_2 \\ -C & & C & \\ | & & | & \\ C & & C & \\ O & OH & N & \end{array}\right]$$

When heated to above its T_g value (about 140°C), but below its decomposition temperature (about 240°C), cyclisation with imidisation occurs. In addition the volatiles formed produce a cellular polymer which has good solvent and heat resistance.

POLYMETHACRYLONITRILE

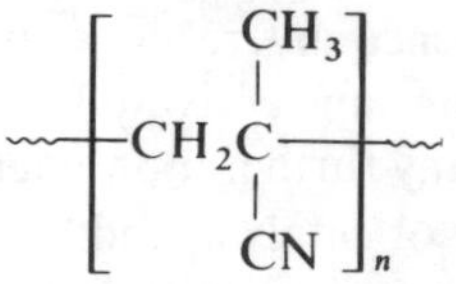

$$\left[\begin{array}{c} CH_3 \\ | \\ -CH_2C- \\ | \\ CN \end{array}\right]_n$$

Produced by free radical polymerisation of methacrylonitrile in a similar way to acrylonitrile. However, methacrylonitrile polymerises less readily and may be more readily polymerised in emulsion. Anionic polymerisation, e.g. with Grignard reagents, can give stereoregular polymer. The T_g value is 120°C so the polymer softens at a lower temperature than polyacrylonitrile, but nevertheless the polymer does discolour at the temperature required for moulding.

POLYMETHYLACRYLATE

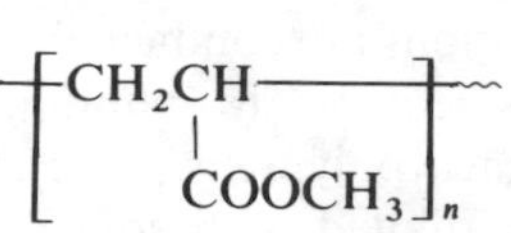

$$\left[\begin{array}{c} -CH_2CH- \\ | \\ COOCH_3 \end{array}\right]_n$$

Tradenames: Acryloid, Plexigum M. Produced by free

radical polymerisation in solution or emulsion of methyl acrylate. The T_g value is about 10°C, so the polymer is tough and leathery at ambient temperatures, but not rubbery. It is also water sensitive. However, it has found uses as a textile size and leather finish.

POLY-(N-METHYL-α-ALANINE)

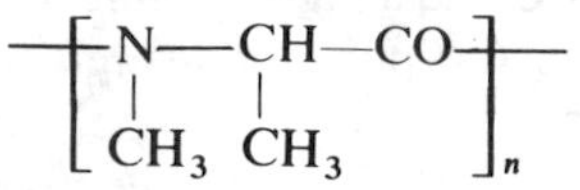

$$\left[\begin{array}{c} N-CH-CO \\ | \quad | \\ CH_3 \;\; CH_3 \end{array} \right]_n$$

Synthesised by polymerisation of its N-carboxy-anhydride. It exists in solution as a threefold right-handed helix with the peptides being all *trans*.

POLY-2-METHYL-1,3-BUTADIENE Alternative name for polyisoprene.

POLY-(3-METHYLBUTENE-1)

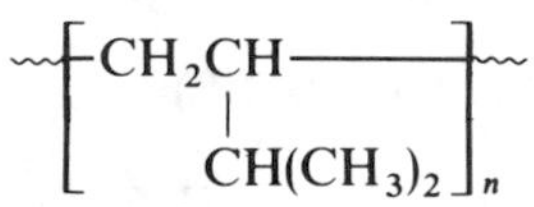

$$\sim\left[\begin{array}{c} CH_2CH \\ | \\ CH(CH_3)_2 \end{array} \right]_n\sim$$

Prepared by Ziegler–Natta polymerisation of 3-methyl-butene-1 in isotactic form; the polymer crystallises in a 3_1-helix having a T_m value of 310°C and a T_g value of about 50°C. Its density is about 0·92 g cm^{-3}.

POLY-(METHYL-α-CHLOROACRYLATE)

$$\left[\begin{array}{c} COOCH_3 \\ | \\ CH_2C \\ | \\ Cl \end{array} \right]_n$$

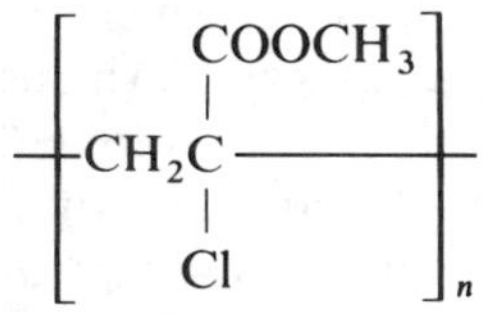

Tradename: Gafite. Prepared by free radical polymerisation of methyl-α-chloroacrylate, which very readily polymerises. The polymer has been of some interest as a harder and higher softening point (T_g value about 140°C) sheet material than polymethylmethacrylate. However, the unpleasant nature of its monomer and its tendency to yellow by thermal degradation have limited its interest.

POLYMETHYLENE

$$\text{-}[CH_2]_n\text{-}$$

Prepared by polymerisation of diazomethane,

$$n\,CH_2{=}N{\equiv}N \rightarrow [CH_2]_n + n\,N_2$$

the polymer nominally has an identical structure to linear polyethylene, but is much more linear. It predated polyethylene but is now largely only of historical interest. High molecular weight polymer may also be obtained by the reduction of carbon monoxide with hydrogen using a ruthenium catalyst under about 100 atm pressure and at about 200°C.

POLYMETHYLENEDIAMIDE A polymer of structure $[CORCONHCH_2NH]_n$, synthesised by reaction of an aldehyde, often formaldehyde, with a dinitrile in sulphuric acid. When higher aldehydes are used a substituted polymer results. With formaldehyde:

$$2\,CH_2{=}O + NCRCN \rightarrow [CORCONHCH_2NH]_n$$

Soluble in the usual polyamide solvents. Polymer T_m values are usually 250–300°C.

POLY-(METHYLENEDIPHENYLENE OXIDE) The aromatic polyether

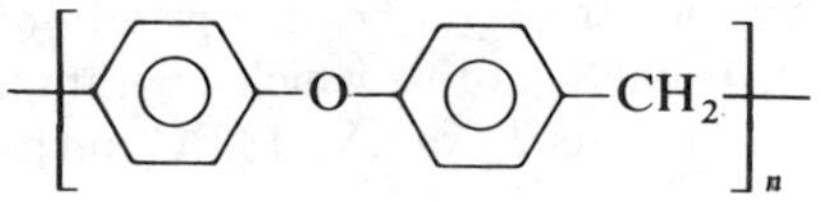

$$\left[\langle\bigcirc\rangle - O - \langle\bigcirc\rangle - CH_2 \right]_n$$

A phenol–aralkyl polymer. Tradename: Doryl. Useful as a high temperature resistant electrical insulation material.

POLYMETHYLENEPHENYLENE (Polybenzyl) (Poly-tolylene) A polymer of the type

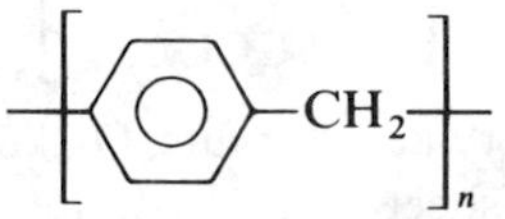

$$\left[\langle\bigcirc\rangle - CH_2 \right]_n$$

the *para*-linked polymer, poly-(p-methylenephenylene), being of greatest interest. Polymerisation of benzyl halides under Friedel–Crafts conditions yields highly branched, amorphous, low molecular weight products of this type, with a softening point of about 75°C. Blocking of the ring positions, as with durylmethyl chloride

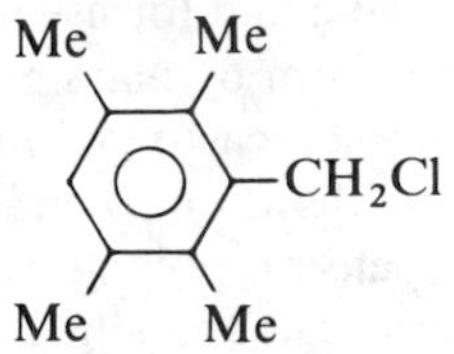

$$\begin{array}{c} Me \quad\;\; Me \\ \langle\bigcirc\rangle - CH_2Cl \\ Me \quad\;\; Me \end{array}$$

makes it possible to prepare linear, crystalline, high molecular weight polymers with T_m values of about 270°C. The Xylok resins may be considered to be broadly of this polymer type.

POLY-(γ-METHYL-α-L-GLUTAMATE)

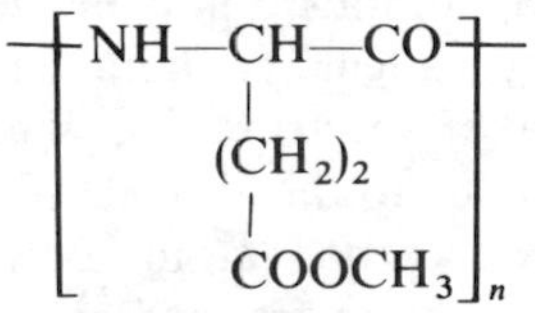

$$\left[\begin{array}{c} NH-CH-CO \\ | \\ (CH_2)_2 \\ | \\ COOCH_3 \end{array} \right]_n$$

Synthesised by polymerisation of the N-carboxyan-hydride of γ-methyl-α-L-glutamic acid. In its widely studied solution and solid state properties and conformations, it behaves similarly to poly-γ-benzyl-α-L-glutamate.

POLY-N-METHYLGLYCINE Alternative name for polysarcosine.

POLY-(4-METHYLHEXENE-1)

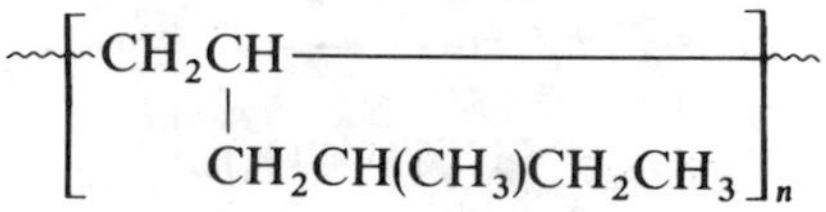

$$\sim\left[\begin{array}{c} CH_2CH \\ | \\ CH_2CH(CH_3)CH_2CH_3 \end{array} \right]_n\sim$$

Prepared by Ziegler–Natta polymerisation of 4-methyl-hexene-1 in isotactic form. It crystallises with a T_m value of 188°C as a 7_2-helix. It is nevertheless transparent. Its T_g value is about 0°C.

POLY-(5-METHYLHEXENE-1)

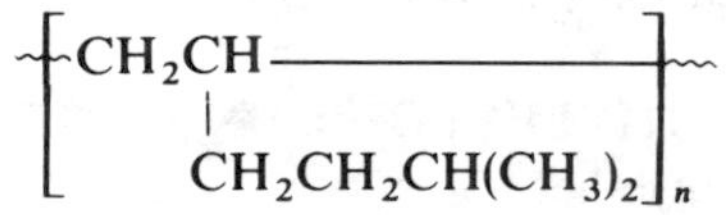

$$\left[CH_2CH \underset{\underset{CH_2CH_2CH(CH_3)_2}{|}}{} \right]_n$$

Prepared by Ziegler–Natta polymerisation of 5-methyl-hexene-1 in isotactic form, which is transparent even though crystalline. Its T_m value is 130°C and its T_g value is −15°C.

POLYMETHYLMETHACRYLATE (PMMA)

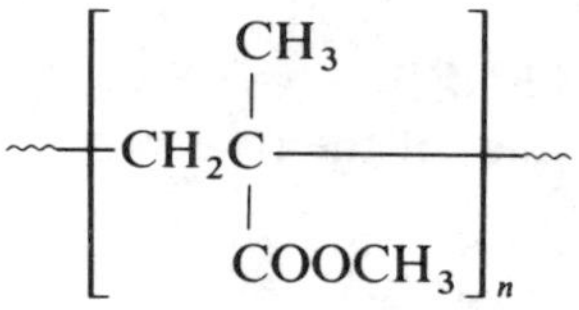

$$\left[CH_2\underset{\underset{COOCH_3}{|}}{\overset{\overset{CH_3}{|}}{C}} \right]_n$$

Tradenames: Acrylace, Acrylite, Diakon, Elvacite, Lucite Vedril and, in sheet form, Asterite, Oroglas, Perspex, Plexiglas. An important commercial thermoplastic material. Produced by free radical polymerisation of methyl methacrylate using the usual peroxide or azo initiators, or by thermal or photochemical initiation.

The polymer produced is atactic and amorphous and has a high transparency. It is therefore commercially important for glazing and for use in signs, for which it is produced in sheet form by mass polymerisation in sheet moulds by monomer casting. Mass polymerisation is prone to autoacceleration. Bead polymer is readily produced by suspension polymerisation.

The main transition is at about 110°C but a β-transition due to side-chain motion occurs at 20°C. Tactic polymers are synthesised by use of anionic catalysts in solution, e.g. organolithium compounds or Grignard reagents. Isotactic polymers tend to be produced in non-polar solvents and syndiotactic polymers in polar solvents.

Isotactic polymer has a T_g value of 45°C and a T_m value of 150–160°C. Syndiotactic polymer has a T_g value of 115°C. Tacticity is easily characterised by nuclear magnetic resonance spectroscopy, since the spectrum is relatively simple due to the absence of a hydrogen atom on each alternate carbon atom. Both diad and triad sequences are readily determined as are configurational sequence lengths. The polymer has the best resistance to ultraviolet light degradation, and hence the best weathering resistance, of all the commoner plastic materials. However, it has only limited impact resistance (0·2–0·4 J $(12·7\,mm)^{-1}$), but this may be improved either by copolymerisation, e.g. with butyl acrylate or acrylonitrile, or by blending with a rubbery polymer such as poly-(n-butylacrylate).

POLY-(4-METHYLPENTENE-1)

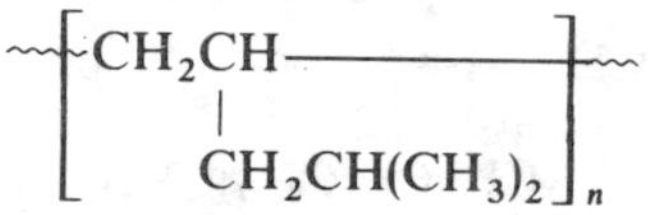

$$\left[CH_2CH \underset{\underset{CH_2CH(CH_3)_2}{|}}{} \right]_n$$

Tradename: TPX. Produced by Ziegler–Natta polymerisation of 4-methylpentene-1, as the isotactic polymer. It crystallises readily but, unusually for a crystalline polymer, it is highly transparent due to the similar densities (very low at 0·83 g cm^{-3}) and hence refractive indices of the crystalline and amorphous regions. The crystalline conformation is a 7_2-helix. A very high T_m value of 235–240°C and a high T_g value of about 40°C (for a polyolefin) give the polymer a high softening point. Mechanical behaviour is similar to polypropylene with a tensile modulus of 1500 MPa, a tensile strength of 30 MPa and an Izod impact strength of about 0·5 J $(12·7\,mm)^{-1}$. It is a useful plastic where high softening point, optical clarity and chemical inertness are required, as in sterilisable medical ware and chemical laboratory apparatus.

POLYMETHYLSTYRENE Alternative name for polyvinyltoluene.

POLY-(α-METHYLSTYRENE)

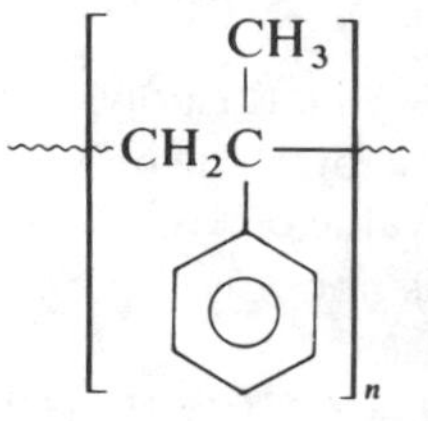

$$\left[CH_2\underset{\underset{\bigcirc}{|}}{\overset{\overset{CH_3}{|}}{C}} \right]_n$$

Tradename: Resin 18. The polymer cannot be produced by conventional free radical polymerisation due to its exceptionally low ceiling temperature (20°C at a monomer concentration of 1M) and hence its tendency to depolymerise. Therefore low temperature (below 0°C) must be used. High molecular weight polymer is a hard, clear material with a T_g value of 168°C. Low molecular weight liquid polymers have been used as plasticisers.

POLY-(METHYLVINYL ETHER) (Poly-(vinylmethyl ether)) (PVME) (PVM) $\{CH_2CH(OCH_3)\}_n$.
Tradenames: Gantrez M, Lutonal, Lutonal M. Prepared by bulk or solution cationic polymerisation of methylvinyl ether, with a Friedel–Crafts catalyst such as BF_3. The polymer can vary from a viscous, readily water-soluble liquid to a stiff rubber depending on molecular weight. The T_g value is −34°C and crystalline polymer melts at 144°C. The polymer is useful as a rubber plasticiser and tackifier and due to its compatibility in aqeuous solution with gums etc., it is also useful in textile, leather, adhesive and paper materials.

POLY-(METHYLVINYL ETHER-co-MALEIC AN-HYDRIDE) Alternative name for methylvinyl ether–maleic anhydride copolymer.

POLY-(METHYLVINYL KETONE)

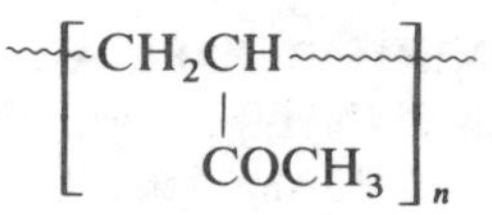

$$\left[CH_2CH \underset{\underset{COCH_3}{|}}{} \right]_n$$

Produced by free radical or ionic polymerisation of methylvinyl ketone. Although featured in many early studies of synthetic polymers it has found little commercial use due to its poor thermal and photochemical stability. It cyclises with dehydration on heating above 250°C by an intramolecular condensation between pairs of ketone groups, isolating unreacted groups. It can be readily converted by polymer analogous reactions to many functional derivatives. Its softening range is 40–60°C.

POLYNONAMETHYLENEUREA   Alternative name for nylon 91.

POLYNONANOAMIDE   Alternative name for nylon 9.

POLYNORBORNANAMIDE   Tradename: Hostamid. A cycloaliphatic polyamide which is the copolymer obtained by melt polymerisation of the bis-(aminomethyl)norbornenes

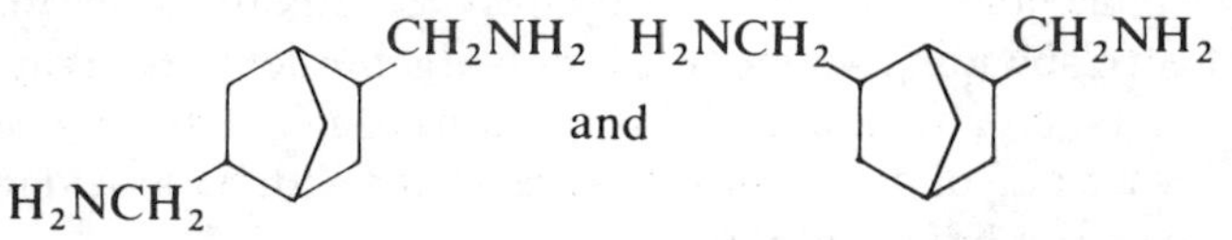

(and possibly other diamines) with a dicarboxylic acid and possibly also caprolactam. The commercial product is amorphous and transparent and has a T_g value of about 150°C, a water absorption maximum of 1–1·5% and a tensile strength of 15 and 91 MPa (dry and wet respectively).

POLYNORBORNENE

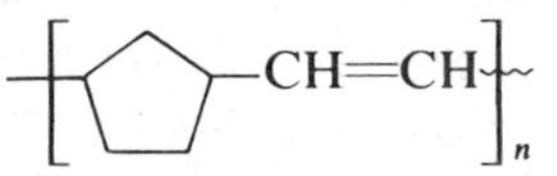

Tradenames: Norsorex, Telene. The polymer obtained by ring-opening polymerisation of norbornene. Both *cis* and *trans* structures may result. The commercial product has a T_g value of 35–45°C, but by incorporation of a mineral oil extender the T_g value may be considerably lowered to −45 to −60°C, thus giving useful rubbery properties, including very soft compositions. Vulcanisation can be by conventional accelerated sulphur vulcanisation. The rubbery polymers are useful as vibration and noise damping materials.

POLYNOSIC RAYON   Tradenames: Toramomen, Tufcel, Vincel, Zantrel. A viscose rayon with a high ratio (about 0·75) of wet to dry strength, low water absorption and low alkali solubility compared with normal rayon. This is achieved by omitting the ageing and ripening process of the viscose, by dissolving the xanthate in water rather than alkali and by spinning into a very dilute acid bath—modifications of the normal viscose process. The product has a higher DP than normal viscose rayon (about 500 compared with about 250), a multifibrillar structure and properties much more like cotton. Tenacity

is typically about 3·3 g denier^{-1}(dry) and 2·5 g denier^{-1}(wet), elongation at break is 9%(dry) and 12%(wet) with about 3% increase in diameter when wet and high initial wet modulus (3% at 0·5 g denier^{-1}). Fabrics have better dimensional stability, will withstand mercerising and handle more like cotton.

POLYOCTANOAMIDE   Alternative name for nylon 8.

POLYOCTENAMER

$$-\!\!\left[CH\!\!=\!\!CH(CH_2)_6\right]_n$$

Tradename: Vestenamer. The alkenamer formed by ring-opening polymerisation of cyclooctene,

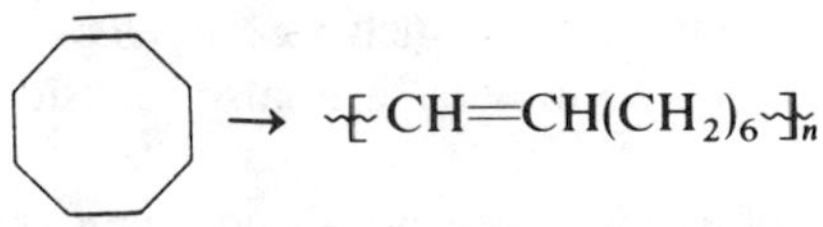

by Ziegler–Natta polymerisation. Polymers with *cis* isomer contents of 75–80% and 40–50% are potentially useful rubbers. The high *cis* polymer has a T_m value of 18°C. The polymers exhibit high tensile strength possibly due to the small number of chain ends, which is due in turn to many polymer molecules existing as macrocyclic rings. Low molecular weight polymers are highly compatible with many other polymers and are useful as rubber additives.

POLY-(*n*-OCTYLMETHACRYLATE)

A rubbery polymer having a T_g value of −20°C or −70°C, produced by free radical polymerisation of *n*-octyl methacrylate, which has found some use as a textile and leather finish.

POLYOL   A molecule containing two or more hydroxyl groups. The term is used to describe both the low molecular weight hydroxy compounds with two hydroxyls (diols or glycols such as ethylene glycol and 1,4-butane diol) and polymeric molecules with hydroxyl end groups. The latter are the prepolymers, usually of a molecular weight of a few thousand, which by reaction with a diisocyanate form a polyurethane:

$$n\,HO\!-\!R'\!-\!OH + n\,OCN\!-\!R\!-\!NCO \rightarrow$$
$$-\!\!\left[R'O\!-\!CONH\!-\!R\!-\!NHCO\!-\!\!-\!O\right]_n$$

They are either hydroxy-terminated polyesters (polyester polyols) or hydroxy-terminated polyethers (polyether polyols) and are used for the formation of polyurethane foams and elastomers.

POLYOLEFIN   A polymer whose repeat unit structure is of the type $-\!\!\left[CH_2CR^1R^2\right]_n$, where R^1 and R^2 are saturated alkyl or cycloalkyl groups or hydrogen atoms, i.e. they are polymers of olefin monomers which contain a single double bond. Usually R^1 is hydrogen, i.e. the polymer is a poly-α-olefin, so that most polyolefins are also vinyl polymers. Sometimes a wider group of polymers is included in the term, covering polydienes and polymers in which R^1 is an aromatic hydrocarbon group, such as in polystyrene. However, these are normally excluded from the group, although those polymers where R^1 is cycloaliphatic, e.g. cyclohexyl (as in poly-(vinyl-cyclohexane)), are included. Several very important commercial plastic materials—polyethylene, poly-propylene, polybutene-1, poly-(4-methylpentene-1) and polyisobutene, and some important rubbers, e.g. ethylene–propylene rubber and butyl rubber are all poly-olefins. Frequently copolymers in which the olefin monomer predominates, such as ethylene–vinyl acetate copolymer and ionomers, are also considered to be polyolefins.

Mostly, polyolefins are produced using stereospecific catalysts (often by Ziegler–Natta polymerisation) and hence are isotactic and therefore crystallise. There is no question of tacticity in polyethylene or polyisobutene, although polyethylene readily crystallises due to its structural regularity. Being saturated hydrocarbons, polyolefins are chemically inert (although susceptible to oxidation), electrically non-polar and highly insulating with low loss. When crystalline, they are highly insoluble and relatively hard and stiff.

POLY-α-OLEFIN A polyolefin of structural type $-\!\!\!-\!\!(CH_2CHR)\!\!-_n$ where R is an alkyl or cycloalkyl group. Thus the important commercial polymers polypropylene, polybutene-1 and poly-(4-methylpentene-1) are poly-α-olefins. Other higher poly-α-olefins such as polyhexene-1 and polyoctene-1 have also been investigated.

POLYORGANOALUMINOSILOXANE A polymer containing siloxane and aluminoxane units in the polymer chain, the aluminium also being substituted with organic groups. Such polymers may be synthesised by the reactions

$$RSiCl_3 + NaOH + Al_2(SO_4)_3 \xrightarrow{H_2O}$$
$$(RSi(OH)_2O)_3Al + Na_2SO_4 + NaCl$$

or

$$RSi(OH)_2ONa + Al_2(SO_4)_3 \rightarrow$$
$$(RSi(OH)_2O)_3Al + Na_2SO_4$$
$$\mathbf{I}$$

In the presence of water further reaction occurs

$$RSi(OH)_2ONa + H_2O \rightarrow RSi(OH)_3$$

and

$$RSi(OH)_3 + \mathbf{I} \rightarrow$$

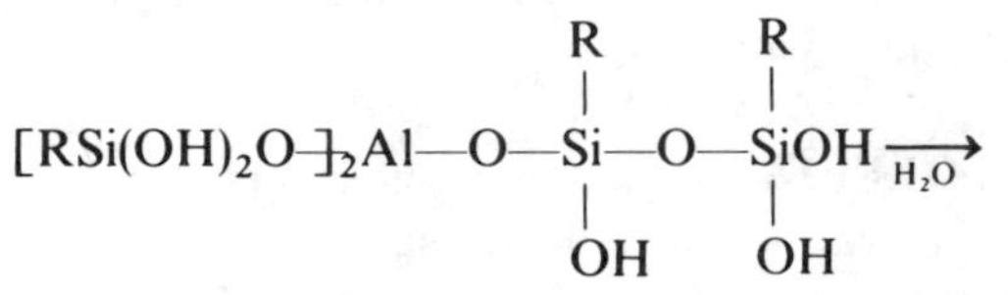

etc. → polymer

When R is phenyl and with one aluminium to every four silicons, the polymer is soluble in organic solvents, forms brittle films and is highly infusible to about 500°C. However, the polymers have poorer thermal stability than unmodified polysiloxanes.

POLYORGANOBOROSILOXANE (Borosiloxane polymer) (Polyborosiloxane) A polymer containing both siloxane and boron atoms in the chain. Such polymers may be prepared by any of a variety of condensation reactions between di- or trifunctional organosilicon compounds (such as acetates, alkoxides or halides) and boron compounds (such as alkoxides or boric acid) e.g.

$$n\,B(OR)_3 + n\,R'_2Si(OCOCH_2)_2 \rightarrow$$

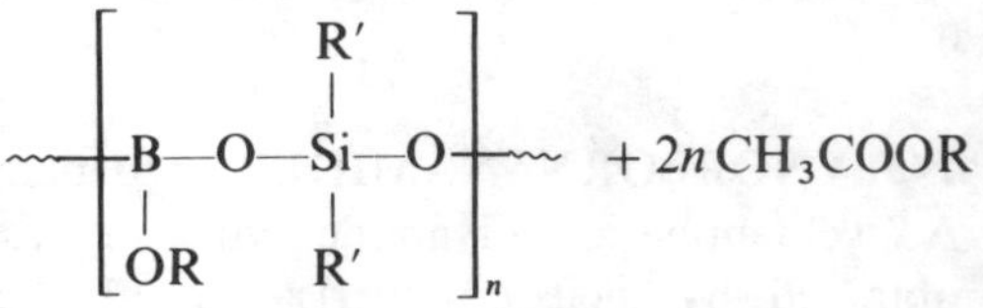

The polymers are in general lower melting, less thermally stable and have poorer hydrolytic stability than the corresponding unmodified polyorganosiloxanes. Polymers containing similar linkages are thought to be formed when polydimethylsiloxanes are heated with boric acid to produce bouncing putty.

POLYORGANOMETALLOSILOXANE A poly-metallosiloxane in which the metal atoms carry organic groups. When these are attached by metal–carbon bonds, the polymer is truly organometallic. However, many metals do not form stable metal–carbon bonds. Nevertheless organic groups may be attached as alkoxide or organosiloxy groups, as in polyorganosiloxy-metalloxanes. Examples include polyorganoalumino-siloxanes, polyorganoborosiloxanes, polyorganophos-phosiloxanes, polyorganostannosiloxanes and poly-organogermanosiloxanes. The incorporation of organic groups enables polymers with a variety of properties, such as solubility in organic solvents, to be prepared.

POLYORGANOMETALLOXANE A polymetallox-ane in which organic groups are attached to the metal atoms. These may be true organometallic compounds with metal–carbon bonds, as in, for example,

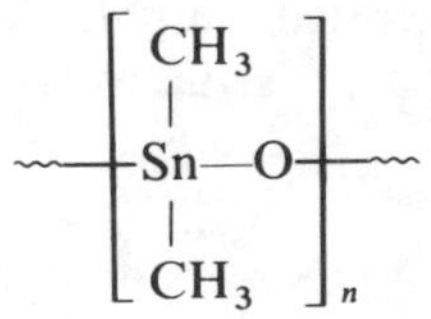

(polymeric dimethyltin oxide) or, as is more often the case, the substituents may be alkoxy groups, as in polyalkoxy-titanoxanes for example.

POLYORGANOSILOXANE A polymer containing

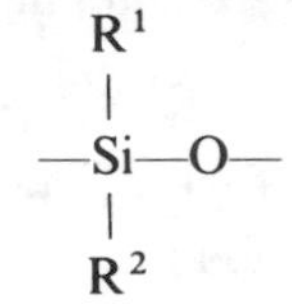

units in which R^1 and/or R^2 are organic groups. The most important type of polysiloxane, these polymers forming the basis of the commercially important silicone fluids, rubbers and resins. Most commonly R^1 and R^2 are alkyl or aryl groups, the most common polymer being polydimethylsiloxane ($R^1=R^2=CH_3$); however, polymers in which some R^1 groups are phenyl are also of commercial interest (the phenylmethylsilicones). Several copolymers containing proportions of other methylsiloxane units are also of interest, e.g. methylvinylsilicone, nitrilesilicone and fluorosilicone rubbers. Many polymers with metal atoms in the chain, the polyorganometalloxanes, have also been widely studied as have ladder polymers such as polyphenylsilsesquioxane. The polymer structures are sometimes represented by the symbols M, D, T and Q for mono-, di-, tri- and quadra- (i.e. tetra-) functional units in the chain. Thus, for example

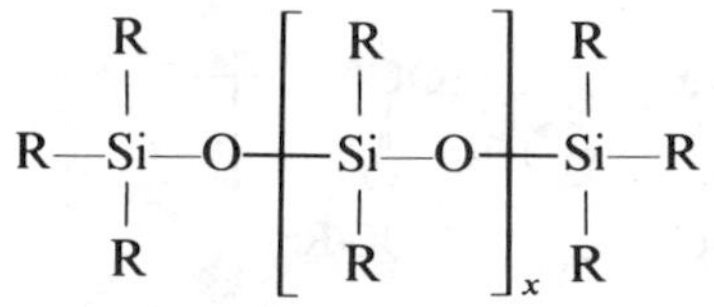

$$
\begin{array}{ccccc}
R & & R & & R \\
| & & | & & | \\
R-Si-O- & -Si-O- & -Si-R \\
| & & | & & | \\
R & & R & & R
\end{array}\Big]_x
$$

would be designated MD_xM.

In general the polymers are synthesised by aqueous hydrolysis of chlorosilanes, being formed by condensation of the unstable intermediate silanols, catalysed by the hydrochloric acid produced. Thus a linear polymer results from hydrolysis of a dichlorosilane:

$$x\,R_2SiCl_2 \rightarrow x\,R_2Si(OH)_2 + 2x\,HCl \rightarrow$$

$$\text{-}\!\!\left[\,R_2SiO\,\right]_x + x\,H_2O$$

Use of a proportion of a trichlorosilane or of tetrachlorosilane results in branched T units or Q units respectively. A monochlorosilane acts as a chain terminator (providing M units). Hydrolysis is often carried out with aqueous hydrochloric acid and results in the formation of a mixture of linear polymer and cyclic oligomers, especially cyclic tetramer. By equilibration by heating with a suitable catalyst, such as $FeCl_3$ or KOH, a higher yield of high molecular weight polymer may be achieved. Replacing dichlorosilane with diethoxysilane also favours formation of linear polymer. However, to produce the very high molecular weight polymer required for silicone rubber, the cyclic tetramer is isolated and polymerised by equilibration with alkali. Use of either a miscible or immiscible organic solvent with the water favours cyclic oligomer formation.

The polyorganosiloxanes are characterised by high thermal stability (to about 400°C) due to the strength of the Si—O bond and thermo-oxidative stability (to about 200°C). They have high chemical resistance, except to acid or base catalysed chain cleavage, as occurs in equilibration. They have a strongly hydrophobic nature and useful non-stick properties.

POLYORGANOSTANNOSILOXANE A polymer containing both siloxane and stannoxane units in the polymer chain, the tin atoms also carrying organic groups. Prepared, for example, by

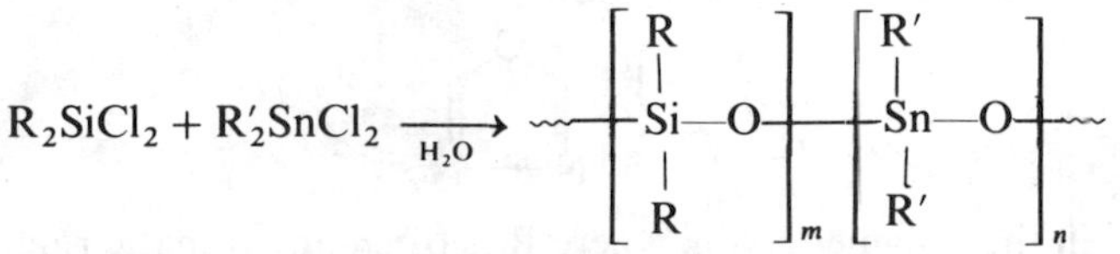

$$R_2SiCl_2 + R_2'SnCl_2 \xrightarrow{H_2O} \sim\!\!\left[\begin{array}{c} R \\ | \\ Si-O \\ | \\ R \end{array}\right]_m\!\!\left[\begin{array}{c} R' \\ | \\ Sn-O \\ | \\ R' \end{array}\right]_n\!\!\sim$$

by reaction of organotin oxides with silanols or by reaction of organotin chlorides with alkali metal silicates.

POLYORGANOTITANOSILOXANE A polymer containing both siloxane and titanoxane units in the polymer chain, the titanium atoms also carrying organic groups. Several polymers of this type have been synthesised, e.g. by

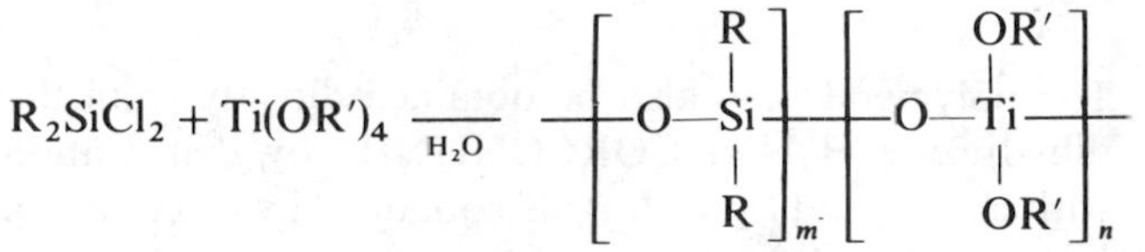

$$R_2SiCl_2 + Ti(OR')_4 \xrightarrow{H_2O} \left[\begin{array}{c} R \\ | \\ O-Si \\ | \\ R \end{array}\right]_m\!\!\left[\begin{array}{c} OR' \\ | \\ O-Ti \\ | \\ OR' \end{array}\right]_n$$

or by reaction of acetoxysilanes with alkoxy titanium compounds

$$(AcO)_2Si(CH_3)_2 + (i\text{—}PrO\text{—})_2Ti\text{—}(OSi(CH_3)_2)_2 \rightarrow$$

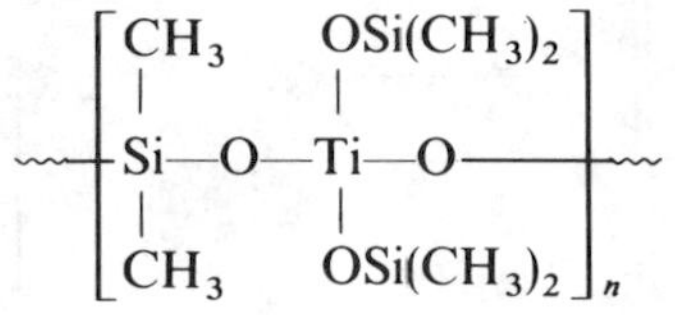

$$\sim\!\!\left[\begin{array}{cc} CH_3 & OSi(CH_3)_2 \\ | & | \\ Si-O-Ti-O \\ | & | \\ CH_3 & OSi(CH_3)_2 \end{array}\right]_n$$

The polymers are of lower thermal stability than unmodified polysiloxanes.

POLYOX Tradename for high molecular weight polyethylene oxide, useful as a water-soluble film, and, in solution, as a thickening agent and as a size.

POLY-(1,2,4-OXADIAZOLE) A polymer containing the ring

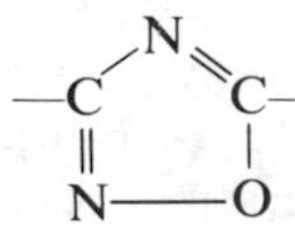

$$
\begin{array}{ccc}
 & N & \\
 & \diagup\ \diagdown & \\
-C & & C- \\
\parallel & & | \\
N & —— & O
\end{array}
$$

in the polymer chain. Synthesised by thermal cyclodehydration of a poly-(acylamide oxime) (obtained from reaction of a diamide oxime

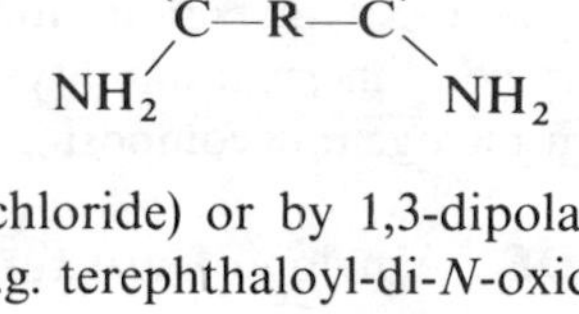

$$
\begin{array}{ccc}
HON & & NOH \\
\diagdown & & \diagup \\
C-R-C \\
\diagup & & \diagdown \\
NH_2 & & NH_2
\end{array}
$$

and a diacid chloride) or by 1,3-dipolar addition of a nitrile oxide, e.g. terephthaloyl-di-N-oxide,

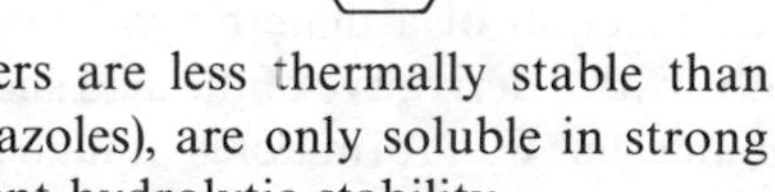

$$O\leftarrow N\equiv C\text{—}\langle\bigcirc\rangle\text{—}C\equiv N\rightarrow O$$

The polymers are less thermally stable than the poly-(1,3,4-oxadiazoles), are only soluble in strong acids and have excellent hydrolytic stability.

POLY-(1,3,4-OXADIAZOLE) A polymer having the structure

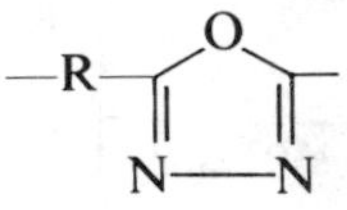

$$-R-\underset{\underset{N-N}{\|}}{\overset{O}{\diagup \diagdown}}-$$

in the polymer chain, where R is often an aromatic ring, e.g. m- or p-phenylene. Most commonly synthesised by dehydration of a polyhydrazide, itself obtained from a dihydrazide and a diacid chloride or from hydrazine and a diphenyl ester

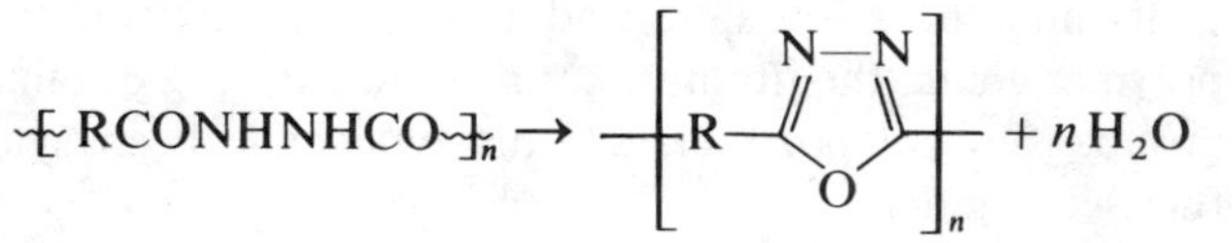

$$\fbox{$\cdots$}RCONHNHCO\fbox{$\cdots$}_n \rightarrow \left[R-\overset{N-N}{\underset{O}{\diagup \diagdown}} \right]_n + nH_2O$$

I

The polymer **I** may also be obtained directly from the dihydrazide $H_2NNHCORCONHNH_2$ by dehydration with H_2SO_4 and P_2O_5. It is also obtained by reaction of a bistetrazole with a diacid chloride

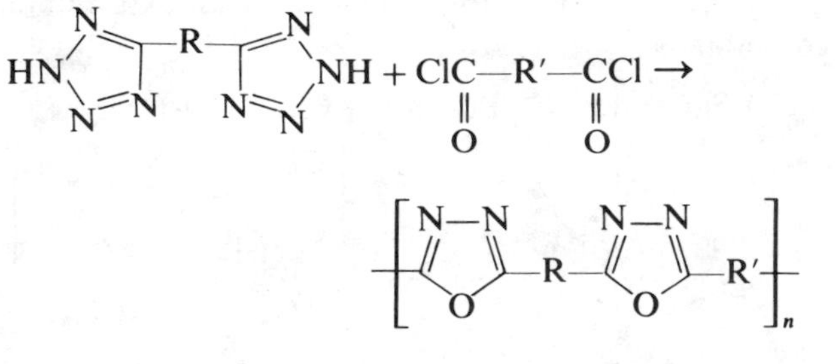

$$HN-\overset{N=N}{\underset{N-N}{\diagdown}}-R-\overset{N=N}{\underset{N-N}{\diagup}}-NH + ClC-R'-CCl \rightarrow$$

$$\left[\overset{N-N}{\underset{O}{\diagup \diagdown}}-R-\overset{N-N}{\underset{O}{\diagup \diagdown}}-R' \right]_n$$

and by heating a polyacylamidrazone with a strong organic acid

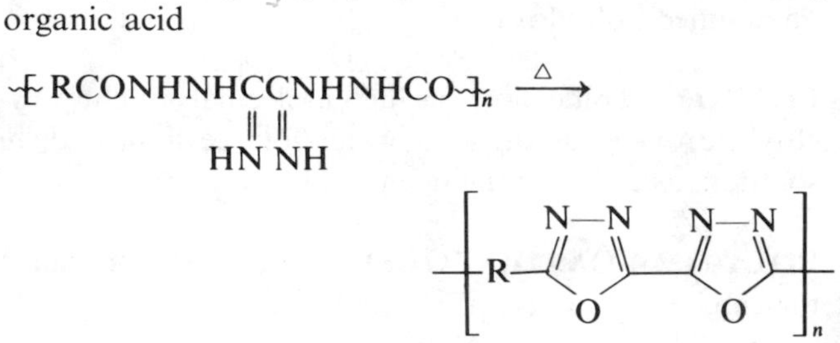

$$\fbox{$\cdots$}RCONHNHCCNHNHCO\fbox{$\cdots$}_n \overset{\triangle}{\longrightarrow}$$
$$\underset{HN\ NH}{\overset{\|\ \|}{}}$$

$$\left[R-\overset{N-N}{\underset{O}{\diagup \diagdown}}-\overset{N-N}{\underset{O}{\diagup \diagdown}} \right]_n$$

The polymers are readily prepared in high molecular weight form and have been widely investigated as potential high temperature resistant fibres (by dehydration of the polyhydrazide in fibre form). They show no change in properties to 200°C and retain about 60% of their mechanical property values to over 300°C. They are soluble only in strong acids and have very good hydrolytic stability in both acidic and basic media. Aliphatic polymers have T_g values of about 50°C and T_m values of 200–300°C. Aromatic polymers decompose before melting, but have very good thermal stability (being almost as good in air as in nitrogen) decomposing at 450–500°C.

POLYOXAMIDE A nylon of structure

$$\fbox{$\cdots$}NHRNHCOCO\fbox{$\cdots$}_n$$

where R is an aliphatic or aromatic hydrocarbon group. Prepared by reaction of a diamine with oxalic acid or, more usually, with a derivative of the acid such as the acid chloride or an ester. Can form fibres of high strength and low moisture sensitivity, but usually with an incon-

veniently high T_m value. Therefore copolymers are of greater interest, especially when prepared from branched or very long chain diamines, which considerably lowers the T_m value.

POLY-(1,3-OXAZA-2,4-DIONE) A polymer containing the ring

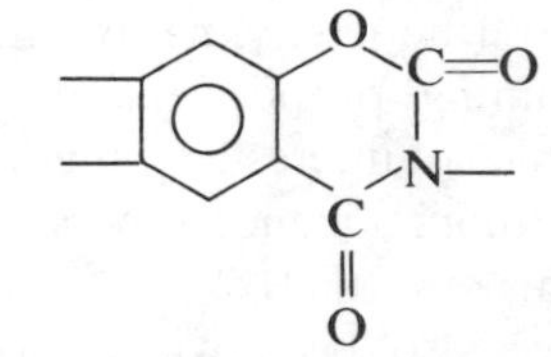

$$\left[\text{ring structure} \right]$$

in the polymer chain. Synthesised by reaction between a diester and a dihydroxyacid, e.g. the dimethyl ester of 2,4-dihydroxyphthalic acid, and an aromatic diisocyanate (OCNArNCO). Reaction proceeds via an intermediate ester–polyurethane (**I**)

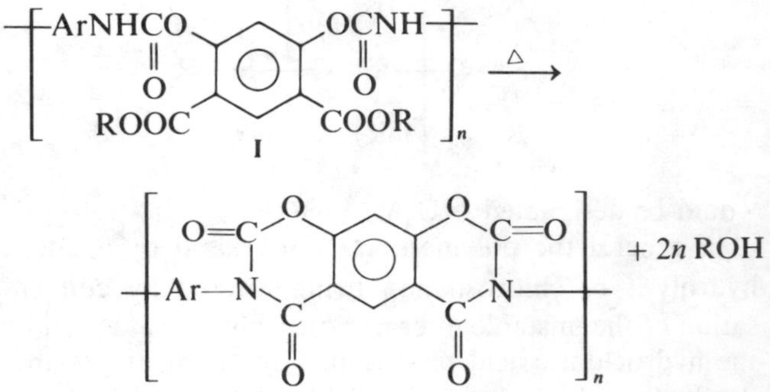

$$\left[\underset{ROOC}{\overset{ArNHCO}{}} \underset{COOR}{\overset{OCNH}{}} \right]_n \overset{\triangle}{\longrightarrow}$$

I

$$\left[Ar-N \underset{}{\overset{}{}} \right]_n + 2nROH$$

The polymers have reasonable high temperature resistance.

POLYOXAZOLIDONE A polymer containing the structure

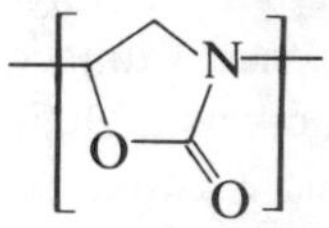

$$\left[\underset{O}{\overset{N}{}} \underset{O}{\diagup \diagdown} \right]$$

in its repeat unit. Prepared by 1,3-cycloaddition of a diglycidyl ether to a diisocyanate

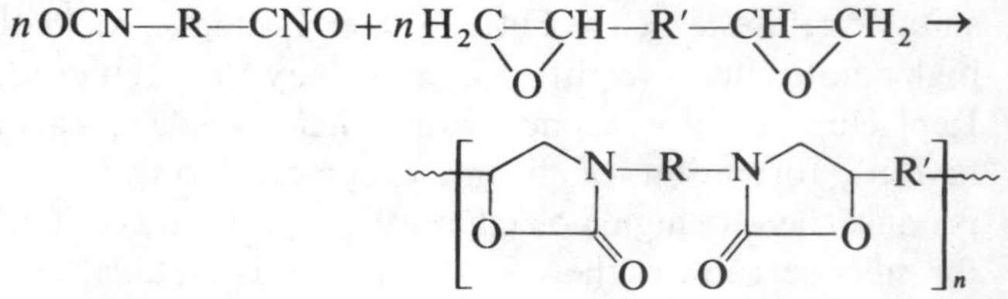

$$nOCN-R-CNO + nH_2C\overset{O}{\diagup \diagdown}CH-R'-CH\overset{O}{\diagdown \diagup}CH_2 \rightarrow$$

$$\left[\overset{N-R-N}{\underset{O\ O}{}}R' \right]_n$$

Commercial polymers using this reaction have been prepared, e.g. for increasing the thermal stability of epoxy resins and polyisocyanurates. A polyoxazolidone elastomer is produced commercially by reaction of bisphenol A diglycidylether with isocyanate end-capped polytetramethylene glycol.

POLYOXETANE (1) $\fbox{$\cdots$}CH_2CH_2CH_2-O\fbox{$\cdots$}_n$. Prepared by ring-opening polymerisation (e.g. using boron trifluoride) of oxetane (oxacyclobutane)

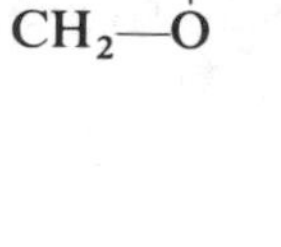

$$\underset{CH_2-O}{\overset{CH_2-CH_2}{\underset{|}{\overset{|}{}}}}$$

The polymer readily crystallises but has a low T_m value (35°C) and therefore, unlike the related poly-(3,3-bis(chloromethyl)oxacyclobutane) it is not of commercial interest.

(2) Alternative name for oxetane polymer.

POLYOXYACETYL Alternative name for polyglycolide.

POLY-(p-OXYBENZOYL) Alternative name for poly-(p-hydroxybenzoic acid).

POLYOXYETHYLENE GLYCOL (Poly-(ethylene oxide) glycol) H$-$[OCH$_2$CH$_2$]$_n$OH. A polyether formed by polymerisation of ethylene oxide in the presence of a diol and initiated using an alkali. The polymer crystallises and has a T_m value of 50–60°C. It is of interest as a prepolymer, of molecular weight 600–3000, for the formation of high strength polyurethane elastomers, but it is rather hydrophilic and so the polyurethanes have poor resistance to hydrolysis.

POLYOXYMETHYLENE (POM) (Polyformaldehyde) (Polyacetal) (Acetal) $-$[CH$_2$O]$_n$. Tradenames: Alkon, Celcon, Delrin, Duracon, Hostaform, Kemetal, Teral, Ultraform. A polymer of formaldehyde. A low molecular weight polymer (paraformaldehyde) may be made simply by evaporation of an aqueous formaldehyde solution. High molecular weight polymers are obtained either by anionic polymerisation in solution using, for example, an amine, phosphine or alkoxide in a hydrocarbon solvent, or by cationic polymerisation, usually of the cyclic trimer of formaldehyde (trioxane), with a Lewis acid. The high molecular weight polymers have a much higher thermal stability than paraformaldehyde. However, in commercial polymers they are further stabilised against their tendency to depolymerise, either by end-capping (by conversion of the unstable hemi-acetal end groups (HO$-$CH$_2$$-O\sim$) to stable acetate groups), or by making a copolymer containing ethylene oxide (or higher alkylene oxide) units which stop the depolymerisation zippers. The high molecular weight polymers crystallise to about 80% with a T_m value of 175°C (lowered to about 163°C in the copolymer) and with T_g type transitions at -17°C and -75°C. The polymers are highly insoluble in all solvents at room temperature. They have good mechanical properties with respect to stiffness (flexural modulus being typically about 2500 MPa), fatigue endurance and creep resistance and they have a reasonably high impact strength. They are therefore widely used as engineering plastics, often competing with nylons.

POLYOXYPROPYLENE GLYCOL (PPG) (Polypropylene glycol) H$-$[OCH(CH$_3$)CH$_2$]$_n$OH. A linear polyoxypropylene polyol prepared by polymerisation of propylene oxide in the presence of a glycol, e.g. ethylene glycol, and hence having a functionality of two. Useful in the formation of polyurethane elastomers by reaction with a diisocyanate.

POLYOXYPROPYLENE POLYOL A polyether, terminated with hydroxyl groups, i.e. a polyol, prepared by polymerisation of propylene oxide with an alkali initiator in the presence of a monomeric polyol. If the latter is a diol, e.g. ethylene glycol, then a linear polymer, polyoxypropylene glycol results:

$$H-[OCH(CH_3)CH_2]_n$$
$$-OCH_2CH_2O-[(CH_2CH(CH_3)O]_mH$$

Use of a triol, e.g. glycerol or trimethylolpropane, gives a branched polymeric triol. The polymers are rather irregular in structure, having both head-to-head and head-to-tail structures and are atactic. Therefore they do not crystallise as readily as other aliphatic polyethers and are liquids.

They are important polyol prepolymers for the formation of polyurethanes by reaction with diisocyanates, both for polyurethane elastomers and for flexible foams. For foams, copolymers with ethylene oxide of molecular weight about 2000 are used for the prepolymer process, but for the one-shot process molecular weights of about 3000 are preferred. The secondary end-group hydroxyls are not as reactive with diisocyanates as primary hydroxyls, and frequently capped or tipped polyols are used, in which ethylene oxide units cap the ends to increase reactivity. Later, ethylene oxide–propylene oxide copolymers, usually with an ethylene oxide content of 10–20% and a molecular weight of 3000–6000, became available for different types of foam production.

POLYOXYTETRAMETHYLENE Alternative name for polytetramethylene ether glycol.

POLYOXYTETRAMETHYLENE GLYCOL (PTMG) Alternative name for hydroxy-terminated low molecular weight polytetrahydrofuran, i.e.

$$HO-[(CH_2)_4-O]_nH$$

useful as a polyol prepolymer for polyurethane production.

POLYPARABANIC ACID (Poly-(2,4,5-triketoimidazoline)) (Poly-(1,3-imidazoline-2,4,5-trione)) (PPA). Tradenames: Tradlon, Tradlac. A polymer containing the heterocyclic ring

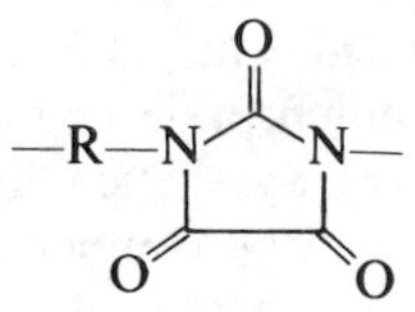

in the polymer chain, and hence closely related to the polyhydantoins. Synthesised by reaction between a dicarbamoyl cyanide and a diisocyanate to form a precursor polyimidoimidazolidione, which may be hydrolysed to the polyparabanic acid. The polymers are usually crosslinked (unless HMDI is used as the diisocyanate) and are of interest commercially as high temperature resistant, high T_g (200–300°C), electrical insulators as lacquers or films. Polymers with R = PhXPh, where X = —CH$_2$— or —O—, have been developed commercially, as have polymers formed by reaction between oxamide esters and blocked diisocyanates.

POLYPELARGONAMIDE Alternative name for nylon 9.

POLYPENTADIENE Alternative name for poly-piperylene.

POLYPENTENAMER $\{CH=CH(CH_2)_3\}_n$. The polyalkenamer product of ring-opening polymerisation of cyclopentene. The *trans* isomer (transpolypentenamer) is a relatively useful commercial rubber and is produced by Ziegler–Natta polymerisation using an aluminium triethyl/tungsten hexachloride catalyst. The *cis* isomer may be obtained using aluminium diethylchloride/molybdenum pentachloride. On polymerisation both linear chains and macrocyclic rings are formed. The T_g values are very low being -114°C (*cis* isomer) and -97°C (*trans* isomer), whilst the T_m values are -41°C (*cis* isomer) and $+18$°C (*trans* isomer). The *cis* isomer shows no tendency to crystallise, whereas the *trans* isomer crystallises on stretching thus giving high tensile strengths. The *trans* polymer can be vulcanised using conventional accelerated sulphur vulcanisation and will accept large amounts of oil extender and carbon black. The vulcanisates, even when highly oil extended, show good strength and abrasion resistance with good ozone and ageing resistance as well. Thus the *trans* isomer is a potentially useful general purpose rubber.

POLYPEPTIDE A polyamide which can be considered, formally, to be formed by the linking together of α-amino acid monomers through elimination of water between amino acid carboxyl groups, although in reality formation of polypeptides does not occur in this way. Thus it is a polyamide consisting of α-amino acid residues (—NHCHR—CO—) linked through peptide bonds. Simple proteins are sequential copolymers consisting of polypeptide molecules with, usually, the twenty different standard amino acid residues linked in a precise sequence and typically containing 100–1000 residues. More than one chain may be present, the chains frequently being linked through disulphide links or, in oligomeric proteins, being merely associated with each other. Conjugated proteins contain a polypeptide component associated with a prosthetic group. Synthetic polypeptides are also important, although largely as models for the study of protein behaviour. Sometimes use of the term polypeptide is restricted to the synthetic polymers (also referred to as poly-(α-amino acids)) and occasionally it is even further restricted to synthetic polypeptide homopolymers.

These are the most readily synthesised, usually by treatment of the α-amino acid N-carboxyanhydride with base. Simple heating of α-amino acid monomers to cause polycondensation usually results in dehydration, decarboxylation or cyclisation to diketopiperazines. Random copolymers may be similarly made. However, synthesis of copolymers with α-amino acid residues in a precise sequence (as in proteins) is much more difficult. Until recently only very small polymers, i.e. peptides, had been synthesised and even these required highly repetitive and time consuming methods. However, by the technique of solid phase synthesis, sequential polypeptides of a hundred or more residues have now been synthesised quite rapidly. The main problem in synthesis is the necessity of blocking all the reactive groups in the monomers except the pair required for peptide bond formation. This is achieved using a protecting group, e.g. *t*-butyl, oxycarbonyl or carbobenzoxycarbonyl, attached to the group it is desired to block, and which, after peptide bond formation, is readily removed.

Polypeptides frequently adopt a regular conformation in both the solid state and solution. One of the main motivations for study of synthetic polypeptides as models for proteins is to gain an understanding of the relationship of conformation to polymer chain structure, and hence of biological activity in proteins. The regular conformations found are the α-helix and the β-conformations. The residues Ala, Leu, Phe, Trp, Met, His, Glu and Val stabilise the α-helix, whereas Tyr, Cys, Asp, Ser, Ile, Thr, Glu, Asp, Lys, Arg and Gly destabilise the α-helix.

The regular α-helix may be disrupted in solution by heating or by change of solvent to yield a random coil—the helix–coil transition. This results in denaturation in proteins. Conformational structure is studied by optical rotatory dispersion, circular dichroism, nuclear magnetic resonance, viscosity properties and electronic spectroscopy. Theoretical calculations of polypeptide conformational stability have often been made, the results being presented frequently as a Ramachandran plot.

POLYPEPTIDYL PROTEIN A protein on to which poly-α-amino acid side chains have been grafted by polymerisation of the amino acid N-carboxyanhydride, thus forming a multichain poly-α-amino acid. Both the α- and ε-amino groups of the protein will initiate N-carboxyanhydride polymerisation under mild (approximately neutral, low temperature, aqueous) conditions. In this way the biological and physicochemical properties of the protein may be modified in interesting ways.

POLYPERFLUOROALKYLVINYL ETHER Alternative name for perfluoroalkoxy polymer.

POLY-(PERFLUOROMETHOXY-1,1-DIHYDROPERFLUOROPROPYL ACRYLATE) (Poly-2F4) (Poly FMFPA)

$$\left[\begin{array}{c} CH_2CH \\ | \\ COOCH_2CF_2CF_2OCF_3 \end{array} \right]_n$$

A fluoroacrylate rubber.

POLY-(PHEN-as-TRIAZINE) A poly-(as-triazine) phenyl substituted in the triazine ring, i.e. containing the structure

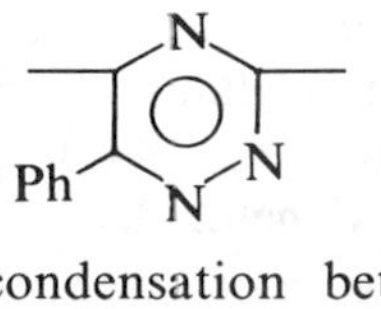

Synthesised by polycondensation between a phenyl substituted bisglyoxal (a bibenzil) and a diamidrazone. The polymers have better thermo-oxidative stability and are more readily soluble than the unsubstituted polymers.

POLYPHENYL Alternative name for polyphenylene.

POLYPHENYLENE (Polyphenyl) A polymer of the type

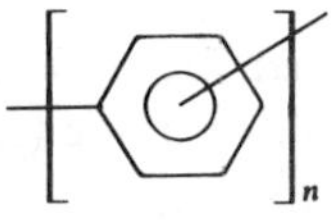

Difficult to synthesise in high molecular weight form without mixed isomeric linking. The pure *para*-linked polymer should, theoretically, have exceptionally high heat stability (to >800°C) but would also be highly intractable. Many syntheses have been attempted. Dehydrogenation (best with chloranil) of poly-(1,3-cyclohexadiene) gives amorphous polymers of low T_m value. Treatment of 1,4-dichlorobenzene with sodium or copper gives polymer with a $T_m > 500$°C. However, these polymers are dark coloured and intractable. Oxidative coupling by dehydropolycondensation of benzene with a Lewis acid catalyst (e.g. $AlCl_3$) and an oxidising agent (e.g. $CuCl_2$ or MnO_2) gives high molecular weight crystalline polymers, either linear and soluble or branched, stable to >500°C in air or to about 700°C in nitrogen. The branched polymers can be fused to simple shapes at about 450°C. Mixed *para*- and *meta*-linked polymers result from treatment of mixed terphenyls with benzene-*m*-sulphonyl chloride and are used in high temperature resistant composites (e.g. with asbestos). Friedel–Crafts polymers produced from terphenyls are slightly soluble and after crosslinking (e.g. with glycols) have been used in rocket motors. Thermosetting branched polyphenylene oligomers (H-resin), crosslinkable using organometallic catalysts, are useful as high temperature and chemically resistant coatings. Many substituted polyphenylenes have also been synthesised, especially polyphenylphenylenes.

POLY-(PHENYLENE...) Prefix for a polymer of the type

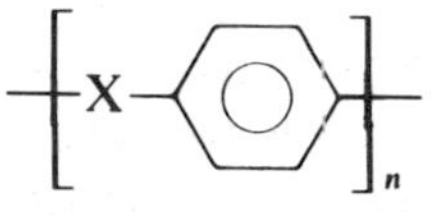

where X is a linking atom or group. Most frequently linking is through the *para* position as this gives the stiffest chains with the highest T_g values, softening and melting temperatures. However some *para*-linked polymers are too intractable, so *meta*-linking is used, e.g. in aromatic polyamides. Mixed isomeric linking may be present, arising from mixed substitution during polymer synthesis,

especially if Friedel–Crafts polymerisation is used. Polymers with X absent (the polyphenylenes) are very intractable, hence the need for the linking X group in useful commercial polymers to provide chain flexibility and to enable the polymer to be melt or solution processed. These polymers form the largest group of aromatic polymers and often have such good thermal stability and high softening points that they are considered as high temperature resistant polymers. Such commercial materials are useful as plastics moulding and laminating resins, fibres and coatings and include poly-(2,6-dimethyl-1,4-phenylene oxide), poly-(phenylene sulphide), polyethersulphones, aromatic polyamides and aromatic polyesters.

POLYPHENYLENEAMINE Alternative name for aromatic polyamine.

POLYPHENYLENE ETHER (PPE) Alternative name for poly-(2,6-dimethyl-1,4-phenylene oxide).

POLY-(m-PHENYLENEISOPHTHALAMIDE)

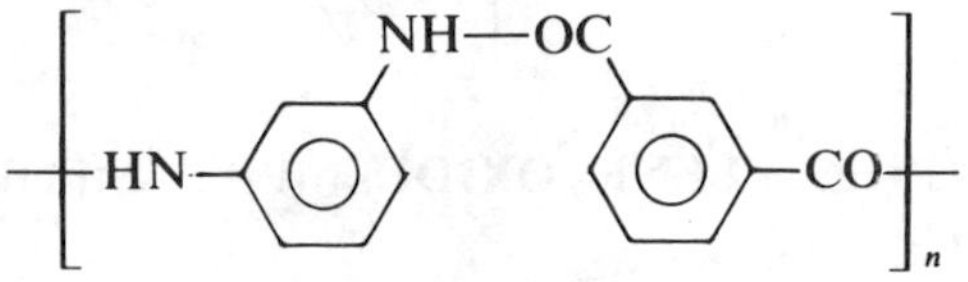

Tradenames: Nomex (previously HTI), Conex, Phenilon. A well-known wholly aromatic polyamide, widely used in fibre, film and paper form. Synthesised by solution polymerisation between *m*-phenylenediamine and isophthalolyl chloride. It does not melt ($T_m > 400$°C) and may be used continuously up to 250°C or to >400°C in the short term, being thermo-oxidatively very stable and also self-extinguishing (LOI is 28). The fibre retains 50% of its 20° strength at 285° but fabrics do shrink markedly on exposure to flame. A recently introduced replacement (HT-2) does not have this disadvantage. Applications include textiles in space and racing drivers' suits, whereas papers are used as high temperature insulation layers, in transformers and are often coated with polyimide for heat sealability. Nomex paper is also fabricated into honeycomb for use as a layer in high performance composites.

POLYPHENYLENE-1,3,4-OXADIAZOLE (POD) A polymer of structure

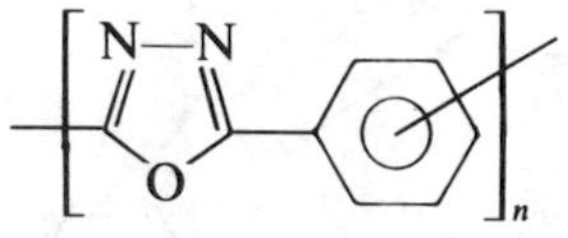

produced by condensation of a phthalic acid with hydrazine. Produced commercially as a film with high temperature resistance almost approaching that of the polyimides and useful for electrical insulation purposes. The *p*-phenylene polymer produced by reaction between terephthalic acid/dimethyl terephthalate and hydrazine sulphate in fuming sulphuric acid is also of commercial interest as a high performance tyre cord reinforcing fibre. The polymer becomes partially *N*-methylated during reaction, the *N*-methyloxadiazole groups then ring-opening to give *N*-methylhydrazide units:

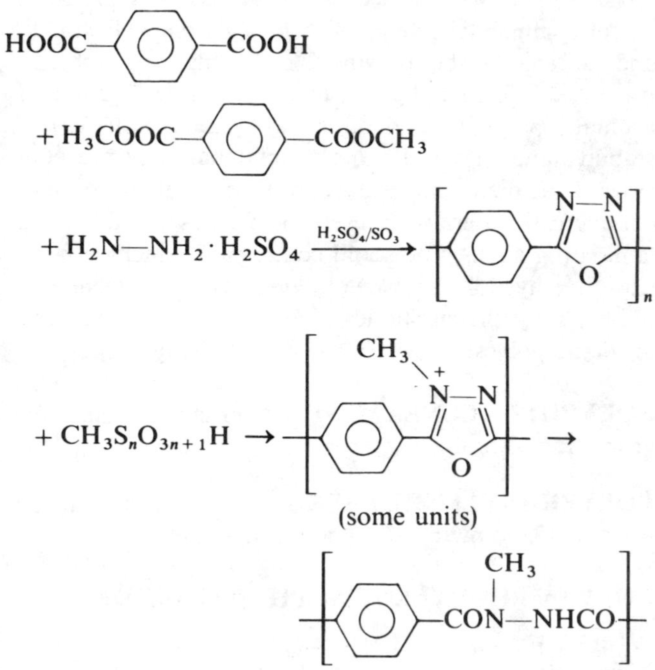

(some units)

POLY-(PHENYLENE OXIDE) (PPO) Strictly, the polymer

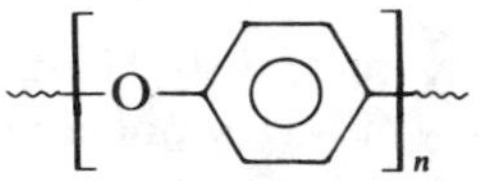

but also used to describe any similar polymer with a substituted benzene ring (often at the 2 and 6 positions) typically with alkyl groups or halogen atoms. Linking is usually through the *para* positions, although it may be mixed depending on the method of synthesis. 2,6-Disubstitution blocks the *ortho* positions and results in solely 1,4-linking, and these polymers are the best known. Synthesised by oxidative coupling of the appropriate phenol using oxygen and a catalyst, e.g. a cuprous salt and pyridine. High molecular weight polymer (**I**) readily forms, although dimerisation to a diphenoquinone (**II**) may compete:

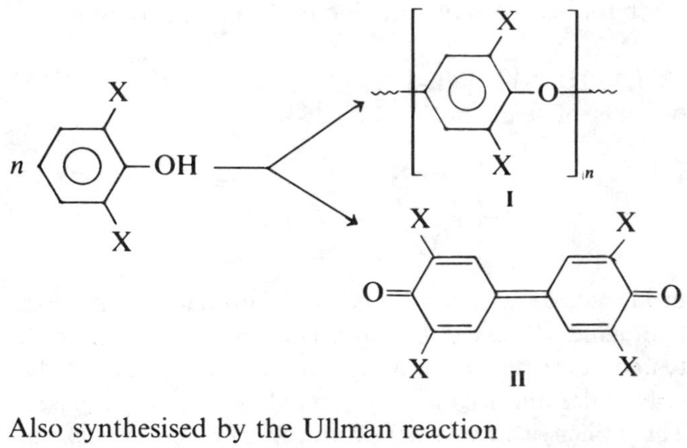

Also synthesised by the Ullman reaction

$$n \, MOArBr \rightarrow \text{---}[OAr]_n \text{---} + n \, MBr$$

where M is a metal.

The polymers have reasonable thermal stability and some, e.g. with X = H, CH_3, phenyl or Cl, have relatively high T_g and T_m values. The unsubstituted polymer has a T_m value of about 250°C and is very thermally stable, but is too intractable and insoluble to be fabricated even into films or fibres. The important commercial polymer has X = CH_3, i.e. it is poly-(2,6-dimethyl-1,4-phenylene oxide).

POLY-(1,4-PHENYLENEPYROMELLITIMIDE)

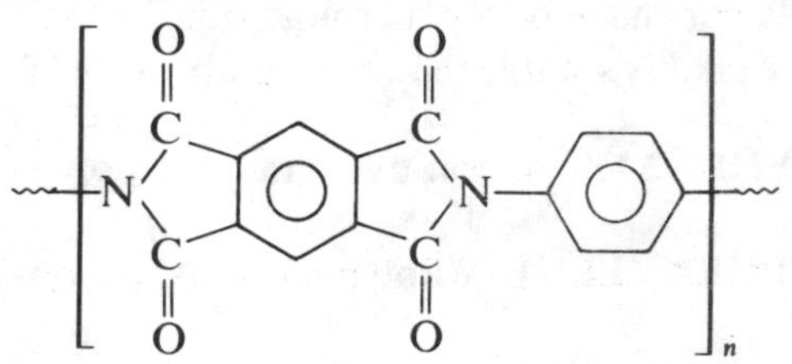

The polyimide formed by reaction of pyromellitic dianhydride with *p*-aminoaniline (1,4-diaminobenzene).

POLY-(*p*-PHENYLENE SULPHIDE) (PPS)

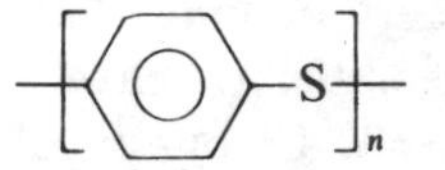

Tradenames: Fortron, Ryton, Tedur. Synthesised by self-condensation of a metal salt of a *p*-halothiophenol

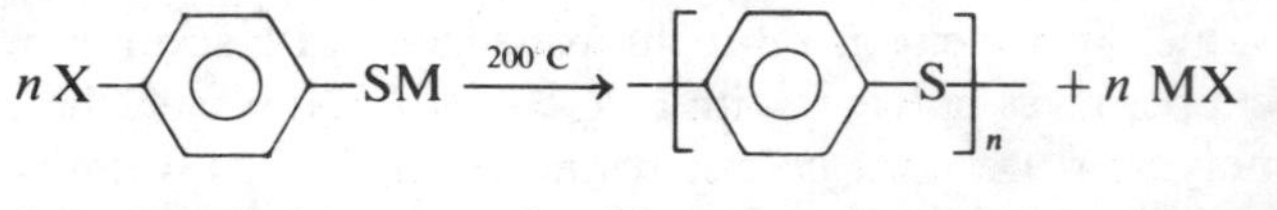

where X = halogen, preferably Br, and M = a metal (Na, Li, K, but preferably Cu). The commercial synthesis involves reaction between *p*-dichlorobenzene and sodium sulphide in a polar solvent, e.g. NMP

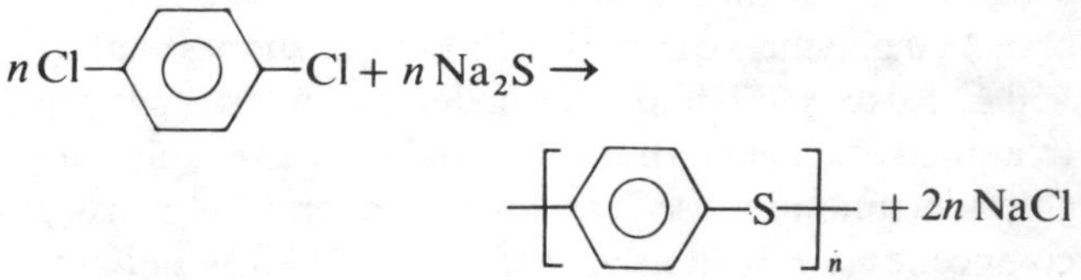

The polymer is highly crystalline with a T_m value of about 285°C. The T_g value is somewhat uncertain, being quoted as about 200°C. It is thermally stable and may be used to above 200°C in air. It is insoluble in all solvents below 200°C and is fire resistant, having an LOI of 44. It crosslinks on heating in air to 400°C and may be melt processed at about 350°C, usually in a slightly crosslinked form. It is mainly used as a chemically and high temperature resistant coating and as a moulding material, especially if filled, e.g. with glass, for electrical connectors, lamp holders, pump parts and valves. It is also useful as a bearing material with low friction. Doping of PPS with AsF_5 considerably raises its electrical conductivity from $10^{-16} \, S \, cm^{-1}$ to up to about $200 \, S \, cm^{-1}$ and doped polymers are of interest as conducting polymers.

POLY-(*p*-PHENYLENESULPHONE)

Synthesised by any of the methods used for aromatic polysulphones in general. Its high T_m value (about 520°C) makes it very intractable, but related ether-linked polymers (polyethersulphones) have lower softening points and several commercial products of this type are available.

POLY-(p-PHENYLENETEREPHTHALAMIDE)

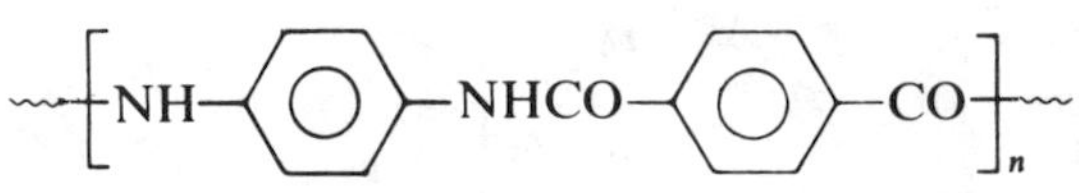

Tradenames: Arenka, Kevlar, Twaron, Vniivilon. Synthesised by solution polymerisation of p-phenylenediamine with terephthaloyl chloride in hexamethylphosphoramide or N-methylacetamide solvent. A wholly aromatic polyamide originally developed as a high modulus (about 130 GPa), high strength (tensile strength about 3 GPa) fibre for use as a tyre cord reinforcement to compete with steel and E-glass. The fibres are wet spun from concentrated sulphuric acid solution. Solutions of the polymer show lyotropic liquid crystalline behaviour. Hot drawn fibres, e.g. Kevlar 49, have a higher modulus than the tyre cord types, e.g. Kevlar 29, and are useful as lightweight reinforcing fibres in plastics composites. Compared with glass and carbon fibres, the fibres have a higher elongation at break (about 3%) and so are tougher. Thus fabrics are useful for bullet-proof clothing for military use.

POLY-(m-PHENYLENE-1,3,5,7-TETRAKIS-(3,3,3-TRIFLUOROPROPYL)TETRASILOXANYLENE)

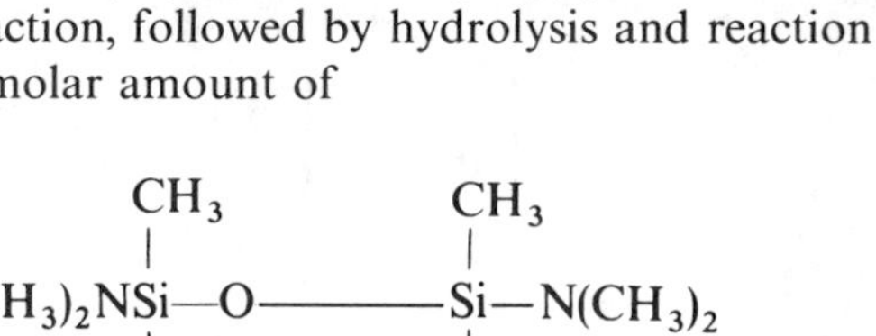

The best known poly-(fluoroalkylarylenesiloxanylene), produced by coupling of m-dibromobenzene with diethoxymethyl-(3,3,3-trifluoropropyl)silane via a Grignard reaction, followed by hydrolysis and reaction with an equimolar amount of

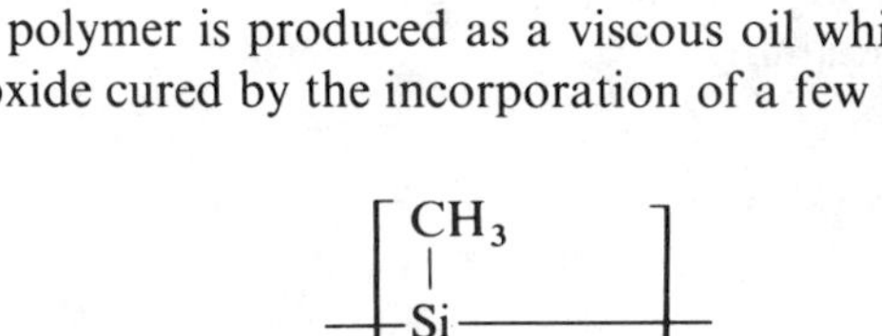

The polymer is produced as a viscous oil which may be peroxide cured by the incorporation of a few per cent of

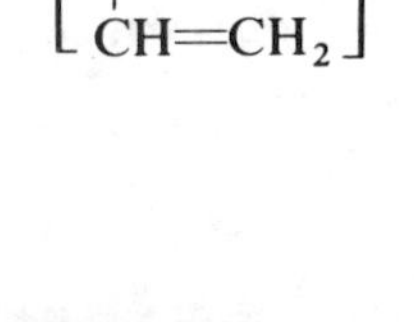

units. Vulcanisates are useful as fuel and high temperature resistant sealants (to 260°C) for the fuel tanks of supersonic aircraft.

POLYPHENYLENETHIOOXYTETRAMETHYLENE

Alternative name for poly-(thio-1,4-phenylenetetramethylene).

POLY-(PHENYL ISOCYANATE)

The polymer produced by anionic polymerisation of phenyl isocyanate

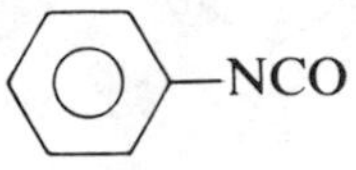

for example with a lithium alkyl catalyst:

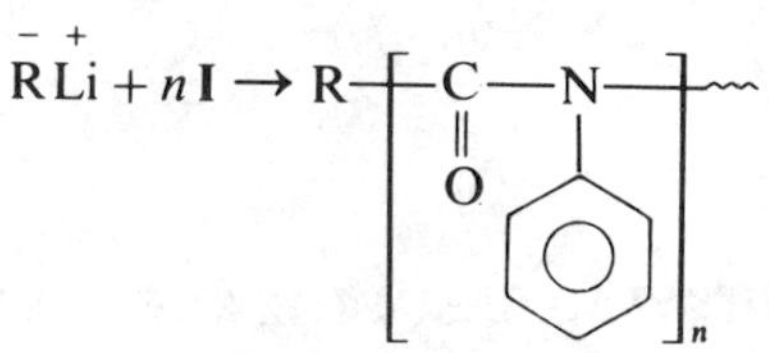

Such polymers have low thermal stabilities, reverting to isocyanate on heating. The polymer is an example of the unusual alternating carbon–nitrogen system (the polycarbazanes).

POLYPHENYLOXADIAZOLE

A polymer containing both

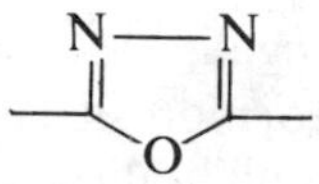

and

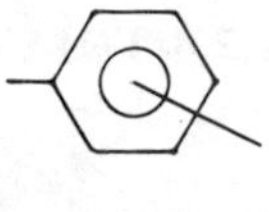

groups in the polymer chain. Synthesised by heating a poly-(N-acyloxamidrazone) (I) with dichloroacetic acid:

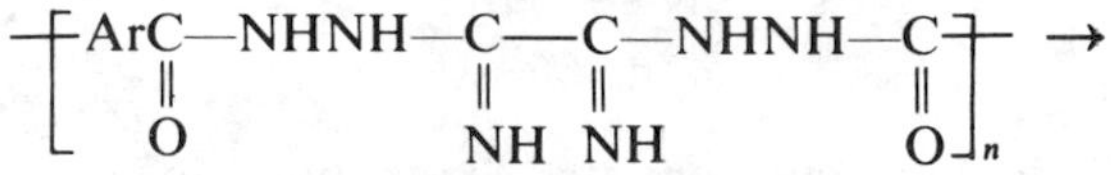

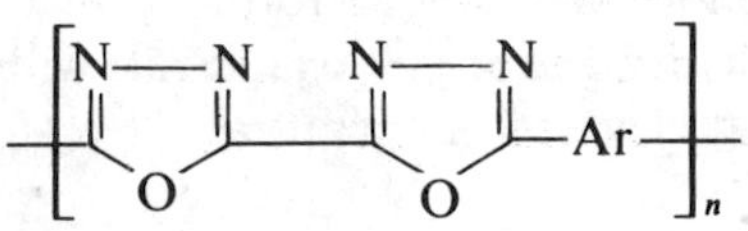

POLYPHENYLPHENYLENE

(Phenylated polyphenylene) A polyphenylene highly ring substituted with phenyl groups. Substitution increases solubility but does not lower thermal stability. Synthesised by inverse Diels–Alder polymerisation of a bistetraphenylcyclopentadienone (a tetracyclone) with a diethynylbenzene:

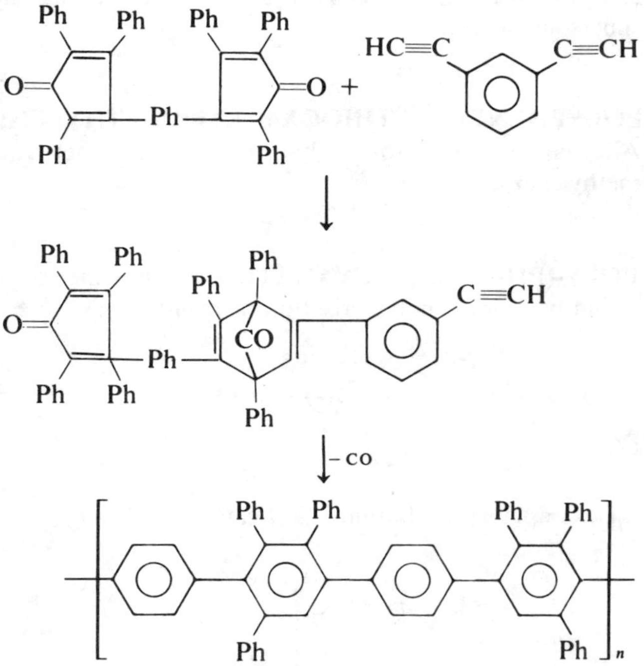

POLYPHENYLSILSESQUIOXANE (Polysilsesquioxane) A ladder polysiloxane of structure

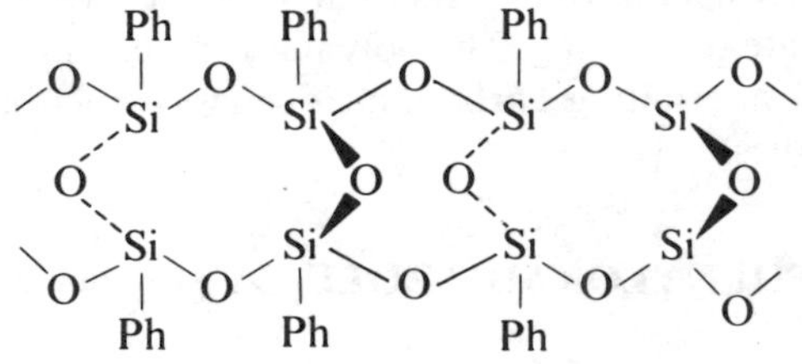

prepared in stereoregular (*cis*-syndiotactic) form by hydrolysis of phenyltrichlorosilane in toluene with dilute alkali, followed by equilibration. The polymer has high thermal and hydrolytic stability.

POLY-(4-PHENYL-1,2,4-TRIAZOLE) A triazole polymer of structure

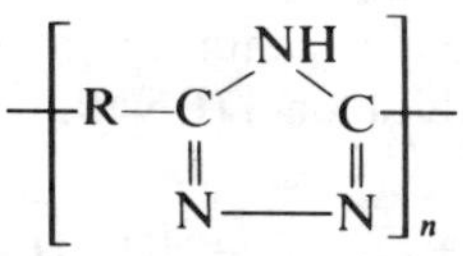

where R is usually aromatic. Synthesised in high molecular weight form by reacting an aromatic poly-hydrazide or polyoxadiazole with aniline in PPA solvent at elevated temperature:

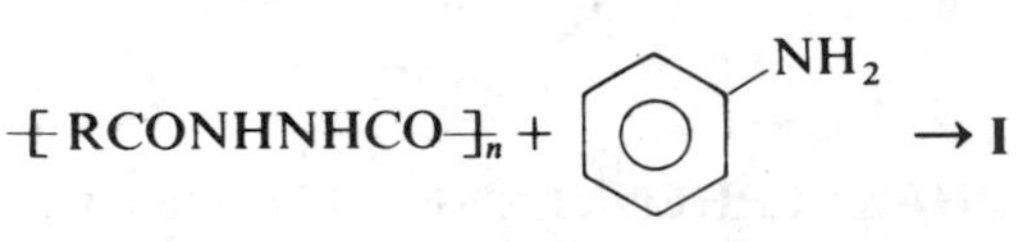

Soluble only in strong acids, having a high T_m value and of good thermal and thermo-oxidative stability and excellent hydrolytic stability.

POLYPHOSPHATE (Polymetaphosphate) A polymer of structure

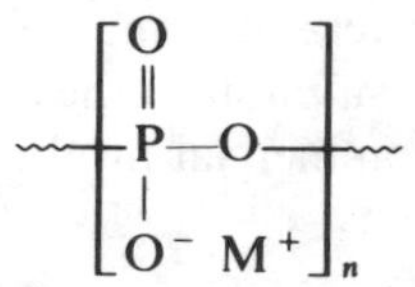

(linear polyphosphate) but which may also contain crosslinking units,

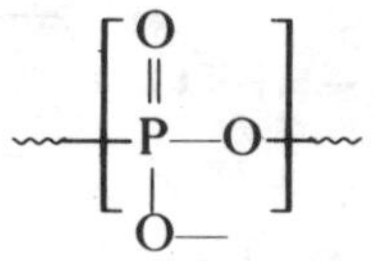

to give a network polyphosphate (or ultraphosphate glass). Linear polymers are obtained by heating metal dihydrogen phosphates to high temperatures: $n\,MH_2PO_4 \rightarrow \{PO_3M\}_n + n\,H_2O$. For the sodium compound temperatures of $> 340°C$ are needed, temperatures below $475°C$ giving one crystalline form (Maddrell's salt) and temperatures of $550-650°C$ giving another form (Kurrol's salt). The potassium salt only gives Kurrol's salt when heated to $> 230°C$. Sodium polyphosphate may be quenched from the melt (at $> 650°C$) to an amorphous glass, readily soluble in water, whereas the crystalline forms are highly insoluble. Polyvalent metal poly-phosphates may be obtained by precipitation from sodium or potassium solutions, initially producing rubbery materials which become hard and glassy as they dry out. Similar glassy polymers are obtained by heating the metal dihydrogen phosphates. Quaternary am-monium salts may also be obtained by precipitation from aqueous sodium polyphosphate. Polymers substituted with long alkyl chains are useful as greases. When metal dihydrogen phosphates are heated with an excess of phosphorus over metal (greater than 1:1 stoichiometry) then the useful crosslinked ultraphosphate glasses are formed.

POLYPHOSPHAZENE Alternative name for phosphazene polymer.

POLYPHOSPHIMATE (Polyphosphonate amide) A polymer of structure

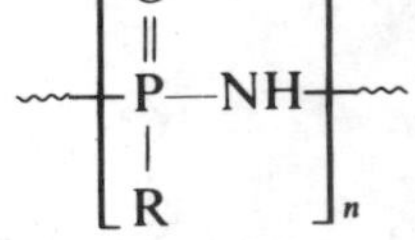

The polymer with $R = NH_2$ is prepared by heating phosphoric acid triamide $((NH_2)_3P{=}O)$. On heating the polymer to about $600°C$ a highly insoluble and inert product containing

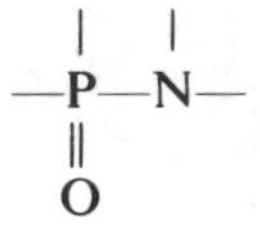

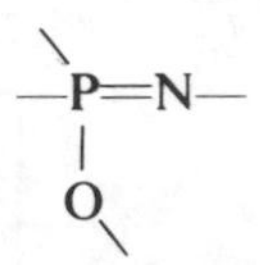

units is obtained, sometimes called PON polymer.

POLYPHOSPHONAMIDE A polymer of general structural type $-(NHRNHPO_2)_n-$, prepared by reaction of a diamine with a phosphorus halide or derivative.

POLYPHOSPHONATE AMIDE Alternative name for polyphosphimate.

POLYPHOSPHONITRILIC CHLORIDE Alternative name for polydichlorophosphazene.

POLYPHOSPHORIC ACID (PPA) A solution of phosphorus pentoxide ($\sim 80\%$) in phosphoric acid, consisting largely of oligomers of phosphoric acid. Since it is a good solvent for many organic monomers and polymers, and reacts with water without loss of acidity, it is widely used as a solvent for cyclopolycondensation reactions in the synthesis of many aromatic and heterocyclic polymers.

POLYPHTHALOCYANINE (Phthalocyanine polymer) A coordination polymer with a structure

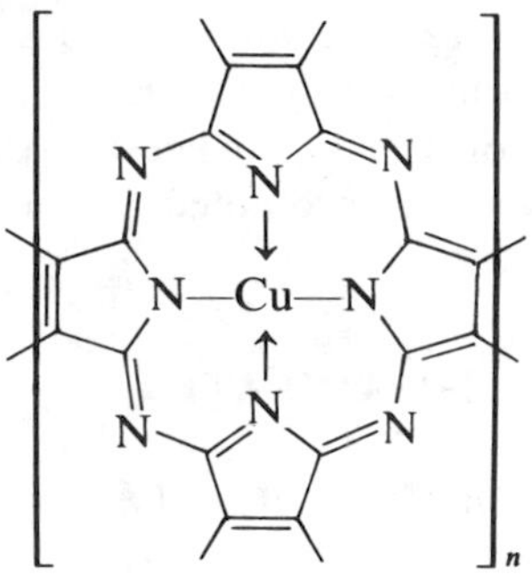

or similar phthalocyanine structure. It may be synthesised by heating tetracyanoethylene with copper or copper salts. Interest in such polymers stems from the good thermal stability of low molecular weight phthalocyanines. However, the polymers have proved disappointing in this respect. The polymers may have high electrical conductivity. Similar polymers may be synthesised with zinc or iron replacing copper.

POLYPIPERIDONE Alternative name for nylon 5.

POLYPIPERYLENE (Polypentadiene)

$$-(CH_2CH=CHCH(CH_3))_n-$$

The polymer can exist in several isomeric forms. In the 1,4-isomer, which is the isomer shown above, both the *cis* and the *trans* forms contain an asymmetric centre and can exist in the regular syndiotactic and isotactic as well as the irregular atactic forms. In 1,2-polypiperylene

$$-(CH_2CH(CH=CHCH_3))_n-$$

similar possibilities of isomerism exist. 3,4-Polypiperylene

$$-(CH(CH_3)CH(CH=CH_2))_n-$$

is ditactic and therefore can exist in any of three regular stereoisomeric forms—disyndiotactic, erythrodiisotactic and threodiisotactic. *cis*-1,4-polypiperylene is produced by Ziegler–Natta polymerisation with a soluble cobalt catalyst such as cobalt acetylacetonate with an organic aluminium chloride, which gives the predominantly syndiotactic polymer. It has a T_m value of 50–55°C. The isotactic form is similarly produced (from either *cis* or *trans* monomer) with an aluminium alkyl/titanium alkoxide catalyst. Its T_m value is 43–46°C. The *cis* polymers yield vulcanisates similar in properties to other diene hydrocarbon rubbers. Isotactic *trans*-1,4-polypiperylene is obtained from either the *cis* or *trans* monomer with vanadium trichloride/aluminium triethyl catalyst. It has a T_m value of 95°C. Syndiotactic 1,2-polypiperylene is also obtained using the same method as for the syndiotactic *cis*-1,4-polymer but using an aromatic solvent or a low Al/Co ratio. It is non-crystalline when unstretched but crystallises on stretching giving a T_m value of 10–20°C.

POLYPIVALOLACTONE

$$\left[CH_2C(CH_3)_2CO \atop \| \atop O \right]_n$$

Prepared by the ring-opening polymerisation of pivalolactone using a weakly basic substance, such as a tertiary amine or a phosphine as the catalyst. The polymer melts at 245°C and depolymerises above the T_m value. It can crystallise in three different crystalline forms, usually with the formation of large spherulites. As moulded, typical crystallinity is about 75%. Fibres with high tenacity may be melt spun. The T_g value is about $-10°C$.

POLYPRO Tradename for polypropylene.

POLYPROLINE

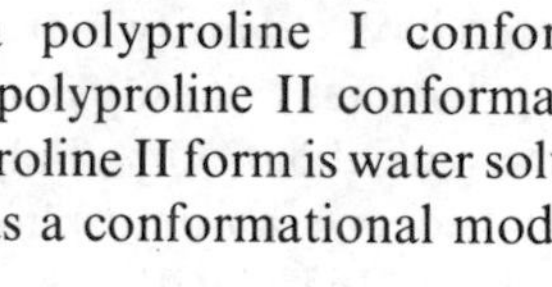

$$\left[N\!-\!CH\!-\!CO \atop | \quad\;\; | \atop CH_2 \;\; CH_2 \atop \diagdown\;\; \diagup \atop CH_2 \right]_n$$

A poly-imino acid polypeptide which, unlike the poly-α-amino acid polypeptides, having no amide hydrogen, cannot form hydrogen bonds. Poly-L-proline is synthesised by polymerisation of its N-carboxyanhydride. It can exist in a polyproline I conformation which mutarotates to polyproline II conformation in organic acids. The polyproline II form is water soluble. It has been widely studied as a conformational model for collagen.

POLY-L-PROLINE I CONFORMATION A helical conformation adopted by poly-L-proline obtained, as normal, by N-carboxyanhydride polymerisation in pyridine, which is stable in aliphatic alcohol solutions but mutarotates (from dextrarotation to laevorotation) on dissolving in water and organic acids, to give the poly-L-proline II conformation. The peptide bonds, unusually, exist in the *cis* form, the helix being right-handed, with a residue repeat of 1·9 Å and 3·3 peptides per turn. The angles of rotation are $\phi = -83°$, $\psi = 158°$ and $\omega = 0°$.

POLY-L-PROLINE II CONFORMATION A helical polypeptide conformation with three peptides per turn (a 3_1 helix), a residue repeat of $3.12\,\text{Å}$ (double that of the α-helix and 87% of that of the fully extended chain). Identical to the polyglycine II helix and responsible for the supercoiling in tropocollagen molecules. Unlike the polyproline I conformation it is a left-handed helix with *trans* peptide bonds. The angles of rotation are $\phi = -78°$, $\psi = +149°$.

POLY-β-PROPIOLACTONE

$$\left[\begin{array}{c} CH_2CH_2CO \\ \| \\ O \end{array} \right]_n$$

Prepared by the ring-opening polymerisation of β-propiolactone. The polymer has a T_m value of 75–85°C.

POLYPROPYLENE (PP)

$$\left[\begin{array}{c} CH_2CH \\ | \\ CH_3 \end{array} \right]_n$$

Tradenames: Carlona P, Daplen, El-Rex, Escon, Hostalen PP, Lacqtene P, Luparen, Moplen, Napryl, Novolen, Polypro, Profax, Propathene, Royalene, Stamylan P, Vestolen P (for plastic materials) and Herculon, Meraklon and Ulstron (for fibres). The polymer produced by the polymerisation of propylene and of major importance as a plastic and fibre forming material. Attempts to polymerise propylene by a free radical mechanism merely lead to low molecular weight oligomers since the monomer acts as a powerful chain transfer agent. Polymerisation to high molecular weight polymer is achieved by Ziegler–Natta polymerisation. The polymer so produced is highly stereoregular being isotactic. Small amounts of atactic polymer are also produced but are separated by their solubility in *n*-hexane.

Commercial polypropylene is 90–95% isotactic and has a number average molecular weight of 40 000–60 000 with a polydispersity of 6–12. It crystallises with the molecules in a 3_1 helix. Several different types of spherulite may be formed on crystallisation. The crystalline regions have a density of $0.94\,\text{g cm}^{-3}$ and the amorphous regions have a density of $0.85\,\text{g cm}^{-3}$, so that overall, polypropylene has a density of about $0.90\,\text{g cm}^{-3}$ with a crystallinity of about 50%. Polypropylene melts may be quenched to give an amorphous polymer. The maximum T_m value is 165°C with a narrow (about 5°C) melting range, so that polypropylene softens at a considerably higher temperature than polyethylene. Its major amorphous transition is at about 0°C so that the polymer embrittles markedly on cooling. Commercial block copolymers with about 5–15% ethylene blocks, referred to as propylene copolymer, have somewhat lower brittle points.

Polypropylene is stiffer than polyethylene, having a tensile modulus of 1000–1300 MPa and a tensile strength of 25–35 MPa. Its elongation at break is 50–300% and its impact strength (Izod) is 0.3–$4.3\,\text{J}$ $(12.7\,\text{mm})^{-1}$. Polypropylene is highly solvent resistant and environmental stress cracking resistant. It has a very high electrical resistivity. However, the presence of a tertiary hydrogen on each repeat unit makes if very susceptible to oxidative degradation. Although usually opaque, thin film, especially if biaxially oriented, may show a sparkling degree of clarity. Uniaxially oriented film may be readily split in the direction of orientation to give tape, twine or fibres (fibrillation). As a plastics material polypropylene is mainly used as an injection moulding material.

α-POLYPROPYLENE The normal crystalline form of polypropylene with a monoclinic unit cell of dimensions $a = 6.65\,\text{Å}$, $b = 20.96\,\text{Å}$, $c = 6.50\,\text{Å}$ and of density $0.936\,\text{g cm}^{-3}$. This cell has four 3_1-helical chains passing through it.

β-POLYPROPYLENE A polymorphic form of polypropylene sometimes found in melt cooled polymer, when its presence results in the formation of some spherulites of a different type than normal. It probably has a hexagonal unit cell, with $a = 12.74\,\text{Å}$ and $c = 6.35\,\text{Å}$. This could be formed from normal α-polypropylene by a slight lateral rearrangement of the 3_1 helices. Formation of β-polypropylene may be promoted by the use of certain nucleating agents.

POLYPROPYLENE ADIPATE

$$\left[\begin{array}{c} OCHCH_2OC(CH_2)_4C \\ | \qquad\quad \| \qquad\quad \| \\ CH_3 \qquad O \qquad\quad O \end{array} \right]_n$$

A polyester formed by the esterification of propylene glycol with adipic acid. It has a T_m value of $-25°C$. It is useful as a polymeric plasticiser for polyvinyl chloride and as a polyol precursor polymer for polyurethanes.

POLYPROPYLENE AZELATE

$$\left[\begin{array}{c} OCHCH_2OC(CH_2)_6C \\ | \qquad\quad \| \qquad\quad \| \\ CH_3 \qquad O \qquad\quad O \end{array} \right]_n$$

A polyester formed by the esterification of propylene glycol with azelaic acid, having a T_m value of $-46°C$.

POLYPROPYLENE GLYCOL (PPG) Alternative name for hydroxy-terminated polypropylene oxide of low molecular weight. Useful as a polyol prepolymer for polyurethane formation.

POLYPROPYLENE GLYCOL MONOPHENYL ETHER

$$\bigcirc\!\!\!\!\!\!-O(CH(CH_3)CH_2O)_xH$$

A plasticiser for cellulose acetate. Also compatible with other cellulose esters and ethers.

POLYPROPYLENE GLYCOL SEBACATE (PPS)

$$H\text{-}[OCH_2CH(CH_3)\text{---}OOC(CH_2)_8CO\text{-}]_nOH$$

A polymeric plasticiser for polyvinyl chloride and nitrile rubber with good permanence and imparting good low temperature flexibility.

POLYPROPYLENE OXIDE

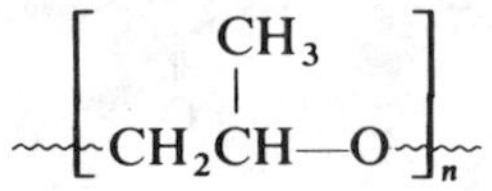

$$\left[\begin{array}{c} CH_3 \\ | \\ \sim\text{-}CH_2CH\text{---}O\text{-} \end{array}\right]_n$$

Low molecular weight polymers may be prepared by polymerising propylene oxide using an alkali in propylene glycol solution, giving a polymer with hydroxyl end groups, i.e.

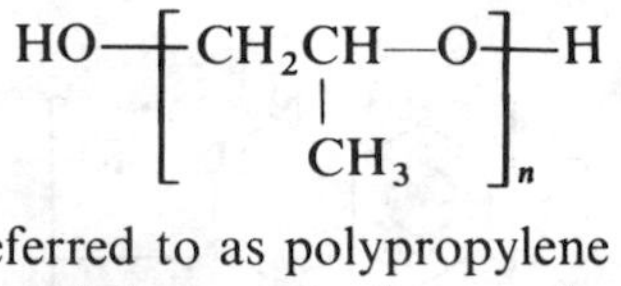

$$HO\text{-}\left[\begin{array}{c} CH_2CH\text{---}O \\ | \\ CH_3 \end{array}\right]_n\text{-}H$$

which is also referred to as polypropylene glycol. These polymers are useful as polyol prepolymers for the formation of polyurethanes, especially when copolymerised with a triol such as 1,1,1-trimethylolpropane or 1,2,6-hexanetriol, to give some crosslinking in the polyurethane. Normally polymers with molecular weights of 3000–3500 are used for flexible foams and those with molecular weights of about 500 for rigid foams. High molecular weight polymers may be prepared either by using an alkaline earth (Ca, Ba or Sr) metal compound, e.g. oxide, carbonate or alkoxide, or by using an organometallic compound such as zinc or aluminium alkyl usually with a co-catalyst (water or an alcohol). Isotactic polymer is often produced using these catalysts, but it is optically inactive. If optically active, i.e. either D- or L-propylene oxide, is used then optically active polymer can be formed. Both forms crystallise with a T_m value of 74°C. Amorphous, atactic polymers are useful especially as copolymers, as propylene oxide rubbers.

POLY-(PROPYLENE OXIDE-b-ETHYLENE OXIDE)

(Propylene oxide–ethylene oxide block copolymer) Tradenames: Pluronic, Tetronic. Prepared by anionic polymerisation of propylene oxide, e.g. with sodium alkoxide initiator, at about 120°C, followed by addition of ethylene oxide. Commercial polymers are triblock, having the structure.

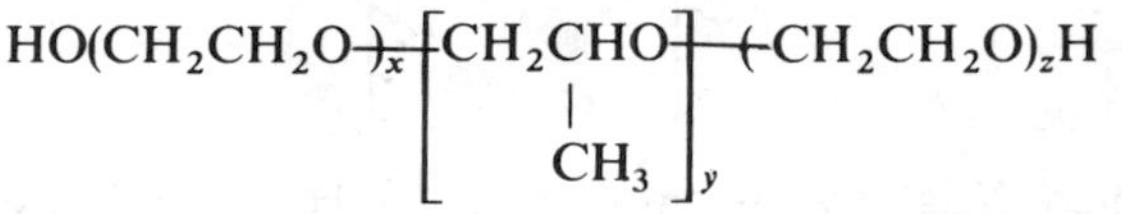

$$HO(CH_2CH_2O)_x\left[\begin{array}{c} CH_2CHO \\ | \\ CH_3 \end{array}\right]_y(CH_2CH_2O)_zH$$

with a molecular weight of 3000–8000. They are useful as non-ionic detergents since the ethylene oxide units are water soluble but the propylene oxide units are not. More complex copolymers, the Tetronics, are prepared by the polymerisation of propylene oxide in the presence of ethylenediamine, followed by the addition of ethylene oxide. Their structure is $X_2NCH_2CH_2NX_2$ where X is

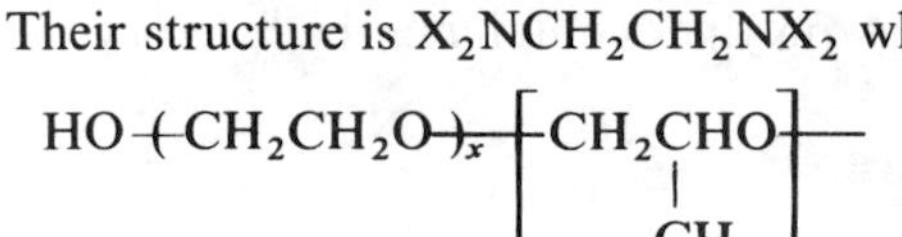

$$HO\text{-}(CH_2CH_2O)_x\left[\begin{array}{c} CH_2CHO \\ | \\ CH_3 \end{array}\right]_y\text{-}$$

POLYPROPYLENE SEBACATE

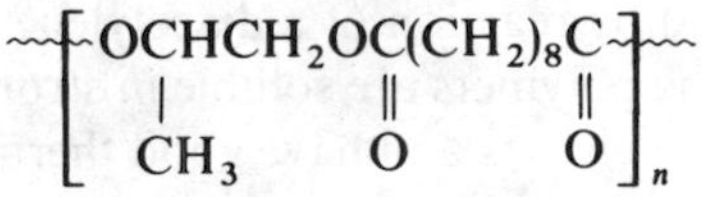

$$\sim\left[\begin{array}{c} OCHCH_2OC(CH_2)_8C \\ | \quad\quad \| \quad\quad \| \\ CH_3 \quad\ O \quad\quad O \end{array}\right]_n\sim$$

A polyester formed by esterification of propylene glycol with sebacic acid. It has a T_m value of −34°C and is useful as a polymeric plasticiser for polyvinyl chloride.

POLYPROPYLENE SULPHIDE

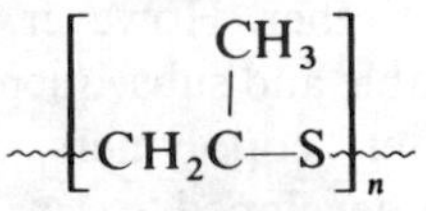

$$\sim\left[\begin{array}{c} CH_3 \\ | \\ CH_2C\text{---}S \end{array}\right]_n\sim$$

Prepared by ring-opening polymerisation of propylene sulphide, using an ionic catalyst, preferably an organometallic catalyst such as zinc diethyl with water as a co-catalyst. The polymer may be amorphous or partially crystalline, with the T_m value varying with the molecular weight and tacticity. The T_g value is possibly about −40°C. The copolymers with allylthioglycidyl ether form the basis of propylene sulphide rubbers.

POLYPUFF Alternative name for poly-(p-ethynylbenzene).

POLYPYRAZINE A polymer containing the pyrazine ring

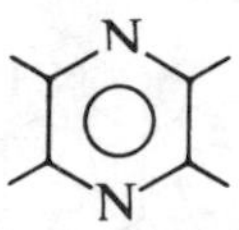

in the polymer chain, most frequently fused to a benzene ring, as in a polybenzopyrazine or polyquinoxaline.

POLYPYRAZOLE A polymer containing the heterocyclic ring

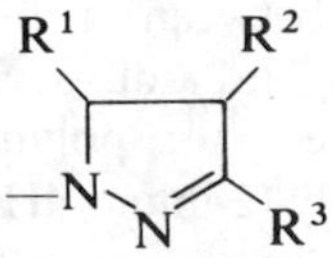

where R^1, R^2 and/or R^3 may be either an alkyl group, a hydrogen atom or the link in the polymer chain. Synthesised by polycondensation between a dihydrazine or dihydrazide and a bis-(β-diketone) either in the melt or by a two-stage solution condensation/cyclisation. Polymerisation probably takes place via a polyhydrazone:

$$H_2NNHCOR^4CONHNH_2 + R^5COCH_2COR^6 \rightarrow$$

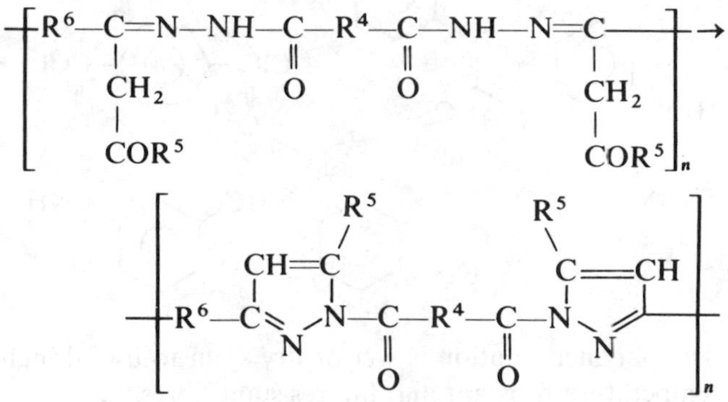

$$\left[\begin{array}{c} R^6\text{---}C=N\text{---}NH\text{---}C\text{---}R^4\text{---}C\text{---}NH\text{---}N=C \\ | \quad\quad\quad\quad \| \quad\quad\quad \| \quad\quad\quad\quad | \\ CH_2 \quad\quad\quad O \quad\quad\quad O \quad\quad\quad CH_2 \\ | \quad\quad\quad\quad\quad\quad\quad\quad\quad\quad\quad\quad\quad | \\ COR^5 \quad\quad\quad\quad\quad\quad\quad\quad\quad COR^5 \end{array}\right]_n \rightarrow$$

Also formed by 1,3-dipolar addition of a sydnone to a diacetylene and by reaction of a diacetylene and a diazo compound. The polymers are soluble in strong acids and in highly polar solvents and have good thermal stability.

POLYPYROMELLITIMIDE An aromatic polyimide obtained by polycondensation between pyromellitic dianhydride and a diamine. The earliest type of polyimide to be developed commercially, where the diamine used is 4,4'-diaminodiphenyl ether. However, these polyimides are not melt processable and subsequently many different modified types, often copolymers, which are more tractable, have been developed.

POLYPYRROLIDONE Alternative name for nylon 4.

POLYPYRROLONE Alternative name for poly-imidazopyrrolone.

POLYPYRRONE Alternative name for polyimidazopyrrolone.

POLYQUINAZOLINEDIONE A polymer of structure

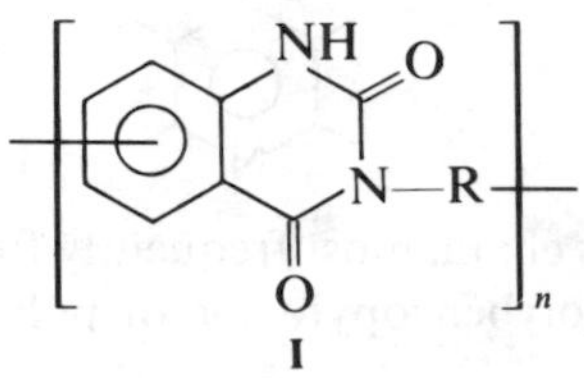

I

where R is usually an aromatic group. Synthesised by polycondensation between a bis-(aminobenzoic acid) (bisanthranilic acid) and a diisocyanate. Reaction occurs via a tractable precursor polyurea-acid which cyclodehydrates to a poly-(2-imino-4H-3,1-benzoxazin-4-one), which rearranges. The polymers have the good thermal stability typical of heterocyclic polymers. A commercial polymer (AFT-2000), is an aromatic quinazolinedione–amide copolymer and has been developed from reaction of a diaminoquinazolinedione monomer and isophthaloyl chloride in DMAc or NMA solvent:

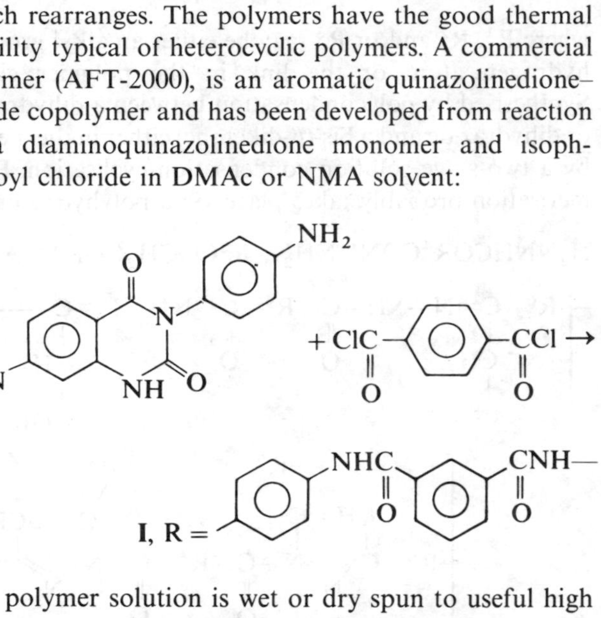

I, R =

The polymer solution is wet or dry spun to useful high temperature resistant and fire resistant fibres.

POLYQUINAZOLINEQUINAZOLONE A ladder polymer containing the heterocyclic ring system:

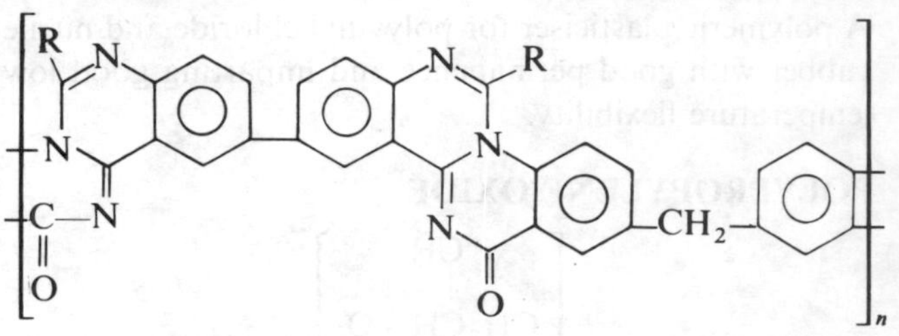

Synthesised by reaction between a bisbenzoxazinone and an aromatic o-bis-(amino-amide).

POLYQUINAZOLONE A polymer containing the heterocyclic ring

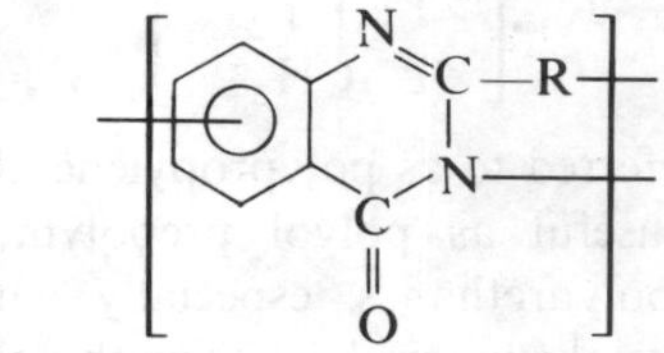

I

in the polymer chain. Synthesised by reaction between a bisbenzoxazinone and an aromatic diamine:

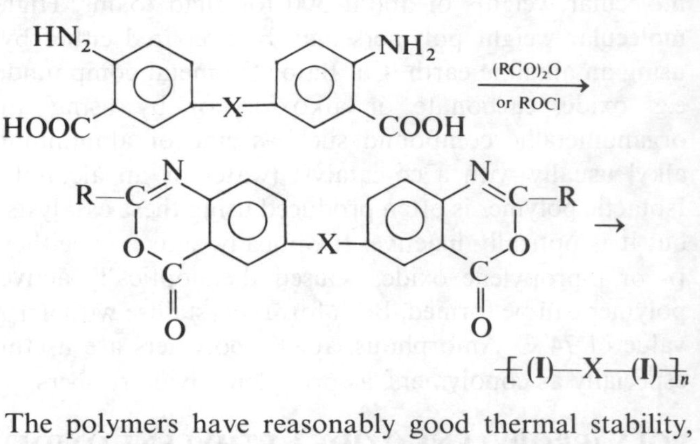

The polymers have reasonably good thermal stability, especially if R is phenyl rather than methyl.

POLYQUINONE A ladder polymer of structure

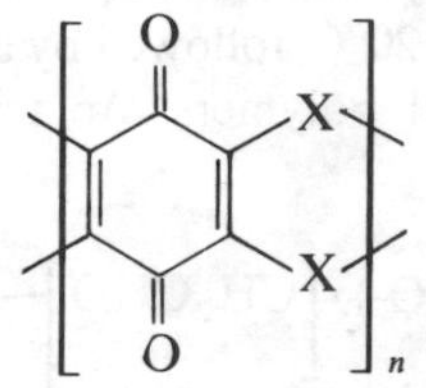

where X is oxygen, sulphur or —NH—. Synthesised by reaction of chloranil with tetrahydroxybenzoquinone, sodium sulphide or ammonia respectively. Dark coloured polymers with reasonably good thermal stability are formed.

POLYQUINOXALINE (Polybenzopyrazine) (Polypyrazine) A polymer containing the heterocyclic ring in the polymer chain. Synthesised by reaction of an aromatic bis-(*o*-diamine) and an aromatic diglyoxal in the melt (at about 375°C) or in solution, e.g. in HMP:

$$n\,H_2N-\bigcirc-\bigcirc-NH_2\ +$$

$$n\,OHC-\underset{O}{\overset{}{C}}-\bigcirc-\underset{O}{\overset{}{C}}-CHO\ \rightarrow$$

$$\left[\ \text{(quinoxaline ladder structure)}\ \right]_n + 2n\,H_2O$$

The polymers are thermally stable in air but rapidly oxidise above about 350°C. Low molecular weight polymers are soluble in, for example dioxan and DMAC, but fully cyclised high molecular weight polymers are very insoluble and infusible. Tractability is improved by having flexible ether links in the polymer chain. Phenyl substituted polymers (phenylated polyquinoxalines) are obtained using an aromatic dibenzyl, e.g.

$$PhCOCO-\bigcirc-O-\bigcirc-COCOPh$$

p,p'-oxydibenzyl, and have improved solubility and oxidative stability. Ladder polyquinoxalines have also been synthesised.

POLYSACCHARIDE (Glycan) A carbohydrate polymer that may be considered as a step-growth polymer of a monosaccharide in which the monosaccharide (sugar) units are linked through glycoside bonds formed by reaction of a hemi-acetal (C-1) hydroxyl on one monomer with any hydroxyl of another, with elimination of water. Polymers with a DP of <10 are oligosaccharides, but most polysaccharides have a DP of ≫10, typically several tens to several hundreds.

They occur widely naturally, generally as either energy reserve materials, e.g. starch and glycogen, or as the structural material in plants, e.g. cellulose, or animals, e.g. chitin; they also occur as bacterial cell-wall, connective tissue and blood-group substances and as gums. They often form the basis of important commercial materials, especially with cellulose, but also with starch and many gums.

They may be classified according to their biological function or chemical structure. Homopolysaccharides contain only one type of monosaccharide unit, heteropolysaccharides contain more than one type of unit. A homopolysaccharide is named as a glycan by adding the suffix -an to the stem of the name of the parent monosaccharide, e.g. a glucan is a homopolymer of glucose. Such homoglycans include glucans (e.g. starch, cellulose, dextran, glycogen, lichenan, pullulan), fructans (e.g. inulin, levan), mannan, galactan, arabinan, xylan, galacturonan and glucosaminan. The heteropoly-

saccharides (heteroglycans), cannot be so simply classified, but most frequently contain only two types of sugar unit, and are then named using the stems of the sugar names, the dominant sugar being written last, e.g. galactomannan.

Typical monosaccharide units are the aldohexoses D-glucose, D- and L-galactose and D-mannose, the ketohexose D-fructose, the aldopentoses D-xylose and L-arabinose, the hexuronic acids D-glucuronic, D-galacturonic, D-mannuronic, L-guluronic and L-iduronic, the 6-deoxyhexoses, L-rhamnose and L-fucose and the hexosamines D-glucosamine and D-galactosamine. Further polysaccharide groups are classified according to the presence of a particular functional group, e.g. aminopolysaccharides and sulphated polysaccharides. Groups classified according to biological function are the reserve deposit (e.g. starch, glycogen), plant cell wall (e.g. cellulose), bacterial (e.g. dextran, levan) and connective tissue (including the mucopolysaccharides, e.g. chondroitin, heparin) polysaccharides and more complex polymers in which the carbohydrate chains are attached to non-carbohydrate polymers, e.g. glycoproteins and lipopolysaccharides.

The name of a polysaccharide must also show the positional linking of the sugar units, e.g. 1,2'-, 1,3'-, 1,4'-, 1,5'- and 1,6'-, the configuration of the anomeric carbon, i.e. α- or β-, and the particular ring form. Thus the full name of cellulose is poly-(1,4'-β-D-glucopyranose) or more simply 1,4'-β-D-glucan. In polysaccharides, names are often abbreviated by using only the first three letters of the sugars involved, e.g. gal- for galactose, except that the single letter G is used for glucose. The suffix A is added for a uronic acid, e.g. GalA for galacturonic acid, the suffix N is added for an aminosugar, e.g. GN for glucosamine and the suffix NAC is added for an *N*-acetylated aminosugar, e.g GalNAC for *N*-acetylgalactosamine. A lower case p or f is added to denote the pyranose or furanose ring form. Thus, for example, cellulose is poly-(1,4'-β-D-Gp).

POLYSALT (Polyelectrolyte complex) (Polyion complex) The network structure produced as a result of mixing an anionic with a cationic polyelectrolyte solution, thus precipitating a salt in which the opposite charges carried by each type of polymer effectively crosslink the network through ionic bonds. A 1:1 complex is often readily formed, e.g. by reaction of a sulphonate containing polymer with a quaternary ammonium containing polymer with the elimination of a low molecular weight salt. Such complexes are hard and brittle, but can form hydrogels, which are useful as ultrafiltration dialysis and desalination membranes and soft contact lenses.

POLYSAR Tradename for styrene–butadiene rubber.

POLYSAR BUTYL Tradename for butyl rubber.

POLYSARCOSINE (Poly-(*N*-methylglycine))

$$\left[\begin{array}{c}N-CH_2-CO\\|\\CH_3\end{array}\right]_n$$

A poly-imino acid. The L-isomer is synthesised by polymerisation of L-sarcosine N-carboxyanhydride. Unusually for a poly-amino acid, the polymer exists in a random conformation containing a random distribution of *cis* and *trans* peptide bonds. Unlike other poly-α-amino acids (apart from glycine) it has no asymmetric α-carbon atom and therefore it cannot be studied by optical rotatory dispersion and circular dichroism.

POLY-(SCHIFF BASE)   Alternative name for poly-azomethine.

POLYSEBACAMIDE   A nylon polymer formed by reaction of sebacic acid with a diamine and having the general structure

$$\text{-[NHRNHCO(CH}_2)_8\text{CO-]}_n$$

Nylon 610 is the most important example.

POLYSEMICARBAZIDE   A polymer of structure

$$\text{-[NHCONHNHRNHNHCONHR'-]}_n$$

where R is an alkyl group. Formed by reaction of a diisocyanate with a bishydrazine in a polar solvent. When —NHRNH— is derived from N,N-diaminopiperazine and R' is aromatic, the polymers have T_m values of 200–300°C and show a capability of forming chelate complexes with metals, such as nickel and copper.

POLYSERINE

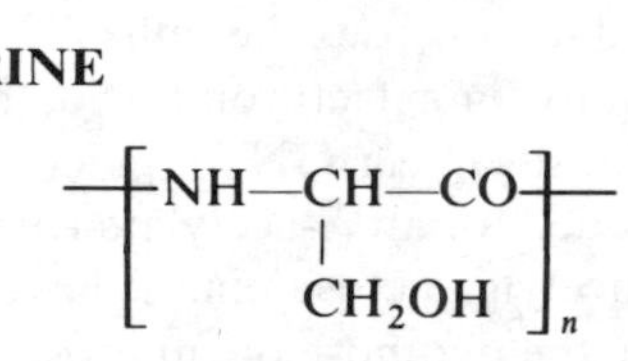

$$\left[\text{NH—CH—CO}\atop\text{CH}_2\text{OH}\right]_n$$

Optically pure poly-L-serine is obtained by polymerisation of O-benzyl-L-serine N-carboxyanhydride with $\overline{\text{O}}$Me, followed by removal of the benzyl protecting group. It is insoluble in water and many organic solvents but dissolves in aqueous lithium bromide solutions, in which it is in a random coil conformation. In the solid state a β-conformation is favoured. In concentrated acids it undergoes an N → O acyl shift in which an amide bond is converted to an ester bond. It is a useful model for certain proteins, e.g. silk fibroin, and for the active sites of serine enzymes, e.g. trypsin.

POLYSILASTYRENE

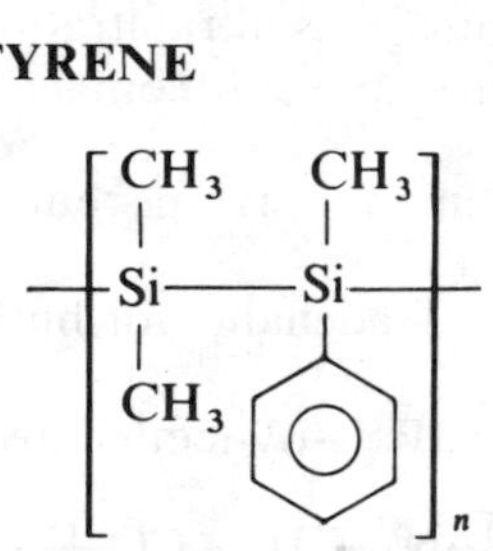

I

Produced by reacting a mixture of dimethyldichlorosilane

and phenylmethyldichlorosilane with metallic sodium dispersed in toluene:

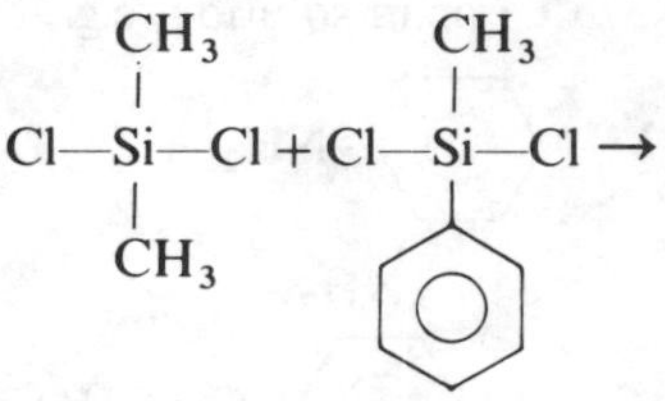

The polymer may be converted to a carbasilane on heating to above 1200°C. It is useful as a matrix material for silicon nitride composites of improved strength, since on heating to about 1400°C the silastyrene is converted, via the carbasilane, to β-silicon carbide.

POLYSILAZANE   A polymer of repeat unit structure:

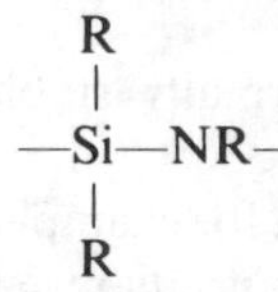

$$-\text{Si—NR—}\atop{\displaystyle\mathop{}^{R}_{R}}$$

Formed by condensation of aminosilanes, themselves formed by reaction of chlorosilanes with ammonia, followed by heating of the largely cyclic low molecular weight silazanes formed. The polymers have excellent thermal stability (unlike the polysiloxanes they do not tend to depolymerise at about 400°C) but have poorer hydrolytic stability. Hydrolytic stability is improved by N-substitution.

POLYSILICATE   Alternative name for silicate polymer.

POLYSILOXANE   A polymer whose chain structure is of the type

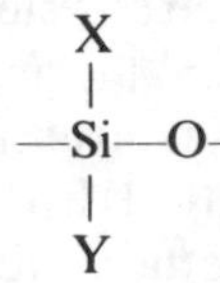

$$-\text{Si—O—}\atop{\displaystyle\mathop{}^{X}_{Y}}$$

The substituents X and Y are nearly always organic groups (frequently alkyl) when the term polyorganosiloxane is used. Such polymers form the basis of the commercially important silicones.

POLYSILOXANECARBORANE   Alternative name for polycarboranesiloxane.

POLYSILSESQUIOXANE   Alternative name for polyphenylsilsesquioxane.

POLYSOAP   A polyelectrolyte which also contains large hydrophobic groups, such as long alkyl chains, which tend to aggregate in aqueous solution through the formation of intramolecular hydrophobic bonds. A widely studied example is:

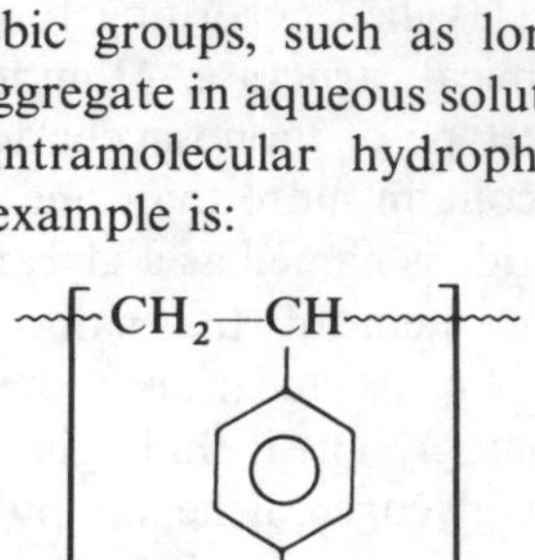

POLYSPIROKETAL A spiro polymer formed by condensation between a cyclic dione, e.g. 1,4-cyclohexadione, and pentaerythritol:

$$O=\text{(cyclohexane ring)}=O + \begin{array}{c} HOCH_2 \quad CH_2OH \\ \diagdown C \diagup \\ HOCH_2 \quad CH_2OH \end{array} \rightarrow$$

$$>C\left[\begin{array}{c} OCH_2 \quad CH_2O \\ \diagdown C \diagup \\ OCH_2 \quad CH_2O \end{array} C \right]_n$$

The polymer is of high crystallinity and has higher thermal stability than the similar non-spiro polyether.

POLYSTANNOSILOXANE A polymer containing both siloxane and stannoxane links in the polymer chain. Polymers may be formed by simply heating organotin oxides (R_2SnO) with dimethylsiloxane ring or chain compounds with silanediols, or by reacting a difunctional tin compound, e.g. an organotin dihalide or diacetate, with a difunctional silane monomer. The tin atom can act as an acceptor from an oxygen in a neighbouring chain and thereby increases viscosity or causes the polymer to behave like the borosilicone bouncing putty.

POLYSTANNOXANE A polymer with chains of alternating tin and oxygen atoms. The polymers formed by dehydration of $RSn(OH)_3$, for example, have high thermal stability, similar to the analogous polysiloxanes:

$$n\,RSn(OH)_3 \rightarrow \left[\begin{array}{c} R \\ | \\ Sn-O \\ | \\ OH \end{array}\right]_n$$

POLYSTYRENE (PS)

$$\sim\!\!\left[CH_2CH\right]_n\!\!\sim$$
(phenyl substituent)

Tradenames: Afcolene, Bextrene, Carinex, Distrene, Dylene, Edistir, Erinoid, Fostarene, Gedex, Hostyren, Lacqrene, Lorkalene, Lustrex, Polystyrol, Restirolo, Sicostyrol, Stiroplasto, Stymer, Styron, Styvarene, Vestyron. A polymer of major importance both as a commercial thermoplastic material and in fundamental studies of polymerisation and polymer structure. One of the few polymers that can be produced by free radical, anionic or cationic polymerisation of its monomer, styrene. Commercially, free radical polymerisation is always used so the polymer employed as a plastic is atactic and amorphous. Styrene polymerisations have been very widely used as models for chain polymerisation in general. Any conventional free radical initiator may be used, e.g. an organic peroxide or azo compound. The polymerisation may be performed as a mass, suspension, solution or emulsion polymerisation. Styrene also polymerises readily on heating (thermal polymerisation) without added initiator or by exposure to ultraviolet or high energy radiation. Anionic polymerisation, e.g. with a metal alkyl catalyst, can yield a monodisperse polymer by the living polymer technique more readily than most other monomers, and such polymers are widely used as polymer molecular weight standards. Ziegler–Natta polymerisation can yield isotactic polymer, although this material is not produced commercially. Normal atactic polystyrene has a T_g value of about $100°C$, although this value may be lower due to the presence of residual styrene monomer or ethylbenzene, the latter often occurring as an impurity in the monomer. The polymer is transparent and has excellent clarity and brilliance. It has very high electrical resistivity and low dielectric loss. It is soluble in a wide range of organic solvents, e.g. aromatic hydrocarbons, ketones and chlorinated hydrocarbons. It is a hard, stiff, glassy material with a typical tensile strength of $50\,MPa$, an elongation at break of 1–2% and a tensile modulus of $4\,GPa$. However, it is a brittle material with a typical Izod impact strength of about 0.1–$0.3\,J$ $(12.7\,mm)^{-1}$. Therefore as a plastic, polystyrene is often toughened, e.g. by the incorporation of a rubber (as in high impact polystyrene), by copolymerisation (as in styrene–acrylonitrile copolymer) or by both methods (as in acrylonitrile–butadiene–styrene copolymer). Unmodified polystyrene is often called crystal polystyrene. Isotactic polystyrene is crystalline and has a T_m value of about $260°C$ and a T_g value of $97°C$. It crystallises in spherulitic form and is even more brittle than atactic polystyrene and is therefore not useful as a plastic. Polystyrene may be converted to various derived polymers by electrophilic substitution on the benzene ring, as in chloromethylated polystyrene and sulphonated polystyrene. It is also widely used as a foamed plastic (expanded polystyrene).

POLY-(STYRENE-co-ACRYLONITRILE) Alternative name for styrene–acrylonitrile copolymer.

POLY-(STYRENE-co-MALEIC ANHYDRIDE) Alternative name for styrene–maleic anhydride copolymer.

POLYSTYROL Tradename for polystyrene.

POLYSTYRYLPYRIDINE (PSP) A high temperature resistant thermosetting resin, whose prepolymer is formed by reaction between an aromatic dialdehyde and 2,4,6-trimethylpyridine (collidine):

$$n \begin{array}{c} CH_3 \\ \text{(pyridine ring)} \\ H_3C \quad N \quad CH_3 \end{array} + n\,OHC-\text{(phenyl)}-CHO \rightarrow$$

$$\left[\begin{array}{c} CH_3 \\ =HC \quad \text{(pyridine ring)} \quad CH=CH-\text{(phenyl)}-CH= \\ N \end{array}\right]_n$$

It becomes crosslinked on heating to above $150°C$.

POLYSULPHIDE A polymer of general structure $+R-S_x+_n$, where x is either unit or a small number, e.g. 2–4. When $x = 1$ then the polymer is the sulphur analogue of the corresponding polyether and is, in general, made by similar methods. The most important types are the polysulphide rubbers (which are polyalkylene polysulphides with $x > 1$). Simple polyalkylene polysulphides, e.g. polyethylene sulphide and polypropylene sulphide, have been investigated, the latter being of interest as a rubber. Polyphenylene sulphide is also a useful commercial product as a high temperature resistant polymer.

POLYSULPHIDE RUBBER (TR) Tradenames: Novaplas, Thiokol A, Thiokol B, Thiokol ST, Thiokol FA, Thiokol N, Perduren G, Perduren H, Vulcaplas, GR-P. Any of a series of rubbery polymers of general structural type $+R-S_x)_n$, where $x = 2$–4 and R is

$$-CH_2CH_2-, \quad -CH_2(OCH_2CH_2)_2-$$

or, as is often the case, a mixture containing a branching point such as

$$\sim CH_2CHCH_2\sim$$

The polymers are made by reaction, in suspension, of a di- or polychlorocompound with hot aqueous sodium polysulphide:

$$n\,ClRCl + n\,Na_2S_x \rightarrow +RS_x+_n + 2n\,NaCl$$

However, the reaction is not a simple polycondensation, since interchange reactions, ring formation (especially if R contains 4–5 carbon atoms in a chain) and desulphurisation (removal of sulphur from polysulphide linkages) may occur. Thus the sulphur rank (the value of x in $-R-S_x-$) may change during polymerisation. In early polymers x was about four but in most modern materials x is usually in the range 2–2·2. High molecular weight materials are useful rubbers, but so are liquid low molecular weight materials (with molecular weights of 500–10 000) formed by reductive cleavage of the high molecular weight polymers by treatment with sodium hydrosulphide.

The polymers are vulcanised by reaction of terminal groups, which are either —OH or —SH groups. Thus linear polymers only undergo chain extension and the 'vulcanisates' have high compression set. By use of a trifunctional monomer, e.g. 1,2,3-trichloropropane, branched polymers are formed which give networks on vulcanisation. Hydroxyl-terminated polymers are vulcanised with a zinc compound, e.g. zinc hydroxide, whereas thiol-terminated polymers, including the liquid polymers, are vulcanised with lead, copper or zinc peroxides. Liquid polymers may be cured with epoxy or phenolic resins or with diisocyanates.

The vulcanisates have poor physical properties, often with an unpleasant odour, but have outstanding oil, solvent and chemical resistance. In particular compression set is high especially at high temperatures, probably due to a disulphide interchange reaction occurring. The rubbers are useful for fuel hose and tubing and as binders and for printing rollers. Liquid polymers are useful as caulking compounds and sealants and as castable rubbers, e.g. for encapsulation.

POLYSULPHONAMIDE A polymer containing the group —RSO$_2$NHR′— in the repeat unit of the polymer chain. When R and/or R′ are aromatic the polymer is an aromatic polysulphonamide and these types are of most interest. Usually prepared by solution or interfacial polymerisation at low temperature by reaction of a diamine with a disulphonyl chloride,

$$n\,H_2NRNH_2 + n\,ClO_2SR'SO_2Cl \rightarrow$$
$$+HNRNHSO_2R'SO_2+_n + 2n\,HCl$$

usually in the presence of an HCl acceptor. Polymers from aliphatic diamines and aromatic sulphonyl chlorides, e.g. poly-(hexamethylene-1,3-benzenedisulphonamide), are the most readily prepared. Aromatic diamines have much lower activity and aliphatic sulphonyl chlorides are very sensitive to hydrolysis and their polymers have low solubility. The polymers dissolve in highly polar solvents such as m-cresol, trifluoroacetic acid and concentrated sulphuric acid and in many basic solvents, even in aqueous alkali. They melt at lower temperatures than the corresponding polyamides.

POLYSULPHONATE A polymer of the type $+RSO_2O+_n$. The aromatic polysulphonates, where R is an aromatic ring, are of most interest.

POLYSULPHONE (1) A polymer of the type $+RSO_2+_n$. Polymers where R is an aromatic ring structure (aromatic polysulphones) are of most interest, particularly when ether links are also present (polyethersulphones). Aliphatic polysulphones may be synthesised by alternating copolymerisation of a vinyl monomer with sulphur dioxide via a 1:1 complex. They have low ceiling temperatures and have been used in the classical studies of ceiling temperature behaviour. Polysulphones are also formed by oxidation of polysulphides.

(2) Alternative name for the polyethersulphone obtained by reaction of bisphenol A with 4,4′-dichlorodiphenylsulphone (Udel).

POLY-(4,4′-SULPHONYLDIBENZAMIDE) A polyamide of general structure

$$\sim \left[NHRNHCO-\bigcirc-SO_2-\bigcirc-CO \right]_n ,$$

formed by reaction of a diamine with 4,4′-sulphonyldibenzoyl chloride. The best known example is poly-(hexamethylenesulphonyldibenzamide). The polymers are lower melting than the terephthalamides and are more soluble.

POLY-(SULPHUR NITRIDE) $+S_2N_2+_n$. When synthesised by a topotactic polymerisation of di-(sulphur nitride), a relatively defect free, electron deficient polymer is obtained. However, although more disorder is present than with topotactically prepared polydiacetylenes, the electrical conductivity is high along the chains, being about $1 \times 10^3 \, S \, cm^{-1}$, i.e. metallic conductivity is observed. At very low temperature, below $1.4 \, K$, the polymer becomes superconducting.

POLYTEREPHTHALAMIDE A polyamide synthesised by reaction of a diamine with terephthalic acid or one of its derivatives and therefore having the structure

$$\sim \left[NHRNHCO - \langle O \rangle - CO \right]_n \sim$$

Examples include nylon 6T and TMDT. The presence of the *para*-linked benzene ring causes the polymer to have a high T_m value, so much so that if the diamine has less than seven methylenes, the polymer decomposes before it melts, and cannot be made by conventional melt polymerisation.

POLYTEREPHTHALOYLOXAMIDRAZONE (PTO)

$$\left[\langle O \rangle - \underset{O}{C} - \underset{NH_2}{\overset{NH-N}{C}} - \underset{NH_2}{\overset{N-NH}{C}} - \underset{O}{C} \right]_n$$

A polyacylamidrazone synthesised by reaction between terephthaloyl chloride and oxalamidrazone (obtained from reaction of oxalyl nitrile and hydrazine). The polymer can exist in both keto and enol forms. It may be spun from an alkaline solution into an acid bath to form fibres. The fibres, or even woven fabrics, form coloured chelate compounds (chelated PTO) with an ammoniacal solution of any of several metal hydroxides, e.g. Ca, Zn, Sr. The fabrics have been produced as useful fire resistant textiles.

POLYTEREPHTHALOYL-2,5-TRANSDIMETHYL-PIPERAZINE (Poly-(2,5-dimethylpiperazineterephthalamide)). A widely studied piperazine polyamide, synthesised by low temperature solution polymerisation by reaction of *trans*-2,5-dimethylpiperazine with terephthaloyl chloride in a chlorinated hydrocarbon solvent. Has a T_m value of greater than $440°C$ and is highly insoluble even in formic acid and dimethylformamide.

POLYTETRAFLUOROETHYLENE (PTFE) $\sim[CF_2CF_2]_n\sim$. Tradenames: Algoflon, Fluon, Hostaflon TF, Polyflon, Soreflon, Teflon, Tetraflon. The most important fluoropolymer, which is a plastic material. Prepared by free radical polymerisation of tetrafluoroethylene in aqueous systems with persulphate or peroxide initiators to give granular or dispersion polymer. The monomer may also apparently spontaneously polymerise with violence. The polymers are extremely linear and crystallise (to about 95%) with the chains adopting a slightly twisted extended chain zig-zag conformation. The T_m value is $327°C$. Typically, molecular weights are in the millions. Like other fluoropolymers, polytetrafluoroethylene has exceptionally high thermal and thermooxidative stability and is completely solvent resistant, only certain fluorinated solvents dissolving it at temperatures near its T_m value. It is a tough, relatively flexible material, of tensile strength about $20 \, MPa$ and an elongation at break of 200–300%. Its tensile modulus is about $0.5 \, GPa$. It has outstandingly good electrical insulation properties and an unusually low coefficient of friction. In addition to T_m, a small transition at about $19°C$ is observed, but even above T_m the material does not flow, the melt viscosity being $10^{10}–10^{11} \, P$ at $350°C$. Therefore it cannot be melt processed in the normal way. Products are made by sintering or by special extrusion techniques. It is widely used, due to its chemical inertness, for pumps, seals, valves and gaskets, and due to its low friction, for non-stick coatings, e.g. for cooking equipment. It is also useful as a high temperature and fire resistant electrical insulation material.

POLY-(TETRAFLUOROETHYLENE-co-ETHYLENE) Alternative name for tetrafluoroethylene–ethylene copolymer.

POLY-TETRAFLUOROETHYLENE-co-HEXAFLUOROPROPYLENE) Alternative name for tetrafluoroethylene–hexafluoropropylene copolymer.

POLY-(TETRAFLUOROETHYLENE-co-PERFLUOROMETHYLVINYL ETHER) Alternative name for tetrafluoroethylene–perfluoromethylvinyl ether copolymer.

POLY-(TETRAFLUOROETHYLENE-co-PERFLUOROPROPYLVINYL ETHER) Alternative name for tetrafluoroethylene–perfluoropropylvinyl ether copolymer.

POLY-(TETRAFLUOROETHYLENE-co-SULPHONYLFLUORIDEVINYL ETHER) Alternative name for tetrafluoroethylene–sulphonylfluoridevinyl ether copolymer.

POLYTETRAHYDROFURAN (PTHF) (Polytetramethylene oxide) (Polyoxytetramethylene glycol) $\sim[(CH_2)_4 - O]_n\sim$. Tradename: Teracol. Prepared by cationic ring-opening polymerisation of tetrahydrofuran, using carbenium ion salts, such as $(C_6H_5)_3C^+SbCl_6^-$, or oxonium ion salts, such as $(C_2H_5)_3O^+BF_4^-$. Low molecular weight polymers terminated with hydroxyl groups, are useful as polyol prepolymers for block copolymer formation and have been termed polytetramethylene ether glycols. The polymer crystallises with a T_m value of 58–60°C.

The polymers are much less hydrophilic than polyoxyethylene glycols and are widely used as prepolymer polyols of molecular weight 600–3000, for the preparation of polyurethane elastomers by reaction with diisocyanates. One of the earliest polyether polyols used in producing flexible polyurethane foams.

POLYTETRAMETHYLENEADIPAMIDE Alternative name for nylon 46.

POLY-(TETRAMETHYLENE ADIPATE) (PTMA)

$$-[(CH_2)_4OOC(CH_2)_4COO]_n-$$

Synthesised by simple polycondensation between 1,4-butanediol and adipic acid. Occasionally used as the polyester prepolymer for the formation of polyurethane elastomers. It has a lower T_g value than poly-(ethylene adipate) and therefore the polyurethane formed from it has a low 'freeze' temperature, but it has a greater tendency to cold-hardening.

POLYTETRAMETHYLENE ETHER GLYCOL (PTMEG) (Polyoxytetramethylene) (Polytetrahydrofuran) $HO-[(CH_2)_4O]_n-H$. A widely used prepolymer for providing the soft segments in block copolymer thermoplastic elastomers. Thus in polyether/esters, hydroxy-terminated prepolymer may be incorporated into the polymer chain by esterification and in segmented polyurethanes by reaction with diisocyanates.

POLY-(TETRAMETHYLENE OXIDE) Alternative name for polytetrahydrofuran.

POLYTETRAMETHYLENE TEREPHTHALATE (PTMTP) Alternative name for polybutylene terephthalate.

POLY-(TETRAMETHYLENETEREPHTHALATE-b-POLYOXYTETRAMETHYLENETEREPHTHALATE)
(Tetramethyleneterephthalate—polyoxytetramethylene-terephthalate block copolymer) (Polyether/ester block copolymer) Tradenames: Arnitel, Hytrel. A commercial thermoplastic elastomer, synthesised by ester interchange between dimethylterephthalate and a mixture of 1,4-butanediol and polyoxytetramethylene; therefore it contains hard segments of tetramethyleneterephthalate units of structure

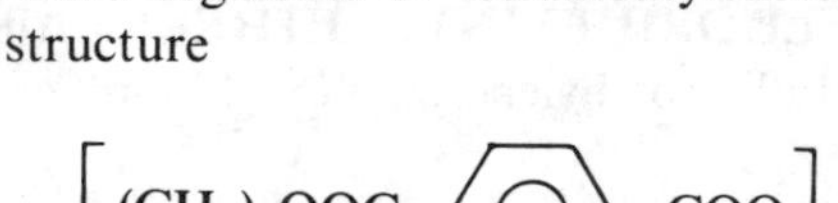

(about 60 wt%) and soft segments of polyoxytetra-methyleneterephthalate units

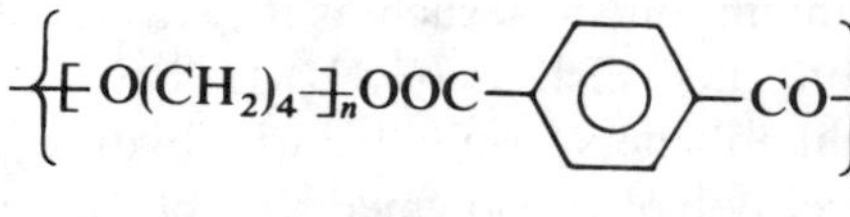

(about 40 wt%), the polyether having a molecular weight of about 1000. Each of the hard segments has more than ten repeat units and can therefore crystallise. The material phase separates on cooling from the melt, the hard segments crystallising and acting as physical crosslinks. The copolymers are useful elastomers, capable of operating from −55°C to +150°C, with good retention of mechanical properties and also with good oil resistance.

POLY-s-TETRAZINE Alternative name for poly-(1,2,4,5-tetrazine).

POLY-(sym-TETRAZINE) Alternative name for poly-(1,2,4,5-tetrazine).

POLY-(1,2,4,5-TETRAZINE) (Poly-(sym-tetrazine)) (poly-(s-tetrazine)) A polymer containing the heterocyclic rings

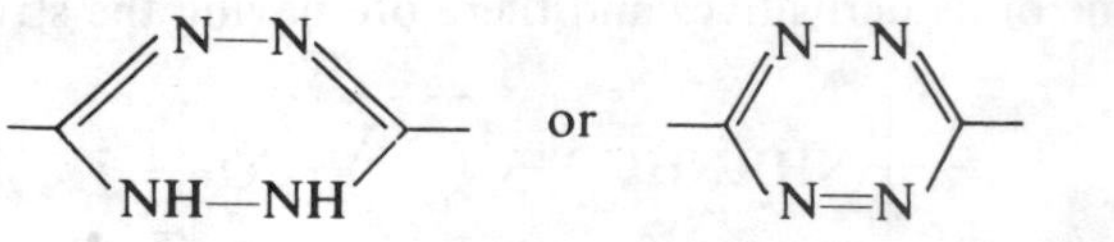

in the polymer chain. **I** is formed by condensation between hydrazine and a diiminoester,

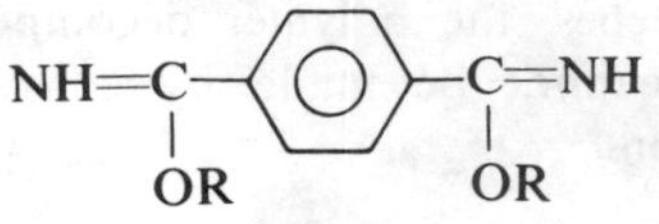

and may be reduced to **II**. **I** is also formed by self-polymerisation (by dipolar addition) of the bisdipolar ion from a bis-(aromatic hydrazide chloride):

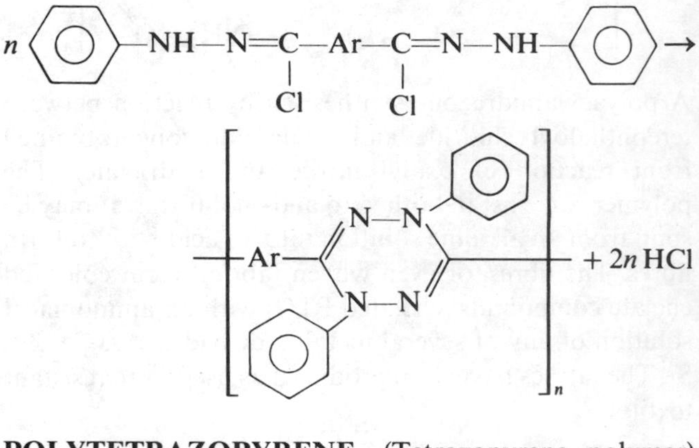

POLYTETRAZOPYRENE (Tetrazopyrene polymer) A polymer of structure

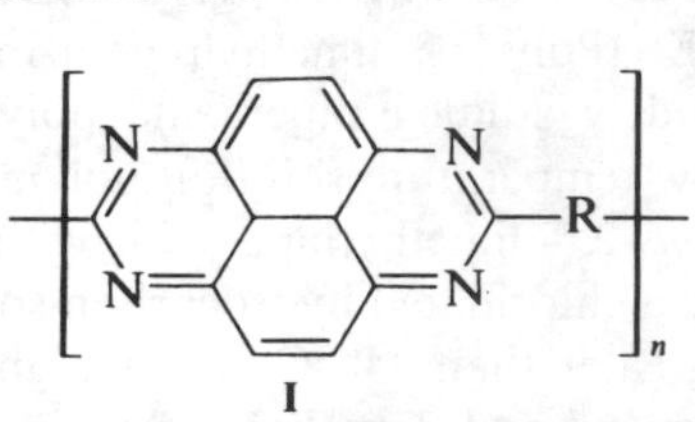

where R is an aromatic group, often a phenylene ring. Synthesised by condensation between 1,4,5,8-tetra-aminonaphthalene and a diphenyl ester under oxidising conditions:

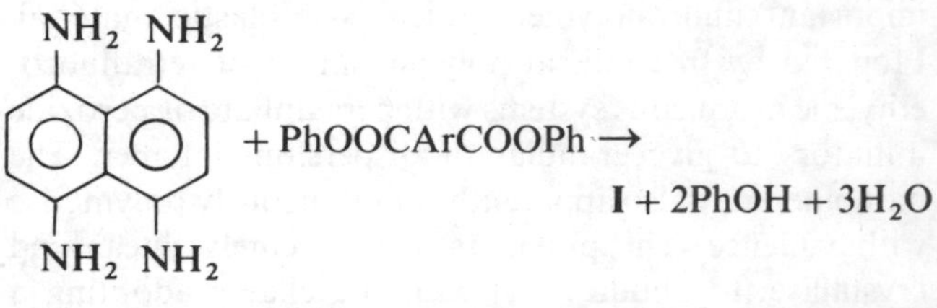

$$I + 2PhOH + 3H_2O$$

The polymers are dark coloured, melt above 400°C and have the good thermal and thermo-oxidative stability of a heterocyclic polymer, although they do not perform as well as some others.

POLYTHENE Alternative name for polyethylene, especially low density polyethylene.

POLY-(1,3,4-THIADIAZOLE) A polymer containing the heterocyclic ring

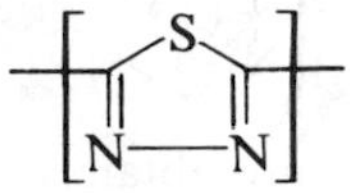

in the polymer chain. Synthesised either by cyclo-dehydration of a precursor polyoxathiahydrazide, or by cyclodehydrosulphurisation of a precursor polydithia-hydrazide. The precursor polymers are obtained by reaction between a tetrathiodiester and a dihydrazide or dithiahydrazide or, preferably, by reaction of phosphorus pentasulphide with a polyhydrazine in pyridine (py):

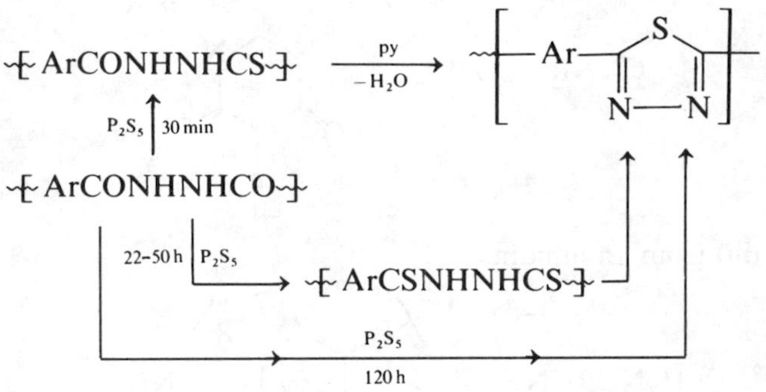

The polymers have the excellent hydrolytic, thermal and, particularly, thermo-oxidative stability of their oxygen analogues (the polyoxadiazoles) and have been studied as potential high temperature reistant fibres. They melt at $>375°C$, are soluble in strong acids and, unlike many heterocyclic polymers, can have little colour.

POLYTHIAZOLE A polymer containing the heterocyclic ring

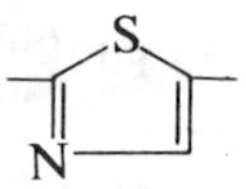

in the polymer chain. Synthesised by reaction between a dithioamide and a bis-(α-haloketone) in refluxing solvent (DMF or acetic acid)

$$n\,H_2NCSR^1CSNH_2 + n\,BrCH_2COR^2COCH_2Br \rightarrow$$

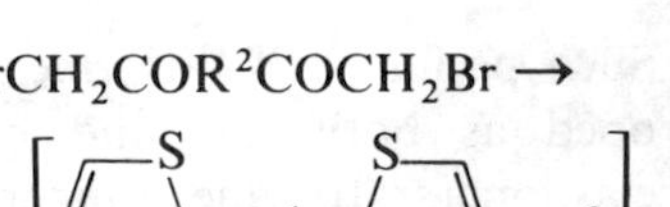

where R^1 and R^2 may be aliphatic, alicyclic or aromatic hydrocarbon groups. The wholly aromatic polymers have melt temperatures of $>400°C$, are crystalline, coloured and soluble only in strong acids. Their good thermal stabilities in nitrogen (to $500°C$ by TGA) and in air (to $\sim350°C$) are typical of a heterocyclic polymer.

POLYTHIAZOLINE A polymer containing the heterocyclic ring

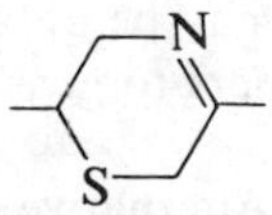

in the polymer chain. Synthesised by heating a poly-thiourea containing pendant hydroxyl groups in PPA:

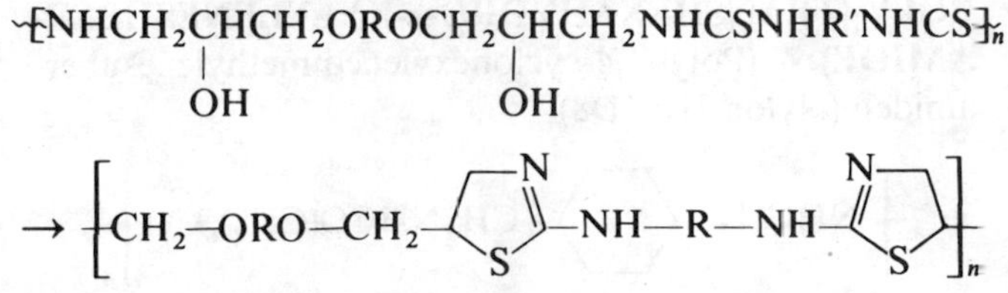

$$\text{-[NHCH}_2\text{CHCH}_2\text{OROCH}_2\text{CHCH}_2\text{NHCSNHR'NHCS]}_n$$
$$\qquad\quad | \qquad\qquad\qquad\quad |$$
$$\qquad\quad OH \qquad\qquad\qquad OH$$

POLYTHIAZYL Alternative name for sulphur nitride polymer.

POLYTHIOCARBONYL FLUORIDE

$$\text{-[CF}_2\text{S-]}_n$$

Prepared by anionic polymerisation or by free radical polymerisation (with BEt_3/oxygen initiator at about $-80°C$) of thiocarbonyl fluoride ($CF_2 = S$), itself obtained, for example, by the high temperature reaction of sulphur with chlorodifluoromethane. The initially rubbery polymer slowly crystallises, but crystallisation is inhibited by copolymerisation with about 2% allyl chloroformate. Useful rubbers may be obtained by vulcanisation with, for example, zinc oxide.

POLYTHIODIETHANOL Tradename: Cymax

$$\text{-[CH}_2\text{CH}_2\text{SCH}_2\text{CH}_2\text{O-]}_n$$

The polymer is prepared by the self-condensation of thiodiethanol and has a T_g value of $-65°C$. Hence it is rubbery if it is non-crystalline. Thus the commercial polymer also contains other comonomers to suppress crystallisation and to provide crosslinking sites for sulphur vulcanisation. The vulcanisates behave similarly to nitrile rubber and have excellent resistance to alcohol based fuels.

POLY-(THIO-1,4-PHENYLENEOXYTETRA-METHYLENE) (Polyphenylenethiooxytetramethylene).

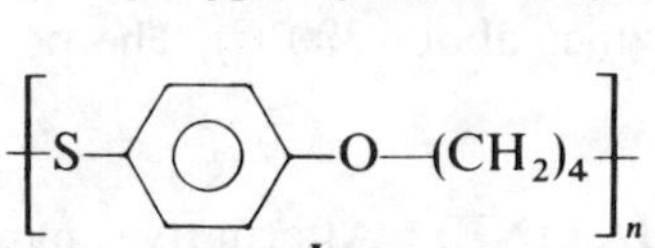

The polymer and its ring substituted derivatives (e.g. with CH_3 and Cl) are synthesised by ring-opening polymerisation of aryl cyclic sulphonium zwitterions with loss of charge (death-charge polymerisation):

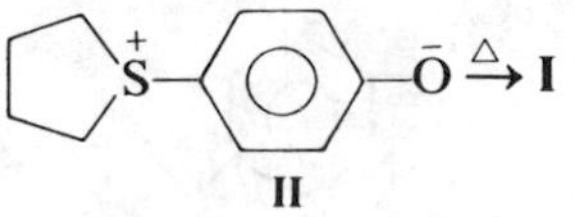

Polymerisation can occur in aqueous solution to yield water-resistant coatings. Polymers from such simple monomers as **II** are soft, e.g. dichloro-**I** has a T_m value of 150°C and a T_g value of about 10°C. Hard coatings with high softening points may be obtained by using similar tetrafunctional monomers to give crosslinked polymers.

POLYTOLYLENE Alternative name for poly-methylenephenylene.

POLY-(TRANSHEXAHYDRO-*p*-XYLYLENESUBER-AMIDE) (Poly-(1,4-cyclohexylenedimethylenesuber-amide)) (Nylon HPXD8)

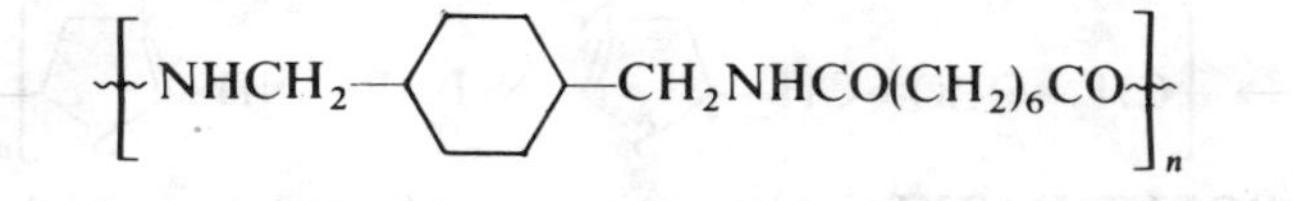

$$\left[NHCH_2 \text{—} \bigcirc \text{—} CH_2NHCO(CH_2)_6CO \right]_n$$

A partially aromatic polyamide obtained by reaction of transhexahydro-*p*-xylylenediamine and suberic acid. Has a T_g value of 86°C, a T_m value of 295°C, a high hydrolytic stability and is a potentially useful plastic moulding material.

POLY-(as-TRIAZINE) A polymer containing the heterocyclic ring

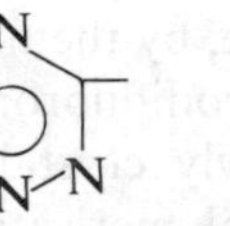

in the repeat unit. Synthesised by cyclopolycondensation between a diamidrazone and a bis-(1,2-dicarbonyl) compound:

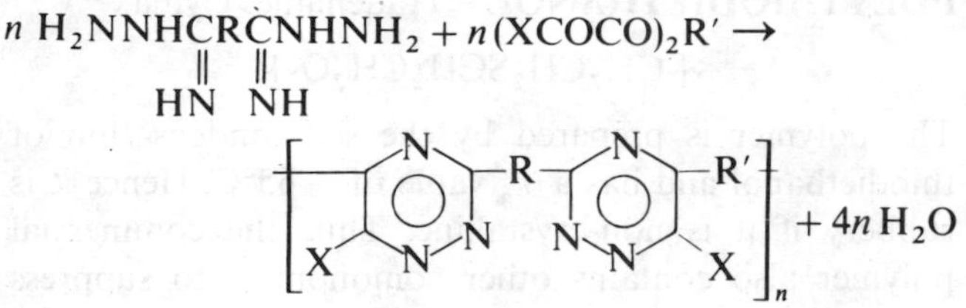

$$n\ H_2NNHCRCNHNH_2 + n(XCOCO)_2R' \rightarrow$$
$$\underset{HN\ \ NH}{\overset{\|\ \ \|}{}}$$

$$+ 4n\,H_2O$$

Typically R is pyridyl or phenyl or is absent and the bis-(dicarbonyl) monomer is an aromatic bisglyoxal (X = H) (as the hydrate). R' is an aromatic ring or a substituted bisglyoxal (X = phenyl) (a dibenzyl), leading to the poly-(phenyl-as triazines), by solution polymerisation in *m*-cresol. The substituted polymers have better solubility and thermo-oxidative stability. T_g values are high (for phenyl-as-triazines about 300°C). The polymers do not crystallise.

POLY-(s-TRIAZINE) Alternative name for poly-(sym-triazine).

POLY-(sym-TRIAZINE) (Poly-(s-triazine)) Trade-names: NCNS polymer, Triazin A. A polymer containing the heterocyclic ring

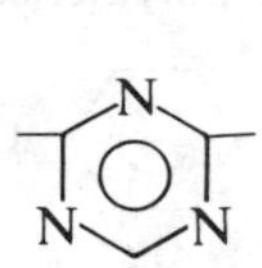

in the polymer chain. Several different polymers containing other linking groups have been synthesised.

s-Triazinylpolyethers are formed by reaction of cyanuric chloride with an aromatic dihydroxy compound:

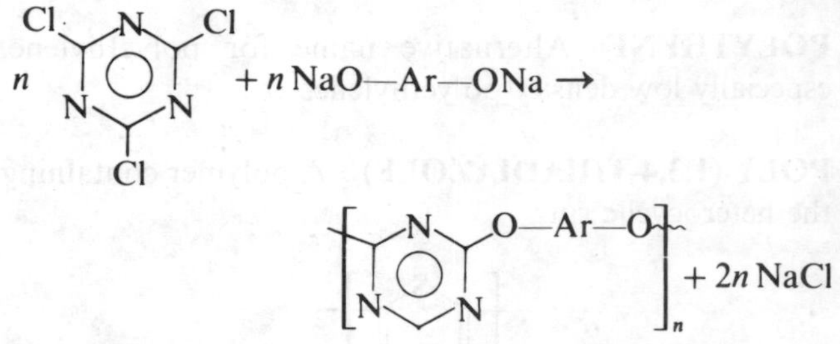

$$+ n\,NaO\text{—}Ar\text{—}ONa \rightarrow$$

$$+ 2n\,NaCl$$

Crosslinked polymers are obtained by:
(i) polymerisation of a nitrile, e.g. terephthaloyl nitrile:

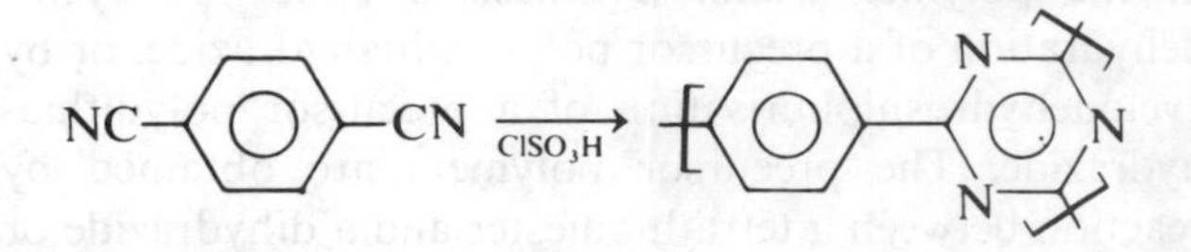

(ii) isocyanate trimerisation:

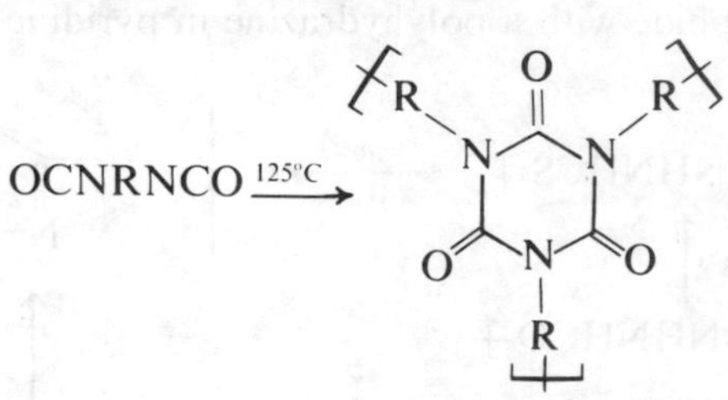

$$OCNRNCO \xrightarrow{125°C}$$

(iii) from an amidine

$$\underset{HN\ \ NH}{\overset{\|\ \ \|}{H_2NCRCNH_2}} \xrightarrow{120°C}$$

$$+ NH_3$$

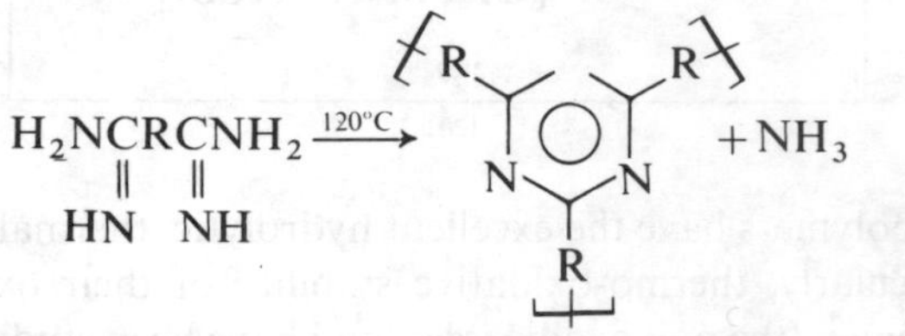

Linear polymers are obtained from an intermediate iminoylamidine polymer (**I**), itself obtained from reaction of a dinitrile with a bisamidine at room temperature

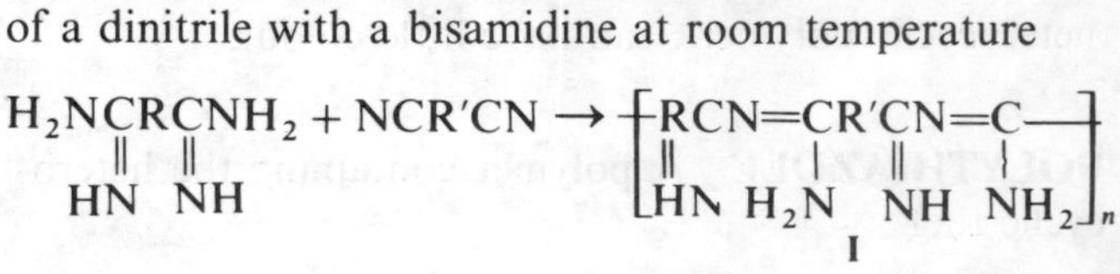

$$\underset{HN\ \ NH}{\overset{\|\ \ \|}{H_2NCRCNH_2}} + NCR'CN \rightarrow \left[\underset{HN\ \ H_2N}{\overset{}{RCN}}=\underset{}{CR'}\underset{NH\ \ NH_2}{\overset{}{CN}}=C \right]_n$$
$$\mathbf{I}$$

where R and R' are often fluorinated hydrocarbon groups.

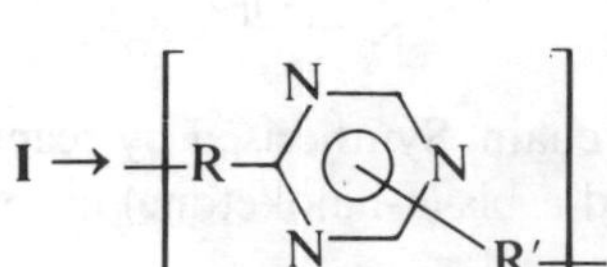

$$\mathbf{I} \rightarrow$$

Such perfluoroalkyltriazine polymers have been developed as both low and high temperature resistant elastomers (triazine rubber) with excellent chemical resistance. Commercial *N*-cyanosulphonamide polymers are also triazine polymers, being crosslinked through triazine rings. s-Triazine–polyamide polymers have also been synthesised from terephthalic acid and melamine or a guanamine.

POLYTRIAZOLE Alternative name for triazole polymer.

POLY-(1,2,3-TRIAZOLE)   A polymer containing the

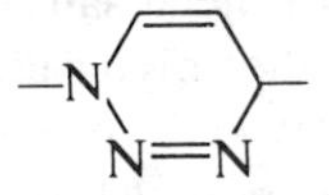

ring in the polymer chain. Synthesised by dipolar addition of an azide to an acetylene. The polymer formed also contains 1,4- and 1,5-disubstituted triazole units.

POLY-(1,2,4-TRIAZOLE)   A polymer containing the heterocyclic ring

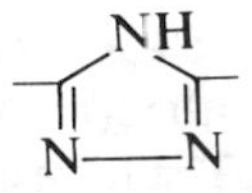

in the polymer chain. Synthesised by heating a poly-(N-acylamidrazone) or a diamidrazone precursor, in refluxing pyridine or NMP

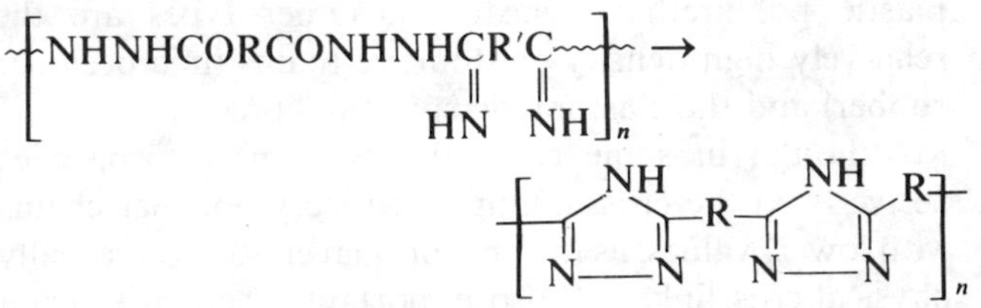

where R and R′ are alkyl or aryl groups, R′ being frequently aryl. Alternatively synthesised by reaction of a diamidrazone with a diacid chloride, particularly with an oxalamidrazone or with all aromatic monomers, when useful high temperature resistant polymers result:

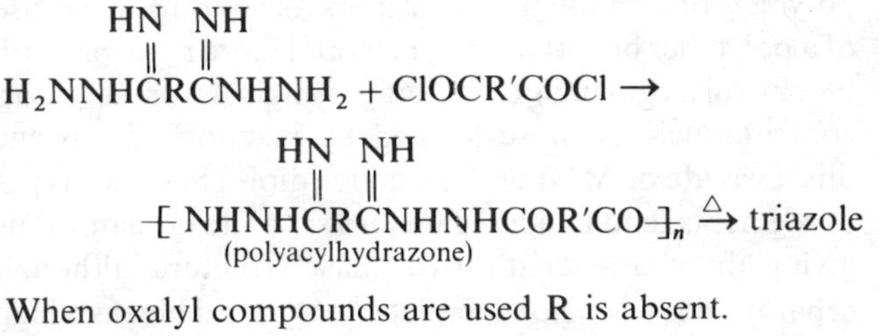

When oxalyl compounds are used R is absent.

POLYTRIAZOLINE   A polymer containing the heterocyclic ring

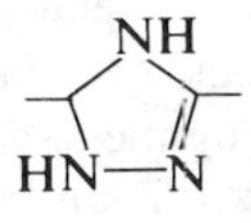

in the polymer chain. Synthesised by reaction of a diamidrazone with a dialdehyde

$$H_2NNHCArCNHNH_2 + OHCAr'CHO \rightarrow$$

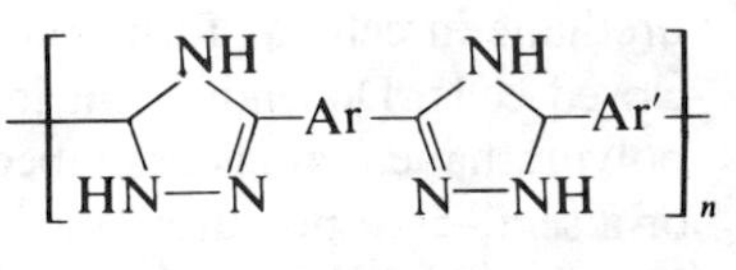

where Ar and Ar′ are aromatic ring structures.

POLYTRIDECAMETHYLENETRIDECANOAMIDE
Alternative name for nylon 1313.

POLYTRIDECANOAMIDE   Alternative name for nylon 13.

POLY-(2,4,5-TRIKETOIMIDAZOLINE)   Alternative name for polyparabanic acid.

POLYTRIMETHYLHEXAMETHYLENETEREPHTHALAMIDE   (TMDT) Tradename: Trogamid T.

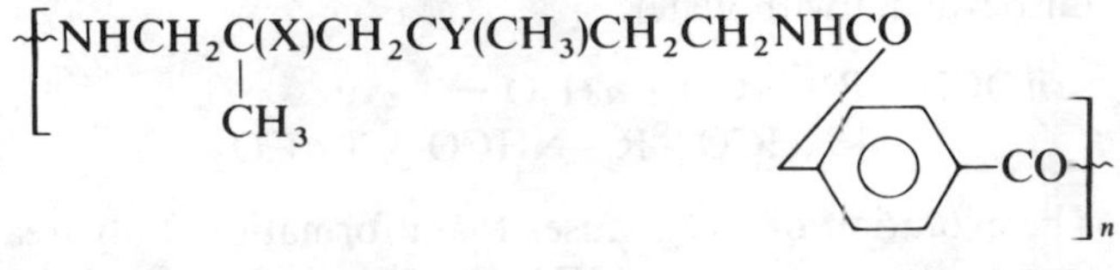

where either X = CH$_3$ and Y = H or X = H and Y = CH$_3$. The two isomers are present in equal amounts in the commercial polymer, which is made from a mixture of 2,4,4- and 2,2,4-trimethylhexamethylenediamines (obtained from acetone via its trimer isophorone) and terephthalic acid. An amorphous, transparent polymer with high yield strength (85 MPa), tensile modulus (3000 MPa) and impact strength. Therefore it is useful as an engineering plastic. It has a T_g value of 145°C and retains its mechanical properties almost to this temperature.

POLYTRIPEPTIDE   A sequential polypeptide consisting of sequences of tripeptide units, i.e.

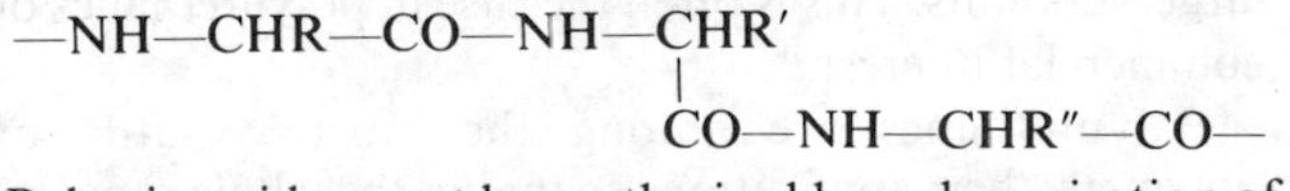

$$—NH—CHR—CO—NH—CHR'$$
$$CO—NH—CHR''—CO—$$

Polytripeptides must be synthesised by polymerisation of the tripeptide monomer, e.g as an active ester, since the N-carboxyanhydride route is unavailable. Several such polymers have been widely studied as models for collagen, e.g. poly-(gly–pro–pro) or silk, e.g. poly-(gly–gly–ala).

POLYTYROSINE

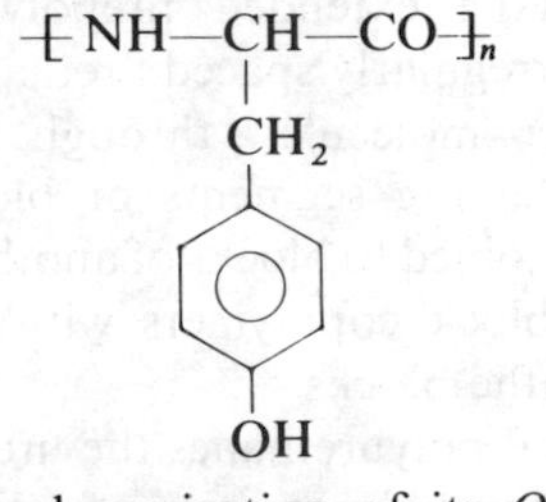

Synthesised by polymerisation of its O-carbobenzoxy-protected N-carboxyanhydride. It is insoluble in water but soluble in alkali pH solutions. It exists as the α-helix in the solid state and in solution, but undergoes helix–coil transition as solvent polarity or alkalinity is increased.

POLYUNDECANOAMIDE   Alternative name for nylon 11.

POLY-(11-UNDECANOAMIDE)   Alternative name for nylon 11.

POLYURAZOLE   A polymer of structural type

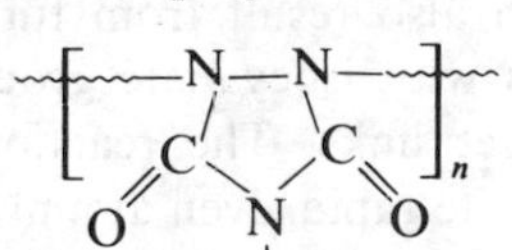

A commercial copolymer containing urazole and epoxide groups has been developed.

POLYUREA A polymer containing —NHCONH— groups in its repeat unit. Formed by reaction of a diisocyanate with water:

$$n\,OCN—R—NCO + n\,H_2O \rightarrow$$
$$—[NHCO—R—NHCO]_n + n\,CO_2$$

The evolution of CO_2 causes foam formation. Polyurea foams often based on MDI, have been developed for thermal insulation purposes as an alternative to urea–formaldehyde foam, since their use does not involve the evolution of formaldehyde.

POLYURETHANE (PU) (PUR) (Urethane) (Polycarbamate) A polymer which contains urethane (—NHCOO—) groups in the polymer chain. Like polyesters and polyamides, a polyurethane may contain the urethane group as part of every repeating unit. However, even polymers in which the urethane groups are only present in minor amounts are also termed polyurethanes. Frequently other functional groups such as ester, ether, amide or urea groups are present in much larger amounts. This is often the case in polyurethanes of commercial interest.

Polyurethanes are among the most versatile of polymers, their applications spanning the whole range of polymeric products—as polyurethane elastomers, polyurethane fibres, rigid, flexible and semi-rigid polyurethane foams, solid plastics, coatings and adhesives.

They range in their structural types from regular polyurethanes, consisting of regularly repeating urethane groups, e.g. of the type $—[ROCONHR'NHCOO]_n$, through urethane extended prepolymers containing relatively few irregularly spaced urethane groups linking the prepolymer molecules, through segmented polyurethanes containing segments or blocks of urethane repeating units joined to blocks of another type of unit, to non-urethane block copolymers with urethane groups merely linking the blocks.

In commercial polyurethanes the urethane groups are nearly always formed by reaction of a diisocyanate with a diol (or polyol): $R—NCO + HO—R' \rightarrow RNHCO—OR'$. Thus linear polymers result from reaction with a diol:

$$OCN—R—NCO + n\,HO—R'—OH \rightarrow$$
$$—[O—CONH—R—NHCO—O—R']_n$$

This is a step-growth polymerisation, but unlike many other such polymerisations, no small molecule is eliminated, so that it cannot be described as a condensation polymerisation. Sometimes this type of polymerisation is called a polyaddition or rearrangement polymerisation. If a polyol containing more than two hydroxyl groups is used, then a branched or crosslinked polymer is formed. Crosslinking can also result from further reactions of urethane groups with isocyanate groups to form allophanate or biuret links. The reactions of isocyanate groups can be quite rapid, even at ambient temperature, but frequently, especially in polyurethane foam forma-

tion, catalysts such as tertiary amines, e.g. 1,4-diazabicyclo-2,2,2-octane, or metal salts, e.g. stannous octoate, are used. Frequently, the polyols used are themselves low molecular weight polymers, often hydroxy-terminated polyesters (polyester polyols) or hydroxy-terminated polyethers (polyether polyols). Another common polyurethane forming reaction (not used commercially) is that between a bischloroformate and a diamine:

$$n\,ClOCOROCOCl + n\,H_2NR'NH_2 \rightarrow$$
$$—[NHOCOROCONHR']_n + 2n\,HCl$$

POLYURETHANE ELASTOMER (Polyurethane rubber) (Urethane rubber) A polyurethane with elastomeric properties. The term covers a very wide range of polyurethanes, classified according to the method of manufacture. The main types are cast polyurethane elastomer, millable polyurethane elastomer and thermoplastic polyurethane elastomer. Other types are the relatively high density elastomeric foams (microcellular rubber) and the elastomeric spandex fibres.

Although in some cases the elastomeric properties derive from the crosslinking of rubbery polymer chains with low T_g values, as in conventional elastomers, usually physical crosslinks are also important. These arise from the two phase nature of most polyurethane elastomers, which consist of block copolymer molecules (i.e. segmented polyurethane molecules) containing elastomeric, or soft, blocks (or segments) of either aliphatic esters units (e.g. derived from the use of a polyester polyol prepolymer) or of aliphatic ether units (derived from the use of a polyether polyol prepolymer) and more rigid, or hard, blocks (or segments) containing a high urethane group content, such as those formed by reaction of tolylene diisocyanate or MDI with an active diol. These two types of segments tend to aggregate into separate domains thus giving the characteristic two phase structure, although separation is often not complete. In the hard blocks a high density of urethane groups exists, which form intermolecular hydrogen bonds, thus stiffening the structure and acting as physical crosslinks. In some linear thermoplastic polyurethane elastomers, all 'crosslinking' is through the hard blocks. Polyurethane elastomers are noted for their high toughness, good strength and wear properties and good oil and oxidation resistance. However, they do suffer from limited resistance to hydrolysis.

POLYURETHANE FOAM A crosslinked polyurethane in cellular form, the cells being either open or closed cells. Depending on its mechanical behaviour, a polyurethane foam is described as being a flexible, a rigid or a semi-rigid polyurethane foam. The foams are made using both polyester and polyether polyols, although usually the latter are used, especially for rigid foams. They are produced by the reaction of the polyol with a diisocyanate in the presence of either water, which produces carbon dioxide to blow the foam by reaction with excess diisocyanate,

$$2\sim NCO + H_2O \rightarrow \sim NH—CO—NH\sim + CO_2$$

or with a blowing agent, such as a chlorofluoromethane. Catalysts are usually also used to give the right balance of reaction rates so that the gas bubbles are trapped in the liquid polymer as the viscosity increases due to polymerisation and cross-linking. A surfactant, usually a silicone–polyether block copolymer, is also often present to control cell morphology. One-shot processes are used with polyester and polyether polyols, but prepolymer and quasi-prepolymer processes are also used with polyether polyols.

POLYURETHANE RUBBER Alternative name for polyurethane elastomer.

POLYUREYLENE Alternative name for polyurea.

POLYURONIDE Alternative name for glycuronan.

POLYVALENT ENZYME An allosteric enzyme which contains more than one binding site for effector or effectors.

POLY-(δ-VALEROLACTONE)

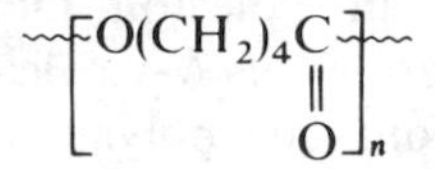

$$\sim\left[O(CH_2)_4 \underset{\underset{O}{\parallel}}{C}\right]_n$$

Prepared by the ring-opening polymerisation of δ-valerolactone. An aliphatic polyester with a T_m value of 50–60°C.

POLYVINYL ACETAL (1) A polymer containing a substantial proportion of repeat units of the general structure

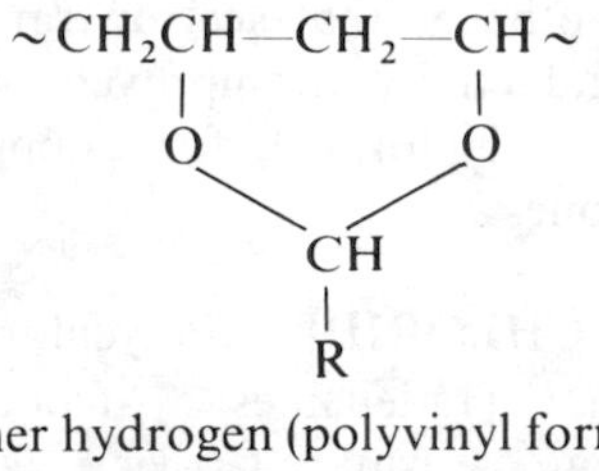

$$\sim CH_2CH - CH_2 - CH \sim$$
$$\underset{O}{|} \qquad \underset{O}{|}$$
$$\underset{CH}{\diagdown \diagup}$$
$$|$$
$$R$$

where R is either hydrogen (polyvinyl formal) or an alkyl group, e.g. CH_3 (polyvinyl acetal itself) or n-butyl (polyvinyl butyral). Such polymers may be produced by reaction of polyvinyl alcohol with the appropriate aldehyde RCHO. Since acetal ring formation involves reaction of neighbouring pairs of hydroxyl groups, statistically some hydroxyls must remain as isolated unreacted groups.

(2) Tradename: Alvar. The particular polyvinyl acetal obtained by reaction of polyvinyl alcohol with acetaldehyde. It has found some use in enamels and lacquers, but less so compared with polyvinyl formal and polyvinyl butyral.

POLYVINYL ACETATE (PVA) (PVAc)

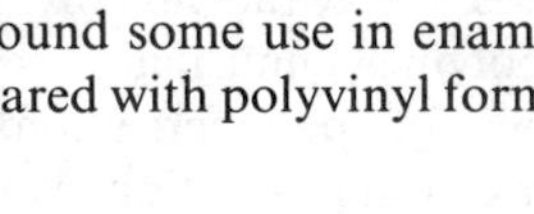

$$\sim\left[\underset{\underset{OOCCH_3}{|}}{CH_2CH}\right]_n$$

Tradenames: Calatac, Elvacet, Emultex, Epok V, Gelva, Mowilith, Texicote V, Texilac, Vandike, Vinalak, Vinavil,

Vinamul, Vinnapas. Produced by free radial polymerisation of vinyl acetate. This is frequently performed by emulsion polymerisation since the product is often used as a latex for surface coatings. Solution, suspension and mass polymerisation, however, are also used. The product of solution polymerisation, especially in methanol as solvent, is useful for conversion to polyvinyl alcohol by alcoholysis. Suspension polymer is also produced for this purpose. Polymerisation is subject to chain transfer to both monomer and to polymer, the latter leading to the characteristic long-chain branching of polymers formed at high conversions.

Polyvinyl acetate varies from a gummy solid at low molecular weight to a leathery material at high molecular weight. Its T_g value is 29°C. The polymer is usually atactic and amorphous and can give clear films, but exhibits considerable cold flow. It tends to decompose to give acetic acid and becomes discoloured at temperatures above about 100°C and cannot satisfactorily be melt processed. It is soluble in alcohols, chlorinated solvents, ketones, esters and aromatic hydrocarbons.

The homopolymers and a variety of copolymers are widely useful commercially as film forming materials in emulsion paints (for which a plasticiser or suitable comonomer may be used to improve film formation and flexibility), in adhesives and as a size and binder in textiles and papers.

POLYVINYL ALCOHOL (PVA) (PVAl)

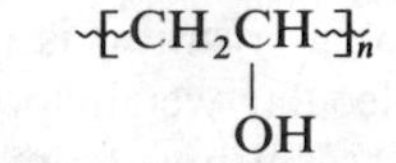

$$\sim\left[\underset{\underset{OH}{|}}{CH_2CH}\right]_n$$

Tradenames: Alcotex, Clarene, Elvanol, Gelvatol, Mowiol, Pevalon, Polyviol, Vinarol, Vinavilol, Vinol, and as films, Cremona, Kanebiyan, Kuralon, Mewlon, Vinal, Vinylon. The monomer vinyl alcohol does not exist, so the polymer is produced by the hydrolysis (or, better, alcoholysis) of polyvinyl acetate, especially of polymer produced by solution polymerisation in methanol as solvent:

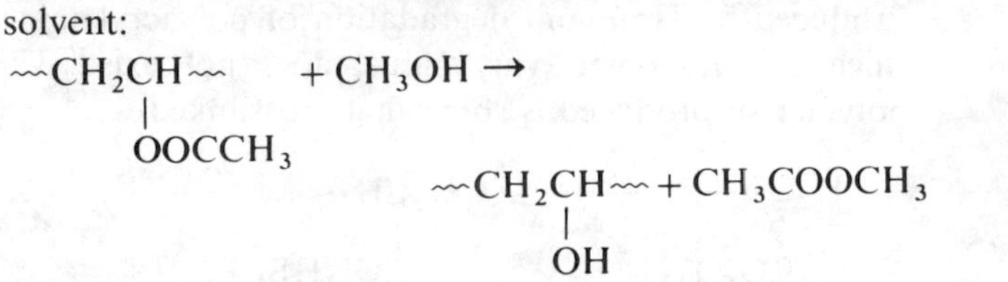

$$\sim CH_2CH \sim \quad + CH_3OH \rightarrow$$
$$\underset{OOCCH_3}{|}$$
$$\sim CH_2CH \sim + CH_3COOCH_3$$
$$\underset{OH}{|}$$

The reaction is usually catalysed by acid or base, e.g. CH_3O^-. Commercial polymers are often about 87% hydrolysed, at which stage they have maximum water solubility, dissolving readily in cold water. Polymers of about 100% hydrolysis are much less water soluble, requiring heating to about 85°C for dissolution. Polymers have good resistance to organic solvents, especially hydrocarbons, unless the solvent is capable of hydrogen bonding as, for example, with formamide and diethylenetriamine. Water and other alcohols, such as glycerol, have a plasticising effect.

Despite being atactic, polyvinyl alcohol can crystallise, especially if highly hydrolysed, since the hydroxyl group can be accommodated in place of a hydrogen atom in the crystal lattice of polyethylene. The T_m value is about 230°C. Stretched fibres can have about 50% crystallinity and high tensile strength. Commercial fibres are frequently insolubilised by crosslinking by reaction with formaldehyde. Film is useful for the production of water-soluble packaging and aqueous solutions are used as paper and textile sizes and as adhesives.

Ethylene–vinyl alcohol copolymers have very low permeability to a wide range of penetrant molecules and are useful in film form as a barrier layer in coextruded film laminates for packaging. Only plasticised homopolymer may be extruded and moulded since the polymer decomposes above about 120°C, eliminating water and producing coloured polyene structures in the polymer chain. The polymer may be converted by reaction with aldehydes or ketones to polyvinyl acetals or polyvinyl ketals respectively.

POLY-(VINYLALKYL ETHER) (Poly-(vinyl ether)) (Vinyl ether polymer) A polymer of general structure

$$\left[CH_2{-}CH \atop {\;\;\;\;|\atop OR} \right]_n$$

where R is an alkyl group. Several examples have found use commercially, e.g. polyvinylmethyl, polyvinylethyl and polyvinylisobutyl ethers. Such polymers are usually prepared by low temperature cationic polymerisation of the vinylalkyl ether monomer with a Friedel–Crafts catalyst. Free radical polymerisation is not very successful, giving only low molecular weight polymers. Crystalline isotactic polymers resulting from low temperature polymerisation have been known for a long time.

POLYVINYLAMINE

$$\left[CH_2CH \atop {\;\;\;\;|\atop NH_2} \right]_n$$

Produced by Hofmann degradation of polyacrylamide, since the monomer, vinylamine, does not exist. The polymer so produced is somewhat crosslinked

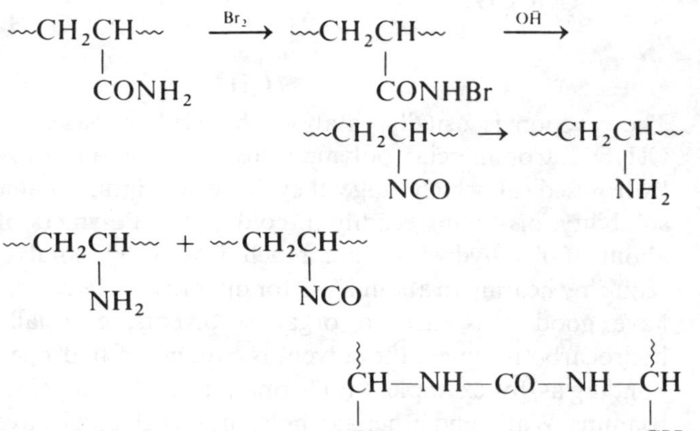

~CH₂CH~ $\xrightarrow{Br_2}$ ~CH₂CH~ $\xrightarrow{OH^-}$
| |
CONH₂   CONHBr

~CH₂CH~ → ~CH₂CH~
| |
NCO   NH₂

~CH₂CH~ + ~CH₂CH~
| |
NH₂   NCO

CH—NH—CO—NH—CH
| |
CH₂   CH₂

POLYVINYL BUTYRAL Tradenames: Butvar, Butacite, Pioloform B. The polyvinyl acetal produced by

reaction of polyvinyl alcohol with n-butyraldehyde. It has repeat units of the type

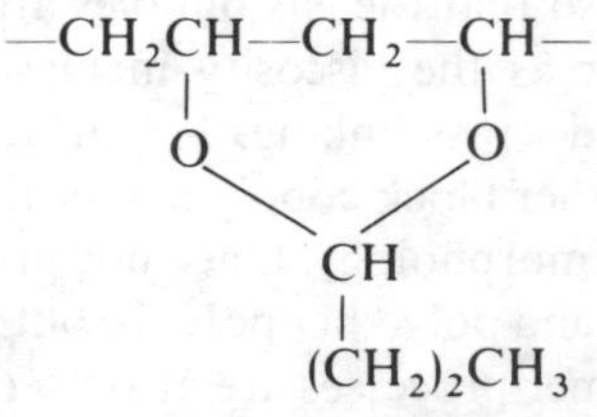

—CH₂CH—CH₂—CH—
| |
O   O
\   /
CH
|
(CH₂)₂CH₃

Commercial polymers contain 10–20% of unreacted hydroxyl groups. It is softer, has a lower T_g value and a lower softening point than polyvinyl formal. Its major use is as the interlayer in laminated safety glass, when the polymer is plasticised with about 30% of, for example, dibutylsebacate plasticiser.

POLY-(N-VINYLCARBAZOLE) (PVK)

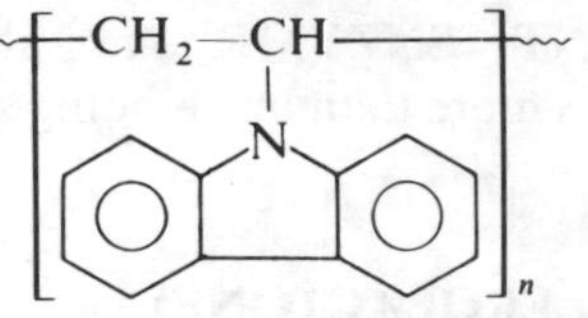

$$\left[CH_2{-}CH \right]_n$$

Tradenames: Luvican, Polecton. Obtained by the free radical polymerisation of N-vinylcarbazole, often by mass polymerisation. The polymer has a T_g value of 208°C, which is unusually high for a vinyl polymer. It is an excellent electrical insulating material, although it is somewhat brittle. It has found use as a dielectric in capacitors. In film form its photoconductive properties are utilised in the xerographic reproduction process. Its high photoconductivity is due to its adopting a helical conformation such that the aromatic groups stack parallel to each other, so that electron transfer is relatively easy. Photoconductivity is improved when a charge-transfer complex is formed, for example with 2,4,7-trinitrofluorenone.

POLYVINYL CHLORIDE (Polychloroethene) (PVC)

$[CH_2CHCl]_n$. Tradenames: Afcodur, Afcoplast, Afcovyl, Breon, Carina, Corvic, Dacon, Ekavyl, Gedevyl, Hostalit, Lacqvyl, Ladene, Lekavyl, Lonzalit, Lonzavyl, Lucolene, Lucorex, Lucovyl, Mavlan, Marvylan, Mipolam, Norvyl, Opalon, Quervil, Ravinil, Rhodopas, Rhovylite, Scon, Sicron, Solvic, Trosiplast, Ultryl, Varlan, Vestolit, Vinatex, Vinoflex, Vinylite, Vinnol, Vipla, Viplast, Viplavil, Vybak (plastics materials) and Fibravyl, Isovyl, Khlorin, Leavil, PCU, Retractyl, Rhovyl, Thermovyl (fibres). A major thermoplastic material. Produced by free radical polymerisation of vinyl chloride. The polymer, unusually, is insoluble in its monomer and therefore polymerisation is heterogeneous, except when carried out in solution, which is rarely done. Thus auto acceleration occurs from the start and the kinetics are non-standard. In addition chain transfer, mostly to monomer, is important and polymer molecular weight is independent of initiator concentration and is controlled by polymerisation temperature. Organic peroxides and azo compounds are used for initiation, but to achieve high molecular weight, low polymerisation temperatures and especially active initiators are needed. Thus peroxides such as *t*-butylperpivalate and peroxydicarbonates are used.

The polymer is produced by mass, suspension and emulsion polymerisation techniques. Much of the latter type of polymer, of fine particle size of about 1 μm, is used for producing pastes (mostly plastisols) and is called paste polymer (or resin). Suspension and mass polymer, with porous particles of about 100 μm in size, readily absorbs plasticiser to give a dry blend and is known as easy processing (or EP) resin. Typical polymer molecular weights for commercial polymers are $\bar{M}_n = 30\,000$ to 70\,000. Technologically, molecular weight is assessed by the dilute solution viscosity as either a *K*-value or as the ISO viscosity number.

Although essentially atactic, the polymer contains syndiotactic sequences which result in some short-range ordering of the molecules, corresponding to about 10% 'crystallinity'. The 'crystalline' regions only melt at about 250°C and persist during normal melt processing. This may account for the very high pseudoplasticity of the melt and the difficulty of fusing the melt during processing. The T_g value is about 80°C.

The polymer is very sensitive to thermal degradation which occurs above 100°C by dehydrochlorination, leading to the formation of conjugated polyene sequences in the polymer chains:

$$+CH_2CHCl+_x \rightarrow +CH=CH+_x + x\,HCl$$

If $x > 5$ the polyenes absorb light and the polymer is coloured. Normally $x = 1$–30 so the polymer becomes very discoloured (white $\rightarrow$ pink $\rightarrow$ brown $\rightarrow$ black) at the processing temperatures used of > 150°C. In addition, the liberated hydrochloric acid is corrosive to metal processing machinery. Therefore the polymer always contains a thermal stabiliser for processing. A very wide range of materials are used for this purpose, including basic lead compounds, metal soaps and organotin compounds at levels from 0·2 to 8%. Lubricants are also required to reduce friction during processing.

Unmodified polymer is a hard, stiff material, typically with a tensile modulus of about 2·5 GPa, a tensile strength of about 50 MPa and with only a moderate impact strength, e.g. an Izod impact strength of 0·7–2·0 J $(12\cdot7\,mm)^{-1}$. The latter is highly dependent on the adequacy of fusion during processing and may be considerably improved, e.g. to 10 J $(12\cdot7\,mm)^{-1}$, by the use of an impact modifier, such as 10–20% of a rubbery EVA, MBS or ABS. However, mechanical behaviour, notably flexibility, can be vastly modified by plasticisation.

Plasticisers are more frequently used in polyvinyl chloride than in any other polymer, usually at the 30–100 phr level. Plasticised, or flexible PVC (PPVC) is to be distinguished from the unplasticised, or rigid, PVC (UPVC). Unusually, small amounts of plasticiser cause a stiffening and embrittlement of the polymer, whereas when more than about 120 phr is used the material is too soft and weak for most uses. Thus typical amounts used are 50–70 phr, when mechanically the PVC is like a stiff rubber with a tensile modulus of about 0·015 GPa and a tensile strength of about 15 MPa. Technologically, flexibility is often measured and expressed by a softness measurement, such as the British Standard Softness number.

Being atactic PVC is intrinsically transparent. Its electrical properties are relatively poor compared with most other plastics, with a volume resistivity of about $10^{13}\,\Omega$ cm and a power factor of about 0·01–0·15. Nevertheless, its properties are good enough for use as a low frequency insulation material and it is widely used as such due to the wide range of flexibilities obtainable by plasticisation. It shows an outstanding resistance to corrosive chemicals, particularly to strong acids, bases and oxidising agents. It is unaffected by some solvents (especially alcohols and aliphatic hydrocarbons) and is only slightly swollen by aromatic hydrocarbons, esters and ketones. It is truly dissolved by relatively few solvents—cyclohexanone, tetrahydrofuran and ethylene dichloride being notable examples. For an organic polymer its resistance to burning is very good, with a limiting oxygen index of about 45%, but this is considerably reduced by plasticisation.

In use, in addition to the incorporation of stabiliser and lubricant (which are always needed for processing purposes) plasticiser and (in UPVC) an impact modifier, fillers are also frequently incorporated. Typically 15–30 phr, often of calcium carbonate or clay, may be used in order to keep materials costs low rather than for the more usual reason of reinforcement. Other additives sometimes used include fire retardants, process aids and blowing agents (for cellular PVC).

Thus due to its all-round reasonably good properties and to the exceptionally wide range of additives that may be employed to considerably modify these properties PVC may be regarded as the most widely used plastic material. Important applications include extruded pipe, tube, wire and cable coverings, calendered film and sheet, and blown bottles. In addition, several important products are produced by processes using the PVC in liquid, rather than melt, form as a paste (usually a plastisol), e.g. playballs, leathercloth and other coated substrates. Modified PVCs are also of commercial importance. These include a wide range of copolymers, especially with vinyl acetate, vinylidene chloride or acrylonitrile. Chlorinated PVC is also of commercial significance.

POLY-(VINYL CHLORIDE-co-ACRYLONITRILE)
Alternative name for vinyl chloride–acrylonitrile copolymer.

POLY-(VINYL CHLORIDE-co-PROPYLENE) Alternative name for vinyl chloride–propylene copolymer.

POLY-(VINYL CHLORIDE-co-VINYL ACETATE)
Alternative name for vinyl chloride–vinyl acetate copolymer.

POLYVINYL CHLOROACETATE

$$\sim\!\!\left[\!\!\begin{array}{c} CH_2CH \\ | \\ OOCCH_2Cl \end{array}\!\!\right]_n\!\!\sim$$

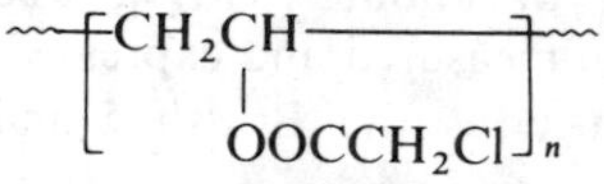

Tradename: Viplavil. Produced by free radical polymerisation of vinyl chloroacetate. Its T_g value is about 35°C and it was at one time used as a lacquer resin.

POLYVINYL CINNAMATE

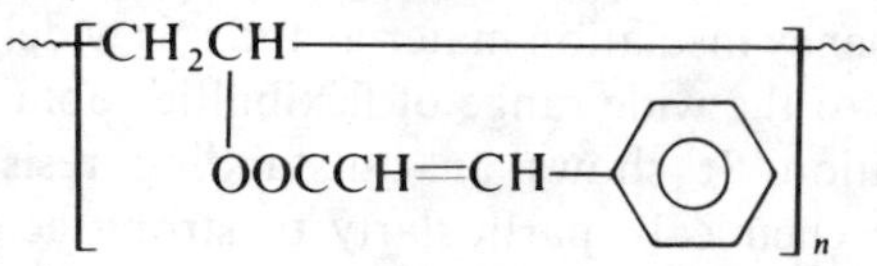

$$\left[\!\!\begin{array}{c} CH_2CH \\ | \\ OOCCH=\!CH-\!\bigcirc \end{array}\!\!\right]_n$$

Polymerisation of vinyl cinnamate yields intractable crosslinked polymers, so polymers containing vinyl cinnamate units are prepared by partial esterification of polyvinyl alcohol by reaction with cinnamyl chloride,

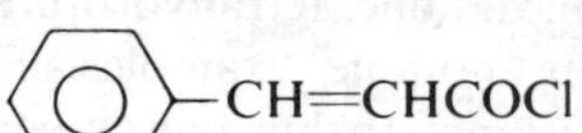

$$\bigcirc\!\!-\!CH=\!CHCOCl$$

in alkaline solution (Schotten–Baumann method). Solutions of the resultant polymer, e.g in dichloromethane, are then used to form a photosensitive coating on a substrate, as the basis of a photoresist. On exposure to ultraviolet light the polymer becomes crosslinked and hence insolubilised. With a photosensitiser, crosslinking may also be induced by visible light. Crosslinked structures include cyclobutane rings:

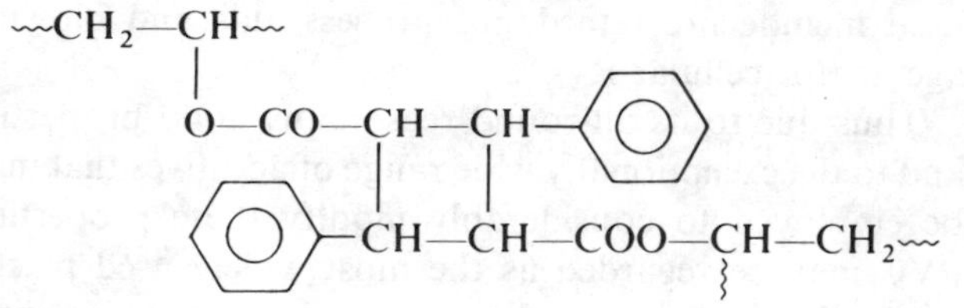

$$\sim\!CH_2-CH\!\sim$$
$$|$$
$$O-CO-CH-CH-\bigcirc$$
$$\bigcirc-CH-CH-COO-CH-CH_2\!\sim$$

POLYVINYLCYCLOHEXANE

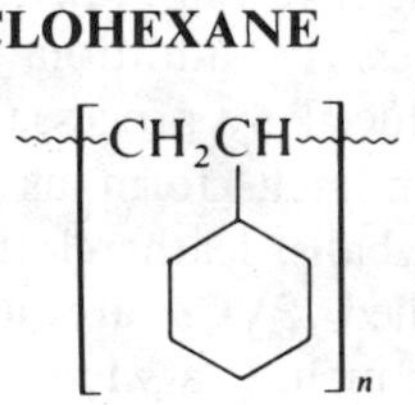

$$\sim\!\!\left[\!\!\begin{array}{c} CH_2CH \\ | \\ \bigcirc \end{array}\!\!\right]_n$$

Prepared by Ziegler–Natta polymerisation of vinylcyclohexane in isotactic form. It crystallises in a 4_1-helical conformation with a T_m value of 385°C and a T_g value of about 80°C. Its density is about $0.95\,g\,cm^{-3}$.

POLY-(2-VINYLDIBENZOFURAN)

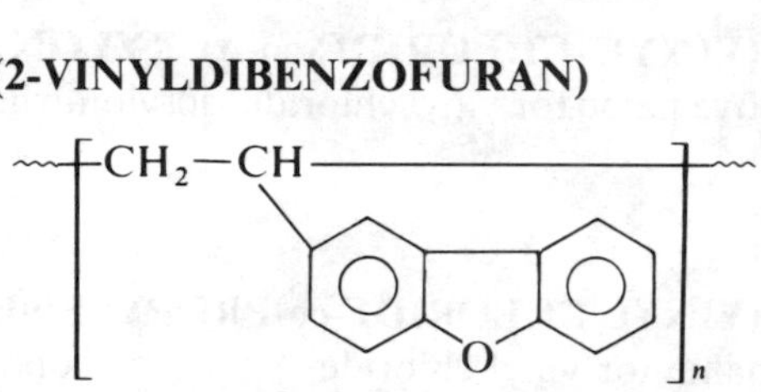

$$\sim\!\!\left[\!\!\begin{array}{c} CH_2-CH \\ | \\ \end{array}\!\!\right]_n$$

Tradename: Lectrofilm. Prepared by free radical or thermal polymerisation of 2-vinyldibenzofuran. It has a high softening point but is rather brittle and at one time was used as a capacitor dielectric.

POLYVINYL ESTER
A polymer of general structure

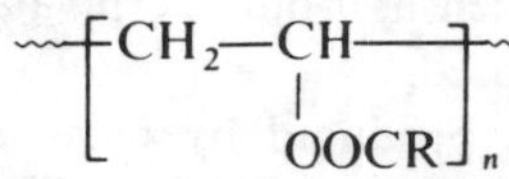

$$\sim\!\!\left[\!\!\begin{array}{c} CH_2-CH \\ | \\ OOCR \end{array}\!\!\right]_n\!\!\sim$$

where R is an alkyl group, i.e. an ester of polyvinyl alcohol. The most important polymer is polyvinyl acetate (where R is CH_3). Other polymers include polyvinyl chloroacetate, polyvinyl formate, polyvinyl propionate and polyvinyl stearate.

POLY-(VINYL ETHER)
Alternative name for poly-(vinylalkyl ether).

POLY-(VINYLETHYL ETHER)

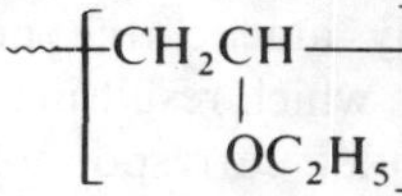

$$\left[\!\!\begin{array}{c} CH_2CH \\ | \\ OC_2H_5 \end{array}\!\!\right]_n$$

Tradenames: Lutonal, Lutonal A. The viscous liquid or rubbery polymer obtained by the cationic polymerisation of vinylethyl ether using a Friedel–Crafts catalyst. It has a T_g value of -42°C and is of interest as a pressure sensitive adhesive, especially due to its solubility in a wide range of solvents.

POLYVINYL FLUORIDE
(PVF) $\{CH_2CHF\}_n$. Tradename: Tedlar. Prepared by high pressure (1000atm) free radical polymerisation of vinyl fluoride in an aqueous system. Unlike the closely related polyvinyl chloride, it crystallises, even though atactic, due to the small size of the fluorine atom. It has a T_m value of 200°C and a T_g value of -20°C. It has better heat resistance than polyvinyl chloride but does eliminate hydrogen fluoride at fairly low temperatures. It has exceptional weathering resistance, which has led to its use as a glazing film for outdoor use.

POLYVINYL FORMAL
Tradenames: Formvar, Pioloform F, Redux. The polyvinyl acetal with repeating units

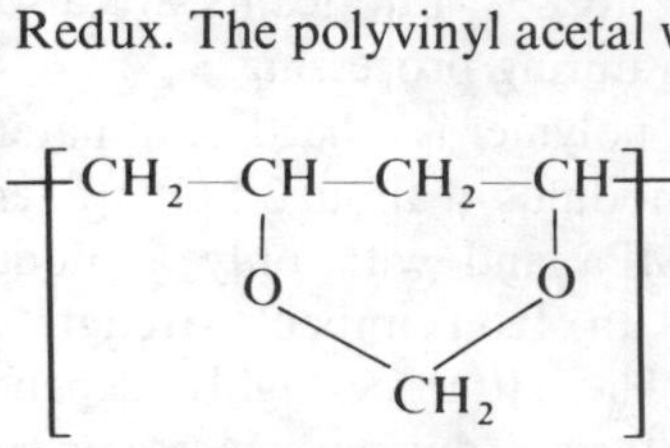

$$\left[\!\!\begin{array}{c} CH_2-CH-CH_2-CH \\ | \quad\quad | \\ O \quad\quad O \\ \backslash \quad / \\ CH_2 \end{array}\!\!\right]$$

produced by reaction of polyvinyl alcohol with acidified aqueous formaldehyde at about 70°C. Commercial polymers can contain up to about 50% of residual acetate groups, which increases solubility in, for example, ketones and esters. The polymer is useful as an electrical insulating wire enamel in conjunction with a resole.

POLYVINYL FORMATE

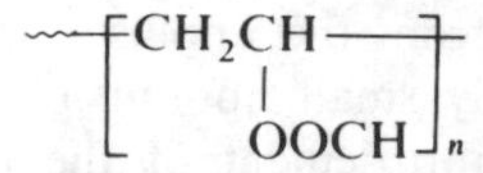

$$\sim\!\!\left[\!\!\begin{array}{c} CH_2CH \\ | \\ OOCH \end{array}\!\!\right]_n$$

Produced by free radical polymerisation of vinyl formate. The polymers have a lower head-to-head content than polyvinyl acetate and therefore on alcoholysis or hydrolysis produce a more stereoregular polyvinyl alcohol. The polymer has therefore been of some interest for this purpose. It has a T_g value of about 35°C.

POLYVINYLIDENE CHLORIDE (PVDC)

$$-\!\!\left[CH_2CCl_2\right]_{\!n}$$

Tradenames: Ixan, Saran, Viclan (plastics), Krehalon, Permalon, Tygan, Velan (fibres). The polymer is readily produced by free radical polymerisation of vinylidene chloride by mass, suspension, emulsion or solution polymerisation. Commercially most copolymers are produced by suspension polymerisation. The homopolymer readily crystallises with a T_m value of about 220°C. It is highly insoluble and is impermeable to gases and has a high density of about $1.9\,\text{g cm}^{-3}$. Commercially, vinylidene chloride–vinyl chloride and vinylidene chloride–acrylonitrile copolymers are produced, containing about 15% of the comonomer, since the homopolymer is difficult to melt process owing to its ready dehydrochlorination, like polyvinyl chloride, at processing temperatures:

$$-\!\!\left[CH_2CCl_2\right]_{\!x} \rightarrow -\!\!\left[CH\!=\!CCl\right]_{\!x} + x\,HCl$$

The copolymers are useful as monofilaments and as moisture and gas barrier films and coatings.

POLY-(VINYLIDENE CHLORIDE-co-ACRYLONITRILE) Alternative name for vinylidene chloride–acrylonitrile copolymer.

POLY-(VINYLIDENE CHLORIDE-co-VINYL CHLORIDE) Alternative name for vinylidene chloride–vinyl chloride copolymer.

POLYVINYLIDENE CYANIDE $-\!\!\left[CH_2C(CN)_2\right]_{\!n}$.
Very readily formed by polymerisation of vinylidene cyanide by either free radical or anionic polymerisation, the latter being catalysed by even very mild bases such as water, alcohols, ketones or amines. Fibres may be spun from, for example, dimethylformamide solution. However, the commercial fibres that have been produced are copolymers with vinyl acetate.

POLYVINYLIDENE FLUORIDE (PVDF)

$$-\!\!\left[CH_2CF_2\right]_{\!n}$$

Tradenames: Dyflor, Foraflon, Kynar, Solef. Prepared by high pressure free radical polymerisation in aqueous systems. It readily crystallises having a T_m value of 171°C and a T_g value of −35°C. It has moderate tensile and impact strengths and good chemical and solvent resistance. It has unusual electrical properties, being a piezoelectric polymer as a stretched film and also being pyroelectric. Its applications make use of these specialised properties as in loudspeakers and pyroelectric detectors. Rubbery copolymers, especially with hexafluoropropylene, are important fluoroelastomers.

POLY-(VINYLIDENE FLUORIDE-co-CHLOROTRIFLUOROETHYLENE) Alternative name for vinylidene fluoride–chlorotrifluoroethylene copolymer.

POLY(VINYLIDENE FLUORIDE-co-HEXAFLUOROPROPYLENE) Alternative name for vinylidene fluoride–hexafluoropropylene copolymer.

POLY-(VINYLIDENE FLUORIDE-co-1-HYDROPENTAFLUOROPROPYLENE) Alternative name for vinylidene fluoride–1-hydropentafluoropropylene copolymer.

POLY-(VINYLISOBUTYL ETHER) Alternative name for poly-(isobutylvinyl ether).

POLYVINYL ISOCYANATE

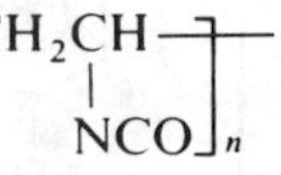

$$\left[\begin{array}{c}CH_2CH\\ \ \ \ |\\ NCO\end{array}\right]_{\!n}$$

Produced by free radical polymerisation using ultraviolet light or peroxide initiation, of vinyl isocyanate. The polymer does not soften on heating but darkens at about 300°C. When heated with ethanol some urethane groups are formed. When heated with amines or ammonia the polymer becomes water soluble, possibly by formation of urea groups.

POLYVINYL KETAL A polymer containing a substantial proportion of repeating units of the type:

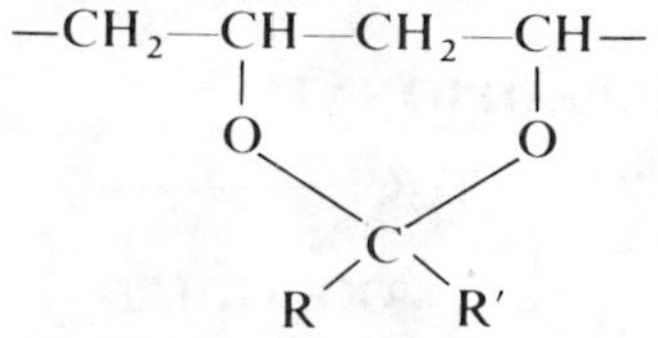

$$\begin{array}{c}-CH_2-CH-CH_2-CH-\\ \quad\ \ |\qquad\qquad |\\ \quad\ \ O\qquad\qquad O\\ \quad\ \ \ \diagdown\ \ C\ \ \diagup\\ \quad\ \ \ R\ \ \ \ \ R'\end{array}$$

Produced by reaction of polyvinyl alcohol with a ketone

$$\left(\begin{array}{c}R\\R'\end{array}\!\!>\!CO\right)$$

under acidic conditions. Such polymers have not proved as useful as the similar polyvinyl acetals.

POLY-(VINYLMETHYL ETHER) Alternative name for poly-(methylvinyl ether).

POLY-(VINYLMETHYL ETHER-co-MALEIC ANHYDRIDE) Alternative name for methylvinyl ether–maleic anhydride copolymer.

POLY-(1-VINYLNAPHTHALENE)

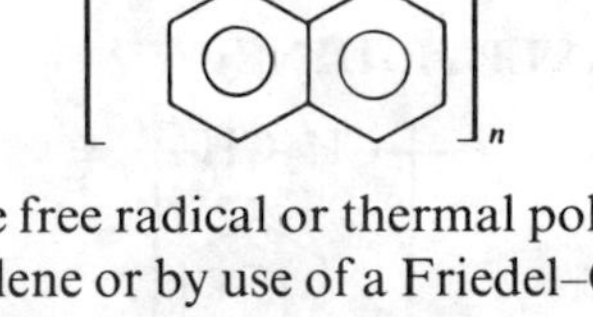

Obtained by the free radical or thermal polymerisation of 1-vinylnaphthalene or by use of a Friedel–Crafts catalyst. The polymer has a T_g value of 159°C and is transparent and brittle.

POLY-(2-VINYLNAPHTHALENE)

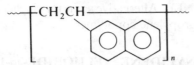

Obtained by the free radical or thermal polymerisation of 2-vinylnaphthalene. The polymer has a T_g value of 151°C and is similar to polystyrene in properties, i.e. it is transparent and brittle.

POLY-(p-VINYLPHENOL)

Tradename: Resin M. Produced by thermal polymerisation of vinylphenol, itself produced by dehydrogenation of p-ethylphenol. Polymers of low molecular weight (about 4000) are useful as hardeners for epoxy resins, giving cured laminates with superior thermal and dimensional stability to conventionally cured laminates. The laminates are useful for printed circuit board manufacture, since they will withstand soldering temperatures of up to 370°C for short periods of time. A brominated resin, Resin MB, is also produced.

POLYVINYL PROPIONATE

Tradename: Propiofan. Produced by free radical polymerisation of vinyl propionate. Its T_g value is about 10°C and it has found use as an emulsion for coatings with relatively high alkali resistance.

POLY-(2-VINYLPYRIDINE)

Obtained by the free radical polymerisation of 2-vinylpyridine. The polymer has a T_g value of 104°C. Although the homopolymer has been used as a photographic film anti-halation backing due to its rapid solubility in dilute aqueous acid, the copolymers, e.g. with acrylonitrile and in styrene–butadiene rubber are more important.

POLY-(4-VINYLPYRIDINE)

Obtained by the free radical polymerisation of 4-vinylpyridine. The polymer has a T_g value of 142°C and has been used as a basis for the preparation of cationic polyelectrolytes by reaction with alkyl halides to yield polyquaternary ammonium halides. It forms donor–acceptor complexes with tetracyanoquinodimethane and other electron acceptors which have semiconducting properties.

POLY-(N-VINYLPYRROLIDONE) (PVP)

Tradenames: Albigen A, Kollidon, Periston, Plasdene, Polyclar. A water soluble polymer prepared by the free radical polymerisation of N-vinylpyrrolidone in aqueous solution with, for example, hydrogen peroxide initiator. The polymer has a T_g value of about 85°C and is very readily soluble in water. Aqueous solutions have been used as blood plasma substitute. The polymer has a great ability to associate, and to form 'complexes', with a wide variety of organic materials and it has many specialised applications as a result of this property, e.g. as a dyestuff stripper or to improve dyeability, in cosmetics and in hair lacquers.

POLYVINYL STEARATE

Produced by free radical polymerisation of vinyl stearate. It is a waxy polymer of low T_g value (about −50°C) which has been used as an oil viscosity modifier and for its lubricating properties.

POLYVINYL SULPHONIC ACID   (Poly-(ethylenesulphonic acid)) (Poly-(ethenesulphonic acid))

Prepared by free radical polymerisation of vinyl sulphonic acid. The polymer and its salts are water soluble and their possible applications utilise their polyelectrolyte nature. The sodium salt is useful as a blood anticoagulant.

POLYVINYLTOLUENE (Polymethylstyrene)

The vinyl toluenes may be polymerised to clear glassy polymers by similar means as used for styrene polymerisation, for the ortho, meta or para isomers. Commercial vinyltoluene is usually a 60:40 mixture of the meta and para isomers. It is used as a comonomer in the crosslinking of unsaturated polyesters and in styrenated alkyd paints and varnishes. Little use has been found for the homopolymers, which have somewhat lower T_g values than polystyrene.

POLYVIOL Tradename for polyvinyl alcohol.

POLY-(2,6-XYLENOL) Alternative name for poly-(2,6-dimethyl-1,4-phenylene oxide).

POLY-(1,4′-β-D-XYLOPYRANOSE)

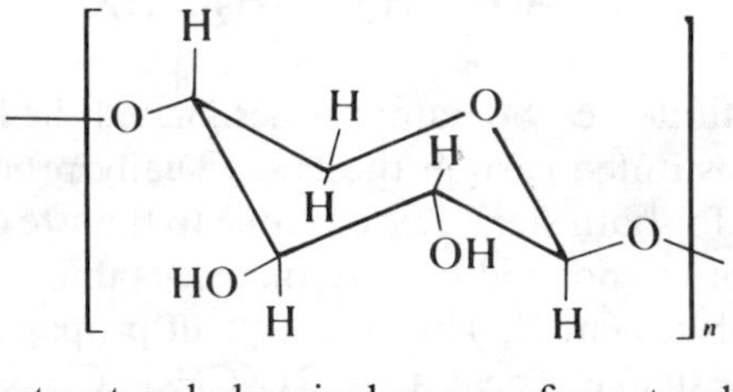

The full structural chemical name of most xylans.

POLYXYLYLENE A polymer of the type

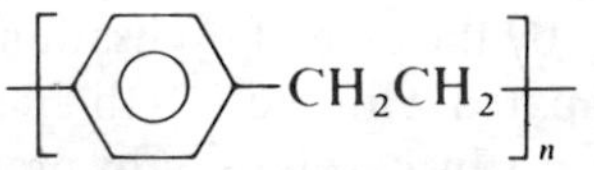

Most interest is with the *para* isomer, since the *meta* and *ortho* isomers have low T_m values. Many polymers substituted in either the ring or alkyl group, e.g. —CF_2—CF_2— linked, have been synthesised, but the unsubstituted polymer (poly-*p*-xylylene) is the best known. This is best synthesised by pyrolysis of 1000°C of *p*-xylene. This produces the monomer, *p*-xylylene,

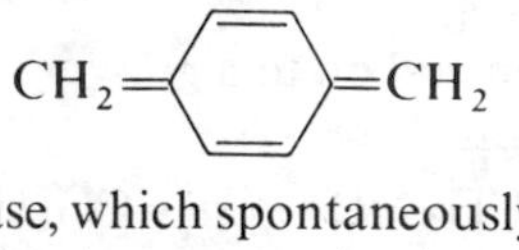

in the vapour phase, which spontaneously polymerises by a free radical mechanism on condensing on a cool surface, forming a coating. Pyrolysis of the cyclic dimer of *p*-xylene, 2,2′-*para*-cyclophane, at about 600°C is favoured in the production of the commercial polymer (Parylene N). This polymer is crystalline with a T_m value of >400°C, a T_g value of about 65°C, and is highly insoluble. It can only be used in air to about 100°C due to low thermo-oxidative stability, but is good in inert atmospheres. It is useful as an electrical insulation film, especially in compact capacitors. A ring substituted chloro-polymer (Parylene C), with a T_m value of 280–300°C and a T_g value of 80–100°C is also produced and may be used to higher temperatures (about 20°C higher).

POLY-(m-XYLYLENEADIPAMIDE) (Nylon MXD6)

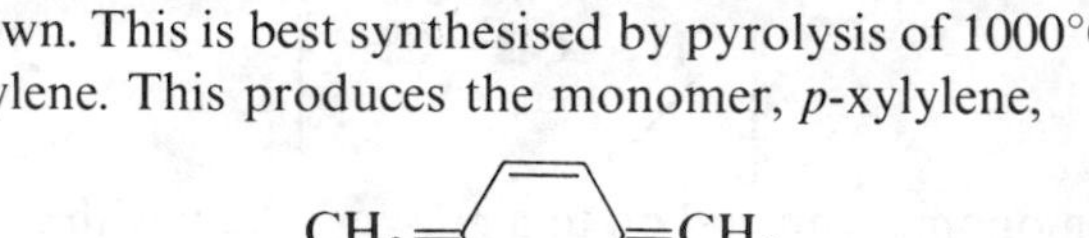

Tradename: Ixef. Synthesised by melt polymerisation of *m*-xylylenediamine with adipic acid. Has a T_m value of 243°C and a T_g value of about 90°C. Although its property values (e.g. modulus of 50 g denier^{-1} and yield

stress of 110 MPa in fibres) are considerably higher than for nylon 66, its water sensitivity has precluded a wider usage. It can be readily obtained in either an amorphous (and clear) or crystalline (and opaque) form depending on cooling conditions from the melt. It is useful as a high stiffness (tensile modulus 12 GPa), high softening point (HDT under 1·8 MPa load, 230°C) engineering plastic.

POLY-(p-XYLYLIDENE)

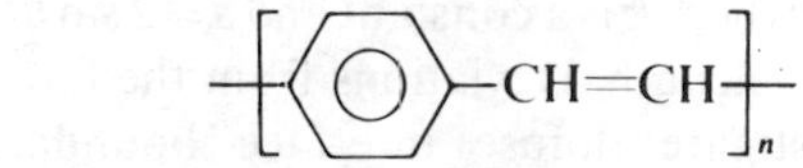

Only low molecular weight polymers, e.g. decamers, have been synthesised, e.g. by treatment of α,α′-dichloro-*p*-xylene with sodium in liquid ammonia. These polymers have T_m values >400°C are intensely yellow and show a thermal stability in air to about 400°C by TGA. They are insoluble in all solvents. The polycyanoterephthalylidene polymers are closely related.

POLY-(p-XYLYLYNE) Alternative name for poly-ethynylbenzene.

POLYZOTE Tradename for expanded polystyrene.

POM Abbreviation for polyoxymethylene.

PON POLYMER The highly insoluble and inert material obtained by heating the polyphosphinate

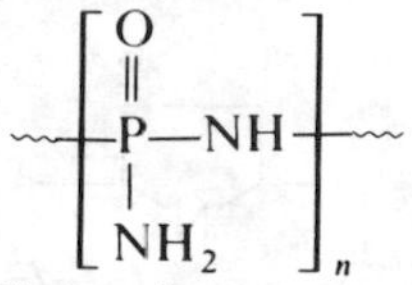

to about 600°C. The polymer possibly contains

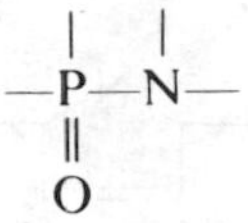

and

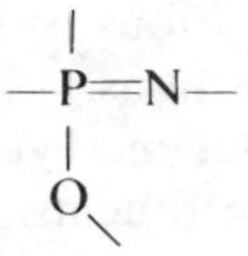

crosslinks.

PONTIANAK A semi-fossil copal natural resin, which is compatible with a wide variety of other varnish ingredients, especially after thermal processing. It softens at 90–135°C.

POPCORN POLYMERISATION (Proliferous polymerisation) The separation during a free radical polymerisation of small, opaque polymer nodules (or popcorns) which, once formed, proliferate rapidly to yield a crosslinked insoluble product of larger volume than the original monomer. Once a crosslinked nucleus has formed, local termination is very low due to the low mobility of the radicals in the popcorn, so that the polymerisation rate is high. Usually confined to the polymerisation of diene monomers except, notably, with methyl acrylate

POROD–KRATKY CHAIN Alternative name for worm-like chain.

POROD'S LAW An expression for the scattered intensities of radiation, $I(s)$, at the high angle (θ) end ($2\theta > 1$–$2°$) of a small angle X-ray or neutron scattering curve. It is, $I(s) = k/s^{-4}$, where k is a constant and $s = 2\sin\theta/\lambda$, with λ being the wavelength. Deviations from the law can be used to investigate diffuse interface boundaries or electron density variations between phases in two phase materials.

POROUS DISC METHOD A method, based on vapour pressure lowering, for the determination of the number average molecular weight of polymers of low molecular weight. Sufficient pressure is applied to a polymer solution to prevent isothermal distillation of solvent from a reservoir to the solution. The reservoir is a column of solvent contained in a tube, closed at the top by a funnel containing the porous disc. This is surrounded by the solution in a closed cell.

PORPHIN The basic ring compound from which porphyrins can be considered to be derived. It consists of four pyrrole rings joined through —CH= groups to form the larger ring:

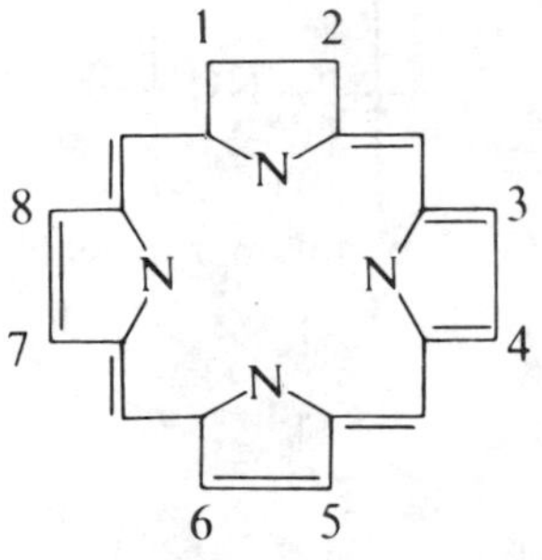

In the porphyrins various substituents replace the hydrogen atoms at positions 1–8. The important protoporphyrin contains four methyl groups, two vinyl and two propionic acid substituents.

PORPHYRAN A sulphated polysaccharide obtained from certain red seaweeds, similar in structure to agarose, but having some D-galactose units present as the 6-methyl ether and the L-galactose units present as both the 6-sulphate and 3,6-anhydride.

PORPHYRIN The nitrogen ring structure which, together with an iron atom to which it is coordinated through four nitrogen atoms, forms haem—the prosthetic group of haemoproteins. A derivative of porphin in which the hydrogens on the four pyrrole rings have been substituted. The important porphyrin found in the haem of most haemoproteins is the protoporphyrin (with four methyl, two vinyl and two propionic acid substituents) isomer, protoporphyrin IX.

POSITIONAL ISOMERISM A type of isomerism which arises through the two possible ways in which a monomer unit in a vinyl polymerisation may add to a growing active centre to produce a head-to-head (h–h) or head-to-tail (h–t) link:

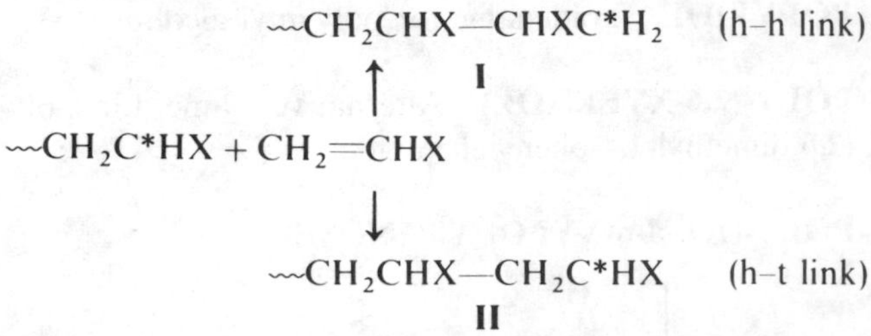

$$\sim CH_2CHX—CHXC^*H_2 \quad \text{(h–h link)}$$
$$\mathbf{I}$$
$$\uparrow$$
$$\sim CH_2C^*HX + CH_2{=}CHX$$
$$\downarrow$$
$$\sim CH_2CHX—CH_2C^*HX \quad \text{(h–t link)}$$
$$\mathbf{II}$$

The substituted carbon atom is designated the head and the unsubstituted carbon the tail. Usually product **II** is favoured for both steric reasons (due to the size of X) and because of the possibility of resonance stabilisation of **II** by the substituent X. Thus subsequent propagation of **I** would most likely give a —CHXCH$_2$—CH$_2$C*HX or tail-to-tail (t–t) link. Typically only about 1% of h–h/t–t links are formed.

Sometimes the extent of h–h links may be determined chemically, e.g. by the ease of cyclisation of the 1,3- (h–t) structures compared with the 1,2- (h–h) structures, or by fragmentation, e.g. by pyrolysis or by preferential scission of the one type of unit over the other. A classic case of the last type is the ready cleavage by oxidation with HIO_4 of the 1,2-glycol h–h structures in polyvinyl alcohol, whereas the h–t 1,3-structures are stable. In certain ring-opening polymerisations, such as of propylene oxide

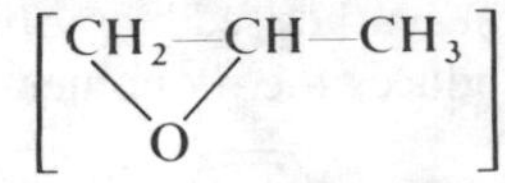

$$\left[\begin{array}{c} CH_2{-}CH{-}CH_3 \\ \diagdown\,\diagup \\ O \end{array}\right]$$

the monomer can add on to a growing chain with a sense of direction, i.e. as

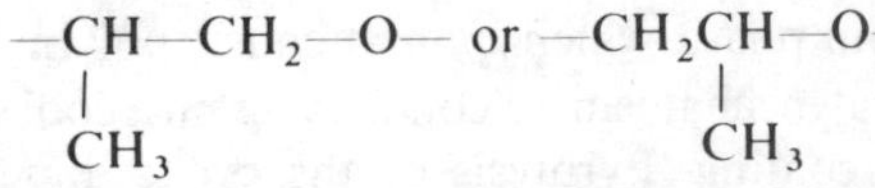

$$-CH{-}CH_2{-}O— \quad \text{or} \quad -CH_2CH{-}O—$$
$$\quad | \qquad\qquad\qquad\qquad | $$
$$\quad CH_3 \qquad\qquad\qquad\qquad CH_3$$

thus giving rise to four different types of dyad. In some fluoropolymers, such as polyvinyl fluoride $\{CH_2CHF\}_n$ and polyvinylidene fluoride $\{CH_2CF_2\}_n$, due to a lack of steric and resonance effects, an unusually high h–h/t–t linking occurs, as shown by ^{19}F and ^{13}C NMR spectroscopy. Some copolymers contain larger than usual amounts of h–h/t–t links, as in vinyl chloride–vinylidene chloride and ethylene–propylene copolymers. Positional isomerism is also possible as a result of the 1,4-polymerisation of substituted dienes ($CH_2{=}CH—CX{=}CH_2$), but not with butadiene itself.

POSITIVE COOPERATIVITY The increase in affinity for binding a second molecule caused by the binding of the first molecule. It occurs with allosteric enzymes when binding of one molecule of substrate or effector increases the affinity for a second substrate or effector.

POSITIVE SPHERULITE A spherulite in which the refractive index is greater across the polymer chains than along them. Owing to the tangential molecular orientation, the greater refractive index is also radial.

POSITRON ANNIHILATION TECHNIQUE (PAT)
A method for the study of polymer structure in which positrons (positive electrons) from a source, such as ^{23}Na, undergo annihilation by interactions with electrons in the polymer with emission of photons as γ-particles. Study of the energy distribution of the photons can give structural information, as can the angular distribution and study of positron lifetimes.

POST EFFECT The continuance of grafting reactions in radiation induced graft copolymerisation after irradiation has ceased. This is due to free radical chain ends becoming occluded and hence having long lifetimes and is favoured by increased crystallinity and low temperatures.

POST VULCANISATION Continued formation of sulphur crosslinks in sulphur vulcanised rubbers during storage subsequent to vulcanisation at ambient temperatures. It occurs especially when basic accelerators are used which promote the formation of polysulphide crosslinks during vulcanisation. The sulphur from these redistributes itself on storage, to give a larger number of short crosslinks. This results in an increase in stiffness and a loss in rubber elasticity.

POTATO VIRUS X An early studied virus with a rod-shaped virion, like TMV, of size 540×10 nm and of estimated molecular weight of 35×10^6. The geometric arrangement is also similar to TMV, consisting of a single RNA molecule forming a helical core and surrounded by a coat of 650 protein subunits each consisting of 463 amino acid units having a molecular weight of about 52 000. Unlike TMV its disaggregated subunits do not reaggregate.

POWELL–EYRING MODEL A model for non-Newtonian flow behaviour, but which predicts Newtonian behaviour in both the limiting cases of very high and very low shear rates. In practice this behaviour is often found for polymer systems, but the upper Newtonian viscosity (η_∞) does require much higher shear rates than are usually generated in polymer processing. One form of the equation is, $\eta = \mu_0[\text{arcsinh}\,(\mu_0\dot{\gamma}/B)/(\mu_0\dot{\gamma}/B)]$, where η is the apparent viscosity, $\dot{\gamma}$ is the shear rate, μ_0 is the zero shear rate Newtonian viscosity and B is a material constant.

POWER FACTOR The cosine of the phase angle (ϕ) when a dielectric is subject to a sinusoidally varying field. For a perfect (no loss) dielectric, $\phi = \pi/2$, so that $\cos\phi = 0$. However, when dielectric loss occurs, $\phi = (\pi/2 - \delta)$, where δ is the loss angle, and hence the power factor equals $\sin\delta$. It is the ratio of the power loss to the product of the root mean square voltage applied and the root mean square current passing through the material. In practice the loss tangent ($\tan\delta$) is frequently referred to as the power factor. However, any confusion is usually not important since for low δ, $\sin\delta$ approximately equals $\tan\delta$.

POWER LAW EQUATION (Ostwald–de Waele equation) An empirical constitutive equation which describes the non-Newtonian flow behaviour of fluids for which the viscosity (an apparent viscosity, symbol η) is a function of the rate of shear ($\dot{\gamma}$). The equation is, $\tau = K(\dot{\gamma})^n$, where τ is the shear stress and K and n are the power law indices, called the consistency index and the flow behaviour index respectively. In logarithmic form the equation is, $\log\tau = \log K + n\log\dot{\gamma}$ and a log–log flow curve plot is linear. Since $\eta = \tau/\dot{\gamma}$ we also have $\eta = K(\dot{\gamma})^{n-1}$. Thus if $n = 1$ and $K = \mu$, the equation represents Newtonian behaviour, whilst if $n > 1$ shear thickening results and if $n < 1$ shear thinning results. For most polymer melts at usual shear rates, power law behaviour is observed at least over a certain, about two decades, shear rate range. It has therefore been very widely used, in place of Newton's law, in the analysis of polymer melt flow behaviour in many polymer processing operations and in melt rheometry. However, the equation is mathematically inconvenient since the units of K depend on the value of n. To overcome this difficulty, K may be eliminated by relating η to η_r, the viscosity at some standard shear rate (often taken to be $1\,\text{s}^{-1}$), as: $\eta = \eta_r|\dot{\gamma}|^{n-1}$.

POWER LAW FLUID A fluid which obeys a power law equation. Many polymeric fluids (both melts and solutions) do so at least over a restricted range of shear rates.

POWER LAW INDEX Either one of the two constants of a power law equation, $\tau = K(\dot{\gamma})^n$, where τ is the shear stress, and $\dot{\gamma}$ is the shear rate, with K (the consistency index) and n (the flow behaviour index) being the power law indices. Sometimes the term is used as an alternative for the flow behaviour index alone.

PP Abbreviation for polypropylene.

PPA Abbreviation for polyphosphoric acid and for polyparabanic acid.

PPE Abbreviation for polyphenylene ether.

PPG Abbreviation for polypropylene glycol.

PPO Abbreviation for polyphenylene oxide and for poly-(2,6-dimethyl-1,4-phenylene oxide).

PPQ Abbreviation for phenylated polyquinoxaline.

PPS (1) Abbreviation for polypropylene glycol sebacate.
(2) Abbreviation for poly-(p-phenylene sulphide).

PPVC Abbreviation for plasticised polyvinyl chloride.

PRANDTL–EYRING MODEL A model for non-Newtonian flow behaviour which results in the relationship, derived from the kinetic theory of liquids, $\eta = \eta_0\,\text{arcsinh}\,B\dot{\gamma}/B\dot{\gamma}$, where η is the shear rate dependent apparent viscosity, $\dot{\gamma}$ is the shear rate and B is a material constant. The model predicts Newtonian behaviour in the limit of zero shear rate (with viscosity η_0), but not at high shear rates, unlike the Powell–Eyring model. It also only predicts pseudoplastic behaviour.

PRD-49   Early tradename for Kevlar.

PREALBUMIN   A blood plasma protein, with a small (about 1·5%) carbohydrate content and with a molecular weight of about 60 000, which migrates electrophoretically even faster than the serum albumin. Its presence is revealed by more sophisticated electrophoresis, e.g. in a gel, than the usual method. It may function in the blood as a thyroxine carrier.

PRECIPITATED SILICA   Alternative name for colloidal silica.

PRECIPITATION CHROMATOGRAPHY   Alternative name for Baker–Williams fractionation.

PRECIPITATION POLYMERISATION   A bulk or solution polymerisation in which the polymer is insoluble in the polymerisation medium and precipitates during polymerisation. Polyvinyl chloride and polyacrylonitrile are among the few of the more common polymers that are insoluble in their monomers, so they always precipitate unless a special polymer solvent is used. In the presence of polymer particles in free radical polymerisation, the growing radicals may become occluded, thus reducing termination, so that autoacceleration occurs at low conversions.

PRECIPITIN   The three-dimensional lattice structure which results from the interaction of an antibody with its complementary antigen and which precipitates from the blood serum.

PRECURSOR POLYMER   A polymer which is subsequently converted by a chemical reaction to a different polymer. Precursor polymers are useful in the synthesis of polymers which are difficult or impossible to make directly by polymerisation of the appropriate monomer. Thus polyvinyl acetate is a precursor polymer for polyvinyl alcohol. Precursor polymers are also useful when the derived polymer is difficult to shape or process. Thus a polymer product may be shaped from a more tractable precursor polymer then the polymer converted to the desired polymer. An important example is the use of a polyamic acid as a precursor for a polyimide. However, the chemical conversion of one polymer to another by the usual methods of a polymer analogous reaction or by cyclisation, is often not complete, so that the final polymer is a copolymer containing both converted and unconverted repeat units.

PREDEFORMATION   A prior straining of a material to improve its mechanical properties by inducing strain hardening, as, for example, in the drawing of fibres.

PRE-EFFECT   In a free radical polymerisation, the period following an abrupt increase in the rate of initiation, e.g. at the beginning of a photo- or radiation-induced polymerisation, during which the radical concentration is increasing, so that the steady state assumption is not valid. Its magnitude is often given by ΔM_{pre}, the difference between the amount of reaction that has occurred and that which would have occurred if the steady state concentration of free radicals had been produced instantaneously at time $t = 0$. Given by $\Delta M_{pre} = k_p[M] \ln 2/k_t$ where k_p and k_t are the propagation and termination rate constants respectively and $[M]$ is the monomer concentration.

PREKERATIN   A fibrous precursor, soluble in acidic buffers, to insoluble, crosslinked keratin. It is rich in sulphhydryl groups but contains no disulphide bridges. It is obtained from certain specialised cells of keratinising tissues, e.g. the epidermin from certain skin tissues. Typically it has a solution molecular weight of about 600 000.

PREMELTING   A transition often observed in crystalline polymers, usually occurring at about $(0\cdot8–0\cdot9) \times T_m$, where T_m is the melting temperature. It is often the α-transition since it is the highest transition, apart from the melting transition. It is also sometimes referred to as the α_c-transition. The molecular interpretation is of a molecular rotation about the chain axis, for chain folded polymer crystals, with an expansion in one or both dimensions of the crystal lattice perpendicular to the chain direction. At T_{α_c} the chain fold length begins to increase. For polyethylene the transition is often resolved into two components, the α'- and the α''-transitions (also called the α_1- and α_2-transitions) one of which may have its origin as described above and is sometimes the only one referred to as premelting. The other component may be associated with interlamellar shear or slip or with c-shear.

PREPARATIVE FRACTIONATION   Fractionation of whole polymer into fractions which are separately isolated. Such fractions of narrow molecular weight distribution are then often used to study the molecular weight dependency of polymer properties. If the molecular weights of individual fractions are characterised then the molecular weight distribution of the whole polymer may be characterised. However, the method is more time consuming than analytical fractionation methods for this purpose. Preparative techniques include fractional precipitation and fractional solution methods, gel permeation chromatography and other chromatographic methods including gradient elution chromatography, ultrafiltration, thermal diffusion and zone melting.

PREPOLYMER   A polymer, often of low molecular weight, i.e. a few hundred or thousand, which is subsequently to be converted to a higher molecular weight polymer. This is achieved either by chain extension through reaction of functional end groups, e.g. polyols extended by reaction with isocyanate groups during polyurethane formation, or through the more precisely distributed functional groups of a structoset polymer either as end groups (structoterminal prepolymer) or as pendant functional groups (structopendant prepolymer). Use of a prepolymer enables a thermoset (network) fabricated product to be made by fabrication, e.g. by moulding, in the fusible, mouldable prepolymer form, followed by crosslinking to the hard, infusible thermoset product.

PREPOLYMER PROCESS Formation of a polymer from a precursor polymer by chain extension and/or crosslinking reactions. The term is used especially for the formation of polyurethanes where, in contrast to the one-shot process, polyurethane formation is by a two stage process. In the first stage a polyester or polyether polyol is reacted with a diisocyanate to form an isocyanate-terminated prepolymer, which is then chain extended and/or crosslinked in the second stage. This latter is by reaction with a diol, amine, water or further isocyanate groups for polyurethane elastomer formation, or with water, catalysts and other ingredients for polyether based flexible polyurethane foam. The prepolymer process was originally the main one used here because of the low reactivity of polyether polyols. However, with the discovery of more active catalyst systems, the one-shot process became possible and is now the more important process. Rigid polyurethane foams, especially those using tolylene diisocyanate, are also produced by the prepolymer process. A disadvantage, apart from having an extra step compared with the one-shot process, is that the isocyanate-terminated prepolymers often have a limited stability, due to reaction with atmospheric water. Also the prepolymers may have high viscosities, thus making mixing difficult in the second stage.

PREPROINSULIN A polypeptide precursor for pro-insulin, itself a precursor for insulin. It consists of a single polypeptide chain of 104–109 amino acid residues, which is converted to proinsulin by removal of a block of 23 residues from the N-terminal end.

PRESCOLLAN Tradename for a cast polyurethane elastomer similar to Vulkollan.

PRESSURE FLOW (Poiseuille flow) Flow brought about by the application of an external pressure to a fluid and in which the boundaries of the fluid are fixed and rigid, as opposed to a drag flow where flow is caused by movement of the boundaries. The classical example is flow through a circular cross-section pipe by the application of a pressure difference across its ends. Analysis of this case for a Newtonian fluid, leads to the Poiseuille equation for the volumetric flow rate. For a non-Newtonian fluid, as with most polymer melts and solutions, the power law equation often provides a good model. Pressure flows are frequently found during flow in polymer processing in moulds and dies.

PRESSURE SENSITIVE ADHESIVE An adhesive, based on a rubbery polymer, which will flow sufficiently on application of low stress, as pressure, to wet the adherend. A tackifier is usually incorporated to lower the viscosity and relaxation time of the polymer.

PREVENTIVE ANTIOXIDANT An antioxidant which works by preventing initiation of oxidative chain reactions by removing thermally labile, free radical-forming groups. Usually a peroxide decomposer. Metal-ion deactivators (e.g. chelators such as oxamide derivatives) also function in this way by preventing catalysis of peroxide decomposition by metals such as copper.

PREVEX Tradename for a poly-(2,6-dimethyl-1,4-phenylene oxide) blend with high impact polystyrene.

PREVULCANISATION Alternative name for scorch.

PREVULCANISATION INHIBITOR (PVI) A type of retarder of vulcanisation which is therefore useful in preventing scorch, but which does not affect the rate of crosslink formation during subsequent vulcanisation at higher temperatures, unlike other types of retarder. *N*-Cyclohexylthiophthalimide is a well-known example.

PRI Abbreviation for plasticity retention index.

PRIMARY ACCELERATOR An accelerator which is either used alone or as the major component of a mixed accelerator in combination with a secondary accelerator.

PRIMARY ACETATE Alternative name for cellulose triacetate.

PRIMARY CHAIN The main polymer chain in a crosslinked polymer, from which the network can be considered to be built up by formation of the crosslinks between the primary chains.

PRIMARY CHARGE EFFECT The retardation of the macroions and acceleration of the counter-ions that occurs during ultracentrifugation of a polyelectrolyte as a result of the establishment of a sedimentation potential.

PRIMARY CRYSTALLISATION The main part of the crystallisation of a polymer, that takes place according to the Avrami equation and is associated with the growth of spherulites. It may be followed at longer times by secondary crystallisation.

PRIMARY NUCLEATION The formation of a primary nucleus for crystallisation by formation from an amorphous material of an incipient crystal above a certain critical size. This occurs by fluctuations in the local order of the polymer molecules but has a positive free energy (the surface free energy of the crystal). Nuclei smaller than the critical size are subcritical or embryos and above this size (but still having a positive free energy of formation) are supercritical. Larger nuclei with negative free energies of formation are stable nuclei or small crystals.

PRIMARY PLASTICISER A plasticiser that has a high solvation power for a polymer and therefore is efficient in increasing polymer flexibility. A primary plasticiser is generally miscible with the polymer in all proportions. The most important primary plasticisers are the phthalates, especially in polyvinyl chloride.

PRIMARY RADICAL (Initiating radical) The free radical generated by thermal or photochemical decomposition of an initiator or by other means, e.g. redox initiation, that reacts with a monomer molecule to produce an activated monomer.

PRIMARY RADICAL TERMINATION In a free radical polymerisation, termination by bimolecular mutual destruction of the free radical active centre with a primary radical ($R^{\cdot}$) from the initiator:

$$\sim\!\!M_n^{\cdot} + R^{\cdot} \rightarrow \sim\!\!M_n\!\!-\!\!R$$

It can occur at high rates of initiation with stable primary radicals, at low monomer concentration with unreactive monomers, or when termination reactions are restricted, e.g. due to high viscosity. It considerably alters the polymerisation kinetics, the polymer molecular weight and molecular weight distribution.

PRIMARY STRUCTURE (1) The amino acid sequence structure of the polypeptide chain or chains of a protein. The primary structure is indirectly responsible for the biological function of a protein through its influence on the specific conformation adopted by the protein in its native state, i.e. the regular secondary structure and overall folding (tertiary structure).

The determination of the primary structure (sequencing) is of fundamental importance in protein studies and is usually a complicated procedure, performed in several stages. Firstly if the protein contains more than one polypeptide chain the chains must be separated, any disulphide bridges must be cleaved and the resulting sulphhydryl groups must be alkylated (to prevent reformation of disulphide bridges). A sample of each chain is then completely hydrolysed and its amino acid composition is determined in an automatic amino acid analyser. On another sample the N-terminal residue is determined using fluorodinitrobenzene, dansyl chloride or another similar method and the C-terminal residue is determined. A further sample is cleaved enzymatically with a proteinase, e.g. pepsin or chymotrypsin, or chemically, e.g. with cyanogen bromide or by mild hydrolysis, to yield a series of smaller peptide fragments. These are separated by ion-exchange or paper chromatography or by electrophoresis and their amino acid sequences are determined. The latter is often achieved by the Edman method, usually in an automatic protein sequenator. The process is repeated by producing a different series of peptides by cleavage of the polypeptide in a different manner. Their sequences are also deter-

mined. By comparing the amino acid sequences of the two sets of peptides, especially when overlapping peptides are involved, i.e. those peptide fragments from the first hydrolysis which span the cleavage points of the second hydrolysis, the complete sequence of the original peptide can be deduced.

The primary structures are now known for many proteins even those with several hundred amino acid residues. In general for each individual protein polypeptide, all molecules have identical primary structure. However, in the same protein from different species some minor differences in the primary structure may be found. These differences are the greater the less closely related are the species. Indeed determination of the sequences of the same protein from many different species can be used to establish evolutionary relationships. The differences may be only in one or a few amino acid residues out of several hundred. In some cases differences in primary structure have been found in samples of the same protein from different individuals of the same species. Although the primary structure is responsible, ultimately, for biological activity, certain segments of the polypeptide chain may not be essential and considerable variation in the sequences in these segments may be found. However, for the biologically required sequences, replacement of a single amino acid residue by a different one may completely alter biological activity as, for example, in the haemoglobins A and S. For these active sequences therefore, it is not surprising that considerable sequence homology exists between species.

For globular proteins as a whole there is little regularity in the primary structure, i.e. no tendency for two or three or more specific sequences to occur repeatedly. In fact nearly all possible sequences of two or three amino acid residues have been found. Also, except for the rarer proteins such as protamines which contain very large amounts of a single amino acid, there is no tendency for sequences of a single amino acid residue to be found. However, in fibrous proteins specific short sequences often occur repeatedly, these usually involving glycine, proline and hydroxyproline in collagen and involving alanine, glycine and serine in silk fibroin.

(2) In carbon black, the clusters, frequently chain-like, of particles of black formed by the permanent fusion of the particles during the manufacturing process. A black rich in these structures may have the suffix -HS (for high structure) after its abbreviated name. The primary particles may form looser aggregates termed secondary structure.

PRINCIPAL AXES OF STRAIN The three mutually perpendicular directions in a body along which the principal strains act. They remain mutually perpendicular during the deformation. Their directions also remain unchanged if no rotation takes place.

PRINCIPAL AXES OF STRESS The three mutually perpendicular axes in a body normal to the principal planes, along which the principal stresses act.

PRINCIPAL EXTENSION RATIO Symbol λ_n. An extension ratio in one of the three principal directions. If the principal directions are chosen as the x, y and z axes then the extension ratios are designated λ_1, λ_2 and λ_3 respectively.

PRINCIPAL PLANE One of the three mutually perpendicular planes on which the principal stresses act and on which the shear stresses are zero.

PRINCIPAL STRAIN The maximum values of the strain components obtained when the strain tensor is transformed to symmetrical axes, when the shear components become zero. The principal strains are designated e_{xx}, e_{yy} and e_{zz} (or e_1, e_2 and e_3 respectively) and act along the principal axes.

PRINCIPAL STRESS The limiting values of the normal stresses acting at a point in a body. The three planes on which they act are the principal planes, which are mutually perpendicular, and the axes, normal to these planes, along which the principal stresses act, are the principal axes. Such a set of principal axes may always be found for which the shear stresses on the principal planes are zero. Thus any state of stress may always be represented by the three principal stresses, with no shear stress components, when referred to the principal axes. The values of the principal stresses are given by the lengths of the axes of the stress ellipsoid. The principal stresses are usually designated σ_1, σ_2 and σ_3 (or p_1, p_2 and p_3) with $\sigma_1 \geq \sigma_2 \geq \sigma_3$, which are the real roots of the stress cubic.

PRINCIPLE OF CORRESPONDING TEMPERATURES A statement of the general observation that linear amorphous polymers have modulus values of $10^{5.5}$–$10^{6.5}$ Pa in the rubbery region and 10^9–10^{10} Pa in the glassy region and that the shapes of the modulus–temperature curves are similar. Thus it may be stated that such polymers are approximately equivalent at corresponding temperatures. The corresponding temperature often used for comparison is the reduced temperature T_r, given by $T_r = T/T_g$. Such correspondence is only found for polymers with similar molecular weights and molecular weight distributions and similar chain stiffnesses and side chain mobilities.

PRO Abbreviation for proline.

PROBIT METHOD A particular statistically determined procedure useful for carrying out the falling weight impact test, in which a fixed number of specimens are subject to testing at each of a number of different impact energies and the fraction of fractures at each energy is recorded. For polymers, the distribution of failures is often Gaussian and the plot of impact energy versus fraction broken, plotted on a normal probability scale, is linear. This method of collecting and handling impact data is more informative than the 50% failure rate method (obtained by the staircase technique), in particular in predicting the impact energy required to cause a small fraction of failures.

PROCESS OIL A rubber oil when used in relatively small amounts as a physical, rather than a chemical, softener to aid processing.

PROCOLLAGEN The precursor of collagen in its biosynthesis. It is larger than collagen, which undergoes cleavage at the N- and C-terminal ends, subsequent to its synthesis, to give tropocollagen. Any hydroxyproline or hydroxylysine residues are formed by hydroxylation after incorporation of proline or lysine into the chain and may subsequently attack carbohydrates at galactose or glucosyl–galactose residues.

PRODEGRADANT A material capable of promoting degradation. Its effect may be desirable, as in the use of a photosensitiser in a degradable polymer or detrimental, as with a pro-oxidant.

PROENZYME Alternative name for zymogen.

PROFAX Tradename for polypropylene.

PROGLUCAGON An inactive precursor polypeptide for glucagon containing an additional eight amino acid residues (at the C-terminal end) to glucagon, which are cleaved to form the active glucagon hormone.

PROINSULIN An inactive precursor polypeptide for the hormone insulin, consisting of a single polypeptide chain of 81–86 residues, which contains the insulin. It consists of a chain sequence of amino acids at the carbonyl-terminal end separated from the insulin B-chain at the amino-terminal end by a C-chain of 34 amino acid units. Three intramolecular disulphide bridges are also present in the A- and B-chain segments in the same position as in insulin. Thus cleavage of the A–C and C–B chain junctions yields the active insulin. Proinsulin itself is formed from preproinsulin.

PROLACTIN A protein hormone secreted by the anterior pituitary gland which stimulates the mammary glands. In consists of a single polypeptide chain of molecular weight 21 500 containing 191 amino acid residues.

PROLAMIN A food storage protein, found in the largest amounts in the seeds of plants. It is characterised by its solubility in the lower aliphatic alcohols, especially ethanol. It is usually isolated as a gluten which is molecularly heterodisperse, but may be fractionated by electrophoresis. Examples include gliadin (from wheat) and zein (from corn). It is relatively rich in glutamic acid, proline, asparagine and glutamine, but it is deficient in lysine. It is often considered to be a globulin.

PROLIFEROUS POLYMERISATION Alternative name for popcorn polymerisation.

PROLINE (Pro) (P)

$$HN-CH-COOH$$
$$CH_2CH_2$$
$$CH_2$$

M.p. 220°C (decomposes).

A non-polar α-imino acid, although frequently classified as an α-amino acid, widely found in proteins in especially large amounts in collagen. Unlike the other α-amino acids it is readily soluble in alcohols and gives a yellow, rather than a purple, colour in the ninhydrin reaction. Proline residues in proteins are not cleaved by chymotrypsin and are often associated with hairpin loops. Its pK' values are 1·99 and 10·96, with the isoelectric point at 6·30.

PROLOY Tradename for an ABS/bisphenol A polycarbonate blend.

PROMOTOR A chemical species capable of increasing the activity of the main catalyst or other species bringing about a polymerisation or other reaction. Thus it is an alternative name for activator in the context of polymerisation. The term is used particularly in cationic polymerisation as an alternative name for co-catalyst especially when a reactive cyclic ether, such as an epoxide (e.g. epichlorhydrin) or oxacyclobutane, is used to promote the polymerisation of relatively unreactive cyclic ethers, such as tetrahydrofuran, with Lewis acid catalysts.

PROOF STRESS The stress required to produce a specified amount of plastic deformation. It is determined by drawing a straight line parallel to the initial elastic portion of the stress–strain curve, but offset by a specified (often 1%) strain. It is the stress at which this line intercepts the stress–strain curve. Often used for metals in which the initial elastic response is linear, but in polymers its use is more doubtful since the initial response is often viscoelastic and the stress–strain curve can be curved in the region where the strain is fully recoverable.

PRO-OXIDANT A promoter of oxidation. Many metal compounds are pro-oxidants, especially those of transition metals which can function by virtue of their variable valencies. Some such compounds are antioxidants under certain conditions, e.g. at a particular temperature or even merely at a particular concentration, other than those when they operate as pro-oxidants.

PROPAGATION (Chain propagation) The series of reaction steps in a chain polymerisation in which many monomer molecules rapidly add on in succession to the active centre, causing rapid formation of a polymer molecule once an activated monomer molecule has been formed. A single propagation step may be represented as, $\sim\!\!\sim\!M^* + M \rightarrow \sim\!\!\sim\!M^*$, the first steps being,

$$I-M^* + M \rightarrow I-M-M^* \xrightarrow{M} I-M-M-M^* \xrightarrow{M} \text{etc.}$$

where I—M* is the activated monomer. In general:

$$I\!-\!(M)_n\!M^* + M \rightarrow I\!-\!(M)_{n+1}\!M^*$$

The process continues as a kinetic chain until the active centre is lost by termination or is transferred to another molecule.

Usually many propagation steps (typically 10^2–10^4) follow from a single act of initiation. For high molecular weight polymer to be formed, the propagation rate must be fast compared with termination or transfer. The rate constant (k_p) typically has values of 10^2–10^4 litre $(\text{mol s})^{-1}$. The rate of propagation (R_p) is given by $k_p[M][M^*]$ where [M] and [M*] are the monomer and active centre concentrations respectively.

Since little monomer is consumed in initiation, the rate of monomer disappearance ($-\mathrm{d}[M]/\mathrm{d}t$) is the same as the rate of polymerisation, which is also R_p. By using the steady state assumption, R_p is shown to be $k_p[M](R_i/2k_t)^{1/2}$ where R_i is the rate of initiation and k_t is the termination rate constant (defined as the rate constant for $\sim\!\!\cdot + \sim\!\!\cdot \xrightarrow{k_t}$ dead polymer).

Propagation in vinyl polymerisation is predominantly by head-to-tail addition and in free radical polymerisation there is little preference for a monomer to add on in either the same or opposite configuration, so that largely atactic polymer results. However, there is a tendency to syndiotacticity as polymerisation temperature is lowered. In ionic polymerisation, especially when the counter-ion is closely associated with the active centre, stereoregulation can result in the formation of tactic polymer. Coordination polymerisation usually produces isotactic polymer. In diene polymerisation several different propagation steps are possible, to produce differently linked repeating units. Thus 1,2-, 3,4- and 1,4-addition and cis–trans isomerism are all possible.

PROPANE-1,2-DIOL Alternative name for propylene glycol.

PROPANE-1,2,3-TRIOL Alternative name for glycerol.

n-PROPANOL (n-Propyl alcohol)

$$CH_3CH_2CH_2OH \qquad \text{B.p. 97·2°C.}$$

A solvent for some natural resins, some phenol–formaldehyde resins and urea–formaldehyde resins and useful as a diluent for lacquers. Miscible with water in all proportions.

PROPAN-2-OL Alternative name for isopropanol.

PROPAN-2-ONE Alternative name for acetone.

PROPATHENE Tradename for polypropylene.

PROPENE Alternative name for propylene.

PROPIOFAN Tradename for polyvinyl propionate.

β-PROPIOLACTAM Alternative name for 2-azetidinone.

β-PROPIOLACTONE

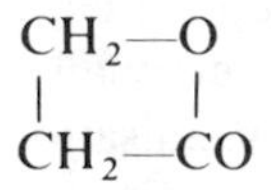

$$
\begin{array}{cc}
CH_2 & O \\
| & | \\
CH_2 & CO
\end{array}
$$

B.p. 62°C/20 mm.

A cyclic lactone that readily polymerises by ring-opening polymerisation to poly-(β-propiolactone), due to its highly strained four-membered ring. It may be polymerised using either an acidic or basic catalyst.

PROPORTIONAL LIMIT The point on a stress–strain curve beyond which it deviates from linearity and hence beyond which Hooke's law is no longer obeyed.

n-PROPYL ALCOHOL Alternative name for *n*-propanol.

PROPYLENE (Propene)

$$CH_2{=}CHCH_3 \qquad \text{B.p. } -47.7°C.$$

Most propylene is produced commercially as a by-product in the manufacture of ethylene or gasoline by the cracking/reforming of various hydrocarbon feedstocks obtained from crude oil or natural gas. About 50% is used as a component in petrol (gasoline) after being either alkylated with isobutene or oligomerised using phosphoric acid. Propylene is converted to a large number of derivatives, especially isopropanol, C_7–C_{12} oligomers, propylene oxide, oxo-products, cumene and glycerol. Various monomers are produced from propylene, including acrylonitrile, isoprene, 4-methylpentene-1 and epichlorhydrin, although, of course, large quantities of propylene are polymerised directly to isotactic polypropylene as well as to ethylene–propylene rubbers, using Ziegler–Natta catalysts.

PROPYLENE COPOLYMER Although this term could refer to any propylene copolymer, it usually refers specifically to the ethylene–propylene block copolymers containing up to about 15% ethylene to improve the low temperature impact strength of polypropylene.

PROPYLENE GLYCOL (Propane-1,2-diol)

$$
\begin{array}{c}
CH_3CHCH_2OH \\
| \\
OH
\end{array}
$$

B.p. 189°C.

Prepared by high temperature/high pressure hydration of propylene oxide. The preferred diol for the preparation of unsaturated polyester resins, since the products are miscible with styrene and show little tendency to crystallise. A solvent for some natural resins.

PROPYLENE OXIDE

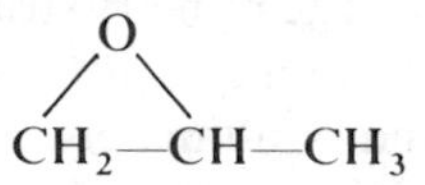

$$
\begin{array}{c}
O \\
/\backslash \\
CH_2{-}CH{-}CH_3
\end{array}
$$

B.p. 34°C.

Produced from propylene, via propylene chlorohydrin, by reaction with chlorine and water, followed by dehydro-chlorination by treatment with calcium hydroxide:

$$CH_2{=}CH{-}CH_3 \rightarrow ClCH_2CH(OH)CH_3 \rightarrow$$

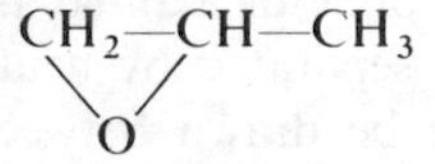

$$
\begin{array}{c}
CH_2{-}CH{-}CH_3 \\
\backslash / \\
O
\end{array}
$$

Direct oxidation of propylene may be used, but gives relatively low yields of propylene oxide due to oxidation of the methyl group which yields large amounts of acrolein. Useful as the monomer for polypropylene oxide and also for the production of propylene glycol.

PROPYLENE OXIDE–ETHYLENE OXIDE BLOCK COPOLYMER Alternative name for poly-(propylene oxide-b-ethylene oxide).

PROPYLENE OXIDE RUBBER (PO) (GPO) Tradenames: Parel 58, Dynagel XP-139. A rubbery polymer of propylene oxide containing a small amount of allyl glycidyl ether units for sulphur vulcanisation. Although not outstanding in any particular respect, the vulcanisates have good physical properties, resembling natural rubber in their low temperature and dynamic properties, reasonable heat resistance and good ozone, weathering and oil resistance.

PROPYLENE SULPHIDE

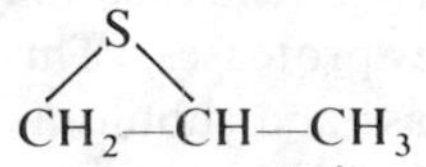

$$
\begin{array}{c}
S \\
/\backslash \\
CH_2{-}CH{-}CH_3
\end{array}
$$

B.p. 75–77°C.

Produced by reaction of thiourea or potassium thiocyanate with propylene oxide. It may be polymerised to polypropylene sulphide by ring-opening polymerisation using anionic or cationic catalysts.

PROSTHETIC GROUP The non-polypeptide part of a conjugated protein, which may or may not be covalently bound to the polypeptide part. Thus in haemoproteins the prosthetic group is an iron–porphyrin complex, in glycoproteins it is a polysaccharide, in nucleoproteins it is a nucleic acid, in lipoproteins it is a lipid and so on. The location of the prosthetic group with respect to the polypeptide component is determined by X-ray diffraction, especially if it contains a metal, or by partial hydrolysis followed by identification of the peptide–prosthetic group fragments.

PROTAMINE A highly basic protein, occurring as a nucleoprotein, bound to DNA, i.e. a nucleoprotamine, in the sperm of many fish and lower animals and usually isolated from fish sperm. Salmine from salmon and clupeine from herring have been particularly intensively studied. Protamines are small proteins, or large peptides, of molecular weight 2000–5000 and usually contain an unusually high content of arginine, or sometimes of another basic amino acid. These basic amino acids neutralise the charge on the DNA and thereby possibly help packing in the sperm head. Although preparations are inhomogeneous, the different molecular species are very similar. Protamines are devoid of many amino acids, e.g. cysteine, tryptophane and aspartine. The N-terminal amino acids are proline or alanine, the sequences being basic amino acids separated by neutral units. A sharp distinction cannot be drawn between protamines and histones, but the protamines are more basic and of lower molecular weight.

PROTEASE (Proteolytic enzyme) An enzyme that catalyses the hydrolysis of peptide linkages in the polypeptide chains of proteins. Therefore they are not only important as enzymes biologically, but are also useful in structural elucidation of protein polypeptides, by their ability to break down the long chain molecules to smaller peptides that can be identified. They are often synthesised biologically as inactive precursor zymogens, to prevent attack on cellular proteins and are only activated when required. A protease may specifically hydrolyse only one type of bond, e.g. the bond at the carbonyl of a lysine or arginine residue hydrolysed by trypsin, or it may not be quite so specific, e.g. chymotrypsin, pepsin and thermolysin. An enzyme which hydrolyses only terminal peptide bonds is an exopeptidase, e.g. carboxypeptidase. An endopeptidase can attack its specific peptide link anywhere in the chain. These are monomeric enzymes and in many of them a serine residue is involved at the active centre, so they are called serine enzymes or serine proteases. This group includes chymotrypsin, elastase and subtilisin. In a thioprotease, e.g. papain, a thiol (—SH) group is similarly involved.

PROTECTING GROUP (Blocking group) A chemical group (Z) which is attached to a reactive chemical group (A) to protect it from chemical reaction whilst the molecule to which A is attached is subjected to other chemical reactions. After these reactions, A may be regenerated by removal of the protecting group. Thus in most general terms:

$$Z + A{\sim}R' \xrightarrow[\text{protection}]{} Z{-}A{\sim}R' \xrightarrow[\text{chemical reaction}]{}$$

$$Z{-}A{-}R'' \xrightarrow[\text{removal}]{} A{-}R'' + Z \text{ (or other products)}$$

Use of protecting groups is sometimes needed during the formation of polymers and is best known in the formation of peptide bonds in the stepwise synthesis of peptides and proteins. Other examples occur in the protection of active hydrogens from reaction with isocyanates in polyurethane formation, and in the blocking of the positions of benzene rings when polymers are being produced by electrophilic substitution when *ortho/para* directing substituents are present, since normally only *para* substitution is required. In this latter case the blocking group (as it is more usually called) is not removed. In the case of peptide and protein synthesis, in the formation of a new peptide bond from precursor amino acids or peptides, both the amino group and the carboxyl groups not involved in peptide formation must be protected. These groups are those at the ends of the precursor, or are side group when acidic or basic amino acids or amino acid residues are involved. Many different protecting

groups are available. They must be capable of being readily removed after peptide bond formation and, since the peptides involved are usually of the L-enantiomer, they must not suffer racemisation. *t*-Butyloxycarbonyl and benzyloxycarbonyl groups are most frequently used for amine group protection, whilst ester formation, e.g. benzyl, methyl or ethyl, protects carboxyl groups.

PROTEIN One of a very large group of biopolymers of outstanding importance, since proteins are present to the extent of 50% or more in the cells of all living matter—animals, plants, viruses and bacteria. Proteins consist of ordered, sequential copolymers of α-amino acids since they are polypeptides, the amino acid residues being joined through peptide bonds. Twenty different amino acids (the standard amino acids) are found in proteins, often all, or most of these, being present in any given protein. Thus the chain structures are very complex and offer an endless number of different structural possibilities. Unlike most synthetic copolymers, the different amino acids are joined in the same precise sequence in all molecules of a given protein, which is therefore monodisperse. Molecular weights vary from a few thousand to several million.

Natural polypeptides of molecular weight less than about 5000 are often referred to as peptides, rather than as proteins, although the dividing line is generally not precisely defined. Proteins which contain only a polypeptide chain (or sometimes more than one chain in association—an oligomeric protein) are simple non-conjugated proteins. A complex or conjugated protein contains, in addition, non-polypeptide as a prosthetic group.

Proteins may be classified in several different ways. Traditionally, simple proteins are grouped according to their solubility as albumins, globulins, prolamins, glutelins protamines, histones and scleroproteins. Conjugated proteins are classified according to the nature of the prosthetic group as nucleoproteins, lipoproteins, glycoproteins (or mucoproteins), chromoproteins (often also metalloproteins), phosphoproteins and flavoproteins. Alternatively, proteins may be grouped as being globular or fibrous, or according to their biological function as enzymes, storage proteins, transport proteins, contractile proteins, immune proteins, hormones or structural proteins. An individual protein may belong to several of these different classes.

A protein is isolated from the tissues in which it occurs by a variety of techniques. Simple extraction, e.g. with an appropriate aqueous solution, may be sufficient. However, a variety of purification techniques are employed to isolate what it is hoped is a molecularly homogeneous material from the extracts. Often such a goal is not achieved.

The structure of a protein may be characterised at several different levels. The primary structure refers to the precise sequence of amino acid residues along the polypeptide chain and is ultimately responsible for the biological activity of the protein. There are often small differences in primary structure of samples of the same protein from different species, or even from individuals of the same species. The primary structure may be represented in general as,

$\sim$NHCHRCONHR$_1$CONHR$_2$CONHR$_3$CO$\sim\cdots$, etc.

where R$_1$, R$_2$, R$_3$,..., etc. are the different substituents on the α-amino acid residues. In proteins it is always the L-enantiomer that is found, so the prefix L- is usually understood and is omitted from structures written in the normal shorthand form using standard abbreviations for the different amino acids. Conventionally the structure is written starting with the N-terminal residue at the left-hand side, e.g. Gly–Val–Thr–Pro...Gly–Ala–Lys. The N- and C-terminal ends may be emphasised by writing H–Gly–Val–Thr–Pro...Gly–Ala–Lys–OH. If part of a sequence is undetermined, it is enclosed in parentheses and the abbreviated residues are separated by commas, as in Gly–Val–(Thr, Pro, Gly, Ala)–Lys–Ser.

A major part of protein science involves the necessarily complex and lengthy procedures required to determine the primary structure. The strategy and techniques for the determination of the primary structure by sequencing are those normally used for polypeptides or peptides. The number of polypeptide chains is first determined by reaction of the N-terminal amino group(s) with fluorodinitrobenzene or dansyl chloride, followed by hydrolysis and separation, and identification of the dinitrofluorobenzene or dansyl amino acids by paper chromatography or electrophoresis. The number of C-terminal ends can also be determined by hydrazinolysis or by cleavage with carboxypeptidase. If more than one chain is present then the chains must then be separated. If they are not covalently joined, this may be achieved by using solubilities in acid, base or denaturing agent. However, if as is often the case, the chains are joined through disulphide bridges of cystine units, these may be cleaved to cysteine groups by reduction with mercaptoethanol. The chains are then separated by gel permeation chromatography or by ion-exchange chromatography. Next, each chain is cleaved at different points and the sequences of the shorter peptides produced are determined, after their separation by ion-exchange chromatography or electrophoresis. Cleavage may be by partial acid hydrolysis or by a reagent which cleaves at a specific residue, e.g. cyanogen bromide at a methionine residue. Alternatively a protease such as trypsin or chymotrypsin is used to cleave at a specific residue. The Edman method is then used, especially in an automatic sequenator, to sequence the peptides. Small peptides can also be sequenced by mass spectrometry. The sequence of the original polypeptide chain may then be deduced by studying the overlapping of peptide sequences. Finally, the positions of disulphide links are determined in the original protein, by hydrolysing it after oxidation of the disulphides to cysteic acid, followed by separation and identification of the peptides, e.g. by using the diagonal method.

The secondary structure refers to the local chain conformation adopted by the polypeptides. Regular conformations may be identified, as the α-helix or the β-conformation. These are very evident in fibrous proteins, but only persist for short segments of the polypeptide chains in globular proteins. Tertiary structure refers to the irregular, but precise larger scale folding of the chains, e.g. due to hairpin loops, enabling the molecule to form the compact structures typical of globular proteins. The quaternary structure of a conjugated protein refers to the arrangement in space of the polypeptide chains and the prosthetic group and the type and site of bonding between them.

PROTEINASE   Alternative name for endopeptidase.

PROTEIN SEQUENATOR   An apparatus for carrying out successive Edman degradations on a peptide or protein automatically, thus enabling the amino acid sequence to be determined. Fifty or more residues may be so determined.

PROTEOLYTIC ENZYME   Alternative name for protease.

PROTHROMBIN   A blood plasma protein occurring in the α$_2$-globulin electrophoretic fraction. A glycoprotein with about 5% carbohydrate and with a molecular weight of 63 000. The zymogen of thrombin, which is formed from prothrombin in the presence of a thromboplastin and Cu^{2+} ions. It is therefore involved in the blood clotting process.

PROTOFIBRIL   A fibrillar subunit of the microfibrils of cellulose, about 35 Å wide. Possibly consists of flat ribbons of 100% crystalline single cellulose molecules.

PROTOHAEM   Alternative name for haem.

PROTOMER   Alternative name for subunit in an oligomeric protein or supermolecular protein complex.

PROTON MAGNETIC RESONANCE SPECTROSCOPY   (PMR spectroscopy) Nuclear magnetic resonance spectroscopy of the protons in molecules. Classically, most NMR spectroscopy of organic molecules, including polymers, has been of protons. However, more recently ^{19}F and ^{13}C NMR have become of increasing importance.

PROTOPORPHYRIN   A porphyrin which contains four methyl, two vinyl and two propionic acid groups substituted for hydrogen on the pyrrole rings of porphin. The most important isomer is protoporphyrin IX with methyls at the 1,3,5 and 8 positions, vinyls at the 2 and 4 positions and propionic acid at the 6 and 7 positions. It is this porphyrin which is found in the haem of haemoglobin, myoglobin and cytochrome C.

PS   Abbreviation for polystyrene.

PSEUDO-AFFINE DEFORMATION   Deformation of an amorphous or crystalline polymer, which is assumed to consist of transversely isotropic units whose symmetry axes rotate on deformation in the same manner as lines joining points in the bulk material. Useful in the description of the crystalline orientation of polymers.

PSEUDO-ASYMMETRIC CENTRE An alternative name for the substituted carbon atoms in a vinyl polymer chain P_x—CH_2C^*HX—P_y, rather than calling them simply asymmetric centres. This term is preferred since although these carbons do have four different groups attached, two of them are the polymer chain lengths P_x— and P_y—, which are so similar in the vicinity of the C* centre that these centres do not give rise to optical activity in stereoregular polymers. According to strict organic chemical use of the term, C* is only pseudo-asymmetric when it is at the precise centre of the chain, i.e. P_x—= P_y—. In an isotactic polymer, inversion of configuration at this centre then occurs, so that half the C* centres are of the *d*-configuration and half of the *l*-configuration.

PSEUDOCATIONIC POLYMERISATION Polymerisation with a cationic initiator, but in which propagation may occur through a non-ionic species. The most studied system is the polymerisation of styrene with perchloric acid initiator, where propagation may take place via the neutral ester

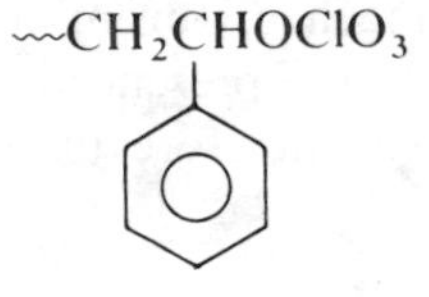

$$\sim\!\!CH_2CHOClO_3$$

although an ion pair may also be the active centre:

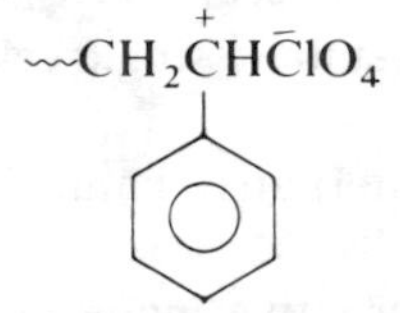

$$\sim\!\!CH_2\overset{+}{C}H\bar{C}lO_4$$

PSEUDOEQUILIBRIUM METHOD Alternative name for Archibald method.

PSEUDOPLASTICITY Alternative name for shear thinning.

PSP Abbreviation for polystyrylpyridine.

PSU Abbreviation for polysulphone.

PTC DERIVATIVE Abbreviation for phenylthiocarbamoyl derivative.

PTFE Abbreviation for polytetrafluoroethylene.

PTH Abbreviation for phenylthiohydantoin.

PTHF Abbreviation for polytetrahydrofuran.

PTMA Abbreviation for poly-(tetramethylene adipate).

PTMEG Abbreviation for polytetramethylene ether glycol.

PTMG Abbreviation for poly-(oxytetramethylene) glycol.

PTMTP Abbreviation for polytetramethylene terephthalate.

PTO Abbreviation for polyterephthaloyloxamidrazone.

PU Abbreviation for polyurethane.

PULLULAN An α-D-glucan produced when cultures of *Pullularia pullulans* are grown on sucrose. The glucose units are 1,4'- and 1,6'-linked in the ratio of approximately 2:1.

PULSED FOURIER TRANSFORM NUCLEAR MAGNETIC RESONANCE SPECTROSCOPY Alternative name for Fourier transform nuclear magnetic resonance spectroscopy.

PULSE INDUCED CRITICAL SCATTERING (PICS) A technique for studying phase separation processes in polymer systems, e.g. determining phase diagrams by cloud point measurements, whereby laser light scattering is measured after the polymer, usually in solution, has suffered a small rapid temperature change.

PUR Abbreviation for polyurethane.

PURE GUM Alternative name for gum rubber.

PURELY VISCOUS FLUID (Inelastic fluid) A fluid which does not show any elastic effects, so that if held at any constant shape the stress instantaneously becomes isotropic or zero. Conversely, if the stress is made instantaneously isotropic or zero then the fluid remains at constant shape. Hence an appropriate model and constitutive equation for such a fluid need not be a viscoelastic one. In general, its behaviour may be described by $\tau = \eta\Delta$, where Δ is the rate of deformation tensor, τ is a stress tensor and η is a viscosity. Roughly, if the recoverable shear, i.e. the ratio of elastic to shear stress, is less than unity, then a purely viscous model and equation will be adequate to describe fluid flow. However, polymers show stress relaxation and normal stress effects in simple shear flow. Nevertheless the purely viscous fluid model has been widely used as the basis of several other models for describing polymer flows, such as the power law, Ellis, Prandtl–Eyring and Powell–Eyring models.

PURE SHEAR A special case of plane strain. For pure shear about the *z*-axis, for example, the only strain components are ε_{21} ($=\varepsilon_{12}$) referred to principal axes. It results, for example, when a sheet is clamped along two edges (which are long compared with the unclamped edges) and is strained by moving the clamps apart but keeping them parallel.

PURE STRAIN A deformation in which the three orthogonal lines chosen as the system of coordinates are not rotated by the deformation and for which the shear strain components are zero. Thus if the lines are chosen as a set of cartesian coordinates, they are the principal axes of strain.

PUSTULAN A 1,6′-linked β-D-glucan found in certain lichen.

PVA Abbreviation for polyvinyl acetate or polyvinyl alcohol.

PVAc Abbreviation for polyvinyl acetate.

PVAl Abbreviation for polyvinyl alcohol.

PVC Abbreviation for polyvinyl chloride.

PVDC Abbreviation for polyvinylidene chloride.

PVDF Abbreviation for polyvinylidene fluoride.

PVF Abbreviation for polyvinyl fluoride.

PVI Abbreviation for prevulcanisation inhibitor.

PVK Abbreviation for poly-(N-vinylcarbazole).

PVM Abbreviation for poly-(vinylmethyl ether).

PVME Abbreviation for poly-(vinylmethyl ether).

PVMQ Abbreviation for a phenylsilicone elastomer containing some vinyl groups.

PVP Abbreviation for poly-(N-vinylpyrrolidone).

PVSI Older alternative for the abbreviation PVMQ.

PYRALIN Tradename for prepregs of the polyimide obtained from pyromellitic dianhydride and 4,4′-diaminodiphenylether, used in high temperature resistant structural applications.

PYRAMIDAL CRYSTAL (Hollow pyramidal crystal) The usual form of a polymer single crystal as initially crystallised. However, in the preparation of such a crystal for electron microscopy, the pyramid usually collapses to a flat lozenge-shaped lamella. The original pyramid edges are then sometimes still evident by the sectorisation of the crystal. The hollow form results from the nature of the chain folding at the crystal surface.

PYRANOSE The ring form of a monosaccharide or monosaccharide unit in a carbohydrate that contains six ring atoms and is therefore considered to be derived from pyran:

Most hexoses and pentoses exist in this form in the solid state, but are often in equilibrium with the furanose form in solution. A sugar in pyranose form is named by adding the suffix -pyranose to its name, e.g. β-D-glucopyranose. A glycoside in pyranose form is a pyranoside.

-PYRANOSIDE Suffix added to the name of a glycoside to show it is in the pyranose ring form.

PYRAN POLYMER Alternative name for divinylether—maleic anhydride copolymer.

PYRE ML Tradename for a solution of the polyimide obtained from pyromellitic dianhydride and 4,4′-diaminodiphenylether, used for a high temperature resistant wire enamel for electrical insulation purposes.

PYREX Alternative name for a borosilicate glass with about 60% SiO_2 and 12% B_2O_3.

PYROELECTRIC POLYMER A polymer whose polarisation changes on heating. Polymer electrets show pyroelectric effects, as in their thermally stimulated discharge.

PYROGENIC SILICA (Fumed silica) (Thermal silica) Tradenames: Aerosil, Cabosil. A synthetic fine particle silica prepared by reaction of silicon tetrachloride with hydrogen in a flame. Particle sizes are typically 0·01–0·05 μm. Useful as a thickening agent and as a white reinforcing filler in rubbers, especially silicone rubbers.

PYROLYSIS Thermal degradation performed at a high temperature in an inert atmosphere (vacuum or inert gas). Results in the formation of volatile low molecular weight fragments, which are conveniently analysed by gas–liquid chromatography (pyrolysis–GLC). Monomer is a major product when the polymer is liable to depolymerise, but more often random scission and rearrangement reactions result in a complex mixture of volatile products. Polymers liable to crosslinking or cyclisation may give a carbonaceous char. The precise product composition depends not only on the individual polymer but also on the size and shape of the sample and the pyrolysis temperature.

PYROLYSIS GAS CHROMATOGRAPHY (Pyrolysis–GLC) A technique of polymer characterisation whereby a small sample of polymer coated on a wire is pyrolysed by heating the wire electrically in the gas stream of a gas–liquid chromatograph. Precise pyrolysis temperature control is best achieved in Curie-point pyrolysis. Chromatographic separation and identification of the products is useful for the structural characterisation of polymers, especially the determination of copolymer composition.

PYROMELLITIC DIANHYDRIDE (PMDA)

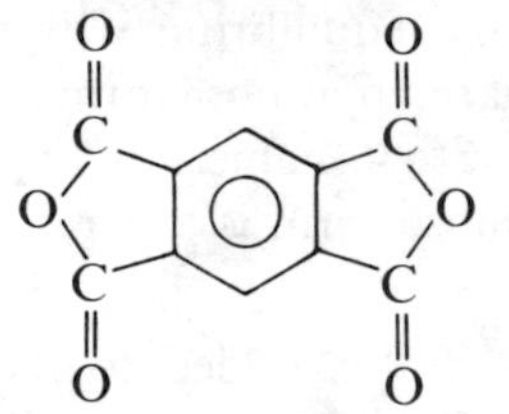

M.p. = 286°C.

Synthesised by oxidation of durene,

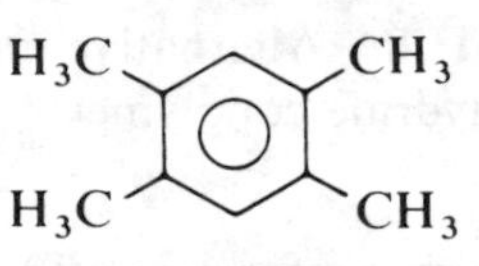

The cheapest and therefore the most widely used tetrafunctional monomer for the synthesis of fused ring aromatic, high temperature resistant polymers, particularly for polyimides by reaction with an aromatic diamine. Also used for curing of epoxy resins to give products with higher heat distortion temperatures than phthalic anhydride cured resins, owing to the higher degrees of crosslinking obtainable.

PYROPOLYMER A polymer produced by pyrolysis of a precursor polymer, e.g. pyrolysis of PAN, 1,2-polybutadiene or cellulose, frequently in fibre form. Often also a ladder polymer, as a result of cyclisation during pyrolysis. When the fibre is oriented, e.g. by stretching during pyrolysis, morphologically regular pyropolymer of high modulus may be produced, e.g. as in the graphitisation process in the formation of carbon fibres.

PYROXENE A linear, crystalline silicate polymer, consisting of SiO_4^{2-} tetrahedra linked so that there are two tetrahedra for each repeating unit joined through common oxygen atoms:

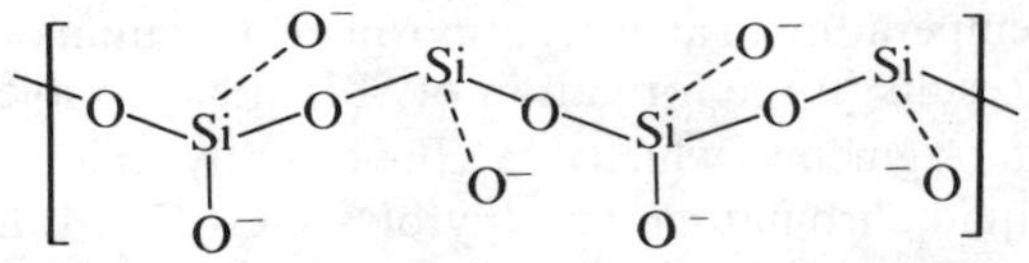

The counter-ions in most naturally occurring pyroxene minerals are combinations of magnesium, calcium and iron, so that these polymers are copolymers with the empirical formula $Mg(Ca, Fe)(SiO_3)_2$. The counter-ions link the chains laterally and can be differently arranged. Important examples of pyroxenes are the diopsides and spodumene.

PYROXYLIN A solution of cellulose nitrate, useful as a lacquer.

PYRROLIDINONE Alternative name for 2-pyrrolidone.

2-PYRROLIDONE (Pyrrolidinone) (γ-Butyrolactam)

$(CH_2)_3CONH$

M.p. 24°C. B.p. 124–126°C/11 mm.

The monomer from which nylon 4 is formed by anionic ring-opening polymerisation. Prepared by reaction of the corresponding lactone with ammonia. The lactone is formed by cyclodehydration of 1,4-butanediol.

PYRRONE POLYMER Alternative name for poly-imidazopyrrolone.

Q

Q (1) Symbol for a tetrafunctional unit in a polyorgano-siloxane, Q being used for quadri- rather than T for tetrafunctional, to distinguish it from the symbol T used for a trifunctional unit.

(2) The letter used in rubber materials to denote a silicone rubber (QR).

(3) Symbol for glutamine.

Q–e **SCHEME** Alternative name for Alfrey–Price *Q–e* scheme.

Q-**FACTOR** Alternative name for dynamic amplification factor.

Q-FILM Tradename for poly-(ethylene-2,6-naphthalenedicarboxylate).

QIANA Tradename for fibres of poly-(bis-(4-cyclohexyl)methanedodecandiamide).

QUACORR Tradename for a furan resin.

QUANTUM YIELD (1) Symbol ϕ. The ratio of the number of molecules reacting (or of product formed) to the number of light quanta (photons) absorbed in a photochemical process. The usual quantitative measure of the efficiency of the process, but can also give information on the mechanism, e.g. if $\phi > 1$ then a chain reaction is occurring. Thus ϕ_{XL} may refer to quantum yield of crosslinks for example.

(2) Symbol ϕ. For free radical polymerisation, the quantum yield (ϕ_{fr}) is the number of free radicals produced for each photon of light absorbed. In a photopolymerisation ϕ_{fr} controls the rate of initiation and includes contributions from the quantum yields for formation of excited species, energy transfer yield and efficiency of forming initiating species.

QUARTZ The most important crystalline form of silica which consists of a three-dimensional network of six-membered Si–O rings (formed by fusion of three SiO_4 tetrahedra) linked such that every six rings enclose a twelve-membered Si–O ring.

QUASI-PREPOLYMER PROCESS (Semi-prepolymer process) A process, intermediate between the prepolymer and the one-shot processes for the formation of polyurethane foam, where the polyol (polyester or polyether) is first reacted with a large excess of diisocyanate, to produce a low viscosity prepolymer. This is then reacted with more polyol, or sometimes a low molecular weight compound such as ethylene glycol, with catalysts and other ingredients to produce the foam. Both rigid and flexible polyurethane foams are produced by this process.

QUATERNARY STRUCTURE The arrangement of the subunits of an oligomeric protein in space to form the complete protein 'molecule'. Like the tertiary structure, quaternary structure is determined by X-ray diffraction, but this is even more difficult due to the large molecular size of the protein and the existence of different polypeptides in different conformations.

QUENCHER (Quenching agent) A molecule capable of quenching photo-induced excited states and therefore of acting as an ultraviolet stabiliser in a polymer. Many transition metal chelates, especially of nickel, e.g.

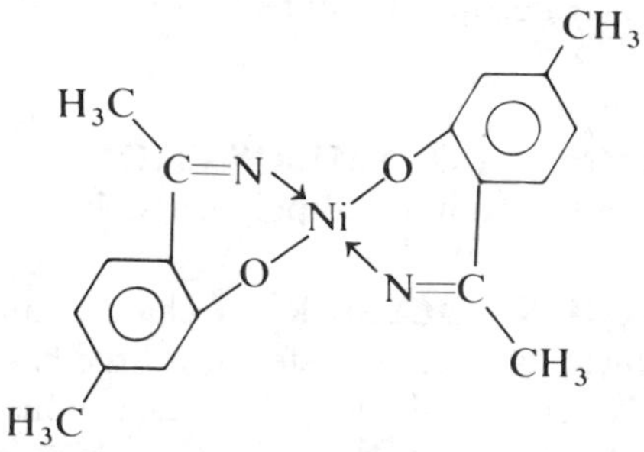

are thought to be effective UV stabilisers by acting at least partly as quenchers. Other UV stabilisers originally thought to work solely due to their UV absorber activity, are now believed to act partly as quenchers.

QUENCHING The process by which the excess energy of the excited state in a photo-irradiated molecule or group is lost by transfer to a second molecule (the quencher) which dissipates the energy harmlessly, usually as heat. Thus the excited state molecule cannot undergo bond dissociation and in the case of polymers use of a quencher is a method of stabilisation against photo-degradation.

QUENCHING AGENT Alternative name for quencher.

QUINOL Alternative name for hydroquinone.

QUINONE Alternative name for p-benzoquinone.

p-QUINONEDIOXIME

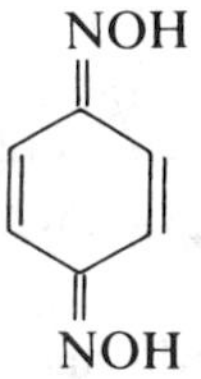

M.p. 240°C (decomposes).

A vulcanisation agent for rubbers, particularly useful for butyl rubber. Used in conjunction with an oxidising agent when it is converted to p-dinitrosobenzene, which then reacts with unsaturated sites on the rubber. The vulcanisates show better ageing and thermal stability than sulphur vulcanisates but are coloured, have poorer mechanical properties and are susceptible to scorch.

QUINONEMETHIDE A conjugated structure of the general type

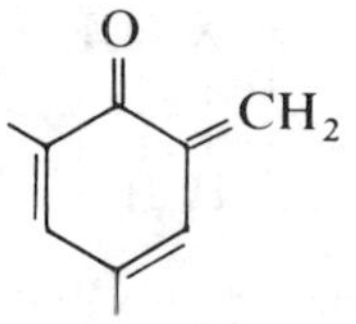

which may be formed during the curing of phenol–formaldehyde polymers by the high temperature decomposition of dibenzyl ether structures present in the prepolymers. Such structures can contribute to the dark colour often found in cured products.

QUIRVIL Tradename for polyvinyl chloride.

QX-13 Tradename for a polyimide formed by reaction between benzophenonetetracarboxylic dianhydride and diacetylmethyleneaniline

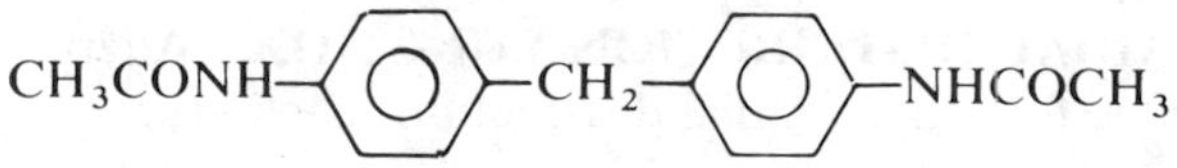

$$CH_3CONH-\!\!\langle O \rangle\!\!-CH_2-\!\!\langle O \rangle\!\!-NHCOCH_3$$

R

R Symbol for arginine.

RABINOWITSCH CORRECTION A correction applied to the viscosity data obtained from a capillary rheometer to allow for the non-Newtonian character of a fluid. It changes the wall shear rate ($\dot{\gamma}_w$) from $4Q/\pi R^3$ to $1/\pi R^3[3Q + \Delta P \, dQ/d\Delta P]$ according to the Rabinowitsch equation, where Q is the volume flow rate, R is the die radius, ΔP is the pressure drop across the capillary and $dQ/d\Delta P$ is the rate of change in Q with P. $\dot{\gamma}_w$ may alternatively be written in the Metzner form of the Rabinowitsch equation for a power law fluid.

RABINOWITSCH EQUATION (Weissenberg–Rabinowitsch–Mooney equation) An equation for the wall shear rate ($\dot{\gamma}_w$) in capillary rheometry in terms of the experimentally measurable quantities—the die radius (R), the volume flow rate (Q) and the pressure difference across the capillary ends (ΔP). One form of the equation is

$$-\dot{\gamma}_w = 1/\pi R^3[3Q + \Delta P \, dQ/d\Delta P]$$

or

$$-\dot{\gamma}_w = \tfrac{3}{4}(4Q/\pi R^3) + (\tau_w/4) \, d(4Q/\pi R^3)/d\tau_w$$

where τ_w is the wall shear stress. An alternative form using the flow behaviour index of the power law equation is the Metzner form.

RACEMIC DYAD (Syndiotactic dyad) Symbol r. A dyad in a polymer capable of exhibiting stereochemical configurational isomerism, in which the configurations of the two asymmetric, or rather pseudo-asymmetric, centres are opposite. A racemic dyad in a vinyl polymer may be represented by a Fischer projection formula as

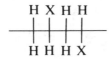

Thus the two methylene protons can be seen to be equivalent and so give a single resonance in their NMR spectrum.

RADEL Tradename for a polyethersulphone, probably of similar structure to Victrex PES and with a T_g value of 204°C.

RADIAL BANDING Alternative name for ringed spherulite.

RADIAL BLOCK COPOLYMER Alternative name for star block copolymer.

RADIAL DISTRIBUTION FUNCTION Alternative name for atomic radial distribution function.

RADIAL POLYMER Alternative name for star polymer.

RADIATION CROSSLINKING Crosslinking brought about by irradiation by high energy radiation, i.e. γ- or X-rays or proton, electron or neutron beams. Irradiation causes the formation of high local concentrations of free radicals, which may combine, forming crosslinks, as in polyethylene, polystyrene and diene rubbers, or which may initiate degradation as in cellulose, polymethylmethacrylate and polytetrafluoroethylene. Usually both processes occur simultaneously but with one or the other predominating.

RADIATION-INDUCED DEGRADATION High energy radiation-induced degradation. Degradation resulting from exposure of a polymer to any of a wide range of high energy radiations, including X-rays, γ-rays, electron, proton or neutron beams. The energy involved in all cases is so high that indiscriminate bond scission occurs, resulting in chemical and physical changes largely independent of the type of radiation involved.

The usual sequence of events is ionisation, formation of excited states and bond dissociation. Very high local concentrations of free radicals are formed in the path of the radiation. The radicals may either undergo β-scission or combine to form crosslinks; the balance between crosslinking and scission depends on the precise con-

ditions of exposure. Analysis of the two processes is performed by determination of the sol and gel contents of the irradiated polymer. Radiation crosslinking (e.g. of polyethylene) is a useful method of raising the stiffness and softening point of a polymer without the need for complex chemical vulcanisation.

RADIATION-INDUCED POLYMERISATION (Radiation polymerisation) (Radiolytic polymerisation) Chain polymerisation initiated by irradiation of the monomer with any form of high energy radiation—electron, neutron or α-particle beams, X- or γ-rays. Interactions of these radiations with the monomer, as with any organic compound, initially produces unstable ions, e.g. by ejection or capture of electrons, which subsequently dissociate to free radicals. Usually polymerisation is initiated by these free radicals, but at low temperature the ions may be stable enough to initiate ionic polymerisation. The method is particularly useful in solid state polymerisation since the energetic radiation readily penetrates into the interior of a solid monomer sample.

RADIATION POLYMERISATION Alternative name for radiation-induced polymerisation.

RADICAL-ION POLYMER A charge-transfer polymer in which complete transfer of charge has occurred. Such a polymer has enhanced electrical conductivity. The most widely studied type is that in which a polymer containing a quaternised nitrogen, such as in quaternised poly-2- or poly-4-(vinylpyridine), is complexed with a tetracyanoquinodimethane radical-ion counter-ion. Conductivities of up to about $1 \times 10^{-2}\,S\,cm^{-1}$ may be reached, but the polymers are dark coloured and oxidise readily in air. Similar salts with ionenes have also been studied.

RADICAL LIFETIME Alternative name for lifetime of a kinetic chain.

RADICAL POLYMERISATION Alternative name for free radical polymerisation.

RADICAL SCAVENGER (Free radical scavenger) (Free radical trap) (Radical trap) A molecule capable of chemically reacting with free radicals to form either non-radical products or radical products with low reactivity. A scavenger molecule may be a free radical itself, which forms a covalent bond by pairing with the unpaired electron of the radical of concern, e.g. diphenylpicrylhydrazyl (DPPH). Such a scavenger is used to test the existence of a free radical reaction. Many molecules forming fairly stable free radicals, e.g. phenols, aromatic amines, quinones and carbon black, are frequently used as chain-breaking antioxidants.

RADICAL TRAP Alternative name for radical scavenger.

RADIOLYTIC POLYMERISATION Alternative name for radiation-induced polymerisation.

RADIUS OF GYRATION Symbol R_G. A parameter characterising the size of a polymer random coil. It is defined as

$$R_G^2 = \sum_i m s_i^2 \bigg/ \sum_i m = \sum_{i=1}^n s_i^2/n$$

where the polymer chain consists of n segments, each of mass m, located at distance s from the centre of gravity of the coil. Thus it is the second moment of mass distribution about the centre of gravity. When all the possible conformations of the chain are considered then an average value of R_G, the mean square value $\langle R_G^2 \rangle$ is used. It is more accessible to experimental determination than the related end-to-end distance (r), to which it is related by $\langle R_G^2 \rangle_0 = (1/6)\langle r^2 \rangle_0$, where the subscripts 0 indicate the unperturbed dimensions. The value of R_G may be obtained from light scattering or ultracentrifuge measurements on a polymer solution, or by using small angle neutron scattering for a solid polymer. For branched chains, the chain dimensions decrease and R_G decreases with the extent of branching and various theories provide relationships between R_G and the branching index. Many relationships involving R_G only hold for the unperturbed dimensions with $\langle R_G^2 \rangle_0$ being used. In other than a theta solvent, $\langle R_G^2 \rangle = \alpha_R \langle R_G^2 \rangle_0$, where α_R is the expansion factor.

RAMACHANDRAN MAP (Ramachandran plot) (Steric map) An energy contour diagram, which plots the atomic interaction energies in polypeptide chains as a function of the torsional angles ϕ and ψ with ϕ as abscissa and ψ as ordinate. Each polypeptide has its own characteristic diagram, but generally they are all rather similar. The energy contours indicate which particular conformations are allowed. Since each regular conformation (such as the α- or other helix) has the same values for ϕ and ψ at each amino acid residue, it is represented by a point on the diagram. In the simplest treatment only those conformations are allowed where the interatomic distances are greater than those of the minimum distances found from X-ray crystallographic data of organic compounds. In more sophisticated calculations all pairwise interactions are summed using '6–12' potentials for van der Waals' and dispersion forces, plus torsional potentials, barriers to rotation, dipole and hydrogen bond interactions. The allowed regions are found to contain the regular conformations (α-helix, β-conformation, etc.) found in polypeptides.

RAMACHANDRAN PLOT Alternative name for Ramachandran map.

RAMAN SCATTERING SPECTROSCOPY Alternative name for Raman spectroscopy.

RAMAN SPECTROSCOPY (Raman scattering spectroscopy) A spectroscopic technique in which the sample is irradiated with monochromatic, usually laser, light and the scattered radiation, which is usually of longer wavelength (Stokes emission) is measured at 90° or 180°. The scattering is inelastic Raman scattering, whereby emission occurs from an excited, very high vibrational state to lower vibrational states, but not to the ground state. Thus a distribution of wavelengths is emitted, corresponding to the different vibrational modes in the molecules of the sample. For a transition to occur, the polarisability of the bond involved must not change during the excitation of the vibration. In general, this means that Raman and infrared spectra are similar. However, homonuclear vibrations, particularly C—C vibrations, are Raman but not infrared active. This is of importance for the observation of the vibrations, e.g. the longitudinal acoustic mode, in carbon–carbon chain polymers. Further advantages of the technique are that only a small sample area can be investigated and that the sample can be in any of many forms, including in aqueous solution, which is not possible with infrared spectroscopy since water strongly absorbs. Experimentally, a laser illuminates a small area of the sample and the scattered light is dispersed with a double grating monochromator and detected with a photomultiplier. Applications of the technique have expanded considerably since the development of lasers (laser Raman spectroscopy).

RAMIE A natural bast fibre obtained from the subtropical nettles *Boehmeria nivea* (China grass) or *Boehmeria tenacissima* (rhea). The degummed and bleached fibres are almost pure α-cellulose and have variable lengths of up to about 50 cm. They are very strong, durable and lustrous but lack elasticity.

RANDOM CHAIN PROCESS Alternative name for Bernouillian process.

RANDOM COPOLYMER (Ideal copolymer) A copolymer in which the different repeat units are arranged randomly along the polymer chain, e.g. for the most common type, a binary copolymer of units A and B:

$$\sim\!\!AABABBBAABBAAAABABBBBAABB\!\!\sim$$

A perfectly random arrangement results from chain copolymerisation by simultaneous polymerisation of monomers A and B, in which the product of the monomer reactivity ratios r_A and r_B is unity—ideal copolymerisation. Some free radical and many ionic chain copolymerisations are ideal, but in the former case copolymerisation behaviour is often intermediate between ideal and alternating. Step-growth copolymerisation will yield random copolymers if the reactivities of the monomers are the same, as is usually the case. Furthermore, often step-growth block copolymers become randomised through the occurrence of interchange reactions. Mixtures of step-growth homopolymers may similarly become randomised in copolymer structure.

A random copolymer formed from monomers whose homopolymers are amorphous will have properties intermediate between those of the homopolymers. A copolymer formed from monomers, at least one of which forms a crystalline homopolymer, may have properties completely different from those of both of the corresponding homopolymers. In particular, random copolymerisation, by increasing structural irregularity, reduces crystallinity, so that large amounts (typically about 25%) of a second comonomer may completely prevent crystallisation. A notable example is found with the random copolymers of ethylene and propylene, which are rubbery, whereas both the homopolymers are crystalline and plastic. Random copolymers are often named by linking the two monomer names by the term -co-, as, for example, in poly-(styrene-co-methylmethacrylate).

RANDOM DISTRIBUTION Alternative name for Schultz–Flory distribution.

RANDOM FLIGHT CHAIN Alternative name for freely jointed chain.

RANDOM ORIENTATION Orientation in which all directions of orientation are equally likely, when the value of the orientation function becomes zero. This produces a circular wide angle X-ray diffraction pattern similar to a powder diagram usually produced in bulk and moulded specimens.

RANDOM PREPOLYMER A prepolymer obtained by reaction of a mixture of monomers (one of which is usually of functionality > 2) by step-growth polymerisation, but in which the reaction has been stopped, usually by cooling, before network formation and gelation have occurred. In this form the polymer is still soluble and fusible (the A stage) or close to the gel point (B stage) and capable of fabrication. Subsequent network formation is brought about merely by continuing the reactions started in the first stage, usually simply by heating. Examples are phenol–formaldehyde resoles, alkyd resins and urea– and melamine–formaldehyde methylol prepolymers.

RANDOM RE-ENTRY Alternative name for switchboard model.

RANDOM SCISSION Polymer chain scission which occurs with equal probability at any repeating unit of the polymer molecule. The result of many polymer degradations, especially when a linking bond (interunit link) between repeat units is capable of being broken by chemical attack, e.g. as in the hydrolysis of ester links. The proportion of links broken is the degree of degradation. Rapid molecular weight reduction is the major result.

RAOLIN Tradename for chlorinated rubber.

RATE OF CURE The rate at which the stiffness of a rubber compound increases during vulcanisation due to crosslink formation. The rate of cure determines the cure time.

RATE OF DEFORMATION TENSOR Symbol Δ_{ij}. The tensor notation for the rate of deformation of a material, where in cartesian coordinates $\Delta_{ij} = (\partial u_i / \partial x_j + \partial u_j / \partial x_i)$ using index notation and the summation convention, where u_i are the velocity components in directions x_i. For most materials the stress components depend only on the Δ_{ij} components so that the latter provide the connection between the dynamics and the kinematics of flow of a fluid.

RATE OF POLYMERISATION Symbol R_p. In chain polymerisation, the rate at which monomer (M), of concentration [M], is converted to polymer, i.e. $-d[M]/dt$. Since propagation occurs hundreds of times more frequently than initiation, rates of polymerisation and propagation are the same and the symbol R_p is used for both. Since propagation is $\sim\!\!\text{M}^* + \text{M} \rightarrow \sim\!\!\text{MM}^*$, $R_p = k_p[M][M^*]$, where k_p is the propagation rate constant and $[M^*]$ is the active centre concentration. By use of the steady state assumption the unknown $[M^*]$ value can be eliminated, since the rate of initiation is $R_i = 2k_t[M^*]$ and the rate of initiation = rate of termination at the steady state, and hence $R_p = k_p[M](R_i/2k_t)^{1/2}$ where k_t is the rate constant for termination defined as the rate constant for

$$\sim\!\cdot + \cdot\!\sim \xrightarrow{k_t} \text{dead polymer}$$

In free radical polymerisation initiated by thermal decomposition of initiator I with rate constant k_d and initiator efficiency f, $R_p = k_p[M](fk_d[I]/k_t)^{1/2}$. Deviations from this behaviour can occur due to autoacceleration, transfer, primary radical termination, inhibition or when R_i is dependent on monomer concentration.

RAVIKAL Tradename for acrylonitrile–butadiene–styrene copolymer.

RAVINIL Tradename for polyvinyl chloride.

RAYFLEX Tradename for a high tenacity viscose rayon filament.

RAYLEIGH–BRILLOUIN SCATTERING Alternative name for Brillouin scattering.

RAYLEIGH EQUATION An equation giving the intensity of scattered light (I_θ) at an angle θ to the incident light (intensity I_0) direction at a distance r from an anisotropic scattering particle whose size is small compared with the wavelength of the light (λ). It is $I_\theta r^2 / I_0 = 16\pi^4 \alpha^2 \cos^2 \theta / \lambda^4$ for polarised light and

$$I_\theta r^2 / I_0 = (8\pi^4 \alpha^2 / \lambda^4)(1 + \cos^2 \theta)$$

for unpolarised light, where α is the polarisability of the particle. The equation forms the basis of much of the theory of light scattering, including that useful for the determination of polymer molecular weights.

RAYLEIGH LINE BROADENING (Rayleigh linewidth spectroscopy) The frequency broadening of the monochromatic line in Rayleigh scattering due to interaction of the incident light with processes involving dissipation of fluctuations of energy, entropy or temperature. The broadening is therefore especially dependent on diffusion and relaxation processes. The effect has been most widely used to determine the translational diffusion coefficient of polymers, especially of proteins, in solution. The shape and linewidth of the Rayleigh peak, at a 90° scattering angle, is measured using a photomultiplier either together with a frequency analyser or by measuring residual photon counts at fixed time intervals. In the light beating spectroscopic technique, the scattered light is analysed together with light from the incident beam, to produce beats of frequency equal to the frequency difference between the two components.

RAYLEIGH LINEWIDTH SPECTROSCOPY Alternative name for Rayleigh line broadening.

RAYLEIGH RATIO Symbol R_θ. In light scattering defined as $R_\theta = I_\theta r^2/I_0$, where I_θ is the scattered light intensity at a point at distance r from the scattering centre and for which the angle of the scattered beam to the incident beam (of intensity I_0) is θ. It enables the scattering intensities to be described independently of various instrumental parameters. The reduced scattering intensity (symbol also R_θ) is similar, being defined as

$$R_\theta = I_\theta r^2/I_0(1 + \cos^2 \theta)$$

and is numerically equal to the Rayleigh ratio when $\theta = 90°$. Therefore the use of the term is sometimes restricted to measurements at $\theta = 90°$. Widely used in the analysis of light scattering data for determination of the weight average molecular weight of a polymer by the Zimm plot method.

RAYLEIGH SCATTERING The elastic scattering of radiation by a random dispersion of independent point scatterers whose size is small compared with the wavelength of the radiation (λ)—commonly taken to be less than $\lambda/20$. The scatterers may be atoms or molecules and since they are independent, the scattered radiation from different scatterers is incoherent and interference effects cancel each other out. The theory was first developed for the scattering of light by an ideal gas. For non-polarised incident light of intensity I_0, the intensity of scattered light (I_θ) at an angle θ to the incident beam and a distance r from the scattering centre is

$$I_\theta = I_0(n^*/r^2)(8\pi^4\alpha^2/\lambda^4)(1 + \cos^2 \theta)$$

where n^* is the number of scattering centres of polarisability α.

To eliminate parameters merely dependent on the experimental arrangement, notably r, the scattered light is expressed as the Rayleigh ratio (R_θ) as $R_\theta = I_\theta r^2/I_0$ or as the reduced scattering intensity (also R_θ) as

$$R_\theta = I_\theta r^2/I_0(1 + \cos^2 \theta)$$

which is independent of θ. For plane polarised light R_θ is doubled. If a dilute solution is similarly considered as a collection of point scatterers, then $Kc/R_\theta = 1/M$, where M

is the solute molecular weight and K is a constant equal to $(2\pi^2 n^2/\lambda^4 N_A)(dn/dc)^2$ (where n is the refractive index, dn/dc is the refractive index increment and N_A is Avogadro's constant).

RAYON Generic name for the regenerated man-made fibres of cellulose from cuprammonium cellulose, viscose or cellulose acetate. The term originally included cellulose acetate itself, but this is now referred to as acetate.

RBZ THEORY Abbreviation for Rouse–Bueche–Zimm theory.

REACTION MOULDED POLYURETHANE ELASTOMER Alternative name for cast polyurethane elastomer.

REACTIVITY RATIO (Monomer reactivity ratio) Symbol r. In chain copolymerisation, the ratio of the rate constants for the addition of one monomer (A) to a growing active centre terminated by an A unit, to that for addition of the other monomer (B) to the same type of active centre. Thus in binary copolymerisation there are four possible propagation reactions, with associated rate constants (k):

$$\sim\!A^* + A \xrightarrow{k_{AA}} \sim\!AA^*$$

$$\sim\!A^* + B \xrightarrow{k_{AB}} \sim\!AB^*$$

$$\sim\!B^* + B \xrightarrow{k_{BB}} \sim\!BB^*$$

$$\sim\!B^* + A \xrightarrow{k_{BA}} \sim\!BA^*$$

Thus the reactivity ratio for monomer A (r_A) is k_{AA}/k_{AB} and the reactivity ratio for monomer B (r_B) is k_{BB}/k_{BA}. The value of r is therefore a measure of the preference for an active centre to add on a monomer of its own type rather than to add on a monomer of the opposite type.

The values of r_A and r_B, together with the monomer feed composition, therefore determine the composition of the copolymer, as expressed in the copolymer composition equation. Furthermore, the product $r_A r_B$ determines the distribution of the monomer units in the copolymer. When $r_A r_B = 1$, then ideal copolymerisation results, leading to a truly random copolymer. When both r_A and r_B separately equal unity, both monomers show equal reactivity to both propagating species and the copolymer composition is the same as that of the monomer feed. When $r_A = r_B = 0$, alternating copolymerisation results. Most commonly $1 > r_A r_B > 0$, so that copolymerisation behaviour is between the ideal and alternating types. When both r_A and r_B are < 1, then at some feed composition azeotropic copolymerisation occurs. If $r_A \gg r_B$, then the tendency is to consecutive homopolymerisation of the two monomers separately. The case of both r_A and $r_B > 1$ is comparatively rare and leads to block copolymer formation. If both r_A and r_B are $\gg 1$, then simultaneous homopolymerisation of both monomers occurs. This is very rare.

In free radical copolymerisation the tendency is towards alternation and the reactivity ratios are comparatively insensitive towards reaction conditions such as solvent, temperature, pressure and monomer concentration. In ionic copolymerisation, the tendency is to ideal copolymerisation, so that it is difficult to prepare copolymers containing appreciable amounts of both monomers. In many cases the product $r_A r_B$ is >1. Furthermore, in contrast to free radical copolymerisation, the values of the reactivity ratios are more dependent on reaction conditions.

RECIPROCAL SCATTERING FUNCTION The function Kc/R_θ, used in the light scattering method of determination of polymer weight average molecular weight, where K is a constant, R_θ is the Rayleigh ratio and c is the polymer solution concentration. Kc/R_θ is plotted against $Ac + B\sin^2\theta/2$ in the Zimm plot used for analysis of the data, where A and B are constants and θ is the scattering angle.

RECLAIMED RUBBER A scrap vulcanised rubber which has been heat treated, often in the presence of 'reclaiming chemicals', typically at up to 200°C. This breaks down the network structure, increasing plasticity and hence enables the rubber to be reprocessed. A typical reclaim contains about 45% rubber, the rest being carbon black, softeners, etc. In tyre reclaim the reinforcing fibre may be removed by chemical or mechanical treatment.

RECOMBINATION Alternative name for geminate recombination.

RECOVERABILITY The ability of a body to recover its original dimensions after deformation when the deforming forces are removed. Although, unlike metals, polymers may recover completely after stressing beyond the proportional limit, when stressed beyond the yield point, permanent deformation will have occurred.

RECOVERABLE SHEAR For an elastic fluid, the ratio of the first normal stress difference $(\sigma_{11} - \sigma_{22})$ to twice the shear stress (σ_{12}) at the same shear rate for steady simple shear. It is thus a measure of the relative importance of elastic to viscous effects. It is useful for the calculation of die swell recovery in extrusion. In one theory, based on the assumption that die swell is an elastic effect, the swelling ratio (χ) is given by $(1 + \frac{1}{2}S_R^2)^{1/6}$, or approximately $\chi = S_R^{1/2}$, where S_R is the recoverable shear.

RECOVERY The decrease of strain on removal of stress. Recovery is complete and instantaneous in perfectly elastic materials; however, it may be delayed and only partial in polymers, when it may be called creep recovery since it is the reduction in the creep strain on removal of, or reduction in, the applied stress. In particular, if a stress σ_0 is applied at zero time and removed at time t_1, the recovery at a longer time t is the difference between the creep that would have occurred at time $t_1 + t$, had σ_0 not been removed, and the actual creep.

As a consequence of the Boltzmann Superposition Principle this is identical to the creep response, had a stress σ_0 been applied at time t_1, i.e. the creep and recovery responses are identical in magnitude. For the commonly observed non-linear behaviour, recovery is even more complex than creep since it is controlled by three variables, the duration of the original creep, the original creep strain and the duration of recovery. Since, like creep, non-linear recovery is important in engineering design with plastics, characterisation over a wide range of conditions is necessary. The task may be simplified by using the parameters fractional recovery and reduced time.

RECOVERY FACTOR The product of the elastic recovery and the work recovery multiplied by 100. The elastic recovery is defined as the fractional length of a fibre recovered from its initial extension after removal of the stressing force, and the work recovery is the area of the stress–strain curve recovered from that required for the initial extension, i.e. the stored energy in deformation. Recovery factor is an important factor in textile behaviour.

RECRYSTALLISATION Crystallisation of partially molten or dissolved material with the remaining crystals acting as a substrate and providing the primary crystallisation nuclei.

REDISTRIBUTION An interchange of the substituents on chlorosilane monomers, in which the less useful components of the mixed chlorosilane produced by the direct process are converted to more useful products. Thus heating a mixture of trimethylchlorosilane and methyltrichlorosilane at 200–400°C with aluminium chloride produces the more useful dimethyldichlorosilane:

$$(CH_3)_3SiCl + CH_3SiCl_3 \rightarrow 2(CH_3)_2SiCl_2$$

REDON Tradename for a polyacrylonitrile fibre.

REDOX INITIATION Initiation of free radical polymerisation by the free radicals generated by an oxidation–reduction (redox) reaction between two components, which by themselves are quite stable. The advantage is that the free radicals may be generated at suitable rates at low temperature (down to -50°C). Most systems are water-soluble and are particularly suitable for emulsion polymerisation. A wide variety of oxidants and reducing agents have been used. Examples of redox reactions are:

$$HO{-}OH + Fe^{2+} \rightarrow HO^- + Fe^{3+} + HO\cdot$$
$$RO{-}OR + Fe^{2+} \rightarrow RO^- + Fe^{3+} + RO\cdot$$
$$RO{-}OH + Fe^{2+} \rightarrow RO^- + Fe^{3+} + HO\cdot$$

Other variable valency metals may be used, such as Co^{2+} (as a naphthenate or octoate as used in the crosslinking of unsaturated polyesters). Peroxide decomposition may also be accelerated by amines:

$$PhNR_2 + R'CO\!-\!OCR' \rightarrow R'CO\cdot + R'CO^- + PhNR_2^+$$

Another common component is the persulphate ion, as in:

$$^-O_3SO\!-\!OSO_3^- + Fe^{2+} \rightarrow Fe^{3+} + SO_4^{2-} + SO_4^-\cdot$$

or

$$^-O_3SO\!-\!OSO_3^- + S_2O_3^{2-} \rightarrow SO_4^{2-} + SO_4^-\cdot + S_2O_3^-$$

REDOX INITIATOR A polymerisation initiation system capable of producing free radicals at low temperatures and causing redox initiation of free radical polymerisation.

REDOX POLYMER Alternative name for oxidation–reduction polymer.

REDOX POLYMERISATION A free radical polymerisation initiated by redox initiation.

REDUCED OSMOTIC PRESSURE Symbol Π/c. The osmotic pressure (Π) of a solution divided by its concentration c. In the membrane osmometry method of determination of polymer number average molecular weight ($\bar{M}_n$), Π/c is plotted against c and the intercept at $c = 0((\Pi/c)_{c\to 0})$ is used to calculate $\bar{M}_n$, since

$$\Pi/c = RT/\bar{M}_n + Bc + Cc^2 + \cdots$$

where B and C are the second and third virial coefficients respectively. Sometimes $(\Pi/c)^{1/2}$ is plotted against c instead, when the value of C is not zero.

REDUCED SCATTERING INTENSITY Symbol R_θ. In light scattering, it is defined as $R_\theta = I_\theta r^2/I_0(1 + \cos^2 \theta)$, where I_θ is the intensity of scattered light at a point at distance r from the scattering centre and for which the scattered beam is at an angle θ to the incident beam whose intensity is I_0. It enables the scattering to be described independently of the parameters r and θ. R_θ also equals $8\pi^4 n\alpha^2/\lambda^4 V$, where α is the polarisability of the scattering particle, λ is the light wavelength and V is the volume in which n scattering centres are present. Closely related to the Rayleigh ratio (symbol also R_θ), defined as $R_\theta = I_\theta r^2/I_0$, and therefore numerically equal to it at $\theta = 90°$. Useful in the analysis of light scattering data for the determination of weight average molecular weight of a polymer by the Zimm plot method.

REDUCED SPECIFIC VISCOSITY Alternative name for viscosity number.

REDUCED TIME In the presentation of the recovery data of non-linear viscoelastic materials, the ratio of the recovery time to the preceding creep time. A plot of fractional recovery versus reduced time can bring data from measurements made under different conditions into coincidence on a master curve, thus simplifying the use of data for engineering design purposes.

REDUCED VISCOSITY Alternative name for viscosity number.

REDUCING SUGAR A carbohydrate which contains a potential aldehyde group which is capable of acting as a reducing agent, e.g. reducing Fehling's solution to copper. Reducing monosaccharides are the aldoses. Reducing disaccharides are those in which the interunit glycoside bond is between the anomeric (reducing) centre (i.e. the hemi-acetal hydroxyl) of one monosaccharide unit and an alcoholic hydroxyl of the other. Thus 1,4′-linked disaccharides, e.g. cellobiose and maltose, are reducing Polysaccharides may similarly contain reducing end groups.

REDUX Tradename for polyvinyl formal used in conjunction with a phenolic resin as an adhesive.

REFLECTION–ABSORPTION SPECTROSCOPY Infrared spectroscopy in which the sample is spread as a thin film on a highly reflective metal surface. By having two such layers parallel to each other many such reflections can be obtained, as in the MIR version of attenuated total reflection spectroscopy. However, in contrast to this technique, the reflections occur external to the sample after the beam has passed through it. The technique is useful for the investigation of polymer films on metal surfaces and for studying polymer to metal adhesion.

REFRACTIVE INDEX INCREMENT (Specific refractive increment) Symbol dn/dc. The rate of change of refractive index (n) of a solution with concentration of solute (c). For polymer solutions very precise values are required for the analysis of light scattering data since the term $(dn/dc)^2$ occurs in the equations relating scattered light intensity to weight average molecular weight. Precise dn/dc values are obtained using a differential refractometer.

REFRACTIVITY INTERCEPT A measure of the aromaticity of a rubber oil. Defined as refractivity intercept $= n_D^{20} - 0.5d^{20}$, where n_D^{20} is the refractive index at 20°C for the sodium D-line and d^{20} is the density at 20°C. Values may vary from about 1·045 for a paraffin oil and 1·05 for a naphthenic oil to about 1·07 for an aromatic oil.

REGAIN Alternative name for moisture regain.

REGENERATED CELLULOSE Cellulose which has been converted to a cellulose derivative to solubilise it, to enable it to be spun as a fibre or cast as a film. After spinning or casting the cellulose derivative is reconverted to cellulose (regenerated). Viscose rayon is the most important example, although cuprammonium rayon and denitrated cellulose nitrate (Chardonnet silk) are other examples. Regeneration makes it possible to produce fibres of a wide range of deniers and other characteristics and enables a cheap cellulose source (wood pulp) to be used. The regeneration process usually causes some degradation of the cellulose molecular chains, resulting in a lowering of DP, e.g. from about 1000 to about 250 in viscose rayon.

REGENERATED FIBRE A man-made fibre made from a naturally occurring polymer, which is dissolved in a suitable solvent and wet or dry spun to produce a fibre. Although the natural polymer may itself have been fibrous, and hence the fibre may be said to be regenerated, this is not so with man-made protein fibres. Usually the polymer is cellulose, either cellulose itself, as with viscose and cuprammonium rayon production, or a cellulose derivative, especially cellulose acetate. The earliest man-made fibres were of this type, but synthetic fibres have become much more important.

REGENERATED PROTEIN A protein which has been dissolved in an aqueous solution and regenerated as a fibre by a wet spinning process. Several commercial regenerated protein fibres have been developed, but are now no longer of importance. They were based on acid-casein, the groundnut proteins arachin and conarachin, the soya bean protein glycinin and the maize protein zein.

REGIONS OF VISCOELASTICITY Alternative name for five regions of viscoelasticity.

REGULATOR (1) Alternative name for modifier. (2) Alternative name for effector.

REINFORCEMENT The enhancement of some aspect or aspects of the mechanical behaviour of a material by the incorporation of a second, reinforcing, component thus producing a composite material. The particular property being enhanced depends on the application of the composite. Most frequently reinforcement refers to an increase in strength properties, often with an accompanying increase in modulus. Particular aspects of strength reinforcement of interest are tensile strength, impact strength, abrasion resistance and tear resistance, although enhancement of other properties are also referred to as reinforcement. For strength and modulus increases, the reinforcing agent must itself be stronger and stiffer than the polymeric matrix. Fibrous reinforcing fillers are the most frequently used fillers for highly reinforced composites, especially glass, but also carbon, boron and stiff organic fibres, in unsaturated polyester and epoxy resin matrices. They may be used as continuous filaments or as short fibres. Some fibrous fillers, such as cellulose, asbestos and whisker materials are only available as short fibres. The performance of many fillers, especially glass, is improved by the use of coupling agents. Reinforcement with particulate fillers is also possible, especially the use of carbon black and silica in rubbers. Particulate fillers, such as talc and other mineral fillers, may reinforce plastics, but usually only if they are of small particle size. Short cellulose fibre fillers are widely used for the reinforcement of phenol–, urea– and melamine–formaldehyde polymers.

REINFORCEMENT FACTOR The ratio of the Young's modulus of a composite to that of its matrix polymer.

RELATIVE MOLAR MASS Alternative name for molecular weight when this is expressed as a dimensionless quantity.

RELATIVE PERMITTIVITY (Permittivity) (Dielectric constant) (Specific inductive capacity) Symbol ε. The ratio of the capacitance of a capacitor filled with the dielectric of interest to its capacitance with a vacuum between its electrodes. Thus ε is a measure of the ability of the dielectric to store electric charge through its polarisation when an electric field is applied. It is sometimes also simply called the permittivity, but this term may also refer to the absolute permittivity. ε is equal to the ratio of the absolute permittivity of the dielectric to that of a vacuum (free space). In a static field a static value of the relative permittivity will be observed. In a varying field, the value will depend on the frequency. A complex relative permittivity (ε^*) is then often used to represent dielectric behaviour. This is given by $\varepsilon^* = \varepsilon' - i\varepsilon''$, where ε' is also called the relative permittivity, ε'' is the loss factor and i is $(-1)^{1/2}$. At a frequency in the region of a dielectric relaxation, the value of ε' goes through a step increase as a particular polarisation mechanism comes into play with decreasing frequency. In a similar way ε' is temperature dependent, increasing temperature causing step increases in ε' again as particular additional molecular motions come into play, for example at T_g. In general, the more polar a polymer, the greater its relative permittivity will be. Thus hydrocarbon polymers have values of 2–3, whilst polar polymers can have values of up to about 10. However, in some polymers the orientations of the dipoles may tend to cancel each other out, as in polytetrafluoroethylene. Measurement of the relative permittivity, as with other dielectric properties (such as loss factor), utilises many different techniques. Frequency domain methods are used at high frequencies, such as the Schering and transformer ratio arm methods (10–10^5 Hz), resonance methods (10^5–10^8 Hz) and wave transmission methods (10^9–10^{11} Hz). Time domain methods are used at 10^{-4}–10^6 Hz.

RELATIVE VISCOSITY Symbol η_{rel}. Alternative name for viscosity ratio.

RELAXATION The time-dependent return to equilibrium, of a system, or property of a system, which has been displaced from equilibrium by an applied constraint. In the context of polymer relaxation, the constraints may be an electric field (dielectric relaxation), a magnetic field (nuclear magnetic relaxation), visible or ultraviolet radiation (luminescence depolarisation or dynamic birefringence) or, most importantly, mechanical stress or strain (mechanical relaxation). In an ideal relaxation the rate of return to the equilibrium value P_{eq} of the property (P) of interest is proportional to the distance from equilibrium and is given by

$$(P - P_{eq}) = (P - P_{eq})_0 \exp(-t/\tau)$$

where $(P - P_{eq})_0$ is the initial value, t is the time and τ is the relaxation time. If the constraint is applied as a pulse then the decay of the material response is followed. Alternatively, the applied constraint may be periodic usually having the sinusoidal waveform $c = c_0 \sin \omega t$, where ω is the frequency and c_0 is the maximum amplitude. For an ideal relaxation, $(P - P_{eq})$ is a simple exponential function. P can also be considered as a complex quantity (P*) which may be resolved mathematically into a real (P') and an imaginary part (P'') as $P^* = P' + iP''$. P' is the in-phase (or storage) component and P'' is the out-of-phase (or loss) component, with

$$P' = P/(1 + \omega^2\tau^2) \quad \text{and} \quad P'' = \omega\tau P/(1 + \omega^2\tau^2)$$

where ω is the frequency, τ is the relaxation time and P is the value of the property at zero frequency, i.e. the static value.

In polymers, relaxations are particularly important since they are manifestations of molecular motions. These motions are particularly restricted in polymers, not only in the solid state but also in dilute solution, due to the long chain nature of polymer molecules. Thus relaxation times may be very long and energy losses high as a result of viscous forces opposing molecular motion. Observations of relaxations using dynamic methods, especially dynamic mechanical and dielectric spectroscopy, are useful in characterising the associated transitions in polymers.

RELAXATION BIREFRINGENCE Birefringence arising during stress relaxation of a deformed material. It is used to test the ideality of the Gaussian elasticity of rubbers by use of the stress–optical law.

RELAXATION IN THE ROTATING FRAME Nuclear magnetic relaxation of nuclear spin magnetisation parallel to the applied field, when the field is varying at radio frequencies. This enables the spin–lattice relaxations to be observed at much lower frequencies (about 10^3 Hz) than is usual with nuclear magnetic measurements.

RELAXATION MODULUS Symbol E_r or $E_r(t)$ for tensile measurements and G_r or $G_r(t)$ for shear measurements. The time-dependent modulus observed during stress relaxation defined as $E_r(t) = \sigma(t)/\varepsilon(0)$, where $\sigma(t)$ is the time-dependent stress, which is decaying at a fixed strain $\varepsilon(0)$ applied at time zero. The value of E_r is, like other moduli, temperature dependent, especially in the region of T_g. To compare values at different temperatures, the time must be held constant. This may be done by performing stress–strain experiments at different temperatures and extracting values of the secant modulus at the same time from the family of curves. Thus, if 10 s was chosen as the time (which is the appropriate time scale for many quasi-static T_g value determinations), then E_r would be given the symbol $E_r(10)$.

RELAXATION SPECTRUM Alternative name for dynamic mechanical spectrum and for relaxation time spectrum.

RELAXATION STRENGTH (1) For a viscoelastic material, the difference in the strains between the initial glassy, unrelaxed (i.e. short time) state and the final, relaxed (i.e. long time) state, usually divided by the initial strain. Thus it is also equal to $(G_u - G_r)/G_r$, where G_u and G_r are the unrelaxed and relaxed moduli respectively.

(2) The change in the real component of the shear modulus (G') in a dynamic mechanical behaviour as the relaxation region is traversed either as a result of the variation of temperature or of frequency. Thus it is the dispersion in G'. It is given by $G' = \int dG'/dT \, dT$. In the case of cyclohexyl derivatives of acrylic polymers, a correlation between $\Delta G'$ and the free energy conformational changes of the cyclohexyl ring has been established.

(3) For a dielectric relaxation it has the symbol $\Delta\varepsilon$ and is the difference between the relative permittivity of a dielectric measured at high frequency (ε_∞) and that measured at low frequency—the static value (ε_s).

RELAXATION TIME (1) Symbol τ. A measure of the speed of a relaxation process. It is defined as the time required for the value of a property P, or more precisely its displacement from the equilibrium value $(P - P_{eq})$, to decay to $1/e$ (i.e. 0·368) of its original value. Under dynamic conditions, with a sinusoidally varying constraint of frequency ω, the real component (P') (the complex form of P) of P* is high when $\omega < \tau^{-1}$, since polymer molecular motions then readily take place, whereas when $\omega > \tau^{-1}$ little motion is possible and P' and P'' are low. When $\omega = \tau^{-1}$, a maximum in P'' occurs, i.e. there is a maximum in the loss. Values of relaxation times can vary from hours or even days (in the case of mechanical relaxations of some polymers when subject to a pulsed applied strain) to as low as 10^{-10} s for nuclear magnetic relaxations.

(2) Alternative name for dielectric relaxation time.

RELAXATION TIME SPECTRUM (Relaxation spectrum) Symbol $H(\tau)$. The distribution of relaxation times $f(\tau)$ of a viscoelastic material whose behaviour can be represented by a model of a large number of Maxwell elements joined in parallel. Thus the stress, $\sigma(t)$, during stress relaxation may be represented as

$$\sigma(t) = G_r e + e \int_0^\infty f(\tau)\exp(-t/\tau)\,d\tau$$

where G_r is the relaxed modulus, $f(\tau)d\tau$ is a weighting function giving the concentration of Maxwell elements with relaxation times between τ and $\tau + d\tau$ and e is the fixed strain. In practice a logarithmic time scale is more convenient and a different relaxation time spectrum $H(\tau)$ is used, where $H(\tau)d(\ln\tau)$ gives the contributions to the stress relaxation associated with relaxation times between $\ln\tau$ and $(\ln\tau + d\ln\tau)$. The stress relaxation modulus $G(t)$ is then given by

$$G(t) = G_r - \int_{-\infty}^\infty H(\tau)\exp(-t/\tau)\,d\ln\tau$$

The relaxation spectrum can be calculated from the experimentally determined $G(t)$ values by Fourier or

Laplace transforms or from the real or imaginary components of the complex modulus.

RELAXED COMPLIANCE Alternative name for equilibrium compliance.

RELAXED MODULUS Alternative name for equilibrium modulus.

RENATURATION The restoration of the native conformation and biological activity of a protein that has been denatured by removal of the cause of denaturation, e.g. by cooling, change in pH or removal of the salt acting as denaturing agent.

RENEKER DEFECT A combined dislocation and disinclination in which the former has a displacement (translation) vector parallel to the polymer molecule and the latter has a rotation vector of 180° with the axis of rotation, which is the molecular axis. Propagation of this defect from a fold or chain end in a folded chain crystal can account for the increase in fold length during annealing.

RENYL Tradename for nylon 6.

REPEATING UNIT Alternative name for repeat unit.

REPEAT UNIT (Repeating unit) (Constitutional unit) The smallest structural unit of a polymer chain and the most important feature in determining the polymer properties. In a simple linear homopolymer the polymer structure may be represented as $-\!\!\!-[M]_n$, where M is the repeat unit. Thus, for example, in polypropylene

$$-\!\!\!-[CH_2CH]_n \quad \text{the repeat unit is} \quad -\!CH_2CH-$$
$$\quad\quad |\quad\quad\quad\quad\quad\quad\quad\quad\quad\quad\quad\quad |$$
$$\quad\quad CH_3 \quad\quad\quad\quad\quad\quad\quad\quad\quad\quad CH_3$$

Usually the repeat unit is the same as the mer, although it may be smaller, as in polyethylene where the repeat unit is $-CH_2-$ but the mer is $-CH_2CH_2-$. However, in such cases the mer is also frequently called the repeat unit. The repeat unit may also be larger than the mer. Thus in AABB-type step-growth polymers the repeat unit consists of two mers derived from the two monomers AA and BB, e.g. in nylon 66 the repeat unit is

$$-NH(CH_2)_6NHCO(CH_2)_4CO-$$

consisting of two mers

$$-NH(CH_2)_6NH- \quad \text{and} \quad -CO(CH_2)_4CO-$$

derived from the diamine and diacid used to form the polymer. In copolymers more than one repeat unit is present. In highly crosslinked network polymers it is often not possible to recognise a repeat unit due to the complex nature of the structure.

REPLICATION A method of specimen preparation for electron microscopic examination of its surface. The specimen surface is shadowed with a heavy metal to increase contrast in the microscope and an impression is made on another material deposited on the shadowed surface. The original material is then dissolved or stripped away leaving a replica of the adhering shadow.

REPTATION The snake-like sliding motion of a polymer chain that enables it to undergo long-range conformational changes by escaping from the imagined tube that surrounds the contorted contour of the polymer chain. The tube places a constraint on the motions of the polymer molecule, modelling the constraints imposed by entanglement couplings.

RESAMIN Tradename for urea– and melamine–formaldehyde polymers.

RESERVE DEPOSIT POLYSACCHARIDE (Food deposit polysaccharide) (Reserve food polysaccharide) (Reserve polysaccharide) A polysaccharide whose main biological function is to act as a reserve food supply. The main plant reserve foods are starch and inulin; in animals glycogen serves this function. When the food energy is required the polymer is broken down to monosaccharide sugars by enzymatic hydrolysis in plants or by enzymatic phosphorolytic fission in animals.

RESERVE FOOD POLYSACCHARIDE Alternative name for reverse deposit polysaccharide.

RESERVE POLYSACCHARIDE Alternative name for reserve deposit polysaccharide.

RESIDUAL PROTEIN A protein fraction, tightly bound to nucleic acid, obtained from cell nuclei (e.g. calf thymus gland) after the histones have been removed by treatment with 1M hydrochloric acid at pH 2·9 and the DNA has been removed, e.g. by treatment with deoxyribonuclease. Unlike the histones, residual protein contains tryptophane and is insoluble at neutral and acidic pH values, i.e. it is less basic.

RESIDUAL STRAIN The time-dependent strain remaining after a material has been subject to creep for time t_1 followed by a period of recovery. Thus for a linear viscoelastic material $e_r'(t) = e_c(t_1) + e_c(t - t_1)$, where $e_r'(t)$ is the residual strain and $e_c(t_1)$ and $e_c(t - t_1)$ are the creep strains at times t and $t - t_1$ respectively.

RESIDUE REPEAT The distance travelled along the length, i.e. the axis, of a helical conformation per polymer chain repeat unit. For polypeptide helices, the α-helix, for example, has a residue repeat of about $1·50\,\text{Å}$.

RESILIENCE The ratio of the energy imparted to a material by deformation caused by an object striking it, to the energy returned to the object. Thus, for example, when the object is a ball, the resilience may be determined simply from the rebound height, i.e. the rebound resilience.

RESILIN A structural protein found in the exoskeleton of many insects and having rubber-like properties, similar to elastin. However, unlike elastin, it has a reasonable (about 25%) proportion of polar amino acids and its crosslinks, one every 100 residues, are not through desmosine residues, but possibly involve di- or tri-tyrosine bridges.

RESIMENE Tradename for urea- and melamine–formaldehyde polymers.

RESIN Originally the term referred to the polymeric exudations of certain plants and trees, especially after these had hardened on exposure to air or after long-term burial in the ground. Examples are rosin, copal and damar. Certain other similarly hard and brittle natural polymers, such as lac, are also referred to as resins. These resins are used as components in surface coating materials, especially varnishes. Since the rise of synthetic polymers as commercially important materials, as plastics with somewhat similar properties to the natural 'resins', the term resin is also used, at least technologically, interchangeably with the term polymer. In addition, the term resin is also used to mean certain liquid prepolymer products, such as unsaturated polyester and epoxy prepolymers, which are subsequently crosslinked to hard, somewhat brittle thermoset polymers, also, confusingly, referred to as resins.

RESIN 18 Tradename for poly-(α-methylstyrene).

RESIN M Tradename for poly-(p-vinylphenol).

RESIN MB Tradename for brominated poly-(p-vinylphenol).

RESINOGRAPHY (Polymerography) The study of polymers and polymer compositions, largely at the morphological level, using microscopic and allied techniques. Early work emphasised simple visual and optical microscopic examination of natural resins such as shellac and rosin, but the term now encompasses all polymeric materials and hence the term polymerography is also used.

RESINOX Tradename for phenol–formaldehyde polymer.

RESIT The final crosslinked network polymer formed from a resole (via a resitol) by crosslinking the prepolymer, after neutralisation, by further heating.

RESITOL The intermediate B-stage in the conversion of a resole to its final fully cured C-stage.

RESOLE A phenol–formaldehyde polymer formed by reaction of a phenol (usually phenol itself) with a molar excess of formaldehyde (often with a phenol:formaldehyde ratio of $1:1.5$–2.0) under alkaline conditions in aqueous solution. Use of sodium hydroxide gives polymers with greater solubility in water than those obtained using ammonia, but of a slightly deeper yellow colour. The resole may be isolated as a hard, brittle solid by removal of the water by vacuum distillation. Typically it has a melting point of 45–50°C.

The phenol, or strictly the phenoxide ion, becomes substituted at the *ortho* and *para* positions by electrophilic attack by formaldehyde to give methylolphenols. These are even more reactive to formaldehyde than is phenol itself and therefore become further substituted to di- and trimethylolphenols:

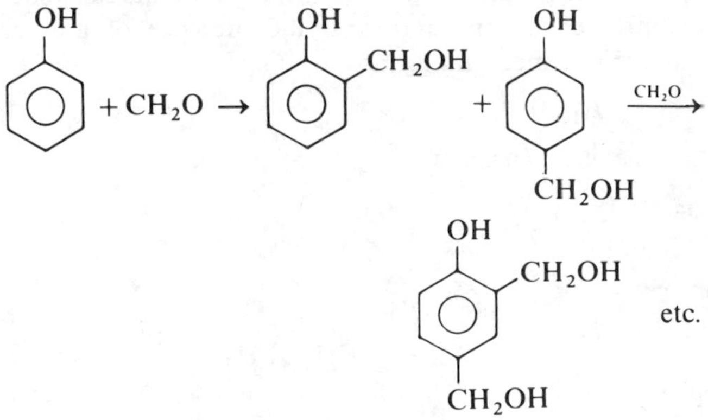

The relatively stable methylolphenols undergo self-condensation to form polynuclear phenols with the rings linked by methylene bridges:

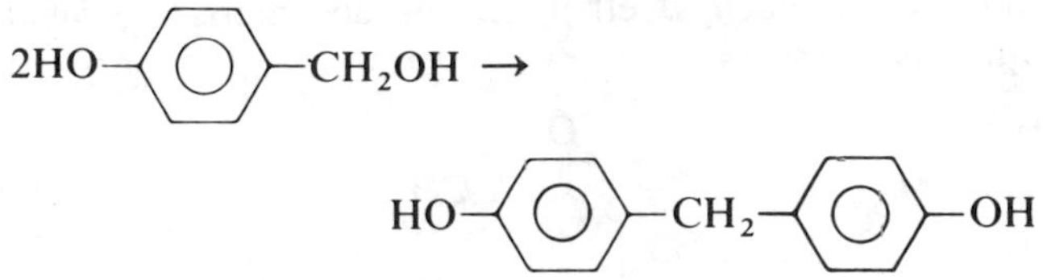

A typical resole consists of a complex mixture of mono-, di- and polynuclear phenols with an average of 2–4 rings per molecule. Most links are through the *para* position with *o*-methylol groups attached to the rings. Such soluble, viscous liquid or fusible solid resins are the A-stage resins.

These prepolymers may be crosslinked simply by further heating often at about 150°C, and usually after neutralisation to a slight acidity. The crosslinking reaction passes through a rubbery B-stage (the resitol) to a final highly crosslinked, infusible network C-stage resin (the resit). A very complex series of reactions is involved in crosslinking. The main reactions are the formation of further methylene bridges by reaction of methylol groups with free *ortho* or *para* hydrogen positions and the formation of ether links by reaction of two methylol groups. The former reaction is favoured under more acidic conditions since benzylic carbonium ions are formed from methylols and it is these which attack the *ortho* and *para* hydrogens:

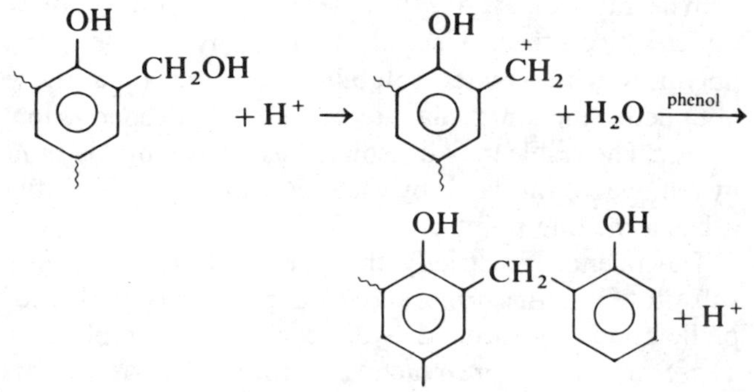

This reaction is also favoured by the presence of a large number of unsubstituted *ortho* and *para* positions. Conversely, ether link formation is favoured under neutral conditions and with the presence of a large number of methylol groups:

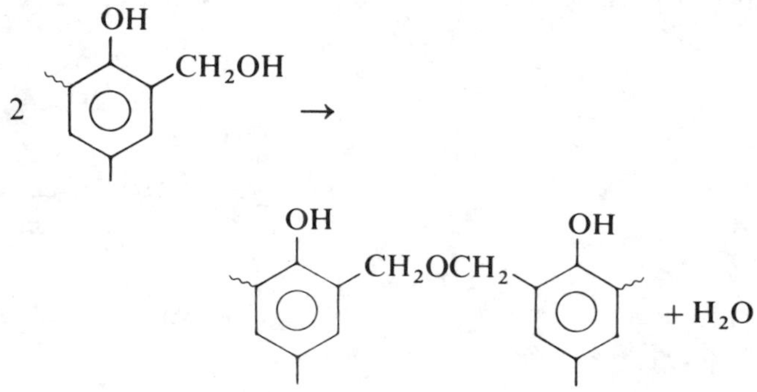

In commercially cured resole products conditions of crosslinking are such that the majority of links are probably methylene. Furthermore, at the high temperatures used dibenzyl ether groups decompose to yield quinonemethides:

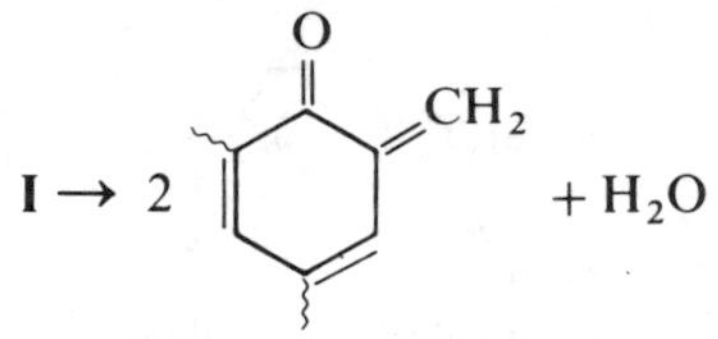

which can further react to give conjugated and other products probably responsible for the dark colour of the cured products.

RESONANCE DISPERSION   The frequency dependence of the components (J' and J'') of the complex compliance, found with some crystalline polymers in which the loss compliance goes through a sharper maximum than if it had only a single retardation time of ordinary viscoelastic behaviour. The storage compliance goes through a maximum and then a minimum with increasing frequency in the same frequency range. Such behaviour may be modelled by a mechanical model containing an inertial unit as well as the usual springs and dashpots.

RESONANCE METHOD   (1) One of several methods for the determination of the dynamic mechanical behaviour in the frequency region of approximately 10^3 Hz. This corresponds to a stress wave of wavelength similar to the sample dimensions, i.e. about 10 cm. Under forced vibration a system of standing waves can be set up with resonance in the sample and from the half-width and half-power points of the resonance peak the storage and loss moduli can be calculated. The most popular technique is the vibrating reed method.

(2) A method for the determination of the dielectric properties of a dielectric in the frequency range 10^5–10^9 Hz, by use of a resonant circuit. This basically consists of an inductance and a capacitance in parallel. Such a circuit is frequency sensitive, especially near one particular frequency. Thus introduction of the sample dielectric causes a change in resonance frequency, which is compensated for either by a change in frequency, by use of a Q-meter, or by a change in capacitance, as in the Hartshorne and Ward method.

RESONANCE PEAK HALF-WIDTH   Alternative name for half-width of the resonance peak.

RESONANCE RAMAN SPECTROSCOPY   Raman spectroscopy in which the exciting laser light has a frequency close to that of a low lying electronic transition. This results in a large increase in the intensity of some of the vibrational bands, such as conjugated double bonds and biopolymer chromophores. It has proved especially valuable in the study of polydiacetylenes and Raman labelled biopolymers. If an ultraviolet light laser is used then the direct study of polypeptides and nucleic acids is possible in this way.

RESORCINOL   (1,3-Dihydroxybenzene)

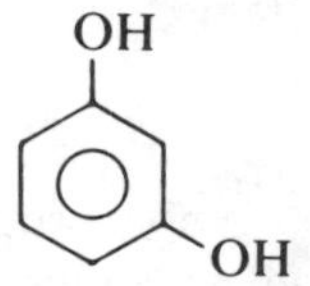

M.p. 110°C. B.p. 281°C.

Produced by alkali fusion of *m*-benzenesulphonic acid, which is obtained by the sulphonation of benzene. It is useful in the formation of phenol–formaldehyde polymer, especially cold-setting adhesives, since the 4- and 6-positions are much more reactive to formaldehyde than is phenol itself.

RESORCINOL–FORMALDEHYDE   POLYMER   Tradename: Aerodux. A phenol–formaldehyde polymer where the phenol used is resorcinol.The presence of a second hydroxyl in resorcinol activates the phenol to reaction with formaldehyde to such an extent that reaction occurs readily at ambient temperatures without a catalyst. However, for commercial use as cold-setting adhesives, especially for wood, a prepolymer is first made using a low formaldehyde to phenol ratio (<1) with an alkaline catalyst. In use this is crosslinked by the addition of further formaldehyde either as a solution or, more usually, as paraformaldehyde.

RESTICELL   Tradename for expanded polystyrene.

RESTIL   Tradename for styrene–acrylonitrile copolymer.

RESTIRAN   Tradename for acrylonitrile–butadiene–styrene copolymer.

RESTIRAN M   Tradename for a blend of polyvinyl chloride and acrylonitrile–butadiene–styrene copolymer.

RESTIROLO   Tradename for polystyrene.

RESTRAINED LAYER THEORY   A mechanism proposed to account for the action of coupling agents in improving the mechanical performance of fillers, especially fibres, in polymer composites. The coupling agent is thought to increase the modulus of the polymer in the region of the fibre by 'tightening up' its structure. This provides an interfacial region of polymer of modulus intermediate between that of the fibre and the polymer, which helps uniform transfer of stress between matrix and fibre.

RETARDATION   The reduction in the rate of a free radical polymerisation due to the presence of a retarder.

RETARDATION SPECTRUM   Alternative name for retardation time spectrum.

RETARDATION TIME   Symbol τ'. A time parameter characterising the delayed elasticity of a viscoelastic material as observed in creep. The faster the molecular response, the shorter the retardation time, i.e. times are short in rubbers and long in rigid plastics. Amorphous polymers exhibit the five regions of viscoelastic behaviour at different temperatures or on different time scales and the observed behaviour on the time scale depends on τ'. τ' generally falls in the middle of the time scale of behaviour, i.e. in the transition zone. For Voight model behaviour, τ' is the ratio of the dashpot viscosity to spring modulus and is the time for the creep strain to reach $(1 - 1/e)$ or 0.632 of its final equilibrium value. In practice, polymer behaviour is not so simple and consideration of a retardation time spectrum is necessary to represent creep.

RETARDATION TIME SPECTRUM   (Retardation spectrum) Symbol $L(\tau')$. The distribution of retardation times $f(\tau')$ of a viscoelastic material whose behaviour can be represented by a model with a large number of Voight elements joined in series. The strain during creep can be represented by

$$e(t) = J_u + \sigma \int_0^\infty f(\tau') \exp(1 - t/\tau')\,d\tau'$$

where σ is the stress, t is the time, J_u is the unrelaxed creep compliance and $f(\tau')$ is a weighting function giving the concentration of Voight elements with retardation times between τ' and $\tau' + d\tau'$. In practice, a logarithmic time scale is more convenient and a different retardation time spectrum $L(\tau')$ is used, where $L(\tau')\,d\ln\tau'$ gives the contributions to the stress retardation associated with retardation times between $\ln\tau'$ and $\ln\tau' + d\ln\tau'$. The creep compliance is then given by

$$J(t) = J_u + \int_{-\infty}^{\infty} L(\tau')(1 - \exp(-t/\tau'))\,d\ln\tau'$$

The retardation spectrum may be determined from the experimental values of $J(t)$ by a Fourier or Laplace transform or from the real or imaginary components of the complex compliance.

RETARDED ELASTICITY   Alternative name for delayed elasticity.

RETARDED ELASTIC STATE   (Leathery region) (Leathery state) (Transition zone) (Viscoelastic state) The state of a polymer between the glassy and rubbery states in which, with increasing temperature, movement of molecular chain segments becomes possible. Thus the state spans the temperature interval of about 20°C either side of the glass transition temperature. Mechanical behaviour changes very rapidly with temperature in this region. As cooperative molecular motions increase, more energy is dissipated as heat and less is stored elastically, i.e. damping increases. However, with further temperature increase, rubber-like response to stress occurs, so damping passes through a maximum in this region. This may be characterised as a maximum in the loss modulus or $\tan\delta$. The moduli decrease by a factor of about 10^3 on passing through this region.

Overall behaviour may be described as leathery. Since retarded elasticity is most pronounced in this state, it is the most characteristic state for observing viscoelastic behaviour. Viscoelastic materials are subject to time–temperature superposition so that this state may also be traversed (from glass-like to rubber-like behaviour) with increasing time of observation or with decreasing frequency in dynamic mechanical situations, when the term transition zone is often preferred.

RETARDER   (1) A substance capable of reducing the rate of a free radical polymerisation by acting as a radical scavenger. However, its effectiveness in reacting with free radicals is not high enough to completely stop polymerisation (inhibition). Thus a retarder differs from an inhibitor only in degree and not in kind. Indeed many compounds that are inhibitors for one monomer are retarders for another. Aromatic nitro-compounds are typical examples, possibly reacting by

Many metal salts, notably ferric chloride, are also retarders.

(2) An additive used in rubbers to prevent premature vulcanisation (scorch) during mixing and shaping of the raw unvulcanised rubber compound. Early materials were acidic, such as benzoic acid and phthalic anhydride, and not only lengthened the vulcanisation induction period (i.e. the scorch time) but unfortunately also reduced the rate of vulcanisation in the vulcanisation stage. More recently various phthalimide derivatives, such as N-cyclohexylphthalimide, have been used which do not affect the rate of crosslinking at vulcanisation temperatures. These are known as prevulcanisation inhibitors.

RETENTION VOLUME Symbol V_R. In chromatography, the volume of solvent that passes through the chromatographic column before the species of interest is eluted from the column.

RETICULATED FOAM A cellular material in which the cell walls have collapsed or otherwise disappeared, so that the material consists merely of a network of struts. Reticulated polyurethane foams are formed when a suitable solvent, such as dimethylformamide, is incorporated in the foam formulation.

RETRACTYL Tradename for polyvinyl chloride staple fibre.

RETROGRADATION The irreversible deposition of an insoluble crystalline form of amylose from its concentrated solutions.

RETROINHIBITION Alternative name for feed-back inhibition.

REUSS MODEL A model for a composite material in which the components are connected in series, so that the modulus of the composite (M_c) is given by the inverse law of mixtures, i.e. $1/M_c = \phi_a/M_a + (1 - \phi_a)/M_b$, where M_a and M_b are the moduli of the components a and b respectively, and ϕ_a is the volume fraction of component a.

REVERSED PHASE CHROMATOGRAPHY A variation of partition chromatography in which an inert support material carries an organic liquid as the stationary phase and the eluting moving phase is aqueous. The technique is particularly useful for separations of transfer and ribosomal ribonucleic acids.

REVERSE ORDER PRECIPITATION (Reversion of fractions) In fractional precipitation, the initial separation of lower molecular weight fractions, rather than of the higher molecular weight fractions by precipitation. This is often caused by the use of too high an initial polymer concentration with certain solvent/non-solvent pairs.

REVERSE OSMOSIS The passage of pure solvent from a solution through a semi-permeable membrane to the solvent side of the membrane, caused by the application of a pressure to the solution which exceeds the osmotic pressure. Thus pure solvent may be separated from the solution. The technique is particularly useful for the desalination of brackish waters. Suitable polymer membranes are of cellulose acetate. Hollow fibres of, for example, polyamides may also be used, as may sulphonated aromatic polyethersulphones.

REVERSIBLE POLYMERISATION Alternative name for equilibrium polymerisation.

REVERSION (1) A reduction in the modulus and tensile strength of a rubber vulcanisate as a result of prolonged thermal ageing. It occurs in sulphur vulcanised rubbers containing polysulphide crosslinks, mostly with polyisoprenes (both with natural rubber and synthetic), either when vulcanisation times are too long (overcure) or during other exposures to temperatures of above about 150°C. Reversion is due to a breakdown in the network structure, probably as a result of the crosslinks forming cyclic structures.

(2) The recombination of the hydrolysis products of a polysaccharide. Especially applied to the D-glucose and maltose produced from starch, which can recombine to form linkages not present in the starch.

REVERSION OF FRACTIONS In fractionation, the separation of fractions in the reverse order of molecular weights to that expected. Specifically in fractional solution, in the last few fractions which are obtained by extraction with the most powerful solvents, the highest molecular weight fractions are obtained first. This may be due to hold-up of low molecular weight material in cavities in the column. In fractional precipitation, reversion is termed reverse-order precipitation.

REVINEX Tradename for styrene–butadiene rubber.

REXENE Tradename for low density and high density polyethylenes.

REYNOLDS NUMBER Symbol Re. A dimensionless parameter which is a measure of the importance of inertia effects relative to viscous flow in flow problems. In general, Re is given by $\rho UL/\eta_0$, where ρ is the fluid density, U is a characteristic average velocity of flow, L is a characteristic length along which the velocity varies and η_0 is the Newtonian viscosity (the coefficient of viscosity). Thus for the particular example of flow along a pipe of diameter D, Re $= \rho UD/\eta_0$, where $U = 4Q/\pi D^2$ with Q as the volume flow rate, so that Re $= 4\rho Q/\eta_0 D$. In general, a flow with Re less than 2100 is a laminar flow and with Re > 2100 it is a turbulent flow. Most polymer melt flows have very low Re values (10^{-3}–10^{-4}) and hence are laminar.

L-RHAMNOSE (6-Deoxy-L-mannose) The ring form is α-L-rhamnopyranose:

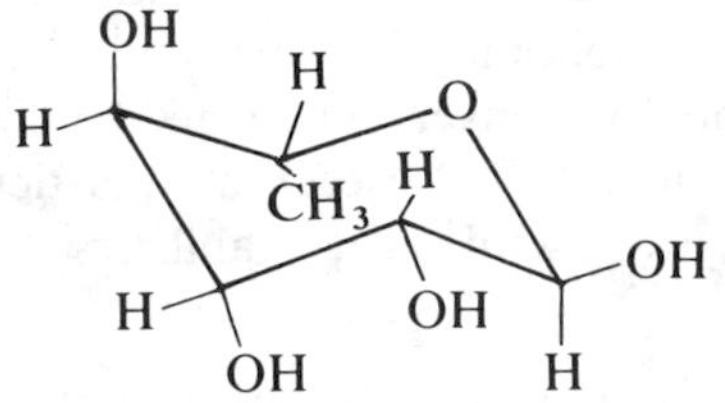

M.p. 93–94°C (monohydrate). $\alpha_D^{20,H_2O} + 8.2°$.

A deoxyaldohexose found in many natural glycosides and in polysaccharide gums, e.g. gum arabic, pectic substances, mucilages and bacterial polysaccharides. It crystallises from water as the monohydrate α-anomer.

RHENOFLEX Tradename for chlorinated polyvinyl chloride.

RHEODESTRUCTION Alternative name for rheomalaxis.

RHEOGONIOMETER A rheometer which measures both the normal forces and the tangential forces in a sheared fluid, either in rotation or in oscillation. The best known type is the cone and plate Weissenberg rheogoniometer.

RHEOLOGICAL EQUATION OF STATE Alternative name for constitutive equation.

RHEOMALAXIS (Rheodestruction) The irrecoverable reduction in viscosity that occurs on shearing. This is usually due to some change in the molecular structure of the fluid as, for example, in the mastication of rubbers.

RHEOMETER An instrument for the determination of the rheological properties of a fluid. If interest is solely in the determination of the fluid viscosity, then the instrument is being used as a viscometer. Many different types of instrument are used for polymer fluids, for both melts and solutions. For melts, capillary rheometers are very popular, but only give comparative data. Precise data may be obtained from cone and plate or couette-type rheometers, although these are only suitable for lower viscosity melts or solutions. Parallel plate viscometers are also used. Several dynamic techniques are used, including eccentric rotating discs, oscillating discs and torsion methods. All these methods refer to shearing flows. Extensional flow properties may be studied by simple elongation or extrudate drawing methods.

RHEOMETRY The experimental determination of the rheological properties of materials, particularly fluid materials. An apparatus for doing this is called a rheometer. For Newtonian fluids the entire rheological behaviour is determined by the viscosity alone. Determination of viscosity is called viscometry. However, for non-Newtonian fluids, rheological behaviour is also shear rate dependent and a flow curve is often determined. For elastic fluids, as with most polymeric systems, the elastic properties of the fluid also need to be determined to characterise the flow behaviour. Most commonly, this involves the determination of the first and second normal stress differences of the fluid.

RHEO-OPTICAL EFFECT The production of an optical effect as a result of deformation, either in the solid state or on flow of a fluid. Examples include photoelastic effects (as in photoelastic stress analysis), stress-induced birefringence and the observation of the variations in the small angle light scattering and X-ray diffraction patterns on stressing a material.

RHEOPEXY (1) An increase in the rate of viscosity increase of a thixotropic material on gentle movement (rather than just an increase in the viscosity) on cessation of shearing.
(2) Alternative name for antithixotropy.

RHEOVIBRON A dynamic rheometer in which the sample is clamped between two strain gauges and subjected to a sinusoidally varying tensile strain at a fixed frequency. A direct reading is obtained of tan δ. The instrument operates at 3·5, 11, 35 or 110 Hz.

RHODEFTAL Tradename for a polyamideimide useful as a high temperature resistant electrical wire varnish and adhesive, and produced by reaction of trimellitic anhydride with an aromatic diamine.

RHODIALITE Tradename for cellulose acetate.

RHODOPAS Tradename for polyvinyl chloride.

RHOPLEX Tradename for acrylate and methacrylate ester polymer emulsions useful as textile finishes and sizes.

RHOVYL Tradename for polyvinyl chloride continuous filament.

RHOVYLITE Tradename for polyvinyl chloride.

RIBBED SMOKED SHEET (RSS) The raw gum natural rubber obtained by coagulation of the natural rubber latex diluted to a rubber content of about 15% and formed into sheets which are subsequently dried in a smokehouse. The most important form of natural rubber.

RIBBON POLYMER A polymer in which the bonds in the polymer chain are all double bonds or in which the chain consists entirely of conjugated ring structures. The repeat units are therefore +R+, where R is usually either

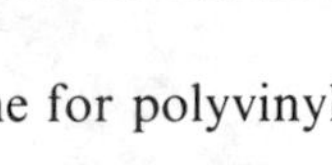

RIBLENE Tradename for low density polyethylene.

RIBONUCLEASE An enzyme, the most widely studied being that obtained from the intestine, e.g. bovine pancreatic ribonuclease I, which hydrolyses ribonucleic acid at those 5'-linkages in which the 3'-linkage is attached to a pyrimidine nucleotide. The products of hydrolysis are thus pyrimidine and nucleoside-3'-phosphates and oligonucleotides terminating in pyrimidine-3'-phosphate residues. It consists of a single polypeptide chain of 124 residues containing four intrachain disulphide crosslinks. It was the first enzyme for which the amino acid sequence was determined. It is also the first protein for which it was demonstrated that the amino acid sequence determines the tertiary structure. Thus, if it is denatured with 5M urea and the disulphide bridges are cleaved, then catalytic activity is lost. However, on removal of urea, catalytic activity is regained, demonstrating that the polypeptide spontaneously refolds and the correct disulphide bridges, of which there are 105 possibilities for four bridges, are reformed to produce the catalytically active conformation. The active centre is situated in a large cleft at which the residues His 12 and His 119 and Lys 41 are involved in catalysis. The enzyme has been synthesised, as has a two-chain form which is also active. The two-chain form can be produced by the action of subtilisin on the one-chain form.

RIBONUCLEASE B   An enzyme glycoprotein isolated from commercial ribonuclease, similar in carbohydrate composition to ovalbumin, consisting of a single carbohydrate group N-glycosidically linked to the asparagine residue at position 34 in the protein. A similar material is isolated from bovine pancreatic juice but the carbohydrate has six mannose units to two 2-acetamido-2-deoxy-D-glucose units.

D-RIBOSE

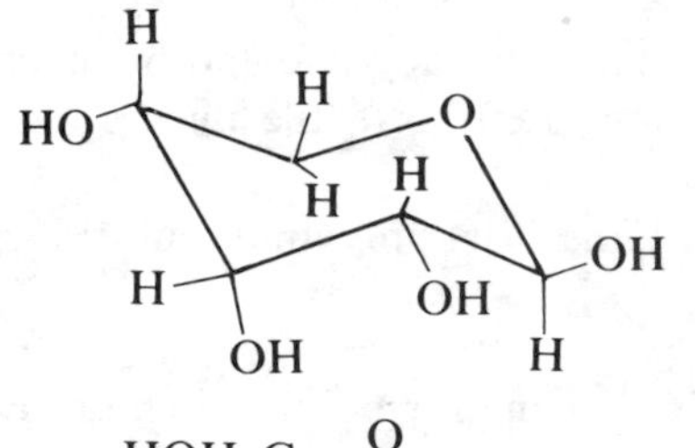

(β-pyranose form)

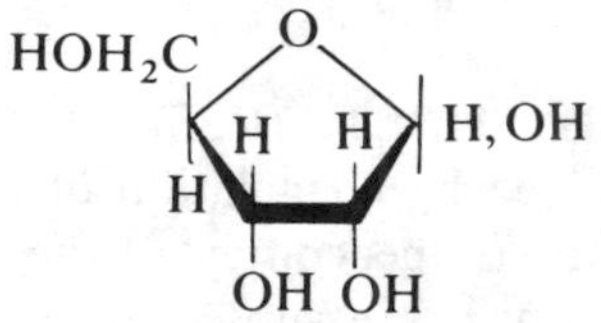

(furanose form)

M.p. 87°C. $\alpha_D^{20, H_2O} - 23.7°$.

An aldopentose which, together with the related 2-deoxy-D-ribose, is the carbohydrate constituent of nucleic acids, in the furanose form. Prepared by the stepwise hydrolysis of yeast nucleic acid. Solutions may also exist in the furanose form.

RICINOLEIC ACID   (12-Hydroxy-cis-9-octadecenoic acid)

$$CH_3(CH_2)_5CHCH_2CH{=}CH(CH_2)_7COOH$$
$$OH$$

M.p. 5°C. B.p. 228°C/10 mm.

Comprises about 85% of the acid residues in castor oil. Commercially, the term refers to the mixture of acids obtained from the hydrolysis of castor oil. Useful as a source of the monomer 11-aminoundecylenic acid for nylon 11 and for conversion to alkyl ricinoleates, e.g. butylacetyl ricinoleate, useful as plasticisers. Metal ricinoleates are useful heat stabilisers for polyvinyl chloride.

RIGID BODY DISPLACEMENT   A uniform displacement of the particles in a body in one direction, so that their relative distances are unaltered and the body is undistorted.

RIGIDEX   Tradename for high density polyethylene.

RIGIDITY MODULUS   Alternative name for shear modulus.

RIGID POLYURETHANE FOAM   A rigid, cellular crosslinked polyurethane, usually with closed cells, formed by reaction of a diisocyanate (sometimes containing components of a higher functionality) and often MDI or polymeric MDI, with a polyester or, more usually, a polyether polyol. Foaming may result from the incorporation of water, which reacts with isocyanate groups to form carbon dioxide, but usually is the result of using a chlorofluorocarbon blowing agent, sometimes in combination with water. The foams achieve their rigidity, compared with flexible foams, by being more heavily crosslinked. This is achieved by the use of polyols, usually poly-(oxypropylene)glycols, of low molecular weight (e.g. about 500) which are highly branched by the incorporation of higher functionality comonomers such as sorbitol or, pentaerythritol. Prepolymer processes and quasiprepolymer processes are used with tolylene diisocyanate to reduce the toxic hazard of this material. One-shot processes are used with MDI and polymeric MDI with polyethers and, normally, a tertiary amine and/or organotin catalyst system is used, together with a silicone surfactant as with flexible foams. The major application of the foam is for thermal insulation as it has exceptionally low thermal conductivity, especially when a fluorocarbon is trapped in the cells, as this itself has a low thermal conductivity. Conductivity can be as low as about $0.2 \, W \, m^{-1} \, K^{-1}$. Typical densities of these foams are about $0.02 – 0.1 \, Mg \, m^{-3}$. However, higher density foams with cell sizes of $0.2 – 1.0 \, mm$, obtained using less blowing agent, can have densities in the range $0.16 – 1.0 \, Mg \, m^{-3}$ and are also useful as structural foams.

RIGISOL   A plastisol which contains a polymerisable plasticiser which polymerises during gelation. Thus the gelled product is rigid, as opposed to being soft and flexible in a normal plastisol product.

RILSAN   Tradename for nylon 11.

RILSAN A   Tradename for nylon 12.

RILSAN B   Tradename for nylon 11.

RING-CHAIN EQUILIBRIUM The equilibrium which is often established between low molecular weight (usually monomeric, sometimes oligomeric) cyclic species and linear high molecular weight polymeric species in step-growth or ring-opening polymerisation. The position of the equilibrium depends upon the stability of the cyclic species, which in turn depends largely upon the ring size. In general, in AB-type step-growth polymerisation the following equilibria are involved:

$$n \text{ A—B} \rightleftharpoons \text{~[A—B~]}_n$$
$$n \text{ A—B} \quad \text{①}$$

e.g.

$$n \text{ H}_2\text{N—R—COOH} \rightleftharpoons \text{~[NH—R—CO~]}_n + n \text{ H}_2\text{O}$$
$$n \text{ NH—R} \quad + n \text{ H}_2\text{O}$$
$$\text{C=O}$$

Similarly, in ring-opening polymerisation, equilibrium ① will be involved. Six-membered rings are the least strained, and hence are the most stable, and most of these cyclic monomers do not polymerise readily, e.g. α-piperidine and oxepane. In a similar way AB monomers capable of forming five- and six-membered rings usually do so rather than forming linear polymer. However, anionic or cationic polymerisation of five- and six-membered cyclic monomers may often be successfully achieved at the low temperatures at which the equilibrium favours the linear polymer.

RINGED SPHERULITE (Banded spherulite) (Radial banding) A spherulite which, when viewed between crossed polars in a polarising microscope, appears as a series of alternating concentric light and dark rings. These arise from the extinction due to zero birefringence of the crystallites whose optic axes are parallel to the direction of the light. The crystallites are radially aggregated into fibrils, which are regularly twisted, and give rise to this pattern. The period of twist corresponds to the spacing of the rings.

RING-OPENING POLYMERISATION Polymerisation by the ring-opening of a cyclic monomer, giving a polymer in which the repeat units are joined together by similar links to those in the monomer, but now linearly. Thus, in general,

$$n \text{ R—X} \rightarrow \text{~[R—X~]}_n$$

where X is usually a functional group, but may also be a hydrocarbon. Polymerisation results from the action of a catalyst which is usually an ionic species, but may be a neutral molecule, e.g. water. A wide variety of monomers may be ring opened, including cyclic ethers (e.g. epoxides, such as ethylene and propylene oxides), cyclic amides (i.e. the lactams, such as ε-caprolactam), cyclic amines (i.e. the

imines, such as ethyleneimine), cyclic sulphides (including sulphur itself—an S_8 ring compound) and inorganic ring compounds such as cyclic siloxanes and phosphonitrilic chlorides. Typically the catalyst combines with the monomer to form an ionic monomer–catalyst species which attacks a further monomer molecule with ring opening, giving a propagating species which frequently contains an ionised heteroatom. Thus, for example, in the cationic polymerisation of cyclic ethers, the active centre is an oxonium ion:

$$\text{Cat}^+ \bar{\text{Y}} + \text{R—O} \rightarrow \text{Cat—}\overset{+}{\text{O}}\text{—R} \xrightarrow{\text{R—O}} \text{Cat—O—R—}\overset{+}{\text{O}}\text{—R}$$

Frequently the polymerisation has some of the characteristics of both step-growth and chain mechanisms. Thus the propagation step involves successive additions of monomer molecules (as in chain polymerisation) but the polymer molecular weight continuously increases as in step-growth polymerisation. Polymerisability is largely governed by the monomer ring size; the smallest rings are ring strained and therefore readily opened, so that polymerisability decreases in the order three-, four-, five-, six-membered rings. Usually six-membered rings are too stable to be ring opened (except notably in δ-valerolactone). As ring size is further increased reactivity again increases in the order six-, seven-, eight-membered rings, etc. As with classical step-growth polymerisation, sometimes a significant ring-chain equilibrium is established.

RIPPLE DECORATION The epitaxial growth of folded chain lamellae on an extended chain crystal fracture surface of the same polymer, by melting the surface, followed by crystallisation. The lamellae have their chain axes aligned with the substrate and improve the ease of examination of surface morphology.

RIVER MARKING Elongated markings, often branching at narrow angles, present on a fracture surface and resulting from the fracture crack propagating along a parallel array of cleavage planes at different levels. The lines are steps in the surface. Thus they are hackle markings when the crack has propagated for a considerable distance along the same level.

RIVLIN–SAUNDERS EQUATION (Mooney–Rivlin–Saunders equation) A relationship between the strain energy function (W) and strain invariants I_1, and I_2 of the form $W = C_1(I_1 - 3) + f(I_2 - 3)$, where f is a decreasing function of I_2 and C_1 is a constant. An improvement on the Mooney equation, especially in describing the behaviour of rubbers in two-dimensional deformation, e.g. as sheets.

RIVLIN THEORY A theory of rubber elasticity based on the most general form possible for the strain energy function (W) and assuming that the rubber is incompressible and isotropic in the unstrained state so that W must be symmetrical with respect to the principal extension ratios λ_1, λ_2 and λ_3. It also assumes that since W is unaltered by a change in sign of the λ_i values then it must depend only on even powers of λ_i. The three simplest functions of this type are the strain invariants $I_1 = \lambda_1^2 + \lambda_2^2 + \lambda_3^2$, $I_2 = \lambda_1^2\lambda_2^2 + \lambda_2^2\lambda_3^2 + \lambda_3^2\lambda_1^2$ and $I_3 = \lambda_1^2\lambda_2^2\lambda_3^2$. $I_3 = 1$ for an incompressible material and hence $I_2 = 1/\lambda_1^2 + 1/\lambda_2^2 + 1/\lambda_3^2$. W may be expressed as

$$W = \sum_{i=0, j=0}^{\infty} C_{ij}(I_1 - 3)^i(I_2 - 3)^j \qquad (1)$$

where the C_{ij} parameters are constants. If only simple forms are considered then $i = 1$ and $j = 0$, and

$$W = C_{10}(I_1 - 3) = C_{10}(\lambda_1^2 + \lambda_2^2 + \lambda_3^2 - 3)$$

The Mooney equation is a form of eqn. (1) in which $i = 1$ and $j = 1$.

RMS END-TO-END DISTANCE Abbreviation for root mean square end-to-end distance.

ROCKITE Tradename for phenol–formaldehyde polymer for moulding.

ROCKWELL HARDNESS An arbitrary number expressing the hardness of a polymer material as measured by a standard indentation test involving the application of first a minor (10 kg) load followed by a major (60–150 kg) load. In some procedures a reading is taken with the major load operating and in others after the major load has been removed. Use of different major loads and indentor diameters gives rise to different Rockwell scales (R, LM, E and K) covering different but overlapping scales of hardness.

ROD-CLIMBING EFFECT Alternative name for Weissenberg effect.

ROHACELL Tradename for polymethacrylimide.

RONFALIN Tradename for acrylonitrile–butadiene–styrene copolymer.

RONFALOY Tradename for a bisphenol A polycarbonate/ABS blend.

ROOM TEMPERATURE VULCANISING SILICONE ELASTOMER (RTV silicone elastomer) A silicone elastomer that may be vulcanised without heating. The unvulcanised gum is based on a low molecular weight polydimethylsiloxane, or similar fluoro or methylphenyl containing polymer, containing reactive end groups. Several different methods of cure may be used, employing different end groups. In condensation cure, silanol end groups in branched polymers condense to give a network in a reaction catalysed by an organic base. Alternatively, condensation involving alkoxy end groups may be catalysed by a tin soap. Condensation curing can

also occur due to atmospheric moisture, e.g. with an acetoxy-terminated polymer in which the acetoxy groups are hydrolysed to silanol groups which then condense.

ROOT MEAN SQUARE END-TO-END DISTANCE (RMS end-to-end distance) The square root of the mean square end-to-end distance, i.e. $(\langle r^2 \rangle)^{1/2}$.

ROPET Tradename for a polyethylene terephthalate/acrylic polymer blend.

ROSIN (Colophony) One of the most important natural resins, obtained by distilling the oleoresin from many species of pine tree. The essential oils (turpentine) volatilise leaving the rosin behind. Wood rosin is obtained from the stumps of old trees. Rosin consists mostly (about 90%) of abietic acid and its isomers, together with esters of these acids and resenes. The abietic acid is formed mostly by the heat treatment during distillation, the original oleoresin containing other similar and often isomeric acids, such as laevopimaric acid. It is a brittle solid which softens quite sharply at about 80°C and is soluble in most organic solvents. Although it is useful in varnishes its properties are improved, e.g. the softening point and hardness are raised, by esterification to form an ester gum. It may also be usefully modified by reaction with maleic anhydride or phenol–formaldehyde resin. Rosin esters with monohydric alcohol, e.g. methyl abietate, are useful plasticisers, whilst dihydric alcohol esters, e.g. with ethylene glycol, are useful as plasticisers and adhesives. Esters with glycerol and pentaerythritol (ester gums) are useful in adhesives and lacquers.

ROSSI–PEAKES FLOW TESTER An apparatus for assessing the melt flow behaviour of a polymer, whereby the polymer melt, held in a chamber, is forced by a known pressure through a capillary, and the length of material that has flowed is measured by a following rod. The test is repeated at various temperatures and the temperature is noted at which the length of flow is 25·4 mm.

ROTATING SECTOR METHOD (Sector method) A technique which enables the individual rate constants for propagation (k_p) and termination (k_t) to be obtained for a free radical polymerisation, through the determination of the average radical lifetime, τ_s. The method involves the determination of the polymerisation rate in a photopolymerisation as a function of the cycle time for alternating light and dark periods of illumination. This is achieved by placing a rotating sector disc between the light source and the polymerisation vessel. A plot is made of the rate ratio $R_p/(R_p)_s$ against $\log t$, where R_p and $(R_p)_s$ are the rates of polymerisation measured under alternate (non-steady-state) and continuous (steady-state) illumination conditions respectively and t is the length of the light period. From this a value of τ_s can be found. Finally, values of k_p and k_t can be calculated from the relations

$$\tau_s = k_p[M]/2k_t(R_p)_s \quad \text{and} \quad R_p = k_p[M](R_i/2k_t)^{1/2}$$

where R_i is the rate of initiation for steady illumination and $[M]$ is the monomer concentration.

ROTENE Tradename for high density polyethylene.

ROUSE–BUECHE–ZIMM THEORY (RBZ theory) A molecular theory of viscoelastic behaviour based on the bead–spring model of a polymer chain. On stressing, the coordinated segmental movements of the chain are treated as modes of cooperative motion, each mode having a characteristic relaxation time, giving the Rouse distribution of relaxation times. Relationships can then be derived giving the coefficients of viscosity of a polymer in dilute solution under sinusoidal shear. The theory can be extended to consider an undiluted polymer above its glass transition temperature, by considering the system to be represented by polymer segments dissolved in a matrix of other polymer segments and replacing viscosity by a monomeric friction coefficient (ξ_0). This enables the continuous relaxation time spectrum to be calculated as

$$H(\tau) = (\rho N_A/2\pi M)(\langle r^2 \rangle_0 NkT\xi_0/6\tau)^{1/2}$$

where $\langle r^2 \rangle_0$ is the unperturbed mean square end-to-end distance of a chain, M is the molecular weight, ρ is the density, N_A is Avogadro's number, k is Boltzmann's constant, T is the temperature and τ is the relaxation time. The retardation time spectrum is also calculated as

$$L(\tau') = (2M/\pi\rho N_A)(6\tau'/\langle r^2 \rangle_0 NkT\xi_0)^{1/2}$$

where τ' is the retardation time. The model only applies to intermediate τ values but is reasonably successful in the low modulus regions of viscoelastic behaviour.

ROUSE DISTRIBUTION The distribution of relaxation times used as a model for viscoelastic behaviour in the Rouse–Bueche–Zimm theory. Given by $\tau_i = \tau_0/p^2$, where $p = 1, 2, 3, \ldots, N$ (where N is a function of molecular weight) and τ_0 is the longest relaxation time corresponding to translational movement of a complete polymer chain.

ROUSE THEORY A molecular theory of polymer solution viscosity (η) which is based on a bead–spring model, like the Bueche and Zimm theories, and the free-draining coil assumption for dynamic conditions, for which the energy dissipation and storage are calculated from considerations of the energy decrease for non-random configurations. This results in an expression for the storage and loss shear moduli, $G'(\omega)$ and $G''(\omega)$ respectively, at angular velocity ω as

$$G'(\omega) = nkT \sum_{p=1}^{N} \omega^2\tau_p^2/(1 + \omega^2\tau_p^2)$$

and

$$G''(\omega) = \omega\eta_s + nkT \sum_{p=1}^{N} \omega\tau_p/(1 + \omega^2\tau_p^2)$$

where N is the number of 'beads' per polymer molecule, n is the number of polymer molecules per cubic centimetre, ω is the angular velocity for a sinusoidally varying deformation and τ_p is the relaxation time which can be written as

$$\tau_p = 6(\eta - \eta_s)/\pi^2 p^2 nkT$$

where η_s is the solvent viscosity, k is Boltzmann's constant and T is the temperature. These results are the same as for the Zimm theory provided that hydrodynamic interactions are negligible.

ROVEL Tradename for acrylonitrile–ethylene/propylene rubber–styrene copolymer.

ROVING A number of strands of fibre (from 2 to 100) wound parallel with no twist. Used particularly for glass fibre, especially when the fibre is to be used for glass reinforced plastics. Rovings, often comprising 60 strands (or ends), are often used directly in spraying techniques when they are chopped and blown on to a mould together with the resin. They are sometimes woven into heavy woven roving fabrics.

ROW NUCLEATION The formation of linearly related crystallisation nuclei during the crystallisation of a strained melt, e.g. in film blowing or fibre spinning. The nuclei contain extended chain molecules; however, secondary (epitaxial) nucleation on the surface of such row structures produces chain folded lamellae oriented perpendicular to the strain direction with their molecules uniaxially oriented in the strain direction. Eventually, parallel cylindrical spherulites (cylindrites) are formed. Row nucleation in solutions produces shish-kebab morphology.

ROYALAR Tradename for a partially crosslinked thermoplastic polyurethane elastomer, based on a polyester or polyether polyol/MDI prepolymer and chain extended.

ROYALENE Tradename for ethylene–propylene rubber, polyethylene and polypropylene.

ROYALITE Tradename for acrylonitrile–butadiene–styrene copolymer.

RSS Abbreviation for ribbed smoked sheet.

RTV SILICONE ELASTOMER Abbreviation for room temperature vulcanising silicone elastomer.

RUBBER An amorphous polymer existing somewhat above its glass transition temperature, so that considerable segmental motion is possible. Rubbers are thus relatively soft (typical modulus of about 3 MPa) and deformable. Since linear polymeric rubbers are rather too soft and weak to be useful, in commercial rubber products the molecules are crosslinked by vulcanisation. This crosslinking raises modulus, strength, softening point and solvent resistance. The term elastomer is often used interchangeably with the term rubber, but is often the term preferred when referring to vulcanisates. Although uncrosslinked gum rubbers show considerable elasticity, true rubber elasticity (high elasticity), with instantaneous recovery from high strain of up to about 1000% elongation, only occurs in vulcanised rubbers. Some rubbery polymers, notably natural rubber, also have the useful property of crystallising on stretching which can considerably raise the modulus and strength.

Although most amorphous linear polymers exhibit rubbery properties at temperatures well above their T_g values, materials commonly known as rubbers are those existing as rubbers at ambient temperatures. Useful rubbers will have a T_g value of well below 0°C, whilst useful plastics will have a T_g value of above about 60°C. Often the term rubber refers specifically to natural rubber which is still of major importance despite the development of many synthetic rubbers. Apart from natural rubber, the other general purpose rubbers are styrene–butadiene rubber (the largest tonnage synthetic rubber), polybutadiene, synthetic polyisoprene, ethylene–propylene rubber and EPDM, and butyl rubber. In addition, there are many other lower tonnage special purpose rubbers, which generally have higher heat and solvent resistance. These include nitrile, polychloroprene, acrylic, fluoro- and chlorosulphonated polyethylene and silicone rubbers. Other speciality rubbers are polyurethane, polyether and polysulphide rubbers and nitrosorubbers.

For the production of useful rubber products, the raw rubber is usually compounded with the following materials: sulphur and vulcanisation accelerator or other vulcanisation agents, filler (usually carbon black) and plasticiser or extender (as in oil extended rubber) as a softener, antioxidant and antiozonant. For compounding and processing the rheological behaviour of the raw rubber is important. This is often assessed as the plasticity or viscosity (especially as the Mooney viscosity). The polymer molecular weight and hence viscosity may be reduced by a prior mastication stage, particularly with natural rubber, which may be aided by the use of a peptiser. The compounding stage may have to be of limited duration due to rapid scorch. Once compounded and shaped, e.g. by a suitable moulding, extrusion or calendering process, vulcanisation is performed by heating.

The major rubber products are tyres; however, hose, footwear, belting, cable covering, mechanical parts such as vibration dampers and seals are also important. Rubber has long been of great scientific interest owing to its unusual elastic and thermoelastic properties. A culmination of this interest is manifest in the highly developed statistical molecular theory of rubber elasticity.

RUBBER CHLORIDE Alternative name for chlorinated rubber.

RUBBER ELASTICITY The elastic behaviour of polymers well above their glass transition temperatures, i.e. in the rubbery region of viscoelastic behaviour, where they show elasticity to very high strains of up to several hundred per cent. If the material is perfectly elastic then all the work done will be stored as strain energy (described by a strain energy function), at least for an isothermal deformation. Molecularly, the phenomenon is due to an uncoiling of the randomly coiled polymer molecules on stressing, and is therefore entropic in origin. In conse-

quence a highly successful statistical molecular theory, known also as the kinetic theory of rubber elasticity, has been developed which closely describes many of the experimentally determined features of rubber behaviour.

RUBBER–GLASS TRANSITION Alternative name for glass transition.

RUBBER–GLASS TRANSITION TEMPERATURE Alternative name for glass transition temperature.

RUBBER HYDROCHLORIDE Tradenames: Pliofilm, Tensolite, Ty-ply. Natural rubber which has been reacted with hydrochloric acid either in solution at about 10°C or as a latex. Addition of the hydrochloric acid across the double bond occurs, to give the approximate theoretical chlorine content of 33·9% for the structure

$$-\!\!\left[CH_2C(CH_3)ClCH_2CH_2\right]_n$$

plus some cyclised structures. A crystalline material melting at about 115°C with low permeability to gases and to water, low flammability and high chemical resistance. However, it tends to dehydrochlorinate on heating and exposure to ultraviolet light, which produces isorubber when this reaction is performed in a controlled way. It has been used as a packaging film and adhesive.

RUBBER OIL (Petroleum oil) A high boiling point petroleum fraction obtained after the lower boiling point fractions have been removed from the petroleum by fractional distillation. Important as plasticisers for natural and synthetic rubbers, when they are classified according to their contents of aromatic, naphthenic and paraffinic components. The exact molecular constitution of these oils is not known in detail due to the complexity of the mixtures and of the molecules themselves. The latter are largely fused ring compounds, individual molecules containing aromatic and cycloaliphatic (naphthenic) rings with side chains—the paraffinic portion. A typical molecule is

Thus aromatic oils contain mostly aromatic rings, whilst naphthenic oils contain more cycloaliphatic rings and paraffinic oils have more side chains. The composition of individual oils is indicated by analytical separations obtained from adsorption on to silica gel or clay, or by chemical means. In this system an oil may be separated into four components: the heterocyclic and polar compounds, which typically comprise from about 1% (in a paraffinic oil) to about 25% (in an aromatic oil); aromatic molecules, consisting largely of two and three ring compounds with paraffinic side chains, which are highly compatible with many rubbers; saturated molecules, which are ring compounds with side chains comprising 20–80% of the molecular structure and finally, waxes (up to about 1% asphaltene may also be present). An alternative method of analysis leads to classification by the types of carbon atoms present according to some measured physical property. The viscosity gravity constant and refractivity intercept have been widely used in this way. These plasticising oils are considered to act as process aids, i.e. as process oils, in amounts of up to about 20 phr and as extenders when used above this amount.

RUBBERY FLOW The region of viscoelastic behaviour of a linear amorphous polymer (following, with increasing temperature, the rubbery state) in which the modulus starts to drop relatively sharply with further increase in temperature, after the plateau of the rubbery state. Typically shear modulus is about 10^5 Pa. In addition to the instantaneous elastic response, significant viscous flow response occurs at longer times. Rubbery flow will not occur in crosslinked rubbers or in crystalline polymers.

RUBBERY PLATEAU (Plateau zone) The relatively flat part of the modulus (or log modulus) temperature curve of a polymer in the rubbery state. Sometimes used as an alternative to the term rubbery state.

RUBBERY STATE (Rubbery plateau) (Plateau zone) The viscoelastic state of an amorphous polymer, extending on the temperature scale from about 30°C above T_g to about 80°C above T_g, i.e. the next state above the retarded elastic state. In this temperature range the polymer exhibits rubber elasticity, typically having a modulus of about 10^6 Pa. The alternative term rubbery plateau indicates the relatively constant value of the modulus in this region, compared with the rapid changes in the preceding zone and also in the following rubbery flow zone. By the time–temperature superposition principle, rubbery state behaviour may be observed in the appropriate time or frequency domains at other temperatures or frequencies, when the alternative term plateau zone is often used.

RUBBONE Tradename for oxidised rubber.

RUBI-CONJUGATED POLYMER A conjugated polymer in which delocalisation of electrons is limited by the presence of structural defects, such as non-linearity, non-planarity or atacticity. This results in a much lower electronic conductivity than in eka-conjugated polymers. Examples include polyphenyls and polyacetylenes.

RUHEMANN'S PURPLE

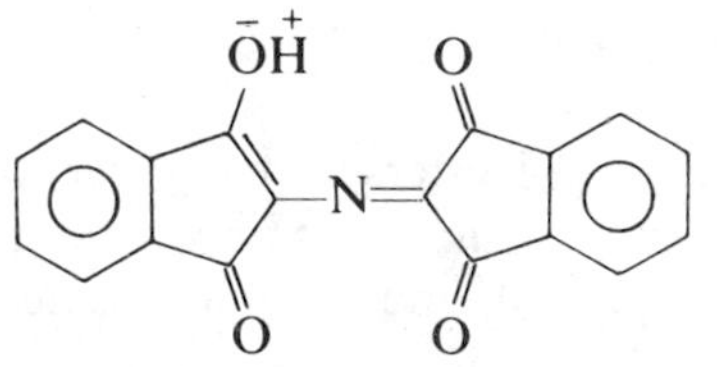

The indandione-2-N-2'-indane enolate formed by reac-

tion of ninhydrin with a free α-amino group. Formation of this coloured product is useful, therefore, in the quantitative determination of amino acids and peptides.

RULE OF MIXTURES Alternative name for law of mixtures.

RUN NUMBER The number of uninterrupted monomer sequences per 100 monomer units in a copolymer. Thus, for example, in a copolymer molecule of A and B units

AABBBABABBBBAAAAABAABAABBBAABAABABB

there are 18 runs (each one being separately underlined) in 35 units so the run number (R) is $18 \times 100/35 = 51.4$. R is also given by

$$R = 200/(2 + r_A f_A/f_B + r_B f_B/f_A)$$

where f_A and f_B are the mole fractions of the monomers A and B in the feed during copolymerisation and r_A and r_B are their reactivity ratios.

RUPTURE Alternative name for fracture, used especially for ductile fracture and for the fracture of rubbers.

RUPTURE FACTOR The ratio of the brittle strength in flexure to that in tension. It often has a value of about 1.5, the higher flexural strengths probably being due to the greater difficulty for crazes and cracks to grow down the stress gradient in flexure.

RUTILE TiO_2. A particular crystalline form of titanium dioxide, which is the most widely used white pigment in polymer materials. Unlike the other crystalline form, anatase, it does not act as a photosensitiser; indeed it is often considered to act as a stabiliser against polymer photodegradation.

RYNITE Tradename for polyethylene terephthalate moulding material.

RYTON Tradename for poly-(p-phenylene sulphide).

S

S Symbol for serine.

SAF Abbreviation for super abrasion furnace carbon black.

SAFFLOWER OIL The oil of safflower seeds, whose triglycerides contain mostly 9,12-linoleic acid residues. It is therefore a semi-drying oil. It is useful in alkyd resins where it combines the good properties of linseed oil (drying) and the non-yellowing nature of soya bean oil.

SAINT VENANT BODY A material which behaves as a Saint Venant model.

SAINT VENANT MODEL A rheological model of a material that is rigid for stresses below a yield value, but which flows at higher stresses. Often it is schematically represented by a solid body (acting as a slider) resting on a flat surface between which two components of frictional force can develop.

SAINT VENANT PRINCIPLE Stresses due to two statically equivalent loadings applied over a small area are significantly different only in the vicinity of the area on which the loadings are applied. At distances large compared with the area concerned, the effects due to these two loadings are the same.

SAINT VENANT SOLID Alternative name for Bingham Solid.

SALIGENIN (*o*-Hydroxymethylphenol)

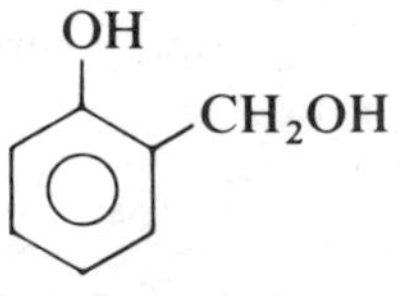

M.p. 86°C.

One of the initial products of the reaction of phenol and formaldehyde to form phenol–formaldehyde polymers, as in the formation of novolacs. In acid conditions it is unstable forming a carbenium ion which attacks a further phenol molecule. In basic conditions, as in resole production, formation of saligenin structures activates the rings to further attack by formaldehyde so that di- and trimethylolphenols are formed.

SALMINE One of the most widely studied protamines, isolated from salmon sperm heads. It contains about 90% arginine (the only basic amino acid present) as short sequences separated by single neutral amino acid residues of only a restricted range of about six amino acids. It has proline as the N-terminal group and arginine as the C-terminal group and contains about 25 arginine units per chain. The total molecular weight is about 5000.

SALOL Tradename for phenylsalicylate.

SALS Abbreviation for small angle light scattering.

SALTING IN The increase in solubility of an electrically charged biopolymer, particularly a protein, in aqueous solution with an increase in ionic strength. This is due to a decrease in the attractive forces between different polymer molecules as the ions come between the molecules, shielding charges of opposite signs on different molecules. Salts with divalent ions, e.g. $MgCl_2$, $(NH_4)_2SO_4$, are especially effective. It is useful as a basis of separation of mixtures of proteins by fractional precipitation, since different proteins will show different solubilities as ionic strength is varied.

SALTING OUT The decrease in solubility of an electrically charged biopolymer, particularly a protein, in aqueous solutions of high ionic strength with an increase in ionic strength. It is possibly due to dehydration of the protein, reducing its water solubility. It is useful as the basis for the separation of mixtures of proteins by fractional precipitation, since the solubilities of different proteins respond differently to ionic strength. Ammonium sulphate is often preferred for this due to its high water solubility.

SAN Abbreviation for styrene–acrylonitrile copolymer.

SANDARAC A natural resin obtained from certain coniferal trees of North Africa. A soft pale-coloured resin soluble in organic solvents and useful in card, paper and leather varnishes.

SANS Abbreviation for small angle neutron scattering.

SANTOGARD PVI Tradename of *N*-cyclohexylthio-phthalimide.

SANTOPRENE Tradename for a polypropylene/ethylene–propylene–diene monomer copolymer blend in which the rubber is crosslinked in the presence of the polypropylene. It is a thermoplastic polyolefin rubber.

SANTOWHITE Tradename for a range of antioxidants which include Santowhite CI (*N*,*N*′-di-*β*-naphthyl-*p*-phenylenediamine), Santowhite powder (4,4′-butylidene-bis-(6-*t*-butyl-*m*-cresol)), Santowhite crystals (4,4′-thiobis-(6-*t*-butyl-*m*-cresol)) and Santowhite TNPP (tris-(*p*-nonylphenyl)phosphite).

SARAN Tradename for vinylidene chloride copolymers with, for example, vinyl chloride or acrylonitrile.

SARCOSINE CH_3NHCH_2COOH. M.p. 210°C (decomposes). An amino acid with pK' values of 2·21 and 10·20. The monomer for the formation of polysarcosine, from its *N*-carboxyanhydride.

SARILLE Tradename for a crimped viscose staple fibre.

SATURATED POLYESTER A polyester which does not contain any $C{=}C$ bonds in the chain, as opposed to an unsaturated polyester which does. Such polymers may be linear, branched or network. Linear polymers of importance are the aliphatic polyesters of low molecular weight useful as polymeric plasticisers and as polyol prepolymers for polyurethanes. Important partially aromatic polymers are polyethylene terephthalate, polybutylene terephthalate and bisphenol A polycarbonate. The alkyd resins are network saturated polyesters. Wholly aromatic polyesters have high melting points and good high temperature resistance and include poly(hydroxybenzoic acid) and the polyarylates.

SAXS Abbreviation for small angle X-ray scattering.

SB Abbreviation for styrene–butadiene copolymers rich in styrene and hence thermoplastic as distinct from the styrene–butadiene rubbers.

SB POLYMER Tradename Solprene. Abbreviation for styrene–butadiene block copolymer, as distinct from the more common random copolymer. Unlike the triblock SBS polymer, this diblock polymer does not attain high stiffness and strength as the styrene domains do not act as physical crosslinks for the butadiene matrix owing to the fact that only one styrene block is attached to each butadiene chain. Radial block copolymers of high styrene content (K-Resin) are also referred to as SB polymers.

SBR Abbreviation for styrene–butadiene rubber.

SBS Abbreviation for styrene–butadiene–styrene block copolymer.

SCANNING ELECTRON MICROSCOPY (SEM) Electron microscopy which utilises the secondary emission of electrons from a surface when bombarded with an electron beam. The main advantage over traditional EM is the great depth of field that may be obtained, although at some loss of resolution. Ideally suited for examination of surface morphology, e.g. fracture surfaces.

SCARAB Tradename for urea–formaldehyde polymer based moulding powder.

SCATTERING CROSS-SECTION Symbol σ. The ratio of the scattered neutron flux to incident flux in neutron scattering. It is related to the scattering length (b) by: $\sigma = 4\pi b^2$. The coherent scattering is given by, $\sigma_{coh} = 4\pi \bar{b}_c^2$, and the incoherent scattering by, $\sigma_{incoh} = 4\pi(\overline{b^2} - \bar{b}^2)$. For a polymer in dilute solution, the scattering cross-section is given by the probability that a neutron will be scattered into the solid angle (Ω) as $(d\sigma/d\Omega)_{coh} = (b_p - b_s)S(q)$, where b_p and b_s are the coherent scattering lengths per unit volume of the polymer and solvent respectively and equal

$$N_A \sum_n b_{p,n}/V_p \quad \text{and} \quad N_A \sum_n b_{s,n}/V_s$$

respectively, where N_A is Avogadro's number and $b_{p,n}$ and $b_{s,n}$ are the atomic scattering lengths of the atoms of the polymer repeat unit and solvent molecules and V_p and V_s are the partial molar volumes. $b_p - b_s$ is thus the contrast factor, analogous to the refractive index increment in light scattering. $S(q)$ is the scattering law. The scattering cross section is identical to the Rayleigh ratio of light scattering and is related to the experimentally measured incident and scattered neutron intensities, I and I_0 respectively, by $I/I_0 = kN\,dT(d\sigma/d\Omega)_{coh}$, where k is an instrument constant, N is the number of scatterers per unit volume, d is the sample thickness and T is the transmission.

SCATTERING ENVELOPE A polar diagram which plots the intensity of scattered light as a function of angle of the scattered beam to the incident beam. When the

scattering particles are less than about $\lambda/20$ (λ being the wavelength of the light) the envelope is symmetrical in the sense that the scattered intensity at angle θ in the forward direction is the same as that scattered at $\pi - \theta$ in the backward direction. When the scatterers are larger, then internal interference occurs and the forward scattering is more intense giving an unsymmetrical envelope, or dissymmetry.

SCATTERING FACTOR Alternative name for particle scattering factor.

SCATTERING LAW (Structure factor) Symbol $S(\mathbf{q})$ or $S(\kappa)$. The density fluctuation correlation function in Fourier space, thus giving the variation in intensity of scattered radiation with scattering vector $\mathbf{q}$ or κ.

SCATTERING LENGTH Symbol b. In neutron scattering, the amplitude of the scattered wave characterising the scattering of a nucleus, since the probability of scattering (the scattering cross-section) is $4\pi b^2$. The value of b varies from one nucleus to another, in particular ^1H and ^2H (protons and deuterium atoms) have very different values. This leads to the usefulness of carrying out small angle neutron scattering experiments by the labelling of some molecules with ^2H.

SCATTERING POWER A measure of the ability of a two phase material to scatter X-rays in small angle X-ray scattering. It is the mean square fluctuation in electron density and is given by,

$$\overline{(\rho - \bar{\rho})^2} = (\rho_2 - \rho_1)^2 \phi_1 \phi_2$$

where ρ_1 and ρ_2 are the electron densities and ϕ_1 and ϕ_2 are the volume fractions of the continuous and dispersed phases.

SCATTERING VECTOR Symbol $\mathbf{h}$ or $\mathbf{q}$. A function of the angle of scatter (θ) of radiation used in small angle X-ray and small angle neutron scattering and defined as, $\mathbf{h} = 4\sin(\theta/\lambda)$, where λ is the radiation wavelength. It is also a measure of the momentum transfer during the interaction of the radiation with the material.

SCF Abbreviation for superconductive furnace carbon black.

SCHARDINGER DEXTRIN A group of cyclic oligosaccharides obtained by the action of the enzyme *Bacillus macerans* on starch. Six and seven α-1,4'-linked D-glucose units are present in the α- and β-dextrins respectively. These are also known as cyclohexaamylase and cycloheptaamylase respectively. Owing to their cyclic structures they are resistant to enzymatic hydrolysis and form complexes with many inorganic compounds.

SCHATZKI MECHANISM (Crankshaft motion) A type of molecular motion thought to occur in polymers which contain at least four $—CH_2—$ groups in a repeat unit. It involves the simultaneous rotation about two in-chain C—C bonds separated by four —CH_2— groups. In a randomly oriented chain such bonds are at least partially collinear, so this motion resembles that of a crankshaft. Thus the onset of such a motion as the temperature is raised is thought to occur at about $-105°C$ and is responsible for the transition occurring at about this temperature in those polymers, such as polyethylene and its copolymers, containing —$(CH_2)_4$— groups in the chain. In a crystalline polymer such a transition may be the dominant amorphous transition and it is then often referred to as the glass II transition or even as T_g.

SCHERING BRIDGE An electrical bridge circuit (similar to a Wheatstone bridge for measuring resistance) which forms the basis of a commonly used method for the determination of dielectric properties in the frequency (ω) range $10–10^6$ Hz by determination of the capacitance (C_x) and conductance (G_x) of the dielectric of interest in a condenser. With alternating current bridges there is a problem in the elimination of stray capacitances. In the Schering bridge, which consists of two capacitative and two resistive arms, this is achieved by the use of a substitution method, the bridge being balanced with and without the unknown capacitance (C_x) connected across a low loss calibrated standard capacitance (C_1). In the opposite resistive arm (resistance R_4), a calibrated capacitance (C_4) is in parallel with the resistance. If the changes in C_1 and C_4 with sample out and with sample in the circuit are ΔC_1 and ΔC_4, then $C_x = \Delta C_1$ and

$$\tan \delta_x = G_x/\omega C_x = \omega C_1' \Delta C_4 R_4 / \Delta C_1$$

where C_1' is the initial value of C_1. Stray capacitances are further reduced by careful screening of the bridge components and by use of a Wagner earth.

SCHERRER EQUATION A relationship between the mean dimension of crystallites of a polycrystalline sample, measured perpendicular to the hkl planes (L_{hkl}) and the broadening of the hkl reflection expressed as the integral breadth, or breadth at half-maximum intensity (β_0)

$$L_{hkl} = k\lambda/\beta_0 \sin \theta$$

where k is a constant usually taken as unity, λ is the wavelength of radiation and θ is the diffraction angle. The relationship is rather unreliable as a measure of crystallite size since broadening may also be due to lattice distortions.

SCHIFF BASE CHELATE POLYMER A polymer prepared by coordination of a substituted Schiff base polymer with a metal ion such as zinc, cadmium, calcium or nickel. An example is the polymer from bis-(salicylaldehyde) and o-phenylenediamine coordinated to a metal (M) to give:

However such polymers have poorer thermal stability than the uncoordinated polymers.

SCHIFF BASE POLYMER Alternative name for polyazomethine.

SCHLEROPROTEIN Alternative name for structural protein.

SCHLIEREN METHOD Alternative name for Schlieren optics.

SCHLIEREN OPTICS An optical system used in an ultracentrifuge for the determination of the solute concentration gradient at any point in the cell. A parallel beam of light from a slit passes through the cell and an image of the slit is obtained which is deflected by different amounts along its length by the radially directed refractive index gradient in the cell. The image is focussed on a suitable phase plate. An image of this is photographed to give a two-dimensional plot in which the abscissa corresponds to the distance from the end of the cell and the ordinate is proportional to the refractive index gradient and hence also to the concentration gradient.

SCHOB PENDULUM An instrument for the determination of the rebound resilience of a material, usually used for rubbers. A pendulum consisting of a rod pivoted at the upper end with a striker attached to the lower end is allowed to fall from a fixed height so that the striker hits a vertically mounted slab of the rubber. The rebound resilience is the ratio of the rebound height to the drop height.

SCHULTZ–BLASCHKE EQUATION An empirical relationship between the specific viscosity (η_{sp}) and concentration (c) of a polymer in dilute solution, of the form, $\eta_{sp}/c = [\eta] + k[\eta]\eta_{sp}$, where $[\eta]$ is the limiting viscosity number. Often valid over a wider range of values of c than either the Huggins or Kraemer equations.

SCHULTZ DISTRIBUTION (Zimm distribution) A molecular weight distribution (originally derived from chain polymerization kinetics with constant rate of initiation and termination by reaction with monomer) having a differential weight distribution function of $W(r) = (-\ln p)^2 r p^r$, where p is the ratio of rate of propagation to rate of termination and r is the measure of the molecular size, e.g. degree of polymerisation or molecular weight, of each species. Later generalised to $W(r) = (-\ln p)^{k+1} r^k p^r / \Gamma(k+1)$, or for the corresponding number distribution, $N(r) = (-\ln p)^{k+1} r^{k-1} p^r / \Gamma(k)$, where k is the degree of coupling and Γ is the gamma function. When $k = 1$ this is the Schultz–Flory distribution and when $k = 2$ it is the distribution for simple chain polymerisation with termination by combination. $W(r)$ has a maximum at $\bar{x}_n$ and $N(r)$ has a maximum at $\bar{x}_n/k$. Also $\bar{x}_n = -k/\ln p$, $\bar{x}_w = -(k+1)/\ln p$ and $\bar{x}_z = -(k+2)/\ln p$ where $\bar{x}_n$, $\bar{x}_w$ and $\bar{x}_z$ are the number, weight and z-average degrees of polymerisation respectively. Thus $\bar{x}_n : \bar{x}_w : \bar{x}_z$ are as $k : (k+1) : (k+2)$. When $k = 2$ the polydispersity is 1·5 and for higher values of k the distribution is even narrower and is therefore used to describe fractionated samples.

SCHULTZ–FLORY DISTRIBUTION (Most probable distribution) (Geometric distribution) (Exponential distribution) (Random distribution) In early literature also sometimes called the normal distribution, although this term is now an alternative for the Poisson distribution. A molecular weight distribution which results from several types of polymerisation, whose distribution function may be derived from the polymerisation kinetics. It results from chain polymerisation with constant rate of initiation, constant monomer concentration, with and without transfer and with termination by disproportionation, from linear step-growth polymerisation where there is equal reactivity of functional groups or with random interchange occurring, or from random scission degradation. A special case of the more general Schultz distribution with a degree of coupling of $k = 1$.

The differential number distribution function of molecular sizes i, expressed as molecular weights M_i, is, $N(M_i) = p^{i-1}(1-p)$ and the differential weight distribution function is $W(M_i) = ip^{i-1}(1-p)^2$, where p is the extent of reaction in a step-growth polymerisation or the ratio of the probability that a chain will propagate to the probability that it will terminate in chain polymerisation. In this latter case

$$p = v_p/(v_p + v_t) = k_p[M]/[k_p[M] + (2R_i k_t)^{1/2}]$$

where v_p and v_t are the rates of propagation and termination respectively, k_p and k_t are the corresponding rate constants, $[M]$ is the monomer concentration and R_i is the rate of initiation.

The number, weight and z-average degrees of polymerisation ($\bar{x}_n$, $\bar{x}_w$ and $\bar{x}_z$ respectively) are $1/(1-p)$, $(1+p)/(1-p)$ and $(1+4p+p^2)/(1-p)(1+p)$, and for high molecular weight polymer they are in the ratios 1:2:3. Alternatively, it can be shown that the weight distribution function of molecular sizes is $W(M_i) = (i/\bar{x}_n^2)[\exp(-i/\bar{x}_n)]$, hence the names exponential and geometric distributions. A plot of $N(M_i)$ versus M_i falls rapidly from a high initial value, whereas a plot of $W(M_i)$ versus M_i has a maximum at $\bar{x}_n$ and an inflection point at $\bar{x}_w$.

SCHWEITZER'S REAGENT Alternative name for cuoxam.

SCLAIR Tradename for a linear low density polyethylene based on octene-1 comonomer.

SCMK Abbreviation for S-carboxymethylcysteine-kerateine.

SCON Tradename for polyvinyl chloride.

SCORCH (Prevulcanisation) The premature vulcanisation of a raw rubber compound whilst being heated during the compounding or processing (shaping) stages of rubber product manufacture. Scorch prevents adequate mixing or shaping before the final vulcanisation stage. Thus a rubber compound must have an adequate scorch time (scorch resistance) to allow these processes to be performed. The scorch resistance is highly dependent on the vulcanisation system used. Thiazoles and especially dithiocarbamates promote scorch. Scorch resistance is determined by the time required for the onset of crosslinking, as measured by an increase in the viscosity of the compound, e.g. in a Mooney viscometer. Scorch time may be increased by the use of a retarder.

SCORCH TIME Alternative name for Mooney scorch time.

SCREW DISLOCATION (Burger's dislocation) A dislocation generated by a slip of one part of a crystal relative to the rest, with a dislocation line vector marking the line of motifs terminating the slip.

SDS Abbreviation for sodium dodecyl sulphate.

SDS–GEL ELECTROPHORESIS A method of gel electrophoresis which is now the most widely used method for determining protein polypeptide chain molecular weight and for checking protein homogeneity, due to both its rapidity and the requirement for only simple equipment. The protein is treated with sodium dodecyl sulphate (SDS) which associates with the polypeptide chain to form a rod-like structure in which the SDS molecules coat the polypeptide with their charged sulphate groups on the outside and their hydrocarbon chains form hydrophobic bonds with the protein. Usually such complexes have an SDS:protein weight ratio of 1·4:1. If the original protein is oligomeric, the SDS disassociates the protein to its subunits. If the original protein contains interchain disulphide bonds these must first be destroyed. On electrophoresis in a polyacrylamide gel, the rate of migration is determined by the mass of the complex and hence by the protein molecular weight (M). Calibration with proteins of known molecular weight (marker proteins) shows that the distance travelled is proportional to $\log M$.

SEBACIC ACID

$$HOOC(CH_2)_8COOH \qquad \text{M.p. } 134°C.$$

Produced by alkaline high temperature hydrolysis of ricinoleic acid (a major component of castor oil). A monomer for the formation of nylon 610 and nylon 610 copolymers. Sometimes used in place of phthalic anhydride in the formulation of unsaturated polyester resins and in alkyd resins for giving cured products with increased flexibility.

SEBS Abbreviation for styrene–ethylene/butylene–styrene block copolymer.

SEC Abbreviation for size exclusion chromatography.

SECANT MODULUS The slope of the line on a stress–strain plot drawn from the origin to a point on the stress–strain curve at some defined strain value. The line is thus the secant to the curve. It is mostly used for characterising the Young's modulus for materials which show non-linear stress–strain behaviour, when the values of the secant modulus at 0·2, 1·0, 100 or some other percentage elongation may be quoted.

SECONDARY ACCELERATOR An accelerator used as the minor component (10–20% of the total) in a mixed accelerator, in combination with a primary accelerator, to activate the latter and to improve vulcanisate properties.

SECONDARY ACETATE Alternative name for secondary cellulose acetate.

SECONDARY BUTYL ALCOHOL (Butan-2-ol)

$$CH_3CH_2CH(CH_3)OH \qquad B.p.\ 99·5°C.$$

The industrial material is a mixture of the two optical isomers. A solvent for many natural resins, ethyl cellulose, polyvinyl butyral and some phenol–formaldehyde resins.

SECONDARY CELLULOSE ACETATE (Secondary acetate) Cellulose acetate with a DS of 2–2·5, obtained by hydrolysis of cellulose triacetate dope by addition of water. This method of partial hydrolysis gives a more evenly substituted product than the product of partial acetylation of cellulose. Earlier a major fibre, often referred to as cellulose acetate rayon, or simply acetate, obtained by dry spinning from acetone solution. Also a useful film-forming material for photographic film and recording tape being less inflammable than the earlier used cellulose nitrate. Previously widely used, when plasticised with, for example, dimethyl phthalate, triacetin or triphenyl phosphite for easy processing, as a plastic material.

SECONDARY CRYSTALLISATION A slow crystallisation process occurring towards the end of the earlier main (primary) process. Frequently observed in an Avrami analysis as a deviation of the log–time plot at long times, continuing beyond the normal 100% crystallisation value. Probably involves reorganisation of the crystallites, producing an increase in interfold distance and larger and more perfect crystallites.

SECONDARY FLOW A circulatory flow in which the flow lines cross the direction of flow, as opposed to laminar flow. Secondary flow always occurs with non-Newtonian fluids except for certain simple flow geometries, e.g. for a circular cross-section tube or infinite slit.

SECONDARY FRACTURE Fracture occurring separately from the main fracture, but in the neighbourhood of the main fracture. It arises from the fracture stress being exceeded in the material surrounding the main fracture front due to stress concentrations from inhomogeneities (such as voids, filler particles or other hard inclusions) in the material. Interaction of secondary fracture, which spreads circularly in a plane parallel to the main fracture, with the main fracture, leads to the formation of conic and other types of markings on the fracture surfaces.

SECONDARY NUCLEATION Nucleation on a smooth crystal surface by the deposition and growth of the motif on the surface. Since one face of the growth unit is in contact with the preformed crystal face, the process has lower free energy than primary nucleation.

SECONDARY PLASTICISER A plasticiser which is of limited compatibility with the polymer concerned and which can therefore only be used in limited quantities. If the compatibility limit is exceeded the excess plasticiser will separate out as an oily surface layer on the polymer. Generally, secondary plasticisers are not as efficient at increasing flexibility as primary plasticisers but often impart good low temperature flexibility.

SECONDARY STRUCTURE (1) The conformational structure of proteins referring to the occurrence of a regular conformational structure. Such a regular conformation may persist over a considerable length and involve a major part of a polypeptide chain, as in most fibrous proteins, or it may occur only over a part of a polypeptide chain, as in many globular proteins.

The regular conformations adopted by polypeptide chains are usually the α-helix or the β-conformation. Occasionally other regular conformations, such as the π-helix and the polyproline I or II conformations, are found. The secondary structure is frequently stabilised by intramolecular (as in the α-helix) or intermolecular (as in the β-pleated sheet) hydrogen bonding. Theoretical predictions may be made of the stability of the various possible conformations for a known sequence of amino acid residues, according to the Ramachandran map. The actual occurrence of the secondary structure is largely determined by X-ray diffraction.

(2) In carbon blacks, the loose aggregates of clusters or chains of primary structure particles.

SECONDARY TRANSITION In an amorphous polymer, a transition of smaller magnitude than the main transition, i.e. the glass transition. It nearly always occurs at a lower temperature than T_g and may be labelled as the β-, γ-transition, etc., T_g being labelled as the α-transition. Most amorphous polymers exhibit at least one such transition, often a very broad one extending over about 100°C. Occasionally a higher temperature (liquid–liquid) transition also occurs. In crystalline polymers, the term secondary transition may refer either to the same type of transition as in an amorphous polymer or to a secondary crystalline transition such as premelting or a crystal–crystal transition. Secondary transitions are usually not sufficiently large to be detected by the static methods, such as dilatometry, useful for T_g determination. They are usually characterised by dynamic mechanical, dielectric or nuclear magnetic resonance relaxation measurements.

SECOND ORDER MARKOV CHAIN PROCESS A Markov chain process in which the occurrence of an event is dependent on the last two previous events. An example is copolymerisation which shows a penultimate effect, when knowledge of the tetrad sequence distributions is required to test for the process. Also in the generation of stereochemical configurational sequences, a penultimate effect is a second order Markov chain process and requires a knowledge of pentad sequence distributions in order to be identified.

SECOND ORDER TRANSITION A thermodynamic transition which shows a sharp discontinuity of the second derivative of the Gibbs free energy (G) with respect to temperature. Hence there is also a sharp discontinuity in the specific heat (C_p), the volume expansion coefficient (α) and the compressibility (κ) with respect to temperature. In addition there is a sharp discontinuity in the first derivative of the entropy, enthalpy and volume with respect to temperature. Polymer glass transition temperatures are often thought to be second order transitions since they exhibit such discontinuities. However, many consider that they should not be regarded as true second order thermodynamic transitions since they do not fulfil the requirement that thermodynamic equilibrium should exist on both sides of the transition. Furthermore, the value of T_g is rate dependent and can also be regarded as a kinetic phenomenon and the values of C_p, α and κ are smaller below T_g than above.

SECOND VIRIAL COEFFICIENT Symbols A_2, B and Γ_2. The coefficient of the most important term of the virial equation which accounts for the non-ideality of behaviour of a system, in particular of the colligative and other properties of dilute solutions. Such non-ideal behaviour arises out of binary and other higher order interactions between solute particles. Hence determination of the value of the virial coefficient provides a measure of the strength of these interactions. Generally the virial equation is of the form,

$$P = RT(c_2/M_2 + A_2c_2^2 + A_3c_2^3 + A_4c_2^4 + \cdots, \text{etc.})$$

where P is the colligative property, c_2 is the solute concentration, M_2 is the solute molecular weight and A_2, $A_3, A_4, \ldots$, etc. are the second, third, fourth, $\ldots$, etc. virial coefficients. Often non-ideality is adequately represented by the c_2 and c_2^2 terms only and rarely are terms higher than c_2^3 needed. Alternatively,

or

$$P = RTc_2/M_2 + Bc_2^2 + Cc_2^3 + \cdots, \text{etc.}$$
$$P = (P/c_2)_0(c_2 + \Gamma_2c_2^2 + \Gamma_3c_2^3 + \cdots, \text{etc.})$$

where $(P/c_2)_0$ is the value of P/c_2 extrapolated to infinite dilution and the solution behaves ideally. Thus the second virial coefficient may be represented by $B(=RTA_2)$ or $\Gamma_2(=A_2M_2)$.

For polymer solutions A_2 is related to the excluded volume (u), e.g. by the expression $A_2 = Nu^2/2M_2^2$, where N is Avogadro's number, and to the polymer solvent interaction parameter (χ) by $A_2 = (\frac{1}{2} - \chi)V_1\rho_2^2$, where V_1 is the solvent molar volume and ρ_2 is the polymer density. Since u increases as solvent power increases, so does A_2, which should also decrease as molecular weight increases. At the theta temperature A_2 will be zero. Thus A_2 provides a measure of the thermodynamic quality of the solvent.

A_2 may be obtained experimentally from colligative property measurements, especially of the osmotic pressure (π), by plotting P/c_2 against c_2. A linear plot will result, assuming $A_3 = A_4 = \cdots$, etc. $= 0$, with a slope of A_2. Alternatively A_2 may be obtained from the light scattering Zimm plot, since it is the slope of the $\theta = 0$ line, where θ is the scattering angle.

SECTORISATION The division of a polymer single crystal into four sectors by diagonal lines, sometimes observed by electron microscopy. A result of the collapse of the original hollow pyramid form of the crystal during the sample preparation required for electron microscopy.

SECTOR METHOD Alternative name for rotating sector method.

SEDIMENTATION The movement of particles, which may be dissolved molecules—especially if polymeric, through a suspending liquid in the direction of an applied gravitational field. High fields may be applied by ultracentrifugation, for the determination of the molecular weight of a dissolved polymer, especially for aqueous solutions of biopolymers. The sedimentation velocity (dr/dt) is given by the relation, $S = (1/\omega^2r)(dr/dt)$, where r is the distance from the centre of rotation, ω is the angular velocity and S is the sedimentation constant. S is related to the molecular weight by the Svedberg equation.

SEDIMENTATION COEFFICIENT Alternative name for sedimentation constant.

SEDIMENTATION CONSTANT (Sedimentation coefficient) Symbol S. The rate of sedimentation in a unit centrifugal field. Important in the sedimentation velocity method of polymer molecular weight determination by ultracentrifugation. Related to the molecular weight (M) by the Svedberg equation. Given by $S = (1/\omega^2r)(dr/dt)$, where ω is the angular velocity, r is the distance from the centre of rotation and dr/dt is the velocity of the sedimenting polymer molecule. S is concentration dependent and its value must be extrapolated to infinite dilution (S_0). This is often complicated, especially for synthetic polymers, which are best studied in a theta solvent. For proteins and nucleic acids the concentration dependence of S is minimised more readily by control of ionic strength and pH of the solution. The unit of the sedimentation constant is the Svedberg (S) (10^{-13} s). Values for polymers are in the range 1–200 S.

SEDIMENTATION–DIFFUSION AVERAGE MOLECULAR WEIGHT
Symbol $\bar{M}_{SD}$. The molecular weight average obtained by the sedimentation velocity method of ultracentrifugation when the weight average values of the sedimentation and diffusion constants are used. Ill-defined theoretically, but in practice lies between the number and weight average values of molecular weight.

SEDIMENTATION EQUILIBRIUM METHOD
(Equilibrium centrifugation) A method for the determination of the molecular weight (M) of a polymer by ultracentrifugation of a dilute solution at relatively low speeds (10 000–20 000 rpm). The rate of sedimentation is then balanced by the rate of diffusion of the polymer molecules and a concentration gradient (dc/dr) is formed. The gradient is measured using an optical system, often the Schlieren method, which yields values of dc/dr as a function of r, the distance from the centre of rotation.

The method is preferred over the sedimentation velocity method for synthetic polymers, since it is more amenable to an exact theoretical treatment for polydisperse polymers. Also it does not need separate measurements of other parameters to be made, but the time required to reach equilibrium can be very long, e.g. several days. M is given by $(1/c)(dc/dr) = (\omega^2 r/RT)(1 - \bar{v}_2\rho)(M)$, where the concentration c is obtained by integration of the Schlieren curve, ω is the angular velocity, $\bar{v}_2$ is the polymer partial specific volume and ρ is the solvent density. On integration

$$\ln c = [(1 - \bar{v}_2\rho)/2RT] - \omega^2 r^2 M = \text{constant}$$

so M may be obtained from the slope of the plot of $\ln c$ versus r^2.

The apparent value of M obtained is concentration dependent and must be extrapolated to infinite dilution. For a polydisperse polymer the weight average value ($\bar{M}_w$) is obtained. However, since $\bar{M}_w$ varies in the cell, depending on r, the above plot will be curved.

A better relationship, obtained by integration over the whole cell volume is,

$$2RT(c_b - c_m)/\omega^2(1 - \bar{v}_2\rho)c_0 = \bar{M}_w(r_b^2 - r_m^2)$$

where c_b, c_m, r_b and r_m are the concentrations and distances from the centre of rotation at the bottom of the cell and at the meniscus respectively and c_0 is the initial solution concentration. In principle the number, z- and higher average molecular weights ($\bar{M}_n$, $\bar{M}_z$, etc.) may be calculated. However, in practice $\bar{M}_n$ is only obtainable if c falls to zero somewhere in the cell and the other averages higher than $\bar{M}_z$ are rather unreliable. The molecular weight distributions may also be obtained in principle. Usually it is difficult to determine c_m accurately, but the meniscus depletion method may be used, when c_m is zero.

SEDIMENTATION POTENTIAL
The potential gradient that is set up when a solution of a polyelectrolyte sediments in an ultracentrifuge. It arises because the macroions sediment faster than the counter-ions, and it acts to slow down the macroions and to accelerate the counter-ions—the primary charge effect.

SEDIMENTATION VELOCITY METHOD
(Boundary sedimentation) A method of ultracentrifugation at such high speeds that the rate of sedimentation of polymer molecules in solution may be measured. As the molecules sediment, a layer of pure solvent is formed, whose refractive index (n) differs from that of the solution. In the boundary region a sharp change in n occurs and the motion of the boundary may be followed.

The molecular weight (M) of the polymer may be related to the sedimentation velocity (dr/dt) by the Svedberg equation. For the absolute determination of M, this requires a knowledge of the diffusion coefficient (D), the sedimentation coefficient (S), which is difficult to obtain, and the partial specific volume ($\bar{v}_2$) of the polymer. Instead, frequently a relationship of the form $S = kM^a$ is assumed, where k and a are constants for the particular polymer/solvent/temperature combination under study. They are determined using polymer fractions of known M.

For a polydisperse polymer the particular average of M obtained depends on the type of average of S and D used. If weight average values of S and D are used then the sedimentation–diffusion average molecular weight ($\bar{M}_{SD}$) is obtained. However, since the type of average of S and D usually measured is uncertain, the method is best restricted to monodisperse (or nearly monodisperse) polymers such as proteins and nucleic acids. For synthetic polymers the sedimentation equilibrium method is more suitable.

SEEDED CRYSTALLISATION
Alternative name for self-seeding.

SEEDED POLYMERISATION
A polymerisation in which preformed polymer species act as 'seeds' for further polymerisation. The seeds may activate the monomer, or polymer active centres may have been preformed by the addition of a small amount of monomer to the initiator. This latter is particularly possible in anionic polymerisation, where living polymer active centres can be used. Subsequently polymer chains can grow from the seeds by the addition of further monomer. Alternatively, in emulsion polymerisation, preformed latex particles may be used as seeds for further polymerisation by growth of these particles.

SEED HAIR FIBRE
A natural fibre which is found attached to the seed of a plant. Cotton, coir and kapok are important examples.

SEE-THROUGH CLARITY
Alternative name for clarity.

SEGMENTED COPOLYMER
(Segmented polymer) In general an alternative name for block copolymer, but usually refers to a multiblock polymer, especially one consisting of relatively long blocks of one type separated

by short blocks of a second type. Its structure may be represented by $-(A_xB_y-)_n$, where A_x and B_y are the blocks of repeat units A and B. The term is applied particularly to segmented polyurethanes where the hard blocks may be very short (only a few repeat units) but where, nevertheless, phase separation with domain formation still occurs.

SEGMENTED POLYMER Alternative name for segmented copolymer.

SEGMENTED POLYURETHANE A polyurethane block copolymer with relatively short blocks or segments. Most polyurethane elastomers are segmented copolymers containing alternating hard and soft blocks or segments, i.e. they are multiblock polymers. The soft blocks are usually aliphatic polyester or polyether units of molecular weight of a few hundred to a few thousand and with a low T_g value, so that they are elastomeric. The hard blocks contain a high proportion of urethane or urea groups, formed by reaction of a diisocyanate with a diol or diamine, and are intermolecularly hydrogen bonded as well as being stiffer molecules. The two types of segments tend to aggregate separately into domains, the hard segment domains acting as physical crosslinks for the soft segment domains.

SELF-ADHESION Alternative name for tack.

SELF-CONSISTENT METHOD A theory for predicting the values of the elastic properties of composite materials which uses as a model a particular shape of the filler phase embedded in a matrix whose elastic properties are those of the composite as a whole, as in the Hill theory. Other more complex models leading to the Kerner and van der Poel equations are also sometimes referred to as self-consistent.

SELF-EXTINGUISHING POLYMER A polymer which will burn in air when exposed to a source of ignition, but which ceases to burn when the source is removed.

SELF-INITIATION POLYMERISATION Alternative name for thermal polymerisation.

SELF-NUCLEATION Alternative name for self-seeding.

SELF-PROPAGATION In chain copolymerisation, propagation in which an active centre adds on a monomer of the same type, i.e. $\sim A^* + A \rightarrow \sim AA^*$, rather than one of the opposite type (cross-propagation). The relative tendency of self- to cross-propagation is given by the monomer reactivity ratio.

SELF-SEEDING (Self-nucleation) (Seeded crystallisation) The primary crystallisation by small crystalline regions of the polymer that have survived the melting or dissolution step. These 'seed' crystals often persist to temperatures well above the normal melting range, but

are progressively destroyed as temperature is raised, as evidenced by the number of spherulites subsequently nucleated on cooling.

SELF TRANSFER Alternative name for back biting.

SEM Abbreviation for scanning electron microscopy.

SEMICONDUCTING POLYMER A polymer having an electrical conductivity in the range $10^{-10}–10^2$ S cm^{-1}. Most polymers have conductivities below 10^{-12} S cm^{-1} and hence are insulators, but polymers with exalted electrical properties may be synthesised. These are mostly semiconducting, charge-transfer polymers, such as poly-(acene quinone radical). Only a few polymers are truly conducting, i.e. having conductivities of greater than 10^2 S cm^{-1}. Many polymers that are often referred to as being conducting are strictly only semiconducting.

SEMI-CRYSTALLINE POLYMER A polymer sample which is only partially crystalline. Apart from polymer single crystals all crystalline polymer samples are partially crystalline. The degree of crystallinity is usually 30–80%.

SEMI-DRYING OIL An oil, such as safflower, soya bean or dehydrated castor oil, which air dries more slowly than a drying oil, such that a film of the oil only becomes tacky after about 7 days.

SEMI-INTERPENETRATING POLYMER NETWORK (Semi-IPN) An interpenetrating polymer network in which one of the polymers is linear rather than crosslinked. Two types may be characterised depending on the method of formation. In a semi-IPN of the first type, a monomer is polymerised to a linear polymer (II) in the presence of network I, or alternatively the network II may be formed in the presence of polymer I. The second type is a semi-IPN which is a graft copolymer with one of the polymers being crosslinked.

SEMI-IPN Abbreviation for semi-interpenetrating polymer network.

SEMI-LADDER POLYMER Alternative name for a partial ladder polymer.

SEMI-PERMEABLE MEMBRANE A membrane which is only permeable to the solvent molecules of a solution and not to the solute. In membrane osmometry, for the determination of a polymer number average molecular weight ($\bar{M}_n$), it provides the membrane separating the solution and solvent compartments of the osmometer. The membrane is the greatest source of uncertainty and limitation of the technique since most available membranes are permeable to polymer solutes of $\bar{M}_n$ less than about 10 000. Even if the polymer sample has a much higher $\bar{M}_n$, as a result of the usual distribution of molecular sizes, the low molecular weight species will diffuse through the membrane, giving a falsely low osmotic pressure and hence a high $\bar{M}_n$ value.

For work at or near ambient temperature with organic solvents, regenerated cellulose membranes (e.g. gel cellophane or deacetylated cellulose acetate) are usually used. Bacterial cellulose is best for aqueous solutions. For high temperature use, porous glass and polychlorotrifluoroethylene membranes have been employed.

SEMI-PREPOLYMER PROCESS Alternative name for quasi-prepolymer process.

SEMI-REINFORCING BLACK (SRF) A type of furnace carbon black of particle size about 70 nm, moderately reinforcing, giving compounds of high elongation and resilience and low compression set. Useful in mechanical goods, footwear, inner tubes, hose and floor mats.

SEMI-RIGID POLYURETHANE FOAM A polyurethane foam, intermediate in its mechanical properties between a rigid and a flexibile polyurethane foam. Formed by reaction of a synthetic polyether triol (with a moderate molecular weight) or a polyester polyol (with a moderate degree of branching) with a diisocyanate.

SEMI-VINYL COPOLYMERISATION Copolymerisation of a vinyl monomer with a non-vinyl monomer. In most cases the latter does not homopolymerise but forms a 1:1 alternating copolymer with the vinyl monomer. Examples of non-vinyl monomers are CO, SO_2 and O_2.

SEPHADEX Tradename for a crosslinked dextran, widely used as a gel in gel permeation chromatography in aqueous systems, especially for proteins and nucleic acids. It is prepared by crosslinking the dextran by reaction with epichlorhydrin:

$$\sim OH + CH_2\!-\!CH\!-\!CH_2Cl \xrightarrow{\bar{O}H}$$

$$\sim OCH_2\!-\!CH\!-\!CH_2Cl \xrightarrow{\bar{O}H} \sim OCH_2\!-\!CH\!-\!CH_2$$

$$\xrightarrow{\sim OH} \sim OCH_2\!-\!CH\!-\!CH_2\sim$$

Derivatives of Sephadex, e.g. diethylaminoethyl Sephadex are useful as ion-exchange resins for ion-exchange chromatography of biopolymers.

SEPHAROSE Tradename for an agarose gel suitable for use as the column packing material for gel permeation chromatography in aqueous systems. Derivatives are made by reaction of some of the hydroxyls and are also widely used in ion-exchange chromatography and in affinity chromatography of biopolymer mixtures.

SEQUENATOR An instrument for the determination of the amino acid sequence in proteins by automatic analysis using the Edman method of repeated sequential removal and identification of the N-terminal amino acid residue of the polypeptide chain.

SEQUENCE HOMOLOGY The existence of identical amino acid residues in identical positions in the polypeptide chains of different proteins which often also means the existence of identical sequences of residues. The homology arises when different proteins are homologous proteins, functionally similar proteins, e.g. haemoglobin and myoglobin, or the different subunits of an oligomeric protein, such as the α- and β-chains of haemoglobin.

SEQUENCE LENGTH The number of repeat units of a specific type joined to each other contiguously in a polymer chain. The specific units may be one type of monomer unit in a copolymer or one type of stereoisomer in a polymer capable of stereoregular configurational isomerism. Since the formation of these polymers is a statistically controlled process, with Bernouillian statistics, the polymer molecules contain a distribution of sequence lengths. For a whole polymer chain the average sequence length may be calculated. For a random copolymer, for example, this is given by $l_A = (1 + r_A[A]/[B])$ and $l_B = (1 + r_B[B]/[A])$, where l_A and l_B are the average sequence lengths of the comonomers A and B respectively, r_A and r_B are their reactivity ratios and [A] and [B] are their mole fractions in the feed. For an alternating copolymer $l_A = l_B = 1$ and for a block copolymer l_A and l_B are very large. Experimentally, sequence lengths are difficult to determine. In the case of stereoisomeric sequences, and to a more limited extent for copolymers, high resolution NMR spectroscopy can give limited information.

SEQUENCING The determination of the sequence of amino acid residues, i.e. the primary structure, of a peptide or polypeptide chain of a protein. This is often performed by repeated application of the Edman method most conveniently using a sequenator.

SEQUENTIAL COPOLYMER A copolymer in which the different comonomer units are joined together in a precisely defined sequence along the length of the polymer chain, the sequence being the same in all polymer molecules. This is the case in the important biopolymers, the proteins and the nucleic acids. In the former case the precise sequential structure is essential for the protein to perform its specific biological function. In the latter case, the precise sequential arrangement contains genetic information and the polymer may also be described as an informational macromolecule.

SEQUENTIAL POLYMERISATION A polymerisation in which a polymer is formed with repeat units of different structure and arranged in a precisely defined sequence along the polymer chain. This is only possible by very careful control over the course of the polymerisation. Usually the structural differences refer to different comonomer units. Examples range from the relatively simple cases of alternating and block copolymerisations to the much more complex cases of protein and nucleic acid synthesis. In the case of proteins, usually about 20 different monomers have to be used to construct polymer chains in which the same long complex sequence running the whole length of the molecule has to be formed for every polymer molecule.

SEQUENTIAL POLYPEPTIDE A polypeptide which contains repeating sequences of amino acid residues. Since some important fibrous proteins, such as silk and collagen, contain many such sequences, the study of synthetic sequential polymers to model these polymers, in order to investigate the influence of sequence type on conformation, is of interest. The polymers are more difficult to synthesise than simple homopolymers, since it is necessary to synthesise a di- or tripeptide and to subsequently polymerise it, as the N-carboxyanhydride route is not available. The simplest such polymers are the polydipeptides, such as poly-(gly–ala), and poly-tripeptides, such as poly-(gly–gly–ala). Polymers with longer sequences are increasingly difficult to prepare, but polymers up to polyhexapeptides have been made. Some sequential polypeptides have been synthesised as models for enzyme active sites, e.g. poly-(ser–his–leu–leu–leu) and poly-(leu–his–leu–leu–ser–leu) models for chymotrypsin.

SER Abbreviation for serine.

SERACETA Tradename for cellulose acetate rayon.

SERICIN (Silk gum) A protein of silk (accounting typically for about 20% of the silk thread) as a coating on the two fibroin filaments which binds them together. In commercial silk, the sericin is removed (degumming) by extraction with hot soap solution. Sericin contains a very high serine content (about 35%) together with other polar amino acid residues, e.g. aspartic and glutamic acids, and much fewer non-polar residues, compared with fibroin. It also contains about 0·5% cystine/2 and is amorphous, acting as a supporting matrix for the fibroin.

SERINE (Ser) (S)

$$H_2NCHCOOH$$
$$|$$
$$CH_2OH$$

M.p. 223°C (decomposes).

A non-polar α-amino acid, widely distributed in proteins, and present in a particularly large amount in fibroin. Its pK' values are 2·21 and 9·15, with the isoelectric point at 5·68. In the hydrochloric acid hydrolysis stage of amino acid analysis of proteins, serine is slowly destroyed. Thus the serine content is determined by conducting hydrolyses for different lengths of time and extrapolating the serine content to zero time.

SERINE ENZYME An enzyme in which a serine residue is present at the active site and which is essential for the catalytic activity of the enzyme. It facilitates the reaction being catalysed by forming a covalent bond with an appropriate specific group on the substrate. Thus trypsin with an acyl group forms an ester bond, giving an acyl enzyme and chymotrypsin with phosphoric acid forms a phosphoenzyme. Many proteinases are serine enzymes.

SERUM ALBUMIN A blood plasma protein comprising about 55% of the total blood protein, and occurring to an extent of 3–4·5 g per 100 cm³ of plasma. On plasma electrophoresis, it is the fastest moving component due to its high negative charge and low isoelectric point, pH 4·7. It is highly water soluble, e.g. 40% at pH 7·4. It has a molecular weight of 68 000; the globular protein molecules are ellipsoidal, about 38 Å by 150 Å, but are more symmetrical than other blood proteins. It is one of the most widely studied proteins due to its ready availability in a homogeneous form. It consists of a single polypeptide chain with one N-terminal aspartic acid and one C-terminal threonine residue. The chain contains large amounts of both acidic and basic amino acids. This explains its behaviour with respect to ion-binding ability, molecular expansion on changing pH and titration behaviour, which have all been widely studied. It is the most important protein influencing osmotic regulation in blood, exerting an unexpectedly high osmostic pressure, but it also has an important role in the transport of fatty acids and other water insoluble substances in the blood.

SERUM PROTEIN One of the large number of proteins found in blood serum, i.e. the clear yellow fluid remaining after the blood has been allowed to clot and the clot removed. The plasma has the same composition, including protein composition, as blood serum except that fibrinogen is absent.

SET The deformation remaining after removal of the deforming stress, i.e. the non-elastic part of the deformation. Most of this deformation may be recoverable in time, but permanent set resulting from viscous flow may remain even after very long times. Set is often measured for rubbers either by compression (compression set) or by tension tests.

SFS Abbreviation for sodium formaldehyde sulphoxylate.

S-GLASS A high strength glass whose oxide weight per cent composition is: SiO_2, 64·3; Al_2O_3, 24·8; Fe_2O_3, 0·2; MgO, 10·3; Na_2O, 0·3. Sometimes used as a fibre reinforcement in polymer composites when the highest strength and stiffness are required. Its fibres have a tensile strength of 4·5 GPa, Young's modulus of 86 GPa, density of 2·48 g cm⁻³ and a refractive index of 1·523. However, it is difficult to draw into fibres.

SHARP FOLDING MODEL A model for chain folding at the boundary of a polymer crystal in which the polymer chain folds back on itself over a short length to give adjacent re-entry of the chain from the surface into the crystal.

SHEAF A morphological feature of the crystallisation of polymers formed during the early stages of spherulite formation. On crystallisation, fibrils initially grow from a nucleated centre and bend over as they grow to form a sheaf-like structure. At a later stage the fibrils become doubled back on themselves to form spherulites.

SHEAR A deformation such that the separation between particles on a line in a body does not alter but their separation from particles on an adjacent line increases. The amount of shear is denoted quantitatively as the shear strain.

SHEAR BAND (Deformation band) Thin planar regions of high shear strain generated on stressing an amorphous polymer beyond the yield point. They arise due to the presence of flaws or stress concentrations and as a result of strain softening in these regions. The band propagates in the direction of maximum resolved shear stress, which is at 45° to the direction of the applied stress, or at a somewhat larger angle if the material dilates. The bands are birefringent and are therefore best observed by transmitted polarised light.

SHEAR CREEP COMPLIANCE Symbol $J(t)$. The creep compliance for deformation in shear. One of the most frequently measured creep functions, especially as it can often be related to the molecular parameters underlying the deformation more easily than can the tensile creep compliance, since in shear no changes in volume occur. However, with many polymers, with Poisson's ratio close to 0·5, the tensile compliance $D(t)$ may be simply related to $J(t)$ by $D(t) = J(t)/3$. Many different sample geometries have been used for measurement of $J(t)$, e.g. a sandwich of material in shear or, for a stiffer material, a bar may be subject to torsion. For viscoelastic fluids, rotation between coaxial cylinders or torsion between parallel plates or between a cone and a plate may be used.

SHEAR FLOW A flow caused by a shearing deformation rather than a tensile or bulk deformation. The most common type of deformation geometry met with in flow. Simple shear flow may be induced either by pressure (pressure flow) or by the movement of surfaces (drag flow). For polymer melts simple shear usually results in shear thinning. Shear flow behaviour is characterised by the shear viscosity, usually simply called the viscosity.

SHEAR-INDUCED CRYSTALLISATION Crystallisation induced in a polymer melt by the application of high shear rates at temperatures just above the melting point. This sometimes leads to the formation of shish-kebab morphology.

SHEARING DISC VISCOMETER Alternative name for Mooney viscometer.

SHEAR LOSS COMPLIANCE Symbol J''. The loss compliance for shear deformation.

SHEAR LOSS MODULUS Symbol G''. The loss modulus for shear deformation.

SHEAR MODULUS (Modulus of rigidity) Symbol G. The elastic modulus for a shear deformation, i.e. the ratio of a shear stress to the strain it produces. The shear modulus is related to the Young's modulus (E) by, $G = E/2(1 + v)$, where v is the Poisson ratio. For many rubbers v is about 0·5, so $G \approx E/3$. For many plastics v is about 0·3 so $G \approx E/2·6$. G may be determined directly for a rubber by deforming a rubber block bonded between two parallel metal plates by stressing one of the plates parallel to its plane. For a plastic, G may be determined by measuring the twist on deforming a beam of the plastic in torsion. G may also be determined by various types of dynamic mechanical measurements.

SHEAR STORAGE COMPLIANCE (Dynamic shear compliance) Symbol J'. The storage compliance for shear deformation.

SHEAR STORAGE MODULUS (Dynamic shear modulus) Symbol G'. The storage modulus for shear deformation.

SHEAR STRAIN Symbol γ. The change in the original right angle between two axes in a body on stressing, usually measured in radians. This is mathematical (or tensor) shear strain; engineering shear strain is half this value. The shear strain is positive if the right angle between two positive directions of the two axes decreases, so its sign depends on the coordinate system.

SHEAR STRESS Symbol τ. A stress acting within a plane of a material so as to cause a shear deformation. In double suffix notation there will be, in general, six shear stress components acting on the coordinate planes passing through any point: τ_{xy}, τ_{yx}, τ_{xz}, τ_{zx}, τ_{yz} and τ_{zy}. Thus the state of stress at any point within a body is determined by three shear stress components (since at equilibrium $\tau_{xy} = \tau_{yx}$, $\tau_{xz} = \tau_{zx}$ and $\tau_{yz} = \tau_{zy}$) as well as the three normal components.

SHEAR THICKENING (Dilatancy) Non-Newtonian flow behaviour in which the apparent viscosity increases with increasing rate of shear. In some cases this is accompanied by a volume increase, i.e. the material dilates. It occurs particularly with suspensions of irregularly shaped particles, e.g. with certain polyvinyl-chloride pastes, and occasionally with polymer melts, e.g. when they crystallise during flow.

SHEAR THINNING (Pseudoplasticity) Non-Newtonian flow in which the apparent viscosity (η) decreases with increasing shear rate. Most polymer melts and many polymer solutions are shear thinning, especially at other than low shear rates, e.g. in the range 10^2–10^3 s^{-1}. This is the usual range for many melt processing operations. Indeed the large reductions in η that occur, typically by about two orders of magnitude, can be said to make melt processing feasible without the need for unrealistically powerful machines. This behaviour can be understood for the highly molecularly entangled polymer chains, which become disentangled on shearing, the more so the higher the shear rate. Conversely extensional viscosity is usually not so dependent on strain rate. Many theoretical and empirical constitutive equations have been suggested to describe this behaviour, but, in general, they only apply over a restricted range of conditions, especially over only a narrow range of shear rate. In addition, the equations are often only soluble with difficulty when applied to other than the simplest types of flow. The most widely used equation is the power law equation, in which for a shear thinning material, the flow behaviour index has a value of less than unity.

SHEAR YIELDING   Distortion of the shape without change in volume, on stressing a material beyond the yield point. One of the two main modes of yielding, the other being craze formation. In a crystalline polymer it may occur by slip along slip planes by dislocation glide. In amorphous polymers it may consist of either a diffuse shear yielding throughout the whole of the stressed region, or it may be localised to give clearly defined shear bands in highly strained samples.

β-SHEET   Alternative name for pleated sheet.

SHEET POLYMER   Alternative name for parquet polymer.

SHELLAC   A natural resin produced by refining the insect secretion lac. A complex mixture of esters mostly based on aleuritic acid. Compared with other natural resins it is hard, has good abrasion resistance and adheres well to metals. It is soluble in alcohols and other solvents but not in hydrocarbons. It is useful in varnishes, but although widely used at one time as a thermoplastic moulding material, it is now little used. Shellac melts at about 75°C, but prolonged heating at above 100°C hardens the melt due to internal polycondensation between free hydroxyl and carboxyl groups.

SHIFT FACTOR   Symbol a_T. The amount by which log(modulus) or log(compliance) versus time (or frequency) curves, obtained at different temperatures, have to be shifted along the time (or frequency) scale to bring them together to form a master curve for a particular reference temperature T_r, as required by the time–temperature superposition principle. The original empirical WLF equation expressed the dependency of a_T on temperature as, $\log a_T = C_1(T - T_r)/(C_2 + (T - T_r))$, where C_1 and C_2 are constants, having universal values of -17.44 and 51.6 K respectively if T_r is taken to be the T_g value.

SHISH-KEBAB   A morphological form of a partially crystalline polymer which consists of crystallites arranged in linear arrays along a central thread of polymer. Formed during crystallisation of rapidly sheared polymer solutions (e.g. by stirring). The central thread of extended chain crystals is produced by elongational flow and forms a nucleus for overgrowths of chain folded lamella as row structures. The model has subsequently provided a useful basis for discussion of the morphologies obtained from strained polymer melts.

SHORE HARDNESS   An arbitrary number expressing the hardness (or softness) of a polymer material, as measured by a simple indentation test involving the application of a force to a vertical indentor using a calibrated spring. The scale is such that a reading of 100 corresponds to pressing the indentor onto a sheet of plate glass and a reading of zero corresponds to the indentor meeting no resistance. Two different indentors are used: type A for soft materials and type D for hard materials.

SHORT CHAIN BRANCHING   The occurrence of branching in a polymer with branches of up to only a few repeat units. It is most frequently found in vinyl polymers, where polyethylene and polyvinyl chloride are the most studied examples. In these cases branching is usually the result of transfer reactions, including back biting, during polymerisation and may be characterised by infrared spectroscopy (for polyethylene) or ^{13}C NMR (for polyethylene and polyvinyl chloride).

SHORT FIBRE COMPOSITE   (Chopped fibre composite) A fibre reinforced material in which the fibres are not continuous but whose length can vary from a few centimetres, as in chopped glass and carbon fibre composites, down to 10^{-6} m in whisker reinforced materials. Since fibre strength and stiffness is usually much greater than for polymers, the potential for reinforcement is great. Furthermore, economic methods for the production of the composite products, such as injection moulding, can be used.

SHORT OIL ALKYD RESIN   An alkyd resin which contains less than about 50% of an oil, and hence is highly aromatic and soluble only in aromatic solvents. Coatings based on such resins are usually crosslinked at elevated temperatures to give very hard, glossy finishes.

SHORT STOP   A free radical trapping agent added (at about the 0.1% level) to an emulsion polymerisation to terminate the polymerisation. Frequently used in styrene–butadiene rubber production. Examples include hydroquinone and sodium dimethyldithiocarbamate.

SI   An older alternative for the abbreviation MQ.

SIALIC ACID   (*N*-acetylneuraminic acid)

M.p. 185–187°C. α_d^{22} $-32°$.

An acetylated aminosugar frequently found as a monosaccharide component of glycoproteins.

SiB Abbreviation for poly-(*m*-carboranesiloxane).

Si–B POLYMER Abbreviation for polycarboranesiloxane.

SICOFLEX Tradename for acrylonitrile–butadiene–styrene copolymer.

SICRON Tradename for polyvinyl chloride.

SILANE One of a group of organosilicon compounds which by analogy to the hydrocarbon compounds termed alkanes, can be considered as the parent compounds from which other organosilicon compounds are derived. Silane itself is SiH_4, but generally the term refers to compounds of the type $H_3Si\text{(}SiH_2\text{)}_n SiH_3$, which are disilanes ($n = 0$), trisilanes ($n = 1$), etc. Compounds containing organic groups replacing hydrogen are organosilanes. If the organic group is chemically reactive then the compound is an organofunctional silane, e.g. vinyldichlorosilane, where the vinyl group is the functional group. Organosilanes with a functional group attached directly to silicon are silicon functional silanes. Such groups are usually readily hydrolysed to silanols and include halogen, usually chlorine (as in chlorosilanes), alkoxy (alkoxysilanes) and amine (aminosilane) groups. These compounds, especially the chlorosilanes, form the monomers from which polyorganosiloxanes are synthesised.

SILANE COUPLING AGENT A chemical compound of the general type $R_{(4-y)}SiX_y$ which acts as a coupling agent for reinforcing fillers in polymer composites. Usually $y = 3$ and the X groups are either chlorine, acetoxy or other alkoxy. The X groups are hydrolysed to silanol groups on application of the coupling agent to the filler from aqueous solution: $RSiX_3 \rightarrow RSi(OH)_3 + 3HX$. The silanol groups, which may subsequently polymerise, provide adhesion to the filler, especially to glass, possibly by covalent or hydrogen bond formation. In early silanes X was chlorine; this gives unstable compounds and the hydrochloric acid produced on hydrolysis is undesirable. In more recent silanes X is usually either $—OCH_3$ or $—OC_2H_5$. The R group provides adhesion to the polymer matrix often by covalent bonding by reaction with the matrix during its crosslinking. Specific R groups are chosen for specific use with specific resins. Thus γ-methacryloxypropyltrimethoxysilane is commonly used with unsaturated polyesters and γ-aminopropyltriethoxysilane is used with epoxy resins. About $0.1–0.5\,wt\%$ of the resin is used.

SILANOL A compound containing a hydroxyl group bound to silicon. Silanol itself is H_3SiOH, whilst other silanol compounds are named by adding the suffix -ol to the name of the parent silane, e.g. $(CH_3)_2Si(OH)_2$ is dimethylsilanediol. Such organosilanols are formed as intermediates in the synthesis of polyorganosiloxanes by the hydrolysis of silicon functional silanes, especially chlorosilanes. The silanols so formed readily condense to form siloxanes with the elimination of water, e.g.

$$(CH_3)_2SiCl_2 \longrightarrow (CH_3)_2Si(OH)_2 \longrightarrow \{Si(CH_3)_2—O\}$$

Silanol groups are often present on the surface of inorganic silicates such as glass and silica. The —OH groups may then participate in reactions, e.g. with a coupling agent, which increase adhesion to a polymer matrix and hence also increase reinforcement.

SILASTIC Tradename for a range of silicone elastomers.

SILASTOMER Tradename for a range of silicone elastomers.

SILICA SiO_2. The most abundant substance in the earth's crust, occurring in the free state as nearly pure silica, e.g. as quartz, or as more impure forms such as sand and flint in sandstone, feldspar and other rocks. Highly pure silica sand is used as a filler in polymers, especially in epoxy resins, to provide good thermal and electrical conductivity, low thermal expansion and high abrasion resistance. Silica is also produced synthetically as pyrogenic silica and as colloidal silica. A three-dimensional network polymer consisting of linked SiO_4 tetrahedra in which every oxygen is shared between two silicons. Crystalline silica is an outstandingly important natural mineral, occurring in three different crystalline forms, quartz (the most common), tridymite and cristobalite, in which the tetrahedra are linked in different ways. On cooling a silica melt, amorphous, or vitreous, silica is produced.

SILICATE POLYMER (Polysilicate) In terms of abundance in the earth's crust, the most important group of polymers found naturally. They may be considered as salts of polymers of silicic acid, H_4SiO_4. In aqueous solution silicic acid will polymerise by polycondensation, forming polysilicic acid

$$\sim\!\!\left[\begin{array}{c} OH \\ | \\ Si—O \\ | \\ OH \end{array}\right]_n\!\!\sim$$

Silicate polymers are usually complex networks and may be either amorphous, as in the synthetic silicate glasses, or vitreous silica. However, naturally occurring silicate polymers are crystalline, comprising linear polymers (e.g. pyroxenes and Woolastonite), ladder polymers (e.g. amphiboles), two-dimensional sheet-like polymers (such as the layer polymers chrysotile, kaolinite and talc) or, most importantly, three-dimensional networks (as found in silica (most commonly found as quartz), the feldspars and zeolites). The polymers may be considered to consist of SiO_4 tetrahedra which may be joined in different ways, with the sharing of one, two or three oxygen atoms giving linear, layer or network structure polymers respectively.

SILICONE An alternative name for polyorganosiloxane, being more frequently used when a polymer of this type is used as the basis of a useful commercial product. Such products span a diverse range from low molecular weight linear or branched polymers useful as silicone oils and greases to high molecular weight crosslinkable polymers useful as rubbers and to the crosslinked products useful as silicone resins. The name arises from the assumed, but erroneous, analogy between the structure of silicones (R_2SiO) and ketones ($R_2C{=}O$).

SILICONE ELASTOMER (Silicone rubber) Trade-names: Silastic, Silastomer, Silopren. A high molecular weight polyorganosiloxane rubber, the basic polymer being polydimethylsiloxane (dimethylsilicone elastomer—abbreviation MQ) with a molecular weight in the range 300 000 to one million. Such polymers are produced by ring-opening polymerisation of octamethylcyclotetrasiloxane. Several copolymers are also of interest, notably those containing about 0·5% methylvinylsiloxane groups (the vinylsilicone elastomer VMQ) which are more readily vulcanised and give vulcanisates with better compression set resistance. Copolymers with 5–15% of the methyl groups replaced with phenyl groups (phenylsilicone elastomers—PMQ) give vulcanisates with better low temperature properties, whilst related fluorosilicone elastomers have improved solvent resistance. Nitrilesilicone elastomers have also been of interest.

The unvulcanised polymer (dimethylsilicone gum) is normally crosslinked to a useful elastomer by heating with an organic peroxide, such as benzoyl peroxide or 2,4-dichlorobenzoyl peroxide. Special low molecular weight polymers with reactive end groups may be room temperature vulcanised (RTV silicone elastomer).

Although silicone elastomer vulcanisates have relatively poor mechanical properties compared with other rubbers, for example having a typical tensile stength of only 4–9 MPa, these properties are retained over a very wide temperature range. The rubbers have excellent high temperature resistance, typically with a useful life of about two years at 150°C, and retain their flexibility to about -50°C or even lower in the phenyl substituted polymers. They retain excellent electrical properties under extremes of temperature and moisture, but have poor abrasion resistance. They swell moderately in oils, fuels and in many solvents, although the fluorosilicones are much better in this last respect.

SILICONE FLUID A liquid polyorganosiloxane, usually polydimethylsiloxane, but sometimes also containing phenylsiloxane units, of relatively low molecular weight ranging from a few hundred to about 25 000. Produced by aqueous hydrolysis of dichlorodimethylsilane, often with dilute hydrochloric acid, which produces a mixture of linear polymer and cyclic oligomers, as an oil. This is separated and subject to equilibration to homogenise the molecular size distribution. The addition of hexamethyldisiloxane as a chain stopper at the equilibration stage is used to control the reaction and to limit the polymer molecular weight. The fluids have low volatility and high thermal stability, to about 150°C for long periods in air, and up to about 200°C under inert conditions for methylsilicone fluids. For methylphenylsilicone fluids the temperature limit is about 100°C higher. They have chemical inertness but are soluble in hydrocarbon and chlorinated hydrocarbon solvents. The viscosities of the fluids are remarkably temperature and shear rate independent. The dimethylsiloxane fluids are useful as release agents, lubricants, greases and hydraulic fluids, whilst silicone fluids containing Si—H bonds are also of interest as textile finishes since they crosslink on heating with alkali. The methylphenylsilicones, with 10–45% phenyl groups, are useful as heat stable lubricants and heat transfer fluids.

SILICONE RESIN A three-dimensional polyorganosiloxane formed by crosslinking the highly branched polymers produced by hydrolysis of dichloro- and trichlorosilanes. Commercial resins are mostly methylphenylsiloxane polymers produced by hydrolysis of blends of methyltrichlorosilane, phenyltrichlorosilane, dimethyldichlorosilane, diphenyldichlorosilane and methylphenyldichlorosilane. Resins are often classified according to their R/Si ratio, where R represents the methyl and phenyl groups combined. This ratio will be two for a polymer from a dichlorosilane and one for a polymer from a trichlorosilane. In practice ratios of 1·2 to 1·6 are used. The chlorosilane blend in a hydrocarbon solvent such as toluene, is stirred with water, the organic layer is separated and concentrated to about 80% solids.

The highly branched polymer contains silanol end groups through which network formation is achieved by heating with a suitable catalyst such as zinc or cobalt octoate. The resins have very good high temperature resistance (which improves with an increase in phenyl content), excellent water repellency and non-stick properties. They are used in their precursor branched form in solution as binders in laminates and composites with glass and asbestos for high temperature electrical insulation products. They are also useful as components for co-polymerisation with alkyd resins and other resins for surface coatings for imparting heat resistance and water repellency.

SILICONE RUBBER Alternative name for silicone elastomer.

SILK The continuous fibrous material secreted by certain insects and spiders. Most silks are composed of proteins but a few are chitin based. The most studied is that produced by the caterpillar of the silk moth, *Bombyx mori*, which is cultivated for commercial silk production. The fibre is produced as two filaments of the protein fibroin cemented together by a surface layer of water soluble sericin. The sericin comprises about 20% of the thread, which is typically 15–25 μm in diameter. The fibre is soft and lustrous, giving fabrics with a smooth and attractive hand. The fibre has a tenacity of 3·5–5·0 g denier^{-1}. Its elongation at break is about 20% and it has a moisture regain of 11%, imparting good comfort properties.

SILK GUM Alternative name for sericin.

SILOPREN Tradename for a range of silicone elastomers.

SILOXANE A compound containing one or more Si—O linkages. The parent series of siloxanes contains unsubstituted silicons, i.e. $H_3Si-[OSiH_2-]_nOSiH_3$. Compounds with $n = 0$ are disiloxanes (e.g. $(CH_3)_3Si—O—Si(CH_3)_3$ is hexamethyldisiloxane), compounds with $n = 1$ are trisiloxanes, etc. Polymers with large values of n are polysiloxanes, or if substituted with organic groups they are polyorganosiloxanes, often commonly being referred to as silicones. Siloxanes are frequently synthesised by hydrolysis of chlorosilanes via the unstable silanols.

Cyclosiloxanes $[Si(R_2)—O-]_n$, e.g. octamethylcyclotetrasiloxane ($R = CH_3$, $n = 4$), are also known.

SIMHA EQUATION A modified form of the Einstein equation, describing the effect of suspended particles on the coefficient of viscosity (η) of a fluid, when the particles are other than spherical. Of the general form, $\eta = \eta_0(1 + \lambda\phi)$, where η_0 is the coefficient of viscosity of the pure liquid and ϕ is the volume fraction of the suspended particles. λ is the constant whose value depends on the particle shape. Dumbell-shaped particles are of particular interest in the development of theories of the viscosity of dilute polymer solutions, since hydrodynamically they behave similarly to the pearl-necklace model of a polymer chain. For dumbells, $\lambda = 3L^2/2a^2$, where $2L$ is the length of the dumbell and a is the radius of its spherical ends.

SIMPLE ELONGATIONAL FLOW (Simple extensional flow) A model simple flow which may be defined by the rate of deformation tensor having the form,

$$\dot{\varepsilon}\begin{pmatrix} 2 & 0 & 0 \\ 0 & -1 & 0 \\ 0 & 0 & -1 \end{pmatrix}$$

where $\dot{\varepsilon}$ is the principal extension rate.

SIMPLE EXTENSIONAL FLOW Alternative name for simple elongational flow.

SIMPLE LAW OF MIXTURES Alternative name for law of mixtures.

SIMPLE PROTEIN (Unconjugated protein) A protein consisting entirely of a polypeptide, unlike a conjugated protein which contains a prosthetic group. The main types of simple protein, classified according to biological function, are the structural proteins, hormones, many enzymes, some transport proteins (e.g. serum albumin), some storage proteins (e.g. ovalbumin), contractile proteins and protective proteins (e.g. antibodies and thrombin). They may also be classified according to

solubility as the albumins, globulins, protamines, glutelins, schleroproteins, prolamins and histones.

SIMPLE SHEAR Shear in which the deformation is identical with that of a pure shear plus a rotation. Produced by applying only a shear stress, i.e. only the shear components of the stress tensor have non-zero values. Thus it is a deformation in which there is a family of parallel material planes (the shearing planes) which move parallel to each other along their own planes in a straight line, their separations remaining constant at constant volume. The relative displacement of a pair of planes divided by their separation has the same value for all pairs.

SIMPLE SHEAR FLOW (Viscometric flow) A model shear flow for which the rate of deformation tensor is

$$\dot{\gamma}\begin{pmatrix} 0 & 1 & 0 \\ 1 & 0 & 0 \\ 0 & 0 & 0 \end{pmatrix}$$

where $\dot{\gamma}$ is the shear rate which may not be constant, direction 1 is the flow direction, direction 2 is the direction of velocity variation and direction 3 is the neutral direction. Experimental set-ups may be designed to achieve simple shear flow in the laboratory from which the fluid viscosity may be determined, when the flow is then called a viscometric flow. This occurs when there are no end effects in capillary flow, during flow in a cone and plate viscometer and flow in the annulus between two rotating concentric cylinders. For a simple fluid under steady state conditions, three material constants characterise the flow. These are the coefficient of viscosity (given by $\tau_{12}/\dot{\gamma}$, where τ_{12} is the shear stress) and the first and second normal stress functions. Plane couette flow is the simplest example. Simple shear flow together with simple elongational flow provide the basis for the description and analysis of flow in polymer melts and for solutions of practical importance.

SIMULTANEOUS INTERPENETRATING POLYMER NETWORK (SIN) An interpenetrating polymer network produced by the simultaneous formation of the two types of network. This is in general possible by forming one polymer by a step-growth polymerisation and one chain polymerisation, e.g. epoxy resin/polyacrylate SIN.

SIN Abbreviation for simultaneous interpenetrating polymer network or for semi-interpenetrating polymer network.

SINGLE CRYSTAL An isolated, microscopic-sized polymer crystal with no intercrystalline molecular connections. Formed only by very slow crystallisation from dilute solution (at $<0.1\%$ concentration). Typically it consists of a lozenge-shaped lamella about $100\,Å$ thick often resulting from a collapsed pyramidal crystal. The polymer molecules (typically about $10\,000\,Å$ in length) have been shown by electron diffraction to lie oriented across the thickness of the crystal and hence must be chain folded. Melt crystallised polymer crystallites are also thought to have chain folded molecules. The interfold distance (and hence crystal thickness) depends on crystallisation temperature and may be increased by annealing. Enthalpy measurements suggest that the crystals have a significant amorphous content. This is present on the surface and is associated with disordered folding produced by the emergence of a polymer molecule from the crystal and its re-entry.

SINGLE CRYSTAL MAT A stack, in flat form, of single crystal lamellae formed by allowing a suspension of the lamellae to settle on a flat surface. Use of such a mat as a kind of diffraction grating in small angle X-ray scattering, enables the long spacings to be determined.

SINGLE CRYSTAL PATTERN The X-ray or electron diffraction pattern obtained from a single crystal. It consists of a regular lattice of spots which represents a projection of the reciprocal lattice onto the photographic plate. Analysis of the patterns yields the structure of the unit cell and hence the crystal structure.

SINGLE-POINT METHOD A method for the determination of the limiting viscosity number ($[\eta]$) of a polymer in dilute solution by making the flow–time measurements in a capillary viscometer at a single concentration. The Kraemer, Huggins, Schultz–Blaschke or Martin equation may then be used to calculate $[\eta]$ provided the appropriate constant in the equation is known. The method, although rapid, is less reliable than the usual method of determining the viscosity ratio at several concentrations and extrapolating to zero concentration.

SINGLE STRAND POLYMER A linear polymer consisting of only a single chain (or strand) of atoms, as opposed to a double strand polymer. Thus linear polymers, that are not ladder polymers, are single strand.

SINGLET OXYGEN Molecular oxygen in an excited singlet state, designated $^1O_2$. It can be produced in photochemical reactions by the quenching of excited states of another molecule by oxygen. It may be involved in the photo-oxidation of polymers by abstraction of polymer hydrogen atoms or by the addition to double bonds, forming peroxides.

SINTERING Heat treatment of separate particles or grains of a material which leads to an agglomeration and compaction.

SINVET Tradename for bisphenol A polycarbonate.

SIREL Tradename for styrene–butadiene rubber.

SIRFEN Tradename for phenol–formaldehyde polymer.

SIRTENE Tradename for low density polyethylene.

SIS Abbreviation for styrene–isoprene–styrene block copolymer.

SISAL A leaf fibre from the plant *Agave sisalana*, consisting of stiff strands 50–100 cm in length, which can be creamy white in colour. It is a cellulosic fibre containing about 6% lignin. It is widely used for twines and cordage.

SIZE (Dressing) A surface treatment applied from solution to glass fibres to lubricate the surface and hence prevent the abrasive damage which reduces fibre strength. A binder is also usually present to improve packing of the single filaments into strands. When a finish is not to be applied to the fibre the size may also contain a coupling agent. Plasticised polyvinyl acetate is often used as a size of this type since it is compatible with polyester and epoxy resins. If the coupling agent is to be applied subsequently as a finish, the size is usually first removed by solvent treatment or by burning it off the fibres. A typical size of this type is dextrinised starch.

SIZE EXCLUSION CHROMATOGRAPHY (SEC) Alternative name for gel permeation chromatography.

SKF Tradename for vinylidene fluoride–chlorotrifluoroethylene or vinylidene fluoride–hexafluoropropylene copolymer.

SKIM RUBBER A form of natural rubber containing a large amount of non-rubber components (15–25%) which increase its rate of vulcanisation. Obtained from the skim latex, which is the by-product from the concentration of the latex by centrifugation.

SKYBOND Tradename for a range of foamable polyimides and polyimide laminating resins, produced by reaction of benzophenone tetracarboxylic dianhydride and 4,4′-diaminodiphenylether, and having the structure:

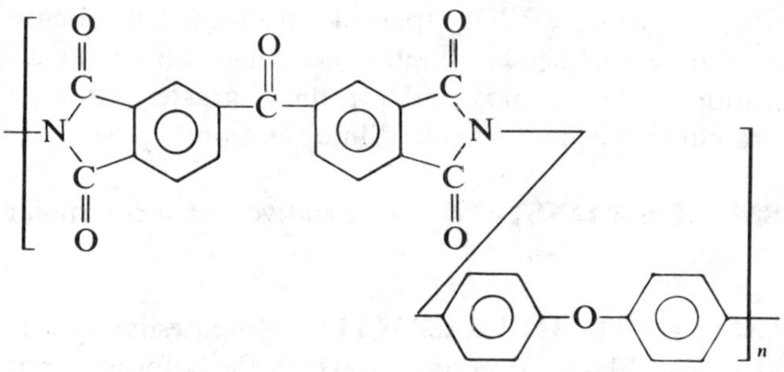

SKYGARD 700 Tradename for a polyimide solution used for a high temperature resistant laminating resin.

SKYPRENE Tradename for polychloroprene.

SLIP PLANE The plane in which a material deforms by slip of one molecular layer past another. In a crystalline polymer the planes are parallel to the polymer chain axes. Slip is propagated by the propagation of dislocations in the slip plane. This motion can be by either glide or climb.

SMALL ANGLE LIGHT SCATTERING (SALS) The scattering of light of optical wavelengths over angles of less than about 5° caused by particles of several hundred ångströms in size. In thin films of solid polymers the appropriate particles are the spherulites and measurement of the radial distribution of the scattered light characterises the spherulite size distribution. Experimentally a laser is used as light source and with an analyser and polariser and film in the beam, the pattern is recorded simply photographically on flat film. With polariser and analyser crossed, a clover-leaf pattern (H_v pattern) is obtained and with polariser and analyser parallel the V_v pattern results.

SMALL ANGLE NEUTRON SCATTERING (SANS) Neutron scattering, usually elastic and coherent, where the scattering angles are $<10°$. At the wavelengths used (about 1 nm), scattering corresponds to the dimensions of polymer molecules. The technique is widely used for the determination of the radius of gyration in polymer solutions, but also especially in melts and solid rubbers and glasses, due to the high contrast possible by selective deuteration, since deuterium and hydrogen have widely different scattering lengths. This is especially valuable since light scattering and small angle X-ray scattering can only be used in dilute solution. The technique has been used to confirm some of the basic postulates of polymer behaviour, e.g. the assumption that in bulk amorphous polymers the molecular dimensions are the unperturbed dimensions (as observed in a theta solvent) and the affine deformation assumption of the theory of rubber elasticity.

SMALL ANGLE X-RAY SCATTERING (SAXS) X-ray scattering at angles with respect to the incident beam of usually $<2°$. Both diffuse and discrete scattering patterns may be obtained. It is especially useful for the determination of long period spacings (and hence crystallite thickness) in single crystal mats, for characterisation of microvoids and for the characterisation of protein shapes and sizes in dilute solution.

SMALL'S CONSTANT Alternative name for molar attraction constant.

SMALL STRAIN ELASTICITY (Infinitesimal strain elasticity) Elastic behaviour in which the deformation is small enough (often typically considered to be $<1\%$) so that the material exhibits perfectly elastic behaviour, described by the classical theory of elasticity. It describes the behaviour of many metals, often to quite high stresses (but below the yield point), but it is often only an approximation to polymer behaviour which can exhibit large strains before yielding. In terms of displacements of points in a body on deformation, the small strain assumption amounts to ignoring the higher order terms such as $(\partial u/\partial y)^2$, $(\partial u/\partial y \cdot \partial u/\partial z)$, etc., which are involved in the finite strain components, where u, v and w are the displacements and x, y and z are the coordinates of a point in the undeformed body. Unlike the finite strain case, it is immaterial whether the components of the stress are referred to the deformed or undeformed body, since to a first approximation the body's dimensions are unaffected by the deformation.

SMECTIC PHASE A liquid crystalline mesophase in which the mesogens show both long-range orientational order and one- or two-dimensional positional order, i.e. they have a layered structure. Since the positional ordering may be of different types, several different phases and phase transitions may be observed.

SMITH DEGRADATION A sequence of chemical reactions used on a polysaccharide for structural elucidation, in particular for the determination of the type of glycoside link present (i.e. 1–6, 1–4, etc.). It consists of the sequence—periodate oxidation, borohydride reduction and mild hydrolysis with dilute acid, giving degradation to C_2, C_3 and C_4 fragments characteristic of the type of glycoside linkage. Particularly useful in identifying 3-linked units which are not oxidised by periodate.

SMITH–EWART THEORY A quantitative treatment of the kinetics of emulsion polymerisation based on the qualitative Harkins model. It shows how the number of latex particles and the rate of polymerisation depend on the emulsifier type and concentration and on the initiator concentration. Three cases are considered. In case 1 the probability of the radicals escaping from a polymer/monomer particle is high thus giving, on average, a low number of radicals per particle. This is thought to apply to monomers which have a high monomer chain transfer constant, such as vinyl chloride and vinyl acetate. In case 2—the most widely considered, radicals do not escape from the particle. For small particles, if a second radical enters, mutual termination with the radical already present will occur rapidly, so that each particle contains either one or two radicals. This leads to the well-known rate expression for interval II of an emulsion polymerisation, $-d[M]/dt = k_p[P][M]/2N$, where k_p is the propagation rate constant, $[P]$ is the particle concentration, $[M]$ is the monomer concentration and N is Avogadro's number. This is the situation for the polymerisations of styrene and other highly insoluble monomers. In case 3 for large particles, the average number of radicals per particle is $>\frac{1}{2}$.

SMOKE SUPPRESSANT An additive which reduces the amount of smoke produced when a material burns. Excessive smoke production is a problem in fires involving polymers, particularly with PVC. Metal borates, such as zinc borate and molybdenum oxide are particularly effective in PVC.

SMR Abbreviation for standard Malaysian rubber.

S/N CURVE   Alternative name for fatigue curve.

SNIAFIL   Tradename for a viscose rayon filament.

SNIAMID   Tradename for nylon 6 and nylon 66.

SOC   Abbreviation for stress–optical coefficient.

SODA CELLULOSE I   The hydrated 'complex', possibly of formula Cell.NaOH.3H$_2$O formed when cellulose is treated with 12–19% aqueous sodium hydroxide.

SODA CELLULOSE II   The 'complex', possibly of formula Cell.NaOH, formed when cellulose is treated with 21–45% aqueous sodium hydroxide.

SODA GLASS   Alternative name for soda-lime glass.

SODA-LIME GLASS   (Soda glass) The commonest type of commercial glass of approximate composition: SiO$_2$, 70%; Na$_2$O, 13%; CaO, 12%. It is easily fabricated due to its low softening temperature (745°C) at which its melt viscosity is about 10^7 N s m^{-2}. Its specific gravity is 2·45, its refractive index is 1·51 and it has a tensile modulus of 70 GPa. However, the glass has a high coefficient of thermal expansion, and therefore has a low thermal shock resistance compared with borosilicate glasses and vitreous silica.

SODIUM CARBOXYMETHYLCELLULOSE   A cellulose ether in which some cellulose hydroxyl groups have been replaced by —OCH$_2$COO$^-$ Na$^+$ groups. Prepared by reaction of alkali cellulose with chloracetic acid or its sodium salt. A water soluble polyelectrolyte. Commercial products have a DS of 0·5–1·2 and are used as thickeners in foods, in detergents and as textile and paper sizes.

SODIUM CELLULOSE XANTHATE   (Cellulose xanthate) (Xanthate)

$$Cell—O—\underset{\underset{S}{\|}}{C}—SNa$$

A cellulose derivative produced by reacting alkali cellulose with carbon disulphide. Soluble in dilute aqueous sodium hydroxide to form viscose, from which viscose rayon is produced.

SODIUM DODECYL SULPHATE   (SDS)

$$CH_3(CH_2)_{10}SO_4^- Na^+$$

A detergent which is sometimes used as a denaturing agent for proteins. It destroys the natural conformation of the polypeptide chain by associating through hydrophobic bonding, forming complexes which are ridged and rod-like. The formation of such complexes is useful in SDS-gel electrophoresis. SDS will also dissociate oligomeric proteins into their subunits.

SODIUM FORMALDEHYDE SULPHOXYLATE (SFS)

$$Na^+HSO_2^-.HCHO$$

A reducing agent commonly used in the production of cold styrene–butadiene rubber by emulsion polymerisation. Its function is to convert the ferric iron of the redox initiator back to ferrous iron by:

$$2 Fe^{3+} + HSO_2^-.HCHO + 3 OH^- \rightarrow$$
$$2 Fe^{2+} + SO_2^{2-}.HCHO + H_2O.$$

SODIUM POLYACRYLATE

$$\left[\underset{CH_2CH}{\overset{\overset{COO^-Na^+}{|}}{}}\right]_n$$

Tradenames: Acrysol, Plexileim, Syncol, Texigel. The sodium salt of polyacrylic acid, useful as a thickening agent, e.g. for latices. Prepared either by hydrolysis of polyacrylonitrile or polymethylacrylate with sodium hydroxide or by neutralisation of polyacrylic acid.

SODIUM RUBBER   An early polybutadiene produced from butadiene using sodium metal as the polymerisation catalyst.

SOFT BLOCK   Alternative name for soft segment.

SOFTENED RUBBER   (Peptised rubber) Natural rubber to which a peptiser has been added to the latex before coagulation. The peptiser may cause softening either during drying of the coagulum or during subsequent mastication.

SOFTENER   In general an alternative name for plasticiser, but used particularly to describe the use of plasticisers (usually rubber oils) in rubbers in relatively small amounts (5–20 phr) as process aids. This method of softening is referred to as physical softening, as opposed to the chemical softening caused by the use of peptisers.

SOFTENING POINT   The temperature at which, on heating, a polymer, usually a plastic, softens appreciably, as measured by a particular test. The most frequently used tests are the Vicat softening point and the heat distortion temperature tests. For amorphous polymers the value obtained is usually just a little below the T_g value. For crystalline polymers it is between the T_g and the T_m values, being closer to the T_m value the higher the degree of crystallinity. The softening point may be raised considerably by the use of a filler, especially in the case of some crystalline polymers which show only moderate crystallinity, such as nylons and thermoplastic polyesters. The softening point is often used as a guide to the maximum working temperature of a plastic material.

SOFTNESS NUMBER   An arbitrary number expressing the softness of a polymer material, used particularly for plasticised polyvinyl chloride, as measured by a particular indentation test. The test involves the measurement of the vertical position of a vertical plunger terminated in the indentor employing a dial gauge, after pressing it using a minor load (294 mN force), followed by a major load (5·25 N force), against a horizontal flat sheet specimen of the polymer.

SOFT SEGMENT The blocks in a block copolymer thermoplastic elastomer which have a T_g value (and a T_m value if they are crystallisable) well below ambient temperature and so retain good molecular flexibility at ambient temperature. These segments therefore impart rubbery characteristics to these materials. Thus, for example in polyether/ester block copolymers the soft segments are provided by the polyoxymethylene-terephthalate blocks. In SBS, SIS and SB polymers the polybutadiene blocks are the soft segments.

SOL That part of a partially crosslinked or network polymer that is not part of the network, i.e. it consists of linear or branched molecules, and is soluble in solvating solvents.

SOLEF Tradename for polyvinylidene fluoride.

SOLID PHASE SYNTHESIS (Merrifield synthesis) A technique for the synthesis of peptides and proteins in which the peptide bonds are built up in the usual stepwise manner, but whilst the growing peptide is covalently bonded to an insoluble synthetic resin. In this way the insoluble peptide–resin material may be readily washed free of excess reagents and reaction products and filtered at each stage. This method speeds up the synthesis and results in high yields of coupled product. Typically the first amino acid (as its t-butyloxycarbonyl (Boc) derivative) is attached to a chloromethylated polystyrene–divinyl benzene resin:

$$t\text{—Bu—O—C—NHCHRCOOH} +$$
$$\overset{\|}{\text{O}}$$

$$\text{ClCH}_2\text{—}\langle\bigcirc\rangle\text{—resin} \rightarrow$$

$$t\text{—Bu—O—C—NHCHRC—O—CH}_2\text{—}\langle\bigcirc\rangle\text{—resin}$$
$$\quad\quad\overset{\|}{\text{O}}\quad\quad\overset{\|}{\text{O}}$$

The Boc group is removed and a dipeptide is formed by coupling with a further Boc-protected amino acid using N,N'-dicyclohexylcarbodiimide. The process is then repeated sequentially using the desired Boc-amino acid at each stage. At the end of the synthesis the required peptide or protein is cleaved from the resin using hydrogen bromide/trifluoroacetic acid or hydrofluoric acid. The whole process may be mechanised and automated so that large polypeptides, e.g. the 124 residue ribonuclease, may be synthesised in a reasonable time.

SOLID-STATE POLYMERISATION Polymerisation with the monomer in the solid crystalline state. Despite the lack of mobility of the monomer and the low propagation rate constant, the polymerisation may be faster than in the bulk liquid phase due to the high concentration of active centres resulting from a lack of termination reactions. Most work has been performed on vinyl monomers initiated by high energy radiation. Many solid-state polymerisations are topochemical, a few also being topotactic. Canal polymerisation is a special type of solid-state polymerisation.

SOLIMIDE Tradename for a foamable polyimide produced by reaction of benzenetetracarboxylic dianhydride, 2,6-diaminopyridine and 4,4′-diaminodiphenylmethane. Available as a polyamic acid precursor which loses ethanol and water on imidisation when heated, thus producing a foam. Useful as a low flammability replacement for polyurethane foam for aircraft seats.

SOLITHANE Tradename for a cast polyurethane system based on a prepolymer system cured with 3,3′-dichloro-4,4′-diaminodiphenylmethane or a diol or triol.

SOLPRENE Tradename for low cis-polybutadiene, for solution styrene–butadiene rubber and for styrene–butadiene star, radial and block copolymers.

SOLTAN S Tradename for a nitrile resin barrier polymer.

SOLUBILITY PARAMETER Symbol δ. The square root of the cohesive energy density of a material. The solubility parameter can be a useful guide to the miscibility of two materials, one or both of which may be polymers. Since the enthalpy of mixing is given by, $\Delta H_m = (\delta_1 - \delta_2)^2 V\phi_1\phi_2$, where δ_1 and δ_2 are the solubility parameters of the materials, ϕ_1 and ϕ_2 are their volume fractions and V is the volume of the mixture, it will be at a minimum when δ_1 and δ_2 are close in value or equal. Therefore similar values of the two solubility parameters indicates mutual miscibility and good compatibility. However, the above relationship predicts that ΔH_m is always positive and therefore cannot be used when specific interactions, such as hydrogen bonding, occur between the components. In these cases the hydrogen bonding ability may be allowed for by using the hydrogen bonding index. For polymer solubility in a solvent, $\delta_1 - \delta_2$ should be about $3{\cdot}0\,(\text{MJ m}^{-3})^{1/2}$, where $(\text{MJ m}^{-3})^{1/2}$ or $(\text{cal cm}^{-3})^{1/2}$ are the usual units used. For polymer–polymer miscibility this difference must be much lower, about $0{\cdot}2\,(\text{MJ m}^{-3})^{1/2}$ or less.

The solubility parameters of liquids may be determined from their heats of vaporisation, but polymers are involatile. Polymer δ values may be obtained by finding the liquid which causes maximum swelling of a slightly crosslinked network of the polymer or which gives a maximum limiting viscosity number. This solvent δ value is then the closest to that of the polymer. Alternatively, δ may be calculated from the group molar attraction constants. Values of δ for polymers (in $(\text{MJ cm}^{-3})^{1/2}$) vary from 16·3 (polyethylene) and 18·7 (polystyrene) for non-polar polymers, to 19·4 (polyvinyl chloride) and 21·8 (polyethyleneterephthalate) for moderately polar polymers, to 27·8 (nylon 66) and 28·7 (polyacrylonitrile) for highly polar polymers.

SOLUTION POLYBUTADIENE Polybutadiene produced by solution polymerisation. The commercially important polybutadiene rubbers are mostly produced by solution polymerisation either by Ziegler–Natta polymerisation or by using lithium alkyl catalysts. They are stereoregular polymers having high *cis*-1,4-contents.

SOLUTION POLYMERISATION Polymerisation with the monomer dissolved in a solvent as a diluent. If the polymer is insoluble in the solvent then a precipitation polymerisation results. In chain polymerisation, by dilution with solvent, many of the problems of bulk polymerisation, i.e. dissipation of the heat of polymerisation, autoacceleration and chain transfer to polymer, may be alleviated. However, the solvent must be chosen with care, in particular it must not be active in chain transfer itself nor must it contain impurities that are active. In step-growth polymerisation, solvents are used either to homogenise the reaction mixture (e.g. by simply dissolving the monomer if crystalline, and by maintaining the polymer in solution) or to moderate the reactivity of extremely reactive monomers, such as diacid chlorides. Disadvantages of solution polymerisation include the difficulty of removing all the solvent from the polymer at the end of the reaction and, on a large scale, the inconvenience and expense of solvent removal. Thus for commercial polymer production, solution polymerisation is generally not used unless the polymer can be easily separated, e.g. by cooling as in many polyethylene processes, or unless the polymer is required in solution, e.g. for use as a lacquer or adhesive.

SOLUTION SBR Styrene–butadiene rubber produced by solution polymerisation.

SOLUTION STYRENE–BUTADIENE RUBBER Styrene–butadiene copolymer produced by solution polymerisation using a lithium alkyl catalyst. By variation of the polymerisation conditions, many polymer structural variations may be achieved. In particular use of ethers and amines as randomisers can result in high 1,2-polymerisation of the butadiene units, whilst coupling of living polymer chains can give branched polymers of improved processability. In use in tyres some polymers show a useful combination of low rolling resistance and good wet grip.

SOLUTION VISCOMETRY (Dilute solution viscometry). The simplest and most widely used method for estimation of polymer molecular weights. The method involves the measurement of the flow times of a series of dilute polymer solutions, and of pure solvent, through the capillary of a capillary viscometer, calculation of the limiting viscosity number and estimation of the viscosity average molecular weight using the Mark–Houwink equation. For reliable results the viscometer must be mounted vertically, in a thermostat of carefully controlled temperature and must be scrupulously clean. The solution must be carefully filtered and its concentration, conventionally expressed in gram per decilitre, should not exceed 1–2 wt %. The viscometer capillary should be such that the flow times are neither so short that a kinetic energy correction is necessary, nor inconveniently long—100–200 s is ideal. It may be necessary to apply an end-effect correction. Only rarely is a Rabinowitz correction required. Laminar flow, required by the theory, is ensured by having a Reynolds number of less than about 100.

SOLVENT CRACKING Alternative name for environmental stress cracking.

SOLVENT GRADIENT CHROMATOGRAPHY Alternative name for column extraction.

SOLVENT NAPHTHA Alternative name for naphtha.

SOLVENT STRESS CRACKING Environmental stress cracking when the liquid causing the cracking also has a solvating effect on the polymer. This can occur with many rigid plastic materials such as polystyrene and polymethylmethacrylate.

SOLVIC Tradename for polyvinyl chloride.

SOMATOTROPIN (Growth hormone) A protein hormone secreted by the anterior pituitary gland which promotes various aspects of growth, e.g. of bone tissue, protein synthesis, metabolism of carbohydrates and fats. A polypeptide with a molecular weight of 21 000, containing 191 amino acid residues, although only part of the molecule appears to be necessary for activity.

SOMEL Tradename for thermoplastic polyolefin rubber.

SONIC MODULUS The tensile modulus measured at high frequencies (10^3–10^4 Hz) by a wave propagation technique. It may be determined by using a sample in monofilament form attached to a diaphragm (e.g. of a loudspeaker) to excite a stress wave and a quartz piezoelectric crystal to detect the amplitude and phase of the response at variable distances along the sample length.

SORBITOL

$$HOCH_2CHCHCHCHCH_2OH$$
$$\quad\ \ |\ \ \ |\ \ \ |\ \ \ |$$
$$\quad\ \ OH\ OH\ OH\ OH$$

M.p. 96·7–97·7°C.

Produced by the hydrogenation of glucose. Occasionally used in the production of alkyd resins, when it has an effective functionality of four rather than six due to the low reactivity of some hydroxyls.

SOREFLON Tradename for polytetrafluoroethylene.

SOVITHERM Tradename for chlorinated polyvinyl chloride.

SOVPRENE Tradename for polychloroprene.

SOYA BEAN OIL The glyceryl esters (obtained from soya beans) of a mixture of acids whose approximate composition is 14% stearic, 23% oleic, 55% linoleic and 8% linolenic. It is thus approximately glyceryl oleate dilinoleate

CH$_2$OR where R is
|
CHOR′ —(CH$_2$)$_7$CH=CH(CH$_2$)$_7$CH$_3$
| and R′ is
CH$_2$OR′ —(CH$_2$)$_7$CH=CHCH$_2$CH=CH(CH$_2$)$_4$CH$_3$

It may be converted to epoxidised soya bean oil. Useful in alkyd resins as a semi-drying oil for good colour retention.

SPACE-CHARGE POLARISATION Alternative name for interfacial polarisation.

SPACE NETWORK Alternative name for a network polymer existing in three dimensions (as is usual) rather than as a planar network.

SPANDEX A polyurethane in fibre form, strictly one containing a thermoplastic polyurethane elstomer with at least 85% polyurethane content. Many commercial examples have been marketed, including Lycra, Vyrene, Dorlastan, Spanzelle and Glospan. Compared with the only other major elastomeric fibre, from natural rubber, spandex has higher tensile strength and modulus, as well as superior resistance to oxidation and fire. Fine denier fibre may be made, and the major use is in stretch fabrics.

SPANZELLE Tradename for a spandex polyurethane elastomeric fibre.

SPECIAL BOILING POINT SPIRITS Petroleum hydrocarbon fractions, largely similar to petroleum ethers, but having wider and generally higher boiling point ranges in the overall range 40–160°C. Useful in extractions and in rubber adhesives and when rubber solutions are required, e.g. for dipping and proofing.

SPECIFIC ACTIVITY The number of enzyme units per milligram of protein, or, in newer units, the number of katals per kilogram of protein. A measure of the purity of an enzyme preparation.

SPECIFIC DAMPING CAPACITY (Specific loss) A measure of the damping of a material when its dynamic mechanical behaviour is considered. Defined as the ratio of the energy dissipated per cycle of the alternating stress field to the maximum energy stored, i.e. the elastic energy.

This equals $2\tan\delta$, where $\tan\delta$ is the tangent of the loss angle.

SPECIFIC INDUCTIVE CAPACITY Alternative name for relative permittivity.

SPECIFIC LOSS Alternative name for specific damping capacity.

SPECIFIC REFRACTIVE INCREMENT Alternative name for refractive index increment.

SPECIFIC VISCOSITY Symbol η_{sp}. Defined as η_{rel} − 1, where η_{rel} is the viscosity ratio. Used in place of η_{rel} since this has a limiting value of unity. It is the fractional increase in the viscosity of a solvent resulting from dissolution of a polymer.

SPHERULITE An aggregation of crystallites as a spherical cluster, consisting of fibrillar crystalline lamellae (the fibrils) radiating from the centre of the spherulite, or arising as branches. Amorphous polymer is present between the fibrils. Spherulites are the most noticeable morphological feature of most crystalline polymers, varying in size from about $0.1\ \mu m$ to several millimetres, usually being in the range $1\ \mu m$ to $0.1\ mm$. They are birefringent and are most conveniently observed using a polarising microscope when a characteristic Maltese cross pattern may be observed. Another extinction pattern sometimes observed is a series of concentric rings (ringed spherulite). Other special morphologies, e.g. dendritic spherulite, are also known. The molecular chains lie perpendicular to the fibril axis (and therefore to the spherulite radius), the extinction patterns arising from a regular twisting of the fibrils. When the refractive index is greater across than along its axis, the spherulite is positively birefringent (a positive spherulite). Conversely the spherulite may be a negative spherulite.

SPILAC Tradename for an acrylic polymer laminating resin.

SPIN-DECOUPLING Alternative name for spin–spin decoupling.

SPIN LABELLING The technique of covalently attaching a molecular group, which contains a stable free radical, to a polymer chain. Analysis of the electron spin resonance spectrum of the label can give useful information on the molecular motions and orientation of the polymer, in particular of that part of the polymer chain to which it is known to be attached, which is often the chain end. Most labels are based on the nitroxide radical and are attached by carrying out an appropriate chemical reaction on the polymer.

SPIN–LATTICE RELAXATION (Longitudinal relaxation) In nuclear magnetic relaxation, it is the relaxation of the nuclear spin magnetisation in the direction of the applied field. The relaxation originates from the interaction of a nucleus with the rapidly fluctuating local field due to movements in the neighbouring magnetic nuclei. When the relaxation is observed in the presence of an applied radio frequency magnetic field, it is called relaxation in the rotating frame.

SPIN–LATTICE RELAXATION TIME Symbol T_2. The half-life of the spin–spin relaxation process that occurs in wide-line nuclear magnetic resonance spectroscopy. Thus at resonance, the populations of the two magnetic energy states for the nucleus are not the same due to energy transfer from the upper state to the surroundings (spin–lattice relaxation). The relaxation time is also the time in which a non-equilibrium state completes $(1 - 1/e)$, i.e. 0·632, of its return to equilibrium. It is obtained from $v_{1/2} T_2 \approx 1/\pi$, where $v_{1/2}$ is the width of the resonance peak at half-maximum intensity.

SPINNING (1) The formation of a fibre by extruding a polymer either in solution into a coagulating bath (wet spinning) or into an air drying medium (dry spinning) or by extruding a polymer melt into a cooling atmosphere when solidification occurs (melt spinning). All man-made fibres are made by one of these methods.

(2) The formation of a yarn from short fibres, i.e. staple, by arranging the fibres parallel into a bundle, pulling them out and thereby thinning the bundle, and at the same time also applying a twist.

SPINODAL The line on a temperature versus composition phase diagram of a mixture of two components, which separates the two phase region from a metastable single phase region between the binodal and spinodal. For a system in which miscibility increases with increasing temperature it lies just beneath the binodal. For polymer/polymer mixtures it is possible to quench a homogeneous mixture through the metastable region into the unstable two phase region, where phase separation occurs by spinodal decomposition.

SPINODAL DECOMPOSITION A process of phase separation that occurs when the temperature of a homogeneous mixture of two components is rapidly changed, so that the system is brought to a state in which both the spinodal and binodal have been crossed. This is much easier to achieve in polymeric mixtures than with low molecular weight mixtures. Phase separation occurs due to fluctuations in composition gradually sharpening rather than by nucleation and growth, where sharp changes in composition are present from the earliest stages of phase separation at the phase boundaries.

SPIN–PHASE RELAXATION Alternative name for spin–spin relaxation.

SPIN PROBE A technique, similar to spin labelling, but in which the stable free radical is merely physically mixed with the polymer. An electron spin resonance study of the tumbling behaviour of the probe can give information on polymer molecular motions (as with spin labelling).

SPIN–SPIN COUPLING (Coupling) In nuclear magnetic resonance spectroscopy, the coupling of the nuclear spins of adjacent nuclei. Since nuclei may be in different spin states, this causes splitting (spin–spin splitting) on the nuclear magnetic resonance peaks by an amount given by the coupling constant (J). If a nucleus has n coupled neighbours, then its resonance peak will be split into $n + 1$ spin states. The splitting can reveal additional structural information, especially with polymers, since J for protons or saturated carbon atoms depends on the dihedral angle and hence the spectrum gives information on the molecular conformation. Complex spectra resulting from splitting may be simplified by spin–spin decoupling using the double resonance technique.

SPIN–SPIN DECOUPLING (Decoupling) (Spin decoupling) In nuclear magnetic resonance spectroscopy, decoupling of the spin–spin coupled nuclei using the double resonance technique. The irradiating frequency, corresponding to the resonance frequency of one nucleus, causes such rapid transitions between its spin states that each state has too short a lifetime to couple with the second nucleus, the spin–spin splitting thus disappearing. This causes a great simplification of the spectrum, thus aiding in its interpretation. However, additional information is also lost. In ^{13}C spectroscopy, decoupling of proton–^{13}C interactions is particularly important in improving resolution of individual carbon atom resonances and in improving sensitivity.

SPIN–SPIN RELAXATION (Spin–phase relaxation) A nuclear magnetic relaxation in which relaxation of nuclear spin magnetisation occurs transverse to the applied field. It originates from the mutual interaction between two coupled spins.

SPIRAL GROWTH (Growth spiral) An interconnected stack of lamellae arranged in a spiral manner, occurring on a polymer single crystal grown from solution. It is due to the formation of a permanent step or wedge on the single crystal surface, which serves as the nucleus for the growth of new lamellae by folding against the exposed face.

SPIRO-LADDER POLYMER Alternative name for spiro-polymer.

SPIRO-POLYMER (Spiro-ladder polymer) A ladder polymer in which the rings are joined through two single bonds arising from the same ring atom:

Many inorganic coordination polymers are of this type, but only a few such organic polymers are known, the polyspiroketals being rare examples. The spiro structure confers increased thermal and chemical stability on the polymer.

SPODUMENE A linear, crystalline, naturally occurring mineral silicate polymer of the pyroxene type, of empirical formula $LiAlSi_2O_6$, i.e. with the linear silicate chains being linked laterally by lithium and aluminium atoms.

S-POLYMER A commercial copolymer of isobutene and styrene, useful for forming transparent plastic sheets with excellent barrier properties.

SPONGE RUBBER A cellular rubber made from solid starting materials, but unlike expanded rubber, its cells are interconnected. Such a material is produced either by using a low viscosity rubber and nitrogen gas or by employing a chemical blowing agent for expansion.

SP RUBBER Abbreviation for superior processing (natural) rubber.

SRF Abbreviation for semi-reinforcing furnace carbon black.

STABILISATION Reduction or elimination of degradation or its effects on properties. Achieved by the incorporation of a stabiliser as an additive, by copolymerisation or by chemical reaction of the polymer. The last two methods are used particularly to prevent depolymerisation (e.g. end-capping of polyacetal), although the former is much more widely used.

STABILISER (Antidegradant) An additive used to reduce or eliminate degradation or its effects on properties. The type of stabiliser used depends on the type of chemical reaction it is desired to prevent. Usually only small quantities (often <1%) adequately protect the polymer. A thermal stabiliser eliminates the effects of thermal degradation and is most commonly required in chloro-polymers, especially polyvinyl chloride. When oxidative degradation occurs, an antioxidant is used. A chelator will reduce the effects of metal ion catalysis on degradation. Photochemical degradation may be prevented by the use of either a screening agent, an ultraviolet absorber or a quencher—all examples of photostabilisers. Chemically induced degradation is often a hydrolysis reaction and therefore may be reduced by the use of an acid or basic stabiliser to eliminate the base or acid catalysis respectively. Stabilisers are often used in combination, when synergism is particularly desirable.

STABIN OSMOMETER Alternative name for Pinner–Stabin osmometer.

STABILITE WHITE Tradename for butylated 4,4′-isopropylidene-diphenol.

STAFLEX Tradename for high density polyethylene.

STA-FLOW Tradename for vinyl chloride–propylene copolymers.

STAINING ANTIOXIDANT An antioxidant whose oxidation products are coloured (often yellow), and hence it imparts a colour to the polymer which it is protecting. Their use is therefore restricted to dark materials—mainly rubbers. Most chain-breaking aromatic amine antioxidants, e.g. *N*,*N*-diphenyl-*p*-phenylenediamine, are staining.

STAMYLAN HD Tradename for high density polyethylene.

STAMYLAN LD Tradename for low density polyethylene.

STAMYLAN P Tradename for polypropylene.

STAMYLEX Tradename for linear low density polyethylene with octene-1 as the comonomer.

STANDARD AMINO ACID One of the twenty α-amino acids commonly found in proteins, all of which are present in most individual proteins. They are glycine, alanine, valine, leucine, isoleucine, serine, threonine, aspartic acid, glutamic acid, asparagine, glutamine, lysine, arginine, methionine, cysteine, phenylalanine, tyrosine, tryptophane, histidine and proline.

STANDARD LINEAR SOLID (General linear solid) A mechanical model of linear viscoelastic behaviour which contains three elements, not just the single spring and dashpot elements of the Maxwell and Voight models. The elements, consisting of either two springs and a dashpot or of two dashpots and a spring, may be arranged in one of four different series or parallel arrangements. The model is thus capable of representing both creep and stress relaxation behaviour, by the differential equation,

$$a_0\sigma + a_1\,d\sigma/dt = b_0 e + b_1\,de/dt$$

where a_0, a_1, b_0 and b_1 are functions of the spring moduli and dashpot viscosities, σ is the stress and e is the strain. Solution of the equation predicts an exponential response with a single relaxation or retardation time. For many polymers, at least over the transition zone, this is a good representation. However, a better fit to real behaviour is given by even more complex models, e.g. four element models, but the equations are difficult to handle. Usually a model consisting of a large number of Maxwell elements in parallel or of Voight elements in series, is used. This leads to the relaxation and retardation time spectra.

STANDARD MALAYSIAN RUBBER (SMR) A grade of raw natural rubber with a particular technical specification based mainly on its plasticity retention index. Crepe, ribbed smoked sheets and Heveacrumb types can all be classified in this way.

STANDARD OIL PROCESS A low pressure process for the polymerisation of ethylene similar to the Phillips process but using a reduced molybdenum oxide catalyst. It produces a highly linear polymer and therefore high density polyethylene.

STANNOUS OCTOATE $Sn(C_7H_{15}COO)_2$. A useful catalyst in the formation of polyether based flexible polyurethane foams. Usually used in combination with an amine catalyst component, such as 1,4-diazabicyclo-2,2,2-octane. It is susceptible to hydrolysis so is sometimes replaced by dibutyltin dilaurate when water mixes are used.

STANYL Tradename for nylon 46.

STAPLE (Staple fibre) Short fibre as opposed to continuous filaments. All natural fibres, apart from silk, are staple fibres with lengths varying from 2–4 cm in cotton to 7–15 cm in wool to several hundred centimetres in jute and hemp. Man-made fibres are produced as continuous fibres, but are frequently chopped to form staple, to enable different fibres to be blended.

STAPLE FIBRE Alternative name for staple.

STAR BLOCK COPOLYMER (Radial block copolymer) A star polymer in which the arms of the star consist of block copolymer chains. Commercial products of butadiene and styrene of this type are manufactured. These include certain Solprenes, Europrene Sol T and the BDS-K resin. The first product is a four-armed star and in the last product a multi-modal molecular weight distribution is found which results in good processability. BDS-K resin is made using a two-stage initiator with a polyfunctional epoxidised linseed oil.

STARCH The principal reserve deposit polysaccharide in plants. A mixture of two glucans: (1) amylopectin (usually comprising 70–85% of the starch) which swells in hot water, but is insoluble; (2) amylose which is soluble in hot water. Starch occurs in plant cells as granules of 0·01–1 mm in size, and is extracted by grinding plant cells rich in starch (containing about 80% starch), e.g. from corn, potatoes and arrowroot, so that the granules pass into an aqueous extract from which they may be centrifuged. About 60% of the starch produced is converted to syrup and dextrose. The remainder is used in food, textile, paper and laundry industries, about half being chemically modified, e.g. as dextrins.

STARK RUBBER Natural rubber that has crystallised during storage at ambient temperature over long periods of time, typically years. The rubber becomes hard and inelastic. The crystals melt at a higher temperature than usual for natural rubber apparently because they are oriented even though no external force has been applied.

STAR POLYMER (Radial polymer) A polymer with several polymer chains attached to the same centre, which may be a single atom or a chemical group. Synthesised by polymerisation with a multifunctional initiator or by reaction of a preformed polymer with reactive end groups with a multifunctional molecule. For example, the reaction of lithium-terminated polystyrene with $SiCl_4$ gives four branched polystyrene and with 1,2,4-(trichloromethyl)benzene gives three star polystyrene. The rheological properties of star polymers have been studied in detail since theoretical relationships for the behaviour of star-branched molecules are more easily derived than for other types of branching.

STATE OF CURE The development of a particular physical property, especially the modulus, of a rubber compound during vulcanisation (or cure). Measured using a curemeter.

STATIC DIELECTRIC CONSTANT Alternative name for static relative permittivity.

STATIC FATIGUE (Creep fracture) (Stress rupture) The fracture that occurs after long term loading under a steady load, as opposed to the fast fracture that occurs under the more usual direct loading type of fracture, where the load is continuously increased until fracture. This type of fracture is highly time dependent, greater times needing lower stresses to produce fracture. Also as temperature increases, fracture occurs more quickly. Since even direct loading must take place over a certain time interval (although much shorter) there must also be a static fatigue component in operation here. Both brittle and ductile fracture may be observed, the former being more likely at higher stresses. In brittle static fatigue, progressive stress crazing is frequently observed.

STATIC RELATIVE PERMITTIVITY (Static dielectric constant) Symbol ε_s. The maximum value of the relative permittivity of a dielectric. This is observed after a constant electric field has been applied for a sufficient time to allow maximum polarisation to occur. Over this long period all orientations of the dipoles have had time to develop.

STATIONARY STATE ASSUMPTION Alternative name for steady-state assumption.

STATISTICAL CHAIN ELEMENT A hypothetical unit of a polymer chain, of length l_s, which is used in place of the actual primary units (the individual covalent bonds) to produce a statistically equivalent chain of N_s such units. In this way the various factors contributing to chain stiffness, and hence chain coil size in solution, can be represented by a suitable choice of element size. Thus the mean square end-to-end distance $\langle r^2 \rangle_0$ is given by $N_s l_s^2$. This way of representing a polymer chain may provide a more useful, although physically less realistic, model since steric hindrance parameters are often not easily calculated. The value of l_s may be calculated from the contour length L, as $L = N_s l_s$ and the experimentally determined mean end-to-end distance.

STATISTICALLY EQUIVALENT CHAIN (Equivalent chain) A hypothetical model for a real polymer chain in which the various factors (bond angle restriction, steric hindrance, bond length) influencing overall chain dimensions are represented by a freely jointed chain of statistical chain elements, each of length l_s, rather than by a larger number of real chain chemical bonds.

STATISTICAL MOLECULAR THEORY OF RUBBER ELASTICITY Alternative name for kinetic theory of rubber elasticity.

STAUDINGER EQUATION The earliest empirical relationship between polymer molecular weight (M) and the polymer's limiting viscosity number ($[\eta]$). It is, $[\eta] = kM$, where k is a constant dependent on the particular polymer and solvent. Of limited validity and now replaced by the more general Mark–Houwink equation.

STAUDINGER INDEX Alternative name for limiting viscosity number.

STEADY-STATE ASSUMPTION (Stationary state assumption) The basic assumption in free radical and certain other chain polymerisation kinetics that enables the rate equation to be readily solved. The assumption is that the concentration of active centres ($[M^*]$) remains constant and hence their rate of formation by initiation (R_i) must equal their rate of disappearance by termination (R_t). Now $R_t = 2k_t[M^*]^2$ and $R_i = 2fk_d[I]$ when initiation is by thermal homolysis of a free radical initiator I, where k_t and k_d are the rate constants for termination and initiator decomposition respectively and f is the initiator efficiency. Substitution in the relationship for rate of polymerisation, $R_p = k_p[M][M^*]$, where k_p is the rate constant for propagation, gives

$$R_p = k_p[M](fk_d[I]/k_t)^{1/2}$$

The assumption is generally valid, with $[M^*]$ rapidly rising to its very low steady-state value of $\sim 10^{-8}$ M after an induction period of only a few seconds. The assumption is not valid during the pre- and after effects.

STEADY-STATE COMPLIANCE Symbols J_0^e (in shear) and D_e^0 (in tension). The compliance associated with the recoverable elastic component of a deformation of a fluid-like viscoelastic material when at long times a steady-state flow is achieved.

STEARIC ACID (Octadecanoic acid)

$$CH_3(CH_2)_{16}COOH$$

M.p. 70°C.

Obtained from the glycerides found in many animal and vegetable fats. Widely used as a lubricant in polymer compositions. An essential ingredient in most rubber sulphur vulcanisation recipes. Metal stearates are useful as thermal stabilisers for chlorinated polymers, especially PVC, and also as lubricants. Some stearate esters have found use as plasticisers.

STEM (Fold stem) The parallel regularly packed segments of polymer chains in chain folded polymer crystals that run between the crystal surfaces across the small dimension of the (usually) lamellar crystals. The stems have a regular conformation, e.g. extended chain for polyethylene, and many stems may belong to one molecule, being connected by chain folds at the crystal surfaces.

STEP-GROWTH POLYMER A polymer formed by step-growth polymerisation. Since such polymer forming reactions involve the formation of interunit links by reaction between functional groups containing heteroatoms, these polymers are also usually heteropolymers. One of the two main classes of polymers, when they are classified according to their mode of formation—the other being chain or addition polymers. Since most step-growth polymerisations involve the elimination of a small molecule, step-growth polymers, unlike chain polymers, do not contain all the atoms present in the monomer. Such polymers are often referred to as condensation polymers. However, a classification of polymers based on polymerisation mechanism can lead to difficulties, since the same polymer may be capable of being formed by both chain and step-growth mechanisms. A notable example is found in the case of monadic nylons which can be produced by the step-growth polymerisation of ω-amino acids or by ring-opening chain polymerisation of lactams. Linear step-growth polymers may be formed from monomers in which the two types of reactive group (A and B) are contained in the same monomer, thus producing AB polymers, or from two monomers each containing either 2A or 2B groups—giving the AABB polymers. In this latter case there is some uncertainty as to what should be considered as the repeat unit. These polymers may be considered formally as alternating copolymers of units derived from the AA and BB monomers. However, usually the whole AABB structure is regarded as the repeat unit of a homopolymer.

STEP-GROWTH POLYMERISATION (Stepwise polymerisation) One of the two main types of polymerisation, the other being chain polymerisation. It occurs by means of normal organic chemical reactions between reactive (often called functional) groups (A and B) on a monomer or monomers containing at least two such groups. If both types of reactive group are present in a single monomer, then AB polymerisation results, as with, for example, formation of a polyester by self-condensation of a hydroxyacid:

$$n\,HO—R—COOH \rightarrow \pm O—R—CO\mp_n + n\,H_2O$$

If one monomer contains the A groups and another monomer the B groups, then an AABB polymerisation results, for example, the reaction of a diol with a diacid to form a polyester:

$$n \text{ HO—R—OH} + n \text{ HOOC—R'—COOH} \rightarrow$$
$$\text{+O—R—OCO—R'—CO+}_n + 2n \text{ H}_2\text{O}$$

If only difunctional monomers are used then only linear polymer is formed. If the monomer has a functionality of >2, then initially branched polymer is formed, which at a sufficiently high conversion of the functional groups, will form a highly crosslinked polymer, giving an infinite network.

It is assumed, usually correctly, that all reactive groups are equally reactive, so that growth of polymer chains occurs slowly. Monomers initially react to form dimers, dimers reacting with themselves to form tetramers or with more monomer to form trimer and so on. Consequently polymer molecular size increases slowly in a series of distinct steps and monomer disappears rapidly. This contrasts with chain polymerisation, in which a monomer molecule, once activated, grows extremely rapidly to high polymer.

In linear step-growth polymerisation, the increase in average molecular size with conversion is given by the simple Carothers equation. To produce polymer with a high enough molecular weight to have useful mechanical properties requires extremely high conversion. Thus, for example, a conversion of 99% only gives a degree of polymerisation of 100. This implies that a successful polymerisation requires a stoichiometric balance of A and B groups (and hence highly pure monomers), no preferential loss of one type of monomer by side reactions or volatilisation, the absence of monofunctional monomers and the choice of reaction type and reaction conditions capable of proceeding to very high conversions. Although many linear polymerisations are step-growth polymerisations, only a few reaction types can be continued to high enough conversions to produce really high molecular weight polymer. These include polyesterification by ester interchange, polyamidation, polyurethane formation and interfacial polymerisations using acid chlorides. In such cases the polymer molecular weight distribution is the most probable distribution.

Linear step-growth polymerisation may be complicated by the establishment of ring–chain equilibria or by the occurrence of interchange reactions. In non-linear polymerisations, involving the use of a monomer with a functionality of >2, above a critical conversion (which may be predicted using a modified Carothers equation, or by the branching coefficient) a network is formed and gelation occurs. Many useful thermoset polymers are formed in this way, but it is often essential to halt the polymerisation before network formation has occurred, i.e. at some prepolymer stage, so that the polymer may be fabricated prior to a final crosslinking, which is often induced by heat. Prepolymers of this type are random prepolymers and are often crosslinked merely by continuing the same polymerisation reaction used in their formation, as, for example, with resoles and glycerol polyesters. In structoset prepolymers, the second stage crosslinking proceeds by a different reaction.

In many step-growth polymerisations a small molecule, such as water, hydrochloric acid, ammonia or carbon dioxide, is eliminated, so such polymerisations are also termed condensation polymerisations. Indeed often the terms step-growth and condensation polymerisation are used synonymously. However, some step-growth polymerisations, notably polyurethane formation, do not eliminate a small molecule, so are not condensations, but instead are sometimes termed polyadditions.

STEP-LADDER POLYMER Alternative name for partial ladder polymer.

STEPWISE POLYMERISATION Alternative name for step-growth polymerisation.

STEREOBLOCK COPOLYMER A block copolymer in which the blocks consist of sequences of monomer units of the same stereochemical configuration.

STEREOBLOCK POLYMER A block copolymer in which the block differences are due to differences in the configurations of the repeat units within them. The differences are often in tacticity, either with blocks of different tactic types or with one tactic block and one atactic block. However, it is generally considered that the blocks must be of reasonable length, e.g. sufficient for crystallisation to occur, for the polymer to qualify as stereoblock. Polypropylene has been the most widely studied example. *Cis–trans* isomerism may also form the basis of a stereoblock polymer, as in *cis*-1,4-polybutadiene with sequences of *trans* units.

STEREOELECTIVE POLYMERISATION A stereoregular polymerisation in which an optically active catalyst causes incorporation of only one of the two stereoisomers of a racemic monomer mixture into the polymer chain. For example, a Ziegler–Natta catalyst of $TiCl_3$ and optically active 2-methylbutyllithium stereoelectively polymerises only one stereoisomer of 3-methylpentene-1. This contrasts with stereoselective polymerisation where a non-optically active catalyst is used.

STEREOREGULARITY In polymers, an alternative name for tacticity.

STEREOREGULAR POLYMER (Tactic polymer) A polymer that exhibits tacticity, i.e. one that has regularity in the stereochemical configurations of its repeat units. For a monotactic polymer, two perfectly regular types are possible—isotactic polymer and syndiotactic polymer. For ditactic polymers other regular forms, the *erythro* and *threo* forms, are possible. Stereoregularity is a result of an atom in the polymer repeat unit being asymmetric and therefore capable of existing in either the D- or L-configurations. Examples are the C* atoms on poly-α-amino acids (polypeptides),

$$\text{+C*HR—CO—NH+}_n$$

and polypropylene oxide

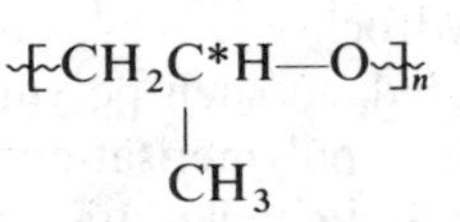

$$-[-CH_2C^*H—O-]_n$$
$$|$$
$$CH_3$$

In vinyl polymers, $-[-CH_2—C^*HX-]_n$, the C* atom is a pseudo-asymmetric atom since according to the classical organic chemist's view, given the similar nature of the polymer chains attached to C* on either side of C*, it is a centre of symmetry. Strictly, it is only the atom at the centre of the polymer chain that is pseudo-asymmetric and the classical viewpoint leads to an illusory reversal of configuration at this central unit. To avoid these inconsistencies, stereoregular polymers are best described in terms of the relative configurations of neighbouring units, as being either *meso* (m) or racemic (r) dyads. Analysis of stereoregularity in these terms is most readily performed experimentally by NMR spectroscopy.

The three-dimensional regularity of structure may be represented, e.g. in the isotactic case, by drawing the all-*trans* extended zig-zag chain for a vinyl polymer as:

and for polypropylene oxide as:

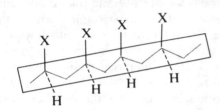

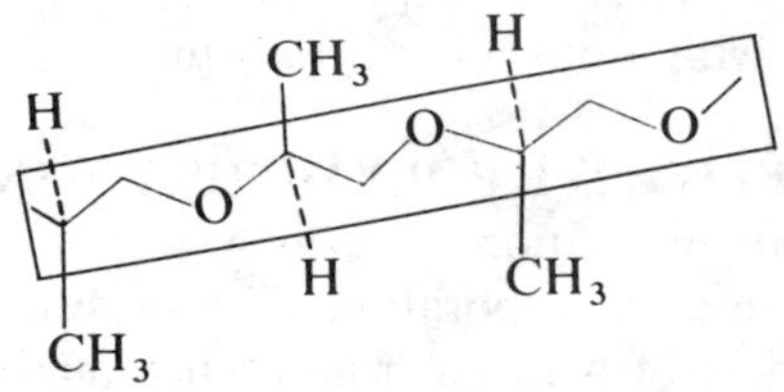

In this second case, as with polypeptides, with three chain atoms per repeat unit, the substituent is on alternate sides of the extended zig-zag chain. In the Fischer projection these structures are shown as

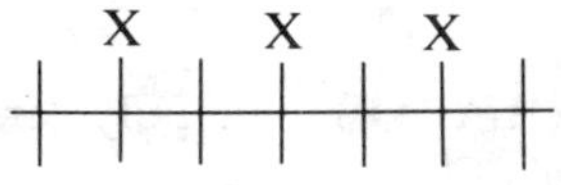

and

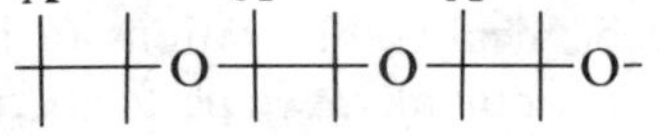

respectively, in which for both types of polymer the substituent is on the same side. Most commonly the three-dimensional structures are represented in two dimensions as:

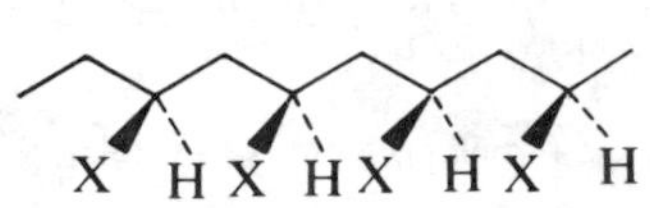

and

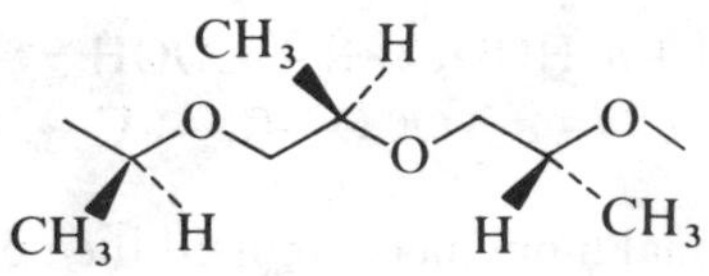

Stereoregular polymers of vinyl and diene monomers are produced by stereoregular polymerisation often by an ionic or coordination (e.g. Ziegler–Natta) polymerisation mechanism, where the catalyst exerts steric control. Free radical polymerisation usually results in largely atactic polymer. Chiral (asymmetric) monomers, such as propylene oxide and α-amino acids, will give isotactic, but never syndiotactic, polymer, simply by polymerising the D- or L-monomer under conditions that do not cause racemisation. Polymerisation of the racemic monomer can give isotactic polymer by stereoelective polymerisation, or atactic polymer. Polymerisation of chiral vinyl monomers of the type

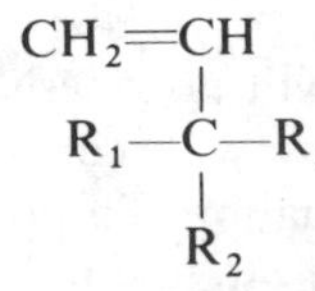

$$CH_2=CH$$
$$|$$
$$R_1—C—R_3$$
$$|$$
$$R_2$$

can also give stereospecific polymer by stereoselective polymerisation.

STEREOREGULAR POLYMERISATION (Stereospecific polymerisation) A polymerisation which produces a significantly stereoregular polymer structure. In vinyl polymerisation stereospecificity is manifest as tacticity. Free radical polymerisation usually produces atactic polymer, although there is a tendency towards syndiotacticity as the polymerisation temperature is lowered. Ionic polymerisations are frequently stereospecific. In both anionic and cationic polymerisation, as the degree of association of the counter-ion–active centre ion pair decreases (with increasing counter-ion nucleophilicity and solvent polarity), so stereoregulation increases. The most powerful stereospecific effects are found in Ziegler–Natta polymerisations, which frequently produce highly isotactic polymers. In 1,3-diene polymerisation stereospecificity is possible in the sense of highly regular 1,4-, 1,2-, 3,4-, *cis* or *trans* polymerisations. Again Ziegler–Natta and anionic polymerisations can produce highly isomerically pure polymers.

STEREORUBBER A rubbery polymer produced by stereospecific polymerisation and therefore one in which one of the several possible stereoisomers predominates. Several important commercial rubbers are of this type. They include high, medium and low *cis*-1,4-polybutadiene and synthetic *cis*-1,4-polyisoprene.

STEREOSELECTIVE POLYMERISATION A stereoregular polymerisation in which only one of the two stereoisomers of a racemic monomer mixture is incorporated into the polymer chain even though a non-optically active catalyst is used (as opposed to the use of an optically active catalyst in stereoelective polymerisation). For example, using Ziegler–Natta catalysts, racemic α-olefin monomers (such as 4-methylhexene-1

$$(CH_2\!=\!CH\!-\!CH_2\overset{*}{C}C_2H_5)$$
$$\qquad\qquad\qquad |$$
$$\qquad\qquad\quad CH_3$$

where the $\overset{*}{C}$ atom is asymmetric) can be polymerised in this way.

STEREOSPECIFIC POLYMERISATION Alternative name for stereoregular polymerisation.

STERIC EXCLUSION The probable main mechanism of molecular size separation in gel permeation chromatography (GPC) at least at low solvent flow rates when diffusion effects are absent. The separation is due to the ability of the smaller molecules in a polydisperse sample to penetrate the pores of the porous gel through which a solution of the polymer is flowing. Thus they are held up far longer in a column of the gel than the larger molecules. On assuming this mechanism, a universal calibration method for molecular weight determination from a GPC experiment may be developed.

STERIC HINDRANCE PARAMETER Symbol σ. A measure of the stiffness of a polymer chain in dilute solution, brought about by a restriction in the number of conformations the chain can adopt due to restricted chain rotations caused by steric interactions between substituents. It is defined by,

$$\sigma^2 = (1 + \langle\cos\theta\rangle)/(1 - \langle\cos\theta\rangle)$$

where $\langle\cos\theta\rangle$ is the mean value of the cosine of the conformational angle of rotation. For a freely rotating chain $\langle\cos\theta\rangle = 0$ and for an all-*trans* chain $\langle\cos\theta\rangle = 1$. Hence σ is a measure of the freedom of rotation in the chain. It increases with size of substituents and varies from about 1·5 to about 3.

STERIC MAP Alternative name for Ramachandran map.

STERNITE Tradename for phenol–formaldehyde and urea–formaldehyde polymers, acrylonitrile–butadiene–styrene copolymer and polystyrene.

STERN–VOLLMER EXPRESSION A mathematical relationship expressing the efficiency of a quenching process of a photochemically excited state. One form is

$$\phi_0/\phi_Q = 1 + \tau k_Q[Q]$$

where ϕ_0 and ϕ_Q are the quantum yields for a particular photochemical change in the absence and presence of quencher respectively, τ is the lifetime of the excited state in the absence of quencher, k_Q is the rate constant for the quenching reaction and [Q] is the quencher concentration.

STICK-SLIP A fracture surface marking consisting of a series of parallel markings, parallel to the fracture propagating front. It is caused by alternate acceleration and slowing due to an alternating development and breakdown of crazes ahead of the crack. As the crack opens further, the craze strands break and more load is placed upon other strands. This catastrophic rupture occurs until the crack reaches undrawn material when the crack slows. Subsequently new crazed material is formed again.

STIFFNESS CONSTANT Alternative name for elastic modulus.

STILAN Tradename for an aromatic polyetherketone.

STIROPLASTO Tradename for polystyrene.

STOCHASTIC PROCESS A process which is determined by the occurrence of certain probabilities of events making up the process. Polymerisation is such a process and various features of polymerisation, especially copolymerisation, and the structural features of the polymers produced, may be analysed by consideration of the probabilities of the individual events occurring. This is done using the appropriate statistics (Bernouillian or Markov) for the chain molecular model which is thought to apply.

STOCKMAYER–FIXMAN EQUATION An equation relating the limiting viscosity number ($[\eta]$) to molecular weight (M) and the unperturbed radius of gyration (S_0) of a polymer molecule. Widely used for the determination of unperturbed chain dimensions by measurement of $[\eta]$ for a series of fractions of the polymer. The equation is,

$$[\eta]/M^{1/2} = k + 0{\cdot}5BM^{1/2}\Phi$$

where k is a constant equal to $\Phi 6^{3/2}(\langle S_0^2\rangle/M)^{3/2}$, Φ is a universal constant (the intrinsic viscosity parameter) and B is another constant whose value depends on the particular polymer, solvent and temperature. A plot of $[\eta]/M^{1/2}$ versus $M^{1/2}$ should give a straight line with an intercept equal to k.

STOKES Symbol St. The c.g.s. unit of kinematic viscosity. It has the dimensions of $10^{-4}\,m^2\,s^{-1}$.

STOKES DERIVATIVE Alternative name for the substantial derivative.

STOKESIAN FLUID (1) Alternative name for a Newtonian fluid, often used instead of this term when considering types of flow other than shear flow.
(2) A generalised fluid which is inelastic, isotropic and homogeneous, i.e. a purely viscous fluid for which the constitutive equation may be written as,

$$\tau_{ij} = \eta_1 e_{ij} + \eta_2 e_{ik}e_{kj}$$

where η_1 and η_2 are the generalised and cross viscosities respectively, the τ_{ij} parameters are the shear stresses and e_{ij}, e_{ik} and e_{kj} are the shear strains.

STORAGE COMPLIANCE Symbols C' (in general), J' (in shear), D' (in tension) and B' (in bulk). In dynamic mechanical behaviour, the ratio of the strain in phase with the stress to the stress. The component of the complex compliance (C^*) which represents the elastic response and hence determines the stored energy, the other component being the loss compliance (C''). C' is usually much lower than C'' having a value of about $10^{-8}\,\text{Pa}^{-1}$ for a stiff polymer or about $10^{-5}\,\text{Pa}^{-1}$ for a rubber. C' is frequency dependent, the value decreasing with increasing frequency (ω) in the transition zone. Mathematically, for a Voight model in shear, $J' = J_v/(1 + \omega^2\tau^2)$, where J_v is the model compliance and τ is the retardation time. For a more general model, more representative of real polymer behaviour,

$$J'(\omega) = J_u + \int_{-\infty}^{\infty} [L(\tau)/(1 + \omega^2\tau^2)]\,\text{d}\ln\tau$$

where $L(\tau)$ is the retardation time spectrum and J_u is the unrelaxed compliance. Unlike the complex compliance, C' is not simply the inverse of the storage modulus M', but is related to it by $C' = (1/M')/(1 + \tan^2\delta)$ where $\tan\delta$ is the tangent of the loss angle.

STORAGE MODULUS Symbols M' (in general), G' (in shear), E' (in tension) and K' (in bulk). The component of the complex modulus (M^*) in dynamic mechanical behaviour which is in phase with the strain, i.e. the ratio of the stress in phase with the strain to the strain. M' thus determines the elastically stored energy and hence is called the storage modulus. Usually M' is much larger than the loss modulus (M''), and hence M^*, which equals $M' + iM''$, approximately equals M'. M' is frequency dependent, having a value for a rubbery polymer of about $10^5\,\text{Pa}$, which increases rapidly as frequency increases to a value of about $1\,\text{GPa}$ in the high frequency glassy region. Mathematically, for a Maxwell model, the frequency (ω) dependence is given by

$$M' = E_m\omega^2\tau^2/(1 + \omega^2\tau^2)$$

where τ is the relaxation time and E_m is the modulus of the spring component of the model. A better more generalised model gives

$$M'(\omega) = M_r + \int_{-\infty}^{\infty} [H(\tau)\omega^2\tau^2/(1 + \omega^2\tau^2)]\,\text{d}\ln\tau$$

where $H(\tau)$ is the relaxation spectrum and M_r is the relaxed modulus, which is more realistic of actual polymer behaviour. M' is not the inverse of the storage compliance (C') but is related to it by $M' = (1/C')/(1 + \tan^2\delta)$, where $\tan\delta$ is the tangent of the loss angle. M' may be determined experimentally by any of the usual dynamic mechanical methods, e.g. torsion pendulum or forced vibration non-resonance methods.

STORAGE PROTEIN A protein whose biological function is the storage of amino acids as nutrients. Examples include casein, ovalbumin, gliadin and zein. Some storage proteins store materials other than amino acids, e.g. iron in ferritin. Some transport proteins such as haemoglobin and myoglobin may also be classed as storage proteins since they provide short term storage of oxygen.

STORED ENERGY FUNCTION Alternative name for strain energy function.

STRAIN Symbols ε and e. The deformation brought about by the imposed stress on a body due to the displacement of points in the body relative to adjacent points. Since the displacement is relative, it is often expressed as a dimensionless function, such as l/l_0, where l and l_0 are the undeformed and deformed lengths of a small element in the body. Several such functions are used: tensor (or mathematical) strain (ε) engineers strain (e), extension ratio (λ), natural strain (ε) and Green's strain (ε') are examples. In double suffix notation the tensor strain components are, $\varepsilon_{xx} = \partial u/\partial x$, $\varepsilon_{yy} = \partial v/\partial y$ and $\varepsilon_{zz} = \partial w/\partial z$ (the normal strain components, corresponding to expansion or contraction along the x, y and z axes respectively) and

$$\varepsilon_{yz} = \tfrac{1}{2}(\partial w/\partial y + \partial y/\partial z)$$
$$\varepsilon_{zx} = \tfrac{1}{2}(\partial u/\partial z + \partial w/\partial x)$$

and

$$\varepsilon_{xy} = \tfrac{1}{2}(\partial v/\partial x + \partial u/\partial y)$$

where ∂u, ∂v and ∂w are the infinitesimally small relative displacements of two close points whose coordinates are x, y, z and $x + \partial x$, $y + \partial y$, $z + \partial z$. Historically, when engineering strains have been used they have been written without the factor $\tfrac{1}{2}$, which can lead to confusion. The six strain components may be written as a strain tensor, whose elements ε_{ij} are given by

$$\varepsilon_{ij} = \tfrac{1}{2}(\partial u_i/\partial x_j + \partial u_j/\partial x_i)$$

where i and j can have the values 1, 2 and 3, using the notation $x_1 = x$, $x_2 = y$ and $x_3 = z$ and $u_1 = u$, $u_2 = v$ and $u_3 = w$.

STRAIN BIREFRINGENCE Birefringence related to the strain in a sample produced as a result of stressing it.

STRAIN ENERGY The elastic energy stored in a body when in a state of strain and resulting from the work done in deforming the body. For a perfectly elastic material it is fully and instantaneously recoverable on removal of the stress. Often expressed as the strain energy function.

STRAIN ENERGY FUNCTION (Elastic potential) (Stored energy function) A relationship between the work done (W) on deforming a body as a function of the deformations (strains) produced. Since the work done must be independent of the loading history, it must be capable of being represented in terms of the strain invariants (I_1, I_2 and I_3). Thus in the most general way the strain energy function may be represented as $W(I_1, I_2, I_3)$. An example is found in the Mooney equation.

STRAIN ENERGY RELEASE RATE The strain energy released when a crack grows during fracture per unit area of crack. The usual symbol is G, but often only the crack opening mode (mode 1) is considered, when the symbol used is G_1. The critical value of G_1 for fracture to occur is G_{1c}.

STRAIN HARDENING (Work hardening) The increase in the slope of a stress–strain curve at high strains. This can be due to a strain induced crystallisation at very high extensions, as in natural rubber, or to orientation producing greater resistance to extension in the direction of drawing.

STRAIN INVARIANT Any of the three combinations of strain components which do not vary with direction of the cartesian coordinate axes chosen. They are e_{ii}, $e_{ij}e_{ji}$ and $e_{ij}e_{jk}e_{ki}$ in index notation. Thus they have 3, 9 and 27 terms respectively, for infinitesimal strains. For finite strains it is more convenient to define three alternative invariants as

$$I_1 = 3 + 2e_{rr}$$
$$I_2 = 3 + 4e_{rr} + 2(e_{rr}e_{ss} - e_{rs}e_{sr})$$

and

$$I_3 = |\delta_{rs} + 2e_{rs}|$$

where δ_{rs} is the Kronecker delta. These have been widely used in discussion of the strain energy function of rubbers when they are usually recast, for homogeneous pure strain, in terms of the principal extension ratios λ_1, λ_2 and λ_3 as,

$$I_1 = \lambda_1^2 + \lambda_2^2 + \lambda_3^2$$
$$I_2 = \lambda_1^2\lambda_2^2 + \lambda_3^2\lambda_1^2 + \lambda_3^2\lambda_2^2$$

and

$$I_3 = \lambda_1^2\lambda_2^2\lambda_3^2 = 1$$

if there is no change in volume on deformation, as with many rubbers.

STRAIN SOFTENING (Work softening) The reduction in the slope of the stress–strain curve (or even of the stress itself) when a polymer is stressed beyond the yield point. This may be simply due to the geometrical effect of reduction in cross-sectional area giving a reduced load, when stress is plotted as load, but it may also be an intrinsic strain softening since it can be observed on a time–stress plot. Adiabatic heating, especially at high strain rates, can also be a cause of strain softening.

STRAIN TENSOR Symbol ε_{ij} in index notation. A notation for representing the components of strain, which are given by the relation

$$\varepsilon_{ij} = \tfrac{1}{2}(\partial u_i/\partial x_j + \partial u_j/\partial x_i)$$

where i and j can have the values 1, 2 or 3. In this notation $x_1 = x$, $x_2 = y$ and $x_3 = z$ and $u_1 = u$, $u_2 = v$ and $u_3 = w$, where the x_i parameters (and hence x, y, z) are the coordinate axes directions and the u_i parameters (and hence u, v, w) are the displacements. The strain tensor is written as

$$\varepsilon_{ij} = \begin{bmatrix} \varepsilon_{xx} & \varepsilon_{xy} & \varepsilon_{xz} \\ \varepsilon_{yx} & \varepsilon_{yy} & \varepsilon_{yz} \\ \varepsilon_{zx} & \varepsilon_{zy} & \varepsilon_{zz} \end{bmatrix} \text{ or as } \begin{bmatrix} \varepsilon_{11} & \varepsilon_{12} & \varepsilon_{13} \\ \varepsilon_{21} & \varepsilon_{22} & \varepsilon_{23} \\ \varepsilon_{31} & \varepsilon_{32} & \varepsilon_{33} \end{bmatrix}$$

or as

$$e_{ij} = \begin{bmatrix} e_{xx} & \tfrac{1}{2}e_{xy} & \tfrac{1}{2}e_{xz} \\ \tfrac{1}{2}e_{yx} & e_{yy} & \tfrac{1}{2}e_{yz} \\ \tfrac{1}{2}e_{zx} & \tfrac{1}{2}e_{zy} & e_{zz} \end{bmatrix}$$

in terms of engineering strain components. When referred to principal axes, the shear components vanish and then

$$e_{ij} = \begin{bmatrix} e_{xx} & 0 & 0 \\ 0 & e_{yy} & 0 \\ 0 & 0 & e_{zz} \end{bmatrix}$$

where e_{xx}, e_{yy} and e_{zz} are now the principal strains.

STRAND The wound-up collection of individual filaments, usually of glass fibres (bound together by the size), resulting from the collection of the filaments from the base of the crucible bushing from which they are drawn during manufacture. The bushing contains 100–800 holes, frequently 204 or a multiple thereof, so that a strand contains this number of filaments. Strands may subsequently be twisted or plied together to form a yarn.

STREAMLINE FLOW Alternative name for laminar flow.

STRESS Symbol p or σ. The intensity of an internal force in a body, i.e. the force per unit area on a given plane passing through the point being considered. The term stress is used to refer to a stress vector, stress component or stress tensor. A stress vector or stress tensor is often analysed in terms of its normal stress components (symbol σ) and tangential (or shear) stress components (symbol τ). In general there are three normal components and six shear components of stress, since for rotational equilibrium the shear components on opposite faces of a cube of material must be equal and opposite (the complementary shear stress condition). With reference to the usual cartesian coordinate system, the normal stress components are designated σ_{xx}, σ_{yy} and σ_{zz} and the shear stress components are τ_{xy}, τ_{yz} and τ_{zx}. The normal components of stress on any plane are given by,

$$\sigma_{p,N} = l^2\sigma_x + m^2\sigma_y + n^2\sigma_z + 2mn\tau_{yz} + 2nl\tau_{zx} + 2lm\tau_{xy}$$

where l, m and n are the direction cosines of the vectors drawn normal to the plane from the coordinate system origin. For any state of stress, a set of cartesian axes may always be drawn (the principal axes) such that the shearing stresses on the principal planes are zero, the resultant normal stresses being the principal stresses $(\sigma_1, \sigma_2$ and $\sigma_3)$. These are the roots of the stress cubic,

$$\sigma^3 - I_1\sigma^2 + I_2\sigma - I_3 = 0$$

where $\sigma = \sigma_1/l = \sigma_2/m = \sigma_3/n$ and I_1, I_2 and I_3 are the stress invariants.

STRESS COMPONENT The component of a stress acting on a body, resolved in a particular direction. The directions chosen are the normals to the planes of a hypothetical small cube of material with edges parallel to the coordinate axes x, y and z, giving the normal components of stress $(\sigma_{xx}, \sigma_{yy}$ and $\sigma_{zz})$, and the two perpendicular directions within each face of the cube which contains the planes referred to above giving the shear components of stress $(\tau_{xy}, \tau_{yz}, \ldots,$ etc.). In the absence of body torques, $\tau_{xy} = \tau_{yx}, \ldots,$ etc., and there are thus six independent components of stress, three normal and three shear. The state of stress at a point is thus defined when these six components are specified for any plane through the point.

STRESS CONCENTRATION FACTOR The measure by which stress is increased in the vicinity of a hole, sharp corner or other such feature present in an object. It is the ratio of the maximum stress in the vicinity of the feature to the stress if the feature were absent. In the simplest case of a circular hole, the factor is three, since the maximum stress is three times the applied stress at the sides of the hole in the equatorial positions. For an elliptical hole of major axis length a and radius of curvature at the tip r, the factor is $(1 + 2(a/r)^{1/2})$.

STRESS CORROSION CRACKING Alternative name for environmental stress cracking.

STRESS CRAZING Alternative name for certain types of environmental stress cracking where many small surface cracks or crazes are formed when a glassy polymer is exposed to certain organic solvents whilst under stress. The stress may be internal frozen-in stress, so that an externally applied load is not needed. The stress involved is less than that required to produce crazing in the absence of solvent. Sometimes the term is used to include those cases where no solvent is required. Polymers which are particularly susceptible are polystyrene, polymethylmethacrylate and polycarbonate.

STRESS CUBIC An equation in σ (the principal stress acting on a principal plane) whose roots are the three principal stresses σ_1, σ_2 and σ_3. It is,

$$\sigma^3 - I_1\sigma^2 + I_2\sigma - I_3 = 0$$

where I_1, I_2 and I_3 are the three stress invariants. The equation allows the principal stresses to be calculated if the stress components for three coordinate planes are known. If l, m and n (the direction cosines of the principal planes with respect to the cartesian axes x, y and z) are known, then the directions of the principal axes can be calculated.

STRESS DEVIATOR Alternative name for deviatoric stress.

STRESS ELLIPSOID The ellipsoid in which the distance from the origin to the surface is proportional to the normal stress, when the axes of the ellipsoid are the principal directions of stress. The principal values of stress are given by the roots to the equation of the ellipsoid, which are the distances from the ellipsoid centre along the axes to its surface.

STRESS INTENSITY FACTOR Symbol K. A factor relating the magnitude of the stress components in the vicinity of a crack to the crack and specimen geometry and the overall stress (σ). It is given by, $K = \sigma(\pi a)^{1/2} Y$, where $2a$ is the crack length, and Y is a geometrical factor, sometimes called the finite plate correction factor, since for an infinite plate containing an embedded crack, $Y = 1$ and hence K is $\sigma(\pi a)^{1/2}$. Fracture will occur if the value of K exceeds the critical stress intensity factor K_c. For the usually considered crack opening mode (mode 1), K is called K_1. The value of Y, and therefore of K depends on the size, shape and location of the crack and on the size and shape of the specimen.

STRESS INVARIANT A function of the stress components which is independent of the coordinate system chosen, i.e. it is invariant with the choice of axes. The stress invariants are the coefficients of the stress cubic,

$$\sigma^3 - I_1\sigma^2 + I_2\sigma - I_3 = 0$$

where σ is a principal stress and I_1, I_2 and I_3 are the stress invariants. These have the values,

$$I_1 = \sigma_{xx} + \sigma_{yy} + \sigma_{zz}$$
$$I_2 = \sigma_{xx}\sigma_{yy} + \sigma_{yy}\sigma_{zz} + \sigma_{zz}\sigma_{xx} - \sigma_{xy}^2 - \sigma_{yx}^2 - \sigma_{zx}^2$$

and

$$I_3 = \sigma_{xx}\sigma_{yy}\sigma_{zz} + 2\sigma_{xy}\sigma_{yz}\sigma_{zx} - \sigma_{xx}\sigma_{yz}^2 - \sigma_{yy}\sigma_{zx}^2 - \sigma_{zz}\sigma_{xy}^2$$

The stress invariants are particularly useful in the analysis of yielding.

STRESS–OPTICAL COEFFICIENT (SOC) The constant of proportionality between the stress in a material and the birefringence resulting from the molecular orientation produced, as expressed in the stress–optical law.

STRESS–OPTICAL LAW A relationship between the stress (σ) in a material and the birefringence (Δn) resulting from orientation: $\Delta n = SOC \times \sigma$, where SOC is the stress–optical coefficient and equals

$$(2\pi/45kT)((\bar{n}^2 + 2)^2/\bar{n})(b_1 - b_2)$$

(where $\bar{n}$ is the average refractive index and b_1 and b_2 are the polarisabilities of the molecular segment parallel and perpendicular to the segment). The relationship provides the basis for the method of stress analysis in complex structures known as photoelastic stress analysis.

STRESS OVERSHOOT (Overshoot) The production of a stress maximum on the start-up of flow at constant rate, in an apparatus producing steady shear flow (e.g. a cone and plate rheometer) before the stress settles down to a steady equilibrium value. Such a phenomenon may be due to an elastic effect or to an instrumental artefact.

STRESS RELAXATION The relatively slow decay of the stress when a viscoelastic material is held at a constant strain after being rapidly stressed initially. Characterised by the time-dependent stress relaxation modulus, defined as the ratio of the stress $\sigma(t)$ to the fixed strain e. Usually the stress decays exponentially and at sufficiently long times may become zero if viscous flow takes place. However, if the material retains some rigidity then eventually a constant finite stress level is reached, corresponding to an equilibrium or relaxed stress relaxation modulus. Like creep, stress relaxation is one of the most commonly performed types of viscoelastic experiment. Quantitatively, the stress decays in a similar manner to the increase in creep strain with time and shows similar temperature dependence and time–temperature superposition. The rate of the stress decay is determined by the relaxation time (τ), which is similar, but not identical to, the retardation time of creep.

The simplest model of linear viscoelasticity that can usefully represent stress relaxation is the Maxwell model, for which the stress decays exponentially as $\sigma(t) = \sigma_0 \exp(-t/\tau)$. However, often real behaviour is more complex and the stress may not decay to zero at infinite times. This behaviour may be represented by more complex models, which may involve a stress relaxation spectrum. Stress relaxation measurements are often easy to carry out, particularly with soft materials, such as lightly crosslinked rubbers. The results of such measurements may be related to the theory of rubber elasticity and are also useful in studying network breakdown.

STRESS RELAXATION MODULUS Symbols $G(t)$ in shear, $E(t)$ in tension and $K(t)$ in bulk. The ratio of the time-dependent stress ($\sigma(t)$) to the fixed strain (e) during stress relaxation of a viscoelastic material. Unlike an elastic modulus, the corresponding compliance is not the inverse of the modulus due to the different time dependencies involved. One of the most frequently determined viscoelastic parameters, especially of rubbers, for which measurements are comparatively simple, yielding useful results for the study of network breakdown. The stress relaxation modulus is an exponentially decreasing function of time. Thus for a Maxwell model of viscoelastic behaviour, which provides the simplest representation of stress relaxation, $G(t)$ is given by $G(t) = (\sigma_0/e)\exp(-t/\tau)$, where σ_0 is the initial stress and τ is the relaxation time. However, such a simple model, giving a single relaxation time, is often not realistic and a relaxation time spectrum must be involved. Like creep, the response to multiple incremental strains can be represented by the Boltzmann Superposition Principle. Furthermore, the stress relaxation modulus can be considered to be made up of three components. Firstly,

the unrelaxed modulus corresponding to very short times, i.e. during the initially rapidly applied strain (although this elastic modulus is not usually considered as a relaxation component). This is followed by the time-dependent component representing the delayed elasticity. Finally, at very long times, if viscous flow occurs, then the stress relaxation modulus becomes infinite, but if flow does not occur then eventually an equilibrium or relaxed modulus results. This may be related to the network parameters of a crosslinked rubber via the theory of rubber elasticity. Typical values are about 1 GPa and 0·1 MPa for the unrelaxed and relaxed moduli respectively.

STRESS RUPTURE Alternative name for static fatigue.

STRESS SOFTENING The smaller stress required to strain a material to a certain strain, after a prior cycle of stressing to the same strain followed by removal of the stress. Mostly observed in filled rubbers (when it is known as the Mullins effect), where it results from the detachment of some polymer molecules from filler particles in the first cycle and which therefore cannot support the stress on subsequent straining to the same strain.

STRESS TENSOR Symbol σ_{ij}. A notation for the stress components given by arranging them as:

$$\sigma_{ij} = \begin{bmatrix} \sigma_{xx} & \sigma_{xy} & \sigma_{xz} \\ \sigma_{yx} & \sigma_{yy} & \sigma_{yz} \\ \sigma_{zx} & \sigma_{zy} & \sigma_{zz} \end{bmatrix}$$

or

$$\sigma_{ij} = \begin{bmatrix} \sigma_{11} & \sigma_{12} & \sigma_{13} \\ \sigma_{21} & \sigma_{22} & \sigma_{23} \\ \sigma_{31} & \sigma_{32} & \sigma_{33} \end{bmatrix}$$

using double suffix notation. For rotational equilibrium $\sigma_{ij} = \sigma_{ji}$, so that there are six independent components. On transformation to principal axes σ_{ij} becomes

$$\begin{bmatrix} \sigma_{xx} & 0 & 0 \\ 0 & \sigma_{yy} & 0 \\ 0 & 0 & \sigma_{zz} \end{bmatrix}$$

since the shear components ($i \neq j$) become zero, and σ_{xx}, σ_{yy} and σ_{zz} are now the principal stresses.

STRESS VECTOR The stress acting on a given plane of a material. When several stresses act on the same plane then the resultant is found by vector addition. Since a different vector is associated with any given plane (even passing through the same point), stress does not behave as a vector but as a second order tensor quantity. The state of stress at a point within a body is known if the stress on any two given planes passing through the point is known (when the stress on any other plane is also known).

STRESS WHITENING The appearance of white regions in a material when it is stressed, due to local changes in the refractive index in the stressed regions. This may be the result of the formation of microvoids or, especially in heterogeneous materials such as rubber reinforced plastics, the formation of microcrazes.

STRETCH Alternative name for deformation ratio.

STRETCHING FLOW Alternative name for elongational flow.

STRUCTOPENDANT PREPOLYMER A structoset prepolymer which may be crosslinked to a network polymer by reaction of functional groups placed along the polymer chain. Examples are epoxy resins with hydroxyl groups, novolacs and unsaturated polyesters.

STRUCTOSET PREPOLYMER A prepolymer of more precisely defined structure than a random prepolymer, which contains functional groups either at the chain end (structoterminal prepolymer) or along the chain (structopendant prepolymer), through which the polymer may be crosslinked to produce a network polymer. This is normally achieved via a reaction different from that which produced the prepolymer and is accomplished using a catalyst or an added crosslinking agent.

STRUCTOTERMINAL PREPOLYMER A structoset prepolymer in which the functional groups are at the chain ends. Examples are hydroxyl-terminated polyol polyethers used in polyurethane formation and epoxy resins with terminal glycidyl groups.

STRUCTURAL IRREGULARITY A chemical group, different from that of the repeating unit, present in a polymer molecule. Most polymers contain such groups but usually only to the extent of a few groups per thousand repeating units. Nevertheless they may have a significant effect on some properties, notably in their ability to act as weak links at which polymer degradation is initiated. They are the result either of reactions other than propagation during polymerisation, or of some subsequent chemical change in the polymer. Common irregularities are chain-end unsaturation (from transfer to monomer or disproportionation during polymerisation), branch points (from transfer to polymer), head-to-head links and oxygenated structures, e.g. hydroperoxides or carbonyl groups.

STRUCTURAL PROTEIN (Schleroprotein) A protein which functions biologically as a structural component of the organism in which it occurs. Structural proteins are also mostly fibrous proteins. Necessarily they are highly water insoluble, which can make structural studies difficult. Solubilisation destroys the often extensive intermolecular bonding—disulphide bridges or hydrogen bonding. Several major proteins are structural proteins. Especially important are keratin, collagen and fibroin. Others include sclerotin and elastin. The contractile proteins myosin and actin are also sometimes considered

as structural proteins. Certain glycoproteins have a structural function.

STRUCTURE The extent of aggregation of the primary particles of carbon black. The spherical particles may be fused together into clusters or chains (the primary structure) which survive compounding into a rubber matrix, or may form loose aggregates (the secondary structure) which are broken down during compounding. High structure blacks give unvulcanised compounds with lower nerve and higher stiffness than do low structure blacks.

STRUCTURE FACTOR (1) Symbol F_{hkl}. The structure factor describes the ideal intensity of a diffracted ('reflected') beam in X-ray diffraction, from a set of hkl planes. Defined by

$$F(hkl) = \sum_{n=1}^{N} f_n \exp\left[2\pi i(hx_n + ky_n + lz_n)\right]$$

where x, y, z are fractional coordinates of the unit cell edges a, b, c and N is the number of atoms in the unit cell, of which the nth atom is located at x_n, y_n, z_n with atomic scattering factor f_n. The ideal intensity of a reflection is $|F(hkl)|^2$ but a number of experimental factors modify this to the actual intensity $I(hkl)$.

(2) Alternative name for scattering law.

STYMER Tradename for polystyrene.

STYMER S Tradename for styrene–maleic anhydride copolymer salt.

STYPOL Tradename for an unsaturated polyester resin.

STYRAMIC HT Tradename for poly-(2,5-dichlorostyrene).

STYRENATED ALKYD RESIN An alkyd resin in which the unsaturated oil part of the resin has been copolymerised with styrene by heating with styrene and a peroxide initiator. The reaction is possibly via a 1,4-addition:

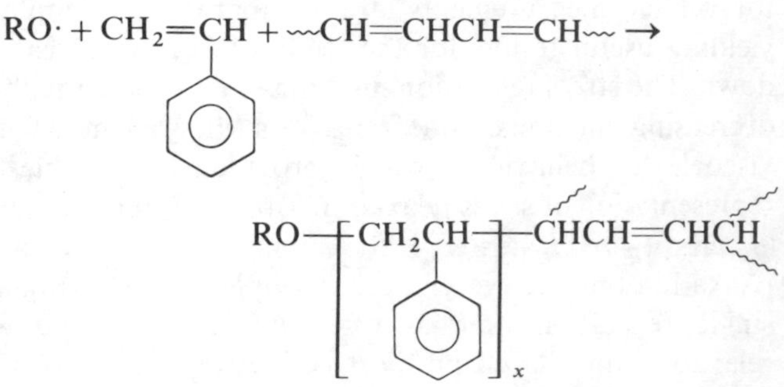

Coatings based on these resins have good water and alkali resistance and colour retention, but poor solvent resistance.

STYRENE (Phenylethylene) (Vinylbenzene)

$$CH_2{=}CH{-}\langle\bigcirc\rangle$$

B.p. 145°C.

An important monomer for the production of several polymers useful as plastics, rubbers and coatings. It has also been widely used in academic studies of the mechanisms (mostly free radical) of chain polymerisation. It is produced almost exclusively by the dehydrogenation of ethylbenzene by thermal cracking at about 600°C in about 90% yield. It is stored containing *t*-butyl catechol as an inhibitor to prevent premature polymerisation. It is unique as a monomer in that it readily homopolymerises by all types of chain polymerisation mechanism—free radical, cationic, anionic and Ziegler–Natta, although the first mechanism is always used for commercial production. Polystyrene is a major thermoplastic, but the monomer is also of great use due to its ability to copolymerise readily with a wide variety of comonomers. Important products include the random copolymers styrene–acrylonitrile and styrene–butadiene rubber, whilst the random/graft copolymer acrylonitrile–butadiene–styrene, the block copolymers styrene–butadiene–styrene and styrene–isoprene–styrene and the alternating copolymer styrene–maleic anhydride are also of commercial significance. Free radically, styrene may be polymerised with any conventional initiator, but it is frequently simply thermally polymerised without an added initiator. It is also used as a copolymerising crosslinking agent in the curing of unsaturated polyester resins and in styrenated alkyd resins. Overall, therefore, styrene is one of the most versatile monomers.

STYRENE–ACRYLONITRILE COPOLYMER (SAN)

(Poly-(styrene-co-acrylonitrile)) Tradenames: Bexan, Fostacryl, Kostil, Lacqsan, Luran, Lustran A, Novodur W, Restil, Styvacril, Tyril, Vestoran. A random copolymer of styrene and acrylonitrile, the commercial materials having an acrylonitrile content of 10–35%, usually in the range 20–30%. The polymers are produced as plastic materials by mass polymerisation at about 150°C. They have better solvent resistance, higher impact strength and higher softening point than polystyrene, these properties improving as the acrylonitrile content increases. However, their tendency to yellow and 'burn' during processing also increases as the acrylonitrile content increases. Typical properties of a copolymer with a 30% acrylonitrile content are a T_g value of 106°C, an Izod impact strength of 0·25–0·35 J $(12·7\,mm)^{-1}$ and a tensile modulus of 3700 MPa. Acrylonitrile is the more reactive monomer in copolymerisation (the reactivity ratios being 0·4 and 0·04 for styrene and acrylonitrile respectively) and thus there is a tendency to alternation in the copolymers. Copolymers with much higher acrylonitrile contents (70–80%) are often called high nitrile polymers.

STYRENE–BUTADIENE COPOLYMER (Poly-(styrene-co-butadiene)) (SB polymer) One of several different types of copolymer containing styrene and butadiene units. Most important are the styrene–butadiene rubbers which are random copolymers containing about 25% styrene. Similar polymers with about 60% styrene are known as high styrene resins. Block copolymers are also well-known commercial rubbers, especially the triblock styrene–butadiene–styrene thermoplastic rubbers. Radial block copolymers are also of interest.

STYRENE–BUTADIENE RUBBER (SBR) (GR-S) (FR-S) Tradenames: Ameripol, Baytown, Buna Huls, Buna S, Butakon, Cariflex S, Cisdene, Copo, Darex, Duranit, Europrene, Gentro, Intex, Intol, Jetron, Krylene, Krynol, Naugapol, Naugatex, Philprene, Plioflex, Pliolite, Polysar, Revinex, Sirel, Solprene, Synpol, Ugitex S, Uridene. A random copolymer usually containing about 25% styrene and 75% butadiene units. The most important commercial synthetic rubber. Mostly produced by free radical polymerisation in emulsion (emulsion SBR), although solution SBR is also significant. The earliest type, produced in Germany, was known as Buna S. During World War II SBR production was rapidly built up under US Government auspices, the product being known as GR-S (government rubber styrene) using the mutual recipe.

Two main types of emulsion SBR are produced. Hot rubber is produced by emulsion polymerisation at about 50°C using persulphate as initiator. Cold rubber is produced by emulsion polymerisation at about 5°C using redox initiation. A typical redox initiation system comprises cumene or *p*-menthane hydroperoxide with ferrous sulphate, sodium formaldehyde sulphoxylate and sodium ethylenediaminetetraacetic acid. The latter complexes with the iron, whilst the sulphoxylate reduces ferric back to ferrous ion. A mercaptan, e.g. dodecyl mercaptan, is used as a modifier to limit molecular weight, whilst polymerisation is terminated at about 70% conversion by the use of a short stop. Typical reactivity ratios are 0·5 and 1·6 for styrene and butadiene respectively, so that their product has a value of about unity, indicating that ideal copolymerisation with random distribution of the comonomer units takes place. The butadiene units have mixed stereoisomerism, typically consisting of about 20% *cis*-1,4-, 60% *trans*-1,4- and 20% 1,2-units, the exact composition being dependent on polymerisation temperature. Hot rubber contains branched molecules and macrogel, which whilst sometimes assisting processing, does yield vulcanisates with inferior properties.

Solution SBR is produced by anionic polymerisation using lithium alkyl catalysts. In non-polar solvents the tendency is to produce block copolymers, but in polar solvents random copolymers are formed. These polymers are linear and may suffer from cold flow. This can be eliminated by linking the polymer chains together to form a star copolymer.

SBR may be vulcanised in a similar way to natural rubber, but generally needs less sulphur and more accelerator due to its lower unsaturation. One advantage of SBR over natural rubber is that it can tolerate much larger quantities of oil extenders without deterioration of vulcanisate properties. Furthermore, it does not need mastication and has better abrasion resistance and ageing resistance than natural rubber but poorer dynamic mechanical properties. Its major use is in tyre treads.

STYRENE–BUTADIENE–STYRENE BLOCK COPOLYMER (SBS) Tradenames: Cariflex TR, Europrene Sol T, Kraton, Thermolastic. One of the best known block copolymers. It has been the subject of major study of phase separation and domain morphology in block copolymers and is a significant commercial product, useful as a thermoplastic elastomer. Usually prepared by sequential monomer addition to the living polymer produced by anionic polymerisation of styrene, e.g. with butyl lithium catalyst. Typically in the commercial material, the styrene blocks have a molecular weight of 10–50 000 and the butadiene blocks have a molecular weight of 30–100 000. As is usual with block copolymers the different blocks are incompatible and phase separate into styrene domains dispersed in a butadiene matrix. The morphology has been widely studied by electron microscopy which is aided by staining the butadiene regions with osmium tetroxide. The morphology is dependent on block length and sample preparation conditions. Spherical, lamellar and cylindrical morphologies have all been observed. Extremely regular arrays of dispersed styrene domains, resembling large scale single crystal structures, may be formed. The glassy styrene domains act as physical crosslinks as they tie down both ends of the butadiene chains. Thus the stiffness and strength of a crosslinked rubber may be achieved without the necessity of vulcanisation, i.e. the material is a thermoplastic elastomer.

STYRENE–ETHYLENE/BUTYLENE–STYRENE BLOCK COPOLYMER (SEBS) Tradenames: Elexar, Kraton G. A hydrogenated version of styrene–butadiene–styrene block copolymer, in which unsaturation is absent, and the polymer has much improved oxidation and weathering resistance. The mixed centre block arises from the presence of moderate amounts of 1,2-butadiene units in the original SBS polymer.

STYRENE–ISOPRENE–STYRENE BLOCK COPOLYMER (SIS) Tradenames: Cariflex TR1107, Europrene Sol T. A triblock polymer with two styrene end blocks and a central isoprene block. Similar in behaviour to the corresponding styrene–butadiene–styrene block copolymer, except that the isoprene blocks are less compatible with the styrene blocks than are butadiene blocks. Hence phase separation occurs and the physical crosslinking effect is apparent at lower styrene contents.

STYRENE–MALEIC ANHYDRIDE COPOLYMER (Poly-(styrene-co-maleic anhydride)) Tradenames: Cadon, Dylarc, Lytron, Stymer S. An alternating copolymer produced by free radical polymerisation of the mixed comonomers and thus having the structure:

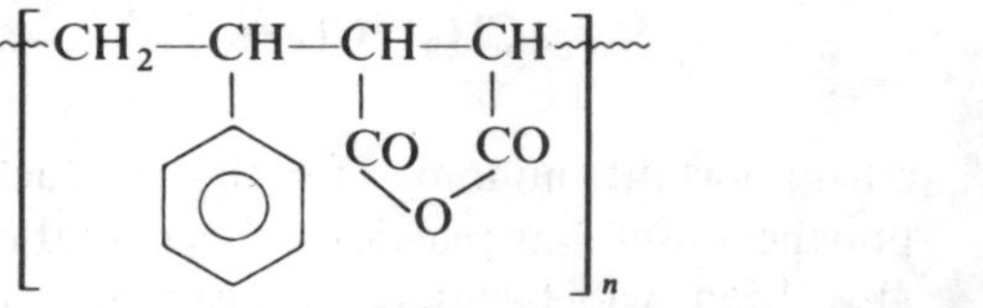

Hydrolysis of the anhydride groups occurs in water, but the polymer dissolves only when converted to a salt. It is useful as an emulsifier and textile size. It can have a higher T_g value (up to 122°C) than polystyrene and forms the basis of some ABS materials with high softening points.

STYROCELL   Tradename for expanded polystyrene.

STYROFOAM   Tradename for expanded polystyrene.

STYROGEL   Tradename for the highly crosslinked porous polystyrene gels widely used for gel permeation chromatography of synthetic polymers in non-aqueous systems.

STYRON   Tradename for polystyrene and high impact polystyrene.

STYROPOR   Tradename for expanded polystyrene.

STYVACRIL   Tradename for styrene–acrylonitrile copolymer.

STYVARENE   Tradename for polystyrene.

SUBCELL   Sometimes used to mean the crystallographic unit cell, when this term used to be employed to describe the lamellar crystals of a crystalline polymer. At one time the subcells were thought to be crystallographically regular and the true unit cell was regarded as a subcell of the lamella.

SUBCRITICAL NUCLEUS   Alternative name for an embryo.

SUBERIC ACID

$$HOOC(CH_2)_6COOH$$

M.p. 143°C.

Synthesised by oxidation of cyclooctane (obtained by hydrogenation of 1,5-cyclooctadiene—a dimer of butadiene), first to a cyclooctanol/cyclooctanone mixture with air, then to suberic acid with nitric acid. Only occasionally used as a monomer for nylons, e.g. nylon HPXD8, despite its high thermal stability (to 350°C) and hence suitability for melt polymerisation.

SUBSIDIARY LAMELLA   A lamellar crystal formed during the later stages of melt crystallisation in the cavities left in the spherulites between the previously formed dominant lamellae.

SUBSTANTIAL DERIVATIVE (Stokes derivative) A mathematical operator useful in continuum mechanics when Lagrangian coordinates are used, which is written as D/Dt and defined as

$$\mathrm{D/D}t = \partial/\partial t + \sum_i v_i\, \partial/\partial x_i$$

where v_i are velocity components and x_i are the directions for rectangular cartesian coordinates. Thus for a scalar variable P:

$$\mathrm{D}P/\mathrm{D}t = \partial P/\partial t + \sum_i v_i(\partial P/\partial x_i)$$

SUBSTRATE The molecule on which an enzyme exerts its catalytic action. Many enzymes are named by adding the suffix-ase to the name of the substrate, e.g. phosphatase catalyses hydrolyses of phosphate esters.

SUBTILISIN A bacterial enzyme, obtained from *Bacillus subtilis*, which is a protease. Enzymes of different genetic origin have quite different amino acid sequences, but nevertheless have similar specific functions. In total it has 270 amino acid residues and a molecular weight of 27 500, but lacks cysteine and cystine residues and the sequences are quite different from animal proteases, e.g. trypsin and chymotrypsin. It causes hydrolysis mainly at amino acid residues containing the carboxyl group whose substituent is aromatic or aliphatic, but is not very discriminating. It is sometimes used for further cleavage of the peptides produced by the initial tryptic or other action on a protein polypeptide for sequencing. At its active site, a serine residue is involved in hydrolysis, i.e. it is a serine enzyme. Since it is stable to alkali and to temperatures of about 65°C it is used commercially in enzyme detergents.

SUBTRACTIVE EDMAN METHOD A variation of the Edman method of N-terminal amino acid sequence determination of peptides and proteins, in which the peptide–thiazolinone (formed by a reaction of phenylisothiocyanate with the α-amino group) is removed and is identified by determination of the amino acid composition of the remaining shortened peptide. Its nature is then known by difference.

SUBUNIT (Protomer) (Monomer) The individual polypeptide chain of an oligomeric protein or supermolecular assembly (such as a virus). The subunit adopts the precise but irregularly folded structure typical of a globular protein and is tightly, although non-covalently, bound to other subunits, or commonly, to the prosthetic group in the case of conjugated proteins, e.g. nucleic acid in viruses, the haem group in haemoproteins. Denaturation will disrupt the quaternary structure of the associated complex, but provided that it is not too severe, it will not disrupt the tertiary structure of the individual subunits, which may then be studied.

SUFFIX NOTATION Alternative name for index notation.

SUGAR A sweet, crystalline, water soluble, low molecular weight carbohydrate, comprising the monosaccharides and their lower oligomers (oligosaccharides), i.e. the di- and trisaccharides. The term is often used interchangeably with the term monosaccharide, disaccharide, etc.

SUGAR ACID A monosaccharide (or sugar) or monosaccharide unit in which either an aldehyde group has been replaced by a carboxyl group (aldonic acid), or a terminal hydroxyl has been replaced by a carboxyl group (uronic acid), or in which both end groups have been so replaced (saccharic acid).

SULPHATED GALACTAN A polymer of galactose in which some of the hydroxyl groups are esterified as half esters of sulphuric acid. They occur widely in red seaweeds as agar, carrageenan and porphyran. They contain galactose units either in the D- form or as both D- and L-enantiomorphs, as well as derived units such as the sulphate esters, 3,6-anhydride and 6-methyl ether.

SULPHATED POLYSACCHARIDE A polysaccharide in which some of the monosaccharide hydroxyl groups are converted to the half esters of sulphuric acid. Found widely among seaweeds. Examples include agarose, carrageenan and fucoidan. Sulphated mammalian tissue polysaccharides which are also aminopolysaccharides, include chondroitin sulphate A and heparin.

SULPHATE PROCESS Alternative name for Kraft process, so named since sodium sulphate is added to replace the sodium sulphide consumed in the process.

SULPHENAMIDE A compound containing the structure —SNR′R″, where R′ may be hydrogen or an alkyl group and R″ is an alkyl group. Several important accelerators for the sulphur vulcanisation of rubbers are of this type, e.g. *N*-cyclohexylbenzothiazylsulphenamide and *N*-oxidiethylbenzothiazylsulphenamide.

SULPHITE PROCESS A process for the production of cellulose pulp (as cellulose I and containing the hemicelluloses) from wood by solubilising the lignin as lignosulphonic acids. The wood is heated with a bisulphite solution (calcium, ammonium, magnesium or sodium salts) at 120–150°C and a pH of 1–5.

SULPHOLANE (Tetramethylene sulphone) (Tetrahydrothiophene-1,1-dioxide)

B.p. 278·7°C (decomposes).
A dipolar aprotic solvent useful for dissolving many polar

polymers and especially useful for the synthesis of polysulphones.

SULPHONATED EPDM (Ethylene–propylene–ethylidenenorbornyl sulphonate) Tradename: Thionic. An ethylene–propylene–diene monomer copolymer containing about 55% ethylene units and about 5% ethylidenenorbornyl units which have been sulphonated by reaction with acetyl sulphate, thus becoming units of:

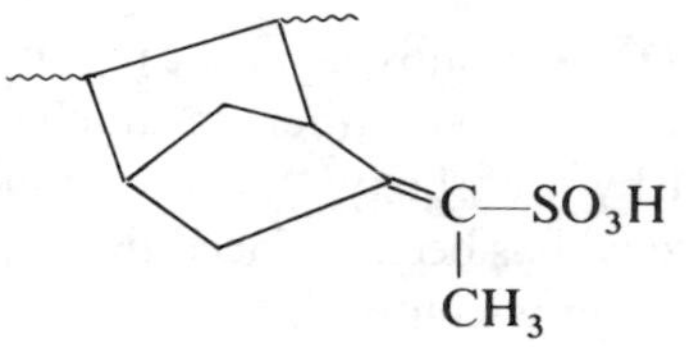

Sulphonate salts are then formed by reaction with metal (especially zinc) compounds. These provide ionic crosslinks between the polymer chains. When quantities of zinc stearate are used, the material processes as a thermoplastic elastomer. The polymer has good heat, weather and ozone resistance and is useful as a sheet material for outdoor use and for hoses and coatings.

SULPHON I Tradename for a sulphone linked poly-(p-benzamide) of repeat unit structure

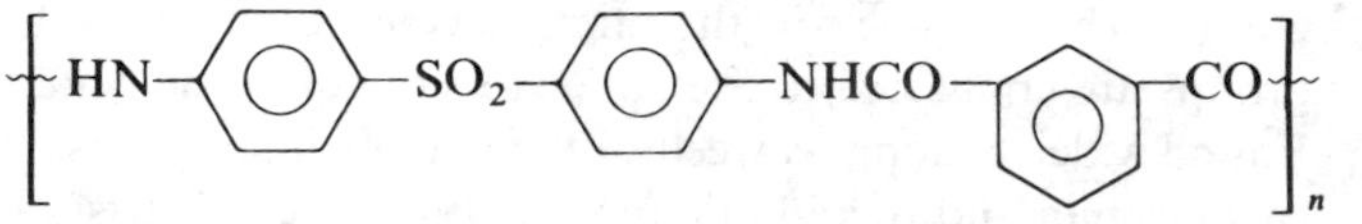

SULPHON T Tradename for a sulphone linked poly-(p-phenylene terephthalamide) of repeat unit structure

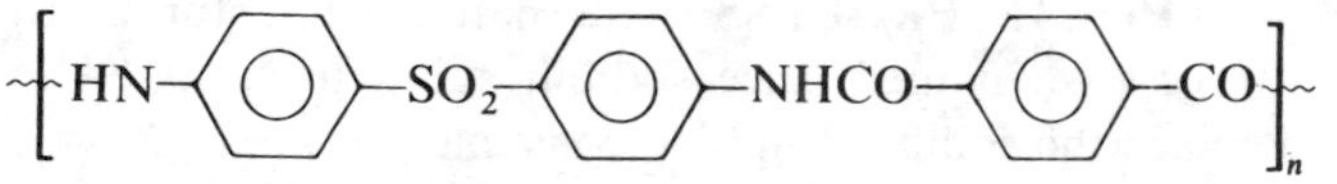

SULPHURLESS VULCANISATION Vulcanisation of a diene rubber through formation of sulphur crosslinks but not involving the use of elemental sulphur. The sulphur is derived solely from the accelerator. This is possible by the use of thiuram disulphides in conjunction with zinc oxide. The vulcanisate crosslinks are largely monosulphidic, and as a consequence reversion is absent, so the vulcanisates have good thermal ageing properties. However, they do have lower tensile strengths than sulphur vulcanisates.

SULPHUR MODIFIED POLYCHLOROPRENE Polychloroprene obtained by the polymerisation of chloroprene in the presence of about 1% sulphur. The polymer contains S_x units, where x is 2–6, in the polymer chain. The modified polymer is more readily vulcanised, can be readily masticated, especially by using a peptiser and is softer and tackier in the raw unvulcanised state than normal polychloroprene.

SULPHUR-NITRIDE POLYMER (Polythiazyl) $\leftarrow(SN)\rightarrow_n$ Prepared by passing cyclic tetrasulphur tetranitride vapour over silver and condensing the resultant unstable disulphur dinitride on a cool surface where it spontaneously polymerises. The polymer is deposited as fibrous bundles having a golden metallic lustrous appearance. The polymer has metallic conductivity and becomes superconducting below 1 K. The linear polymer chains are nearly planar with all S—N bonds being of approximately the same length.

SULPHUR RANK The average value of x in the repeat unit structure of polysulphide rubbers, i.e. in $\leftarrow R—S_x\rightarrow_n$.

SULPHUR VULCANISATION Vulcanisation of a diene rubber by heating to about 150°C with sulphur. In practice, a vulcanisation accelerator and activators are used in addition to sulphur. Without accelerator about 8 phr of sulphur is required but vulcanisation is very slow; about 5 h for natural rubber. It is also of low efficiency in that much sulphur is used to form polysulphide crosslinks and cyclic structures, whereas for the development of useful vulcanisate properties short (mono- and disulphide) crosslinks are required. Use of an accelerator not only speeds up vulcanisation giving times down to a few minutes, but also increases the proportion of mono- and disulphide crosslinks, e.g. from 40–50 sulphur atoms to 10–15 per crosslink. Typically about 2 phr of sulphur plus 2–4 phr of accelerator are used. Use of higher amounts of accelerator (to about 6 phr) and less sulphur (0·5–2 phr) can give efficient vulcanisation, requiring only 2–5 sulphur atoms per crosslink. Sulphur vulcanisation is the commonest method of vulcanisation for a wide range of diene rubbers including natural, isoprene, ethylene–propylene, styrene–butadiene, butadiene and nitrile rubbers, but is too slow for butyl rubber and does not work with polychloroprene. A typical vulcanisation recipe consists of rubber (100 parts), sulphur (2 phr) MBT (about 1 phr), stearic acid (2 phr) and zinc oxide (3 phr).

The mechanism of the chemical crosslinking process is complex but involves formation of a zinc perthiomercaptide (XSS_xZnS_xSX) by reaction of sulphur with previously formed XSZnSX. This is obtained from zinc oxide and the accelerator XSH, XSSX or $XSNR_2$ (thiazole, sulphenamide or thiuram disulphide respectively), where X is

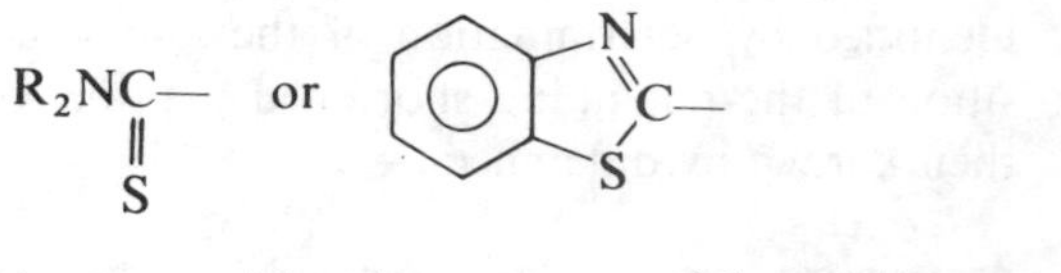

Zinc perthiomercaptide then reacts with rubber at positions α to the double bond to give RS_xH species, where R is the rubber hydrocarbon. Crosslinks are formed by the reaction

$$RS_xH + RH \rightarrow RS_{x-1}R + XSH$$

where RH is a further rubber molecule.

SUMMATION CONVENTION In index notation, the convention is that the product of several variables, written, for example, as $x_{ij}y_iz_j$, represents the sum of each of the products of every possible combination of the repeated indices. Often a repeated index which is to be summed is given a Greek letter and is called a dummy index. A suffix repeated in a term means summation of the term over the range of values of the suffix. Thus e_{ii} means $e_{11} + e_{22} + e_{33}$.

SUMMATIVE FRACTIONATION (Summative precipitation) A method of fractionation in which separate samples of a polymer solution are mixed with a large quantity of precipitant, which precipitates more polymer as fractions of lower molecular weight, each time the solvent power of the precipitant is decreased. Each time the weight of polymer remaining in solution is determined on an aliquot and its average molecular weight is also determined. In this way a summative molecular weight distribution curve is obtained.

SUMMATIVE PRECIPITATION Alternative name for summative fractionation.

SUNN (Indian hemp) A natural bast fibre obtained from the plant *Crotalaria juncea*. The fibres occur in bundles as strands of up to 150 cm in length. They are light in colour and of similar strength to hemp. The material is useful for cordage, sacking and fishing nets.

SUPER ABRASION FURNACE BLACK (SAF) A type of furnace carbon black with particle size about 20 nm. A highly reinforcing black useful in special tyre treads and in products for highly abrasive service conditions.

SUPERCONDUCTING POLYMER A polymer exhibiting electrical conductivity of $\sim 10^{20}\,\mathrm{S\,cm^{-1}}$. It has been postulated that a linear polyene chain containing highly polarisable side groups, e.g. dyes such as diethylcyanine iodide, could be superconducting in one dimension. Poly-(sulphur nitride) is superconducting at below 1·4 K.

SUPERCONDUCTIVE FURNACE BLACK (SCF) A type of furnace carbon black with a low number of surface groups and high porosity. This gives a highly electrically conducting black suitable for conducting products, e.g. anti-static products.

SUPERCOOLING The cooling of a polymer melt to below the melting point without crystallisation. It is also the temperature difference between the melting point and the temperature of crystallisation, when it is also called the undercooling. In polymers, high supercoolings (typically 30–50 K) are necessary for crystallisation to occur at a reasonable rate. Crystallisation rate increases with supercooling, reaching a maximum, then decreases due to increasing viscosity of the melt.

SUPERCRITICAL NUCLEUS A crystallisation nucleus for primary nucleation of positive free energy of formation, but larger than the critical size required for crystal growth.

SUPER DYLAN Tradename for high density polyethylene.

SUPERFLEX Tradename for high impact polystyrene.

SUPERIOR PROCESSING NATURAL RUBBER (SP Rubber) Natural rubber obtained by coagulating a mixture of a normal latex with a latex of a vulcanised rubber. Commonly the ratio is 80/20. The polymer has improved extrusion and moulding behaviour including reduced die swell due to elastic retraction.

SUPERMOLECULAR STRUCTURE Polymer structural features observable at a level above that of individual polymer molecules. These include crystal structure, crystallite, multilayer crystalline, fibrillar, spherulitic and fibrous morphologies.

SUPERPOSITION The summation, in a linear manner, of the individual strains (given by the appropriate Hooke's law relationship) produced by several forces acting on a body. Usually valid as long as deformations are small and do not affect the action of the external forces.

SUPERPOSITION PRINCIPLE Alternative name for Boltzmann Superposition Principle.

SUPRAMOLECULAR COMPLEX An association of individual polypeptide or protein molecules (the subunits) that behaves as a single, separate molecular unit due to the strong binding (not necessarily covalent) between the subunits, so that the complex may be isolated in homogeneous and often crystalline form. Sometimes it is also classed as an oligomeric protein. Examples range from the multienzyme complexes, such as fatty acid synthetase containing seven different enzyme subunits, to the viruses whose molecular weights or, perhaps better termed, particle weights are usually several million, e.g. 40 million in tobacco mosaic virus.

SURFACE FORCE An external force which acts on the surface of a body, such as hydrostatic pressure or pressure of one body on another.

SURFACE RESISTIVITY The electrical resistance between two electrodes forming the opposite sides of a square in or on the surface of a material. The size of the square is immaterial and the unit used is the ohm. Surface resistivity is hardly a material property since it usually depends on the presence of contaminating materials, especially moisture, on the surface of the material concerned.

SURLYN Tradename for the ionomers produced by the neutralisation of the methacrylic acid units in ethylene–methacrylic acid copolymers.

SUSPENDED-LEVEL VISCOMETER Alternative name for Ubbelohde viscometer.

SUSPENSION AGENT An ingredient in suspension polymerisation which, together with agitation, maintains the monomer as a dispersion of droplets in the suspension medium, i.e. it prevents coalescence. Finely divided mineral powders may be used, but normally a water soluble polymer is employed. A wide variety of water soluble polymers is used, including hydrolysed polyvinyl acetate, acrylic and methacrylic polymers and copolymers, cellulose derivatives, gelatin and pectin. The suspension agent is normally found at the droplet/water interface and therefore can prevent coalescence mechanically. However, it may also act by modifying surface tension and local viscosity. Typically, it is used in the range of 0·1–1·0% of the water.

SUSPENSION POLYMERISATION (Bead polymerisation) (Pearl polymerisation) Polymerisation, usually chain polymerisation of vinyl monomers, in which a water insoluble monomer is dispersed in water, or occasionally in another liquid, as droplets using a combination of agitation and a suspension agent. In this way the problem of dissipation of the heat of polymerisation, most acute in bulk polymerisation, can be alleviated since the water acts as a heat sink. Typically up to about 50% monomer in water may be used. A conventional monomer soluble initiator is employed, so that each monomer droplet acts as an isolated bulk polymerisation and the standard free radical polymerisation kinetics apply. The suspension agent must prevent coalescence of the droplets, especially at intermediate conversions when they may be sticky. Two types of suspension agent are used, either a finely divided powder, such as kaolin or magnesium silicates, or, more commonly, a water soluble polymer. Typically about 1% of the suspension agent is used and the droplet size, and hence the final polymer particle size, is in the range 0·1–1 mm. The product is easily separated from the suspension and dried, and is convenient to handle. It may, however, contain suspension agent as a contaminant, since it becomes grafted onto the polymer. In the usual case when the polymer is soluble in the monomer, e.g. polystyrene and polymethylmethacrylate, the product is in the form of small, clear, glass-like beads. If the polymer is insoluble in the monomer, as in the case of polyvinyl chloride, the product particles are granular with internal structural features such as porosity, which are dependent on the choice and amount of suspension agent.

SVEDBERG Symbol S. The unit for sedimentation constants. One Svedberg has a value of 10^{-13} s. Values for polymers are in the range 1–200 S. Proteins are frequently characterised by their sedimentation constant in aqueous solution at 20°C, when the symbol S_{20} or $S_{20,w}$ is used.

SVEDBERG EQUATION An equation used in the sedimentation velocity method for determination of molecular weight (M), relating M to the sedimentation (S_0) and diffusion (D_0) constants at infinite dilution as, $M = [RT/(1 - \bar{v}\rho)](S_0/D_0)$, where T is the absolute temperature, R is the universal gas constant, $\bar{v}$ is the

partial specific volume of the polymer and ρ is the solvent density. S_0 is given by $(1/\omega^2 r)(dr/dt)$ where ω is the angular velocity and r is the distance between the pure solvent boundary and the centre of rotation. The diffusion constant is given by $D = kT/f$, where f is the frictional coefficient given by $f = 6\pi\eta R'$, for spherical molecules of radius R' in a solvent of viscosity η.

SWELLING RATIO (1) The ratio of extrudate diameter to die diameter in extrusion. For polymer melts this ratio is usually greater than unity due to die swell and may have values up to about 3·0, but is typically about 1·5. The ratio is difficult to determine accurately due to shrinkage on cooling or crystallisation and due to difficulties in avoiding draw down.

(2) The ratio of the swollen to unswollen volumes of a lightly crosslinked polymer swollen by a solvent. Measurement of the equilibrium swelling ratio provides a method for the determination of the network parameter through the use of the Flory–Rehner equation.

SWIRL MAT Alternative name for continuous strand mat.

SWITCHBOARD MODEL (Random re-entry) A model for the surface structure of a chain folded polymer crystal in which the chain emerges from the crystal, forming a relatively large loop, and re-enters the crystal more or less randomly rather than adjacently. Such re-entry is thought to predominate in melt crystallised polymers.

SYNCOL Tradename for polyacrylic acid and its salts.

SYNDIOTACTIC DYAD Alternative name for racemic dyad.

SYNDIOTACTIC 1,2-POLYBUTADIENE A butadiene polymer produced commercially by Ziegler–Natta polymerisation. More recent materials have been produced using a cobalt halide complex/aluminium trialkyl/water catalyst and contain about 90% 1,2-linked units, with about 60% in the syndiotactic form, i.e. as

$$\sim\!\!\left[\begin{array}{c}CH_2CH\!\!-\!\!\!\!\\ |\\ CH\!\!=\!\!CH_2\end{array}\right]\!\!\sim$$

units. The polymer is partially crystalline, the crystallinity (20–30%) and T_m value (75–90°C) depending on the degree of structural regularity. The material is a soft thermoplastic, processing like a thermoplastic elastomer. It is useful for the production of packaging films of outstanding toughness and high gas and water vapour permeability.

SYNDIOTACTIC POLYMER A stereoregular polymer in which at least one chain atom of the repeat unit is capable of exhibiting stereochemical configurational isomerism and in which the configuration alternates between neighbouring isomeric centres. In terms of dyad relative configurations, the structure is

$$-mrmrmrmrmr-,$$

where —m— is a *meso* dyad and —r— is a racemic dyad. In vinyl polymers of the type $-\!\!\left[CH_2CHX\right]_m$ with the polymer in its fully extended chain conformation, the substituents point alternately in opposite directions from the plane of the chain:

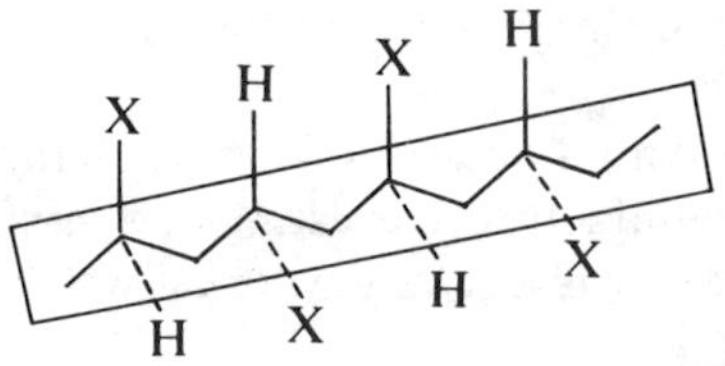

Syndiotactic polymers are often more difficult to produce than isotactic polymers, but may sometimes be made by anionic or Ziegler–Natta polymerisation. Since steric interaction between X groups is less than in an isotactic polymer, the polymers often exist as extended chains with better packing and higher T_m values. A tendency to syndiotacticity is frequently found in free radical polymerisation, especially at low temperatures.

SYNDIOTACTIC POLYPROPYLENE Polypropylene in which the asymmetric carbon atoms have opposite configurations on each alternate asymmetric centre. The polymer may be synthesised by the use of certain Ziegler–Natta catalysts, such as $VCl_4 + AlEt_2Cl$ at $-78°C$. The polymer crystallises with a T_m value of 183°C (or 134°C) with a 7_3- or 4_1-helix and has a T_g value of about 0°C.

SYNERGISM The production of a combined effect due to several components of a mixture, whose magnitude is greater than the sum of the individual effects of the separate components. Thus a synergistic stabiliser mixture provides greater protection for a polymer against the effects of degradation than expected from the individual stabiliser materials.

SYNOLITE Tradename for an unsaturated polyester resin.

SYNPOL Tradename for styrene–butadiene rubber.

SYNRES Tradename for an unsaturated polyester resin.

SYNTHAPRET Tradename for polycarbamoyl-sulphonate.

SYNTHETIC FIBRE A man-made fibre produced by spinning (wet, dry or melt) from a synthetic polymer. Now the most important fibre type for textile and other purposes. The most important examples are nylons, polyester (i.e. polyethylene terephthalate) and polyacrylonitrile, although many other polymers are used, including polyvinyl chloride, polyvinylidene chloride, polypropylene, polyurethanes and polyvinyl alcohol.

SYNTHETIC POLYMER A polymer produced by polymerisation of its monomer or monomers by a chemical reaction controlled by man, as opposed to a polymer produced in nature by biosynthesis giving a natural polymer. The most important commercial materials that are based on polymers (plastics, rubbers, fibres, coatings and adhesives) use synthetic polymers, although natural polymer fibres are still important. Natural polymers which have been chemically modified to different polymers for subsequent use are man-made polymers. Synthetic polymers for use as plastics are sometimes referred to as synthetic resins.

SYNTHETIC RESIN (Resin) Alternative name for a synthetic polymer which is to be used as the basis for a plastic material.

SYNTHOMER Tradename for emulsion styrene-butadiene rubber.

T

T (1) Symbol for threonine.

(2) Symbol for a trifunctional unit in a polyorganosiloxane, i.e. for the branched unit

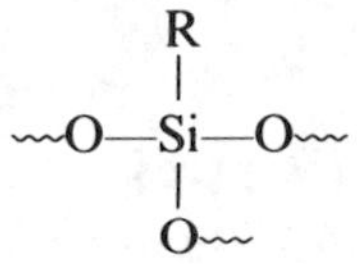

Its presence thus provides the potential for forming crosslinks. Derived from the use of a trichlorosilane monomer.

TACK (1) The ability of a material to stick to a surface on momentary contact and then to resist separation. Unvulcanised rubbers frequently show high tack. The rubber must be of sufficiently low viscosity, and hence of low molecular weight, to flow and wet the adherend under the low pressure applied. However, viscosity must be high enough so that there is some elastic resistance to separation. Tack is commonly assessed subjectively simply by applying finger pressure.

(2) (Self-adhesion) (Autohesion) The tendency of a rubber to stick to itself rather than to other surfaces. High tack is useful in building up rubber products from several layers (or plies) which must adhere to each other. Natural rubber and chloroprene rubber generally show high tack, especially if of low molecular weight. Tack may be increased by the addition of a tackifier such as coumarone–indene polymer, rosin or a phenol–formaldehyde resin.

TACRYL Tradename for a polyacrylonitrile fibre.

TACTICITY (Stereoregularity) The occurrence of a regularity of configurational isomeric units in a polymer whose repeat unit is capable of exhibiting stereochemical configurational isomerism. This is possible when the repeat unit contains an asymmetric, or pseudo-asymmetric, centre, as in a vinyl polymer $-\!\!\left[CH_2C^*HX\right]_n$ where C^* is asymmetric. Two types of regularity are recognised, one in which the sequences have the same configuration (isotactic polymer) and one in which the configuration alternates (syndiotactic polymer). Polymers with no such regularity are atactic polymers. Such polymers do not usually have completely regular structures, but rather they have long sequences of regularity. The tacticity may be analysed experimentally by the use of NMR spectroscopy in terms of the relative configurations of dyads, triads, tetrads or pentads. Since tacticity is an aspect of configurational isomerism, it is determined by the bond making processes during polymerisation and is highly dependent on the polymerisation conditions, especially on the type of polymerisation initiator or catalyst used.

TACTIC POLYMER Alternative name for stereoregular polymer.

TAIL EFFECT The presence of appreciable amounts of low molecular weight components in the fractions obtained by polymer fractionation. It is more serious in fractional precipitation than in fractional solution and arises from the intrinsic solubilities of the different molecular weight species, rather than as a result of the experimental method.

TAIL-TO-TAIL STRUCTURE A structure of the type —CHX—CH$_2$—CH$_2$—CHX— in a vinyl polymer or related polymer. It is one of the possible positional isomers, whose formation must follow the formation of a head-to-head link during chain polymerisation.

TAKA-AMYLASE A crystalline enzyme glycoprotein obtained from the mould *Aspergillus aryzae*. The carbohydrate is linked to aspartic acid protein units and consists largely of D-mannose, with some D-xylose, 2-amino-2-deoxy-D-glucose and D-galactose units.

TAKANAYAGI MODEL A model useful in explaining the mechanical behaviour of composite materials, analogous to the spring and dashpot mechanical relaxation models. In the simplest models the two phases comprising the composite are considered to be combined in series (the iso-stress model) or in parallel (the iso-strain model). Combinations of series and parallel elements often better represent the composite behaviour, e.g. for incompatible polymer blends. The models have been applied to other types of composites including crystalline (and therefore two phase) polymers and semi-compatible blends.

TAKTENE Tradename for high or medium *cis*-polybutadiene.

TALC A crystalline layer silicate mineral of approximate formula $Mg_6(Si_4O_{10})_2(OH)_4$. It consists of a sheet of brucite $(Mg(OH)_2)$ sandwiched between two layers of SiO_4 tetrahedra joined as rings, such that two of every three hydroxyls are replaced by oxygen atoms of the tetrahedra. The interlayer forces are weak so the material is a soft mineral. It is useful as a filler in some plastic materials, especially polypropylene.

TALL OIL An oil obtained from the Kraft process of wood pulp manufacture, consisting of a mixture of rosin and fatty acids. It is a semi-drying oil which is useful in alkyd resins.

TAN δ (1) Alternative name for tangent of the loss angle.
(2) Alternative name for dielectric loss tangent.

TANGENT OF THE LOSS ANGLE (1) (Tan δ) (Damping factor) (Dissipation factor) (Loss tangent) The tangent of the angle by which the strain lags behind the stress in dynamic mechanical behaviour. $\tan \delta = M''/M'$, where M'' and M' are the loss and storage moduli respectively. It is therefore a measure of the ratio of the energy dissipated to the energy stored during a complete cycle of loading and unloading, i.e. it is a measure of damping. Tan δ has low values at low and at high frequencies, passing through a maximum in the transition zone. For a Maxwell model material, $\tan \delta = 1/\tau\omega$, where τ is the relaxation time and ω is the frequency. This model predicts, incorrectly, that tan δ falls continually with temperature. More complex, models, e.g. standard linear solid models, predict actual behaviour more correctly. At low frequencies tan δ is conveniently determined by the torsion pendulum method, where it is given by $\tan \delta = (\Delta/\pi)/(1 - \Delta^2/4\pi^2)$, where Δ is the logarithmic decrement. At higher frequencies tan δ is determined by forced resonance vibration methods, such as the vibrating reed method, or by a non-resonance method.
(2) Alternative name for dielectric loss tangent.

TAPERED BLOCK COPOLYMER Alternative name for gradient block copolymer.

TAPPI HOLOCELLULOSE A holocellulose produced by a standard method of isolation from plant tissue consisting of alternate treatments with chlorine gas and extractions with methanolamine.

TAYLOR–OROWAN DISLOCATION Alternative name for edge dislocation.

TBA Abbreviation for torsional braid analysis.

TBLS Abbreviation for tribasic lead sulphate.

TBP Abbreviation for tributyl phosphate.

TCNQ Abbreviation for tetracyanoquinodimethane.

TCP Abbreviation for tricresyl phosphate.

TCR Abbreviation for technically classified rubber.

TDI Abbreviation for tolylene diisocyanate.

TDI INDEX The tolylene diisocyanate content of a polyurethane formulation. A TDI index of 100 means that there is an exact stoichiometric equivalence of hydroxyl groups and isocyanate groups in the system. A 5% excess of isocyanate groups, which is often used to take account of isocyanate side reactions, would have a TDI index of 105.

TEARING ENERGY Alternative name for tear strength.

TEAR STRENGTH (Tearing energy) A measure of the energy required to propagate a tear in a flexible material such as a thin plastic film or elastomer. Somewhat analogous to the impact strength of brittle polymers. For its determination, typically a cut is made in an edge of a specimen and the force required to propagate the cut is measured by means of a tensile test. The sample is a strip or may be in the form of a pair of trousers (the trousers tear test). The tear strength is usually expressed as the ratio of the maximum load to the thickness of the specimen.

TECHNICALLY CLASSIFIED RUBBER (TCR) A grade of raw natural rubber classified according to its rate of vulcanisation. This can vary according to the conditions of coagulation of the rubber from the latex. Differences in cure rate are largely due to the variation in the non-rubber hydrocarbon content.

TECHNOFLON FOR Tradename for vinylidene fluoride–hexafluoropropylene copolymer.

TECHNOFLON S Tradename for vinylidene fluoride–1-hydropentafluoropropylene copolymer.

TECHNOFLON T Tradename for vinylidene fluoride–tetrafluoroethylene–1-hydropentafluoropropylene terpolymer.

TECHNYL Tradename for nylons 6, 66 and 610.

TECHSTER E Tradename for a polyethylene terephthalate moulding material.

TECHSTER T Tradename for polybutylene terephthalate.

TEDLAR Tradename for polyvinyl fluoride.

TEDUR Tradename for poly-(p-phenylene sulphide).

TEFLEX Tradename for tetrafluoroethylene–hexafluoropropylene copolymer.

TEFLON Tradename for polytetrafluoroethylene.

TEFLON EPE Tradename for tetrafluoroethylene–perfluoropropylvinyl ether copolymer, but of lower T_m value (295°C) than Teflon PFA.

TEFLON FEP Tradename for tetrafluoroethylene–hexafluoropropylene copolymer.

TEFLON PFA Tradename for tetrafluoroethylene–perfluoropropylvinyl ether copolymer.

TEFZEL Tradename for tetrafluoroethylene–ethylene copolymer.

TEGO Tradename for phenol–formaldehyde polymer.

TEGOFAN Tradename for chlorinated rubber.

TEICHOIC ACID A polymer of glycerol phosphate or ribitol phosphate. The former types are found in cell walls and membranes of certain Gram-positive bacteria and have the polyol residues linked through phosphodiester groups and may contain D-alanine side groups and glycosidically linked sugar side groups, e.g. glucopyranose in certain strains of *Streptococcus*. The sugar units may also be present in the polymer chain, e.g. *N*-acetyl-D-glucosamine units in *Micrococcus*, 1,3-linked to glycerol through phosphodiester bonds. The ribitol teichoic acids are similar to the glycerol acids but only occur in cell walls and always contain glycosidically bound carbohydrate side groups, e.g. the single 4-*O*-β-D-glucopyranose units in the acid from *Bacillus subtilis*.

TEKLAN Tradename for a modacrylic fibre consisting of 50% acrylonitrile and 50% vinylidene chloride.

TELCAR Tradename for a thermoplastic polyolefin rubber.

TELECHELIC POLYMER A polymer with deliberately introduced end groups of a specific desired type. Step-growth polymers which have reactive end groups arising from the normal monomer functional groups, are not considered as telechelic polymers. Telechelic chain polymers may be prepared by reacting, and therefore quenching, a living polymer with a specific reagent. Thus a carboxy- or hydroxy-terminated polybutadiene is formed by reaction of living polymer with carbon dioxide or ethylene oxide respectively. They are also formed using an initiator containing the desired group, or by chain transfer using an appropriate chain transfer agent. Such polymers may undergo subsequent further reaction through their end groups. Thus chain extension can lead to block copolymer formation. Reaction with tri- or polyfunctional reagents causes crosslinking. By using this reaction suitable low molecular weight liquid polymers can be employed as liquid rubbers. The best known materials of this type are the carboxy-, hydroxy- and thiol-terminated polybutadienes. Such polymers are best prepared by anionic living polymerisation with a difunctional catalyst. The resultant two living polymer chain ends are then reacted with an appropriate reagent to obtain the desired functional groups.

TELENE   Tradename for copolymers of norbornene.

TELESCOPIC FLOW   Flow of a fluid through a straight tube with a circular cross-section, such that each infinitesimally thin liquid cylinder of radius r coaxial with the tube moves rigidly parallel to the tube axis with a velocity (v) which depends only on r and is zero at the tube wall and at a maximum at the tube axis. If v depends linearly on r^2 then the flow is Poiseuille flow. If v is independent of r over some range of r then the flow is plug flow.

TELOGEN   An active chain transfer agent which is used in the formation of telomers by telomerisation. Fragments of the telogen become incorporated as end groups in the telomer. Commonly used telogens are carbon tetrachloride and chloroform.

TELOMER   An oligomer formed by chain polymerisation in which extensive chain transfer has occurred due to the presence of a chain transfer agent, so that the telomer contains fragments of the transfer agent as end groups. A widely studied example is the formation of ethylene telomers by free radical polymerisation of ethylene in the presence of carbon tetrachloride:

$$\dot{R} + xCH_2{=}CH_2 \rightarrow R{-}(CH_2CH_2)_{x-1}CH_2\dot{C}H_2 \xrightarrow{CCl_4}$$
$$R{-}(CH_2CH_2{-})_x Cl + C\dot{C}l_3$$

$$\dot{C}Cl_3 + yCH_2{=}CH_2 \rightarrow$$
$$CCl_3{-}(CH_2CH_2{-})_{y-1}CH_2\dot{C}H_2 \xrightarrow{CCl_4}$$
$$CCl_3(CH_2CH_2)_yCl + \dot{C}Cl_3$$

Typically x and y have values of 1–6. If the telomers can be separated, then they form useful intermediates for the synthesis of α,ω-disubstituted hydrocarbons. These can be converted to other useful products, e.g. to α,ω-amino acids useful as nylon monomers. Similar telomers of vinyl acetate provide useful coating materials.

TELOMERISATION   Polymerisation carried out in the presence of a large amount of an active chain transfer agent (the telogen) so that only low molecular weight polymers (the telomers) are produced, which have end groups that are fragments of the telogen. Such end groups may subsequently be chemically converted to give useful α, ω-disubstituted molecules. The classic example is the free radical telomerisation of ethylene with carbon

tetrachloride as the telogen:

$$\dot{R} + xCH_2{=}CH_2 \rightarrow R{-}(CH_2CH_2)_{x-1}CH_2\dot{C}H_2 \xrightarrow{CCl_4}$$
$$R{-}(CH_2CH_2)_x Cl + \dot{C}Cl_3$$

$$\dot{C}Cl_3 + yCH_2{=}CH_2 \rightarrow$$
$$CCl_3{-}(CH_2CH_2)_{y-1}CH_2\dot{C}H_2 \xrightarrow{CCl_4}$$
$$CCl_3{-}(CH_2CH_2)_yCl + \dot{C}Cl_3$$

Typically x and y have a distribution of values in the range 1–6, so that separation of pure compounds is difficult. However, reaction conditions can be optimised for the production of a particular telomer. The telomers of ethylene can be converted to α,ω-amino acids useful as monomers for nylons. Similar telomers of vinyl acetate are useful as surface coatings.

TEMPERATURE GRADIENT FRACTIONATION Alternative name for Baker–Williams fractionation.

TEMPERATURE–TIME SUPERPOSITION   Alternative name for time–temperature superposition.

TEMPLATE POLYMERISATION (Matrix polymerisation) The polymerisation of monomer (or monomers) attached in some ordered fashion to a template to produce a polymer with a precisely determined structure, (e.g. order of comonomer units or stereoregularity) complementary to that of the template. Frequently, enhanced rates of polymerisation also result. Biologically, protein synthesis is a template polymerisation, the template being the DNA. However, synthetic template polymerisations have also been studied, e.g. the polymerisation of α-amino acid N-carboxyanhydrides on polypeptide templates, 4-vinylpyridine on polystyrene-4-sulphonic acid, and methylmethacrylate on polymethylmethacrylate.

TENACITY   The tensile strength of a fibre or yarn expressed as the breaking load in grams per denier or decitex. The tenacity and conventional tensile strength differ by the factor of the fibre density. Typically tenacities fall in the range 1–10 g denier^{-1}.

TENASCO   Tradename for a high tenacity rayon.

TENAX   Tradename for poly-(2,6-diphenyl-1,4-phenylene oxide).

TENITE   Tradename for cellulose acetate, cellulose nitrate and polybutylene terephthalate.

TENITE BUTYRATE   Tradename for cellulose acetate butyrate.

TENITE POLYETHYLENE   Tradename for low density polyethylene.

TENITE POLYTEREPHTHALATE   Tradename for a polyethylene terephthalate moulding material.

TENITE PROPIONATE Tradename for cellulose acetate propionate.

TENSILE CREEP COMPLIANCE (Extensional creep compliance) (Young's creep compliance) Symbol $D(t)$. The creep compliance for deformation in tension. Often very conveniently measured by means of a simple tensile test, but not of such great theoretical significance as the shear creep compliance, since tensile deformation occurs with a change in volume as well as with a change in shape. However, when Poisson's ratio is near 0·5 or at small deformations, the volume change may be negligible and the shear and tensile compliances are related by $D(t) = J(t)/3$.

TENSILE FLOW Alternative name for elongational flow.

TENSILE IMPACT STRENGTH The impact strength of a material measured using tensile loading, with a tensile testing machine operating at a very fast rate.

TENSILE LOSS COMPLIANCE Symbol D''. The loss compliance for tensile deformation.

TENSILE LOSS MODULUS Symbol E''. The loss modulus for tensile deformation.

TENSILE MODULUS Alternative name for Young's modulus.

TENSILE STORAGE COMPLIANCE (Dynamic tensile compliance) Symbol D'. The storage compliance for tensile deformation.

TENSILE STORAGE MODULUS (Dynamic tensile modulus) Symbol E'. The storage modulus for tensile deformation.

TENSILE STRENGTH (Ultimate tensile strength) Symbol σ_B or $\hat{\sigma}$. The tensile load at fracture divided by the initial, undeformed cross-sectional area. For brittle and elastomeric materials fracture occurs at the maximum applied load, without significant plastic flow. The tensile strength in brittle fracture is often called the brittle strength. However, for tough materials, yielding will precede fracture and a maximum load may be at the yield point rather than at break.

The tensile strength is both temperature and strain rate dependent. As temperature increases so tensile strength decreases and as strain rate increases so does tensile strength; the effects are greater the more ductile the material. Tensile strength also increases with increasing molecular weight. For polydisperse polymers, it is not certain as to which average molecular weight should be used to express this dependency. Correlations with viscosity are often better, indicating that the role of entanglements may be significant. For crystalline polymers the molecular weight effect is less.

For rubbers, an increase in the crosslink density initially increases the tensile strength, but beyond a maximum value it subsequently decreases. Theories of vulcanised rubbers predict that $\sigma_B \propto v_e^{2/3}$, or $\sigma_B \propto v_e$ for low degrees or crosslinking; often however, $\sigma_B \propto v_e^{0.5-0.6}$ (where v_e is the effective number of crosslinked chains per unit volume).

Crosslinking of rigid polymers has little effect on σ_B, unless the initial, uncrosslinked molecular weight is low, as it often is in thermosetting polymers. Orientation can considerably increase σ_B in the direction of orientation but σ_B is reduced perpendicular to this direction. Extreme orientation, especially in highly drawn fibres, can produce enormous increases in σ_B. In the most favourable cases strengths approaching the theoretical strengths for perfectly formed crystals have recently been approached, in particular, in ultradrawn polyethylene.

For many polymers, σ_B is approximately 300 MPa at very low temperatures, i.e. in the brittle state. Although this is a reasonable value for the rupture of intermolecular forces, it more likely arises from the scission of chain bonds, since molecular motions will be severely restricted at these very low temperatures. However, calculations of the theoretical value of σ_B for homogeneous materials suggest values of about $E/10$ (where E is the tensile modulus) giving somewhat higher values (300–1200 MPa) for σ_B. Values can approach the theoretical values in highly crosslinked or oriented samples. Typically, thermoplastic (linear) polymers have values of the order of 5–50 MPa and thermoset polymers have values of 50–150 MPa. For rubbers σ_B is around 1 MPa.

TENSILE VISCOSITY Alternative name for elongational viscosity.

TENSOLITE Tradename for rubber hydrochloride.

TERACOL Tradename for polytetrahydrofuran.

TERAL Tradename for polyoxymethylene copolymer.

TEREBEC Tradename for a polyester–imide synthesised by polycondensation between a hydroxy-terminated polyester and a iacid diimide from trimellitic anhydride and m-phenylenediamine.

TEREPHTHALIC ACID

$$HOOC-\langle\bigcirc\rangle-COOH$$

Sublimes at 300°C.

Prepared by the high temperature oxidation of p-xylene, with air or nitric acid. The monomer for the preparation of terephthalate polyesters. Although polyethylene terephthalate can be prepared by the esterification of ethylene glycol, ester interchange with dimethyl terephthalate is preferred, owing to its easier purification and greater miscibility with the glycol.

TERGAL Tradename for a polyethylene terephthalate fibre.

TERITAL Tradename for a polyethylene terephthalate fibre.

TERLENKA Tradename for a polyethylene terephthalate fibre.

TERLON Tradename for poly-(p-benzamide).

TERLURAN Tradename for acrylonitrile–butadiene–styrene copolymer.

TERMINAL CHAIN In a crosslinked network polymer, particularly a vulcanised rubber, a polymer chain bound to the network at a crosslink at one of its ends, but not at the other end, i.e. it has a free end. Thus on deformation of the network, although the conformation of the terminal chain may become temporarily altered, after relaxation it will contribute nothing to the elastic recovery of the network on removal of the deforming force.

TERMINAL RELAXATION TIME The dominant relaxation time in the terminal zone of the frequency scale of the regions of viscoelastic behaviour. It is thus the time required to achieve steady flow under constant stress or for recovery after removal of stress. In terms of the Rouse–Bueche–Zimm theory it is given by, $\tau_1 = 6\eta_0 M/\pi^2 \rho RT$, where η_0 is the steady state viscosity, M is the molecular weight, ρ is the density, R is the universal gas constant and T is the temperature.

C-TERMINAL RESIDUE The α-amino acid residue at the end of a peptide, polypeptide or protein chain containing the free carboxylic acid group, i.e. the group ⏤NHCHRCOOH usually existing as ⏤NHCHRCOO⁻. It is identified and determined by hydrazinolysis, which converts all other residues to hydrazides, but leaves the C-terminal residue unchanged. It is then separated by ion-exchange chromatography and identified by reaction with dinitrofluorobenzene or dansyl chloride. It may be determined enzymically using carboxypeptidase, which removes amino acids from the C-terminal residue sequentially. Thus the first amino acid to appear is the original C-terminal residue.

N-TERMINAL RESIDUE The α-amino acid residue at the end of a peptide, polypeptide or protein chain containing the free amino group, i.e. the group $H_2NCHRCO$⏤ which usually exists as $H_3\overset{+}{N}CHRCO$⏤. It may be determined by conversion to a derivative using dinitrofluorobenzene, dansyl chloride or the cyanate method, followed by complete hydrolysis and identification of the free amino acid derivative by electrophoresis or paper chromatography. The Edman degradation method may also be used. In a protein the number of such derivatives identified must equal the number of polypeptide chains originally present. However, sometimes the N-terminal residue is present in the protein as a derivative, e.g. it is acetylated, and does not react with the reagent. Furthermore, functional groups other than the N-terminal NH_2 may form derivatives with the reagent.

TERMINAL ZONE Alternative name for the viscous flow region of viscoelastic behaviour.

TERMINATION (Chain termination) In chain polymerisation, the reaction step in which an active centre disappears (termination of a kinetic chain) and growth of the polymer molecule ceases—it becomes a dead polymer molecule. Also sometimes used to mean merely the termination of growth of a polymer molecule with the active centre persisting after transfer to another molecule. Termination of a kinetic chain occurs by interactions between two active centres. Since such interactions are unlikely on electrostatic grounds in ionic polymerisation, termination in this sense in these polymerisations is often absent. In free radical polymerisation, termination is either by combination or, less commonly, by disproportionation between two polymer active centres (or it can be by a mixture of both processes). Values of k_t (the termination rate constant) are typically 10^6–10^8 litre $(\text{mol s})^{-1}$. At high conversion the termination rate may decrease due to an increase in viscosity or as a result of occlusion, causing an increase in the rate of polymerisation (autoacceleration). Primary radical termination may occur when high initiator concentrations are used.

TERMONOMER One of the three monomers involved in a terpolymerisation.

TERNARY COPOLYMER Alternative name for terpolymer.

TERPENE RESIN A low molecular weight hydrocarbon resin obtained by heating a terpene (α- or β-pinene or dipentene) with $AlCl_3$ in xylene at about 40–50°C. The resins are useful in pressure sensitive adhesives, hot-melt adhesives, sealants and coatings.

TERPOLYMER · (Ternary copolymer) (Tripolymer) A copolymer which contains three different repeat units. It is comparatively rare compared with the enormous number of binary copolymers studied, but some important commercial copolymers are of this type, including acrylonitrile–butadiene–styrene and ethylene–propylene–diene monomer copolymers.

TERPOLYMERISATION A copolymerisation involving three different comonomers, yielding a terpolymer.

TERTIARY NUCLEATION Nucleation in which the initial step is to start a new row of crystal growth along an edge, so that two faces of the growth unit are in contact with the preformed crystal.

TERTIARY STRUCTURE The overall three-dimensional folded shape of a protein polypeptide. Tertiary structure is most evident in globular proteins where compact, roughly spherical, folded molecular shapes occur. Fibrous proteins usually occur as elongated molecular shapes which result from the fact that their polypolypeptides exist almost completely in regular conformations. The tertiary structure adopted is ultimately determined by the amino acid sequence of the polypeptide chain, i.e. the primary structure, through the operation of a combination of ionic, hydrogen and hydrophobic bonding forces and sometimes disulphide bridges. Hydrophobic bonding is now thought to be the most important factor determining tertiary structure.

A protein in its native conformation, in its usual aqueous environment, has most of its hydrophobic α-substituents located in the interior of the molecular coil, whereas the hydrophilic basic and acidic amino acid α-substituents are on the outside of the coil. The operation of these stabilising bonding forces is finely balanced so that denaturation readily occurs when conditions alter to upset this balance (e.g. changes in pH, temperature or salt concentration).

The precise tertiary structural conformation adopted is essential for the biological activity of the protein. Although this conformation is a precise one, and the same conformation is adopted by all protein molecules of a given type in their natural environment, it is not regular; however, segments of the polypeptide often adopt a regular conformation such as the α-helix or the β-conformation. These segments are separated by randomly coiled segments to give the overall folded three-dimensional structure.

The complete tertiary structure may be determined by X-ray diffraction, although the analysis of the diffraction pattern is difficult due to the irregular molecular shape. Nevertheless over 50 protein structures have been so determined. Notable examples are myoglobin (the first to be determined, although not considered to be typical due to its high α-helix content), haemoglobin (the first oligomeric protein structure to be determined), lysozyme, ribonuclease, chymotrypsin, carboxypeptidase A and cytochrome C. X-ray analysis is considerably assisted by introducing heavy metal atoms into the protein to provide reference points. By using such methods a resolution of structure to the 0·2 nm level is possible.

TERYLENE Tradename for a polyethylene terephthalate fibre.

TET Abbreviation for triethylenetetramine.

TETD Abbreviation for tetraethylthiuram disulphide.

TETORON Tradename for a polyethylene terephthalate fibre.

TETRAALKYLTITANATE Tradename: Tyzor. A type of coupling agent, of chemical structure $(RO)_4Ti$, for glass and other reinforcing fillers in polymer composites. Often R is butyl or isopropyl. In use the titanate is probably hydrolysed and the resultant hydroxyl groups react with, or hydrogen bond to, the glass surface to improve adhesion to the polymer.

TETRABROMOPHTHALIC ANHYDRIDE

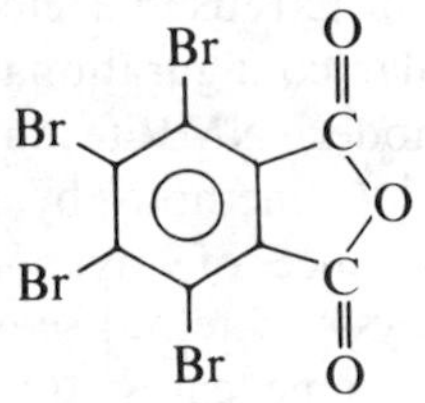

M.p. 280°C.

Useful, in place of phthalic anhydride, in the formation of unsaturated polyester resins with improved fire resistance.

TETRACARBOXYLIC DIANHYDRIDE Alternative name for dianhydride.

TETRACHLOROPHTHALIC ANHYDRIDE

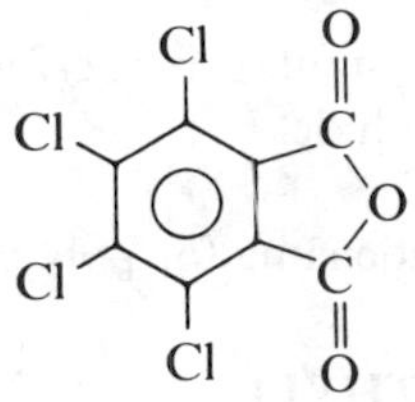

M.p. 225°C.

Useful, in place of phthalic anhydride, in the formation of unsaturated polyester resins with improved fire resistance.

TETRACYANOQUINODIMETHANE (TCNQ)

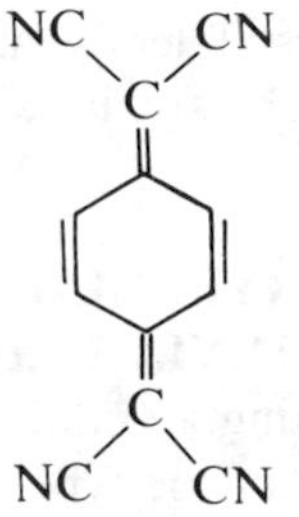

I

A very strong electron acceptor molecule forming a radical anion:

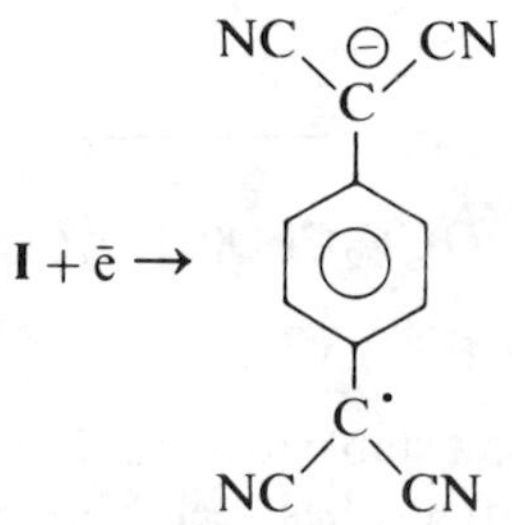

$$I + \bar{e} \rightarrow$$

It forms highly electrically conducting radical-ion salts with organic cations and is the counter-ion for the most widely studied radical-ion polymers.

TETRACYCLONE Alternative name for tetraphenyl-substituted biscyclodieneones, such as a phenylated biscyclopentadienone.

TETRAD A sequence of four repeat units in a polymer molecule. Some aspects of polymer microstructure may be analysed in terms of tetrads in a similar way to dyads and triads. In particular, configurational tetrad sequences may be revealed by modern NMR techniques such as high field (220 MHz) or ^{13}C methods by observing the α-methylene group resonances of vinyl polymers of the type $-\!\!\!\!\left[\text{CH}_2\text{CHX}\right]_n$. Analysis of tetrad sequences in copolymers has shown, in some cases, that propagation in copolymerisation depends on higher order Markov chain statistics.

TETRADECANOIC ACID Alternative name for myristic acid.

TETRAETHYLTHIURAM DISULPHIDE (TETD)

$$(\text{C}_2\text{H}_5)_2\text{N}\underset{\underset{\text{S}}{\|}}{\text{C}}\text{SS}\underset{\underset{\text{S}}{\|}}{\text{C}}\text{N}(\text{C}_2\text{H}_5)_2 \qquad \text{M.p. } 70°\text{C.}$$

An ultraaccelerator similar in performance to tetramethylthiuram disulphide.

TETRAFLON Tradename for polytetrafluoroethylene.

TETRAFLUOROETHYLENE

$$\text{CF}_2\!\!=\!\!\text{CF}_2 \qquad \text{B.p. } -76\cdot3°\text{C.}$$

Prepared by the thermal cracking at about 700°C of chlorodifluoromethane, itself produced by reaction of hydrogen fluoride with chloroform. The monomer for the production of polytetrafluoroethylene and its copolymers, by free radical polymerisation. It very readily polymerises, sometimes violently, and must therefore be stored and handled with care, usually with the aid of an inhibitor.

TETRAFLUOROETHYLENE–CARBALKOXYPER-FLUOROALKOXYVINYL ETHER COPOLYMER
A terpolymer containing tetrafluoroethylene and perfluorovinyl ether units of the type

$$\sim\!\!\left[\begin{array}{l}\text{CF}_2\text{—CF—}\\ \quad\quad|\\ \quad\quad\text{O(CF}_2)_3\text{COOCH}_3\end{array}\right]$$

and

$$\left[\begin{array}{l}\text{CF}_2\text{—CF—}\\ \quad\quad|\\ \quad\quad\text{OCF}_2\text{CFO(CF}_2)_3\text{COOCH}_3\\ \quad\quad\quad\quad|\\ \quad\quad\quad\quad\text{CF}_3\end{array}\right]$$

The polymer is heat stable to about 320°C and is useful as a film material (after it has been converted to its sodium salt) in a similar way to Nafion, as a permselective membrane for electrolytic cells, especially for the chloralkali process of brine electrolysis.

TETRAFLUOROETHYLENE–ETHYLENE COPOLYMER (ETFE) (Poly-(tetrafluoroethylene-co-ethylene)) Tradenames: Aflon, Hostaflon ET, Tefzel. A largely alternating copolymer of tetrafluoroethylene and ethylene, similar in properties to polytetrafluoroethylene but with higher impact and tensile strengths. It is also melt processable. Typically, it crystallises to about 50–60% with a T_m value of 270°C, but has a lower use temperature at about 170°C. It is used largely as an electrical insulation material.

TETRAFLUOROETHYLENE–HEXAFLUOROPROPYLENE COPOLYMER (FEP) (Poly-(tetrafluoroethylene-co-hexafluoropropylene)) (Fluorinated ethylene–propylene copolymer) Tradenames: Fluon FEP, Neoflon, Teflex, Teflon FEP. A random copolymer similar in properties to polytetrafluoroethylene, but with a T_m value of about 290°C and melt processable at 300–380°C. It has a somewhat better impact strength but a lower service life temperature than polytetrafluoroethylene.

TETRAFLUOROETHYLENE–PERFLUORO-METHYLVINYL ETHER COPOLYMER (PFE) (Poly-(tetrafluoroethylene-co-perfluoromethylvinyl ether)) (Perfluorinated elastomer) Tradenames: Kalrez, Hostaflon TFA. A fully fluorinated fluoroelastomer having about 60% —CF_2CF_2— and 40%.

$$\begin{array}{l}\text{—CF}_2\text{CF—}\\ \quad\quad|\\ \quad\quad\text{COCF}_3\end{array}$$

groups. It has one of the best thermo-oxidative stabilities of all the fluoroelastomers with a continuous use temperature in air of up to 260°C, or with intermittent use, to 315°C. It also has excellent resistance to swelling by a wide variety of liquids and good resistance to chemical attack. It has good high temperature compression set resistance. Its T_g value is about $-10°$C. The copolymers also contain about 2% of another fluorinated comonomer for vulcanisation with a diamine or bisphenol.

TETRAFLUOROETHYLENE–PERFLUOROPROPYLVINYL ETHER COPOLYMER (Poly-(tetrafluoroethylene-co-perfluoropropylvinyl ether)). Tradenames: Teflon PFA, Teflon EPE. A perfluoroalkoxy polymer, which, being fully fluorinated, has even higher thermal and thermo-oxidative stabilities than the otherwise similar tetrafluoroethylene–hexafluoropropylene copolymers.

TETRAFLUOROETHYLENE–PROPYLENE COPOLYMER Tradenames: Aflas, Xenox. Produced by free radical redox polymerisation or by γ-irradiation. The polymer is an amorphous elastomer, crosslinked by peroxide curing, which is enhanced using triallyl cyanurate. Vulcanisates have good mechanical properties and are usable to about 200°C, being useful for seals, diaphragms and tubes especially in hot, corrosive environments.

TETRAFLUOROETHYLENE–SULPHONYLFLUORIDEVINYL ETHER COPOLYMER

(Poly-(tetrafluoroethylene-co-sulphonylfluoridevinyl ether)) Tradename: XR-resin. A copolymer of tetrafluoroethylene with about 10–20 mol% of

$$\sim\left[CF_2CF-\underset{\underset{CF_3}{|}}{O}(CF_2CFO)_n CF_2CF_2SO_2F\right]\sim$$

units. The copolymer can be melt processed in the normal way (unlike polytetrafluoroethylene). It is hydrolysed to the acid, i.e. by converting the $-SO_2F$ groups to $-SO_2H$ groups by treatment with base followed by acidification. The acid form of the polymer (tradename Nafion) is produced commercially as a permselective film for cations, but not for anions, for use in electrochemical processes.

TETRAHYDROFURAN (THF)

B.p. 65°C.

Prepared from furfural (itself obtained by hydrolysis and cyclisation of the aldopentoses occurring in various agricultural wastes, such as corncobs) which is pyrolysed to yield furan, which is then hydrogenated:

$$HOCH_2CH(OH)CH(OH)CH(CHO)OH \rightarrow$$

$$\underset{\underset{O}{\diagdown}}{\overset{CH-CH}{\underset{CH}{\|}} \overset{\|}{CH-CHO}} \xrightarrow{-co} \underset{\underset{O}{\diagdown}}{\overset{CH-CH}{\underset{CH}{\|}} \overset{\|}{CH}} \rightarrow \square$$

Alternatively it may be obtained by cyclodehydrogenation of butane-1,4-diol. It gives polytetrahydrofuran by ring-opening polymerisation, e.g. with carbenium ion salts. A good solvent for a wide range of polymers and hence a commonly used solvent for gel permeation chromatography. Especially useful as a solvent for polyvinyl chloride. It is soluble in water and readily forms explosive peroxides.

TETRAHYDROFURFURYL ALCOHOL

B.p. 178°C.

A solvent for cellulose esters and ethers and for many natural resins.

TETRAHYDROFURFURYL OLEATE

$$CH_3(CH_2)_7CH=CH(CH_2)_2COOCH_2$$

B.p. 240°C/5 mm.

A secondary plasticiser for polyvinyl chloride and its copolymers. Also compatible with cellulose esters and ethers, polystyrene and acrylic polymers.

TETRAHYDRONAPHTHALENE

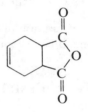

B.p. 200–209°C.

Tradename: Tetralin. Produced by the vapour phase partial hydrogenation of naphthalene. Readily oxidises in air. A solvent for many natural resins, rubbers and polyvinyl chloride. At elevated temperatures (above 80°C) also a solvent for polyethylene.

TETRAHYDROPHTHALIC ANHYDRIDE

M.p. 103–104°C.

Prepared by the Diels–Alder addition of maleic anhydride to butadiene. Useful in the formation of unsaturated polyester resins that may be crosslinked by atmospheric oxidation, in a similar manner to the air drying of alkyd resins. Also useful for curing epoxy resins, especially for producing light coloured products. Unlike phthalic anhydride it does not sublime.

TETRAHYDROTHIOPHENE-1,1-DIOXIDE Alternative name for sulpholane.

TETRAKISHYDROXYMETHYLPHOSPHONIUM CHLORIDE

(THPC) $(HO-CH_2)_4\overset{+}{P}\overset{-}{Cl}$. One of the most widely used phosphorus compounds for imparting a durable flame resistant finish to cellulosic materials, especially cotton. Usually used in combination with an amido compound, e.g. methylolmelamine, with which it forms polymers which are deposited in the cellulose substrate.

TETRALIN Tradename for tetrahydronaphthalene.

TETRAMETHYLENEGLYCOL DIMETHACRYLATE

$$\underset{\underset{CH_3}{|}}{CH_2=CCOO(CH_2)_4OCOC}\overset{CH_3}{\underset{|}{=}CH_2}$$

A typical dimethacrylate monomer useful as the basis of anaerobic adhesives.

TETRAMETHYLENESULPHONE Alternative name for sulpholane.

TETRAMETHYLENE TEREPHTHALATE–POLY-OXYTETRAMETHYLENE TEREPHTHALATE BLOCK COPOLYMER   Alternative name for poly-(tetramethylene terephthalate-b-polyoxymethylene terephthalate).

TETRAMETHYLOLMETHANE   Alternative name for pentaerythritol.

TETRAMETHYLTHIURAM DISULPHIDE   (TMTD)

$$(CH_3)_2NCSSCN(CH_3)_2$$
$$\overset{\parallel}{S}\quad\overset{\parallel}{S}\qquad\text{M.p. }137°C.$$

An ultraaccelerator frequently used in small amounts (about 0·5 phr) as a secondary accelerator. Also used as a vulcanisation agent in sulphurless vulcanisation.

TETRAMETHYLUREA   (TMU)

$$(CH_3)_2NCON(CH_3)_2\qquad\text{B.p. }175°C.$$

A dipolar aprotic solvent useful for the synthesis of aramid polymers and other aromatic and heterocyclic polymers.

TETRAZOPYRENE POLYMER   Alternative name for polytetrazopyrene.

TETRONIC   Tradename for tetrablock copolymers of poly-(propylene oxide-b-ethylene oxide), useful as anionic detergents.

T-EVEN PHAGE   One of the most widely studied groups of phages of the bacterium *Eschericia coli*. The individual members (T2, T4 and T6 being the best known) are closely related serologically. The virion consists of an outer coat (capsid) of protein consisting of a head in which the DNA is encased and a long tail, so that it is often said to be tadpole shaped. The DNA, comprising about 60%, occurs in double helix form and has a molecular weight of about 130 million with about 5000 nucleotides. The DNA is adenine/thymine rich, with about 32–35% of each, and the cytosine is replaced by 5-hydroxymethylcytosine. The hydroxymethyl group is glucosylated in an unusual way, which protects the DNA from enzymes in the infected cell. The protein may consist of about 50 different types of molecule possibly of about 2000 subunits and probably of molecular weight about 40 000, but readily forming dimers. The head contains most, about 90%, of the protein.

TEVILON   Tradename for a polyvinyl chloride fibre.

TEX   A measure of the coarseness of a fibre, yarn or tow, often used in particular for synthetic fibres. The weight in grams of 100 m of fibre, yarn or tow.

TEXICOTE V   Tradename for polyvinyl acetate.

TEXIGEL   Tradename for polyacrylic acid and its salts.

TEXILAC   Tradename for polyvinyl acetate.

TEXIN   Tradename for a partially crosslinked thermoplastic polyurethane elastomer, based on a polyester or polyether polyol/MDI prepolymer and chain extended.

TEXTURE   Alternative name for texture orientation.

TEXTURE ORIENTATION   (Texture) The non-random orientation of crystallites in a polycrystalline sample.

TFE   Abbreviation for tetrafluoroethylene.

TGA   Abbreviation for thermogravimetric analysis.

THERBAN   Tradename for hydrogenated nitrile rubber, of higher heat stability than normal nitrile rubber, but requiring curing by peroxide or high energy radiation.

THERMAL ANALYSIS   One of a range of techniques for determining the temperature dependence of certain properties of materials. Owing to both the commercial and scientific importance of this dependence in polymers, the use of these techniques is often dominated by polymer applications. Some techniques were specially developed for polymer work. The techniques include thermogravimetric analysis, differential thermal analysis, differential scanning calorimetry, thermal volatilisation analysis, dynamic mechanical spectroscopy, torsional braid analysis and thermal optical analysis.

THERMAL BLACK   A type of carbon black produced by the thermal decomposition of natural gas at about 1300°C in the absence of air. Thermal black has little structure and the particles are generally large (150–500 nm) as in the most common FT and MT types. Useful as a rubber reinforcement, although thermal types have less reinforcing effects than furnace blacks. However, it may be used at high loadings. It is used in linings, inner tubes, V-belts and general mechanical goods.

THERMAL BREAKDOWN   Dielectric breakdown resulting from Joule heating (due to the presence of conducting species) or dielectric loss. If sufficient heat is produced then a catastrophic rise in temperature occurs, since the rise in temperature increases conductivity even further and also increases dielectric relaxations. Breakdown follows either as a result of a decrease in electric strength or due to melting or chemical decomposition.

THERMAL DEGRADATION   Degradation induced by exposure to an elevated temperature. For thermally sensitive polymers such as polyvinyl chloride, the temperature involved may be as low as 100°C, but for many other polymers temperatures in the region of 250°C and above are of interest. For most organic polymers rapid degradation occurs well below 500°C, although the better high temperature resistant polymers show little change on prolonged heating to these temperatures. Degradation is often initiated at structural irregularities which lower the thermal stability. High temperature degradation in an inert atmosphere is pyrolysis, whereas in an oxygen-containing atmosphere the degradation becomes thermo-oxidative and more rapid due to oxygen participation in the free radical reactions often involved. Chemically, simple depolymerisation with evolution of monomer may occur, but more frequently polymer molecule breakdown gives a complex mixture of degradation products. The course of thermal degradation may be followed by any of the wide range of thermal analysis techniques.

THERMAL DIFFUSION FRACTIONATION A fractionation method in which a solution of polymer is placed in a temperature gradient between two vertical surfaces, the space between the surfaces being connected to an upper and a lower reservoir. Thermal circulation occurs which causes the polymer to migrate to the lower reservoir, the longer molecules diffusing faster.

THERMAL DIFFUSIVITY A parameter determining the rate of heat exchange of a material with its surroundings, when rapid temperature changes are involved, so that the relevant static parameter (the thermal conductivity) cannot be used. It is defined as the ratio of the thermal conductivity to the product of the density and the specific heat. Polymer melts typically have values of about $10^{-7} \, m^2 \, s^{-1}$.

THERMAL FAILURE Failure during dynamic fatigue caused by the development of a sufficiently high temperature internally in a specimen due to the high mechanical loss over certain ranges of frequency and temperature. Owing to the low thermal conductivity of polymers, the temperature rise can be sufficiently high to raise the specimen temperature above the glass transition temperature, when failure can result from excessive permanent deformation.

THERMAL GRADIENT ELUTION Alternative name for Baker–Williams fractionation.

THERMALLY STIMULATED CURRENT Alternative name for thermally stimulated discharge.

THERMALLY STIMULATED DISCHARGE (Thermally stimulated current) A technique for the study of dielectric relaxation processes by heating a polymer sample that is permanently polarised, i.e. an electret. The sample, contained between the electrodes used for its polarisation, is short circuited through an electrometer, then heated and the discharge current is recorded as a function of temperature.

THERMAL OPTICAL ANALYSIS (TOA) A group of techniques based on the measurement of the variation in light intensity, transmitted by a sample, with temperature. Often polarised light is used with advantage as in thermal depolarisation analysis and depolarised light intensity analysis. The techniques are very sensitive to the detection of melting and glass transitions and crystallisation.

THERMAL POLYMERISATION (1) Free radical polymerisation initiated by thermal homolysis of an added initiator, usually a peroxide or azo compound, to yield free radicals. For peroxides temperatures of 50–90°C are usually suitable, although the more stable peroxides may be used at 100–140°C. Azobisisobutyronitrile is the only commonly used azo compound, usually at 50–70°C.

(2) (Self-initiation polymerisation) A purely thermally initiated polymerisation, without the use of an added initiator and usually occurring by a free radical mechanism. The initiating free radicals are formed by the action of heat on the monomer itself. Often impurities are responsible for initiation, but some monomers, notably styrene, do polymerise slowly in this way even when highly purified. The mechanism is unclear but may involve the bimolecular formation of free radicals:

$$2PhCH=CH_2 \rightarrow Ph\dot{C}H-CH_3 + CH_2=\dot{C}Ph$$
$$or \rightarrow Ph\dot{C}HCH_2-CH_2\dot{C}HPh$$

THERMAL SILICA Alternative name for pyrogenic silica.

THERMAL SOFTENING The softening that occurs due to a rise in temperature on cyclically stressing a material, as a result of hysteresis. It may lead to fatigue failure.

THERMAL STABILISER A stabiliser additive to reduce or eliminate thermal degradation or its effects. Most commonly used with chlorinated polymers (notably polyvinyl chloride) which are particularly thermally sensitive due to their tendency to dehydrochlorinate. Here such stabilisers have the two-fold role of absorbing the hydrochloric acid evolved and of reducing the resulting discolouration of the polymer. A wide range of materials are used for this purpose, including basic lead compounds, metal soaps and complexes and organotin compounds. Other polymer types also sometimes require stabilisation against purely thermal degradation, as opposed to the more commonly encountered thermo-oxidative degradation, e.g. polyacetals require the use of basic compounds to reduce thermal depolymerisation.

THERMAL VOLATILISATION ANALYSIS (TVA) (Evolved gas analysis) A technique for the analysis of the volatile thermal degradation products of a polymer by noting the response of a Pirani gauge, flame ionisation detector or thermal conductivity detector to their presence. By using a series of detectors and traps for condensing the volatiles at different temperatures, mixtures of products may be analysed in terms of fractions boiling within certain temperature ranges. The information obtained is similar to that available from thermogravimetric analysis but the technique has higher sensitivity and is cheaper.

THERMID 600 Tradename for a thermosetting polyimide produced by reaction between benzenetetracarboxylic dianhydride, an aminophenylacetylene and α,α′-dimethoxy-*p*-xylene.

THERMOBALANCE The instrument used for thermogravimetric analysis. Consists of an analytical balance, one arm of which is attached to the sample container by a quartz rod such that the sample is situated in an oven remote from the balance. Alternatively, the sample may be suspended in the oven from a quartz spring balance. The oven may be operated isothermally or at a programmed heating rate.

THERMOELASTIC EFFECT The related effects of temperature (and enthalpy) and elasticity of a material. In the case of rubbers the effects are anomalous compared with other materials. The temperature of a rubber increases on stretching and decreases on retraction. Furthermore if the temperature of a stretched rubber is increased, the rubber contracts, whereas expansion occurs if the rubber is cooled, i.e. the linear coefficient of expansion is negative except at low elongation, where thermoelastic inversion is said to occur.

THERMOELASTIC INVERSION The decrease in tensile force with increase in temperature necessary to maintain a constant length of a rubber sample under tension. Arises only at low (< 10%) elongations; at higher elongations thermoelasticity is observed. Caused by thermal expansion of the rubber, which increases the length in the unstrained state, and thereby reduces the effective elongation.

THERMOELECTRIC METHOD Alternative name for vapour pressure osmometry.

THERMOGRAVIMETRIC ANALYSIS (TGA) A thermal analysis technique whereby the weight of the sample is continuously recorded whilst it is being heated in a thermobalance. Data may be obtained isothermally (static TGA) or at a constant heating rate (dynamic TGA). Widely used in polymer thermal degradation studies—a classical application for which the theory of TGA has been developed in detail to aid analysis of degradation kinetics. Of more limited use in other weight loss applications, e.g. solvent evaporation studies.

THERMOLACTYL Tradename for a blended yarn of polyvinyl chloride and polyacrylonitrile.

THERMOLASTIC Early tradename for a styrene–butadiene–styrene block copolymer.

THERMOLYSIN A heat-stable bacterial protease, which preferentially hydrolyses peptide bonds at which the amino group is contributed by the non-polar amino acids leucine, isoleucine and valine. Therefore it is useful in the sequencing of polypeptides containing no arginine or lysine and also in the sequencing of the protamines, which contain so much arginine and lysine that cleavage by trypsin gives no useful sequence information.

THERMOMECHANICAL ANALYSIS (TMA) A technique for the investigation of the deformation, with a non-oscillating load, of a small sample of a material as a film, fibre or thin disc, as a function of temperature. Several different deformation modes are usually available, such as penetration, expansion, bending, shear flow or fibre torsion. The results are displayed as a thermomechanical. spectrum.

THERMOMECHANICAL SPECTRUM A plot of the variation of a mechanical property against temperature. If the temperature range includes the transition zone of viscoelastic behaviour then the term viscoelastic spectrum may be used. Although any mechanical property may be plotted, since it is frequently the dynamical mechanical behaviour which is determined as a function of temperature, plots are often of storage or loss moduli or of tangent of the loss angle against temperature.

THERMOPLASTIC A plastic material, i.e. one which is either a polymer or based on a polymer, existing below the T_g value (if amorphous) or between the T_g and T_m values (if crystalline), in which the polymer is a linear polymer. The material therefore softens on heating and hardens on cooling and the softening/cooling cycle may be repeated many times. This is in contrast to the other main group of plastic materials, the thermosets, which do not soften on heating. The basis for most processing methods for the production of thermoplastic products thus involves heating the material to soften it, followed by the application of pressure to fuse the material and to shape the product by viscous flow. The majority of plastics materials are thermoplastic, including the largest tonnage 'commodity' plastics polyethylene, polypropylene, polystyrene, polyvinyl chloride, and most 'engineering' plastics.

THERMOPLASTIC ELASTOMER (Thermoplastic rubber) An elastomer which displays the typical high elasticity of a rubber, but without needing to be crosslinked through covalent bonds by vulcanisation. The rubbery polymer chains are tied together instead by physical crosslinks. These crosslinks may usually be destroyed by heating, when the material behaves as a linear polymer, i.e. as a thermoplastic melt. The 'crosslinks' reform on cooling. Thus it is possible to shape a rubber product by simple thermoplastic shaping techniques rather than having to perform the somewhat less convenient normal rubber processing methods, including vulcanisation. The crosslinks are said to be thermoreversible crosslinks. Such tying together of the rubbery chains occurs since the materials are two phase systems in which a stiff phase, consisting of a glassy or crystalline polymer, is dispersed in a rubbery matrix. Most simply the stiff phase is a crystalline polymer merely physically blended with a rubbery polymer. Polypropylene/EPDM blends provide well-known examples of this type. Other thermoplastic elastomers are block copolymers in which hard blocks (or segments) aggregate into domains to provide the crosslinking effect. The hard block may be a high T_g plastic block, such as styrene in SBS, or it may consist of a crystallisable block in a multiblock polymer, as with the tetramethyleneterephthalate blocks in polyether/ester block copolymers. In segmented polyurethane block copolymers hydrogen bonding provides the crosslinking effect.

THERMOPLASTIC POLYAMIDE ELASTOMER Alternative name for polyether block amide.

THERMOPLASTIC POLYETHERESTER Alternative name for polyetherester block copolymer.

THERMOPLASTIC POLYOLEFIN ELASTOMER Alternative name for thermoplastic polyolefin rubber.

THERMOPLASTIC POLYOLEFIN RUBBER (Thermoplastic polyolefin elastomer) (TPO) Tradenames: Dutral TP, Dutralene, ET polymer, Kelburon, Kelprox, Keltan TP, Lavaflex, Moplen, Nordel TP, Santoprene, Somel, Telcar, TPR, Vestolen EM, Vestopren TP, Vistaflex, Uneprene. A type of thermoplastic elastomer usually based on polypropylene blends with ethylene–propylene rubber.

THERMOPLASTIC POLYURETHANE (TPU) A linear, or only slightly crosslinked, polyurethane which may therefore be softened on heating and processed by melt processing methods such as injection moulding. Most TPUs are also elastomeric (thermoplastic polyurethane elastomers) and are segmented polyurethanes consisting of hard and soft segments which phase separate into domains. However, some early polyurethane products (e.g. Durethan U, the linear polymer obtained by reaction of poly-(ethylene adipate) with hexamethylene-diisocyanate are stiff and do not show elastomeric properties, but have been used as plastics and fibres.

THERMOPLASTIC POLYURETHANE ELASTOMER (TPU) Tradenames: Daltomold, Desmopan, Elastollan, Estane, Jectothane, Royalon, Texin. A linear or only slightly crosslinked polyurethane elastomer, which is also a segmented polyurethane in which the hard blocks, consisting of segments rich in urethane or urea groups, are the main source of crosslinks (or the sole source in the case of some products such as Estane), as physical crosslinks. Since the hard blocks soften on heating, the polymers may be processed by injection moulding, extrusion and other melt processing methods. The polymers are formed by reaction of a polyester (e.g. polycaprolactone) prepolymer with a diisocyanate, usually MDI, followed by chain extension with a glycol. Proportions close to stoichiometric equivalence of hydroxyl and isocyanate groups are used so that both high molecular weight polymer and a minimum of allophanate crosslinks (from excess isocyanate) are formed. These materials have the high tear and tensile strengths and good oil and abrasion resistances of the cast polyurethane elastomers, but suffer from high compression set.

THERMOPLASTIC RUBBER (TPR) Alternative name for thermoplastic elastomer.

THERMOREVERSIBLE CROSSLINK A physical crosslink which is destroyed on heating and which reforms again on cooling. Present in thermoplastic elastomers, which enables them to be processed as thermoplastic melts, but which exhibit properties typical of a vulcanised elastomer when cooled.

THERMOSET A polymer that is so extensively crosslinked that on heating it does not significantly soften, unlike a thermoplastic polymer. Moulded plastic products are produced by forming the polymer (or plastic material based on the polymer) as a thermosetting prepolymer which does soften and flow and which on prolonged heating becomes crosslinked to the final crosslinked thermoset structure. Common examples include phenol–formaldehyde, melamine–formaldehyde and urea–formaldehyde polymers, some polyurethanes, unsaturated polyesters and epoxy resins. In the last two cases the final crosslinked structure may be formed using catalysts at ambient temperature, rather than using heat. Nevertheless the products are still regarded as thermoset polymers. However, crosslinked polymers which are not relatively hard and rigid, notably vulcanised rubbers, are not referred to as thermoset polymers.

THERMOSETTING POLYMER A polymer, frequently of low molecular weight, which may be crosslinked by the application of heat or by use of a suitable catalyst (or a combination of both) to a thermoset polymer. Thus rigid thermoset plastic products may be made by forming a heat softened, or liquid, thermosetting prepolymer into the shape required, followed by crosslinking.

THERMOTROPIC POLYMER A polymer which exhibits a transition from a glassy melt or crystalline state to a liquid crystalline state at a characteristic temperature without being diluted with a solvent.

THERMOVYL Tradename for a polyvinyl chloride staple fibre.

THETA POINT The condition that a liquid exists as a theta solvent for a polymer. These theta conditions may be arrived at either by changing the solvent composition, e.g. by adding more good solvent to a binary solvent mixture, or by changing the temperature.

THETA SOLVENT (Flory solvent) A solvent which, at the particular temperature being considered, is at its theta temperature.

THETA TEMPERATURE (Flory temperature) (Ideal temperature) Symbol Θ. The temperature at which, for a given polymer–solvent pair, the polymer exists in its unperturbed dimensions. Under these conditions, the long-range forces between polymer molecular segments which cause contraction are just balanced by the polymer–solvent interactions which cause the polymer molecular coil to expand. At the theta temperature the second virial coefficient becomes zero and the Flory–Huggins interaction parameter is also zero. It is also the critical solution temperature for polymer of infinite molecular weight. Below Θ the polymer chain segments attract one another and the polymer will eventually phase separate. Since there are no interactions between segments at Θ, deviations from ideal solution behaviour also disappear. The theta temperature may be determined by determining the second virial coefficient, e.g. by a colligative property measurement, at temperatures near to Θ for samples of different molecular weight. For each sample the plot of A_2 against temperature will give a curve, and the curves will all intersect at the theta temperature.

THF Abbreviation for tetrahydrofuran.

THIACRYL Tradename for acrylic elastomers.

THIAZOLE (Benzothiazole) A compound containing the structure

where X is a further atom or group. Thiazoles form one of the most important groups of accelerators for the sulphur vulcanisation of rubbers. Examples include mercaptobenzothiazole and dibenzothiazole disulphide.

THIAZOLINONE The intermediate of structure

formed during the Edman method for protein sequence determination, by the hydrolysis of a phenylthiocarbamoyl.

THIIRANE Alternative name for episulphide.

THIN LAYER CHROMATOGRAPHY (TLC) A chromatographic technique in which the stationary phase is spread as a thin layer on an inert carrier, such as a glass plate or thin plastic sheet, rather than being packed into a column. This increases the rate of separation. The method is particularly suitable for the separation of small samples for analytical purposes, but small-scale preparative separations may also be made. It is usually a variation of either the adsorption chromatographic technique (involving the use of, for example, hydroxyapatite for protein separations) or of partition chromatography (using a layer of silica gel or cellulose for separations of fragments of proteins and nucleic acids). It is only rarely used for separations of synthetic polymers.

4,4′-THIO-BIS-(6-t-BUTYL-m-CRESOL) (Bis-(2-methyl-4-hydroxy-5-t-butylphenyl)sulphide)

M.p. 150°C.

Tradename: Santowhite crystals. A chain-breaking, and possibly also a preventive, antioxidant, often used synergistically with carbon black. Widely used in polyolefins.

2,2′-THIO-BIS-(4-METHYL-6-t-BUTYLPHENOL)

M.p. 82–88°C. B.p. 180°C/3 mm.

Tradenames: CAO-6, CAO-4. A chain-breaking, and possibly also a preventive, antioxidant, which forms synergistic mixtures with carbon black. Widely used in polyolefins.

THIOCARBANILIDE

M.p. 149°C.

An early rubber vulcanisation accelerator, now little used due to its low rate of vulcanisation, fast onset of cure and the relatively poor mechanical properties of the vulcanisates.

THIOKOL A Tradename for an early polysulphide rubber based on 1,2-dichloroethane as monomer and with a sulphur rank of four, i.e. the polymer has the strucutre $+CH_2CH_2-S_4+_n$.

THIOKOL B Tradename for a polysulphide rubber made using di-2-chloroethyl ether, thus giving a polymer of repeat unit structure $+CH_2CH_2OCH_2CH_2-S_x+_n$.

THIOKOL FA Tradename for polysulphide copolymer rubber formed from di-2-chloroethyl formal $(CH_2(OCH_2CH_2Cl)_2)$ and ethylene dichloride.

THIOKOL LP Tradename for a range of liquid polysulphide rubbers obtained by cleavage of Thiokol FA or Thiokol ST.

THIOKOL N Tradename for an early polysulphide copolymer rubber formed from ethylene dichloride and propylene dichloride as comonomers.

THIOKOL ST Tradename for a branched polysulphide rubber formed from di-2-chloroethyl formal $(CH_2(OCH_2CH_2Cl)_2)$ with about 2% 1,2,3-trichloropropane as trifunctional branching units.

THIOL ENZYME (Cysteine enzyme) An enzyme in which a cysteine residue is present at the active site and which is essential for catalytic activity of the enzyme. The sulphydryl group forms a thioester bond by reaction with a specific acyl group of the substrate forming an acyl-enzyme. An example is papain.

THIOL-TERMINATED POLYBUTADIENE Tradename: PM polymer. A low molecular weight polybutadiene with thiol end groups. Prepared by anionic polymerisation of butadiene using a difunctional catalyst to produce a living polymer with two living ends. These are then reacted with sulphur, cyclic disulphide or episulphide to give two thiol ends. Useful as a liquid rubber which may be crosslinked through the end groups by reaction with a multifunctional reagent.

THIONIC Tradename for a sulphonated EPDM zinc salt which is a thermoplastic elastomer.

THIOUREA

$$SC(NH_2)_2 \qquad \text{M.p. } 181\text{–}182^\circ C.$$

Prepared either by heating ammonium thiocyanate at about $140^\circ C$, $NH_4CNS \rightarrow SC(NH_2)_2$, or by reaction of cyanamide with ammonium sulphide. It is useful in the formation of thiourea–formaldehyde polymers.

THIOUREA–FORMALDEHYDE POLYMER Tradenames: Beetle, Cibanoid. A polymer formed by reaction of thiourea with formaldehyde in aqueous solution, in a similar manner to the formation of urea–formaldehyde polymer. Initially, water-soluble methylolthioureas are formed, which on further heating condense becoming hydrophobic. Finally, crosslinking occurs to produce a network thermoset polymer. The polymers have higher water resistance but are more brittle than the similar urea–formaldehyde polymers. Some early commercial aminopolymers were urea/thiourea–formaldehyde polymers.

THIRD VIRIAL COEFFICIENT Symbols A_3, C and Γ_3. A coefficient of the virial equation for dilute solution colligative property behaviour, expressing higher order interactions between solute molecules than given by the second virial coefficient (A_2). Frequently it may be neglected in dilute ($<1\%$) solutions, but in more concentrated solutions it may be significant. For polymer solutions a non-zero value causes curvature in the π/c versus c plots, where π is the osmotic pressure and c is the polymer concentration. A value of A_3 is then needed to obtain a linear plot. Frequently A_3 is taken to be about $0.25A_2$.

THIURAM SULPHIDE A compound of structure

$$\underset{\underset{S}{\overset{\|}{}}}{(R_2NC)_2S_x}$$

useful as a vulcanisation accelerator (as an ultra-accelerator) but with short scorch times and rather slow curing. The thiuram disulphides ($x = 2$) may be used as vulcanisation agents in the absence of sulphur itself. Examples include tetramethyl- and tetraethylthiuram disulphide, dipentamethylene tetrasulphide and tetramethylthiuram monosulphide. Frequently used in small quantities 'about 0.5 phr) as a secondary accelerator.

THIXOTROPY Time-dependent fluid behaviour in which the apparent viscosity decreases with the time of shearing and in which the viscosity recovers to, or close to, its original value when shearing ceases. The recovery may take place over a considerable time. This may sometimes occur with polymer systems, when molecular disentanglement increases with time of shearing.

THORNEL Tradename for carbon fibres from rayon, polyacrylonitrile and pitch fibre precursors.

THPC Abbreviation for tetrakishydroxymethylphosphonium chloride.

THR Abbreviation for threonine.

THREODIISOTACTIC A ditactic polymer in which the configurations of both types of stereoisomeric centre (the C* atoms in the polymer $-\!\!+\!\!C^*HX\!\!-\!\!C^*HY\!\!-\!\!\}_n$ for example) are the same in all the repeat units, but in which one type of centre is of opposite configuration to the other. On a Fischer projection the substituents will appear on opposite sides of the chain:

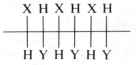

This is more frequently written:

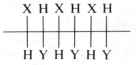

THREONINE (Thr) (T)

$$H_2NCHCOOH$$
$$|$$
$$CHOH$$
$$|$$
$$CH_3$$

M.p. 253°C (decomposes).

A polar α-amino acid found in most proteins. Its pK' values are 2·71 and 9·62, with the isoelectric point at 6·16. In the hydrochloric acid hydrolysis stage of protein amino acid analysis, threonine is slowly destroyed. Thus the threonine content is determined by conducting hydrolyses for various times and extrapolating the threonine value to zero time.

THREOPOLYMER
A ditactic polymer in which the configurations of the two types of stereoisomeric centre are different.

THRESHOLD STRESS
The stress above which ozone cracking of a stressed diene rubber occurs very rapidly.

THROMBIN
An enzyme formed in blood from its zymogen prothrombin, which catalyses the conversion of serum-soluble fibrinogen to the insoluble fibrin, which is mainly responsible for the clotting of blood.

THYROGLOBIN
A glycoprotein found in the thyroid, probably consisting of four peptide chains of molecular weight about 660 000. Most samples have a similar carbohydrate composition, typically containing D-galactose, D-mannose, N-acetamido-2-deoxy-D-glucose, L-fucose and sialic acid in the approximate ratios 1:2:2:0·5:1.

THYROTROPIN
A protein hormone secreted by the anterior pituitary gland, which acts on the thyroid gland to stimulate thyroxine production. It is a glycoprotein of molecular weight 28 300, containing 220 amino acid residues in two polypeptide chains.

TIE MOLECULE
The intervening section in a polymer molecule which has started to crystallise independently in two different crystals. The intervening section cannot crystallise in either crystal. Situated in the amorphous layers separating the lamellar crystallites in the same stack. When strained, the tie molecules may crystallise and provide strong interlamellar fibrillar links. When drawn, additional tie molecules are formed by partial unfolding of the lamellae and they become incorporated into microfibrils.

TIME-DOMAIN METHOD
(DC step-response method) A method for the determination of the dielectric properties of a material in which a step change in voltage (V) is made across the sample and the resulting transient current, which changes as polarisation takes place, is monitored. The measurements may be transformed into

frequency-domain terms by a Fourier transformation. However, this involves considerable computation. Often an approximate method, such as the Hamon approximation, is used. In practice a step voltage of 1–500 V is applied across the sample held in a normal guarded electrode and the resulting changing current is measured by the voltage it produces across a standard resistor. The method has classically been used for frequencies of 10^{-1}–10^{-4} Hz, but more recently higher frequency (up to 10^6 Hz) determinations have been made using an electrometer amplifier to measure the time integral of the current. Even higher frequency measurements are possible by time-domain spectroscopy.

TIME-DOMAIN SPECTROSCOPY
A time-domain method for the determination of dielectric properties at very high frequencies (10^8–10^{12} Hz). The sample in a waveguide is subjected to very short radiation pulses with fast rise times and the reflected or transmitted pulses are observed with an oscilloscope. The time-domain readings may be converted to the frequency domain by a Fourier transform.

TIME–TEMPERATURE SUPERPOSITION
(Temperature–time superposition) The observation that the effects of different time scales of measurement (or correspondingly different frequencies) and of different temperatures can have equivalent effects on the viscoelastic properties of a material. Thus the viscoelastic behaviour at one temperature can be related to that at another by a change in the time scale only. For example, a series of plots of log modulus or log compliance (especially creep compliance or relaxation modulus) versus log time at various temperatures can, by shifting the curves horizontally by different amounts (by the shift factors a_T), be combined to give a single master curve for a particular reference temperature. The shift factor is found to be temperature dependent only and the WLF equation expresses the observation that a_T is approximately the same for all amorphous polymers. The principle can be used to predict creep or stress relaxation behaviour at temperatures or times other than those for which data are available.

TINUVIN
Tradename for a range of 2-hydroxyphenyltriazole ultraviolet absorbers. The best known is Tinuvin P, 2-(2-hydroxy-4-methylphenyl)benzotriazole,

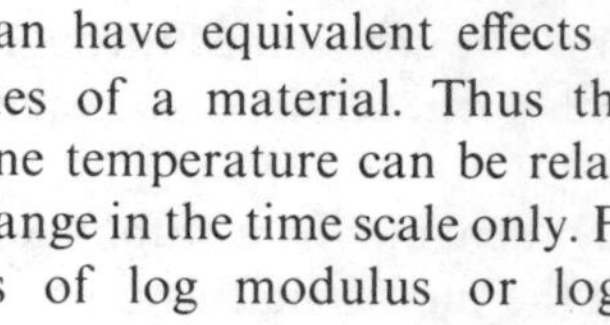

M.p. 132°C.

TIOTM
Abbreviation for triisooctyl trimellitate.

TIPPED POLYOL
(Capped polyol) A polyoxypropylene polyol in which the polymer chains are terminated in ethylene oxide units, so that the hydroxyl

end groups are primary rather than secondary and hence are more reactive with diisocyanates when the polyether is used in the formation of polyurethane foams.

TLC   Abbreviation for thin layer chromatography.

TMA   (1) Abbreviation for thermomechanical analysis. (2) Abbreviation for trimellitic anhydride.

TMD   Abbreviation for trimethylhexamethylenediamine.

TMDT   Abbreviation for poly-(trimethylhexamethylene-terephthalamide).

TMIB   Abbreviation for 3,3,5-trimethylpentane-1,3-diol-diisobutyrate.

TMTD   Abbreviation for tetramethylthiuram disulphide.

TMU   Abbreviation for tetramethylurea.

TMV   Abbreviation for tobacco mosaic virus.

TNF   Abbreviation for 2,4,7-trinitrofluorenone.

TOA   Abbreviation for thermal optical analysis.

TOBACCO MOSAIC VIRUS   (TMV) One of the most studied viruses. Each virion consists of a single helical coil of ribonucleic acid, accounting for about 5%, embedded in a coat of 2130 identical protein subunits to form an overall supermolecular complex nucleoprotein with a rod shape about 300 nm long and 10 nm wide. The overall particle molecular weight is about 40 million. The protein subunits, or 'monomers', have a molecular weight of 17 530 comprising 158 amino acid residues whose sequence is known. The proteins are terminated by threonine at the carboxyl end and an acetylated serine at the amino end and have a single cysteine group and no histidine. The denatured protein will refold to its original conformation even though no disulphide links are involved. The native protein tends to aggregate in weakly alkaline media (often as trimers called the A protein) as the pH decreases and/or the temperature or ionic strength increases. Larger aggregates form until the virus-like rod results. If ribonucleic acid is also present the reconstituted virions are much more readily formed and are much more stable, e.g. over a pH range of 3–9, and have similar biological activity to the original virus. The ribonucleic acid molecular weight is about 2 million with about 6340 nucleotides. The ribonucleic acid alone is usually considered to be sufficient for infectivity.

TOF   Abbreviation for trioctyl phosphate.

TOHALON   Tradename for a cellulose acetate fibre produced by acetylating viscose rayon after spinning. This allows the production of a cellulose acetate fibre using an aqueous, rather than an organic, spinning medium.

TOLUENE

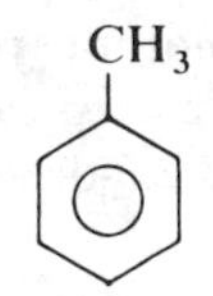

B.p. 110·6°C.

An important solvent for many polymers, often being preferred commercially (e.g. in adhesives and coatings) to benzene due to its lower toxicity. A solvent for many rubbers, natural resins, acrylic polymers, polystyrene, polyvinyl acetate, epoxy, amino and alkyd resins. A non-solvent for polyvinyl chloride, cellulose acetate, polyethylene and nylons. Usually obtained from the distillation of petroleum when it may contain some methylcyclopentane (b.p. 117–120°C).

TOLUENE-2,4-DIISOCYANATE

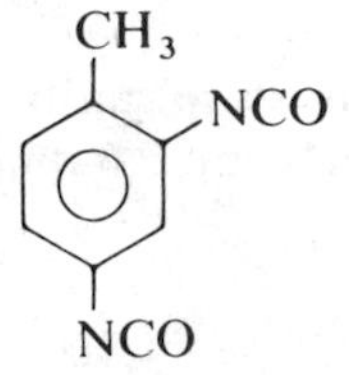

M.p. 22°C. B.p. 120°C/10 mm.

A particular isomer of tolylene diisocyanate. It is used (instead of the more frequently employed mixed isomers) as the tolylene diisocyanate for polyurethane formation when a higher reactivity is required.

o/p-TOLUENESULPHONAMIDE

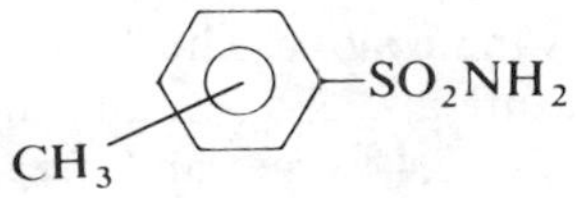

M.p. 105°C. B.p. 360°C/760 mm.

A plasticiser for phenol–, urea– and melamine–formaldehyde resins, polystyrene and polyvinyl acetate.

p-TOLUENESULPHONYL GROUP   (Tosyl group)

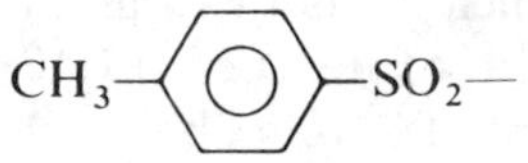

A protecting group useful for protecting amino groups of amino acids during peptide and protein synthesis. It is often used particularly for protecting the α-amino group of lysine and the guanidino group of arginine. It is introduced by reaction of the amino group with *p*-toluenesulphonyl chloride

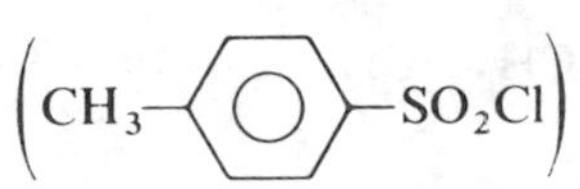

It is removed by reaction with sodium in liquid ammonia.

TOLYLENE DIISOCYANATE (TDI) The commonest diisocyanate used in the preparation of polyurethanes by reaction with a diol or polyol. Three different commercial products are used each containing different amounts of the 2,4- and 2,6-diisocyanate isomers

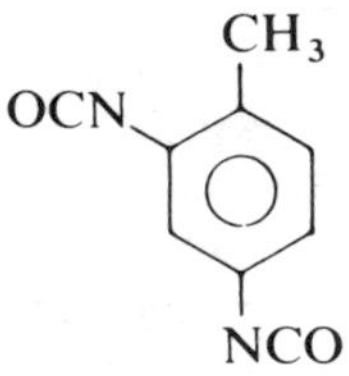

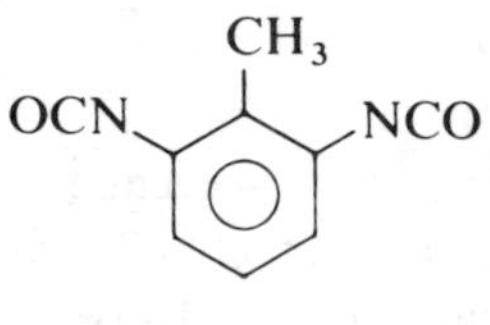

They are made by nitration of toluene which initially gives a 60/40 mixture of the 2- and 4-nitrotoluenes

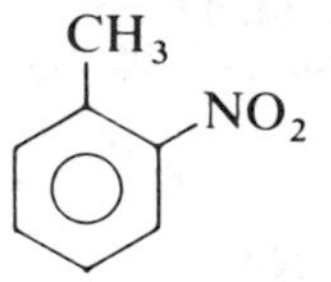

respectively. This mixture can be further nitrated to an 80/20 mixture of 2,4- and 2,6-dinitrotoluenes

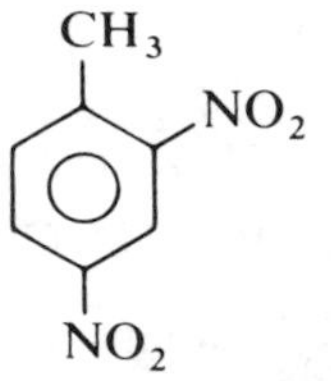

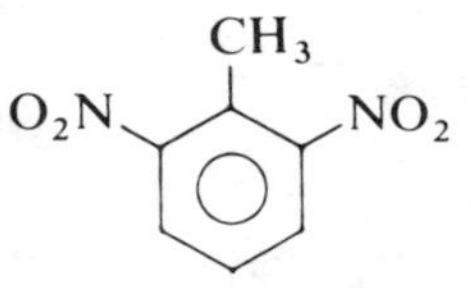

which when reduced to the similar diaminotoluenes and then reacted with phosgene gives an 80/20 mixture of 2,4- and 2,6-toluenediisocyanates. A similar reaction sequence performed on separated pure 2-nitrotoluene gives a 65/35 mixture of 2,4- and 2,6-toluenediisocyanates. A similar reaction sequence performed on 4-nitrotoluene gives 2,4-toluenediisocyanate alone. The 2,4-isomer melts at 22°C, the 65/35 mixture at 3·5–5·5°C and the 80/20 mixture at 11·5–13·5°C. The 2,4-isomer and the mixtures all boil at about 120°C/10 mm. The 80/20 mixture is the most widely used, especially for flexible foams. The 65/35 mixture is used where lower reactivity is required and the pure 2,4-isomer is used where higher reactivity is required, the 4-isomer group having high reactivity.

TOMATO BUSHY STUNT VIRUS An early studied plant virus which is polyhedral in shape. It has a ribonucleic acid content of 15–17% and a virion molecular weight of 8–10 million with each virion being about 28 nm in diameter. The protein subunits have a molecular weight of about 60000.

TOPANOL CA Tradename for the phenolic antioxidant 1,1,3-tris-(2-methyl-4-hydroxy-5-t-butylphenyl) butane.

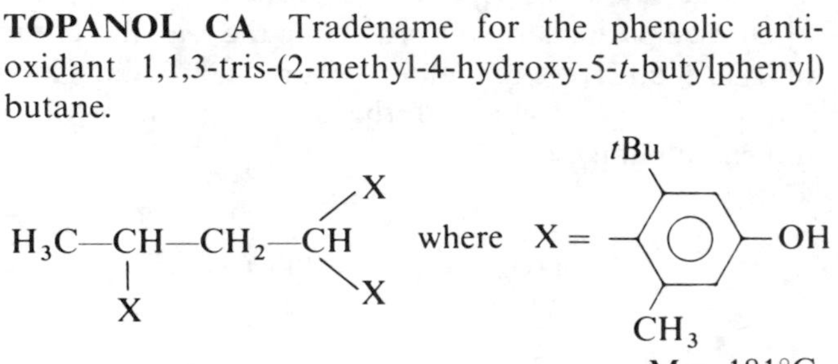

M.p. 181°C.

A non-staining, chain-breaking antioxidant widely used, but especially in polyolefins, being very effective when synergised with a peroxide decomposer such as DLTDP.

TOPEL Tradename for a crosslinked viscose rayon fibre with properties similar to Corval.

TOPOCHEMICAL POLYMERISATION Polymerisation of a crystalline monomer in which the polymerisation is influenced by the crystalline structure of the monomer. Thus it is a solid state polymerisation in which the kinetics of polymerisation and the crystalline or other structural features of the resultant polymer are all influenced by the crystal structure of the monomer. Well studied examples are the polymerisation of trioxane to polyoxymethylene and the polymerisation of substituted di-ynes to polyacetylenes. In the particular case when the polymer crystal structure is formed from the monomer crystal structure without an intermediate loss of order, the polymerisation is topotactic.

TOPOTACTIC POLYMERISATION Topochemical polymerisation in which a crystalline monomer is converted to a crystalline polymer without any intermediate loss of order. Many topochemical polymerisations start by being topotactic, but order is lost as propagation proceeds beyond the formation of oligomers. Polymerisations of some substituted di-ynes provide rare examples of topotactic polymerisations.

TORAMOMEN Tradename for a polynosic rayon.

TORAYCA Tradename for polyacrylonitrile based carbon fibre.

TORLON Tradename for a polyamide–imide obtained by reaction of trimellitic anhydride with an aromatic diamine. It has the structure

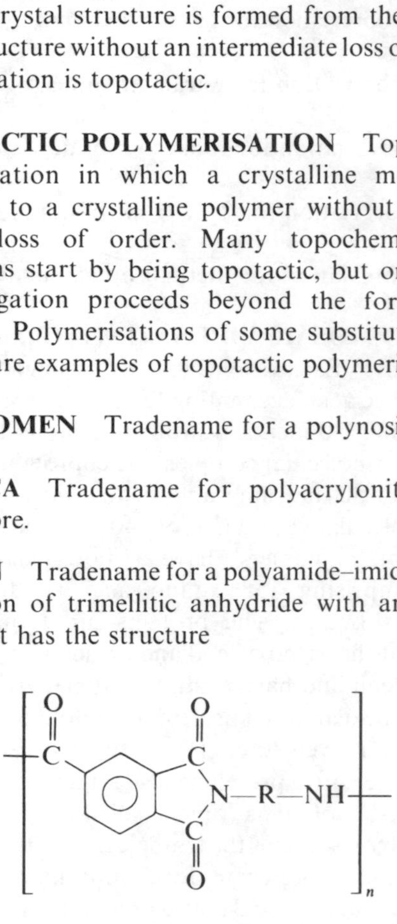

and is useful as a high temperature resistant (to 260°C) injection moulding engineering plastic.

TORSIONAL ANGLE (Angle of rotation) Symbol ϕ or ψ. An angle characterising the amount of rotation that has taken place about a chemical bond from some arbitrary base angle. Thus for a carbon–carbon single bond, the angle ϕ (in this case called the dihedral angle)

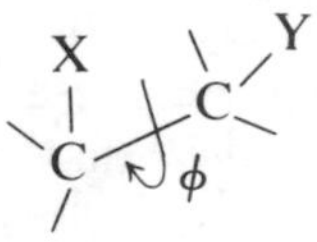

could be defined as the angle of rotation about the carbon–carbon bond from the position shown, i.e. with groups X and Y eclipsed. The sum of all such angles in a molecule thus defines the conformation. In polymers, even when certain values for ϕ are preferred, an extremely large number of conformations may be possible. In the particular case of polypeptides, three such angles (ϕ, ψ and ω) are defined according to the presently accepted convention as shown

However, several other conventions for describing conformational rotation have been used in the past. Since the peptide bond has considerable double bond character, little rotation is possible and the value of ω is fixed, by convention being assigned the value $+180°$ for the usual *trans*-peptide form (or $\omega = 0°$ for the rarer *cis*-peptide form). The convention for ϕ and ψ is that they are both assigned values of $+180°$ for the fully extended chain conformation shown, and that their values increase positively in the direction shown, i.e. for clockwise rotations when the chain is viewed from the N-terminal end. In polypeptides, steric and other interactions restrict the permissible values of ϕ and ψ. The interaction energies have been subject to much theoretical calculation and may be plotted as a function of ϕ and ψ on a Ramachandran plot. The occurrence of specific preferred values for ϕ and ψ leads to the adoption of a preferred conformation for the polypeptide, which is a regular conformation, e.g. the α-helix or the β-conformation, if values of ϕ and ψ are the same for each amino acid residue.

TORSIONAL BRAID ANALYSIS (TBA) An extension of the torsional pendulum technique for the determination of the dynamic mechanical behaviour of polymers, useful when only small amounts of polymer are available, or when the polymer will not support its own weight. A glass braid substrate is impregnated with the polymer (from solution) and the solvent is evaporated. The braid is then subject to torsion and the damped oscillations are followed as in a conventional torsion pendulum experiment. Thus the loss modulus (as given by the logarithmic decrement) and the storage modulus (often simply as a relative rigidity parameter of $1/p^2$, where p is the period of oscillation) can be determined, often as a function of temperature, when the results are displayed as a thermomechanical spectrum.

TORSION MODULUS Alternative name for shear modulus.

TORSION PENDULUM A technique for the determination of the dynamic mechanical properties, whereby a sample is subject to free vibration torsional oscillations.

The oscillations are damped by virtue of the mechanical energy dissipated, i.e. by the loss. The polymer, in the form of a strip or filament, is attached at one end to a torsion wire held in tension. The other end is rigidly clamped. Also attached to the wire is a torsion arm with a known large moment of inertia, usually consisting of a bar to which weights (often moveable) are attached. In operation the arm is displaced and the damped oscillations are followed by recording the amplitude of successive swings, e.g. by deflection of a light beam from a mirror attached to the wire, or electrically. From a tracing of the oscillations, the period and the diminution in amplitude (as the logarithmic decrement (Δ)) due to damping may be determined. The shear storage modulus (G') is given by $G' = (2lM\omega^2)/\pi r^4$, where l is the length of the strip or filament with a cylindrical cross-section of radius r, M is the moment of inertia of the torsion arm and ω is the frequency. The shear loss modulus (G'') is given by $G'' = \omega^2 \Delta 2lM/\pi^2 r^4$ and hence $G''/G' = \tan\delta = \Delta/\pi$ for small damping. For a specimen with a rectangular cross-section $\pi r^4/2l$ is replaced by ab^3/l, where a is the width and b is the thickness of the specimen. Torsion pendulum measurements cover the low frequency range of $0 \cdot 01–50\,Hz$.

TOSYL GROUP Abbreviation for *p*-toluenesulphonyl group.

***N*-TOSYL-L-PHENYLALANYLCHLOROMETHYL KETONE** (TPCK)

An affinity label for chymotrypsin. It irreversibly alkylates the histidine-57 residue, as well as binding to the active site in the normal manner of a substrate. However, hydrolysis by the enzyme does not occur. Thus histidine-57 is shown to be close to the active site and to be involved in normal catalytic activity.

TOTAL BUTYL Tradename for butyl rubber.

TOUGH–BRITTLE TRANSITION Alternative name for brittle–ductile transition.

TOUGH FRACTURE Alternative name for ductile fracture.

TOUGHNESS In general, the ability of a material to withstand fracture. In particular, sometimes used as an alternative name for critical strain energy release rate.

TOW A loose rope of several thousand (tens to hundreds of thousands) filaments. Typically a tow may be of 200 000 denier consisting of 40 000 filaments each of 5 denier.

TPCK Abbreviation for *N*-tosyl-L-phenylalanylchloromethyl ketone.

TPE Abbreviation for thermoplastic elastomer.

TPO Abbreviation for thermoplastic polyolefin rubber.

TPP Abbreviation for triphenyl phosphate.

TPR (1) Abbreviation for thermoplastic rubber.
(2) Tradename for a thermoplastic polyolefin rubber.

TPU Abbreviation for thermoplastic polyurethane and for thermoplastic polyurethane elastomer in particular.

TPX Tradename for poly-(4-methylpentene-1).

TR Abbreviation for polysulphide rubber.

TRACKING The progressive formation of an electrically conducting carbonised pathway across the surface of an insulator by surface discharges. The process is considerably accelerated by the presence of surface films of moisture and dirt. The film causes a surface current to flow between points at different potential, which may generate sufficient heat to evaporate the water. Sparking can then occur across the gap through the surface layer causing carbonisation of the polymer. Phenol–formaldehyde polymers and polylvinyl chloride are particularly prone to tracking.

TRACTION VISCOSITY Alternative name for elongational viscosity.

TRADLAC Tradename for a polyparabanic acid resin material.

TRADLON Tradename for polyparabanic acid.

TRAGACANTHIC ACID The major polysaccharide component of gum tragacanth. It consists of chains of 1,4′-linked D-galacturonic acid (about 40%) with β-D-xylose, α-L-fuco-1,2′-β-D-xylose and β-D-galacto-1,2′-β-D-xylose side chains 1,3′-linked to the main chain.

TRANSACETALISATION An interchange reaction between acetal groups (analogous to transesterification), such as occurs during copolymerisation of trioxane with ethylene oxide, whereby the initially formed block copolymer structure becomes randomised by the interchange, thus producing a random copolymer. In general the reaction may be represented as:

$$\sim\!O\!-\!R\!-\!O\!-\!R\!-\!O\!\sim + \sim\!O\!-\!R'\!-\!O\!-\!R'\!-\!O\!\sim \rightarrow$$
$$\sim\!O\!-\!R\!-\!O\!-\!R'\!-\!O\!- + \sim\!O\!-\!R'\!-\!O\!-\!R\!-\!O\!\sim$$

TRANSAMIDATION (Amide interchange) The reaction of one amide in a polyamide with another amide such that the lengths of the polymer molecules become altered,

$$P_x NHCOP_y + P_n NHCOP_m \rightarrow$$
$$P_x NHCOP_m + P_n NHCOP_y$$

where P_x, P_y, P_n and P_m are the remaining polymer chains of different lengths. Unlike ester interchange, amide interchange is slow, even at the high temperatures (up to about 300°C) employed in nylon synthesis and processing. However, in nylon blends some redistribution of amide repeat units may be expected.

TRANSANNULAR POLYMERISATION Chain polymerisation of a non-conjugated, cyclic (often bicyclic) diene in which isomerisation of the active centre occurs involving both double bonds such that linking of the monomer units is across the ring. This occurs particularly in the cationic polymerisation of norbornenes, e.g. of 2,5-norbornadiene to polynortricyclene:

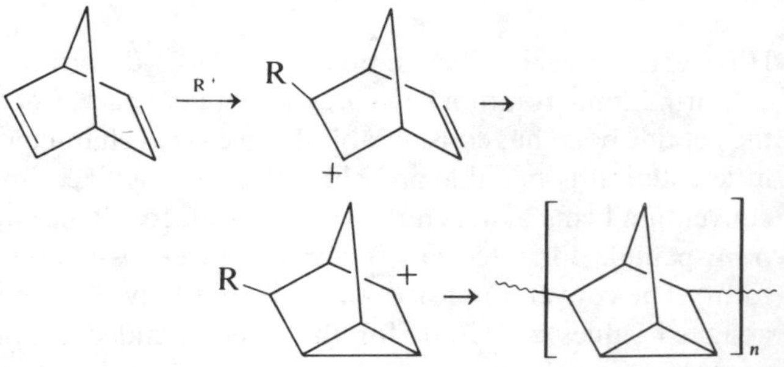

***TRANS* CONFIGURATION** The particular geometric isomeric form of a repeat unit of a polydiene (or other repeat unit containing an in-chain double bond or saturated ring) in which the attachment of the polymer chain residues is *trans* to the double bond or ring:

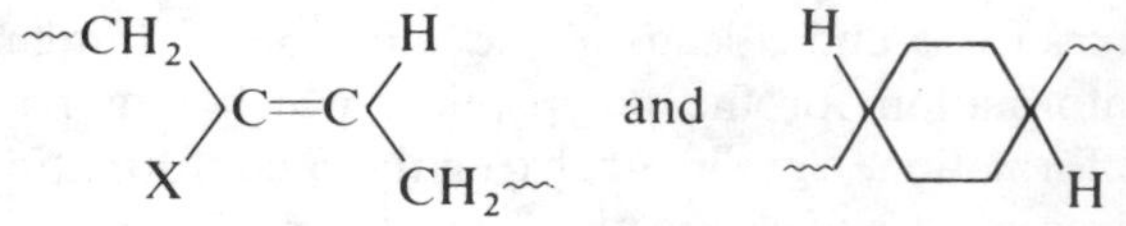

Polymers with high *trans* isomeric content have higher T_g and T_m values than high *cis* polymers.

***TRANS* CONFORMATION** Symbol t. The conformation of a 1,2-disubstituted organic compound in which rotation about the bond joining a pair of adjacent atoms (i.e. atoms 1 and 2) is such that the substituents are staggered *trans* to each other. For the simplest example, *n*-butane, the 'saw-horse' representation is

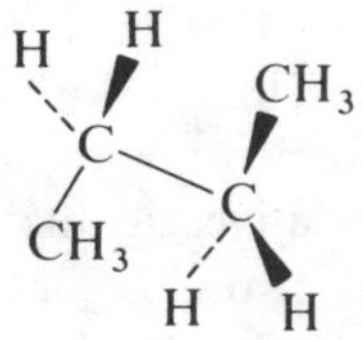

and the Newman representation is

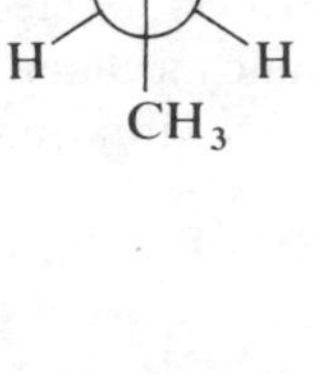

In such simple cases this conformation is energetically the most favoured and has a high population. In carbon–carbon polymer chains sometimes all the chain bonds are *trans* (all-*trans* or —ttttttttt—) as in some crystalline polymers such as polyethylene.

TRANSCRYSTALLISATION Growth of crystalline fibrils (not developed into spherulites) from crystallisation in the melt or on a surface, such that the nuclei are located in a plane surface. Nucleation is caused by the surface of the containing vessel or by straining the melt.

TRANSESTERIFICATION Alternative name for ester interchange.

TRANSFER Alternative name for chain transfer.

TRANSFER AGENT Alternative name for chain transfer agent.

TRANSFER CONSTANT Alternative name for chain transfer constant.

TRANSFER GRAFTING A method of producing graft copolymers by forming free radical sites on a polymer chain (by hydrogen abstraction) in the presence of the monomer for a second polymer, which becomes grafted onto the chains by polymerisation at these sites.

TRANSFER LENGTH The length of a short fibre, or of a broken continuous fibre, in a fibre reinforced composite over which stress transfer from the matrix to the fibre occurs. The ends of the fibre do not support any load but suffer a shearing force due to the tendency of the fibres to pull away from each other and due to the elastic/plastic response of the matrix. These shearing forces build up stress in the fibre, which increases with distance from the fibre end. Above the critical fibre length the stress can build up sufficiently to exceed the tensile strength of the fibre which may therefore fracture.

TRANSFERRIN (β-Metal-binding globulin) A blood plasma protein found in the β-globulin fraction on electrophoresis. It contains about 5% carbohydrate, probably as two units attached to aspartic acid protein units, and has a molecular weight of 85 000. It can be obtained in a sufficiently homogeneous form to be crystalline. It is responsible for the transport of iron, which it binds maximally at pH 7·0 in the presence of carbon dioxide.

TRANSFER TO INITIATOR Alternative name for induced decomposition.

TRANSFER TO MONOMER Chain transfer in a chain polymerisation where the propagating polymer active centre abstracts an atom from a monomer molecule, e.g. in the free radical polymerisation of a vinyl monomer:

$$\text{---CH}_2\dot{\text{C}}\text{HX} + \text{CH}_2\text{=CHX} \rightarrow$$
$$\text{---CH}_2\text{CH}_2\text{X} + \text{CH}_2\text{=}\dot{\text{C}}\text{X}$$

Typical values of the transfer constant (C_M) are 10^{-4}–10^{-5}, except, notably, for vinyl acetate and vinyl chloride ($C_M \sim 10^{-3}$), so that transfer does not seriously limit the polymer molecular weight. In cationic polymerisation transfer to monomer is common and often controls polymer molecular weight. It frequently involves transfer of the catalyst–co-catalyst complex to monomer involving the loss of a proton and leaving the chain end unsaturated. Transfer can also occur by hydride ion abstraction from the monomer.

TRANSFER TO POLYMER Chain transfer in a free radical polymerisation by abstraction of an atom (usually hydrogen) from a dead polymer molecule by the propagating active centre:

$$\text{---}\dot{\text{M}}_n + \overset{\text{X}}{\underset{|}{\text{------}}} \rightarrow \text{---} + \text{---M}_n\text{---X}$$

The new active centre can continue to grow by normal propagation steps to produce a branched polymer molecule:

$$\text{---} + \text{M} \rightarrow \overset{\dot{\text{M}}}{\underset{|}{\text{------}}} \xrightarrow{x\text{M}} \overset{\dot{\text{M}}}{\underset{\underset{|}{\text{M}_x}}{\text{------}}}$$

The reaction is only significant at high conversions of monomer to polymer but can cause a large increase in the molecular weight of the polymer formed in the later stages. Determination of the transfer constant (C_P) is difficult since it requires accurate determination of the number of branches formed. Typical values are $\sim 10^{-4}$. The extent of branching can be calculated, as the branching density, as a function of C_P and conversion. Transfer to polymer is higher in polymerisations with active centres of high activity, such as ethylene, vinyl acetate and vinyl chloride. It is particularly important in the free radical high pressure polymerisation of ethylene.

TRANSFORMER RATIO-ARM BRIDGE An electrical bridge circuit used for the determination of dielectric properties over the frequency range 50–10^4 Hz. The secondary winding of a transformer forms the bridge arms, the other two arms containing the unknown and standard impedances. Stray impedances are measured in this circuit, and since modern transformers can be very accurately tapped, only a few cheap standard capacitances are needed, compared with the expensive variable capacitances required for the Schering bridge.

TRANSHEXAHYDRO-*p*-XYLYLENEDIAMINE

$$\text{H}_2\text{NCH}_2\text{---}\langle \bigcirc \rangle\text{---CH}_2\text{NH}_2$$

Synthesised by hydrogenation of *p*-xylylenediamine, which is obtained by hydrogenation of terephthalonitrile, itself derived either from terephthalic acid or *p*-xylene. Useful as a monomer for poly-(transhexahydro-*p*-xylylenesuberamide).

TRANSITION The pronounced changes in the properties of a material that occur at a certain temperature (the transition temperature) or over a range of temperature. The commonest transition is melting. Melting is typical of a first order transition, i.e. one in which a discontinuity in the intensive properties, such as enthalpy, entropy and volume, occurs. Melting is a very sharp transition in low molecular weight compounds, but in crystalline polymers it may extend over 10–20°C, giving a melting temperature range rather than a sharp melting point. A polymer melting temperature usually refers to the highest temperature of the range.

However, in polymers, several other transitions, having the characteristics of second order transitions, are also usually observable. These are associated with the onset of particular modes of molecular motion, i.e. relaxations, as the temperature is raised. They may be characteristic of motions occurring in an amorphous material (a glass-forming polymer or, more rarely, a low molecular weight substance), or in the amorphous regions of a crystalline polymer, or the motions may be in a crystalline phase or in both phases. Usually the most important of these non-melting transitions is the glass transition, with its associated glass transition temperature (T_g). It is nearly always the highest temperature transition and is therefore referred to as the α-transition in amorphous polymers, which of course have no melting transition. Lower temperature transitions are often labelled as the β-, γ-, δ-transitions, etc., with decreasing temperature. The associated relaxation processes are correspondingly labelled the α-, β-, γ-, δ-processes or relaxations. The T_g value usually extends over several degrees, but the lower transitions are often much broader, e.g. the β-transition often extends over about 100°C.

A crystalline polymer may exhibit the range of amorphous transitions, whose strength will be less, the lower the degree of crystallinity. In addition, there may also be some secondary crystalline transitions, such as premelting and crystal–crystal transition.

The most important changes that occur in the properties over the transition region are in the mechanical properties. Indeed mechanical tests are frequently used for the characterisation of transitions. Thus over the region of the T_g value, the moduli typically change by a factor of about 10^3.

For many transitions, but especially the glass transition, the value of the transition temperature or temperature range, is dependent on the rate of testing, since the transition is dynamic in its molecular origin. Furthermore, the transitions, or perhaps rather the relaxations, may be observed at constant temperature by testing over a range of frequencies as in dynamic mechanical spectroscopy and dielectric spectroscopy. These two techniques are frequently used for characterising transition behaviour.

α-TRANSITION The highest temperature transition of an amorphous polymer, apart from any liquid–liquid transition that may be observed. Usually it is the most important transition and may be identified as the glass transition. Lower temperature transitions are therefore regarded as secondary transitions. In crystalline polymers the α-transition is the next highest transition to the melting transition (T_m) and is often referred to as premelting. In some cases, notably polyethylene, it can be resolved into several components—the α'- and α''-transitions for example.

α'-TRANSITION (α₁-Transition) The lower temperature component of the α-transition in polyethylene. It is possibly a crystalline transition attributed variously to the mosaic block boundaries in the crystal lamellae and to interlamellar slip. It is also sometimes called the α-transition when the α''-transition is called the α'-transition.

α''-TRANSITION (α₂-Transition) The higher temperature component of the α-transition in polyethylene. A crystalline transition probably due to *c*-shear. It may be obscured by the α'-transition. It is also sometimes called the α'-transition, when the α'-transition, is referred to as the α-transition. Sometimes it is referred to as the premelting transition.

α₁-TRANSITION Alternative name for α'-transition.

α₂-TRANSITION Alternative name for α''-transition.

β-TRANSITION The next highest temperature transition to the α-transition. In amorphous polymers this is often observed as a broad transition, sometimes extending over about 100°C, often as a shoulder on the main T_g transition. At high frequencies the transition may merge into the glass transition and disappear. The transition is often thought to be associated with local chain motion relaxation, possibly crankshaft motion or side-group motions. Polymers in which the transition is relatively large, often retain their toughness to well below their T_g value into the region of this transition. In these cases the temperature range between the α- and β-transitions is sometimes known as the working range. In crystalline polymers, the β-transition is usually the glass transition.

γ-TRANSITION The next highest temperature transition to the β-transition. In amorphous polymers this occurs well below the T_g value and its origin is thought to be due to any of several processes, e.g. side-group motions, end-group motions, or even quantum mechanical tunnelling of side-group motions. In crystalline polymers it corresponds to the β-transition of the amorphous polymer and is sometimes resolvable into two or three components, one probably involving a crystalline relaxation.

δ-TRANSITION The lowest temperature transition, sometimes observed in amorphous polymers and having origins similar to the γ-transition. In crystalline polymers, its origins may be similar to those of the γ-transition in amorphous polymers, or it may be due to a crystalline relaxation, e.g. as a result of crystal defects.

TRANSITION MOMENT A measure of the change in dipole moment that occurs as a result of a molecular vibration. It is a vector quantity whose magnitude and direction depends on the mode of vibration and the orientation of the molecule involved. Absorption of infrared radiation, inducing such a vibration, only occurs if the transition moment is non-zero. This provides the basis for infrared dichroism.

TRANSITION ZONE Alternative name for the retarded elastic state. However, it is often restricted to describing this state when observed in the appropriate time domain or, in the case of dynamic mechanical behaviour, frequency domain, rather than in the temperature domain.

TRANSMISSION METHOD A method for the measurement of birefringence by the determination of the retardation (R) of a sample. The polarised light intensity (T) transmitted through the sample when viewed between crossed polarisers is determined, when $T \sim \sin^2(\pi R)$ and $R = (d/\lambda)\Delta n$, where d is the sample thickness, λ is the light wavelength and Δn is the birefringence. It is especially useful when high speed deformation of the sample is involved.

TRANSPIP Tradename for *trans*-1,4-polyisoprene.

TRANS-1,4-POLYBUTADIENE

One of the stereoisomeric forms of polybutadiene. Produced by Ziegler–Natta polymerisation with an aluminium triethyl/vanadium trichloride or other vanadium catalyst. It has a T_g value of $-106°C$ and a T_m value of $145°C$ and therefore, unlike the *cis*-1,4-isomer, is not a useful rubber due to its crystallisation.

TRANS-1,4-POLYISOPRENE

Tradename: Transpip. Occurs naturally as the main constituent of gutta-percha and balata but is also produced synthetically by Ziegler–Natta polymerisation using, for example, an aluminium tributyl/vanadium trichloride catalyst, which gives a polymer with virtually 100% pure *trans*-1,4-structure. The polymer has a T_g value of about $-60°C$ and a T_m value of about $65°C$,

considerably higher than for the *cis* isomer (natural rubber). The polymer is normally crystalline and is therefore stiffer and harder than the *cis* isomer, having a higher tensile strength even when unvulcanised. However, sulphur vulcanisation is carried out to improve high temperature properties and solvent resistance. The synthetic polymer is very similar in properties to the natural product.

TRANSPOLYOCTENAMER Tradename: Vestenamer. A polyoctenamer which is of high *trans* isomer content. It is highly compatible with many other polymers and is useful as an additive to improve rubber processing.

TRANSPORT PROTEIN A protein whose biological function is the transport of substances in an organism. Important examples are the haem proteins for transport of oxygen, the lipoproteins for transport of lipids, serum albumin for the transport of fatty acids and cytochrome C for electron transport.

TRANSVERSE MODULUS (Lateral modulus) The elastic tensile modulus of a uniaxially drawn and oriented fibre or film in a direction perpendicular to the draw direction. If the draw direction (fibre axis in the case of fibres) is the z-direction then the transverse modulus is $E_1 = E_2 = 1/S_{11}$. Alternatively it is the elastic modulus of a uniaxial long fibre composite in the direction perpendicular to the fibres.

TRAVIS Tradename for a vinylidene cyanide/vinyl acetate copolymer fibre.

TREEING The formation of a tree-like system of channels in a dielectric due to the occurrence of internal discharges. These can arise from the use of sharp, as opposed to plane, electrodes, or from voids in the dielectric. Dielectric breakdown occurs in a series of steps producing the channels until eventually complete breakdown occurs when a continuous path has formed between the electrodes. The voltage at which the discharges start is the discharge inception voltage.

TREMOLITE A fibrous amphibole asbestos of composition $Ca_2Mg_5Si_8O_{22}(OH)_2$.

TRESCA YIELD CRITERION The yield criterion that yielding occurs when the maximum shear stress (σ_s) reaches a critical value. Thus if the principal stresses are σ_1, σ_2 and σ_3 with $\sigma_1 > \sigma_2 > \sigma_3$, then the criterion is $\frac{1}{2}(\sigma_1 - \sigma_3) = \sigma_s$. For a tensile test with $\sigma_1 =$ applied stress and $\sigma_2 = \sigma_3 = 0$, then $\sigma_s = \sigma_1/2 = \sigma_y/2$, where σ_y is the yield stress in tension. Although applicable to some metals, the criterion is not very useful for polymers. A modified Tresca criterion may apply to polymers which form shear bands.

TREVIRA Tradename for a polyethylene terephthalate fibre.

TRIACETATE Generic name for a cellulose triacetate fibre in which at least 92% of the cellulose hydroxyl groups are acetylated.

TRIACETIN Alternative name for glyceryl triacetate.

TRIAD A sequence of three repeat units in a polymer molecule, useful in representing some features of polymer microstructure. Thus stereochemical configurational isomers may be analysed in terms of iso-, syndio- and heterotactic triads. The frequency of occurrence of each type of triad may often be determined by analysis of the NMR resonances (e.g. of the α-substituent of a vinyl polymer $-(CH_2CHX-)_n$) which are different for each triad. The above-named triads will have relative configurations of the dyads of mm, rr and mr respectively. They may be represented by a Fischer projection as:

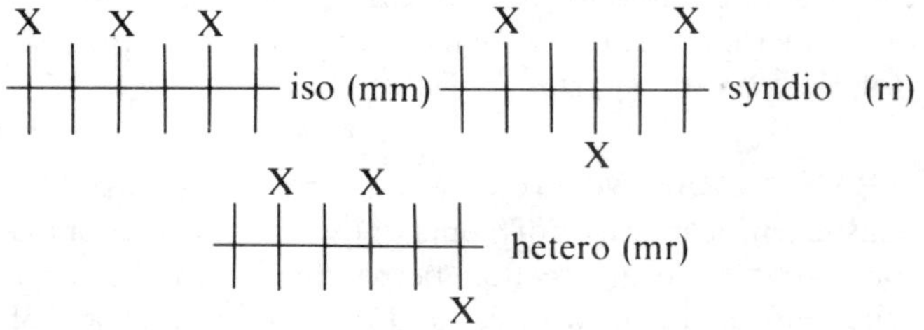

If the probabilities of an m or an r placement during polymerisation are the same, giving a purely random or atactic polymer, then the triads are in the ratios mm:mr:rr of 1:2:1 if the triad statistics are Bernouillian. If the probabilities are not the same then the ratios are $p_m^2:2p_m(1-p_m):(1-p_m^2)$, where p_m is the probability of a *meso* (m) (or isotactic) placement. If the frequencies of the triads are not in these ratios then p_m is dependent on the configuration of the previous unit in the polymer chain, i.e. Markovian chain statistics apply. Triad sequences may also be analysed in copolymers.

TRIALLYL CYANURATE

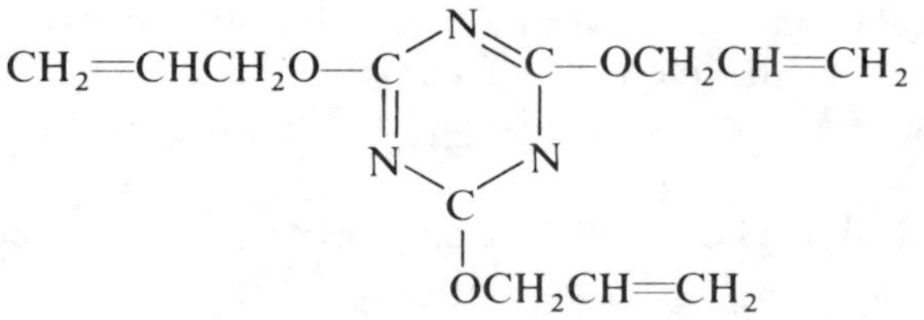

M.p. 27·3°C. B.p. 162°C/2 mm.

Prepared by the reaction of cyanuric chloride with excess allyl alcohol. Useful in place of styrene in the crosslinking of unsaturated polyester resins, to give cured products with higher softening temperatures.

1,3,5-TRIAMINO-2,4,6-TRIAZINE Alternative name for melamine.

TRIANGULAR FRACTIONATION A method of fractional precipitation in which sufficient non-solvent is added to precipitate about half the polymer, the precipitated gel (or coacervate) is separated and then redissolved and the process is continually repeated. Similarly, enough non-solvent is added to the supernatant solution from each stage to precipitate about half the remaining polymer. The method can produce sharper fractions, take a shorter time and keeps solution volumes to a minimum, compared with other precipitation methods.

TRIAX Tradename for an ABS/nylon blend.

TRIAZIN A Tradename for an s-triazine polymer produced by reaction of bisphenol A with cyanuric chloride to yield a prepolymer which is cured by further heating at 170–200°C to give a cyclised and crosslinked s-triazine structure. It is useful in place of epoxy resins where higher temperature performance, especially with respect to electrical properties, is required:

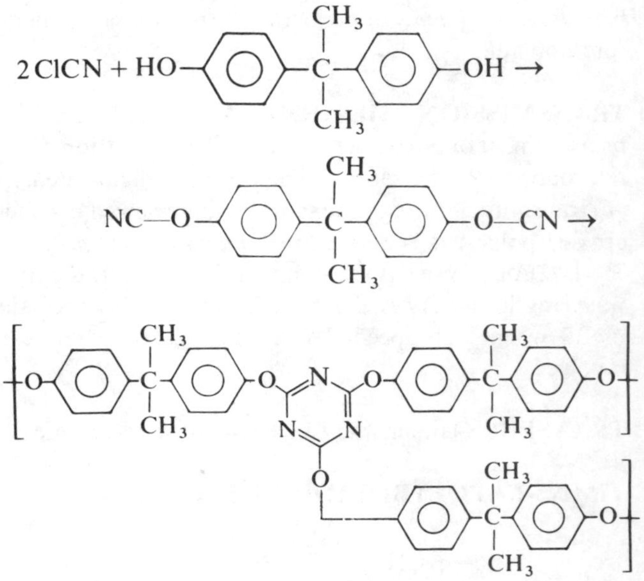

s-TRIAZINE POLYAMIDE A polymer of structure

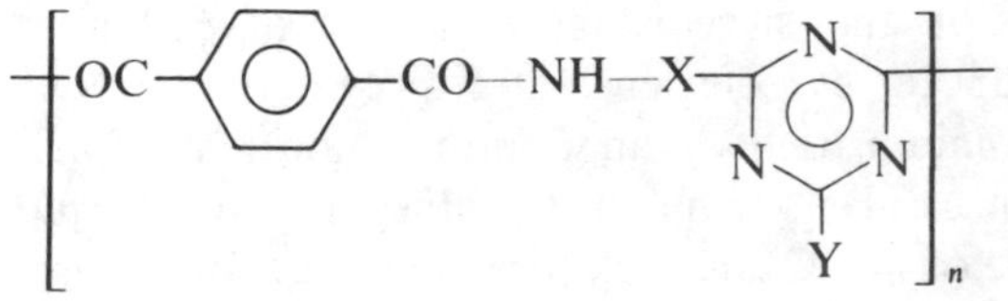

with either X absent and Y = NH₂ (polymer I), X absent and Y = phenyl (polymer II), or X = m-phenylene and Y = H (polymer III). Synthesised by reaction between terephthalic acid and melamine (yields I), phenylguanamine (yields II) or aminophenylguanamine (yields III). Only III has been synthesised in pure form. The polymers have unexpectedly poor thermal stability.

TRIAZOLE POLYMER (Polytriazole) A polymer containing a triazole ring in the polymer chain. Several different isomeric ring polymers, also linked differently in the chain, have been synthesised. These include poly-(1,2,3-triazoles), poly-(1,2,4-triazoles), poly-(4-amino-1,2,4-triazoles) and poly-(4-phenyl-1,2,4-triazoles).

TRIBASIC LEAD SULPHATE (TBLS)

$$3PbO.PbSO_4.H_2O$$

A widely used basic lead stabiliser for the thermal stabilisation of polyvinyl chloride. It is effective and cheap, having good electrical properties. Its toxic hazard in use is not as great as that of basic lead carbonate.

TRIBLOCK POLYMER (ABA block copolymer) A

block copolymer consisting of two terminal blocks of A repeating units and a central block of B units. If the outer blocks are stiff and the central block is rubbery, then due to phase separation and domain formation, the stiff polymer domains can act as physical crosslinks for the rubbery domains which comprise the matrix. The material then behaves as a thermoplastic elastomer. Styrene–butadiene–styrene block copolymer is the best known example of this type of material.

TRIBUTOXYETHYL PHOSPHATE

$$[CH_3(CH_2)_3OCH_2CH_2O\text{-}]_3PO$$

B.p. 220°C/5 mm.

A flame retardant plasticiser for polyvinyl chloride and its copolymers, synthetic rubbers and cellulose esters.

TRIBUTYL CITRATE

$$(CH_3(CH_2)_3OOCCH_2)_2C(OH)COO(CH_2)_3CH_3$$

B.p. 294°C/760 mm, 150°C/3 mm.

A plasticiser for a wide range of cellulose esters and ethers. Also compatible with polyvinyl chloride and its copolymers and with polystyrene. It has poor permanence.

TRIBUTYL PHOSPHATE (TBP)

$$[CH_3(CH_2)_3O\text{-}]_3PO$$

B.p. 289°C/760 mm.

A plasticiser for cellulose acetate and nitrate. Also compatible with polyvinyl chloride and its copolymers and with phenol–formaldehyde resins. It is rather volatile but has low flammability.

TRIBUTYRIN Alternative name for glyceryl tri-butyrate.

TRICEL Tradename for a cellulose triacetate fibre.

1,1,1-TRICHLOROETHANE

$$CH_3CCl_3 \qquad B.p. 74.1°C.$$

A low flammability solvent having much lower toxicity than many other chlorinated hydrocarbon solvents. It is therefore often preferred as a replacement for these. A solvent for hydrocarbon rubbers, silicone oils, polyvinyl acetate, polystyrene and acrylic polymers.

TRI-(2-CHLOROETHYL) PHOSPHATE (Tri-β-(chloroethyl) phosphate)

$$[ClCH_2CH_2O\text{-}]_3PO \qquad B.p. 194°C/5 mm.$$

A flame retardant plasticiser, mainly for cellulose esters, but also used in other polymers, especially polyurethane foams.

TRICHLOROFLUOROMETHANE (CFM 11)

$$CFCl_3 \qquad B.p. 23.8°C.$$

The most useful physical blowing agent for polyurethane foams, especially rigid foams. In closed cell foams, it only diffuses out of the cells slowly, and due to its low thermal conductivity, it can usefully contribute to such a foam acting as a thermal insulation material.

TRICRESYL PHOSPHATE (Tritolyl phosphate) (TCP)

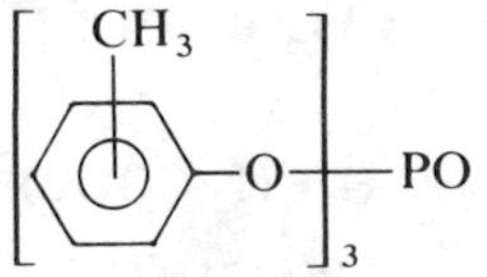

B.p. 420°C/760 mm (decomposes).

Usually a mixture of *meta* and *para* isomers. Commercial materials derived from cresols from coal tar may contain some of the toxic *ortho* isomer. Therefore petroleum derived cresols are preferred. A widely used flame retardant plasticiser for cellulose ethers and esters, polyvinyl chloride and its copolymers, rubbers, alkyd and phenolic resins.

TRIDYMITE A crystalline silica mineral which con-

sists of sheets of SiO_4 tetrahedra linked to form fused six-membered rings, the sheets being joined to form a network through the fourth oxygen, which points alternately above and below the planes of the sheets.

TRIETHYL CITRATE

$$(CH_3CH_2OOCCH_2)_2C(OH)COOCH_2CH_3$$

B.p. 294°C/760 mm, 150°C/3 mm.

A plasticiser for a wide range of cellulose esters and ethers. Also compatible with polyvinyl chloride and its copolymers and with polystyrene, but of poor permanence.

TRIETHYLENEDIAMINE Alternative name for 1,4-diazabicyclo-2,2,2-octane.

TRI-(β-CHLOROETHYL) PHOSPHATE Alter-native name for tri-(2-chloroethyl) phosphate.

TRIETHYLENE GLYCOL

$$HO(CH_2CH_2O\text{-}]_3H \qquad B.p. 287°C.$$

A solvent for cellulose acetate. Miscible with water in all proportions.

TRIETHYLENE GLYCOL DICAPRYLATE

$$[CH_3(CH_2)_6COOCH_2CH_2OCH_2\text{---}]_2$$

B.p. 212–254°C/5 mm.

Commercial mixtures also contain other fatty acid esters. Compatible with a wide variety of natural resins, polyvinyl chloride and rubbers. Occasionally used as a plasticiser in organosols.

TRIETHYLENE GLYCOL DI-2-ETHYLBUTYRATE

$$[CH_3CH_2CH(CH_2CH_3)COOCH_2CH_2OCH_2\text{---}]_2$$

B.p. 202°C/5 mm.

A plasticiser used mostly for polyvinyl butyral but sometimes for polyvinyl chloride and its copolymers and for cellulose esters, especially for low temperature flexibility.

TRIETHYLENE GLYCOL DI-2-ETHYLHEXOATE
(Triglycol dioctoate)

$$[CH_3(CH_2)_3CH(CH_2CH_3)COOCH_2CH_2OCH_2\text{---}]_2$$

B.p. 215°C/5 mm, 370°C/760 mm.

A plasticiser for polyvinyl butyral and synthetic rubbers for good low temperature flexibility.

TRIETHYLENEGLYCOL MONOETHYL ETHER
(Trioxitol) (Ethoxytriglycol)

$$CH_3CH_2(OCH_2CH_2\text{---})_3OH$$

A solvent, similar to diethylene glycol monoethyl ether, but of lower hygroscopicity and higher boiling point.

TRIETHYLENETETRAMINE (TET)

$$H_2N(CH_2CH_2NH)_2CH_2CH_2NH_2$$

B.p. 227°C.

A pungent liquid produced by reaction of ammonia with 1,2-dichloroethane. Useful as a curing agent for epoxy resins; its behaviour is similar to diethylenetriamine.

TRI-2-ETHYLHEXYL PHOSPHATE (Trioctyl phosphate) (TOF)

$$[CH_3(CH_2)_3CH(CH_2CH_3)CH_2O\text{---}]_3PO$$

A plasticiser for polyvinyl chloride and synthetic rubbers, imparting good fire resistant properties.

TRIGLYCERIDE
A triester of glycerol. i.e. having the structure

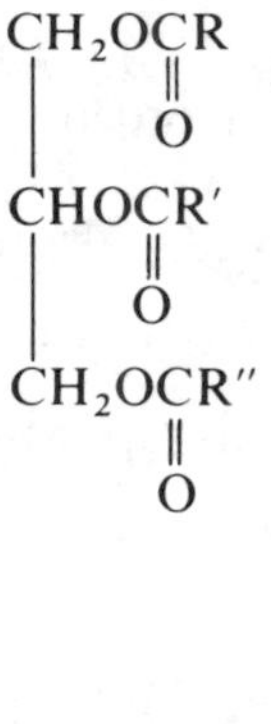

$$
\begin{array}{l}
CH_2OCR \\
\quad \| \\
\quad O \\
CHOCR' \\
\quad \| \\
\quad O \\
CH_2OCR'' \\
\quad \| \\
\quad O
\end{array}
$$

where the R, R' and R" groups may be the same. The main components of many natural plant oils, which are complex mixtures of triglycerides, e.g. linseed, soya bean, castor, tung, safflower, tall, cottonseed and coconut oils. The R, R' and R" groups are long alkyl chains (16 or 18 carbon atoms). Thus the oils are fatty acid triglycerides, often with the alkyl groups being unsaturated. The oils therefore harden by the process of air drying. This property and the toughness they impart, make them essential ingredients in alkyd resins.

TRIGLYCOL DIOCTOATE
Alternative name for triethylene glycol di-2-ethylhexoate.

TRIHYDRAZINOTRIAZENE

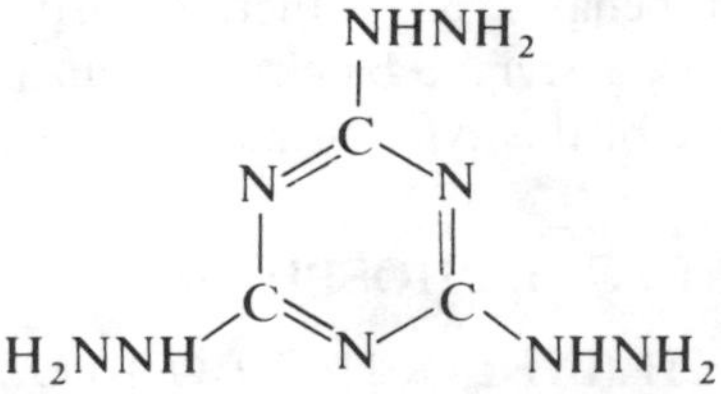

A chemical blowing agent that decomposes at the relatively high temperature of around 275°C to yield nitrogen and ammonia. It is therefore useful for producing cellular polymers from materials which require high melt processing temperatures such as nylons and bisphenol A polycarbonate.

TRIISOOCTYL TRIMELLITATE (TIOTM)

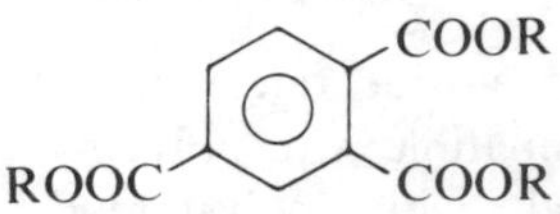

where R is the isooctyl group. A commonly used trimellitate plasticiser for polyvinyl chloride and its copolymers having low volatility and low extractability.

TRIKETOHYDRINDENE HYDRATE
Alternative name for ninhydrin.

TRIMELLITATE PLASTICISER
A plasticiser which is an ester of trimellitic acid, having the structure

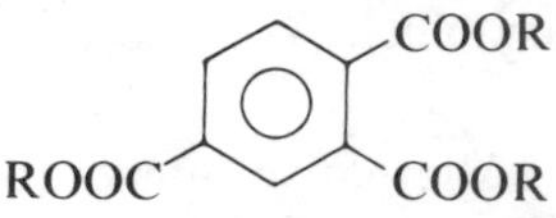

Usually R is the octyl (i.e. 2-ethylhexyl), isooctyl or mixed 7–9 carbon group. Useful plasticisers for polyvinyl chloride for low extractability and low volatility.

TRIMELLITIC ANHYDRIDE (TMA)

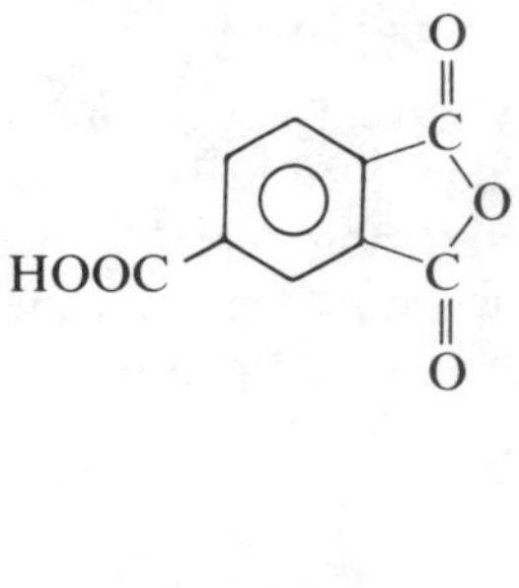

M.p. 168°C.

A useful monomer for the synthesis of polyamide–imides, other polyimides and trimellitate plasticisers. Also useful as an epoxy resin curing agent, giving cured products with high heat distortion temperatures.

TRIMER
A molecule consisting of three mers or repeat units, i.e. an oligomer with a degree of polymerisation of three. Trimers may be formed on degradation of much larger polymer molecules. Thus the extensive hydrolysis of polysaccharides and proteins produces some trisaccharides and tripeptides respectively. Characterisation of trimer structure provides useful information about the structure of the polymer from which it was derived.

TRIMETHYLCHLOROSILANE

$$(CH_3)_3SiCl \qquad \text{B.p. } 58°C.$$

A chlorosilane produced, as a minor product, by the direct process of alkylation of silicon. On hydrolysis an unstable silanol is formed which readily dimerises with loss of water to hexamethyldisiloxane:

$$(CH_3)_3SiCl \rightarrow (CH_3)_3SiOH$$
$$\rightarrow (CH_3)_3SiOSi(CH_3)_3$$

2,3,6-TRI-O-METHYLGLUCOSE

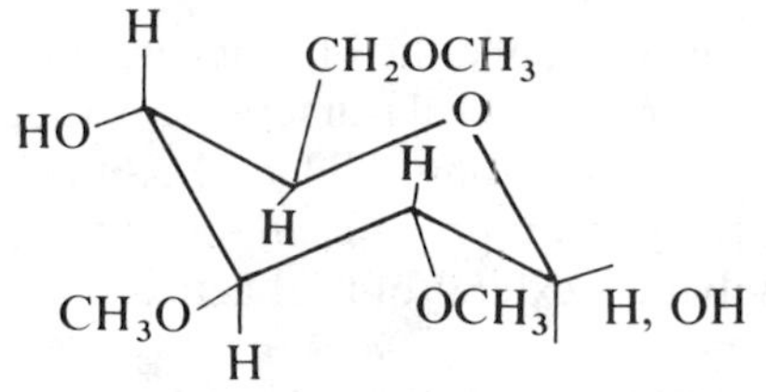

$$\text{M.p. } 124°C. \; \alpha_D^{20} \; +118°.$$

The product of hydrolysis of methylated cellulose, which establishes the cellulose structure as that of a 1,4'-linked glucan. Direct hydrolysis with aqueous mineral acid only gives a 10% yield, but when the methylated cellulose is converted to the trimethylglucoside this may then be hydrolysed to about 90% yield of 2,3,6-tri-O-methyl-glucose.

TRIMETHYLHEXAMETHYLENEDIAMINE (TMD)
A 1:1 mixture of the 2,2,4- and 2,4,4-trimethyl isomers,

$$H_2NCH_2C(CH_3)_2CH_2CH(CH_3)CH_2CH_2NH_2$$

and

$$H_2NCH_2CH(CH_3)CH_2C(CH_3)_2CH_2CH_2NH_2$$

useful as the diamine mixture for the synthesis of poly-trimethylhexamethyleneterephthalamide). Synthesised by trimerisation of acetone to isophorone, followed by hydrogenation and oxidation to trimethyladipic acids, conversion to the dinitriles and hydrogenation to the mixed diamine.

ε-N-TRIMETHYLLYSINE

$$H_2NCHCOOH$$
$$|$$
$$(CH_2)_4$$
$$|$$
$$^+N(CH_3)_3$$

A rare amino acid found in small amounts in myosin

1,1,1-TRIMETHYLOLPROPANE

$$CH_3CH_2C(CH_2OH)_3$$

M.p. 58·8°C. B.p. 295°C.

Produced by the condensation of formaldehyde with butyraldehyde. Sometimes used as a comonomer for the production of slightly branched polyoxypropylene triols for use in flexible polyurethane foam and used for alkyd resin manufacture.

2,2,4-TRIMETHYLPENTANE-1,3-DIOL DIISOBUTYRATE (TMIB)

$$(CH_3)_2CC(CH_3)_2CH_2OOCCH(CH_3)_2$$
$$|$$
$$OOCCH(CH_3)_2$$

A plasticiser for polyvinyl chloride with low staining properties and giving plastisols with low viscosity and good viscosity stability.

2,4,7-TRINITROFLUORENONE (TNF)

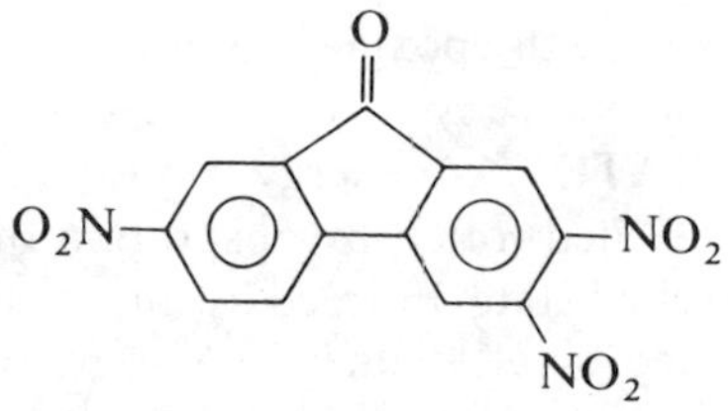

An electron acceptor molecule capable of considerably enhancing the photoconductivity of poly-(N-vinyl-carbazole) by forming charge transfer complexes with it.

TRIOCTYL PHOSPHATE (TOF)
Alternative name for tri-2-ethyl phosphate.

TRIOXANE

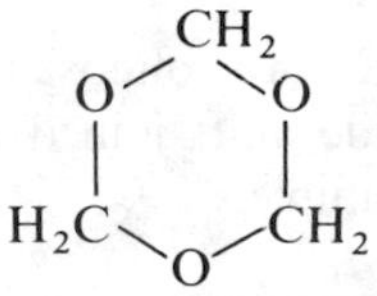

M.p. 62–64°C. B.p. 115°C.

A cyclic formal trimer of formaldehyde, prepared by heating a concentrated (about 55%) aqueous solution of formaldehyde with 2% sulphuric acid. The monomer for the formation of polyacetal copolymers by cationic polymerisation and sometimes used as a source of formaldehyde in the curing of phenol–formaldehyde polymers.

TRIOXITOL   Tradename for triethylene glycol mono-ethyl ether.

TRIPEPTIDE   A peptide containing three α-amino acid residues and therefore two peptide bonds. In general of structure:

$$H_2NCHRCONHCHR'CONHCHR''COOH$$

A well-known naturally occurring example is glutathione whose full name is γ-glutarylcysteinylglycine. Tripeptides are also produced on extensive hydrolysis of proteins.

TRIPHENYL PHOSPHATE   (TPP)

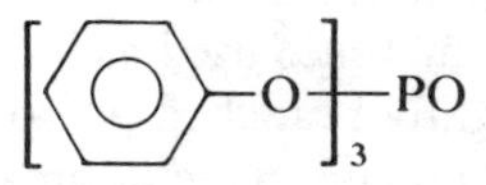

B.p. 220°C/10 mm. M.p. 48·5°C.

A plasticiser mainly for cellulose esters, used in conjunction with another primary plasticiser to prevent its crystallisation. Not compatible with polyvinyl chloride.

TRIPHENYL PHOSPHITE

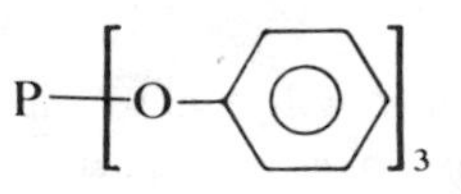

B.p. 360°C, 235°C/18 mm.

Useful as a reactive diluent for epoxy resins, where it reacts with hydroxyl groups, liberating phenol, which itself then reacts with epoxide groups.

TRIPLET STATE   An energy state of a molecule energetically excited (often by absorption of ultraviolet light) to an excited state. Formed by intersystem crossing from an excited singlet state. The spins of a pair of electrons are unpaired in the triplet state. The lowest excited triplet state (T_1) is the most important, especially in photodegradation, when carbonyl groups are involved, as in many hydrocarbon polymer photo-oxidations. Formation of carbonyl triplet states may be pictured as opening of the $>C=O$ bond to form the diradical $>\dot{C}-\dot{O}$.

TRIPOLYMER   Alternative name for terpolymer.

TRISACCHARIDE   An oligosaccharide containing three monosaccharide units joined by glycoside bonds. Cellotriose is an example.

TRIS(1-AZIRIDINYL)PHOSPHINE OXIDE   (APO)

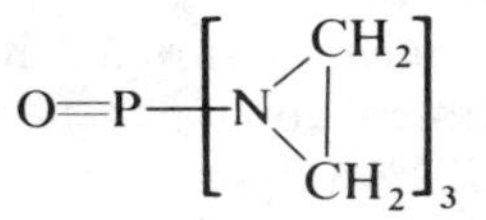

M.p. 41°C.

Formed by reaction of phosphorus oxychloride and ethyleneimine. Widely used for imparting durable flame resistant properties to cellulosic materials. Used either alone, when it chemically reacts with the cellulose hydroxyl groups as well as polymerising, or used in combination with a polyfunctional amine, e.g. ethylenediamine, or amido compound, e.g. urea, with which it forms phosphorus containing polymers deposited in the cellulose substrate. However, APO treatment results in the loss of tensile strength and yellowing of the fabric.

1,1,3-TRIS-(2-METHYL-4-HYDROXY-5-t-BUTYL-PHENYL)BUTANE   Tradename: Topanol CA.

TRIS-(p-NONYLPHENYL)PHOSPHITE

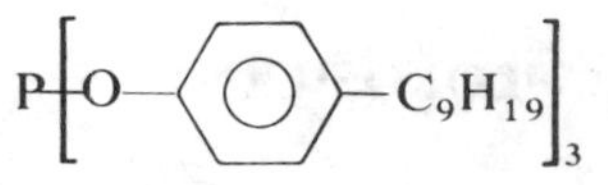

Tradenames: Santowhite TNPP, Polygard. A versatile ultraviolet absorber. Commercial materials are often mixtures of mono- and disubstituted phenyl compounds.

TRITACTIC POLYMER   A polymer in which there are three centres of stereochemical configurational isomerism in each repeat unit. This arises most notably in the 1,4-polymerisation of 1,4-disubstituted butadienes. The polymers have the structure

$$-[CHY-CH=CH-CHX]_n-$$

Such a polymer has one site of geometric isomerism and two sites of stereochemical isomerism associated with the asymmetric carbon atoms. Unlike the situation in vinyl polymers, these last types of centre are truly asymmetric and the polymers exhibit optical activity.

TRITOLYL PHOSPHATE   Alternative name for tricresyl phosphate.

TRIXYLYL PHOSPHATE   (TXP)

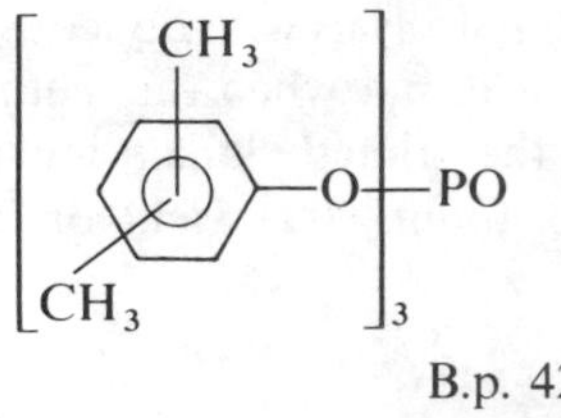

B.p. 420°C/760 mm.

Commercial materials usually contain a minimum of the toxic *ortho* substituted phosphates, but these may be present in materials derived from coal tar xylenols, hence petroleum derived materials are preferred. A flame retardant plasticiser for polyvinyl chloride and its copolymers and for cellulose esters.

TROGAMID   Tradename for nylon 66.

TROGAMID T   Tradename for poly-(trimethylhexamethyleneterephthalamide).

TROLIT   Tradename for cellulose acetate.

TROLITAN Tradename for phenol–formaldehyde polymer.

TROLIT F Tradename for cellulose nitrate.

TROMSDORFF EFFECT Alternative name for auto-acceleration.

TROMSDORFF–NORRISH EFFECT Alternative name for autoacceleration.

TROPOCOLLAGEN The characteristic triple helix structure adopted by the polypeptide chains of collagen. Each tropocollagen molecule consists of three poly-peptide chains wound round each other, with three residues per turn. The triple helix is held together by hydrogen bonds between the —NH— of glycine and a $>$C=O of another chain. This is only possible if every third residue on each chain is glycine. Furthermore interchain crosslinks are also present. Some crosslinks are formed by reaction between two lysine groups, which can be enzymically deaminated to aldehydes, which then condense by an aldol reaction:

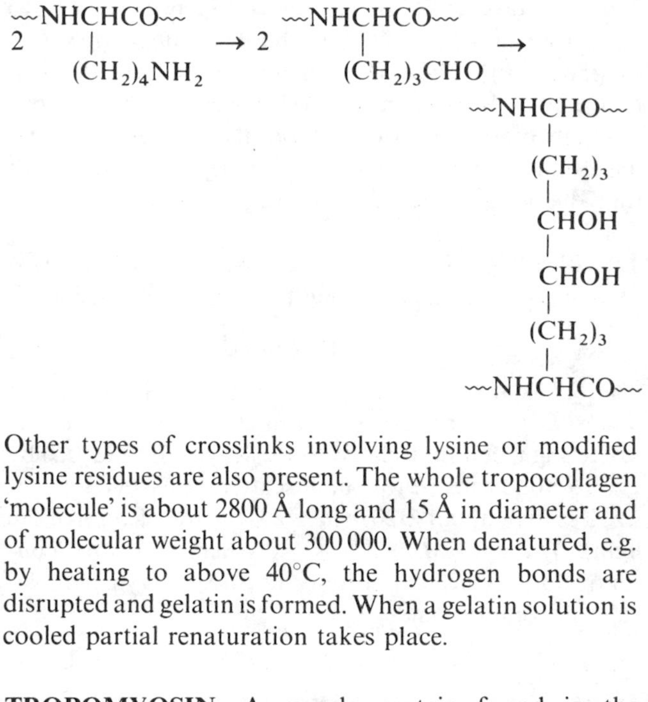

$$2 \underset{\underset{(CH_2)_4NH_2}{|}}{\sim\!\!\sim NHCHCO\sim\!\!\sim} \rightarrow 2 \underset{\underset{(CH_2)_3CHO}{|}}{\sim\!\!\sim NHCHCO\sim\!\!\sim} \rightarrow$$

$$\begin{array}{c} \sim\!\!\sim NHCHO\sim\!\!\sim \\ | \\ (CH_2)_3 \\ | \\ CHOH \\ | \\ CHOH \\ | \\ (CH_2)_3 \\ | \\ \sim\!\!\sim NHCHCO\sim\!\!\sim \end{array}$$

Other types of crosslinks involving lysine or modified lysine residues are also present. The whole tropocollagen 'molecule' is about 2800 Å long and 15 Å in diameter and of molecular weight about 300 000. When denatured, e.g. by heating to above 40°C, the hydrogen bonds are disrupted and gelatin is formed. When a gelatin solution is cooled partial renaturation takes place.

TROPOMYOSIN A muscle protein found in the filaments of striated muscles of animals. It is water soluble, being extracted by hot, slightly alkaline water, after the major proteins myosin and actin have been removed. It consists of a double stranded coil of α-helical chains about 40 nm long, which fit inside the grooves of the F-actin. Its molecular weight is about 70 000 and it contains a large excess of acidic amino acid residues (about 24%) over the basic residues (16%); thus it has a very high zwitterion content. It functions in regulating actin–myosin cross-bridges. The usual form is sometimes called tropomyosin B, whereas a somewhat different form, tropomyosin A, occurs in some invertebrate muscles, e.g. 'catch' muscles of molluscs.

TROPOMYOSIN A (Paramyosin) A special form of tropomyosin found in the 'catch' muscles of molluscs and, unlike the common tropomyosin B, it is water insoluble It has a molecular weight of about 220 000 and is α-helical with a double stranded coiled-coil structure.

TROPOMYOSIN B The almost universally found form of tropomyosin.

TROPONIN A muscle protein found in the thin filaments of striated muscles of higher animals. A globular protein consisting of three polypeptide subunits, each having different functions. Troponin C (TN-C), also called troponin A, which binds Ca^{2+} ions, has a molecular weight of 18 000. Troponin I (TN-I) inhibits the formation of myosin–actin cross-bridges by binding actin and has a molecular weight of 23 000. Troponin T binds tropomyosin.

TROSIPLAST Tradename for polyvinyl chloride.

TROUSERS TEAR TEST A particular testing geome-try for measuring the tear strength of a soft polymer material, such as a soft plastic or, more particularly, a rubber. A uniform cut is made in the edge of a thin sheet or film of the material and the two edges on either side of the cut are pulled apart in opposite directions. The force required to do this may be measured using a tensile testing machine and the tearing energy may be determined.

TROUTON VISCOSITY Alternative name for elongational viscosity. However, often the term is more specifically used to mean the elongational viscosity of an incompressible Newtonian fluid (which is three times the shear viscosity) or of a non-Newtonian fluid for which the same is true at low or zero rates of elongation (which can occur with polymer melts).

TRP Abbreviation for tryptophane.

TRUE STRAIN The integral of nominal strain, i.e.

$$\int_{l_0}^{l_1} dl/l = \ln(l_1/l_0)$$

where l_1 and l_0 are the final and initial strains. Identical to nominal strain at low strains.

TRUE STRESS The load divided by the instantaneous (i.e. deformed) cross-sectional area of the sample on which it acts. It is of more basic significance than nominal stress, when considering intrinsic properties of a material.

TRYPSIN An enzyme found in the digestive tract where it acts as a protease. It is very specific in its action, causing, almost exclusively, hydrolysis at the carboxyl group of the basic amino acids lysine and arginine. Because of its specificity, it is also very useful for the initial cleavage of polypeptide chains of other proteins in their sequencing. The active site is probably similar to that of chymotrypsin, except that the hydrophobic pocket contains an aspartic acid residue which aids in the binding of the basic side groups of lysine or arginine. It is thus a serine enzyme forming an acyl-enzyme as an intermediate in the hydrolysis. It is first produced biologically as its inactive zymogen trypsinogen.

TRYPSINOGEN The zymogen precursor for trypsin. It is converted to trypsin by the loss of a hexapeptide from the N-terminal end, caused by the action of the enzyme enterokinase.

TRYPTOPHANE (Trp) (W)

$$H_2NCHCOOH$$
$$CH_2$$
$$C$$
$$CH$$
$$NH$$

M.p. 281°C (decomposes).

An aromatic α-amino acid found widely in proteins; however, unlike most other standard α-amino acids, it is sometimes absent. Its pK' values are 2·38 and 9·39, with the isoelectric point at 5·89. It has a strong ultraviolet light absorption, which is useful for its quantitative determination, free or combined in a protein. It is destroyed by the normal hydrochloric acid hydrolysis stage of amino acid analysis of proteins, and therefore it is determined by a separate hydrolysis either with mercaptoethanesulphonic acid or by an alkaline hydrolysis. Tryptophane residues are cleaved by chymotrypsin.

TTP Abbreviation for tritolyl phosphate.

TUFCEL Tradename for a polynosic rayon staple fibre.

TUNG DISTRIBUTION An empirical molecular weight distribution which is a particular form of the generalised exponential distribution of molecular sizes (r) expressed, for example, as degrees of polymerisation or molecular weight. Usually given as the integral distribution function,

$$\int_0^r W(r)dr = 1 - \exp(yr^m)$$

where m is an adjustable parameter and $y = -\ln p$, where p is the extent of reaction in a step-growth polymer or the ratio of propagation to termination rates in a chain polymer. In differential form this is

$$W(r) = ym\exp(-yr^m)r^{m-1}$$

In effect it is a modified Schultz distribution.

TUNG OIL An oil obtained from the tung tree, whose triglycerides contain predominantly eleostearic acid residues (about 80%). It is therefore a drying oil and is useful in alkyd resins for its fast drying and water resistant properties.

TURBIDIMETRIC TITRATION A technique of analytical fractionation in which the polymer is continuously precipitated from solution, the higher molecular weight fractions first, by progressive additions of non-solvent. The amount precipitated is measured by the increase in optical density of the suspension, which increases as the turbidity (τ) increases. Assuming that τ depends on the weight of polymer precipitated, and that the volume fraction of the polymer precipitated (P), for incipient precipitation, depends on the concentration c of the particular polymer species about to precipitate and on its molecular weight, then $P = k\log c + f(M)$ has been found to hold. Furthermore, an increase in P (ΔP) causes an increase in $\tau(\Delta\tau)$ due to precipitation of molecules of a certain molecular weight M, and also $\Delta\tau/\tau_\infty = w_i/w$, where τ_∞ is the turbidity at complete precipitation and w_i is the weight of polymer precipitated from a total weight w, corresponding to species i. In addition,

$$w_i = c_i/c = (\Delta\tau/\tau_\infty)/\{1 - [(1 - P)/(1 - P - \Delta P)]\}(10^{\Delta P/k})$$

k and $f(M)$ are determined by calibration with fractions of known molecular weight. Then for any chosen P, the right-hand side of the above equation can be calculated and the corresponding value of $\Delta\tau/\tau_\infty$ is read off from the experimental curve of $\Delta\tau/\tau_\infty$ versus P. Hence c_i/c ($= w_i/w$) is ascertained, whence $f(M_i)$ is obtained from $f(M_i) = P - k\log[c(1 - P)]$. Then M_i is inferred from the empirical $f(M_i)$ versus M_i relationship. Apart from the assumptions made in the theory, the method also has experimental uncertainties, notably that the turbidity often varies with time due to agglomeration of the precipitate.

TURBIDITY Symbol τ. A parameter sometimes used to express the intensity of light scattering, defined as

$$\tau = (1/l)[\log(I_0/I)]$$

where I is the intensity of a beam of light, of initial intensity I_0, after passing through a length l of the scattering medium. Related to the Rayleigh ratio (R_θ) by $\tau = (16/3)R_\theta$, whence for small scattering particles, i.e. of size less than about $\lambda/20$ (where λ is the wavelength of the light), $H'c/\tau = 1/\bar{M}_w + 2Ac + \cdots$, where c is the concentration of the scattering centres (e.g. polymer molecules in solution) and

$$H' = (16/3)[2\pi^2 n_0^2(dn/dc)^2]/[\lambda^4 N_A(1 + \cos^2\theta)]$$

for a dilute polymer solution of refractive index increment dn/dc, where n_0 is the solvent refractive index, N_A is Avogadro's number and θ is the scattering angle. Since I_0/I is usually nearly unity for polymer solutions, turbidity values are very small and are rarely measured directly. Instead the Rayleigh ratio or the reduced scattering intensity are used. However, turbidity is useful in turbidimetric titration and in detailed theoretical treatments of light scattering.

TURBULENT FLOW Flow in which eddies are present, i.e. flow which is no longer laminar. Turbulence occurs when a critical value of the Reynolds number is reached and depends on the flow rate. Thus in extrusion, die turbulence may occur as a result of interruption of the flow lines at the die entry or within the die. This can lead to melt fracture.

β-TURN A type of hairpin loop found in polypeptides and proteins in which the polymer chain turns back on itself over a span of four amino acid residues, forming locally at least, an anti-parallel β-structure. Three different such turns have been identified in the folds of globular proteins. Such turns may also be present in fibrous proteins and synthetic polypeptides adopting the anti-parallel β-sheet structure and in cross-β-structures.

TURNIP YELLOW MOSAIC VIRUS (TYMV) A small isometric plant virus with a diameter of 280 Å comprised of about 35% RNA of molecular weight about 2 million and containing 5770–6500 nucleotides. The particle (i.e. virion) molecular weight is about 5·5 million. The virion contains 180 identical protein subunits as a coat thus making a nucleoprotein supermolecular complex with icosahedral symmetry. The protein has a molecular weight of about 20 000 and is folded in the native virus into a prolate ellipsoid of revolution of about 30 Å and length 60–70 Å.

TURNOVER NUMBER Alternative name for molar activity.

TURPENTINE A hydrocarbon solvent of varying composition obtained by the steam distillation of either the resin exudations of various coniferous trees (giving gum spirit turpentine) or of the stumps or chips remaining after the trees have been cut (giving wood turpentine). The latter may also be obtained by destructive distillation of the stumps or chips. Also obtained as a by-product of wood pulp manufacture, giving sulphate or sulphite turpentine of variable composition. The boiling range is typically about 150–170°C. In its main use as a solvent thinner in paints and varnishes, it has largely been replaced by white spirit.

TUSSAH SILK A wild silk produced in India and China.

TVA Abbreviation for thermal volatilisation analysis.

TWARON Tradename for an aramid fibre similar to Kevlar.

TWINNED CRYSTAL Two crystals joined microscopically in a symmetrical fashion, so that one is a mirror image of the other. It is due to a disturbance in the regular placement of the motifs during the packing of the initial nucleus.

TWO-DIMENSIONAL CHROMATOGRAPHY (Two-dimensional method) A variation of the paper chromatography technique, widely used for the separation of complex mixtures of peptides (from protein hydrolyses) and amino acids, in which the mixture is first chromato-graphed in one direction and the paper is dried and then the mixture is chromatographed with a different solvent in a direction at right angles to the first. In this way a two-dimensional 'map', e.g. a peptide map, of the separated components is formed, with much better separations than are possible in ordinary, i.e. one-dimensional, paper chromatography.

TWO-DIMENSIONAL METHOD A method of separation of peptides or amino acids on paper by paper chromatography or by paper electrophoresis in one direction, followed by chromatography or electrophoresis in a second direction at right angles to the first. Thus a two-dimensional peptide or amino acid 'map' is obtained with much better separations than with the conventional one-dimensional process.

TWO STAGE POLYMER Alternative name for two stage resin.

TWO STAGE RESIN (Two stage polymer) A linear prepolymer, i.e. an A-stage resin, which can only be crosslinked to a network polymer by reaction involving an added crosslinking agent. The term is applied particularly to novolac phenol–formaldehyde polymers, which require the addition of hexamethylenetetramine or another source of formaldehyde for crosslinking.

TXP Abbreviation for trixylyl phosphate.

TYBRENE Tradename for acrylonitrile–butadiene–styrene copolymer.

TYGAN Tradename for a vinylidene chloride copolymer fibre spun from Saran.

TYMV Abbreviation for turnip yellow mosaic virus.

TYPE 8 NYLON Alternative name for N-methyl-methoxynylon.

TY-PLY Tradename for rubber hydrochloride.

TYR Abbreviation for tyrosine.

TYRIL Tradename for styrene–acrylonitrile copolymer.

TYRIN Tradename for chlorinated polyethylene.

TYROSINE (Tyr) (Y)

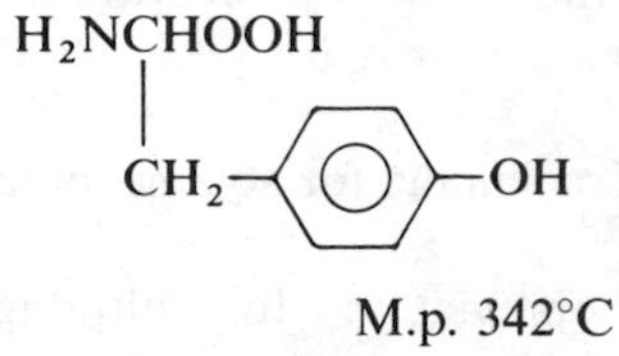

$$H_2NCHOOH$$
$$CH_2 - \langle O \rangle - OH$$

M.p. 342°C (decomposes).

An aromatic α-amino acid found widely in proteins. Its pK' values are 2·20 and 9·11, with the isoelectric point at 5·66. Its strong ultraviolet light absorption at 280 nm is used for its quantitative determination or indeed for the estimation of a tyrosine containing protein in solution. Tyrosine protein residues are cleaved by chymotrypsin and are often associated with the β-conformation. Its very low water solubility is useful for its isolation from protein hydrolysates.

TYZOR Tradename for tetraalkyltitanate.

U

UBBELOHDE VISCOMETER (Dilution viscometer) (Suspended level viscometer) A modified Ostwald U-tube capillary viscometer which does not require a fixed volume of liquid for its use. Therefore dilution of a polymer solution may be carried out in the viscometer by successive additions of solvent without emptying and cleaning it each time, as is required with the Ostwald type. It is therefore more convenient to use when determining the limiting viscosity number of a polymer/solvent pair in the dilute solution viscosity method of polymer molecular weight estimation. A third vertical tube is attached just below the bottom of the capillary so that the liquid emerging from the capillary flows down the walls of the tube below the capillary, due to the formation of a suspended level at the capillary exit.

UCARDEL Tradename for a polysulphone/SAN polymer blend.

UCAR FLX Tradename for very low density polyethylene.

UCARSIL Tradename for polycarboranesiloxane.

UCST Abbreviation for upper critical solution temperature.

UDEL Tradename for the polyethersulphone synthesised from bisphenol A and 4,4'-dichlorodiphenylsulphone. Also called Bakelite sulphone or simply polysulphone.

UF Abbreviation for urea–formaldehyde polymer.

UFORMITE Tradename for urea–formaldehyde polymer.

UGIKRAL Tradename for acrylonitrile–butadiene–styrene copolymer.

UGITEX-S Tradename for styrene–butadiene rubber.

UHMWPE Abbreviation for ultrahigh molecular weight polyethylene.

UKAPON Tradename for an unsaturated polyester resin.

ULDPE Abbreviation for ultralow density polyethylene.

ULSTRON Tradename for polypropylene fibre.

ULTEM Tradename for a polyetherimide.

ULTIMATE ELONGATION Alternative name for elongation at break.

ULTIMATE TENSILE STRENGTH Alternative name for tensile strength.

ULTRAACCELERATOR An accelerator for the sulphur vulcanisation of rubbers which is particularly reactive and hence provides very rapid vulcanisation. Dithiocarbamates and thiuram disulphides are the best known types.

ULTRACENTRIFUGATION The technique of subjecting a dilute polymer solution (or suspension of small particles) to a high gravitational field so that the molecules (or particles) undergo sedimentation. This is performed by spinning a cell containing the solution at a high angular velocity ω in an ultracentrifuge. The variation of the concentration of solute with position in the cell either at equilibrium (equilibrium centrifugation) or with time (sedimentation velocity method) is determined by optical means in an analytical ultracentrifuge. From the data obtained, the weight average molecular weight ($\bar{M}_w$) may be evaluated. Used especially with biopolymers, e.g. proteins, since these are often monodisperse which makes the data easier to interpret than with synthetic polymers. Information on other molecular weight averages and molecular weight distribution may also be obtained. Mixtures of biopolymers may also be analysed, especially by density gradient ultracentrifugation. Preparative ultracentrifugation is also used to achieve separation of components on the gram or milligram scale for biopolymer mixtures.

ULTRACENTRIFUGE An instrument in which a small cell containing a solution or suspension is mounted in a rotor, which is spun at very high angular velocities, so that the cell contents are exposed to high gravitational fields—ultracentrifugation. In this field the particles, which may be dissolved polymer molecules, sediment. The sedimentation equilibrium method requires speeds of several thousand revolutions per minute, but the sedimentation velocity method requires speeds of up to 10^5 rpm, corresponding to a gravitational field of up to $10^6 g$. The rotor chamber is evacuated to reduce air friction. In an analytical ultracentrifuge sedimentation may be studied by measuring the solute concentration distribution in the cell.

For polymer solutions this is a valuable method of determining the weight and z-average molecular weights, particularly of proteins and nucleic acids. The cell is of a truncated sector shape whose (theoretical) apex is at the centre of rotation. The rotor and cells contain windows so that the concentration of solute as a function of distance from the centre may be followed by passing a beam of light parallel to the axis of rotation through the sample during rotation. Several different optical systems are used to detect and record solute concentration. With absorption and interference optics, the output is proportional to the concentration at the point in the cell being monitored. In the Schlieren method the concentration gradient is determined.

ULTRADUR Tradename for polybutylene terephthalate.

ULTRAFILTRATION (Membrane filtration) A speeded-up version of dialysis in which centrifugation or the application of pressure to the polymer solution is used to force small molecules through the membrane. Since solvent, as well as small solute, molecules can be removed, the process can be used for concentrating as well as for desalting polymer solutions.

ULTRAFORM Tradename for polyoxymethylene copolymer.

ULTRAHIGH MOLECULAR WEIGHT POLYETHYLENE (UHMWPE) A linear polyethylene produced by Ziegler–Natta polymerisation, with a weight average molecular weight in the range 1–5 million. The polymers, although linear, crystallise with difficulty (owing to the extremely large size of the molecules) to a solid with a density of about $0.94 \, \mathrm{g \, cm^{-3}}$. They thus have a slightly lower stiffness and yield strength than HDPE, but have improved impact strength, environmental stress cracking resistance and creep resistance.

ULTRALOW DENSITY POLYETHYLENE (ULDPE) An ethylene copolymer with a higher olefin comonomer such as butene-1 or octene-1, with a density in the range $0.860–0.900 \, \mathrm{g \, cm^{-3}}$. These polymers approach ethylene–propylene–diene monomer rubber in properties.

ULTRAMICROTOMY Microtomy in which extremely thin slices are produced. Often necessary for samples which are to be viewed by electron microscopy.

ULTRAMID Tradename for nylons 6, 66 and 610.

ULTRAMID A Tradename for nylon 66.

ULTRAMID B Tradename for nylon 6.

ULTRAMID K Tradename for a transparent polyamide produced by reaction of adipic acid, hexamethylenediamine and

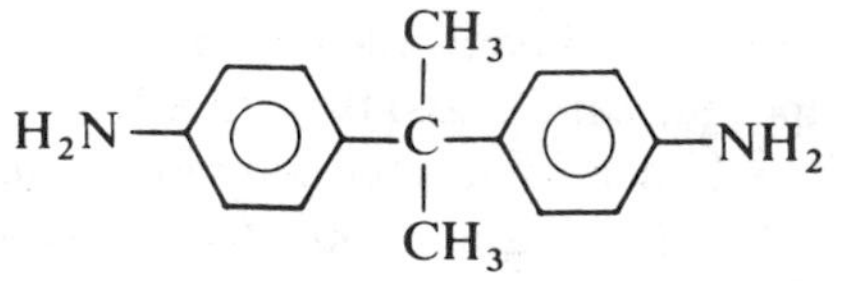

ULTRAMID S Tradename for nylon 610.

ULTRAPAS Tradename for melamine– and melamine/urea–formaldehyde polymers.

ULTRAPEK Tradename for an aromatic polyetherketone.

ULTRAPHOSPHATE GLASS A polyphosphate in which, compared with the linear polymer, pairs of $-\mathrm{O^-M^+}$ ions have been replaced by an $-\mathrm{O}-$ crosslink, to give a network polymer, the crosslinking units being

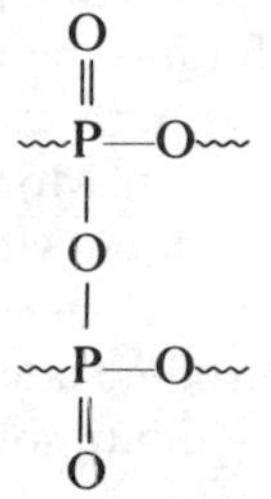

The polymers are made by heating a metal dihydrophosphate ($\mathrm{MH_2PO_4}$) with additional phosphoric acid or with a precursor to phosphoric oxide, such as ammonium dihydrogen phosphate, to provide the crosslinking sites. Glasses of complex composition may be made by heating mixtures of metal oxides or carbonates with phosphoric acid or ammonium phosphate. Up to about 500°C, linear polymers are formed containing acidic hydrogens. These crosslink on further heating to above 600°C. The T_g values of these amorphous polymers can vary considerably depending on both the crosslink density and the metals present. The latter participate to different extents in 'tightening' the structure through ionic forces. Quite low values of T_g, giving low softening point polymers, are possible, e.g. lead/alkali metal polymers have T_g values in the range 150–250°C. The polymers are subject to hydrolysis,

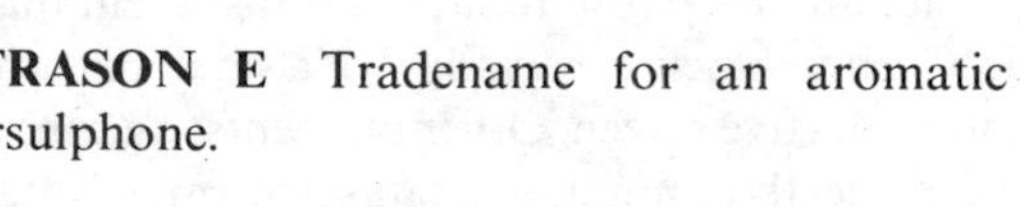

the rate being dependent on the metals present and the crosslink density. The alkali metal containing polymers are attacked even at room temperature; other salts, such as lead, give more durable polymers. However, the related borosilicate glasses are even more durable.

ULTRASON E Tradename for an aromatic polyethersulphone.

ULTRASONIC DEGRADATION A type of mechanochemical degradation occurring when a polymer solution is irradiated with ultrasonic radiation. Cavitation (rapid collapse of regions of low pressure) occurs and the resulting very high local shearing forces can cause mechanochemical chain scission of the polymer, thus reducing molecular weight.

ULTRASONIC RELAXATION Relaxation occurring as a result of ultrasonic irradiation. Irradiation results in a periodic longitudinal compression and rarefaction (i.e. a pressure fluctuation) of the material. This may be resolved into an isotropic and a shear component. With appropriate sample geometry $L^* = K^* + G^*/3$, where L^*, K^* and G^* are the complex moduli for longitudinal waves, bulk and shear deformations respectively. The response of a sample to the two components may be evaluated so that effectively such measurements amount to the determination of the moduli at higher frequencies than in other mechanical relaxation methods. Usually measurements are made at a fixed frequency over a temperature range, so that transitions may be observed at higher frequencies than is normal. In addition the passage of a wave may cause adiabatic heating and cooling with consequent effects on wave propagation. Monitoring of this response yields thermodynamic data which can be related to specific conformational changes in the polymer. In practice the elucidation of the rates and energetics of these changes is restricted to dilute solutions.

ULTRASON S Tradename for an aromatic polysulphone.

ULTRATHENE Tradename for ethylene–vinyl acetate copolymer.

ULTRAVIOLET ABSORBER (UV absorber) A type of ultraviolet stabiliser, soluble in the polymer it is protecting, which is a powerful absorber of ultraviolet radiation in the wavelength range which causes photodegradation of the polymer. Usually only 0·01–0·1 wt% provides adequate protection against the ultraviolet component of sunlight responsible for weathering.

Ideally an absorber must be sufficiently compatible with the polymer so that it does not separate or 'bloom' out. It must be stable to ultraviolet light itself, it must not be coloured or give coloured products and it must have low volatility and good thermal stability. The main types are the 2-hydroxybenzophenones, the 2-hydroxyphenylbenzotriazoles and various derivatives of salicylic acid, e.g. phenyl salicylate. Ultraviolet absorbers are capable of dissipating the ultraviolet energy harmlessly as heat by internal conversion.

ULTRAVIOLET SCREEN (UV screen) (UV screening agent) A type of ultraviolet stabiliser that is insoluble in the polymer and therefore renders a transparent material opaque. Carbon black, especially if of a small particle size, is the most effective screen. Other pigments, e.g. zinc oxide and the rutile (but not the anatase) form of titanium dioxide, have lesser screening power and act by reflecting rather than absorbing the ultraviolet radiation.

ULTRAVIOLET STABILISER (Photostabiliser) An additive capable of inhibiting ultraviolet light induced degradation (photodegradation), which is especially important in polymers used out-of-doors where exposure to sunlight results in weathering. Both ultraviolet screens and absorbers are widely used. More recently, quenching agents, capable of acting as energy transfer agents for the excess energy of the polymer excited states, have been developed.

ULTRAX Tradename for an aromatic polyester thermotropic liquid crystalline polymer.

ULTRYL Tradename for polyvinyl chloride.

UNCONJUGATED PROTEIN Alternative name for simple protein.

UNDERCOOLING (1) Alternative name for supercooling.
(2) The temperature difference between the melting point and crystallisation temperature of a polymer.

UNEPRENE Tradename for a thermoplastic polyolefin rubber.

UNIAXIAL ORIENTATION (Axial orientation) (Fibre orientation) (Monoaxial orientation) Orientation with cylindrical symmetry about a characteristic direction (the principal axis). It is present in drawn and spun fibres, when the polymer molecules (or crystallites) are oriented in the long direction of the fibre, which is also the principal axis, or in this case, the fibre axis.

UNIAXIAL STRAIN A strain whose only component is in a single direction, this direction often being designated the z-direction, so that $\varepsilon_{zz} \neq 0$, $\varepsilon_{xx} = 0$ and $\varepsilon_{yy} = 0$. A common strain situation, e.g. found in a simple tensile test, when the sides of the sample are constrained from expanding or contracting.

UNIAXIAL STRESS A stress whose only component is in a single direction, this direction often being designated as the z-direction, so that $\sigma_{zz} \neq \subset 0$, $\sigma_{xx} = 0$ and $\sigma_{yy} = 0$. Commonly found in a simple tensile test. It produces a strain $\varepsilon_{zz} = \varepsilon_{zz}/E$, where E is the Young's modulus, a contraction along the x and y axes accompanied by a dilatation, $\Delta = (1 - 2v)\sigma_{zz}/E$, and since $v < \frac{1}{2}$ this is a volume increase.

UNIPLANAR ORIENTATION Orientation in which the oriented elements, e.g. polymer chains, are parallel to a particular plane. It usually applies to polymer films in which the chains are parallel to the film surface.

UNIPOLYMER Alternative name for homopolymer.

UNIT CELL The basic unit for describing the ordered arrangement of atoms in a crystal. More exactly, the smallest parallelepiped which can generate the crystal lattice by repeated translations along the axes of the lattice. It is characterised by the lengths of the sides of the parallelepiped (a, b and c) and the angles between them (α, β and γ). In polymer crystals the commonest unit cells are monoclinic ($a \neq b \neq c$; $\alpha = \gamma = 90°$, $\beta \neq 90°$), e.g. isotactic polypropylene and nylon 6, orthorhombic ($a \neq b \neq c$; $\alpha = \beta = \gamma \neq 90°$), e.g. polyethylene, polyisobutylene and syndiotactic PVC, triclinic ($a \neq b \neq c$; $\alpha \neq \beta \neq \gamma \neq 90°$), e.g. polyethyleneterephthalate and nylon 66, trigonal ($a = b = c$; $\alpha = \beta = \gamma \neq 90°$), e.g. isotactic polystyrene and tetragonal ($a = b \neq c$; $\alpha = \beta = \gamma = 90°$), e.g. many isotactic vinyl polymers with large side groups. The motifs are not usually located at the lattice points. The lattice axes chosen are those giving the simplest analytical description. The unit cell structure (the crystal structure) is usually determined by wide angle X-ray diffraction, but is more difficult to determine with polymer crystals than with non-polymer crystals since the former are often not so well developed. When only small polymer single crystals are available electron diffraction may be more useful.

UNIT ELONGATION   Alternative name for deformation ratio.

UNIVERSAL CALIBRATION   A method of calibration in gel permeation chromatography (GPC) which enables the chromatogram for any polymer/solvent combination to be converted from a concentration–retention volume (V_R) plot to a concentration–molecular size plot. This is possible on the assumption that the separation of different molecular sizes is solely on the basis of their molecular size and not on the basis of their chemical structure. Assuming further that the equivalent sphere hydrodynamic volume may be taken as the size of the polymer coils in solution, then according to the Flory-Fox equation, the product $[\eta]M$ is proportional to the volume of the hydrodynamically equivalent sphere, where $[\eta]$ is the limiting viscosity number and M is the molecular weight of the polymer coil. Hence a plot of log ($[\eta]M$) versus V_R, obtained with standard polymers, usually polystyrene, provides a universal calibration curve. To convert the $[\eta]M$ values obtained for the sample being analysed requires further calibration of $[\eta]$ versus V_R. This is often conveniently carried out by collecting fractions from the GPC eluate and by determining $[\eta]$ by dilute solution viscometry, especially if a continuously recording viscometer is built into the GPC instrument. Alternatively published values of the Mark–Houwink constants may be used.

UNPERTURBED DIMENSIONS   The dimension of a polymer coil in dilute solution at the theta temperature. Under these conditions the long-range interactions between segments of the polymer chains, causing the chain to contract, are just balanced by the solvation forces. The polymer chain conformation is then solely determined by short-range forces through bond angles, bond distances and bond rotations. The chain then assumes a shape as though the solvent were not present. The average mean square end-to-end chain distance in the unperturbed state ($\langle r^2 \rangle_0$) may be calculated for various models of an isolated polymer chain, e.g. the freely jointed chain, the freely rotating chain and the chain with restricted rotation. Recently small angle neutron scattering measurements have confirmed the earlier postulate

that in an amorphous polymer the molecules exhibit their unperturbed dimensions.

UNRELAXED COMPLIANCE   (Glassy compliance) Symbols J_u (or J_g) in shear, D_u (or D_g) in tension and B_u (or B_g) in bulk. The limiting compliance of a viscoelastic material when measured at very short times, thus corresponding to the purely elastic response characteristic of the polymer if it was in the glassy state. Experimentally it is often difficult to measure reliably.

UNRELAXED MODULUS   (Glassy modulus) Symbols G_u (or G_g) in shear, E_u (or E_g) in tension and K_u (or K_g) in bulk. The inverse of the unrelaxed compliance. The limiting modulus of a viscoelastic material when measured at very short times.

UNSATURATED POLYESTER   (UP) Tradenames: Alpolit, Atlac, Beetle, Cellobond, Gabraster, Grilesta, Hetron, Leguval, Norsodyne, Stypol, Synolite, Synres, Ukapon, Vestopal, Ugikapon. A polyester which contains repeating units having carbon–carbon double bonds in the main chain. These may enter into free radical chain copolymerisation with an added unsaturated monomer, thus crosslinking the polyester chains. This is the basis for their commercial importance as unsaturated polyester resins for the manufacture of glass reinforced plastic products by the techniques of laminating, hand and spray lay-up and dough and sheet moulding.

Resins are made by the high temperature (150–200°C) melt polymerisation of a diol (usually propylene glycol) with an unsaturated acid (maleic acid as maleic anhydride) together with a modifying acid (often phthalic acid as phthalic anhydride). A typical molar ratio would be 1·2:0·67:0·33 respectively. However, a wide variety of other monomers are also used. Other glycols may be diethylene or neopentylene glycols as well as bisphenol A. Fumaric acid is sometimes preferred to maleic acid. Similarly phthalic acid may be replaced by isophthalic acid, adipic or sebacic acid, or endomethylenetetrahydrophthalic anhydride. The crosslinking monomer is usually styrene at about 35 wt% of resin, which when added to the viscous resin dilutes it and so lowers the viscosity. Sometimes other vinyl monomers, such as methyl methacrylate, diallyl phthalate or triallylcyanurate are used. If especially fire resistant resins are needed, then halogenated monomers, such as chloromaleic acid, tetrachloro- or tetrabromophthalic anhydride, chlorendic acid or 2,5-dichlorostyrene may be used.

The linear polymer, typically with a molecular weight of 1000–2000 is a viscous liquid and is crosslinked (or cured) to a hard tough product by copolymerisation with the added vinyl monomer by free radical polymerisation. This is initiated either at high temperatures (about 100°C) with a peroxide, or at ambient temperature by use of an initiator together with an activator or accelerator—cold curing. Commonly methylethyl ketone peroxide or cyclohexanone peroxide together with a soluble metal salt such as cobalt naphthenate or octoate are used.

Sometimes cold curing is performed with an amine and peroxide system.

The very wide variety of crosslinked products that are manufactured can have a wide variety of mechanical properties. In general they are hard and relatively tough, with reasonable heat resistance (to 100–150°C) and solvent resistance. In general, chemical resistance is also good, but the ester links are susceptible to hydrolysis.

UNZIPPING A rapid sequence of chemical reaction steps progressing along a polymer chain, once the first step has initiated the process in the particular polymer molecule being considered. Encountered in polymer degradation reactions particularly in the unzipping loss of monomer molecules that occurs in depolymerisation. The tendency to unzip is given by the zip length.

UP Abbreviation for unsaturated polyester resin.

U-POLYMER Tradename for a polyarylate copolymer of terephthalic acid and isophthalic acid with bisphenol A. An amorphous polymer with a T_g value of about 180°C, typical engineering polymer properties, good impact strength and electrical properties.

UPPER CRITICAL SOLUTION TEMPERATURE (UCST) The critical solution temperature (T_c) above which a binary mixture exists as a homogeneous, single phase solution no matter what the composition, and below which phase separation can take place. It occurs near the maximum on the cloud point curve. It is found in the majority of mixtures for which solubility increases with temperature due to the decrease in the Flory–Huggins parameter (χ) with temperature. At the UCST the quantities $\partial(\Delta G_m)/\partial x_2$, $\partial^2(\Delta G_m)/\partial x_2^2$ and $\partial^3(\Delta G_m)/\partial x_2^3$ are all zero (ΔG_m is the free energy change on mixing and x_2 is the solute (polymer) mole fraction). Differentiation of the expression for ΔG_m from the Flory–Huggins theory gives, $\phi_{2c} = 1/(1 + x^{1/2})$ and $\chi_c = \frac{1}{2}(1 + x^{-1/2})$ at the critical point, where χ_c is the critical value of χ, x is the polymer degree of polymerisation and ϕ_{2c} is the value of the polymer volume fraction (ϕ_2) at the critical point. For infinite molecular weight ($x \rightarrow \infty$) χ reaches 0.5, corresponding to the theta point, i.e. T_c is the theta temperature. For some systems a lower critical solution temperature exists as well, but, confusingly, this is at a higher temperature.

UPPER NEWTONIAN VISCOSITY The coefficient of viscosity of a fluid at very high shear rates, where Newtonian behaviour is observed, although the fluid is non-Newtonian at lower shear rates. This is often true for polymer melts since the molecules become fully elongated and disentangled above a critical shear rate, so that above this rate the resistance to flow does not change. However, this usually only happens at higher shear rates than those achieved in rheometers or during processing. Since polymer melts are usually shear thinning, the value of the upper Newtonian viscosity is usually lower than the value of the apparent viscosity at low shear rates, and several orders of magnitude lower than the 'static' Newtonian viscosity.

UPVC Abbreviation for unplasticised polyvinyl chloride.

URAC Tradename for urea–formaldehyde polymer.

URANOX Tradename for an epoxy resin.

UREA

$$CO(NH_2)_2$$

M.p. 132–136°C.

Produced by reaction of liquid carbon dioxide with ammonia at 135–200°C and 70–230 atm pressure. Initially ammonium carbamate is formed,

$$2\,NH_3 + CO_2 \rightarrow NH_2COONH_4$$

which subsequently decomposes to urea and water. Urea is highly water soluble and is weakly basic, forming salts with some acids. On heating, ammonia is liberated and biuret ($NH_2CONHCONH_2$) and cyanuric acid (1,3,5-trihydroxy-2,4,6-triazine) are formed. Urea is useful as the monomer for the production of urea–formaldehyde polymers and for conversion to melamine by heating with ammonia.

UREA–FORMALDEHYDE POLYMER (UF) Tradenames: Aerolite, Beckamin, Beetle, Cellobond, Cibanoid, Epok, Gabrite, Iporka, Kaurit, Mouldrite, Nestorite, Plaskon, Pollopas, Resamin, Resimene, Scarab, Sternite, Uformite, Urac. An aminopolymer formed by reaction of urea with formaldehyde at about a 1:3 ratio in aqueous solution, initially at neutral or slightly alkaline pH. This produces a mixture of mono- and dimethylolureas:

$$H_2NCONH_2 + HCHO \rightarrow H_2NCONHCH_2OH \rightarrow$$
$$HOCH_2NHCONHCH_2OH$$

The mixture is then acidified, when condensation occurs between methylol and unreacted amide groups to form methylene bridges:

$$H_2NCONHCH_2OH + H_2NCONHCH_2OH \rightarrow$$
$$\sim\!\sim\!NHCONHCH_2NHCONHCH_2\!\sim\!\sim$$

Further reaction of the $\sim\!\sim\!NH\!\sim\!\sim$ groups produces pendant methylol groups:

$$\sim\!\sim\!NH\!\sim\!\sim + HCHO \rightarrow \sim\!\sim\!N\!\sim\!\sim$$
$$|$$
$$CH_2OH$$

Crosslinking can then result by further reaction of methylol groups with other methylols, to give ether links, or with $\sim\!\sim\!NH\!\sim\!\sim$ groups, to give methylene links if more than two such reactions occur in each molecule. In practice reaction is stopped at an intermediate stage before sufficient crosslinking has occurred to cause gelation.

Such prepolymers find many uses as the basis for moulding materials after compounding with filler (usually α-cellulose), acidic crosslinking catalysts (often called hardeners or accelerators) and other ingredients. In moulding such a compound at 125–160°C, the acid produced from the hardener causes extensive crosslinking to occur through formation of further methylene and ether links. The resultant thermoset plastic product is not coloured, unlike phenol–formaldehyde thermosets, and is of low cost. The mechanical properties of a typical α-cellulose filled material (with a 2:1 polymer:filler ratio) are: tensile strength 50–80 MPa and Izod impact strength 0·17–0·24 J (12·7 mm)$^{-1}$. Dielectric strength is 120–200 V per 0·001 m and volume resistivity is 10^{13}–10^{15} Ω m. For adhesive use, especially with wood, a urea:formaldehyde ratio of about 1:2 is used and the prepolymer is used as an aqueous solution after mixing with a hardener, often ammonium chloride, which liberates hydrochloric acid by reaction with excess formaldehyde.

Urea–formaldehyde polymers may be modified by reaction with an alcohol, especially n-butanol giving butylated urea–formaldehyde with methylol groups:

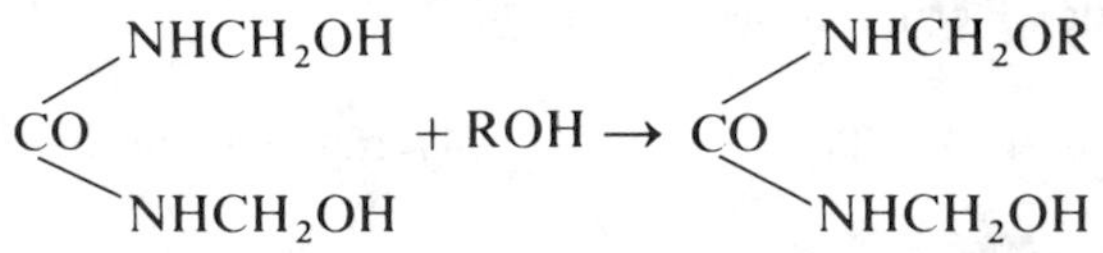

$$CO\begin{array}{c}NHCH_2OH\\ \\NHCH_2OH\end{array} + ROH \rightarrow CO\begin{array}{c}NHCH_2OR\\ \\NHCH_2OH\end{array}$$

The etherified resins are more soluble in organic solvents and are used with alkyd resins in stoving enamels. Water solubility may be increased by the formation of ionic groups on the polymer, either by reaction of methylol groups with sodium bisulphite, or by reaction with organic bases followed by acidification. Such ionic resins are useful in improving the wet strength of paper. Methylolurea ethers, especially the methyl ether, and polymers with various cyclic ureas are useful as textile finishes.

UREPAN 600 A millable polyurethane elastomer, based on poly-(ethylene adipate) and tolylene diisocyanate dimer, cured using an excess of diisocyanate. Formerly Desmophan A.

UREPAN 640 A millable polyurethane elastomer, based on a polyester and MDI and cured with peroxides.

URETHANE A compound of the type

R—O—CONH—R'

Often the term is also used as an abbreviation for polyurethane. The urethane group, from which urethanes get their name, is formed by reaction between a hydroxyl group and an isocyanate group. Therefore polyurethanes result from reaction of a di- or polyisocyanate with a diol or polyol. This reaction occurs readily at 25–50°C with primary alcohols, but less readily with secondary and tertiary alcohols. Aromatic isocyanates react more readily than aliphatic isocyanates in the absence of steric effects.

URETHANE OIL A product of the reaction of the mixed partial esters (derived from drying oil fatty acids

with one or more polyhydric alcohols) with a diisocyanate. Widely used in wood varnishes and other surface coatings applications. A generalised structure is

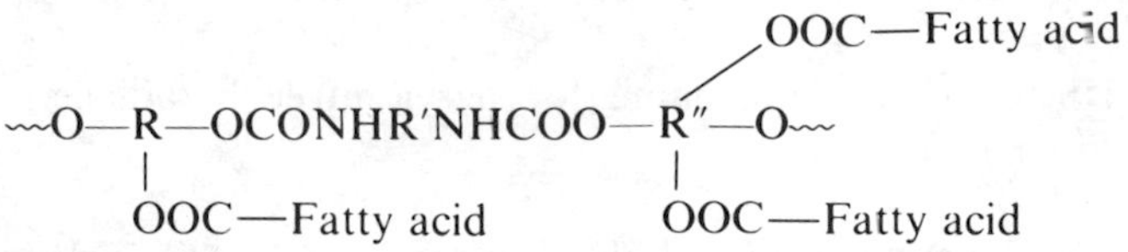

~~O—R—OCONHR'NHCOO—R''—O~~ with OOC—Fatty acid groups

where R(OH)$_3$ and R''(OH)$_4$ are the original polyols and OCN—R'—NCO is the original diisocyanate. They are therefore analogous to oil modified alkyd resins, with the urethane units derived from the diisocyanate replacing phthalic acid units. Long oil length urethane oils are made from pentaerythritol or triols such as glycerol or trimethylolpropane, but for short oil lengths (below about 60%) the mean polyol functionality must be below three. This is achieved using a diol either of low molecular weight or a polyoxypropylene diol of molecular weight about 1000. Such oils are known as chain extended urethane oils. Drying oils used in ester formation are linseed, soya bean, safflower, sunflower and tall oils, the last three (as with alkyds) giving less yellow products. Tolylene diisocyanate is often used as the diisocyanate. However, coatings based on this isocyanate yellow on exposure to sunlight. Those based on cycloaliphatic diisocyanates show much less yellowing. Coatings from these oils dry by air-drying (through the unsaturation in the fatty acid chain) and yield products superior to alkyd resins in resistance to marring, abrasion, water and dilute aqueous solutions.

URETHANE RUBBER Alternative name for polyurethane elastomer.

URETIDINE DIONE Alternative name for isocyanate dimer.

URIDENE Tradename for solution styrene–butadiene rubber.

URON A compound containing the ring structure

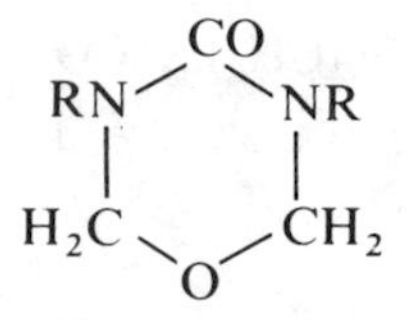

where R is hydrogen in uron itself. The compound with R = —CH$_2$OCH$_3$ (N,N'-bis-(methoxymethyluron)) is useful as a cellulose textile crease resistant finish.

URONIC ACID A sugar acid in which the terminal hydroxyl group of a monosaccharide has been replaced by a carboxyl group, i.e. OHC(CH$_2$OH)$_n$COOH, usually with n = 4 (hexuronic acid). Common examples are D-glucuronic, D-galacturonic, D-mannuronic, L-guluronic and L-iduronic acids. Often found as a component of plant gums and pectic substances.

URTAL Tradename for acrylonitrile–butadiene–styrene copolymer.

URYLON Tradename for nylon 91 fibre.

USCOLITE Tradename for acrylonitrile–butadiene–styrene copolymer.

UV ABSORBER Abbreviation for ultraviolet absorber.

UV SCREEN Abbreviation for ultraviolet screen.

UV SCREENING AGENT Alternative name for ultraviolet screen.

V

V Symbol for valine.

VAL Abbreviation for valine.

VALANIS–LANDEL HYPOTHESIS In the theory of rubber elasticity, the hypothesis that the strain energy function (W) can be represented as the sum of three separate functions of the three principal extension ratios λ_1, λ_2 and λ_3, i.e. as, $W = W(\lambda_1) + W(\lambda_2) + W(\lambda_3)$, where $W(\lambda_i)$ are of identical form. This is consistent with the theory of a Gaussian network, with the simple forms of a non-Gaussian network and with Ogden's theory. The hypothesis makes it possible to describe the elasticity of a rubber over a wide range of strains in terms of the comparatively simple problem of finding the form of the function W of a single variable. Also it is possible to obtain the form of $W(\lambda_i)$ from a single experiment in pure shear.

δ-VALEROLACTAM Alternative name for 2-piperidone.

δ-VALEROLACTONE

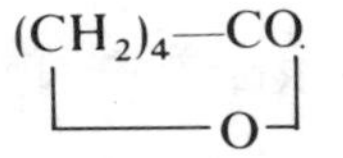

$$(CH_2)_4—CO$$
$$\underline{\qquad\qquad O}$$

M.p. 220°C.

Prepared by the oxidation of cyclopentanone. It readily polymerises to poly-(δ-valerolactone) by ring-opening polymerisation, especially with acid catalysts.

VALINE (Val) (V)

$$H_2NCHCOOH$$
$$|$$
$$CH$$
$$/ \ \backslash$$
$$CH_3 \ \ CH_3$$

M.p. 315°C (decomposes).

A non-polar α-amino acid present in most proteins, but usually only in small amounts. Its pK' values are 2·32 and 9·62, with the isoelectric point at 5·96. It is often associated with the β-conformation in proteins. In the hydrochloric acid hydrolysis step of protein amino acid analysis, valine is only slowly released and therefore its occurrence is obtained by extrapolation to infinite hydrolysis time.

VALOX Tradename for polybutylene terephthalate.

VAMAC Tradename for an ethylene–methylacrylate–carboxy monomer terpolymer rubber.

VAN DER POEL EQUATION A theoretical expression for the elastic properties of a composite, derived on the basis of a self-consistent model, in which a spherical filler particle is embedded in a shell of matrix which, in turn, is embedded in a material whose properties are those of the composite as a whole. The shear modulus relationship is considerably more dependent on composition and is much more complex than in the similar Kerner equation.

VANDIKE Tradename for polyvinyl acetate.

V and PM NAPHTHA Abbreviation for varnish and paint makers naphtha.

VANILLIN

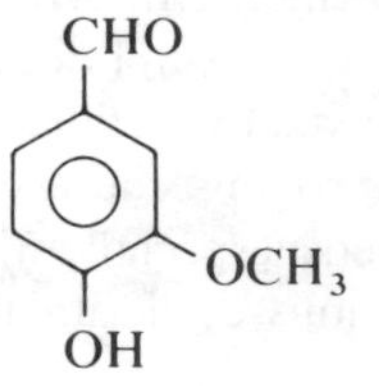

M.p. 81°C.

One of the main products of oxidative degradation of lignin.

VAPOUR PRESSURE LOWERING One of the colligative properties of a solution and the basis of a method for the determination of the molecular weight of a solute, since for a dilute solution, the lowering of the solvent vapour pressure, $(p_0 - p)/p_0 = x_2$, where p_0 and p are the vapour pressures of the pure solvent and the solution respectively and x_2 is the mole fraction of solute. However, direct measurement of the very small pressure differences involved is difficult, especially for polymer solutions, e.g. a 1% (w/v) solution of solute of molecular weight of 20 000 in benzene gives a relative lowering of 2×10^{-5}, corresponding to a change of only 0·5 Pa. Several indirect methods have been used more successfully, based on the isothermal distillation of solvent. These include the isopiestic distillation and porous disc methods. However, these earlier methods have been largely replaced by the thermoelectric method (vapour pressure osmometry). This is one of the most popular methods for polymers with number average molecular weights of less than about 20 000.

VAPOUR PRESSURE OSMOMETRY (VPO) (Thermoelectric method) A technique, utilising the principle of vapour pressure lowering, in which isothermal distillation of solvent occurs from solvent to a solution of a solute whose molecular weight is required. The distillation is detected and measured by the heat of condensation of the solvent on a drop of solution placed in an atmosphere saturated with solvent. The temperature of the drop rises until the consequent increased solution vapour pressure is the same as that of the pure solvent.

One of the most widely used techniques for the determination of the number average molecular weight ($\bar{M}_n$) of polymers with $\bar{M}_n$ less than about 20 000. In practice, the temperature increase of the drop is measured by placing a solution and a solvent drop on each of two thermistor beads, both mounted in an atmosphere saturated with solvent vapour and the whole apparatus is closely thermostatically controlled. The temperature difference (ΔT) between the two beads is proportional to ΔR, the change in the balancing resistance of a Wheatsone bridge circuit containing the thermistors. For an ideal solution, $\Delta R = -k\ln a$, where k is a constant for a given solvent and temperature and also takes account of heat losses and the constant of proportionality between ΔT and ΔR. For a non-ideal polymer solution, of concentration c, the appropriate virial equation is, $\Delta R/c = k/\bar{M}_n + Bc + Cc^2 + \cdots$, where $B, C, \ldots$, etc. are the virial constants, and where k can be calculated, but is best determined as a calibration constant using a solute of known molecular weight.

VARCUM Tradename for phenol–formaldehyde polymer.

VARIATION METHOD A method useful in developing theoretical relationships between the mechanical properties, especially moduli, of a composite and those of its components, by using the theorems of minimum complementary energy and minimum potential energy. Although more sound theoretically, the method only yields bounds to the values of the mechanical properties and these may be very far apart, as in the case of bounds obtained from the simple and inverse laws of mixtures. However, sometimes closer bounds may be obtained, as in the Hashin–Shtrikman equations.

VARLAN Tradename for polyvinyl chloride.

VARNISH AND PAINT MAKERS NAPHTHA (V and PM naphtha) A petroleum hydrocarbon solvent, similar to white spirit, but somewhat more volatile, having a boiling range of about 100–160°C.

VAROX Tradename for 2,5-di(*t*-butylperoxy)-2,5-dimethylhexane.

VAT POLYMER A conjugated polymer capable of existing in a reduced (leuco) form, or in a non-conjugated form, analogous to a vat dye. Highly insoluble in the oxidised form but soluble in aqueous alkali in the reduced form. Also a type of oxidation–reduction polymer. Examples include polyindigo and a poly-(5,5-bisisatyl-thiopheneindophenone).

VC Abbreviation for vinyl chloride.

VCM Abbreviation for vinyl chloride monomer, often used instead of the simpler abbreviation VC for vinyl chloride.

VECTRA Tradename for an aromatic polyester thermotropic liquid crystalline polymer, based on naphthalene ester repeat units linked through the 2,6-positions:

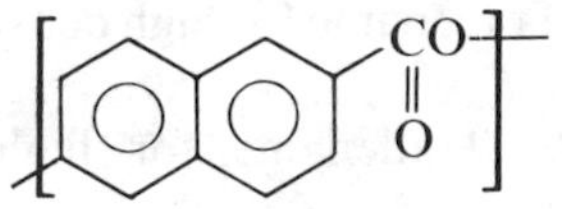

VEDRIL Tradename for polymethylmethacrylate moulding material.

VEGETABLE FIBRE A natural fibre obtained from a vegetable, rather than an animal, source. The main types are seed hair, bast and leaf fibres.

VELAN Tradename for a vinylidene chloride–vinyl chloride copolymer fibre.

VEREL Tradename for a modacrylic fibre.

VERMICULITE A crystalline layer silicate mineral of similar structure to talc. However, some silicon atoms are replaced with aluminium, producing a negative charge that is neutralised by interlayer cations, mostly magnesium. The ionic forces bind the layers together strongly. Water molecules in the layers are hydrogen bonded to oxygens of the silicates. On rapid heating to about 300°C the steam produced causes separation of the layers, giving an expansion (exfoliation) of up to 30-fold.

VERSAMID Tradename for fatty polyamides.

VERSEL Tradename for polybutylene terephthalate.

VERY LOW DENSITY POLYETHYLENE (VLDPE) Tradenames: Norsoflex, Ucarflex. An ethylene copolymer with a higher olefin comonomer such as butene-1 or octene-1 with a density in the range 0·900–0·915 g cm^{-3}. Commercial materials are useful as stretch films, in film laminates and in flexible containers, often competing with ethylene–vinyl acetate copolymer.

VESPEL Tradename for moulded compositions based on the polyimide obtained from pyromellitic dianhydride and 4,4'-diaminodiphenyl ether.

VESTAMID Tradename for nylon 12.

VESTAMID X 504

VESTAMID X Tradename for a polyether amide based on lauryllactam, 1,10-decanedicarboxylic acid and polytetrahydrofuran.

VESTAMID X4308 Tradename for a transparent polyamide produced by reaction of lauryllactam, the monomer for nylon 12, and other comonomers.

VESTAN Tradename for poly-(cyclohexane-1,4-dimethyleneterephthalate).

VESTENAMER Tradename for polyoctenamer.

VESTODUR Tradename for polybutylene terephthalate.

VESTOLEN Tradename for high density polyethylene.

VESTOLEN A Tradename for high density polyethylene.

VESTOLEN BT Tradename for polybutene-1.

VESTOLEN EM Tradename for a polypropylene/ethylene–propylene rubber thermoplastic polyolefin elastomer.

VESTOLEN P Tradename for polypropylene.

VESTOLIT Tradename for polyvinyl chloride.

VESTOPAL Tradename for an unsaturated polyester resin.

VESTOPREN TP Tradename for a polyolefin/ethylene–propylene–diene monomer rubber thermoplastic elastomer.

VESTORAN Tradename for a poly-(2,6-dimethyl-1,4-phenylene oxide) blend with high impact polystyrene and for styrene–acrylonitrile copolymer.

VESTYRON Tradename for polystyrene.

VGC Abbreviation for viscosity gravity constant.

VIBRATHANE Tradename for a cast polyurethane elastomer system based on a polyester or polytetramethylene glycol polyether, cured with 3,3′-dichloro-4,4′-diaminodiphenylmethane.

VIBRATHANE 5000 Tradename for a millable polyurethane elastomer cured with peroxides.

VIBRATING REED METHOD The most popular forced vibration method for the determination of the dynamic mechanical properties. A small, thin strip of polymer is firmly fixed to a vibrator head, the other end being free, i.e. the strip is arranged as a cantilever. The amplitude of the vibration of the free end is recorded, e.g. by optical/electrical means using a photocell. Results are obtained over a range of frequency which includes the resonance frequency (f_r). From these results the elastic modulus in tension (E) is given by $E = cL^4 \rho f_r^2/D^2$, where c is a numerical constant, L is the free length, ρ is the density and D is the thickness. Tan δ is given by f/f_r, where f is the half-width of the resonance peak. The method is appropriate in the audio frequency range of $10–10^4$ Hz.

VICABAR Tradename for a nitrile resin barrier polymer.

VICARA Tradename for a man-made protein fibre from the maize protein zein.

VICAT SOFTENING POINT (Vicat temperature) A measure of the temperature at which a polymer softens appreciably, as indicated by the penetration of a loaded needle with a cross-sectional area of 1 mm². The lower plane surface of the needle presses on a sheet of the polymer whilst it is being heated at 50°C h⁻¹. The needle is loaded to a force of 9·81 or 49 N. The temperature at which the penetration of the needle is 1 mm is the Vicat softening point. Sometimes a 1/10 Vicat softening point is measured, corresponding to 0·1 mm penetration. For an amorphous polymer, the value is about 5–10°C below the T_g value. For a partially crystalline polymer, the value depends on the degree of crystallinity, but is between the T_g value and the melting temperature.

VICAT TEMPERATURE Alternative name for Vicat softening point.

VICLAN Tradename for a vinylidene chloride–acrylonitrile copolymer.

VICTREX 720P Tradename for a polyethersulphone containing both

units and having a T_g value of about 250°C.

VICTREX PEEK Tradename for polyetheretherketone.

VICTREX PEK Tradename for polyetherketone.

VICTREX PES Tradename for an aromatic polyethersulphone.

VINAL Generic name for a fibre consisting of a polymer with at least 50 wt% of vinyl alcohol units and in which the total of these, together with any acetal units, account for at least 85% of the fibre.

VINALAK Tradename for polyvinyl acetate.

VINAMUL   Tradename for polyvinyl acetate.

VINAROL   Tradename for polyvinyl alcohol.

VINATEX   Tradename for polyvinyl chloride.

VINAVILOL   Tradename for polyvinyl alcohol.

VINCEL   Tradename for a polynosic rayon.

VINNAPAS   Tradename for polyvinyl acetate.

VINNOL   Tradename for polyvinyl chloride.

VINOFLEX   Tradename for isobutylvinyl ether–vinyl chloride copolymer and for polyvinyl chloride.

VINOL   Tradename for polyvinyl alcohol.

VINYL ACETATE

$$CH_2\!=\!CH$$
$$|$$
$$OOCCH_3 \qquad B.p.\ 72.5°C.$$

Produced by any of several different important methods. The reactions of acetylene and acetic acid in the liquid phase at about 80°C with a mercury catalyst or, more commonly, in the vapour phase at about 210°C with a zinc or cadmium catalyst, are both widely used:

$$CH\!\equiv\!CH + CH_3COOH \rightarrow CH_2\!=\!CH$$
$$|$$
$$OOCCH_3$$

Alternatively, ethylene may be oxidised to a mixture of acetaldehyde and acetic anhydride which are subsequently reacted together to yield ethylene diacetate:

$$CH_2\!=\!CH_2 \rightarrow CH_3CHO + (CH_3CO)_2O$$
$$\rightarrow CH_3\!-\!CH(OOCCH_3)_2$$

This may then be pyrolysed to vinyl acetate. Often this sequence of reactions is conducted in a single stage process using a mixture of ethylene, palladium chloride and sodium acetate.

The monomer for the production of polyvinyl acetate by free radical polymerisation and also very useful as a comonomer for the production of a wide range of copolymers, e.g. with ethylene, vinyl chloride, acrylates and dialkyl maleates and fumarates. In general, however, it does not readily enter into copolymerisation. Polymerisation is subject to retardation by several impurities which may be present in the monomer, such as acetaldehyde and crotonaldehyde. Some monomers, e.g. styrene and acrylonitrile, have a strong inhibiting effect on its polymerisation.

VINYLAL   Generic name for polyvinyl alcohol fibres.

VINYLBENZENE   Alternative name for styrene.

VINYL BENZOATE

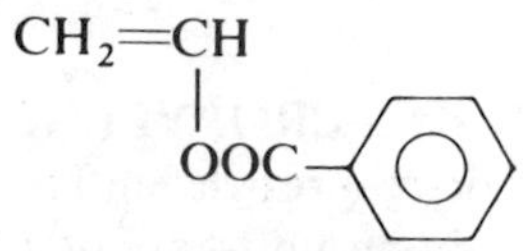

$$CH_2\!=\!CH$$

B.p. 80.5°C/9 mm, 203°C/760 mm.

Produced either by transvinylation of vinyl acetate with benzoic acid or by reaction of acetylene with benzoic acid. It has been used as a comonomer with vinyl acetate for water based emulsion paints.

N-VINYLCARBAZOLE

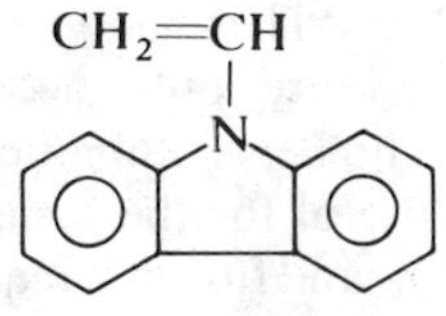

$$CH_2\!=\!CH$$

M.p. 67°C.

Prepared by the reaction of carbazole, itself obtained from coal tar, with acetylene. It may be polymerised thermally, free radically or cationically to poly-(N-vinylcarbazole).

VINYL CHLORIDE   (Chloroethene) (VC) (VCM)

$$CH_2\!=\!CHCl \qquad B.p.\ -14°C.$$

The monomer for the production of polyvinyl chloride and its copolymers. Produced commercially either by the addition of hydrochloric acid to acetylene or, more commonly, by the chlorination of ethylene to ethylene dichloride, followed by its dehydrochlorination. Usually the hydrochloric acid by-product so produced is converted back to chlorine, these reactions being conducted in a single stage (oxychlorination) process:

$$2\,CH_2\!=\!CH_2 + Cl_2 + \tfrac{1}{2}O_2 \rightarrow 2\,CH_2\!=\!CHCl + H_2O$$

The gaseous monomer is readily liquefied at a few atmospheres pressure and is normally stored, handled and polymerised as a liquid. It is considered to be a toxic hazard, so human exposure must be kept to a minimum and considerable efforts are made to achieve a very low (a few parts per million) levels of residual monomer in polyvinyl chloride. It may be readily polymerised, but only by a free radical mechanism, to polyvinyl chloride. Usually peroxide or azo initiators are employed at temperatures of 30–80°C using mass, suspension or emulsion polymerisation techniques.

VINYL CHLORIDE–ACRYLONITRILE COPOLYMER   (Poly-(vinyl chloride-co-acrylonitrile)) Tradenames: Dynel, Vinyon N (fibres). The copolymers with about 40% acrylonitrile have increased solubility, compared with polyvinyl chloride and polyacrylonitrile, especially in acetone, and have higher softening points than polyvinyl chloride. They form the basis of useful fibres with low flammability, but still have rather low softening points even with 40% acrylonitrile content. The

fibres are useful for non-clothing applications such as for blankets, carpets and filter cloths.

VINYL CHLORIDE–PROPYLENE COPOLYMER
(Poly-(vinyl chloride-co-propylene)) Tradename: Sta-Flow. Copolymers containing up to about 10% propylene are produced commercially and are claimed to have higher heat stability and easier processability than vinyl chloride homopolymer.

VINYL CHLORIDE–VINYL ACETATE COPOLYMER
(Poly-(vinyl chloride-co-vinyl acetate)) Tradename: Vinidur. The most important vinyl chloride copolymer, typically containing up to about 20% vinyl acetate in commercial products. The copolymers have lower melt viscosities (and therefore only need low processing temperatures) and higher solubilities than polyvinyl chloride. They are useful for the formation of products where better melt flow and fusion is required than is easily possible with the homopolymer. Products made from copolymer include calendered sheet, for flooring and thermoforming, and gramophone records. Because of their higher solubility, they are also used in coating applications.

VINYL CHLOROACETATE

$$CH_2{=}CH$$
$$\quad\ \ |$$
$$\quad OOCCH_2Cl$$
B.p. 44–46°C.

Produced either by transvinylation of vinyl acetate with chloroacetic acid or by reaction of acetylene with chloroacetic acid. The monomer for the formation of polyvinyl chloroacetate.

VINYLCYCLOHEXENE DIOXIDE

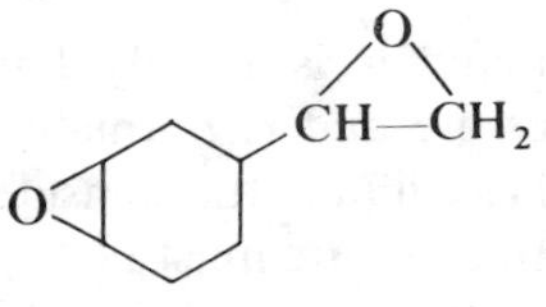

B.p. 227°C.

A liquid epoxy resin, formed by epoxidation of vinyl-cyclohex-3-ene, itself obtained by Diels–Alder dimerisation of butadiene. Useful as a cycloaliphatic epoxy resin and as a reactive diluent in diglycidyl ether bisphenol A epoxy resins.

2-VINYLDIBENZOFURAN

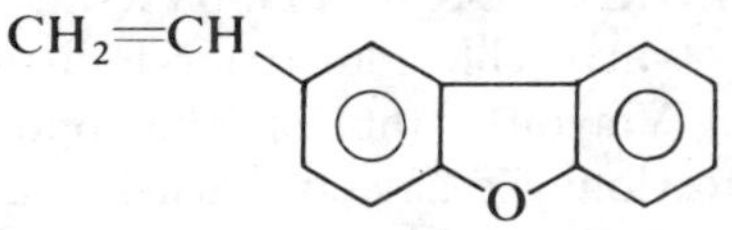

B.p. 128°C/1·5 mm. M.p. 93°C.

Prepared from dibenzofuran by acetylation, reduction to the carbinol and then dehydration. It polymerises thermally or free radically to poly-(2-vinyldibenzofuran).

VINYL ESTER RESIN
Tradenames: Atlac, Corezyn, Corrolite, Derakane, Diacryl, Epocryl, Silmar, Spilac. Low molecular weight prepolymers and the network polymers resulting from their crosslinking, prepared by reacting an epoxy resin prepolymer with an unsaturated carboxylic acid, such as acrylic or methacrylic acid. This results in the formation of unsaturated end groups on the epoxy resin giving polymers of the structure:

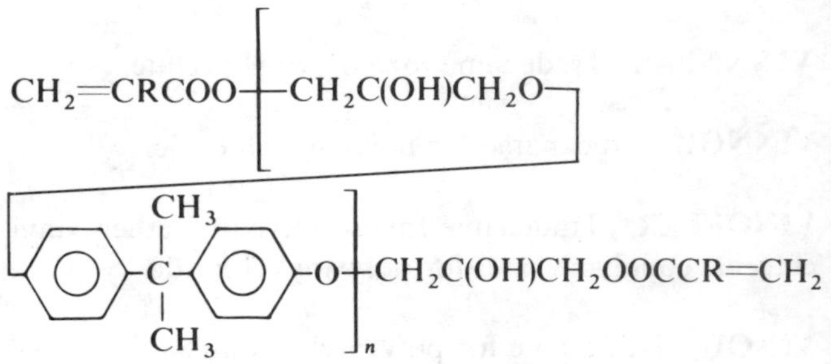

The resins have properties intermediate between those of epoxy resins and unsaturated polyester resins. Like the latter they may be rapidly crosslinked through the unsaturated groups by copolymerisation with a vinyl monomer, which is often styrene. Like epoxy resins they have good thermal and mechanical properties when cured and, especially when compared with unsaturated polyester resins, they have very good chemical resistance. The main use of the resins is in the production of chemically resistant plant and in transport products.

VINYL ETHER POLYMER
Alternative name for poly-(vinyl-alkyl ether).

VINYLETHYL ETHER

$$CH_2{=}CH{-}OC_2H_5 \qquad \text{B.p. } 35\cdot5°C.$$

Prepared either by the vinylation of ethanol by reaction with acetylene in the presence of potassium ethoxide or by reaction of acetylene with ethanol via the acetal. It polymerises by cationic polymerisation to poly-(vinylethylether).

VINYL FLUORIDE

$$CH_2{=}CHF \qquad \text{B.p. } -72°C.$$

Prepared by dehydrogenation of fluoroethane (obtained by the addition of hydrogen fluoride to ethylene), by the addition of hydrogen fluoride to acetylene or by the pyrolysis of trifluoropropylene. The monomer for the preparation of polyvinyl fluoride by free radical polymerisation.

VINYL FORMATE

$$CH_2{=}CH$$
$$\quad\ \ |$$
$$\quad OOCH$$
B.p. 46·6°C.

Produced either by transvinylation of vinyl acetate with formic acid or by reaction of acetylene with formic acid. The monomer for the formation of polyvinyl formate.

VINYLIDENE CHLORIDE

$$CH_2{=}CCl_2 \qquad \text{B.p. } 31{\cdot}6°C.$$

Produced either by reaction of 1,1,2-trichloroethane ($CH_2ClCHCl_2$) with calcium or sodium hydroxides, or by iron catalysed thermal dehydrochlorination of 1,1,1-trichloroethane (CH_3CCl_3). A toxic, volatile monomer which readily polymerises by free radical polymerisation to polyvinylidene chloride even on standing (unless inhibited). It is useful commercially for the production of a variety of vinylidene chloride copolymers, mostly vinylidene chloride–vinyl chloride and vinylidene chloride–acrylonitrile copolymers.

VINYLIDENE CHLORIDE–ACRYLONITRILE CO-POLYMER (Poly-(vinylidene chloride-co-acrylonitrile))

Tradenames: Saran, Viclan, Teklan (fibre). These modified vinylidene chloride polymers, containing 5–15% acrylonitrile (Saran and Viclan) are useful for low moisture and gas permeability coatings and packaging films (usually as part of a laminated structure). They are prepared by emulsion polymerisation and a coating may be applied from the polymer latex. Fibres are produced using an approximately 50/50 copolymer and have high burning resistance.

VINYLIDENE CHLORIDE–VINYL CHLORIDE COPOLYMER (Poly-(vinylidene chloride-co-vinyl chloride))

Tradenames: Saran, Velan. Commercial copolymers with about 15% vinyl chloride have similar properties to polyvinylidene chloride, i.e. they crystallise and have low gas permeability, but they are much more readily melt processable. They are useful for films with high durability and chemical resistance and as films and coatings with low gas and water vapour permeability.

VINYLIDENE CYANIDE

$$CH_2{=}C(CN)_2 \qquad \text{B.p. } 50{\cdot}5°C.$$

Prepared by pyrolysis of 1,1,3,3-tetracyanopropane (obtained from the condensation of formaldehyde with malononitrile), by pyrolysis of 1-acetoxy-1-dicyanoethane, or by pyrolysis of 4,4-dicyanocyclohexane. It very readily polymerises to polyvinylidene cyanide by free radical or anionic mechanisms, even such mild bases as water, alcohols, amines or ketones acting as catalysts for the anionic mechanism. It has been used as a comonomer for the formation of fibres as a copolymer with vinyl acetate.

VINYLIDENE FLUORIDE

$$CH_2{=}CF_2 \qquad \text{B.p. } -82°C.$$

Obtained either by the pyrolysis of trifluoropropylene,

$$CH_2{=}CH \rightarrow CH_2{=}CHF + CH_2{=}CF_2$$
$$\underset{\displaystyle CF_3}{|}$$

or by the pyrolysis of chlorodifluoroethane,

$$CH_3CF_2Cl \rightarrow CH_2{=}CF_2 + CH_2CFCl$$

The monomer for the preparation of polyvinylidene fluoride and its copolymers by free radical polymerisation.

VINYLIDENE FLUORIDE–CHLOROTRIFLUORO-ETHYLENE COPOLYMER (CFM) (Poly-(vinylidene fluoride-co-chlorotrifluoroethylene))

Tradenames: Kel-F, SKF, Voltalef. A fluoroelastomer copolymer containing —CH_2CF_2— and —CF_2CFCl— units. It does not have such good heat stability as the vinylidene fluoride–hexafluoropropylene copolymers, but has better resistance to oxidising acids.

VINYLIDENE FLUORIDE–HEXAFLUOROPRO-PYLENE COPOLYMER (FKM) (Poly-(vinylidene fluoride-co-hexafluoropropylene))

Tradenames: Dai-El, Fluorel, SKF, Technoflon FOR, Viton A. A random copolymer of

$$-CH_2CF_2- \quad \text{and} \quad -CF_2CF-$$
$$\underset{\displaystyle CF_3}{|}$$

units, which does not crystallise and has a low T_g value. It has exceptional heat, chemical and solvent resistance as a fluoroelastomer. Terpolymers which also contain some tetrafluoroethylene units are commercial rubbers (tradenames Dai-El, Viton B). Prepared by emulsion polymerisation at high temperatures and pressures. Typically it contains 60–85% vinylidene chloride units. The polymers are vulcanised by heating with diamines, via dehydrofluorination, forming double bonds which subsequently react with further amine. This leads to vulcanisates with high compression set and possibly also high porosity due to water produced during vulcanisation. Improved vulcanisates are obtained with bisphenols (activated with a quaternary ammonium or phosphonium compound) or with peroxides (and a coagent such as triallylisocyanurate), when a copolymer containing a termonomer which is attacked by free radicals is used.

VINYLIDENE FLUORIDE-1-HYDROPENTAFLUO-ROPROPYLENE COPOLYMER (Poly-(vinylidene fluoride-co-1-hydropentafluoropropylene))

Tradename: Technoflon S. A fluoroelastomer copolymer containing units:

$$-CH_2CF_2- \quad \text{and} \quad -CF_2CH-$$
$$\underset{\displaystyle CF_3}{|}$$

A terpolymer (tradename Technoflon T) which also contains some tetrafluoroethylene units is also produced. The copolymers are similar in properties to the vinylidene fluoride–hexafluoropropylene copolymers but have a poorer solvent resistance.

VINYLIDENE MONOMER A monomer of the type $CH_2{=}CX_2$, where X is any atom or simple group. Polymerises readily by free radical polymerisation to

$\{CH_2CX_2\}_n$. Although not many examples are known, they include vinylidene chloride (X = Cl), vinylidene fluoride (X = F) and vinylidene cyanide (X = CN).

VINYLISOBUTYL ETHER

$$CH_2{=}CH{-}OCH_2CH(CH_3)_2$$

B.p. 83°C.

Prepared in a similar way to vinylethyl ether from isobutanol and acetylene. It polymerises by cationic polymerisation to poly-(vinylisobutyl ether). Copolymers with vinyl chloride are useful as coatings.

VINYL ISOCYANATE

$$CH_2{=}CH$$
$$|$$
$$NCO$$

B.p. 32°C/660 mm.

Prepared by reaction of sodium azide with acrylyl chloride. Polymerised by free radical polymerisation to polyvinyl isocyanate.

VINYLITE   Tradename for polyvinyl chloride.

VINYL LAURATE

$$CH_2{=}CH$$
$$|$$
$$OOC(CH_2)_{11}CH_3$$

B.p. 142°C/10 mm.

Produced by transvinylation of vinyl acetate with lauric acid or by reaction of acetylene with lauric acid. It has been used as a comonomer (to act as an internal plasticiser) in copolymers with vinyl acetate and vinyl chloride.

VINYLMETHYL ETHER   Alternative name for methylvinyl ether.

VINYL MONOMER   Strictly, a monomer containing the vinyl group $CH_2{=}CH{-}$, i.e. one of the type $CH_2{=}CHX$, where X is an atom or chemical group. However, the term, as often used, includes all monomers of the type $CYZ{=}CWX$, where W, X, Y and Z are atoms (including hydrogen) or groups. Monomers containing the structural unit $C{=}C{-}C{=}C$, i.e. 1,3-dienes, are also vinyl monomers, but are usually considered separately as diene monomers. Since vinyl monomers contain a $C{=}C$ double bond they are susceptible to chain polymerisation (vinyl polymerisation) to polymer (vinyl polymer). However, 1,2-di-, tri- or tetrasubstituted monomers, i.e. with substituents on both carbon atoms, usually do not polymerise unless the substituents, W, X, Y and Z are small, e.g. fluorine atoms, and are therefore of little interest.

Monomers with both Y = Z = H and W = X are vinylidene monomers. Most interest is with monomers of

the type $CH_2{=}CWX$, where W = H or CH_3. These form the largest group of monomers and are the basis of many important commercial polymers. The most common type has W = Y = Z = hydrogen and some have the word vinyl as part of their trivial name by which they are best known, although according to strictly correct nomenclature they should be named differently. Thus, for example, vinyl chloride, strictly named, is 2-chloroethene. Examples are legion and include the α-olefins, where X is an alkyl group (with W = Y = Z = hydrogen) such as propylene (X = CH_3), butene-1 (X = C_2H_5) and 4-methylpentene-1 (X = $CH_2CH(CH_3)_2$); X may be aryl as in styrene (X = C_6H_5) and its derivatives, or a halogen as in vinyl chloride (X = Cl). Acrylic monomers, e.g. acrylic acid (X = COOH) and its esters, acrylonitrile (X = CN) and methacrylic monomers (W = CH_3), e.g. methyl methacrylate (W = CH_3, X = $COOCH_3$), are also important. Many others of less importance include N-vinyl carbazole where

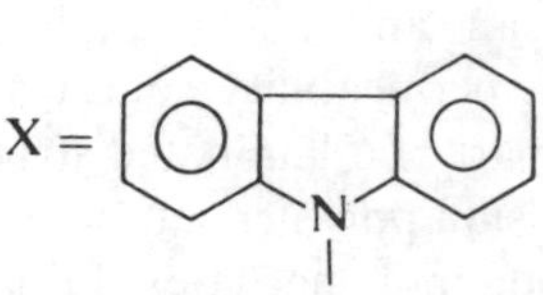

X =

vinyl pyridine where

X = —N

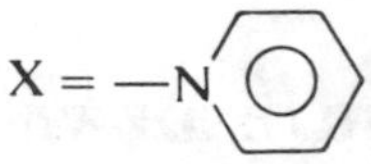

and N-vinyl pyrrolidone where

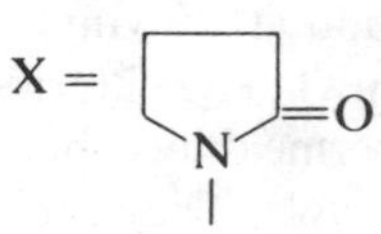

X =

Vinyl polymerisation of these monomers occurs by chain polymerisation by a wide variety of mechanisms. Susceptibility to a particular mechanism, radical, anionic, cationic or coordination, depends on the electronic demands of the substituents X, W, Y and Z.

1-VINYLNAPHTHALENE

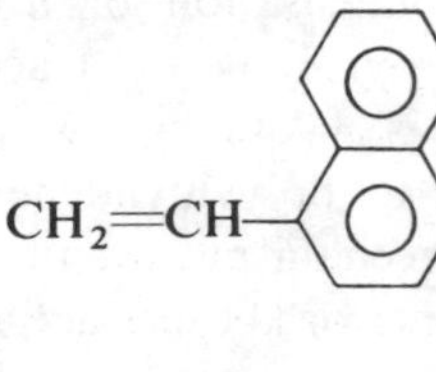

$CH_2{=}CH{-}$

B.p. 125°C.

Obtained from naphthalene by alkylation to mixed 1- and 2-ethylnaphthalenes, followed by dehydrogenation, or by oxidation followed by dehydration. The monomer for poly-(1-vinylnaphthalene).

2-VINYLNAPHTHALENE

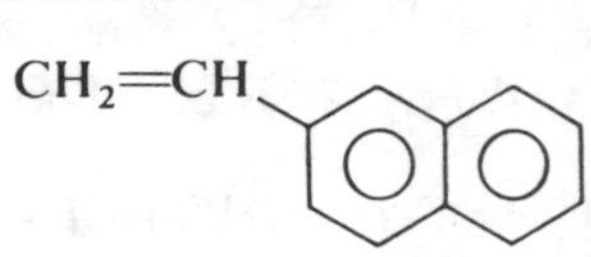

$CH_2{=}CH$

B.p. 82°C/2 mm. M.p. 63°C.

Obtained by the alkylation of naphthalene to mixed 1- and 2-ethylnaphthalenes followed by dehydrogenation, or by oxidation followed by dehydration. The monomer for poly-(2-vinylnaphthalene).

VINYLON Tradename for a polyvinyl alcohol fibre crosslinked by reaction with formaldehyde to enhance water resistance.

VINYL POLYBUTADIENE A polybutadiene with a high proportion of 1,2-linkages (giving vinyl groups), together with the more usual 1,4-linkages. The polymers show higher reversion resistance than the normal 1,4-polybutadiene, so may be vulcanised at higher temperatures. They are also capable of giving a good combination of wet grip and low rolling resistance in tyres.

VINYL POLYMER A polymer of a vinyl monomer, i.e. strictly of the type $-\!\!\lbrack CH_2\!-\!CHX\rbrack_n$ where X is an atom or simple group. However, the term is often used to mean any polymer of the more general type $-\!\!\lbrack CWX\!-\!CYZ\rbrack_n$ where W, X, Y and Z are either hydrogen (H) or substituent atoms or groups. 1,2-Disubstituted polymers are not very common (unless X = Y = fluorine) since the corresponding monomers do not readily polymerise. Most interest is with polymers of the type $-\!\!\lbrack CH_2\!-\!CWX\rbrack_n$ where W = H (hydrogen) or CH_3. When W = X, the polymer is a vinylidene polymer, although this term is not used when X = H or alkyl.

Most often W = H, i.e. the polymer is of the simple type $-\!\!\lbrack CH_2\!-\!CHX\rbrack_n$. This group forms the most important group of commercial plastic materials as well as being the most widely studied synthetic polymers scientifically. Some polymers contain the word vinyl in their name, e.g. polyvinyl chloride; these are sometimes called the vinyls in plastics technology.

Important polymers are the polyolefins, e.g. polyethylene (X = H), polypropylene (X = CH_3), poly-(butene-1) (X = C_2H_5) and poly-(4-methylpentene-1) (X = $CH_2CH(CH_3)_2$), several halogen polymers, e.g. polyvinyl chloride (X = Cl) and fluoride (X = F), polystyrene (X = C_6H_5) and its derivatives, polyvinyl acetate (X = $OOCCH_3$) and its derivatives, e.g. polyvinyl alcohol (X = OH) and the polyvinyl acetals, polyacrylic acid (X = COOH) and its esters (X = COOR) and polyacrylonitrile (X = CN). Examples of some less important polymers are poly-(N-vinyl carbazole) where

X =

poly-(N-vinyl pyrrolidone) where

X =

and the poly-(vinyl ethers) (X = OR).

Polymers of the type $-\!\!\lbrack CH_2CWX\rbrack_n$ where W ≠ H, commonly have W = CH_3, as in poly-(methacrylic acid)

(W = CH_3, X = COOH) and its esters (X = COOR), e.g. polymethylmethacrylate and poly-(α-methylstyrene) (X = C_6H_5, W = CH_3).

The polymers are prepared by chain polymerisation of the appropriate vinyl monomer, most often by free radical polymerisation, although some are best prepared by ionic or coordination polymerisation. The polymers are predominantly head-to-tail linked and, if prepared free radically, they are largely atactic, although with some tendency to syndiotacticity when low polymerisation temperatures are used. Largely isotactic or syndiotactic polymers can be prepared using ionic or coordination polymerisation. The polymers are usually linear, although sometimes branching is present to a significant extent, e.g. in low density polyethylene and polyvinyl acetate. They may be produced in crosslinked form by copolymerisation with a divinyl monomer or by subsequent crosslinking of the preformed polymer.

VINYL POLYMERISATION Polymerisation of a vinyl monomer, i.e. strictly one of the type $CH_2\!=\!CHX$. However, as usually used, the term includes all monomers of the type $CWX\!=\!CYZ$ where W, X, Y and Z are hydrogen atoms or other substituent atoms or groups. 1,2-Disubstituted monomers show little tendency to polymerise unless W and X are small, such as fluorine atoms. Monomers with two vinyl groups (diene monomers) are said to undergo either divinyl polymerisation, when the vinyl groups are not conjugated, or diene polymerisation when they are conjugated. Polymerisation is by chain polymerisation to a vinyl polymer, $-\!\!(CH_2\!-\!CHX)_n$. Any of a wide range of initiators may be used, the electronic demands of the substituents determining whether a particular monomer will undergo free radical, anionic, cationic or coordination polymerisation.

Most vinyl monomers (but not α-olefins) will undergo free radical polymerisation since the substituent X provides some resonance stabilisation of the growing active centre $\sim\!\!\sim\!CH_2\!-\!\dot{C}HX$. Cationic polymerisation is favoured by electron-donating substituents, e.g. W = X = CH_3 (isobutene) and X = H, W = $COCH_3$ (methyl vinyl ketone), and anionic polymerisation is favoured by electron-withdrawing substituents, e.g. X = COOR (acrylic and methacrylic esters) and X = CN (acrylonitrile). α-Olefin monomers are exclusively polymerised by coordination polymerisation, usually with Ziegler–Natta catalysts.

Since polymerisation involves conversion of a high energy C=C π-bond to a lower energy C—C σ-bond the reaction is exothermic. The values of the high negative heat of polymerisation can range from 50 to 130 kJ mol^{-1} and the polymerisation method must be chosen to adequately dissipate this heat. Propagation is almost exclusively by head-to-tail addition, usually $\sim 1\%$ head-to-head groups being formed. With monomers of the type $CH_2\!=\!CXY$ with X ≠ Y, the resulting polymer contains asymmetric (chiral) carbon atoms and the possibility of tacticity arises. In free radical polymerisation there is little preference for iso- or syndiotactic placement during propagation, so the polymers are predominantly atactic but with a tendency to syndiotacticity as polymerisation temperature is lowered. In ionic polymerisation both iso- and syndiotactic polymers may often be made by a judicious choice of reaction conditions. In coordination polymerisation isotactic polymers are frequently formed.

VINYL PROPIONATE

$$CH_2{=}CH$$
$$\mid$$
$$OOCC_2H_5$$

B.p. 95°C.

Produced either by transvinylation of vinyl acetate with propionic acid or by reaction of acetylene with propionic acid in the vapour phase. The monomer for the formation of polyvinyl propionate, but also useful as a comonomer for emulsion polymer coating latices.

2-VINYLPYRIDINE

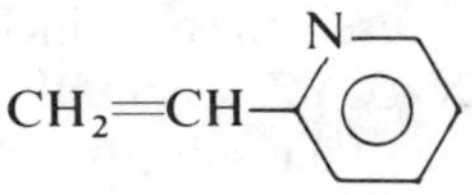

$$CH_2{=}CH{-}$$ (pyridine ring)

B.p. 159°C.

Obtained by the reaction of 2-methylpyridine (2-picoline) with formaldehyde to give 2-(hydroxyethyl)pyridine, followed by dehydration. Polymerises by free radical polymerisation to poly-(2-vinylpyridine). Although the homopolymer is not important, the monomer is useful for several commercial copolymers, e.g. to improve the dyeability of polyacrylonitrile fibres and in polybutadiene and styrene–butadiene rubbers to improve adhesion of the rubber to tyre cords (vinylpyridine rubbers).

4-VINYLPYRIDINE

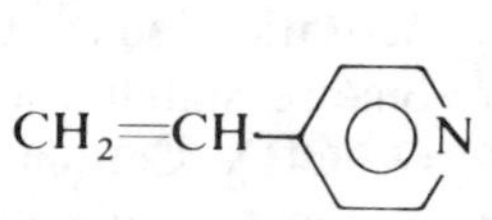

$$CH_2{=}CH{-}$$ (pyridine ring)

B.p. 65°C/15 mm.

Prepared in a similar way to 2-vinylpyridine but from 4-methylpyridine (4-picoline). It polymerises to poly-(4-vinylpyridine) by free radical polymerisation.

VINYLPYRIDINE RUBBER A butadiene–vinylpyridine copolymer, with comonomers in the ratio of about 75:25, at one time widely investigated as a potentially useful rubber. A terpolymer also containing some styrene is useful in latex form for rubber to textile bonding.

N-VINYLPYRROLIDONE

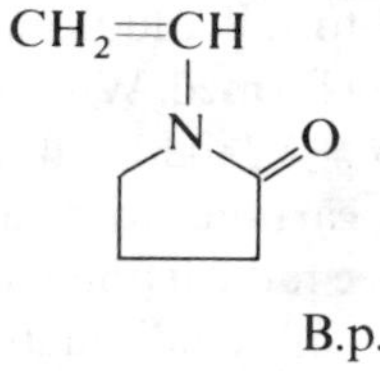

$$CH_2{=}CH$$ (pyrrolidone ring with N and O)

B.p. 96°C. M.p. 13·5°C.

Prepared by vinylation of α-pyrrolidone with acetylene, the α-pyrrolidone being obtained from butane-1,4-diol which is cyclised to γ-butyrolactone and then treated with ammonia. It is polymerised, usually in aqueous solution, with a free radical initiator to poly-(N-vinylpyrrolidone). Copolymers with vinyl acetate and graft copolymers with other vinyl comonomers are commercial products.

VINYLSILICONE ELASTOMER (Methylvinylsilicone elastomer) (VMQ) A silicone elastomer based on poly-dimethylsiloxane in which some of the methyl groups (less than 0·5%) have been replaced by vinyl groups, i.e. the polymer contains some methylvinylsiloxane units. This enables the gum to be vulcanised at higher rates and the vulcanisates to have reduced compression set compared with dimethylsilicone elastomer. Peroxide crosslinking agents are used for vulcanisation, including dicumyl peroxide, ditertiarybutyl peroxide and 2,5-dimethyl-2,5-bis-(t-butylperoxy)hexane.

VINYL STEARATE

$$CH_2{=}CH$$
$$\mid$$
$$OOC(CH_2)_{16}CH_3$$

M.p. 35–36°C. B.p. 187–188°C/43 mm.

Produced either by transvinylation of vinyl acetate with stearic acid or by reaction of acetylene with stearic acid. The monomer for the production of polyvinyl stearate, and also used as a comonomer, acting as an internal plasticiser, in copolymers of vinyl acetate and vinyl chloride.

VINYL SULPHONIC ACID (Ethylene sulphonic acid) (Ethene sulphonic acid)

$$CH_2{=}CH$$
$$\mid$$
$$SO_2OH$$

B.p. 125°C/1 mm.

Prepared by the elimination of sulphuric acid from ethionic acid, itself obtained by reaction of ethanol with sulphur trioxide:

$$C_2H_5OH + SO_3 \rightarrow C_2H_5OSO_2OH$$
$$\rightarrow HOSO_2CH_2CH_2SO_2OH$$
$$\rightarrow CH_2{=}CHSO_2OH + H_2SO_4$$

It readily polymerises by a free radical mechanism to polyvinyl sulphonic acid. Its salts polymerise even more readily. The monomer and its salts are also useful as comonomers, e.g. to provide vulcanisation sites in rubbers, to improve dyeability in fibres and to stabilise the copolymer in emulsion.

VINYLTOLUENE (o-, m-, p-Methylstyrene)

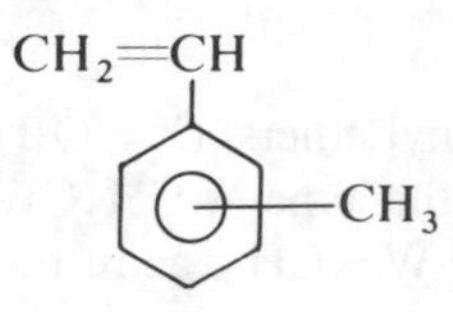

$$CH_2{=}CH$$ (benzene ring with CH_3)

B.p. 172°C.

A monomer, sometimes used as a replacement for styrene. Produced by dehydrogenation of ethyltoluenes, which are formed by the alkylation of toluene with ethylene at 500–600°C. Commercial samples consist of a 60:40 mixture of *meta* and *para* isomers. It readily thermally polymerises to polyvinyltoluene but is mostly used as a comonomer in coatings and in unsaturated polyester resins.

VINYLTRICHLOROSILANE

$CH_2=CHSiCl_3$

An early silane coupling agent for glass fibre reinforced unsaturated polyester (GRP) composites. Rather unstable, producing undesirable hydrochloric acid and now superseded by vinyltriethoxysilane.

VINYLTRIETHOXYSILANE

$CH_2=CHSi(OC_2H_5)_3$

Tradename: Garan. A silane coupling agent for glass fibre reinforced unsaturated polyester resin (GRP) composite.

VINYON Generic name and tradename for a vinyl chloride/vinyl acetate copolymer fibre, which contains at least 85 wt% of vinyl chloride units.

VINYON HH Tradename for a staple fibre of a copolymer of vinyl chloride (88%) and vinyl acetate (12%).

VINYON N Tradename for a vinyl chloride–acrylo-nitrile modacrylic copolymer fibre.

VIPLA Tradename for polyvinyl chloride.

VIPLAST Tradename for polyvinyl chloride.

VIPLAVIL Tradename for polyvinyl chloride and polyvinyl chloroacetate.

VIRION The complete virus particle containing both the nucleic acid and the protein coat.

VIRTUAL CROSSLINK Alternative name for physical crosslink.

VIRUS A biological particle capable of reproducing itself in the presence of a specific host cell and therefore having some of the attributes of a living organism. A nucleoprotein, composed principally of nucleic acid (of one or two chains) and many protein subunits. Viruses can infect plant, animal and bacterial cells. In the latter case the virus is called a bacteriophage. Many diseases of man, e.g. poliomyelitis and influenza, are transmitted by viruses.

The virus particle (called the virion) molecular weight is usually in the range of a few to a hundred million. Many viruses have been isolated in a molecularly homogeneous form and have been crystallised. The viruses vary considerably in particle size and shape from long (300 nm) rod-like particles of the tobacco mosaic virus and icosahedral particles of tomato bushy stunt plant viruses to the 'tadpole' shape of many bacteriophages.

The nucleic acid is RNA in plant viruses, which contain 5–20% nucleic acid, but may be RNA or DNA in animal (5–30% nucleic acid) or bacterial viruses (25–70% nucleic acid). The nucleic acid may have from a few thousand to several hundred thousand nucleotide units, the polymer chain often being helical or circular. The protein units are arranged in a regular way around the nucleic acid core, forming a coat, and thus giving a supermolecular complex, held together by a combination of ionic, hydrogen and hydrophobic bonding forces. The most widely studied viruses are the tobacco mosaic viruses, with only one type of protein subunit, and the bacteriophages of the bacterium *Escherichia coli*, with many different protein subunits. Other widely studied examples are the tomato bushy stunt, tobacco necrosis, turnip yellow mosaic, influenza, papilloma and adeno-viruses.

VISCOELASTIC FLUID (Elasticoviscous fluid) A fluid, which although it exhibits predominantly viscous flow behaviour, also exhibits some elastic recovery of the deformation on release of the stress. To emphasise that viscous effects predominate, the term elasticoviscous is sometimes preferred; the term viscoelastic is reserved for solids showing both elastic and viscous behaviour. Most polymeric systems, both melts and solutions, are viscoelastic due to the molecules becoming oriented on shear of the fluid, but retaining their equilibrium randomly coiled configuration on release of the stress. Elastic effects are manifest in die swell, melt fracture and frozen-in orientation in many polymer melt processing operations. Although viscoelastic constitutive equations may be derived, they often do not adequately describe real behaviour, or are very difficult to handle in solving flow problems. The Maxwell fluid model has been extensively used to provide a mathematical description of viscoelastic fluid behaviour, in particular in the White–Metzner modified form. Use is also often made of the Oldroyd or Jaumann derivatives.

VISCOELASTICITY Mechanical behaviour having the characteristics of both classical elasticity and viscous flow. However, with some materials, particularly polymeric solids, since there is a delayed deformation, which is nevertheless often completely recoverable, viscous flow cannot be said to have occurred. Thus viscoelasticity is perhaps better viewed more broadly as mechanical behaviour in which the relationships between stress and strain are time dependent, as opposed to classical elastic behaviour in which deformation and recovery both occur instantaneously on application and removal of stress respectively.

On stressing a viscoelastic material, three deformation responses may be observed—an initial elastic response, followed by a time-dependent delayed elasticity (also fully recoverable) and, finally, a viscous, non-recoverable, flow component. Most polymer containing systems (solid polymers, polymer melts, gels, dilute and concentrated solutions) exhibit viscoelastic behaviour due to the long-chain nature of the constituent molecules.

The viscoelastic deviations from ideal elasticity or purely viscous flow depend on both the experimental conditions (particularly temperature and magnitudes and rates of application of stress or strain) and on polymer structure (particularly molecular weight, molecular weight distribution, crystallinity, crosslinking and branching).

For a high molecular weight glassy polymer, i.e. an amorphous polymer well below its T_g value, very few chain motions are possible and hence the material largely behaves elastically, with a very low value for the creep compliance of about 10^{-9} Pa^{-1}. On the other hand, well above the T_g value, i.e. for a rubbery polymer, the creep compliance is about 10^{-4} Pa^{-1}, since considerable segmental rotation can occur. Sometimes the intermediate temperature region, which corresponds to the region of the T_g value, is referred to as the viscoelastic region, the leathery region or the transition zone. Well above the T_g value is the region of rubbery flow followed by the region of viscous flow. In this last region flow occurs owing to the possibility of slippage of whole molecular chains occurring by means of coordinated segmental jumps. These five temperature regions give rise to the five regions of viscoelastic behaviour. Light crosslinking of a polymer will have little effect on the glassy and transition zones, but will considerably modify the flow regions.

A constitutive relationship between stress and strain describing viscoelastic behaviour will have terms involving strain rate as well as stress and strain. If there is direct proportionality between the terms then the behaviour is that of linear viscoelasticity, described by a linear differential equation. Polymers may exhibit linearity but usually only at low strains. More commonly, complex non-linear viscoelastic behaviour is observed.

Viscoelasticity is characterised by dependencies on temperature and time, the complexities of which may be considerably simplified by the time–temperature superposition principle. Similarly the response to successively applied loadings can be simply represented using the Boltzmann Superposition Principle.

Experimentally viscoelasticity is characterised by creep (quantified by creep compliance for example), stress relaxation (quantified by stress relaxation modulus) and by dynamic mechanical response (as given by the complex moduli or compliances or the storage and loss moduli and compliances).

Theoretically, constitutive relations may be derived on the basis of models for the material behaviour. These may be simple mechanical models of springs and dashpots, e.g. Maxwell, Voigt and standard linear solid models, leading to a differential representation of viscoelasticity, or the models may be molecular as in the Rouse–Bueché–Zimm theory, leading to an integral representation.

VISCOELASTIC LIQUID A viscoelastic material for which, at long times of applied stress, e.g. in a creep experiment, a steady flow is eventually achieved. Thus in a generalised Maxwell model, all the dashpot viscosities must have finite values and in a generalised Voight model one spring must have zero stiffness. For materials with very long relaxation times the distinction between a viscoelastic liquid and a viscoelastic solid becomes uncertain.

VISCOELASTIC RELAXATION Relaxation occurring in a viscoelastic fluid, usually when it is subject to a dynamic constraint. The influence of frequency on viscoelastic properties is often of interest. When the fluid is subject to a periodically varying shear at low frequencies, the molecular motions responsible for viscous flow can occur. However, with increasing frequency, the molecules have less time to undergo translation so the viscosity decreases and elastic behaviour predominates. At frequencies close to $1/\tau$ (τ being the relaxation time), maximum energy loss occurs so that G'' (the loss modulus) is at a maximum and the real part of the complex viscosity is at a minimum. In dilute solution, viscoelastic relaxation has been modelled and explained by such theories as the Rouse–Bueché–Zimm theory (RBZ theory) based on the normal modes of motion of a flexible chain. However, these theories break down at higher frequencies or with stiff chains. Nevertheless, for perfectly rigid rods the theory of the frequency dependency of the complex viscosity is well understood, but unfortunately really stiff polymers, such as poly-(α-amino acids) and nucleic acids, are intermediate in behaviour. For polymer melts no successful theory based on molecular motions has been developed. However, at high frequencies real behaviour is fitted to that predicted by macroscopic models, involving the addition of the complex shear admittances (the reciprocal impedances) for a Hookean solid and a Newtonian fluid. This leads to a complex compliance predicting a wider dispersion in viscosity and shear modulus than that predicted by a simple Maxwell model. In a further refinement, a contribution from the retardational compliance has also been considered. For low frequencies, real behaviour can be fitted to the model by adding the RBZ normal modes.

VISCOELASTIC SOLID A viscoelastic material for which, at long times of applied stress, e.g. in a creep experiment, an equilibrium is eventually achieved, i.e. $G(t) \rightarrow G_e$ as $t \rightarrow \infty$, where G_e is the equilibrium modulus. In terms of viscoelastic models, one viscosity of a generalised Maxwell model must be infinite and all the springs of a generalised Voight model must have non-zero stiffnesses.

VISCOELASTIC SPECTRUM A thermomechanical spectrum whose temperature range includes the transition zone of viscoelastic behaviour.

VISCOELASTIC STATE In general the state of a material in which it exhibits viscoelasticity. Since this is

VISCOMETER Alternative name for capillary viscometer.

VISCOMETRIC FLOW (1) Alternative name for simple shear flow.

(2) A flow, which may be either a pressure flow or a drag flow, developed in an apparatus for the determination of the viscosity or the apparent viscosity of a fluid. Such flows are often simple shear flows but may not necessarily be so. Examples include flow in a pipe of circular cross-section, flow between concentric cylinders or discs, one of which is rotating, or flow in a cone and plate apparatus.

VISCOMETRY The experimental determination of the viscosity of a fluid, as opposed to the more complete characterisation of its rheological properties implied by the term rheometry. It is only for inelastic fluids that viscometry can give a complete characterisation of rheological behaviour.

VISCOSE The viscous solution of sodium cellulose xanthate in alkali from which viscose rayon fibres are spun. Sometimes used to mean the viscose rayon itself.

VISCOSE PROCESS A process for solubilising cellulose, spinning it into fibres and regenerating the cellulose. Used to produce viscose rayon. Alkali cellulose is treated with carbon disulphide to form sodium cellulose xanthate

$$(\text{Cell}-\text{O}-\overset{\displaystyle\overset{\text{S}}{\|}}{\text{C}}-\text{SNa})$$

This is then dissolved in dilute sodium hydroxide to form a viscous solution (the viscose) in which the cellulose is partially regenerated. The viscose rayon fibres are wet sp in from viscose into sulphuric acid which completes the regeneration. Modifications of the process are used to produce high tenacity rayon and polynosic rayon.

VISCOSE RAYON Tradenames: Avlin, Avril, Avron, Corval, Evlan, Fibro, Lanusa, Sarille, Sniafil, Topel. The regenerated cellulose fibre (as yarn or fabric) produced by the viscose process. Sometimes it is simply referred to as viscose, although this term also often refers only to the solut on from which the fibres are spun. Ordinary viscose rayons have a tenacity of about 2·6 g denier^{-1} (dry) and about 1·4 g denier^{-1} (wet) with an elongation at break of about 15% (dry) and 25% (wet) and a moisture content at 65% RH of 12%. The specific gravity is 1·52. The fibres and staple are readily dyed and are widely used in all textile applications except where excessive biodegradative or chemical agencies come into contact. Similar modified rayons are the high tenacity and polynosic rayons.

VISCOSITY In general, the ability of a fluid to resist flow, although the term may be used for any deformation of any material. Quantitatively the term may sometimes be specifically used as an alternative name for coefficient of viscosity or apparent viscosity. Many other types of viscosity may be used to describe the flow behaviour of a fluid. These include coefficient of viscous traction, complex viscosity, dynamic viscosity, elongational viscosity, inherent viscosity, intrinsic viscosity, kinematic viscosity, logarithmic viscosity number, plastic viscosity, reduced viscosity, relative viscosity and specific viscosity.

For a Newtonian fluid, quantitatively, it is the constant (η) relating the shear stress (τ) to the shear rate ($\dot{\gamma}$) (= dγ/dt, where γ is the shear strain) in the relation $\tau = \eta\dot{\gamma}$. η is the coefficient of viscosity, often simply called the viscosity, or sometimes the dynamic viscosity. For a non-Newtonian fluid, η is sensitive to $\dot{\gamma}$, increasing with increasing $\dot{\gamma}$ in dilatant fluids and decreasing in pseudoplastic (or shear thinning) fluid. The viscosity is also temperature dependent, the dependence being given by the Arrhenius relation, $\eta = A e^{-B/RT}$, where A and B are constants for the fluid. The shear rate for a Newtonian fluid through a tube is given by the Poiseuille equation. The unit of viscosity is the poise (P) (units of 10^{-1} Pa s) or the centipoise (cP) (10^{-2} poise).

Ordinary liquids and dilute polymer solutions typically have a viscosity of about 1 cP. Concentrated polymer solutions and polymer melts have very high viscosities (often > 10^5 P), which are also highly dependent on molecular weight. Melt viscosities are non-Newtonian, the shear rate dependence of η often being expressed by a power law equation. Dilute solution viscometry provides the most convenient method of estimation of polymer molecular weight, using capillary viscometry. Viscosities of more viscous polymer systems, such as melts, dispersions and concentrated solutions, are evaluated in capillary and cone and plate rheometers or in couette flow.

VISCOSITY AVERAGE MOLECULAR WEIGHT Symbol $\bar{M}_v$. The molecular weight average obtained from the limiting viscosity number of a dilute polymer solution, using the Mark–Houwink equation. Defined as,

$$\bar{M}_v = \left[\sum_{i=1}^{\infty} N_i M_i^{1+a} \bigg/ \sum_{i=1}^{\infty} N_i M_i\right]^{1/a}$$

where N_i is the number of molecular species i of molecular weight M_i and a is the exponent of the Mark–Houwink equation. Since the value of a is dependent on the solvent and temperature used, there is no unique value to $\bar{M}_v$ for a particular polymer sample.

Determination of the limiting viscosity number by dilute solution viscometry provides the most convenient method of polymer molecular weight determination and since it yields $\bar{M}_v$, this is the most frequently measured average. Values of a fall in the range 0·5–0·8 and therefore $\bar{M}_v$ is closer to the weight average ($\bar{M}_w$) than the number average ($\bar{M}_n$) molecular weight. Therefore in the determination of a by calibration with samples of known molecular weight, it is preferable to use $\bar{M}_n$ rather than $\bar{M}_w$, unless monodisperse samples are available. By determination of $\bar{M}_v$ in both a poor and a good solvent, it is possible to get a measure of the polydispersity of the sample. $\bar{M}_v$ values of copolymers cannot be reliably obtained.

VISCOSITY COEFFICIENT Alternative name for coefficient of viscosity.

VISCOSITY GRAVITY CONSTANT (VGC) A measure of the aromaticity of a rubber oil. Defined as, VGC = $[10G - 1.0752\log(V-38)]/[10 - \log(V-38)]$, where G is the specific gravity at 15°C and V is the Saybolt viscosity at 38°C. As VGC increases so does aromaticity, having a value close to 1·0 for a highly aromatic oil, about 0·90 for a naphthenic oil and about 0·80 for a paraffinic oil.

VISCOSITY NUMBER (Reduced specific viscosity) (Reduced viscosity) Symbol η_{sp}/c. The ratio of the specific viscosity of a solution to the concentration (c) of the solute. In a dilute polymer solution c is usually expressed as grams of polymer per decilitre (dl) of solution. A measure of the ability of a polymer to increase the viscosity of a solvent. Even in dilute polymer solutions, intermolecular interactions occur between polymer molecules, so η_{sp}/c is concentration dependent. Therefore to assess the effect of isolated polymer molecules it is necessary to extrapolate to infinite dilution to obtain the limiting viscosity number (or intrinsic viscosity), $(\eta_{sp}/c)_{c \to 0}$.

VISCOSITY RATIO (Relative viscosity) Symbol η_{rel}. The ratio of the viscosity of a solution (η) to the viscosity of pure solvent (η_0). In capillary viscometry of a dilute polymer solution $\eta_{rel} = \eta/\eta_0 = t/t_0$, where t and t_0 are the measured flow times of the solution and solvent respectively.

VISCOSITY STABILISED NATURAL RUBBER Alternative name for constant viscosity natural rubber.

VISCOUS FLOW Flow occurring in a fluid or solid material resulting from shear, bulk or tensile deformational forces. The term therefore excludes particle flow. For a viscoelastic material it is the final region of viscoelastic behaviour of a linear amorphous polymer at the highest temperature or lowest frequency (or longest time) range, beyond the rubbery flow region. For a crosslinked or crystalline polymer below its T_m value, viscous flow does not occur. In this region, the modulus is typically about 10^5 Pa and decreases as the temperature increases, the polymer behaving as a viscous liquid with very little elastic recovery.

VISTAFLEX Tradename for a polypropylene/ethylene-propylene thermoplastic polyolefin rubber.

VISTALON Tradename for ethylene-propylene rubber.

VISTANEX Tradename for polyisobutene.

VITELLIN An egg yolk protein which exists in the egg as the lipoprotein, lipovitellin. A phosphoprotein, containing about 1% phosphorus, comprising about 15% of the egg yolk protein. It is precipitated when the yolk is diluted with water. It contains an unusual amount of arginine (8%) as well as high serine (11%) and glutamic acid (11%) contents.

VITELLININ An avian yolk protein, which occurs naturally as the lipoprotein, lipovitellinin. It is very similar to vitellin except for a lower phosphorus content of about 0·3%; however, some doubt exists as to the nature of the phosphorus content, which is variable and may not be phosphoprotein.

VITON A Tradename for vinylidene fluoride-hexafluoropropylene copolymer.

VITON B Tradename for vinylidene fluoride-hexafluoropropylene copolymer which also contains some tetrafluoroethylene termonomer units.

VITON G Tradename for a quaterpolymer of vinylidene fluoride, tetrafluoroethylene, hexafluoropropylene and a crosslinking monomer. Different variations are Viton GH, Viton GLT and Viton GF.

VITON GF Tradename for a fluoropolymer which is a quaterpolymer of vinylidene fluoride, tetrafluoroethylene, hexafluoropropylene and a crosslinking monomer.

VITON GH Tradename for a fluoropolymer which is a quaterpolymer of vinylidene fluoride, tetrafluoroethylene, hexafluoropropylene and a crosslinking monomer.

VITON GLT Tradename for a fluoropolymer which is a quaterpolymer of vinylidene fluoride, tetrafluoro-ethylene, hexafluoropropylene and a crosslinking monomer.

VITREOUS SILICA A three-dimensional network polymer of SiO_2, which can be considered to be made up of SiO_4 tetrahedra joined at their vertices by the sharing of each oxygen atom between two silicons, giving:

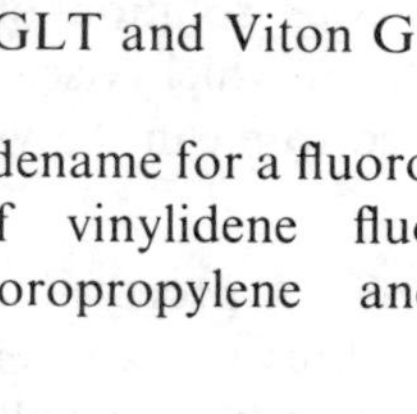

$$\left[\begin{array}{c} | \\ O \\ | \\ \sim\!Si\!-\!O\!\sim \\ | \\ O \\ | \end{array}\right]$$

Essentially the structure of the network is random due to a distribution of Si—O—Si bond angles about the mean value of about 153°, giving a variable orientation of the tetrahedra. However, some short-range order, extending

VLDPE Abbreviation for very low density polyethylene.

VMQ Abbreviation for vinylsilicone elastomer.

VNIILON Tradename for poly-(p-phenyleneterephthalamide).

VOID VOLUME The volume of a gel permeation chromatographic column not occupied by the gel. This is also the same as the retention volume for molecules eluted above the exclusion limit.

VOIGT ELEMENT (Kelvin element) A Voigt model which is a component, together with other Voigt or Maxwell model components, of a more complex viscoelastic model system, such as the standard linear solid.

VOIGT MODEL (Kelvin model) A simple mechanical model for viscoelastic behaviour which consists of a spring of Young's modulus E and a dashpot of coefficient of viscosity η connected in parallel. It is thus an isostrain model, the stress being the sum of the individual stresses on the spring and dashpot. This leads to a differential representation of linear viscoelasticity as $\sigma = E\varepsilon + \eta\, d\varepsilon/dt$, where σ is the stress and ε is the strain. The model is useful in representing simple creep behaviour; thus integration of the above equation gives $\varepsilon = \varepsilon_0 \exp(-t/\tau')$, where ε_0 is the initial strain and τ' is the retardation time and equals E/η. For stress relaxation the model predicts purely elastic behaviour and so cannot be used as a model for this aspect of viscoelastic behaviour. Such a model is often referred to as a Voigt solid model, in contrast to the Maxwell fluid model, since the deformation is recovered completely on removal of the stress, given sufficient time. Thus at long times the model behaves as a Hookean solid.

VOIGT SOLID MODEL Alternative name for Voigt model.

VOLAN Tradename for methacrylatochromic chloride.

VOLTALEF Tradename for polychlorotrifluoroethylene and for vinylidene fluoride–chlorotrifluoroethylene copolymer.

VOLUME RESISTIVITY The electrical resistance across the opposite faces of a unit cube of a material. The usual units are ohm centimetre, the size of the cube being 1 cm in this case. Materials with a resistivity of $> 10^8\ \Omega$ cm are electrical insulators and those with values of $< 10^2\ \Omega$ cm are electrical conductors. Materials with intermediate values are semiconductors. Most polymers are insulators, having resistivity values ranging from 10^{11}–$10^{19}\ \Omega$ cm. The volume resistivity of insulators may be measured on flat sheets of material sandwiched between specially designed circular electrodes with low contact resistance. The current flowing through the material is measured, often with a voltage of 500 V being applied for 1 min.

VOLUMETRIC STRAIN Alternative name for dilatation.

VON MISES YIELD CRITERION A yield criterion which states that yielding occurs when the shear strain energy reaches a certain value. This may be represented as $(\sigma_1 - \sigma_2)^2 + (\sigma_2 - \sigma_3)^2 + (\sigma_3 - \sigma_1)^2 =$ a constant, where σ_1, σ_2 and σ_3 are the principal stresses. For simple tension $\sigma_2 = \sigma_3 = 0$ and the constant is $2\sigma_y^2$, where σ_y is the yield stress. For pure shear $\sigma_1 = -\sigma_2$ and $\sigma_3 = 0$ and $\sigma_1 = \sigma_y/(3)^{1/2}$, hence the shear yield stress is $1/(3)^{1/2}$ times the tensile yield stress. Many metals obey this criterion, but many polymers do not, although their yield behaviour is closer to that predicted by the Von Mises criterion rather than by the Tresca criterion. For polymers the criterion has to be modified to take account of the hydrostatic dependence of yield.

VPO Abbreviation for vapour pressure osmometry.

VSI Older alternative for the abbreviation VMQ.

VULCANISATE (Vulcanised rubber) A rubber which has been lightly crosslinked by the process of vulcanisation. Compared with the unvulcanised rubber it has higher stiffness, strength, rubber elasticity and resistance to solvent swelling and to set.

VULCANISATION The process of conversion of a raw rubber composed of linear molecules to a lightly crosslinked network—the vulcanisate. Most rubbers in useful commercial rubber products have been vulcanised since this considerably improves their properties. Thus most raw rubbers are soft, low modulus materials for which vulcanisation increases stiffness, reduces permanent deformation (set) on stressing and reduces sensitivity to solvent swelling. The commonest method of vulcanisation, applicable to most diene rubbers, such as natural rubber, styrene–butadiene, polybutadiene, ethylene–propylene and nitrile rubbers is sulphur vulcanisation, which involves heating with sulphur at about 140°C. Heating with sulphur alone (when about 8 phr may be used) is very slow, requiring several hours, and much sulphur is used inefficiently. Use of various metal oxides, especially zinc oxide, speeds up the process. However, organic vulcanisation accelerators are almost universally used in addition since they not only give faster rates of vulcanisation but also ensure that the sulphur is used more efficiently; generally only 1–3 phr of sulphur and

over 1–2 nm, may exist. The material is prepared by heating a naturally occurring crystalline silica, e.g. quartz, and cooling. It has a T_g value of about 1200°C and a low coefficient of thermal expansion making it useful as a thermal shock-resistant glass. The glass prepared from silica sand is opalescent due to locked in air bubbles. Closely related are the alkali metal silicate glasses such as soda glass.

vulcanisation times of a few minutes are required. The use of ultraccelerators makes vulcanisation at ambient temperatures possible. In some cases only a very low amount of sulphur (about 1 phr) is required (low sulphur vulcanisation) or even no sulphur at all (sulphurless vulcanisation). The vulcanisation activators, zinc oxide and stearic acid, are usually also used.

Chemically, sulphur vulcanisation involves the formation of sulphur crosslinks, $-S_x-$, between the rubber molecules. For best mechanical ageing properties the crosslinks should be short, i.e. with $x = 1$ or 2. In the absence of accelerator much sulphur is used in forming inefficient long sulphur crosslinks with x greater than two, and also cyclic and pendant sulphur groups. On average about 40 sulphur atoms may be required for each crosslink formed. In accelerated vulcanisates only about 10–15 sulphur atoms are required and the figure may be as low as 4–5 for efficient vulcanisation systems. The structure of the crosslinks is determined experimentally by their reactions with appropriate chemical probes. Typically about one crosslink for each 100 monomer units gives optimum properties.

Non-sulphur vulcanisation of diene rubbers may be carried out with sulphur chloride (cold curing) or with a thiuram disulphide or bismorpholine disulphide. Butyl rubber may be vulcanised with quinonedioxime or phenolic resins. Peroxide vulcanisation can be almost universally applied and is used particularly with ethylene–propylene rubber when no termonomer is present and with silicone rubber. Polychloroprene does not vulcanise, but is vulcanised by zinc oxide or a thiourea. Acrylic rubbers containing a halogenated comonomer are vulcanised with polyamines and fluororubbers with a mixture of a metal oxide and amine.

The progress of vulcanisation is usually followed in an instrument recording the increase in modulus of the vulcanisate using a curemeter. In very active vulcanisation systems it is sometimes necessary to guard against premature vulcanisation during the compounding and shaping stages, to prevent prevulcanisation or scorch. This tendency may be reduced by using a retarder. Too long a vulcanisation time (overcure) can, especially with natural rubber, lead to a diminution in tensile strength or modulus (reversion). The efficiency of use of sulphur is sometimes expressed as the efficiency parameter. Efficient vulcanisation systems use very low sulphur (less than 1 phr) but high accelerator (about 3 phr) contents.

VULCANISED RUBBER Alternative name for vulcanisate.

VULCANITE Alternative name for ebonite.

VULCAPLAS Tradename for a polysulphide rubber made from 1,3-glycerol dichlorohydrin,

$$CH_2Cl$$
$$|$$
$$CHOH$$
$$|$$
$$CH_2Cl$$

as the monomer.

VULCAPRENE Tradename for an early polyurethane elastomer, made by reaction of hexamethylene di-isocyanate with a polyester–amide (itself made by reaction of adipic acid and monoethanolamine) and crosslinked by formaldehyde or isocyanates. It has good ozone, oxygen and oil resistance, but poor mechanical properties and is mainly used for lacquers and adhesives. Subsequently similar polymers called Daltolac were developed.

VULCUP Tradename for di-t-butylperoxydiisopropylbenzene.

VULCOLLAN Tradename for a cast polyurethane elastomer, with good tear and abrasion resistance and high load bearing capacity. Usually based on a poly-(ethylene adipate) polyester prepolymer, of molecular weight about 2000, reacted with naphthalene-1,5-di-isocyanate to form a further prepolymer with limited shelf life. This is further reacted with a diol, e.g. 1,4-butanediol, and crosslinked by heating at about 100°C.

VYBAK Tradename for polyvinyl chloride.

VYCOR A borosilicate glass containing 55–70% SiO_2 and 20–40% B_2O_3. The boron compound forms a disperse phase which grows on heating to about 600°C forming a continuous network in a silica matrix. It may subsequently be leached by treatment with dilute hydrochloric acid to give a porous glass with pores of 0·2–0·5 nm in diameter.

VYCRON Tradename for a polyethylene terephthalate fibre modified by the incorporation of some isophthalate units.

VYDAX Tradename for telomers of tetrafluoro-ethylene.

VYDYNE Tradename for a mineral filled nylon 66.

VYRENE Tradename for a spandex polyurethane elastomeric fibre formed by reaction of either a polyester (from adipic acid and a mixture of ethylene and propylene glycols) or a polyether polyol with excess MDI. This isocyanate-terminated prepolymer is then reacted with water to give some amine end groups. The polymer is then spun to a fibre and heated, when amine and isocyanate groups react to form urea groups.

W

W Symbol for tryptophane.

WAGNER EARTH An electrical circuit used in a Schering bridge to reduce stray capacitances. It consists of a voltage divider connected across the generator, so that the bridge detector terminals may be adjusted to earth potential.

WALLNER LINE A periodic rippled appearance to a fracture surface due to the occurrence of lines which are the loci of intersections of the main fracture front with transverse secondary waves released during fracture. Alternatively a periodic stress wave may be deliberately generated in order to specifically study the Wallner effect, by applying external ultrasonic radiation.

WALS Abbreviation for wide angle light scattering, often simply called light scattering. Usually it refers to the use of light scattering from polymer molecules in dilute solution as a method of determination of the weight average molecular weight.

WASTED CROSSLINK In a crosslinked rubber network, a crosslink joining two terminal chains to the network and therefore not contributing to the elasticity of the network.

WATER EXTENDED POLYESTER A crosslinked unsaturated polyester which contains up to 90% water dispersed as fine droplets of about 2–5 μm in diameter. Useful as a plaster and wood replacement and as an ablative plastic.

WAVE TRANSMISSION METHOD A method for the determination of dielectric properties in the high frequency region of 10^8–10^{11} Hz. Electromagnetic waves are propagated down a waveguide or coaxial line terminated by a slab of the sample dielectric filling the end space. A standing wave is established in the air filled portion and a probe, inserted in a slot in the line, monitors the field intensity as a function of position. The ratio E_{min}/E_{max} (the ratio of wave amplitudes at node and antinode positions), the distance (x_0) of the first node from the specimen surface and the wavelength (λ_1) in the air filled portion are all measured. This enables the complex propagation factor (γ^*) of the dielectric to be calculated, from which the relative permittivity (ε') and loss factor (ε'') are obtained from.

$$\varepsilon^* = \varepsilon' - i\varepsilon'' = [(1/\lambda_c^2)^2 - (\gamma^*/2\pi)^2]/[(1/\lambda_1^2)^2 + (1/\lambda_c^2)^2]$$

where λ_c is the cut-off wavelength of the waveguide. λ_c is infinity for a coaxial line.

WAXS Abbreviation for wide angle X-ray scattering.

WEAK LINK A structure present in a polymer chain or a side group, as a structural irregularity, which is of lower stability than the normal repeat unit structure and therefore at which polymer degradation is likely to be initiated. Such weak links are difficult to characterise since they may be present to less than one per polymer molecule. Examples of possible weak links are branch points, head-to-head structures, unsaturated structures, copolymerised impurities, initiator and other end groups.

WEATHERING The deterioration of a material when exposed out-of-doors to the degradative agencies of the weather. The most important agency is the long wavelength (300–400 nm) ultraviolet component of sunlight reaching the earth's surface. This is absorbed by most polymers and is sufficiently energetic to cause bond dissociation of most organic chemical bonds and to initiate photodegradation.

The overall weathering effects are very complex, depending on many factors which include intensity and wavelength of radiation, the presence of atmospheric oxygen (weathering is essentially a photo-oxidation process), pollutants (e.g. oxides of nitrogen and sulphur), ozone and atmospheric moisture, rainfall, the day/night and seasonal cycles, temperature and the geometry and orientation of the polymer sample. Many undesirable changes in properties can occur, including embrittlement, discolouration and surface crazing (often appearing as a whiteness). Failure can be rapid, e.g. in polyolefins, if the polymer is unprotected by an ultraviolet stabiliser. Accelerated weathering is used to rapidly evaluate the performance of polymers and stabilisers.

WEIGHT AVERAGE MOLECULAR WEIGHT Symbol $\bar{M}_w$. A polymer molecular weight average which is the mean value of the weight distribution of molecular sizes, defined as,

$$\bar{M}_w = \sum_{i=1}^{\infty} N_i M_i^2 \bigg/ \sum_{i=1}^{\infty} N_i M_i = \sum_{i=1}^{\infty} M_i w_i \bigg/ \sum_{i=1}^{\infty} w_i$$

where N_i, M_i and w_i are the number of molecules, the molecular weight and the total weight of the molecular species i. Since the term M_i^2 occurs in the numerator, $\bar{M}_w$ is always larger than $\bar{M}_n$, but its value is smaller than the higher z and $z+1$ averages.

Compared with $\bar{M}_n$, small amounts of high molecular weight polymer make a larger contribution to $\bar{M}_w$ and can considerably raise its value, whereas low molecular weight species contribute little due to their low weights. $\bar{M}_w$ is therefore the relevant average to use when discussing a property, such as melt viscosity, which is most dependent on high molecular weight species. $\bar{M}_w$ may be determined by any method in which the property being measured depends on the weight of the species present. Light scattering and sedimentation in an ultracentrifuge are the most widely used methods.

WEIGHT DISTRIBUTION Alternative name sometimes used for frequency distribution when this is describing the variation in molecular size i in a polymer sample as a function of the weight (or mass) of each species i present. Sometimes also used for the differential weight distribution.

WEISSENBERG EFFECT (Rod-climbing effect) The movement of a fluid up a rod or cylinder whilst being rotated in the fluid, rather than (as is more usual) the formation of a depression or vortex. The effect occurs

WEISSENBERG NUMBER Symbol W_s. A dimensionless parameter which is a measure of the importance of elastic relative to viscous effects in flow problems. In general, for viscometric flows, W_s is given by the product of the relaxation time (λ) of the fluid and the rate of deformation, i.e. $W_s = U\lambda/L$, where U is a characteristic velocity, and L is a characteristic length, e.g. diameter of a pipe, in the direction along which the velocity varies.

WEISSENBERG–RABINOWITSCH–MOONEY EQUATION Alternative name for Rabinowitsch equation.

WEISSENBERG RELATION The equality of the two normal forces, σ_{22} and σ_{33}, for the flow of a non-Newtonian fluid. This is often approximately true and is indeed predicted by many constitutive equations.

WEISSENBERG RHEOGONIOMETER A cone and plate rheometer capable of operating over a wide range of shear rates and temperatures and of evaluating elastic effects, such as normal stresses, during flow.

WENZEL EQUATION A relationship between the change in the contact angle of a liquid on a solid surface and the roughness factor r, which is the ratio of the true surface area to the apparent surface area. It is: $\cos\phi/\cos\theta = r$, where ϕ is the apparent contact angle and θ is the intrinsic contact angle on a smooth surface. Thus if θ is less than 90° roughening decreases ϕ and hence promotes wetting. Thus in adhesion, roughening will often improve the strength of the adhesive bond.

WESSLAU DISTRIBUTION A particular form of the generalised logarithmic-normal distribution of molecular weights, whose differential weight distribution function is,

$$W(r) = \exp\left[-(\ln r - \ln r_m)^2/2\sigma^2\right]/r\sigma r_m(2\pi)^{1/2}$$

where r is the size of each individual molecular species, e.g. degree of polymerisation, r_m is its mean value and σ is the standard deviation of the distribution. Although an empirical distribution, not derivable from any particular polymerisation mechanism, it does describe some real systems, such as polyethylene produced by Ziegler–Natta polymerisation.

WET SPINNING A process for the production of man-made fibres in which a solution of a polymer is extruded through a multi-orifice die (a spinneret) into a liquid in which the polymer is insoluble and therefore in which it is coagulated. Viscose and cuprammonium rayons, polyvinyl alcohol and polyacrylonitrile fibres are produced in this way.

WHISKER A small fibre up to a few micrometres in diameter and with an aspect ratio of 10–100, which is a perfect single crystal and which therefore has exceptionally high tensile strength and Young's modulus. Therefore whiskers have great potential as reinforcing fibres in composites, including polymer composites, particularly as they are somewhat flexible (2–4% elongations) and therefore unlike glass and other rigid fibres, do not break during composite processing. Usually grown by high temperature vapour deposition reactions, requiring a high degree of purity in the precursor materials, so ensuring structural perfection in the crystalline product. The most common commercial whisker material is β-silicon carbide which can have values for its moduli and strength close to those theoretically calculated from interatomic forces. Such whiskers can have a tensile strength of up to about 4000 MPa. However, the use of whiskers has been severely limited due to their high cost.

WHITE ASBESTOS Alternative name for chrysotile.

WHITE LEAD Alternative name for basic lead carbonate.

WHITE–METZNER MODEL A modification of the Maxwell model of fluid behaviour, utilising the Oldroyd derivative (d/dt), giving the relation,

$$\Delta(\dot\gamma)\,\eta(\dot\gamma)\,d/dt + \tau$$

where τ is a shear stress, Δ is the rate of deformation tensor, λ is the relaxation time and η is the viscosity. The last two are both functions of the shear rate ($\dot\gamma$) and are related through a 'modulus' (G) as $\lambda = \eta/G$. This leads to the normal stress components $\Psi_{12} \sim \eta^2$ and $\Psi = 0$, which is often approximately true in polymeric fluids. Consequently this model has been widely used for modelling polymer flows.

WHITE SPIRIT (Mineral spirit) A petroleum hydrocarbon fraction, largely similar to the special boiling point spirits, but of even wider and higher boiling point ranges (being typically of about 50°C in the range 140–250°C) and containing more aromatics (about 15%). An important paint and varnish solvent and thinner.

WHITING Particulate calcium carbonate resulting from the grinding of chalk. Sometimes also refers to synthetic precipitated calcium carbonate.

WHOLE POLYMER A polymer which is heterogeneous with respect to molecular structure, before it has been separated into more homogeneous fractions by fractionation. The usual structural heterogeneity is the molecular weight (the whole polymer is then polydisperse), although tacticity and copolymer composition variations are sometimes of interest.

WHOLLY AROMATIC POLYAMIDE (Aramid) A polyamide in which the polymer chain consists entirely of aromatic rings and amide groups. The polymers have very high T_m values and are almost exclusively synthesised by low temperature solution methods, often from the appropriate diamine and diacid chloride. Those from a phthaloyl chloride and o-phenylene diamine have a T_m values which are too low to be of interest, whereas those from iso- or terephthaloyl chloride and m- or p-diamines, are rather intractable. Thus the important polymers poly-(p-phenyleneterephthalamide) and poly-(m-phenylene-isophthalamide) are not melt processable, but must be processed in solution.

WHOLLY AROMATIC POLYMER An aromatic polymer which does not contain any aliphatic hydrocarbon groups in the polymer chain. Thus, for example, polyoxybenzoyl

$$\sim\!\!\left[CO\!-\!\langle\bigcirc\rangle\!-\!O\right]_n\!\!\sim$$

is wholly aromatic whilst polyethylene terephthalate

$$\sim\!\!\left[OC\!-\!\langle\bigcirc\rangle\!-\!COOCH_2CH_2\right]_n\!\!\sim$$

is not—it is only partially aromatic.

WIDE ANGLE LIGHT SCATTERING (WALS) Alternative name for large angle light scattering. Commonly simply called light scattering.

WIDE ANGLE X-RAY SCATTERING (WAXS) The scattering of X-rays by the regular arrays of atoms in a lattice of a crystalline material in which the scattering angles are given by the Bragg Law at angles below about 90° (generally around 20–50°). The resultant diffraction pattern is usually recorded photographically. The film may be either flat or as a cylindrical strip surrounding the sample, as in the Debye–Scherrer method. For a melt crystallised polymer, which consists of randomly oriented crystallites, a pattern similar to a powder pattern, consisting of a series of concentric rings, results. When recorded by the Debye–Scherrer method these appear as lines whose positions give the d-spacings between planes (hkl) of atoms. For uniaxially (fibre) oriented specimens (e.g. drawn synthetic or natural fibres) the pattern is similar to that obtained by rotating a single crystal (rotation photograph), and is known as the fibre diagram. This consists of a series of spots (reflections) lying on lines (the layer lines) perpendicular to the direction of the X-ray beam. The spots are often diffuse, or elongated to arcs, due to the small size of the crystallites, the presence of lattice imperfections or departures from ideal fibre orientation. By correctly assigning the hkl indices to the reflections, the dimensions of the unit cell may be calculated. A major success has been the determination of the chain conformations of a variety of helical biological polymers, including DNA. The degree of crystallinity may be estimated from the variation of intensity with diffraction angle. The value obtained is usually lower than that found using other methods and is not absolute due to difficulties in accounting for diffuse

scattering and in defining degree of crystallinity. Crystalline orientation can be characterised by measurement of the variation of intensity around the diffraction axis in a fibre diagram, with the uniaxial axis tilted with respect to the X-ray beam by varying amounts. Such data is obtained using a pole-figure analyser.

WIDE-LINE NUCLEAR MAGNETIC RESONANCE SPECTROSCOPY (Broad-line nuclear magnetic resonance spectroscopy) Nuclear magnetic resonance spectroscopy in which broad resonances are observed and analysed. Broadening results from relatively slow molecular motions, such as those that occur in polymeric solids, giving a distribution of local magnetic fields. As the motion of a resonating nucleus increases, the variations in magnetic field are smeared out and all nuclei experience the same field, thus giving a narrower resonance peak. Thus a measure of molecular motion is given by the width of the peak. The variation in field arises from spin–spin interactions which produce a spread of frequencies given by the spin–spin relaxation time. Also, at resonance, the populations of the two energy states are not the same due to spin–lattice relaxation, causing a loss of energy from the upper state to the surroundings. This is a maximum when molecular motions occur near the frequency of the resonance. Wide-line NMR has been extensively used in the investigation of molecular transitions and relaxations in polymers, which are frequently characterised by the spin–lattice relaxation times.

WILD RUBBER Natural rubber obtained from Hevea brasiliensis trees which grow wild rather than being cultivated on a plantation (plantation rubber). It is subject to large variations in its nature and properties and is now little used.

WILLIAMS–LANDEL–FERRY EQUATION (WLF equation) An equation relating the value of the shift factor (a_T) (associated with time–temperature superposition) required to bring log modulus (or log compliance) versus time (or frequency) curves for different temperatures into a single master curve, for a particular reference temperature (T_0). The equation is empirical in origin and has the form,

$$\log a_T = C_1(T - T_0)/[C_2 + (T - T_0)]$$

where the constants C_1 and C_2 are approximately identical for all amorphous polymers. The values of C_1 and C_2 were originally determined by choosing T_0 as 243 K for polyisobutylene. If T_0 is taken as the static (dilatometric) glass transition temperature (T_g), then C_1 and C_2 take on the universal values of -17.44 and 51.6 K respectively. Subsequently the WLF equation was given theoretical justification from a consideration of the free volume theory of T_g. This leads to a form of the equation as

$$\log a_T = \log(\eta_T/\eta_{T_g})$$
$$= (b/2.303 f_g)(T - T_g)/[(f_g/\alpha_t) + T - T_g]$$

where b is a constant of the Doolittle equation, f_g is the fractional free volume at T_g and α_f is the coefficient of expansion of the free volume.

WILLIAMS PLASTIMETER A plastimeter in which a preheated cylindrical rubber sample is compressed between two parallel plates by a specified load for a specified time. The resulting thickness of the sample is the plasticity number.

WILLIAMS THEORY A molecular theory of the non-Newtonian flow behaviour of polymeric fluids, which unlike the Bueche and Graessley theories, does not rely on any specific mechanism of interaction between molecules or segments of molecules. Instead it assumes that a force of interaction exists and may be expressed by a potential function and a segment distribution function, which varies with the shear rate ($\dot{\gamma}$) in an assumed way. It leads to a similar result, i.e. that η/η_0 is a function of $\lambda\dot{\gamma}$, where η is the coefficient of viscosity, η_0 is the coefficient of viscosity at zero shear rate and λ is a relaxation time.

WLF EQUATION Alternative name for the Williams–Landel–Ferry equation.

WOLCRYLON Tradename for a polyacrylonitrile fibre.

WOLLASTONITE A linear, crystalline, naturally occurring silicate mineral polymer, in which three SiO_4^{2-} tetrahedra (each with two shared oxygen atoms) comprise the repeating unit. The chains are linked laterally through calcium ions, the empirical formula being $CaSiO_3$, although iron, magnesium and manganese may also be present. Available in both fibrous and particle forms, and useful as a filler in polymer composites. It may act as a reinforcing filler especially when the particles have a high aspect ratio and are treated with a coupling agent. Wollastonite also confers good electrical properties with low oil and water absorption.

WOOL A natural protein fibre obtained mainly from the outer coat of various sheep. Raw wool contains 10–20% grease which is removed by scouring with a soap solution. The fibres vary from 2–15 cm in length and from 10–50 μm in diameter and have a natural crimp. They are composed of the protein keratin. The fibres are relatively weak, having a tenacity of about 1·5 g denier^{-1} (dry) and about 1·0 g denier^{-1} (wet), but they are very durable, having an elongation at break of 25% and an elastic recovery of 99%. Wool is the most hygroscopic fibre, having a moisture regain of 13–18%.

WORKING RANGE In general, the temperature range over which a polymer shows useful properties as a material, especially as a plastic. In particular, the temperature range between the brittle–ductile transition and the softening point. For an amorphous polymer this may roughly correspond to the temperature interval between the β-transition and the T_g value.

WORK HARDENING Alternative name for strain hardening.

WORK SOFTENING Alternative name for strain softening.

WORK OF FRACTURE Alternative name for critical strain energy release rate.

WORK OF COHESION Symbol w_c. The energy that must be expended in creating unit area of new surface from a homogeneous material by cleavage or fracture. Conversely it is the energy liberated when two surfaces of the same material coalesce. Thus $w_c = 2\gamma$, where γ is the surface tension of the material.

WORK OF ADHESION Symbol w_A. The energy that must be expended to separate two surfaces which meet at an interface, due to intermolecular forces of attraction. w_A is thus given by the Dupré equation, $w_A = \gamma_1 + \gamma_2 - \gamma_{12}$, where γ_1 and γ_2 are the surface tensions of the two components and γ_{12} is the interfacial tension. Experimentally w_A is determined by measuring the wetting angle (θ) for incompletely wetting liquids on solids using the Young equation to give, $w_A = \gamma_1(1 + \cos\theta)$, where γ_1 is the surface tension of the liquid. If wetting is complete then $\theta = 0$, $\cos\theta = 1$ and $w_A = 2\gamma_1$.

WORM-LIKE CHAIN A model for a polymer chain consisting of n segments each of length b, with adjacent bonds forming an angle ϕ to each other and with free rotation. A persistence length a is defined as the projection of the end-to-end vector of an infinitely long chain on the first bond and $a = b/(1 + \cos\phi)$. The worm-like chain is generated in the limit $b \to 0$ and $\phi \to \pi$, whilst keeping a and the contour length $L\ (=nb)$ fixed. This yields a mean square end-to-end distance

$$\langle r^2 \rangle = 2a^2[(L/a) - 1 + \exp(-L/a)]$$

which reduces in the limit of $L \gg a$ to $\langle r^2 \rangle = 2La$ and in the limit of $a \gg L$ to $\langle r^2 \rangle_0 = L^2$. The former case describes flexible chains and the latter rigid, rod-like chains. Thus the model can describe the whole spectrum of chains from small rods to long coils, but it does ignore the thickness of the chains.

WOVEN ROVING A heavy fabric produced by weaving rovings, especially of glass fibres. It is heavier than usual glass fibre fabrics but does not produce such high strengths in composites. However, woven roving does not need desizing since the coupling agent is applied with the size.

X

X-500 Tradename for an aromatic polyamide-hydrazide fibre of structure

produced by reaction of p-aminobenzhydrazide and terephthaloyl chloride.

XANFLOOD Tradename for xanthan.

XANTHAN Tradenames: Actigum, Kelzan, Keltron, Xanflood. A random copolymer polysaccharide produced by the bacterium *Xanthomonas campestris* grown on dextrose. The polymer consists of units of D-glucuronic acid, D-mannose and D-glucose irregularly linked. The linear chains contain single monosaccharide side chains of either D-mannose or 4,6-O-(1-carboxyethylidene)-D-glucose units:

HO H OH H — CH_2 — O — C — COOK
H_3C — C — O — H

About 8% of the hydroxyl groups are acetylated and the polymer is isolated and used as its potassium salt. The polymer is water soluble forming helices in solution, which can become ordered forming a cholestic phase at high concentrations. At very low concentrations (a few parts per million) use of an aqueous solution can cause considerable friction reduction in turbulent flows and the polymer is therefore useful in oil recovery. It is also used as a dispersing agent and thickener.

XANTHATE Alternative name for sodium cellulose xanthate.

XDI Abbreviation for xylylene diisocyanate.

XENON ARC LAMP A light source frequently used in artificial, accelerated weathering devices, owing to the similarity of the spectral distribution of energy of its light (when suitably filtered) to that of natural sunlight.

XENOX Tradename for a tetrafluoroethylene-propylene copolymer.

XENOY Tradename for a bisphenol A polycarbonate/polybutylene terephthalate blend.

XPI 182 Tradename for a soluble and melt processable polyimide produced by reaction of 3,4-dicarboxy-1,2,3,4-tetrahydronaphthalene-succinic acid dianhydride (itself produced by reaction of styrene with 2 mol of maleic anhydride) with a diamine.

XPS (1) Abbreviation for expanded polystyrene.
(2) Abbreviation for X-ray photoelectron spectroscopy.

X-RAY PHOTOELECTRON SPECTROSCOPY (XPS) Alternative name for electron spectroscopy for chemical application.

XR-RESIN Tradename for tetrafluoroethylene-sulphonylfluoridevinyl ether copolymer.

XT-POLYMER Tradename for methylmethacrylate-butadiene-styrene or methylmethacrylate-acrylonitrile-butadiene-styrene copolymer.

XYDAR Tradename for an aromatic polyester thermotropic liquid crystalline polymer.

XYLAN A polymer of D-xylose. Found in association with cellulose in the lignified tissues of land plants. It amounts to about 20–25% of wood fibres and cereal straw cell walls and 15–20% of grass stems. Frequently the major component of the hemicellulose in association with glucomannans in coniferous woods. Most xylans consist of linear chains of 1,4'-linked β-D-xylose units, but with single or double sugar side groups. Those from woody tissues frequently have single α-D-glucuronic acid side groups 1,2'-linked to the main chain; sometimes these groups are methylated (the glucuronoxylans). Cereal flour xylans are relatively simple having randomly distributed 1,3'-linked unit α-4-arabinofuranose side groups amounting to up to 30–40% of the polymer. Stem and straw xylans (commonly called arabinoxylans) have complex structures containing α-4-arabinofuranose, β-D-xylose and β-(D- or L-)galactopyranose units in the side groups of 1–3 units. Algal xylans, found in certain seaweeds, are 1,3'- and/or 1,4'-linked β-D-xylose polymers.

XYLENE

The three different isomers boil at 144.2°C (*ortho*), 139.1°C (*meta*) and 138.8°C (*para*). Commercial xylenes, obtained from petroleum or coal tar, are a mixture of all three isomers. An important solvent; the solvent properties are not significantly affected by the isomeric composition. Impurities, such as toluene and trimethylbenzenes, may affect the properties. A solvent for many natural resins and rubbers, fluid silicones, alkyd resins, melamine- and urea-formaldehyde resins, polystyrene and polyisobutene. A non-solvent for cellulose esters, polyester resins and nylons. Somewhat swells polyvinyl chloride, polyethylene and polyvinyl acetate. Widely used in coatings, especially as it is a good solvent for alkyd resins, and wire enamels. Also useful as an intermediate for the production of terephthalic acid.

XYLENOL (Dimethylphenol) There are six possible isomers of which 3,5-xylenol

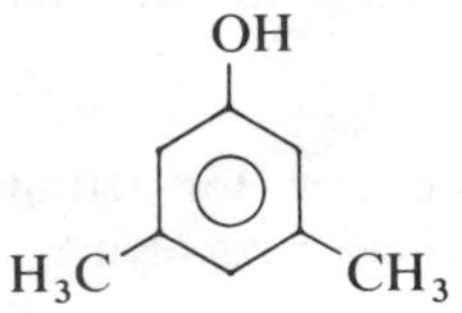

[structure: 3,5-xylenol — benzene ring with H_3C and CH_3 substituents and OH]

(M.p. 62·5°C, b.p. 226°C)

is useful for the production of phenol–formaldehyde polymers with improved chemical resistance and increased oil solubility for surface coating uses. The 2,6-isomer

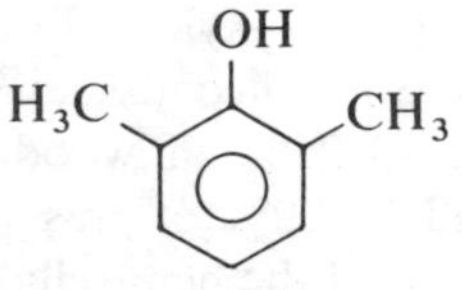

[structure: 2,6-xylenol — benzene ring with H_3C and CH_3 substituents and OH]

(M.p. 49°C, b.p. 212°C)

is produced in high purity form, by the alkylation of phenol with methanol at about 350°C, for use as the monomer for the production of polyphenylene oxide.

2,6-XYLENOL Alternative name for 2,6-dimethyl-phenol.

XYLOK RESIN Tradename for a range of phenol–aralkyl Friedel–Crafts polymers which are condensation products of aralkylethers or halides and phenols:

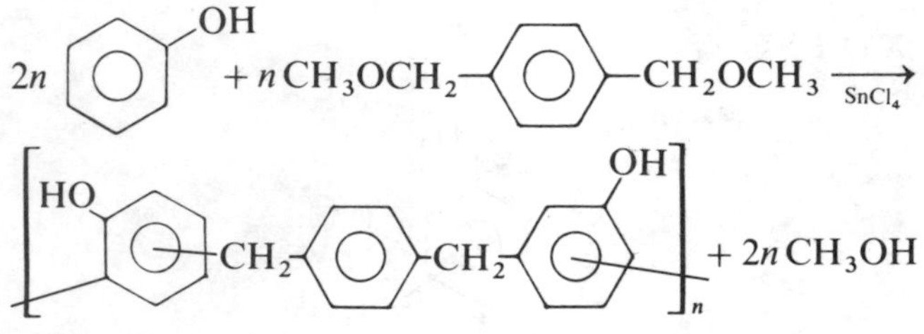

[reaction scheme: $2n$ (phenol, OH) $+ nCH_3OCH_2$—C_6H_4—CH_2OCH_3 $\xrightarrow{SnCl_4}$ polymer ($-CH_2$—C_6H_4—CH_2— ... with OH, HO) $+ 2n\,CH_3OH$]

The polymers can be cured with hexamine or epoxy resins to hard, intractable products, which may be considered as high temperature resistant types of phenolic resin. They resist deformation to about 250°C and are especially useful as composites with glass, asbestos or carbon fibres.

XYLONITE Tradename for celluloid.

β-D-XYLOPYRANOSE The usual anomeric ring form of D-xylose units in a carbohydrate which are therefore more fully described by this name.

D-XYLOSE Ring form (α-D-xylopyranose):

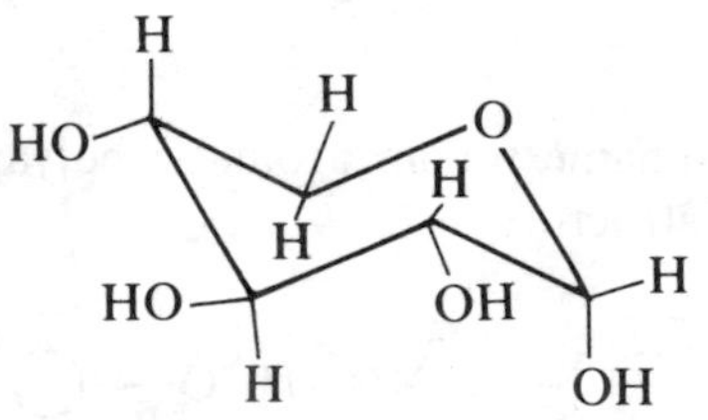

[structure: pyranose ring with OH, H, HO, O labels]

M.p. 145–148°C, $\alpha_D^{20,\,H_2O}$ +18·8°.

An aldopentose, widely found in the polysaccharides of many woody plant materials, e.g. straw and seed hulls, either as an essentially homopolysaccharide (xylan) or as part of a heteropolysaccharide, e.g. glucuronoxylan, arabinoxylan, gum tragacanth.

XYLYLENE DIISOCYANATE (XDI)

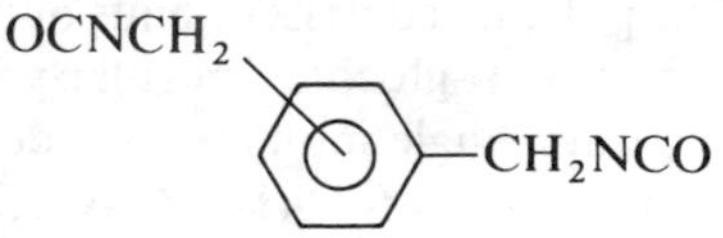

[structure: benzene ring with $OCNCH_2$ and CH_2NCO substituents]

A diisocyanate, useful in the production of polyurethanes, especially coatings, with a good resistance to yellowing on thermal or photo-oxidation. Synthesised by ammon-oxidation of xylene, followed by hydrogenation and phosgenation. The commercial materials consist of a 70/30 mixture of the *meta* and *para* isomers.

XYRON Tradename for poly-(2,6-dimethylphenylene oxide) grafted with styrene.

Y

Y Symbol for tyrosine.

YARN An assemblage of fibres laid or twisted together which is the form from which textiles are subsequently spun or woven. Filament yarn is made from long continuous fibres and is usually only slightly twisted and relies mainly on the fibre strength for its performance. Staple yarn is made by twisting (by spinning) short, i.e. staple, fibres and its strength depends more on the cohesiveness of the fibres and hence on the spinning process. The number of filaments in a yarn may vary from one (a monofilament) to several hundred, but is usually in the range 15–100.

YIELD CRITERION A function of the general stress system which must reach some critical value for yielding to occur. The criterion is independent of the mode of stressing (tension, shear, compression, etc.). For an isotropic material the criterion is a function of the invariant of the stress tensor and may be formulated in terms of the principal stresses. Originally, yield criteria

YIELDING Classically considered to be the onset of permanent (plastic) deformation on stressing a material beyond a certain critical value—the yield stress—in contrast to the completely reversible elastic deformation which occurs at stresses below the yield stress. Such behaviour, although approximately followed by metals, is not so closely followed by polymers. Thus polymers not only show some irreversible deformation before the yield point, but may also show considerable recoverability on stressing beyond the yield point, or even complete recoverability, e.g. if annealed. However, amorphous polymers well below the T_g value, i.e. in the glassy state, do yield approximately in the classical manner and the theory of plasticity, involving considerations of yield criteria and plastic flow, has been usefully applied. Polymers which yield are ductile, and yielding is accompanied by necking, which is then followed either by a relatively rapid ductile fracture or by cold drawing.

YIELD POINT The point on a stress–strain curve beyond which deformation is not completely recoverable, i.e. plastic flow has occurred and permanent deformation has resulted. Very clear with metals, where two deform-ation mechanisms operate—lattice distortion in the elastic region before yield and dislocation motion in the plastic flow region after yield. However, with polymers the distinction between elastic (recoverable) and plastic (permanent) deformation is not so clear and the yield point is perhaps not of such significance. The yield point also often occurs at a maximum on the true stress versus strain curve, when it is sometimes called the intrinsic yield point. The yield stress and the yield strain are the values of the stress and strain at the yield point. A further definition of yield point, sometimes called the extrinsic yield point, is given by Considère's construction and is the point at which deformation becomes unstable and necking occurs, the sample also carrying maximum load at this point.

YIELD STRESS (Yield value) The stress beyond which yielding occurs. On a stress–strain curve, the stress at the yield point. This is usually a maximum on the curve, beyond which the load falls as the strain is increased further. Strictly, the yield stress is the true stress at yield, but since yielding usually occurs at relatively low elongations, the nominal stress is adequate. Sometimes, especially in shear, no maximum is observed on the stress–strain curve and the yield stress may be taken as the point at which the tangents to the stress–strain curve before and after yielding intersect. The yield stress is often a more useful guide (than the tensile strength) to the limit to which a polymer may be stressed, since often after yielding the material has deformed so much that a component made from it will have lost its function.

were developed for metals, for which yielding is independ-ent of hydrostatic pressure and hence only depends on the deviatoric components of stress. Widely used criteria are the Tresca, Von Mises and Coulomb criteria.

YIELD VALUE Alternative name for yield stress, but used particularly for fluids rather than solids. Thus it is the shear or tensile stress for a Bingham model at which flow begins.

YOUNG–DUPRÉ EQUATION Alternative name for Young equation.

YOUNG EQUATION (Young–Dupré equation) A relationship between the surface tensions of a solid in equilibrium with the vapour of a liquid (γ_{SV}) and of a liquid (γ_L) and the contact angle of a drop of the liquid resting on the solid. The relationship is, $\gamma_{SV}\gamma_{SI} = \gamma_L\cos\theta$, where γ_{SI} is the interfacial surface tension. This provides a convenient way of measuring the work of adhesion of the liquid to the solid since γ_S and γ_{SI} cannot be easily measured separately.

YOUNG'S CREEP COMPLIANCE Alternative name for tensile creep compliance.

YOUNG'S MODULUS (Tensile modulus) (Modulus of elasticity in tension) Symbol E. The elastic modulus for uniaxial extension. The most frequently and easily measured of the elastic constants and the most commonly used parameter characterising material stiffness. Young's modulus is usually measured by a simple tensile test as the slope of the stress–strain (elongation) curve. However, for many polymers a simple stress–strain plot is not linear. E is then often taken to be given by the slope of a secant, e.g. at 0.2% or 1% elongation, giving a secant modulus. In other respects modulus behaviour for polymers is different from other materials, especially in its tempera-ture sensitivity. Typically for an amorphous polymer, E changes by about three orders of magnitude in the region of the T_g value, from about 10^3 MPa in the glassy region well below T_g, to about 1 MPa for a rubbery polymer well above T_g. Furthermore the value of E obtained in a test also depends on the rate of testing, decreasing with decreasing test speed, since a slow speed allows more time for molecular disentanglement. E is increased if the polymer crystallises due to the increased intermolecular forces of attraction. A calculation of E for a completely extended chain crystalline polymer yields a theoretical modulus approximately equal to that of steel, e.g. that for polyethylene is calculated as 250 GPa. In practice moduli are about a hundred times lower. However, careful preparation of highly oriented (e.g. ultradrawn) samples can give material with a modulus approaching the theoretical value. E is related to the other elastic moduli. For plastics at small strains, E is approximately the same as the compressive modulus. It is related to the shear modulus (G) by, $E = 2G(1 + \nu)$, where ν is Poisson's ratio. For many rubbers $\nu \sim 0.5$, so $E \sim 3G$.

YPHANTIS METHOD Alternative name for men-iscus depletion method.

Z

Z Symbol for glutamine and/or glutamic acid in a protein.

ZANTREL Tradename for a polynosic rayon.

Z AVERAGE MOLECULAR WEIGHT Symbol $\bar{M}_z$. A higher molecular weight average defined as,

$$\bar{M}_z = \sum_{i=1}^{\infty} N_i M_i^3 \bigg/ \sum_{i=1}^{\infty} N_i M_i^2$$

where N_i and M_i are the number of molecules and molecular weight respectively of each molecular species i. Also equal to the ratio of the third to the second moments of the distribution. Its value is higher than that of the weight average molecular weight ($\bar{M}_w$) and even more sensitive to the higher molecular weight species. It is therefore possibly the better average to use when describing the molecular weight dependency of certain properties, such as melt viscosity, which are also very dependent on the higher molecular weight species. Methods of determining $\bar{M}_z$ are few; sedimentation in an ultracentrifuge and gel permeation chromatography are the main methods.

(Z+1) AVERAGE MOLECULAR WEIGHT Symbol $\bar{M}_{z+1}$. A higher molecular weight average defined as,

$$\bar{M}_{z+1} = \sum_{i=1}^{\infty} N_i M_i^4 \bigg/ \sum_{i=1}^{\infty} N_i M_i^3$$

where N_i and M_i are the number of molecules and molecular weight respectively of each molecular species i. Also equal to the ratio of the fourth to the third moment of the molecular weight distribution. Its value is even higher than that of $\bar{M}_z$ and is extremely sensitive to high molecular weight species. The values of $\bar{M}_{z+1}$ and the higher averages ($\bar{M}_{z+2}$, $\bar{M}_{z+3}$,..., etc.) can only be determined experimentally via a detailed knowledge of the molecular weight distribution, as obtained, for example, by gel permeation chromatography, and even then they are very uncertain.

ZDC Abbreviation for zinc diethyldithiocarbamate.

ZEFRAN Tradename for a polyacrylonitrile fibre, containing a dye receptive component, possibly poly-N-vinylpyrrolidone.

ZEIN The protein fraction of maize seeds, obtained by extraction with 70-80% ethanol, i.e. the prolamine component. Its molecular weight is about 50 000 and it is rich in leucine and proline, but is deficient in glycine, tryptophane and lysine. It has been used for the production of man-made fibres.

ZEOLITE A complex network aluminosilicate polymer, similar to feldspar, but with a much more open structure, so that the sodium, potassium and calcium ions, as well as water, can move freely through the interconnected cavities. Thus the zeolites can exhibit base exchange properties and absorb water, by intercalation, as well as small organic molecules. Zeolites are therefore important as molecular sieves and for chemical separation processes, e.g. they can separate linear from branched chain paraffin hydrocarbons.

ZERO ORDER MARKOV CHAIN PROCESS Alternative name for Bernoullian process.

ZETAFIN Tradename for ethylene-ethyl acrylate co-polymer.

Z-GROUP Abbreviation for benzyloxycarbonyl group.

ZIEGLER-NATTA CATALYST A catalyst for Ziegler-Natta polymerisation. In general, it consists of two components. One component is an alkyl or hydride of a group I-III metal, most commonly an aluminium alkyl, such as AlEt₃, Al(isobutyl)₃ or AlEt₂Cl, other examples being n-butyllithium, diethylzinc or a Grignard reagent. The second component is a transition metal salt (usually a halide) of a group IV-VIII transition metal, particularly of titanium, such as TiCl₄ or TiCl₃, or of vanadium, such as VCl₄, VCl₃ or VOCl₃. The catalyst components when mixed, usually in a hydrocarbon solvent, may form a homogeneous catalyst or a heterogeneous catalyst, either by precipitation or by use of an insoluble transition metal compound, e.g. TiCl₃. Particularly important examples of the former are AlEt₂Cl/VCl₄, which can produce syndiotactic polypropylene, and (cyclopentadienyl)₂TiCl₂/AlEtCl₂, which is very active for ethylene polymerisation. The most widely studied heterogeneous catalyst is TiCl₃/AlEt₃, which is highly stereoregulating, particularly for producing isotactic polypropylene. In this type of catalyst the surface titanium centres become alkylated by the aluminium component forming titanium–carbon σ-bonds. For stereospecific polymerisation of conjugated dienes, catalysts based on AlR₃ with cobalt, nickel, titanium and vanadium compounds are used. 1,4-Cis polymerisation predominates with metals of groups IV and VIII and 1,4-trans polymerisation predominates with group IV and V halides. Vinyl type 1,2-polymerisation is favoured with group VI metal compounds. More recently second generation catalysts of much higher activity have been developed. These have either been specially solvent (e.g. with an ether) or heat treated, or a third component has been involved, e.g. Bu₂Mg. Alternatively activity is enhanced using an active supporting medium such as Mg(OH)₂.

ZIEGLER-NATTA POLYMERISATION (Ziegler polymerisation) Chain polymerisation using a Ziegler-Natta catalyst. Such polymerisations are remarkable in that olefins may be polymerised under mild conditions to high molecular weight polymer and also high stereoregulating effects are possible. The catalysts consist of two react with the catalyst or with the active centres and hence kill the polymerisation.

A wide variety of monomers may be polymerised by Ziegler–Natta catalysts. However, their importance is largely due to their ability to polymerise α-olefins, especially propylene, to high molecular weight isotactic polymers, and to polymerise dienes, especially butadiene and isoprene, to stereoregular forms. Heterogeneous catalysts are more stereoregulating and are formed by reacting (for example) AlR_3 or AlR_2Cl (where R is an alkyl group, often ethyl) with crystalline $TiCl_3$, so that the catalytic activity resides at surface crystal sites which have been alkylated by the AlR_3. Many mechanisms of propagation have been suggested, reflecting the wide variety of catalysts that have been investigated. However, typically, propagation involves the insertion of a monomer molecule in the σ-bond between the active chain end and a titanium atom, whilst simultaneously complexing the monomer with the titanium at a vacant lattice site on the surface. Thus the active centre is coordinated to the catalyst and Ziegler–Natta polymerisations are a type of coordination polymerisation. Furthermore, the active centre is often anionic in nature so the polymerisation is also referred to as an anionic-coordination polymerisation. Both monometallic and bimetallic mechanisms having been suggested. The former seem more likely in heterogeneous systems and the latter in homogeneous systems. The complexing of the monomer is an important factor in stereoregulation. Non-polar monomers are not so strongly complexed and need a heterogeneous catalyst, possibly with surface ligands that can additionally help to sterically direct the monomer, for stereoregulation. More polar monomers which complex more strongly, can be polymerised to tactic polymers with homogeneous catalysts. Monomers of very high polarity may inactivate the catalyst by complexation which is too strong or by reaction with the catalyst.

ZIEGLER POLYMERISATION Alternative name for Ziegler–Natta polymerisation.

ZIG-ZAG CONFORMATION Alternative name for extended chain.

ZIMM DISTRIBUTION Alternative name for Schultz distribution.

ZIMM–MEYERSON OSMOMETER A particularly simple capillary osmometer, consisting of an open-ended cylindrical solution cell, to the side of which a vertical filling tube and measuring capillary are attached. A membrane is clamped across the open ends and the cell is immersed in a bath of solvent, which constitutes the solvent cell.

ZIMM PLOT A widely used method of analysis of the data obtained from light (or other radiation) scattering measurements on a dilute polymer solution of concentration c, whereby a plot is made of Kc/R_θ versus $\sin^2(\theta/2) + kc$, for different values of c at different scattering angles θ, where K is an instrumental constant, R_θ is the reduced scattering intensity (if R_θ is the Rayleigh ratio then K is replaced by $K(1+\cos^2\theta)$) and k is an arbitrary constant, chosen to give a reasonable spread of points. The plotted points form a grid (the Zimm plot itself) on which a double extrapolation is performed to zero concentration and to zero angle. The extrapolated points form the lines $\theta = 0$ and $c = 0$, whose common intercept on the Kc/R_θ axis has the value $1/M_w$, where M_w is the weight average molecular weight of the polymer. The initial slope of the line of $c = 0$ yields the second virial coefficient A_2 and that of $\theta = 0$ yields the mean square z average radius of gyration $\langle s^2\rangle_z$, since the equation describing the scattering is:

$$\frac{Kc}{R_\theta} = 1/M_w + 1/M_w \times \left[(16\pi^2/3\lambda^2)\langle s^2\rangle_z \sin^2(\theta/2) + 2A_2c + \cdots\right]$$

The method utilises the fact of zero internal interference at zero angle and no assumption needs to be made about the shape of the polymer molecules, in contrast to the dissymmetry method.

ZIMM THEORY A molecular theory of the dynamic (in the oscillatory sense) viscosity of a polymer solution, based on the bead–spring model of subchain units, as in Bueche theory. It leads to an expression for the complex viscosity (η^*) as,

$$\eta^* = \eta_s + cRT/M \sum_{p=1}^{N'} \lambda'_i/(p^2 + i\omega\lambda'_i)$$

where η_s is the solvent viscosity, c is the polymer concentration, R is the universal gas constant, T is the temperature, λ'_i is a terminal relaxation time, ω is the angular frequency of stressing and N' is the number of subunits per polymer chain. λ'_i cannot be obtained except in the following limiting cases. (1) The free-draining coil (no hydrodynamic interaction) when $\lambda'_i = \langle r^2\rangle \zeta N^2/6\pi^2 kT$, where ζ is the monomer friction coefficient, $\langle r^2\rangle$ is the mean square chain end-to-end distance, k is Boltzmann's constant and N is Avogadro's number. This leads to $\lambda'_i = 6M\eta_s(\eta_0 - \eta_s)/c\pi^2 RT$ which is a result identical with that obtained by Rouse theory. (2) For a dominant hydrodynamic interaction when $\lambda'_i = M(\eta_0 - \eta_s)/2.369cRT$, where M is the molecular weight.

ZINC DIETHYLDITHIOCARBAMATE (ZDC)

$$[(C_2H_5)_2NCSS]_2Zn \qquad \text{M.p. } 178°C.$$

An ultraaccelerator for the sulphur vulcanisation of rubbers, especially for lattices.

ZINC ISOPROPYLXANTHATE (ZIX)

$[(CH_3)_2CHOCOS]_2Zn$

M.p. 145°C (decomposes).

An ultraaccelerator for the sulphur vulcanisation of rubbers.

ZIP LENGTH The ratio of the rate of depropagation to the sum of the rates of termination and transfer in a polymer undergoing depolymerisation. Hence a measure of the tendency to depropagation by unzipping of the polymer.

ZIX Abbreviation for zinc isopropylxanthate.

ZONAL CENTRIFUGATION Alternative name for density gradient centrifugation.

ZONE CENTRIFUGATION Alternative name for band centrifugation.

ZONE ELECTROPHORESIS Electrophoresis in which the polymer solution (usually protein) is held in a rigid hydrophobic matrix across which the electric field is applied. The sample is confined to a narrow zone, unlike free boundary electrophoresis where the sample can suffer convective and mechanical disturbance. The protein components become separated into zones, which are visualised by staining, whose positions may be determined and whose optical density is proportional to the amount of protein. The section of the support containing the desired protein may be cut out and the component extracted. It is therefore adaptable for preparative as well as analytical separations. The technique requires only small quantities of sample, is simpler and has a higher resolution than free boundary electrophoresis which it has largely replaced. The support may be paper or cellulose acetate strips (paper electrophoresis) or may be a granular material such as a block (block electrophoresis) or a column (column electrophoresis). The support, in the form of a gel (gel electrophoresis), such as starch, polyacrylamide or agarose, may also separate on the basis of molecular size to give even better resolution. More refined variations of gel electrophoresis are disc electrophoresis and isoelectric focussing which can give even higher resolution and require extremely small sample quantities. SDS gel electrophoresis is widely used for protein molecular weight determinations.

ZONE MELTING A method of polymer fractionation in which a polymer sample is placed on top of a column of solid (frozen) solvent. The top solvent layer is melted and dissolves the polymer and is then re-frozen. The next lower layer is then melted and the cycle is repeated down the column. Individual polymer species move down the column at different speeds depending on their molecular weight and become differentially distributed throughout the length of the column, which is finally cut into sections and the solvent is sublimed.

ZONE SEDIMENTATION Alternative name for band sedimentation.

ZYMOGEN (Proenzyme) A biologically inactive precursor of an enzyme, from which the active enzyme is produced by cleavage of a peptide fragment or fragments. For example, chymotrypsin is the zymogen for trypsin and is initially produced in cells as trypsin would attack the cells in which it is produced. It is activated by cleavage of an arginine–leucine bond followed by loss of two dipeptides. Many monomeric enzymes are first produced as zymogens.

ZYTEL Tradename for nylons 6, 66, 610, 612 and their copolymers.

ZYTEL ST Tradename for rubber toughened nylon 66.

APPENDIX A: UNITS

The SI system of units is now very widely used and is used throughout this dictionary. It is based on a few basic units, together with various derived units. Certain other non-SI units are permitted.

Basic SI units

Quantity	Symbol	Unit	Symbol
Length	L	metre	m
Mass	M	kilogram	kg
Time	T	second	s
Electric current	I	ampere	A
Temperature	θ	kelvin	K
Luminous intensity	I_v	candela	cd
Amount of substance	n	mole	mol

Supplementary units

Quantity	Symbol	Unit	Symbol
Plane angle	α, β, etc.	radian	rad
Solid angle	Ω	steradian	sr

SI derived units (with special symbols)

Quantity	Unit	Symbol	Special symbol
Force	newton	$kg\ m\ s^{-2}$	N
Energy	joule	$N\ m$	J
Power	watt	$J\ s^{-1}$	W
Pressure	pascal	$N\ m^{-2}$	Pa
Stress	pascal	$N\ m^{-2}$	Pa
Frequency	hertz	s^{-1}	Hz
Luminous flux	lumen	$cd\ sr$	lm
Temperature	degrees Celsius		°C
Electric charge, electric flux	coulomb	$A\ s$	C
Electric potential	volt	$J\ C^{-1}$ $W\ A^{-1}$	V
Electric resistance	ohm	$V\ A^{-1}$	Ω
Electric conductance	siemen	$A\ V^{-1}$	S
Electric capacitance	farad	$C\ V^{-1}$ $A\ s\ V^{-1}$	F

Other derived units

Area	square metre	m^2
Volume	cubic metre	m^3
Density	kilogram per cubic metre	$kg\ m^{-3}$
Concentration	mole per cubic metre	$mol\ m^{-3}$
Dynamic viscosity	pascal second	$Pa\ s$
Entropy	joule per kelvin	$J\ K^{-1}$
Heat capacity	joule per kelvin	$J\ K^{-1}$
Thermal conductivity	watt per metre per kelvin	$W\ m^{-1}\ K^{-1}$
Luminance	candela per square metre	$cd\ m^{-2}$

APPENDIX B: CONVERSION FACTORS

Prefixes

The following prefixes may be used to construct decimal multiples of units.

Prefix	Symbol	Multiple	Prefix	Symbol	Multiple
tera	T	10^{12}	deci	d	10^{-1}
giga	G	10^{9}	centi	c	10^{-2}
mega	M	10^{6}	milli	m	10^{-3}
kilo	k	10^{3}	micro	μ	10^{-6}
hecto	h	10^{2}	nano	n	10^{-9}
deca	da	10	pico	p	10^{-12}

Quantities

To convert to SI units multiply by the relevant conversion factors given below.

	SI unit	Other units	Conversion factor
Length	m	micrometre (μm)	10^{-6}
		inch (in)	0·0254
		foot (ft)	0·3048
		ångström (Å)	10^{-10}
		mil ($=0·001$ in)	$2·54 \times 10^{-5}$
Area	m^2	square inch (in^2)	$6·452 \times 10^{-4}$
		square foot (ft^2)	$9·290 \times 10^{-2}$
Volume	m^3	litre (l)	10^{-3}
		gallon (UK) (gal)	$4·456 \times 10^{-3}$
		gallon (US) (gal)	$3·785 \times 10^{-3}$
		barrel (oil) (bbl) (42 gal)	0·1589
Angle	rad	degree (°)	0·01745
Mass	kg	tonne (SI) (t)	10^{3}
		ton (UK) (t)	$1·016 \times 10^{3}$
		ton (US) (t)	$9·072 \times 10^{2}$
		pound (lb)	0·4536
Temperature	K	x degrees Celsius (°C)	$x + 273·15$
	K	x degrees Fahrenheit (°F)	$0·556(x + 459·67)$
	[°C]	x degrees Fahrenheit (°F)	$0·556(x - 32)]$
Density	kg m^{-3}	g cm^{-3}	10^{3}
		lb ft^{-3}	16·02
Force	N	dyne	10^{-5}
		kilogram-force (kgf)	9·807
		pound-force (lbf)	4·448
Pressure, stress	Pa	kgf cm^{-2}	$9·807 \times 10^{4}$
		atmosphere (atm)	$1·013 \times 10^{5}$
		bar	10^{5}
		pounds per square inch (psi) (lbf in^{-2})	$6·895 \times 10^{3}$
		torr (mm Hg)	$1·333 \times 10^{2}$
		dyne cm^{-2}	10^{-1}

Quantities—*contd.*

	SI unit	Other units	Conversion factor
Energy (or work)	J (= N m)	calorie (cal) (international)	4·187
		erg	10^{-7}
		British thermal unit (Btu)	$1·055 \times 10^3$
		foot pound-force (ft lbf)	1·356
		litre atmosphere (l atm)	$1·013 \times 10^2$
		kilowatt hour (kW h)	$3·6 \times 10^6$
Power	W	horsepower (hp)	$7·457 \times 10^2$
		foot pound-force per second (ft lbf s^{-1})	1·356
		calorie per second (cal s^{-1})	4·187
		British thermal units per hour (Btu h^{-1})	0·2931
Viscosity (dynamic)	Pa s (N s m^{-2})	poise (P)	10^{-1}
Viscosity (kinematic)	m^2 s^{-1}	stoke (St)	10^{-4}
Surface tension	N m^{-1}	dyne cm^{-1}	10^{-3}
Impact strength	N m^{-1}	pound-force per foot (lbf ft^{-1})	14·59

APPENDIX C: PHYSICAL CONSTANTS

Atmospheric pressure (standard)	P_0	$1 \cdot 013 \times 10^5$ Pa
Avogadro constant	N_A	$6 \cdot 022 \times 10^{23}$ mol^{-1}
Base of natural logarithms	e	$2 \cdot 718$
Boltzmann constant	k	$1 \cdot 381 \times 10^{-23}$ J K^{-1}
Charge on an electron	e	$1 \cdot 602 \times 10^{-19}$ C
Gas constant	R	$8 \cdot 314$ J K^{-1} mol^{-1}
Gravitational acceleration	g	$9 \cdot 807$ m s^{-2}
Molar volume of ideal gas (under standard conditions)	V_0	$2 \cdot 241 \times 10^{-2}$ m^3 mol^{-1}
Pi (ratio of circumference to radius of a circle)	π	$3 \cdot 142$
Planck constant	h	$6 \cdot 626 \times 10^{-34}$ J s
Velocity of light *in vacuo*	c	$2 \cdot 998 \times 10^8$ m s^{-1}

APPENDIX D:
RELATIVE ATOMIC MASSES (ATOMIC WEIGHTS)

Element	Symbol	Relative atomic mass	Element	Symbol	Relative atomic mass
Aluminium	Al	26.98	Manganese	Mn	54.94
Antimony	Sb	121.8	Mercury	Hg	200.6
Arsenic	As	74.92	Molybdenum	Mo	95.94
Barium	Ba	137.3	Nickel	Ni	58.71
Bismuth	Bi	209.0	Nitrogen	N	14.01
Boron	B	10.81	Osmium	Os	190.2
Bromine	Br	79.90	Oxygen	O	15.99
Cadmium	Cd	112.4	Palladium	Pd	106.4
Calcium	Ca	40.08	Phosphorus	P	30.97
Carbon	C	12.01	Platinum	Pt	195.1
Chlorine	Cl	35.45	Potassium	K	39.10
Chromium	Cr	52.00	Rhodium	Rh	102.9
Cobalt	Co	58.93	Silicon	Si	28.09
Copper	Cu	63.55	Silver	Ag	107.9
Fluorine	F	19.00	Sodium	Na	22.99
Gold	Au	197.0	Sulphur	S	32.06
Hydrogen	H	1.008	Tin	Sn	118.7
Iodine	I	126.9	Titanium	Ti	47.90
Iron	Fe	55.85	Tungsten	W	183.9
Lead	Pb	207.2	Vanadium	V	50.94
Lithium	Li	6.941	Zinc	Zn	65.37
Magnesium	Mg	24.31			

APPENDIX E: THE GREEK ALPHABET

A	α	alpha	N	ν	nu
B	β	beta	Ξ	ξ	xi
Γ	γ	gamma	O	o	omicron
Δ	δ	delta	Π	π	pi
E	ε	epsilon	P	ρ	rho
Z	ζ	zeta	Σ	σ	sigma
H	η	eta	T	τ	tau
Θ	θ	theta	Y	υ	upsilon
I	ι	iota	Φ	ϕ	phi
K	κ	kappa	X	χ	chi
Λ	λ	lambda	Ψ	ψ	psi
M	μ	mu	Ω	ω	omega